THE IMMUNE SYSTEM AND MENTAL HEALTH

HYMIE ANISMAN
Department of Neuroscience, Carleton University, and The Royal's Institute of Mental Health Research, Ottawa, Canada

SHAWN HAYLEY
Department of Neuroscience, Carleton University, Ottawa, Canada

ALEXANDER KUSNECOV
Department of Psychology, Rutgers University, New Brunswick, New Jersey, United States

ELSEVIER

ACADEMIC PRESS
An imprint of Elsevier

Academic Press is an imprint of Elsevier
125 London Wall, London EC2Y 5AS, United Kingdom
525 B Street, Suite 1650, San Diego, CA 92101, United States
50 Hampshire Street, 5th Floor, Cambridge, MA 02139, United States
The Boulevard, Langford Lane, Kidlington, Oxford OX5 1GB, United Kingdom

Notices
Knowledge and best practice in this field are constantly changing. As new research and experience broaden our understanding, changes in research methods, professional practices, or medical treatment may become necessary.

Practitioners and researchers must always rely on their own experience and knowledge in evaluating and using any information, methods, compounds, or experiments described herein. In using such information or methods they should be mindful of their own safety and the safety of others, including parties for whom they have a professional responsibility.

British Library Cataloguing-in-Publication Data
A catalogue record for this book is available from the British Library

Library of Congress Cataloging-in-Publication Data
A catalog record for this book is available from the Library of Congress

ISBN: 978-0-12-811351-6

For Information on all Academic Press publications
visit our website at https://www.elsevier.com/books-and-journals

  Working together
to grow libraries in
developing countries

www.elsevier.com • www.bookaid.org

Publisher: Nikki Levy
Acquisition Editor: Joslyn Chaiprasert-Paguio
Editorial Project Manager: Timothy Bennett
Production Project Manager: Anusha Sambamoorthy
Cover Designer: Miles Hitchen

Typeset by MPS Limited, Chennai, India

THE IMMUNE SYSTEM AND MENTAL HEALTH

Contents

Preface

It is difficult to dance concurrently at two weddings. And yet, this is what we tried to do in this book. Broadly, it was our intention to provide an overview of the relationship between inflammatory immune processes and several mental illnesses and neurological disorders. This entails introducing behavioral scientists and clinicians to fundamental aspects of immune functioning and how these processes are linked to hormonal, neurotransmitter, neurotrophic, and microbial factors. Simultaneously, we aimed to introduce basic neuroscientists to the chief characteristics of diverse mental illnesses, and some of the factors that moderate these conditions. In so doing, we explored the processes by which the brain and immune system, separately and conjointly, contribute to various psychiatric and neurological states.

If it seems that we strayed or overlooked some things, it was more by design, and less due to ignorance—although the breadth and depth of the field imposes its own form of humility at what there is to know and learn. In fact, our goal was to provide sufficient clarification to what can be quite confusing for the newcomer. We omitted coverage of numerous physical illnesses that likely have their roots in inflammatory processes, and which account for why physical, mental, and neurological disorders are so frequently comorbid with one another. Dedicated chapters on heart disease, cancer, and stroke did not make it into this book—although we do discuss them at relevant times. In calibrating our focus, we were squarely concerned with knowing what we can say about the impact of immune processes on brain and behavioral functions, and the destabilization that might result from this impact. We hope we were successful.

Several basic chapters kick things off, dealing with diverse biological processes, including neurons and glial cells, hormones, neurotransmitters, and neurotrophins, as well as immune functioning and microbiota. This is followed by a description of the ways by which various life-style factors (eating processes, exercise, and sleep), stressor experiences, prenatal, and early postnatal experiences, come to influence (and are influenced by) neurobiological systems that have been implicated in mental disorders. Having described these processes, we then relate them to the most frequent mental illnesses, including depressive illnesses, anxiety disorders, PTSD, and schizophrenia, autism spectrum disorders, chronic pain conditions, and finally, neurodegenerative disorders. Throughout, we offer views concerning how immunity and brain processes are interlinked, with an eye toward the identification of biomarkers relevant for the development of illness and predicting treatment responses, as well as the potential for novel treatment targets. Many of these mental health conditions are comorbid with one another, and with many physical conditions. Therefore, we also provide a description of the links between mental illness, and several comorbid illnesses, primarily heart disease, stroke, and diabetes, although throughout the book many other comorbid illnesses are covered.

It was not our intent to cover any single topic in great depth. There are many books that are dedicated for this purpose, but none that provide an overview of how the immune system has fully integrated itself into psycho-pathology. We aimed to strike a balance: Provide sufficient detail for mental health experts to gain an understanding of those biological processes with which they are less familiar, and give those with expertise in biological processes an appreciation of the complexities of mental health issues. At present, there is certainly the need for an integrated approach to this complex issue, and the present effort represents a stab at an intriguing topic. Most certainly, we may not have done justice to all fields equally (after all, it is difficult to dance concurrently at two weddings), so we encourage more informed readers to forgive us our trespasses, and invite the curious to enjoy, delight, and marvel at how far we have come since the days when evil spirits were released with the tap of a blunt chisel and faith in a world well-imagined—much like the one we describe in the following pages.

Hymie Anisman, Shawn Hayley, and
Alexander Kusnecov
February, 2018

Acknowledgments

There are many people we ought to thank for their help in many facets of this volume. In fact, too many to mention them all without exceeding our page limits. However, Hymie would be remiss not to thank Zul Merali, Kim Matheson, and Steve Ferguson for their many thoughtful suggestions in relation to the precision medicine approaches, the impact of challenges in vulnerable populations, and the complex molecular biological underpinning of disease. As well, several former graduate students, now colleagues at other institutes, specifically Sheena Taha, Amy Bombay, Opal McInnis, and Robyn McQuaid, were exceptionally generous with their ideas pertaining to social neuroscience and psychosocial determinants of health. Rob Gabrys was so very kind in taking on the gruesome editing process, and we're very grateful for his help. As always, Simon, Rebecca, Max, and Maida were immensely supportive. Shep and Oefa were also helpful to "grumpy" in several ways (I think "grumpy" is a variant on "grandpa," but I might be wrong).

Shawn thanks all the wonderful graduate students that actually conducted the research that underpins many of the concepts talked about within this book, particularly the chapters on neurodegeneration. Most notably, Emily Mangano (now Rocha), Darcy Litteljohn, Zach Dwyer, Chris Rudyk, and Kyle Farmer have all helped fuel the many neuroimmune studies conducted over the years that have helped keep me excited about research. My colleagues have also been tremendously supportive of my work, with a special shout out to David Park and Michael Schlossmacher who serve as models of research excellence and are also wonderful friends. Of course, a hearty thank you also goes out Olivia and Louise for providing a shining light to come home to each day!

Alex would like to thank his friends and colleagues—especially George Wagner, John Pintar, and Gleb Shumyatsky—who tolerated long absences and deferrals due to constantly shifting deadlines, and provided favors and even advice during this long process. In addition, our intellectual lives would collapse without the inspiration and encouragement from our present and former graduate students. In particular, I thank Dr Ruthy Glass, Sara Norton, Audrey (Qing) Chang, and Janace Gifford, for keeping the embers of neural-immune cogitation glowing bright. Their efforts empirically, as well as on the theoretical front, are always appreciated. A field like this sparks the imagination, and theirs was on fire. And of course, Wendy Kusnecov, and all the "little" Kusnecov's—Saskia, Aidan, and Simone—now in their early 20s, deserve a special thank you for tolerating the late nights, surly behavior, and drinking all the coffee! It took a while, but it's done!

We are indebted to April Farr, our first editor at Elsevier who got us into this, and Nikki Levy and Timothy Bennett who continued in this capacity. Their patience and helpfulness throughout the various phases of this project is appreciated.

Some of our own research, mentioned in various chapters (sometimes, too often), was made possible by the generosity of the Canadian Institutes of Health Research, the

Natural Sciences and Engineering Research Council, the Canada Research Chairs Program, the National Institutes of Mental Health, National Institutes of Drug Abuse, and the Dana Foundation. We can only hope that their generosity will be still greater over the coming year to support the dedicated young researchers who continue to search for ways to diminish and prevent illnesses, thereby enhancing the lives of very many.

Multiple Pathways Linked to Mental Health and Illness

BROAD IMPACT OF MENTAL ILLNESS

Despite the best efforts of scientists and clinicians, mental illnesses are often resistant to pharmacological and to behavioral/cognitive treatments. Moreover, even when a positive response is obtained, residual features of the illness may persist, and illness recurrence is notably high. Making matters worse, mental illnesses are frequently comorbid with other psychological problems, as well as numerous physical disorders. In some instances, the comorbid conditions may be reciprocally perpetuating, whereas in other cases they might be promoted by common elements. These include growth factors and inflammatory immune processes that have been implicated in various mental illnesses and neurological disorders, as well as heart disease, diabetes, stroke, cancer, and an array of immune-related conditions. Treatments for many physical illnesses have been advancing, even if a bit slowly, and the potential for still better outcomes have been bolstered by the recognition that a precision (personalized) medicine approach may be instrumental in determining which patients will be most responsive to particular treatments. This perspective opened the door to precision public health that could potentially predict the most efficacious interventions to benefit whole communities or populations. With new perspectives, new approaches, and new technologies, the modes of treatment and their effectiveness are becoming progressively better.

Yet, there's another side to mental illnesses that needs to be addressed. Despite many treatment advances, and the understanding that psychiatric disturbances are medical conditions and not a personal failing, attitudes concerning mental illness have considerable stigma attached to them. This is all the more remarkable given that mental illness affects more than 20% of people at some time in their life, which means that an awful lot of families are affected by these illnesses. It's hard enough on individuals to deal with their mental illness, but when this is compounded by various forms of stigma, as well as insufficient or ineffective social support, maintaining a positive quality of life becomes much more difficult (Corrigan,

The Immune System and Mental Health
DOI: https://doi.org/10.1016/B978-0-12-811351-6.00001-2

(cont'd)

Rafacz, & Rüsch, 2011). If nothing else, reducing the stigmatization of individuals with mental health problems will increase the likelihood that those at risk will seek treatment early.

Aside from public stigma, it isn't unusual for self-stigma to occur, frequently accompanied by shame and humiliation, social devaluation, internalization of a negative self-concept, a need to maintain secrecy (Corrigan, Rafacz, & Rüsch, 2011) and a decline of help-seeking. If this weren't sufficiently diminishing, individuals with mental health issues frequently have to deal with "structural stigma" which is manifested as inordinately extended waits for treatment as well as a variety of biases when they actually do seek help. Diminished care for other physical illnesses is not uncommon, being misattributed to the manifestation of mental illness. Is it any wonder that life span of patients with mental illness is reduced?

Over the past decade, there has been a decline in overt stigma, possibly reflecting political correctness, but the problem is still extensive. Attempting to "educate" people about mental illness has not been especially effective in eliminating stigma (Adams, Lee, Pritchard, & White, 2010). Even health professionals aren't immune from maintaining stigmatizing attitudes concerning mental illness, and they are also likely to self-stigmatize when they experience symptoms of a mental health problem. It isn't unusual for health professionals to have serious concerns about exposing themselves to the judgment of their peers, and they may also be reluctant to seek treatment (Givens & Tjia, 2002). Given that health professionals are affected in this manner, it's hardly surprising that negative attitudes occur in so many others.

A GENERAL PERSPECTIVE OF MENTAL ILLNESS

Considerable efforts have been devoted to the identification of specific brain regions and neuronal processes that are responsible for the provocation or inhibition of particular behavioral phenotypes. There is merit to this approach, as critical neuronal pathways and brain regions have been identified that contribute to cognitive and emotional functioning, and a variety of behaviors, as well as to mental and neurological illnesses. At the same time, complex behaviors involve complicated neural circuits that won't readily be discerned by simply assessing single brain regions (Northoff, 2013). Consistent with this perspective, a systems approach is frequently adopted that involves analyses of the functional connectivity among neural circuits that subserve normal

behaviors and that of psychological illnesses. It has become increasingly apparent that the activity and functioning of neuronal circuits within the brain are influenced by peripheral and brain hormones, sympathetic nervous system activity, immune factors within the periphery, and immune-like factors that act within the brain (e.g., released by microglia) as well as processes associated with gut bacteria (e.g., Cryan & Dinan, 2012; Dantzer, O'Connor, Freund, Johnson, & Kelley, 2008). In essence, there has been a push toward analysis of mental illnesses within the context of a broader systems-based approach.

This opening chapter introduces multiple systems that may contribute to the evolution of physical, neurological, and psychological disorders, and why some individuals may be more vulnerable (or resilient) to such illnesses. As a particular illness may come about owing to any of several factors, it may be understandable that a treatment that is effective for one individual may not be equally effective for a second. Accordingly, the concept of personalized (precision medicine) treatment strategies are introduced as these have been used for several physical illnesses, and are being incorporated in the treatment of mental health disturbances.

In considering mental illnesses, it is essential to assess not only the diverse processes and mechanisms that are operative within the central nervous system (CNS) itself, but also those apparent across several other systems. The focus in this book is on brain-immune system interactions as they pertain to health, which is consistent with the literature of the past decade or so, wherein inflammatory factors have moved from being a marginal consideration, to one that is a primary player in a range of brain conditions. Indeed, it is now evident that inflammatory immune cells, such as macrophages and microglial cells, and the inflammatory messengers they produce (cytokines), can influence neurotransmission, hormonal release, and even neuronal survival. The involvement of these systems (and their interactions) will be considered in relation to mental health processes as well as comorbid physical illnesses. Finally, genetic, prenatal, and early life experiences, environmental challenges and life-style factors all feed into these illness-related biological processes, and thus need to be considered within any systems-based approach.

It is generally accepted that psychosocial factors, such as early negative life experiences (e.g., parental neglect, abuse), living in poverty, or other types of psychologically toxic environments, together with a failure to cope effectively, would be at or near the top of the list of damaging experiences that can lead to mental illness (Shonkoff, Boyce, & McEwen, 2009). In concert with these psychological stressors are the many other "physical" (neurogenic), systemic, and environmental stressors that affect well-being, including chemical toxicants, industrial contaminants, food additives, and bacterial or viral agents. Given this spectrum of differing challenges, it is easy to get lost in all the specifics, which we wanted to avoid. Thus, we focus on some of the mechanistic commonalities of these varied challenges insofar as they pertain to mental and neurological illness.

Inherent in a systems approach is that in addition to pharmacologically based strategies, psychosocial influences ought to be considered in dealing with at least some psychiatric illnesses (Cruwys, Haslam, Dingle, Haslam, & Jetten, 2014). Moreover, while the treatments of neurodegenerative diseases are obviously not amenable to psychological approaches alone, their comorbid features include depression and anxiety, which may be affected by these treatments. Accordingly, in considering the development and recovery from mental illness, "systems" not only encompass intrinsic biological processes, but also those related to the individual in the context of their social and physical environment, their experiences, and the influence of their ethnic (including cultural) background (Matheson, Bombay, & Anisman, 2018)

In evaluating the multiple factors that are involved in mental illnesses, it is also important to rely on a constellation of "omics" that might be used to decipher how diverse systems come together to produce different phenotypes. Piecing together the interactions of multiple biological networks, at different levels of analysis, can provide a fuller picture of what's going on in the body (see Table 1.1). Unlike the research conducted a decade or two earlier, data sets are now immense and becoming progressively larger. There are some 3 billion chemical coding units that form an individual's DNA. The proteome exceeds 30,000 proteins, and 40,000 metabolites have so far been identified. Huge numbers of epigenetic marks and gene mutations appear on genes, any of which have the potential to

TABLE 1.1 So Many Omes

Ome	What Is It Made of?	What Does It Show?
Genome	DNA: The complete set of instructions for building an organism	Genes for building proteins and other molecules
Transcriptome	RNA: Copies of protein-building instructions	Determines which genes are turned on and off and their activity level
Proteome	Proteins	How the genetic instruction manual is actually applied
Metabolome	Small molecules (metabolites)	Chemical reactions happening inside an organism
Epigenome	Molecules that modify the genome to influence when genes are turned on and off	How environmental and experiential factors influence the activity of genes
Phenome	Outcomes determined by the layers of biological processes	Traits and diseases

From Hamers (2016).

influence health risks. As well, the presence of inflammatory factors and microbiota (bacteria everywhere in and around our bodies) contribute to illnesses, including psychiatric disturbances. All these elements only begin to describe the complex interplay between the multiple processes that determine our phenotype. At every level of analysis, including prenatal and perinatal experiences, psychosocial, experiential, and environmental factors have a say on the expression of specific outcomes. Piecing together the interactions of multiple biological networks can provide a fuller picture of what's going on in the body and within the brain. Biological determinants of health are considered at several levels of analysis.

Analyses of the processes leading to or underlying particular illnesses necessarily imply that these illnesses can be accurately identified. But, this can be far more difficult than the uninitiated might think. Concluding that certain symptoms denote the presence of a particular mental illness is complicated by the fact that many different illnesses share common features. Conversely, a given illness assessed across a population may comprise diverse symptoms to the extent that individuals might present with very few common features. Moreover, an individual's symptom profile may vary over time, and may even morph from one condition to another, as in the case of unipolar depression transforming to bipolar illness. Aside from the difficulty of differentiating between the symptoms of various illnesses, deciding how illnesses should be treated involves yet another level of complexity. This is attested to by the ongoing debate concerning the usefulness of the Diagnostic and Statistical Manual of Mental Disorders (5th ed.; American Psychiatric Association, 2013) and the approach promoted by the National Institutes of Mental Health (NIMH) concerning the use of the Research Domain Criteria (RDoC) in relation to mental illness (Insel, 2014). These issues obviously have considerable bearing on the well-being of affected individuals, but as we'll see, many factors contribute to the difficulties associated with diagnosis and treatment. We still have limited knowledge concerning the workings of the brain, as well as how and why brain functioning is sometimes disrupted, hence leading to mental illness and cognitive deterioration, let alone how to cure a malfunctioning brain.

THE IMMUNE SYSTEM AND MENTAL HEALTH

MULTISYSTEM COORDINATION

Although each of our organs has its own unique functions, it's essential that they communicate with one another and with the brain. This is not only essential so that the left leg doesn't trip over the right, but also because the brain may respond to, and conversely, influence the functioning of various hormonal systems, immune activity, and gut-related processes that can affect other organs, such as the liver or kidney (e.g., Vaikunthanathan, Safinia, Lombardi, & Lechler, 2016) and the brain itself.

Just as maintaining general well-being is dependent upon the smooth operations of multiple intersecting systems, the development and emergence of various pathological conditions may reflect a disturbance in any of several nodes within and between these systems. What this implies is that in adopting prevention or treatment strategies for illnesses, a systems approach might be ideal. As we'll see in several ensuing chapters, this includes attention to psychosocial process in the emergence and treatment of mental and physical illnesses. This by no means implies that a serious or chronic illness (e.g., heart disease) can be cured by altering the functioning of other systems or focusing solely on brain changes. However, it does suggest that amelioration of some illnesses or symptoms can be facilitated by adjunctive (auxiliary) treatments that deal with other systems. Further to this, diminishing the adverse effects of illness on psychological states can enhance the effectiveness of treatments and might also influence treatment compliance and the course of recovery. In later chapters in which we'll be dealing with illnesses comorbid with psychological disturbances, such as heart disease and diabetes (Chapter 16: Comorbidities in Relation to Inflammatory Processes), this perspective will be considered in greater depth. For the moment, suffice it that a sizable portion of the population experiences multiple illnesses that may or may not be related to their primary mental disturbance (chronic pain or illnesses for which the source cannot be identified), making it that much more difficult to treat disease conditions. Without a diagnosis or target for treatment, effective ways of dealing with the illness, or even masking symptoms adequately, may not be attained.

Sequential or Concurrent Influences on Mental Illnesses

The development of illnesses often reflects an insidious process that progressed over months or years, even though individuals mistakenly come to believe that it occurred suddenly ("One day I was fine, and then bam, I wasn't"). This may be the case for viral illnesses, but the development of most chronic illnesses occurs in progressive phases. For instance, type 2 diabetes may develop owing to a combination of eating the wrong foods, not exercising, being overweight, poor sleep quality, and genetic factors, which may lead to metabolic syndrome, eventually culminating in diabetes. Some of these very same factors contribute to the buildup of plaque that produces heart disease, and diabetes itself is a good predictor of later heart disease. In some instances, the presence of particular factors may be especially pertinent among individuals who are "at risk" owing to genetic influences, specific experiences, or the presence of other illnesses.

Researchers in several fields have adopted the view that the development of some illnesses involves "two-hits" or "multiple hits". For instance, genetic constituency may be a first hit that placed people *at risk*, but actual illness only occurred with a second hit. This may comprise the actions of a second gene, or it may stem from unhealthy behaviors or the presence of particular environmental factors. The second hit can also include the cumulative actions of *toxicants* (foreign materials from man-made sources, such as insecticides or industrial chemicals released into the environment, such as herbicides, insecticides, fungicides, rodenticides, and even food additives) and *toxins* (substances that come from a biological source, such as living cells or organisms that could result in illnesses).

First, second and multiple hits aren't simply composed of current stressful events or even repeated exposure to negative stimuli, but instead may reflect events that occurred prenatally or during early life. Such experiences can profoundly influence the developmental trajectory related to neurobiological, behavioral, and emotional development, so that illnesses materialize many years later. It is likewise possible that negative early life experiences prime (sensitize) biological processes so that the organism becomes more reactive when these or other challenges are encountered again much later. As we'll see, negative experiences can promote the suppression of particular gene actions (i.e., through epigenetic changes) and hence affect the phenotypes that would otherwise be apparent. As it happens, some of these epigenetic changes can be passed from one generation to the next so that individuals can be affected by their parent's experiences, and they, in turn, can pass these features on to their own children (the sins of the father).

In the context of multiple hits, the view has been expressed that random mutations and epigenetic changes accumulate with age, and may contribute to many illnesses. Genes related to neurobiological functioning in some brain regions may be more vulnerable to mutations and hence may preferentially contribute to pathology. For instance, with age, mutations normally increase at a higher rate in genes controlling functioning within the dentate gyrus of the hippocampus than in the prefrontal cortex, possibly because the former cells have the potential of dividing (Lodato et al., 2018). Throughout this book, we will repeatedly see that life-style factors are important ingredients in determining pathological conditions. However, at times it may simply be a matter of bad luck.

WINNING THE WRONG LOTTERY

Some illnesses occur as a result of specific genes or gene mutations, or the presence of certain genes coupled with a second hit in the form of an environmental trigger or a stressor. Other illnesses simply occur owing to the accumulation of mutations that appear with ageing. There's little question that some forms of cancer have a high heritability rate, and thus the development of cancer may be linked to inherited genes. Being BRCA1 or BRCA2 positive are well known genetic risk factors in relation to breast cancer, but many other genes, amounting to more than 180, also contribute in this regard (Michailidou et al., 2017). Mutations may also occur on a random basis, although numerous agents (cigarettes, sun rays) can encourage development of mutations (mutagenesis). Cancers can certainly develop owing to specific lifestyles or environmental factors, but not everyone who smokes develops lung (or other) cancers, and not every sun worshipper develops a malignant melanoma. With each cell division, the chance of a mutation occurring increases, and thus, cancerous cells should be more frequent in rapidly reproducing cells than in those that are slower. An analysis of 31 tumor types, in fact, revealed that the lifetime risk for cancers was linked to the rate of total stem cell division (Tomasetti & Vogelstein, 2015). The more rapidly reproducing cells might also be more likely to be affected by environmental triggers, and thus again more readily turned into cancer cells. To a significant extent, individuals who develop cancer are the unfortunate winners of the wrong lottery.

(cont'd)

Various gene mutations have been identified that predict the subsequent occurrence of cancer, irrespective of family history of this illness (Natrajan et al., 2016), possibly suggesting that the mutations occurred on a random basis or were engendered by environmental factors. However, if cancer occurs owing to random mutations, then why is it that many patients may end up being affected by more than a single type of cancer? What are the odds of winning a lottery twice? As it happens, individuals who are affected by several types of cancer are more likely to carry a particular genetic marker, the KRAS-variant. Indeed, about 25% of those with cancer carry this mutation, and more than 50% of those carrying this marker develop more than one cancer.

A common approach to understand processes that are linked to illnesses comprise epidemiological analyses that ask what life-style, experiential, and genetic factors predict disease occurrence. When systematic individual differences are detected they sometimes provide fundamental clues regarding the processes leading to illness, point to interventions to preclude the development of illness, or means by which to treat an ongoing disease. In other instances, the question can be turned on its head. Why do some people who ought to be at high risk seem not to develop illnesses? Why is the occurrence of cancer among people with Alzheimer's disease about half that expected based on simple probabilities. This curious relation wasn't initially attributable to people dying of one disease before the other had an opportunity to emerge, and the focus on one illness didn't result in symptoms of the other condition being obscured. Yet, dementia patients receive fewer cancer screening tests, making it less likely that both illnesses will be detected in the same patients. As well, people with cancer that also suffer dementia, die sooner than those with cancer without dementia. It seems that physicians are less likely to recommend aggressive cancer therapy for patients with dementia, hence their shorter survival time.

Methods of Evaluating Biological Substrates of Mental and Neurological Illnesses

Potential links between specific neurochemical processes and depressive behaviors have been addressed using several approaches. Each has considerable merit, and the triangulation of methods has been especially useful. As a whole, these studies have pointed to the complexity of trying to identify the mechanisms responsible for depression, especially as depressive illnesses are biochemically and behaviorally homogeneous. The data have also pointed to the contribution of several organismic variables (genetic, age, and sex) and experiential factors (ongoing stressors, early life trauma or neglect, previous stressful experiences) in the provocation and the maintenance of depression, as well as its recurrence following successful treatment. The research approaches adopted to determine the processes associated with mental illnesses have included studies comprising

(1) genetic analyses, including whole genome analyses, assessment of polymorphisms, and epigenetic changes in relation to the appearance of illness;
(2) evaluation of hormone and neurochemical factors in blood and cerebrospinal fluid of patients before and after therapy;
(3) imaging studies (e.g., PET, fMRI) that assessed either functional changes in specific brain regions in patients versus

healthy controls (either by assessing blood flow indicative of cellular activity in certain brain regions, or assessing activity or density of particular neurotransmitter receptors), as well as studies that imaged the size of particular brain sites;

(4) determination of brain chemicals and receptors in postmortem tissues among depressed individuals that had died of suicide in comparison to those that had not been depressed and died through causes other than suicide (e.g., sudden cardiac arrest, accidents);

(5) studies in humans that assessed the effectiveness of drug treatments, as well as studies that compared the relative efficacy of several treatments (alone or in conjunction with other agents); and

(6) analyses involving animal models in which biochemical effects of treatments were assessed and analyses performed regarding the effectiveness of pharmacological treatments in attenuating behaviors reminiscent of mental illness.

GENETIC CONTRIBUTIONS TO ILLNESS

In a diploid organism, such as humans, the contributions of dominant alleles largely determine the phenotype. However, the *expressivity* (the extent to which a gene influences the phenotype) varies appreciably between individuals even though they might carry the same genotype. A related concept, *penetrance*, refers to that proportion of individuals carrying a particular gene who show the expected phenotype. In essence, carrying a particular gene does not guarantee a particular phenotype, but instead is influenced by interactions with other factors.

There are a number of conditions that are determined by single genes, but most of the illnesses that we'll be considering involve polygenic actions. These genes could operate in an

additive fashion in which each gene contributes to the phenotype, some perhaps a bit more than others. Alternatively, the actions of particular genes could potentially interact with one another in relation to phenotypic outcomes. Environmental factors, experiences, and the presence of other genes can moderate the phenotypic actions of a particular gene (epistatic interaction). It also seems that a particular gene or set of genes can have more than a single phenotypic outcome (pleiotropy), thereby contributing to several illness comorbidities.

Increased gene diversity enhances the ability of organisms to deal with many environmental influences that can have negative consequences. However, the price for this is that gene variants (mutations) also appear that favor the development of disease states (Lenz et al., 2016). One would think that with so many generations of selection for ideal traits, many of the genes that favor disease would have been selected against and hence genetically related illnesses would be far less frequent than they actually are. No doubt, there has been some selection of this nature, but at the same time, over the course of evolution in which genes for desirable traits were selected, some of these genes may have had downstream consequences that were maladaptive. The classic example of this is that being heterozygous for a particular gene may prevent malaria, but has the unfortunate side effect of producing sickle cell anemia. For those living in Africa, being heterozygous for the sickle cell anemia gene may be advantageous, but it has little value for someone living in North America.

The common view for many decades had been that for better or worse, the genes carried by an individual were immutable, and thus what they were born with was what they were stuck with. Indeed, some of the debates in the 1960s and 1970s seemed to be about how much of the *phenotypic variance* (P_v) stemmed from our genes (G_v), that produced by the environment (E_v) and the variance attributable to the gene x environment interaction (the latter was often ignored or valued at close to zero). Efforts directed at uncovering the contribution of these elements to a phenotype wasn't conducted merely to obtain an understanding concerning how these processes came together, but in some cases had serious implications for government policy.

Twin Research

A common method to evaluate the extent to which a trait is genetically inherited (*heritability*) is through twin studies. Given that *monozygotic* (identical) twins largely have identical genotypes whereas *dizygotic* (nonidentical; fraternal) twins share only half their genes, it would be expected that for phenotypes primarily determined by genes, the correlation between monozygotic twin pairs would be greater than that evident among dizygotic twins. In theory, that's correct, but random mutations can affect gene expression and thus reduce the monozygotic similarities. As well, relative to monozygotic twins, dizygotic twins could have more and greater differences in their experiences. Indeed, monozygotic twins are sometimes treated more alike than are dizygotic twins, in a sense being interchangeable, perforce creating greater similarities between monozygotic pairs. The solution to dissociating environmental from genetic contributions was to assess the phenotypes of identical twins that had either been raised together versus those twins that had been raised apart, thereby limiting environmental similarities. This said, the environment of twins reared apart was not entirely different, as most adopted twins ended in relatively good middle or upper class environments. Furthermore, the prenatal (intrauterine) environment of identical twins may be more similar than that of fraternal twins.

Twin studies have been used to assess heritability in relation to many phenotypes, including physical pathologies, such as heart disease, various types of cancer, diabetes, addictions (alcohol, smoking), and a host of many other physical and psychological disorders. Many studies also explored heritability of various social attitudes (e.g., socialism, abortion, gay rights, and racial segregation) and dispositional (personality) features (e.g., neuroticism, agreeableness, religious fundamentalism, and right-wing authoritarianism). For the most part, there hasn't been appreciable agreement concerning the contribution of genetic factors in regard to these personality and ideological features.

Molecular Genetic Approaches

Humans have 23 pairs of chromosomes each comprising about 6 m of DNA coiled into a small packet that can fit into the nucleus of a cell. The DNA is made up several million nucleotides that come in 4 flavors: guanine (G), adenine (A), cytosine (C), and thymine (T), which in sets of 3 form amino acids. The double stranded DNA, and the many genes present, are the template for "transcription" or RNA formation, which in most respects is identical to the DNA (thymine is replaced by uracil in RNA). The strand of RNA transcribed from DNA, referred to as messenger RNA (mRNA) travels through small openings of the nucleus to enter the cell's cytoplasm. Thereafter, another form of RNA, transfer RNA (tRNA), operating with ribosomes, serve in the "translation" process wherein lengthy chains comprising amino acids form a protein. Depending on the specific sequence of amino acids, a range of different hormones and enzymes, and other complex neurobiological factors are formed, which determine behavioral and psychological outputs.

Only a small fraction of the amino acids of a DNA strand form genes containing the information that contributes to the formation of our phenotypes. These genes, serve like a set of drawings that provide the instructions for a building being constructed. Another sequence of amino acids form promoter or regulatory regions that provide instructions to the primary gene, essentially telling it when to turn on or off, as well as how and when to interact with other genes. What makes these regulatory processes especially interesting is that psychosocial, experiential, and environmental triggers, affect a genes' influence on neurobiological processes and behavioral phenotypes. Contrary to views that had been held only a couple of decades ago, it seems that although the functions of particular genes are more or less fixed, social and environment influences may affect whether these genes will actually reach their potential or be subverted.

Social and Environmental Moderation of Gene Actions

It is curious that although the debates concerning the gene (nature) versus environment (nurture) contribution to phenotypes was fairly tempestuous, scant attention was given to the role of gene x environmental interactions, and for some reason was thought to account for little of the variance. As we now know, the environment in which an organism is raised or tested can markedly influence the expression of a gene. Inbred strains of mice, which had been developed through successive brother x sister matings so that every mouse of a given strain is identical to every other mouse, share many behavioral phenotypes. However, even small environmental differences (e.g., where animals were bred) could result in appreciable differences in the appearance of behavioral phenotypes (Crabbe, Wahlsten, & Dudek, 1999). Likewise, a pregnant mom's experience could affect the expression of genes among her pups, and the early life of rodent pups can affect the later phenotypic expression of their genes (e.g., Szyf, 2011).

MUTATIONS IN HOSTILE ENVIRONMENTS

Given the frequency of cell multiplication and genes being duplicated it can reasonably be expected that some errors will occur, a nucleotide (or string of nucleotides) is deleted, replaced by another nucleotide, or a small string is duplicated (amplification), or translocations may occur in which a gene fragment from one location appears elsewhere. Some mutations that occur might have negative consequences, and indeed, the term mutation may conjure up expectancies of creatures from the dark lagoon (or Ninja Turtles). Some mutations, such as those caused by environmental toxicants could result in illnesses, such as some forms of cancer, but others may reside on portion of the DNA strand that don't seem to have much bearing on phenotypic outcomes.

Importantly, the occurrence of mutations could be a fundamental component of evolution. When environmental changes pose a challenge for organisms, those with mutations that are helpful in dealing with these conditions might be more likely to survive and pass on their genes, whereas those without the mutation would be less likely to survive and commensurately less likely to pass on their genes. Studies in bacteria indicated that there's still more to it than this. Upon being challenged by conditions that comprised potential starvation or the presence of an antibacterial agent (antibiotic) that could kill them, the rate of E. coli bacteria exhibiting gene mutations (mutagenesis) increased appreciably. Evidently, in the context of severe challenges mutagenesis increases, possibly in an effort for some of the bacteria to survive. To be sure, most bacteria will die in response to an antibiotic assault, but given sufficient bacterial diversity, a very small number may have

---

(cont'd)
---

developed just the right mutation(s), which allowed them to survive and form the core of a bacterial colony, thereby contributing to the formation of more resistant bacteria should the antibiotic again be present.

Although it is usually thought that bacterial mutations within DNA occur on a random basis, it seems that a mutation can also occur within an RNA strand, particularly when the cell is stressed. The mutated RNA gives rise to proteins with enhanced functions, and could potentially influence the development of DNA with this mutation in subsequent bacterial generations. This retromutagenesis has been considered as a way for bacteria to escape destruction by antibiotics (Morreall et al., 2015). Mutations could similarly increase in humans, especially in relatively hostile environments, and might be preserved within the gene pool, and passed down across generation. Although this could be very advantageous for humans, when it happens in bacteria, it may contribute to bacteria becoming resistant to antibiotics, which is very bad news for us.

Gene Polymorphisms

It has been estimated that approximately 10 trillion cells are formed over the average life span, and the sequence of nucleotides ought to be the same in each of these. When DNA is transcribed to mRNA, the sequence of bases should be faithfully reproduced, and to help ensure that this occurs, a built-in proofreading process exists to limit errors. Understandably, however, with so much complexity and activity, some errors will occur.

Acquired mutations may be present owing to environmental influences, such as environmental toxicants, and should the mutations occur within sperm or egg, they can be transmitted across generations, and thus are considered to be inherited mutations. A mutation that appears in a significant portion of the population (more than 1%) is referred to as a polymorphism, and if this mutation comprises a change of only a single nucleotide, it is referred to as a single nucleotide polymorphism (SNP; pronounced snip). Just as changing one letter of a word or one word of a sentence can entirely alter its meaning, a change of a single nucleotide at an important portion of a gene can have profound effects on a phenotype. By example, some perturbations may result in certain enzymes, hormones, or hormone receptors not operating as they should. This has provided a means of assessing whether the presence of a SNP (or several SNPs) on a given gene (e.g., coding for a specific hormone receptor) are predictive of the appearance of specific pathological conditions or behavioral styles. As very many mutations occur within every individual, and several SNPs may even be present within a single gene, linking specific polymorphisms to particular phenotypes is often difficult, especially considering that complex behaviors or pathological conditions typically involve multiple genetic contributions. There had been reports indicating relations between certain SNPs and behaviors, but the findings of many early studies, especially those with small sample sizes, weren't readily replicated, although greater success was achieved in subsequent studies that included large cohorts of individuals.

The difficulty linking particular SNPs to phenotypic outcomes was made more complicated with the realization that the occurrence of a given SNP may vary dramatically across cultures. By example, among Euro-Caucasians a particular SNP on the gene coding for the oxytocin receptor appears at a rate of 15%—20%, but this same SNP occurs in about 80% of Asian people (Kim et al., 2011). If this weren't sufficiently complicated, the phenotypic correlates of a particular SNP might vary in the presence of other genetic factors, as well as in the context of particular earlier experiences. Essentially, certain psychological or physical illnesses might only occur when "multiple hits" are experienced, and while genetic constitution may represent one or several hits, a psychologically toxic environment and negative experiences may reflect additional hits that culminate in pathology. This multihit notion has been fairly well accepted in relation to certain types of cancer, Parkinson's disease, stroke recovery, and likely contributes to the emergence of at least some psychological disorders, including depression and schizophrenia.

Epigenetics

Although it was eventually accepted that both genes and environment, as well as their interactive effects, contributed to the expression of numerous phenotypes, it had hardly been considered that experiential and environmental factors could actually influence how the genes were expressed. Following the lead of cancer researchers, it was demonstrated that in response to particular toxicants, or even some foods, as well as stressors, including those that occurred prenatally (Cao-Lei et al., 2017; Szyf, 2015), epigenetic changes could be engendered which might influence psychological well-being. These epigenetic modifications constitute alterations of gene expression in which the actions of some genes or their promoters (regulatory elements) may be suppressed (or activated), but without the DNA sequence being altered. In essence, a change of phenotype can occur without a corresponding change of genotype. Epigenetic modifications are evident in

cell differentiation (e.g., in the formation of different types of cells, such as skin cells versus brain cells over the course of prenatal development), and may be responsible for the development of altered immunity (Hoeksema & de Winther, 2016). The emergence of diseases, such as cancer, immune-related illnesses, neuropsychiatric disorders, pediatric disorders, as well as several developmental disorders (e.g., Fragile X, Rett syndrome, Beckwith-Weidman, Prader-Willi, and Angelman syndromes) may likewise occur owing to epigenetic actions (Szyf, 2009). Epigenetic alterations may occur on genes coding for hormones and hormone receptors, growth factors, and immune processes, which have been implicated in several psychological conditions, including depression, posttraumatic stress disorder (PTSD), and drug addiction (Maze et al., 2010; Mehta et al., 2013).

Epigenetic modifications may come about in several ways, with the most common being DNA methylation, histone modification, and noncoding RNA-associated gene silencing, each of which influences the way in which genes are expressed. Methylation refers to the addition of methyl groups to the DNA at specific sites, and as a result they are less transcriptionally active. The second method involves chromatin modification. Chromatin comprises the complex of proteins (histones) and DNA that is tightly bundled to fit into the nucleus. The complex itself is modifiable through numerous ways, including the incorporation of acetyl groups (acetylation), enzymes, and forms of RNA, specifically small interfering RNAs and microRNAs. As a result, the chromatin structure is altered, resulting in gene expression being affected. When chromatin is tightly folded it tends to result in gene expression shutting down, whereas open chromatin tends to be more functional or expressed. Over the past years the number of epigenetic marks discovered has increased exponentially, and in 2015 the National Institutes of Health (NIH) Roadmap Epigenomics provided global maps of regulatory elements, their presumed activators and repressors, which were linked to cell types that might be relevant for the analysis of the molecular basis of many disease conditions (Kundaje et al., 2015). Incidentally, epigenetic changes had long been thought to be restricted to DNA, but have now been seen on RNA, and not only occur at adenine bases within DNA, but can also appear at cytosine bases. The trick now is to determine their phenotypic actions, and whether and how these epigenetic can be erased or preserved (Willyard, 2017).

It is particularly important that if the epigenetic changes occur within germline cells (sperm or egg), these changes can be recapitulated through cell divisions and across multiple generations (Bird, 2007). To assume that a transgenerational effect is occurring and is not due to common factors that lead to epigenetic changes uniquely within each generation, the epigenetic effects need to be demonstrated across at least three (preferably more) generations. There is, in fact, compelling evidence that the experiences of one generation, through epigenetic alterations, can affect phenotypic outcomes in ensuing generations (Franklin et al., 2010). Indeed, transgenerational epigenetic actions may be fundamental for the fine tuning of gene activation (and inactivation) processes during the course of embryogenesis (Zenk et al., 2017).

Over the course of development, epigenetic changes may play a critical role in determining the different types of cells being formed from the original zygote. As about 200 different cell types will eventually be developed, machinery needs to be present to erase genetic marks so that a cell can return to its "naïve" state, and then develop into yet another specialized cell. Most genes escape the reprograming process, but about 1% are "imprinted" and thus these become permanent features of the organism (e.g., Youngson & Whitelaw, 2008) (Fig. 1.1).

It had at one time been thought that epigenetic changes were uncommon events, and thus when an epigenetic change was detected congruent with a behavioral alteration, these were taken to be linked in some fashion. It is now certain that epigenetic changes occur exceedingly frequently (Koziol et al., 2016), making it difficult to identify a one-to-correspondence with a particular phenotype, let alone assume a causal connection between the two. For instance, 366 separate epigenetic marks were apparent in the hippocampus of individuals who died of suicide (273 hypermethylated and 93 hypomethylated) (Labonté et al., 2013), and no doubt epigenetic changes would be apparent in other brain regions (e.g., hypothalamus, prefrontal cortex) that have been aligned with depressive illnesses.

The difficulty of linking epigenetic effects and specific phenotypes is still more difficult than first thought. Epigenetic effects are not only exceptionally common, but they can be transient, changing through the lifespan in response to environmental events, such as smoking, exercise, and what we eat, or conversely in response to fasting. In fact, epigenetic changes can occur even after an hour of exercising. Obviously, it is necessary to determine whether an epigenetic change is transient or permanent, and if the latter, then what processes make them permanent.

Although some of epigenetic changes may play a causal role in the emergence of a given pathology, others may evolve as a result of the illness (or distress associated with the illness), or they might simply be bystanders that are neither causally related nor biomarkers that predict illness onset. As repeatedly mentioned, aside from the sheer number of epigenetic marks that exist, any given illness likely involves a large number of genes and thus it is questionable how significant a single epigenetic mark would actually be. To complicate analyses further, epigenetic patterns are dynamic and changeable over time (as already

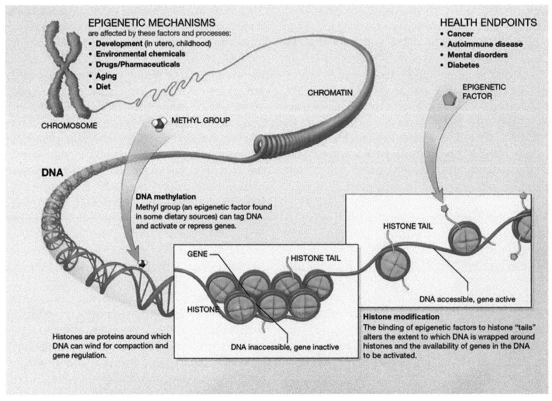

FIGURE 1.1   Epigenetic changes can be elicited by in utero and early childhood experiences, aging, environmental chemicals, drugs, and diet. DNA methylation and histone modification, each of which alters how genes are expressed, do so without altering the underlying DNA sequence, and when epigenetic changes occur within germline cells, they may persist through cell divisions for the duration of the cell's life, and may also occur over multiple generations. In the figure, epigenetic effects are depicted as occurring as a function of methylation of particular DNA sites or through histone modification wherein DNA wraps around histones tightly so that gene expression is prevented. These epigenetic actions have notable effects on cancer processes, autoimmune disease, mental disorders, diabetes, and other illnesses. *Source: From National Institutes of Health—http://commonfund.nih.gov/epigenomics/figure.aspx, Public Domain, https://commons.wikimedia.org/w/index.php?curid = 9789221.*

said, only a portion of epigenetic changes that occur are permanent), and thus links to particular pathological conditions will be still more difficult to discern and the probability of false positives will increase commensurately (Birney, Smith, & Greally, 2016). Longitudinal designs may diminish some of the problems in linking epigenetic marks to the subsequent appearance of certain pathologies (Birney et al., 2016), but even these types of studies won't necessarily reveal whether a link is causal or simply correlational, especially when the epigenetic profile and the pathology occur concurrently. Despite these limitations, identifying epigenetic markers could be used to determine who might be vulnerable to illness and perhaps what treatments might be most efficacious for treatment in any given individual (Mullins & Lewis, 2017; Szyf, Tang, Hill, & Musci, 2016).

Genetic Links to Mental Illness

The picture seems rosy with respect to the use of genetic factors for the purpose of illness identification, but skepticism in this context would be highly appropriate, especially in light of the long track record of less than staggering successes. Considerable hope was placed on genetic influences in relation to mental illnesses, and in being able to identify suitable markers for illnesses and their treatments. Perhaps we were fooled by the fact that there are illnesses that can be ascribed to a single major gene (e.g., cystic fibrosis, sickle cell disease, Tay Sachs, Fragile X syndrome, muscular dystrophy, Huntington disease, and Thalassaemia), but these tend to be relatively uncommon. Indeed, a large number of core genes, say a few dozen, regulate most illnesses. But, if one considers the vast array of gene networks in which the actions of one gene interacts with those of a second, third, and fourth gene, it will be clear that most complex illnesses are likely influenced by hundreds or thousands of genes (Boyle, Li, & Pritchard, 2017). The very same thing can be said of epigenetic effects that occur in remarkably large numbers, and their contribution could interact with other genetic actions.

Neuroimmune Links

It had long been considered that the immune system was independent of brain functioning, which might have contributed to mental illness not being recognized as being tied to immunologically related processes. This chasm was bridged by several findings. Immune functioning was found to be subject to classical conditioning (Ader, Felten, & Cohen, 1990), and immune activity was influenced by and could similarly promote neuroendocrine, neurotransmitter and other aspects of brain functioning, and could thus influence disease progression (Reiche, Nunes, & Morimoto,

2004). Evidence supporting a link between immune factors and mental illness came from several sources, including reports that activation of the inflammatory immune system (e.g., by immunogenic agents or by stressors) could instigate psychological disturbances, and in some individuals illness (e.g., depression) could be attenuated by anti-inflammatory treatments. At the same time, complex behavioral traits and states, including mental illness and immune-related disorders, involve multiple interactions with other factors, such as innate and adaptive immune functioning (Mangino, Roederer, Beddall, Nestle, & Spector, 2017). Thus, even if immune factors were related to mental illnesses, this would likely involve complex interactions with other factors.

We're only beginning to understand other connections between the immune system and mental illnesses, including the contribution of gut bacteria (microbiota) that affect inflammatory process (Cryan & Dinan, 2012; Sherwin, Rea, Dinan, & Cryan, 2016). The appreciation that multidirectional processes are at work in affecting brain processes and immune functioning, and may contribute to gut-related disorders (Eisenstein, 2016), has led to still other novel approaches and targets being developed in the treatment of mental and physical disorders.

BIOLOGICAL PROCESSES

Autonomic Nervous System

The autonomic nervous system (ANS) is involved in the regulation of body organs and processes over which we don't have voluntary control (e.g., heart, gut, and stomach). This does not imply that the ANS is independent of the CNS. Regulation of the ANS occurs through the medulla oblongata, and other brain regions involved in this capacity include the hypothalamus, which influences energy regulation (e.g., eating), and the amygdala, which contributes to basic emotions, particularly fear and anxiety, that directly or indirectly affect ANS functioning.

The sympathetic component of the ANS, involving the release of epinephrine (adrenaline) and the parasympathetic nervous system (in which acetylcholine is the primary neurotransmitter) act together in the functioning of various organs. These systems are ordinarily in balance with one another, but occasionally environmental triggers will instigate changes so that sympathetic activity predominates, as observed in response to emotionally arousing events that produce an increase of blood pressure and heart rate. In other instances, the compensatory antagonistic system may be overly active, and thus blood pressure and heart rate may become inordinately low. These systems are readily manipulated through pharmacological means, although the focus has been more on modifying epinephrine and norepinephrine activity than that of altering acetylcholine.

Sympathetic activation, together with parasympathetic responses, also contribute to the shunting of blood away from those organs that are not necessary to deal with threats, whereas blood flow increases to organs involved in survival (e.g., those required for elevated physical actions). As we'll see in greater detail later, the release of epinephrine also increases immune reactivity, setting in motion cells from primary immune organs, such as the spleen (Padro, McAlees, & Sanders, 2013) so that they are out doing the work they are meant to do.

The Central Nervous System and Neurotransmission

Experts in cardiovascular, endocrine, or immune system functioning might say that their system is particularly important, unique, and glamorous. Although they're all special and necessary for survival, most people likely see the CNS, which comprises the brain and spinal cord, as the "first among equals". Through billions of neurons that are able to communicate with one another, the brain is responsible for sensory and motor processes, judgments and decision-making, sleep and wakefulness, memory, as well as fundamental drives, such as eating, sleeping, defense, and sexual behaviors. As well, the brain influences activity within the sympathetic and parasympathetic components of the ANS, thereby influencing endocrine and immune functioning.

Our 80 billion or so neurons have numerous dendritic branches, with each having many synapses (points at which one neuron makes contact with another). At birth, each neuron has approximately 2500 connections, increasing to about 15,000 by 2 or 3 years of age, so that more than 1000 trillion synaptic connections are present within the brain. Specific experiences can instigate the formation of synapses, which are strengthened as these experiences are repeated, and then serve to maintain memories of the events. Those synapses that are fired often are strengthened, whereas those that are hardly activated will be pruned (literally, use it or lose it).

Various brain regions serve specific functions, but as we've already said, complex behaviors or pathological conditions typically are determined by neuronal circuits (or systems) comprising several brain regions operating sequentially or in parallel. For example, a syndrome such as PTSD may involve brain regions governing fear (aspects of the amygdala), memory of the trauma (hippocampus and amygdala), and judgment and appraisals (aspects of the prefrontal cortex). Likewise, addictions may involve areas associated with anxiety (aspects of the amygdala), cognitive processes and impulsivity (prefrontal cortex), and reward processes (nucleus accumbens). Indeed, even seemingly basic functions, such as eating or sex, are influenced by systems involving interconnections between multiple brain regions and multiple neurotransmitters (and hormones), although disruption of any single neurochemical can undermine these behaviors.

Approximately 100 different neurotransmitters have been identified, including gaseous substances, which as a group stimulate several different receptors that can lead to diverse consequences. Ordinarily, each neuron can release only one type of transmitter, essentially meaning that it sends messages in only one language. However, a given neuron can be stimulated (activated) by a variety of different neurochemicals (i.e., it understands many languages). The message sent is not only affected by the specific neurotransmitter released and the particular receptors triggered in specific brain regions, as the rate of neuronal firing can have different meanings for an adjacent receiving neuron. Of the many neurotransmitters present within the mammalian brain, the most abundant is glutamate, which serves to activate neurons that they impinge upon. With so many excitatory neurons abounding and being at play concurrently, considerable organization is needed to assure that messages are passed along in a coordinated fashion as well as to diminish noise. To this end, other neurotransmitters have an inhibitory effect on adjacent neurons, with gamma-aminobutyric acid (GABA) neurons being the most abundant.

It is often stated that this or that neurotransmitter has this or that action. In fact, however, most neurotransmitters and their receptors are present across multiple brain regions and may serve in diverse capacities. When triggering neuronal activity within the prefrontal cortex, a given neurotransmitter, such as norepinephrine, may contribute to appraisals and decision-making, whereas the same neurotransmitter in the hippocampus might be involved in the circuitry linked to memory processes, and in still other areas, such as the amygdala, it may dispose individuals toward vigilance and anxiety. Likewise, dopamine in the nucleus accumbens may contribute to reward processes, but in nuclei of the hypothalamus it may promote or interact with several hormones to influence energy regulation, feeding, and sexual behavior. For that matter, when a drug is administered to increase a neurotransmitter in one brain area, as in the case of l-DOPA being administered to increase dopamine in the substantia nigra in order to reduce parkinsonian symptoms, neuronal activation also occurs in other brain regions, such as the nucleus accumbens, which had been operating appropriately all along. The excessive dopamine activity in the latter region may increase the value of rewarding stimuli, and may thus encourage the emergence of reward-related disturbances, such as gambling problems. As it happens, dopamine also plays a fundamental role in decision-making, and manipulating nigrostriatal dopamine levels as animals were engaged in particular tasks could trigger a change in their choice behaviors (Howard, Li, Geddes, & Jin, 2017). As such, dopamine could play a role in both good and bad decisions that are made (e.g., in the case of addictions, including gambling). In considering the actions of various neurotransmitters, it's obviously necessary to distinguish between the multiple brain regions in which they are acting, despite the earlier caution regarding analyses of systems rather than specific brain sites.

Receptor Functions

Upon being released from axon terminals, the neurotransmitter can activate postsynaptic receptors on the adjacent neuron. For any given transmitter, several receptor subtypes can be triggered, each of which can produce different actions and outcomes. Receptors may also be present at the presynaptic end of axons (autoreceptors) that released the neurotransmitter. Upon being stimulated, these autoreceptors serve to tell the neuron to slow down their production of the neurotransmitter. As the amount of neurotransmitter released into the synapse increases, the likelihood of autoreceptors being triggered is increased, and hence neurotransmitter production is self-regulated.

Once a postsynaptic receptor has been triggered, it sets in motion a cascade of intracellular changes. The first messenger, the neurotransmitter, typically doesn't enter the cell (although steroid hormones do) and so a second messenger system is essential for the message initiated by the neurotransmitter to be propagated within the neuron. When a receptor is bound by a ligand (e.g., a neurotransmitter) conformational changes occur (e.g., G-protein activation), giving rise to second messenger production, leading to activation of intracellular processes. Several different second messenger systems have been identified, including a cyclic adenosine monophosphate (cAMP), a phosphoinositol, as well as the arachidonic acid system. The prime function of each second messenger is basically the same, primarily involving sequential activation of intracellular proteins through the process of phosphorylation (addition of a phosphate group, which favors the protein adopting a structural state conducive to signaling). Yet, the presence of particular biochemical features can distinguish one pathway from one another. As well, the actual impact of any particular pathway being activated depends upon the tissue and system in which it is embedded. Thus, they can potentially serve as drug targets for a wide range of conditions (Dhami & Ferguson, 2006). In fact, most drugs used to treat depression, schizophrenia, and Parkinson's disease target elements of the key monoamine pathways, hence they involve many similarities at the intracellular signaling level.

Turnover and Reuptake

Once a neurotransmitter has been released into the synaptic cleft it ought to stimulate nearby receptors, after which it is normally eliminated to preclude excessive receptor stimulation. Enzymes present in the synaptic cleft can degrade some of the neurotransmitter that is present, but many transmitters are eliminated by being transported back into a neuron (to be recycled) through a specialized transporter mechanism. This ecologically friendly process, referred to as reuptake, allows for the transmitter to be available for later reuse. The longer the neurotransmitter remains viable in the synaptic cleft, the greater the opportunity to stimulate a receptor. Pharmacological means of altering the efficiency of the neurotransmitter have been developed wherein the time spent in the synaptic cleft is altered. Degrading enzymes can be inhibited so that transmitter won't be destroyed, or the reuptake of the transmitter back into the neuron can be diminished [e.g., selective serotonin reuptake inhibitors (SSRIs) and selective serotonin and norepinephrine reuptake inhibitors (SNRIs) that are commonly used in the treatment of depression]. Such manipulations are among the obvious ways by which treatments could have their positive effects, but the actual beneficial effects of the drug treatments may arise owing to downstream actions that occur. In fact, the long latency between drug administration and the onset of clinical benefits of antidepressants resulted in the proposition that some time-dependent cellular effects, independent of the levels of monoamines, must be involved. In this regard, neuroplastic changes downstream of the initial actions of SSRIs might underlie symptom remediation in cases of depression or similar stress disorders, and it was thus proposed that drug treatments ought to target growth factors responsible for the neuroplastic alterations (Duman & Li, 2012).

Glial Cells

In addition to neurons, the brain is replete with glial cells, largely comprising astrocytes (astroglia), microglia, oligodendrocytes, and ependymal cells. For the longest time, research attention focused almost exclusively on neurons, whereas glial cells were considered secondary players that acted primarily as support cells for neurons, providing nutrients and taking away debris ("the help"). However, this view turned out to be far too narrow.

Microglia develop from myeloid progenitors in the yolk sac, entering the CNS during early embryonic development. They have several fundamental roles in brain development and during adulthood. Aside from just providing nutrients to neurons, and maintaining ion balances within the fluid outside of the brain cells, microglia promote growth and branching of dendrites. Beyond these actions, they are involved in surveillance, programed cell death, and clearance of apoptotic neurons, and neuronal plasticity (Salter & Stevens, 2017). They are also essential in the repair of neurons within the brain and spinal cord, and through the release of some neurotransmitters (e.g., GABA) they can communicate with neurons. As well, glial cells contribute to the clearance of neurotransmitters from the synaptic cleft, thereby limiting the damaging effects that would otherwise occur owing to a build-up of some transmitters, such as glutamate. Fortunately, unlike neurons that die off with age, microglia renew themselves several times over the course of a lifetime, thus the number of such cells remain relatively steady (Askew et al., 2017).

As much as glial cells have multiple positive effects, they can also engender neurotoxic actions through excessive inflammatory immune processes stemming from excessive cytokine release. The microglia respond to a signal coming from the immune system, notably the complement pathway (Stephan, Barres, & Stevens, 2012), which prompt them to adopt a phenotype that allows them to engage in defensive behaviors, clear pathogens and cellular debris, and also allows them to prune or get rid of unneeded synapses. A disturbance in relation to signaling involving microglia, so that too few (or too many) synapses are pruned, could lead to both developmental and degenerative disorders, including schizophrenia and Alzheimer's (Chung, Welsh, Barres, & Stevens, 2015; Hong et al., 2016). Fig. 1.2 describes some of the many functions that are attributable to microglia.

The discovery that microglia could release inflammatory immune molecules that had traditionally been viewed as being involved in signaling between peripheral immune cells, gave rise to new perspectives concerning processes related to depressive disorders (Rivest, 2009). These inflammatory factors were also implicated in recovery from stroke as well as neurodegenerative disorders, such as Alzheimer's and Parkinson's disease (Litteljohn & Hayley, 2012), and as we'll see in later chapters (Chapter 14: Inflammatory Roads to Parkinson's Disease and Chapter 15: A Neuroinflammatory View of Alzheimer's Disease), inflammatory immune molecules have also been implicated in schizophrenia and in developmental disorders (Chapter 12: Autism and Chapter 13: Schizophrenia). The findings regarding microglia have opened new avenues in understanding illnesses and comorbidities that exist between various disorders, and holds promise for effective treatments becoming available that target inflammatory processes (Salter & Stevens, 2017).

Protecting the Brain (Sometimes)

Astrocytes are able to coax neurons to increase their production of a protein C1q, which is fundamental for pruning to occur (Christopherson et al., 2005). The presence of C1q is essential for early life neural development, and may be essential for synaptic plasticity in the aging brain (Stephan et al., 2013). Reducing C1q or hindering the ability of this

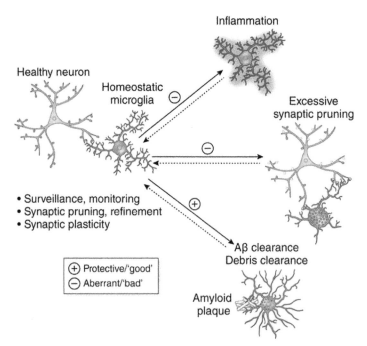

FIGURE 1.2   Microglia constantly survey the brain and spinal cord for potential threats in the form of plaques, infectious agents, damaged neurons, and unnecessary synapses. As microglia are exceptionally sensitive to perturbations within the brain, they are effective in preventing damage. Aside from microglia, glial cells come in three other varieties, each with their own function in relation to neuronal processes (Allen & Barres, 2009). Ependymal cells are not typically considered as microglia, but are included in our description because of their importance in the trafficking between the brain and cerebral spinal fluid.

Astrocytes, the most abundant type of glial cell, are involved in maintaining ion balances within fluid outside of brain cells, and are fundamental in the repair of brain and spinal cord neurons. They are also able to communicate with neurons through the release of particular neurotransmitters (e.g., GABA, glutamate). As well, these cells may be important for repair of brain damage caused by stroke, and may even contribute to the formation of neurons (Kokaia & Lindvall, 2012).

Oligodendrocytes are involved in myelination of neurons (myelin forms the sheath around axons), which is essential for the rapid propagation of electrical signals down the axon.

Ependymal cells (not shown in the figure) line the brain ventricles and the central canal of the spinal cord and contribute to the production and regulation of cerebrospinal fluid (CSF). Junctions between ependymal cells allow for the fluid release across the epithelium so that some exchange occurs between CSF and nervous tissue of brain and spinal cord. Thus, analyses of CSF obtained through a spinal tap, can serve as an observation post of the CNS. *Source: From Salter & Stevens (2017).*

protein can limit the initiation of the complement cascade, which comprises a series of small proteins that upon activation can cascade to activate one another, and contribute to the clearance of pathogens by working together with innate immune cells or antibodies. When this occurs, immune processes are affected as is the age-related decline of cognitive and memory in aging mice (Shi et al., 2015). In line with animal studies, the *C4A* gene, which serves to encode a complement protein downstream of C1q, was linked to synaptic loss and to schizophrenia (Sekar et al., 2016).

Following brain injury or the presence of disease, astrocytes changed dramatically (reactive astrocytosis), and a variety of genes were upregulated (Zamanian et al., 2012). It had been debated whether this reflected a destructive or protective effect, given that astrocytes, like microglia, adopt various plastic phenotypic states (Heppner, Ransohoff, & Becher, 2015). Indeed, a subset of astrocytes were deemed to be destructive as they were associated with the upregulation of genes linked to the inflammatory cascades that were known to disturb synapses. These astrocytes, which are induced by microglia secreting interleukin-1β (IL-1β), tumor necrosis factor-α (TNF-α), and C1q following CNS injury, develop neurotoxic functions, destroying neurons as well as mature oligodendrocytes. In contrast, another subset of astrocytes was associated with enhanced neurotrophic processes, and was thus taken as being neuroprotective (Liddelow et al., 2017). In effect, it seems that at low levels inflammatory cytokines may have neuroprotective effects, but at higher concentrations they may be neurodestructive (Ekdahl, Kokaia, & Lindvall, 2009). These negative actions could come about through the promotion of excessively strong inflammatory responses or through the production of free oxygen radicals that promote neurodestruction, thereby giving rise to neurodegenerative disorders (Litteljohn & Hayley, 2012).

Some exogenous and endogenous factors that could potentially act in a pathogenic capacity might not readily reach the brain and spinal cord owing to the presence of endothelial cells that comprise part of the blood brain barrier (BBB). In this regard, cytokines in the periphery that are released when specific immune cells are activated might not reach the brain readily owing to their large size. However, cytokines can reach some aspects of the brain where the BBB is absent or porous, such as the area postrema, or they might access the brain through an active transport system (Banks, 2016). Under some conditions (such as in response to traumatic brain injury, or perhaps in response to stressors) the BBB may be compromised, allowing for greater entry of cytokines, which could potentially have damaging effects.

Neurotrophins

One of the most remarkable attributes of brain neurons is their plasticity (i.e., their ability to modify their connections, essentially amounting to the rewiring of the brain). Dendritic branching (arborization) and the formation of synapses (synaptogenesis) is promoted by experiences, and acquiring new stimulus—stimulus or stimulus—response associations is due to new synapses being formed, and then strengthened with use. In addition to being involved in learning and memory processes, neural plasticity is essential in recovery from stroke and head injury, and contributes to emotions and disturbed mood states, such as depression (Duman, 2014). For synaptic plasticity to occur, neurotrophins (growth factors) are necessary. Particular attention in this regard has focused on brain derived neurotrophic factor (BDNF), although other neurotrophins, such as fibroblast growth factors (FGF-2) and vascular endothelial growth factor (VEGF) may also be involved (Kirby et al., 2013). It is also well accepted that neurtrophins can stimulate neurotransmitter release, and interact with cytokines in promoting mental illness (Audet & Anisman, 2013). Table 1.2 provides a listing of several neurotrophins and their functions.

Hormones

Much like neurotransmitters, hormones can also act as signaling molecules within the brain and in the periphery. In the periphery, hormones are manufactured by a variety of glands and transported to distant organs to regulate physiology and behavior. Some hormones are released directly into the blood stream (endocrine hormones), whereas others are secreted into a duct and flow either into the bloodstream, or spread from cell to cell by diffusion (exocrine hormones). As well, neuropeptide

TABLE 1.2　Neurotrophins and Their Function

Neurotrophin	Biological Effect	Outcome
Brain-derived neurotrophic factor (BDNF)	Support survival of neurons; encourages growth and differentiation of new neurons; promotes synaptic growth	Influences memory processes, stress responses, mood states
Basic fibroblast growth factor (bFGF or FGF-2)	Involved in neuroplasticity; formation of new blood vessels; protective actions in relation to heart injury; essential for maintaining stem cell differentiation	Contributes to wound healing; neuroprotective; diminishes tissue death (e.g., following heart attack); related to anxiety and depression
Nerve growth factor (NGF) and family members Neurotrophin-3 (NT-3) and Neurotrophin-4 (NT-4)	Contributes to cell survival; growth and differentiation of new neurons. Fundamental for maintenance and survival of sympathetic and sensory neuron; axonal growth	Survival of neurons; new neuron formation from stem cells; related to neuron regeneration, myelin repair, and neuro-degeneration. Implicated in cognitive functioning, inflammatory diseases, in several psychiatric disorders, addiction, dementia as well as in physical illness, such as heart disease, and diabetes
Insulin-like growth factor 1 (IGF-1)	Secreted by the liver upon stimulation by growth hormone (GH). Promotes cell proliferation and inhibits cell death (apoptosis)	Secreted by the liver upon stimulation by growth hormone (GH). Promotes cell proliferation and inhibits cell death (apoptosis)
Vascular endothelial growth factors (VEGF)	Signaling protein associated with the formation of the circulatory system (vasculogenesis) and the growth of blood vessels (angiogenesis)	Creates new blood vessels during embryonic development, encourages development of blood vessels following injury, and creates new blood vessels when some are blocked. Muscles stimulated following exercise. Implicated in various diseases, such as rheumatoid arthritis, and poor prognosis in relation to breast cancer

hormones, such as β-endorphin and dynorphin, can be secreted by neurons and activate receptors present within the CNS.

Hormones are fundamental in metabolic processes, eating and energy balances, stress reactions, cell growth and cell death (apoptosis), sexual characteristics and behaviors, and they can have profound actions on brain processes, such as neurogenesis (Mahmoud, Wainwright, & Galea, 2016). Beyond these actions, hormones are involved in cognitive functioning, including learning and memory processes, and may contribute to various mood and motivational states (de Kloet, 2014), and

contribute factor to sex differences that have been related to illnesses, such as heart disease and autoimmune disorders.

It is of particular significance, that several different hormones influence various biological and behavioral processes as well as aspects of immune functioning (e.g., Frank, Watkins, & Maier, 2015), and drugs that affect hormone functioning have been used to treat a variety of immune-related illnesses. In later chapters the influence of various hormones will be discussed in more detail, including their roles in a variety of stressor and immune-related illnesses.

IMMUNE PROCESSES

The primary job of the immune system, as discussed in Chapter 2, The Immune System: An Overview, is to recognize what's part of us and what isn't, and to protect us from foreign particles that could create harm. During prenatal development the immune system learns "what us comprises" and by extension, what isn't part of us. In addition to this innate immunity, over the course of postnatal development the organism will also acquire immunity, in which it learns through experiences about foreign particles, including bacteria and viruses. Thus, the immune system ought to protect us from numerous invaders, although the fact that we get sick with many illnesses, including those in which the immune system turns on the self (autoimmune disorders), speaks to the limitations of these systems. Still, with the different types of immune cells present (neutrophils, monocytes, macrophages, natural killer cells, T helper cell, cytotoxic T cells, and B cells) and the ability to destroy foreign particles floating around in the body or those that have gained entry to healthy cells, we ought to be pretty well set to deal with multiple challenges that can be encountered. As we'll see though, bacteria and viruses don't sit around passively, and instead go about looking for ways to get through our natural defenses. In fact, under certain conditions, bacteria can subdivide into two distinct populations, one of which is especially virulent, and the "memory" related to this hyper-virulent state can be maintained for several generations (Ronin, Katsowich, Rosenshine, & Balaban, 2017).

Contrary to earlier misconceptions, we know that immune signaling factors (cytokines, chemokines) can also affect brain functioning, and by virtue of several hormonal and neurotransmitter alterations, may influence mood disorders, psychosis, and neurodegenerative disorders. Conversely, many hormones and neurotransmitters are known to affect immune functioning (Blalock, 2005). Given the multidirectional exchanges that involve the immune system, it follows that factors that influence immune activity (e.g., life-style, obesity, sleep) may come to affect psychological state.

The prototypical stress hormone cortisol (corticosterone in mice and rats) can have profound actions on immune functioning, which can affect physical and mental well-being. At physiological levels (those ordinarily seen in the absence of a pharmacological treatment), corticoids can limit immune activation (e.g., preventing excessive immune activity in response to an acute stressor), and at pharmacological levels (which are markedly elevated through exogenous factors, such as drug treatments) immune functioning can be drastically suppressed. The latter treatments have been used to act against illnesses in which the immune system attacks the self (autoimmune disorders, such as rheumatoid arthritis, lupus erythematosus). In addition to glucocorticoids, through their actions on immune functioning, estrogen and progesterone can likewise affect autoimmune disorders (Hughes & Choubey, 2014).

As we'll see in Chapter 6, Stress and Immunity, numerous psychological factors can affect immune activity, and hence might influence physical illnesses. Beyond these peripheral actions of stressors, brain cytokine variations can also be engendered by a variety of stressors, and may interact with immunogenic agents in synergistically increasing brain cytokine functioning (Gibb, Al-Yawer, & Anisman, 2013). These brain cytokine (e.g., IL-1β) changes, such as those that occur in the hippocampus, have been implicated in fear learning processes (Jones et al., 2018), whereas variations of IFNγ may be related to memory processes (Litteljohn, Nelson, & Hayley, 2014). It also appeared that by undermining BBB integrity, particularly among stress-sensitive mice, access of cytokines to the brain parenchyma was increased, hence influencing the development of illnesses, such as depression (Menard et al., 2017).

MICROBIOTA

Microbiota that colonize the gut appear to be regulators of immune processes, and various microbial species can have dramatic effects on brain functioning (some direct and some through their interactions with immune constituents). The involvement of gut microbiota in health and disease has quickly become the darling of the neuroscience world having been implicated in numerous disease states. More than half the published reports have been conducted in animals, and as in many health-related fields, translation from the mouse to the human can be difficult, especially given several differences related to mucosal and systemic immunity and disease pathogenesis across species. This aside, several studies indicated that the links between the microbiome and health are not simply correlational, as manipulating gut bacteria can have health benefits in rodents, although comparable studies in humans are limited.

Enteric Nervous System

Until very recently, little attention was devoted to the connections between the gut and the brain, even though the "enteric nervous system" (the nervous system associated with the gut) shares several transmitters that are present in brain. The cells and neuronal composition of the gut is highly dynamic, and like brain cells, it is subject to neurogenesis as well as apoptosis (Rao & Gershon, 2017). The enteric system, which courses through tissues that line the colon, small intestine, stomach, and esophagus, can influence brain functioning, just as the brain can influence gut

functioning. Communication between the gut and brain can occur through stimulation of the vagus nerve, which extends from the viscera to the brain stem. As well, enterochromaffin cells that compose 1% of the gut epithelium, serve as chemosensors that respond to metabolic and homeostatic signals, which through the release of serotonin, inform the CNS of changes that are occurring (Bellono et al., 2017). The enterochromaffin cells are exquisitely sensitive to stimulation, and might be one of the first line defenses against threats and injury. These cells may be overly sensitive, and may contribute to inflammation present among individuals experiencing inflammatory bowel disease. In this regard, serotonin within the gut can stimulate 5-HT$_4$ and 5-HT$_7$ receptors, which promote anti and proinflammatory actions, respectively (Spohn & Mawe, 2017). In light of the many common elements between the enteric nervous system and the brain, and the bidirectional communication that occurs between them, the position was adopted that the enteric nervous system might also contribute to several neurological disorders (Rao & Gershon, 2016).

Gut Microbiota

Functioning of the enteric nervous system can be affected by the presence of bacteria present within the gut, and can profoundly influence brain process that determine psychological well-being (Hyland & Cryan, 2016). Likewise, microbiota related to nutrition markedly influence metabolic processes, and could thereby affect physical health (Sonnenburg & Bäckhed, 2016). A variety of microorganisms (e.g., bacteria, archaea, fungi, viruses and eukaryotes, collectively referred to as microbiota) are present in huge numbers, several trillion, in fact, comprising hundreds of species, are present within the gut, as well as on our skin, in nasal passages, mouth, and between our teeth. Cooperation occurs across bacterial species (Rakoff-Nahoum, Foster, & Comstock, 2016), and as these microbes colonize different parts of the body, new species evolve, but behave harmoniously with one another (Silverman, Washburne, Mukherjee, & David, 2017).

There is the view that gut microbes have been fundamental in human evolution. The microbial communities that exist within the gut and the rest of the body are largely dependent upon or influenced by the resources available at these sites (Fisher, Mora, & Walczak, 2017). When environments change, so do aspects of the microbiome, which can then influence the host. Those aspects of the microbiome that are most useful for us will be maintained and those least useful will fade away (Shapira, 2016). Of the many types of gut bacteria present, some can have positive actions, but others may negatively influence health. In general, microbiota can encourage resistance to colonization by pathogenic species, and when these gut bacteria are eliminated or diminished, such as when mice are bred in a sterile environment or are treated with antibiotics, susceptibility to pathogenic bacteria is elevated (Sassone-Corsi, & Raffatellu, 2015). Other forms of bacteria, in contrast, promote the expansion and virulence of pathogens (Cameron & Sperandio, 2015).

Co-evolution occurs wherein we influence our microbiome, and our microbiome reciprocally affects what is carried forward across generations of bacteria. It seems we've made a deal with our microbiota; we provide them with a place to live and they help us in multiple ways, ranging from digestion, eliminating toxic xenobiotics (molecules that are foreign to the body), and influence immune functioning. There is still a great deal that isn't known concerning the interactions that occur between microbiota and xenobiotics, but it is fairly certain that gut microbes contribute significantly to the modification of a variety of dietary components and thereby affect health. When more is understood regarding the functioning of gut microbiota and the genes that control them, it may provide new arsenals to deal with disease vulnerabilities.

In view of gut bacteria being fundamental for metabolic processes and efficient immune functioning, alliances form between the many microbial species that inhabit the gut. This balance is necessary so that tolerance can develop to innocuous antigens (e.g., foods that are not contaminated), thereby limiting unnecessary immune activation (Belkaid & Hand, 2014; Hyland & Cryan, 2016). Gut bacteria are at once vulnerable to being disturbed and hardy in the face of challenges. When bacteria suffer the loss of essential enzymes, others seem to take on the load, even if this means creating a patchwork of multiple components.

The microbiome composition varies considerably across individuals, likely being determined by the host's (us) genetic composition, together with hormonal, metabolic, and immune factors. As well, microbiota are subject to change with environmental influences, including foods consumed, exercise, and other life-style factors, and is even affected by changes of our circadian clock. For that matter, some gut bacteria have their own circadian clocks, which are influenced by melatonin variations (Paulose, Wright, Patel, & Cassone, 2016). Yet, following microbial changes, this system seems to reset itself, returning to the state it had been in prior to the perturbations encountered.

There has been an explosion of studies suggesting that microbial disturbances may contribute to numerous diseases. So many, in fact, it has become difficult to link particular illnesses with specific bacterial elements, as well as whether causal connections exist between bacterial agents and illnesses (and the directionality of these relationships). It is supposed that various gut bacteria need to be in balance with one another, and when these balances are disturbed (dysbiosis) illnesses may evolve for a variety of reasons, including the disruption of the tight coordination that exists between the microbiome and immune processes (Honda & Littman, 2016; Thaiss, Zmora, Levy, & Elinav, 2016). Not only is dysbiosis connected to greater vulnerability to the impact of pathogens, but it might also contribute to autoimmune disorders, such as multiple sclerosis (MS). Specific bacterial taxa that were elevated in MS patients were able to increase inflammatory responses in peripheral blood mononuclear cells, whereas those that were diminished in MS patients, activated processes that served in an anti-inflammatory capacity (Cekanaviciute et al., 2017). Moreover, when certain gut microbiota (*Akkermansia*) from twins discordant for MS were transferred to mice, signs of MS, including disturbed immune regulatory factors, were apparent only in those that had received the bacteria from the twin with MS (Berer et al., 2017).

Aside from the gut bacteria-immune link, it is clear that dietary elements are converted by gut bacteria into signals that affect adipose tissue, and when dysbiosis exists, functioning of the intestine, liver, lungs, cardiovascular system and brain can be affected, and pathology can develop at these sites (Schroeder & Bäckhed, 2016). Some of these disturbances are shown in Fig. 1.3 and are described in greater detail in several excellent reviews (Lynch & Pedersen, 2016; Schroeder & Bäckhed, 2016). In view of the links between gut microbiota and foods consumed, dietary interventions could be expected to have health benefits through their actions on gut bacteria (David et al., 2014).

Although gut bacteria have been implicated in many diseases in humans, the available data have largely been correlational, and there have been only a few syndromes that have been causally tied to microbial factors (e.g., C. difficile related diarrhea; inflammatory bowel disease). To some extent, the analysis of connections between microbiota and illnesses is complicated by the fact that there is some uncertainty as to what constitutes a healthy microbiome, especially given the pronounced variability that exists across healthy individuals. There is also evidence indicating that diversity of the gut microbiome is appreciable across communities and cultures, and differs widely between Euro-Americans and American-Indians (Sankaranarayanan et al., 2015), and thus developing specific remedies for particular illnesses could prove difficult. At the same time, core bacteria need to be present for well-being to be maintained (Flint, Scott, Louis, & Duncan, 2012), and deep sequencing of gut microbiota have revealed some of the "normal" factors that were linked to well-being (Zhernakova et al., 2016). Moreover, genome guided microbial communities can be established that could act against specific pathogens (Brugiroux et al., 2016). This said, at the moment it is unlikely that proper diagnoses will be obtained based *solely* on a bacterial profile, and this variable would be best suited as one of many that is incorporated in the diagnoses and selection of treatment strategies. Assessment of microbiota dysbiosis among critically ill patients is another thing entirely as it could be linked to organ failure and could provide clues in the development of intervention strategies (Jacobs, Haak, Hugenholtz, & Wiersinga, 2017). Furthermore, as we'll see, microbiota and immune system functioning reciprocally influence one another, and are tied to a huge range of illnesses involving brain functioning and several physical illnesses, and like inflammatory processes, microbial factors might contribute to the comorbidity that is so often apparent with several disease conditions.

PERSONALIZED TREATMENT STRATEGIES (PRECISION MEDICINE)

Regardless of the illness being considered, some individuals tend to be highly vulnerable

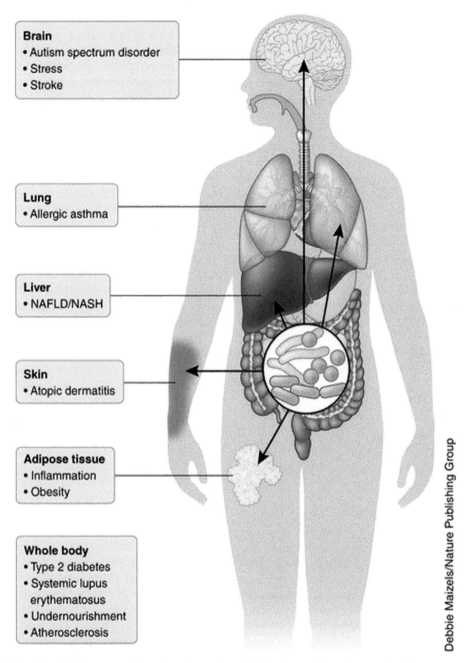

Brain
- Autism spectrum disorder
- Stress
- Stroke

Lung
- Allergic asthma

Liver
- NAFLD/NASH

Skin
- Atopic dermatitis

Adipose tissue
- Inflammation
- Obesity

Whole body
- Type 2 diabetes
- Systemic lupus erythematosus
- Undernourishment
- Atherosclerosis

Debbie Maizels/Nature Publishing Group

FIGURE 1.3 Alterations in composition, diversity, and metabolites derived from the gut microbiota are associated with diseases affecting different organs of the human body. Evidence for a causative role of the gut bacteria is strongest in metabolic disease. Not shown in the figure is that gut dysbiosis has also been linked to inflammatory processes and to numerous mental illnesses. *Source: From Schroede & Backhed (2016).*

(high risk), whereas others seem to be relatively resilient, being able to overcome symptoms or might actually not develop the illness at all. Many individuals may likewise encounter events or environments (e.g., chronic stressors) that could promote illnesses, but the nature of the illness experienced varies broadly. Individuals differ with respect to their "weak links" that might have been acquired through experience or through inherited genes, and the nature of the illness that emerges will vary based on these risk factors. Marked individual differences are also seen with respect to the efficacy of treatments. Although some individuals overcome illnesses readily following treatment, there are those who seem to recover slowly, if at all. Once again, numerous factors might contribute to these individual differences, including genetic constitution that affects other biological processes, together with psychosocial influences and previous experiences in dealing with illness.

Treatments Based on Endophenotypic Analyses

In the case of some illnesses, a physician can pretty well count on a positive response occurring with a particular treatment, but in other instances, things aren't nearly as predictable. Individuals can have similar illness symptoms, yet might be differentially responsive to treatments. Some individuals being treated for depressive illness, for instance, may not respond positively to a particular drug, but exhibit diminished symptoms in response to an alternative treatment. A second individual, in contrast, might respond to the first treatment, but might not have responded positively to the alternative. Moreover, individuals may respond differently over time following some prescribed treatment. In the case of Parkinson's, some patients respond well for

many years to dopamine replacement using l-DOPA, whereas others experience the loss of drug efficacy within a mere three or 4 years. Further to this point, symptom clusters that co-occur can also influence treatment regimens differently. Indeed, about half of Parkinson's patients are afflicted with comorbid depression and similarly, many MS patients experience comorbid depression and anxiety and again, these symptoms can evolve differently over time. Still in other instances people may receive superficially similar diagnoses (e.g., at one time all breast cancers were viewed as being the same as one another), but the nature of these conditions may actually be very different from one another (e.g., with respect to genetic factors and/or whether they are hormonally responsive). Predictably, the effectiveness of treatments will also differ based on such considerations. Accordingly, if the characteristics of the illness are defined appropriately, more effective treatment strategies can be applied.

A treatment approach was developed some time ago in response to the repeated failures to treat patients effectively. This method, the endophenotypic approach, involved connecting specific genes to disease conditions or specific features of an illness (Gottesman & Gould, 2003). The endophenotypes involved comprise the "measurable" aspects that link genetic factors and illness symptoms. These genetic contributions may themselves be linked to endocrine, neurotransmitter, immunological, microbial, neuroanatomical, neuropsychological, cognitive, or behavioral factors. Moreover, if these factors are also tied to specific treatments, then the selection of effective treatments can be facilitated. Theoretically and practically, genetic or neurobiological factors can serve as biomarkers for illness development, the recurrence of illness, and treatment approaches. There are occasions where an endophenotypic approach may not be possible or practical. The alternative then might be one of focusing on dimensions of illness in broader terms, especially as particular symptoms often fall in clusters (e.g., sets or clusters of either neurovegetative or cognitive features). Both approaches could also incorporate several risk factors (premorbid conditions as well as stressor experiences) and biomarkers of illness in the endophenotypic analysis (Fig. 1.4).

Linking specific genes to psychological disturbances, which then determines treatment strategies, is not an entirely new notion. Precisely this approach was used in treating some forms of cancer and heart disease, but only recently was this proposed for use in mental illnesses. The search for biomarkers of mental illness has recently been going on vigorously, encompassing genetic markers as well as other biological substrates and specific symptoms presented that offer clues for the nature of the disturbance. The personalized or precision medicine approach has taken on increasing allure, and has posed a significant challenge to traditional methods related to diagnosis and treatment.

The American Psychiatric Association had over a number of years released several iterations of the *Diagnostic and Statistical Manual of Mental Disorders* (DSM), which were meant to provide a detailed and systematic description of the symptoms of various psychiatric disorders in the hope of facilitating diagnosis and treatment of a broad set of mental illnesses. Despite the importance of having psychiatrists work from a common playbook, the Fifth Edition of the DSM, published in 2013, was met with considerable resistance and criticism in some quarters, and those who viewed classification systems of illnesses as inherently counterproductive were especially vigorous in their dissent, pushing for a personalized treatment approach.

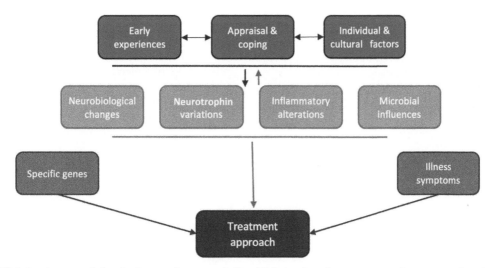

FIGURE 1.4 An expanded endophenotypic approach. In addition to focusing on genes and gene mutations in relation to illness symptoms, this approach considers the influence of stressors, early experiences and cultural factors in moderating neurobiological processes, as well as epigenetic factors, that culminate in disease processes and predict their most effective treatment.

If you talk to God, you are praying; If God talks to you, you have schizophrenia. If the dead talk to you, you are a spiritualist; If you talk to the dead, you are a schizophrenic
(Szasz, 1973).

The usual approach to diagnosis, including the use of the DSM criteria, have for many years encountered stubborn resistance, and some early theorists found the entire classification system troublesome. Thomas Szasz, who made a mark in his time, but has since been relegated to the attic, argued against the entire notion of diagnosing "illnesses that people have" as opposed to "symptoms people exhibit." Szasz went so far as to suggest that the labels that are applied to patients, such as "schizophrenia," represent an unprincipled concept, something used for convenience, rather than having any benefits for the affected individual. For that matter, Szasz even maintained that psychiatry was a pseudoscience with aspirations to become a genuine medically based scientific entity. In his 1961 book, The Myth of Mental Illness, Szasz described mental illness as a myth, suggesting that individuals get labeled, marginalized by society, and then preyed upon by a coercive system monopolized by drug companies.

While Szasz was a key figure of the anti-psychiatry movement of the 1960s and 1970s,

his was not an isolated struggle. Others also rejected the medical model of mental illness, and questioned the use of pharmacological approaches in the treatment of schizophrenia and were strong in the belief that illnesses were born through experiences (Laing, 1960). During those days, many opponents of psychiatry saw it as controlling and repressive, ascribing certain conditions to the category of mental illness, stigmatizing large numbers of people, only to reverse themselves years later. During this period, pharmacotherapy was making the first inroads into psychiatric treatments, and there were many opponents of the psychiatric/medical model, including Cooper who published Psychiatry and Anti-psychiatry (1970), as well as other notables, such as Lacan, Basaglia, Lidz, Arieti, and Foucault. These theorists, however, didn't hold a candle in relation to the negative impact of Kesey's book, and subsequent movie "One Flew over the Cuckoos nest" in promoting distaste for the then current medical approaches. The attack on the medical model in psychiatry and drug therapies had abated for some time, but was rejuvenated with repeated reports of failed drug treatments, and it was even suggested that this science, if it is a science, was entirely unsound.

Dissatisfaction with the DSM, and the approach to treatment that it encompassed, prompted the NIH to offer a new multilevel framework for the diagnosis and treatment of mental illnesses. This comprised the RDoC, which followed from the endophenotypic approach (Cuthbert & Insel, 2013). The RDoC was conceived as being a dimensional system, comprising several layers of analysis ranging from neuronal through to behavioral processes. The very nature of the diagnostic process was,

in fact, turned on its head. Instead of beginning by defining an illness and then attempting to treat it, the RDoC approach begins by defining the behavior (symptoms) presented by patients, and then linking these to brain process in an effort to identify links to pathological conditions, which could inform treatment. The RDoC framework, provided in Table 1.3, comprises five domains:

(1) Negative Valence Systems (i.e., response to aversive threats or events)

TABLE 1.3 The RDoC framework, listing the units of analysis (X axis) and the behavioral domains (Y axis) provides a matrix that can potentially use for predicting psychological disorders, and based on the matrix of features can also serve to predict the efficacy of treatments and illness recurrence

The RDoC Matrix[a]								
	Unit of Analysis							
	Gene	Molecules	Cells	Circuits	Physiology	Behavior	Self-Report	Paradigms
Domain (Construct)								
Negative Valence Systems								
Acute threat								
Potential threat								
Sustained threat								
Loss								
Frustrative nonreward								
Positive Valence Systems								
Approach motivation								
Initial responsiveness to reward								
Sustained responsiveness to reward								
Reward learning								
Habit								
Cognitive Systems								
Attention								
Perception								
Declarative memory								
Language behavior								
Cognitive control								
Working memory								
Systems for Social Processes								
Affiliation and attachment								
Social communication								
Perception and understanding of self								
Perception and understanding of others								
Arousal and Regulatory Systems								
Arousal								
Circadian rhythms								
Sleep and wakefulness								

[a] See: http://www.nimh.nih.gov/research-priorities/rdoc/nimh-research-domain-criteria-rdoc.shtml#toc_matrix
Front Hum Neurosci. 2014; 8: 435. Published online Jun 20, 2014.

(2) Positive Valence Systems (responses to stimuli or events that are perceived to be rewarding)
(3) Cognitive Systems that entail those related to cognitive control, attention, perception, memory, response selection
(4) Social Processes Systems (attachment, social communication, self-perception, perception, and understanding others)
(5) Arousal/Regulatory Systems (arousal, sleep)

On the other axis are the diverse units of analysis that may be tied to the behavioral signs. These include genes, molecules, cells, neural circuits (determined through imaging methods or well-established circuits associated with particular behaviors), physiological processes (e.g., cortisol, heart rate, startle reflex), behaviors, and self-reports (interviews, questionnaires). As in epidemiological analyses, this matrix can provide clusters that link particular attributes or symptoms of a syndrome to specific units of measurement, such as genes, hormones or circuits. Should the features of an illness map onto the efficacy of different treatment methods, then this would ultimately prove to be ideal in predicting which of several treatment options would be best for individual patients. In this regard, many illnesses are comorbid with one another (e.g., depression predicting later heart disease), early signs of an illness could also serve as the proverbial canary in the coal mine, informing physicians of further health risks so that interventions are initiated (Anisman & Hayley, 2012a).

The RDoC approach has received considerable support in many quarters, but it has also met resistance. Among other things, psychiatric diagnoses aren't as stable as one would hope, and as indicated earlier, one condition may morph into another (e.g., the conversion of major depression into bipolar disorder). As well, because of its complexity there may be too much room for measurement error, thereby affecting diagnoses. Others have argued that too much emphasis was placed on biological measures, and that some behavioral phenotypes were not necessarily a direct result of biological dispositions. Moreover, even if they were, they might vary with many other variables, so that identifying all of these might be virtually impossible.

Considerable attention has been devoted to identifying specific genes that might serve as causal factors or markers of many disease conditions, but many problems are likely to be encountered, even with diseases that are seemingly less complicated than those involving the brain (Joyner, 2016). The assumption that genetic-based research approaches will take us many steps forward in treatment is in some instances perfectly reasonable, and it could drive precision medicine and identification of new targets for treatment. The cornerstone of precision medicine has been that large population studies might be able to identify specific associations between genotypes and phenotypes, but with only a few exceptions, these association studies have a limited capacity to predict specific phenotypes in any given individual (Khoury & Galea, 2016). Indeed, as more and more information has come forward regarding the link between genes and disease, it has become eminently clear that the initial hopes might have been a tad optimistic, although this doesn't necessarily imply that precision medicine can't take forward steps. In this regard, a large-scale analysis based on whole exome sequencing (in which sequencing is performed of all the expressed genes in a genome) which was linked to individual's electronic health records, was able to identify some gene variants that might be tied to illnesses (Abul-Husn et al., 2016; Dewey et al., 2016). Yet, as indicated earlier, given that most disease states involve multiple gene variants or gene x environment interactions, these approaches can go only so far. Added to this is that several distinct gene variants can share common biological functions, and thus affect the same disease conditions (Li et al., 2016).

Each of hundreds of genes may contribute to disease conditions, individually accounting for a very small portion of the variance, making it exceptionally difficult to target specific pathways for treatment or even as useful genes for biomarkers. Aside from complex illnesses involving multiple genetic contributions, many illnesses may vary in relation to epistatic interaction, referring to one gene having effects on the actions of a second gene or that environmental and social factors may interact with the actions of a gene. We'll see that this is precisely the case for many disorders, and understandably, identifying these interactions is unwieldly even in large scale studies, making a personalized approach orders of magnitude more difficult. By example, using a large cohort of individuals who self-reported whether or not they had been depressed, 17 genetic variations were identified that might contribute to the illness (Hyde et al., 2016). But, the contribution of some genes to depression may be negligible if interactions with varied experiences, such as current or early life stressful experiences, aren't considered. The very same issues are faced in dealing with other disorders, such as addictions, obesity, hypertension, some cancers, and most certainly, other mental illnesses. It ought to be mentioned, as well, that even leaving genetic factors aside, personalized treatments can still have enormous benefits. For instance, features of illness, such as sex and body-mass index, could be used to predict the efficacy of antidepressant treatments (Green E et al., 2017).

Yet another issue that warrants consideration is that many genes may have pleiotropic actions. That is a particular gene can influence two or more phenotypic traits. These pleiotropic actions can occur in parallel or in series. In the latter instance, a given gene can have a particular effect that leads to biological outcomes and several downstream changes that culminate in a pathological condition. Thus, treatments that target one or more steps on this string of biological factors could potentially modify the pathological conditions. However, a parallel pleiotropy could be present, in which a gene comes to affect different neurobiological pathways directly or indirectly and hence provoke more than a single pathology, but these pathologies may be independent of one another. Once more, this makes it difficult to identify which treatments may be best to treat one or another condition.

It is widely accepted, of course, that prevention of illness may often be more effective than treating illnesses once they have occurred. As indicated by Khoury and Galea (2016), a fundamental benefit of a precision medicine approach is that it might lead to detailed descriptions of risk factors for diseases, and hence programs can be created to prevent illnesses. Here, however, we face the problem that even when individuals are aware of the problems they could encounter (e.g., as a result of smoking, exposure to sun), they often do not adopt behavioral changes, even when they are part of a high-risk group (Hollands et al., 2016).

This brings us to a critical factor that is incorporated into the RDoC approach that focuses on mental illness, but is often overlooked in some aspects of precision medicine involving other disease conditions. As we'll see repeatedly, for many illnesses, the amount of variance accounted for by genetic factors is considerably smaller that the contribution of behavioral, social, and experiential factors. Too often, however, these are not considered adequately in precision medicine. To what extent do diet, stressful experiences, as well as prenatal or early postnatal factors, and many other psychosocial variables, come to affect disease conditions, and does knowing the answers to this have implications for illness prevention and treatment? As we'll see in ensuing chapters, it most certainly does, and consequently having this information may inform best practices.

There are, indeed, a constellation of factors beyond those described in Table 1.3 that are important considerations in the evolution of an illness and the effectiveness of treatments. Kirmayer and Crafa (2014) made the important point that the RDoC approach fails to consider culturally and historically based phenotypes. It was argued that "normality" in relation to mental illness is defined on the basis of what is statistically common or average in a population. Yet, what might be considered "normal" in one culture may not be in a second. They also indicated that the interactions that occur between biological, developmental, and psychosocial processes that define normal versus pathological conditions are driven by cultural factors and may be dependent on the context in which measurements are made (e.g., in relation to Indigenous Peoples vs Euro Caucasians).

In considering treatment strategies, it is also important to ask whether brain neurobiological processes are reciprocally related to social support and trust, and thus will contribute to psychological disturbances and the response to treatment. For instance, do Indigenous People trust non-Indigenous physicians or do African American patients have greater trust in African American physicians relative to Caucasian health providers. Clearly, numerous issues are relevant to how patients fare, but often these considerations are given short-shrift.

CULTURE IN RELATION TO PERSONALIZED MEDICINE

Culture, which can be influenced by social norms and values, geographical conditions, and environmental exposures, may influence the development and course of disease, and may have a profound impact on the efficacy of treatments. It is broadly thought that culture is rooted in ancestry (systems of knowledge), the collective history (including trauma), and evolving environmental and social contexts (climate change; colonization; migration). In essence, culture is a dynamic construct that changes with the passage of time (cultural evolution), being modified by a constellation of factors, such as social structures, relationships, beliefs, activities, diet, multiple environmental factors, and the adaptations that these require. The significance of culture varies based on the extent to which individuals see themselves as reflecting their cultural norms. Importantly, as well, is their sense of belonging to the group, their feelings of pride in their group, and collective esteem it provides (vs, shame, resentment, or anxiety). Together, these aspects of their identity will determine whether individuals adopt their culture as a means to deal with adversity (e.g., in seeking social support in the mobilization of collective actions). In essence, cultural factors are instrumental in shaping an individual's core identity and hence affect resilience in the face of numerous challenges. Thus, in considering a personalized approach to deal with pathologies, cultural factors ought to be a key ingredient (Matheson, Bombay, Dixon, & Anisman, 2018).

The notion of personalized medicine, despite its presumed limitations, has advanced several steps in recent years. The "high-definition medicine approach,", relies on newly developed technological methodologies (e.g., DNA sequencing, advanced imaging, physiological and environmental monitoring), behavioral tracking, and information technologies, to promote sophisticated treatment strategies (down the road, these might include genome editing, cellular reprograming, tissue engineering) (Torkamani, Andersen, Steinhubl, & Topol, 2017). This effort is obviously creative, but it will require concerted government funding for this to be implemented. Getting simple precision medicine off the ground has been enormously difficult, this next step will be much more of a callenge.

Yet another suggestion that goes beyond straight forward precision medicine comprises P4 medicine, which entails four components, predictive (indicating early predictors illness), preventive (intervening before an illness appears), personalized (identifying markers that are most effective in treatment), and participatory (such that patients have a voice in their treatments, as do physicians and health workers) so that policy and implementation of strategies are enhanced (Hood & Auffray, 2013).

Aside from such efforts, the precision medicine approach, which focuses on individual patients, has been extended to apply to whole communities through the application of public health interventions (Khoury, Iademarco, & Riley, 2016). The medical problems stemming from industrialization, overcrowding, various pollutants, new emerging diseases, and many other factors, have grown far too large and too broad to be effectively handled with current approaches. Thus, much greater attention ought to focus on prevention (rather than just treatment), and health efforts need to be addressed at the community level.

Despite the current limitations, with the use of a personalized medicine approach (together with the modifications offered by Khoury, Topol, and Hood), it may be possible at some point to develop appropriate treatment strategies for each individual. Important to this understanding is that genes likely don't link with one another in any simple fashion. In addition to a regulatory gene influencing the actions of a second gene, networks exist that

may be orchestrated by a common element or set of elements, and diverse regulatory elements may interact with one another (Waszak et al., 2015). Mental illnesses and their links to inflammatory processes might represent the interactions among multiple gene networks that control varied symptoms whose presence might provide clues for effective treatments. These gene networks, of course, are likely connected to multiple psychosocial determinants (including context and environmental contributions) that additively or interactively operate with genetic factors in determining the development of various illnesses and influence the best treatment strategies to adopt. This optimism, however, needs to be tempered given the complexity of the issues being addressed. It may simply be that at the moment the issues are too complex (and too expensive to adopt), and is still not workable within the primary care context.

CONCLUDING COMMENTS

Mental illnesses are typically viewed as those that involve disturbances regarding how a person feels, thinks, perceives, or acts. These disorders range from disturbances of mood, anxiety disorders, PTSD, schizophrenia, personality, sleep, eating, and sexual disturbances, as well as developmental and conduct disorders. Mental illnesses involve multiple mechanisms; glial and neuronal processes, and there's little question that hormonal, neurotransmitter, neurotrophic, immunological, and microbiome-related processes play into these disorders. There are ample data indicating that mental illnesses are frequently accompanied by organizational changes within the brain, and serious mental illnesses may be accompanied by variations of connectivity as well as structural brain changes, including variations of the hippocampus, prefrontal cortex, and amygdala. These conditions are most often treated through drugs or various cognitive and behavioral therapies, although other options, such as transcranial magnetic brain stimulation, electroconvulsive shock, and deep brain stimulation (electrical pulses to brain regions in which electrodes have been surgically placed) have been used for some of these conditions.

The diverse features of mental illness are reminiscent of several characteristics that might be evident in "neurological disorders" that involve structural, biochemical, or electrical abnormalities in the brain, spinal cord, or other nerves. The brain changes can be accompanied by symptoms such as altered levels of consciousness, confusion, paralysis, seizures, poor coordination, muscle weakness, pain, and loss of sensation. Impairments of brain and of mind both comprise neuronal disturbances, and are modified still further by psychological and social processes. Likewise, traumatic brain injury can give rise to psychological disturbances, such as the fear and anxiety associated with PTSD, and these processes can be linked to dementia.

Many scientists coming from either a neurological or a psychiatric domain, working with animal models or with patients, have focused on epilepsy, Alzheimer's disease, Parkinson's disease, stroke, brain trauma, amyotrophic lateral sclerosis (ALS), brain tumors, migraine (and other headache disorders), multiple sclerosis, lupus erythematosus, autism and attention deficit hyperactivity disorder (ADHD), as well as neurological disorders that occur as a result of malnutrition. Clearly, there is appreciable overlap regarding the research conducted within neuroscience, neurology, psychology, and psychiatry, making is especially curious that, to a considerable extent, they have existed in independent silos. Moreover, cross-talk and collaborations occur less frequently than one

would have expected, although there have been repeated calls for better integration between these fields (Insel et al., 2010; Martin, 2002).

It was only about 50 years ago that one of the major debates in psychology involved the role of nature versus nurture in relation to intelligence, even debating whether the split was 80:20 or 50:50? Once it was discovered that the position of the nature side might have been tainted by some questionable methods, and even falsified data, the debate petered out, and it wasn't long before it was realized that the debate was actually pointless from the outset. Irrespective of proportion attributed to nature vs nurture, there is little question that nurture need to be fostered in order to obtain the most from each person's nature. Genetic contribution, variations of neurotransmitters, hormones and growth factors, together with microbial and immune alterations all influence one another, and they are all modified by experiential and environmental factors.

The actions of inflammatory processes and microbial factors on mental illnesses are late comers in the game, and it was certainly the case that immune functioning as a player in this regard met considerable resistance, although looking from the outside in, it seems that the microbial contributions were more readily accepted. Perhaps the ice had been broken with the widespread perspective of inflammation and disease, coupled with the demonstration that microbiota and immune functioning reciprocally innervated one another, and both affect brain processes, making for an easier transition to mental illnesses. Of course, the zeitgeist regarding (w)holistic medicine may have contributed to the acceptance of complex interactions in relation to physical and mental illnesses.

2

The Immune System: An Overview

PROTECTION IN A SCARY WORLD

Immunology as a discipline is a relatively young field. However, given that the principle of immunology is protection of the host against nefarious biological forces—germs—humans understood well before white blood cells captivated our imagination, the power of contamination, and the incapacitation that arises from spoiled foods, sickly animals, and febrile children. After all, human civilizations came and went not merely because of the expansionist behaviors of emperors and kings, but because of plagues that decimated millions of people and altered the geopolitical landscape wherever infection arose. The conquest of central and South America owed as much to the brutality of the Spanish armies led by Cortes, as they did to the foreign microbes they brought from Europe. The lack of immunity among the indigenous peoples, led to thousands dying weekly, thereby weakening resistance against the conquistadors. Such historic examples inspired authors like HG Wells to utilize this trick of nature in his classic novel, War of the Worlds, in which humanity was rescued by earthly microbes to which the alien invaders had no acquired immunological defense. Of course, the converse did not apply, as it was evident that the Martians did not harbor any extraterrestrial viruses or bacteria that might have vanquished human life on earth (Wells made sure to emphasize that the Martians were "clean," having rid their planet of the pesky microbial underworld). Interestingly, and quite wisely—for all our sakes—NASA has strict decontamination procedures before bringing astronauts back to earth. And, of course, modern travel is strictly regulated to ensure that infectious agents do not cross borders. Such was not the case in the past, when the passengers on ships were disease carrying rats and mice.

But let us focus less on populations and more on individual behavior, and on the natural connection between the immune system, behavior, and emotions. One of the strongest human emotions is disgust. It is accompanied by a distinct facial expression in which the nose and mouth are seen to be compressed, and perhaps the tilt of the head and neck has an avoidant slant, as if to distance the person from the source of disgust. The reasons are obvious, as the communicative and personal feelings generated by disgust involve rejection. Much has been written about the biological basis for this emotion, and evolutionary thinking has attributed fear of contamination as a primary factor. Odious smells

(cont'd)

and tastes elicit the tell-tale arrangement of facial muscles and feelings associated with disgust, as if the offending stimulus is being expunged. And it is a rapid learning experience; basic one-trial learning of place and/or food aversion. The lesson for the individual is that contact and interaction with the inciting stimulus is biologically dangerous. Live snakes incite fear. But a dead snake, decomposing and crawling with maggots, incites disgust. Here, therefore, is the essence of immunological protection, operating in behavioral terms.

Humans and other organisms evolved what seem to be innate behavioral responses to unpleasant odors. When food spoils, the nose is a reliable ally, an expert judge of palatability. Not surprisingly, the olfactory system and connected circuitry in the brain are strongly linked to emotion-generating neural substrates (e.g., piriform cortex and amygdala). Therefore, olfaction,

as a sensory system that detects the chemical components of the environment, is a prominent mechanism for avoiding potentially dangerous substances. In performing this function, the olfactory system is now recognized to perform pattern recognition, very similar to the innate cells of the immune system, which also discriminate between various pathogens based on recognition of specific molecular patterns on "germs." We know this now, but what we understood behaviorally for centuries—that meat which smells "off" can kill you—did not extend to a deeper biological scrutiny until the 19th century. Analysis of what it is that's "off" and how those that consume tainted food came to survive, engaged a variety of intellectuals who sowed the seeds for the greater discoveries of the 20th century regarding macrophages, lymphocytes, antibodies, and cytokines. So we turn now to this team of players who protect and defend against certain illness.

INTRODUCING IMMUNITY

The immune system is a network of blood-derived cellular operations that orchestrates protection against the microbial world[1]. The mechanisms involved in conferring this protection constitute *host defense*, the "host" being the organism in which the immune system resides. Most vertebrate and invertebrate organisms have a system of defense against outside invaders, with the vertebrate immune system providing the complexity and sophistication that emerges in highly evolved mammalian species, such as man. However, rodents, and especially mice, have served as the main model for our

understanding of immunity, even though recent evidence suggests that translating our knowledge of the murine immune system to humans is not always a simple matter (Beura et al., 2016). Nonetheless, whether we are looking at the immune systems of a mouse or a human, what we see is an intricate network of cells, with considerable heterogeneity of function and a strategic anatomical organization. Our overview will focus on the essential elements necessary to appreciate how immunity works. A major goal is to emphasize key principles of immunological functioning, introduce the cast of cellular actors, and offer various details in (hopefully) measured and digestible amounts that will inform later

[1] In fact, the term "immunity" is derived from the latin word, *immunitas*, which referred to the protection from prosecution (and most likely persecution) that Roman senators enjoyed in ancient times. In legal circles, this term is still used today to signify the pardons and/or light sentences that witnesses may receive for their testimony.

chapters and render them meaningful. To the newcomer, the "stuff" of immunology may seem like a confusing array of different types of cells and molecules that interact in specific locations and for all sorts of different reasons. If at times it seems bewildering, rest assured it is! A diffuse cluster of cells circulates through the body at least once each day, in search of microbial trespassers. This part is easy to comprehend. But as with anything, the challenge is in the details.

The immune system is very much a "dynamic" system, a term often used to emphasize a state of recurring functional change. Within this dynamism, there are different levels of functioning. Where a microbial invasion has taken place, the nearest location of available immunological reserves, and over time, the makeup of the cellular and molecular elements that make up an "immune response," are all determinants of a highly coordinated network of signals and cellular interactions. And not just between the cells of the immune system. Movement and migration of cells is also dependent on the local tissue environment and the molecules expressed therein. It is only now that we have come to understand that for this to work normally, the immune system has to rely on communication from neural and hormonal sources. In Chapter 6, Stress and Immunity, in which we discuss the impact of stress on immune function, we explain this interplay of neural-immune interactivity a little further.

This sense of the immune system as a dynamic set of cells working its way around the body in search of danger stands in contrast to the vision we have of the brain. The complexity of this hallowed organ is unquestionable; but it is fixed, the key cells (neurons) trapped in a network of synaptic connections and molecular signals, often compared to a computational system and a switchboard of lights and circuits. To follow the activity of these circuits is no cakewalk, and we are still figuring out how they all operate simultaneously to give rise to ongoing conscious life. Still, information received by the brain is bounced around quickly (in milliseconds or less) and results in rapid decisions and actions. In contrast, the immune system processes information on the run, as cells circulate through the blood, exit into tissue (i.e., extravasate), traverse the tissue, and then return to the blood via specialized vessels that constitute the lymphatic system. A circulating lymphocyte literally has to hang on for its life—extending cellular spindles that latch on to adhesion molecules like a rock climber seeking a foothold—as it rides the turbulent waves and shear stress of pumped blood, and seeks to hang back and enter particular tissues to sniff out a vagrant microbe (Abadier et al., 2017)[2]. It is all in a day's work, as the immune system tries to contain infection and keep the host healthy.

In making these introductory points, two important issues come to mind. First, the immune system is—like the brain—an *information processing* system. The second is that by virtue of its inability to sit still, the success of the immune system is based on mobile *surveillance*. As such, it gains access to every nook and cranny of the body[3], and as with any good

---

[2] A recent study (Abadier et al, 2017) demonstrated that within blood vessels, mature T cells display a remarkable and highly energetic ability to swing out membraneous protrusions like lassoes or slings, which tether the cell to endothelial adhesion molecules (e.g., selectin). This enables the cells to roll in stable fashion while the blood rushes past within a blood vessel. This operation is similar also to another cell type, the neutrophil, which grapples the "rapids" in similar fashion. The end result is that immune cells can enter nonlymphoid tissue and explore the area for signs of damage or infection.

[3] Historically, there were some exceptions to this rule, in which immune cells were thought to ignore the brain altogether (thereby giving the brain the apparent distinction of being "immune privileged"). But, as we have known for some time—and this book points out—that is simply not true. Leukocyte infiltration of the brain is quite common, both under pathological and normal conditions.

information-processing system, it acts on local information in a judicious and effective manner. To this extent, the immune system has to generate action (or more accurately, reaction), which immunologists refer to as "effector" function. The main effector functions are cytokine production, cell killing (i.e., cytotoxicity), and antibody production. The cells that mediate these functions are simply called *effector cells*, although each class of effector cell belongs to a particular lineage of immune cells that, on activation, differentiate into a particular effector type. For example, B lymphocytes make antibody; however, when a B cell has achieved a fully differentiated effector status it is called a *plasma* cell. In order for immune cells to evolve into an effector state, they require that the information which triggers this transformation fit a particular profile. This information needs to be a specific type of stimulus, one that is cellular and/or molecular in nature, and is also identified as being nonself or foreign (i.e., not part of the host). Generally, we think of this stimulus as a microbe—a virus or bacterium. However, in the parlance of immunology, such a stimulus is referred to as an *antigen*, which we discuss in more detail shortly. It is the cognate stimulus for the immune system, and elicits the classical type of immune response, in which there is cellular proliferation, development of effector function, and eventually, a denouement and return to a smaller cadre of memory cells.

This is the classical view of how an immune response progresses. It is induced through antigenic stimulation, and is part of what is called the *acquired* or adaptive immune system, which culminates in the production of billions of antibody molecules that have highly specific molecular targets on microbial entities (each of these molecular targets is an antigen). An alternate immune repertoire, known as the *innate* immune response, involves less refined and more global effector arms, and is less concerned with specific antigens, than it is with molecular categories (or molecular signatures). One such category includes the so-called PAMPs (pathogen-associated membrane patterns) (Medzhitov, 2009). Recognition of PAMPs on bacteria engages cells of the innate immune system, which automatically generate molecular and cytotoxic responses that can initially decimate the number of invading microbes, but in the process, will recruit cells of the acquired immune system—the T and B lymphocytes. Even if an infection is largely headed off by a robust initial response by the innate immune arm, adding lymphocytes to the fray will result in the making of probably the main effector molecule with which the immune system is associated: Antibody. In the absence of antibody, infection is likely to be prolonged and unresolved. Antibody attaches to viruses and bacteria that induced them, which tags or flags them for destruction. Without this tagging, pathogens can continue to circulate and multiply, evading the lethal effects of cells in the innate compartment of the immune system. However, with antibody bound to the microbe, these innate immune cells can attach to the antibody, and use this to immobilize the microbe long enough for it to be destroyed. In a scenario like this, the antibody is said to have contributed to "opsonization."

More formal descriptions of acquired and innate immune processes will be provided in the coming sections. As we will see, the multiple types of cellular and molecular interactions that each involves (in a nonmutually exclusive manner) follows a set of commands and instructional guidelines that are determined by genetics, cell-surface molecules, secreted signaling proteins and various other nonimmune cells (e.g., endothelial and epithelial cells) that can flag down and/or facilitate immune responses. While the key players of the immune system originate from stem cells in the bone marrow (see Fig. 2.1), important functions can be attributed to the stromal and

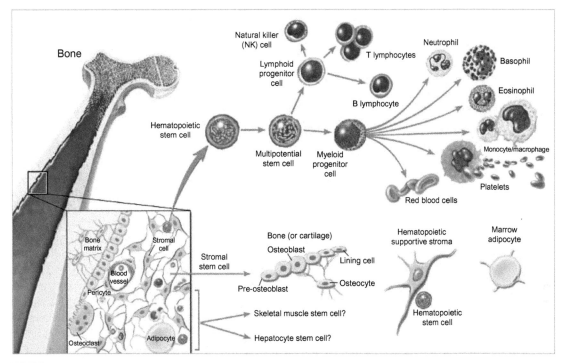

FIGURE 2.1   Formation of cells of the immune system from hematopoietic stem cells. As explained in the text, leuko-cytes originate from a bone marrow-derived multipotent stem cell, that itself was derived from a hematopoietic stem cell. The multipotent stem cell differentiates into two sets of progenitor cells that constitute the lymphoid and myeloid cell lineages. The latter gives rise to cells of the innate immune system, while lymphoid cells develop into T and B lympho-cytes, as well as Natural Killer cells (large granular lymphocytes), which can kill virus-infected cells and tumor cells. *Source: From https://stemcells.nih.gov/info/Regenerative_Medicine/2006Chapter2.htm. © 2001 Terese Winslow (assisted by Lydia Kibiuk).*

epithelial cells of tissues, and the endothelial cells of blood vessels. These represent key sources of information that can guide and recruit immune cells to sites of damage and infection. At these locations, the presence of immune cells forms the basis of tissue-specific inflammatory responses, a major consequence of immunological activity.

THE IMMUNE SYSTEM CONSISTS OF WHITE BLOOD CELLS: LEUKOCYTES

The main constituents of the immune sys-tem are leukocytes ("white cells"). In an adult human, a liter of blood typically contains 7−8 billion white blood cells that are in transit, coming from all parts of the body and recircu-lating back to the same sites as part of a mas-sive and unceasing surveillance operation that persists throughout an individual's lifetime. The word, "leukocyte," is a collective term for a heterogeneous group of cells: Lymphocytes, monocytes, granulocytes (e.g., neutrophils), dendritic cells (DCs), macrophages, and many more (these cells are described further in Fig. 2.1). To say that all leukocytes do the same thing, because they are involved in defending against infection, is to miss the unique contri-bution that each type of cell brings to the fight against microbial invaders. As a collective, however, they work together in a highly organized and coordinated manner to defend the body against infection. If there are any other purported functions, these are ancillary to this key purpose of protecting the tissues and organs against damage from microorganisms.

In thinking about the immune system, the metaphor of a military operation often comes to mind. Leukocytes are the sentinels of the body, alert to any form of disruption to the host; and when this happens, they can mobilize quickly and mount an intense campaign through sophisticated forms of molecular signaling. This generates a rich armamentarium of degradative enzymes, highly specific antibody molecules, a plethora of regulatory cytokines, and exponential rates of cellular replication that will ensure that these molecular tools are in abundant supply. These events are not without their collateral impact on the health of the host organism. However, the efficiency with which the immune response is generated and ultimately winds back to a quiescent state ensures recuperation, recovery, and a return to a healthy state.

The evolution and refinement of such a system resulted from pressures that arose due to the manner in which mammalian organisms interact with their environment. There are multiple portals of entry that microorganisms can slip through, and given our need to eat and procreate, oral and genital surfaces are placed at increased risk of microbial encampment and opportunistic entry into the internal milieu for more extensive colonization. Anatomical surfaces that serve sensory functions are also vulnerable, including the eyes and ears; while the skin and its regular cycles of shedding and regeneration limits any major breach, absent a cut or wound that provides a tempting opportunity for airborne and topical (present on the skin) microbes to transcend the epidermal surface and feast on the dermal matrix of cells and fluids. But it is here that classic descriptions of inflammation emerge, describing the redness and swelling indicative of the early stages of

an immune response. Tissue damage engages local cells to generate chemokines, signaling molecules that flag down the leukocytes circulating in blood. The cells extravasate (leave the blood) and roll into the damaged or infected tissue and begin the process of engaging any extant pathogens, while also helping to rebuild the damaged site. These events are highly orchestrated, and in the case of, say, damaged skin, ensure that restoration of the epidermis and the underlying dermal and subcutaneous tissue progresses quickly and efficiently so that the risk of infection is eliminated.

The best way to appreciate the nature and function of the immune system is to consider where leukocytes congregate. After all, most biological systems operate within an organ. Gastrointestinal (GI) cells are part of the GI tract, alveolar cells the lung, and myocardial cells, the heart, which drives the cardiovascular system and the network of blood vessels that permeate all parts of the body. The latter is critical to our understanding of the immune system and immunity, since irrigation of the body with blood not only ensures that all cells will receive molecules necessary for energy (e.g., glucose), but immune cells will be transported to most regions of the body where they are needed to perform defensive and reparative functions. For this reason, the stem cells that originate in the bone marrow, and which are destined to become immune cells, are referred to as *hematopoietic* (blood-forming) cells.

TYPES OF IMMUNITY

As mentioned earlier, immunity from infection is accomplished through twin avenues of cellular reactivity: (1) natural or *innate* immunity and (2) *acquired* (or adaptive) immunity. These are implemented in staggered fashion, but in reality form a cooperative network of interactions that ensure thorough and complete elimination of invading pathogens. The innate (sometimes also called *native*) form of immunity is performed by established defense mechanisms that generally have a low threshold for discriminating between different types of infectious agents, and do not show changes in magnitude nor intensity of responsiveness with repeated exposure to the same infectious microorganism. However, these notions of discrimination and unvarying stability of performance should be qualified. For the most part, innate immune responses respond to broad categories of potentially infectious microorganisms. That is, while more refined and surgical forms of attack are the role of cells that are part of the acquired/adaptive immune response, the innate immune system was always perceived as attacking any old microbial organism. However, this notion has been rewritten, as we now know that innate immune cells do possess selective recognition capabilities.

A related concept is memory. Inherent in the term "immunity" is the notion of continued remembrance of events that have passed, for which a form of protection is owed (as befits the original use of the latin word *immunitas*). The innate immune system does not bother with memory formation: It typically sees things in the same way every time (although see discussion that follows). Like someone with short-term amnesia, a conversation can be recycled repeatedly, as fresh and alive as it was the first time. So, to some degree, to an innate immune cell, one bacterial cell looks like any other bacterial cell, although in more recent years we have come to understand that the innate immune system has a categorical manner of recognition, through recognition of PAMPs. Thus, the responses are consistently the same, kill and destroy. The same goal is evident for cells of the adaptive immune system, although their role is to be more selective about who gets the chop, and who should have known better about coming back. These cells are bouncers with strong cognitive capacities; they learn

and remember (although they're not as beefy as innate immune cells, which tend to be 50% larger than adaptive immune cells). We will have more to say about innate immune cells and their microbe recognition capacity later.

The alternative view of acquired immunity as *adaptive*, reinforces the notion of some form of biological improvement over time and experience with particular challenges. It is change for the better. With acquired immunity there is specific recognition of a given foreign molecule (antigen) by unique cellular and molecular components of the immune system (viz., receptors), and this recognition (or information) is stored for later retrieval should this foreign molecule ever return. This is a cornerstone of immunity in vertebrate organisms. The chief cells and molecules involved in this specific form of recognition are the *T and B lymphocytes* (also called T and B cells), and the antibody molecules produced by B cells, which we mentioned earlier. It is this form of immunity that reveals the vertebrate immune system to be closely aligned with that of the central nervous system: Both are responsible for information processing, storage, consolidation of that information, as well as its retrieval. For the adaptive immune system, this information is molecular. When we say that the T and B cells recognize a specific molecule, we must remember that, for example, each bacterial cell will present with a multitude of molecules (or antigens). The hundreds of millions of lymphocytes that are present in the vertebrate organism actually have the ability to differentiate between each of these. It is as if the bacterial cell has become pixelated, and each lymphocyte has the task of recognizing a given unique pixel. In this way, lymphocytes can identify given microbial entities, whether they are viruses or bacteria, and differentiate among them in ways the innate immune cells cannot. This specific form of recognition is what allows lymphocytes to pounce on a returning virus or bacterium much faster and with greater force. We will discuss this notion of specificity shortly and introduce the clonal selection theory, which served to explain the existence of this fine-grained nature of differentiation.

WHAT ACTUALLY INDUCES IMMUNITY?

Generally speaking, immunity is formed against (potentially) infectious microorganisms. But, in practice, as already introduced earlier, the immune response is generated against specific molecules (antigens from the term antibody-generating) that are expressed by microorganisms. In most cases, the chief target molecule needs to be a protein. These molecules possess configurations (or strictly speaking—amino acid sequences or short peptides) that the immune system recognizes as foreign (nonself). A single bacterial cell or virus may express many hundreds or thousands of different antigens, against which the (adaptive) immune system can respond. However, for convenience, when speaking about a given immune response, it is common practice to refer to a bacteria or virus as an "antigen" or antigenic stimulus. For this reason, discussion of antigens can further be broken down into *antigenic determinants* or *epitopes*—specific portions of the conceptual antigen (the typical bacterial cell or virus) that can be bound by an antibody.

THE CELLS AND ORGANS OF THE IMMUNE SYSTEM

We noted earlier that the cells of the immune system are a heterogeneous group of cells called leukocytes. As Fig. 2.1 shows, these cells all originate in the bone marrow from a common progenitor cell called the pluripotent hematopoietic stem cell (HSC). These cells differentiate under the influence of locally produced soluble factors into various progenitor cells, two of which give rise to the *lymphoid* and *myeloid* cell lineages. The latter gives rise to cells of the innate immune system, and among these, the most prominent are neutrophils, monocytes, and macrophages. These cells will mature in the bone marrow before entering the blood and localizing in various tissues (e.g., the tissue macrophages).

The lymphoid lineage emerges from the lymphoid progenitor cell and gives rise to T and B lymphocytes, as well as additional cells, such as the large granular lymphocytes (LGLs) that are associated with cytotoxicity functions, and have come to be called Natural Killer (NK) cells (see Fig. 2.1). Of the lymphocytes, only the B cell matures in the bone marrow (and also initially in the fetal liver). In contrast, the T cell initially develops in the bone marrow, but at an early immature state exits to enter the circulation, whereby it is sequestered in the thymus (a gland that is nested between the lungs and above the heart).

In the thymus, the cells encounter various influences that drive them to a fully differentiated and mature state, whereby they are able to leave the thymus and populate other lymphoid organs. In their mature state, T lymphocytes are equipped with the molecular tools to allow recognition of the host, which allows them to discriminate between the molecular information of self and nonself. Maturity, therefore, is in part the capacity to respond aggressively only to that which is foreign. Failure to accomplish this results in autoimmune disease. Consequently, events in the thymus are critical to the development of a fully functional T cell repertoire.

The Innate Immune System

We have already provided a sense of what the innate immune system represents. This is a collective term for a system of immediate defenses against the microbial world, and strictly speaking can begin with actual physical barriers, such as the skin and mucous secretions (e.g., secretions of the eyes and nose, the phlegm of the lungs, and the mucus lining of the inner intestine). However, beyond the "walls" of these initial defenses, are the cells of the hematopoietic system, that are largely confined to the myeloid lineage, with LGLs or NK cells (developed within the lymphoid lineage), also being included as part of the innate immune system. The myeloid cells that chiefly comprise the innate system are granulocytes (most commonly neutrophils), monocytes, and macrophages (see Fig. 2.1). Each is typically associated with mediating an inflammatory response by virtue of their ability to contribute to the build-up of regulatory molecules, fluids, and recruitment of other immune cells (e.g., lymphocytes) to a site of infection where they are typically engaged in a cleanup process ("phagocytosis;" literally, "eating up"), and are thus sometimes referred to as phagocytes. When they are engaged in phagocytosis, they are ingesting necrotic tissue and bacterial cells, and throughout this process liberating humoral mediators of innate immunity, which includes lysozymes and complement, proteins that have cytotoxic capabilities (as we shall learn later in our discussion of antibody).

Additional soluble mediators are secreted which serve to amplify and expand the immune response by alerting the cells of the adaptive immune system. These mediators are cytokines and chemokines, which will both enhance further killing of bacteria and recruit (via the chemokines) other cells (lymphocytes and monocytes) to the site of activity. For example, consider the irritation of a splinter wedged in the skin. If the individual is unable to remove the splinter, the innate immune system will, over time, dissolve the foreign material. There will be redness and swelling, some pain and discomfort, but it will eventually be degraded. This is due to the infiltration of circulating monocytes, as well as activation of any resident macrophages in the dermal tissue, which will create an increased flow of plasma from the blood through the increased gaps between the endothelial cells of the blood vessels. The end result is an edema (increased volume of interstitial fluid), and the reason why an observable area of the body can enlarge dramatically (the swollen finger, foot or knee) after an injury or infection. Therefore, when we speak of inflammation, we can refer to the clinical description (redness and swelling), but in actuality, the underlying basis of this is an innate immune response that has initiated an immunologic defense and increased leukocyte migration to the site of infection or damage. Moreover, inflammation need not be in external regions of the body, but is an ongoing event taking place in the lungs of a smoker, an atherosclerotic cardiovascular system, a cirrhotic liver, an irritable bowel, or an arthritic knee. In all these cases, and many more, the increased presence of innate immune cells and their soluble products is an indication of inflammation.

Specificity and Memory in the Innate Immune System

Two major issues have preoccupied immunologists regarding innate immunity: Specificity and memory. The first pertains to whether the innate immune system possesses the same specificity of lymphocytes. The traditional dogma was that innate immunity was nonspecific. However, there was always a nagging question: Just how did neutrophils and other granulocytes recognize a bacterial cell as something worth engaging and destroying? Moreover, when necrotic tissue needs to be removed, how did the innate immune cells manage to recognize that this is self-tissue worth disposing? The answer to this was arrived at more than 20 years ago when it was proposed—and eventually proven—that innate immune cells expressed pattern recognition receptors (PRRs) for a variety of molecules expressed by microbial agents. These molecules include complex polysaccharides, glycolipids, and lipoproteins that collectively are referred to as PAMPs, although PRRs also recognize nucleotides and nucleic acids (Riera Romo, Perez-Martinez, & Castillo Ferrer, 2016).

The second question that has been posed relates to memory. Do innate immune cells possess memory? In contrast to the adaptive immune compartment, the innate immune system has often been described as having little (if any) memory for previous encounters with microbial agents. Each bacterial cell, whether from *Escherichia coli* or *Staphylococcus aureus*, is dealt with equally, and if infection recurs, the innate immune system will attack these bacteria with the same intensity and speed as it did the first time. At least that had been the long-held view. More recently, a growing body of evidence suggests that innate immune cells, such as neutrophils and NK cells, can respond to PAMPs and generate a short-lived memory for the encounter through epigenetic modification. The form of this memory (which was referred to as "trained memory") is metabolic, to the extent that innate immune cells can respond more strongly and efficiently to later infection, whether it is the same or a related microorganism. The research on innate memory is still on exploratory ground, but at the very least, there is a sense that neutrophils and NK cells may be trained to be better prepared for future encounters with pathogens (Netea, Latz, Mills, & O'Neill, 2015). This may not be the same as the highly specific responses and memory cell formation shown by lymphocytes, but it does suggest that learning is not exclusive to the adaptive immune compartment.

Sterile Inflammation

The induction of inflammation by innate immune cells is part of the normal immune response. And while it is often thought to be something to attenuate and eliminate, there is the alternative view that inflammatory cells are also trying to correct areas that induced their activity in the first place. This is a recurring question that keeps popping up in the neural-immune literature, and especially in relation to the "inflammatory" events occurring in the brain. How much is an effort to help, and how much help is eventually a burden? Inflammation is essentially the double-edged sword problem, and one that has yet to be resolved. However, it was shown that neutrophils were instrumental in rebuilding damaged tissue incurred by thermal hepatic injury, a procedure that produced *sterile inflammation* (Wang, Hossain et al., 2017). The neutrophils in this case were critical to the reconstruction of blood vessels, which were compromised if these cells were prevented from accessing the damaged tissue. This finding inverted the long-held notion that sterile inflammation, and the presence of neutrophils and macrophages, can delay or interfere with tissue recovery. As such, it reinforces the idea that anti-inflammatory treatment may not always be the best option.

Sterile inflammation is a concept introduced to account for cases when there is an accumulation of inflammatory cells and inflammation-related molecules in the absence of actual microbial organisms or infection (Chen & Nunez, 2010). The conditions for induction of sterile inflammation can include a range of factors, including physical injury (e.g., trauma), cancer, autoimmune tissue damage, ischemia, atherosclerosis, and exposure to toxins. Since inflammation is aroused in the absence of PAMPs, it was confirmed that innate immune cells have a recognition mechanism for detecting cells undergoing distress and/or death. The collection of signals generated by damaged tissue is now referred to as danger associated molecular patterns (DAMPs), and these are recognized by PRRs on neutrophils and other innate immune cells. As can be expected with necrotic cells, liberated intracellular molecules function as DAMPs (e.g., chromatin-associated proteins, heat shock proteins, and purinergic molecules, such as ATP), as well as molecules associated with the extracellular matrix and fragments of enzymatically degraded proteins and certain proteases normally required for tissue repair (Chen & Nunez, 2010). Many of the studies investigating the impact of immune processes in psychiatric patients invoke the concept of inflammation, and in so doing, may be directly or indirectly alluding to a state of sterile inflammation. However, in light of the findings by Wang, Hossain et al (2017), it may be prudent to exercise caution in just how we interpret studies showing elevations in pro-inflammatory cytokines and/or circulating neutrophils and monocytes.

In having discussed PAMPs and DAMPs, we should specify precisely the nature of the receptors for these molecules. For some time, macrophages, neutrophils, and monocytes were known to respond to a molecule called lipopolysaccharide (LPS), which is derived from the cell walls of gram-negative bacteria, such as E.coli; as such, LPS is also referred to as *endotoxin* (endogenous toxin). This form of innate immune cell activation is a prominent model of immune activation in neural-immune investigations. In recent years, it has been discovered that LPS binds a molecule called Toll-like receptor 3 (TLR3). When this receptor is stimulated, innate immune cells show activation of a transcription factor, NF-kappa B (NF-κB), which results in the further transcription of genes for a variety of pro-inflammatory cytokines. This explains why LPS is a potent inducer of cytokines such as IL-1β, TNF-α, and IL-6, three prominent pro-inflammatory cytokines, produced by innate immune cells[4]. The family of TLRs, of which 10 were identified in humans, are now widely recognized as the major recognition receptors for PAMPs and DAMPs.

Further discussion of the innate immune compartment will be incorporated into our description of the adaptive immune compartment. It will be recognized that when innate immune cells engage and phagocytose microbial agents, their immunologic importance does not stop there. For instance, they serve an important and necessary role in ensuring activation of T cells, and they will be important in capitalizing on the accumulation of antibody molecules generated by B cells. This highlights the fact that segregation of the immune system into innate and adaptive immune components is an arbitrary and temporary measure that helps to explain the workings of the immune system more effectively. However, the immune system in and of itself comprises both compartments working together in complementary fashion to rid the body of unwanted microbial invaders.

THE LYMPHOID SYSTEM

Lymphocytes and certain cells of the innate immune system (e.g., monocytes), circulate throughout the body performing what are largely surveillance functions. This patrolling behavior is more formally referred to as cell trafficking or migration, and is driven by unique anatomical and soluble factors. Although many immune cells are constantly in transit, they nonetheless localize in well-defined and widely distributed organs, the lymph

---

[4] Interestingly, the name Toll-like receptor is based on the Toll protein that was first identified and named in the fruit fly Drosophila. This protein was initially thought to be critical for the development of Drosophila, but eventually it was recognized to be just as important in helping them fight off infection. When the mammalian homolog of the Toll protein was discovered, and similarly identified with host defense, the name Toll-like receptor was adopted, ostensibly to honor the protein which inspired the discovery.

nodes. These nodes drain the local area that surrounds them, and as such, we have nodes located in various strategic locations (see Fig. 2.2). For example, the walls of the gut contain the *peyer's patches*, which are actually small lymphoid nodules that line the walls of the small intestine, and the cells in these regions travel (or drain) to the mesenteric lymph nodes (MLN). These comprise a pearl-like arrangement of lymph nodes that form a daisy-chain of interconnections and are linked to the gut by the mesentery, a fan-like arrangement of connective tissue and capillaries that contain fluid and lymphocytes (this mix of fluid and lymphocytes is called *lymph* and the capillaries from the gut wall to the MLN are the lymphatic vessels). The upper respiratory tract also contains lymphatic portals to a daisy-chain arrangement of

lymph nodes, as well as the arm pits, the pelvic and abdominal regions (inguinal lymph nodes), and also the rear of the knee (the popliteal lymph nodes). For example, stepping on a rusty nail that might contain tetanus toxin, will engage local immune cells that can drain to the popliteal lymph nodes. Experimentally, these nodes may swell in size after injection of antigen into the footpad of mice—a measure of delayed type hypersensitivity (DTH) (Smith & White, 2010). This swelling or expansion is due to activation of lymphocytes, resulting in further activation and proliferation (cell division) of the cells in the draining node. The nodes, therefore, represent regions where information gathered by immune cells in the surrounding area can be passed on to cells in the lymph node. This transmission of information (regarding the nature of

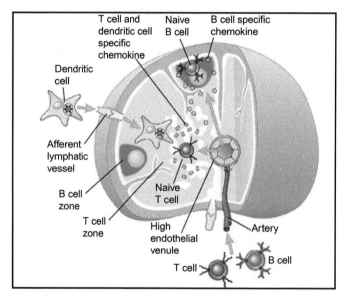

FIGURE 2.2   Lymph Nodes. The lymphatic system drains via the thoracic duct into the vena cava and into the general blood circulation. Lymphatics drain into and away from lymph nodes, which are distributed in gastrointestinal, abdominal and cervical regions; additional nodes can be found around the tonsils, under the arms, regions emanating from the small intestines, and behind the knee (so-called popliteal nodes). The figure shows a typical node which displays a point of entry for an *afferent* lymphatic vessel entering the left side, and which transports lymphocytes and other cells into the node. Within the lymph node, B and T cells are segregated into distinct locations. Lymphocytes can exit from arteries by reaching a high endothelial venule (HEV) which allows for easy extravasation, and then migrate to different areas of the node as a result of attractive chemicals called chemokines, and which are produced in different areas of the node. Note also the migration of dendritic cells (DCs) through the afferent lymph, which captures antigens that enter via the lymphatics. Once the DCs enter the node they move to the T cell-rich areas of the node, where they present the antigen and induce T cell activation. *Source: From Abbas et al. (2007). Cellular and Molecular Immunology. 6th edition. Elsevier press.*

THE IMMUNE SYSTEM AND MENTAL HEALTH

the antigen) is carried out by a professional antigen-presenting cell (APC) referred to as a DC. These cells encounter antigen in the tissues, and then present this on their surface to lymphocytes. The close confines or packed arrangement of cells in the lymph nodes optimize interactions between lymphocytes and DCs, thereby leading to lymphocyte activation. Both T and B lymphocytes are activated.

In addition to the lymph nodes, the spleen is another major lymphoid organ. The spleen does not represent a drainage site for circulating lymphocytes in the tissues. Rather, it filters the blood and traps antigens and foreign cells that are circulating in the blood. The liver also performs the same function, the difference being that most of the immune cells in the liver belong to the innate component of immunity (viz., macrophages called Kuppfer cells). The spleen, however, contains a full complement of macrophages, DCs, and T and B lymphocytes. Because cells of the spleen are activated in response to an antigen that gains access to the circulation, this lymphoid organ is commonly used in experimental animal studies. This is also due to the fact that it yields a large number of immune cells that can be assayed for multiple parameters. Lymph nodes generally do not yield as many cells, especially when there has been no antigenic stimulation. As noted earlier, if the popliteal lymph node draining the foot of a mouse were collected without any antigen injection into the footpad, the experimenter would be hard pressed to locate this node, let alone isolate enough cells to conduct useful assays. The node from an immunized mouse, however, would be plump and ripe with millions of lymphocytes. This very much describes the clinical presentation of swollen lymph nodes below and around the

jaw in someone with an upper respiratory tract infection. Infection in the throat drains antigen to the local tonsillar lymph nodes.

The *primary* immune organs are the bone marrow and thymus, since cells at early stages of development are found there. Once they develop and achieve a fully differentiated state of functional maturity, they will leave the bone marrow and thymus, and circulate throughout the blood and spleen, and lymphatic vessels and lymph nodes. The spleen and lymph nodes are considered to be *secondary* lymphoid organs, since they contain mature circulating leukocytes. In this secondary physical terrain, the cells carry out the immune surveillance that ensures protection of the body.

Circulation of lymphocytes throughout the body is continuous. A given lymphocyte, for example, can circulate and return to its point of origin at least once over a 15–48 hour period, depending on retention rates in select organs (Ganusov & Auerbach, 2014). The trafficking of cells follows a particular pattern, whereby cells in the blood of major arteries gradually move into smaller blood vessels, such as arterioles, which irrigate the tissue, and then eventually into the vessels of the lymphatic system, a network of capillaries containing lymph[5] (Miyasaka & Tanaka, 2004). When the cells extravasate, they migrate through the tissue, sampling and responding (if necessary) to any signals present on or produced by tissue cells or other immune cells. The cells will then be carried forward into a separate (nonvascular) system of small capillaries that drain tissue fluids and push this uni-directionally through a one-way valve system into larger networks of capillaries (Liao & von der Weid, 2015). These networks of tissue-draining capillaries comprise the lymphatic system, since they are filled

---

[5] The latin *lympha* means "clear water," and is derived from interstitial fluid and blood plasma to give it it's more translucent appearance. Interestingly, however, lymph collected from the digestive system can have a milky appearance due to high fat content, and which is called chyle. The lymph, of course, is not just fluid, but filled with lymphocytes and other leukocytes.

only with interstitial fluid and leukocytes. Therefore, once in the lymphatic capillaries, the leukocytes have essentially left the tissue parenchyma. In the lymphatics (i.e., lymphatic capillaries), the cells will enter progressively larger lymphatic vessels, some of which will drain directly into lymph nodes. The lymphatic vessel that brings cells to a lymph node is called an *afferent* lymphatic, and the lymph carried forward through this is simply referred to as afferent lymph. To exit the lymph node, the cells are carried forward in efferent lymph, and do so through an *efferent* lymphatic. When lymph nodes are daisy-chained (arranged in sequence, as shown in Fig. 2.2), each efferent lymphatic is essentially draining the cells toward the afferent lymphatics of the next lymph node in the chain. In time, the cells will leave the lymph nodes and drain via the efferent lymphatics into the thoracic duct, a large lymph-filled vessel that drains into the superior vena cava and directly into the heart and general circulation.

To summarize, the lymphatic system drains mostly lymph nodes. However, the spleen and bone marrow, lack a lymphatic system. Interestingly, it was shown that the brain contains lymphatic vessels in the meninges (Louveau et al., 2017). This, however, appears to drain the contents of brain interstitial fluid or CSF, which also occurs through a *glymphatic* system, so-called because it involves glial cells orchestrating efflux of interstitial content out of the brain parenchyma (Plog & Nedergaard, 2017). Historically, this may be the basis for demonstrations of effective immunization against benign proteins, such as ovalbumin, that were delivered into the brain (Gordon, Knopf, & Cserr, 1992). Interestingly, no evidence of an afferent lymphatic input to the brain has been documented, although cells and soluble factors can enter the brain via areas that have a weak blood-brain barrier or are actively transported through endothelial cells (Banks, 2015).

THE ACQUIRED IMMUNE RESPONSE

There Are Two Major Classes of Acquired Immune Responses

Acquired immune responses are mediated by lymphocytes. As we've seen, there are two main types of lymphocytes: The B cells, which produce antibody; and the T cells, which serve regulatory and accessory or helper-like functions, as well as direct cell-killing functions. Traditionally, the two forms of immunity conformed to an adaptive or acquired modality. The first—and the one that enjoyed the most attention during the early, pre-World War II period of immunology's history (Silverstein, 2003)—is *humoral* immunity. Choice of the term "humoral" was used to highlight the notion that protection is mediated by blood-borne soluble factors. In fact, in 1901, the first Nobel Prize in Physiology and Medicine acknowledged the importance of this notion by awarding the prize to Emil Von Behring, who discovered that when serum from animals immunized with tetanus toxin was injected into naïve (nonimmune) recipient animals, these animals (rabbits, actually) survived the challenge with the toxin. Hence was born the notion of passive transfer of immunity (or just passive immunity) (Graham & Ambrosino, 2015). The fact that it was produced by serum from immunized animals, revealed that the humoral space (i.e., blood compartment) contained protective elements of immunity. This was subsequently found to be a product of B lymphocytes, namely, antibody molecules that are released into the general circulation (lymphatic and blood).

The second type of adaptive immunity, *cellular immunity*, was not recognized until the late 1950s and early 1960s—in spite of several earlier important demonstrations that cells—like antibodies—were capable of providing protection against infection and/or hypersensitivity (Silverstein, 2003). If there was a presiding cellular theory of immunology, it was beholden to Metchnikoff's phagocyte theory of host protection (to which we will come later). However, this essentially took a back seat to the immunochemical tradition pioneered by Behring and Erlich that revolved around the antibody molecule. Eventually, however, recognition of the importance of the T cell, and it's quite special—"helper"—relationship to the B cell (Crotty, 2015), solidified the notion that immunity is very much a cellular affair. The fact that cells mediate protection was demonstrated in a similar manner to that of humoral immunity using passive transfer of lymphocytes from immunized animals to naïve recipients. In this case, no soluble factors or fluids were transferred, only the cells of the immune system. The reason this is attributed to T lymphocytes is because these transferred cells were derived from the thymus, and shown to help naïve B lymphocytes (of the recipient) produce antibody to the antigen (against which transferred cells were sensitized or immune). Or, they directly killed bacteria or host cells that contain viral antigens, when recipient cells were experimentally incapacitated by irradiation and other procedures. Consequently, cell-mediated immunity is generally synonymous with T cell-mediated immune responses.

Hallmark Features of the Adaptive Immune Response

The chief, distinguishing features of any immune response are (1) specificity, (2) diversity, (3) memory, (4) self-limitation, and (5) self-/nonself discrimination. These hallmark features are shared by T and B lymphocytes (see Text Box and Fig. 2.3), and some of these features were introduced earlier (e.g., specificity and memory). Here we will elaborate a little further on these distinguishing characteristics of acquired immunity.

Specificity

As already mentioned, antibody molecules are made against unique elements of a foreign protein molecule (antigenic determinants or epitopes). An antibody made against one particular part of a protein, say region X_{ag}, will not bind another part of the protein, say region Y_{ag}; conversely, the latter region Y_{ag}, will have elicited a response from a subset of B cells that make antibody that will bind this region, but will not bind region X_{ag}. Therefore, within the total pool of B cells in the body, there are cells that are specific to only one antigenic determinant, and no other. Such a group of B cells is called a clone. And if each antigenic determinant—no matter what protein it is derived from—has a unique antibody molecule by which it is bound, then it follows that there must be as many B cell clones as there are antigenic determinants that the immune system encounters. This reasoning is part of the clonal selection theory proposed by Sir McFarlane Burnet, and for which he received the Nobel Prize (Hodgkin, Heath, & Baxter, 2007). Space does not allow for further elaboration of this well-established feature of the acquired immune response, but suffice it that not only does this principle of one-antibody/one-epitope matching apply to B cells, but it also extends to T lymphocytes, which possess antigen receptors that are called T cell receptors (TCRs). As for B cells, the population of T cells is proposed to consist of a multitude of clones, each of which recognizes with its TCR only one specific antigenic determinant. Therefore, when a B cell clone recognizes epitope X_{ag}, there is a corresponding T cell clone that also recognizes epitope X_{ag}, and neither clone will recognize any other epitope.

This shared specificity by the X_{ag}-specific B and T cell clones, is mutually beneficial. That is, T cells that recognize the same antigen as B cells, can serve to help these B cells produce antibody against the antigen. More practically, consider for example, a mouse that is exposed to a given virus, say herpes simplex 1 (HSV-1). This mouse will generate a T cell and B cell response to the virus. However, if the mouse is exposed to a second virus, say cytomegalovirus (CMV), there will also be a T and B cell response to the virus, but the population of T and B cells will be different from that which responded to HSV-1. Each virus will have a separate group of clones that possess specific recognition to the various antigenic determinants on each unique virus. To be sure, there may be some overlap—or cross-reactivity—to some epitopes shared by each virus, but for the most part each virus will recruit different clonally specific T and B cells.

Diversity

This concept can be readily appreciated in light of the foregoing discussion of specificity and clonal selectivity. In being able to generate highly specific responses against the many different antigenic determinants encountered by the immune system, it is evident that there is considerable diversity in the potential reactive cells in the B and T cell populations. Such diversity allows for a large protein molecule (a toxin, perhaps), as large as it may be, to be responded against by multiple clones, each of which will target different components of the protein. Such diversity of targets is matched by the diverse instructions given to each clone. If all targeted the same region, the chances of destroying the foreign protein might be minimal, if that section does not happen to be a weak point. In essence, there is strength in diversity, and therefore, the reason why the adaptive immune response is so powerful. In addition to this, the same antigen is being bound by antibody from B cells, and the TCR of cytotoxic T cells (CTLs). In this way, diverse mechanisms of inactivation are being initiated against the same antigen.

Memory and Self-Limitation

When lymphocytes respond to their respective antigens, they undergo mitotic activity, undergoing many rounds of cell division (or what is sometimes referred to as *transformation*) in order to expand the original population of antigen-specific T and B cell clones. This expansion or proliferation of clones increases the number of cells that are responding against the specific antigen, and once the antigen is eliminated, the expanded lymphocyte population is culled through the generation of signals that promote programmed cell death, or apoptosis. This is an important aspect of the notion of self-limitation—although self-limitation is a concept that can apply during the exponential phase of an immune response, during which some cells are being restrained and redirected in their behavioral profile. The end result is that the response is sufficiently fine-tuned to allow for necessary and sufficient numbers of cells and the molecules that they produce. To ensure that this progresses in a manner that does not threaten the organism (e.g., development of leukemia or self-directed immune responses), important regulatory mechanisms are in place to ensure that some optimal level of responding occurs. The regulation of the immune response is a major area of immunological research, and much can be said about this. We will mention some of the regulatory processes that are mediated by cytokines as well as T cells (the so-called Tregs or regulatory T cells).

As the peak proliferative phase has been reached and apoptosis initiated, a small population of the reactive clones are retained and remain viable as *memory B and T cells*. These memory cells are antigen-specific in the same way as the original clones, but they are now endowed with additional properties that allow them to be more robust responders should the antigen return. Memory cells respond more rapidly, proliferate in greater numbers, and produce greater amounts of antibody. In addition, the antibody produced has greater affinity and avidity for the antigen in question.

Self-/Nonself-Discrimination

As we noted earlier, the thymus is a primary lymphoid organ in which T cells from the bone marrow achieve maturity and then enter the general circulation and take up residence in the secondary lymphoid organs. The cells are said to be mature because they now express cell surface markers that endow them with unique functional properties (e.g., helper or CTLs). More importantly, a T cell remains viable and exits the thymus because it has failed to show reactivity to self-tissue molecules, and recognizes histocompatibility molecules that function to present antigens that are nonself to the TCR (cells that fail to achieve these criteria undergo apoptosis and are removed by phagocytic cells). The mature T cell, therefore, is able to discriminate between molecules that are self and those that are nonself, and in so doing restricts responding against the host. While this intrathymic development and maturation of T cells is one of the most compelling properties of the immune system, it can go wrong, resulting in autoimmune disease.

In summary, these features of the adaptive immune response comprise a high level of plasticity and regulation, and demonstrate highly focused attention to specific molecules (viz., antigens) that are not integral parts of the host. The presence of memory for encounters with antigen ensures accelerated, precise, and expedient elimination of reemerging microbial agents.

THE IMMUNE RESPONSE TO ANTIGEN PROGRESSES THROUGH A DISTINCT SET OF PHASES

The adaptive immune response can be subdivided into three distinct phases or steps. These phases progress along a continuum in which the cellular response is first *induced*, and then the intracellular biochemical machinery of the cell is *activated*, and finally, the effector arm of the response is engaged, generating actions designed to eliminate the microbial antigens. The inductive stage may also be referred to as the *cognitive* phase, since it involves recognition and processing of antigen, while the closely related activation phase involves signal transduction through cell surface receptor stimulation with antigen, as well as signaling by other accessory cell surface molecules expressed on APCs. The activated lymphocyte may be considered "switched on," and will undergo *transformation* and *differentiation*, which includes mitosis and expansion of the original clonal population, changes in individual cell size, and reorganization of intracellular organelles and cytoplasmic space. This latter aspect observes the cells in their differentiated state, when they are fully capable of performing effector functions. Not surprisingly, this is referred to as the effector (or elimination) phase, because the lymphocytes now "effect" their influence on antigens and antigen-expressing bodies; in essence, they perform the neutralizing and cytotoxic functions for which they evolved. However, effector functions can also include the production of cell-signaling molecules, such as the cytokines, whose actions involve the facilitation of proliferation and differentiation of surrounding lymphocytes that bear the same antigenic-specificity. Consequently, during the effector stage, we can have amplification of the number and range of lymphocyte functions, which is designed to provide a quantitative advantage over the microbial agents that triggered the immune response. One can think of effector function as the final blow (or raining of blows), that a given cell is specialized to perform. Whether this is lysis of a target bacterial cell or virus-infected host cell, or production of soluble factors, such as the cytokines or antibodies, the differentiated cell is imposing an immunologic effect.

CELLS OF THE ADAPTIVE IMMUNE SYSTEM

By way of introducing the lymphocytes, we will make a number of declarative points that serve to introduce these cells more fully, but at the same time summarize some points already made. We will then turn directly to the two major classes of lymphocytes: B lymphocytes, which are the precursors of antibody-secreting cells; and T lymphocytes (thymus-dependent), which serve regulatory and cytotoxic functions.

1. Lymphocytes are the preeminent cells in the immune system. Their proper functioning is central to fulfilling its purpose. This is because lymphocytes determine (1) the specificity of immunity and (2) their response orchestrates the effector limbs of the immune system. All other cells (e.g., monocytes/macrophages, dendritic cells,and granulocytes) that interact with lymphocytes serve accessory functions. For example, dendritic cells specialize in the presentation of antigen to lymphocytes, while macrophages, monocytes, and neutrophils produce regulatory cytokines and remove dead and foreign tissue through phagocytosis. The role of these accessory cells in antigen-presentation and innate immunity will be addressed shortly.

2. Each individual lymphocyte, as we've seen, exists as part of a set of unique clones pre-committed to respond to a restricted range or molecular segment of structurally related antigens. This commitment is due to the presence of cell surface receptors (antigen receptors), which recognize antigenic determinants (i.e., epitopes). Given that each lymphocyte membrane is peppered with thousands of antigen receptors (Labrecque et al., 2001), all the antigen-specific receptors on a single lymphocyte clone will recognize the same epitope. The eponymously named

molecule, the TCR, is the antigen-binding receptor on T cells, while the antigen receptor on B cells is actually the antigen-specific antibody molecule embedded in the cell membrane. Since each B cell belongs to a clonal group (cells with the same antigen specificity), the antibody that is expressed on the surface of the cell, is specific to only a single antigenic determinant. Once the antigen stimulates the membrane antibody, the B cell undergoes transformation and differentiation, and in the cytoplasm, actively makes and then secretes antibody that has the same antigen specificity as the membrane-bound antibody.

3. Each lymphocyte clone differs from other lymphocyte clones on the basis of the structure of its receptor binding site. This is the part of the antigen receptor that binds and recognizes the antigenic determinant or epitope. This means that different clones will differ in terms of the epitopes that they will recognize on an antigen. Thus, lymphocytes, as a class of antigen-specific responders are actually a heterogeneous group of cells. Some estimates put the number of possible different epitopes recognized by lymphocytes at around 1 billion, and commensurately a comparable number of different sets of clones are present. This also means that only a fraction of the total number of lymphocytes will be engaged in an immune response against any given virus or other microbial agent.

Fig. 2.3 provides a general schematic of the origin, differentiation, maturation, and ultimate effector role of T and B lymphocytes. After reading through the text, a return to this chart will make clear the general roles played by this rich armamentarium of specialized cells.

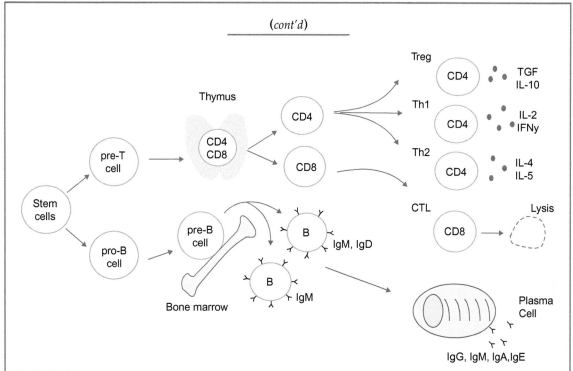

(cont'd)

FIGURE 2.3 *Development of lymphoid cells into effector T and B lymphocytes.* From the lymphoid progenitor stem cell, lymphocytes differentiate into pre-T and pre-B cells, which then separately develop in the bone marrow or thymus into mature lymphocytes. In their mature state, they are typically naïve or "virgin" cells until called upon to act if their particular antigen (as per the clonal selection theory) has entered the body. The development of B cells is marked by surface expression of immunoglobulin molecules (esp. IgM and IgD), which are essentially antigen receptors. Once B cells have been activated, they produce copious amounts of antibody. In this state, they are known as plasma cells, which will eventually revert into a small subset of antigen-specific memory cells that will typically express most of the antibody isotypes. For T cells, the immature state is marked by the double positive CD4/CD8 expressing state of an immature T cell. In the thymus, such cells will lose one or the other molecule to become either CD4 or CD8 positive, but not both (cells that fail to become either one or the other may die and enter programmed cell death). Those that survive and leave the thymus, will circulate as mature T cells. As can be seen, the CD8 T cells will progress into CTLs, which will mainly attack and lyse virus infected cells, and along the way during their fight against infectious antigens, develop into memory cells that can be reactivated more quickly. Similarly, the CD4 T cells differentiate into a variety of T cells that have helper (Th1 and Th2) and regulatory functions (Treg).

B Lymphocytes and Antibody

In this section, we will take a general look at B cells and their main role in the immune system: Antibody production. This is perhaps the preeminent role that the immune system has in relation to eliminating foreign particles. The cytotoxic removal of bacteria and neutralization of viruses is dramatically facilitated by antibody. Moreover, the activation of B cells propagates a highly energetic process that involves the production of billions of antibody molecules, present in such abundance that antigen would have to be deeply sequestered to escape detection and binding by antibody. Failure to develop antibody responses results in serious pathology. Therefore, we will begin first by discussing B cells, and then move on to the antibody molecule and the immunoglobulin class of proteins. Afterward, we will turn to T lymphocytes, the additional line of immunological defense that work to augment B cell responses, as well as promote cytotoxic effector functions that expose viruses and other antigens to the binding functions of antibody.

Development of B Cells

Lymphocytes, whether T or B lymphocytes (we will refer to them interchangeably as lymphocytes or simply "cells"), are derived from self-renewing HSCs (see Fig. 2.1). For B cells, most of this development takes place in the bone marrow, although prior to birth, and early during embryonic development, this commences in the fetal yolk sac [6] and after its retardation persists in the fetal liver. However, the bone marrow is the predominant site for B cell development and continues through life. Prior to becoming a fully mature B lymphocyte, a developing B cell passes through various differentiative stages globally called pro-, pre-, and immature B lymphocyte stages. Key events influence the progression of B cells through these stages. This includes the presence of growth factors, genetic rearrangement of antigen receptors, and selection of cells with appropriate antigen receptors. This will ensure that the mature population of B lymphocytes that enter the circulation and

take up residence in secondary lymphoid organs is fully competent.

B Cell Induction

B cell induction occurs through binding of antigenic determinants to the B cell receptor. This receptor is an immunoglobulin (Ig) molecule[7], a heterodimeric protein that possesses antibody activity. Therefore, when we refer to "antibodies," as a functional class of proteins, we should note that these are Ig molecules. Moreover, when integrated into the cell membrane, the Ig molecule serves the initial function of being an antigen receptor. Once Ig molecules are released subsequent to B cell induction (i.e., antigen binding to the membrane-bound Ig molecule that specifically recognizes it) and activation of the cell, the antigen-specific Ig molecules are released into the extracellular environment at extremely high rates, such that a single B cell may release over 10^5 Ig molecules per minute. Since this is equivalent to the release of antigen receptors into the surrounding environment, Ig molecules are functioning as soluble antigen-binding receptors, or as the German scientist Paul Ehrlich coined them in the late 19th century, *antikörper* (literally, antibody). Interestingly, this principle of releasing soluble "receptors" in order to "mop up" circulating ligands for the receptor, also exists for cytokines. That is, immune cells will release molecules that act as soluble receptors for cytokines. In this case, it serves in a regulatory capacity to reduce or interfere with the function of

---

[6] This provides the initial blood-borne supply of nutrients to the embryo prior to the full development of the placenta; later in gestation, the yolk sac diminishes in size and becomes integrated into the developing gut.

[7] The three main proteins in blood are albumin, globulin, and fibrinogen. Globulin—which received its name for its spherical or global tertiary (three-dimensional) shape has three major types, called alpha, beta, and gamma. The gamma globulin fraction is the immune component, and therefore is referred to as immunoglobulin. It is also the basis for the name given to the basic immunoglobulin unit, IgG. As we will learn, other antibodies are multimeric conglomerates of two or more IgG-like molecules, which resulted in the discovery of other Ig molecules: IgA, IgD, IgE, and IgM.

cytokines that may be in high abundance or no longer needed.

B Cell Activation

The receptor-mediated recognition of antigen results in a complex series of intracellular biochemical events (signal transduction) which cause the B cell to undergo clonal expansion and transformation. This type of activation can either be T lymphocyte dependent (involving cognate T cell help), or T cell independent (cross-linkage dependent).

Cross-linked Activation. Cross-linkage forms of B cell activation require that the same epitope is repeated multiple times on some larger molecule or cell body, such that closely associated Ig receptors on the cell surface can bind (or grasp) each of these epitope exposures at the same time (i.e., cross-link the molecule or cell presenting them). The analogy may be that of the fingers (Ig receptors) of a hand all fitting the five spaces (identical epitopes) provided for their entry into a glove[8]. This occurs with viruses, as they expose multiple copies of proteins thereby repeating identical antigenic determinants at multiple sites on the viral envelope. In addition, common bacteria, such as pneumococci, streptococci, and meningococci, repeatedly express polysaccharides that can bind Ig receptors on B cells. This can serve to anchor a bacterial cell to the B cell, and lead to activation of tyrosine kinase activity and signal transduction through the phosphatidyl-inositol pathway and elevated calcium levels. Membrane cross-linkage activation of B cells is considered to be the major protective immune response mounted against infectious agents. It operates in similar fashion to the PRRs of innate immune cells, such as the neutrophils. As such, it serves to eliminate or neutralize antigen in the absence of specific T cell derived co-stimulatory signals, but does recruit the involvement of other factors, such as complement, a system of circulating proteins with cytotoxic properties that recognize sections of the antibody molecule exposed when it is bound to antigen.

Cognate T cell help of B Cells. After antigen binds to the B cell receptor, it is endocytosed and enzymatically digested into peptide fragments. The peptide fragments are loaded onto *MHC class II* molecules (defined later), which migrate to the membrane to be expressed on the cell surface. The MHC Class II bound peptide interacts with a specific type of T cell (a $CD4^+$ helper T cell) via the TCR on T-helper cells, which has the same specificity for the peptide fragment (antigen) as the B cell Ig surface receptor. This results in the T cell synthesizing and secreting a regulatory substance (a *lymphokine*[9] molecule—but which is now more commonly referred to as a *cytokine*) that binds to a receptor on the B cell. This promotes growth and differentiation of the B cell. In fact, it is of interest to note that IL-6 was once known as B cell growth or differentiation factor. This cytokine is T cell derived, but is also made by cells of the innate immune system, suggesting that these cells can also be recruited to contribute to the activation of B cells. Another cytokine, interleukin-4 (IL-4), is a prominent modulator of B cell antibody production, serving to refine the quality of the antibody produced by B cells, as they progress into the effector stage of responsiveness.

---

[8] Of course, this analogy might work better in reverse—fingers serving as the epitopes (but you get the picture).

[9] In the early days of immunology, soluble mediators derived from lymphocytes were called "lymphokines." This denoted their status as being messengers produced by lymphocytes (lympho). If a soluble messenger came from a monocyte, it was called a *monokine*. Moreover, while we now have the more collective term "cytokine" to refer to a cell-derived messenger or signaling molecule, the original contextually specific terms, such as interleukin ("between leukocytes") have been retained. As a final note, the original Greek meaning from which "kine" was borrowed, relates to movement; therefore, a cytokine is a moving piece of information: a message.

Effector B Cell Response: Plasma Cells

Ultimately, whether through T cell assistance or independently, the activated B cell differentiates into an end stage plasma cell. The B cell is now in the effector stage, and is producing abundant quantities of antibody molecules that specifically bind the antigen that induced and activated the cell. A plasma cell can also be referred to as an *antibody forming cell (AFC)*. In certain older assays of B cell function, single antibody-producing B cells are called plaque-forming cells (PFC) by virtue of their ability to kill (in the presence of complement) sheep erythrocytes or other cells containing antigen against which the B cell is responding.

Memory B Cells

When responding to a specific antigen, during the proliferative or clonal expansion stage (when B cells are differentiating and acquiring the capacity to synthesize and secrete antibody), a fraction of the responding B cells reverts into a resting, dormant state. These cells are the B cell memory cells. When antigen is encountered for the first time, the response induced (whether B or T cell mediated) comprises a *primary* response, and the cells that are responding are said to have been "primed." Since many cells die after the primary response is complete, the reactivation of memory cells by the return of antigen is referred to as a *secondary* response[10]. The memory cells can be referred to as sensitized or primed cells, and this more accurately

reflects the use of these terms, since the cells responding are the surviving cells that possess memory for the initial antigen exposure. Finally, it should be noted that the molecular and cellular profile of the secondary response can be quite different from that seen in the primary response. Relative to the primary response, a secondary antibody response is (1) greater in magnitude, (2) more prompt, (3) the antibody has greater affinity and avidity for the antigen, and (4) certain immunoglobulin classes of antibody (viz., IgG) dominate the response.

Distribution of Antibody

At the cellular level, antibody molecules are found in the endoplasmic reticulum and Golgi apparatus, and are also expressed on the surface of the cell where, as we have already said, they act as receptors for antigen. However, once secreted, antibody is most abundant in the fluid portion of blood (plasma/serum), although as B cells circulate through lymphatic vessels, as well as traffic to lymph nodes, detection of antibody in these areas is expected. However, for evidence of a strong and effective immunization against a given antigen (e.g., flu virus vaccinations), measurement of serum antibody is the clinical standard[11]. In addition, antibody is present in secretory fluids, such as the mucus of the upper respiratory tract and the GI tract. It is also found in a mother's milk after parturition, which confers passive protection to the neonate

---

[10] One should note that subsequent responses to further antigenic reexposure are called tertiary or quarternary responses. However, this becomes cumbersome to discuss, suffice it that responses taking place after the primary response are essentially memory responses. When it comes to vaccines, subsequent memory responses are induced by "booster" shots and these are designed to produce the eponymous boost in antibody production.

[11] We should note that when serum is referred to as *antiserum*, this indicates that it contains antibodies against specific antigens. However, using the term "antiserum" is not synonymous with "antibody," which refers to an individual molecule, part of billions present in the antiserum. From a practical standpoint, however, antiserum possesses antibody, and therefore, antigen neutralizing properties. Many techniques in biological research, as well as clinical testing, rely on the application of serum from immunized animals being used to reveal antigens or block certain biological effects. These are considered immunochemical procedures, since they capitalize on actions of antibody in the "raised" serum.

against pathogens to which the mother has already developed immunity. The main Ig type in the pool of antibody found in these areas—which are part of the common mucosal immune system—is IgA.

With respect to effector functions, antibody can bind to the surface of innate immune cells, such as phagocytic cells (e.g., macrophages and polymorphonuclear leukocytes), as well as non-phagocytic cells such as NK cells and mast cells. Antibody is not produced by these cells, but a receptor for antibody is present on the surface of these cells. This receptor does not bind the portion of the antibody that binds antigen. This capture of antibody by innate immune cells serves to bring them closer to the antigen that antibodies bind, thereby allowing the phagocytic immune cells to engulf and destroy the antigen or microbe that is expressing the antigen. Earlier we mentioned opsonization. In this scenario, antibody is the opsonin, and the phagocytes ingest the antigen-antibody assembly. This is a highly cooperative function between B cells and innate immune cells, which complete the effector function initiated by antibody binding of antigen. Alternatively, a less positive outcome of antibody binding of surface antigens is that involving mast cells. These cells express a receptor for the IgE form of antibody molecule, which results in the release of mast cell histamine and the development of symptoms associated with allergy (runny nose, teary eyes).

THE ANTIBODY MOLECULE: AN IMMUNOGLOBULIN

Serum contains numerous proteins: Albumin, globulins, and other proteins (e.g., acute phase proteins, fibrinogen, and hormones). These proteins can be separated by electrophoresis according to size or charge, with the globulins separating into three distinct zones called alpha, beta, and gamma. The gamma fraction was found to possess immune activity, in that it can bind antigenic molecules. This ultimately revealed that the immune-binding activity of serum was present in gamma globulin proteins, which is where we find antibody in the molecular form that we call immunoglobulin (Ig).

Structure of Immunoglobulin

The basic unit of an immunoglobulin protein consists of two pairs of identical amino acid chains, which make the basic Ig unit a dimeric protein. The chains of one pair are called *Heavy (H) chains*, while the chains of the other pair are called *Light (L) chains* (see Fig. 2.4). These distinctions are based on differences in molecular weight between the H and L chains, the former having the greater number of amino acids. Emphasis on referring to a "basic Ig unit" comes from the fact that immunoglobulins can consist of one or more of these basic units.

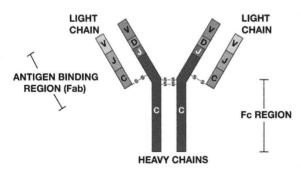

FIGURE 2.4 Basic structure of the immunoglobulin molecule. The antigen-binding component (Fab) of the Ig molecule consists of variable amino acid sequences that show binding affinity and avidity to the peptide antigen that is recognized. This variability in amino acid sequence is dictated by sections of the genome denoted by the letters V, J, D, and C. The amino acid sequence of the Fc region of the molecule remains relatively constant and binds to Fc receptors on macrophages and neutrophils, as described in the text. *Source: From Sompayrac (2016). How the Immune System Works, Fifth Edition. © 2016 John Wiley & Sons, Ltd. Published 2016 by John Wiley & Sons, Ltd.*

Various physicochemical properties (amino acid sequences, size, charge, and conformation) of the H chain determine the identity of an immunoglobulin. Thus, there are five main types (actually called *isotypes*)[12] or *classes* of immunoglobulin: IgG, IgM, IgA, IgD, and IgE. The IgG molecule conforms to the basic unit shown in Fig. 2.4. The H chain of the IgG basic unit will have different physicochemical characteristics from the H chain of the basic Ig units that make up IgM (which has five basic Ig units) or IgA (which has two basic Ig units). That is, what defines an immunoglobulin as IgG, IgA, IgM, and so on, is the composition of the H chain.

The IgG and IgM molecules are the most prominent immunoglobulins released during the first 5 days of a primary antibody response, with IgM being most abundant initially, but followed, within days, by greater production of IgG. However, during a secondary immune response, IgG is the most prominent antibody produced very early after antigen reexposure, with the exception of mucosal regions, such as the gut, lungs, and respiratory tract, where IgA production is prominent. Still, whether systemically or in mucosal areas, IgG is always a key immunoglobulin, persistently refined by activated B cells to have greater affinity and avidity with prolonged or repeated exposure to antigen. This is why it is the main immunoglobulin produced in secondary immune responses.

Why Do Antibodies Possess Specificity?

Within a given Ig class or isotype, there are regions on each H and L chain that are shared (or *constant*) across all Ig molecules of that class. But, there are also regions that are not shared. The regions that differ in amino acid sequence from one immunoglobulin molecule of a given class to another are called *variable* or *V regions*. These variable regions are the binding sites for antigenic determinants. Therefore, if antibody is produced against several different antigenic determinants, within the same class of immunoglobulin secreted against these antigens, there will be differences in the amino acid sequences in the variable regions of each antibody molecule. For example, after immunization with a given virus, a sample of antiserum will have a fraction of IgG molecules specific for one epitope (e.g., sequence A of the viral protein envelope), while another fraction of IgG molecules will be specific for a different epitope (e.g., sequence B of the viral protein envelope). However, collectively all the IgG molecules produced by the B cells responding to the virus will be able to bind the virus. What will be different among the IgG molecules directed to different epitopes of the virus is the amino acid sequence in the V region of the IgG. But, we should note that some IgG molecules with the same epitope specificity may have additional differences that delegate the IgG to different subsets or isotypes (e.g., IgG1, IgG2). Interestingly, this relatively more recent understanding of antibody structure and function also relates to the types of T cells and cytokines that impact antibody-producing B cells. For example, B cells may be releasing IgG1 against a different epitope, but in the presence of a given cytokine produced by helper T cells, there is a shift to the production of IgG2, a process called isotype switching (although one could just as easily say subtype switching). Similarly, B cells switching from IgM production to mainly IgG, is also called isotype class switching.

---

[12] Note that each class of Ig is also referred to as an *isotype*. For example, the statement, "immunization with the protein antigen, ovalbumin, results in the generation of several isotypes of antibody," is merely saying that the immunization has resulted in B cells producing antibody that consists of IgG, IgM, and IgA immunoglobulin forms, with perhaps other forms, such as IgD and IgE, being less abundant in response to the antigen.

The different amino acid sequences in the V regions of an antibody Ig molecule determine the *affinity* and *avidity* of the Ig to various antigenic determinants. Affinity and avidity are terms that refer to the degree of specificity (affinity) and strength of binding (avidity) between the antibody molecule and the antigenic determinant. The V region is found in the antigen-binding segment of an immunoglobulin molecule, and this is called the *Fab fragment* (fragment antigen-binding). The constant region is referred to as the *Fc fragment* (since the constant region was shown to crystallize in solution through self-association).

The Function of Antibody Is to Neutralize Antigen

There are three main processes by which antibody serves to eliminate and/or neutralize antigen. These are described as follows.

1. When protein is liberated from a virus or bacteria or simply ingested (e.g., a toxin) or injected (experimentally), it will be known as soluble antigen, and will circulate through the blood and lymph. This soluble antigen will be bound to antibody and form an antigen-antibody pairing that can further be bound by free portions of the Fab region of antibody that is already paired with this same protein antigen. What eventuates is a network of connections called an *immune complex*. The immune complex can be trapped in tissue where local macrophages can bind the complexed antibody via the Fc region (i.e., via Fc receptors on the cell surface). The immune complex will be endocytosed (incorporated into the inner cell compartment) and the immune complex will be rendered innocuous by enzymatic breakdown.

2. The degradative fate of immune complexes, is aligned with a process called *opsonization*. The term "opsonin" is derived from the Greek, meaning "prepare to eat." Therefore, antibodies are opsonins, in the sense that they bind antigens or surfaces riddled with antigens, and dress the target body for phagocyte consumption. The target body can be bacteria, viruses, and other insoluble particles. For example, when antibody binds to the surface antigens of a bacterial cell, it exposes the Fc portion for binding to the Fc receptor on phagocytic macrophages. Once the macrophage has "docked" at the antibody-bacteria complex, the bacterial cell is ingested and degraded by intracellular enzymes.

3. An alternative to opsonization is activation of the *complement* system. Complement is a set of approximately 20 proteins that are made primarily by the liver, and which are available in blood for a variety of cytotoxic functions. The proteins are modified sequentially to eventually achieve a cytolytic capability. This can be done independently of antibody (in which case, it is part of a humoral innate immune defense), as well as with the aid of antibody. In addition to being synthesized by cells in the liver (hepatocytes), macrophages also generate complement proteins. Overall, complement is an essential and ancient part of immune defense that most likely predated the arrival of antibody-mediated immune responses.

It appears, in summary, that there is a common principle at play when Ag-Ab complexes are formed. Once bound, antigen is summarily degraded and removed by innate cellular (macrophages) and biochemical (complement) mechanisms. In order for this to happen expediently, massive amounts of antibody is required to ensure that detection of antigen is inevitable. This is why failure to generate antibody effectively prolongs infection and promotes morbidity.

Immunoglobulin Class Switching

As we noted earlier, the immune response has primary and secondary characteristics. In a primary antibody response, IgM is the most prominent immunoglobulin secreted. However, with time, specific gene rearrangements begin to generate antibody with greater specificity for the various antigenic determinants that induced the B cell response in the first place. The major class (or isotype) of immunoglobulin in this regard is IgG. However, IgA is the predominant isotype at mucosal surfaces (e.g., GI, respiratory and urogenital tracts).

When antigen is reintroduced to the host, a secondary response is generated (i.e., the memory response), and the predominant class of immunoglobulin will be IgG (if the response is systemic) or IgA (if at the mucosal surfaces). This is not to say that other immunoglobulin classes will not be produced. Rather, the IgG and IgA molecules that are produced in the memory response will be greater in concentration and have greater affinity and avidity, which will allow for faster and more effective binding (immune-complexing). Further, the more robust and prolific production of antibody will ensure that antigen is bound, and subjected to the processes of neutralization and degradation as described earlier.

T LYMPHOCYTES

Development of T Cells

T cells leave the bone marrow in an undifferentiated state and migrate to the thymus gland. Within the thymus, T cells mature into cells which express various distinct molecules (e.g., CD4, CD8, and a TCR) that mediate important T cell functions). Maturation in the thymus is essentially a selection process based on whether T cells possess (1) close affinity for molecules expressed by the major histocompatibility complex (MHC) and (2) the absence of reactivity to molecules that are not encoded by MHC and that are expressed on self-tissue. This latter criterion is the basis of *self-tolerance*. Therefore, a mature T cell will not react against non-MHC self-antigens—which are all the proteins and peptides expressed by endogenous host cells— but will strongly recognize MHC encoded molecules. This optimizes the ability of T cells to recognize foreign antigens (which are not part of the host), and to acknowledge that they are being presented to the T cell by host MHC-encoded molecules. This is the molecular basis of antigen presentation, and why T cells need to develop strong affinity to MHC proteins. As we will learn, this MHC requirement allows for selective engagement of different subtypes of T cells.

THE CD MOLECULE

A large number of CD molecules (several hundred, in fact) have been identified on lymphocytes and other cells of the immune system (and also applied to nonimmune cells). In many cases, cells are almost always described in terms of the cell surface markers they express. In some cases, it can get downright confusing, since a given CD molecule may be expressed on multiple cell types, as for instance, CD2, a cell adhesion molecule found on T cells and NK cells. The term "CD" is derived from the expression "cluster of differentiation," to reflect the identification

of a unique protein through immunohistochemical procedures (i.e., using antibody recognition). In almost all cases, a particular CD molecule is given a number, in order to simplify nomenclature. This is because many CD molecules initially had different names (e.g., CD2 was once called "T cell surface antigen"). It was eventually found to enhance binding of APCs to T cells, but not in an indispensable manner, since CD2 knockout mice were not terribly immunocompromised. In many cases, CD molecules are involved in cell-cell interaction and receptor function, but in

---

(cont'd)

some cases the function is unknown. Nonetheless they do serve as a useful marker of a given cell type, as is the case for T cells, which are all CD3-positive, but as mentioned earlier, not all are CD4- or CD8-positive. Moreover, as described in the text, the CD3 molecule turns out to be more than just a differentiating marker. It forms a complex with the TCR, which ensures antigen recognition. Moreover, most B cells express a marker called CD19, and innate immune cells (macrophages, granulocytes and dendritic cells) express CD14.

Generating antibodies that specifically recognize these various CD molecules allows immunologists to isolate and purify cells that have a particular CD signature. Techniques used to do this are beyond the scope of this chapter, but isolating, for example, $CD3^+/CD4^+/CD8^-$ cells allowed immunologists to determine that when these cells were added to purified CD19 + (B cells) and CD14 + (dendritic cells) cells, antibody production was substantially augmented. From this was born the notion of T cell help, and which was confined to CD4, and not CD8, T cells.

On leaving the thymus, all T cells express a TCR and CD3 molecule expressed on their cell surface. However, they have further identifying characteristics that are functionally important, and segregate T cells into two functional classes that express either the CD4 or CD8 molecule. Hence, mature T cells are designated as either $CD3^+/CD4^+/CD8^-$ or $CD3^+/CD4^-/CD8^+$, and can simply be referred to as CD4 and CD8 T cells. These subtypes mediate distinct functions, which we introduced earlier: Helper and cytotoxic functions.

Induction of a T Cell Response

Induction of a T cell response requires a recognition step in which the antigenic determinant is specifically bound to a receptor expressed on the T cell surface, which is referred to as the *T cell antigen receptor* (or just TCR). Unlike the B cell antigen receptor, the TCR does not disengage from the cell membrane, but serves as one end of a coupling

event between the T cell and the APC, during which the TCR is determining its complementarity to the antigen. This recognition step between antigen and TCR involves a physical interaction between the T cell and another (non-T) host cell that presents the antigen in association with a protein encoded by the MHC. This antigen-presenting non-T cell will also express additional molecules that allow for the induction of a T cell response. These additional molecules are referred to as accessory molecules (e.g., CD28, B7)[13], and these ensure that T cell activation will occur.

In most cases, activation of T cells will induce the transcription and translation of the gene for the cytokine, interleukin-2 (IL-2; see Table 2.1 for a description of cytokine functions), along with other cytokines. For instance, with regard to T-helper cells (which are CD4 positive), it was found that isolated CD4 T cells could be polarized into what came to be called Th1 and Th2 cells, each with their own particular cytokine profile (e.g., Th1: IL-2 and IFNγ;

---

[13] T cells will express CD28, which will bind the B7 molecule (which is a combination of CD80 and CD86) on the antigen-presenting cell (APC). The complete interaction between the TCR on the T cell with the MHC molecule on the APC, along with CD28 on the T cell interacting with B7 on the APC, ensures activation of the T cell. Yes, it becomes a bit dizzying, but see Fig. 2.5 for a schematic illustration of these interactions.

Th2: IL-4). Subsequently, other subtypes were discovered, and these T cells came to be known as Treg cells, owing to their powerful regulatory or suppressive functions on other immune cells. The Treg cell is CD4 + /CD25 + and is defined by the presence of a gene for FoxP3. The Treg cells are thought to be important to blocking autoimmune responses, and in doing so exert powerful suppressive effects on all other T cells (CD4 and CD8 positive), as well as B cells and DCs. This ensures containment of exuberant and/or prolonged inflammatory responses.

As just alluded to, the presentation of processed antigen in association with a Class II MHC molecule is performed by an APC. The major cells that function as APCs are macrophages, B cells, and DCs[14]. Of these cells, the DCs are considered true professional APCs, whose role is exclusively to present antigen. They are prominent in all lymphoid organs, capable of cytokine production, and possess the morphology (i.e., multiple dendritic arborizations) that maximize surface expression of MHC Class II molecules loaded with antigenic peptides.

Overall, *induction* involves a set of cellular and molecular interactions which culminate in an antigenic determinant being specifically recognized by the appropriate T cell clone (the one T cell in a million that will recognize the particular molecular characteristics of the antigenic determinant that is presented). To understand this more fully, it is necessary to consider more closely the MHC, and the TCR.

The Major Histocompatibility Complex

The MHC is a set (or *complex*) of genes found in all vertebrate mammals and located on distinct chromosomes (viz., chromosome 6 in humans and chromosome 17 in the mouse). The human MHC is also referred to as the HLA region, which is an abbreviation for "human leukocyte antigen." In the mouse, the MHC is known as the H2 region. Skipping between these two species may generate some confusing use of terminology, but essentially each MHC region is similarly organized into groups of alleles that code for the different MHC molecules that are involved in antigen presentation. To limit the confusion, when not making reference to any particular species, the literature uses MHC[15], but within a particular species will adopt more specific terms.

The genes (or alleles) of the MHC code for protein molecules that are expressed on the cell membrane and which play a critical role in the presentation of antigen to T cells. The main protein molecules encoded by the MHC are the Class I and Class II molecules. There are also MHC Class III molecules, but they are more secretory in nature, and will not be discussed here.

---

[14] Not only do B cells produce antibody, but they are also capable of expressing MHC II and presenting antigen, this will normally occur in the context of cognate T cell help as was described earlier.

[15] The use of the abbreviation "MHC" sometimes refers to the Class I and II molecules encoded by the MHC. For example, "MHC-peptide complex" refers to an MHC molecule bound by an antigenic determinant (usually a peptide fragment of the original, larger antigenic protein). Further, the phrase, "MHC expression," refers to the production and cell surface appearance of the Class I and II proteins encoded by the MHC. Thus, it is important to remember that MHC refers to a unique set of genes, while MHC "molecules,", "proteins," or "expression," refers to the products of these genes when they are transcribed and subsequently translated into protein molecules.

Class I MHC molecules are expressed on most cells in the body: Cells of the immune system and cells not of the immune system. For example, MHC I molecules are found in the brain, and are important in shaping the development of the brain. Therefore, in addition to cells of the immune system, we can include neurons, pancreatic cells, kidney cells, epithelial cells, heart cells, and other important tissues, as all possessing a functional gene that codes for the production and expression of MHC Class I molecule.

The Class I MHC molecules are glycoproteins which consist of a large alpha-chain that is linked to another protein called beta2-microglobulin. The alpha chain is highly variable (i.e., *polymorphic*) at the extracellular amino-terminal end, and binds the antigenic peptide and components of the TCR. These regions (whether in the mouse or human) all display very high degrees of polymorphism, which essentially means that the peptide-binding region of the MHC I molecule is able to bind many different types of amino acid sequences and with different physicochemical properties (e.g., hydrophobic and acidic/basic properties). Finally, Class I molecules bind only very small peptides (approx. average size 9 amino acids).

Antigen presentation via MHC Class I occurs after intracellular antigen processing that then loads peptide fragments (of the larger protein antigen) onto the alpha chain of the MHC I molecule. This peptide-MHC assembly is then transported to the surface of the cell for extracellular expression. Thus, when a virus infects a host cell, the virus can be internalized, digested and it's "bits" (peptide fragments) expressed by the MHC I molecule on the cell surface. Any local or passing CD8 + T cells can interact with the infected cell, and by virtue of their expression of CD8, will recognize the MHC I molecule (a trick it learned in the thymus), and then bring it's TCR around to determine whether it is the clone for the particular peptide being displayed. If there is no

recognition of the viral peptide, another CD8 cell—the right one perhaps—will eventually come by and make the correct recognition and swing into action to lyse the infected cell and liberate viral particles for immune-complexing and opsonization.

Since viruses can burrow into practically any type of cell, all cells of the body are susceptible to viral infection. Therefore, the advantage of MHC I expression being ubiquitous is to ensure that presence of a virus can be advertised to the cells of the immune system, no matter where the virus has taken up lodging. This does require sacrifice, as infected cells are killed in the process of CD8 (and NK cell) cytotoxicity, as well as antibody-dependent cytotoxicity (i.e., opsonization of host cells), and this is why the process needs to progress quickly before viral spreading compromises organ function.

Class II MHC molecules are expressed in a more restricted manner, being found on B cells, macrophages, DCs, epidermal langerhans cells, thymic epithelial cells, and in the case of humans, activated T cells. The presence of MHC molecules on epithelial cells of the thymus is essential in determining whether mature T cells are selected for release into the peripheral T cell population. In the mouse, Class II molecules are encoded by the *I* (for "immune") region of the MHC, containing two subregions H2-A and H2-E, which code for the various components of the MHC II molecule, said to be I-region-associated (Ia). In humans, there are three sets of genes that code for the Class II molecule. These are the DR, DQ, and DP regions of the HLA complex (also referred to as HLA-DR, HLA-DQ and HLA-DP), each of which codes for the whole MHC Class II protein assembly. This means that there can be polymorphism in the MHC II molecule, as well as differential distribution of these polymorphic forms of MHC II in different cells and different body regions (e.g., gut vs lymph nodes).

The actual MHC II protein assembly consists of two glycoproteins: The alpha and beta chains[16]. Extracellular regions of each chain contain domains that are highly polymorphic (i.e., variable in terms of the consistency of amino acid sequences), since this is where the molecule binds the processed antigenic peptides. In contrast to the Class I molecule, peptides of much larger size can be bound by the Class II molecule (e.g., 13–25 amino acids in length). Furthermore, MHC II interactions are primarily with CD4+ T cells. The CD4 molecule is the mechanism by which the T cell recognizes an MHC II molecule, and which was the purpose of the education received in the thymus. Graduation from the thymus for CD4+/CD8− T cells rested on the condition that they acknowledged MHC II molecules and ignored MHC I molecules. As such, CD4 T cells are the chief mechanism by which the immune response to antigen is amplified and regulated (such as helping B cells).

Summarizing briefly, the major function of MHC encoded Class I and II molecules is the binding and presentation of peptides to T cells whose receptors are capable of recognizing the MHC-peptide complex. The MHC Class I molecules are widespread in most tissue cells, while the MHC II molecule is exclusive to immune cells, which use this in a more "cognitive" manner to direct attention to antigens from foreign entities. As such, professional APCs, such as DCs, will be localized in most areas where interactions with T lymphocytes will be maximized, such as in lymph nodes.

The T Cell Receptor

The T cell antigen receptor (TCR) is a heterodimeric protein (composed of two unique divisions) found on the surface of all T lymphocytes. On most T cells, the TCR consists of the *alpha* and *beta* chains, and is therefore designated the alpha/beta TCR (A gamma/delta TCR is found on a very small percentage of T cells). The TCR heterodimeric protein is usually complexed with several other protein molecules, in particular the CD3 molecule. This TCR-CD3 complex is sometimes referred to as the *functional TCR complex*. Whether a T cell is CD4+ and thereby specialized for recognizing MHC II, or CD8+ and specialized for recognizing MHC I, the peptide antigen being presented on the MHC I or II molecule can be the same and needs to be recognized by the precommitted T cell clone circulating somewhere in the body. Since CD3 facilitates this TCR-mediated presentation, it makes sense for it to be present on all T cells.

The $TCR_{\alpha/\beta}$ heterodimer interacts directly with antigenic determinants via variable regions on the alpha and beta chains. The arrangement of these chains is very similar to that of immunoglobulins (an antigen-binding variable-region and a nonantigen binding constant-region). The purpose of the TCR—as we have been noting—is to recognize specific antigenic peptides and initiate intracellular signals that will result in cell division (mitosis or proliferation) and differentiation toward an effector state. The CD3 molecule and other surface molecules mediate these intracellular signal transduction mechanisms, and activate the T cell.

Antigen Presentation to T Cells

The TCR binds to and recognizes all appropriate antigenic peptides while they are bound to either a Class I or Class II molecule. However, T cells that are CD4+ will recognize antigen only if it is bound by Class II molecules, and T cells that are CD8+ will recognize

[16] We should note the alpha chain in the MHC II molecule is not the same one as for the MHC I molecule. Across different proteins with multiple folding subunits, alpha, beta, gamma, etc., are formal designations for distinct sequences (chains of amino acids) within the larger protein.

antigen only if it is bound by Class I molecules. Cells that present antigen in the presence of Class II MHC molecules are generally considered professional APCs. Strictly speaking this applies to DCs, while macrophages and B cells also readily perform APC functions, while at the same engaging in phagocytic activity and antibody production, respectively.

Macrophages ingest large foreign molecules by phagocytosis or endocytosis. The antigen is then digested by intracellular enzymes, and the antigen fragments are loaded onto the MHC Class II molecule and transported to the cell surface. Macrophages generally do not express MHC Class II molecules constitutively. That is, the MHC II molecule has to be

induced, most commonly by a cytokine called interferon-γ, which is a major inducer of MHC II molecules.

B lymphocytes can bind native antigen (whole and unprocessed) and internalize it by endocytosis. Enzymatically facilitated fragmentation of the antigen takes place, followed by loading onto a MHC II molecule.

DCs are so named because of their highly arborized morphology consisting of an extensive array of thin processes. These serve to both sample their surroundings, as well as make contact with other cells. In many ways, these cells resemble the microglial cells of the brain, also known to have APC function, albeit quite limited. The development of DCs occurs in the

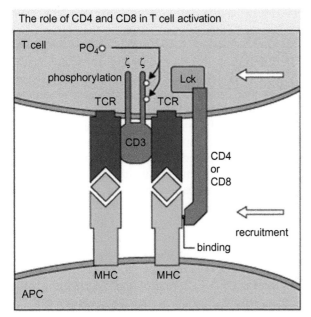
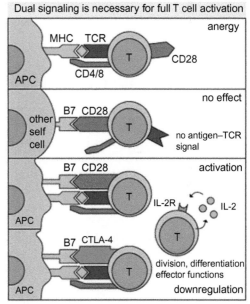

FIGURE 2.5 Interlinked molecular interactions required to produce T cell activation by an antigen-loaded APC. Shown on the left is MHC (whether I or II) presentation of peptide antigen (blue diamond) to the TCR, which is complexed to CD3. If the T cell is CD4 positive, and APC expresses MHC II, recognition will take place, and the kinase Lck will be brought closer to the intracytoplasmic tail of the CD3, which will then be phosphorylated, and which will facilitate signal transduction. Similarly, if the T cell is CD8 positive, and presentation is by MHC I, the result will be the same. On the right is an additional co-stimulatory requirement to ensure that T cell activation will result in IL-2 production and proliferation of the T cell. As mentioned in the text, B7 and CD28 are needed to provide additional signals that will fully activate T cells. Without this co-stimulation, cells can become anergic or unresponsive. Finally, when CD28 is replaced by the surface molecule CTLA-4, T cell activity is reduced or downregulated. *Source: From Male, Brostoff, Roth, & Roitt (2006).*

bone marrow (save for the follicular DCs of lymph nodes) and progresses along the myeloid cell lineage, from a precursor cell that also can redirect differentiation toward the development of monocytes. The alternative switch, to becoming DCs, requires a molecule, Flt3, without which the precursor cell will favor progression toward a monocyte phenotype. The DCs are found in the skin (where they are called Langerhans cells), and in all the various lymphoid organs. As professional APCs, DCs, wherever they may be located, will encounter and ingest foreign proteins, and then seek interaction with lymphocytes to "present" antigenic peptides on MHC II molecules. DCs show considerable initiative in performing this role, as they will actually migrate into the lymph nodes draining the area of antigen exposure, and in a MHC II-dependent manner trigger the lymphocyte response (Fig. 2.5).

T Cell Activation and Effector Function

We have learned that antigen recognition by the TCR is through association with either MHC I or II molecules. This is a necessary condition for activation, but not a sufficient condition. Additional accessory molecules are required to initiate the drive into cell cycle (see earlier discussion regarding T cell expression of CD28). These include CD3, as well the CD4 and CD8 molecules. The CD3 molecule activates signal transduction by means of its close association with the $TCR_{\alpha/\beta}$ heterodimer. The CD4 molecule binds MHC Class II, which contributes to cell signaling. Likewise, the CD8

molecule binds the Class I MHC molecule and is believed to transduce the cell signal. The CD28:B7 interaction completes this set of molecular interactions, and sets up the conditions to initiate T cell activation and movement toward an effector phenotype[17].

T cell effector functions include: (1) helping B cells generate an antibody response (mediated by CD4$^+$ lymphocytes); (2) eliminating virally infected cells (mediated by CD8 T-cytotoxic lymphocytes); and (3) activating macrophages (mediated by T-cell derived cytokines, e.g., IFNγ not only induces MHC II, but also transforms macrophages into a tumoricidal state). Recall that MHC I expression is ubiquitous throughout the body. Given that viruses can infect nonimmune cells, the immune system evolved a mechanism to recognize foreign viral antigens through recognition of MHC I molecules. The CD8 T cell, therefore, has the capacity to attack virally infected cells. This will liberate the virus, and expose it to antibody that may already have been generated by B cells. The drawback of CD8 activity is damage to self-tissue, although if regulated appropriately, the response can limit the spread of the virus and with the aid of antibody, can remove it from the body.

We noted earlier that the CD4 T cell, also called a T-helper cell, provides "help" to B cells through two subtypes of T-helper cells—Th1 and Th2—a distinction that has more or less dominated immunological research over the past decade. Functionally, these subtypes are understood in terms of their cytokine profiles. In more recent years, the T-reg cell (for

---

[17] While interlocking of molecules or "handshakes" between the T cell and APC is the prerequisite for activation, a poorly understood conundrum was how this revved things up in terms of T cell effector function, since simply injecting a benign protein (e.g., egg albumin) into a mouse did not generate much of an immune response. Charles Janeway and colleagues resolved this paradox by showing that adjuvants could kick-start a process that ostensibly was in a stalled state. It was eventually realized that adjuvants induced innate immune responses, and in particular, the production of accessory molecules, which turned out to be cytokines. These accessory molecules stimulated the intracellular machinery of T cells that had recognized their antigen, and onward they progressed into a proliferative and cytokine-producing effector state.

regulatory T cells) has been an active area of T cell research. These cells are primarily CD4 T cells, and serve to suppress or inhibit the responses of CD4 and CD8 T cells. At one time in the history of immunological thinking, there was talk of a T-suppressor cell. The nomenclature has changed, and the precise characteristics of such a cell was not fully understood. We now know that Tregs, as they have come to be known, exert suppressor activity, utilizing specific cytokines, such as IL-10 and TGFβ. Furthermore, an additional T cell, that almost exclusively produces a cytokine called IL-17, has emerged. This T cell, called a T17 helper cell, is induced by the cytokine IL-6, and plays a role in activating innate immune components. Consequently, from the earlier dichotomy of Th1 and Th2 cells, other CD4$^+$ T cell subtypes have emerged that represent important elements of regulation in the adaptive immune response.

CYTOKINES

We have learned that lymphocyte activation with antigen requires cell-to-cell contact with attending membrane molecules, such as the TCR-CD3 complex, MHC I and II, and the B7 and CD28 costimulatory molecules. However, there are millions of lymphocytes and just as many circulating monocytes and resident macrophages, that all need to be tuned in to the events of antigen retrieval, presentation, and ultimate lymphocyte activation. Their recruitment and involvement is part of the necessary amplification of an immune response, in which antigen-specific lymphocytes proliferate and antibody is liberally dispersed to bind and neutralize the bacteria and viruses that express the antigens. For all this to work, there has to be a secondary layer of communication which involves soluble factors that communicate between the individual leukocytes. This secondary layer is the cytokine network. Like neurons

that communicate with each other through neurotransmitters that diffuse across a synapse, so are leukocytes informed and instructed to migrate in a given direction, bind to a particular cell, enter a mitotic phase of cell expansion and differentiation, and perform various effector functions such as cytotoxicity and antibody production. All these functions, and more, are induced by soluble proteins called *cytokines*. The name literally means "cell messenger," and these molecules serve in an autocrine (acting on self—e.g., clonal division) and paracrine (acting on bystanders) manner to mobilize cells toward the goal of finding and eliminating pathogens. Within the immune system, cytokines are the chief effector and regulatory molecules of T cells, monocytes, neutrophils, and macrophages. Similarly, while not commonly recognized as a cytokine producing cell, B lymphocytes also are capable of making and releasing cytokines (Lund, 2008). Furthermore, since their discovery in the immune system, cytokines have been found to be made by other cells in the body, including fibroblasts and endothelial cells, and glial cells (and possibly neurons) in the brain. Therefore, while it is recognized that cytokines are an indispensable element of how the immune system goes about its work, it is also recognized that when immune cells produce cytokines, these may act on cellular targets outside the immune system. This renders cytokines as endocrine signaling molecules.

Within the general family of immune-derived cytokines, there are different categories of signaling and regulatory molecules. Historically, cytokines produced by T cells were called *lymphokines*, whereas those from monocytes and macrophages were simply *monokines*. However, for convenience, and the fact that some molecules were both monokines and lymphokines (produced by monocytes and lymphocytes), the generic term *cytokines* was adopted. However, each individual cytokine has retained a unique name, usually reflective of the original function that was attributed to it, or part of another system of nomenclature. For example, many cytokines are called interleukins (i.e., "between leukocytes") and abbreviated, "IL." Many interleukins are produced by T lymphocytes, but monocytes and macrophages also produce interleukins. As an example, interleukin-1 (IL-1) was one of the first major cytokines to be purified and sequenced at the protein and gene level. However, it was first known under different names, such as endogenous pyrogen and thymocyte-stimulating factor. Both names were based on cell-free solutions from stimulated leukocytes producing pyrogenic effects in animals (i.e., inducing fever) or priming thymocytes for mitotic responses to other factors. Ultimately, whether it was endogenous pyrogen or thymocyte stimulating factor, it eventually became known as IL-1, perhaps the most potent neuromodulatory molecule ever discovered. In fact, the first clue that leukocytes had neural effects was inherent in the pyrogenic effects (a hypothalamic mechanism) of the nonpurified culture supernatant that contained "endogenous pyrogen." No one seemed to catch on, and it was

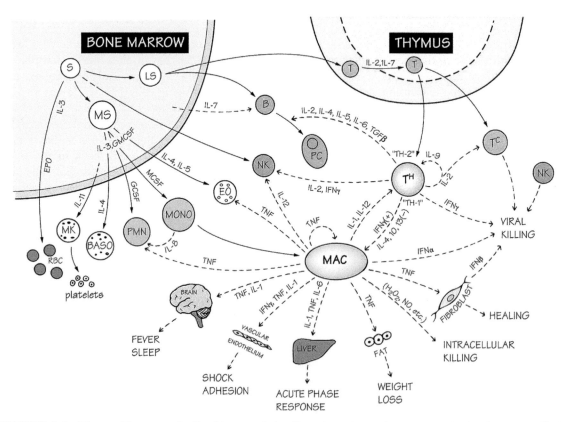

FIGURE 2.6 The cytokine network. Cytokines work in all environments where communication between cells is required at some distance, even if cells are present locally. This is why they are hormone- or neurotransmitter-like. In the bone marrow and thymus, they contribute to the differentiation and maturation of cells. And once immune cells are mature and recirculating between blood and lymphatics, cytokines serve as instructional messages. Note the many cytokines released by T-helper cells (T^H in the figure), which target B cells, and the prominence of TNF and IL-1 as targets for both immune and nonimmune targets. These and other cytokines are central players in neural-immune interactions, and will be mentioned often. *Source: From Playfair and Chain.*

some years before this was more formally demonstrated by Besedovsky and colleagues who demonstrated that IL-1 acted on the hypothalamic-pituitary adrenal axis. In any case, once IL-1 was coined, others followed suit. The next was IL-2, which was originally known as T cell growth factor. This was isolated from cultures of activated T lymphocytes, and was able to promote thymocyte proliferation in the presence of IL-1 (since this allowed for greater expression of IL-2 receptors). Since those early days more than three decades ago, many interleukins have been discovered and the genes cloned. Fig. 2.6 provides a typical description of the cytokine network in operation, and Table 2.1 provides a summary of some of the more prominent cytokines that we will encounter and their basic functions and cellular origins and targets.

The major points to be made about cytokines are as follows:

1. Most are not constitutively expressed and in many cases need to be induced;
2. Their actions are mediated through a specific G-protein coupled receptor synthesized and expressed by the target cell;

3. Many cytokines have overlapping functions (i.e., they are *pleiotropic*), which allows for some redundancy;
4. There is no one role for cytokines—they can cause cells to proliferate (e.g., IL-2), and they can cause cells to die (e.g., TNF promotes apoptosis); their function is instructional, conferring a particular directive that is needed for a given aspect of the immune response to take place (e.g., signaling macrophages to express more MHC Class II molecules, which is one of the directives conveyed by IFNγ, another being to augment the cytotoxic functions of NK cells).

The following table lists some of the major cytokines involved in regulating important immune processes, as well as those that are most commonly studied in psychoneuroimmunology. Immunologists have classified cytokines into superfamilies based on shared structural elements of the receptors they bind. Table 2.1 adopts this grouping, providing descriptions of selected members of (1) the interleukin-1 (IL-1) family, (2) tumor necrosis factor superfamily (3) Type I (e.g.,

TABLE 2.1 Common Cytokines Encountered in Neural-Immune Investigations. Compilation Based on Details from Abbas et al (2015)

Cytokine	Main Immune Cell Origin	Receptor Types	Most Common Function	Target Cells
INTERLEUKIN-1 FAMILY				
IL-1 (IL-1α, IL-1β)	Monocytes, macrophages, and other innate immune cells (incl. DCs)	IL-1R1 IL-1R2	Inflammation, fever	Endothelium, CNS, hepatic cells, T cells
IL-18	As above	IL-18α IL-18β	Inflammation Induces IFNγ by T and NK cells	T cells, NK cells, Monocytes, neutrophils
TUMOR NECROSIS FACTOR SUPER FAMILY				
TNF	Macrophages NK Cells T cells	TNFRI TNFRII	Inflammation, cachexia	Endothelium Neutrophils

(Continued)

TABLE 2.1 (Continued)

Cytokine	Main Immune Cell Origin	Receptor Types	Most Common Function	Target Cells
TYPE I CYTOKINE FAMILY				
IL-2	T cells	IL-2Rα	Mitotic (proliferation) activity	T, B and NK cells
		IL-2Rβ	T cell effector and memory formation	
			Treg generation; NK cytotoxicity	
IL-4	CD4 T cells	IL-4Rα	Isotype switching to IgE, proliferation and differentiation of Th2 cells	B cells
				T cells
	Mast cells			Macrophages
				Mast cells
IL-6	Macrophages	IL-6Rα	Inflammation, acute phase response, B cell differentiation (in plasma cell)	Hepatic cells
		gp130		B cells
	T cells			
IL-12 [IL-12A (p35) IL-12B (p40)]	Macrophages	IL-12Rβ1	Th1 differentiation	T cells
			IFNγ induction	NK cells
	Dendritic cells	IL-12Rβ2	Promotes cytotoxicity	
IL-17 (IL-17A, IL-17F)	CD4 T cells (Th17)	IL-17RA	Increase chemokine and cytokine production by macrophages and endothelial cells	Endothelium
				Macrophages
	Group 3 innate lymphoid cells	IL-17RC		Epithelial cells
TYPE II CYTOKINE FAMILY				
IFNγ	CD4 T cell (Th1)	IFNγR1	Induces MHC II, augments NK cytotoxicity	Macrophages
				NK cells
	CD8 T cell	IFNγR2		
	NK cell			
IL-10	Macrophages	IL-10Rα	Suppression (cytokines, MHC II)	Macrophages
				Dendritic cells
	T cells (esp. Treg)	IL-10Rβ		T cells (esp. Th1)
UNGROUPED CYTOKINES				
Transforming growth factor-β (TGFβ)	T cells (esp. Tregs), macrophages	TGFβR1 TGFβR2 TGFβR3	Suppression (B and T cells; macrophages), Th17 and Treg differentiation, tolerance promotion	T cells B cells Macrophages

IL-2, IL-4), and (4) Type II (IFN-γ and IL-10) cytokines. Only key functions are listed, and of course, one needs to go to the literature to find many other functions and exceptions for each selected cytokines. Overall, there are up to 30 or more major cytokines shown to have important immunological and nonimmunological functions.

Chemokines

The chemokines are a subset of cytokines. Their categorical name comes from a concatenation of "chemo" from chemotactic and "kine" from cytokine. This should indicate the particular role that they play in the recruitment of an immune response. Chemotactic effects have been recognized for a long time, in that the release of certain molecules from a given source act as attractants for cells that are nearby. In principle, the release of a chemokine creates a concentration gradient along which cells will travel or "roll" to reach the area where the chemokine was being released. This can be an effective means to flag down circulating monocytes, for example, and redirect them to a site of infection. The field of chemokine biology has grown dramatically, since these molecules play an important role not only in immune function, but also the development of the nervous system. Axonal guidance, for instance, can now be explained in terms of chemokine release.

A special nomenclature exists for the naming of individual chemokines. These particular cytokines are polypeptides of 8-10 KD in size, and belong to one of four major families. The family category is based on the location of N-terminal cysteine residues in the polypeptide, which determines the name of the particular chemokine. There is the CC family, containing two adjacent cysteine residues; the CXC family, wherein two cysteine residues flank a different amino acid; the CX3C family, in which three amino acids interdigitate between two cysteine residues; and finally, there is a family of chemokines that contain only a single cysteine residue. Like many other cytokines, the chemokines were identified on the basis of the responses they elicited from various cells. The CC and CXC chemokines are produced by multiple cells of the immune system, as well as fibroblasts, endothelial, and epithelial cells. The latter are important sources of signaling to circulating leukocytes, should an infection be taking place in a given tissue location. There are more than 17 different chemokine receptors, each of which is G-protein coupled. Once stimulated, signal transduction will change those elements of a cell that influence motility, including the cytoskeleton actin/myosin polymerization. Although the chemokine receptors are expressed on all leukocytes, a particularly large representation is found on T cells. The relevance of this is obvious, given that T lymphocytes are in continual circulation, and if necessary, can reorient the direction of their movements based on chemokines released by sites of inflammation and injury.

By way of example as to how chemokines can operate, we first note that neutrophils and monocytes are often the first cells recruited to the site of an infection. Local macrophages that have been activated by pathogen can release the chemokine CXCL8 (i.e., CXC family, L for ligand), and since this is a ligand for the chemokine receptors CXCR1 and CXCR2, neutrophils that express these receptors will become modified and "brake" as they pass, and redirect their attention to the area of CXCL8 release. This redirection is driven by molecular changes and expression of adhesion molecules, such as leukocyte adhesion molecule. The neutrophil will then pass between the endothelial cells of a blood vessel, and enter the parenchyma of the infected organ, where it will engage in phagocytic activity and amplification of the immune response through the release of cytokines that act on local and recruited immune cells.

TOLERANCE AND AUTOIMMUNE DISEASE

To this point, we have focused on an overview and description of some of the major elements of the immune system. At the very beginning, we suggested that thinking of the immune system in militaristic terms was useful, since it is essentially a defense system. But, in mounting a defense, it is, like any military campaign, subject to mistakes and considerable calamity and collateral damage. As well, it needs to know who its allies are, and who comprises the enemy. The latter is critical; self-annihilation is not part of the "military" objective. Therefore, when we discussed the thymus and the MHC complex, we noted that recognizing the MHC as part of self was important in order to trust self-reports of an external threat (MHC presenting antigen). Furthermore, a peripheral, mature, extra-thymic T cell already has learned not to react (i.e., become activated and develop effector responses) to any other proteins that the body expresses. The consequence of this self-tolerance is that the organism does not develop autoimmune disease, an immunopathological condition in which the immune system treats certain organs, tissues, and/or molecules expressed by these tissues as essentially foreign or nonself.

There are a numerous debilitating conditions that are defined as autoimmune diseases. Dysregulation of T cell and macrophage function underlies diabetes, thyroiditis and rheumatoid arthritis, while B cell autoantibody responses underlie the immunopathology of myasthenia gravis, systemic lupus erythematosus, and hemolytic anemia. Similarly, multiple sclerosis, a demyelinating disease of the central nervous system, is thought to involve a dysregulated T cell response and monocyte infiltration of the brain and spinal cord. Later we will get to discuss some of the research trying to link autoimmune disease to conditions like autism and schizophrenia.

To conclude, very little is actually known concerning the etiology of many autoimmune diseases. One epidemiological finding is that females are more likely to develop autoimmune disease than men. The reasons for this are unclear, and an obvious hypothesis is sex-hormone regulation. However, for the most part, the reasons why immune cells become auto-aggressive, is still uncertain. To some extent, it provides an opportunity to hypothesize that perhaps certain behavioral abnormalities (like schizophrenia) are the consequence of an autoimmune process.

CONCLUDING COMMENTS

It is hoped that sufficient elemental information has been provided to make the remainder of this book easier to read. Where certain information was omitted, it is highlighted in subsequent chapters. These new pieces of information should make sense, since the purpose of this chapter was to generate some assumption of knowledge, rendering new terms and processes less confusing. We urge the reader to explore further and deeper the details of immunology, and in addition to the citations that have been used, recommend as starting points the texts by Playfair and Chain (2013), Abbas, Lichtman, and Pillai (2015), and Sompayrac (2016). The latter provides a humorous and gentle overview of the immune system, while Playfair and Chain provide excellent illustrations, that offer a pictorial guide of sorts to the immunological landscape. For a more in-depth, and clinical approach, Abbas is an excellent introduction, after which readers should be well-versed to venture farther afield.

Bacteria, Viruses, and the Microbiome

ROCKING THE BOAT

During the early part of the 19th century, postpartum bacterial infections (puerperal infections) were not uncommon and were fatal in as many as 10–15% of cases. As odd as it may seem from today's standards of care, at the time little was understood regarding bacterial infections and proper hygiene within medical settings. Ignaz Semmelweis had the audacity to suggest that postpartum bacterial infections might stem from unhygienic medical staff or by medical equipment that hadn't been cleaned, and he demonstrated that infection frequency could be diminished by having staff wash their hands with chlorinated lime solutions. These suggestions were not well received (weren't the hands of doctors always clean?). With Lister's discovery of antiseptics in 1865 (published in 1867), thanks in part to Pasteurs work related to "germ theory," the practice of maintaining cleanliness became paramount in surgical practice. Arguably, antiseptics and anaesthetics changed surgery and surgical risks forever. However, poor Semmelweis didn't get to see these breakthroughs in medicine, nor was he rewarded for his observations. Instead he was ostracized from others in his profession. He became progressively more depressed, and his battles with institutionalized medicine eventually landed him in a medical asylum. Upon trying to leave, he was beaten by guards, and died two weeks later as a result of internal injuries or gangrene secondary to his injuries.

The situation within hospitals has obviously improved since then, except that hospital acquired infections have been on the rebound for years, including those that are treatment resistant. One terrible condition that can afflict patients is sepsis, comprising infection that spreads to the bloodstream, and the resulting inflammatory cascade can damage various organ systems, eventually leading to death. Early treatment with antibiotics and lots of fluids had been the treatment of choice, but it would be far better to prevent the condition as Semmelweis had done with regard to postpartum bacterial infections. As it happens, simple procedures, such as increased training for staff and the use of a special observation chart to identify early signs of sepsis, can appreciably reduce its occurrence.

Risk of bacterial infection within hospital settings has gone beyond sepsis. Soon after hospital admission patients often lose commensal gut bacteria, whereas pathogenic bacteria might rise

---

(McDonald et al., 2016). This could occur because of the stress of illness, the novel hospital environment, change of sleep pattern, or hospital foods, but whatever the case, this dysbiosis (microbial imbalance) could affect immune functioning, thereby influencing vulnerability to illnesses. With the proviso that resources are available, it may be possible to track individual microbiota that could eventually inform patient vulnerabilities and potentially point to the optimal treatments strategies.

Despite the ignominious treatment Semmelweis received while alive, he was treated better posthumously. He became known as the "saviour of mothers," and the Semmelweis Klinik, a woman's hospital in Vienna, is named after him, as is Semmelweis University in Budapest, and the Semmelweis Hospital in Miskolc (Hungary). The sceptics among us are certainly pleased by the naming of the "Semmelweis reflex," which refers to the reflexive rejection of new knowledge that challenges the old norms and beliefs.

BACTERIAL CHALLENGES

Bacteria, which are thought to be among the first life forms that appeared on earth, are present everywhere, coming in varied sizes and shapes. They typically live in harmony with plants and animals, acting symbiotically, but can also function in a parasitic relationship with other living things. Many of the trillions of bacteria present in animals (and humans) have important beneficial actions, whereas others are less kind, causing a variety of diseases. Under the right environmental conditions, including the temperature and pH, availability of water, oxygen, and a source of energy, bacteria will grow to a particular size and then reproduce asexually through binary fission. Bacteria can double in number every 30 minutes, and some bacteria can do this within 10 minutes. Certain bacteria, such as streptococcus, can stay alive for some time on various external objects (e.g., door handles, toys, and cribs) and thus can represent a fairly persistent threat. Other bacteria, like their viral cousins, are subject to airborne transmission, droplet contact, or direct physical contact. It seems that different strains within the same bacterial species can engender very different immune responses, which may contribute to the diverse outcomes elicited across individuals (Sela, Euler, Correa da Rosa, & Fischetti, 2018).

Infection stemming from bacteria and viruses can also be transmitted indirectly. For example, carrying an infection on unwashed hands, depositing these on a surface, which is then touched by another person, can lead to infection being passed along (fecal-oral transmission). As we know from a number of other illnesses, such as cholera, dysentery, diphtheria, scarlet fever, tuberculosis, typhoid fever, and viral hepatitis, disease agents can also be transmitted through water, ice, food, serum, plasma, or other biological products. In some instances, a disease (e.g., the bacterial infection syphilis, or the parasitic disease toxoplasmosis, as well as viruses, such as HIV and measles) can be passed from a pregnant mother to her fetus. Furthermore, zoonotic diseases in which infection is transmitted from animals to humans, are a constant threat, but can become exceptionally hazardous if they mutate so that they can then be passed between humans.

As we'll see in ensuing chapters, viruses and bacteria, by virtue of inflammatory processes being activated, and the downstream actions on many hormones, brain neurotransmitters, and growth factors, may contribute to several psychological disturbances as well as a great number of physical disorders. To a significant extent, common mechanisms account for these varied conditions, and may be responsible for the frequent comorbidities that are seen amongst illnesses. Increasingly, the significance of infection in relation to mental illnesses has been acknowledged. Methods to prevent or control infection are thus essential in this capacity, but as we'll see, the very best treatments to ameliorate bacterial infection may destroy commensal gut bacteria, leading to the emergence or exacerbation of other illnesses. Clearly, the sword cuts both ways.

Antibiotics

The development of penicillin, and other antibiotics to fight bacteria, was undoubtedly among the most important medical discoveries of the first half of the 20th century. Although Alexander Fleming, who identified penicillin obtained from particular molds, is usually given the credit for antibiotics, infections have been treated with mold extracts for about 2,000 years. In general, antibiotics either kill bacteria (being bacteriocidal) or inhibit their multiplication (bacteriostatic). They do this by either preventing bacteria from building cell walls (e.g., by affecting bacterial ribosomes involved in the creation of cell walls), or breaking down the cell walls of bacteria that already exist. Some antibiotics, such as quinolones, disturb DNA and prevent their repair, so that the bacteria are unable to reproduce and thus die off. Based on the response to a gram stain, and characteristics of the cell walls, bacteria are designated as either gram-positive or gram-negative (the latter being more resistant to antibiotics). When the nature of the bacterial infection is known, a narrow spectrum antibiotic is used, whereas a bacteria that has not been identified is treated with a broad spectrum antibiotic. The former is preferable as they are less likely to create antibiotic resistance. Some antibiotics can produce uncomfortable side effects, and in some instances, allergic reactions can occur that cause anaphylaxis.

Antibiotic Resistance

We grew accustomed to being able to destroy bacteria through treatment with antibiotics, and for some time it had simply been assumed that when one antibiotic failed to do the job effectively, then another could do the trick. Ironically, their very effectiveness contributed to their undoing. As bacteria began to form resistance to antibiotics (reflected by greater difficulty in treating some infectious diseases, lengthier recovery times from infection, and the probability of death increasing), and the first alarms were sounded, a generally cavalier attitude persisted, and most people continued to behave as they had previously. Inevitably, most bacteria followed an effective game plan to get around antibiotics, and hence they all successively became less effective or entirely ineffective.

The factors that generated treatment resistant bacteria comprised the perfect storm. One should never have imagined that bacteria were passive travelers who were simply waiting to be killed by antibiotic agents. Instead, like an opposing army (or groups of terrorists) bent on the host's destruction, some harmful bacteria are clever and vicious, so that with time and experience they develop resistance to the drugs. It was suggested that in response to stressors, such as nutrient deprivation, microbiota respond in a coordinated manner to deal with the insult. Being a new challenge for bacterial communities, an antibiotic might result in bacteria rapidly searching for new methods of dealing with the challenge. Ultimately through a process much like natural selection based on random mutations occurring, bacteria develop resistance to the antibiotic (Jensen, Zhu, & van Opijnen, 2017).

The ability of bacteria to become resistant might have been facilitated by the inappropriate use of antibiotics to fight viruses (e.g., strep throat, bronchitis) for which antibiotics are ineffective. In fact, in the face of a serious threat, such as an antibiotic treatment, bacterial mutation rates increase appreciably, thereby increasing the probability of a mutation occurring that will protect the bacteria from destruction. Furthermore, it was thought that when confronted with an antibiotic, especially if the full course of treatment wasn't adopted (because patients felt better and believed they no longer need the antibiotic, or because they were saving pills in the event that they were needed at some later time), a few hardy bacteria may survive. This may then give rise to similarly resistant clones, so that over successive generations and increased development of evasion methods, the effectiveness of antibiotic agents diminishes[1].

The rate of bacterial mutation increases with a person's age as well as with the social environment in which bacteria find themselves. At the other end of the age spectrum, babies born very prematurely are at increased risk of illnesses developing. As a matter of course, preemies were treated with antibiotics in the mistaken belief that this couldn't cause harm, but this resulted in a marked decline in the diversity of microbiota and simultaneously enhanced survival of bacteria that are resistant to antibiotics. Thus, should a blood infection subsequently arise, a large proportion of these infants will not fair well, especially as resistance to one type of antibiotic also dials up resistance to other antibiotics (Gibson et al., 2016).

The massive use of antibiotics in animals to prevent them from developing infections has contributed to resistant bacteria evolving (Johnson et al., 2016). The antibiotic infested meats end up on our dinner plates, and thus contribute to the development of resistance. As well, some rivers and streams contain *Escherichia coli* resistant to antibiotics. There's a good chance that animal waste, laden with antibiotics and antibiotic-resistant bacteria, leeched into waterways. Air pollution may also affect commensal bacteria, and influence resistance of some bacteria (*Staphylococcus aureas* and *Streptococcus pneumonia*) to antibiotic treatments (Hussey et al., 2017). There has also been the reasonable concern that some household products, such as the disinfectant triclosan, may contribute to antibiotic resistance. As a result, it has been banned from hygiene products, such as hand, skin, and body washes, but triclosan and similar agents appear in numerous other products.

In addition to the mutations that are due to overuse of antibiotics, bacteria have several dirty tricks that they can fall back on. For instance, the genes involved in the development of resistance can be transferred to other cells (conjugation) so that they too will become resistant, although it may be possible to prevent or reverse this action (Lopatkin et al., 2017). Furthermore, in response to an antibiotic, bacteria can go dormant, making them less likely to be attacked (termed persistence). With repeated antibiotic attacks, they essentially "learn" to stay in the dormant state for periods that line-up with the antibiotic's actions, emerging once it seems safe (Fridman, Goldberg, Ronin, Shoresh, & Balaban, 2014). On top of this, bacteria seem to act collectively, coordinating their actions to render maximal toxic

---

[1] The seemingly common sense perspective concerning overuse of antibiotics was almost universally accepted, even though it has been argued that there was actually little evidence supporting this contention. Physicians may be loath to undertreat patients, and thus typically prescribe based on precedent, which could actually reflect *overtreatment*, thereby placing patients at increased risk for antibiotic resistance (Llewelyn et al., 2017). Rather than destroying bacteria entirely, it might be sufficient to simply slow down potentially dangerous bacteria, and in doing so it is less likely that antibiotic resistance will develop (Spellberg, 2016).

effects based on messageing from some external source (quorum sensing), such as the medium in which the bacteria are present (Ng & Bassler, 2009). Bacterial communities can secrete substances, such as β-lactamase, which can proffer passive resistance for other bacteria that are present in that particular environment. Similarly, bacteria that express the resistance factor chloramphenicol acetyltransferase (CAT) can deactivate antibiotics present in that immediate environment. In essence, the response to an antibiotic could be affected by the microbial environment that is present (Sorg et al., 2016).

ANTIBIOTIC RESISTANCE AS AN INCREASING WORLDWIDE THREAT

The WHO has indicated that antibiotic resistance has become among the most pronounced threats to global health and food security. Several bacterial species can cause illnesses by acting as "opportunistic" pathogens. These common threats comprise species such as E. coli and those referred to as the ESKAPE organisms, comprising Enterococcus faecium, Staphylococcus aureus, Klebsiella pneumoniae, Acinetobacter, Pseudomonas, and Enterobacter. Several species and strains of commensal bacteria will readily be destroyed by antibiotics, but may ultimately be replaced by those that are resistant. One of the more recent threats has come out of China, where a hypervirulent form of K. pneumoniae emerged that was multidrug resistant, as well as highly transmissable (Gu et al., 2018).

Staphylococcus aureus (S. aureus or Staph infection), is the best known and most frequent cause of postsurgical infection, but hospital acquired infections have also included bacteremia (bacteria in the blood), endocarditis (inflammation of the inner layer of the heart), sepsis, toxic shock syndrome, meningitis, and pneumonia. We had counted on antibiotics, such as methicillin, in the treatment of such conditions, but a bacterial strain evolved that stopped responding to this agent. It has been estimated that hospital-acquired infections within the US, particularly methicillin-resistant S. aureus (MRSA) and Clostridium difficile (C. difficile) doubled between 2000 and 2010. These infections occurred in about two million patients, leading to between 23,000 and more than 100,000 deaths yearly. Following a hospital stay, one of four seniors had antibiotic resistant bugs on their hands, which they could spread elsewhere (Cao, Min, Lansing, Foxman, & Mody, 2016). Likewise, antibiotic resistance is relatively more common among diabetics using insulin, individuals undergoing chemotherapy, or who have burns, cuts or lesions on the skin, patients undergoing breathing intubation, or who have urinary or dialysis catheters inserted, as well as those with HIV/AIDS or with a weakened immune system owing to still other factors. The immunosuppressive actions of stressors can likewise increase vulnerability to S. aureus infection, especially in vulnerable populations, such as older people.

It is worrisome that MSRA has surfaced outside of hospital environs, becoming a community-acquired infection. It has become increasingly common within individual homes, infecting meat, and poultry. Community-linked infection recently stood at 12%, being attributed to sharing contaminated items, active skin diseases or injuries, poor hygiene, and crowded living conditions. Furthermore, antibiotic resistant bacteria can be "picked up" from other people or foods, as observed among tourists who visit countries where these bacteria are relatively common. The good news, even though it's still a bit limited, is that analyzing the DNA of MSRA can identify those individuals who are at greatest risk of dying as a result of infection, and could potentially facilitate the development of personalized treatment strategies (Recker et al., 2017).

Some illnesses that we hadn't thought about becoming resistant to antibiotics are doing just this. Neisseria gonorrhea, which is responsible for gonorrhea has been showing increased signs of resistance to antibiotics, no longer being responsive to some agents (Unemo et al., 2017). Quinolones, a class of antibiotics that had long been used to treat gonorrhea, have lost their effectiveness, and other drug classes, such as cephalosporins, have also been losing their effectiveness. The last line of defense, the go-to antibiotic colistin, was recently found to have lost its effectiveness in certain cases, possibly owing to a transferable gene mcr-1 that makes it resistant to colistin. This gene can appear in a variety of bacteria and consequently they too could potentially develop resistance.

It's only a matter of time before other threats emerge for which we have little protection or cure. One of these, Shigella currently affects upward of 165 million people worldwide, most often being transmitted through the "the fecal-oral" route. Historically, this highly contagious condition was treated successfully with ciprofloxacin, but its efficacy is now questionable (CDC, 2016). Further, some bacterial illnesses that should have been wiped out years ago, such as tuberculosis (TB), still haunt us. While largely eliminated in Western countries, it is still devastating within parts of Asia and Sub-Saharan Africa, infecting about 9—10 million people in 2015, leading to about 1.5 million deaths, and more than 600,000 individuals have a treatment resistant form of the illness. An antibiotic-resistant form of typhoid has also evolved, infecting large numbers of people within Asia. As typhoid infections ordinarily occur in as many as 30 million people each year, the spread of a treatment resistant strain may be devastating

to an already illness-ridden population. As much as basic health conditions are required to beat various diseases, the development of treatment resistant bacteria, together with the lack of funds or global political will, may limit prevention and treatment of illnesses (WHO, 2016a,b).

Dealing With Antibiotic Resistance

As a first step to combat antibiotic resistance, it might be appropriate to limit the use of these agents for minor bacterial infections. Failing this, alternating doses of antibiotics, and changing the particular antibiotics administered with successive infections might be helpful. Combinations involving several antibiotics administered concurrently that can act synergistically have shown promise in dealing with particular bacteria, and the use of two compounds, one that shreds the shell of bacteria, and the second a potent antibiotic that attacks the exposed bacteria, may be useful in dealing with resistant bacteria (Stokes et al., 2017). As the development of antibiotic resistance has been attributed to the ability of bacteria to limit antibiotic entry into cells, as well as the production of an enzyme, β-lactamase, which is able to destroy antibiotics, β-lactamase inhibitors have been developed to attenuate resistance (Jiménez-Castellanos et al., 2018).

It is also possible to act on bacterial genes to make them more sensitive to antibiotics. A novel compound Teixobactin, which was isolated from microorganisms present in soil, was capable of destroying pathogens effectively, including C. difficile, septicemia, and tuberculosis, without resistance developing. Teixobactin seemed to be effective because it attacks bacteria through multiple methods, and resistance to its effects may not be seen for several decades (Ling, 2015). The discovery of Teixobactin effectiveness was soon followed by several analogs of this compound, and there is a good chance that this is a first step in the development of new antibiotics. Even with so many new treatment in the works to deal with bacterial infection, given the ruthlessness of bacteria in finding ways around our treatments, there is the concern that these treatments will meet the same fate as the antibiotics that are currently available.

Alternatives to Standard Antibiotic Treatments

Severeal novel approaches have been used to eliminate bacteria. Efforts have been directed to treat specific conditions by having bactria turn on one another. Bacteriocins (proteins produced by bacteria to kill their own competitors) could be harnessed to kill pathogens while leaving other microbiota intact. It similarly appears that particular strains of C. difficile are adept at destroying each other (they are competitive strains) by firing a harpoon-like needle through their membrane, which promotes the death of the cell. Thus, the interesting notion was broached that the human microbiota could be used as a potential source for the development of novel ways of dealing with bacteria (Kirk et al., 2017). Using a somewhat different approach, it was demonstrated that resistant bacteria, such as MRSA, could be manipulated by altering ingredients that they need for survival. For instance, MRSA is reliant on folate (vitamin B9), and hence blocking the production of folate can be used as a way of overcoming their resistance to treatments (Reeve et al., 2016).

Viruses have been identified that attack bacteria (termed bacteriophages, or simply phages). These phages appear in mucus, such as in the gums and gut, and may influence immune functioning. The development of viruses that can deal with resistant bacteria, such as MRSA (Green et al., 2017), would be welcome, but it's still very early and the extent to which phages can be used in this capacity isn't altogether certain. In addition to these approaches, nanoparticles have also been developed that can produce chemicals effective in destroying otherwise treatment resistant bacteria (Courtney et al., 2017) and CRISPR-Cas9 could potentially be used to cut out genes from bacteria that show resistance to antibiotic treatments. Appreciable attention has also focused on developing strategies, as well as computer algorithms, that would inform best treatment approaches (Bucci et al., 2016).

VIRAL ILLNESSES

Several viruses, like their bacterial cousins, may contribute to illnesses that have psychological ramifications, which we'll consider in later chapters. Viruses are often said not to be a life form since they are not able to reproduce unless they have the opportunity to use a cells machinery to do so. Upon penetrating a cell, the virus enters into the host cell's genome, and thus uses it in order to replicate. Once sufficient replication has occurred, the virus can force itself through the host cell's membrane, and the viruses that escape will have the opportunity to infect nearby cells. As the virus has its own complement of genes, they can mutate so that new variants of the virus can appear.

Viruses can spread from one person to another through various routes (e.g., through the air or through body fluids), and they can linger for various amounts of time within external environments. In some instances, a virus can lie dormant within the body for extended periods before re-emerging to induce an illness. Viruses and bacteria can also be transmitted to humans through a vector, such as mosquitoes or ticks, leading to illnesses such as malaria, Zika, West Nile virus, dengue fever, and yellow fever, and severe illnesses have spread to humans through birds, pigs, cattle, and rodents. Typically, vector-borne viruses typically don't make the leap to being transmitted between humans. However, these viruses can mutate, and could potentially be transmitted between humans, leading to diseases such as swine flu, HIV/AIDS as well Ebola.

The virulence of a microbe varies so that some create mild symptoms, whereas others can have rapid and powerful consequences. How quickly and broadly a virus can spread within a human population is dependent on several factors; (1) how readily it can be passed one from one person to another, (2) the route by which it is transmitted (e.g., aerobic transmission is more rapid than transmission that involves exchange of fluids), (3) the ability of the virus to penetrate the host's tissues and enter into cells, (4) the capacity of the virus to inhibit the host's immune defenses, and (5) how well equipped it is in obtaining nutrition from the host. Although it is often thought that transmission varies with how quickly the virus kills the host, given that death of the infected person diminishes the opportunity for viral transmission, although passage from one person to the next, as in the case of Ebola, may come about even after death.

In some cases, viruses have nefarious ways of getting around the host's immune defenses. Using particular proteins, they can mask themselves so that they are not readily recognized by immune cells (Holm et al., 2016), and with the assistance of other proteins (neuraminidase), as in the case of influenza virus, for instance, they are able to counter the attack of NK cells that would otherwise destroy the virus. Fortunately, inhibitors of neuraminidase have been developed that enhance the effectiveness of NK cells, and antibodies have been created that act act against proteins that limit NK cell activity (Bar-On, Seidel, Tsukerman, Mandelboim, & Mandelboim, 2014).

PEOPLE NOT TO HANG WITH

People react differently to viruses and to vaccines. Women, in general, seem to be more reactive to vaccinations, possibly because of hormonal factors increasing immune activity. As well, a protein TLR7 which detects viruses and effectively activates immune cells is encoded by genes present on the X chromosome, and hence leads to a greater immune response among women than in men (Karnam et al., 2012). While the greater immune responses among women might seem advantageous, it could also contribute to the greater female disposition toward autoimmune disorders.

Some individuals, often referred to as "superspreaders," seem to be particularly adept in passing on viruses and bacteria. Some feature of their immune response might be responsible for this facility, or they may may have occupations that lead to more social contact either directly or indirectly, or they may simply be especially social, thus coming into contact with a particularly large number of people. Mary Mallon, a cook in the early 1900s seemed to have been a virtuoso in spreading typhoid, despite not presenting with any symptoms herself. She is now best known as "Typhoid Mary" for having infected 51 people, several of whom died. Today's version of Mary Mallon would be far more dangerous owing to larger populations, crowded conditions, and more efficient travel. Indeed, the Middle East Respiratory Syndrome (MERS) that affected South Korea from May to July of 2015, infected 186 individuals, of which 36 died, and caused quarantine of thousands

(*cont'd*)

more. It turns out that an individual who contracted the illness transmitted it to another person, who then reported to hospital with respiratory difficulties. But, as MERS wasn't yet on the radar, he wasn't isolated, and over the course of the next few days he infected 82 others, accounting for 45% of the cases during the outbreak. The Pareto principle, also known as the 80−20 rule, seems to be pertinent to the spread of infection in that 80% of cases transmitted occur through 20% of the people. We can only hope that the potential spreaders choose to be vaccinated, but failing this, we might get lucky and they'll find friends other than us.

Vaccines

For centuries viral illnesses (as well as bacterial infections) decimated human populations, but the discovery of vaccines to prevent illnesses was an obvious game-changer. Using inactivated or dead virus the immune system is primed to respond to similar viruses when they are encountered subsequently, thereby preventing the illness from occurring. Despite the effectiveness of many vaccines, others have been less than perfect, varying across individuals and in some instances their effectiveness diminishes with age. Viruses are also able to mutate so that they won't be recognized. Influenza viruses are notorious in this regard, but since they come in a set number of formats, vaccine makers may (often) be able to anticipate next year's threats. Still, the accuracy of these predictions are variable, such that in some years the vaccine created will be very good, but in other years it has almost no effect. Even if the vaccine is an effective one, individuals vary in the extent to which they are "vaccine responders," possibly owing to whether they produce sufficient antibodies to fight future infection. As well, some flu virus mutations tend to be preferentially effective in infecting immune-compromised individuals. These individuals might be especially sensitive, and they may serve as harbingers of the virus mutations that will be evident during next year's flu season (Xue et al., 2017).

Some vaccines can be developed readily (as in the case of many seasonal flu vaccines, although effective immunization runs around 50−60%), but developing others are more difficult owing to rapid mutations that occurred, as in the case of H7N9 bird flu. This virus spreads from birds to humans and hopefully won't mutate so that the virus spreads between people. However, the CDC has ranked H7N9 at the top of the list of flu strains that could produce a human pandemic, making it essential that new vaccines be available.

It's thanks to mass vaccinations that diseases such as polio have been almost eradicated and measles, mumps, and rubella, which also caused many deaths, have been diminished. Because it takes some time for vaccines to be produced, even when the correct vaccine has been identified, methods are being developed that might be made more rapidly. "Naked DNA" vaccination is one possible approach to this. Administering a viral gene (as opposed to the virus itself) into animals is known to elicit an immune response. Once sections of DNA that encode a viral gene are injected, nearby cells take up the DNA, and will form proteins associated with the virus, to which the immune system ought to mount a response and like other methods of vaccination, a memory of the virus should be maintained.

Given the moderate efficacy of current vaccines, it might be fruitful to develop new approaches to enhance their effectiveness. It has been maintained that gut microbiota may have regulatory actions in relation to influenza, and thus could potentially be harnessed to limit the consequences of influenza infection (Chen , Wu, Kuo, & Shih, 2017). There have been efforts to develop peptide molecules that are able to inhibit a variety of influenza strains by grasping onto common features of a set of influenza A viruses (Kadam, Juraszek, Brandenburg, Buyck, & Schepens, 2017). At some time, a universal vaccine will be developed that acts across a still broader range of influenza viruses (Paules, Marston, Eisinger, Baltimore, & Fauci, 2017).

Just as some individuals may be virus superspreaders, there seem to be those who are particularly susceptible to infection. In this regard, individuals who choose not to be vaccinated (or have their children vaccinated) leave themselves open to illnesses. There are many reasons (or rationalizations) for individuals choosing not to be vaccinated for common illnesses. Frequently, there is mistrust of media and government agencies with respect to recommendations that have been made concerning vaccination (Taha, Matheson, & Anisman, 2014). Alternatively, they may be listening to the sage advice coming from a large cadre of Hollywood types, a few politicians, or friends with strong, albeit fallacious opinions, about the possibility that vaccination is dangerous.

A detailed analysis pointed to a fairly extensive set of factors that were linked to individuals choosing whether or not to be vaccinated (Nowak, Sheedy, Bursey, Smith, & Basket, 2015). Those who opted not to be vaccinated may have based their decisions on earlier experiences, such as having had a negative response to a vaccine, or beliefs that the illness (e.g., flu) is manageable. Resistance to vaccination is also attributable to the belief that recommendations for vaccination actually might be correct for others, but don't apply to them. Some individuals believe that vaccines are often ineffective, or the the misguided notion that one could get the flu (or another illness) from a vaccine. Those railing against vaccinating their children might also not have had the experience of growing up at a time when illnesses like measles, were damageing or killing children, and diseases such as polio were a horrible threat that kept reappearing[2].

Predictably, those ammenable to receiving (flu) vaccination tended to believe that they were flu susceptible, and that vaccines were effective. The propensity to be vaccinated was also elevated among older people or those having an existent chronic health conditions that might be complicated by becoming ill. Having previously experienced a bad flu or a similar illness, favored individual's choosing to be vaccinated, as did easy access to vaccination. As well, intention to vaccinate was particularly high if the recommendation to do so came from a physician[3].

Some viruses, such as measles, are remarkably effective in spreading, so that one person might infect about 90% of people close to them. Other viruses spread less readily, so that one person may infect very few others (say, 0.5

---

[2] In considering the factors associated with depressive disorders (Chapter 8), the work of Kahneman and Tversky related to decision making was raised in the context of how individuals appraise stressful events. Their work indicated that individuals are apt to make some seemingly puzzling decisions in certain situations, and may have much to say about the irrational decision making processes that are common in relation to whether or not people choose to be vaccinated.

[3] In some studies, participants are asked about their "intent" to be vaccinated. While this is reasonable, intent doesn't necessarily translate into action (i.e., actually being vaccinated), and so the data must be interpreted caustiously. It is certainly of interest to determine what factors determine whether intentions become actions.

people), and thus the disease will disappear. Fortunately for individuals who choose not to be vaccinated, when enough people in a population are vaccinated, the source for transmission will be diminished, and thus even potent viruses might not spread (herd immunity). Ironically, this herd immunity also protects the children whose parents refused to have them vaccinated. However, should antivaxxers be successful in their campaign, so that enough people within a population decide not to be vaccinated (or not have their children vaccinated), a "tipping point" will be reached so that herd immunity no longer protects people who are not vaccinated, or those in whom the inoculation was not particularly effective (i.e., vaccine nonresponders). Should an individual become infected with measles (including the children whose parents decided against having them vaccinated), they will have a tough illness to deal with, and they may also experience serious downstream effects. Specifically, following measles infection the immune system may be altered, possibly for as long as 2–3 years, so that the risk for other illnesses may be elevated (Mina, Metcalf, de Swart, Osterhaus, & Grenfell, 2015). Furthermore, if the immune system is not fully developed, as in the case of young children, infection with measles may result in the virus hiding in the body, only to emerge years later to infect the brain.

Bacterial and Viral Challenges Affect Hormonal and CNS Processes

Pathogenic stimuli, such as bacteria and viruses, cause marked effects on glucocorticoids and on central neurotransmitters. Many of these changes are comparable to those usually elicited by both psychological and physical stressors, and thus it was suggested that these systemic challenges were interpreted by the brain as if they were stressors (Anisman & Merali, 2002). In addition to affecting brain neurotransmitters, such as norepinephrine and serotonin (Hayley, Lacosta, Merali, van Rooijen, & Anisman, 2014), immune activating agents may influence the presence of growth factors (e.g., BDNF) as well as proinflammatory cytokines, presumably released by microglia (Audet & Anisman, 2013). As expected, these outcomes vary with sex and age, and at least some of the effects of immune challenge are subject to a sensitization-like effect in that exaggerated responses are evident upon reexposure to a challenge. Moreover, bacterial agents and stressors may act cooperatively in producing brain neurochemical changes that favor the development of depressive disorders (Anisman, 2009).

As described in Chapter 2, The Immune System: An Overview, multidirectional communication occurs between immune, autonomic, microbial, hormonal, neurotransmitter, neurotrophin, and other brain-related processes. These systems are so intimately intertwined that actions in any one may influence the functioning of others. By example, when mature lymphocytes are not present, ordinary behavioral stress responses might be absent, even in mice that are very stress sensitive (Clark et al., 2014). Likewise, manipulating microbial processes may come to influence brain functioning tied to mental health.

MICROBIOTA

For a time, limited attention had been devoted to the links between the brain and the enteric nervous system, other than analyses related to hunger and satiety. In fact, it came as a bit of a surprise that the brain influences gut functioning, and that signaling through processes that line the esophagus, stomach, small intestine, and colon, affect brain functioning (Bravo et al., 2012). Messages from the gut to the brain may occur through stimulation of the vagus nerve, which extends from the viscera to the brain stem, and gut processes can influence hormones, such as ghrelin, that affect brain activity, thereby moderating hunger and obesity. As well, gut functioning may influence immune processes (Hooper, 2012), and by virtue of effects on brain functioning, may influence mood and reward processes (Mayer, Knight, Mazmanian, Cryan, & Tillisch, 2014).

Although most of the research concerning microbiota have focused on those residing within the gut, bacteria that affect us are also found in the mouth and nasal passages, on our skin, between our teeth, and within other body orifices. The trillions of bacteria present ordinarily live harmoniously with one another (i.e., commensal bacteria). Over the course of evolution various microbes adapted and colonized different parts of the body. For their mutual benefit many bacteria behaved cooperatively, although others were parasitic, consuming other bacteria (Silverman et al., 2017). In general, when an imbalance occurs between "good" and "bad" microorganisms (dysbiosis) or in the absence of specific types of bacteria being present, an immense range of physical and mental illnesses may follow. Over the short run, animals with a compromised microbiome can survive, but their ability to do so will be curtailed owing to disturbed immunological functioning. Rodents born entirely germ-free have their immune development hindered, and the balance between various aspects of the immune system is disturbed, thus rendering them more vulnerable to pathologies. Even the response to vaccines, known to be highly variable across individuals, may be dependent on the gut microbiome, which contributes to the shaping of immune responses (Lynn & Pulendran, 2017).

Gut bacteria, in the main, come from four phyla, Bacteroidetes, and Firmicutes, Actinobacteria, and Proteobacteria, whereas others, such as Fusobacteria and Verrucomicrobia phyla appear in lesser abundance. As depicted in Fig. 3.1, gut bacteria and their metabolites can affect immune, neurotransmitter, and hormone systems, and may thus influence inflammatory diseases, neurodevelopmental disorders (e.g., ADHD, autism spectrum disorder),and several mental illnesses (Hyland & Cryan, 2016; Schroeder & Bäckhed, 2016), and may even affect sensitivity to cocaine reward and thus increase the risk for addiction (Kiraly et al., 2016).

What we eat influences our gut microbiota (David et al., 2014), and gut microbiota can influence what we eat. Gut bacteria help to break down foods and contribute to the absorption of otherwise difficult to digest substances. Thus, their presence may help individuals stay lean, and many useful bacteria are themselves strengthened by fiber-rich foods. When fiber isn't available, microbes may die off, or they may feed on the mucus lining, which ordinarily keeps the gut wall healthy. Some foods, such as modest amounts of wine, increase the presence of *Pediococcus pentosaceus* CIAL-86, which sticks to the intestinal wall and fights against bad bacteria (Garcia-Ruiz et al., 2014). Other foods, such as emulsifiers (food additives that serve to stabilize processed foods) can have negative health consequences, possibly by altering the gut microbiota and the induction of inflammation, and these effects may be linked to particular genes (Chassaing et al., 2015).

Given their potential health benefits, gut bacteria have become a target to help individuals deal with obesity. For instance, diets spiked with modified bacteria diminished eating and altered metabolism, thereby lowering adiposity and insulin resistance (Chen et al., 2014). As well, some bacterial families, such as Christensenellaceae, appear in greater number among thin individuals than in people who are heavy, and may causally contribute to this difference. When the bacteria associated with slimness were transferred to other mice, weight gain was diminished relative to mice that hadn't received this transplantation. In theory, Christensenellaceae could be useful in

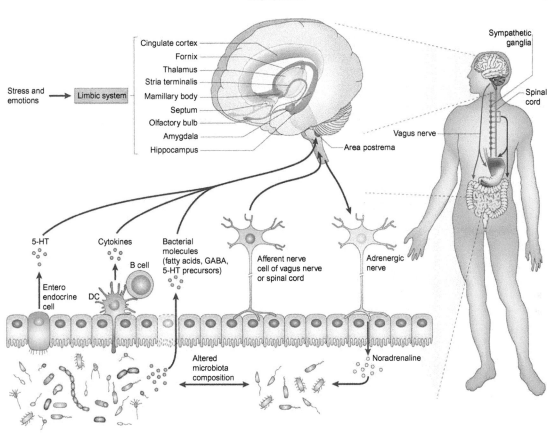

Nature Reviews | Microbiology

FIGURE 3.1 Neural, immunological, endocrine, and metabolic pathways by which microbiota influence the brain, and the proposed brain-to-microbiota component of this axis. Bacteria and bacterial products can reach the brain via the bloodstream and gain passage to the brain though the area postrema. Commensal bacteria may form ligands that activate G-protein coupled receptors (GPCRs), which are fundamental signaling pathways for a majority of neurotransmitters (Cohen et al., 2017). Brain functioning can also be affected by cytokine release from mucosal immune cells, as well as through the release of gut hormones, such as serotonin (5-HT) from enteroendocrine cells, or via afferent neural pathways, including the vagus nerve. Stress and emotions can influence the microbial composition of the gut through the release of stress hormones or sympathetic neurotransmitters that influence gut physiology and alter the microbiota habitat. Moreover, host stress hormones, such as norepinephrine, might influence bacterial gene expression or signaling between bacteria, which can alter the microbial composition and activity. Immune elements, such as antibody (IgM) secreting B cells, also contribute to microbial diversity (Magri et al., 2017). As multidirectional communication occurs between gut bacteria and immune, autonomic and brain processes, factors that affect brain neuronal activity may influence the microbiome and immune functioning. DC, dendritic cell; GABA, γ-aminobutyric acid. *Source: From figure and portions of the text are from Collins, Surette, & Bercik (2012).*

reducing weight, but it's still early to assess this in humans.

The link between the microbiome and obesity is complex and involves multiple steps, and could be subject to genetic differences across individuals (Duranti, Ferrario, van Sinderen, Ventura, & Turroni, 2017). Gut bacteria also differ between males and females, and may be differentially affected by diet. These sex differences may be related to hormonal factors (estrogen, in particular) interacting with gut microbes. It is conceivable that diet in men and women will also have different effects on illnesses related to bacteria, and thus diets meant to treat particular disturbances need to be tailored on the basis of sex as well as other individual difference factors that span several domains.

In retrospect, it isn't surprising that gut bacteria play a prominent role in feeding and energy processes, and may contribute to eating- and gut-related disorders (e.g., Bäckhed et al., 2015; Dinan & Cryan, 2016). In an effort to assess the contribution of the microbiota to a variety of phenotypes, analyses were undertaken to assess mice born and raised in germ-free environments. These mice seemed not to develop in the usual fashion, in that their immune system was deficient, and the gut of these mice had a smaller surface area and hence could not absorb nutrients as readily as in mice raised in a standard germy environment. The germ-free mice also had leaky intestinal walls, and blood vessels that ordinarily supplied food to the gut wall were in short supply. Upon receiving microbiota harvested from the intestines of conventionally raised mice, marked changes occurred in the germ-free mice within 2 weeks. Among other things, their body fat content increased and insulin resistance became apparent even though their food intake was reduced (Bäckhed et al., 2004). Evidently, more food was converted into fat and hence these mice gained weight. More than this, the microbiota was integral in the absorption of monosaccharides from the gut lumen (interior of the gastrointestinal tract), resulting in a process by which fatty acids are produced (lipogenesis) and then stored. Moreover, a type of protein "Fasting-induced adipocyte factor" (Fiaf) was suppressed in the intestinal epithelium, which is essential for triglycerides to be stored within adipocytes. In essence, these studies were among those that led the way in suggesting that the gut microbiota is fundamental in moderating energy being obtained from foods as well as subsequent energy storage.

MICROBIOTA AMONG SUPER-AGERS

The microbiome may contribute to both the physical deterioration that accompanies ageing, as well as to healthy ageing and extreme longevity. In older animals, it is not uncommon for gut dysbiosis to occur, leading to intestines becoming leaky, so that released bacterial products promote inflammation and immune dysfunction. Individuals with high circulating levels of proinflammatory cytokines, particularly TNF-α, are generally more frail and less independent, more vulnerable to some types of infection, and more likely to experience chronic illnesses. To be sure, it is possible that the TNF-α elevations initially came about because of the "wear and tear" associated with ageing, together with greater exposure to infections over the lifetime, which in turn, modified the microbiome, the presence of inflammatory factors, and neuronal processes leading to mental disability.

In older individuals, especially in response to stressors, gut permeability increases, accompanied by elevated circulating proinflammatory cytokine levels. As well, changes occurred in the levels of a particular microbial family Porphyromonadaceae, which has been linked to cognitive decline and affective disorders, and was associated with elevated anxiety in older mice (Scott et al., 2017). It seems that ageing may be accompanied by a shift of the microbial community toward a profile reminiscent of that apparent in inflammatory diseases, and may contribute to the development of behavioral and cognitive disturbances. In fact,

(*cont'd*)

in young rodents that received gut bacteria from old mice, chronic inflammation could be accompanied by elevated leakage of inflammatory bacterial factors into circulation (Fransen et al., 2017).

It is interesting that with normal ageing, certain bacterial species disappear and others become more common, which could contribute to healthy ageing. Specifically, dominant species are replaced with subordinate species and particular bacterial groups (e.g., Akkermansia, Bifidobacterium, and Christensenellaceae) are more prevalent or are enriched. Thus, among extremely healthy individuals who were 100 years or older, their microbiota constituency resembled that of healthy young people (Bian et al., 2017). Cognitive performance among healthy older adults was also linked to the presence of particular gut bacteria (Manderino et al., 2017). To be sure, these findings are simply correlational, but they nonetheless offer interesting hints regarding processes related to extreme longevity.

Among turquoise killifish, which have a relatively short lifespan (4−6 months), several genes located on sex chromosomes, were linked to longevity (Valenzano et al., 2015), which might speak to the greater longevity of females. It was particularly interesting that when older fish of this species consumed the poop of younger fish, they lived longer, raising the possibility that some bacterial factors present in young poop produced benefits for the older fish. In other studies using *Caenorhabditis elegans*, elimination of 29 bacterial genes increased longevity, and limited age-related diseases. These effects seemed to have been related to a substance, colonic acid, which affects the worm's mitochondria, thereby altering energy regulation (Han et al., 2017). These data raise the possibility that the link between bacteria and longevity is a causal one, at least in worms, but it's some distance from worms to humans.

Not long after these initial findings, it was demonstrated that in genetically obese (ob/ob) mice, microbial communities could be distinguished from those that appeared in lean animals (Ley et al., 2005). Most prominently, the firmicutes were increased by 50% and the bacteroidetes were diminished to a comparable extent among those who were obese. When microbes harvested from fat and lean mice were fed to germ-free mice, those that received microbes from obese donors exhibited a much greater increase of fat than did those who received microbes from the lean donors (Turnbaugh et al., 2006), pointing to the causal connection between microbial factors and fat storage. It was later demonstrated that predictable phenotypic changes were provoked by the transplantation of fecal microbiota from adult female human twin pairs discordant for obesity into germ-free mice that were maintained on low-fat chow. Specifically, body and fat mass, together with obesity-associated metabolic phenotypes, varied with the fecal bacteria cultures received (i.e., from the heavy or lean twin). Tellingly, when mice that had received an obese twin's microbiota (Ob) were housed with mice containing the lean co-twin's microbiota (Ln), the increased body mass and obesity-associated metabolic phenotypes in Ob mice was prevented, which was likely because of lateral transmission of microbiota (Ridaura et al., 2013).

Gut bacteria produce spores that can survive in open air, and can be transmitted from one person to another, causing dysbiosis in the second individual (Browne et al., 2016). Ordinarily, the skin microbiome plays a fundamental role in protection from infection, allergies, and the provocation of inflammation, so that when skin microbiome dysbiosis exists, the impact of a parasite can be markedly elevated. Interestingly, in mice the disturbance of the skin microbial community can be transferred to cage-mates (Gimblet et al., 2017), and it is conceivable that people living within the same home, may share a similar microbiome, and thus may share illness vulnerabilities, as well as the propensity for weight gain.

Gut bacteria also exist that can favor weight loss, rather than weight gain. For instance, among lean mice, the bacteria Akkermansia muciniphila is far more common among lean mice than in mice that are prone to diabetes, and when these bacteria were fed to the obese mice they tended to lose weight and the warning signs of type 2 diabetes diminished (Plovier et al., 2017). In humans, prebiotics fed to overweight children, reduced the weight gains that would otherwise appear in growing children, which has important long-term implications given that childhood obesity is often carried into adulthood (Nicolucci et al., 2017). These findings, and others like them, suggest that the gut microbiome might provide a target for obesity treatments, and for the reduction of type 2 diabetes symptoms (Remely et al., 2016). At the same time, as we've already seen, gut bacteria comprise many different subtypes and the specific combinations that are present will dictate different phenotypes.

Factors That Affect Microbiota and Their Implications for Well-Being

The microbial community is negatively affected by poor life-styles, as we saw earlier in relation to food consumption and obesity. Especially harmful effects are elicited by antibiotics that kill useful bacteria along with those

that are not our friends. It was estimated that one in five hospitalized patients experience adverse effects related to antibiotic treatments, including gastrointestinal, renal, or hematologic disturbances. Some common antibiotic treatments (e.g., amoxicillin and azithromycin) taken over a period of just 7 days can have pronounced and long lasting negative effects on gut microbiota diversity (Abeles et al., 2016), and when administered early in life, gut hypersensitivity may persist into adulthood (O'Mahony et al., 2014). Yet, there are instances in which some bacterial species may produce positive side actions. For instance, antibiotics could influence gut bacteria that might otherwise contribute to the development of brain disorders. It is likely that the individual's genetic background, along with exposures to environmental stresses, play a role in shaping the microbiome and hence, determining what long-term repercussions result from its disturbance.

It should be considered that antibiotic treatments can affect the transmission of microbial factors from a pregnant mom to her fetus (vertical transmission), and may thereby disturb protective qualities associated with bacteria (Bäckhed et al., 2015). Hence, offspring can bear the benefits or risks associated not only with the genes passed on to them, but also the microbiota they inherit.

The impression might be gained from Fig. 3.2 that each of the main contributing factors independently influences microbial factors and well-being. Ultimately, however, microbial functioning and intestinal immunity are shaped and maintained by multiple interactive processes. Each of the factors shown in the figure affects others, and additively or interactively, gut bacteria will be affected. It is also the case that commensal microflora can affect and interact with immune processes, which can influence nutrition, including the presence of short chain fatty acids and particular vitamins, which then feedback and affect the

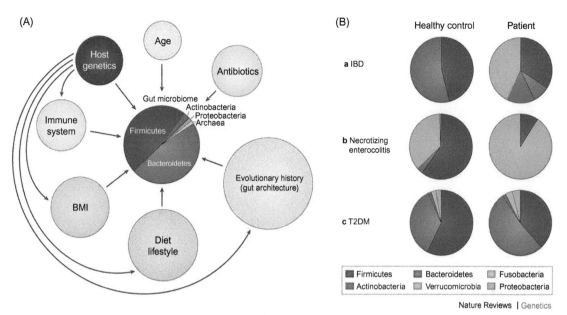

FIGURE 3.2 (A): The gut's physical architecture influences the microbiome constituency. Species that are phylogenetically related typically display more similar microbiomes than species that are only distantly related. Irrespective of age, diet, and geographic location, human gut microbiomes are more similar to one another (despite their own marked variability that stems mostly from diet and lifestyle factors) than they are to the gut microbiomes of other species. In general, the gut microbiome is readily affected by diet, age, genetic factors, host genetic factors, immune system functioning, and age, with the relative contribution of each of these being depicted by the size of their representation in the figure. The use of antibiotics also markedly influences the composition of the gut microbiome, essentially eliminating gut bacteria. The diversity of the gut bacteria in humans also varies with age. Prior to the age of 3 years, the gut microbiome is described as comprising limited species diversity, but takes on a more diverse, adult-like profile after the age of three. (B): Gut microbiome dysbiosis has been linked to many illnesses. As depicted in the figure, this has included inflammatory bowel disease (IBD) (part **a**), necrotizing enterocolitis (part **b**), as well disorders unrelated to the gut, such as type 2 diabetes (T2DM) (part **c**). In the case of each illness, the nature of the gut dysbiosis was very different. *Source: The left panel (A) and the Figure caption are taken from Hall, Tolonen, and Xavier (2017). The right panel (B) comes from Spor, Koren, & Ley (2011).*

microbiota (Spencer & Belkaid, 2012). By example, the presence of microbiota contributes to mast cell functioning following consumption of fat, which can then influence immune activity and instigate particular allergic and inflammatory responses (Sato et al., 2016).

Gut microbiota and the genes involved in regulating them (microbiome) are exquisitely sensitive to psychological stressors, and can be reversed by oral prebiotic treatment (Bharwani et al., 2017). Likewise, microbiota alterations can be provoked by prenatal stressors (Golubeva et al., 2015) and can markedly affect

the development of immune processes. Conversely, intake of specific gut microbes can diminish the impact of stressors, such as strenuous exercise, and can limit abdominal dysfunction and discomfort associated with academic stressors. It also appears that in the absence of an inflammatory inhibitor NLRP12, the presence of beneficial bacteria were reduced and that of disruptive bacteria were elevated, leading to still more inflammation. As expected, increasing the presence of the good bacteria could terminate this negative cycle (Chen et al., 2017).

GETTING AROUND ANTIBIOTIC RESISTANCE TO C. DIFFICILE THROUGH FECAL TRANSPLANTS

The hospital-acquired antibiotic resistant bacterium C. difficile has proven to be particularly able to be transmitted from one individual to the next. If patients were treated with an antibiotic, even using the hospital bed that had recently been occupied by a C. difficle patient increased the risk of contracting this condition (Freedberg, Salmasian, Cohen, Abrams, & Larson, 2016). Not only was the occurrence of C. difficile infection more common among hospital patients treated with antibiotics, but variants of this bacteria emerged that were increasingly destructive owing to their ability to produce a toxin to destroy eosinophils in the gut that ordinarily act in a protective capacity (Cowardin, Buonomo, & Saleh, 2016). Making matters much worse, a new strain (NAP1) associated with multiple recurring C. difficile has been on the rise, doubling between 2001 and 2012.

The approach to deal with C. difficile comprised fecal microbiota being obtained from a healthy donor and then transplanted (in a purified form, most often by colonoscopy or through the nasogastric route, or more recently through acid-resistant capsules) to patients with resistant C. difficile. This results in reestablishment of a beneficial bacterial colony, and abatement of illness. The media has treated fecal transfer as a far-out procedure (probably because of the "yuck" factor), but it is hardly a novel one, as documented by de Groot, Frissen, de Clercq, and Nieuwdorp (2017) in a brief history of this topic. This approach was used as early as the 4th century in China to treat food poisoning and diarrhea, and was used elsewhere over the centuries to treat gastrointestinal problems. The 19th century Russian zoologist Metchnikoff proposed that balances of microbes within the colon, particularly elevated lactic acid bacteria, could lead to gut problems, and Nissle later extended this to include E. coli as a protective agent against Shigella and gastroenteritis. Somewhat later, fecal enemas were found to attenuate a form of colitis, and after a few years, fecal microbiota transplants were used in inflammatory bowel disease (Pigneur & Sokol, 2016). In animals, fecal transplants may also be effective in treating other resistant bacteria, including Eterococcus faecium and K. pneumoniae (Caballero et al., 2015).

Complete fecal transplants might not be necessary to deal with C. difficile, and potentially could be dealt with by transplantation of the bacteria C. scindens and three other bacteria (Buffie et al., 2015). Many factors present in feces, including colonocytes, archaea, viruses, fungi, and protists, may be fundamental to the effectiveness of the treatment and could potentially be enhanced by particular probiotics (Spinler, Ross, & Savidge, 2016). Should the mix of bacterial and nonbacterial components that lead to positive effects be identified, then it will be possible to generate treatments more efficiently and without having to rely on poo donors.

Psychological Functioning Associated with Microbial Changes

The potential involvement of gut microbiota in relation to multiple disease states instigated a great number of studies that traced the links between the microbiome, immune functioning, and brain neuronal changes, which might contribute to psychiatric disorders (Kennedy, Cryan, Dinan, & Clarke, 2017). Some psychological disturbances, such as mood disorders, may appear owing to variations of serotonin formed in the digestive tract, or neurotoxicity brought about owing to increased metabolites of bacterial enzymes, such as D-lactic acid and ammonia (Galland, 2014). There is also ample evidence pointing to the microbiota having an impact on depressive-like features (see Chapter 8, Depressive Disorders). For instance, as adults, mice that had been born germ-free, exhibited altered dendritic morphology in the amygdala and hippocampus that were linked to depressive-like features (Luczynski et al., 2016). Likewise, among nonobese diabetic mice, gut microbiota could drive depressive-like symptoms (e.g., social avoidance), which could be attenuated by antibiotic treatment, and then resurrected through reconstitution of the microbiota from donor mice (Gacias et al., 2016).

Multiple routes have been identified (as shown in Fig. 3.2) by which microbiota affect the brain and thus could promote psychopathology. Being a major pathway between the gut and brain, the vagal nerve was implicated as a player in accounting for anxiety and depressive-like behaviors, possibly acting through anti-inflammatory processes (Borovikova et al., 2000). As we'll see in later chapters, various hormones, neurotransmitters, growth factors, and immune related molecules that have been linked to depression are influenced by the microbiome. Moreover, ss described in Chapter 1, Multiple Pathways Linked to Mental Health and Illness, enterochromaffin cells of the gut epithelium can release serotonin and can activate CNS functioning (Bellono et al., 2017). In addition, peripheral cytokines are altered by gut bacteria, and manipulations of the gut microbiota can affect serotonin levels and specific serotonin receptors, norepinephrine, dopamine, and GABA activity in limbic brain regions, thereby influencing mood states (e.g., Clarke et al., 2013; Stilling et al., 2015). In addition, gut bacteria can influence neuroendocrine factors (e.g., CRH) and neurotrophins (e.g., BDNF) within brain regions that are sensitive to stressors and which have also been tied to the development

and maintenance of depressive disorders. Further to this, strong immunogenic agents engender prounced corticoid responses, which can diminish the effectiveness of the gut barrier (Söderholm & Perdue, 2001). The migration of bacteria from the gut will thereby be facilitated, and these bacteria may promote the production of immune signaling molecules (cytokines), that instigate mood disturbances (Maes, Kubera, & Leunis, 2008).

In Chapter 8, Depressive Disorders, where we'll consider the processes associated with depressive illness, it will become clear that a balance exists amongst commensal bacteria so that some are aligned with the development of illness, whereas others seem to act in a protective capacity by acting against inflammatory processes. In line with this, among rats highly vulnerable to depressive-like states, probiotic treatment may counter behavioral disturbances that may have been provoked by proinflammatory changes instigated by a high fat diet (Abildgaard et al., 2017). As we'll see, as well, antibiotics can affect mood states by altering microbiota, but also affect processes beyond the microbes that they target. This includes their well documented effects on mitochondrial functioning, microglia reactivity, and factors important for neuroplasticity, such as mTOR, which collectively can impact many processes aligned with mental illness. Although antibiotics have been a primary concern in relation to microbiota changes and dysbiosis, such treatments are hardly alone in affecting microbiota within the gut as well as elsewhere. Indeed, nonantibiotic drugs, such as antipsychotics, antidepressants, nonsteroidal anti-inflammatory drugs (NSAIDs), opioids, statins, metformin, and proton pump inhibitors (PPIs) were found to influence a wide range of microbiota classes. The secondary effects of some of these have been suggested to influence the frequent weight gain and visceral fat provoked by antipsychotioc mediction (Le Bastard et al., 2018).

Microbiota and Immunity

The gut is a dirty place, being the recipient of various foods, some of which may be contaminated, and a good portion of our immune cells operate within the gut. Intestinal immunity comprises collaboration between specific types of immune cells and many different cytokines, various nutritional factors (e.g., short chain fatty acids, particular vitamins), and commensal bacteria. Moreover, reciprocal communication occurs between gut microbiota and CNS processes, and consequently gut-level disturbances can instigate psychological, metabolic, and immune-related disorders (O'Mahony, Clarke, Dinan, & Cryan, 2017a).

Microbial factors are fundamentally involved in the development of immunosuppressive responses generated by regulatory T-cells (Treg) to dietary antigens (Kim et al., 2016). This, of course, is critical for the prevention of excessive immune responses being generated in the face of normal dietary intake. It is thought that gut dysbiosis may promote disturbances of Treg cells and imbalances of Th1, Th2 and Th17 lymphocytes, which can promote autoimmunity to antigens derived from the diet and may contribute to autoimmune diseases. Even the regulation of neuron myelination within the prefrontal cortex (Hoban et al., 2016) is affected by the microbiome, and thus might dispose individuals to susceptibility to multiple sclerosis (Mangalam et al., 2017).

Beyond the effects on neuronal processes, the microbiota can affect brain microglia, and through the release of cytokines can affect psychopathological conditions. Furthermore, pre- and probiotic manipulations are capable of altering glycemic dysregulation, such as glucose tolerance and insulin resistance, which

was accompanied by elevated plasma levels of the anti-inflammatory cytokine IL-10. It was thus suggested that microbial manipulations could potentially influence any illness that involves ongoing inflammatory or metabolic disturbances (de Cossío et al., 2017)[4].

An interesting meta-analysis that considered ten diseases indicated that the number of microbiota genera varied with different illnesses. Whereas some illnesses were associated with the presence of numerous genera, others appeared to be associated with a lack of particular microbiota, and many microbiota were linked to multiple illnesses. In effect, some of these conditions are not disease-specific and are general vulnerability factors, so that other elements contribute to the specific illness that emerges. Such findings might be a step in providing information that could be relevant to the use of probiotics and prebiotics (whereas probiotics comprise living micro-organisms, prebiotics are food ingredients that induce the growth or activity of beneficial microorganisms) in relation to specific illnesses.

As much as the findings concerning germ-free mice are interesting, their relevance for neuroimmune processes are, in some ways, not as clear as one might like, especially as the blood—brain barrier (BBB) may be disturbed in these mice (Braniste et al., 2014). Thus, endogenous circulating immune cells and any microbial species that might escape the confines of the gut could potentially access the brain. As well, this would result in exceptional vulnerability to a range of potentially toxic insults that are present in the periphery following environmental exposure. Remarkably, exposing germ-free mice to the normal microbial constituents of the gut obtained from normally housed

---

[4] As we'll emphasize repeatedly, caution should always be exercised in embracing any emerging therapies. This is particularly evident with regards to "hot" areas of research, where there might be the inclination to "jump on the bandwagon." Microbiota have been implicated in a very large number of illnesses, and there may be the concern, as often stated, that when something explains everything, it explains nothing.

mice, reversed BBB deficits. In fact, even treating them with the short chain fatty acid metabolites from normal microbiota appeared to repair the BBB deficits, such that tight junction integrity was increased, which was also associated with the prevention of protein (e.g., IgG) infiltration into the brain parenchyma.

The fact that constituents of the microbiota send signals to the brain that affect BBB functioning and brain homeostasis, has wide ranging implications for virtually all neuronal pathologies. This is illustrated by the finding that the microbiota influences the development of infection following stroke (Stanley, Moore, & Wong, 2018). In fact, when fecal samples from a mouse that experienced ischemic stroke were transferred to other mice, the size of a stroke-induced neuronal infarct was increased in the microbiota receipients (Singh et al., 2016). Conversely, antibiotic treatment reduced stroke damage and diminished the infiltration of inflammatory T cells (Benakis et al., 2016). These findings suggest bi-directional communication between the microbiota and brain, and that the translocation of microbial elements might infiltrate the brain or other organs to influence their functioning.

It is highly likely that the complex interactions between the differing microbial species might contribute to autoimmune disorders. Germ-free mice with impaired BBB integrity were found to be more vulnerable to the development of autoimmune pathology in an experimental autoimmune encephalitis (EAE) model of MS. Furthermore, having the normal gut commensal bacteria was essential for mounting CD4 + T cell and B cell antibody responses in a model of relapse remitting MS (Berer et al., 2011), and thus germ-free mice might be at greater MS vulnerability. In parallel with such germ-free studies, antibiotic-induced microbiota changes also impacted MS pathology. In MS patients marked gut dysbioisis was present, particularly reductions of clostridium and Bacteroidetes species (Miyake et al., 2015). A further interesting aspect related to autoimmune disorder is that fusobacteria increased relapse rate (Tremlett et al., 2016), but successful treatment of MS was associated with elevations of Prevotella and Sutterella species (Jangi et al., 2016). Thus, various microbial species likely have differing effects on brain processes and that their collective impact depends upon a delicate balance between them and the metabolites they excrete.

Physical Illness, Immunity, and Gut Bacteria

The link between microbiota, immune functioning, and disease conditions has been supported by the finding that germ-free mice raised in a sterile environment lived longer following a skin graft, possibly because immune functioning was diminished and hence foreign tissue was not attacked. Conversely, if these mice received microbes from untreated mice, they rejected the skin graft more readily. In essence, these data point to the importance of the microbiota in determining immune functioning and tissue rejection (Alhabbab et al., 2015). Paralleling these findings, tissue transplants involving lungs, skin, and intestines, which had been exposed to external influences, were less successful than transplants of tissues that were less directly affected by external microbial factors (Lei et al., 2016).

Variations in immune function are likely pertinent in the links between the microbiome and development of physical illness such as chronic kidney disease (Nallu, Sharma, Ramezani, Muralidharan, & Raj, 2017). The gastrointestinal tract shares reciprocal connections with the immune system. In Chapter 2, The Immune System: An Overview, we saw that immune cells have membrane pattern recognition receptors (PRRs) that mediate recognition of damage and pathogen-associated molecules (PAMPs), and damage-associated molecular patterns

(DAMPs). These PRRs are rudimentary aspects of the immune system that evolved, in part, as a way of detecting pathogens or other microbial threats, and may operate to enable microbiota to communicate with the immune system (Chu & Mazmanian, 2013).

Generally, PAMPs are initial sensors that allow immune cells to recognize microbial presence, and determine their pathogenic valence (e.g., commensal microbes can often be tolerated). The toll like receptors (TLRs) and NODs (nucleotide-binding oligomerization domain-like receptors) are among the most prominent PAMPs and are found throughout the brain and immune system. They have evolved to recognize specific motifs that characterize bacterial, viral, or fungal invaders. Upon their recognition, very robust intracellular pathways are engaged that give rise to the mobilization of defensive inflammatory (e.g., cytokines), enzymatic, and oxidative (e.g., superoxide) factors, depending upon the nature of the threat. DAMPs act in a similar fashion but in the absence of a microbial constituent, instead becoming active following the detection of specific factors that are released in response to cellular distress. Among these distress signals, adenosine triphosphate (ATP), and other purines, along with mitochondrial and other intracellular factors are released into the extracellular space in the presence of a damaged cell, creating a "sterile" inflammatory reaction.

A distinction has been made between PAMPs and the largely interchangeable term, microbe-associated molecular patterns (MAMPs), which respond to various microorganisms and also act as a bridge between the enteric nervous system and innate immune system. The MAMPs, by virtue of their effect on immune cells, modulate inflammation (Chu & Mazmanian, 2013) and might even influence microbial interactions with TLRs (Zhou et al., 2015). Some of these interactions may not always lead to pathological or inflammatory conditions, which would be in keeping with the symbiotic relationship between the gut microbiota and the host organism.

However, problems may arise when microbial dysfunctions result in improper "sensing" of microbiome signals. If such protective processes are not doing the job effectively, excessive immune activation and chronic inflammation may evolve (Chu & Mazmanian, 2016; Chu et al., 2016). Fortunately, we are blessed with a gene (SIGIRR) that operates to stimulate immune responses that interfere with bacteria forming colonies that would ordinarily have negative health effects. Disruptions of SIGIRR owing to antibiotic treatments, can cause dysbiosis wherein the battle for supremacy moves toward the side of the harmful bacteria (Sham et al., 2013).

As we've seen, microbial dysbiosis has been implicated in a number of diseases that involve gastrointestinal processes and eating disorders, such as anorexia nervosa and bulimia (Chu et al., 2016). Beyond these conditions, gut bacteria have also been linked to metabolic dysfunctions (e.g., insulin resistance), ultimately promoting the development of type 2 diabetes, and it has been proposed that the disturbed balance of gut bacteria might serve as a target in the treatment of this illness. Altered bacterial levels have also been associated with increased proinflammatory activity (e.g., IL-17) that exacerbates autoimmune conditions (López, Rodríguez-Carrio, Caminal-Montero, Mozo, & Suárez, 2016). Still other illnesses might come about because the wound-healing capacity associated with the microbiome might not be operating properly or might not be present. A wide range of other illnesses, which will be discussed in ensuing chapters, have also been linked to immune disturbances that might have their roots in microbial dysbiosis and inflammatory processes. These comprise cardiovascular illnesses, periodontal disease, rheumatoid arthritis, and allergies, as well as seemingly unrelated illness conditions, such as chronic kidney disease, uremic toxicity, multiorgan failure, and several forms of cancer (e.g., Chen et al., 2016). The microbiota has also been associated with the accumulation of amyloid proteins that were linked to neurodegenerative disorders (Chen et al., 2016), which may also be linked to inflammatory processes (see Chapter 14: Inflammatory Roads to Parkinson's Disease and Chapter 15: A Neuroinflammatory View of Alzheimer's Disease).

SICKNESS FEATURES ASSOCIATED WITH SYSTEMIC INFLAMMATION STEMS FROM BRAIN CHANGES

Sickness behaviors in rodents (diminished social interaction, ruffled fur, hunched posture, inactivity, sleepiness) are typically associated with the administration of LPS or cytokines, such as IL-1β, and have been taken to model some of the symptoms associated with depressive disorders. Characteristics of sickness behaviors are frequently apparent among patients experiencing autoimmune disorders, likely reflecting elevated inflammatory immune activation. It seems that in the context of organ inflammation, increased TNF-α levels give rise to monocytes being recruited to the brain, thereby increasing microglial activation and the production of sickness behaviors. The sickness profile associated with liver inflammation in mice can be diminished by a probiotic treatment without affecting severity of the illness, the actual gut microbiota composition, or permeability of the gut, but were tied to diminished microglial activation, and cerebral monocyte infiltration (D'Mello et al., 2015). The sickness behavior and its resolution by probiotics may thus involve brain processes, leading to the possibility that altering systemic immune functioning or microglial activation, as well as limiting recruitment of monocyte-secreting TNF-α within the brain, may diminish some of the features of systemic inflammatory diseases.

Caveats Concerning the Potential for Using Microbiota for Health Benefits

With the increased understanding regarding the contribution of microbiota to illness occurrence, one might be seduced into thinking that we may be on the cusp of being able to target microbiota in order to diminish or prevent illness. Although there have been reports consistent with this perspective (as in the case of rheumatoid arthritis) (e.g., Marietta et al., 2016), in the main, the positive effects observed were modest, and altering microbiota through diets or by probiotics, did not have sufficiently powerful effects to moderate systemic inflammation. Furthermore, it has proven difficult to identify specific bacteria that cause the appearance of illnesses, mainly because so many processes are linked to different pathological conditions. Very many bacterial species exist, each with thousands of genes, and they may interact with numerous hormonal and gut-related processes, and can be modified by multiple environmental and experiential influences. The treatment and prevention of illnesses may also be subject to dynamic processes that are affected by multiple environmental influences, and these vary appreciably across individuals. Accordingly, potential treatments will no doubt have to comprise multiple bacterial changes, rather than any one or two bacteria, and the contribution of treatments may well vary over time (Lynch & Pedersen, 2016).

Even when the processes leading to a disease have been identified, it shouldn't be expected that manipulating these processes would necessarily attenuate the chracteristics of the condition. Once an illness is sufficiently advanced, or particular factors well entrenched, simply altering the microbiome can help in limiting further illness progression, but might be insufficient to reverse the already existing damage. It should be added that aside from the other actions that have been ascribed to microbiota, they are also involved in the production of metabolites that enter into circulation, which then affect various conditions outside of those involving the gut itself. For the moment, firm conclusion regarding the effectiveness of probiotics in the treatment of most illnesses ought to be held in abeyance (Bravo-Blas, Wessel, & Milling, 2016).

These caveats notwithstanding, there are excellent possibilities of being able to capitalize on individual differences in relation to health risks. It has been maintained that individuals can be stratified based on a few dominant bacteria, and thus the broad variability often discussed in relation to microbiota may be somewhat more limited. It may be possible to use these broad classes of bacteria in designing ways (e.g., through prebiotics, probiotics, or synbiotics — the latter contains both pre-and probiotics) to enhance gut bacterial functioning (Cani & Everard, 2016). People who consume the same diet, may nevertheless present with glycemic responses that differ appreciably, possibly owing to individual differences in gut microbial composition. Thus, finding appropriate diets for any given individual might benefit from a personalized approach, which could include glycemic responses, microbial factors, and genetic contributions. (Zeevi et al., 2015). For instance, among young women, socioeconomic factors, specific food choices, such as fat intake, and the presence of a gene variant (DRD4 VNTR), together could predict susceptibility to obesity (Silveira, Gaudreau, Atkinson, Fleming, & Sokolowski, 2016). Obviously, using a personalized approach to deal with diets and obesity would be enormously difficult (and financially constraining), but given the obesity crisis that seems to be escalating, such an approach could have both short- and long-term benefits.

Moderating Variables Concerning Gut Bacteria and Health Outcomes

It cannot be emphasized enough that any positive (or negative) influences of microbial factors may be dependent on a constellation of genetic, experiential, and psychosocial factors (life experiences, trauma, and social learning). The cumulative effects of these varied factors can also affect the effectiveness of intervention and treatment strategies. Just as stressful events can affect microbiota and/or immune functioning, and hence the provocation of disease, exposure to bacteria can affect the subsequent response to a stressor, thereby affecting physical ailments and psychological disturbances.

Microbiota variations are influenced by the presence of particular inherited genes (Goodrich et al., 2016), foods eaten, and epithelial cells that line the various cavities and surfaces of multiple structures. As already indicated, there is a constant battle within the gut between bacteria that try to cooperate and those are antagonistic with one another. Foods eaten fuel these processes so that when the needs of microbes are consistent with the needs of the host, well-being ensues; however, when these needs are at odds with one another (as occurs in response to sugars and fats), the antagonistic relationship may lead to illnesses. In the latter instance, microbes may begin to use nutrients that the body requires (e.g., iron), resulting in the immune system activity increasing to deal with these microbes, potentially leading to inflammation, obesity, and diabetes. In addition, the use of antibiotics have had an enormous impact on disturbing microbial diversity and hence affecting resilience and vulnerability to illness (Belkaid & Hand, 2014).

Beyond these many linkages, gut bacteria may also influence neurogenesis. Specifically, eliminating gut bacteria through antibiotics can diminish hippocampal formation of new neurons and disrupt performance in memory tasks. Interestingly, mice that had exhibited memory disturbances displayed lower white blood cells, primarily monocytes that carried Ly6Chi as a marker, implicating a link between gut bacteria, aspects of immune functioning, and brain neurogenesis. This connection was confirmed by showing that these outcomes could be reversed by reconstitution with normal gut flora provided that mice were also able to exercise (using a running wheel) or given probiotic treatments (Baruch & Schwartz, 2016). In addition to affecting hippocampal neurogenesis, disrupting the microbiome in mice though antibiotics also affects hippocampal glial reorganization, thereby favoring the development of depressive-like features (Guida et al., 2017).

CONCLUDING COMMENTS

Bacteria and viruses have been constant concerns, but in several ways they have become more threatening. Many vaccines aren't as effective as they ought to be (e.g., in the case of influenza vaccines) and the evolution of antibiotic resistant bacteria have become more apparent in relation to a number of existing diseases. Of course, the possibility of new emerging diseases seems to be more a certainty than a possibility. Beside the obvious consequences of infection, it has become apparent that activation of inflammatory processes may promote multiple diseases, including physical and psychological disorders.

Although it had been suspected for well more than a century, it has only recently been established that bacteria and other microorganisms exist throughought the human body, serving to maintain well-being. When dysbiosis occurs, vulnerability to multiple illnesses may occur, and some of the agents, such as antibiotics, which protected us from bacterial infections, may have acted against us by disruption of the gut microbiota. The increased knowledge regarding microbiota has also provided us with tools that could be used to enhance health. Specifically, it is now understood that the human gut is not equipped to digest the many macronutrients that are consumed, such as plant polysaccharides. Thus, commensal bacteria, such as Lactobacillus and Bifidobacterium, are involved in doing the job, ultimately producing short chain fatty acids (SCFAs), such as butyrate, lactate, and propionate. These factors may affect immunity, possibly diminishing potentially pathogenic microbes and augmenting gut barrier function (Slavin, 2013), and they may have neuroprotective effects (Horn & Klein, 2013). The actions of short chain fatty acids on CNS processes are only beginning to be understood, but it seems that some fatty acids (e.g., sodium butyrate) can attenuate stressor-provoked serotonin and BDNF alterations (Sun et al., 2016) and may thereby influence the function of microglia (Erny et al., 2015) and hence affect psychological functioning.

In view of the health benefits (and risks) associated with gut bacteria, there has been an obvious effort to enlist the microbiome to enhance well-being. In this regard, the lifestyles that are often adopted could produce microbial dysbiosis thereby promoting psychological disturbances, which can be attenuated, at least to some extent, by pre- and probiotic consumption. The prebiotics that have been used to modify gut-related disorders, may have their effects owing to multiple changes that evolve. These include the production of antimicrobial compounds, growth substrates, such as vitamins and polysaccharides released into the internal environment, reduction of the luminal pH, prevention of particular microbes from adhering to epithelial cells, augmented barrier functioning, and modulation of immune responses (Power, O'Toole, Stanton, Ross, & Fitzgerald, 2014). Prebiotics, such as certain oligosaccharides in human milk can also inhibit monocytes, lymphocytes, and neutrophils from binding to endothelial cells, and might thereby contribute to the relatively low frequency of inflammatory diseases in milk-fed human infants (Bode et al., 2004).

The microbiota present, as we've seen, varies greatly across individuals, being affected by many environmental factors, including diet, stressors, and environmental toxicants. But, the individual variability that exists rearding microbiota, coupled with the many factors that affect microbiota balances, makes it difficult to discern what reflects a harmful versus a beneficial compliment of bacteria. It

has indeed been suggested that in relation to treatment of illness, the intestinal microbial population might need to be individually tailored by diet and other manipulations (Shoaie et al., 2015). But, it's still uncertain which good bacteria to call upon, how much of it is needed, and how to do battle with bad bacteria. While the actions of some bacteria are known, many others are hardly understood, but their positive (or negative) actions are being uncovered.

Life-Style Factors Affecting Biological Processes and Health

ONCE DOUBT SETS IN, CERTAINTY (AND TRUST) RARELY RETURNS

There's little debate concerning the value of moderate exercise in relation to heart disease, breast cancer, colon cancer, and ischemic stroke. In some instances, as in the case of type 2 diabetes, effective outcomes could be obtained with even modest levels of exercise, and considerable evidence has indicated that regular moderate exercise has multiple psychological benefits. Thus, it's fairly remarkable how little attention most people devote to exercise and instead risky behaviors are maintained. Unless it becomes a habit that is undertaken on a systematic schedule, perhaps facilitated by exercising with others, it's difficult to engage in this behavior on a sustained basis. The sad fact is that "doing nothing is simpler than doing something."

There's also certainty regarding the importance of sleep. We have a basic biological need for sleep, even if it gets in the way of work or play. Failure to get enough sleep will likely undermine physical and psychological health, and some negative physical health consequences may be a result of sleep disturbances that accompany stress and depression (Irwin, 2015). There is such a thing as getting too much sleep, and in some instances sleeping too much might be a reflection of biological disturbances or illnesses being present (Irwin, Olmstead, & Carroll, 2016).

As much as eating well is fundamental to good health, it's not overly surprising that individuals engage in poor eating, especially in the face of so many temptations. What could be more reinforcing than some yummy comfort food to deal with life stressors (OK, for some people alcohol serves as an alternative lousy coping strategy)? At first blush, it might seem surprising that many people barely have an inkling regarding which foods are good for them and which are very bad. Even experts in the field are at odds with one another, frequently providing confusing and contradictory information. In fact, what's healthy and what's not seems to change on a frequent, but irregular basis. Butter used to be a no-no, whereas

---

(cont'd)

margarine was good. Then it was argued that margarine was bad, just a step or two removed from plastic, whereas butter was actually OK. Carbs and fatty foods were the scourge that increased cholesterol levels that led to heart disease, but now it seems that they're not that bad, unless individuals are diabetic or at high risk for heart disease. The business with carbs versus fats, or for that matter whether certain types of fat are worse than others is confusing, although, there is general agreement that we should avoid trans fats. Based on data from large-scale prospective studies, every 2% increase of trans fat intake was associated with a 16% greater chance of premature death, and every 5% increase of saturated fat intake was associated with an 8% increase of early death (Wang, Li et al., 2016).

Adopting a poor life-style can have multiple adverse consequences that culminate in diminished well-being through actions on multiple biological processes. Changes to the inflammatory immune system are no doubt among the key agents responsible for multiple illnesses, including those of a psychological nature, and those that affect our lifespan and our health-span. Among individuals who live very long lives (>95 years of age), the difference from others wasn't so much that these individuals didn't become ill, but rather that their illnesses were compressed into a relatively brief period near the end of life.

RISK FACTORS FOR HEALTH AND DISEASE

Go figure! Actions actually have consequences.

Try telling that to your teenage kid, or to the person carrying a food tray with a double whopper and supersized fries, faithfully offset by a diet Coke. Many illnesses obviously develop because of risky behaviors, particularly poor life-style choices, or engaging in dangerous behaviors or those that are outrageously stupid (thereby making the genes for these traits less likely to reappear in the gene pool). These are all common sins of commission, but some individuals will suffer the consequences that come with sins of omission. These include not receiving recommended screening for colon, breast or prostate cancer (for individuals more than 50 years of age), failing to use sunscreen, choosing not to be vaccinated against an imminent flu epidemic, not having children inoculated for measles, mumps and rubella (MMR), or failing to wear a helmet when skiing, skateboarding, or bicycling.

Our sympathies are much greater for those who encountered illness or accident through no fault of their own. They were unlucky to be born with particular genes or gene mutations or epigenetic modifications that affected their proneness to a disease. Alternatively, they might have been unfortunate to live downwind from a plant spewing pollutants or microparticles that led to cancer, or worked or lived in an environment in which second hand smoke sickened them. Many of us have also been the victim of a martyr who chose to go to work despite having a viral illness that s/he indiscriminately and thoughtlessly spread to others, including me[1].

---

[1] In a slightly different context, the multitalented writer, composer, and singer, Shel Silverstein summed it up well in his song "Don't give a dose to the one you love most" that is still relevant.

BLAMING THE VICTIM

It's unfortunate that there is often a tendency to blame the victims for bad things that happen to them, even if their misfortune was not of their own making. It had been proposed some years ago (Lerner & Montada, 1998) that people have a cognitive bias in which they naively see the world as being just (morally fair), so that positive and noble acts will ultimately be rewarded, whereas evil acts will ultimately be punished. In this context, when people suffer from a negative event occurring, it must have somehow been their own fault.

When it is believed that negative outcomes stem from a person's risky behaviors, irrespective of whether they comprise actions or inactions (sins of commission or omission), the sinners are made to suffer doubly. Will their associates (or medical professionals) treat a smoker and a nonsmoker similarly following a lung cancer diagnosis? The biases and the stigma can be so acute that some nonsmokers will not confide that they have this form of cancer.

We often assume that individuals have free will and might choose to smoke, and hence their illness is self-inflicted. But, is this is entirely true? A 65 or 70 year old person with a smoking related disease, might have been born into a family where smoking was the norm. Alternatively, they may have worked in an environment where everyone smoked, or lived in the poor part of town where smoking was common, and hence they too took up smoking. It is similarly possible that they inherited genes that were linked to impulsivity and high reward satisfaction, both of which favored addiction, making them particularly likely to engage in a risky behavior. In her reviews related to addictions, Volkow et al. (2010) made the important suggestion that drugs, such as alcohol or cocaine, can "overwhelm" cognitive control circuits, thereby impairing rational thinking, thus allowing for addiction to persist. In light of the brain changes that might have occurred in response to these agents, is it reasonable to continue to frown upon patients as if they were sinners?

It isn't particularly unusual to find that policies and procedures that had been adopted with the most positive intentions, turn out to have unintended consequences. National, community, and individual level attempts have been made to promote appropriate behaviors to diminish health risks (campaigns to reduce smoking, encourage weight loss, and children vaccinated against a variety of viral illnesses). These have frequently been successful, but there are instances when unintended consequences act against favorable outcomes. In large measure this occurs because people have a puzzling tendency to adjust their behaviors, essentially adopting more risky actions based upon their appraised risk of negative outcomes being reduced. According to "risk compensation"

theory, when people ordinarily perceive risks to be high, they behave more carefully, but should they perceive personal risk to be low, such as when safety measures have been instituted, they might be less cautious (Peltzman effect). Following the introduction of penicillin to combat sexually transmitted diseases (STDs), such as syphilis, the rates of infection declined precipitously, as those treated were no longer passing on the illness. However, after some time, rates of infection increased once again; because a cure was possible, freewheeling sexual behaviors may have increased and condom use might have declined. Likewise, with the introduction of mandatory use of seat belts, some people drove more recklessly, and the introduction of helmets for skiers resulted in the engagement of riskier behaviors on the hills (e.g., Specht, 2007). In the same fashion, when type 2 diabetic patients find that their blood sugars normalize with medications, they may be less attentive in relation to their diet, and revert to previous bad behaviors that led to their condition.

Unintended consequences may come about for a variety of reasons. For starters, ignorance of the issue results in it being entirely unlikely that individuals (or groups of individuals) are able to predict all potential outcomes that could occur. Further, problems might not be appropriately analyzed, and the solutions adopted may have been more germane to earlier problems, but were unsuitable for the issue being faced at the moment. Individuals and governments may also find themselves adopting policies in which basic values actually interfere with particular behaviors (e.g., the attitude that governments don't have the authority to demand children be vaccinated), which could have unfavorable consequences. On other occasions, government agencies may be reluctant to adopt particular policies or procedures as this could produce fear or panic (e.g., the presence of an imminent threat), and so steps are taken to find alternative strategies, but these could end up having negative consequences. Finally, like governments and other organizations, there is a tendency for individuals to favor short-term gains without appropriate consideration of long-term negative consequences.

In its worst form, proposed solutions can give rise to perverse results in which particular actions can lead to solutions that can make situations worse. When antidepressant agents (SSRIs) were seemingly associated with elevated suicide risk among adolescents, restrictions were imposed so that these drugs were not offered to young people. As a result, however, outpatient visits (ambulatory care) among young people declined, leading to untreated depression and a commensurate increase of suicides (Katz et al., 2008).

There are also times where actions taken can have unintended positive effects, just as certain actions can elicit negative unintended consequences. Unanticipated benefits were frequently realized when drugs developed for one purpose ended up having positive effects for other conditions. For instance, through its anti-inflammatory actions, the pain reliever acetylsalicylic acid (aspirin) may be useful in relation to heart disease, stroke, and other inflammatory conditions. There are many such positive outcomes, but we usually hear more about negative events, possibly because these sell more newspapers.

DIETARY FACTORS AS DETERMINANTS OF HEALTH

Diet Adaptations

Over centuries, as humans moved from one type of environment to another, diets changed, and gene adaptations and microbial variations followed. This might have been particularly notable when humans shifted from the hunter-gatherer mode to become more reliant on farmed foods. The introduction of some foods may have had positive effects in some cultures, but this same food might have been less beneficial, and even harmful, in other groups. Some groups in Northern Canada and Northern Aklaska obtain much more of their calories from fats than do people living much further south, but seem to do well with these diets. In fact, the frequency of death by heart disease is about half that of their southern neighbors (this has been termed the Inuit Paradox). To be sure, the fats obtained from wild animals differs from that of grain-fed farm animals, having different cholesterol content, are free of antibiotics and a variety of other farm-related chemicals, which could account for some aspects of the Inuit Paradox. More than this, in far Northern groups, selection pressures may have resulted in genetic changes that would facilitate dealing with the diet rich in protein, fats, and omega-3 polyunsaturated fatty acids. Indeed, Greenland Inuit Peoples carry particular genetic mutations (that are relatively rare in Europeans), which modulate fatty acid composition, thus contributing to lower low density lipoprotein (LDL) and fasting insulin levels, as well as serving in the regulation of growth hormones (Fumagalli et al., 2015). These data certainly are consistent with the view that personalized diets (like personalized medicine) might be best for determining the foods that ought to be consumed by any given individual.

The adaptation to certain foods based on selection for particular gene mutations obviously occurs over many generations. Often, however, the sudden introduction of new foods into a region or culture may engender adverse outcomes. In this regard, Western diets can be hazardous to the health of some Northern groups (type 2 diabetes is almost endemic in some Northern First Nations communities). Likewise, those living in rural parts of Asia and Africa rely on diets that are relatively unique to their groups, but upon migrating to urban settings and encountering unhealthy diets, obesity and various eating-related disorders may follow. Urbanization in China has similarly resulted in a microbiome reminiscent of that seen in American people, including the loss of microbiota that are thought to be beneficial, which were associated with increase occurrence of Escherichia and Shigella (Winglee et al., 2017). In essence, the impact of specific diets may vary depending on the presence of certain genetic influences that favor or act against the development of particular pathological conditions (Ye et al., 2017).

Beyond issues related to processed versus unprocessed foods, it has often been recommended that we consume foods that contain dietary fiber, heart-healthy oils, and low-fat proteins (the latter obtained from fish, poultry, legumes, nuts, and seeds), and avoid foods replete with preservatives, sugar and sodium.

Consumption of foods high in fiber, for instance, may limit a wide range of age-related health disturbances, including cardiovascular illness, cancer, and a reduction of all-cause mortality (Aune et al., 2016). Conversely, diets high in fats may promote metabolic disorders that alter serotonin functioning, thus leadings to increased risk for psychological disturbances, such as anxiety and depression.

As we've already indicated, confusing and contradictory information has frequently been proffered regarding the foods that should or shouldn't be consumed. The view was frequently advanced that we ought to favor polyunsaturated fats and avoid foods with saturated fats, but it was argued that the available data in this regard aren't all that scientifically convincing (Chowdhury et al., 2014), although polyunsaturated fat consumption was associated with a modestly lower incidence of chronic heart disease. In fact, studies of diet in relation to health are often contaminated by the simple fact that individuals don't accurately recall what they've actually consumed (even using daily eating diaries).

Contrary to the standing dogma, it was reported that saturated fats were not linked to stroke, heart disease or diabetes, although trans-fat was tied to these pathologies. Similarly, in a prospective epidemiological analysis across 18 countries, consumption of total fat and individual types of fat were not related to lower mortality related to heart disease and stroke, whereas high carb consumption was tied to greater risk of mortality (Dehghan et al., 2017). This said, in other well conducted studies, such as those that measured polyunsaturated fats (e.g., linoleic acid) in adipose tissue over a 15 year period, a link was evident in relation to cardiovascular disease and all-cause mortality (Iggman, Ärnlöv, Cederholm, & Risérus, 2016). A recent report from the American Heart Association made sense of the inconsistent findings that have appeared (Sacks et al., 2017). Essentially, it was suggested that research addressing the question needs to look more deeply at the effects of getting rid of saturated fat from the diet, particularly with respect to what replaced these fats. In some instances, saturated fats were replaced by refined carbohydrates and sugars, and even trans fats that were worse for heart health than saturated fats. However, when saturated fats were replaced by polyunsaturated fats, monounsaturated fats, and whole grain carbohydrates, risk of heart disease declined.

Individuals wishing to maintain a proper life-style, including eating well, may become frustrated when diet gurus provide divergent perspectives, and as a result trust in them may be lost. Most diets, including those that focus on reducing carbs (Atkins, South Beach, Zone), fat consumption (Ornish, Rosemary Conley) or moderation of macronutrients (Biggest Loser, Jenny Craig, Nutrisystem, Volumetrics, Weight Watchers) are effective, provided that the individual sticks to it, more so if dieting is accompanied by regular exercise (Johnston et al., 2014). As we'll see shortly, having particular genes may also be fundamental in determining metabolic changes that accompany food ingestion, and thus there are occasions in which adherence to a particular diet might not work out well. As indicated in an editorial in the British Medical Journal, the scientific community desperately needs to offer a consistent and unified message concerning healthy eating.

Just about everyone agrees that sugar from desserts are sadly a definite no-no, and sugar sweetened beverages are just as bad, and perhaps worse. This gave rise to the manufacture of diet drinks spiked with artificial sweeteners, but we shouldn't be misled about all those "diet" drinks. Among other things, their continued consumption was associated with an increase in belly size and cardiometabolic risk (Azad et al., 2017), and by their actions on gut microbes, can encourage metabolic syndrome and obesity, and might thus contribute to the development of type 2 diabetes. Sugary drinks and artificial sweeteners have also been linked to diminished hippocampal volume, poorer cognitive functioning, dementia, and increased risk of stroke (Pase et al., 2017). Once more, these data are only correlational, and the possibility exists that numerous factors, such as preexisting diabetes or prediabetes, influenced outcomes as well.

I CAN RESIST ANYTHING EXCEPT TEMPTATION
Oscar Wilde

The equivalent of a "Western diet" together with a sedentary life-style, increased levels of proinflammatory monocytes, and activity of brain microglia in mice, which can impair neuronal survival. As well, this diet was accompanied by elevated β-amyloid plaque and microglia that express TREM2, which is also observed in Alzheimer's disease (Graham et al., 2016). Similar findings in humans should be sufficient to promote good eating practices. However, as much as people might have earnest intentions to eat properly, when faced with a chocolate éclair, or perhaps some really nice gelato, resisting will be enormously difficult. Even the person affected by diabetes might say "well, just this once, but only a very small bit." Beating temptation is exceedingly difficult, and the response to yummy foods is in many respects reminiscent of an addiction. Indeed, over-eating and drug addiction seems to share some underlying mechanisms, including activation of dopamine brain neurons (Abizaid, 2009). Saturated fats particularly can instigate effects on specific brain dopamine receptors within the nucleus accumbens, thereby affecting feelings of reward (Hryhorczuk et al., 2015), hence promoting further eating.

(cont'd)

The hormone orexin, which plays a significant role in arousal and food intake, may affect addiction processes by provoking dopamine release at the nucleus accumbens and prefrontal cortex (Borgland et al., 2009), and ghrelin, a key component of eating processes, also influences dopamine reward processes (Abizaid, Luheshi, & Woodside, 2014). Fibroblast growth factor (FGF)-21 released from the liver likewise communicates with the brain's reward system to affect food preferences, and might have clinical applications in the management of diabetes (Talukdar et al., 2016). In fact, liraglutide (Victoza), a glucagon-like peptide-1 (GLP-1) antagonist used in the treatment of type 2 diabetes, has the added benefit of diminishing the rewarding value of desirable foods (van Bloemendaal, Ten Kulve, la Fleur, Ijzerman, & Diamant, 2014).

White Fat and Brown Fat in Relation to Obesity

Not all fat is the same, nor is all fat bad. Fat located on the belly (abdominal fat) is considered to be particularly unhealthy as it contains proinflammatory cytokines, which in excess can favor the development of heart disease, diabetes, and other inflammatory diseases. Aside from the location in which fat is found, the nature of the fat itself is important. Brown fat tends to be readily burned energy, and consequently is easily diminished, whereas white fat is an energy storage tissue that is relatively resistant to being modified by exercise or dieting, and is usually considered "bad fat." As infants, humans have appreciable brown fat, which diminishes during adulthood, but it seems that adults have much more brown fat than had initially been thought. Women are more able to activate brown fat than are men, and thinner people have proportionally more brown fat than those who are heavier. There are people who seem to have an extraordinary ability to activate brown fat, but the reasons for this are uncertain, although creatinine clearance might somehow contribute to this (Gerngroß, Schretter, Klingenspor, Schwaiger, & Fromme, 2017). It seems that while thin mice (and perhaps thin people) are able to convert white fat to brown, among obese people the inflammatory response is elevated, and this works against the conversion of white fat to brown (Sanyal et al., 2017). In essence, being overweight has downstream effects that keep people that way.

There has been the recognition that beige fat, an intermingling of white and brown, also burns quickly and like brown fat, serves to maintain our body temperature. If it were possible to convert white fat to brown (or beige) fat, then it might be possible to burn it more efficiently, and thereby diminish obesity and the occurrence of diabetes (Tran et al., 2016). Circulating serotonin has been implicated in eating processes and to adult obesity, perhaps because it contributes to brown fat activity. Consistent with this perspective, mice engineered to be deficient of tryptophan hydroxylase 1, an enzyme necessary for the formation of serotonin, did not become obese when fed a high-fat diet, nor did they become insulin resistant, possibly owing to greater energy expenditure and increased burning of sugar present in brown fat (Crane et al., 2015).

Hypothalamic factors may also contribute to the conversion of white fat to brown fat. By attaching a sugar molecule (O-GlcNAc) to neurons of the hypothalamus that are associated with hunger, it may be possible to influence the conversion of white fat to brown fat, and prevent the negative consequences associated with a high fat diet (Ruan et al., 2014). It was similarly observed that following a meal, hypothalamic processes are activated by circulating insulin that promotes the browning of fat, which can then be used as a rapid source of energy. Following a fasting period, this switch is moved in the other direction so that the brain causes brown fat to be converted to white fat. Among some obese people, the switch that causes white fat to turn brown may be ineffective and hence energy expenditure isn't promoted (Dodd et al., 2017).

Among mice genetically engineered not to produce folliculin, a substance involved in regulating mitochondria in fat cells, the typical weight gain associated with a high fat, junk food diet did not occur and less white fat accumulated (Yan, Audet-Walsh et al., 2016). Further, the administration of specific drugs through nanoparticles that encourage angiogenesis (formation of new blood vessels) can cause white fat-storing cells to turn brown, leading to appreciable weight loss. As an added bonus, these mice also exhibited a decline of cholesterol and triglycerides and improved sensitivity to insulin (Xue, Xu, Zhang, Farokhzad, & Langer, 2016). Preclinical tests with other drugs, such as the anticancer agent bexarotene, also showed promising results, but further studies are necessary to assess and eliminate side effects of the treatment (Nie et al., 2017).

Not to be left out, gut bacteria may play a pivotal role in obesity. Depleting the microbiota (e.g., by antibiotics) had the effect of creating fat burning beige fat within white fat packages, possibly through an increase of certain types of macrophages (Suárez-Zamorano et al., 2015). Moreover, following a meal, gut microbes may themselves produce proteins that signal release of the hormone GLP-1, which inhibits eating (Breton et al., 2015), and in this sense gut bacteria act much like satiety peptides. It also appeared that antibiotic treatment prior to 2 years of age increased the risk for obesity (Scott, Horton et al., 2016), and the view was

offered that antibiotics in early life can dispose individuals to weight change and type 2 diabetes (Cox & Blaser, 2015).

A detailed review of the relevant literature concluded that variations in microbial diversity and composition are tied to obesity, as well as other disorders. This occurs, in part, because certain gut bacteria influence host energy harvesting, regulation of metabolic processes, insulin resistance, fat deposition, and the consequent inflammation, as well as central appetite and food reward signaling. In the latter regard, particular bacterial strains and their metabolites are able to affect brain functioning by stimulating vagal afferents or through immune-neuroendocrine processes, thus favoring the development of obesity (Torres-Fuentes, Schellekens, Dinan, & Cryan, 2017).

Like the involvement of hormones and neurotransmitters, levels of P75 neurotrophin receptors that contribute to neuronal survival and growth, might also be involved in energy regulating processes. Among mice that lacked P75 neurotrophin receptors, weight gain did not occur even if mice were kept on a high fat diet (Baeza-Raja et al., 2016). Finally, one of the strongest biological links to obesity is a gene region referred to as FTO, which might be related to how adipocytes (fat cells) function. Specifically, two genes within this region (IRX3 and IRX5) control whether adipocytes burn versus store fat, and by manipulating features of the gene, fat storage (or burning) could be altered, and thus at some time might be useful in developing anti-obesity treatments (Claussnitzer et al., 2015). It is also possible that finding methods of coaxing stem cells to become brown fat rather than white fat cells could act against obesity (Moisan et al., 2015).

Current research is poised to offer novel treatments to diminish obesity. However, we shouldn't fool ourselves to think that "browning or beiging of white fat," will be an alternative for exercise to stay healthy. In fact, given the many benefits of exercise to heart health and immune functioning, using drug-based shortcuts to weight loss may be counterproductive. This said, for individuals who have been sedentary for years, and who have only seen a gym from a distance, a new drug is in development that burns fat over sugar and was thus able to provide mice with an almost 50% increase of endurance (Fan et al., 2017).

Hormonal Regulation of Eating Processes

The discovery of two hormones, leptin and ghrelin, caused a sea change in how eating and energy regulation were considered. Leptin and ghrelin, released by adipocytes (fat cells) and by cells within the stomach, respectively, enter circulation and then, through their actions on both the brain and peripheral organs, affect food intake, energy expenditure, and adiposity (fat) (Abizaid, Luheshi, & Woodside, 2014). When levels of ghrelin increase, eating is promoted, whereas elevations of leptin that ordinarily occur during and following a meal, signal satiety. As both these hormones influence several brain processes, such as HPA axis activity, as well as dopamine and serotonin functioning within frontal cortical limbic regions (Abizaid, 2009; Fulton et al., 2006), it is reasonable to expect that food consumption and the type of food consumed, may influence mood states. Given that ghrelin can influence dopamine within the ventral tegmentum and nucleus accumbens, which have been linked to reward processes, craving or anticipatory responses to food might be encouraged (Abizaid et al., 2006). Ghrelin stimulation of orexin cells in the lateral hypothalamus may likewise contribute to food craving, including the desire for comfort foods following stressor experiences (e.g., Borgland et al., 2009). In this regard, it has been suggested that craving and the actual consumption of food might involve distinct processes; dopamine within the nucleus accumbens may be associated with craving, whereas orexin may be aligned with consumption itself (Parker et al., 2015). Indeed, nutritional and hedonic value of foods may differentially influence dopamine release from the ventral and dorsal striatum (Tellez et al., 2016). Imaging studies have also confirmed that the links between dopamine receptors and preference for sweet foods evident in lean people were absent in obese individuals, suggesting a disturbance of these connections (Pepino et al., 2016).

In view of their roles in eating processes, it is hardly surprising that ghrelin and leptin could influence psychologically based eating disorders. Elevated ghrelin levels were detected in anorexia and bulimia nervosa patients, whereas decreased ghrelin was associated with binge eating (Méquinion et al., 2013). Consistent with a causal role for ghrelin in eating or in the desire for food, ghrelin administration increased food-related imagery and stimulated feelings of reward, influenced responses to visual and olfactory cues related to food, and enhanced responses to incentive cues (i.e., secondary stimuli that have been associated with reward), thereby affecting craving. It may be significant in this regard that among emotional eaters the typical ghrelin decline was not apparent following food consumption, suggesting that ghrelin functioning was not operating as it should (Raspopow, Abizaid, Matheson, & Anisman, 2014). Paralleling such findings, MRI images taken before and after eating among obese woman, indicated that the neo- and limbic cortices and midbrain regions that were activated when hunger was present, still persisted after a meal was eaten. Again, these data point to poor regulation of satiety processes being associated with obesity.

As craving is a prime promoter of food intake among dieters (and drug consumption among individuals trying to beat the habit), identifying the neurocircuitry underlying the craving process, may help to establish a target to limit obesity (and drug reinstatement). The insular cortex may be involved in this regard given that activity in this region ordinarily increases in anticipation of food, but the functioning of this brain region is disturbed among obese individuals. Using rodent models, it was possible to activate and inhibit neurons within this region, thereby confirming its involvement, and that of hunger-related genes coding for Agouti-related protein (AgRP), in processes related to craving (Livneh et al., 2017).

SET POINT AND FAMINE PREPAREDNESS, AND THE OTHER SIDE OF THE EQUATION

Each of us has a "set point" so that when our weight falls sufficiently, metabolic processes adjust in an effort to limit further weight loss, and activation of related hormones concurrently prompt us to increase eating. When dieters begin to lose weight, their cells go into "famine mode" such that their metabolic rate slows in an effort to sustain energy (the body's cells might not have understood that the person was intentionally dieting and thus responded inappropriately). Neurons within the hypothalamus that respond to Agouti-related neuroptide act as a switch that affects the response to reduced food, serving to determine whether calories should be burned or conserved (Burke, Darwish et al., 2017). Once regular eating is reinstated the "famine mode" may continue so that calories aren't readily burned off, and consequently weight gain may occur despite individuals not over-eating. With repeated efforts to lose weight (yo-yo dieting), metabolic disturbances become accentuated and weight loss becomes progressively more difficult to maintain. Thus, it is questionable whether dieting is, in fact, the best solution to losing weight and keeping it off. Indeed, metabolic changes elicited, even after a missed meal, may produce inflammatory responses that encourage fat storage (Kliewer et al., 2015). These phenotypes may reflect remnants of selection processes that date back to times when hunting food was more difficult than a trip to the grocery store.

Although the harms created by yo-yo dieting are well known, this shouldn't be interpreted as meaning that occasional fast days are counterproductive. In fact, in mice and possibly in humans, periodic cycles of prolonged fasting or a diet that mimics fasting for two or more days can have positive actions, and may even limit some adverse secondary effects associated with chemotherapy. Likewise, these treatments can attenuate and reverse the immunosuppression or immunosenescence associated with both chemotherapy and ageing, possibly by promoting hematopoietic stem cell-based regeneration (Cheng, Adams et al., 2014). Intermittent fasting, or a ketogenic diet, was also reported to preclude inflammation and enhance neuroprotection, thus limiting signs of experimental autoimmune encephalomyelitis (EAE) in an animal model of multiple sclerosis (Kim, Hao et al., 2012). In line with these reports, fasting resulted in the death of destructive autoimmune cells and generation of cells that create myelin that had been destroyed during the course of the disease (Choi et al., 2016). In response to intermittent days of diminished eating, leading to a 25% calorie reduction, reparatory processes may be instigated, including changes of gene expression that have been associated with longevity (Wegman et al., 2015). Similarly, a 25% reduction in daily calorie intake was accompanied by anti-inflammatory actions, which would ordinarily diminish lifespan and increase disease states. It seems that caloric restriction or fasting, enables changes of inflammatory processes, thereby limiting diseases that might otherwise be provoked (Youm et al., 2015). This could come about by promotion of the growth factor VEGF, which contributes to the formation of blood vessels and activation of a subset of immune cells, specifically anti-inflammatory macrophages, which stimulate fat cells so that they burn more fats (Kim, Kim et al., 2017).

Although leptin and ghrelin have been the primary players in research regarding energy regulation and food intake, several other peptide hormones contribute to these processes, as indicated in Table 4.1. For instance, the mammalian analogues of the amphibian hormone bombesin (BB), specifically gastrin-releasing peptide (GRP), and neuromedin B (NMB), contribute to the regulation of eating, acting as a satiety signal, and are also involved

TABLE 4.1   Hormones Related to Energy Regulation and Eating.

Secreted Hormone	Biological Effect	Behavioral Outcome
Leptin	Produced by fat cells. Influences neurons in hypothalamic regions	Reduces food intake and appetite, and increases energy expenditure
Ghrelin	Synthesized in gut. Affects many of the same brain regions as leptin, but in an opposite manner. Influences brain regions associated with reward	Stimulates food intake and appetite, while reducing energy expenditure. Enhances reward seeking behaviors; modulates stress responses
Orexin (hypocretin)	Made by cells in lateral hypothalamus	Involved in the regulation of appetite, arousal, and wakefulness
Insulin	Produced by pancreatic beta cells: Regulates fat and carbohydrate metabolism. Involved in getting glucose from the blood into various body cells and storing it as glycogen	In brain, insulin stimulates hormones that reduce food intake
Bombesin (appears in humans as neuromedin B [NMB] and gastrin releasing peptide [GRP])	Synthesized in gut and in several hypothalamic regions, and is also found in limbic regions	Acts as a satiety peptide (signals when individual is full) and is released in response to stress, thereby promoting anxiety
Neuropeptide Y (NPY)	Produced by the gut and in several brain regions, including the hypothalamic arcuate nucleus	Increases food intake and reduces physical activity; increases energy stored in the form of fat; blocks nociceptive (noxious) signals to the brain; acts as an anxiolytic agent
	Increases vasoconstrictor actions of norepinephrine	
Orexin (hypocretin)	Created within the lateral hypothalamus, but orexin receptors are found throughout the brain	Involved in appetite, as well as stress and reward processes, arousal, and wakefulness
α-Melanocyte stimulating hormone (α-MSH)	Produced in the arcuate nucleus of the hypothalamus. Acts as an agonist of melanocortin (MC-3 and MC-4) receptors in brain, including stress-related regions	Reduces appetite, increases energy expenditure as modulated by leptin
Agouti-related peptide (AGRP)	Produced in the arcuate nucleus (by same cells that produce NPY), and serves as a natural antagonist of MC-3 and MC-4 receptors	Increases appetite and reduces energy expenditure. Modulated by leptin and ghrelin
Glucagon-like peptide-1 (GLP-1)	Manufactured in the gut. Important in stimulating insulin release	Acts in brain to reduce appetite and to diminish functioning of reward processes

THE IMMUNE SYSTEM AND MENTAL HEALTH

in anxiety processes (Merali, Graitson, Mackay, & Kent, 2013). As well, insulin plays a pivotal role in energy processes and may contribute to stress responses, and by causing the release of dopamine within the striatal region of the brain, it may contribute to feelings of reward and pleasure (Stouffer et al., 2015). Beyond this, insulin also acts with or on other hormones that affect energy processes. This includes glucocorticoids, leptin, neuropeptide Y (NPY), and regulatory hormones implicated in the development of obesity and metabolic disturbances associated with chronic stressors.

In addition to these factors, a variant of the gene coding for the neurotrophin BDNF may contribute to the development of obesity, so that boosting BDNF protein levels may diminish appetite and obesity (Mou et al., 2015). Further, among morbidly obese woman, μ-opioid receptors in brain regions associated with reward processing were lower than among normal weight women (Karlsson et al., 2015). Thus, activation of these receptors might subserve reward gained from foods, but when the receptor number is diminished women may eat more as a compensatory response to regain rewarding feelings from food.

In view of the multiple processes associated with eating, a key role for genetic factors in this regard is hardly unexpected, and it might be anticipated that obesity would be linked to genetic processes. A whole genome analysis identified about 90 genes present within subcutaneous fat that could contribute to obesity and hence to diabetes and heart disease (Civelek et al., 2017). In addition, it seems that epigenetic modifications can be present on TRIM28 related genes, which have been implicated in the development of obesity, so that a gene switch is tripped, thereby diminishing obesity (Dalgaard et al., 2016). Genetic polymorphisms also exist that promote eating, but these only express themselves when individuals initiate a diet program (Aberle et al., 2008), and exercise-related epigenetic alterations have been implicated in obesity and type 2 diabetes, which could be expressed transgenerationally (Barrès & Zierath, 2016).

Stress and Eating

Severe stressors typically cause reduced eating. Moderate stressors may have a similar effect in some people, but in others it may increase their propensity to eat, possibly reflecting self-medication through "quick fixes" in an effort to diminish distress. Processes related to stress and to eating are intricately entwined, so that treatments that diminish anxiety, such as benzodiazepines, typically increase food consumption (e.g., Merali, Bédard et al., 2006). As well, several stressor-elicited neurochemical changes, notably variations of dopamine and CRH, have also been related to eating processes, and along with cortisol, may contribute to stress-related food craving (Chao, Jastreboff, White, Grilo, & Sinha, 2017). Predictably, hormones fundamental in eating processes (leptin and ghrelin), have marked effects on stress responses, and chronic social stressors could influence leptin levels (Patterson, Khazall, Mackay, Anisman, & Abizaid, 2013).

From an adaptive perspective, it makes sense that eating and stress experiences would engage antagonistic processes. It would be counterproductive, after all, for an animal under threat to stop for a meal. Yet, in a subset of people, particularly those who use avoidant coping strategies to deal with stressors, negative events can lead to disordered eating, possibly acting as a coping response, much as alcohol might serve in this capacity. The elevated eating is particularly notable among emotional eaters who encounter stressors but receive unsupportive responses from their friends (Raspopow, Matheson, Abizaid, & Anisman, 2013). Among those who adopt eating as a way of coping, the foods selected are typically those that taste good, and individuals may experience cravings for carbs or what is usually referred to as "comfort food." The words "I could really use a celery stick right about now" will never escape the lips of an emotional eater.

Stressor-provoked eating, unfortunately, is linked to elevated glucose, insulin levels, insulin resistance, and incidents of prediabetes and diabetes. Such effects can even develop long after the stressor experience terminated. Childhood trauma, for instance, has been linked to subsequent increased stress reactivity, elevated C-reactive protein reflecting increased inflammation, and compensatory emotional eating (Schrepf, Markon, & Lutgendorf, 2014). Stressful events may also sabotage an individual's ability to make proper decisions and for those already dieting, good intentions may go by the wayside in the face of stressful events. Predictably, ineffective use of eating as a coping mechanism is related to more frequent stress-related pathology, such as PTSD.

Several brain neurochemical changes that accompany stressful experiences contribute to increased eating. Stressor-provoked activation of dopamine neuronal processes, which can stimulate reward processes, could also cause particular stimuli or responses to be more salient (Berridge & Robinson, 2016). Thus, the positive effects of food in diminishing moderate stress responses may become especially significant, thereby leading to increased consumption of tasty foods upon further stressor encounters. By affecting dopamine functioning, hedonic appraisal of foods high in sugars and fat might favor the development of an "addictive-like" condition so that withdrawal of the palatable diet could lead to a stress-like response (Morris, Beilharz et al., 2016). This outcome varies with stressor intensity. If a relatively strong stressor is encountered, circuits involving the hormone, CRH, which has been implicated in stress processing and anxiety, may be activated so that the organism will focus on adopting defensive strategies, rather than those related to reward processes (Lemos et al., 2012).

There has been the view that cortisol elevations associated with stressors contribute to eating and might be an important element in the provocation of obesity. In this regard, cortisol measured in hair samples (hair grows at about 1.25 cm a month, and thus a sample of about 3.75 cm reflects cortisol accumulation over the preceding 3 months) was directly related to body weight, and it was argued that chronic stressors contributed to obesity (Jackson, Kirschbaum, & Steptoe, 2017). In her work linking stressful experiences and eating, Dallman (2010) considered how chronic stressors could have such consequences. From her perspective, in response to stressors, the fast caloric fix provided by sugars and carbohydrates, together with brain neurochemical changes provoked by these comfort foods, provide sufficient energy to facilitate coping with psychological stressors. In essence, eating comfort foods might facilitate the brain's attempt to limit the distress and anxiety created by chronic stressors. Central to Dallman's reasoning was that with continued distress, persistently high concentrations of corticosterone influence CRH activity within the central amygdala, thus engendering strong emotional responses. At the same time, corticosterone may contribute to caloric intake, by increasing the salience of pleasurable or compulsive activities (ingesting sucrose, fat, and drugs), and might thus contribute to obesity. It is of particular significance that glucocorticoids act on the redistribution of fat so that it appears as abdominal fat depots (Peters & McEwen, 2015). As we've seen, abdominal fat acts as a storehouse for inflammatory factors, and as a result may influence several pathological conditions, and it was even suggested that dietary interventions may be useful in limiting the course of some of these conditions, including the cognitive decline associated with ageing.

Early Life Experiences

Stressful events experienced early in life may be associated with elevated adult stress reactivity, anxiety, and a preference for comfort foods, thus contributing to adult obesity. In fact, adult obesity was 30%−50% more frequent among individuals who experienced negative early life events, particularly abuse (Hemmingsson, Johansson, & Reynisdottir, 2014). Aside from proactive effects on appraisal and coping processes, negative early life stressors give rise to disturbed neuroendocrine functioning that feeds into appetite regulation and metabolism, negative thinking, poor emotional regulation, as well as disturbed sleep and cognitive functioning, which also encourage the development of obesity. It is significant, as well, that among mice on a high fat diet, obesity-related inflammatory changes and microglial alterations, seemed to be related to disturbed cognitive functioning, which could be reversed when mice were put on a low fat diet (Hao, Dey, Yu, & Stranahan, 2016). Further, owing to epigenetic modifications within germline cells, a high fat diet that promoted diabetes could lead to a similar outcome in the next generation (Huypens et al., 2016). In this regard, in young animals, an epigenetic change related to a gene associated with impaired glucose metabolism (*Igfbp2*) had ramifications for adult diabetes, and this gene's action was similarly modified in morbidly obese humans with diabetes. Thus, the status of this gene might turn out to be a useful biomarker in predicting the development of diabetes (Kammel et al., 2016)

Stress and Ghrelin

Chronic stressor experiences can increase caloric intake among mice maintained on a low fat diet, as well as diminished leptin levels (Finger, Dinan, & Cryan, 2012), which may act together with ghrelin (Schellekens, Finger, Dinan, & Cryan, 2012). As ghrelin stimulates dopamine functioning in brain regions associated with reward processes, when emotional eaters indulged themselves in comfort foods following a stressor experience, the rewarding feelings might have been potentiated, thus favoring eating when stressors were subsequently encountered. Given that adolescent consumption of a high fat diet resulted in elevated sensitivity of dopamine reward pathways, and such an outcome is subject to a sensitization-like effect, the effects of early eating on subsequent reward-related behaviors may favor later obesity (Naneix et al., 2017).

Obesity and Illness

There have been reports that being somewhat overweight isn't as bad as it had once been thought. It was reported that during the last three decades the age of all-cause mortality increased among individuals who were modestly overweight, so much so, that their life span was longer than that of individuals who were relatively thin. This was the case in the entire population tested, as well as in a subgroup of people who had never smoked and never endured heart disease or cancer (Afzal, Tybjærg-Hansen, Jensen, & Nordestgaard, 2016). A large-scale longitudinal study, surprisingly, revealed that a good number of obese individuals did not present with high blood pressure, insulin resistance, diabetes, low good cholesterol, high bad cholesterol, or high triglycerides, and they were 38% less likely to die early relative to obese individuals who had two or more of these markers (McAuley et al., 2012). It seems that it might be inappropriate to consider obesity from a singular perspective. In fact, overweight individuals who were aerobically fit (measured by cycling until they were fatigued) had a lower risk of death than less heavy individuals who were not aerobically fit (Högström, Nordström, & Nordström, 2014).

Furthermore, energy reserves in heavier people might help them get through sickness, which would not occur among thin people. A somewhat similar conclusion was reached based on a meta-analysis of 97 studies comprising 2.88 million people, indicating that poor health was related to the extent of the person's obesity. Being somewhat overweight or moderately obese body-mass index (BMI) of 25–30 and 30–35, respectively) was not linked to earlier mortality, whereas frank obesity was associated with earlier death (Flegal, Kit, Orpana, & Graubard, 2013).

To be sure, these findings were unexpected, especially as decades of research indicated that being moderately overweight was a risk factor for many diseases as well as their treatment. Simply being overweight during middle age diminished life span by 3.1 and 3.3 years among males and females, respectively, and in obese individuals life span was reduced by 5.8 and 7.1 years (Peeters et al., 2003). Even gaining a moderate amount of weight prior to the age of 55 was associated with an increase of chronic diseases and premature death, which of course, is inconsistent with reports indicating that being modestly overweight may be less harmful than previously thought. For a 5 kg weight gain the risk of type 2 diabetes was increased by 30%, hypertension increased by 14%, cardiovascular disease increased by 8%, obesity-related cancer was elevated by 6%, dying prematurely increased by 5%, and the odds of experiencing healthy ageing was reduced by 17% (Zheng et al., 2017). Likewise, a report on 3.5 million people who were followed for 5.4 years, including many without metabolic problems, indicated that obese individuals were at 49% greater risk for coronary

heart disease, 96% increased risk for heart failure, and 7% increased risk for cerebrovascular disease (Caleyachetty et al., 2017). A meta-analysis of 239 prospective studies conducted across 4 continents confirmed that all-cause mortality was uniformly increased among overweight and obese individuals (The Global BMI Mortality Collaboration et al., 2016).

Considerable data have pointed to obesity being tied to promotion or exacerbation of gastrointestinal (GI), esophageal, breast, renal, and reproductive cancers. At the same time, it is possible that rather than obesity itself, processed foods and sugary beverages, contribute to breast, prostate and colorectal cancers. The American Society of Clinical Oncology (ASCO) indicated that it wouldn't be long before obesity would become the leading preventable cause of cancer, even exceeding that of smoking. In addition to these obviously disastrous outcomes, obesity may be associated with difficulties in the delivery and efficacy of chemotherapy in the treatment of some cancers (Lashinger, Rossi, & Hursting, 2014), and has been linked to increased risk of complications related to surgery and anesthesia and increased incidence of cancer recurrence.[2]

Poor dietary habits and obesity in children predicted adult obesity and contributed to premature development of adult diseases, including heart disease, hypertension, fatty liver disease, osteoporosis, and immune-related disorders, such as multiple sclerosis. As well, the combination of obesity and ageing represents a double whammy for heart disease, as small blood vessels that feed the heart may become damaged under these conditions, likely owing to inflammatory processes (TNF-α and IL-6), ultimately leading to heart failure and diabetes

[2] On average, influenza vaccination is only effective 50%–60% of the time. Following influenza vaccination, individuals who are overweight are twice as likely to develop infuenza relative to normal weight people. This occurs even though they had built up as many antibodies against the influenza virus as normal weight people. It seems that processes related to obesity (e.g., elevated cortisol levels) may undermine immune efficacy, hence making it more likely that illness would occur.

(Dou et al., 2017) as we'll see in Chapter 16, Comorbidities in Relation to Inflammatory Processes.

Abdominal white fat can promote the release of cytokines, particularly IL-6, TNF-α, and monocyte chemoattractant protein 1 (MCP-1), and the resulting chronic inflammatory state can promote the development of depressive illnesses as well as several comorbid conditions (Shelton & Miller, 2010). There is also appreciable evidence suggesting that obesity is associated with increased recruitment of processes that affect CNS functioning, promoting activation of microglia and hence the provocation of neurologic disorders. Indeed, data from the English Longitudinal Study of Ageing (ELSA) indicated that disturbed executive functioning and memory were linked to body mass and systemic inflammation reflected by C reactive protein levels (Bourassa & Sbarra, 2016). Despite the contentions of several naysayers, there is simply too much information pointing to the health risks associated with being overweight, to dismiss the dangers inherent with being obese.

It should also be added that maternal obesity during pregnancy may have profound effects on offspring. Maternal obesity in rodents was accompanied by changes of neuronal functioning within brain reward circuits and disturbed social interactions that were reminiscent of human autism. These effects seemed to be related to a microbial imbalance that could be reversed through gut microbial reconstitution (Buffington et al., 2016). It is significant as well that among obese female mice the presence of metabolic problems can be transmitted through mitochondrial DNA to offspring over three generations, pointing to epigenetic actions (Saben et al., 2016).

The obvious question is how the vastly different outcomes observed across studies be resolved? Some light was shed on this, as it appeared that individuals who were overweight from an early age onward were most likely to suffer early mortality, whereas individuals who were lean throughout life or who gained weight at midlife were less likely to suffer this fate (Song et al., 2016). An analysis of obesity and mortality that included weight history, concluded that sampling weight at one time in a person's life may not provide an accurate reflection of the burden carried, and a better index of how obesity affects health can best be gleaned from the individual's history of being overweight (Stokes & Preston, 2016). In this regard, a 20-year prospective study conducted with 2,500 men and women in the United Kingdom indicated that about a third of obese people were deemed to be in good condition, showing acceptable blood pressure, fasting blood sugar, cholesterol, and insulin resistance. But, even among the seemingly healthy obese individuals, 40% developed risk factors after 10 years, and after 20 years more than 50% fell into the unhealthy obese category (Bell et al., 2015). So, how does one who is overweight know in advance that their current good health will persist over the next 10 or 20 years? Will they be among the lucky ones, or will they become one of those who's intermediate or last years will be spent in a sickly state?

Among obese people, irrespective of whether they were insulin resistant or sensitive, their cells were not behaving like that of nonobese people. When the obese people were challenged with insulin, there were changes in the expression of more than 200 genes, whereas in nonobese people only 2 genes displayed altered expression (Rydén et al., 2016). Findings such as these point to the conclusion that "fat but fit" is likely a misnomer, as these individuals may well be at greater risk for pathology. Rationalizations that obesity doesn't necessarily increase vulnerability to illness in a subset of people might have had the unfortunate consequence of allowing some overweight individuals to take the view that "I may be overweight, but I'm perfectly healthy" giving them license to continue their poor life-styles.

Diet and Immunity

Various balances exist within our neurobiological and metabolic systems, so that one mechanism or process might be fundamental in the promotion of particular responses in a second system, whereas another mechanism may operate in an inhibitory capacity to regulate the excitatory effects of the first. For instance, reactive oxygen species (ROS) and other free radicals, which are by-products of cellular metabolism, will appear at sites of inflammation where they facilitate the death of unhealthy cells. Antioxidant processes are then engaged to limit the actions of ROS once they've done their job. It appears likely that certain foods (e.g., red beans, blueberries, raspberries, cranberries, and artichokes) increase the production of antioxidants, thereby limiting cellular damage. In contrast, a diet rich in animal fats encourages gut bacteria that aren't as beneficial as those that come from a diet rich in plant fibers.

Aside from these actions, dietary factors can also influence microglia, the brain's resident macrophage-like cells. Among mice maintained on a diet rich in fat for 4 weeks, a treatment that increased the number of microglia within the mediobasal hypothalamus, inflammation was promoted along with the tendency to burn fewer calories. The microglia also appeared to recruit immune cells to the mediobasal hypothalamus where they begin to behave like microglia, modulating inflammation. However, if microglia were reduced by experimental drug treatments, mice consumed less and gained less weight (Valdearcos et al., 2017). It was similarly observed that increased hypothalamic glial cells and increased weight engendered by a high-fat diet could be attenuated by genetically influencing glial cell production. Simply increasing inflammation within the hypothalamus, even in the absence of a high-fat diet, altered eating and weight gain, pointing to the glial involvement in these processes (Valdearcos et al., 2017). Thus, interventions that focus on microglia and inflammatory processes might contribute to the development of treatments to diminish obesity.

IT'S OBVIOUSLY NOT FOR EVERYBODY

"I do not like broccoli. And I haven't liked it since I was a little kid and my mother made me eat it. And I'm President of the United States and I'm not going to eat any more broccoli." So said President H.W. Bush in 1990, having become quite the rebel. The problem with broccoli is that it actually has no flavor, and getting people to eat it might be through changing its taste. However, broccoli has powerful antioxidant capabilities owing to their phenolic composition. There are now efforts to enhance the antioxidant capacity of broccoli and other vegetables (e.g., kale and cabbage) by manipulating their genes.

Virtually every wine drinker is likely able to provide a lecture on the health benefits of wine, which stems from the presence of the strong antioxidant resveratrol. However, are antioxidants as useful as they're made out to be? In fact, many studies that examined the effects of antioxidants failed to find a positive effect, a conclusion that was also reached in a large meta-analysis examining the links between antioxidant supplements and a variety of disease conditions (Bjelakovic, Nikolova, Gluud, Simonetti, & Gluud, 2012), including the often touted benefits of moderate drinking in relation to heart disease. So, does this mean that we do away with all the fruits and veggies said to be good for our health? Not at all, they contribute to well-being, but possibly not because of their antioxidant actions. In this regard, at high doses, resveratrol inhibited the T helper cell response elicited by an antigen, whereas at low doses resveratrol caused immune cells to

(cont'd)

be more responsive to an antigen and increased the production of the inflammatory cytokine interferon-γ. Thus, it may be premature to use resveratrol to treat immune-related pathologies, particularly autoimmune disorders (Craveiro et al., 2017).

The widespread belief that dietary factors contribute to immune system functioning isn't based simply on the belief that this ought to be so. Considerable hard data have supported this view and several aspects of digestion were found to affect immunity and immune-related illnesses. The GI tract acts as a barrier that limits potential adverse effects stemming from the foods eaten; essentially the GI tract is responsible for distinguishing between substances which have positive attributes from those that do not. Epithelial cells that line the GI tract, along with protective mucous, minimize the passage of damaging molecules into the body, but at the same time permit the passage of beneficial substances. About 70% of our immune cells are located within the gut so that we are protected from potentially dangerous microbes present on food. A molecule present in the gut (gp96) helps in the regulation of the immune system so that inflammatory overreactions that are associated with disease conditions, such as colitis, do not occur (Hua et al., 2017). It seems that among obese women, caloric restriction comprising a diet of 800 kcal/day, not only produces substantial weight loss, reductions of plasma glucose, insulin, and leptin in adipose tissue, but also enhances the integrity of the gut barrier, and reduces circulating CRP levels, reflecting diminished inflammation (Ott, Skurk, Hastreiter, Lagkouvardos, & Fischer, 2017).

Just as microbiota can affect immune functioning, immunological factors can affect gut bacteria, and immune and commensal bacteria consistently undergo a degree of rebalancing. Too great an immune response can induce harmful inflammation, favoring the development of colitis and inflammatory bowel disease. However, as the processes that provoke these conditions vary considerably across individuals, treatments of gut-related illnesses, like so many other disease conditions, might require a personalized medicine approach (Barroso-Batista, Demengeot, & Gordo, 2015). Related to this, in order that immune functioning progress efficiently, essential vitamins and minerals obtained from foods need to be present, including zinc, selenium, iron, copper, and folic acid, as well as essential fatty acids and mono-unsaturated fats. Omission of foods, such as fibers, can markedly affect the microbial diversity of the gut, and in mice deprived of fiber, many of the constituents of the gut that had been present earlier were not readily reconstituted. It is fascinating that the disappearance of particular gut microbiota develops over generations, speaking to the selection processes that might contribute to cultural differences that are seen in links between diet and health (Sonnenburg et al., 2016).

As we saw in the Chapter 3, Bacteria, Viruses, and the Microbiome, changes in the composition of gut bacteria has been associated with numerous illnesses beyond gut disorders. Given that so many factors can affect the balance of good and bad gut bacteria (dysbiosis), it's rather surprising how little attention is given to these variations in the development of medications to treat existent diseases. For instance, one of the most effective treatment for type 2 diabetes, metformin, produces changes in the composition of gut microbiota, which could potentially be responsible for a portion of the beneficial effects of the treatment (Forslund et al., 2015). Specifically, transplanting fecal samples from metformin treated donors to germ-free mice, enhanced glucose tolerance (Wu, Esteve et al., 2017). Likewise, NSAIDS, such as indomethacin, alter gut microbiota, which then influence drug metabolism, including the effectiveness of the NSAID itself (Liang et al., 2015). Pathways by which microbiota can affect host tissues are provided in Fig. 4.1 (Schroeder & Bäckhed, 2016).

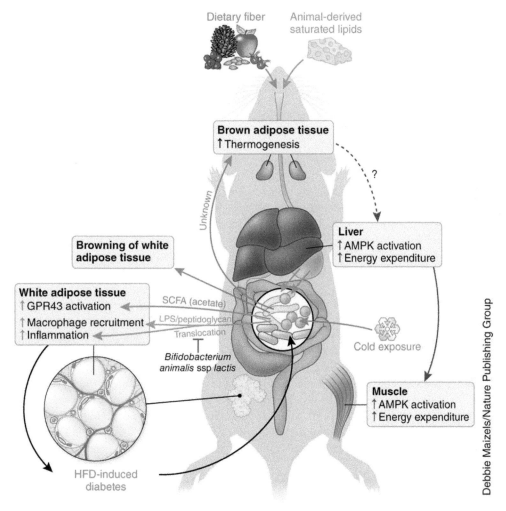

FIGURE 4.1 Pathways through which communication occurs between gut microbiota and host adipose tissue. Saturated lipids (orange arrows) promote the translocation of several bacteria (Gram-negative bacteria and peptidoglycan) into the circulation, which in turn, favors CD14- and NOD1-dependent inflammation within white adipose tissue. This may facilitate immune-related disturbances, such as type 2 diabetes and heart disease. Treatment with *Bifidobacterium animalis* can prevent these negative outcomes. Ordinarily, gut bacteria serve to ferment dietary fibers (green arrows), thereby producing short chain fatty acids (e.g., acetate), which then activate the G-protein coupled receptor 43 (GPR43). The microbial composition can be altered by cold exposure (blue arrows), so that the browning of white adipose tissue ensues, as is the activation of brown adipose tissue. This promotes increased thermogenesis, thereby affecting liver and muscle function through AMP kinase activation, resulting in elevated energy expenditure. *Source: From Schroeder & Bäckhed (2016).*

Immune factors within the gut walk a tight-rope in dealing with bacteria. It needs to be able to destroy anything that is not part of us, yet not destroy beneficial gut bacteria that are also technically not part of us (even if they've been within us for a very long time). To a considerable extent, this sensitive job falls to dendritic cells, which are necessary for an immune response to be mounted, and are needed for immune tolerance to develop by activation of Treg cells. Apparently, under some condition, the dendritic cells that act to suppress immune functioning undergo apoptosis (programed cell death), and thus will no longer be available to limit immune functioning (tolerance), hence allowing for continuous immune activation, inflammation, and pathology (Barthels et al., 2017).

THE SCOOP ON VITAMINS

We had been led to believe that vitamin supplementation was a good idea, especially in children. For individuals with vitamin deficiencies there's little question that supplements are beneficial, and for growing children who weren't consuming a varied diet, obtaining sufficient vitamin A, C, and D through dietary supplements might be useful. But, under other conditions vitamin supplements actually had few beneficial effects. Recently, there's been some buzz about the benefits of vitamin D in preventing upper respiratory infection, although the effect was relatively small (Martineau et al., 2017). However, the inconsistent data reported in this regard can't but produce skepticism. Aside from the involvement of vitamin D in preventing colds, there has been a small swing towards the use of vitamin D in treating depressive disorders. However, based on epidemiological and randomized control trials there doesn't appear to be appreciable support for Vitamin D involvement in depression.

Being a subtype of steroid, vitamin D has many functions, and thus is a good candidate for being involved in, well, just about anything. As indicated by McGrath (2017) "it is a promiscuous biological candidate, likely to be attracted to any passing half-baked hypothesis (it is a risk factor in search of an adverse outcome)." As harsh as this criticism might sound, contrary to the occasional reports suggesting otherwise, the evidence is weak concerning the usefulness of vitamin D supplementation in the prevention of cancer, respiratory infections, rheumatoid arthritis, and multiple sclerosis, and it is unlikely to diminish susceptibility to asthma, atopic dermatitis, or allergies, as too often proclaimed (Manousaki et al., 2017)

Despite the limited effects of vitamins and supplements, even after this was broadly reported in the media, the sales of vitamins and supplements hadn't declined all that much. It may be that individuals feel some uncertainty over dropping a long-held habit regarding their vitamin intake. Individuals who have doubts might still say "well, these vitamins and supplements may or may not improve my health, but taking them surely couldn't hurt," but this assumption might not be entirely accurate. For instance, risk of death was moderately increased among older women taking daily diet supplements of iron, copper, magnesium, zinc, or multivitamins, and excessive vitamin increased intake can cause immune system changes so that infections occur in response to agents to which immunity had previously developed. Calcium supplements have been linked to increased risk of heart attack, and excessive use of supplements was associated with increased cancer risk, as in the case of daily vitamin E, which predicted a 17% greater risk of prostate cancer. In fact, some supplements seemed to have negative health effects with several thousand people ending up in hospital emergency rooms each year. Within the

---

(cont'd)

United States, there were 274,998 such "exposures" between 2000 and 2012.

An interesting piece by Hall (2013) discusses the issue of "uncertainty" in medicine, and in doing so, quotes from Voltaire that "Uncertainty is an uncomfortable position. But certainty is an absurd one." It's remarkable how some naturopaths are as certain as they are that their treatments will definitely heal the patient. In contrast, in making diagnoses and treating patients, uncertainty is not infrequent even among physicians who have been in practice for years.

Omega-3 Polyunsaturated Fatty Acids

Amid the cacophony of voices pointing to one or another food that will prevent illness, are those promoting omega-3 polyunsaturated fatty acids. A major component of the modern Western diet is increased consumption of proinflammatory omega-6 polyunsaturated fatty acids (PUFAs) relative to anti-inflammatory omega-3 PUFAs, and the imbalance may contribute to inflammatory diseases. A broad range of brain changes are promoted by PUFAs, but particular attention regarding the effects of omega-3 has focused on their anti-inflammatory actions. Research emanating from both humans and animals have revealed that omega-3's play a role in the degradation of cell components that are unnecessary or dysfunctional through autophagy (self-eating or degradation of cellular "junk"), which then results in suppression of cytokines, such as IFN-α as well as CXCL-10 secreted by macrophages (Mildenberger et al., 2017). Further, omega-3 fatty acids are converted into cannabinoid epoxides and in this way may promote anti-inflammatory effects (McDougle et al., 2017), thereby limiting several diseases.

Omega-3 PUFAs have received an unusual amount of attention and have reached fad status, and in the form of eicosapentaenoic acid (EPA) and docosahexaenoic acid (DHA) have been assessed in relation to virtually every known psychiatric illness. They have been offered up as adjunctive treatments for major depressive disorders, bipolar disorder, and schizophrenia, neurodevelopmental disorders, as well as age-related cognitive decline. A report from the Canadian Network for mood and Anxiety Treatments recommended omega-3 PUFAs, as well as exercise, S-adenosyl-L-methionine, and yoga, as first or second line treatments for mild to moderate depression (Ravindran, Balneaves, Faulkner, Ortiz, & McIntosh, 2016). However, unlike earlier reports, a large-scale prospective study assessing cognitive decline in older people indicated that these supplements had little effect (Chew et al., 2015).

In addition to omega-3 PUFAs, a diet high in polyphenols (e.g., citrus fruits, cocoa, red wine, tea, and coffee) could affect brain functioning (Letenneur, Proust-Lima, Le Gouge, Dartigues, & Barberger-Gateau, 2007), possibly owing to anti-inflammatory or antioxidant effects. This outcome could also occur through microbial alterations, as most dietary polyphenols accumulate in the large intestine and are then converted by the gut microbiota to less complex metabolites. Several other supplements and natural products could also have health benefits owing to their microbial actions. Black and green tea influence the growth of *Bifidobacterium* species that can have positive actions. But, they have also been implicated in the growth of nasty bacteria, such as *Helicobacter pylori*, *Staphylococcus aureus*, *Salmonella typhimurium*, and *Listeria monocytogenes*, pointing to the need for proper balances within bacterial communities within the gut (Duda-Chodak, Tarko, Satora, & Sroka, 2015). Several plants and mushrooms can similarly influence microbiota and have the benefit of containing fiber and phytochemicals that diminish obesity and type 2 diabetes (Martel et al., 2016).

Meds to Curb Eating

As much as dieting and exercise can be used to reduce weight, it's obvious that too many people want or need something much simpler to attain significant weight loss. The pharmaceutical industry for some time worked diligently to develop profitable weight reducing compounds. The serotonin releasing agent fenfluramine had short lasting effects on weight and was quickly discarded, only to reappear as a composite with another weight reducing agent, phentermine. The combination of these treatments, fenfluramine/phentermine, commonly referred to by the cutesy name fen-phen, caused vital valve dysfunction and pulmonary hypertension, and so was discarded.

Such experiences resulted in many drug companies backing away from the search, but some companies persisted and have developed interesting compounds. Lorcaserin (Belviq), which acts as a $5-HT_{2C}$ receptor agonist seems to be effective, reducing weight in some people by 5%−10% and has received approval as an antiobesity agent. Consistent with the emerging view that eating processes and addiction may involve some overlapping neural circuits, this agent also has potent actions in diminishing some forms of addiction (Higgins, Sellers, & Fletcher, 2013). Another compound, inulin propionate, or a type of fiber, inulin, which influences particular gut bacteria, diminished activity of neurons within the nucleus accumbens and caudate, which have been linked to reward processes. Paralleling the brain changes, the treatment also reduced participant's ratings of the appeal of particular foods, which could be linked to diminished cravings

(Byrne et al., 2016). Still another compound was developed based on the finding that a diet containing fermentable carbohydrates prevented obesity in mice. This protection was lost when mice lacked the relevant receptor (FFAR2), but when present, the peptide YY that signals satiety, was elevated (Greenhill, 2017).

A different strategy for weight reduction is to fool the body into mistakenly thinking that it had already eaten. The drug fexaramine causes bile acid release into the intestine, which ordinarily occurs after people have eaten in order to promote digestion. It also has the effect of reducing blood sugar and cholesterol, and increasing brown fat (Fang et al., 2015).

The involvement of gut bacteria in eating processes and obesity has become increasingly evident, and in mice manipulations of the microbiome has been useful in moderating obesity and inflammation. Once more, speaking to the value of personalized treatments, the ratio between specific bacteria present in the gut (e.g., Prevotella vs Bacteroides) was predictive of the efficacy of certain diets (Hjorth et al., 2018). It can reasonably be expected that it won't be long before strategies involving gut bacteria will be developed to inform the most efficacious diets. In the interim, it should be kept in mind that dietary factors and exercise both influence gut microbiota, and may affect behavior and mood states, but they do so relatively independently of one another.

Natures Little Miracles

There seems to be the belief in some circles that if it grows on trees or bushes, or comes directly from the ground, it must be good for us. As we've seen, some products can have some very strong effects on the body and the brain, and they can have potent medicinal properties. The Pacific Yew Tree was instrumental in giving us potent cancer medications, aloe has been used against burns, and marijuana, deadly night shade, magic mushrooms, and jimson weed have all been implicated as having some medicinal value. Strychnine and digitalis in very low doses can have positive effects for some conditions, but these plant-derived agents can be lethal at somewhat higher doses.

Despite the long-held belief that St John's wort had genuine antidepressant effects, and positive effects were reported in diminishing low and moderate grade depression (Seifritz, Hatzinger, & Holsboer-Trachsler, 2016), studies from the National Center for Complementary and Alternative Medicine (NCCAM) indicated that it was no better than a placebo (Hypericum Depression Trial Study Group, 2002). A systematic review likewise indicated that this herbal treatment had limited effects in large trials, but effects that were more positive were seen in small trials. As in the case of St Johns wort, echinacea has long held the reputation as being effective in treating the common cold, despite supportive data being limited. Based on a meta-analysis that included 24 randomized control trials, Echinacea reduced the risk of catching a cold by 10%–20% (Karsch-Völk et al., 2014).

Most naturalistic treatments aren't being marketed as drugs, and thus haven't been subject to rigorous, expensive, and time-consuming procedures that are necessary for drugs to come onto the market. Not only have most of the natural products available not been assessed in a rigorous scientific fashion, but information is typically not provided concerning the purity, quality, chemical stability, and active constituents of the products. Similarly, the packaging typically provides little information concerning contraindications, such as side effects or potential risks for some people, such as those taking particular medications. The lax policies regarding herbal remedies, as well as comparative studies evaluating natural versus standard medical treatments, have led to the market being saturated by products without genuine value. The simple fact is that even after drugs receive FDA approval, about a third was later found to have safety concerns. How much more dangerous are products that have absolutely no oversight?

Plants often have nasty chemical means of protecting themselves from herbivorous animals, and these chemicals can also affect humans. Coumarin, ephedra, and calamus oil are natural products that have been banned for health reasons, and while sugars and salt are natural, too, in excess they can have bad effects on our health, and can even affect brain functioning. Some mushrooms are poisonous, and poison hemlock, poison ivy, poison oak, and poison sumac haven't been named as they have for no reason. Herbals containing the plant *Aristolochia*, which is widely consumed in some places, can cause kidney disease and cancer in genetically susceptible individuals. As only 5% of the millions of people taking the compound fall ill, and because of the time lapse between consumption and illness occurring, it was difficult for the linkages to be identified (Grollman & Marcus, 2016). In addition to these potential hazards, it seems that risk of bleeding exists in relation to Ginko, and individuals with diabetes should not be taking Asian Ginseng, and there are many substances that should not be used by people undergoing chemotherapy or those with cardiac problems. Some people who swear by natural food supplements are also adherents of other seemingly "natural ways" of obtaining good health, but fortunately some of these fads, such as colon cleansing, a dangerous practice that causes bloating, nausea, electrolyte imbalances, and possible kidney failure, is just about at its end.

There are individuals, as we'll see in the text box, who prefer alternative medicines instead of standard treatments, even if these alternatives are unsound, and potentially dangerous. Perhaps to minimize the dangers of alternative medicine, there has been a push for purveyors of alternative treatments, particularly herbalists, to work side by side with hospital doctors (or at least in an adjacent building or wing). Complementary and alternative medicines (CAM), involves add-on treatments (e.g., relaxation training, acupuncture, or some natural products) to standard medications. It has been argued that having both approaches available allows patients to reap the benefits of both approaches, and to decrease the possibility of patients going to untrained herbalists. There are several advantages to this approach, particularly as some neutraceuticals can serve as effective adjunctive treatments with standard antidepressant agents, as in the case of S-adenosylmethionine (SAMe) and to some extent methylfolate and omega-3 (Sarris et al., 2016). Further, having standard and alternative options concurrently available may help build confidence among people who are entrenched in traditional ways of healing (e.g., in many African countries, as well as Indigenous Peoples in Canada and Australia). Problems arise, however, in dealing with infectious diseases, such as Ebola, when traditional medicines are ineffective, and the most basic forms of quarantine are not maintained.

Although in most instances there is scant evidence attesting to the effectiveness of CAM treatments, it was suggested that there are cases where they might be having some positive effects attributable to their biochemical actions or because they might be powerful placebos, which can, to be sure, be useful. As Deng Xiaoping said in the context of China's economic development, but applicable here as well, "It doesn't matter if a cat is black or white, as long as it catches mice it's a good cat."

THE HOME FOR NATURAL PRODUCTS

Pseudoscience abounds just about everywhere, assaulting us from magazines and are passed around through multiple internet sites. Testimonials are offered about this or that natural compound that has amazing healing powers, and a few physicians have appeared on television peddling various super remedies. Too often, actors somehow are able to influence large numbers of people with the flakiest products ever imagined (Caldwell, 2015). Health food stores peddle books dealing with homeopathy, alternative healing, integrative medicine, acupuncture, the benefits of weekly chiropractic joint adjustments, cupping, organic farming, and reiki therapy (the latter is based on the belief that by touch the therapist can channel energy into the patient, thereby activating natural healing processes within the patient's body). The use of Traditional Chinese Medicines (TCMs) had, for a time, gained momentum, possibly because it has an aura that makes them right for some people. Acupuncture has been in vogue for several years, even though needles stuck in random places yield the same "positive" effects.

So, why is it that people so often resort to alternative medicines and are willing to take the advice of actors or politicians concerning what's healthy and what's not, and disparage the work of physicians and scientists on these same issues? To some extent it may reflect a loss of faith (trust) in physicians and scientists. The sad fact is that standard medicines have had a poor track record for particular illnesses. Antidepressant agents are claimed, in some circles, to be only a shade better than placebos, and opioid-based pain killers obviously have risks attached to them. Many cancer therapies are ineffective (although this may be changing with immunotherapies) and come with distressing side effects. The benefits of some standard heart medications (e.g., beta blockers) have been questioned, and as we've seen, antibiotics have increasingly encountered resistance by bacteria.

(cont'd)

Furthermore, it's well known that iatrogenic illnesses (those brought on by physician error, either those of omission or commission) are not uncommon, and can have severe consequences. If all of this weren't sufficiently discouraging, in some countries patients may encounter long wait lists before they can see a specialist, further encouraging them to seek alternatives. Given the various roadblocks to effective treatment, the movement toward alternative medicines is perfectly understandable, but this shouldn't be misconstrued as being the smart thing to do. Aside from these treatments being ineffective, their use may be accompanied by delays in seeking and initiating potentially effective treatments.

Although it is essential that alternative treatments be assessed in scientifically sound clinical trials, at times, the issues concerning alternative treatments simply becomes silly. How many more trials are necessary to show that particular complementary and alternative medicines don't work (Gorski & Novella, 2014)? Homeopathy, for instance, has been among the most impressive scams going for well more than a century and has repeatedly been shown to be ineffective in treating anything. Nonetheless, trials continue to be undertaken to confirm that the treatment doesn't work. Most of us know the adage that insanity comprises doing the same thing over and over again and expecting different results.

EXERCISE

As important as eating properly might be for those wishing for a healthy life span, exercise in some form is necessary, with aerobic exercises (distance running, bicycling, and jogging), which increases free oxygen use to meet energy demands, thereby strengthening the heart and lungs, being the form that is typically recommended. Certainly, anaerobic exercise to build muscle strength has benefits as well. Most people are probably aware that failing to exercise has adverse effects that are greater than those simply stemming from obesity. Obesity and the development of type 2 diabetes are potent risk factors for further illnesses, including poor bone health, potentially leading to bone fracture and osteoporosis, but exercise attenuates these outcomes. At the same time, even if exercise is undertaken religiously, sedentary behaviors need to be avoided as sitting for extended periods is dangerous to heart

health and has been linked to diabetes and cancer (Biswas et al., 2015).

Engaging in exercise can also diminish vascular and systemic proinflammatory cytokines and might thus have positive actions in relation to other conditions that are promoted by myokines (small cytokines or peptides released by muscle cells; myocytes). As well, exercise increased natural killer cells in breast cancer survivors (Pedersen, Idorn et al., 2016), and was linked to diminished frequency of some forms of cancer, especially when accompanied by reduced alcohol consumption and smoking cessation (Colditz & Sutcliffe, 2015). At the same time, acute aerobic exercise produced a reduction of inflammatory gene expression among patients with lupus erythematosus, (Perandini et al., 2016), supporting the contention that exercise may be useful in diminishing features of autoimmune disorders. As little as 20 min of exercise can act against inflammation, possibly through sympathetic nervous system processes (Dimitrov, Hulteng, & Hong, 2017). Likewise, even single bouts of moderate exercise can result in leukocytosis (an increase of white blood cells) coupled with redistribution of immune cells from storage sites into circulation, likely owing to stimulation by epinephrine and glucocorticoids (Simpson, Kunz, Agha, & Graff, 2015). Thus, even light or moderate daily activity can be beneficial, and can diminish risk of death from any cause even among men already affected by coronary artery disease. Beyond these actions, exercise can have epigenetic effects that are manifested in muscle cell genes that are involved in oxidation and glucose regulation, and thus may be intricately linked to type 2 diabetes (Barrès, Yan, Egan, Treebak, Rasmussen, 2012).

Simply dieting, as we saw earlier, is usually not enough to maintain weight loss, and might likewise be insufficient in the production and maintenance of good physical and psychological health. The combination of diet and exercise can have potent effects on intestinal integrity as well as microbial diversity, varying in thin versus overweight mice (Campbell et al., 2016). It will hardly be surprising that in a 30 year prospective study revealed that exercise and healthy eating, together with other life-style changes, were predictive of fewer occurrence of type 2 diabetes, vascular disease, dementia and all-cause mortality (Elwood et al., 2013).

Cognitive and Affective Benefits

Aside from physical health benefits, fitness has been associated with improved psychological well-being, possibly being related to enhanced synaptic plasticity and may even promote the growth of new neurons within the hippocampus, and increase neuronal connectivity (Voss et al., 2015). Thus, aerobic exercise training may promote multiple cognitive enhancements, including modestly improved attention and information processing speed, executive functioning, and memory. Exercise beginning early, through effects on the microbiome, can engender persistent enhancement of metabolic functioning and brain processes. The effects of exercise might be most apparent among older individuals in whom neuronal functioning ordinarily declines to a moderate degree, but little would be gained among individuals already affected with dementia.

It has been maintained that exercise can reduce mild or moderate anxiety and depression, but tends to be less effective in relation to severe illness. The positive psychological and cognitive effects of exercise may develop through several processes. Most obviously, exercise may act as a way to diminish (cope with) distress that otherwise causes impaired mood and cognitive functioning. As already said, increasing growth factors, including BDNF, IGF-1, VEGF, and GDNF, may have positive effects on neuronal processes, particularly in relation to neurogenesis. Exercise may also influence stressor-related neural circuits so that responses to subsequently encountered stressors are diminished, possibly by facilitating the activation of the inhibitory neurotransmitter GABA (Schoenfeld, Rada, Pieruzzini, Hsueh, & Gould, 2013).

Exercise and Immunity

The positive effects of exercise have been realized in relation to many physical illnesses, but many issues still need to be addressed, such as which individuals will benefit and under what circumstances. Numerous reports have pointed to the impact of exercise on immune functioning and immune-related disorders, but is this always apparent? Animal studies generally indicated that moderate intensity exercise enhanced immune functioning, and wheel running reversed the diminishing production of new neurons in the hippocampus of aged mice (Littlefield, Setti, Prister, & Kohman, 2015). Aside from an exercise regimen enhancing immune functioning, it may also limit the immune system's functional decline with advancing age (immunosenescence) (Simpson et al., 2015). Moderate exercise regimens in humans likewise diminished excessive levels of inflammation, reflected by reduced acute phase proteins, owing to changes of adipose tissue as well as other factors, and might thereby limit the development of illnesses involving activation of inflammatory processes.

The diminished cytokine levels that accompany moderate exercise have positive effects among healthy individuals and in some patients with persistent systemic inflammation. This may arise owing to altered metabolic signals, browning of white fat, and regulation of innate immune functioning, diminished cytokine as well as increased antioxidant activity. It is significant that in relapsing-remitting multiple sclerosis, having patients engage in high-intensity resistance training had positive effects on proinflammatory cytokine levels, and positively influenced fatigue and health-related quality of life (Kierkegaard et al., 2016). Chronic exercise likewise had a protective action in relation to EAE in an animal model of multiple sclerosis. Relative to nonexercising mice, those that received the chronic exercise treatment displayed later onset of the disease and less severe neurological symptoms, notably reduced immune cell infiltration and demyelination in the ventral white matter tracts of the lumbar spinal cord. Yet, in systemic lupus erythematous patients, inflammatory gene expression was diminished immediately after acute aerobic exercise, but was up-regulated during recovery. It also appeared that gene networks were less organized in lupus patients, raising the possibility that in these patients a disturbance existed in the processes needed to trigger a typical exercise-elicited immune transcriptional change (Perandini et al., 2016)[3].

As in the case of other disorders linked to immune functioning, diabetes that was induced in rats through streptozotocin administration, was accompanied by increased inflammatory cytokines (IL-1β, TNF-α, IL-6, and IL-4) and ROS (nitric oxide and malondialdehyde). However, these actions were attenuated by low-intensity exercise training, leading to the suggestion that the positive effects of moderate exercise in protecting against diabetes may stem from the anti-inflammatory effects engendered (Kim et al., 2014).

There have been a large number of reports showing the positive impact of exercise on individuals with illnesses related to ageing, including cancer progression, and might be useful as an add-on treatment (Bigley & Simpson, 2015). This may or may not be correct, but even if it ultimately does not have these positive effects, it might be an excellent coping strategy for cancer patients, thereby diminishing distress and depressive symptoms.

Is It the Case of the More, the Better?

Aside from increasing life span, regular exercise increased health-span, and adopting a proper diet coupled with maintaining a

---

[3] Inflammatory processes have been implicated in the neurodegenerative process associated with Parkinson's disease, which can be attenuated by exercise. Although it is tempting to attribute these effects to reductions of inflammation, the observed neuroprotective effects were just a likely due to activation of the BDNF-TrκB signaling pathway (Wu et al., 2011). In essence, although inflammation may cause problems, its deleterious effects appeared to be modifiable through noninflammatory pathways.

reasonable exercise regimen acted against cellular senescence, essentially protecting individuals against diseases associated with ageing (Schafer et al., 2016). The link between exercise and positive health is not linear, and instead comprises an inverted U shape function, so that the beneficial effects of moderate exercise may be lost in response to excessive exercise. In fact, excessive exercise may be counterproductive, resulting in lymphocyte loss and increased illness (e.g., in response to an influenza virus) (Hoffman-Goetz & Quadrilatero, 2003).

Unlike the positive actions of moderate exercise, single bouts of prolonged exercise can, at least temporarily, disturb multiple aspects of immunity, including disruption of T cell, NK cell, neutrophil functioning, the balance between pro- and anti-inflammatory cytokines, and blunting of the immune responses to specific antigens (Simpson et al., 2015). Furthermore, the "extreme exercise," often engaged by high-performance athletes, can impair NK cell functioning and may reduce immunoglobulin within mucosal secretory (nose and salivary) glands, thus allowing infectious molecules the opportunity to invade. It is puzzling that although high intensity exercise in noncompetitive settings was tied to diminished immune cell proliferation, in a competitive situation cell proliferation was primed, leading to enhanced immune functioning (Siedlik et al., 2016). Why this difference exists is uncertain, but it may have to do with psychological and hormonal processes that accompany competitive situations, thereby overriding the effects of exhaustive exercise alone.

OLD AGE AIN'T NO PLACE FOR SISSIES
Bette Davis

It's unfortunate that age is ordinarily accompanied by the deterioration of various biological systems, hence favoring or even encouraging heart, kidney, liver, and lung diseases, endocrine-related disorders (e.g., diabetes), depressive illness, and dreaded neurodegenerative diseases. Among other things, the capacity of the immune system to protect us declines (immunosenescence) as we become older, which may be exacerbated by stressors encountered. Ageing is associated with low numbers of naïve T cells, disturbed proliferative responses to mitogen challenges, and a poor ratio between CD4 and CD8 cells, although inflammatory factors may be elevated owing to the presence of common viruses. The net response is that with age inflammatory factors may be more damaging, and poor vaccine responses are more common. It is of particular significance, certainly to people who are older, that exercise can enhance immune functioning, and among those who exercised on a regular basis, morbidity and mortality associated with inflammatory or immunologically related illnesses was diminished (e.g., Turner, 2016). In contrast, among extreme exercisers, immune capacity may be undermined and thus these individuals may not enjoy the hoped for healthspan (Simpson & Bosch, 2014).

While we're on the topic of ageing, a few more comments seem in order. Ageing, as almost everyone knows, comes with a variety of psychological challenges. Social activities and contacts tend to decline owing to mobility problems and the presence of illnesses. Social support systems deteriorate as a result of the loss of friends who predeceased them or moved away, and their family members may have dispersed to take opportunities elsewhere. With the individual's social network disintegrating, loneliness may set in, promoting depression and exacerbating vulnerability to other illnesses (Perissinotto, Cenzer, & Covinsky, 2012).

(cont'd)

To make things still more distressing, older individuals may experience a loss of control and independence, and have to rely on others. They may also experience stigmatization and unsupportive social interactions, frequently being patronized, dismissed, and made to feel invisible or even a burden, and seeing signs of elder abuse are not uncommon among neurologist or care workers who come in frequent contact with older people. It's hardly a wonder that the frequency of depression and drug treatments to deal with depression and sleep disturbances have been sky-rocketing in recent years. Although it is unlikely that exercise itself will act as a remedy for these problems, group-based exercise can have distinctive benefits on psychological and cognitive processes (Haslam, Cruwys, & Haslam, 2014).

Exercise and Microbiota

Gut dysbiosis, as we've seen, is predictive of increased vulnerability to inflammatory diseases, but gut immune cell homeostasis and microbiota-immune interactions engendered through exercise, can inhibit illnesses related to microbiota dysbiosis, although they can also operate in the direction of promoting illness (Cook et al., 2016). Among germ-free mice, exercise affected the responses observed when recolonization was undertaken using microbiota from donor mice that had exercised. This treatment also limited the development of chemically induced colitis, possibly through actions on cytokines (Allen et al., 2017). The influence of exercise on gut microbiota also varies with the animal's age. Among juvenile rats, exercise led to changes of several gut phyla, including increased Bacteroidetes and decreased Firmicutes, and produced other profiles associated with adaptive metabolic changes (Cerdá et al., 2016). In this regard, exercise can have multiorgan benefits, which likely stem from microbial factors together with changes of other systems. When exercise was initiated during early life, microbial processes were especially enhanced (Mika & Fleshner, 2016).

The inverted U-shaped relationship between amount and intensity of exercise and immune functioning is also seen in relation to gut microbiota alterations. Whereas mild exercise can be beneficial in relation to some illnesses associated with gut microbiota disturbances (e.g., irritable bowel syndrome and inflammatory bowel disorder), excessive exercise generally promotes intestinal cell damage and leaky gut syndrome may develop (Costa, Snipe, Kitic, & Gibson, 2017). Cross talk occurs between mitochondria and microbiota so that extreme exercise, by affecting production of ROS, may elicit changes of microbiota. As genetic influences exist regarding mitochondrial functioning, the effects of exercise may also interact with these features (Clark & Mach, 2017).

SLEEP

Neurobiological Factors Associated With Sleep

Like eating and drinking, sleep is a basic need that is essential for well-being. Its timing is largely controlled by internal clocks regulated by the suprachiasmatic nucleus located within the hypothalamus, but internal clocks are also present in other brain regions, and have an enormous and long lasting influence on well-being (Hood & Amir, 2017). Disturbances of circadian clock functioning has been associated with impaired hormonal activity and antioxidant production, and has been tied to numerous illnesses, including neurodegenerative conditions, such as Alzheimer's and Parkinson's disease. While such findings do not suggest a causal relationship between clock disturbances and illness, it is nevertheless significant that sleep problems are among the earliest features that emerge in these diseases (Hood & Amir, 2017). Stressful experiences may upset the operation of clock proteins (Al-Safadi et al., 2014), and when individuals were sleep-deprived for an extended time, neuronal functioning in cortical brain regions diminished, essentially overriding the brain's master clock (Muto et al., 2016).

The circadian rhythmicity ordinarily present within cortical regions was greatly affected by sleep loss, whereas activity within subcortical regions were primarily linked to circadian cycle (Muto et al., 2016). In essence, circadian factors and sleep loss might both contribute to the destructive effects of sleep loss on well-being. At the same time, sleep loss can also be part of a symptom cluster observed in immune-related disturbances and mood disorders (Irwin et al., 2014). Aside from these direct actions of sleep loss on the development of illness, sleep disturbances can interfere with treatment adherence for illnesses that might have emerged, thereby increasing morbidity and mortality.

Ordinarily, over the course of the day essential biological resources may be used up and sleep facilitates their recovery. During sleep most of our biological systems are in an anabolic state, which essentially means that skeletal and muscular systems are being rejuvenated, and both hormones and neurotransmitters are being replenished, so that we're ready to deal with the travails encountered the next day. As well, on any given day, and the many events experienced, the number

and strength of synaptic connections increases, which entails the use of considerable energy, and restoration of these processes is necessary. Brain activity declines during sleep, and the number of synapses present may diminish, making recently strengthened synapses relatively more prominent, thereby enhancing memory and performance on the ensuing day. Conversely, lack of sleep may give rise to hippocampal atrophy, reduced length and spine density on dendrites within the hippocampus, as well as reduced connectivity (Havekes et al., 2016). As a result, individuals may experience cognitive impairments, including disturbances of attention, learning, memory, memory consolidation, executive functioning, decision-making, stressor appraisals and emitting appropriate coping responses (e.g., Boyce, Glasgow, Williams, & Adamantidis, 2016). Neuronal disturbances within the prefrontal cortex that stem from disturbed sleep, may also affect appraisals so that rewards are overvalued, whereas losses are undervalued. This, in turn, can influence risky decision making (e.g., in the context of excessive gambling and relapse in relation to other addictions).

Sleep loss has also been linked to increased food intake, possibly because food odors are more enticing under these conditions or it may be that following sleep deprivation the ability to deal with temptation is diminished. The link between sleep and eating was reinforced by reports that individuals who slept poorly were more likely to make poor food choices, perhaps stemming from altered neuronal activity within the frontal and insular cortex. Likewise, disturbances of internal biological clocks have been linked to obesity and insulin sensitivity, possibly owing to variations in the release of eating-related hormones leptin and ghrelin.

SLEEP AND ILLNESS

There has been speculation that during sleep and dreaming brain garbage that accumulated during the day (e.g., unimportant memories or trivial synaptic connections) is removed. In fact, owing to shrinkage of glial cells during sleep, openings between cells become larger, allowing more rapid flow of cerebrospinal fluid within the brain and spinal cord, so that toxins can be removed more readily. For that matter, exogenously administered β-amyloid was more readily eliminated from the brain during sleep periods. Thus, lack of sleep could have important ramification in the provocation of neurodegenerative disorders related β-amyloid accumulation (Mendelsohn & Larrick, 2013). To be sure, although the data are simply correlational, slow wave activity during sleep in humans was linked to the accumulation of β-amyloid in the prefrontal cortex (Mander, Winer, Jagust, & Walker, 2016). As well, neurological disorders have been associated with sleep disturbances and even neuronal connectivity is undermined with sleep deprivation (Kaufmann et al., 2016), and sleep is essential for production of factors, such as myelin, that support brain functioning (Bellesi et al., 2013). Retrospective analyses of medical records also revealed that Alzheimer's and Parkinson's disease were preceded by sleep problems decades earlier (Claassen et al., 2010).

A meta-analysis that included 15 prospective cohort studies indicated that various forms of insomnia (difficulty initiating or maintaining sleep, and nonrestorative sleep) were linked with increased risk for heart attack and stroke, being somewhat more pronounced among women relative to men (He, Zhang, Li, Dai, & Shi, 2017). A further meta-analysis of 153 studies, which evaluated more than 5 million participants, indicated that short sleep was associated with increased occurrence of type 2 diabetes, hypertension, cardiovascular disease, coronary heart disease, stroke, obesity, and earlier death (Itani, Jike, Watanabe, & Kaneita, 2017).

In addition to limiting illness, sleep is also important for healing processes. Ironically, hospitals are exceptionally inhospitable for patients needing recuperative sleep, especially given the many interruptions experienced. Consistent with the expression of sleeping with one eye open, sleeping in a strange place, such as within a hospital setting, is accompanied by one hemisphere of the brain remaining in an alert state.

Sleep is often characterized as comprising periods of REM sleep and four stages of non-REM (NREM) sleep. The switch between sleep and nonsleep states, and the sleep period itself, are governed by about 20 different neuropeptides and neurotransmitters, many of which are controlled by and correspondingly influence circadian cycles (Richter, Woods, & Schier, 2014). As well, movements of gut bacteria over the course of a day may contribute to daily rhythms (Thaiss, Levy et al., 2016) and gut microbes themselves are also subject to diurnal rhythms, which together with eating patterns can have health-related consequences. There has been particular interest in defining the links between circadian changes and sleep loss in relation to disturbed neuroplasticity, which can have wide-ranging cognitive ramifications (e.g., Frank & Cantera, 2014).

Variations of sympathetic activity occur over the course of the sleep cycle. Sleep and sleep depth contribute to sympathetic outflow, and during the transition from wakefulness to sleep, a shift occurs from sympathetic to parasympathetic outflow. During NREM or short-wave sleep, sympathetic activity diminishes, whereas elevated outflow, resembling daytime sympathetic activity occurs during REM sleep. Furthermore, among individuals with insomnia, circulating levels of norepinephrine and epinephrine are elevated, as are other indices of sympathetic functioning (Vgontzas, Fernandez-Mendoza, Liao, & Bixler, 2013), outcomes that were similarly observed in response to experimentally induced sleep deprivation.

It had long been thought that sleep was regulated by serotonin produced in the anterior raphe nucleus, which sends projections to the preoptic portion of the anterior hypothalamus. In addition, dopaminergic neurons of the ventral tegmental region, which has been linked to heightened arousal and reward processes, appears to be a strong modulator of sleep (Eban-Rothschild, Rothschild, Giardino, Jones, & de Lecea, 2016). The production of adenosine, which contributes to energy regulation, is also a major player in relation to sleep processes (Bailey, Udoh, & Young, 2014). Produced from adenosine triphosphate (ATP), adenosine accumulates as neuronal and glial activity increase. With adenosine receptor stimulation, NREM sleep is provoked, and glycogen energy stores recover.

Genetic analyses have suggested that sleep and insomnia may comprise multiple complex mechanisms. A genome-wide analysis based on a cohort of more than 113,000 participants, identified seven key genes linked to insomnia, as well as sleep disorders, such as Restless Legs Syndrome and Periodic Limb Movements of Sleep. The contribution of these genes to insomnia differed between males and females, each having their own predictors, and the genetic correlates of insomnia were also associated with metabolic characteristics and psychiatric disturbances (Hammerschlag et al., 2017).

Sleep Contribution to Emotional Regulation and Mental Health

It's not just babies that are fussy and grumpy when they don't get enough sleep, lack of sleep among adults is associated with poor emotional regulation and emotional intelligence. Lack of sleep was linked to elevated behavioral reactivity in response to a stressor, the amplification of negative appraisals and emotions, elevated impulsivity, as well as increased cortisol levels and amygdala neuronal activity (Guyon et al., 2014). Poor sleep quality among depressed individuals has also been linked to elevated neuronal activity within the dorsal anterior cingulate cortex, as well as to difficulty in a reappraisal task in which individuals were asked to put a positive spin on pictures clearly depicting negative events (Klumpp et al., 2017). In essence, with a lack of sleep the anterior cingulate cortex needs to work that much harder to make sense of situations.

The notion that a lack of sleep, particularly REM sleep, may have psychosocial and mental health consequences or may be a marker for later mental illness has frequently been discussed. The impact of strong stressors on mood states may be mediated by sleep alterations and REM sleep disturbances that follow from these stressor experiences (Kim & Dimsdale, 2007). Furthermore, sleep loss that follows from stressors, may act together with synaptic plasticity, hormonal, and neurotransmitter processes, thereby provoking psychological dysfunctions.

Impact of Sleep and Sleep Loss on Immune Processes

REM sleep has held special interest for many researchers, being linked to psychological well-being. But, as described in Fig. 4.2, both REM and NREM sleep are important for mental health, being related to neurobiological

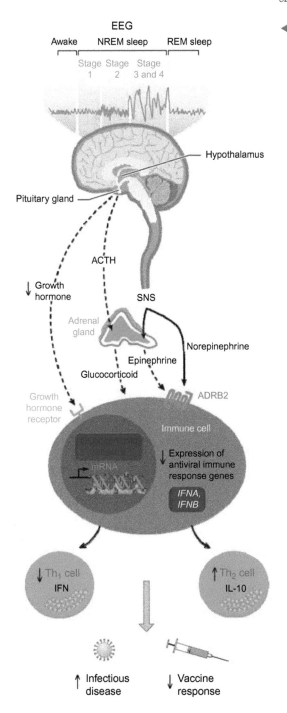

◀ FIGURE 4.2 Sleep disturbance and adaptive immunity are linked to hormonal and immune changes. Sleep disturbances, like stressors, are associated with glucocorticoid release, which alters gene expression at various brain sites. As well, activation of glucocorticoid receptors on leukocytes, promotes sympathetic nervous system activation, which elicits the release of immune factors from primary lymphoid organs.

Early in the sleep period, the secretion of growth hormone enhances T cell proliferation and differentiation and the promotion of cytokine functioning. When sleep is disrupted growth hormone release is diminished, resulting in a shift of immune functioning so that anti-inflammatory factors are predominant over those of a proinflammatory nature. Consequently, susceptibility to infectious disease is increased and the response to vaccines is reduced. The HPA and sympathetic nervous system variations (Vgontzas, Fernandez-Mendoza, Liao, & Bixler, 2013) provoked by chronic sleep loss could promote altered immune functioning (Slavich & Irwin, 2014), and impaired immunological memory (Westermann, Lange, Textor, & Born, 2015). As described by Irwin (2015), glucocorticoid binding interrupts transcription of proinflammatory and antiviral genes, whereas transcription of factor NF-κB promotes the disruption of the inflammatory cascade. As well, variations of this transcription factors, along with AP-1, are disturbed as a result of protein-protein interactions that interfere with inflammatory gene transcription.

In the figure, the dashed lines from the CNS represent release of hormones (i.e., growth hormone, ACTH, glucocorticoid, epinephrine) into systemic circulation, which can affect immune cells. The solid line represents a neural connection (i.e., norepinephrine) that can affect receptors present on immune cells. *Source: From Irwin (2015).*

and immune consequences as a result of altered sleep. Repeatedly waking individuals during Stage 4 NREM sleep causes hormonal dysregulation and altered circadian cycles, and sleep disturbances are tied to aggravation of psychopathological features.

Normal sleep is accompanied by a decline of epinephrine and cortisol, and at the same time growth hormone, prolactin, and melatonin increases occur (Besedovsky, Lange, & Born, 2012). Ordinarily, when individuals sleep, T cell functioning declines after a few hours, as if these cells (along with other hormones) need to rest and rejuvenate. In the absence of restorative sleep the inflammatory marker CRP was elevated (Zhang et al., 2013). Likewise, when individuals were forced to stay awake, their T cell functioning remained high, as if they were on alert (Besedovsky, Dimitrov, Born, & Lange, 2016), and for the moment, the capacity of the immune system to deal with antigenic challenges may be affected. These actions may be moderated by peripheral norepinephrine and circulating cortisol, which are also subject to circadian cycles.

Nocturnal sleep ordinarily favors effective inflammatory immune signaling, whereas the absence of sleep is accompanied by diminished levels of inflammatory cytokines (Redwine, Hauger, Gillin, & Irwin, 2000). In effect, sleep might serve to reallocate energy resources from functions related to wakefulness to processes that facilitate and promote defensive immune responses to deal with infectious challenges (Besedovsky et al. 2012; Motivala & Irwin 2007). Among other things, immune cell division and differentiation that follows antigenic challenges, requires metabolic resources, possibly involving prolactin and other hormones formed and released early in the night.

As observed in relation to several hormonal variations, some immunological changes that occur with sleep are related to circadian factors, whereas others are primarily a reflection of sleep's restorative powers (Besedovsky et al. 2012). For instance, the numbers of leukocytes, granulocytes, and monocytes, as well as the major lymphocyte subsets, including T-helper cells, cytotoxic T cells, activated T cells, and B cells, ordinarily attain their peak in the evening or early portion of the night, and then decline progressively toward the morning, essentially showing the inverse pattern of cortisol. These particular immune variations appear to be tied to a circadian rhythm, rather than sleep itself. However, other changes, notably the production of IL-2, IFNγ, and augmented IL-12 production by dendritic cells and monocytes are affected by nocturnal sleep, rather than circadian factors (Lange, Dimitrov, & Born, 2010).

Thus, sleep loss may affect responses to intracellular viral and bacterial challenges. As in the case of adaptive (acquired) immune responses, innate immunity is altered by nocturnal sleep. Ordinarily, NK cell activity increases over the course of a night's sleep (Kronfol, Nair, Zhang, Hill, & Brown, 1997), but it diminished among individuals with sleep disturbances (Irwin et al., 1994). Other aspects of innate immune functioning, such as the production of IL-6, is affected by both circadian factors and sleep itself (Redwine et al., 2000), whereas the production of the proinflammatory cytokine TNF-α, was promoted by circadian influences (Born et al. 1997). Reductions of hypersomnia may also be related to a decline of IL-1β and BDNF, whereas insomnia was less closely aligned with these factors (Rethorst et al., 2015). Just as sleep loss could affect immune functioning, it is clear that immune alterations, particularly variations of IFN-α, can affect sleep. In rodents, administration of proinflammatory cytokines increased NREM sleep, whereas inflammatory cytokine antagonists and anti-inflammatory cytokines, notably IL-4 and IL-10, elicited the opposite effect (Imeri & Opp, 2009).

Unlike a night or two of fragmented sleep, which may actually have limited effects on circulating inflammatory factors, when sleep is entirely prevented, cytokine variations are instigated, but can be reversed by a subsequent brief period of sleep. Likewise, elevated levels of C-reactive protein and that of the proinflammatory cytokines IL-1β, IL-6, and IL-17 associated with partial sleep deprivation over successive nights (Haack, Sanchez, & Mullington, 2007), may normalize after a night of proper sleep (van Leeuwen et al., 2009). It was similarly observed that cognitive disturbances that occurred with increased inflammation were attenuated among older men who had received adequate sleep. Such effects may arise because sleep loss affects the transcription of IL-6 and TNF-α owing to the alterations of NF-κB, a key transcription control pathway in the inflammatory signaling cascade (Irwin et al., 2008).

When sleep is disturbed, responses to stressors may be inadequate and inefficient, and allostatic overload is more apt to develop, leading to varied illness conditions. By virtue of the endocrine and immune changes imparted, sleep disorders (which are distressing in themselves) have been implicated in a range of illnesses involving inflammatory processes. Likewise, jobs that entail shift work, and hence changes of circadian cycles, were also associated with disturbed gut microbiota and illnesses, such as gastro-intestinal disorders, obesity, disturbed glucose metabolism and metabolic syndrome, cardiovascular diseases. It has been maintained that adequate sleep and proper circadian cycles are essential for effective functioning of biological processes and for adequate responding to stressful events. By virtue of the endocrine and immune changes imparted, sleep disorders have been associated with a range of illnesses involving inflammatory processes, create difficulties in dealing with viral illnesses (Cohen, Doyle, Alper, Janicki-Deverts, & Turner, 2009), and may contribute to some types of cancer. As a proxy for stressor effects on immune functioning to deal with antigenic challenges, the effects of sleep loss was evaluated in relation to the response to vaccination. As vaccines comprise weakened or dead viral particles, they generate an immune response, as described in Chapter 2, The Immune System: An Overview, and thus it is possible to determine whether factors, such as stressors or sleep loss (or the presence of a psychological illness), affects the response to a particular challenge. As expected, the immunologic response to an influenza A vaccination was markedly diminished in response to an experimentally provoked night of partial sleep loss (Spiegel, Sheridan, & Van Cauter, 2002), and a full night of sleep loss likewise provoked a persistent reduction in the response to hepatitis A vaccination. Moreover, relatively brief sleep durations were associated with the protection ordinarily afforded by hepatitis B vaccination being reduced. In essence, the response to vaccines can be diminished by loss of sleep, and might thus increase the risk for infectious diseases.

In other cases, severe sleep disorders may be secondary to other conditions, such as depressive illness, PTSD, or breast cancer. Insomnia has also been linked to shortened telomere length in peripheral blood mononuclear cells of elderly individuals, supporting the view that a lack of sleep exacerbates cellular ageing (Carroll et al., 2016). In this regard, sleep disorders have also been associated with the cognitive decline in Alzheimer's disease and correlated with cortical β-amyloid presence, as well as phosphorylated tau measured in cerebrospinal fluid (Liguori et al., 2017). There is little question that sleep disturbances in older people can be problematic, being associated with a variety of illnesses. However, it is still difficult to determine whether these individuals are in less need of longer sleep, or whether their neurobiological systems fail to create the conditions that allow for more sleep.

During some illnesses, particularly those that involve infection, the release of IL-1β and TNF-α promote or modulate sleep (Imeri & Opp, 2009), likely reflecting an adaptive response to facilitate recuperation. In this regard, as proinflammatory cytokines favor the development of fatigue, should individuals become prone to fatigue and display extended sleep periods, it may well portend something being physically amiss and may predict the occurrence of heart problems (Cappuccio, Cooper, D'Elia, Strazzullo, & Miller, 2011). In this sense, elevations of inflammatory factors might be an illness biomarker, but as inflammation is linked to a wide variety of illnesses, it may initially be difficult to discern which illnesses are in the making.

The fact that immune functioning varies with the circadian cycle has implications for vulnerability to illness as well as its progression. Among mice infected with the herpes virus at the beginning of the day (which is their resting phase), viral replication was an order of magnitude greater than when the viral inoculation occurred prior to the active phase. However, in mice in which the gene controlling circadian cycles (Mmal1) was knocked out, viral replication was elevated irrespective of the time of day (Edgar et al., 2016). Such findings are consistent with the involvement of circadian factors in disease processes. Circadian rhythm is dramatically altered in critically ill patients, which may be related to dysfunction of immune responses (Dengler, Westphalen, & Koeppen, 2015). Not unexpectedly, given the wide ranging influences of altered circadian cycles, these rhythms are disrupted among cancer patients (Altman, 2016). It has even been suggested that a clock gene, Per2 (which is one of many genes involved in controlling circadian rhythms), acts as a tumor suppressor, and disturbances of this gene's functioning encourages uncontrolled cell proliferation, genomic instability, and tumor promoting inflammation. This gene has also been implicated in cell proliferation and apoptosis in oral squamous cell carcinoma (Wang, Lin et al., 2016), as well as in other types of cancer. The idea has also been advanced that epigenetic processes related to clock genes may contribute to cancer development and progression (Masri, Kinouchi, & Sassone-Corsi, 2015).

It might be possible to exploit what is known about clock genes and variations of immune functioning over the course of a day in the development of the most efficacious treatment strategies for illnesses. It has been suggested that cancer treatment delivery (e.g., using programmable-in-time pumps) could be varied with an individual's circadian rhythms in an effort to increase treatment efficacy and tolerability (Innominato et al., 2014). Certain clock genes may also serve as biomarkers in predicting the benefits of some procedures (e.g., neoadjuvant chemoradiation therapy) in the treatment of particular forms of cancer.

Sleep, Diurnal Variations in Relation to Microbiota and Immunity

In view of the link between microbiota and immunity, it might be expected that sleep disorders would influence gut microbiota, and conversely, by affecting cytokines, gut dysbiosis might influence sleep processes. Even though microbiota have been studied extensively in relation to all manner of health conditions, there is still a paucity of information linking microbial factors and sleep disturbances, and illnesses that are secondary to sleep disorders. Nonetheless, there are several reasons to suppose that links exist between sleep alterations, microbial constituency, and health. The microbiome is affected by circadian rhythms, stressors, diet, and exercise, and notable changes occur in the presence of fever responses, likely reflecting or promoting adaptive immune responses to infection. As well, bacterial cell wall factors can trigger sleep, potentially by provoking pathogen-associated molecular pattern recognition receptors, which instigate cytokine release (Krueger & Opp, 2016).

It is known that circadian rhythms can be regulated by diet and time of food being ingested, which can affect microbial communities and metabolic functioning, which then influence immune activity (Voigt et al., 2016), and may also affect enteric infections (Rosselot, Hong, & Moore, 2016). Sleep deprivation alters gut microbiota, perhaps owing to increased white fat accumulation, and the resulting increase of inflammatory responses, and even partial sleep deprivation in humans was linked to microbial changes. There is also evidence that gut dysbiosis contributes to obstructive sleep apnea and hypertension and other cardiovascular conditions that follow (Durgan, 2017). In view of the importance of sleep disturbances to mental illnesses, such as depression, and the emerging evidence linking gut dysbiosis to depression, it would seem propitious to determine the microbial-sleep-depression connections.

CONCLUDING COMMENTS

To a significant extent, physical and mental health are dictated by genetic factors, but as we know, these can be influenced by stressful experiences and all manner of early life challenges. Further, what we eat, whether we engage in exercise, and whether we have recuperative sleep, can interact with stressors and genetic influences in determining well-being. Further, environmental factors can have epigenetic actions that influence well-being, or can directly or indirectly affect the microbiome, which can act on immune processes as well as central nervous system functioning, thereby influencing both physical and psychological well-being and the emergence of illnesses. In light of such findings, it is certainly appropriate to focus on the development of treatment strategies that incorporate multiple dimensions, including genetic, biological and lifestyle factors.

An implicit message that we intended in this chapter is that although there may be some illnesses that simply develop (e.g., owing to random mutations, or being exposed to toxicants that we were unaware of), many diseases fall into the category of "preventable illnesses." In the world of retail sales, a common mantra for success has been 'location, location, location', and the mantra for health and well-being ought to be "prevention, prevention, prevention." As much as it has become a cliché, the expression that "An ounce of prevention is worth a pound of cure" is absolutely correct, but it took quite some time for this to catch on. It's been a century and half since Metchnikoff offered the view that gut-related processes could affect diseases. Likewise, it's been more than two centuries since Pott reported that cancer could come about as a result of environmental factors, having linked soot exposure to scrotal cancer among chimney sweeps, which could be prevented by wearing appropriate attire. These and subsequent demonstrations encouraged the development of fields, such as epidemiology and preventive medicine. The fundamental importance of these fields is that they contribute to the identification of factors linked to disease provocation, which may influence the development of strategies to prevent these illnesses from occurring. With multiple demonstrations pointing to the importance of disease prevention (e.g., through vaccinations or by having people engage in healthier life-styles), it became certain that health approaches that simply focused on cures of illness were insufficient, and thus preventive programs evolved still further. In hindsight, it's remarkable that these efforts were slow to be accepted and in some respects they are still in the process of emerging.

Stressor Processes and Effects on Neurobiological Functioning

HAPPY FAMILIES ARE ALL ALIKE; EVERY UNHAPPY FAMILY IS UNHAPPY IN ITS OWN WAY

This statement from Tolstoy's Anna Karenina gave rise to the "Anna Karenina principle," which is suitable for our exposition regarding the impact of stressors. It can be taken to mean that there is considerable inter-individual variability in relation to bad experiences and how people react to them. However, the Anna Karina principle is usually taken to mean that there are any number of ways through which things can break down, but everything needs to be fully operational for a process to function properly. In order to come out of stressful situations unscathed, it is necessary to identify life challenges accurately, consider every possible eventuality, determine the best ways of dealing with these challenges, and then take appropriate actions while being mindful that the characteristics of stressful situations change over time, and the best ways to deal with some stressors may not be relevant in relation to other challenges.

Diverse stressors may be associated with numerous antecedent events and have many different consequences, thus making comparisons between different stressful events near impossible. As the expression goes, "at night, all black cows look alike." Is a severe debilitating illness equivalent in its impact to the loss of a loved one? Is a business failure more distressing than the shame created by public humiliation? Is PTSD that stems from a car accident and that which occurs as a result of childbirth equally distressing? All stressful experiences are disturbing, and each distressing experience is distressing in its own way. To be sure, some stressors are clearly much worse than others. Yet, whatever strong stressor we are experiencing at the moment might feel as if it's the absolute worst we have ever encountered.

Still, there are several fundamental principles that may be considered in predicting the types of events that are perceived as most distressing and which are associated with the most negative outcomes. Ordinarily, when a stressor is encountered, an impressive set of behavioral, cognitive, and neurobiological

(cont'd)

processes is available that can facilitate an individual's ability to contend with the challenge. These include an array of hormones, neurotransmitters, neurotrophins (growth factors), inflammatory, immune, and microbial factors, all of which may be regulated by gene-related processes. Experiences with traumatic events, especially those encountered early in life, may also cause the sensitization of neuronal mechanisms so that later stressor responses, for better or worse, are exaggerated. Likewise, such experiences may generate epigenetic changes wherein changes of gene expression foster effective or ineffective physiological and behavioral methods of contending with stressors. Conversely, a constellation of processes can render individuals more resilient.

Understanding stressor actions and stress outcomes is complicated by virtue of how many biological systems are concurrently or sequentially affected by stressors and the very great number of variables that can moderate these actions. As well equipped as an individual might be to contend with challenge, in general, there is only so much distress that an individual can handle before stress systems become overloaded, culminating in the emergence of psychological or physical pathology.

VULNERABILITY VERSUS RESILIENCE

Vulnerability, in the context of illnesses, typically refers to the propensity for an individual or a group (or even a whole society) to be at increased risk for physical or psychological problems emerging in response to particular environmental or social challenges. Resilience is not quite the opposite of vulnerability, although it is often considered in this manner. Instead, resilience refers to the ability to recover from illness, but it is also used to describe factors that limit or prevent particular stimuli or events from having negative effects. However, the absence of factors that favor increased vulnerability might not translate as elevated resilience. By example, an individual might carry a constellation of genes that ought to make them relatively nonvulnerable to illness, but it only takes one severe event (e.g., an aneurysm) to turn that on its head. A person can likewise carry a heavy stress load and have few adaptive biological resources, and hence ought to be at increased risk of becoming ill or for an illness to progress. However, the individual may be blessed with a strong social network or family support system, that helps them get through the worst times, and illness may not evolve or symptoms of an existing illness may be tempered. There are also instances in which individuals carry genetic mutations that ought to render them very ill, as in the case of cystic fibrosis, and yet they seem not to have been affected, possibly because they carry yet another genetic mutation that somehow protects them from developing the illness (Chen, Shi et al., 2016).

It's fairly simple for things to break, but may be difficult to repair (the Humpty Dumpty principle). Likewise, it's not overly difficult to identify the constellation of variables that exacerbate the illness-promoting effects of stressors, but it's more difficult to identify the variables that encourage resilience in relation to specific illnesses. Perhaps because of this, most studies that assessed the relationship between stressful events and specific pathologies have focused on identifying the characteristics of individuals or the features of the stressors that are associated with illness, but far fewer have considered the specific factors that promote resilience.

Different views have been offered concerning the elements that lend themselves to resilience, even though it is generally acknowledged that factors or interventions that are effective in limiting distress or illness in one situation may not be equally effective or useful in other situations. Resilience has been described as being related to neural processes underlying reward and motivation (hedonia, optimism), limited responsiveness to negative (fear) situations, and the availability of social behaviors (altruism, bonding, and teamwork) that minimize the impact of stressors (Charney, 2004). Resilience has also been described as stemming from the ability to adapt and to be flexible in the context of changes, the ability to problem solve, and possessing an optimistic outlook on life. Other views of resilience have included acceptance of change, control, and spirituality.

Considerable attention has been devoted to individual difference factors (personality) that operate in relation to resilience. The response to illness, for instance, may vary as function of self-efficacy, self-esteem, self-empowerment, tolerance for uncertainty, optimism, mastery, hardiness, hope, internal locus of control, and acceptance of illness. Being personality traits, these features aren't readily altered (although the can be to some extent), whereas the manner of appraising and coping with illness can be modified (e.g., through cognitive behavior therapy (CBT) or mindfulness training) so that some of the cognitive consequences of severe illness are affected. Fig. 5.1 lists some of the many factors that additively or synergistically contribute to resilience, but these are only a few of the very many factors that can act in this capacity.

ATTRIBUTES OF THE STRESSOR

Most of us encounter an astounding number of different stressful experiences (stressors), each leading to a variety of possible outcomes (stress or stress response). Stressors may comprise isolated events, or they may be experienced in bunches; often, one stressor experience generates others (financial concerns can create family difficulties, which can instigate health problems). Moreover, stressors can morph from being acute events to those that are chronic and intractable, which are more likely to promote illness.

The stressors encountered may comprise those of a processive nature, meaning that they involve cognitive (information) processing, and may comprise psychological (psychogenic) or physical (neurogenic) insults. Another type of stressor that can be encountered is composed of bacterial or viral presence, as well as metabolic changes, collectively referred to as systemic stressors. These challenges can influence brain neurochemical changes, much like those engendered by processive stressors, even though individuals may be entirely unaware of their presence. The brain interprets systemic insults much like it does other stressors, at least at a neurobiological level, and might thus surreptitiously and insidiously contribute to the provocation or maintenance of mood disorders or physical illnesses that are sensitive to stressor actions (Anisman et al., 2008).

STRESSOR APPRAISALS AND REAPPRAISALS

Perhaps the most predictable aspects of stressors are their unpredictability and the individual differences that exist in response to various challenges. One individual may interpret a particular event as stressful, whereas a second might not perceive the stressor in the same way, and even if appraised similarly, the ways in which individuals cope with challenges can differ appreciably. Furthermore, when comparable behavioral and cognitive coping strategies are endorsed, biological responses to contend with the challenge may nonetheless vary across individuals, and

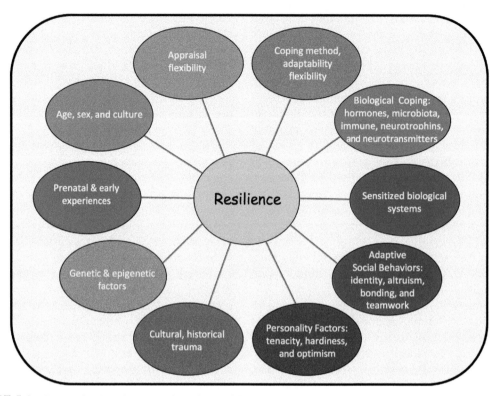

FIGURE 5.1   Stress related resilience can be influenced by a variety of factors. Many of these are outside of the individuals control, but others (e.g., appraisal, coping) are skills and strategies that can be acquired, and thus can be used to act against some of the processes that increase vulnerability to negative stressor effects. The effectiveness of many factors that favor or act against resilience may be context-dependent, varying with different types of challenges, and previous experiences, and can also vary over time as the stressor situation plays out.

consequently they may be differentially vulnerable to negative health outcomes.

As described in Fig. 5.2, upon encountering a potentially challenging situation, individuals make appraisals as to whether it represents a threat, and whether they have the ability, opportunity, and skills necessary to contend with the challenge. These appraisals can take different forms across people. By example, some individuals may appraise (interpret) a stimulus or a condition as a chore that has to be endured. Alternatively, they may see it as a challenge and opportunity that will have positive long-term gains, or prevent some potential negative outcomes from occurring. Still other people may interpret such situations as a threat of harm/loss

(or bring back traumatic memories), giving rise to negative emotions, such as anxiety or anger.

Appraisals and Misappraisals: Role of Heuristics

In their analyses of decision-making processes, Kahneman and Tversky (1996) offered a theoretical model that has implications for stressor appraisal and coping. Inherent in their view is that biases exist concerning decision-making processes, and thus in some instances individuals might not behave like rational actors. Decisions made in various situations, and this would certainly apply to stressful circumstances, are often made based on easily

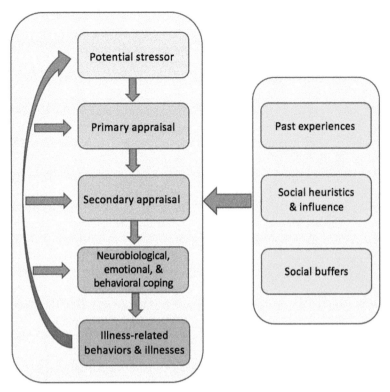

FIGURE 5.2   Upon encountering a potential stressor, individuals may ask themselves whether this represents a threat or risk to well-being. As described by Lazarus and Folkman (1984), following this primary appraisal, individuals ordinarily make a secondary appraisal, assessing whether they have adequate coping resources to meet the demands placed on them. On the basis of these appraisals specific coping responses to deal with the challenge will be adopted. Social heuristics, referring to ways of interpreting events based on similar or dissimilar experiences, can alter stressor appraisals, and social influences can likewise affect the secondary appraisals. The appraisals that are made and coping strategies used may also vary with numerous dispositional (personality) characteristics, such as intolerance for uncertainty, perceived self-efficacy, optimism, hardiness, and locus of control, to name a few. Social buffers (social coping) can also affect other coping methods used. Given accurate appraisals and the adoption of effective coping methods, stressor-related problems ought to be limited. In contrast, appraisal errors or the adoption of poor coping methods may favor the development of pathological conditions.

accessible information that is tied to earlier experiences or to previously established *heuristics* (i.e., rules or shortcuts). These strategies are influenced by several fundamental factors. Decisions can be based on *primed* intuitions that could have developed owing to previous experiences (or second-hand information). Appraisals and decision making are likewise influenced by experiences that are similar to or reminiscent of other events (representative heuristic). As the situation becomes more complex, making it difficult to appraise all the options available, individuals tend to consider the problem in a simpler way, even if this is inappropriate, or they might just go along with the opinion of their in-group, despite the hazards of doing so. In addition, individuals are affected by "associative coherence" in which a stimulus or event is appraised in a particular way simply because it is consistent with preconceived notions. Having formed a perception about one feature of a situation, related attributions (even occurring unconsciously) will be applied to this situation ("attribute substitution"). For instance, in the absence of further information, believing that a person is charitable may also result in the attribution of other positive qualities to that person (e.g., kind, warm, and easy to get along with). Many other factors influence decision making, such as base rate fallacy (in which individuals rely on particular information, while ignoring more general information even if it is more pertinent), ignorance of sample size (in which judgments are made without taking into account the larger picture). These are a few of the great number of biases that affect decision making, only some of which were described by Kahneman and Tversky.

Kahneman brought considerable attention to their views with the lay book "Thinking Fast and Slow" (Kahneman, 2011), in which he discusses dual systems that operate in making decisions. The first comprises an automatic or Fast Thinking system (System 1), whereas the second is a more cognitively based, Slow Thinking system (System 2). Behaviors in decision-making situations are often determined by the automatic, fast thinking system that has been primed to react in a particular manner[1]. There are many occasions, however, where these rapid responses are apt to be inappropriate, but the cognitively oriented slow thinking System 2 can limit blunders associated with System 1.

Individuals may find themselves in situations where decisions need to be made, but their knowledge, experiences, and abilities aren't adequate in relation to the pertinent issue. For instance, how certain are we in our decisions about how we should react to the threat of Zika, H1N1, bird flu, the risk associated with genetically modified organisms (GMOs), or for that matter, the safety of new medications? Many of us know little about these topics and there are more than a few of us who don't have the first clue about what to do in some threatening situations. Thus, in making decisions, we might look for "anchors" that can guide us (e.g., how others have responded to a new medication). Or, we might turn to our ingroup members to determine what they think about vaccine safety or GMOs, and then behave accordingly (even if the group's choices aren't necessarily the best course of action; in fact, group-think tends to favor relatively risky decisions). It's hardly necessary to say that decision-making abilities in stressful situations may be compromised (or augmented) by the mood state engendered by the stressor.

Impact of Stressor Characteristics

Stressors can vary over many dimensions, each of which can influence how they are appraised, and ultimately how coping strategies are affected. Distress increases with the perceived severity of the stressor, and stressors can have cumulative effects so that even day-to-day minor hassles can ultimately have damaging effects. As we'll see in our later discussion of early life experiences, stressors can be seen as being "toxic" challenges or "tolerable" (Shonkoff et al., 2009). The former refers to environments that entail extreme poverty, psychological or physical abuse, neglect, maternal

---

[1] In recent years, the importance of priming has been down-graded as an important component of decision making, largely in response to overstatements or unreliable findings reported by other researchers, particularly in relation to the influence of stimuli or events that weren't consciously perceived. In contrast, conscious experiences may have considerable sway in some stressful situations (irrespective of whether or not this is defined as priming), particularly when appraisals and decisions on how to respond must necessarily occur very quickly.

depression, parental substance abuse, and family violence, which can be particularly damaging and lead to later psychopathology. Modest ("tolerable") stressors, in contrast, can have positive actions, producing an activating effect that can facilitate coping and active responses (e.g., mild anxiety can enhance public performances). As well, modest stressors experienced at an early age may also influence the child learning how to deal with stressors, which can have long-term benefits.

Some stressors that need to be dealt with have already passed (e.g., loss of a loved one), whereas others are anticipatory stressors (e.g., the fear of imminent surgery, or meeting the bully in the schoolyard or workplace). The anxiety created by anticipatory stressors can be exaggerated if the challenge is ambiguous (e.g., the pilot of a plane indicating the need to return to the airport without further explanation), and might be accompanied by disorganized cognitions while the situation plays out.

The stressor characteristic that has received particular attention concerns its controllability, as seen in the work concerning learned helplessness (Maier & Seligman, 1976), which is discussed in more detail in Chapter 8, Depressive Disorders, when considering factors that favor the development of depressive illness. Uncontrollable stressors promote a behavioral profile in which animals cease making attempts to escape from a noxious stimulus, seeming to passively accept the stressor. It was proposed that these animals had essentially learned that "nothing I do matters," leading to a cognitive schema of "helplessness." As we'll see shortly, as an alternative to learned helplessness, it was argued that uncontrollable stressors provoke several neurochemical changes that might be responsible for the behavioral disturbances seen in this paradigm (Anisman et al., 2008).

The predictability of a stressor's occurrence markedly influences stress reactions. Most individuals find the occurrence of unpredictable stressors to be more aversive than those that are predictable, and in animals, unpredictable stressors led to more pronounced neurobiological changes (Baker & Stephenson, 2000). There are also occasions in which it's uncertain whether a threatening event will actually occur, and this uncertainty can create considerable anxiety. This said, in some instances uncertainty may also have beneficial effects. In the context of severe illness that is likely to be fatal, even a small amount of uncertainty may allow a person to maintain a degree of hope so that they can function effectively.

WHY IS THIS STRESSOR DIFFERENT FROM EVERY OTHER STRESSOR?

For animals in the wild, the search for food or approaching the water hole can be fraught with dangers (e.g., from predators awaiting them), requiring some dicey cost-benefit analyses. Thus, interplay exists between the prefrontal cortex responsible for several executive functions and the amygdala that is involved in fear and anxiety. In fact, aspects of the basolateral amygdala interact with the prelimbic cortex in order for fast decisions to be made among rodents placed in an approach-avoidance conflict situation (Burgos-Robles et al., 2017). These regions may likewise be relevant to the bad decision making that sometimes accompanies mental illness (e.g., whether to continue taking a beneficial drug or not).

It is not unusual to encounter risky situations that involve ambiguity or unknowability, which can increase anxiety. In these instances, neuronal activity increases markedly in brain regions involved in decision making, possibly in an effort to make sense of the situation (Bach, Seymour, & Dolan, 2009). Within a laboratory context, knowing that there is a chance, even a

(cont'd)

small one, of receiving a painful stimulus (as part of a computer game) elicited greater anxiety than knowing that the painful stimulus would definitely be experienced (de Berker et al., 2016). Whether uncertainty is linked to specific neurochemical processes has not been studied extensively, but norepinephrine, dopamine and acetylcholine may underlie various facets of uncertainty. Among individuals who can't deal with uncertainty, anxiety symptoms are often prominent, accompanied by increased information seeking in an effort to diminish uncertainty.

In some cases, the information pertaining to a threat's possible occurrence can be ambiguous (e.g., are these symptoms of a heart attack or is it indigestion?). Symptoms of this sort are worrisome, but they might not seem sufficiently coherent to prompt emergency medical help. Too often individuals do nothing in the hope that the vague symptoms will simply disappear, which in this instance is obviously not a reasonable strategy. Under these conditions, being high in intolerance of uncertainty may be advantageous.

As uncomfortable as ambiguity, uncertainty, and unpredictability may be, the fundamental question arises as to where the cut-off lies between feeling "safe" and feeling sufficiently "at risk" to take actions to overcome threats? How unambiguous must symptoms be before an individual decides to take ameliorative action? Knowing an individual's sensitivities in relation to these dimensions can be a useful predictor of how they will respond to certain stressors, and perhaps predict the most effective treatment strategies to reduce their anxieties. However, here again, numerous factors (e.g., previous experiences) moderate the influence of these variables on appraisals and defensive responses.

Although worry and anxiety are troublesome and could promote illness, within limits these cognitive and emotional responses may have adaptive value in that individuals will remain vigilant and ready to respond should a stressful situation escalate. Unfortunately, when these emotions are persistently present owing to inappropriate appraisals and beliefs that a stressor is imminent a good deal of the time, the risk for illness is elevated.

How individuals respond to uncertain and unpredictable events may be tied to when these events are apt to occur. By example, if some horrific event may or may not occur within the next few seconds or minutes, our intolerance of uncertainty is largely irrelevant as things will be moving too quickly for the uncertainty to have significant consequences. In contrast, uncertainties pertaining to somewhat more remote events usually have profound and worrisome effects (will that biopsy show malignancy?), and those who can't deal with uncertainties suffer the most. Finally, the uncertainty of an event in the distant future

might simply fall off the radar as being a potential threat, even though its consequences can be devastating (e.g., predictions of global warming and elevated sea levels within the next 30 or 100 years, or an earthquake hitting in the next 10−20 years). For far off events it's just too easy to adopt an avoidance coping strategy. Only recently have scientists changed their wishy-washy descriptions and predictions of the potential for an earthquake. By making the threat seem more certain (without using scare tactics that often backfire), individuals could possibly be convinced that while they have no control over an earthquake's occurrence, they have some control regarding the consequences, thereby encouraging preparedness. Still better responses might be obtained if an individual was portrayed as part of an in-group, and that their behavior reflects a moral imperative to protect and assist other group members. Likewise, "nudge" approaches can be used so that through indirect suggestions and positive reinforcement, people are slowly moved to the place where they behave in a particular fashion (Thaler & Sunstein, 2008).

A less studied feature of some stressors or stressor situations concerns their volatility. It is not unusual for stressors to increase (or conversely decrease) in intensity, depending on a number of factors. Often these changes occur in a predictable fashion, allowing for preparatory responses to be adopted. However, there are instances in which stressors show volatility, do not follow a predictable trajectory, and expand in just an instant from a few embers to a full-blown conflagration, which is not unusual in the experiences of first responders.

Coping With Stressors

Numerous methods can be used to cope with stressors, which can be condensed to 15 or so general methods that fall into 3 broad classes comprising problem-, emotion-, and avoidant-focused coping. Table 5.1 provides a description of several coping strategies that fall into these three categories. In this table, most coping methods were assigned to one category or another. However, some of these coping methods actually aren't comfortable within any specific category, varying with specific stressor contexts, as we'll see later when we discuss social support.

Problem solving coping methods are often thought to comprise the ideal way to diminish distress and thus preclude pathology, whereas emotion-focused coping methods are frequently considered to be maladaptive. This view, however, is a bit simplistic as the effectiveness of a particular strategy is situation-dependent. Specifically, emotional coping strategies may have beneficial effects if for no other reason

TABLE 5.1   Coping Methods

PROBLEM-FOCUSED STRATEGIES

Problem solving: Finding methods that might deter the impact or presence of a stressor

Cognitive restructuring (positive reframing): Reassessing or placing a new spin on a situation so that it may take on positive attributes. This can entail finding a silver lining to a black cloud

Finding meaning (benefit finding): A form of cognitive restructuring that entails individuals finding some benefit or making sense of a traumatic experience. This might involve emotional or cognitive changes, or active efforts so that others will gain from the experience

AVOIDANT OR DISENGAGEMENT STRATEGIES

Active distraction: Using active behaviors (working out, going to movies) as a distraction from ongoing problems

Cognitive distraction: Thinking about issues unrelated to the stressor, such as immersing ourselves in our work, or engaging in hobbies

Denial/emotional containment: Not thinking about an issue or simply convincing oneself that it's not particularly serious

Humor: Using humor to diminish the distress of a given situation

Drug use: Using certain drugs in an effort to diminish the impact of stressors, including the physiological (physical discomfort) and emotional responses (anxiety) elicited by a stressor. Some individuals may engage in eating in an effort to cope with challenges

(Continued)

TABLE 5.1   (Continued)

EMOTION-FOCUSED STRATEGIES

Emotional expression: Using emotions such as crying, anger and even aggressive behaviors to deal with stressors

Blame (other): Comprises blaming others for adverse events. This is used to avoid being blamed, or as a way to make sense of some situations

Self-blame: Blaming ourselves for events that occurred. Often associated with feelings of guilt or anger directed at the self

Rumination: Continued, sometimes unremitting thoughts about an issue or event, or replaying the events and the strategies that could have been used to deal with events

Wishful thinking: Thinking what it would be like if the stressor were gone, or what it was like in happier times before the stressor had surfaced

Passive resignation: Acceptance of a situation as it is, possibly reflecting feelings of helplessness, or simply accepting the situation without regret or malice ("it is what it is")

RELIGION

Religiosity (internal): A belief in god to deal with adverse events. This may entail the simple belief in a better hereafter, a belief that a merciful god will help diminish a negative situation

Religiosity (external): A social component of religion in which similar minded people come together (congregate) and serve as supports or buffers for one another to facilitate coping

SOCIAL SUPPORT

Social support seeking: Finding people or groups who may be beneficial in coping with stressors. This common coping method is especially useful as it may buffer the impact of stressors as well as serve multiple other functions in relation to stressors, including providing information as to how to best deal with a particular situation. Social support from ingroup members is generally superior to support from outgroup members

Modified from Anisman (2016) and Matheson & Anisman (2003)

than messaging (through verbal and nonverbal ways) the need for help. Further, this coping method might facilitate an individual's coming to terms with their feelings, and consequently diminishing distress.

Problem solving efforts are certainly important in helping individuals diminish or eliminate stressors that are controllable. However, in situations that are beyond the individual's control (e.g., dealing with a terminal illness), these efforts might not be productive, although they may offer the illusion that the individual maintains some control over their destiny. At the same time, if a stressor is severe and uncontrollable, people might encounter difficulty planning and initiating actions, and thus problem coping efforts might be in vain. In fact, rather than adopting straightforward and simple problem-focused approaches, in some

instances exceptionally stressed individuals might engage in excessively complex and unfeasible strategies. The alternative approach, as much as it might be ineffective in other situations, is the adoption of avoidant coping methods (e.g., distraction), allowing the individual to function effectively.

In some instances, nothing can be done to change a situation, and victims need to dig deeply to cope with their challenges. This might comprise a form of cognitive restructuring in which individuals find something positive coming from the trauma. This can emerge in the form of posttraumatic growth (finding meaning or benefit finding). In response to severe stressors (e.g., loss of a loved one through accident, terrorism, and suicide), survivors might try to make sense of the event, and in some instances (e.g., in the case of cancer survivors) some benefit may be derived through these experiences. Having battled through a severe illness or having engaged in a lengthy caregiving routine (e.g., for a parent with a form of dementia) a person might recognize positive implications of their experience (smelling the roses) or that in some ways they were strengthened or feel that they have grown from the experience. In other cases, efforts are made so that others might gain from their own horrid experiences (e.g., creation of organizations to raise funds for specific illnesses, supporting gun control and campaigns against drunk driving). No doubt, these methods of coping are very effective for some individuals, but this isn't always the outcome. Engaging in a search for meaning ("meaning-making efforts") doesn't guarantee that the person will actually discover the meaning of an event ("meaning made") (Park, 2010). Sometimes, there simply isn't anything positive that comes from a tragedy or it isn't in the person's make-up to find meaning in such a situation. Persistent, but unsuccessful efforts to find meaning might be indicative of a person having an unhealthy preoccupation with a tragedy, and is unable to let go as they probably should.

The fact is that in many situations there isn't a single best strategy to deal with challenges, and it's unlikely that a particular strategy will be used in isolation of other methods. The specific cocktail of strategies adopted might determine the "adaptiveness" of the approach in a given situation. High rumination (an emotion-focused strategy) predicts the subsequent development of depression (Nolen-Hoeksema, 1998), depending on the nature of the ruminative content. Depressive symptoms are most apt to occur when rumination is accompanied by other emotion-focused methods, such as self-blame or emotional expression (negative rumination). In contrast, if rumination is accompanied by problem-focused strategies, it might be instrumental in diminishing distress, and negative outcomes might be less likely to develop. Further, having a broad range of coping strategies available, applying these appropriately, and being flexible in both the choice of coping methods and the ability (flexibility) to shift between strategies as the situation demands, may be the ideal approach to deal with stressors. In this regard, the characteristics of stressors may shift over time (e.g., as an illness progresses) and it is important that individuals be able to shift coping methods accordingly. Having a narrow range of coping methods, and being rigid in maintaining them when they ought to be abandoned, will likely be counterproductive (Cheng, Lau, & Chan, 2014).

POOR COPING AT A COST

In desperate situations, people may adopt desperate solutions to get through their travails. In the face of an incurable illness, there may be a temptation to use complementary alternative medicines, even if there isn't scientific evidence to support their use. For that matter, there have been efforts to adopt a "right to try" strategy when nothing else seems to be effective, essentially allowing terminally ill patients to receive treatments (e.g., biologics or drugs) that are at an early phase of testing and haven't received approval from relevant agencies, such as the Federal Drug Administration (FDA). For the patient there is the sense that "there's nothing to lose," but some treatment can promote serious side effects that make things worse.

In the face of severe stressors, there may also be an increase in the propensity of some individuals to turn to religion, even though in some quarters, this is viewed as being primitive. Curiously, the very same individuals might view homeopathy, colonics, acupuncture, and reiki therapy as perfectly reasonable. Religion may be a core component in

(cont'd)

some individuals' identity and may serve as an effective coping strategy (Ysseldyk, Matheson, & Anisman, 2010). Provided that belief and prayer aren't used as a substitute for potentially effective treatments, even if they serve primarily in a palliative capacity, religion can be effective in diminishing distress. It has likewise been suggested that through religion,

individuals might have gained a social support network that serves to provide solace, peace of mind, distraction, and cognitive reinterpretation (Ysseldyk et al., 2010). As well, spirituality has been associated with lower depression and greater neuronal density within the prefrontal cortex and hippocampus (Miller et al., 2014).

Personality and Sex as Moderators of the Stress Response

Personality traits, as we saw in our discussion of resilience, can influence behavioral appraisal, coping processes, and neurobiological functioning, and might consequently affect stressor-related disease conditions. Some individual traits (neuroticism, depressive personality) may favor the development of stress-related pathology, whereas others (optimism, self-efficacy, and hardiness) were linked to resilience (e.g., Carver & Connor-Smith, 2010).

Several illnesses are more common in women than in men (e.g., depression, autoimmune disorders), which might be attributable to women carrying a greater stress load than do men, frequently being the one taking on the responsibility of child rearing, day-to-day tasks within the home, and they disproportionately serve in a caregiving capacity for family members who are ill (e.g., aging parents). Sex-differences also exist in relation to behavioral coping methods, which could influence the ability to overcome stressors. No doubt, hormonal and genetic factors also play a prominent role in this regard, as we'll see when we discuss specific biological stress responses. In some ways, females fare better than men, and in many ways they may be the more resilient sex. Women in industrialized and in nonindustrialized countries

continue to outlive men, although the gap is not as wide as it had been in the past. Aside from estrogen providing women protection against some illnesses (e.g., heart disease), the dimorphism might be due to men being more likely to engage in risky life-styles, smoke and drink more, and eat more cholesterol producing foods, which could limit their lifespan. Pending more research assessing biological differences in males and females, and how these link to disease states, including the contribution of stress and inflammatory processes, it will be difficult to make specific recommendations concerning how to prevent sex-biased illnesses.

Social Support

There is little doubt that social support can serve as one of the most effective ways of contending with negative events. The usefulness of support can take multiple forms and can change over time, which is obvious among individuals dealing with a serious illness. It can serve in a problem solving capacity (e.g., obtaining advice; finding alternative treatment strategies for a severe illness) and can also be combined with emotional coping that can facilitate venting or a shoulder to cry on. Having support might also diminish some of the hormonal changes (e.g., cortisol) associated with stressors, as well as limiting stressor-elicited neuronal activation that was otherwise apparent within the right prefrontal cortex and amygdala (Taylor et al., 2008). These findings attest to the effectiveness of social support as a stress buffer, which can act against the adverse psychological effects that might otherwise occur.

The source of social support can be important in determining its usefulness. Typically, support is most likely to come from friends, family, and other members of our in-group, and their support is typically more effective than that coming from others (out-group members). Those who share our pain also understand what we might be going through, and hence support groups that include others with similar problems (e.g., support groups for families of those that died of suicide or for the parents of children with cancer) are particularly effective in diminishing distress.

POSITIVITY AND WELL-BEING

There has been a movement that focuses on "positive psychology" in the promotion of wellness. From this perspective, the absence of distress doesn't necessarily imply the presence of a positive state. The main goal of positive psychology isn't necessarily to cure mental illness or physical disorders, but instead to use positive experiences and thoughts to enhance well-being (Seligman & Csikszentmihalyi, 2000). Part of this entails the promotion of positive emotions, which can enhance individuals' thought—action repertoires so that psychological and social resources are established and strengthened.

Positive states can act prophylactically to prevent the development of stress-related pathology, such as depression, and may contribute to enhanced recovery from illness. Likewise, relative to individuals with ongoing negative experiences, those with positive emotions displayed superior immune functioning (Barak, 2006) and activation of the reward system is able to enhance both innate and adaptive immune functioning (Ben-Shaanan et al., 2016). As expected, optimism was associated with effective regulation of stress hormones and was accompanied by advantages in the course of cancer progression (Carver et al., 2005). In a 15 year prospective study, positivity was linked to diminished risk for chronic heart disease, and among individuals with an established illness, positive mood was accompanied by lower mortality (Chida & Steptoe, 2008). Even though positive psychology can favor increased survival in the face of chronic illnesses, possibly through effects on life-styles and treatment adherence, it is less likely that it can turn back the clock and produce a cure for chronic, progressive diseases, where damage is already extensive.

Not all researchers believe that positivity is as effective as these reports suggest, and at best might only be useful as an auxiliary treatment for dealing with characteristics related to illness, but not the illness itself. One might also question the meaningfulness of some of the findings reported in relation to positive psychology. There have been indications that positive life experiences are accompanied by modest elevations of immune functioning, but it is questionable whether these limited immune changes have meaningful effects on the ability to fight infection, although there have been reports that positive events can diminish the risk for developing upper respiratory infection and colds. There is also the view that the beneficial effects of positive affect come from acting as a stress buffer (Pressman & Cohen, 2005), and as expected, the relationship between positive affect, well-being, and inflammation was most apparent when perceived stress levels were high (Blevins, Sagui, & Bennett, 2017).

In some circles, there seems to be the belief that we're all supposed to be happy, and negativity isn't tolerated for very long. While not denying that positivity can have some very

---

(cont'd)

good effects, it needs to be acknowledged that there are limits to its benefits, and forcing positive attitudes might not have the intended consequences. Indeed, positive psychology may be fomenting what Held (2002) referred to as the "tyranny of the positive attitude."

Turning Support on Its Head: The Case of Unsupportive Relations

As useful as social support might be, there are limits to its effectiveness, and in some instances, it can actually be counterproductive. Some people prefer to maintain their independence and privacy and consequently might be reluctant to accept support. For others, the act of seeking or obtaining support might reflect weakness (more so in some cultures), or they resist support since it might make them feel indebted. Even if an individual is hesitant to ask for social support, they may find themselves in a situation where their options are limited and so they reach out to a friend or relative, fully expecting that the support will be forthcoming. It is likely that their expectation will be met, but it's possible that the support will be conditional. Worse still, the request might be rebuffed, or they may encounter a response that is entirely unhelpful (e.g., blaming the victim, forced optimism or minimizing the victim's distress, disconnecting or distancing in order to avoid hearing about the problems, and bumbling feebly as a result of not knowing what to say). Not having support can be disappointing, but encountering *unsupportive* interactions can be far more damaging (Ingram, Betz, Mindes, Schmitt, & Smith, 2001). This has been seen among individuals with HIV/AIDS who were abandoned by family members, and among women in abusive dating relationships who were diminished because they stayed in their situation (Matheson, Skomorovsky, Fiocco, & Anisman, 2007).

It's especially unfortunate that occasions of unsupportive actions not only occur at a personal level, but also appear in relation to group behaviors. This is particularly notable when an ongoing injustice is met by silence from witnesses, who might simply "not want to get involved." In the political arena, we've seen this playing out in relation to disadvantaged groups within countries (e.g., Indigenous Peoples within Canada, Australia, the United States, and elsewhere). Diminishing the suffering of the dispossessed and marginalized groups has been a keystone within many philosophies for millenia, but in recent times this was passionately expressed by Nobel laureates Elie Wiesel ("…to remain silent and indifferent is the greatest sin of all") and Reverend Martin Luther King ("In the end we will remember not the words of our enemies, but the silence of our friends").

Social Rejection and Ostracism

Social rejection and targeted ostracism may be especially painful forms of unsupportive relationships. Although members of a group might consider themselves as a unified social entity (entitativity), some group members may be viewed as not representing them as they would like, and as a result, these individuals might be rejected and denigrated (black sheep effect) in order to preserve the groups stature (Eidelman & Biernat, 2003). Social rejection is sufficiently strong that even within a laboratory context, a seemingly benign rejection manipulation can elicit negative ruminative thoughts, alter mood, promote hostility, and elicit elevated cortisol levels (McQuaid, McInnis, Matheson, & Anisman, 2015). Such manipulations also provoked elevated neuronal activity within the dorsal anterior cingulate cortex, much as strong stressors have such an effect (Eisenberger, 2012). The bottom line is that targeted rejection, irrespective of how or when it occurs, can undermine an individual's self-esteem, and can promote or exacerbate depressive mood.

NEUROBIOLOGICAL CHANGES AS AN ADAPTIVE RESPONSE TO STRESSORS

Concurrent with behavioral methods of dealing with stressors, an assortment of biological changes occur that serve in this adaptive capacity. Some of these neurobiological changes operate to enhance vigilance and augment preparedness to deal with impending challenges. Others serve to blunt the psychological and the physical impact of stressors, facilitate appraisal, enhance effective coping, favor adaptation to chronic challenges, and activate neuronal and endocrine processes necessary for survival, while limiting the actions of still other biological processes that could engender negative outcomes (Anisman et al., 2008; Sapolsky, Romero, & Munck, 2000). Psychological factors related to the stressor situation (e.g., stressor controllability, predictability, and ambiguity) may influence neurochemical functioning, thereby affecting emotional and cognitive processes. The behavioral and biological methods of dealing with stressors operate in tandem, and as the effectiveness of behavioral methods diminish, the load placed on biological systems increases. This may not create problems over the short-term, but with chronic stressor experiences, individuals may become more susceptible to pathological outcomes.

TELOMERES: AN INDEX OF GETTING OLD OR CUMULATIVE STRESSOR EXPERIENCES

The impact of chronic strain might be detectable through neurobiological changes that occur, such as variations of daily cortisol rhythms or cortisol levels, but for the most part, simple biological markers that reflect the cumulative effects of chronic stressors are limited. A possible exception to this involves the length of telomeres that comprise a region at the end of each chromosome, made up of repetitive nucleotide sequences. Like the aglet at the end of a shoelace, these telomeres prevent DNA from unraveling.

With each cellular replication that accompanies ageing, the telomeres shorten, and thus their length might serve as index of cellular ageing. Shorter telomere length was also related to age-related illnesses, such as diabetes and heart disease, which were dose-dependently tied to early-life stressors (Price et al., 2013). In some instances, the link between telomere length and age may be more related to illness than ageing itself (Sharifi-Sanjani et al., 2017). In fact, telomere shortening can occur in young individuals experiencing serious health conditions.

Childhood adversities, including lower socioeconomic status, which predicted poorer health, was also associated with shortened telomere length (Cohen et al., 2013), and frequent early life infections were associated with shorter telomere length in adulthood (Eisenberg, Borja, Hayes, & Kuzawa, 2017). Cumulative stressful experiences in children, including maternal violence, verbal or physical assault, or even witnessing domestic disputes, resulted in the shortening of telomere length, and prenatal stressor experiences were likewise linked to shorter telomere length in later adulthood (Entringer et al., 2011).

(cont'd)

As in the case of ageing, stressors and poor life-styles (e.g., smoking and high alcohol consumption) were predictive of telomere shortening. The actions of stressors in this regard were especially notable when accompanied by reduced social support, lower optimism, and elevated physiological stress reactivity (Zalli et al., 2014). Although telomere length can be used as an index of biological aging, it seems that the clock can be turned back (somewhat) as the length of telomeres can be rejuvenated through the enzyme telomerase reverse transcriptase. Exercise, for instance, may diminish the impact of ageing on telomere length and survival of T cells (Silva et al., 2016).

There were indications that telomere length may be linked to psychological illnesses, such as depression and PTSD (Zhang et al., 2014). Conversely, stress reduction procedures, such as meditation, were associated with increased telomere length (Alda et al., 2016). It is uncertain how stressors and mood disorders come to be related to telomere length, but activation of inflammatory processes could potentially contribute in this regard (e.g., Kiecolt-Glaser et al., 2011). As with so many other stressor-related disturbances, changes of telomere length might be an outgrowth of early life and adult stressors and could serve as a biomarker for illness vulnerability, including heart disease, insulin resistance, and type 2 diabetes.

Chronic Stressors and Allostatic Overload

Many biological changes that occur soon after a stressor encounter serve to maintain equilibrium within and between biological systems, and are essential for proper appraisal and behavioral stress responses. As well, they contribute to the distribution of energy resources to assure proper functioning of various organs, including the brain. This process, termed allostasis, is reminiscent of the balancing involved in homeostasis, but because of the urgency related to stressors, it involves a much more rapid mobilization of biological resources. In fact, some of the brain processes that are involved in stress responses may have developed short cuts (the neural equivalent of the fast thinking processes that Kahneman described in relation to decision making), so that rapid responses can be elicited without having to go through the multistep process of appraisal and reappraisal. When an organism's life is on the line, the luxury of spinning its

wheels as to what actions to take simply isn't on the agenda.

Neurobiological systems are remarkably adept at adapting to moderate, predictable stressors that are repeatedly encountered. However, finding adequate ways to deal with repeated challenges that are particularly severe, uncontrollable, and unpredictable, is exceptionally difficult (e.g., chronic illness or chronic pain). Under such conditions, brain regions responsible for dealing with stressors might not have an opportunity to recuperate, as neurobiological processes may be continuously engaged. As a result, neuronal damage can be accrued (allostatic overload), as seen in the case of excessive cortisol release causing damage to hippocampal neurons (McEwen, 2000). As well, when adaptive neurobiological systems becoming overly taxed, hormones or neurotransmitters become less available or receptor sensitivity is altered, leading to a variety of pathological conditions. Stressors can also have cumulative additive or interactive effects that might contribute to the development of illnesses and exacerbation of their progression, especially when these stressors begin during childhood.

In humans experiencing chronic, unremitting stressors (stigmatization, war experiences, and chronic illness), the impact of these challenges may be doubly disastrous, not only because of the continuous nature of the stressor, but also because victims lack safe places that would allow for healing to occur. In most instances, the stressors considered are those that are encountered on an individual basis, but it was proposed that insidious challenges at a broader level (e.g., social disturbances, social conflict, and poverty) can provoke "Type 2" allostatic overload, which requires changes of social structures to prevent the development of pathology (McEwen & Wingfield, 2003). As depicted in Fig. 5.3, the relationship between stressor levels and pathology has been

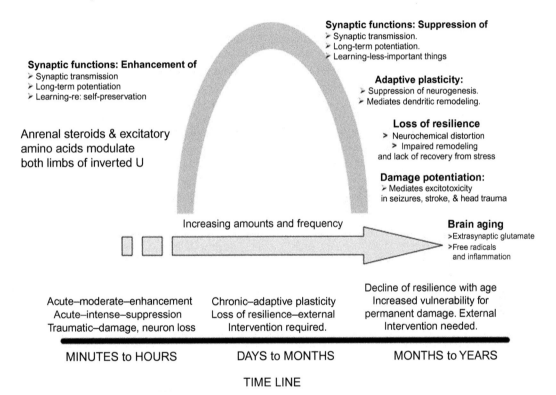

Synaptic functions: Enhancement of
➤ Synaptic transmission
➤ Long-term potentiation
➤ Learning-re: self-preservation

Synaptic functions: Suppression of
➤ Synaptic transmission.
➤ Long-term potentiation.
➤ Learning-less-important things

Adaptive plasticity:
➤ Suppression of neurogenesis.
➤ Mediates dendritic remodeling.

Loss of resilience
➤ Neurochemical distortion
➤ Impaired remodeling
and lack of recovery from stress

Damage potentiation:
➤ Mediates excitotoxicity
in seizures, stroke, & head trauma

Anrenal steroids & excitatory amino acids modulate both limbs of inverted U

Increasing amounts and frequency

Brain aging
➤Extrasynaptic glutamate
➤Free radicals
and inflammation

Acute–moderate–enhancement
Acute–intense–suppression
Traumatic–damage, neuron loss

Chronic–adaptive plasticity
Loss of resilience–external
Intervention required.

Decline of resilience with age
Increased vulnerability for
permanent damage. External
Intervention needed.

MINUTES to HOURS   DAYS to MONTHS   MONTHS to YEARS

TIME LINE

FIGURE 5.3   Acute and chronic stressors influence the risk for illnesses, varying with the chronicity of the stressor. The actions of stressors are mediated by glucocorticoids, glutamate, BDNF, tissue plasminogen activator, CRH, endocannabinoids (eCBs), as well as immune changes and microbiota. With brief stressor exposure, preparatory mechanisms are engaged that might enhance well-being (the left part of the inverted U), although if sufficiently stressful or traumatic, these acute events can be damaging. With continued stressor experiences, a series of further adaptive changes occur to maintain well-being, although under these circumstances social support or other external coping processes are needed to maintain well-being. Eventually, with excessive stressor experiences, these adaptive systems may fail or multiple neurobiological alterations may be instigated that increase risk for illness. Brain plasticity plays a prominent role in determining how effectively individuals are able to deal with stressors, and as early life experiences are fundamental in shaping many neural connections, they may have a marked influence on the impact of stressors that are subsequently encountered. At the same time, neuronal malleability or plasticity also allows for successful interventions that can limit the destructive effects of earlier negative experiences. *Source: From McEwen et al. (2015).*

described as an inverted U-shaped function, being dependent on the severity and frequency of the stressor (McEwen et al., 2015).

NEUROTRANSMITTER ALTERATIONS PROVOKED BY STRESSORS

Before describing the neurobiological effects of stressors, several caveats should be mentioned. There is much to be said about assessing the impact of stressors in both human populations and in laboratory animals (as well as animals in their natural habitats). Human studies provide a richness of information (e.g., concerning the contribution of psychosocial factors, personality, and appraisal-coping processes) that isn't easily obtained through studies using animals. Likewise, important features in humans, such as rumination or the complex cognitive changes that evolve, can't be recapitulated in animals. Although imaging studies allow for analyses of brain region and circuit involvement in response to or anticipation of stressors, animal studies provide the opportunity to examine the brain changes more intensively. As well, it becomes possible to assess the moderating effects of numerous variables (e.g., early life experiences) on the impact of stressors on brain neurochemical functioning within discrete brain sites.

The impact of assorted stressors in animals, generally engage several common neurobiological systems (e.g., increasing HPA activity). Yet, there may also be neurobiological changes that are more readily elicited by some types of stressors than others. For instance, psychological stressors that involve higher-order processing (e.g., stressors that were learned), might not engage the same neural circuits as those elicited by an innate threat. These challenges may also involve different brain networks relative to those activated by neurogenic or systemic insults (Merali et al., 2004). As we'll see when discussing some of the effects of stressors on hormonal changes, some stressors (e.g., those that involve public performances and public scrutiny, and hence more likely to elicit emotions, such as shame or anger) are more likely to cause cortisol changes than are other stressors (e.g., stress of examination) (Dickerson & Kemeny, 2004; Matheson & Anisman, 2012).

Like the behavioral effects of uncontrollable stressors, biological changes may vary over time. Initially, stressors elicit response activation that might reflect anxiety and arousal, but are then replaced by processes that lead to a depressive-like condition. In some instances, the effects of the stressor grow with the passage of time, as does the potential for several neurochemical changes. Aside from pointing to the (dynamic) evolution of stress reactions, these findings raise the possibility that the efficacy of treatments to attenuate stress-related illnesses may also differ over the course of illnesses playing out.

ANIMAL MODELS OF PSYCHOPATHOLOGY

Given the importance of animal models for both physical and psychological disorders, criteria were established to maximize their validity. Yet, it is important to acknowledge that even under the best of conditions, what happens in one mouse strain might not happen in another, and thus it should not be surprising to find that rodent models don't necessarily translate well to humans. To an extent, the validity of an animal model is reminiscent of the validity of a jury trial. A jury might find a particular defendant to be not guilty, but this doesn't mean that the defendant is actually innocent.

(cont'd)

The requirements for an animal model provided here are based on the description provided by Anisman & Matheson (2005):

1. Face validity: The animal model should
   comprise symptoms that are similar to those
   present in the human condition.
2. Predictive validity: Those treatments
   effective in the ameliorating (or preventing)
   symptoms in humans, should be similarly
   effective in animal models. Conversely,
   those treatments that are ineffective in
   attenuating the human disorder should
   similarly not be effective in an animal
   model.
3. Etiological validity: Conditions that provoke
   the pathology in humans should also do so
   in the animal model.

4. Construct validity: The presumed biological
   processes should be similar in the human
   and animal models.

Meeting all these criteria would be ideal, but
often this isn't possible. We might not know, for
instance, which treatments are effective in the
human condition, and thus the predictive validity can't be met in the animal model. Likewise,
the purpose of a study might be to determine
the neurobiological underpinnings of a disorder, and thus the construct validity of the
model is uncertain. It should also be added that
for ethical reasons it isn't possible to assess the
effects of traumatic events in animals, like those
that humans may unfortunately endure. Thus,
once more, there are limits to what can be
gained from animal research.

Researchers prefer to believe that the mouse
and rat may, within limits, be excellent choices
to model anxiety and depression, particularly
given the range of gene manipulations that are
possible. Thus, it's unfortunate that even
among genetically identical mice, subtle laboratory perturbations, or the source from which
they had been obtained, can dramatically alter
the behavioral phenotypes expressed (Crabbe,
Wahlsten, & Dudek, 1999). Added to this is
that the diet of mice or rats, which can vary
across laboratories, can affect the microbiome
and various hormonal processes. There is also
the issue of selecting the strains that are most
appropriate for whatever phenotype is being
examined. Some strains, for instance, might be
considered appropriate to assess anxiety or
depression, whereas others, for a number of
reasons (including their marked hardiness), are
less useful.

As we'll see in considering pathologies other
than depression, such as diabetes and heart

disease, some laboratory manipulations that
were effective in mice are frequently ineffective
in humans. To be sure, there have been reports
showing genetic similarities between mice and
humans (see the ENCyclopedia Of DNA
Elements; ENCODE), and several published
reports have focused on this issue (see the NIH
report Comparing the Mouse and Human
Genomes), but there are also distinct differences across species. At the end of the day,
mice aren't humans (they aren't even pint-
sized rats), and after multiple generations of
inbreeding, it's even uncertain to what extent
they are guaranteed to be valid representations
of typical mice. Added to this is that laboratory
mice are raised in fairly clean environments
and hence they hardly represent the dirty
world of field mice (or that of humans). In this
regard, wild mice that were similar to laboratory mice in many ways exhibited microbiota
that differed markedly from their laboratory
counterparts. When the laboratory mice

received microbiota from wild mice, they experienced reduced inflammation, and enhanced survival in response to an influenza virus (Rosshart et al., 2017). With these caveats in mind, we can now turn to a description of the neurobiological changes that are introduced by stressful experiences.

Norepinephrine, Dopamine, and Serotonin

Upon first perceiving a potential stressor, an animal (or a human) may be uncertain as to whether this actually represents a threat, whether it can be dealt with, or how long the potential stressor will persist. Thus, in response to threats, rapid monoamine responses ought to occur, and then, as characteristics of the stressor are understood (e.g., controllability, chronicity), these neurochemical responses can be adjusted as needed.

In response to stressors, the synthesis and utilization of serotonin, norepinephrine, and dopamine increases across several brain regions, including the hypothalamus, amygdala, hippocampus, and prefrontal cortex, to name a few. As the intensity of the stressor increases so does the rate of neurotransmitter release (utilization), which is met by an increase of synthesis, and thus the availability of these neurotransmitters does not vary appreciably. If the stressor is relatively intense, and coping methods are unavailable (e.g., in the face of an uncontrollable stressor), the coping burden rests more heavily on biological systems. Thus, monoamine utilization increases markedly, eventually outstripping the rate of synthesis, and the absolute level of the neurotransmitter declines (varying across brain regions), rendering the organism less able to contend with immediate challenges (e.g., Amat et al., 2014).

The neurotransmitter changes elicited by stressors are typically short-lived, normalizing within less than an hour. Stressors that comprise social challenges, such as changes in the social environment or defeat by a conspecific, may have somewhat more persistent effects. It has been suggested that the circuitry associated with certain stressful experiences, particularly social threats, or other innately driven challenges, may be hard wired and might be accompanied by particularly profound and persistent neurochemical alterations (Krishnan & Nestler, 2011).

In addition to the altered neurotransmitter turnover (i.e., the relative rates of utilization and synthesis), several receptor changes occur in response to intense stressors, including down-regulated postsynaptic norepinephrine receptor functioning, possibly reflecting a further adaptation to limit excessive activation of neuronal processes that can be detrimental to well-being. Likewise, an uncontrollable stressor regimen which elicited disturbed behavioral performance was paralleled by altered expression of several serotonin receptors ($5-HT_{1A}$, $5-HT_{1B}$, $5-HT_{2A}$, and $5-HT_{2C}$) within the dorsal raphe nucleus, the site of serotonin cell bodies. In the case of $5-HT_{1A}$ receptors (autoreceptors) of the dorsal raphe nucleus, as well as $5-HT_{2C}$ receptors within the basolateral amygdala, the changes were more readily induced by an uncontrollable than a controllable stressor (Rozeske et al., 2011). Similarly, if rats had first learned that a stressor was controllable, then the behavioral disturbances and the serotonin variations elicited by a later uncontrollable stressor were prevented, attesting to the importance of rats perceiving controllability in determining behavioral and biochemical outcomes. It turns out that exercise in the form of running on a wheel also protects animals against the depressive-like effects of a stressor and precludes the serotonin responses at the dorsal raphe (Clark et al., 2015). It is uncertain whether this was a reflection of exercise enhancing resilience, or a consequence of animals having control over their environment.

Given that these neurotransmitter systems affect one another in significant ways, the functioning of norepinephrine, dopamine, and serotonin may need to be integrated for proper behavioral outputs to occur. By example, norepinephrine neurons of the locus coeruleus, through their actions on ventral tegmental dopamine functioning, may influence reward processes, and hence may promote the anhedonia associated with depressive disorders (Isingrini et al., 2016). Likewise, infusion of a 5-HT_{1A} receptor agonist directly into the prefrontal cortex resulted in stressor-elicited dopamine and norepinephrine changes being attenuated. Several studies have also demonstrated that serotonin activity or activation of particular serotonin receptors, could interact with the neurotrophin BDNF in promoting behavioral changes (Jiang, Jin et al., 2017), and GABA alterations could modify the serotonin response to a stressor (Rozeske et al., 2011). Assessing any particular neurotransmitter is important, but as indicated earlier, understanding psychopathology necessitates analyses of system changes.

Chronic Stressor Challenges

It is remarkable how many different adaptive responses develop concurrently or sequentially, accompanied by varied checks and counterchecks in an effort to maintain allostasis. As a stressful experience continues over an extended period, or is encountered repeatedly, the initial neurochemical changes are followed by a sequence of further adaptive changes that may facilitate coping and limit the development of stress-related illness. To meet the demands imposed, a compensatory increase of monoamine synthesis occurs, likely owing to down-regulation of autoreceptors, which signals that more transmitter should be produced.

Thus, the available neurotransmitter normalizes, and frequently exceeds basal (nonstress) levels. This may be an effective adaptive response, at least in the medium term, as it may facilitate the organism's ability to deal with ongoing challenges. Yet, as indicated in our discussion of allostatic overload, when a neurochemical system is overly active, negative secondary effects may develop. Perhaps as a protective response, further postsynaptic receptor changes are engendered, possibly modulating the actions of the high levels of neurotransmitter released. These adaptive neurochemical changes vary across strains of mice, but seem to be less robust in highly stressor reactive mouse strains, and in older mice.

Should adaptation develop in response to a chronic stressor regimen, this may be unique to a particular challenge, and should a novel stressor be encountered, neurochemical systems may be hyper-responsive. Furthermore, if a stressor occurs on a sufficiently protracted basis, an excessive load may be placed on critical systems so that negative outcomes may develop, including the development of several illnesses. It is important to consider that not all neurochemical systems operate identically, so that an adaptation may develop in one system, but this might not occur as readily within a second system or a second brain region.

Chronic stressor that are unpredictable and vary from day to day, are less likely to result in adaptive changes that preclude excessive neurochemical changes and pathology related to allostatic overload is more likely to develop. Although, these outcomes are most likely to occur with relative intense stressors, they have also been witnessed in response to a chronic mild unpredictable stressor regimen. In response to this challenge depressive-like symptoms can be provoked, as can cognitive and learning disturbances measured 1-month later. These actions were accompanied by cell loss within the dorsal raphe nucleus (the site of serotonin cell bodies) and projections to the prefrontal cortex, possibly being mediated by glutamate processes (Natarajan, Forrester, Chiaia, & Yamamoto, 2017). The actions of a chronic mild stressor regimen aren't as reliable as one would like, but the source for the different outcomes isn't certain (Willner, 2016).

The scenario outlined here is meant to provide a general schema as to what occurs in response to stressors, and it is not the case that each of the monoamines behaves identically under all conditions. Nonetheless, such changes have been observed in relation to monoamines in brain regions involved in appraisals and executive functioning (prefrontal cortex), feelings of reward (nucleus accumbens), as well as vegetative or basic life processes (e.g., specific hypothalamic nuclei) (Anisman et al., 2008; Bland, Twining, Watkins, & Maier, 2003). Reiterating the obvious, but worth repeating, the stress burden carried by individuals is moderated by age, gender, genetic, experiential, and psychosocial factors. Moreover, the response to stressors may also be moderated by the individual's life-styles and health behaviors, including diet, physical activity, sleep, and substance abuse.

Previous Stressor Experiences: Stress Sensitization

Even if a stressor experience lasts for only a short time, activation of multiple neuronal systems will follow within minutes, and some of these actions may persist for weeks (Musazzi, Tornese, Sala, & Popoli, 2014). Likewise, the rumination that follows a stressor can have significant and lasting consequences, and stressors may also promote secondary effects (e.g., changes in life-style, employment) that can have downstream consequences, particularly when reminders of the initial trauma reappear. Indeed, some events, such as being publicly shamed or humiliated, could jeopardize

self-esteem and self-efficacy, thereby affecting how subsequent stressors are dealt with. For that matter, the brain's exceptional neuroplasticity that serves us so well in acquiring and maintaining information, can work against us when the stressor is one that undermines our self-perceptions, and is incorporated into one's self-schema.

It is tempting to assume that once neuronal activity has normalized, the stressor's actions will have similarly ended. However, as discussed in relation to behavioral processes, having been exposed to a stressor, the potential for altered neuronal functioning may be altered so that neurotransmitter release or receptor processes are more readily induced (sensitization) by later stressor encounters. This occurs even if the second stressor experience is different from the one initially experienced (Anisman, Hayley, & Merali, 2003). Indeed, "cross-sensitization" effects occur in that a stressor experience augmented responses to later exposure to drugs, such as amphetamine and cocaine that ordinarily affect monoamine functioning (Anisman et al., 2003) and similar actions have been reported in relation to immune challenges (Tilders & Schmidt, 1998). A frequent characteristic of these sensitized responses is that they might not be immediately evident, but appear and grow with the passage of time, at least within a 3 weeks period (Schmidt, Aguilera, Binnekade, & Tilders, 2003). The sensitization associated with a stressor may involve a psychological component. Specifically, paralleling the "immunization" effects observed in behavioral studies, among animals initially trained in a controllable stressor situation, the serotonin changes that would otherwise be elicited by a subsequent uncontrollable challenge were precluded, although this was not the case for the dopamine changes (Bland, Twining, Watkins, & Maier, 2003).

Several stress-related processes are subject to sensitization. For instance, variations of

CRH$_2$ receptors may contribute to the sensitization of serotonin activity within the raphe nucleus, whereas the cross-sensitization effects that have been reported between stressors and amphetamine might be mediated by glutamate functioning. A reasonable possibility is that the sensitization stems from a time-dependent increase of certain types of receptors within the amygdala (notably glutamate receptors), and blocking these receptors at the time of the initial stressor experience diminished the amygdala changes (Yasmin, Saxena, McEwen, & Chattarji, 2016). Further, as already mentioned, cytokine processes likewise appear to be influenced by sensitization processes, which likely involve still other mechanisms, possibly being related to microglial functioning. The essential point is that the impact of stressors may continue long after the initial challenge has passed.

γ-Aminobutyric Acid

With so much information coming in at any given moment, especially under distressing circumstances, the brain needs to distinguish between relevant versus irrelevant information, and even different forms of relevant information. The inhibitory neurotransmitter GABA, it will be recalled, serves as a brake to regulate neuronal activity within other systems, and thus plays an essential function in information gating (Yang, Murray, & Wang, 2016).

Within the hippocampus and amygdala, GABA levels and the conversion of glutamate to GABA (reflected by elevated functioning of the enzyme GAD65) were typically altered in response to stressors. Within the basolateral amygdala, such effects occurred irrespective of whether the stressor was controllable or uncontrollable, but in a portion of the hippocampus, namely the dentate gyrus, GAD65 was differentially affected by stressor controllability (Hadad-Ophir et al., 2017).

Following a chronic mild stressor regimen that engendered depressive-like behavior, GABA functioning was reduced in the cortex. This outcome was accompanied by variations of GAD67, the vesicular GABA transporter, and the GABA transporter-3 (Ma et al., 2016), as well as a reduction in the number of GABA neurons within the orbitofrontal cortex (Varga, Csabai, Miseta, Wiborg, & Czéh, 2017). Beyond these important effects on mood disorders, GABA appears to be associated with other stressor-related conditions, including inflammatory-related illnesses, such as gastrointestinal disorders and visceral pain (Jembrek, Auteri, Serio, & Vlainić, 2017).

The influence of GABA on behavioral outputs varies with the type of GABA receptor that is stimulated (i.e., GABA$_A$ vs GABA$_B$). In the case of stressors, particular attention has focused on changes of GABA$_A$, which are influenced by chronic challenges (Poulter, Du, Zhurov, Merali, & Anisman, 2010). The characteristics of GABA$_A$ receptors differ from that of many other receptors. This receptor comprises a pentameric (5 subunit) protein complex from a set (cassette) of 21 different proteins/genes. Thus, the specific subunits that can make up this complex is enormous, and they can often be distinguished from each other based on their conformation and the resulting docking sites for the neurotransmitter or other binding proteins (e.g., Fritschy & Brünig, 2003). The specific subunits that comprise a given receptor will determine what stimuli will excite it (e.g., benzodiazepines will only affect receptors with particular conformations). Likewise, stressful events may affect subunits that make up the GABA$_A$ receptor (Poulter et al., 2010), and GABA$_A$ receptor subunit mRNA expression can be either up- or down-regulated in several brain regions of depressed individuals (Choudary et al., 2005; Merali et al., 2004). Predictably, treatments that affect specific subunits of the GABA$_A$ receptor can attenuate the behavioral and emotional effects of stressors (Piantadosi et al., 2016).

The characteristics of GABA receptors are also subject to effects stemming from epigenetic and negative early experience. In this regard, early life stressors or those encountered during adolescence, affected GABA functioning measured in subsequent adulthood, thereby influencing the course of anxiety (Jacobson-Pick, Audet, McQuaid, Kalvapalle, & Anisman, 2012). At the same time, it is important to recognize that GABA$_A$ functioning generates a state-dependent fear/anxiety response in that these responses vary with the animal's internal state, which can be influenced by contextual cues (Radulovic, Jovasevic, & Meyer, 2017).

Imaging studies in humans indicated that in response to threats, GABA activity declined, which fits with the finding that GABA concentrations were reduced within the frontal cortex of depressed individuals. In view of the influence of stressors on GABA functioning in animals, and that GABA manipulations can influence both anxiety- and depressive-like behaviors, it has been assumed that the link between GABA functioning and mood states is a causal one. This does not belie the common view that serotonin, corticotropin-releasing hormone (CRH), and other transmitters play a role in these conditions, as GABA may serve in a regulatory capacity in affecting these factors. Furthermore, GABA$_A$ receptor expression, and responses to stressors, can be affected by ovarian hormones, such as progesterone, and thus might be key in accounting for sex differences in mood disorders.

Glutamate

Excitatory glutamate functions are moderated by the inhibition produced by GABA, and the two operate together to maintain behavioral functioning. When an imbalance occurred between glutamate and GABA inputs within the prefrontal cortex, neuronal atrophy and loss of synaptic connections were apparent, thus engendering cognitive disturbances and depression (Ghosal, Hare, & Duman, 2017). Among its other actions, glutamate influences synaptic plasticity and thus affects learning and memory, and likely contributes to the development of schizophrenia and depressive illness (e.g., Sanacora, Treccani, & Popoli, 2012). Stressor experiences readily provoke glutamate release, thereby contributing to the establishment and retention of fear responses as well as stressor-related psychopathologies.

Chronic stressors produced especially marked variations of glutamate and GABA that varied across brain regions. Among rodents that were exposed to chronic social defeat, a depressive-like behavioral profile emerged, together with alterations of glutamate and the glutamate-glutamine cycle (glutamine is catalyzed to glutamate by glutaminase) (Rappeneau, Blaker, Petro, Yamamoto, & Shimamoto, 2016). Following a chronic stressor, glutamate levels and glutaminase activity were reduced in the cortex and cerebellum, but increased in the hippocampus and striatum, whereas GABA levels increased in the cortex and hippocampus, but declined in the striatum (Kazi & Oommen, 2014). These findings are generally in line with the view that stressful events affect GABA-glutamate balances, but these actions are brain-region specific. Based on electrophysiological findings, it was concluded that a chronic stressor regimen produced inhibition of glutamatergic output neurons within the prefrontal cortex, thereby diminishing the contribution of this brain region to stress reactivity (McKlveen et al., 2016). Stressors similarly influence dopamine receptors located on GABA$_B$ interneurons within the prefrontal cortex, thus affecting glutamate functioning (Lupinsky, Moquin, & Gratton, 2017).

Consistent with the role of stressors on anxiety, a restraint stressor influenced glutamate and cytokine expression within the hippocampus of mice, but the observed effects differed between stressor-sensitive and relatively resilient strains (Sathyanesan, Haiar, Watt, & Newton, 2017). Like these strain-specific effects, a sub-chronic stressor that selectively provoked depressive-like effects was also linked to variations in the frequency of particular glutamatergic inputs to the nucleus accumbens, and could thus influence dopamine functioning within this region (Brancat et al., 2017).

Glutamate receptors comprise two broad classes, ionotropic and metabotropic, each of which has several receptor subtypes. Within the ionotropic family, N-methyl-D-aspartate receptor (NMDA) and α-amino-3-hydroxy-5-methyl-4-isoxazolepropionic acid (AMPA) receptors, are activated by stressors (Marrocco et al., 2014), and early life insults persistently alter NMDA-receptors (Ryan et al., 2009). The activation of these receptors can promote opposite effects on neurogenesis and neuronal survival, varying with the synaptic or extrasynaptic concentration of glutamate. Upon exposure to a chronic stressor regimen, glutamate levels rise markedly, and it is conceivable that the depressogenic effects of this treatment stem from excessive stimulation of NMDA receptors (Rubio-Casillas & Fernández-Guasti, 2016). Indeed, at sufficiently high levels, glutamate may be neurotoxic and the resulting cell loss may contribute to psychological disturbances.

These effects of stressors, together with indications that antidepressant medications affect glutamate release (Musazzi, Treccani, Mallei, & Popoli, 2013), encouraged efforts to target glutamatergic processes in the development of treatments for depression. As we'll discuss in Chapter 8, Depressive Disorders, the glutamate antagonist ketamine, can produce rapid antidepressant effects, even among treatment resistant patients (Serafini, Howland, Rovedi,

Girardi, & Amore, 2014). These positive actions were attributed to the downstream effects of the glutamate changes on the neurotrophin BDNF that enhances neuroplasticity (Duman, 2014). If nothing else, these findings indicated that perspectives of depression that strictly focused on monoamines were outdated, and that it would be more appropriate and productive to consider the impact of stressors on neuroplasticity, and perhaps the effects on coping processes and on cognitive flexibility (Zhang et al., 2017). It was indeed suggested that hippocampal neurogenesis influences cognitive flexibility so that individuals are better equipped to deal with changing situations and threats (Anacker & Hen, 2017).

NEUROTROPHINS

As we saw earlier, neurotrophins (growth factors) are fundamental for neuronal plasticity, including the arborization and synaptogenesis that contribute to memory processes and have been connected to stress-related mood disturbances. Acute stressors or even reminders of strong stressors provoked a reduction of BDNF gene expression and protein levels within the hippocampus (Duman & Monteggia, 2006), and were still more prominent in response to a chronic stressor regimen that comprised a series of different challenges (Gronli et al., 2006).

Stressors do not uniformly instigate growth factor reductions across stress-sensitive brain regions, including those that might underlie depression. In fact, several reports indicated that stressful events increased rather than reduced BDNF mRNA expression within portions of the prefrontal cortex (e.g., prelimbic, infralimbic, and anterior cingulate) and the ventral tegmentum, the latter being associated with reward processes. Moreover, unlike several neurotransmitter changes, the increased BDNF expression within the anterior cingulate cortex was more pronounced after controllable than uncontrollable stressors (Bland et al., 2007). These BDNF changes might have reflected the adoption of active coping efforts or to new learning related to the controllability of the stressful situation, rather than directly stemming from the distress created.

That BDNF varies across brain regions is important for understanding its relationship to pathological conditions, and it is possible that stressor-related BDNF changes in different brain sites may be related to diverse emotional and cognitive changes. For instance, hippocampal BDNF might contribute to the memory of negative events, whereas prefrontal cortex BDNF might be related to cognitive changes. It needs to be considered, as well, that other neurotrophins, such as FGF-2, are also influenced by stressful experiences and have been associated with depression (Evans et al., 2004). There is reason to suppose that neural plasticity may be a key factor associated with sensitization effects, much like it influences learning and memory, and might thus contribute to the development of psychopathology. In particular, an initial stressor experience could engender an increase of a growth factor, such as FGF-2, which could influence the neuronal responses elicited upon later stressor encounters (or in response to amphetamine or cocaine), and might thereby encourage psychological disorders (Flores & Stewart, 2000).

There is yet another important consideration that can't go unmentioned. Among mice that had experienced chronic social stress over a 2 week period, neurons born during the last 5 days of the stressor regimen were elevated relative to that which had occurred in nonstressed controls (in fact, the increased neurogenesis was comparable to that seen when mice were raised in an enriched environment). These findings were surprising as chronic stressors typically have been viewed as resulting in diminished plasticity, and diminished neurogenesis might have been expected. However,

in this instance, the analysis comprised cells born during the last 5 days of the social stressor, at which time mice had no longer been attacked, having adopted a submissive position, essentially reflecting adaptation to the otherwise distressing conditions. It was of particular significance, that these new neurons were more likely to be incorporated into the dentate gyrus of the hippocampus during the ensuing 10-week poststress recovery period. When these mice were chronically stressed again, these neurons were more likely to be affected. Among other things, spine density and branching nodes were reduced relative to neurons that had not been born during the stress period, and performance in a spatial learning task was impaired. It was suggested neurons born during a stress period when coping methods were being established, were uniquely adapted to deal with future challenges in the form of social stressors. However, it wasn't certain whether the neurobiological changes rendered animals more vulnerable to psychopathology, or conversely, made animals more resilient (and more able to recover) from further challenges (De Miguel, Haditsch, Palmer, Azpiroz, & Sapolsky, 2018).

HORMONAL CHANGES STEMMING FROM STRESSOR EXPERIENCES

Numerous hormones (see Table 5.2) are responsive to stressors and to immune challenges. As in the case of neurotransmitters, several receptor subtypes can be activated for many of these hormones, and consequently different behavioral outcomes can develop. As life-style factors, such as exercise, can moderate hormonal stress responses, possibly through actions of GABA neuronal activity, it can reasonably be expected that behavioral and physical pathologies will be similarly affected.

As briefly mentioned earlier, numerous hormones influence inflammatory immune

TABLE 5.2   Hormones Related to Stress Responses

Secreted Hormone	Biological Effect	Behavioral Outcome
Corticotropin-releasing hormone (CRH)	Formed in the paraventricular nucleus of the hypothalamus, as well as several limbic and cortical regions. Stimulates ACTH release from the pituitary gland	Involved in stress responses, promoting fear and anxiety, and diminishes food intake and increases metabolic rate
Adrenocorticotropic hormone (ACTH)	Formed in the anterior pituitary gland. Stimulates corticosteroid (glucocorticoid and mineralocorticoid) release from adrenocortical cells	Stress responses elicited are primarily due to actions on adrenal corticoids
Arginine vasopressin (AVP)	Released by both the paraventricular and supraoptic nucleus: promotes water reabsorption and increased blood ACTH	Together with CRH, may synergistically increase stress responses. Influences social behaviors
Cortisol (corticosterone in rodents)	Released from the adrenal gland. Has anti-inflammatory effect, promotes release and utilization of glucose stores from liver and muscle, and increases fat storage	Prototypical stress hormone; influences defensive behaviors, memory processes, caloric intake, and may promote preference for high calorie foods under stressful circumstances (stimulates consumption of comfort foods)
Mineralocorticoids (e.g., aldosterone)	Released from the adrenal gland. Stimulates active sodium reabsorption and passive water reabsorption, thus increasing blood volume and blood pressure	Aldosterone influences salt and water balance. Excessive sodium and water retention leads to hypertension. Low levels of aldosterone lead to a salt-wasting condition evident in Addison's disease.
Epinephrine (adrenaline) and norepinephrine (noradrenaline)	Produced in the adrenal gland (medulla) and within sympathetic neurons; increases oxygen and glucose to the brain and muscles; promotes vasodilation, increases catalysis of glycogen in liver and the breakdown of lipids in fat cells; increases respiration and blood pressure; suppresses bodily processes (e.g., digestion) during emergency responses; influences immune system activity	Elicits fight or flight response; in the brain, EPI and NE have multiple behavioral actions related to defensive behaviors (e.g., vigilance, attention)
Beta-endorphin	Secreted from several sites, such as the arcuate nucleus	Inhibits perception of pain
Oxytocin	Formed in magnocellular neurosecretory cells of the supraoptic and paraventricular nuclei of the hypothalamus	Associated with parental bonding, ingroup bonding, maternal behaviors, and a wide range of prosocial behaviors, as well autism, depression, fear, and anxiety

From Anisman, 2016.

processes, and conversely, immune and cytokine changes markedly affect hormonal processes. For instances, cortisol at physiological levels can limit immune and cytokine elevations that could otherwise be instigated (e.g., by stressors), and at pharmacological levels (i.e., at doses greater than those produced naturally) immune functioning is suppressed. Conversely, treatments that increased circulating IL-1β levels engender marked cortisol elevations. Thus, in considering stressor-provoked pathology, particularly those involving inflammatory processes, it is necessary to consider that multidirectional links exist between diverse processes.

Within a matter of seconds following the appearance of a potential threat, several brain regions, such as the prefrontal cortex, may be activated in order for primary appraisals of the situation to be made, and presumably, secondary appraisals involved in the selection of coping strategies follow quickly afterward. Within moments of stressor onset, autonomic nervous system functioning is set in play so that various organs are altered in preparation for the challenge. Peripheral epinephrine release promotes increased heart rate and blood pressure, consequently providing elevated oxygen to the brain and peripheral organs, and stimulates the release of immune cells from lymphoid organs so that they're in circulation, where they might be needed.

Corticoid Responses

Stressors ordinarily promote the activation of several brain regions, such as the prefrontal cortex and amygdala, which are involved in appraisal of stressful events. When an event is judged to be a threat, the paraventricular nucleus (PVN) of the hypothalamus is stimulated, resulting in the release of CRH from cells at the median eminence (situated at the ventral portion of the hypothalamus). The CRH stimulates the anterior portion of the pituitary gland, provoking the release of adrenocorticotropic hormone (ACTH), which stimulates cortisol (corticosterone in rodents) release from the adrenal gland. Cortisol then enters the bloodstream and comes to affect several of target organs, producing multiple adaptive effects essential for survival. The cortisol also reaches

the brain where neurons in the hippocampus and hypothalamus are stimulated, which inhibits further CRH release.

Among other things, cortisol functions to increase blood sugar that is derived from lactate and glycogen present in the liver and in muscle, and facilitates the metabolism of fats, carbohydrates and proteins, and moderates the functioning of neuronal activity in several brain regions. More than this, cortisol may have preparatory actions, facilitating coping with impending stressors, and may promote neuroplasticity, thereby affecting processes (e.g., sensitization) that are important in dealing with subsequently encountered challenges (Sapolsky, Romero, & Munck, 2000). Cortisol has permissive actions that contribute to the provocation or amplification of other stressor-provoked hormonal changes, and it can also suppress the actions of other hormones and can diminish or limit the immune and inflammatory activity that might otherwise be detrimental to health. For this reason, several medications that affect or act like cortisol (e.g., hydrocortisone, prednisone, prednisolone, and dexamethasone), are widely used to diminish allergic reactions, suppress inflammatory responses involved in autoimmune disorders, limit tissue rejection that can occur following organ transplants, and limit the graft (the transplanted tissue) from attacking the recipient (i.e., graft vs host disease).

Cortisol regulation not only includes being released in response to significant stimuli, but also the appropriate termination of its release once it has served its immediate function. When circulating cortisol reaches the brain, it influences hypothalamic and hippocampal neurons so that further HPA functioning is inhibited. If this self-regulation does not operate appropriately, persistent release of cortisol can occur, which may have detrimental consequences on immune functioning as well as on brain processes. Specifically, the sustained stimulation of glucocorticoid receptors present on hippocampal cells, may lead to cell damage. As the hippocampus is a fundamental component of the negative feed-back loop, HPA activity may persist, leading to still further cell damage, and eventually the development of cognitive disturbances. Given that hippocampal cell loss and elevated cortisol levels ordinarily increase with age, potentially being exacerbated by chronic stressors, the risk for cognitive impairments is that much more pronounced (McEwen & Gianaros, 2011).

Like so many other neurobiological systems, stressors encountered early in life or during the prenatal period may have long-term implications for later stressor provoked corticoid responses. These sensitization-like effects may occur owing to the plasticity of stress-relevant neural systems. Alternatively, such effects may be due to epigenetic processes (e.g., Szyf, 2011). Either way, the lasting glucocorticoid reactivity and receptor sensitivity can promote persistent changes of immune functioning upon reexposure to a stressor.

Impact of Chronic Stressors

With repeated stressor exposure, particularly if the stressor is a homogenous one (i.e., the same stressor is administered on successive days), the extent of cortisol released declines (Anisman et al., 2008). Once again, this seeming "adaptation" should not be misconstrued as reflecting the system being unable to respond to a stressor, but instead reflects a diminished response to specific stimuli. Indeed, after repeated exposure to a given stressor, subsequent exposure to a novel challenge promotes an exaggerated corticosterone response (Uschold-Schmidt, Nyuyki, Füchsl, Neumann, & Reber, 2012). This would seem to be a highly adaptive response. It would, after all, be disadvantageous for an adaptation to develop to all stressors following repeated encounters with one type of stressor. Thus, the

HPA system remains prepared to deal with novel insults, despite the decline in reactivity to a particular stressor.

Aside from the diminished levels of corticosterone that may accompany chronic stressor experiences, it also seems that glucocorticoid receptors may become progressively less sensitive (glucocorticoid resistance) (Avitsur, Stark, Dhabhar, Padgett, & Sheridan, 2002). Adequate glucocorticoid functioning is fundamental in controlling (limiting) autoimmune, infectious, and inflammatory disorders, as well as some psychiatric conditions (e.g., depression), and hence the downregulation of corticoid sensitivity may contribute to aggravation of illness symptoms. In addition to the contribution of chronic stressors, persistent activation of inflammatory cytokines and downstream signaling pathways (e.g., mitogen-activated protein kinases, NF-kB, signal transducers and activators of transcription, and cyclooxygenase) may contribute to the diminished glucocorticoid receptor functioning (Pace, Hu, & Miller, 2007) and could affect illness vulnerability. Findings such as these make it clear that a more detailed understanding is necessary regarding the links between cytokines and glucocorticoid signaling in the hope of developing novel strategies to deal with the impacts of glucocorticoid resistance.

Stressor-Provoked Cortisol Changes in Humans

There is a serious disconnect between the impact of stressors in rodents relative to those apparent in humans. Simply taking a rodent out of its cage and placing it in a novel chamber can elicit a 100% increase of corticosterone, and stronger stressors, can produce an increase of 400%−800%. These strong effects of stressors may have given rise to the erroneous expectation that this would also occur in humans, but the changes observed are typically relatively modest. Academic examinations, for instance, hardly affects cortisol levels, and even some fairly powerful stressors (e.g., in blood samples taken just prior to heart surgery) engendered relatively modest (30%−50%) cortisol elevations (Michaud, Matheson, Kelly, & Anisman, 2008). In fact, diurnal cortisol variations are more pronounced than those associated with many challenges.

In contrast to these anticipatory stressors, pronounced cortisol increases are evident in the "Trier Social Stress Test" (TSST), which comprises a public speaking and an arithmetic challenge. It was suggested that this outcome is generated because the stressor involves a "social evaluative" threat that promotes shame (Dickerson & Kemeny, 2004) or anger (Matheson & Anisman, 2012). As expected, having social support available may limit cortisol changes ordinarily elicited by some stressors, but in the context of social evaluative threat the cortisol changes may occur irrespective of the presence of social support, suggesting that the individual's concerns over being evaluated are especially germane and are not overridden simply by having support present (Taylor et al., 2010). Indeed, in some instances (e.g., when a male partner is present to support a woman) cortisol reactions were more pronounced than in the absence of this support, pointing to the source of support also acting as an evaluative threat.

Cortisol in Relation to Traumatic Experiences

Like the cortisol response that accompanies a chronic stressor, traumatic events that lead to PTSD were associated with cortisol levels that were diminished relative to that of nonstressed individuals or those that had encountered trauma but had not developed PTSD (Yehuda, 2002). The diminished cortisol levels may reflect a protective response to limit the potentially damaging effects that could be elicited by persistent rumination and memories that could readily be rekindled by a variety of triggering cues (see Chapter 10: Posttraumatic Stress Disorder). This seemingly protective action should not be misinterpreted as suggesting that HPA action is down-regulated in all situations. Instead, the diminished HPA response is selective, so that effective activation occurs when this is necessary.

Ordinarily, injection of CRH promotes ACTH release from the pituitary, and the resulting increase of circulating levels of this hormone promotes cortisol release from the adrenal gland. However, among depressed women who had experienced abuse when they were young, the ACTH release elicited by CRH administration was diminished, indicating that HPA functioning was down-regulated. However, when these women experienced a stressor in the form of a social evaluative threat (public speaking), the ACTH release response was exaggerated (Heim, Newport, Mletzko, Miller, & Nemeroff, 2008). It was similarly reported that cortisol levels were low among women who had experienced PTSD symptoms related to psychological and/or physical abuse in a dating relationship. However, upon encountering reminders of their abuse, cortisol levels markedly increased (Matheson & Anisman, 2012). It seems that strong stressors or those experienced when young, can promote down-regulated HPA functioning among individuals with PTSD, which ought to be protective given that hyper-reactivity to environmental cues may result in adverse effects (e.g., on hippocampal cells). At the same time, indiscriminate down-regulated HPA functioning might be maladaptive, as activation of this system might be necessary to deal with challenges. Thus, in response to potentially meaningful stressors, activation of brain regions (e.g., prefrontal cortex or amygdala) involved in stressor-appraisal processes, may over-ride the HPA blunting, and an exaggerated ACTH or cortisol response will be apparent.

DISCRIMINATION AS A CHRONIC STRESSOR

It won't be surprising to anyone that racial discrimination is a powerful stressor that undermines psychological and physical health in adults and children. Likewise, cortisol increases can be elicited by threats to an individual's social identity, as observed in response to acute racial discrimination (Matheson & Anisman, 2012). As with other chronic challenges, chronic racism was accompanied by flattening of the diurnal cortisol profile, much as it does with chronic or severe stressors that lead to PTSD (Adam et al., 2015). Although this cortisol profile could be attributed to a PTSD-like effect, it had been suggested that the acute awareness of racism that occurs in daily life might result in individuals developing protective mechanisms that act against stigmatization and hence the cortisol down-regulation. For instance, attributing the behavior of others to racist attitudes, might protect individuals from concluding that negative behaviors directed at them are due to their own shortcomings.

Aside from cortisol variations, interpersonal racism was accompanied by elevated levels of circulating proinflammatory cytokines, although this was not apparent if individuals had positive racial identities (Brody, Yu, Miller, & Chen, 2015). Perceived discrimination was also accompanied by emotional dysregulation in the form of venting and denial, which were linked to the presence of chronic inflammation reflected by elevated IL-6 and C-reactive protein (Doyle & Molix, 2014). Effects such as these also occurred within the brain, as perceived discrimination among ethnic minority members was associated with activation of the perigenual anterior cingulate cortex and ventral striatum, which are associated with decision-making and reward processes, respectively (Akdeniz et al., 2014).

The pervasive effects of chronic racism, not unexpectedly, are also manifested in the form of hypertension, blood pressure, and cardiovascular reactivity, shortened telomere length (Chae et al., 2014), and increased health risk, particularly mental health problems (Wallace, Nazroo, & Bécares, 2016). As indicted by Berger and Sarnyai (2015), the effects of discrimination are exceptionally powerful, and have effects that are "more than skin deep."

The Morning Cortisol Response

Like many hormones, levels of circulating cortisol vary over the course of the day, being high in the morning, after which a decline occurs over the course of the afternoon and evening, eventually reaching lowest levels at about 2300 hours. Significant cortisol changes also appear soon after awakening. During the first 30 minnute, cortisol levels (usually measured in saliva that participants take on their own at home) rise by about 40%, and then decline rapidly (Schmidt-Reinwald et al., 1999). The initial cortisol rise is more pronounced among individuals who are experiencing ongoing psychological challenges, and the subsequent cortisol decline may occur relatively slowly in those experiencing distress, as well as among individuals with low social positions or experiencing poor health.

The initial enthusiasm in assessing the morning cortisol response in relation to life stressors seems to have waned, as it is procedurally problematic, as participants might not be reliable in taking saliva samples immediately after awakening and again precisely 30 minutes afterward. Still, the studies that have been reported revealed a tie between feelings of distress and the morning cortisol rise. In contrast, morning cortisol levels are diminished among individuals who had been undergoing a chronic stressor or those that had encountered a traumatic stressor that provoked PTSD (Michaud, Matheson, Kelly, & Anisman, 2008). Once again, the flattening of the cortisol curve might reflect an adaptive response to preclude the adverse effects that could otherwise emerge.

Despite the problems inherent in determining the morning cortisol response, interesting data have been obtained through this procedure. For example, the serotonin transporter gene was associated with the variations of the morning cortisol response (Li-Tempel et al., 2016), and were marked in stressed elderly individuals (Ancelin et al., 2017). This particular polymorphism may be relevant to depressive illness, which we'll consider to a greater extent when we examine factors related to depression in Chapter 8, Depressive Disorders.

Corticotropin Releasing Hormone

As described earlier, the release of CRH from the PVN of the hypothalamus sets in motion HPA hormonal variations. With chronic stressor experiences, several further hormonal changes develop, which might serve to maintain integrity of neurobiological processes, although they might also reflect the failure of adequate neurobiological functioning. With repeated variable stressor exposure, arginine vasopressin (AVP) is increased within CRH nerve terminals, and the co-release of CRH and AVP synergistically increases ACTH release (Schmidt et al., 2003). In addition, with a chronic stressor, the density of norepinephrine and glutamate receptors is elevated on CRH cell bodies and dendrites, promoting greater downstream effects (Herman & Tasker, 2016), likely contributing to the sensitization of HPA axis functioning.

Aside from the involvement of CRH in HPA functioning, this peptide is present in several other brain regions that are fundamental for stressor-related behavioral changes. Specifically, CRH activity within prefrontal cortical regions may contribute to stressor-related appraisals and decision making, while amygdala and hippocampal CRH changes may be involved in fear memories. Stressor-related CRH activity can also affect locus coeruleus activity, potentially influencing vigilance in threatening situations. Amygdala variations of CRH are exceptionally well documented, and changes of this peptide in different portions of the amygdala contribute to the acquisition, expression, and extinction of fear responses, as we'll see when we discuss anxiety disorders (LeDoux, 2000) in Chapter 9, Anxiety Disorders. Relatedly, the bed nucleus of the stria terminalis (a portion of the extended amygdala), has been implicated in the development and maintenance of anxiety, possibly by acting together with GABA changes (Partridge et al., 2016).

Cannabinoids

There has been growing interest in the use of cannabis (marijuana) to diminish pain perception, symptoms of physical illness, and nausea resulting from cancer treatments, as well as stress-related psychological disorders, such as PTSD (e.g., Berardi, Schelling, & Campolongo, 2016) and anxiety disorders (Blessing, Steenkamp, Manzanares, & Marmar, 2015). It should, however, be mentioned that cannabis could also provoke adverse actions, such as dizziness, dry mouth, fatigue, somnolence, euphoria, disorientation, drowsiness, confusion, loss of balance, and hallucination (Whiting et al., 2015), as well as more serious problems, ranging from dependence, respiratory problems, psychosis, and even cancer development (Andrade, 2016).

The psychoactive component of cannabis, Δ9-tetrahydrocannabinol (THC), and naturally occurring molecules (eCBs) bind to specific cannabinoid receptors (e.g., CB_1 and CB_2), which are thought to contribute to the medical benefits of THC. When administered directly into the basolateral amygdala, CB_1 agonists diminished stressor-induced HPA changes (Ramikie & Patel, 2012). There have also been indications that CB_1 receptors within the lateral habenula (which receives input from lateral septum, hippocampus, nucleus accumbens, thalamus, lateral hypothalamus, and the raphe nucleus), through actions on acetylcholine and serotonin, might also be involved in the expression of memories involving aversive events (Dolzani et al., 2016). As activation of CB_1 receptors can also influence the release of stress-sensitive neurotransmitters, the position was adopted that CB_1 receptors might represent a novel drug target to attenuate anxiety-related disorders.

The naturally occurring brain eCBs, anandanine (AEA), and 2-arachodonoylglycerol (2-AG), have been implicated as key players in basal and threat-elicited anxiety. While AEA is believed to mediate basal feelings of calmness as well as anxiety associated with stressor experiences, the function of 2-AG, which is activated somewhat later, is to turn off stress responses. From this perspective, stress responses don't simply whither and fade with the passage of time, but seem to involve an active process governed by 2-AG functioning. Treatments that increased 2-AG availability increased stress resilience, even in mice that were genetically engineered to exhibit high levels of anxiety (Bluett et al., 2017), implicating 2AG functioning as a potential target in the treatment of anxiety-related conditions.

Estrogen

Some of the behavioral and neuroendocrine changes elicited by stressors in rodents, such as corticoid elevations, are greater in females than in males, possibly owing to the influence of gonadal hormones (Toufexis, Rivarola, Lara, & Viau, 2014). Consistent with these changes, neuronal functioning in brain regions that govern behavioral and cognitive responses to stressors (e.g., various cortical sites) are decidedly more pronounced in female than in male rodents, and these effects may vary over the estrous cycle. As expected, cognitive impairments elicited by stressors in female rats were abrogated by estrogen applied to the prefrontal cortex (Yuen, Wei, & Yan, 2016). Stressors also have more pronounced effects within the locus coeruleus-norepinephrine system among females, which is accompanied by hyperarousal that may be relevant to the elevated occurrence of PTSD in women relative to men (Bangasser, Wiersielis, & Khantsis, 2016).

To a considerable degree, the greater effects of stressors in female rodents are recapitulated in humans. These stressor actions are dependent on the estrous phase in that stressor-provoked cortisol responses are especially notable during the latter part of the menstrual cycle (luteal phase), a time during which progesterone is particularly high, and varies following menopause. In line with such findings, among women using oral contraceptives the influence of a social-evaluative stressor on cortisol release was diminished (Kudielka & Kirschbaum, 2005).

In addition to the neurobiological differences elicited by stressors in males and females, proceptive behaviors (e.g., courting) are more affected by stressors in females who become less responsive to overtures (receptivity), although male behavior is also influenced, but less so. In females, the stressors may also influence reproduction as a result of disrupted ovulation coupled with uterine alterations that impede implantation of a fertilized egg (Wingfield & Sapolsky, 2003). By promoting the secretion of the opioid peptide β-endorphin, stressors increase the availability of gonadotropin inhibitory hormone (GnIH), which diminishes luteinizing hormone release, consequently promoting infertility. As expected, knocking out the gene for GnIH eliminated the reproductive failure otherwise produced by a stressor (Geraghty et al., 2015). As described in Table 5.3, several other sex-related hormones are affected by stressors, which have considerable bearing on mental and physical health.

Prolactin

Prolactin, which is essential for milk production, is affected by stressors and may disturb

TABLE 5.3   Sex Hormones

Secreted Hormone	Biological Effect	Behavioral Outcome
Testosterone	Male steroid hormone produced in the testis in males and ovaries in females. To a lesser extent it is produced in adrenal glands. Involved in the development and sexual differentiation of brain and reproductive organs; fundamental in secondary sexual features, including body hair, muscle, and bone mass	Associated with sexual behavior and libido. Linked to aggressive and dominant behaviors
Dehydroepiandrosterone (DHEA)	In males, produced in adrenals, gonads, and brain. Acts as an anabolic steroid to affect muscle development	Acts like testosterone. Has been implicated in maintaining youth
Estrogens (estrone, estradiol, and estriol)	Estradiol is the predominant form of the three estrogens produced in the ovaries. It is the principal steroid regulating hypothalamic-pituitary ovarian axis functioning. It is involved in protein synthesis, fluid balances, gastrointestinal functioning and coagulation, cholesterol levels, and fat depositions. It affects bone density, liver, arterial blood flow, and has multiple functions in brain	Influences female reproductive processes, and sexual development; important for maternal behavior, maintaining cognition, as well as anxiety and stress responses
Progesterone	Formed in the ovary; precursor for several hormones; involved in triggering menstruation, and for maintaining pregnancy (inhibits immune response directed at embryo); reduces uterine smooth muscle contraction; influences resilience of various tissues (bones, joints, tendons, ligaments, and skin)	Influences female reproductive processes, and sexual development; affects maternal behavior, disturbs cognitive processes. Has antianxiety actions
Luteinizing hormone (LH)	Produced in the anterior pituitary gland. In females, an "LH surge" triggers ovulation and development of the corpus luteum, an endocrine structure that develops from an ovarian follicle during the luteal phase of the estrous cycle	Behavioral changes associated with estrogen or testosterone are elicited indirectly through actions on other steroids

(Continued)

TABLE 5.3 (Continued)

Secreted Hormone	Biological Effect	Behavioral Outcome
Follicle stimulating hormone (FSH)	Secreted from the anterior pituitary gland; regulates development, growth, pubertal maturation, and reproductive processes. Together with LH, it acts synergistically in reproduction and ovulation	Behavioral changes associated with estrogen or testosterone are elicited indirectly through actions on other steroids
Prolactin	Secreted from the anterior pituitary; regulated by the arcuate nucleus of the hypothalamus	Involved in lactation in mammals, and involved in sexual behavior and gratification, influences levels of estrogen and progesterone. Regulates immune functioning, and acts like a growth-, differentiating- and antiapoptotic factor

maternal behaviors. Under especially threatening conditions, reduced prolactin may limit attention to the offspring, and thus may favor parental survival (Angelier & Chastel, 2009). At the same time, during the period before or a shortly after delivery, changes of prolactin or its receptors may diminish neuroendocrine and behavioral stress responses that would otherwise negatively affect the offspring. In addition to the peripheral actions of prolactin, receptors for this substrate are present within the central amygdala and the nucleus accumbens, supporting the possibility that this hormone contributes to emotional responses.

Oxytocin

Oxytocin has received considerable attention given its proposed contribution to a variety of prosocial behaviors. Among other things, oxytocin has been linked to social and pair bonding, generosity, trust, altruism, attention to positive cues, and mood (Insel & Hulihan, 1995; Taylor et al., 2000). It was also suggested that oxytocin might play a significant role in coping with stressors, especially those that involve social processes (McQuaid, McInnis, Abizaid, & Anisman, 2014; McQuaid et al., 2016), and its administration limited stressor-provoked responses, such as cortisol release (Cardoso, Kingdon, & Ellenbogen, 2014). Conversely, having social support has been associated with stress-buffering actions, which may have been tied to oxytocin functioning (McQuaid et al., 2016). In mice, administration of oxytocin attenuated the behavioral changes induced by a stressor, but this only occurred in male mice, supporting the view that this hormone contributes to sex-dependent behavioral outcomes (Steinman et al., 2016; Taylor, 2000).

It was maintained that oxytocin might diminish the impact of stressors by promoting social coping. In fact, among humans treated with oxytocin, cortisol stress reactivity was reduced in response to an interpersonal stressor (Cardoso et al., 2014) and anxiety was diminished, even in people with low emotional regulatory abilities. Findings such as these have reinforced the view that oxytocin might effectively be used in attenuating stress-related psychological disturbances.

Tend and Befriend Versus Tend and Defend

Oxytocin might play different roles in males and females given that oxytocin was germane to maternal bonding experiences. Specifically, acting together with opioid peptides and gonadal hormones, oxytocin was seen as promoting "tend-and-befriend" characteristics in females, including nurturing behaviors and the development and maintenance of social connections, which serves to augment self-protection (Taylor et al., 2000). The prosocial behaviors attributed to oxytocin are not only apparent in females, but occur in males as well, although there were several distinguishing features evident between the sexes. Males were more likely to engage in an evolutionarily advantageous "tend-and-defend" characteristic. In addition to altruism in which individuals engage in behaviors to facilitate and encourage the well-being of group members, males also display parochial altruism that involves support of ingroup members, coupled with defense (and warning) in relation to outgroup members (De Dreu, 2012). Oxytocin in males thus elicits cooperation, and has also been tied to defense-motivated noncooperative attitudes and behaviors with the aim of supporting members of their own group.

A Social Salience Perspective

A related view concerning the function of oxytocin is that in addition to serving as a prosocial hormone, it acts to enhance sensitivity to social stimuli. Thus, the influence of oxytocin would depend upon the context in which oxytocin changes occurred. When oxytocin levels were elevated, positive social interactions would be seen as especially positive and significant, whereas negative social interactions would be viewed as being more negative (Ellenbogen, Linnen, Cardoso, & Joober, 2013). Accordingly, in the context of negative events, such as early life mistreatment or neglect, the presence of adequate oxytocin functioning might favor damaging outcomes. In fact, among individuals carrying a polymorphism of the oxytocin receptor, which would be accompanied by diminished social sensitivity, the negative effects of early life abusive events were not as pronounced as they might otherwise have been (McQuaid et al., 2014). In line with this suggestion, in the presence of genes

linked to high oxytocin levels, interpersonal stressors predicted elevated depression, whereas a similar outcome was not observed among individuals carrying a polymorphism associated with low levels of this hormone (Tabak et al., 2015). Further to this, among women who had received oxytocin through a nasal spray, forgiving attitudes toward a breach of trust were less prominent, likely because they viewed betrayal as being more profound. Thus, among individuals who had normal levels of oxytocin, administration of this hormone might cause individuals to be overly sensitive to negative social. It might be expected that if oxytocin were not disturbed among depressed individuals, increasing its levels through exogenous administration might aggravate illness features through elevated sensitivity wherein even neutral social stimuli might be more likely to be negatively misinterpreted. From this perspective, although oxytocin could potentially be useful in treating depressive illness, this would depend on the basal levels of this hormone, as well as the functioning of other neurotransmitters (e.g., dopamine) associated with reward and social processes (McQuaid et al., 2014).

Neuropeptide Y

Like many other hormones, Neuropeptide Y has multiple functions, although it is well known for its role as a vasoconstrictor, in diminishing pain perception, affecting circadian rhythms, and in affecting food intake and storage of energy as fat (Abizaid & Horvath, 2008). As well, it serves to diminish anxiety and stress, and is thought to be an important contributor to resilience (Russo, Murrough, Han, Charney, & Nestler, 2012). Anxiety levels were elevated in mice genetically engineered to lack NPY receptors, and in Chapter 10, Posttraumatic Stress Disorder, we'll see that NPY has been implicated as a resilience factor relevant to PTSD.

STRESS AND ENERGY BALANCES

Stressors can either increase or reduce eating, depending on their severity and pattern of appearance. As we described earlier, strong stressors typically reduce eating as it would be dangerous for animals to continue their search for food if a stressor or threat were present. It similarly appears that in response to severe stressors, diminished food consumption occurs among most humans. Mild and moderate stressors, in contrast, may be accompanied by increased food consumption among some people. The linkage between stress and eating is also illustrated by the consistent reports that drugs which reduce anxiety (e.g., benzodiazepines) tend to increase eating, whereas treatments that reduce eating are apt to engender anxiety. In fact, it has been exceptionally difficult to create antianxiety treatments without altering food intake and promoting weight gain. It seems that a set of neurons within the central portion of the amygdala act to encourage eating, and that the receptors involved act in opposition to the stress-activated receptors in this region (Douglass et al., 2017).

Leptin and Ghrelin

Initiation and cessation of eating, as we saw in Chapter 4, Life-Style Factors Affecting Biological Processes and Health, are largely determined by ghrelin and leptin, although other hormones (e.g., neuromedin B and gastrin releasing peptide) may also contribute in this regard. These hormones are affected by stressors and might contribute to symptoms that accompany stress-related disorders, such as anxiety and depression (Andrews & Abizaid, 2014). In this regard, ghrelin activation may diminish anxiety, thereby allowing animals to maintain "appropriate" food-seeking behavior, thus making energy available to contend with challenges (Spencer et al.,

2012). As expected, leptin increases in response to acute stressors, but when this occurs on a sustained basis, "leptin resistance" could develop, so that leptin will fail to signal "stop eating." As cortisol and ghrelin were linked to self-reported anxiety, it was suggested that the interplay between these hormones was responsible for stress-related eating (Patterson & Abizaid, 2013).

Ghrelin or interactions between leptin and ghrelin may also enhance coping and thus act against the development of depressive-like states (Abizaid, Luheshi, & Woodside, 2014). Indeed, in rodents, the depressive-like behavioral disturbances and corticosterone elevations provoked by stressors were largely eliminated by leptin treatment (Stieg, Sievers, Farr, Stalla, & Mantzoros, 2014). This said, in humans, the connection between both leptin and ghrelin and depressive states is less clear (Chuang & Zigman, 2010), particularly as several additional stressor-sensitive hormones, including CRH, cortisol, neuromedin B, and gastrin releasing peptide may act together in affecting mood states. As indicated in discussing CRH_1 and CRH_2 receptors, some hormones may act primarily on the mood features of illness (e.g., eating, sleeping), whereas others might contribute primarily to neurovegetative symptoms (Abizaid et al., 2014).

CRH and Cortisol

Glucocorticoids and CRH are best known for their capacity to deal with stressful experiences, but they are also intimately related to eating and energy processes. The limbic CRH release associated with threats may promote nucleus accumbens dopamine activation, and hence positive stimuli, such as food, may appear more salient or more rewarding. In this regard, stressors may promote a dopamine-mediated craving for comfort foods (a fast caloric fix in the form of sugars and carbs) that might temporarily make individuals feel better. The salience of pleasurable or compulsive activities (ingesting sucrose, fat) is increased in the presence of insulin, and consequently comfort foods might be still more rewarding. These foods may also be instrumental in providing needed energy to facilitate coping with stressors, and the biological changes provoked by comfort foods might diminish anxiety that could otherwise occur. If, in contrast to a mild challenge, a stressor experience is intense, then dopamine levels will decline and the considerable CRH release will not have comparable effects on eating. Instead, a shift away from appetitive responses and toward defensive and vigilant behaviors will be provoked (Lemos et al., 2012).

It was suggested that linkages between stress, cortisol, altered metabolic processes, eating, and the preference for comfort foods, might contribute to obesity as well as the redistribution of stored fat, preferentially appearing as abdominal fat depots (Dallman, 2010). As we'll see in ensuing chapters, the cytokine release from abdominal fat may contribute to inflammatory-related disease conditions, such as diabetes, autoimmune disorders, heart disease, and depression.

STRESS AND IMMUNITY

Stressors Effects in Animals

Stressful events profoundly influence various aspects of immune functioning. These changes are described in detail in Chapter 6, Stress and Immunity, thus only a few key points are introduced here. The impact of stressors on immune functioning vary with characteristics of the stressor, several genetic and experiential factors, and the immune compartment examined (e.g., in blood vs spleen). In general, acute stressors that are moderately intense result in nonessential functions, such as digestion or reproduction, being suppressed,

while immune functioning was enhanced in order for the organism to be able to contend with potential pathogenic risks. However, in response to severe or chronic stressors, it is more likely that both primary and secondary immune functioning will be impaired (Dhabhar, 2009). Such effects are apparent following intense physical injury, and they emerge following psychological challenges (e.g., loneliness, disturbed social stability), indicating that immune disturbances aren't simply a result of tissue damage. For instance, in rodents, disruption of social hierarchies or social defeat, especially if these challenges are encountered chronically, undermine the immune system's capability of mounting an effective response.

In response to acute (or sub-chronic) stressors (e.g., social disruption), the production of circulating proinflammatory cytokines, such as IL-1β and TNF-α, is increased, and several days of this treatment promoted greater cytokine amounts within lymphoid organs (spleen, lung). The initial rise of proinflammatory cytokines is followed soon afterward by elevated anti-inflammatory cytokines, allowing for the balance between different cytokine subtypes to be reestablished. Moreover, under conditions of chronic social instability the influence of immune factors on brain BDNF functioning is increased (Nowacka, Paul-Samojedny, Bielecka, & Obuchowicz, 2014), and cytokine presence in the periphery and brain is altered appreciably (Audet, Mangano, & Anisman, 2010).

In considering the impact of stressors on hormonal processes, continued stressor experiences resulted in glucocorticoid levels being reduced and/or receptor sensitivity diminished, and consequently the immunosuppressive effects otherwise engendered might be absent. In view of the many challenges that animals experience in their natural habitat, a delicate bit of juggling is needed so that glucocorticoid functioning will operate efficiently, and yet not undermine immune functioning.

Brain Cytokine Variations

Stressors can influence concentrations of brain cytokines, although the changes that occur may be distinct from those seen peripherally. Acute stressors provoke an increase of inflammatory cytokine gene expression within the prefrontal cortex, especially if the stressor occurs on the backdrop of an immune challenge (Gibb, Al-Yawer, & Anisman, 2013). Furthermore, when animals are reexposed to a stressor sometime after an initial challenge, the cytokine response may be greatly exaggerated, even when this involved a very different stressor (Johnson et al., 2002). Thus, encounters with a stressor may prime immune or brain microglia processes to respond more vigorously to later challenges, and the excessive cytokine release could potentially lead to pathological outcomes (Anisman et al., 2003).

Once cytokine changes occur within the brain, further downstream effects can be engendered, including the provocation of neurochemical changes that can promote pathophysiological outcomes. As indicted earlier, cytokine challenges elicit brain neurochemical changes that are reminiscent of those engendered by strong stressors. This includes, among other things, increased monoamine utilization in the prefrontal cortex, central amygdala, and hippocampus, GABA and glutamate within limbic and hypothalamic regions, and growth factors in the hippocampus (Audet & Anisman, 2013). Some of these actions, including the elevated serotonin turnover in the prefrontal cortex elicited by the cytokine interferon-α, could be diminished if animals were pretreated with a nonsteroidal anti-inflammatory drug (Asnis et al., 2003).

It was surmised that some of the effects of stressors on central neurotransmitter functioning might actually be governed by an inflammatory response. In this regard, the monoamine changes otherwise provoked by stressors were prevented by inhibiting the actions of IL-1β. Likewise, the behavioral and neuroendocrine effects stemming from a chronic mild stressor, as well as the reduced hippocampal neuroplasticity, were precluded in mice in which IL-1β receptors were genetically deleted (Goshen et al., 2008). Such findings supported the view that the brain interprets inflammatory immune activation much like it does other stressor challenges (Anisman & Merali, 1999), and could thus have similar psychological consequences.

Individual differences are common in response to stressors, and could occur as a result of numerous mechanisms that could impact inflammatory-related illnesses. By example, rats that seemed resilient (reflected by active coping) in the face of social defeat exhibited increased presence of particular microRNAs (miR-455-3p) within the ventral portion of hippocampus, whereas a different microRNA (miR-30e-3p) was elevated in those rats that coped passively. These microRNAs are involved in inflammatory and vascular remodeling pathways, which could influence blood-brain-barrier permeability and could also influence the number of microglia present. Modifying inflammatory processes, either by treating rats with the proinflammatory cytokine vascular endothelial growth factor, or by diminishing inflammation using the nonsteroidal anti-inflammatory drug meloxicam, altered vulnerability to stressor effects. Like other reports, these findings point to the potential use of anti-inflammatory agents in diminishing the consequences of trauma in vulnerable individuals.

Stressor Effects in Humans

Evaluation of stressor effects in humans comprised determinations of circulating immune substrates, or analyses of the actions of *in vitro* challenges (antigens or mitogens) on NK cytotoxicity or lymphocyte proliferation.

Functional outcomes have also been assessed through analyses of susceptibility to viral infection or time for wounds to heal, as well as analyses of the impact of stressors on responses to vaccines, which in essence are like (inactivated) viral threats (Yang & Glaser, 2002). Acute psychological stressors of moderate severity usually increased immune functioning reflected by some or all of these measures, whereas intense or chronic stressors, such as caregiving (e.g., for a partner with Alzheimer's disease), typically disrupted immune activity (Slavich & Irwin, 2014), possibly secondary to hormonal changes that were provoked. Much like other neurobiological responses, immune functioning was disturbed by early life negative experiences (Miller, Chen, & Parker, 2011), and toxic stressors, such as severe poverty, were associated with immune-related epigenetic changes.

Psychological stressors, such as public speaking or exercise, were accompanied by elevated IL-6 and TNF-α together with increased presence of the inflammatory marker C-reactive protein (Zoccola, Figueroa, Rabideau, Woody, & Benencia, 2014). In contrast to the impact of modest stressors, elevated inflammatory markers are not present in response to severe, chronic stressors, such as that experienced by parents of children with cancer (Miller, Cohen, & Ritchey, 2002). Similarly, chronic distress associated with caregiving, was accompanied by dysregulation in the balance between pro and anti-inflammatory cytokines, which could favor the emergence of immune-related disorders (Glaser et al., 2001).

Not unexpectedly, immune system functioning was linked to several personality and emotional factors that were often tied to stressful experiences. Specifically, diminished NK cell activity was reported among individuals who were hostile, especially negative, engaged in high levels of rumination, or expressed a depressive mood (Zoccola et al., 2014). Likewise, trait characteristics, such as hostility, attributional style, and extraversion-introversion, influenced cytokine variations ordinarily provoked by stressors (Segerstrom, 2000). Disturbed cytokine responses ordinarily elicited by stressors were limited among individuals with higher self-esteem, but were more pronounced among those who felt low in social status. Being optimistic, which has been linked to enhanced health and well-being, was also accompanied by altered stressor-provoked cytokine variations (Brydon, Walker, Wawrzyniak, Chart, & Steptoe, 2009). It also appeared that stress-related cytokine changes varied with the emotions elicited by the challenge. The increase of blood IL-6 and the anti-inflammatory cytokine IL-10 provoked in the Trier social stress test was linked to feelings of shame or anger (Danielson, Matheson, & Anisman, 2011).

MICROBIOTA, INFLAMMATORY RESPONSES, AND STRESSORS

Bacterial communities that inhabit the gastrointestinal tract are not only sensitive to diet, antibiotics, and immune activating agents, but are profoundly affected by physical and psychological stressors (Bharwani et al., 2016; Tarr et al., 2015). Considerable diversity exists within the microbiome, which is usually taken to be the preferred state, but in response to stressors microbial dysbiosis occurs, so that an unstable and potentially unhealthy community may develop (Zaneveld, McMinds, &Vega Thurber, 2017). As in so many other contexts, the influence of stressors in rodents vary with sex, being particularly notable in females, which may have implications for the more frequent anxiety and depression seen in human females (Bridgewater et al., 2017).

The microbiota variations were linked to immune and cytokine changes (e.g., plasma IL-6 alterations) associated with stressors, supporting the view that microbiota and inflammatory processes might operate together in affecting stressor outcomes. It also appeared that immune activation reflected by elevated interferon-γ and gut bacteria, especially A. muciniphila, acted together in determining glucose metabolism and thus could be related to metabolic syndrome (Greer et al., 2016).

Multiple factors come together to influence the effects of stressors on microbial processes, which then affect brain functioning. Intestinal bacteria can modulate HPA activity and might thus contribute to stress-related conditions. Among germ-free mice in which basal levels of corticosterone were elevated, stressors provoked an exaggerated corticosterone response (Crumeyrolle-Arias et al., 2014), which could be attenuated by colonization with a specific Bifidobacteria species (Sudo, 2014). Furthermore, the transplantation of microbiota from stressed mice to germ-free mice provoked a marked increase of inflammatory responses following infection (Willing, Vacharaksa, Croxen, Thanachayanont, & Finlay, 2011). Likewise, stressors encountered during early life, influenced the composition of gut bacteria, and even affected the later development of stress responses (De Palma et al., 2015). Microbial variations during early life are readily provoked by stressors, and the absence of microbiota during early life can have pronounced ramifications on functioning of brain regions linked to stress-related behaviors (Stilling et al., 2015). As we'll see in Chapter 7, Prenatal and Early Postnatal Influences on Health, prenatal experiences can also have significant actions in this regard. The influence of stressors on microbial factors is not only pronounced during early life, but with older age the gut becomes more permeable, hence allowing microbiota to enter the blood stream more readily, leading to increased inflammatory factors (e.g., TNF-α) being present, thereby encouraging poor health (Thevaranjan et al., 2017).

Just as stressor-provoked microbial changes can affect central neurochemical functioning, variations of brain functioning can affect the microbiome. Fig. 5.4 describes some of the processes by which multidirectional communication occurs between these processes.

Gut microbiota can have positive or negative effects on neurodevelopmental processes and the balance between good and bad bacteria may be accompanied by psychological disturbances, including elevated risk for depressive illnesses or a PTSD-like condition in animal models (Leclercq, Forsythe, & Bienenstock, 2016). Of practical significance, pretreating mice with particular gut bacteria (*Mycobacterium vaccae*) may diminish stressor effects and inflammation and might thus attenuate trauma actions, such as the development of PTSD in a mouse model (Reber et al., 2016). Evidently, early life variations of the microbiome may have persistent microbial effects that influence psychological well-being, and understanding these relationships might provide important points of intervention to limit the development of disorders (Dinan & Cryan, 2017).

Treatment with prebiotics effectively attenuated anxiety and microbiota changes introduced by stressors (Tarr et al., 2015), and prevented both plasma and brain cytokine elevations that were otherwise provoked (Ait-Belgnaoui et al., 2012). Dietary prebiotics similarly improved sleep and generally limited adverse effects of stressors (Thompson et al., 2017). As predicted, when combined with exercise, the antistress effects of prebiotics were still more pronounced, likely owing to the combined effects on gut microbial processes (Mika, Rumian, Loughridge, & Fleshner, 2016). These changes lend themselves to antidepressant actions, including depression secondary to chronic illness.

Probiotics were similarly reported to diminish anxiety- and depressive-like behaviors in rodents and reduced the hormonal and brain neurochemical alterations ordinarily elicited by stressors (Liang et al., 2015). Probiotic

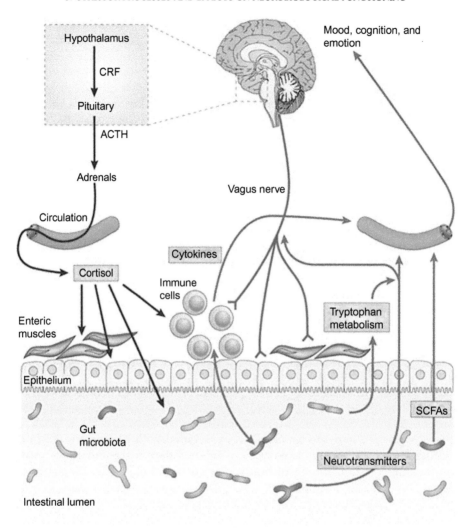

Nature Reviews | Neuroscience

FIGURE 5.4   Direct and indirect pathways exist through which gut microbiota modulate peripheral and central nervous system processes. Through effects on endocrine (e.g., cortisol), immune (cytokines) and neural (vagus and enteric nervous system) pathways, brain processes are affected, leading to behavioral variations. The vagus nerve and functioning of circulating tryptophan levels may contribute to gut-related information being relayed to the brain. In addition, neuroactive bacterial metabolites of dietary fibers, specifically, short chain fatty acids (SCFAs) also affect brain and behavior. In response to stressors, several brain neurotransmitters are affected, as are hormonal processes. Several of the hormone variations, such as corticosterone increases, affect immune cells and cytokine secretion, both in the gut and systemically. Significantly, cortisol can affect gut permeability and barrier function, thereby influencing the composition of gut microbiota. *Source: From Cryan & Dinan (2012).*

treatments also precluded stressor-provoked BDNF changes and limited impaired neurogenesis (Ait-Belgnaoui et al., 2014). It would seem that there's good reason to suppose that manipulating microbial factors could have significant effects on mood states, but the data are not entirely uniform. A meta-analysis confirmed that probiotics seem to have beneficial effects on depressive symptoms, but it was cautioned that more double-blind randomized control trials are still needed (Wallace & Milev, 2017). In this regard, it had been thought that Lactobacillus played a prime role in relation to stressor-related disturbances, but treatment with this microbial factor in humans did not influence cytokine levels, sleep quality, or mood (Kelly, Allen et al., 2017).

CONCLUDING COMMENTS

Stressful experiences engender a range of neurobiological effects that have been linked to numerous pathological conditions. Many of the effects of stressors act in an adaptive capacity, but can nevertheless aggravate existing pathologies. As the load becomes progressively heavier by experiences with severe or chronic challenges, the weakest link in the proverbial chain will give, leading to a disease condition. Once this occurs, which can give rise to further physiological alterations, comorbid conditions are likely to emerge. The weak link in this context may be related to genetic dispositions or the impact of earlier stressor experiences, particularly those encountered during early life, essentially reflecting an initial step of a two-hit or multihit phenomenon.

Several different systems (microbial, inflammatory immune, hormonal, neuronal, and neurotrophic) may additively or interactively govern the impact of stressors, thereby determining the development of pathology. Further to this, the nature and extent of the biological responses that occur may be influenced by a constellation of factors related to the characteristics of the stressor, individual appraisal and coping methods, personality factors, and numerous psychosocial determinants. Clearly, this speaks to the individualistic nature of stressor experiences and outcomes, and a precision medicine approach ought to be considered in determining the optimal treatment strategies for stress-related pathology.

Stress and Immunity

THE BREADTH OF STRESSOR-RELATED IMMUNE DISORDERS

An enormous literature has accrued over the past few decades that addressed the impact of stressors on immunological processes and immunologically-related diseases. Of the latter, the most prominent are the various autoimmune diseases, but in addition, inflammatory processes can also exacerbate asthma, cancer, and cardiovascular disease. Indeed, inflammatory processes may contribute to atherosclerosis, myocardial infarction, and stroke (Courties, Moskowitz, & Nahrendorf, 2014), thus adding the immune system as yet another mediating influence in these conditions. In addition, given that the immune system may encroach on neurodegenerative actions in the brain, it has been implicated in Parkinson's and Alzheimer's disease, and it is possible that stressor-induced alterations of inflammatory processes could exacerbate the progression of neuropathology (Piirainen et al., 2017). Findings such as these point to the importance of assessing how stressors impact the immune system, and whether the resulting immune perturbations contribute to either the onset or progression of these diseases.

Given the preponderance of evidence for a variety of immune alterations attributable to stressful experiences, it would seem that a deeper understanding of this linkage will contribute to preventative and/or therapeutic efforts to treat a wide range of human ailments. It is also pertinent that whereas we once thought "immune-related" diseases were confined to infection, autoimmune disease, and cancer, the range of medical pathologies enveloped by the immune system has expanded. This intrusion of the immune system into unconventional domains of biomedical investigation (for immunologists, at least), reinforces long-held views from psychosomatic research that recognized the strong comorbidity between emotional states like anxiety and depression, and conditions like asthma and heart disease. Since many diseases do not have a clear etiology, the imputation of poorly defined combinations of genetics and environmental experience have been raised as explanatory models. Into such models, it is reasonable to incorporate the immune system as a moderating variable. Moreover, the cells and molecules of the immune system could potentially modify pathological processes within distinct organ systems, exerting either positive (i.e., protective) or negative (e.g., degenerative) influences that are affected by psychosocial factors.

The Immune System and Mental Health
DOI: https://doi.org/10.1016/B978-0-12-811351-6.00006-1

IMPLICATIONS OF STRESSORS FOR IMMUNE-RELATED DISEASES

Immune surveillance occurs broadly, patrolling the entire body, every nook and cranny, but, at one time, it was thought to skip the brain altogether. The brain was too sacrosanct and should be protected against the destabilizing effects of inflammation. However, time only looks forward, and it is now recognized that T cells can infiltrate the nervous system. They can reside in the meninges, as well as extravasate from the blood and enter the parenchymal spine and brain tissue. The functional consequences of this infiltration have been vigorously debated, with some evidence that this serves in a neuroprotective capacity (Filiano, Gadani, & Kipnis, 2017). Alternatively, the presence of T cells may undermine neural repair and recovery. Irrespective of the particular conditions required to optimize the protective role of T cells in neural injury, it is likely that stressors and autonomic nervous system (ANS) activation will influence the migration of T cells into the CNS. Thus, it is conceivable that stressors can affect the homing of T cells to the brain and alter their sensitivity to monoamines and neuropeptides. Similarly, monocytes can infiltrate the brain, especially in multiple sclerosis (MS) (Ashhurst, van Vreden, Niewold, & King, 2014), in which cycles of relapse and remission can be affected by psychologically stressful daily hassles (Ackerman et al., 2002), and which may be due to recruitment of circulating T cells and monocytes that can be a prominent feature of relapse and/or aggravation of disease (Steinman, 2015).

There are additional reasons for us to enter into a discussion about stressor effects on immunity. As we discussed in earlier chapters, stressful events can be a precipitating factor in the development of psychiatric disorders, possibly stemming from the neurochemical changes that occur. Given that the immune system has the capability of altering CNS function—primarily through the influence of proinflammatory cytokines, identification of the relevant stressor-related immune factors that affect the brain are important in determining the links between stressors and mental disorders. In this chapter, we will consider how stress can inhibit (suppress) immune function and what types of immune parameters are affected. As well, we will also consider evidence that stressors can *augment* immune function. This observation materialized after a heavy emphasis in the 1980s and 1990s on the immunosuppressive effects of stressors in both animal and human studies. In hindsight, the augmentation of immune responses makes considerable sense, but created a conundrum, since it was difficult to reconcile with presiding theories of glucocorticoids as a primary mechanism for suppression of the immune system. Nonetheless, as stressors can have both inhibitory and enhancing effects on immune function, this warrants a close assessment of which neural-derived factors impact immune cells, and how they influence the tenor of immune responsiveness to antigen. Moreover, given that stressors can push immune responses beyond some predetermined optimal level of function, raises the possibility that stressors may contribute to an excessive immune-derived assault on the brain, which would then precipitate problems in mood and emotional regulation.

IMMUNE DYSREGULATION IN RELATION TO MENTAL ILLNESS

An important variable to consider in assessing stressor effects is the emotional and cognitive state of otherwise healthy people. The immune changes introduced by stressful events

(cont'd)

vary appreciably based on the background contextual and psychosocial conditions upon which this stressor is superimposed. Considerable evidence has pointed to the involvement of immune-related factors in the evolution of mental illnesses. Conversely, the immune changes associated with stressors can also be moderated by various psychiatric disorders, possibly being related to neuroendocrine and peripheral neurotransmitter alterations that may be present in psychiatric populations. Acknowledging the impact of repeated and/or chronic distress on immunity in such individuals may be relevant to understanding whether perturbation of the immune system through behavioral means results in perpetuation of the diagnosed mental condition.

Stressor effects on immune functioning may result in failure to clear viral and/or bacterial infections, which serve to assault the nervous system, possibly through infectious and inflammatory mechanisms. Bacterial meningitis, rabies, and herpes simplex virus are just some of the CNS-related infections that have to be avoided to ensure stable brain function and mental health. Immunological clearance of these infectious agents is paramount, but may be compromised if individuals experience stressors that are immunomodulatory. Moreover, as we will discuss in several ensuing chapters, immune factors have been related to several mental disorders. In some instances (e.g., depression) illness may arise owing to neurotoxic effects related to inflammatory processes, whereas in the case of schizophrenia, a prominent hypothesis has considered a higher than normal incidence of autoimmune disease. The data in this regard are suggestive, but given the highly stressful nature of schizophrenia, it would not be out of place to hypothesize a role for stress in driving autoreactive immune responses.

DYNAMIC INTERPLAY BETWEEN THE CNS AND IMMUNE SYSTEM

In Chapter 5, Stressor Processes and Effects on Neurobiological Functioning, we learned that stress is a physiological state that draws on many different biological resources to allow the organism to deal with the demands of the environment. Our focus was on neurobiological processes, since our reaction to stressors is based on cognitive and sensory information. In cases, where that information generates alarm, the brain is equipped to initiate a cascade of changes that pass through to the peripheral nervous system and neuroendocrine systems. Stressors promote a rapid increase of norepinephrine and acetylcholine release from the terminal endings of the sympathetic and parasympathetic branches of the ANS. Although the branches of the ANS terminate in virtually all organs of the body, it was the discovery that sympathetic nerve endings terminate in lymphoid organs (see Bellinger, Nance, & Lorton, 2013), that set the field of psychoneuroimmunology alight with excitement. Here was evidence that the immune system was "hard-wired" by the brain. It also suggested that when the brain is stressed, and floods its synapses with neurotransmitter, the end result must be the same as that other important synapse in the periphery: between a sympathetic nerve terminal and a lymphocyte. Indeed, it is now well established that various types of leukocytes express noradrenergic and cholinergic receptors (Sternberg, 2006; Tracey, 2009). This highlights an anatomically based continuity of function, extended not just to cardiac muscle or liver function, but also to immune actions. In essence, were there a change in the immune system as a result of the brain's efforts to engage a psychological or physical stressor, the result would likely be functionally important. Were it not, one would have to question why mammalian systems evolved such an intimate form of communication between the nervous and immune systems. Most surely, we cannot immediately conclude that stressor-induced changes of immune function should be considered as having pathological repercussions. Other circumstances need to come into play.

The same reasoning can be applied to neuroendocrine activation. As with receptors for products of the ANS, there is considerable evidence for the presence of hormone receptors on immune cells. Adrenal-derived glucocorticoids (e.g., corticosterone in rats and mice; cortisol in human and nonhuman primates) primarily serve in an anti-inflammatory capacity (Cain & Cidlowski, 2017). Indeed, in the early years of research on how stressors impact immune function, this served as the major explanatory mechanism for stressor-induced immunosuppression. Nonetheless, in addition to the adrenal, the pituitary represents a source of immunomodulatory hormones (e.g., endorphin, prolactin), which have the ability to bind to immune cells via specific receptors. Moreover, noradrenaline and acetylcholine, which undergo tonic release, even under resting conditions, can influence immune functioning. To add further to this *potpourri* of influences, peripheral neuropeptides, such as substance P and vasoactive intestinal peptide (VIP), can exercise strong functional effects on immune cells, operating through specific receptors (Ganea, Hooper, & Kong, 2015). Indeed, the list of endocrine hormones, peptides, and neurotransmitters that serve as ligands for selective receptors on lymphocytes, monocytes, polymorphonuclear leukocytes, and dendritic cells (DCs), is fairly extensive.

This serves as testimony to the close functional interplay between the brain and the immune system, and reinforces the notion that the immune system evolved to play host to a wide range of products generated by an activated central nervous system (CNS).

Although this conclusion is fairly certain, it is more difficult to answer how immune-related processes impact CNS function, and as a consequence, affect mental illnesses. To answer this, we must also consider whether in affecting the CNS, the immune vector is always the same. Are there behavioral influences that modify the quality and quantity of the immune processes that modify the CNS? For example, if a given cytokine (e.g., IL-1) impacts the brain to induce sickness behavior, is it the same amount and duration of IL-1 release that exerts this effect in all people, or just some? And if the answer is the latter, what accounts for the different parameters of concentration and temporal persistence? Further, is the individual worse off with higher and longer amounts of a given cytokine bombarding their brain?

The CNS and immune system are intertwined, and engage in a dynamic dance in which molecules are exchanged and cellular modifications are made. This is a functional relationship as relevant to physiological actions as the relationship between enzymes and substrates is to the biological functions of a cell. But, when the organism is exposed to psychogenic or neurogenic stressors, what happens to this "dance"? If the CNS and immune system are "holding hands" is the grip tightened or loosened? At the molecular level, the study of stress has revealed a loosening of molecular relationships between enzymes and substrates, or between transcription factors and their promoter regions. Researchers can look more closely at these types of relationships when focusing on specific immune cells. But the higher-order questions concern how this ultimately affects the ability of the immune system to fight off infections, maintain antigenic accuracy, avoid autoimmune pathology, and regulate aberrant neoplastic cells that threaten the development of cancer. This is the traditional rationale for investigating stressor effects on immune function, but newer ones have emerged, in light of suggestions that immune processes may impact depression, anxiety, PTSD, autism, and schizophrenia, as well as several neurodegenerative disorders. Furthermore, how do stressor-related changes in immune function affect neurodevelopment? As well, is an immune system that has been pummeled by life's stressors able to alter the aging brain and promote neurodegenerative disease? These are all big questions, for which the answers are not at all clear.

CONSIDERING THE VALUE OF IMMUNE ALTERATIONS PROVOKED BY STRESSORS

A caveat in any discussion of stress and immunity is that the literature in this area is far from straightforward. At times, it can be downright perplexing. Straight out of the gate we acknowledge that irrespective of whether measurement of immune function is *in vitro* or *in vivo*, it has universally been found that the immune system is perturbed or altered by stressors. There is no argument here. Where the confusion lies is predicting the magnitude, and even the direction of the change, and also determining the clinical significance of some of the changes. Aside from effects on immune functioning, it should be considered whether stressors influence clinical conditions, such as infectious and autoimmune diseases. It is also necessary to reconcile studies showing stressor effects on immune function with evidence that the CNS and immune system are intimately connected at a molecular and at a hormonal level. In effect, stressor-induced immune modulation is hardly an aberrant event, but a naturally evolved phenomenon. Indeed, often the immune changes may be something we need, as opposed to something which requires elimination. Consider that activation of the hypothalamic-pituitary-adrenal (HPA) axis can serve a protective and life-saving capacity during infection, and it may also be necessary to inhibit autoreactive immune responses, thereby limiting the progression of autoimmune disease (del Rey & Besedovsky, 2000).

STRESSOR INFLUENCES ON IMMUNITY: ANIMAL STUDIES

Recognition that stressors impact the immune system predated any major understanding that the immune system was "hardwired" to stress-related processes. However, the relationship of the adrenal to certain immune parameters was recognized many decades ago, although more specific analyses of glucocorticoid receptors present on leukocytes occurred much later. Well before the neural-immune connections were recognized, it was known that the immune system of the laboratory animal reacted strongly to stressful impositions. Exposure of animals to a variety of stressors would shrink the thymus (thymic involution). As it is, the thymus shrinks with age, but this reflection of senescence could be accelerated with stressor experiences. Likewise, the relationship of stress to increased susceptibility to infectious diseases was suspected for some time before being formally demonstrated (Cohen & Williamson, 1991). Based on a very large number of studies on immune functioning in stressed human and infra-human subjects, we are comfortable with the notion that stressors influence immunity, and there is ample reason to believe that these immune changes may in some fashion contribute to the exacerbation (and perhaps provocation) of some illnesses.

Pursuing some measure or indication of "immune function," however, is not a simple matter, given that there are many different facets to what we call immunological activity. In psychology, the same thing is said of "abnormal behavior," wherein by asking what exactly we mean by "abnormal," the tale of the blind men and the elephant is resurrected as a means to emphasize that perspective matters. How one approaches the question of immune function can similarly be a matter of perspective, but it remains necessary to focus on the inherent differences among immune cells. Few would argue that an immune system is not functioning well if an infectious virus is expediently trounced and removed from the body. Conversely, we would conclude poor functioning if someone develops auto-aggressive antibodies against self-tissue and develops organ pathology, or if cancer cells multiply because immune surveillance mechanisms failed to restrict their growth. These examples are clinical indications of functional and dysfunctional immunological activity. And they represent the critical endpoints of stress research on the immune system. Indeed, it is always of fundamental interest to know how a biological system operates under varying environmental conditions (e.g., hot, cold, high altitude, low altitude, and so on), but when we invoke the specter of "stress," the expectation is to consider potential clinical sequelae. So, what we need to ask, in addition to whether stressors modify immune function, is the likelihood that such a modification is clinically relevant. Does a stressed individual's immune system fail to maintain health? The answer to this question is important. As we have already discussed, the connection between the brain and the immune system is natural. If stressors shake up the organism, the way bumps in the road may test

a car's suspension, then any clinical repercussions are likely to be due to faulty engineering. Identifying where that fault may lie is the challenge. However, before we look at this more closely, we ought to attend to how we would assess the consequences of stressor exposure on immune function.

Immune Assessment and Stress

Although it would make research far simpler if stressors affected all aspects of the immune system in the same way, this is hardly the case. The diversity of responses across the various aspects of the immune system is likely beneficial in the sense that as one system is diminished by a stressor, a second system may be unaffected, and even influenced in an opposite manner. In view of the heterogeneity of stress responses that can occur, basic research requires selection of particular parameters of immune function, and perform a detailed assessment of the individual cellular and/or molecular players.

Recall from Chapter 2, The Immune System: An Overview, that the immune system is divided into acquired and innate components, both of which need to be considered in relation to stressor effects. The chief cells of the innate immune system are monocytes, neutrophils, and macrophages, and assays for these parameters include phagocytosis, chemotaxis, and cytokine production. For the acquired or adaptive immune response, we rely on measures of lymphocyte function, and here we further divide our attention between T and B lymphocytes. The most prominent measure of the latter is antibody production or simply immunoglobulin release, if the B cell stimulus is not a specific antigen. In so far as T cell function is concerned, mitogen-induced proliferation has been a prominent measure, although antigen-specific T cell responses have also been investigated; both forms of T cell assessment lend themselves to cytokine measurement

(Note: mitogens are typically plant lectins that non-specifically stimulate T cells and B cells and cause them to divide and generate cytokines and antibody; in essence, mitogens are polyclonal stimuli, with no affinity or selective attraction to a particular antigen receptor).

Many of these assessments have involved *in vitro* assays. In human studies, this has often been the case, although the response to infectious agents and immunization have also been assessed. For the most part, however, animal studies lend themselves better to *in vivo* immune measurements, as well as to the study of infectious disease models and exposure to bacterial toxins. Several animal studies that examined stressor-induced modulation of influenza and herpes simplex virus infections provided an opportunity to consider the immunological mechanisms that have been affected by stressors. For instance, T cytotoxic cells are a critical mechanism with regard to viral infections. However, NK cells also have antiviral functions, as well as providing some insight into surveillance against cancer. Accordingly, in the many studies that examined NK cell function, the clinical implication has often been cancer susceptibility.

Stressor Characteristics

As described in Chapter 5, Stressor Processes and Effects on Neurobiological Functioning, a variety of conditions have served as stressors, which can differentially influence biological and behavioral responses. For example, social disruption through changes in the group composition of group-housed animals has proven to be a strong stressor, while the opposite, social isolation, can also serve as a powerful stressor. Diverse stressors can also instigate some significant differences in the neurochemical pathways activated across brain regions. Different psychogenic stressors (e.g., exposure to a fear-provoking stimulus, restraint, and social defeat) can have diverse effects, which may differ yet again from the actions of neurogenic stressors (e.g., electric shock, exposure to cold water). It is of particular importance in these studies to determine the broad physiological perturbations produced by the stressor (e.g., analyses of hormones or neurotransmitters), and how these may be linked to immune system changes. Moreover, it is essential to identify the factors that act as moderators (e.g., stressor controllability, early life experiences) that are fundamental in determining the nature of the immune responses provoked by stressors.

Aside from the type of stressor, it is important to consider the number and duration of stressor exposures. This is the issue of *acute* versus *chronic* (or repeated) stress. In the current era of stress research, increased attention is being directed to chronic stressors. For the most part, it is understood that the physiological response to a sudden, temporally distinct stressor is adaptive, and unlikely to be of much use in revealing pathological changes, although very severe stressors may have profound effects (but, for ethical reasons researchers typically do not conduct such studies in humans). In fact, from a lay perspective, and the popular media in general, a bad day at work or some form of disappointing news is perceived as a biological threat. However, this alarmist view of things is a gross misunderstanding of the importance of biological plasticity and the role it plays in conferring resilience to everyday stressors. Most creatures, humans included, are made of sterner stuff, although minor day-to-day annoyances can under some circumstances can have appreciable negative action. Unlike the effects of occasional insults, chronic or prolonged stressor experiences are more problematic, often representing a challenge to behavioral and biological adaptation, and are more likely to promote pathology.

Stressors are a ubiquitous aspect of life, and the greatest danger they pose is often linked to their persistence. As we have indicated repeatedly, it is when they sap adaptive resources that their biological impact, and immunopathological potential, is most likely to be realized. To this extent, it has become fashionable to think of chronic stressors as the true problem that we should be addressing, with acute stressors being part and parcel of manageable physiological fluctuations. Of course, given a sufficient number of acute stressor episodes, even if they are distinguishable and unrelated to one another, their cumulative action can be significant. This is indeed the general story with regard to chronic stressor exposure, whereby immune responses that are normally unchanged or modified (e.g., enhanced) by acute stressors, eventually come to be suppressed or impaired in some manner.

Nonetheless, our understanding of any biological system requires that we know how it reacts when first exposed to an unexpected and distressing stimulus—that is, when the organism is acutely stressed. Then it becomes possible to evaluate whether the effects of chronic stressors reflect further adaptive changes or, conversely, the break-down of these processes. In essence, according to this view, what occurs in response to an acute stressor is a defensive change, not an indication of pathology. As an evolutionary advantage, such changes were subject to natural selection and continue to represent useful responses.

We considered earlier why immune cells likely have receptors for glucocorticoids, norepinephrine, and other neurally derived neurochemicals, steroids, and peptides. Several compelling studies have offered suggestions, from protecting against autoimmunity to regulating and improving immune responses (Bottasso, Bay, Besedovsky, & del Rey, 2007). It is plausible that in response to stressor-provoked release of neuroendocrine and ANS monoamines, for which immune cells express receptors, the impact on the immune system is less likely to be one which is designed to compromise immune function. More likely, these changes would serve to facilitate or regulate immune responsiveness.

As we learned in the previous chapter, the brain, as the ultimate information processor, is equipped to appraise and respond to a multitude of stimuli, ultimately producing a stress response. As a result, strategies are implemented that negotiate the stressor and minimize its potential as a biological or psychological threat. This is not always possible, of course, and in the course of "negotiation", tissue damage may be incurred, as might happen when the twin options of fight or flight results in the former. This is evident in the social confrontation stress model, and good evidence exists that animals subjected to various forms of social disruption display pronounce immune changes (Weber, Godbout, & Sheridan, 2017). In any case, responses to stressors require behaviors that generate steps to distance the organism from the stressor either through withdrawal, elimination of the stressor, or in the case of social contexts, acts of diplomacy, which may or may not involve a struggle. With the elimination or attenuation of a stressor, the organism can remain at ease until it is compelled to attend to the next threat or challenge. These challenges come in all shapes and sizes, often comprising events that aren't frequently considered to be stressors from a cognitive perspective, but as described in Chapter 5, Stressor Processes and Effects on Neurobiological Functioning, systemic challenges have effects much like those elicited by psychogenic and neurogenic stressors. In this context, immune functioning may be affected by seasonal changes, temperature fluctuations, and reduced energy intake (e.g., hunger) (Bilbo et al., 2002). When chronic stressors involve compound challenges, some of which occur on an unpredictable basis, the immune system's capacity to ward off infectious agents, and self-regulatory functions may be impaired, thereby resulting in autoimmune disease. As we will address further, in humans, stressors can also modify *in vivo* immune responses to influenza vaccines, and differences in certain cell-mediated immune parameters may exist according to whether a stressor is acute or chronic in nature.

Any instance of chronic stressor exposure bears an inherent acute component (the initial response to challenges of limited duration) and a chronic (or protracted) element, in which that stimulation continues to persist. The acute-chronic dichotomy can be viewed as the equivalent of being thrown an object of heavy weight, which requires the necessary postural and muscular adjustments that absorb the force of the object's impact. However, when this initial acute event is repeated, over and over, it can transform into a persistent state of stimulation that is "chronic," and to which the organism needs to keep mounting the same response it generated upon the first encounter. The question now is how long can the physiological or psychological response of the organism be sustained at an optimal and effective level. With regard to immunity, the question is how long before the initial immune change observed in response to an acute stressor suffers a change that is no longer a reflection of adaptation and of benefit to the organism.

In an interesting study, mice susceptible to UV light-induced skin cancer were stressed for 2 weeks (6 hours/day restraint), and exhibited earlier onset and increased incidence of squamous cell carcinoma. This was associated with reduced cutaneous entry of CD4 + T cells that normally restrict the cancer growth in this model (Saul et al., 2005). In a follow up study, imposition of a shorter, seemingly more "acute" stress regimen (2.5 hours/day for 9 days, which can actually be seen as a subchronic, rather than an acute stressor) augmented the T cell mediated immune response, and reduced the incidence and time to skin cancer onset (Dhabhar et al., 2010). These results, using an immunopathological approach, with well-defined cancer monitoring immune mechanisms, highlights the importance of duration of stressor exposure. Moreover, given that short-term stressor exposure augmented antitumor immunity, emphasizes the notion that the stress response can be viewed as an additional defense mechanism that boosts immunoprotection. Indeed, it was argued that physiological changes induced by exposure to a stressor may operate like a natural adjuvant (Viswanathan, Daugherty, & Dhabhar, 2005)[1].

Aside from duration of stressor exposure, the nature of the stressor may be fundamentally important. In ecologically relevant situations, successful negotiation of a stressor is dependent on the assessment of the situation according to natural implementation of cognitive processes. For instance, acute exposure to a predator (e.g., mice in the presence of rats; or rats in the presence of ferrets), results in activation of the HPA axis and reduction of basic *in vitro* measures of macrophage and NK cell cytotoxicity (Anisman et al., 1997; Lu, Song, Ravindran, Merali, & Anisman, 1998), presumably by triggering genetically predetermined

---

[1] The implications of this notion are quite interesting in the context of other illnesses. There are usually yearly variations in the success of the flu vaccine. Currently (the 2017–2018 winter in North America) the death rate from influenza has tripled relative to previous years, and there is concern that flu vaccines are not especially effective. Were the conditions to be identified by which acute stressors enhance immune responses, perhaps this knowledge could be harnessed and applied as a physiological modulator superimposed on flu shots. If this were possible, natural adjuvant mechanisms might boost initial protection and efficacy against influenza exposure. Subjecting people to stressors is not in the cards—but perhaps an acute bout of exercise—a natural stressor—prior to immunization would do the trick. Just a thought . . . but considering some of the findings related to exercise on antigen-specific memory responses, this is a theoretical possibility.

defense strategies. Assuming the changes are initially adaptive, what might transpire if repetition of the predator and restraint stressors results in different forms of habituation? Indeed, the health outcome following a naturalistic stressor might vary, since such stressors are potentially lethal, militating against the luxury of reduced vigilance. A telling example was provided when mice were subjected to either repeated social disruption or restraint. Those in the social stressor group developed increased rates of mortality in response to an endotoxin (viz., LPS) challenge, whereas sensitization to the inflammatory effects of LPS did not occur in restrained animals (Quan et al., 2001). This effect was further related to the loss of the anti-inflammatory properties of glucocorticoids, that is, glucocorticoid resistance (Engler et al., 2008).

We cannot presume to know what cognitive processes are operating during restraint, but it most certainly appears that whatever cognition and attendant physiological changes (peripherally and in the CNS) are provoked in an acutely restrained animal, appears to translate into less threatening immunobiological effects following chronic restraint. Part of the explanation may lie in how a stressor comes to be perceived. Restraint procedures that utilize a constant duration of immobilization (e.g., 2 hours/day, or 6 hours/day) have somewhat predictable outcomes to which animals can habituate, and learn that release and return to the home cage lies at the end of the experimenter-initiated imposition. It has long been known that measurement of neuronal excitability (as measured by immediate early gene activation—viz., c-fos) in stress-related circuits of the brain, shows a decline with repeated exposures to restraint (Girotti et al., 2006). Natural stressors, in contrast, can elicit very different outcomes. Specifically, rodents exposed to repeated social stressors (daily agonistic encounters that result in social defeat), fail to show significant decline in elevated

patterns of neuronal activation in the brain relative to that induced after the first encounter (Martinez, Phillips, & Herbert, 1998). One way to interpret this is that relaxation of neuronal resources during conflict situations was not possible, since the outcome of the conflict involves possible defeat and death. Therefore, if we return to the different outcomes for endotoxin exposure among animals exposed to social or restraint stressors (Quan et al., 2001), the former was likely more disruptive to normal physiology. Similar findings are emerging for studies of depression-like behavior, such that prolonged social stress (social defeat) imposes a considerable burden on neural adaptation (Laine et al., 2017).

As we have already noted, social stressors can exert different health outcomes than physical restraint. As stressor-provoked glucocorticoid, noradrenergic, cholinergic and neuropeptide changes influence immune function, the immune system responses to these modulating factors may fluctuate with different types of stressors and may vary with respect to the adaptation that could occur with repeated exposure. We do not know enough at present to determine whether slight alterations in the concentration of hormones and neurotransmitters, reflective of different neural "stress" states, are processed by immune cells as packets of information that require different immunological adjustments at the cellular and cytokine level. As we discussed earlier, immunocytes are undoubtedly immersed in an internal milieu that allows them to bind via relevant receptors to various hormones and neuropeptides. These ligands circulate and/or are released at basal levels, likely being receptive to their presence in a manner similar to the balance of salts and various other ions that govern normal cell biology. However, rapid and/or persistent elevations in the concentration of these ligands represent new information to which cells respond. This may result in several different functional changes (e.g., altered cytokine production), many of which have been documented. If there is a lawful relationship between the pattern and concentration of different neurohormonal factors and the threshold sensitivity of immune cells to these stimuli, it has yet to be adequately described.

Stressor Effects on Immune Function

Evaluating the status of the immune system involves a number of *in vitro* and *in vivo* procedures designed to elicit canonical immunological responses, such as lymphocyte proliferation, antibody production, phagocytosis, antigen presentation, and cytokine production. However, the best evaluation of immunity is the retention of stable health in the face of host invasion by viruses and bacteria. Infection with the human immunodeficiency virus offered important lessons and confirmation about the importance of CD4 + T helper lymphocyte function in the fight against opportunistic infections. Since HIV infects and replicates in T cells—most commonly CD4 + T cells—there is a dramatic depletion and destruction of T helper cells. This impairs antigen-specific B cell antibody production, and results in deadly opportunistic infections. As such, the consequences of HIV infection demonstrate the importance of a fully functional immune apparatus. Accordingly, in assessing the relevance of stressor effects on immunity we need to keep in mind whether the immune alterations measured represent a potential breakdown in immunity against infection. Many of the immune parameters assessed under noninfectious conditions are mitogenic stimulation of lymphocyte transformation, natural killer cell activity, and antibody responses to antigen. The clinical relevance of these measures has been established by way of association with additional approaches such as exposure to viral and bacterial antigens, induction of experimental autoimmune disease, and tumorigenesis.

Analyses of stressor effects on immune functioning have involved a variety of different species and often very different stressors. Typically, rats and mice have been the species of choice, but nonhuman primates have been used on occasion. Infrequently, when agricultural applications are needed, pigs, cattle, and fowl have been evaluated. In the main, the effects of stressors on immune functioning have been consistent across species. Yet, as we will see, in some instances, predicting the outcomes, and even the direction of stressor effects on immunity (i.e., suppression or enhancement), is not always straight forward. In this regard, neurophysiological changes provoked by stressors operate against a highly intricate set of intercellular interactions within the immune system. Furthermore, as described in Chapter 2, The Immune System: An Overview, the immune system is heavily compartmentalized, having unique regulatory requirements operating in several anatomical regions of vulnerability (e.g., immune cells in circulation, those present in local lymph nodes, as well as those present within common mucosal system comprising the gut, lung, and urogenital tracts). As a result, unique sets of interactions can be expected between the brain and regionally specific immune processes (Powell, Walker, & Talley, 2017). This is made still more complex as different sets of experimental conditions (e.g., species, stressor) and the type of antigen to which animals are exposed, can instigate very different immunological outcomes.

Cell-Mediated Immune Responses

Measures of cell-mediated immune responses typically reflect assessment of T lymphocyte function. Most often, this is conducted *in vitro*, using mitogen-stimulated lymphocyte proliferation assays, since antigen-specific stimulation does not yield a sufficiently robust stimulation index due to a low number of antigen-specific T cell clones. However, the use of certain toxins and transplantation based antigen-stimulation systems (e.g., mixed lymphocyte reaction) has provided useful information beyond the more artificial conditions of using mitogenic plant lectins. (see Chapter 2: The Immune System: An Overview). The use of mitogens typically revealed suppression of proliferative activity following acute stressor exposure of rats. However, the opposite outcome was elicited by an acute stressor in the analysis of the delayed hypersensitivity (DTH) response, an *in vivo* measure of T cell antigen-specific sensitization and proliferation. This paradigm involves initial sensitization with antigen, followed days to weeks later by challenge with the sensitizing antigen, which results in an inflammatory response characterized by increased redness and swelling of the challenged part of the body (typically the footpad or pinnae of the ear). Comparison of the effects of acute and chronic stressor exposure showed that short-term (acute) stressor exposure enhanced the DTH response, whereas repeated stressor exposure produced suppression of DTH. These findings are in line with the previously mentioned work of Dhabhar and colleagues, who championed the notion that an acute stressor is more likely to enhance immune responses, whereas chronic stressors engender immunosuppressive actions. It is especially interesting that the DTH response could be influenced by stressors that were applied to animals before being sensitized (i.e., given a primary exposure) with the chemical 2,4-dinitrofluorobenzene (DNFB) (Dhabhar, 2014; Dhabhar & McEwen, 1999).

Several older studies have been instructive in establishing potential principles by which stressors modify the direction of antigen-specific immune responses. For example, antigen-specific spleen cell memory proliferation to cholera toxin was enhanced by exposure of rats to acute foot-shock (Kusnecov & Rabin, 1993), which was similar to memory DTH responses being augmented by acute restraint (Flint, Valosen, Johnson, Miller, & Tinkle, 2001). Acute stressors applied at the time of memory induction (i.e., when cells are naïve and are being stimulated with antigen for the first time), very likely exert a suppressive effect, but not once memory cells have formed. Indeed, acute restraint can interfere with antigen sensitization of naïve T cells, resulting in reduced memory DTH or *in vitro* proliferative responses to a recall (i.e., previously encountered) antigen (Flint et al., 2001; Kusnecov & Rabin, 1993; Tournier et al., 2001). Additionally, acute stressor exposure may promote the generation of more effective memory cells from among naïve lymphocyte precursors. In this regard, rats exposed to a single stressor session and immunized with the protein antigen keyhole limpet hemocyanin (KLH) around the time of stressor exposure, showed enhanced proliferative responses by memory spleen cells two weeks later (Wood, Karol, Kusnecov, & Rabin, 1993). Since mitogen-induced proliferation was unaffected, it seems that a stressor at the time of immunization promoted better sensitization and development of a greater pool of memory cells, which was reflected by greater leukocyte infiltration of antigen injection sites. In essence, acute stressor exposure proximal to the time of antigen immunization can augment future reactions to the same antigen. These findings clearly differed from those indicating that exposure to an antigen at the time of stressor exposure had the effect of disturbing antigen-specific T cell memory proliferative responses (Kusnecov & Rabin, 1993; Tournier et al., 2001). It may simply be that different antigens yield different outcomes in response to stressor challenges. Therefore, for theoretical and practical purposes, it may be useful to concurrently evaluate the impact of diverse antigens (benign proteins vs bacterial toxins/viral determinants) in response to stressors.

Unlike the effects elicited by an acute stressor, daily exposure to 6 hours of restraint (combined with shaking) for 3–5 weeks, can suppress DTH responses to a recall antigen (Dhabhar & McEwen, 1996). Clearly, in response to the more prolonged stressor, immune effects were more likely to move in an alternative direction to those observed after an acute stressor. Such a reversal may be in the organism's interests given that the persistent maintenance of a heightened state of immune reactivity is energetically costly, and the risk for immunological dysregulation may be elevated. As a case in point, the reactivity of the immune system to negative feedback regulation by the HPA axis can be affected by persistent social disruption (Stark et al., 2001), thereby allowing for increased inflammation and greater likelihood for the development and/or exacerbation of autoimmune disease. Glucocorticoids are known to be immunosuppressive, and their ability to exert inhibitory restraint on the magnitude of most immune responses is considered important to ensure that inflammation does not result in disease. Consequently, attenuation of immune responses after chronic stress may represent a mechanism to protect against dysregulation both within the immune system, as well as between the immune system and regulatory neuroendocrine mechanisms.

The impact of acute and chronic stressors on immune function, as described in Fig. 6.1, can be deeply embedded in the type of immune response that is being measured. The foregoing discussion was concerned with *in vivo* T cell-mediated inflammatory responses. But as noted earlier, many studies have focused on *in vitro* mitogen-induced lymphocyte responses. These studies have shown that spleen and blood lymphocytes from rats exposed to acute conditioned and unconditioned stressors displayed reduced proliferative capacity to T cell mitogens (Kusnecov & Rossi-George, 2002). In contrast, more prolonged exposure to a variety of different stressors (e.g., foot-shock, immobilization, and isolation) shifts these responses in the opposite direction—enhancement (see Kusnecov & Rossi-George, 2002). For example, when an isolation stressor is imposed for less than one week, blood and spleen lymphocyte proliferation is inhibited; however, if isolation persists for greater than 2 weeks, lymphocyte proliferation can be enhanced[2]. Similarly, varying the amount of exposure to a forced swim stressor can produce similar bimodal effects (acute exposure leading to immunosuppression, with eventual chronic exposure resulting in enhancement).

Together, these observations are in keeping with the principle that the direction of an acute stressor effect is likely to be reversed by chronic or more repeated stressor exposure. Unfortunately, data from other studies have not always cooperated with this perspective, since the imposition of restraint or electric shock can still exert a depressive effect on adaptive functions of lymphocyte proliferation (Kusnecov & Rossi-George, 2002; Maslanik, Bernstein-Hanley, Helwig, & Fleshner, 2012). Therefore, the interpretation of stressor effects on immune function might need to be reevaluated. What may be important is that the appearance of a mitogenic rebound response (from one of initial suppression) is governed by some combination of stressor intensity and frequency of exposure. This said, not all studies used a true chronic exposure regimen (which theoretically should extend over several

---

[2] These actions are likely species-dependent, as protracted isolation in prairie voles, which is a social species (at least with respect to their life partner), produced reduced ability of the innate immune system to kill bacteria, coupled with an increase of agonistic behavior (Scotti, Carlton, Demas, & Grippo, 2015).

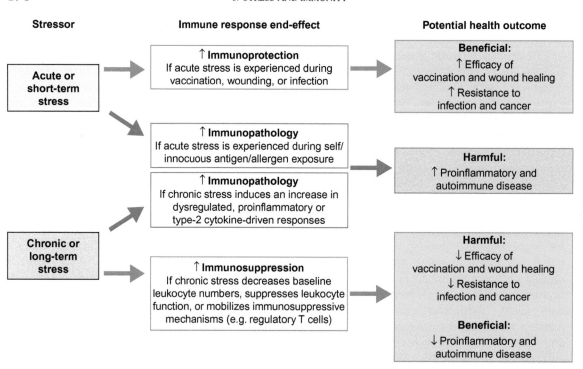

FIGURE 6.1   The relationship among stress, immune function, and health outcomes. Acute stressors experienced during vaccination, wounding, or infection may enhance immunoprotective responses. Acute stressors experienced during immune activation in response to self/innocuous antigens or allergens may exacerbate proinflammatory and autoimmune disorders. Chronic stressor-induced increases of proinflammatory or type-2 cytokine-mediated immune responses may also exacerbate inflammatory and autoimmune disease. Chronic stressor-induced suppression of immune responses may decrease the efficacy of vaccination and wound healing and decrease resistance to infection and cancer. *Source: From Dhabhar (2009).*

weeks, as opposed to several days), and therefore, in cases where repeated exposure is only a few days, insufficient time was allowed to elapse to glimpse some form of immune adaptation. To find studies that utilized a truly chronic level of stressor exposure, one has to turn to a rare, and older, study that exposed rats to 6 months of foot-shock, which revealed that splenic mitogen-induced proliferation continued to be suppressed in younger animals, but not in aged rats, which showed a normalized mitogenic response (Odio, Brodish, & Ricardo, 1987). One should note, however, that given that most rodents live for only about 2 years, a few weeks of stress translates to an

appreciable portion of the life-span. Thus, it is impressive that some age-related adaptation takes place with respect to spleen cell mitogenic function. Whether such adaptations would occur for other immune measures is not known.

Humoral Immune Responses: B Cell Function and Antibody Production

Assessing the effects of stressors on humoral immunity involves the analysis of B cell function and, in particular, the ability of B cells to generate antigen-specific antibody. An important issue in this regard is the timing of stressor application relative to introduction of the antigen to the host. Prior to antigen exposure *in vivo*, naïve B cells are largely in a "resting" or quiescent state. This changes quickly once a novel antigen engages the immune system, initiating a dynamic process of interactions between antigen-presenting cells (APCs), T cells, and B cells. Over the next few days, there is a build-up of antibody produced by antigen-specific B cells, which proliferate and differentiate into plasma cells. The immunoglobulin isotype of the antibody will be primarily IgM and IgG, although at mucosal surfaces, IgA will be dominant. The end result of this antibody accumulation is binding and removal of antigen. This is a critical aspect of the immune response, involving many cell types, and a temporally precise sequence of events that are all potentially susceptible to the neurophysiological consequences of stressor exposure. Accordingly, determining the impact of a stressor on the humoral immune response needs to consider the timing of the stressor relative to immunization, or the impact of a stressor on the resting state of immune cells prior to immunization. Further, once initiated, the impact of stressors on the ongoing immune response also becomes important. In nonimmunized animals, mitogenic responses are reduced by acute stressors, indicating that in this altered context, the arrival of antigen might alter the process, presentation, and induction of antibody-mediated immune responsiveness. In early studies, sheep erythrocytes [i.e., sheep red blood cells (SRBC)] were commonly used as the antigen, mainly because this was the common approach in immunology. When animals received SRBC for the first time and were exposed to an acute stressor, the number of antigen-specific antibody-forming cells (AFC) and the concentration of circulating antibody was reduced (Bhatnagar, Shanks, & Meaney, 1996; Zalcman & Anisman, 1993). These effects were seen even when the stressor was administered several days after the initial (i.e., primary) immunization. This suggested that well after the initial antigen processing by APCs[3], stressors can still impact the initial stages of antibody production and antibody-producing B cell proliferation. In contrast, administering the stressor prior to or at the time of immunization can produce variable effects. In some cases, a one-time application of the stressor at or a day prior to the time of antigen administration, failed to alter the number of antibody-producing B cells (Zalcman & Anisman, 1993). In other cases, a single stressor session immediately prior to or following primary immunization with various antigens (e.g., SRBC, KLH, or tetanus toxoid) either augmented or attenuated antibody production (Kusnecov & Rossi-George, 2002). There were methodological differences that need to be resolved with this type of research. Different laboratories used different species (mouse vs rat), different stressors, and different antigens. Thus, although the B cell antibody response is malleable by stressor perturbation, the outcome is not always predictable. This is no less problematic in the human literature, which we will discuss shortly. Still, one is urged to look for consistencies that could prove useful. For example, in experiments that used KLH as the antigen, enhanced antibody responses were observed after a single, brief (1 hour or less) stressor exposure (Shanks & Kusnecov, 1998). However, inhibitory effects were observed in response to more intense stressors (restraint combined with tail shock), which can also suppress the antigen-specific T and B lymphocyte proliferative response (Gazda, Smith, Watkins, Maier, & Fleshner, 2003). Thus, it seems that primary humoral immune responses may be

---

[3] Recall that APCs can be macrophages, DCs or even B cells. Unless directly investigated, one can only assume that all three types may be involved, although DCs are widely considered the main APCs in lymphoid organs.

refractory to acute stressors, especially when these coincide with antigen exposure. More severe stressors may overcome this resistance or refractoriness, especially if the glucocorticoid response is pronounced (Gazda et al., 2003). Overall, however, the evidence is more aligned with the conclusion that cognate interactions between APCs, B cells, and T cells resist acute stressor effects, such that the antibody response either remains unchanged or exceeds normal levels. This again supports the notion that in the short-term, stressors could actually serve as an immunological adjuvant (Viswanathan et al., 2005). However, this window of stressor-promoted "adjuvanticity" may be narrow, since application of an acute stressor several days after antigen presentation may suppress AFC numbers. This reduction in the number of AFCs is a deviation from the normal course of the B cell response, suggesting that once B cells are activated and begin generating antibody, the elevated concentrations of stressor-induced neurotransmitters and hormones play a more important role in modifying the immune response.

The foregoing comments apply to short-term or acute stressor exposures. When stressor exposure is more prolonged (e.g., mice isolated or exposed to foot-shock (1 hour/day) for 2 weeks), and then immunized with SRBC, the antibody response remained unaffected (Shanks, Renton, Zalcman, & Anisman, 1994). While this suggests some form of physiological adaptation that maintains the stability of the humoral immune compartment, additional manipulations after the chronic stressor experience revealed that this was not necessarily the case. If chronically stressed mice were rested for a few days and then given one more exposure to the foot-shock stressor 3 days after immunization with SRBC antigen, AFC formation was enhanced (Zalcman & Anisman, 1993). As discussed earlier, if naïve (non-stressed) mice were first immunized, and then subjected to a stressor a few days later, the

antibody response was suppressed (Zalcman & Anisman, 1993). In effect, a stretch of time during which the organism is working harder than normal to maintain physiological homeostasis might actually prime the immune system to mount a more pronounced immune response if an additional stressor is experienced after a period of rest. Indeed, for the isolation study mentioned earlier, the initial week of isolation was marked by a suboptimal antibody response, which subsequently normalized (or showed a rebound) after more prolonged isolation. A similar principle operates in rats in that exposure to two daily restraint sessions reduced the anti-SRBC antibody response, but SRBC immunization given after 4 days of restraint, normalized the antibody response (Millan et al., 1996). These findings indicate that adaptation of the humoral immune response occurs in the face of persistent changes in an organism's environmental conditions. This might explain why most organisms survive life's interminable cycles of stress and rest, suggesting that preexposure to stress modifies (and may even reinforce) the effects of subsequent stressor exposure on antigen-specific humoral immune responsiveness. In the same way as the brain learns and adapts to the myriad stimuli to which an organism responds (i.e., displays neuroplasticity), it is likely that this similarly occurs with respect to how the immune system deals with constant fluctuations or "waves" of neurohormonal levels that are by-products of stressors. But, the possibility must also be considered that such adaptations may not occur when stressors are severe, and are naturally meaningful. We talked earlier about the powerful role of stress emanating from social conflict, and, once again, this particular stressor holds an important instructive lesson. Whereas acute exposure of mice to social conflict did not affect the antibody response to KLH, more prolonged social conflict exerted a significant inhibition of the antibody response (Lyte, Nelson, & Thompson,

1990). Thus, while there is every reason to think that some form of plasticity in the dynamic relationship between the brain and the immune system occurs during chronic or persistent exposure to stressors, there will be exceptions. These will rest with the types of stressors with which the organism needs to cope, as well as in relation to genetic and experiential elements.

Stressor Effects on Macrophage Function

Mononuclear phagocytic cells (macrophages) are key elements of the innate component of the immune response. The study of neural-immune interactions is heavily focused on these cells—or the myeloid lineage in general—as their activation is considered proinflammatory and elaborates cytokines (e.g., IL-1) that have potent CNS effects. However, these cells also orchestrate antigen-specific induction and effector mechanisms related to T and B lymphocytes. As the first line of defense against microbial agents, it is important to know how these cells are affected by stressors.

An important aspect of innate immunity is the ability to display chemotactic responses to sites of infection or inflammation. Mice infected with Listeria monocytogenes at the beginning of a 7-day period of daily restraint showed reduced migration of macrophages to the peritoneum where the bacteria were introduced (Zhang et al., 1998). However, reductions in migratory behavior may reflect altered responses to chemotactic factors (e.g., chemokines), which are independent of altered phagocytic function. Indeed, stressors can enhance phagocytic and suppressor functions of peritoneal macrophages (Bailey, Engler, Powell, Padgett, & Sheridan, 2007), which is consistent with the notion that the expression of genes linked to the innate immune system are augmented following stressor exposure (Maslanik et al., 2012). In addition, elimination of macrophages either *in vitro* or *in vivo*, removes stressor-induced suppression of spleen cell mitogenic function, as well as restraint-induced enhancement of antigen-specific AFC numbers (Shanks & Kusnecov, 1998). Evidently, stressor actions on macrophages can influence how T and B cell functions are affected, possibly through changes in macrophage cytokine production. For example, periodic restraint or social disruption increased LPS-induced IL-1β, IL-6 and TNF-α levels in mouse brain, spleen, liver, lung, and peritoneal cells (Engler et al., 2008; Quan et al., 2001). Similarly, increased TNF-α and IL-12 production occurred in mice infected with toxoplasma gondii and exposed to a cold stressor (Aviles & Monroy, 2001), while in rats exposed to neurogenic or psychogenic stressors, IL-1β production by alveolar macrophages was increased in response to LPS (Broug-Holub, Persoons, Schornagel, Mastbergen, & Kraal, 1998). It also appears that HPA activation by inflammatory factors may be mediated by prostaglandin synthesis by perivascular macrophages (Serrats, Grigoleit, Alvarez-Salas, & Sawchenko, 2017). Together, these observations indicate that stressors can elevate proinflammatory cytokines, including those that have the capacity to modify neural and behavioral functions. In addition, these cytokine changes can be instrumental in supporting humoral immune responses to various antigens. In effect, a dynamic interplay between innate and adaptive immune cells may operate during stressor exposure, and dictate the direction of a given antigen-specific immune response.

Stressor Effects on Dendritic Cells

Examining macrophages can provide insight into how they might influence antigen presentation. However, as discussed in Chapter 2, The Immune System: An Overview, DCs represent the chief cells that process and present antigen to T and B cells. A brief review of stressor effects on DC function was provided by Kohman and Kusnecov (2009), although relatively little research has been conducted in this area. As we noted earlier, the application of stressors at the time of immunization with antigen, provides the opportunity to gain an indirect sense of how APC function might be affected. In one study, mice were subjected to 8 hours restraint, after which they were sensitized with fluorescein isothiocyanate (FITC); contact sensitivity to FITC—which primarily recruits a specific CD8 + T cell response—was reduced upon testing 5 days later (Kawaguchi et al., 1997). It was also observed that dermal Ia + Langerhan cells (LCs) sampled immediately after stressor exposure, occupied less cutaneous terrain and displayed a more compact, rounded appearance, with less dendritic branching. The LCs are the primary APCs in skin, and although direct evidence is lacking, it is possible that stressor-induced interference in epidermal DC antigen uptake and processing impaired the antigen-specific T cell response to FITC. In a related study, 2.5 hours of restraint that preceded sensitization with DNFB, enhanced the DTH response to the sensitizing antigen (Saint-Mezard et al., 2003). This effect was dependent on altered DC function, which was demonstrated using two different approaches. In the first, bone-marrow derived DCs were antigen pulsed ex vivo (i.e., in which purified DCs are incubated *in vitro* with antigen for a fixed period of time) and then passively transferred to mice that were stressed. When these mice were challenged with antigen 5 days later, the DTH response in the challenged ear pinna was enhanced. In a second approach, fluorescently labeled DCs that were injected into the footpads of mice just prior to restraint were more likely to migrate into the draining popliteal lymph nodes (those closest to the footpads). It was thus concluded that acute stressor exposure augments T cell mediated contact sensitivity through increased DC

antigen presentation. However, given that the responding T cells were always those of the stressed animal, it cannot be ruled out that a combination of APC and T cell interactions occurred.

Further to such findings, mice exposed to acute restraint, and then injected with the sensitizing agent, DNFB, later displayed greater leukocyte infiltration and higher mRNA levels of chemokines and proinflammatory cytokines (e.g., IL-1β, TNF-α, IL-6, and IFNγ). Moreover, this was associated with a nonselective increase in cell numbers in the dermal tissue and draining cervical lymph nodes, and included mature DCs, macrophages, as well as mature and naïve T cells (Viswanathan et al., 2005). Evidently, in response to an acute stressor, multiple immune factors, including APCs (DCs and macrophages) and T lymphocytes behave in a coordinated manner, thereby facilitating augmented antigen-specific immunoreactivity. The combined actions of the different immune cells likely contribute to the enhanced stressor-induced DTH responses that have been reported (Dhabhar, 2014).

In humans, functional assessments of DC-mediated antigen presentation are uncommon. Nonetheless, it was reported that human circulating DCs, that exhibit a CD11c$^+$/CD14$^-$/CD19$^-$ phenotype, are increased during laparoscopic cholecystectomy surgery or physical exercise, followed by a decline to baseline upon termination of the stressor (Ho et al., 2001). The initial rise in DC numbers may reflect a stress-related demand that extrudes sequestered cells from locations with ready access to the vasculature (e.g., the spleen). Afterward, they undergo redistribution to various tissues in order to maximize their chances of encountering and processing antigen. Unlike these findings, however, a stressful public speaking procedure conducted 24 hours earlier, resulted in a reduction in the number of epidermal Langerhans cells obtained from cutaneous tissue (Kleyn et al., 2008). In this case, one interpretation is that stress promoted emigration of LCs to the lymph nodes that drained the area where skin was biopsied.

Stress and Infection

Analyses of stressor actions on the progression of infectious disease processes have generally revealed elevated susceptibility to replicating viral or bacterial antigens (Kusnecov & Rossi-George, 2002). Further, stressor-provoked suppression of host defense against influenza and herpes virus infections has been well-established in rodents (Sheridan, 1998). In the main, these findings have come from studies assessing the impact of repeated or chronic stressors; however, acute social defeat of an intruder by a resident mouse, reduced lymph node cellularity, and attenuated production of IL-2 and IFNγ in response to primary infection with herpes simplex related pseudorabies virus (de Groot, van Milligen, Moonen-Leusen, Thomas, & Koolhaas, 1999). This finding was in line with an earlier report that impairment of antiviral immunity required a relatively extended period of a restraint stressors (Sheridan, 1998). This said, the pseudorabies strain used in the study by de Groot et al (1999) was avirulent, which was very different from the replicating, virulent strains used by other researchers. Yet, repeated social defeat also generated enhanced immunological memory to influenza virus (Mays et al., 2010). This is obviously at odds with the commonly held view that stressors compromise host defense against microbial targets. To the contrary, as we indicated repeatedly, it seems that there are conditions in which stressors might operate in a protective capacity, thereby facilitating greater resistance against disease.

A fundamental issue that ought to be addressed is whether subclinical doses of infectious agents are affected by stressful experiences so that they fulminate into frank infectious disease. It has been reported that exposure to influenza virus fails to produce symptomatic disease in a large segment of the population, likely owing to the effectiveness of preinfection T cell immunity (Hayward et al., 2014). Clearly, the emergence of illness is not guaranteed to occur simply as a result of a pathogen's presence, and symptoms might be more likely to occur with a second hit. Early studies in rodents had revealed precisely this, in that subclinical infection can be encouraged by the presence of stressors to convert into an infectious illness, presumably owing to reduced immune functioning. Later studies confirmed these findings in human volunteers who were challenged with moderate doses of rhinoviruses, the source of the common cold. These studies revealed that those individuals who reported high levels of perceived distress were likely to become symptomatic and exhibited typical signs of infection, whereas those with low stressor experiences tended to remain asymptomatic (Cohen et al., 1998; Cohen, Tyrrell, & Smith, 1991). As we indicated earlier, at the end of the day, one of the big questions that confront stress researchers is not simply one of whether stressful events undermine specific components or operations of the immune system, but whether these experiences leave the individual at increased risk of becoming ill. Studies such as these speak directly to such questions, and more clinically-relevant studies of this type are needed.

IMMUNOLOGICAL CONSEQUENCES OF STRESSOR EXPOSURE IN HUMANS

There is a substantial human literature reporting on the impact of stressors on basic immune functions, antiviral immunity, and infectious disease, and the general healing effects of the immune system (Fagundes,

Glaser, & Kiecolt-Glaser, 2013; Glaser & Kiecolt-Glaser, 2005). The complexity of the various findings is in some instances no less confusing than the animal studies, which is understandable given the heterogeneity of genetics and experiences that human participants bring to these investigations. A review and meta-analysis of close to 300 studies that assessed healthy human subjects, focused on basic immune parameters derived from peripheral blood leukocytes, since more invasive analyses are typically not possible (Segerstrom & Miller, 2004). In this analysis, the various stressors used were classified as being acute, chronic, naturalistic, and what was referred to as "event sequence" and "distant." Bereavement was an example of an "event sequence" stressor that occurred within a year, and potentially possessed persisting residual effects. Chronic stressors were persistent challenges, such as those faced by caregivers, while distant stressors included trauma experienced 5–10 years prior to immune assessment. The outcome of this analysis led to a number of conclusions which tended to differ between studies that involved acute and chronic stress conditions. Here the findings were very similar to some of the reports from the animal literature discussed earlier. The acute stressors that were considered included public speaking, performing challenging mental arithmetic, or parachuting from an airplane, which cover the social, cognitive and emotional domains of experience. Irrespective of the duration of the stressors, which ranged from 5–100 minutes, there appeared to be some consistency in the types of immune changes that

occurred. The strongest effects were an increase in NK cell numbers or the related large granular lymphocytes (LGLs)[4], which are associated with NK cell function. The functional significance of the increased NK cell numbers was related to increased cytotoxicity, but this was not related to a fundamental enhancement of cytotoxicity at the individual cell level (Segerstrom & Miller, 2004). Peripheral blood lymphocyte proliferation to mitogens that globally stimulate T cells and/or B cells was decreased by acute stressors, and in some studies that examined cytokine production in these mitogen assays, the most consistent effect was increased IL-6 and IFNγ production. This was interesting as reduced proliferative function might be more related to reduced efficiency or production of IL-2. Still, the elevation of IL-6 and IFNγ is notable in that two indices of lymphocyte function—proliferation and cytokine production—generated opposing effects. Moreover, we should note that humoral immune function (e.g., measures of immunoglobulin) was found not to be dramatically affected. This is neither the first nor last time that different parameters of immune function would reveal differential changes in the face of a stressor. Unfortunately, this adds to the confusion regarding how to interpret immune changes.

There are alternative measures of humoral immunity, some of which are sensitive to stressor effects. Short-term natural stressors, such as academic examinations, have commonly been used to assess immune function, and a small study showed that the professors themselves can be impacted by a large crowd

---

[4] LGLs are morphologically distinct cells of the lymphoid lineage (see Chapter 2: The Immune System: An Overview). These cells were originally thought to be primarily NK cells, which target tumor cells and virally infected cells. However, a separate subset of LGLs exist which are thymus-related (hence, T-LGLs), and which are seen in certain forms of leukemia. In the studies reviewed by Segerstrom and Miller, as well as other studies on stress and human immune function, LGL numbers and function is generally interpreted in terms of their cytotoxic functions (i.e., their role as so-called NK cells).

of students. For example, professors teaching to a body of 200 students showed elevated salivary TNF-α, IL-2 and IL-4 concentrations on days when they lectured, relative to off days (Filaire et al., 2011). This reinforces the notion that social stressors and/or public speaking, as an arousing and potentially intimidating influence, (given the presence of public scrutiny) can have pronounced physiological effects. For students taking exams, the stakes are less about social confrontation and shame, but more about academic success and the opportunities this provides. Even though this type of stressor hardly affects cortisol levels, as we have already seen, it does promote reduced *in vitro* IFNγ production, but increased IL-6 and IL-10 production. Interestingly, the changes in IFNγ production may be independent of changes in T cell proliferation. Recall that T cell proliferation tends to be reduced by acute laboratory stressors, and using naturalistic stressors, such as social engagement, the same outcome was observed. This differentiation raises the obvious possibility that some parameters of lymphocyte function may be unrelated to one another. For example, IFNγ is a MHC Class II inducer, as well as an activator of NK cell function. Consequently, to fully assess its function, one needs to examine the cytokine in the context of what it typically is required to accomplish.

In addition to T cell proliferation, other parameters affected by short-term naturalistic stressors include loss of restraint on the normally dormant Epstein Barr Virus (Glaser & Kiecolt-Glaser, 2005). Ordinarily, cell-mediated immune processes, including T cell cytotoxic activity, suppress latent herpes viruses (e.g., Herpes Simplex Virus and EBV). In the case of EBV, large segments of the population may be infected, but the virus remains dormant within B lymphocytes, likely stemming from ideal immunosurveillance and restricted viral proliferation within the memory B cells where it resides. A strong indication that immune functioning has been compromised by weakened T cell control is an increase in circulating anti-EBV antibody titers. During high stress periods (e.g., academic exams for medical students), an increase of IgG antibody specific for the protein shell or capsid of EBV was observed (Glaser & Kiecolt-Glaser, 2005), which was consistent with greater incidence of infectious mononucleosis in other high stress populations (Glaser, Rabin, Chesney, Cohen, & Natelson, 1999). Moreover, the mechanism for the reactivation of the virus was due to impaired cytolytic destruction of proliferating EBV-infected B cells (Glaser & Kiecolt-Glaser, 2005). Other cytotoxic cells are also inhibited by naturalistic stressors. In the Segerstrom and Miller (2004) meta-analysis, naturalistic stressor experiences in seemingly healthy individuals imposed a reduction of NK cell function, which was opposite to the enhancing influence on NK cells by acute laboratory stressors. These variations, which were tied to the type of stressor investigated, exemplify the difficulty in predicting stressor-induced outcomes. Although laboratory stressors allow for more experimental control, they fail to capture the demands that attend the stressor of pending life challenges (like exams), and the preparation and cognitive rumination associated with such events. While such natural phenomena need to account for altered sleep, exercise, and dietary patterns, they do provide a better view into the potential immunological status of the individual living in demanding circumstances.

This brings us to the consequences of chronic stressors in a natural context. Caregivers, for example, have received considerable attention, due to the constant strain of working with people with dementia and other disabilities. For example, chronic stressor conditions impose suppressive effects on T cell proliferation and antibody responses to influenza vaccines (Glaser & Kiecolt-Glaser, 2005). This is compelling, since it implies that the efficacy of an annual vaccine is potentially modified by the psychological state of the individual, and how much hardship they may have endured. Subsequent studies in humans that examined the relationship of psychosocial factors and stressors to antibody production after *in vivo* immunization had reached similar conclusions (Cohen, Miller, & Rabin, 2001). These data, in part, extend what has been found in the animal literature regarding stressor-induced modulation of humoral immune responses to systemically delivered protein antigens. However, the effects in humans are not as dramatic, nor as predictable. Human studies involving immunizations are constrained by methodological compromises. That is, researchers capitalize on established vaccination programs and need to operate within established guidelines for carrying out community-based vaccinations. There is no control of antigen dose, and variations in the composition of the injected vaccine usually occur, such as in the case of influenza studies. Finally, in these studies, it can be difficult to determine the relationship between onset, duration, and frequency of stressful episodes and measures of antibody. As well, the temporal relationship between stressor exposure and antigen delivery (the vaccine) may be an important influence on the ultimate nature of the antibody response. Clearly, it can be difficult to explore with any great precision how stressors impact the immune response to vaccinations. Efforts to explore responses to benign protein antigens are more likely to circumvent this problem.

Despite the constraints, it was suggested that stressors may, in fact, reduce the secondary antibody response to vaccinations and/or other antigens, but is less effective in modifying the primary antibody response (Cohen et al., 2001). However, in healthy volunteers that received an influenza vaccination, increased reports of daily stressors were associated with lower antibody titers (Miller et al., 2004). In this study, stress monitoring occurred over a period of 13

days, prior to, during and after inoculation. Levels of distress for the 2 days prior to antigen delivery did not account for the reduction in the antibody response, but the 10 days during which the immune response was being mounted appeared to be critical in influencing the antibody response. Furthermore, loneliness and a smaller social network was associated with a reduced anti-influenza antibody response (Pressman et al., 2005).

Interestingly, a meta-analysis of the influenza vaccination literature (Pedersen, Zachariae, & Bovbjerg, 2009), concluded that perceived or chronic stressors (e.g., caregiving) was associated with a reduced antibody titer to vaccination, being evident in both older (mean age 70) and younger (mean age 40) participants. In most of the studies reviewed, the inoculation involved a cocktail of different viral strains, and the relationship to stressors existed mainly to the AH1/N1 strain. The reasons for this selectivity are unknown, and may relate to preexisting higher antibody levels for other strains. Nonetheless, it appears that experiencing higher stress levels may result in less than optimal concentrations of circulating antigen-specific antibody. Whether this also results in less protection against infection remains to be determined. Clearly, there is good reason to determine whether psychosocial factors play a role in the protective potential of annual vaccinations. Overall, the pattern of results for healthy human subjects suggests that the longer a stressful situation is experienced, the more likely it is to suppress a given measure of immune function, whether this was determined *in vitro* (e.g., mitogen-induced proliferation) or through vaccination/immunization procedures.

There appears to be somewhat of a disconnect between the animal and human literature with regard to the concordance between *in vitro* and *in vivo* measures. Earlier in our discussion of the animal literature, we saw that spleen cells from rodents exposed to a stressor also displayed suppressed mitogen-induced proliferative response. However, injection of antigen at such a time (when mitogen capacity is inhibited), can actually augment the antibody response (Kusnecov & Rossi-George, 2002). This is not always the case, but we do need to be careful in how we use the outcomes of one immunological assessment to predict those of another. Most certainly, *in vitro* assays are useful as an indication that the behavior of cells has changed, but may not shed new light on how handling of antigens and the response to them transpires *in vivo* before, during, or after some period of stress. Nonetheless, in healthy human participants, the initial wave of research established that T lymphocyte proliferative capacity and NK cell cytotoxicity, as well as virus-specific T cell cytotoxicity, were particularly affected by stressors (Segerstrom & Miller, 2004). Focusing on just these standard immune measures in healthy subjects, however, does not provide immediate insight into the health relevance of these observations. Other research, which focused more on *in vivo* dynamics, provides a better perspective on this question. As discussed earlier, the studies on viral reactivation and impaired cell-mediated control over viral clearance is particularly relevant, as are studies involving more direct investigation of infectious disease and stress in humans.

As already described, susceptibility to various rhinoviruses was found to vary according to the individual's previous psychosocial experiences (Cohen et al., 1998, 1991). Specifically, the greater the incidence of stressor experiences, as well as duration of the stressors, the greater the likelihood that human volunteers showed cold symptoms. Moreover, stressed individuals in these studies showed greater propensity for somatic sensations to immune-mediated processes (Cohen, Janicki-Deverts, & Doyle, 2015). This may be related to augmented immune-mediated effects on afferent or central neuronal systems, since individuals who were exposed to the rhinovirus, exhibited greater intranasal

cytokine responses. Specifically, higher levels of intranasal IL-6 and TNF-α were associated with greater levels of stressor-related threat, as well as greater susceptibility to the cold virus (Cohen, Janicki-Deverts et al., 2012). Furthermore, this was directly correlated with increased glucocorticoid resistance, a notion that is increasingly gaining traction as an explanatory mechanism for enhanced inflammation and autoimmunity.

To conclude this section, the value of many of these studies, and the vaccination studies in particular, is that they extend and provide some useful corroboration for the *in vitro* findings of stressor effects on immune function. However, the relationship between *in vitro* measures and *in vivo* immune effects can be complicated. It has always been a pressing question whether a given immunophenotype (e.g., poor *in vitro* lymphoproliferation) is predictive of impaired responsiveness to *in vivo* antigen exposure. This question has not been adequately answered even within immunology (Kaczorowski et al., 2017). But it is encouraging that the impact of stress on immune function is not strictly an *in vitro* phenomenon. In the next section, we turn our attention more closely to the effects of stress on cytokine production. Here we will also discuss some of the contributions of animal studies to the question of how stressors alter the many different cytokines that both regulate the immune system, as well as potentially influence the brain.

STRESS AND CYTOKINE PRODUCTION

Cytokine synthesis and release is a fundamental property of all leukocytes, and may be especially sensitive to stressor effects. The ability of lymphocytes and macrophages to alter synthesis and rates of cytokine production in response to glucocorticoids, norepinephrine, acetylcholine, and various neuropeptides has been well documented (Olofsson, Rosas-Ballina, Levine, & Tracey, 2012; Sternberg, 2006). This has the potential to alter the series of events that culminate in the effector phase of the immune response, be it antibody production, cytotoxicity, or more cytokine production. As we saw in Chapter 2, The Immune System: An Overview, regulation of T-helper cell cytokines may influence antibody production, as well as the antigen specificity of antibody subtypes. Moreover, the expansion and lytic potential of cytotoxic T cells and NK cells is affected by a variety of T cell and macrophage-derived cytokines, including IFNγ and TNF-α. However, in addition to this local immunoregulatory impact, stressor effects on cytokines may affect how they interact with elements necessary to affect neural and behavioral functions. In this particular domain we have to ask whether stressors will increase cytokine levels to a degree that has greater neural influence, or conversely, precipitates cellular changes that increase migration of monocytes and T cells into the brain. The details of these dynamics are not altogether clear at this time, but are worth considering.

As much as cytokines might be involved in the provocation of psychological or physical disturbances, their actions vary with multiple individual factors, experiences during early and later life, and a constellation of other environmental factors. A schematic representation of such a process is depicted in Fig. 6.2, but as we will see in the ensuing section, the many different pro- and anti-inflammatory cytokines may be differentially influenced by stressors and have diverse consequences.

Stress Effects on Th1 and Th2 Cytokine Responses

The adaptive immune response relies heavily on the balance between Th1 and Th2 cell function. Since a Th2 bias is associated with a drive toward improved support and initiation of humoral immune responses, such as antibody production, it is thought that predominance of Th1 responses would be counterproductive to the promotion of the antibody response. Therefore, a bias toward proinflammatory processes might operate through activation of macrophages (if shifted toward Th1 cytokine responses), whereas greater humoral immune responsiveness would operate through a greater shift toward Th2 cytokine output. This can be relevant to immune-related pathology, since several animal studies have shown that exaggerated and protracted skewing in either direction—Th1 or Th2—can influence infectious and autoimmune disease (Raphael, Nalawade, Eagar, & Forsthuber, 2015). For instance, during the initial stages of an immune response to pathogen, Th1 cell cytokine responses are required to drive increased phagocytic functions, through activation of macrophages and facilitating antibody-dependent opsonization of the pathogen. Over time, this is down-regulated by Th2 cell cytokines, such as IL-10, which can attenuate the damage that a needlessly protracted immune response can impose on local tissue. Therefore, if stressors modify the production of Th1 and Th2 cytokines, this may induce critical imbalances in their mutual counter-regulatory functional relationship, which may result in pathology and inflammatory disease. The key cytokines that have received attention in the context of stress studies include the Th1 cytokines, IL-2, IFNγ, and TNF-α, whereas Th2 function is generally attributed to IL-4, IL-6, and IL-10 production. However, it should be acknowledged that cytokines from either Th cell subtype are synthesized and release by other immune cells. Thus, in assessing circulating cytokines, the cellular source is unknown, and may not originate from T cells.

Many of the effects of stressors on cytokine production have been conducted using *in vitro* approaches, in which harvested leukocytes,

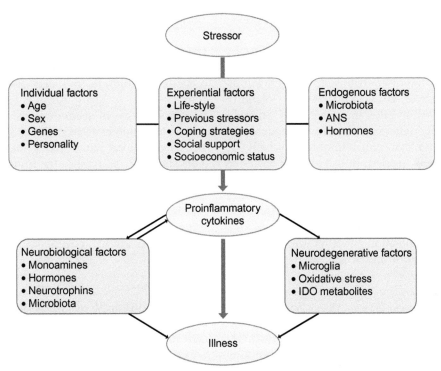

FIGURE 6.2 A schematic representation of the relations between stressors, pro-inflammatory cytokines, and illness, with a particular focus on the individual, experiential, and endogenous factors that may moderate these effects. The capacity of social stressors to promote inflammatory variations that might lead to depression (or other illnesses) may be influenced by the presence of genetic and personality factors, sex, and age. Individuals carrying specific gene combinations or polymorphisms (e.g., variants of IL-6, IL-1β, and TNF-α) may be more vulnerable to the depressive effects of inflammatory activation provided that they also encounter social stressors. Earlier stressor encounters both prenatally and during early life, coping processes (including social support), and gut microbiota variations may influence inflammatory processes and sensitize immune responses to subsequent stressors. The activation of pro-inflammatory processes may directly or indirectly influence depressive states. Elevations of cytokines may influence monoamine (e.g., 5-HT, NE), hormone (e.g., CRH), and growth factor (e.g., BDNF) activity which might favor the evolution of depression (conversely, stressor elicited hormonal and neurochemical functioning may impact cytokine processes). Cytokine elevations may also stimulate the enzyme indoleamine 2,3-dioxygenase (IDO) and promote the release of neurotoxic metabolites, including kynurenic acid, quinolinic acid, or 3-hydroxykynurenine, and cause oxidative stress, culminating in depression. *Source: From Audet, McQuaid, Merali, & Anisman (2014).*

whether from blood or lymphoid organs, are stimulated with mitogens or antigens to which animals or humans had previously been primed or sensitized *in vivo*. There have been studies in animals and humans that assessed *in vivo* elevations in cytokines after stressor exposure (Marsland, Walsh, Lockwood, & John-Henderson, 2017), and it likewise appears that the cytokine variations are more pronounced among animals that had previously been exposed to a stressor, corticosterone treatment, or cytokine exposure (Anisman, Hayley, & Merali, 2003). In addition to sensitization of epigenetic changes, it was suggested that this neuroinflammatory priming involves a signal cascade within the brain that involves DAMPs and the inflammasomes (NLRP3). Whatever the case, it appears that earlier experiences can promote exaggerated neuroinflammatory responses that can promote behavioral disturbances, and might contribute to psychiatric illnesses (Fleshner, 2013; Frank, Weber, Watkins, & Maier, 2015). Like the effects in animals, stressful experiences in humans can also promote marked cytokine alterations, which we will discuss as needed. But, it should be noted that in human studies that have measured circulating cytokine concentrations (in plasma or serum), rarely is the cellular source of the measured plasma cytokine reported. Consequently, it is open to question whether the measured cytokine is of immune origin, or from alternative tissue groups.

Interferon-γ

Exposure of animals to acute stressors was shown to enhance IFNγ production in the DTH immune model of antigen-specific responding (Dhabhar, 2014). This has been shown to apply to *in vivo* measures, but when *in vitro* assessments were conducted, inhibition of IFNγ production was provoked by acute stressors (Kohman & Kusnecov, 2009). Similarly, when stressors were prolonged (Iwakabe et al., 1998) or repeated daily, IFNγ production was attenuated in response to a variety of antigens, including tetanus toxin, herpes simplex and influenza viruses, and less pathogen-associated stimuli, such as ovalbumin and CD3 crosslinking with monoclonal antibody. Importantly, *in vivo* regional lymph nodes displayed inhibited ability to produce IFNγ after exposure of the animal to a stressor (Bonneau, Zimmerman, Ikeda, & Jones, 1998; Dobbs, Feng, Beck, & Sheridan, 1996).

It was noted earlier that in humans, acute stressors are more likely to increase IFNγ production assessed ex vivo, which was confirmed in the meta-analysis of human stress-induced cytokine changes (Marsland et al., 2017). However, the suppression is eliminated after prolonged or more persistent stressor exposure. Under these conditions, stressors either failed to alter the percentage of IFNγ+ CD4 and CD8 T cells (Glaser et al., 2001) or exerted a modest reduction in T cell IFNγ production (He, Gao, Li, & Zhao, 2014). Interestingly, the persistent stress of academic examinations produced an enhancement of IFNγ production by stimulated blood leukocytes (Maes et al., 1998). This stands in contrast to the findings from animal studies, which showed chronic stressors to inhibit IFNγ expression. Of course, animal studies access cells from lymphoid organs, whereas human studies rely on circulating cells, which might account for the differences observed.

Interleukin-2

Stressor-induced suppression of IL-2 production was established some time ago in various rat and mouse studies (Bonneau, 1996). Similarly, psoriasis patients or healthy controls that were exposed to the Trier Social Stress Test, both showed a significant decline in mitogen-induced IL-2 production by peripheral blood leukocytes (Buske-Kirschbaum, Kern, Ebrecht, & Hellhammer, 2007). This is also consistent with research showing that caregivers, who reported high levels of perceived distress, also displayed reduced capacity to generate IL-2 (Bauer et al., 2000). However, a meta-analysis revealed that stimulated IL-2 production was unaffected by social threat (Marsland et al., 2017). Thus, while IL-2 gene activation and cytokine output in humans is susceptible to stressor effects, it is apparent that this is not a guaranteed outcome. In having considered stressor effects on IL-2 and IFNγ, we are in essence gaining a perspective on the response capacity of Th1 cells.

For the most part, commonly employed experimental stressors in animal studies exerted a suppressive influence on Th1 cytokine production, although largely when this was assessed by *in vitro* restimulation methods.

Less research is available on measures of Th1 cytokine production *in vivo*, which may help to determine whether the implications of the *in vitro* studies are relevant to Th1 dependent diseases. As noted earlier, the argument has been made that augmented *in vivo* Th1 responses occur after acute, but not after chronic stressor exposure (Dhabhar, 2014). This implies a requirement for cognate stimulation of T cells (i.e., with antigen), since mice exposed to prolonged restraint (from 1 week to several weeks) failed to show plasma IFNγ elevations (Voorhees et al., 2013). Given that Th1 cytokines generally need to be induced by a cognate stimulus, this finding is not surprising, and suggests that spontaneous elevations of circulating Th1 cell cytokines are less susceptible to the effects of stressors, which was also the case for human studies (Marsland et al., 2017).

Interleukin-4

In considering Th2 type responses, we can begin with IL-4. This cytokine is important in regulating B cell activation and differentiation, but is most prominent in promoting IgE antibody responses, which is involved in allergic reactions. In humans, mitogen-induced production of IL-4 was reduced in response to psychosocial and academic stressors (Buske-Kirschbaum, Gierens, Hollig, & Hellhammer, 2002; Buske-Kirschbaum et al., 2007; Maes et al., 1998; Uchakin, Tobin, Cubbage, Marshall, & Sams, 2001), although in another study that utilized a social stressor (public speaking task), no changes in mitogen-stimulated IL-4 production were observed (Ackerman, Martino, Heyman, Moyna, & Rabin, 1998). It is evident that further research is needed to address important interactions between stressors and immunological mechanisms underlying allergic reactions, which in some cases represent a significant life-risk.

In animal studies, restraint failed to affect IL-4 production induced by T cell mitogens (Iwakabe et al., 1998; Li, Harada, Tamada, Abe, & Nomoto, 1997). However, restraint reduced spleen cell IL-4 responses to herpes simplex virus (Bonneau, 1996), and murine exposure to a social stressor imposed a long-term memory deficiency in IL-4 production to a porcine pseudorabies virus (de Groot, Boersma, Scholten, & Koolhaas, 2002). The latter was selective for mice that exhibited wounds, with nonwounded mice failing to show any effect on IL-4. These data suggest that infection and injury interact with psychogenic stressors to inhibit Th2 cell cytokine production. The consequences may be an extension of Th1 mediated immunity, which is necessary to drive inflammatory and phagocytic processes that are designed to eliminate pathogens. However, this notion is not in keeping with the known suppressive effects of stressors on IFNγ that were mentioned earlier, relegating the significance of changes in IL-4 production as something yet to be confirmed.

Interleukin-10

If there is one cytokine that could be considered the "glucocorticoid" of the immune system, IL-10 would certainly fit this role. It has been referred to as the master anti-inflammatory regulator of the immune system, and there are few elements of immune activity that are not inhibited by this cytokine (Rojas, Avia, Martin, & Sevilla, 2017). As with other cytokines, production of IL-10 is modifiable by stressors. For example, *in vitro* IL-10 production in response to tetanus toxin, influenza virus, and mitogen was suppressed following prolonged restraint or social disruption (Dobbs et al., 1996; Merlot, Moze, Dantzer, & Neveu, 2004; Tournier et al., 2001). However, when IL-10 was induced *in vivo* by an LPS injection, either acute swim or restraint increased elevated IL-10 production by spleen cells (Connor, Brewer, Kelly, & Harkin, 2005; Curtin, Boyle, Mills, & Connor, 2009; Curtin, Mills, & Connor, 2009). In essence, two different scenarios are revealed by observations that focus either on the induction of IL-10 *in vitro* or *in vivo*. The *in vivo* findings are very much in keeping with the notion that acute stressor exposure can enhance key components of an immune response. In this particular case, the source of IL-10 is most likely macrophages, since LPS does not directly stimulate T cells, but does induce IL-10 in cells of the innate immune system. Additional models are required to focus more closely on T cell derived IL-10. Indeed, it was found that the development of IL-10-secreting memory T cells responsive to pseudorabies antigen was inhibited by exposure of mice to social stressors (de Groot et al., 2002).

An increase in perceived stress among elderly individuals that received an influenza vaccine was associated with an augmented *in vitro* IL-10 response to influenza antigen restimulation (Kohut, Cooper, Nickolaus, Russell, & Cunnick 2002). This was in keeping with a more recent study that examined hepatitis B patients, who were partitioned into high and low daily life stress groups. In these patients, the high stress group showed the greatest stimulated IL-10 production (He et al., 2014). In contrast, a social stressor applied in a laboratory setting suppressed *in vitro* IL-10 production by isolated blood leukocytes (Buske-Kirschbaum et al., 2007). This points out that the particular conditions of stressor exposure and/or induction of IL-10 show differential effects on cytokine output.

In some cases, the influence of stressors on IL-10 production was studied in relation to the Th1 cytokines IL-2 and IFNγ, the production of which is reduced by elevations of IL-10. Indeed, high levels of stress among younger caregivers was associated with elevated numbers of IL-10 T regulatory cells, while the number of IL-2 or IFNγ Th1 cells remained unchanged (Glaser et al., 2001). These caregivers had previously exhibited attenuated antibody responses to influenza vaccine, as well as reduced output of IL-2 (Kiecolt-Glaser,

Glaser, Gravenstein, Malarkey, & Sheridan, 1996), which was in keeping with the increase in IL-10-producing T cells. This inverse relationship was replicated among students with increased exam pressure, showing elevated production of IL-10 (Maes, Christophe, Bosmans, Lin, & Neels, 2000), that was associated with reduced IFNγ output (Marshall et al., 1998). However, in other studies, stressor-induced enhancement of IFNγ production (Maes et al., 1998, 2000), a finding commonly observed after exposure to psychosocial stressors in healthy subjects and patients with atopic dermatitis or MS (Ackerman et al., 1998; Buske-Kirschbaum et al., 2007).

It is evident that stressors can modify IL-10 concentrations, which might exert greater inhibitory control over Th1 cytokine production. If such regulation is inappropriate or untimely in the face of a need to mount an effective immune response, stressor-induced elevations of IL-10 could be considered a risk-factor. However, we do not know enough about the role of IL-10 at the interface between stress and disease. To be sure, since IL-2 and IFNγ are not entirely suppressed by stressors, it is possible that some refractoriness to the inhibitory effects of IL-10 may also develop during stressor experiences. At the very least, it should be considered that the observed increases of IL-10 are compensatory changes, serving as a physiological check against excessive inflammation in scenarios that would benefit from an augmentation of Th1 cytokine responses. Most certainly we need to know more about this relationship to gain a better understanding of how stress may benefit or compromise immunologically-mediated protection against disease.

Interleukin-5

As a Th2-derived cytokine, IL-5 has received attention in the development, maturation, and recruitment of eosinophils. These cells are granulocytic leukocytes with antimicrobial cytotoxic effector capability, and are thought to contribute to a range of immunopathologic situations, including respiratory allergies, asthma, and gastrointestinal disease. In addition, they are involved in the coordination of antigen-specific T and B cell responses. As observed in the case of other Th2 cytokines, academic examinations in students with mild forms of asthma induced pulmonary eosinophilia and increased the production of IL-5 in cells harvested from sputum samples (Liu et al., 2002). Interestingly, peripheral blood eosinophilia was noted in stressed subjects who presented with atopic dermatitis, suggesting that in allergic situations, stressors can promote the migration of potentially pathogenic eosinophils (Buske-Kirschbaum et al., 2002).

Conclusions Regarding Th1/Th2 Cytokine Findings

So, how do we interpret these varying observations? It is evident that attempting to predict differential changes in Th1 and Th2 cytokine function in the context of stressful circumstances is not simple. We also struggle with the additional question of determining whether stressors differentially influence cytokine activity in different compartments of the immune system—something that is seldom addressed. For example, the spleen, blood, lymph nodes, and mucosal-associated tissue compartments all contain varying micro-environments that involve tissue-specific regulation of the immune response. We know very little about how stressors impact the immune system in the mucosal-associated tissue, as opposed to the spleen, or in the case of humans, blood. These are important shortcomings that will hopefully be redressed by the current zeitgeist of interest in the gut microbiota, where immune processes are very likely to be important in any theories of gut regulation of brain

and behavioral functions. In this regard, it was suggested that stressors may undermine natural defense barriers, including the blood-brain-barrier, and those that provide commensal microbiota within the gut. Aside from altering the microbiota present, as described earlier, stressful events promote the translocation of microbiota from cutaneous and mucosal surfaces into regional lymph nodes, and may thus produce neuroendocrine changes as well as immune alterations (Bailey, 2016). Consistent with the view that microbiota can influence stressor-related psychological disorders, as indicated earlier, among mice pretreated with Lactobacillus rhamnosus for 28 days, anxiety-like behaviors and disturbed social interactions were brought about by social defeat. This was accompanied by activation of DCs being limited, while increasing the presence of IL-10 regulatory T cells (Bharwani, Mian, Surette, Bienenstock, & Forsythe, , 2017). Social stressors were also capable of promoting the translocation of the rod-shaped, gram positive bacteria (lactobacilli) from the gut into the circulation, where they can augment the production of IL-1 by spleen cells (Lafuse et al., 2017). Although it is odd that lymph nodes draining the gut were not primed by lactobacilli, with the effect operating on a more systemic level, the findings are in keeping with the notion that stressor-induced alteration of the gut microbiota may precipitate the production of factors that alter behavior. Since IL-1 is a powerful regulator of behavior, and implicated in depression, this work has strong implications for mental health. Of course, several related pathways exist between gut microbiota and brain functioning. As described in Fig. 6.3, interactions occur between the microbiota and the gut, which come to affect brain functioning. Stressful events give rise to the release of peripheral norepinephrine, M2 macrophages, which may be the start of a cascade that leads to changes of immune processes. Some of the released cytokines may also trigger changes of

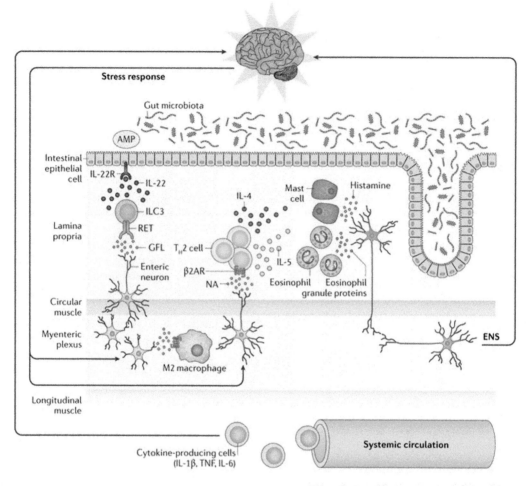

Nature Reviews | Gastroenterology & Hepatology

FIGURE 6.3   There is complex crosstalk between the immune system, the brain and the gut. Noradrenergic neurons in the gut release noradrenaline (NA), which ligates the β_2 adrenergic receptor (β_2AR) on macrophages in the myenteric plexus (supporting differentiation toward the M2 phenotype) and T cells, which limits T helper (T_H)1 differentiation (indirectly favoring T_H2 and T_H17 differentiation). Mast cells and eosinophils are found degranulating next to enteric neurons, providing a mechanism for sensory excitation (which can be perceived by the enteric nervous system (ENS) and central nervous system). Gut immune cells (e.g., $\gamma\delta$ T cells) differentiating in the gut can traffic to the brain under some circumstances (e.g., after brain injury). Blood-borne cytokines generated in the gut can also signal in the brain. Enteric glial cells also produce glial-cell-derived neurotrophic factor family ligands (GFL), which stimulate IL-22 production by innate lymphoid cell (ILC)3 (acting on the specific receptor RET). IL-22 acts on an epithelial-restricted receptor (IL-22R) to stimulate epithelial proliferation, antimicrobial peptide production. IL-4 and IL-5 (potentially generated locally by T_H2 T cells) support activation of mast cells and eosinophils, respectively. AMP, antimicrobial peptides. *Source: From Powell et al. (2017).*

eosinophils and mast cells that come to affect the enteric nervous system, which, in turn, affects CNS functioning (Powell et al., 2017).

Overall, the available data show that stressors can modulate T cell cytokine activity. As we learn more by varying stressor parameters, and probe into critical areas where the host meets the environment (viz, mucosal associated tissues) alternative interpretations will surely arise that will hopefully illuminate how stressors affect Th1 and Th2 cell cytokine output. The extrapolation of animal data to human studies requires conceptual agreement that is sometimes lacking. While the human and animal data are concordant with regard to the mere phenomenon that stressors can modulate the immune system, they diverge with regard to how cytokine production by human immunocytes is affected by stressors. Admittedly, human studies are handicapped by a focus on mainly peripheral blood cells. As well, variations in genetic background, experiential history, and age also need to be addressed (Shields & Slavich, 2017).

Interleukin-6

The attributed functions of IL-6 have grown since its discovery, but for the most part it is recognized as a B cell growth and differentiation factor. It is also involved in fibroblast and neuronal growth, and has been viewed as a neuroprotective factor (Sun et al., 2017; Sun, Xie, Tang, Li, & Shen, 2017), although it is more commonly discussed in the context of various neuronal and systemic pathologies (Rothaug, Becker-Pauly, & Rose-John, 2016). Alterations in the production and/or release of IL-6 have long been recognized, and it has gained attention as perhaps the main and most reliable cytokine to be elevated in plasma after exposure to stressors (Marsland et al., 2017; Steptoe, Hamer, & Chida, 2007). Moreover, increased attention has been directed to IL-6, largely connected with depression in the human literature.

Animal studies demonstrated that simply exposing animals to a series of different stressors would elevate circulating plasma IL-6 (LeMay, Vander, & Kluger, 1990; Zhou, Kusnecov, Shurin, DePaoli, & Rabin, 1993), with more severe stressors producing greater IL-6 elevations. When animals were exposed to a series of stressors that included handling, exposure to novelty, social disruption, restraint, and then electric shock—a sequence that involves increased intensity of threat—the IL-6 concentrations became increasingly higher (Hale, Weigent, Gauthier, Hiramoto, & Ghanta, 2003; Kitamura et al., 1997; Zhou et al., 1993). In earlier animal research, spleen cell IL-6 production induced by a T cell mitogen was increased in lactating (but not nonlactating) female rats exposed to a conditioned psychological stressor (Shanks, Kusnecov, Pezzone, Berkun, & Rabin, 1997). This drew attention to IL-6 elevations being subject to unique physiological states in animals. Similarly, mice exposed to social disruption showed increased *in vitro* IL-6 production to LPS, suggesting heightened macrophage reactivity (Stark, Avitsur, Hunzeker, Padgett, & Sheridan, 2002). Alternatively, stimulation of IL-6 production was reduced in response to influenza and herpes virus antigens following prolonged restraint (Bonneau, 1996). Interestingly, in mice a chronic social stress increased the entry of peripheral IL-6 into the brain, and was suggested as a mechanism for promoting depressive-like behavior (Menard et al., 2017).

In several studies, the effects of stressors applied on a backdrop of an immune challenge were assessed. In this instance, it appeared that in mice that had received LPS, an acute stressor enhanced circulating IL-6 levels measured 1.5 hour after treatment in the hardy C57BL/6By strain of mouse, but had lesser effects in the anxious BALB/cBy strain. In contrast, TNF-α was only modestly affected, whereas IL-10 levels rose dramatically in both strains, although a bit less so in the more anxious strain (Gibb, Hayley, Poulter, & Anisman, 2011). These actions varied appreciably with a chronic stressor, and also with time following the endotoxin treatment. For instance, in the non-inbred CD-1 strain, an acute stressor increased plasma IL-6 measured 1.5 hours after LPS treatment, but by 3 hours the IL-6 levels were reduced. In contrast, both acute and chronic stressors reduced circulating TNF-α levels at both times, whereas IL-10 was markedly increased 1.5 hour after LPS treatment in both acutely and chronically stressed mice (Gibb, Al-Yawer, & Anisman, 2013). Aside from the strain-dependent effects, these data make it clear that what happens in blood is very different from what occurs in the brain in relation to stressors, and the actions in the prefrontal cortex and hippocampus can be distinguished from one another. Analyses focused on brain cytokine changes linked to illnesses, such as depression, might do well to consider these compartment-specific cytokine changes as well as those that occur across brain regions.

The chronic strain experienced by parents of young cancer patients showed a stress-related suppression of IL-6 production (Miller, Cohen, & Ritchey, 2002), whereas for individuals experiencing the more short-term distress of exams, public speaking, or exercise, caused enhanced *in vitro* mitogen-induced IL-6 production (Goebel, Mills, Irwin, & Ziegler, 2000; Maes et al., 1998). In human studies, stressor exposure resulted in similar effects, which was corroborated in a meta-analysis of 29 studies that investigated plasma/serum IL-6 elevations in response to stressors (Marsland et al., 2017). When measuring cytokines in plasma, it is assumed that the cytokine may have emerged from an activated leukocyte. However, an important measure that needs to be obtained simultaneously with circulating cytokines is the activation state of peripheral blood leukocytes. In the absence of confirmation in the same individuals that circulating leukocytes are actively producing and

secreting IL-6, there is the very real possibility that stress-induced plasma IL-6 elevations are emerging from an endocrine-like organ responsive to stressor intensity. This was suggested long ago when it was shown that stressor-induced plasma IL-6 elevations were eliminated in adrenalectomized rats (Zhou et al., 1993). This does not necessarily mean that IL-6 was derived from the adrenal, but it did allude to the possibility that an immune source was unlikely. Other studies with mice showed that an immobilization stressor increased circulating IL-6, and was not associated with the spleen, but was instead linked to increased IL-6 mRNA expression in the liver (Kitamura et al., 1997). This raises the possibility that stressor-induced plasma IL-6 elevations are due to activation of hepatocytes, a long-recognized source of the acute phase inflammatory response.

Interleukin-1 and Tumor Necrosis Factor

Plasma elevations of IL-1 and TNF-α are readily induced by injection of LPS, which targets TLR4 innate immune cells, such as macrophages. In mice challenged with LPS, the IL-1 response was significantly reduced by an acute restraint session, whereas repeated social disruption augmented both spleen and lung IL-1 and TNF-α responses (Goujon et al., 1995; Quan et al., 2001). Similarly, rats exposed to a single session of inescapable tail shock generated greater amounts of IL-1β and TNF-α in response to an *in vivo* LPS challenge (Johnson et al., 2002). Evidently, exposure to stressors varying in intensity and/or stressor duration, results in opposing inflammatory cytokine responses. Moreover, mice exposed to a relatively severe stressor, showed elevated circulating levels of IL-1, TNF-α, and IL-6, in the absence of any LPS challenge (Cheng, Jope, & Beurel, 2015). Whether the source of these cytokines was immunological is not clear. Interestingly, however, stressor-induced elevations in circulating IL-1 and IL-6 can be blocked by immunological activation with LPS 1 week prior to stressor exposure (Hale et al., 2003; Merlot, Moze, Dantzer, & Neveu, 2004), implying a form of desensitization or tolerance at the cellular level. Since LPS targets macrophages, this points to the possibility that such cells are involved in the spontaneous production of pro-inflammatory cytokines following stressor exposure.

There are many more studies on the effects of stressors on immune function and cytokine production in particular. In this section, we highlighted some of the older and newer pieces of information, which hopefully has provided an understanding of the types of changes taking place when individuals are exposed to different types of stressors. Cytokines, such as IL-1β, IL-6, and TNF-α are major activators of the neuroendocrine system, and have become prominent points of discussion in theories of depression. It is uncertain whether stressor-induced elevations in cytokines represent a potential neurotrophic influence that affects normal brain function. Most certainly, the demonstration of increased IL-6 entry into the brains of chronically stressed mice establishes the plausibility of this notion. Finally, there are numerous studies regarding the effects of stressors on cytokine induction in the brain, which are discussed in Chapter 5 Stressor Processes and Effects on Neurobiological Functioning, dealing with glial cell responses and degeneration. Our focus here, however, has been to address how stressors affect peripheral cytokine components. The overall consensus—always with exceptions—is that stressors can increase the concentration of circulating cytokines. However, in what way this might influence mental health, is part of an ongoing investigation in this exciting area.

MODERATION OF CYTOKINE CHANGES BY PSYCHOLOGICAL FACTORS

Much like other biological effects introduced by stressors, cytokine changes can be moderated by cognitive and emotional changes. In a laboratory stress test in which emotion-inducing manipulations elicited an increase of plasma IL-1β, IL-6, and IL-8, these actions were abrogated among individuals who displayed cognitive control over emotional information (Shields, Kuchenbecker, Pressman, Sumida, & Slavich, 2016). Likewise, among women given a scenario of abuse, the rise of plasma IL-6 was linked to the extent of anger and sadness evoked, but not that of shame and anxiety. However, if women had themselves been in an abusive relationship earlier, then IL-6 levels also increased in relation to their expressed shame and anxiety (Danielson, Matheson & Anisman, 2011). It was similarly reported that the IL-6 response in the Trier Social Stress Test was linked to the specific emotions displayed, which could be dissociated from the cortisol response elicited (Moons, Eisenberger, & Taylor, 2010). Specifically, anger reactions provoked by the stressor were directly related to cortisol changes and not to IL-6, whereas fear and anxiety (but not anger) elicited in association with the stressor was related to IL-6 (Moons & Shields, 2015; Moons et al., 2010). These studies, and others like them, make it clear that the impact of stressors on cytokine production are not just linked to characteristics of the stressor, but are moderated by emotional and cognitive control factors that are associated with the challenge.

AUTOIMMUNE DISORDERS

As we've seen throughout this chapter, stressful events have been associated with immune alterations, and it seems that vulnerability to infectious diseases may be affected. Some of these illnesses might arise because stressors weaken immune functioning, thereby allowing viruses to do their unsavory business. However, stressors can also exacerbate disorders in which aspects of the immune system attack the self (i.e., autoimmune disorders). Stressors are unlikely to cause these illnesses, but by instigating immune activation, they may aggravate the symptoms of an illness that is already present.

The influence of stressors has been examined in relation to many autoimmune disorders, with the most common being MS, systemic lupus erythematosus (SLE), rheumatoid arthritis (RA), inflammatory bowel disease (IBD) and Type 1 diabetes. Other autoimmune disorders, such as celiac disease, inflammatory bowel disorder (IBD), psoriasis, and Sjogren's syndrome, are also affected by stressors. The 65 or 70 other autoimmune disorders identified or suspected, have not been examined as extensively in relation to stressor actions, although many can be exacerbated by the same factors that affect MS, SLE, and RA. Here we will focus primarily on the latter three. In Chapter 16, Comorbidities in Relation to Inflammatory Processes, we discuss stressors, inflammatory factors, and microbiota in relation to diabetes and IBD.

It is generally thought that although the processes associated with autoimmune disorders are not identical to one another, they may share common elements (including that they preferentially occur in females), and thus might be affected by many of the same factors. In this regard, autoimmune disorders may be determined by a gene variant that influences T helper cells so that normal cells are attacked

more readily. Alternatively, cells that ordinarily are responsible for immune suppression may not be operating as they should, thus allowing for autoimmune disorders to emerge. It was suggested, as well, that still other genetic variants might determine which specific aspect of the body will be attacked. For instance, it was suggested that inheriting mutations of the "autoimmune regulator" gene (AIRE) may change immune cell competency, thus allowing them to turn on the body's own organs (Oftedal et al., 2015). As we'll see, however, each of the autoimmune disorders also has their own unique contributing agents.

Multiple Sclerosis

MS comprises a disorder in which immune responses are directed toward the myelin surrounding axons of the spinal cord and brain, leading to plaques or lesions. The disorder may present in a form wherein new symptoms appear as discrete attacks, but can be followed by lengthy periods, even years, before further incidents are experienced (relapsing-remitting form). Despite symptoms seeming to be in abeyance between episodes, the neurological disturbances may persist and even progress. In a variant of the disorder, that of primary progressive MS, remission does not occur following the initial symptoms. In secondary progressive MS ("galloping MS"), a variant in which patients had initially been diagnosed with relapsing-remitting MS experience, a progressive neurologic decline occurs between episodes. Typically, definite periods of remission do not occur, although hints and false signs of abatement may appear.

In animal models, such as Theiler's murine encephalomyelitis virus (TMEV) infection, stressors exacerbated several aspects of MS. However, as we saw in relation to stressor effects on immune changes promoted by antigenic challenges, the MS variations were tied to the timing of the stressor relative to infection. For instance, when social defeat coincided with infection, the course of the disease was diminished and the extent of inflammation was limited, whereas a similar outcome did not occur when the stressor preceded infection, possibly owing to the actions of IL-6 changes elicited by the stressor (Meagher et al., 2007). Studies using other animal models also indicated that acute stressors may limit MS symptoms, perhaps owing to the immunosuppressive actions of corticosterone. In contrast, chronic stressors seemed to promote illness exacerbation.

The occurrence of MS flares has been attributed to numerous factors, including medication usage, cardiovascular reactivity, baseline heart rate, and stressful life events, accounting for as much as 30% of the variance of symptoms (Ackerman et al., 2002). In line with aspects of these findings, distressing events were especially common during the period prior to relapse (Ackerman et al., 2002). Of particular interest in the present context is that stressful experiences were accompanied by new brain lesions coupled with the exacerbation of MS symptoms (Mohr, Hart, Julian, Cox, & Pelletier, 2004). Moreover, these actions were particularly common among individuals who focused on their illness through emotion-based coping, but were least notable among individuals who used avoidance/distraction to cope (Mohr, Goodkin, Nelson, Cox, & Weiner, 2002). It is particularly noteworthy that viral infection was also associated with MS relapse, possibly stemming from the immune activation that accompanies infection. As we have already indicated, viral illnesses may be influenced by stressor experiences, thus the combination of psychological and viral insults may contribute to the MS flares. As much as these studies have been informative, they generally had modest effect sizes owing to the use of a small number of participants. As well, the studies involved retrospective self-reports of stressful experiences (recall of these experiences are notoriously biased), making them less convincing than would be liked (Brown, Tennant, Dunn, & Pollard, 2005).

These caveats notwithstanding, a relationship does seem to exist between stressful events and MS flares, and interestingly, the frequency of stressful experiences, rather than the severity of acute stressors, was most closely aligned with the worsening of symptoms. In fact, although modest acute stressors could aggravate symptoms, relapse was diminished if individuals experienced a strong acute stressor, such as major surgery or fractures (Mohr et al., 2004). As we've seen, the neurobiological processes associated with moderate versus severe or chronic stressors differentially influence immune and corticoid functioning, and it is reasonable to hypothesize that the effects of stressors in relation to MS may involve these mechanisms (e.g., a strong stressor might suppress immune functioning and thus inhibit symptoms).

Systemic Lupus Erythematosus

SLE, like the other autoimmune disorders is considerably more prevalent in women than in men, usually commencing anywhere from 15 to 35 years age. Given the sexual dimorphism, it is likely that estrogen contributes to the development of the illness. In addition to genetic influences, environmental factors have been implicated in the development and aggravation of SLE. The environmental triggers have ranged broadly, having been attributed to such factors as cigarette smoke, pesticides, phthalates, and it was suggested that infection, and estrogen intake, might also act in this capacity.

The illness can involve an attack on just about any portion of the body, including the nervous system. Thus, comorbidity with multiple illnesses is common, often being associated with RA as well as osteoporosis, cardiovascular disease, and cancer (Cervera, Khamashta, & Hughes, 2009). Characteristics of the illness also include elevated fatigue, pain, rash, fever, abdominal discomfort, headache, and dizziness, which are often antecedents of flares, and at this point of the disease neurological and psychiatric disturbances may appear. Similarly, at this stage, white matter hyperintensities are detectable within the brain as are lesions, hemorrhage, cellular atrophy, and microstructural abnormalities. Thus, it is hardly surprising that patients present with impaired executive functioning, speed of information processing, and poor memory.

The development and progression of SLE has been attributed to a variety of biological mechanisms. One view has it that the disease arises owing to impaired clearance of debris comprising dead cells, and the development of antinuclear antibodies that attack portions of healthy cells (Lisnevskaia, Murphy, & Isenberg, 2014). Given that blood barrier permeability may be compromised in SLE, autoantibodies could also access brain sites in which damage is present (Williams, Sakic, & Hoffman 2010), hence leading to cognitive disturbances and depressive symptoms (Kowal et al., 2006). Indeed, these auto-antibodies can affect glutamate (NMDA) receptors, thereby promoting cell death (Faust et al., 2010), and behavioral disturbances.

As we will see in Chapter 8, Depressive Disorders, depressive illness has been linked to the 1019 G allele of the 5-HT$_{1A}$ gene, and it appeared that SLE flares associated with stressors were especially notable in those individuals that carried a particular form of this gene, which may influence stressor sensitivity. Thus, this gene polymorphism might account for the comorbidity often seen between SLE and depressive illness. To be sure, receiving a diagnosis of SLE can be devastating and could favor the development of stress-related disorders, such as depressive illness.

As observed among MS patients, symptoms of SLE were aggravated by day-to-day irritations, particularly those associated with social relationships, but were less likely to be exacerbated by major life stressors. The symptom aggravation associated with daily hassles was not only predictive of flares in SLE patients, but was also associated with disturbed cognitive functioning, impaired total attention speed, and poor visual memory performance (Peralta-Ramírez, Jiménez-Alonso, Godoy-García, & Pérez-García, 2004). These outcomes could reflect the immune enhancement elicited by mild and moderate stressors, whereas strong stressors are more likely to induce immunosuppression. It may be significant that SLE patients seemed to be particularly sensitive to stressors, at least with respect to changes of immune and cytokine responses, which is in keeping with the symptom exaggeration elicited by stressors (Jacobs et al., 2001). Indeed, it seems that the way in which individuals cope with the distress also influenced health-related quality of life among SLE patients (Hyphantis, Palieraki, Voulgari, Tsifetaki, & Drosos, 2011). Perhaps not surprisingly, cognitive behavioral therapy effectively diminished depressive symptoms and enhanced quality of life among SLE patients. Somewhat unexpected, however, is that psychosocial and coping factors which can influence depressive symptoms and quality of life, did not result in parallel organ deterioration related to SLE (Bricou et al., 2006).

Rheumatoid Arthritis

RA most commonly appears in middle age, primarily among women, and can also appear in children (juvenile arthritis). Even when the disorder initially manifests in adulthood, systemic inflammation and autoimmunity often begin several years earlier (Demoruelle, Deane, & Holers, 2014). The features of RA comprise an attack and inflammation of the synovial joints, the membrane lining joints, tendon, as well as sheaths, and cartilage, and in some cases ANS disturbance and heart problems may occur (Adlan, Lip, Paton, Kitas, & Fisher, 2014). The distress created by the disease is made worse by the chronic pain experienced, as well as general malaise or feelings of fatigue, weight loss, and poor sleep. As we've seen in other autoimmune disorders, symptoms may wax and wane and patients may even experience lengthy periods during which symptoms diminish, before again recurring. Numerous factors, including environmental and genetic processes likely contribute to the development of RA, and as with other illnesses, there is the view that multiple hits contribute to its occurrence. It has been known for some time that RA is associated with specific antibodies, rheumatoid factors, and citrullinated peptides, often being detectable in advance of clinical signs of the illness. This does not necessarily imply that these are causative agents of RA, but they might be useful biomarkers for disease occurrence (McInnes & Schett, 2007). As in the case of other autoimmune disorders, day-to-day stressful events have been suggested as an important ingredient that exacerbates symptoms, possibly owing to increased cytokine functioning. Likewise, chronic interpersonal stressors were accompanied by worsening of symptoms, together with elevated *in vitro* IL-6 production and diminished effectiveness of glucocorticoids in the inhibition of cellular inflammatory responses (Straub, Dhabhar, Bijlsma, & Cutolo, 2005). Analysis of the everyday changes of pain symptoms among workers, indicated that on days when they encountered stressful work-related events, their pain levels were relatively elevated, and was also related to job strain. Consistent with such findings, chronic stressful experiences aggravated the symptoms of RA, particularly juvenile arthritis. It appeared that under these conditions, changes of cortisol activity and α-norepinephrine

receptor functioning played a role in producing arthritic flares (Straub et al., 2005).

Attempts to understand RA have focused mainly on immune dysregulation, but there are indications that microbial factors contribute to the illness. Specifically, the disorder was associated with increased bacterial antigens within the gut and lung, and even in the tissues surrounding the teeth (Brusca, Abramson, & Scher, 2014). Given that stressors profoundly affect microbiota, which have been linked to immune processes, as described in Chapter 3, Bacteria, Viruses, and the Microbiome, it ought to be considered that linkages exist between stressful experiences and these processes (e.g., Gur & Bailey, 2016).

Although we have focused here on the contribution of stressors to influence the development of autoimmune disorders, there is reason to believe that other life-style factors, many of which were described in Chapter 4, Life-Style Factors Affecting Biological Processes and Health, such as eating and exercise, which affect microbial processes and immune functioning, might also contribute in this respect. In view of the contribution of stressful experiences and other lifestyle factors in the exacerbation of autoimmune disorders, it might be of appreciable benefit for treatment of these conditions to include psychological approaches to diminish distress (especially as these illnesses themselves may create considerable stress) and education to enhance illness management and treatment compliance (Lisnevskaia et al., 2014).

CONCLUDING COMMENTS

The immune system, through its exquisitely designed components and interactions with endocrine processes, typically protects from an assortment of challenges. As such, it is predictable that stressors would set in motion a sequence of adaptive immune changes to deal with imminent or ongoing threats.

However, there are limits to the capacity to deal with such challenges, and with chronic stressful experiences, the immune system's capabilities may be undermined, thereby influencing vulnerability to bacterial and viral infection. As well, immune dysregulation may exacerbate the symptoms of autoimmune disorder and perhaps cancer as well.

On the surface, assessing the impact of stressors on immune functioning should be easily achievable. However, there is no *a priori* reason to assume that all facets of the immune system (e.g., innate vs acquired immunity; different immune compartments; varied immune cells that come into play at different times following antigen presentation; components that stimulate or suppress immune processes; the actions of different cytokines) would operate in precisely the same way. Like so many aspects of neurobiological systems, effective immune functioning is dependent on intricate balances between various components of this system as well as the countervailing forces that keep functioning within an effective range. As well, the actions of the immune system are dictated by multiple hormones, and it is subject to psychological influences that affect these neuroendocrine processes.

In describing the actions of chronic stressors on various immune mechanisms it seemed that inconsistent findings were common, and the effects on some elements of the immune system behaved in strikingly different ways from that evident on other aspects of immunity. At one level, this might be interpreted as reflecting a smorgasbord of outcomes without any apparent overarching schema to assure effective functioning. Alternatively, there may be a master conductor that operates to ensure that balances are maintained within and between components of the immune system, but we have yet to discover how this principal modulator operates.

It is equally possible that the seemingly disparate outcomes provoked by stressors may be more a reflection of our naivety about how the immune system operates, and what subtle aspects of the immune system dictate the outcomes that emerge. As a case in point, it had long been thought that while the adaptive immune system was capable of memory, whereas the innate immune system did not have this capability, this view has been undergoing some modifications. To an extent, cells of the innate immune system may be capable of functional memory in response to particular inflammatory immune signals, reflected by elevated cytokine changes upon reexposure to these stimuli. This "trained innate immunity" may be related to sensitized neuronal processes in the brain (which speaks to the immune system) or epigenetic changes that alter sensitivity to this same challenge at a later time (Salam et al., 2018).

There is yet another essential element that needs to be addressed. As we have seen repeatedly, pronounced inter-individual differences exist in every aspect of human physiology and behaviors, as well as how stressors are managed. Some individuals are more apt to exhibit either excessive immune activity, or alternatively diminished responses, leading to illness. Stressors may produce especially pronounced adverse consequences in older individuals (Fali, Vallet, & Sauce, 2018), and it is clear that women and men respond differently to such challenges. It is also likely that the constellation of variables that affect neuroendocrine and brain processes (e.g., early life stressor experiences, personality factors) may come to indirectly affect immune functioning, and hence vulnerability to psychological and physical diseases. Despite our best efforts, many of these illnesses have been refractory to treatment. It is possible that treatments to diminish distress (along with many other life-style changes) may be useful in disease prevention and treatment. However, even if this turns out not to be effective in the treatment of illnesses, simply diminishing distress may have immeasurable advantages for the affected individual.

Prenatal and Early Postnatal Influences on Health

A SIMPLE TWIST OF FATE

An editorial published in Lancet (2017) made the point that being born into poverty has an enormous impact on health and well-being, and there's no question that income and life expectancy are directly related to one another (Chetty et al., 2016). Where an individual sits on an income gradient, as Michael Marmot indicated some years earlier (2010), determined their life trajectory and life-span. Numerous similar reports indicated that the top 20% of income earners lived a decade longer than those at the bottom of the income ladder. Indeed, children born in poverty generally died 10–15 years before those of the wealthiest 20%, and the interval between birth and death was accompanied by multiple health disparities, including, obesity, elevated emergency department visits, mental health issues, and self-harm. A 5-year prospective study also revealed that death stemming from heart attack or stroke was much more frequent in rural and remote areas than in large cities. Those living in remote areas were less likely to be screened for risk factors related to heart disease, and were more likely to endorse poor life-styles, such as smoking and behaviors that led to obesity, pointing to the importance of preventive care (Tu et al., 2017).

Multiple physiological processes no doubt contribute to increased illness and premature death among individuals who had been raised in an unfavorable socioeconomic climate. Poor diet, together with other early life factors, may have influenced the microbiome in a way that lent itself to disease occurrence (Snijders et al., 2016). As well, the chronic strain associated with low socioeconomic status may have produced a defensive response style that promoted greater resistance to glucocorticoid signaling, which then promoted elevated inflammatory immune functioning. The HPA and immune activation are beneficial in response to acute threats, but early life social stressors may leave a biological "residue" that renders individuals at elevated risk for poor well-being (Miller et al., 2009). Fortunately, the effects of such experiences may be modifiable by later social support from family and friends, as well as overall social connectedness (Cruwys, Haslam, Dingle, Haslam, & Jetten, 2014). Likewise, maternal support during early years can attenuate behavior and brain alterations associated with adverse early life experiences (Luby, Belden, Harms, Tillman, & Barch, 2016). It is notable that events experienced in childhood, including those associated with nutritional

<hr>

(cont'd)

factors, microbial state, and psychosocial stressors, are associated with multiple epigenetic actions detected during adulthood, including those related to inflammation, and thus might be linked to illnesses, such as depression, diabetes, and heart disease (McDade et al., 2017).

EARLY DEVELOPMENT

The prenatal period and early postnatal life are precarious times. Cell duplication and the formation of a diverse number of different cells progresses rapidly, dictated by DNA and mRNA functioning. This process is dynamic in the sense that various endogenous substrates are added as part of the instruction for the developing fetus, and then removed at appropriate times. These modifications necessarily require that particular genes be turned on and off in a well-orchestrated sequence.

Gene transcription that occurs early in embryonic development can instill a stable epigenetic state that has implications for later physiological functioning (Greenberg et al., 2017). Yet, with so many changes occurring during the course of prenatal development, especially those stemming from multiple rounds of cell multiplication, the risk for chance mutations and those that are instigated by toxicants is considerable, and should such an outcome occur, it will be amplified with cell multiplication. At the same time, dynamic epigenetic processes are thought to be important in the regulation of developmental processes, and here again, there is ample room for problems to occur owing to environmental challenges and stressors. During this time, immune system functioning is blunted so that the fetus isn't swarmed by the maternal immune system, and to some extent, components of the immune system are functional by the second trimester (McGovern et al., 2017). Indeed, during pregnancy a mother's immune functioning

may be down-regulated, which may account for the alleviation of autoimmune symptoms women may experience during this period, but this may also leave her at risk for infection, which could impact the developing fetus. In addition to variations of peripheral immune functioning, cytokine changes are diminished in the maternal brain (Sherer, Posillico, & Schwarz, 2017), which can have profound behavioral consequences that are apparent during the postpartum period.

In the context of developmental problems, there was a time when the focus of teratogenic effects involved drugs and environmental chemicals, and to some extent immunologically related illnesses (e.g. Rubella). But, it has become clear that a wide range of factors, including moderate stressors, may affect the developing brain. It also appears that prenatal stressors may interact with genetic factors in determining the occurrence of developmental disorders as well as pathologies that emerge in adulthood. As we'll see in Chapter 8, Depressive Disorders, individuals carrying the short variants of the gene for the 5-HT transporter (5-HTT) may be at increased risk for the development of mood disorders, provided that they also experienced strong early life or adult stressors (although there have been indications that these actions might be less straight forward than initially thought). It likewise seems that in mice, prenatal stressors were more likely to promote depressive-like features among female offspring carrying the short allele of the 5-HTT promoter gene (Van den Hove et al., 2011). Moreover, behavioral effects of prenatal immune challenge could develop through epigenetic actions on the gene coding for the 5-HTT transporter (Reisinger et al., 2016).

PRENATAL CHALLENGES

Consumption of particular drugs during pregnancy, especially during the first trimester, places the fetus at risk for a variety of adverse outcomes (teratogenic effects) ranging from gross physical abnormalities, through to profound or relatively subtle neuropsychiatric conditions and developmental disturbances. In many instances, these challenges might not directly cause pathology, but instead increase the susceptibility to such outcomes.

The agents that increase the propensity for negative outcomes, include several pollutants that are frequently encountered, such as second-hand smoke, polychlorinated biphenyls (PCBs), estrogen-related products that have been showing up in water supplies, as well as common household chemicals. Women who had been exposed to organic pollutants were at four times greater risk of developing gestational diabetes, which could have affected their offspring. As well, being exposed to air pollution during the perinatal period (i.e., shortly before and after birth) was accompanied by increased asthma risk, as well as childhood emotional problems (Margolis et al., 2016).

The Need for Prenatal Resources

Ideally, prenatal development proceeds smoothly with few hazards appearing along the way. However, the fetus can encounter gestational exposure to infection, nutritional challenges, obstetric complications, and stress reactions in the pregnant woman. Each of these events can have lasting effects on the offspring's cognitive, emotional and social behavior, as well as on processes related to neuroplasticity. Additional risks to the fetus can be encountered if mom isn't knowledgeable about prenatal care (e.g., use of drugs).

Experiences tied to poverty within Western countries (e.g., living in distressed neighborhoods, high rate of unemployment, uninsured status, and low education level), and an array of maternal factors (young maternal age, nonmarital status, planned pattern of prenatal care, and late recognition of pregnancy) have been associated with negative outcomes for the fetus, so that postnatal life may begin with numerous physiological or psychological disadvantages. There's no question that mothers and offspring in wealthier countries have it better than do those in poorer nations. However, obesity and gestational diabetes, which are not uncommon in wealthier nations, have also been associated with neurodevelopmental conditions, such as autism spectrum disorder and other health problems related to immune dysfunction (Macpherson, de Agüero, & Ganal-Vonarburg, 2017).

Studies of Prenatal Stress

A wide range of stressors mom experiences can affect the developing fetus, such as extreme challenges, war-time experiences, or natural disasters, as well as more common challenges (e.g., including work-place stressors, discrimination, and bereavement). These experiences can have profound effects that become apparent soon after birth, and may also affect the developmental trajectory so that pathological outcomes appear in adulthood. The negative consequences of these psychological challenges are greater with more intense experiences, but are particularly marked in the offspring of women who encountered multiple adverse experiences (Robinson et al., 2011).

The risk of premature delivery and low birth weight is appreciably increased among women with high anxiety or depression, as well as women who encountered strong stressors during pregnancy. Chronic prenatal stressors are accompanied by elevated circulating CRH, possibly of placental origin, along with variations of other hormones that pass through the placenta, such as cortisol and the endogenous opioid met-enkephalin. These factors may contribute to fetal development, and could precipitate preterm labor and reduced birth weight (Ghaemmaghami, Dainese, La Marca, Zimmermann, & Ehlert, 2014).

Unfortunately, these preterm infants are at elevated risk of endocrine or metabolic disorders (e.g., diabetes) and illnesses that develop as a result of diabetes (Paz Levy et al., 2017). As well, several hormonal alterations that were associated with prenatal stressor experiences and the accompanying premature delivery predicted delayed fetal neuromuscular and nervous system maturation, and diminished gray matter volume (Sandman, Davis, Buss, & Glynn, 2011). Not surprisingly, these conditions were also accompanied by neurodevelopmental disorders, cognitive disturbances (e.g., delayed language development; attention deficit hyperactivity disorder), and emotional problems (e.g. fearfulness and anxiety) that often carried into adulthood (Glover, 2011). Prenatal stressors that led to maternal PTSD was accompanied by lower birth weights of their offspring in comparison to offspring of nonstressed mothers, and also those of mothers who experienced trauma, but did not develop PTSD (Seng, Low, Sperlich, Ronis, & Liberzon, 2011). In effect, the outcomes may not simply be a reaction to the stressor, but instead developed owing to the mother's psychological or physical responses to these challenges.

Prenatal stressor experiences have been associated with diminished adult levels of cortisol and ACTH (Entringer, Kumsta, Hellhammer, Wadhwa, & Wüst, 2009), consistent with that seen in adult patients diagnosed with PTSD. Moreover, treatments that simulate some stressor actions, notably the administration of the synthetic glucocorticoid betamethasone (used in an effort to promote lung maturation in fetuses at risk of preterm delivery), provoked persistent effects on subsequent infant temperament and behavioral reactivity in response to stressors (Davis, Waffarn, & Sandman, 2011). Likewise, among rodents, prenatal administration of the synthetic corticoid dexamethasone, rather than a psychological or physical insult, promoted hypertension and increased basal plasma corticosterone levels in offspring, as well as altered hippocampal glucocorticoid receptors.

Epidemiological studies demonstrated that prenatal stressor experiences were associated with numerous physical illnesses, such as disturbed cardiovascular regulation, type 1 diabetes, and metabolic syndrome that might foreshadow adult type 2 diabetes. Offspring were also at elevated risk of immune-related disorders, such as allergies and asthma, and a greater likelihood of contracting infectious diseases (Avitsur, Levy, Goren, & Grinshpahet, 2015). Aside from these effects, prenatal stressors have been linked to psychological disturbances, such as depression, schizophrenia, drug addiction, and eating-related disturbances (e.g., binge eating).

Numerous studies conducted with rodents have indicated that stressful events encountered by a pregnant mom can subsequently affect the behavior of her pups when they became adults. It is tempting to attribute these actions to neurobiological disturbances promulgated by the prenatal stressor, but the stressor experiences can also disturb maternal behaviors that could have long-term effects on the offspring. Yet, behavioral disturbances, including altered sexual behaviors, could be attenuated or reversed by neonatal administration of testosterone (Pereira, Bernardi, & Gerardin, 2006), speaking to the importance of sex hormones in determining the persistent actions of prenatal stressors.

Some of the actions of the prenatal stressor were not apparent during the normal course of development, but were only manifested when animals encountered a further stressor (e.g., in the case of eating disorders), again supporting the double-hit perspective regarding pathological outcomes. Significantly, the occurrence of binge eating could be attenuated by exposing the offspring to a methyl-balanced diet during adolescence (Schroeder et al., 2017).

The breadth of the psychological and physical illnesses that can accompany prenatal stressors may reflect the diversity of biological changes introduced by these challenges. Alternatively, it is possible that a core set of processes, such as particular hormones, and the increase of inflammatory immune factors, play a key role in these disorders, many of which involve inflammatory processes.

THE IMPACT OF MOM'S PRENATAL AND PERINATAL DIET

Commensal bacteria and their numerous metabolites that originate within mom influence the fetus. Thus, what mom eats, the drugs consumed, toxins encountered, and the stressors experienced, which can affect her microbiota, may indirectly affect the fetus (Macpherson et al., 2017). Rats that consumed the equivalent of a high fat/high sugar Western diet (junkfood) during pregnancy had offspring who were inclined to gain excessive weight during the suckling period, and were likely to develop diabetes. In fact, such a diet during gestation and lactation disposed mouse pups toward an epigenetic profile associated with diabetes. However, this could be reversed if mice were maintained on a low-fat diet following weaning (Moody, Chen, & Pan, 2017). The maternal diet during fetal development has also been associated with obesity-related epigenetic changes that can even appear in later generations (Drummond & Gibney, 2013).

A Western diet in pregnant rodents influenced the subsequent brain reward circuitry of their offspring so that they were more attuned to rewards, which could have contributed to their elevated eating and weight gain (Paradis et al., 2017). This diet was also associated with anxiety in adolescence, possibly by affecting glucocorticoid signaling (Sasaki, de Vega, Sivanathan, St-Cyr, & McGowan, 2014). Predictably, maternal prenatal and early childhood nutrition may affect facets of inflammatory immune functioning, thereby contributing to neurodevelopmental disorders and immune-related disturbances.

Pregnant women who partake in a high-fat, high-sugar diet were more likely to have children with conduct problems and with attention deficit/hyperactivity disorder (ADHD) symptoms, which were linked to epigenetic changes on the gene coding for IGF-2 within the cerebellum and hippocampus (Rijlaarsdam et al., 2017). Likewise, maternal diets that were rich in omega-6, but low in omega-3 (creating a proinflammatory bias), resulted in offspring that had a smaller brain and displayed disturbed emotional responses in adulthood (Sakayori et al., 2016), and were at increased risk for developing schizophrenia-like symptoms (Maekawa et al., 2017).

An interesting natural experiment was conducted in Gambia, which ordinarily has a rainy season in which food is available and a dry season when food is scarcer. Children born during the latter portion of the food-scarce season, when their mothers had been food deprived, were generally smaller (at the age of 2) than children born during the rainy season. Importantly, the growth failure persisted for several subsequent generations (Eriksen et al., 2017). It's hard to know whether these outcomes reflected microbial factors or epigenetic effects, but it's clear that nutritional stress can have persistent consequences.

Timing of Prenatal Stressors

The teratogenic effects of most compounds are greatest during the first trimester of pregnancy, and to an extent this gestational stage is likewise most susceptible to the negative effects of stressors. Reduced birth weight was more apparent in response to a stressor encountered during the first trimester (Sandman et al., 2011). However, stressors experienced as late as the third trimester of pregnancy were still accompanied by psychopathology as well as disruption of some aspects of immune functioning. As we've seen, prenatal stressors can produce a diverse range of effects, and it is conceivable that stressors encountered at different stages may lead to varied outcomes.

Stressors in humans promote dynamic behavioral and neurobiological consequences that vary over time. Thus, it might not be profitable to attempt to isolate specific times when stressors in pregnant women would have their most pronounced actions. For that matter, a stressor encountered just prior to pregnancy was linked to shorter gestation and reduced birth weights, speaking to the proactive effects of stressors (e.g., rumination) on fetal well-being (Talge, Neal, & Glover, 2007).

Although it seems fairly clear that stressful prenatal experiences proactively affect the well-being of the offspring that may carry on throughout life, the impact of the prenatal trauma may be confounded with postnatal rearing conditions or other maternal factors. It is similarly possible that prenatal stressors interact with genetic contributions in determining outcomes within the offspring. In an effort to differentiate these confounded factors, the influence of stressors was evaluated in pregnant women who were either genetically related to their child or were genetically unrelated to the fetus, having received the fertilized egg through *in vitro* fertilization (IVF) (Rice et al., 2010). It turned out that regardless of whether the offspring was genetically related or unrelated to the mother, the offspring's birth weight and antisocial behavior was associated with prenatal stressor experiences. Interestingly, postnatal maternal anxiety/depression was linked with offspring anxiety, whereas the link to offspring attention deficit hyperactivity disorder was most apparent in genetically related mother-offspring pairs. Evidently, genetic, prenatal, and postnatal factors can all influence the offspring, but their specific contributions depend on the phenotype of interest.

Biological Correlates of Prenatal Stress in Humans

Prenatal stressors, as we've seen, can lead to numerous neurobiological alterations, which can affect psychological and physical functioning. A general representation of some of these variations is provided in Fig. 7.1. Psychosocial stressors experienced by a pregnant woman (or rodent) lead to changes of cytokines, tryptophan, cortisol, reactive oxygen species (ROS), monoamines, and gut bacteria in mom. In addition, prenatal stressors influenced placental cytokines, neurotrophins, and commensal bacteria, which were linked to adult anxiety among the offspring, together with BDNF alterations in the adult amygdala (Gur et al., 2017). Clearly, stressors experienced by a pregnant female can be transmitted to the fetus, which can then have consequences that persist throughout life.

Glucocorticoid Variations

Exposing a pregnant female to a stressor gives rise to elevated glucocorticoid levels within both the mom and the fetus, hence influencing the programing of fetal neurons. Among other things, this may be accompanied by structural changes within stress-sensitive brain regions, which can then affect behavioral

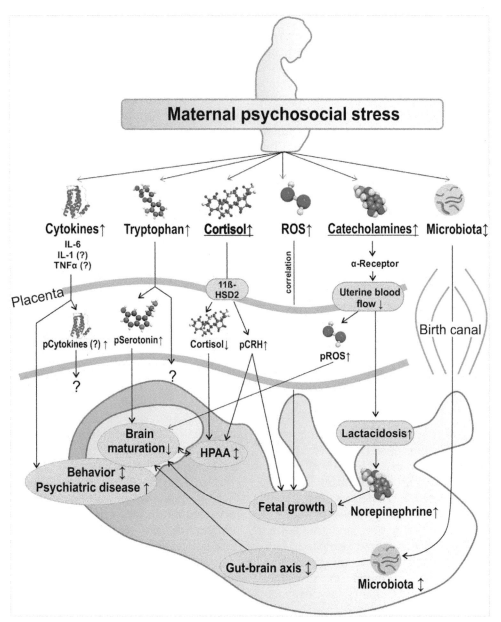

FIGURE 7.1 Multiple neurobiological processes are affected by prenatal stressors, which may affect the fetus and thereby influence postnatal development, even affecting adult behavior and physical processes. The involvement of HPA functioning and placental CRH have received particular attention in this regard, although other processes, notably inflammatory mechanisms, monoamine functioning, placental reactive oxygen species (pROS) and microbial factors, all have pronounced consequences on fetal development and postnatal functioning. These consequences of prenatal stressors are fairly broad, but represent only a portion of the changes that occur. *Source: From Rakers et al. (2017).*

and cognitive processes (e.g., attention and learning deficits), and favor the development of anxiety and depression. These actions, like so many others, may be moderated by genetic factors, and by the postnatal environment, including the maternal care received. Yet, not every neurobiological change stemming from prenatal stressors should be taken as promoting adverse outcomes. For instance, the adult offspring of rats that had been stressed during the second half of pregnancy, exhibited elevated corticosterone reactivity to stressors, increased firing of norepinephrine and dopamine neurons, but reduced firing of serotonin neurons. However, these neurobiological changes were not accompanied by the presence of anxiety or depression, raising the possibility that the neurobiological variations reflected adaptive responses that facilitated effective functioning in offspring (Oosterhof, El Mansari, Merali, & Blier, 2016).

Sex-Dependent Effects of Prenatal Stressors

Females are far more vulnerable than males to conditions, such as anxiety, depression, and autoimmune disorders. Numerous factors could account for this dimorphism including the impact of prenatal challenges. The anxiety and depression seen among female offspring of rodents stressed while pregnant could stem from estrogen changes and elevated HPA responsiveness, whereas learning deficits, which were more common in prenatally stressed males, might be due to sex-dependent reductions of hippocampal neurogenesis and cortical dendritic spine density (Weinstock, 2007). Essentially, prenatal stressors can have effects in both sexes, but the phenotype detected may be dependent on the specific hormonal processes that were altered and when during the course of prenatal development these occurred.

When encountered early in prenatal development, stressors can also instigate demasculanization in males, which was also seen in male descendants of prenatally stressed mice (Morgan & Bale, 2011). A sensitive period seems to exist in early gestation during which epigenetic programing of the male germ line occurs most readily, thereby affecting specific sex-related phenotypes. Stressors experienced during this period of prenatal development can promote a surge of gonadal hormones, potentially resulting in brain functioning being biased in a sexually dimorphic manner. During the course of postnatal development, these gonadal hormones might promote neuronal changes that come to affect the expression of sex-specific phenotypes.

In rodents, a chronic stressor experienced prior to pregnancy was accompanied by reduced GABA and glutamate within the right hippocampus of female offspring, but upon further exposure to a stressor during adolescence males reacted more strongly (Huang, Shen et al., 2016). Why these sensitization-like effects were apparent in a sexually dimorphic fashion is uncertain, but nonetheless point to the different stressor vulnerabilities in the two sexes. In this context, the presence of high levels of the inflammatory cytokine TNF-α, coupled with low levels of anti-inflammatories during pregnancy were associated with increased adult depression in males, but not in females despite the greater incidence of depression in females (Gilman et al., 2016). Once more, the source for this dimorphism isn't certain, but could be related to hormonal factors, and TNF-α was associated with dimorphic brain circuitry that occurs over the course of prenatal development.

Neurotrophins

Just as adult stressors can affect neurotrophins, it seems that prenatal stressors can have such actions. The variations of BDNF were accompanied by reduced brain cell proliferation, and as we've seen with respect to other biological processes, these actions of prenatal stressors could be exacerbated by further stressor experiences during adulthood (Fumagalli, Bedogni, Perez, Racagni, & Riva, 2004). Paralleling the BDNF gene changes, when a pregnant dam encounters a stressor, FGF-2 gene expression may be affected in her unborn pups, and might thereby influence subsequent stressor responses (Fumagalli, Bedogni, Slotkin, Racagni, & Riva, 2005). Conversely, treatments to alleviate depressive symptoms, such as repeated antidepressant or electroconvulsive shock treatment, increased FGF-2 in cortical and hippocampal sites. As expected, administering FGF-2 directly into the brain attenuated signs of depression within animal models of this disorder. It seems that prenatal stressors could set the groundwork for later depressive disorders, and other pathological conditions that involve growth factors. Beyond these actions, FGF-2 has also been implicated in disturbances associated with hypoxia (reduced oxygen) that follows from very premature birth, and was also associated with mild to moderate cognitive delay and increased incidence of anxiety, attention deficit hyperactivity, and autism spectrum disorders (Salmaso, Jablonska, Scafidi, Vaccarino, & Gallo, 2014).

Studies in humans linking prenatal stress, neurotrophins, and later pathology are obviously difficult to conduct, but data collected in "unnatural" conditions, have been instructive. For instance, among pregnant women that experienced war-related trauma, BDNF methylation was elevated in maternal venous and umbilical cord blood, as well as in placental tissue, supporting the possibility that prenatal epigenetic changes can be promoted by stressors (Kertes et al., 2017). The implications of these particular effects for later well-being are uncertain, especially as many other effects of these stressor experiences can have downstream consequences.

Inflammatory Factors

Prenatal stressful experiences were associated with elevated levels of circulating pro- and anti-inflammatory cytokines (Entringer et al., 2008), microglial activation markers, and elevated inflammatory cytokines within the prefrontal cortex and hippocampus of offspring (Posillico & Schwartz, 2016). Although IL-6 blockade attenuated the microglial changes, the behavioral alterations were unaffected, pointing to the possible involvement of other (or multiple) processes in determining the impact of prenatal stressor challenges (Gumusoglu, Fine, Murray, Bittle, & Stevens, 2017). More relevant to the action of prenatal stressors might comprise the morphological changes of brain glial cells accompanied by elevated release of proinflammatory cytokines and chemokines (Ślusarczyk et al., 2015).

In light of the many neurobiological and behavioral disturbances that have been associated with prenatal stressors, it is exceedingly difficult to link specific neurobiological factors to particular physical or behavioral disturbances that are associated with prenatal stressors. The view was advanced that owing to the extent and breadth of biological changes associated with prenatal stressors, a "general susceptibility" to pathology is created by these prenatal experiences, and the specific pathologies that emerge vary with genetic influences as well as postnatal experiences (Huizink, Mulder, & Buitelaar, 2004).

Despite the pronounced adverse actions of prenatal stressors, this should not be misconstrued to suggest that once the dye has been cast, it isn't modifiable. Specifically, the potential negative consequences engendered by having a depressed mother could be diminished if the infant received appreciable contact comfort during the initial weeks of life. Similarly, although the elevated prenatal cortisol levels detected in amniotic fluid (at 17 weeks of gestation) predicted subsequent cognitive impairment, this was primarily apparent among children expressing insecure attachment, but not in children with secure attachment[1].

Impact of Prenatal Infection

There's little question that infection during pregnancy can have widespread and long-lasting repercussions on the offspring. Most studies in rodents indicated that prenatal infection was accompanied by reduced cortical volume, and by anxiety- and depression-like behaviors (Depino, 2015), and in humans perinatal inflammation has been associated with these same psychological disorders, as well as schizophrenia and autism (Depino, 2017).

Like the effects of prenatal stressor challenges, it seems that prenatal immune activation may have sex and strain dependent actions on HPA functioning, anxiety- and depressive-like behaviors. The depressive and anxiety behaviors associated with prenatal LPS treatment were seen across generations (Walker, Hawkins, Sominsky, & Hodgson, 2012), raising the possibility that such effects could be linked to epigenetic changes, although they could also be attributable to altered maternal care that the pups received following LPS treatment (Penteado et al., 2014).

In addition to mood disturbances, an endotoxin administered to pregnant rodents induced a schizophrenia-like profile in offspring, accompanied by decreased myelination of neurons within cortical and limbic brain regions, more so in males than in females (Wischhof, Irrsack, Osorio, & Koch, 2015). Prenatal administration of moderate doses of a bacterial endotoxin (LPS) in rodents also promoted white matter abnormalities and typically was accompanied by reduced volume of brain regions linked to psychological disorders. It has also been observed that treatment of pregnant rhesus monkeys with an endotoxin similarly resulted in offspring being more stress-reactive, and promoted an attentional profile thought to reflect an autism-like condition (although it might also be seen as a schizophrenia-like profile). Curiously, at 1-year of age, gray and white matter volume was increased (rather than reduced) in parietal, medial temporal, and prefrontal cortices (Willette et al., 2011). This clearly stands in contrast to the impact of such challenges in rodents, but it is uncertain what might have been responsible for these differences. Given that these investigators had observed that influenza virus infection produced a fairly persistent decline of gray matter in frontal and parietal brain regions (Short et al., 2010), it seems likely that the outcomes may have been related to the nature of the prenatal infection experienced.

There's no shortage of factors that might contribute to the impact of infection during pregnancy on the behavior of offspring.

---

[1] There are instances in which prenatal stressors, like those encountered in early life, can have beneficial effects. Prenatal insults, for example, can instigate epigenetic changes so that oxytocin receptors are more readily activated, thereby facilitating maternal bonding with the infant and elevated social coping (Unternaehrer et al., 2016). It was also reported, however, that prenatal stressors reduced amygdala oxytocin expression, and disturbed social interactions with conspecifics. Postnatal administration of oxytocin directly into the amygdala attenuated the negative effects of the prenatal stressor (Lee, Brady, Shapiro, Dorsa, & Koenig, 2007), suggesting a causal connection between the hormone and social interactions. Similar outcomes, however, were not produced by having pups of the stressed dams fostered by mothers that had not been stressed. Thus, although the effects of prenatal stressors can be altered, simply having a mother that had not been stressed was not sufficient to attenuate the effects of prenatal insults.

Prenatal LPS treatment that increased anxiety and elevated stress reactivity in adulthood, was accompanied by altered DA and serotonin within the prefrontal cortex and hippocampus, and reduced adult BDNF and neurogenesis, and many of these effects could be attenuated by repeated antidepressant (fluoxetine) administration (Lin et al., 2014). Prenatal immune activation may also interact with stressors encountered in adulthood, leading to altered corticoid responses and activation of N-methyl-D-aspartate receptor (NMDA) receptors, which then affect behavioral stability (Burt, Tse, Boksa, & Wong, 2013). It seems likely, as well, that prenatal immune activation may have its behavioral consequences by compromising neural maturation and survival owing to deficiencies of neurotrophins, such as vascular endothelial growth factor (VEGF) (Khan et al., 2014).

Mental Illnesses Stemming From Prenatal Viral and Bacterial Challenges

Viral challenges during prenatal development in animals consistently revealed negative effects that were manifested throughout the postnatal period and into adulthood. For instance, influenza infection during pregnancy among rhesus monkeys, resulted in gray matter volume being reduced throughout the brain of offspring assessed at 1 year of age (Short et al., 2010). Further still, influenza infection might disturb neuronal migration during development resulting in disordered architecture of the cortex. For obvious reasons, controlled studies are more difficult to conduct in humans, but data have been available related to naturally occurring (or illness related) immune disturbances.

Fetal exposure to syphilis, rubella, and herpes simplex, and other viruses can engender varied morphological disturbances, as well as psychological problems (e.g., schizophrenia)

and learning disabilities (Brown, 2011). Given how often and how readily pregnant women can contract viral or bacterially related illness, it is obviously of considerable importance that the effects observed, and how these evolved, be better understood. This was highlighted with the recent Zika outbreak, but it's been an issue for some time in relation to other infections, such as influenza. Understandably, analyses in humans focused on linking prenatal viral infection to the later development of pathology in the offspring can be difficult, particularly if the disturbances, as in the case of schizophrenia, might not emerge (or at least be detected) until mid or late adolescence.

These limitations aside, the confluence of data obtained using several approaches have supported the view concerning the links between prenatal infection and the emergence of pathological conditions. In some instances, the effects observed, as in the case of heart defects and facial deformities, may stem from fever experienced early in pregnancy (Hutson et al., 2017), which speaks to the possible usefulness of pregnant women using acetaminophen to reduce fever (but, see Avella-Garcia et al., 2016).

Viral epidemics in humans (Rubella, influenza, toxoplasma gondii, and herpes simplex virus type 2) were associated with a 500%–700% increase in the birth of children who subsequently developed schizophrenia. In fact, more than one-fifth of individuals that had been exposed to rubella prenatally later received a diagnosis of either schizophrenia or a schizophrenia spectrum disorder (Brown, 2011). Such an outcome was not uniquely tied to Rubella. Among women exposed to influenza during the first trimester of pregnancy, a 700% increase in risk of schizophrenia was reported in offspring, while the intracellular parasite Toxoplasma gondii was associated with a 200% increase of schizophrenia as well as a variety of other neuropsychiatric difficulties. To be sure, these are correlational findings and numerous other factors could have been responsible for the apparent link between prenatal inflammation and later schizophrenia in offspring. While viral insults and the resulting inflammatory changes, including brain cytokine elevations, could potentially affect neuronal processes that provoke schizophrenia, febrile (fever) responses associated with viral insults, and fetal hypoxia secondary to infection, could have lasting effects (Brown, 2011). In addition to schizophrenia, prenatal infection was also associated with autism, cerebral palsy, epilepsy, and even Parkinson's disease. Once more, prenatal infection might create a general vulnerability for neurodevelopmental disorders, but still other factors dictate which specific pathologies might arise (Harvey & Boksa, 2012).

Given the difficulties of forming one-to-one connections between prenatal infection and later pathological conditions in humans, there have been many attempts made to do this through animal models. Of course, there is no certainty that animal models of schizophrenia actually reflect the human disorder, particularly as some symptoms can't be adequately assessed (e.g., hallucinations). Nonetheless, prenatal infection in rodents produced disturbed social interaction in pups, together with attentional disturbances and altered exploratory behaviors. These behavioral phenotypes were accompanied by various brain alterations that have been linked to schizophrenia, such as diminished size of the cortical and hippocampal regions, coupled with an increase in cortical pyramidal cell density and more tightly packed pyramidal cells (Meyer & Feldon, 2009).

As dopamine and glutamate neuronal functioning may be key ingredients in the emergence and maintenance of schizophrenia, it is significant that prenatal infection in rodents was accompanied by dopamine receptors variations and activity (Meyer & Feldon, 2009), as well as changes of glutamate signaling in offspring (Meyer, Nyffeler, Yee, Knuesel, & Feldon, 2008). The dopamine changes occurred regardless of whether pups were raised by their biological mother (that had been infected during pregnancy) or a surrogate mother that cared for them following birth, indicating that the prenatal infection (as opposed to postnatal maternal care) was fundamental for the neurochemical alterations. Significantly, paralleling what occurs in humans, the behavioral disturbances among pups that experienced the prenatal infection only emerged during the adolescent period (Meyer et al., 2008), and could be diminished by antipsychotic medications that attenuate dopamine functioning.

Not every prenatal infection leads to adverse outcomes. Ordinarily, a small number of maternal cells get through the placenta and embed themselves in the organs of the developing fetus (microchimerism). This may be important for the fetus and mom's immune system to become compatible and hence not attack one another. In countries where malaria is endemic, the offspring of mothers that had malaria during pregnancy carried more of mom's DNA than did offspring of moms that had not been infected. Although they were more likely to be infected with malaria (showing a positive blood smear), they were less likely to exhibit malaria symptoms (Harrington et al., 2017). How this resistance comes about may be related to mom's immune system teaching the offspring how to deal with infection.

AUTISM AND LINKS TO IMMUNITY

The research concerning autism in relation to immunity, and particularly to vaccination, has had a shady history, thanks to the controversial reports by Wakefield suggesting that vaccination with MMR (mumps, measles, Rubella) led to Autism. We needn't go through this sketchy

---

(cont'd)

story as it's likely well known to readers. Suffice it that it has had enormous adverse effects as parents stopped immunizing their children, thus leaving them vulnerable to diseases. It may also have hampered legitimate research evaluating the impact of viral factors on the provocation of autism and other conditions that may be linked to inflammatory processes.

It's likely that a single factor doesn't lead to all instances of autism, although prenatal or early life experiences (e.g., very premature birth and the hypoxia that goes with it) might contribute to this disorder, as might epigenetic changes. A credible view of autism has emerged that immune-related disturbances in pregnant women might be at the root of autism in a substantial number of cases, and may also contribute to schizophrenia (Estes & McAllister, 2015). These immune-related alterations include major histocompatibility complex class I molecules, microglia, complement factors, and pathways downstream of cytokines. Among women who had experienced a viral infection during the first trimester of pregnancy, a 300% increase of autism was later detected, and when the infection occurred in the second trimester, a 40% increase occurred. Much as previously reported for schizophrenia, it was proposed that it's not the virus itself that causes the problem, but the strong immune responses on the part of the mother that produces the collateral damage in the offspring.

IMPACT OF EARLY LIFE CHALLENGES

Besides the effects of prenatal stressors, negative events encountered during the early postnatal period may also have pronounced long-term ramifications on well-being. Among other things, such negative early life experiences can alter the course (trajectory) of stress relevant neurobiological processes so that stressor reactivity is exaggerated in later adulthood, and the occurrence of psychological disturbances is elevated.

Not all early life adversities have negative long-term repercussions. Moderate ("tolerable") stressors could serve to facilitate learning how to appraise and cope with such events (Shonkoff, Boyce, & McEwen, 2009). Similarly, tolerable stressors might prime stress-relevant biological systems so that a later stressor experience will give rise to moderate neurochemical changes that operate to enhance coping efforts (Santarelli et al., 2017). Moderate challenges during early life and adolescence might also prime growth factor functioning, and the resulting enhanced synaptic plasticity could potentially facilitate the ability to contend with subsequently encountered challenges. This raises the question as to whether prenatal stressors, like those encountered early in life, can in some way enhance resilience in response to postnatal adversities that might be encountered[2].

In contrast, many stressors that children encounter comprise "toxic challenges," such as psychological, sexual or physical abuse, neglect, withholding affection, family disturbances and

---

[2] This type of change is not unique to stressors. "Hormesis" is said to occur when positive outcomes emerge following exposure to low levels of a toxicant, which at high doses as negative effects. The low dose can also have positive effect in protecting against the adverse consequences that might later be engendered by a high dose. Similar effects are seen in other contexts. Low level brain stimulation by external means, for instance, can protect against the damaging effects that might otherwise be provoked by a subsequent large seizure.

parental problems (e.g. mental illness, addiction), and extreme poverty. These childhood experiences profoundly shape behavioral, neurobiological, immunological, and microbial trajectories, and thus may have considerable sway on the appearance of later behavioral and physical pathologies (e.g., depression, anxiety), and may even be linked to earlier death. These toxic early life experiences can also program later behavioral and emotional responses so that responses to new stressor encounters are exaggerated. These experiences can also undermine behavioral styles so that individuals have difficulties forming social relations, become less trusting, experience disturbed self-regulation, and adopt poor lifestyle choices, any of which could aggravate already disturbed adaptive neurobiological processes. In addition, strong early life stressors have been associated with impaired decision-making, increased stress reactivity, and resistance to fear responses being extinguished, as well as the development of depression, anxiety, risk for PTSD, and elevated vulnerability to physical illnesses. Given the multiple negative consequences associated with early life abuse, it probably comes as no surprise that the depression that occurs in adulthood among these individuals, tends to be especially resistant to positive actions of antidepressant medications (Williams, Debattista, Duchemin, Schatzberg, & Nemeroff, 2016). Still, among children brought up in low socioeconomic conditions and are more prone to develop chronic illnesses, such as metabolic syndrome, this outcome is less likely to occur if children received adequate maternal nurturance (Miller, Chen, & Parker, 2011).

Passage of Poor Appraisal and Coping Methods

Traumatic experiences encountered during early life could affect the way individuals appraise the world around them, including how they appraise and interpret later stressful encounters. Such experiences may also be associated with warped internal attributions (i.e., self-blaming and self-criticizing) and cognitive distortions linked to appraisal of safety, a preoccupation with danger, as well as excessive focus on being able to have control over their own lives. These characteristics, particularly the exaggerated perceptions of future harm, and maintaining this sense of threat and unpredictability about the future, may contribute to the emergence of anxiety, and the generally negative worldview might favor later depressive illnesses. The seemingly warped (negative) perspective about future harms is not entirely without merit, as individuals who experienced trauma are, in fact, at elevated risk of encountering further stressors. As adults, children who experienced childhood abuse were at elevated risk of re-victimization experiences (e.g., domestic violence, partner abuse, and rape). This concept of "stress generation" (Liu & Alloy, 2010) has been known for some time, even appearing in the work of the philosopher Rashi (1040–1105), who asserted that "one misfortune invites another."

It is unfortunate that toxic stressor encounters during early life frequently aren't isolated incidents, but may comprise one of a set of distressing events within a constellation of adverse experiences. It was thus suggested that victimization should not simply be viewed as "an event," but should instead be considered as "a condition" that entails ongoing distress (e.g., Pratchett & Yehuda, 2011). Predictably, perhaps, these stressor experiences all too often extend over the course of the life span and even across generations (Bombay, Matheson, & Anisman, 2011). As such, when evaluating the impact of childhood stressors on the subsequent emergence of pathology, it might not be productive to limit analyses to particularly severe stressors, and instead, it might be appropriate to consider the cumulative effects of multiple lifetime stressors that might have been endured.

Inappropriate appraisals may give rise to poor coping abilities, and children who experienced traumatic events, often employ ineffective coping strategies to deal with ongoing challenges. Children and adolescents who experienced community violence, sexual abuse, or maltreatment tended to use emotion-focused and avoidant coping strategies (as well as risk-taking, confrontation, and the release of frustration), rather than problem-oriented coping that might be more effective in eliminating the stressor. Ordinarily, the resources and experiences to appraise stressors accurately are not well formed among children, and the coping strategies available to them are limited. Early life stressors may further diminish cognitive flexibility and may stunt coping development, and the poor ability to adopt new and more effective strategies are carried into adulthood. It is also thought that chronic or repeated adversities may give rise to the child attempting to understand why abusive or neglectful experiences are happening to them. These children might attribute the stressor to aspects of themselves, and they may internalize the belief that the adverse events are due to their own shortcomings. Thus, these experiences may influence individual characteristics (e.g., self-perceptions, self-blame, mastery, and self-esteem), which may foster the development of negative cognitive styles, further undermining the use of effective coping strategies to deal with later stressors.

How a child behaves might not always provide information concerning the actual impact of very adverse experiences. Even if children that had experienced family discord and violence displayed little general anxiety, this didn't imply that the potential for disturbances weren't present. Among children who had experienced strong early life stressors, even if they didn't show outward signs of distress, depictions of angry faces provoked increased neuronal activity within the amygdala and the anterior cingulate cortex, much like that associated with strong threat responses in other individuals. These brain responses were not a reflection of general reactivity, as depiction of sad faces did not elicit similar neuronal changes (McCrory et al., 2011).

As discussed earlier, stressful experiences may promote neuronal sensitization so that later responses to stressors are exaggerated, and this may be especially notable in children. It should be said that despite the considerable evidence supporting the view that early life adverse events may influence the response to later challenges, there have been reports suggesting that early and midlife stressors can act independently in affecting several indices of inflammation measured peripherally, although what occurs within the brain may be something else entirely. Indeed, resting frontal EEG asymmetry was linked to the presence of inflammation, but this was most apparent among individuals who had experienced relatively extensive levels of childhood maltreatment (Hostinar et al., 2017).

Impact of Parenting

Poor parental behaviors come in several forms, such as neglect, disengagement and disorganization, hostility, coercion, and low positive parent-child interactions, sometimes stemming from parental drug addiction. Regardless of the form, these behaviors may have profound repercussions on children, including the development of depressive symptoms, suicidality, PTSD, and interpersonal difficulties, often mediated by negative perceptions of the self and the future. It is still more unfortunate, as we've known for at least 50 years, that poor parenting styles in rhesus monkeys may be recapitulated when these offspring have their own children. In rodents, as in monkeys, having an attentive and supportive mom leads to positive outcomes, whereas having an ineffective and distant mom leads to negative outcome. These long-term actions may stem from persistent stressor-related neuroendocrine alterations, such as CRH or glucocorticoids, or those associated with attachment, such as oxytocin. As we'll see shortly, stress reactivity and health outcomes that stemmed from early life neglect influenced the behavior of mice when they subsequently became mothers. In humans, poverty during early life (and the multiple hardships that come with poverty), can engender epigenetic effects that influence neuroendocrine and immune functioning in adulthood, which can also be passed on to their offspring. In this regard, genetic factors and parenting interacted to predict the parenting style that offspring later expressed (Beaver & Belsky, 2012). In essence, these experiences within a given generation, together with genetic influences, comprise the ingredients that govern the recapitulation of negative parenting and stressor experiences that appear in the next generation.

Many illnesses, such as type 2 diabetes and heart disease, take years to develop. While these conditions are usually associated with aging, increasing evidence has suggested that stressors in early life or even in utero set the stage for the expression of these illnesses (Eriksson, Räikkönen, & Eriksson, 2014). By example, prenatal psychosocial stressors were accompanied by insulin resistance in young adults, and risk for diabetes was increased among the children of women who lost a loved one during pregnancy (Li, Olsen, Vestergaard, Kristensen, & Olsen, 2012). Childhood adversities were also predictive of a doubling of subsequent gestational diabetes and a 500% increase in the occurrence of postpartum depression (Madigan, Wade, Plamondon, Maguire, & Jenkins, 2017). These experiences, in turn, were linked to negative health outcomes in their children, including anxiety and depression.

Although early life stressful events can have long-term ramifications owing to the reprograming of neuroendocrine or neurochemical stress responses, such outcomes might also disturb mom, who then affects her offspring. Using a cross fostering procedure (i.e., pups were transferred from their biological mother to one that displayed either high or low maternal care) and other related paradigms, it was demonstrated that some behaviors of the offspring can be "inherited" from their nursing mother, rather than from just their biological mother (Anisman, Zaharia, Meaney, & Merali, 1998).

Early Life Stressors in Relation to Hormonal Changes

Early life stressors experiences, like those encountered prenatally, can alter the developmental trajectory of several neurobiological processes, including glucocorticoid functioning, thereby affecting adult behavioral and biological reactivity in response to further stressful experiences. Increased glucocorticoid responses elicited by stressors in early life may also promote glutamate release at the prefrontal cortex, amygdala and hippocampus, which could favor anxiety in adulthood if these glutamate variations persist. These actions could come about owing to the sensitization of neurobiological processes so that further challenges promote exaggerated responses. As well, epigenetic alterations can develop in relation to glucocorticoid receptor functioning, hence affecting adult responses to stressors (e.g., Szyf, 2011).

Based on early studies, it was assumed that the infant period in rats, spanning 4–14 days, is exceptionally sensitive to stressors, such as separation from the dam, especially when this occurs for an extended period, such as 24 hours. It was subsequently reported that separation from mom for as little as 3 hours a day on 7–10 successive days was sufficient to produce elevated anxiety in pups (e.g., Plotsky & Meaney, 1993), even though moms under naturalistic conditions may leave the nest for such periods in order to forage for food. As much as the notion concerning the effects of maternal separation (for 3 hours each day) is intuitively appealing, other investigators found that separation for extended periods did not uniformly have the negative effects initially reported, although several sex-dependent effects were observed in relation to some phenotypes (McIntosh, Anisman, & Merali, 1999).

It also turned out that the very early life period (postnatal days 4–14) was not unique for adverse effects of stressors to appear. The window from 10–20 days postnatally was as sensitive to prolonged separation from the dam relative to that seen at earlier periods. Moreover, the stress sensitivity seen at later times seemed to be related to genes linked to reward processes involving the ventral tegmentum, which could confer life-long sensitivity to further stressors (Peña et al., 2017). Using an inflammatory challenge, a window from 14–21 days following birth was also reported in relation to the provocation of anxiety response, independent of any actions related to maternal influences (Spencer, Martin, Mouihate, & Pittman, 2006). Still other studies pointed to the juvenile period (days 28–35 postpartum in rats) as having particularly pronounced effects on GABA functioning and anxiety (e.g., Jacobson-Pick, Elkobi, Vander, Rosenblum, & Richter-Levin, 2008). Evidently, there are multiple windows during which stressor experiences can affect different phenotypes, although they might not involve identical processes.

Despite the inconsistencies and the different maternal separation paradigms used, among germ-free mice, separation from the dam for 3 hours a day was sufficient to alter HPA functioning and colonic cholinergic neural regulation. Upon bacterial recolonization, anxiety behaviors were introduced along with a microbial profile distinct from that seen in control animals (De Palma et al., 2015). Evidently, separation from the mom can serve as a stressor that renders mice vulnerable to bacterial dysbiosis, which favors anxiety.

SELECTIVITY OF THE HPA RESPONSE

The cortisol rise associated with acute stressors has multiple adaptive attributes, but as we discussed earlier, sustained cortisol release may have negative effects, including loss of hippocampal neurons. Thus, there may be benefits for the glucocorticoid response to be blunted under certain conditions, such as in response to chronic challenges. Yet another instance in which HPA functioning is blunted is in relation to the corticoid response seen in pregnant moms and their pups. During the last trimester of pregnancy, and during lactation, the HPA response to stressors is diminished in rodent dams. The reduced cortisol response may be attributable to any of several neurobiological processes, including alterations in the function of other hormones, such as oxytocin, prolactin, and opioid peptides, as well as a down-regulation in the ability of norepinephrine to promote hypothalamic CRH and arginine vasopressin release (Tu, Lupien, & Walker, 2005; Walker, 2010). It seems that during lactation, moms are protected from biological changes ordinarily associated with anxiety (although postpartum depression is an obvious deviation from these findings). The suppression in the dam prevents excessive corticosterone from reaching the fetus or pups (the latter through the mother's milk). There is no doubt a fundamental evolutionary need for stress hormone responses to be inhibited during the lactation period, but at the same time, there is also a need for selectivity to exist regarding the effects of different stressors. Thus, in contrast to the cortisol suppression characteristic of threats made toward the dam herself,

---

(cont'd)

exaggerated corticosterone release may occur in the mom in response to threats directed toward the pups. While still nurturing pups, animals are able to "filter" relevant from irrelevant stimuli at least in terms of their offspring's well-being. Accordingly, a threat directed at her offspring, provokes a profound change of cortisol levels, presumably to provide resources essential for her to maintain the well-being of her pups. Messing with momma is one thing, targeting her pups is another thing entirely.

The glucocorticoid changes associated early life stressors are certainly fundamental to well-being, but other processes are also affected by these experiences. As we saw earlier, stressors experienced early in life can have enduring effects on sexually related behaviors that are manifested during adulthood, perhaps stemming from persistent changes of estrogen receptors. It was likewise reported that mild postnatal infection reduced receptivity among females, as well as behavioral responsiveness to estradiol and progesterone in adulthood (Ismail, Garas, & Blaustein, 2011). Early life stressful events may also have marked effects on prolactin levels. Upon its release from the anterior pituitary, this hormone promotes lactation, and contributes to sexual behavior and sexual pleasure, eating processes, pain perception, and responses to emotional stressors. In the latter regard, prolactin receptors are present within brain regions that control emotional responses (e.g., central amygdala, the bed nucleus of the stria terminalis, and the nucleus accumbens), and stressor-elicited prolactin functioning may contribute to HPA activation.

As we discussed earlier, oxytocin has been linked to social bonding and attention to socially salient environmental triggers, and oxytocin levels and functioning can be influenced by psychosocial stressors. After 6 weeks of isolation, beginning just after weaning, prairie voles displayed high levels of anxiety coupled with enhanced mRNA expression of oxytocin, CRH, and AVP (Pournajafi-Nazarloo et al., 2013). Moreover, the maternal care female rats received during the early postnatal period predicted the subsequent appearance of oxytocin receptors in adulthood (Francis, Young, Meaney, & Insel, 2002).

Paralleling the findings in animals, stressful experiences in humans affected oxytocin functioning, and negative early life experiences, including those related to maternal care and emotional abuse, had persistent repercussions on oxytocin throughout adulthood (Heim, Newport, Mletzko, Miller, & Nemeroff, 2008). Congruent with the oxytocin changes, early life stressor experiences can have lasting consequences on prosocial behaviors, thereby affecting the organism's ability to contend with further psychosocial challenges. There have been indications that such outcomes were due to epigenetic effects, but as already indicated, other actions related to maternal care could ostensibly elicit these effects.

Given oxytocin's presumed role in social interactions, it may be significant that levels of this hormone were diminished among women who had experienced childhood abuse and then subjected to a social stressor, but a different form of distress, specifically, that of childhood cancer, did not produce equally pronounced effects (Pierrehumbert et al., 2010). Thus, it is tempting to surmise that the persistent effects on oxytocin are limited to those that involve social stressors. However, it seems that chronic physical illnesses had long-term ramifications on the development of anxiety and depression (Secinti, Thompson, Richards, & Gaysina, 2017). Furthermore, in rodents exposed to a prenatal stressor, the disturbed oxytocin levels and disrupted social interactions seen in adulthood could be attenuated by oxytocin administered directly to the amygdala, making it likely that the connections between prenatal stressors, oxytocin changes, and social interactions were of a causal nature (Lee et al., 2007).

BREAST FEEDING—MORE THAN JUST AN OXYTOCIN HIT

Breast feeding may facilitate bonding between the offspring and mom owing to the release of hormones, such as oxytocin. Beyond this very important aspect of breast feeding, breast milk contains nutrients such as fat, protein, vitamins and minerals, and long chain polyunsaturated fatty acids that contribute to the development of cognitive and motor processes (Jonas & Woodside, 2016). Breast milk also contains growth factors and other hormones that come to affect neurodevelopment, and this form of nutrition also contains immunoglobulins that may be important to preclude some infections. Thus, breast fed infants were less likely to develop gastrointestinal tract problems, respiratory tract infections, obesity, diabetes, and childhood leukemia and lymphoma, relative to their bottle-fed peers (e.g., Ahmadizar et al., 2017), possibly operating through inflammatory processes. There is evidence indicating that the mother's breast milk, through the presence of particular microRNAs, can provide preterm infants with a metabolic boost that enhances their ability to thrive (Carney et al., 2017).

Breast milk not only contains fats and lactose, but also more than 200 different human milk oligosaccharides (complex sugars) that are an exceptionally rich energy source. These oligosaccharides, which aren't easily digested, travel to the gut where they feed good bacteria, which then act to create adhesive proteins that seal the gut so that microbes are kept out of the bloodstream. In addition, there is the belief that breast milk acts against bacteria that promote disease. Among premature infants, some protective intestinal bacteria may not be present and if they are unable to engage in breast feeding, which is often the case, they might be at increased risk for infection. However, lactoferrin, a protein ordinarily obtained in breast milk, can diminish these adverse effects (Sherman, Sherman, Arcinue, & Niklas, 2016).

Aside from the elevated anxiety and depressive features associated with early life stressor experiences, chronic physical illnesses were frequently evident in adulthood, which were particularly potent in the promotion of later anxiety and depression (Secinti et al., 2017). Early life adversities may also be linked to chronic pain in adulthood (Burke, Finn, McGuire, & Roche, 2017), which may be tied to multiple neurochemical alterations promoted by early life stressors, including elevated inflammation, although this could also be secondary to the depressive illness that evolved. Significantly, the effects of earlier stressors on mood disturbances are modifiable by antidepressant treatments alone, and in combination with omega-3 polyunsaturated fatty acids (Pusceddu et al., 2015).

Morphological Brain Changes Tied to Early Adversity

The volume of both the amygdala and hippocampus was reduced among individuals who had endured early life abuse or neglect, and these outcomes were especially marked with cumulative stressor experiences (Frodl & O'Keane, 2013). Negative early life challenges that were linked to variations of brain structure and volume were also associated with internalizing symptoms, such as depressive affect and anxiety, as well as impulse control and bipolar disorder.

The brain morphological changes tied to previous distressing events appeared to be moderated by genetic influences, although it has been reported, at least in relation to PTSD and depression, that preexisting hippocampal disturbances promoted increased risk for illness. The diminished hippocampal volume was not unique to depression or PTSD, having been seen in patients with schizophrenia. Thus, small hippocampal size may be a core element for an assortment of mental illness, but still other factors determine whether and which pathologies will emerge. Specifically, some of the effects of stressors on brain morphology were primarily apparent in the presence of particular gene mutations, such as those related to BDNF, serotonin reuptake, or an enzyme involved in degrading norepinephrine (COMT) (Rabl et al., 2014), each of which has been implicated in the development of depressive illnesses or high stressor reactivity. Hippocampal and other brain changes were also linked to diminished social support, disturbed coping, as well as to epigenetic effects (Mehta et al., 2013).

Being empathetic, some humans are affected simply by witnessing others experiencing a distressing event (even if they might not take actions to diminish this). It seems that this also occurs in rodents as well as in other species so that when an animal is stressed, nearby conspecifics display stress responses, although in rodents this might reflect a stress reaction brought about by pheromones, rather than empathy for a buddy. These reactions are fairly powerful, in that the offspring of pregnant rats who witnessed another rat being stressed, exhibited marked depressive-like behaviors, accompanied by more frequent epigenetic

changes, an altered gene expression profile, and brain morphological variations (Mychasiuk et al., 2011). The brain changes were characterized by diminished dendritic arborization, neuronal and glial cells being reduced within the prefrontal cortex and hippocampus, and limited maturation of hippocampal granule cells (a small type of neuron that is subject to neurogenesis), which persisted into adulthood. These stressor conditions influenced cortical brain regions that subserve attention, executive functioning, as well as information processing and memory, so that the behavior of the offspring was greatly disturbed.

Early Life Stressor and Brain Neurochemical Variations

Adverse experiences early in life elicit pronounced brain neurochemical changes. Separation of pups from their mom led to social anxiety when these pups were adults, accompanied by reduced 5-HT$_{1A}$ receptor expression in the dorsal raphe nucleus, the site of 5-HT cell bodies (Franklin et al., 2010). When previously stressed mice encountered a stressor during adulthood, tryptophan hydroxylase mRNA expression and that of 5-HTT was elevated (Gardner, Hale, Lightman, Plotsky, & Lowry, 2009). Early life social isolation likewise downregulated dopamine activity within the nucleus accumbens and increased kappa opioid receptor reactivity, which could have implications for later pathology (Karkhanis, Rose, Weiner, & Jones, 2016).

Early life stressors engendered especially marked and lasting alterations of GABA$_A$ receptor activity, particularly if rodents were reexposed to the stressor in adulthood (Skilbeck, Johnston, & Hinton, 2010). Similarly rats that had been stressed during adolescence exhibited more pronounced GABA variations within the lateral amygdala (Zhang & Rosenkranz, 2016) and GABA$_A$ subunit variations were more pronounced in response to later stressors relative to that evident in nonstressed rodents. Predictably, based on the GABA$_A$ subunit conformational variations stemming from the stressor treatment, these animals were also more sensitive to the anti-anxiety agent diazepam (Jacobson-Pick et al, 2008), and might have contributed to elevated anxiety that developed subsequently (Salari & Amani, 2017). Evidently, by influencing the developmental trajectory regarding GABA functioning, adverse experiences in early life can determine later basal and stressor elicited levels of anxiety. It also seemed that diminishing inflammatory responses (e.g., through the administration of a cyclooxygenase inhibitor), reduced the behavioral signs linked to depression, and altered the GABA and dopamine D2 receptor changes, otherwise provoked by an early life stressor (Lukkes, Meda, Thompson, Freund, & Andersen, 2017).

Growth Factors in Relation to Early Life Experiences

Early life stressors can disturb hippocampal synaptic plasticity, more so in males than females (Derks, Krugers, Hoogenraad, Joëls, & Sarabdjitsingh, 2016), and can diminish the hippocampal neurogenesis ordinarily engendered by exercise in male mice (Abbink, Naninck, Lucassen, & Korosi, 2017). Moreover, in line with the presumed involvement of early life experiences in the development of depression, growth factors, such as BDNF, as well as cytokines within the brain, were disturbed in association with stressors encountered early in life (Bilbo & Schwartz, 2012). Such effects, including those produced by prenatal stressors, may have arisen owing to epigenetic actions (Fumagalli et al., 2004). The persistent suppression of regulatory elements for genes that code for BDNF, might promote or exacerbate the development of depressive illness in

adulthood. Female rhesus macaques who experienced maternal deprivation developed a depressive-like profile that was accompanied by reduced plasma BDNF levels. These outcomes could be diminished if monkeys were reared by peers, but this was least apparent if they carried a particular BDNF polymorphism (Cirulli et al., 2009).

Similar studies are obviously difficult to accomplish in humans, but depressive symptoms among pregnant women predicted reduced BDNF methylation in male and female infants (Braithwaite, Kundakovic, Ramchandani, Murphy, & Champagne, 2015), although the significance of these findings for later mental health remain uncertain. This said, in humans who experienced negative early life events and who also carried the BDNF mutation were likely to exhibit biases in favor of recalling negative stimuli, which was linked to the development of depression (Aguilera et al., 2009). These actions were still more prominent among individuals with a double polymorphism comprising BDNF and the gene for the serotonin transporter (5-HTT) (Carver, Johnson, Joormann, Lemoult, & Cuccaro, 2011).

Early life abuse and the presence of the BDNF mutation was accompanied by diminished gray matter within the subgenual anterior cingulate cortex, but not in other brain regions that have been implicated in depression (hippocampus, prefrontal cortex). Thus, this region might comprise a mechanistic tie between stressful early life experiences, the BDNF polymorphism, and adult depression. It seems, as well, that early life abuse that was associated with later depression and suicide, was accompanied by altered myelination of axons that linked the cingulate cortex and both the amygdala and nucleus accumbens, and might thus affect emotions and reward processes (Lutz et al., 2017). As well, among individuals who had experienced early life emotional abuse, the amygdala neurons responded very strongly to threats (an anticipatory response), and neuronal processes within a greater expanse of brain regions were consequently affected (Klumpers, Kroes, Baas, & Fernández, 2017).

The neurotrophic changes elicited by stressors and their implications for depression had received considerable support, but findings in both rodents and humans have suggested that the stress-BDNF-depression linkages are not as straightforward as first believed. Specifically, under normal conditions, individuals ought to be responsive to environmental stimuli so that positive experiences would enhance later well-being and negative experiences would have the opposite experience. However, in the presence of the polymorphism for serotonin and BDNF, an individual's malleability (synaptic plasticity) would be reduced so that the usual influence of environmental or experiential factors, whether positive or negative, would not be realized (Caldwell et al., 2013). Thus, the positive effects of a stimulating environment would not occur, but neither would the adverse actions of negative experiences. Of course, the BDNF and 5-HTT interactions with life stressors are only two among many others that have been reported, including polymorphisms of the FKBP5 gene that influences glucocorticoid receptor sensitivity (Grabe et al., 2016). Just as a more "nuanced" role for stressor-elicited neurotrophic factors should be considered in relation to learning and memory processes, the link between stressor-elicited neurotrophic factors and mood states need to be evaluated in conjunction with other neurobiological changes that occur (Bondar & Merkulova, 2016).

IMMUNITY AND MICROBIAL FACTORS

Early Life Stress and Immune Functioning

Stressful early life experiences can have a lasting influence on immune functioning, and might thereby affect well-being throughout the life span (Cattaneo et al., 2015). The negative effects of early life stressors were observed in response to psychological as well as systemic insults (e.g., among rodents challenged with a bacterial endotoxin or to an influenza virus), and were moderated by adult stressor experiences, social support systems, and coping processes (Fagundes, Glaser, & Kiecolt-Glaser, 2013). For example, a stressor in the form of maternal separation diminished cytokine gene expression and might thus have affected later well-being (Dimatelis et al., 2012). Early isolation of rat pups, which led to a depressive behavioral profile, was also accompanied by microglial activation and elevated proinflammatory cytokine expression within the hippocampus, and these outcomes could be prevented by the antibiotic minocycline (Wang, Huang et al., 2017). Supporting the importance of cytokines, treatment with an antidepressant during adolescence reduced the anxiety and depressive features provoked by maternal separation, and concurrently enhanced immune regulation, notably diminished levels of proinflammatory cytokines (IL-1β) and increased levels of the anti-inflammatory IL-10 (Wang, Dong et al., 2017).

A systematic review indicated that a variety of childhood stressors could increase circulating proinflammatory cytokines and acute phase proteins (CRP, fibrinogen), and this chronic inflammatory state might lend itself to a variety of disease conditions (Coelho, Viola, Walss-Bass, Brietzke, & Grassi-Oliveira, 2014; Walker, Nakamura, & Hodgson, 2010). These range from airway inflammation through to the development of depression (Jonker, Rosmalen, & Schoevers, 2017), and can modulate the actions of genes related to bipolar disorder, so that the illness is more apt to emerge (Scaini et al., 2017). Once again, many neurobiological routes are present through which such outcomes may evolve, including epigenetic changes. As well, early life trauma, such as abuse, can affect lifestyles that come to promote both inflammatory processes and depression, so that when controlling for lifestyle, the link between abuse and later depression disappears (e.g., Jonker et al., 2017). This doesn't mean that early life trauma isn't tied to depression, but instead suggests that life-style factors act as an intermediary in this regard.

A prospective analysis similarly revealed that early life social adversity predicted elevated inflammation in adulthood, which may be a biomarker of heart disease (Slopen et al., 2015). Analyses within a unique sample of participants that could be followed for an extended period, revealed that microbial and psychosocial experiences in infancy and childhood predicted epigenetic changes linked to inflammation. These epigenetic marks were predictive of later heart disease and other illnesses tied to inflammation (McDade et al., 2017), even affecting cellular functioning and successful ageing (Ambeskovic, Roseboom, & Metz, 2017).

Studies conducted in humans have been consistent with the view that early life adversities may result in biological changes in which a proinflammatory bias developed, thereby increasing the risk for adult pathology emerging. Early life sexual abuse, for instance, was accompanied by chronic inflammation characterized by elevated circulating CRP and IL-6 in adulthood (Bertone-Johnson, Whitcomb, Missmer, Karlson, & Rich-Edwards, 2012). Moreover, among adolescent girls who had experienced early life adverse events the *in vitro* production of IL-6 from stimulated monocytes was elevated and at the same time glucocorticoid sensitivity to challenges was reduced (Ehrlich, Ross, Chen, & Miller, 2016).

Just as stressors affected immune functioning, among rat pups that were treated with an endotoxin, subsequent adult stressor responses were altered, including greater suppression of lymphocyte proliferation and NK cell functioning (e.g., Shanks et al., 2000) and increased central cytokine responses elicited by stressors encountered during adulthood (Walker et al.,

2010). Inflammation provoked during early life similarly increased amygdala neuronal functioning and resulted in impaired auditory fear extinction in adulthood (Doenni, Song, Hill, & Pittman, 2017), a feature seen in PTSD. Once again, these actions were not only apparent in response to immune challenges during the very early postnatal period, as immune activating agents administered 14 days following birth also engendered lasting effects on emotional reactivity in rodents (Dinel et al., 2014). Pathogenic encounters early in life likewise elevate HPA functioning, and might thereby dampen immune activity. Such actions can carry over into adulthood, provided that the initial challenge occurred during a critical window during prenatal or early postnatal development (Mouihate, 2013). Consistent with the reports in rodents indicating that early life stressors affect adult psychopathology, neonatal immune activation promoted anxiety and depressive symptoms, coupled with HPA hyperactivity, and exaggerated hippocampal TNF-α and IL-1β presence upon stressor re-exposure in adulthood (Majidi, Kosari-Nasab, & Salari, 2016). Likewise, early life LPS administration could promote anxiety during adolescence, accompanied by elevated expression of NLRP3 inflammasome proteins within the CNS (Lei, Chen, Yan, Li, & Deng, 2017).

Persistent cytokine variations related to early life stressors can undermine physical health and promote psychological disturbances, and could potentially render individuals at increased jeopardy for disorders promoted by inflammatory processes. To the point, childhood adversity may increase reactivity to adult stressors, accompanied by increased perceived stress, fatigue, depressive symptoms, poor quality of life, together with diminished NK cell activity and elevated levels of IL-6. Pronounced effects of this nature were observed among women who experienced greater childhood emotional neglect/abuse, and could affect overall quality of life, as well as promoting lasting immune disturbances that may place individuals at elevated risk for diverse types of illness (Janusek, Tell, Albuquerque, & Mathews, 2013). In view of these linkages, it was suggested that new strategies ought to be explored to limit the consequences of childhood trauma prior to clinical symptoms being present, including anti-inflammatory interventions and alterations of adaptive immunity (Danese & J Lewis, 2017), although such interventions may pose other problems.

ADOLESCENCE AND EARLY ADULTHOOD

Adolescence is accompanied by reorganization of neuronal and hormonal systems, which may be especially sensitive to stressors. During these times, stressors are apt to produce disturbance of nerve cell growth within the hippocampus, altered CRH_1 and CRH_2 receptor functioning, changes of certain brain neurotransmitters, such as GABA and its receptors, and persistent alterations of glutamate NMDA-receptors (Yohn & Blendy, 2017). These neurochemical changes occur in the context of socialization processes, and it might be expected that social instability (and other stressors) at this time could engender pronounced and lasting effects, manifesting as increased adult anxiety and depression. Fear and anxiety responses established during adolescence are difficult to overcome, and it is thought that the majority of adult fear- and anxiety-related disorders have their roots in moods that developed at these and at earlier ages. Transitions from one phase of life to another often requires the ability to adapt to new people and circumstances, the development of new social networks and new social identities, and for college-aged individuals this includes the formation of an adult-like identity, which entails a need for social, economic, and emotional independence (which may clash with the continued need for parental support). Most often this transition occurs seamlessly, but among some individuals it is destabilizing, distressing, and lonely, breeding high levels of depression and anxiety. Aside from the actions of the transitions themselves, some individuals may experience bullying at various stages of adolescence, which can have ramifications throughout life, favoring the development of inflammatory-related diseases (depression, heart disease) in later years (Espelage, Hong, & Mebane, 2016), as well as elevated drug use (Earnshaw et al., 2017).

Gut and Immunity

Events experienced early in life that influence the microbiome can have pronounced consequences that carry-on through adulthood. For example, if bacteria normally present within or on tadpoles were disturbed, enduring effects were apparent in that they were affected by more parasites as adults relative to frogs that didn't have this early life experience. In contrast, if the initial bacterial challenge occurred in adult frogs, then the subsequent negative effects did not occur, attesting to the unique actions of the early life encounter (Knutie, Wilkinson, Kohl, & Rohr, 2017). It was similarly observed that fecal bacteria collected from infants at 1-year of age predicted cognitive performance measured when they were 2-years of age. Although it is often thought that the greater the bacterial diversity the better, in this particular case, children with less diversity and high levels of the genus Bacteroides performed better relative to that apparent in the presence of greater diversity and lower Bacteroide levels. Thus, there may be an optimum with respect one or both of these variables (bacterial diversity and the presence Bacteroides) in predicting later cognitive outcomes (Carlson et al., 2017).

Stressors affect the gut microbiota, and early life stressors may be particularly adept in this regard. Simply disturbing a rodent's nest promotes erratic maternal behaviors and instigates excessive corticosterone levels in pups, coupled with diminished microbial diversity (Moussaoui et al., 2017). Likewise, separating pups from their mom can provoke gut bacterial alterations (O'Mahony, Hyland, Dinan, & Cryan, 2011), which could be responsible for increasing susceptibility to later psychopathological outcomes (Dinan & Cryan, 2017) as well as immune-related disorders, such as

inflammatory bowel syndrome (O'Mahony, Clarke, Dinan, & Cryan, 2017b).

Given the sensitivity of the gut at this time, it can be expected that the diet consumed during early life can affect an organism's microbiome, and might thus affect CNS processes. As expected, microbial manipulations, notably supplementation with B. pseudocatenulatum CECT7765 was able to reverse the immune and neuroendocrine disturbances, as well as elevated anxiety associated with early life maternal separation (Moya-Pérez et al., 2017). In line with such reports, a probiotic (*Lactobacillus fermentum*) could be used to diminish intestinal barrier dysfunction stemming from early life maternal separation, possibly through immune alterations that increased IFNγ and reduced IL-4 (Vanhaecke et al., 2017).

NOTHING TO SNEEZE AT

Among neonates, the frequency of infections is relatively high, which has typically been attributed to immune functioning not being sufficiently mature to ward off bugs that happen to be around. This may be correct, but there's more to it than that. Clostridia, a bacteria present in the gut, protects neonatal mice from at least two potentially lethal pathogens, and a set of anaerobic, spore-forming bacteria (clostridia) protect neonatal mice against diarrhea-causing pathogens. It has similarly been reported that byproducts of gut microbiota, together with two types of fungi, are associated with increased inflammation, leading to asthma and allergies (Fujimura et al., 2016), and children deficient of several gut bacteria at 3 months of age were at increased risk for developing asthma by age 5 (Arrieta et al., 2015). As the pathophysiological profile of those with allergies is in some ways similar to that which accompanies chronic stressors,

it is possible that stressful events encountered early in life may contribute to the development of allergies (Schreier & Wright, 2014).

Microbial and immune processes are functionally related, and the two mature in concert with one another, so that perturbations of one system may affect the other. Even small changes in the intestinal environment can affect microbiota, which will affect immune processes. Likewise, immune changes may profoundly affect microbiota functioning. In a detailed review, it was suggested that prenatal and early postnatal modulators of microbiota can have profound and lasting effects on illness vulnerability in adulthood (Amenyogbe, Kollmann, & Ben-Othman, 2017). Indeed, early life use of antibiotics has been implicated in the formation of allergies (Ahmadizar et al., 2016). Thus, targeting this phase of development (for example, through pre- or probiotic supplements) may be a way of preventing later illness (Shaaban et al., 2017).

Consistent with the finding that diet manipulations can have marked consequences, supplementation with the prebiotic BGOS during the neonatal period influenced BDNF protein levels during young adulthood (Williams, Chen et al., 2016), and micronutrient supplements provided early in life limited later stressor-provoked corticosterone changes and cognitive disturbances that would otherwise be apparent (Krugers et al., 2017; Naninck et al., 2017). Likewise, early life diets that included prebiotics and bioactive milk fractions affected stress-elicited mRNA expression of dorsal raphe nucleus 5-HT$_{1A}$ receptors and largely precluded behavioral disturbances otherwise provoked by uncontrollable stressor exposure (Mika et al., 2017). Among germ-free mice, hippocampal levels of serotonin and its primary metabolite 5-hydroxyindoleacetic acid, were appreciably elevated, primarily in males, suggesting a stress-like profile. Although later colonization with microbiota did not restore serotonin functioning, the elevated anxiety that was otherwise present was diminished, suggesting that these behaviors were independent of microbially related serotonin changes (Clarke et al., 2013).

Among rodents born entirely germ-free, postnatal development of immune processes are undermined; lymphoid disturbances may be present, and balances between various aspects of the immune system can be disturbed, resulting in elevated vulnerability to immune-related disorders. Likewise, among pregnant women, good gut bacteria can positively affect the fetus, and the use of antibiotics by a pregnant woman negatively affects both mom and her fetus. The vertical transmission of microbial protection comes with vaginal birth when the emerging infant meets a variety of bacteria within the mom's birth canal, whereas Caesarean delivery may render offspring more vulnerable to illnesses, such as inflammatory and metabolic diseases, including allergies and asthma, and other chronic childhood immune disorders (Bäckhed et al., 2015; Gensollen, Iyer, Kasper, & Blumberg, 2016).

To accommodate the loss of bacteria associated with Caesarean birth, there were efforts made to provide newborns with bacteria by applying vaginal swabs (vaginal seeding) from their birth mother (Dominguez-Bello et al., 2016). Although the mode of delivery was associated with altered abundance and diversity of gut bacteria at 3 months following birth, these effects were no longer apparent at 6 months (Rutayisire, Huang, Liu, & Tao, 2016). Likewise, although there had been suspicions that Caesarean section might be associated with an adverse cardiometabolic risk profile early in life, analyses of infants at 1 year of age did not confirm these concerns (Haji et al., 2014). There continues to be debate regarding the value of vaginal seeding, but the American College of Obstetricians and Gynecologists recently indicated that they do not recommend or encourage the procedure given that the data available are not strong enough to support a causal connection between seeding and well-being. Moreover, the procedure could have the undesirable effect of transmitting pathogens to the infant.

Ordinarily, organisms such as Lactobacilli and Bifidobacteria appear in the gut within a few hours of birth, typically being obtained from mother's milk. As well, low numbers of other nonpathogenic bacteria like Streptococcus, Micrococcus, Staphylococcus, and Corynebacterium soon colonize the gut. In an effort to improve health and limit allergies and hypersensitivities, prebiotics and probiotics, as well as a combination of the two, have been included in infant formulas, but a systematic review concluded that their usefulness is uncertain (Mugambi, Musekiwa, Lombard, Young, & Blaauw, 2012). There have been reports in which dietary factors were manipulated in an effort to alter the microbiota of children born by Caesarean section. As expected, synbiotic treatment (a combination of prebiotics and probiotics) normalized gut microbiota (Chua et al., 2017), but once again, the effectiveness of this in relation to other phenotypes is less clear. While probiotics may diminish the odds of eczema developing, the evidence related to other allergic conditions is generally weak (West, Jenmalm, Kozyrskyj, & Prescott, 2016).

Modifying gut bacteria, as expected, can have marked ramification on some illness conditions. By example, upon being treated with normal gut bacteria from an adult mouse, female mice that were genetically at risk for type 1 diabetes exhibited a marked decline in the occurrence of this illness. It was also reported that in rural portions of India in which neonatal sepsis is a major problem, a synbiotic mix could reduce this condition by as much as 40%.

EVEN IF "CLEANLINESS IS NEXT TO GODLINESS," IT DOESN'T NECESSARILY LIMIT DISEASE

Disturbances of immunoregulatory processes may be attributable, at least in part, to the failure of encountering microorganisms from of our evolutionary past. Not meeting with these "old friends" has made humans less able to develop optimal immune functioning and hence more vulnerable to disease states. Typically, when the Old Friends hypothesis is considered, the focus has been on immune related processes, but environmental microbial factors are also important players in this respect (Lowry et al., 2016). Not adapting adequately to microbes may result in exaggerated inflammatory immune responses, thereby influencing vulnerability to psychiatric illnesses. In fact, some of our old friends, including helminths and bacteria, have been almost entirely eliminated from the urban environment, thus elevating our dependence on microbial factors obtained from our mothers, other people, animals, and specific features of the environment.

According to the "hygiene hypothesis," protecting children from every bug can render their immune system less practiced and hence more vulnerable to infection. Likewise, early experiences with environments that are too sterile can limit the development of immune tolerance to foreign substances, thus allowing for exaggerated responses to some harmless elements. In fact, the increased frequency of allergies (and asthma) in recent decades has been attributed to excessive cleanliness (Strachan, 2000), as have other immune-related illnesses, such as multiple sclerosis, inflammatory bowel disease, and depressive disorders (Okada, Kuhn, Feillet, & Bach, 2010). Conversely, being raised in a rich microbial environment (e.g., living on farms) was associated with reduced occurrence of inflammatory bowel disorder during adulthood, although the greater use of pesticides within such environments may have increased vulnerability to neurodegenerative disorders (Mostafalou & Abdollahi, 2017).

Tight regulation of immune functioning during the prenatal and early postnatal period is necessary to prevent excessive inflammation that can have adverse actions. For this to occur, young children need early, regular, and frequent exposure to harmless microorganisms (old friends) that have been present over the course of evolution and thus are readily recognized by the immune system (Rook, Lowry, & Raison, 2015). These experiences begin in the uterus and are marked during the perinatal period. Among other experiences, Caesarean delivery, antibiotic abuse, and migration from poor to richer environments in which a change of immunoregulatory organisms are encountered, can all affect the microbiome, thereby influencing later health.

The original hygiene hypothesis suggested that people frequently became ill because they hadn't encountered potentially harmful bugs, but to be properly trained, the immune system also needs to encounter a diversity of bugs, not just infectious pathogens, but also friendly

(cont'd)

bugs. Encounters with a diverse set of bacteria serves to train Treg cells that are important for immune regulation, and preventing excessive immune activity that favors the development of allergies and asthma, as well as autoimmune conditions. Clearly, the hygiene hypothesis is broader than initially envisioned, and the processes that are presumed to be involved in creating barriers to infection have been reconsidered (Bloomfield et al., 2016).

The increasing incidence of allergies has been attributed to new external factors (possibly related to changing diets, increased use of antibiotics, or increased presence of pollutants) that cause disturbances related to balancing of immune system responses (Belkaid & Hand, 2014). Foods consumed are, in a sense, an antigen that should cause the immune system to react against them. However, the mucosal membrane within the digestive system, along with particular immune factors, limit these immune reactions, and permit the development of tolerance to the food antigens. Yet, there are occasions in which the barrier system is disturbed and hence problems may arise. Certain foods eaten during early life can influence gut bacteria that favor the development of allergies and inflammatory illnesses, and the microbiome of individuals with allergies are distinguishable from those without allergies. For that matter, gut bacteria present during the first few postnatal months could predict food sensitization that appeared at 1 year of age (Azad et al., 2015). Aside from the severe allergic responses that can lead to anaphylactic shock (e.g., in response to peanuts and related products), even minor allergies can have profound effects. Possibly owing to elevated inflammatory functioning, these allergies were associated with the risk of adult heart disease doubling as well as earlier onset of this condition (Silverberg, 2016).

EPIGENETICS AND INTERGENERATIONAL ACTIONS

Stressor-Related Epigenetic Effects

As we've seen throughout this chapter, early life experiences, whether good or bad, may affect the developmental trajectory of neurobiological, behavioral, and emotional processes, rendering people more or less vulnerable or resilient to pathology, even decades later. Furthermore, we can be affected by the experiences of our parents as well as the toxicants to which they had been exposed, and we, in turn, might pass these characteristics on to our own children.

Environmental and experiential factors that promote epigenetic changes that cause the silencing (or activation) of particular genes can occur at any time of life, but may be especially pertinent if they occur prenatally or early in development (Szyf, 2011). The most pertinent influence within a pup's life is their mom, and thus a dam's behavior toward her pups during early development (exhibiting good attention to pups vs being neglectful) may influence particular genes, including those that regulate HPA functioning (Champagne, 2010). Indeed, poor maternal care during early postnatal development was accompanied by elevated methylation of the promoter for the gene regulating hippocampal glucocorticoid receptors, and persistently elevated stressor reactivity (Szyf, 2011). It was similarly observed in mice that repeated prenatal stressor experiences encountered during the first trimester of pregnancy reduced DNA methylation of the gene promoter for CRH and sex-dependently increased the methylation of the promoter region of the gene for glucocorticoid receptors (Grundwald & Brunton, 2015). Of potential practical relevance, the impact of early life stressors can be altered by treating rats with either a methyl donor (methionine) or a histone deacetylase inhibitor (trichostatin A) (Weaver et al., 2005).

As much as reports of epigenetic changes of hormonal and behavioral processes were important in popularizing this line of research, as we indicted earlier, simply because early life maternal neglect causes both HPA-related epigenetic changes and behavioral alterations, doesn't necessarily mean that the two are connected. In fact, periodic maternal deprivation can produce a change in behavior coupled with elevated plasma corticosterone and hippocampal nerve growth factor, without provoking epigenetic changes within the promoter region of the gene for the glucocorticoid receptor. While not dismissing the potential importance of epigenetic changes associated with stressors, toxicants, and other external influences, there have been problems in identifying which specific epigenetic changes contribute to the effects of adverse early life or prenatal experiences. Many epigenetic marks are present within the genome, any of which could be linked to a given behavioral phenotype. By example, childhood abuse was associated with 997 differentially methylated gene promoters relative to that evident among individuals that had not experienced early life abuse (Suderman et al., 2014). Aside from the epigenetic modifications related to glucocorticoid receptors, early life events in monkeys were linked to DNA methylation of the gene coding for the serotonin transporter, which might contribute to the emergence of depression and other illnesses (Kinnally et al., 2010). Similarly, in female rodents, poor maternal care was accompanied by methylation of the gene

promoter of estrogen receptor alpha (ERα) in the hypothalamus (Champagne, 2010). As this receptor is fundamental for the functioning of estrogen and oxytocin, which contribute to maternal behaviors, it is possible that the silencing of the gene for ERα might contribute to the poor maternal behaviors stemming from impoverished early life care.

As described at the outset of this chapter, negative early experiences in humans were associated with an increase of various epigenetic marks related to inflammatory processes relevant to pathologies that appear in later life (McDade et al., 2017). Thus, linking any given epigenetic effect to particular phenotypes is difficult, certainly in studies with a small number of participants, as is the case in most research involving animals. Furthermore, most of the research that assessed the link between epigenetic changes and specific behavioral outcomes has comprised association studies, precluding conclusions regarding causal connections. Research is obviously needed to assess the effects of reversing epigenetic changes in order to determine the causal connections between epigenetic effects and particular phenotypes (Szyf & Bick, 2013).

Epigenetics in Relation to Environmental Toxicants

The marked cellular proliferation and differentiation that occurs during fetal development makes this a particularly sensitive period for environmental toxicants and stressors to turn genes on or off (Champagne, 2010). It was suggested that environmental toxicants (e.g., pesticides and fungicides, dioxin, jet fuel, and plastics) may produce epigenetic effects that can be transmitted across generations, just as trauma and maternal stressors can have such effects (Manikkam, Haque, Guerrero-Bosagna, Nilsson, & Skinner, 2014). A number of prenatal challenges (e.g., methyl mercury, diesel fumes, the androgenic fungicide vivclozolin, the estrogenic peptide methoxychlor, and the endocrine disruptor bisphenol-A) were also accompanied by epigenetic changes within genes coding for BDNF, as well as several immune factors (IFNγ and IL-4) that have been linked to psychological and physical disturbances (Champagne, 2010). Prenatal bisphenol A treatment produced epigenetic modifications in expression of the genes encoding estrogen receptors within the cortex and hypothalamus of juvenile mice, which was coupled with sex-dependent effects on social anxiety (Kundakovic et al., 2013).

Like so many other environmental toxins, bisphenol has been implicated as a health risk that could affect behavioral processes. Even at very low doses administered to pregnant rats, marked hormonal variations were evident in offspring, as was amygdala expression of receptors coding for estrogen, androgen, oxytocin, and vasopressin, which could affect later stressor responses (Arambula, Jima, & Patisaul, 2017). Consistent with these findings, the pups of rats that had been treated with bisphenol A during both pregnancy and lactation, exhibited an anxiety-like profile that may have stemmed from glutamate receptor alterations (Zhou, Chen et al., 2015). It is especially interesting that prenatal bisphenol A treatment in pregnant mice produced lasting epigenetic changes in the BDNF gene within the hippocampus and blood of mouse pups, supporting the view that blood BDNF changes introduced by a prenatal challenge might be an effective marker of hippocampal changes (Kundakovic et al., 2015).

Even relatively low concentrations of bisphenol could affect genes that are regulated by estrogen and could thus influence estrogen-related conditions, such as infertility, endometriosis, endometrial cancer, osteoporosis, prostate cancer, obesity, and breast cancer (Jorgensen, Alderman, & Taylor, 2016). It may be especially significant from an intervention perspective that the impact of toxins, like that of stressors, can be attenuated by increasing the presence of folate in the dam's diet (McGowan & Szyf, 2010), which has its positive effects by increasing the availability of methyl donors to limit methylation that had occurred earlier. It is interesting in this context that the use of folic acid (and multivitamin) supplements before and during pregnancy predicted a marked reduction in the occurrence of autism (Levine et al., 2018).

Intergenerational and Transgenerational Epigenetic Actions

Environmental triggers can elicit biological and behavioral effects that may be passed on across generations (transgenerational actions) provided that the epigenetic changes occurred in a germline cell (Franklin et al., 2010). What mom eats, the stressors she (or dad) encountered, the pesticides to which they had been exposed (some of which have profound effects on hormonal systems), can engender adverse effects that are recapitulated over several generations (Manikkam et al., 2014). Likewise, maternal immune activation through the viral mimic poly I:C, induced depressive-like effects that spanned several generations (Ronovsky et al., 2017)[3]. Thus, the transgenerational effects of systemic insults yield effects much like those engendered by neurogenic and psychogenic

---

[3] Animal models have strongly supported the view that prenatal inflammation may contribute to the later development of schizophrenia. Prenatal treatment of poly I:C, which mimics the actions of a virus, but without multiplying, has been one of the favorite methods of assessing the infection-schizophrenia link (Dickerson & Bilket, 2013).

insults. It has, indeed, been surmised that the progressive increase of some diseases witnessed over the past few decades could be due, in part, to environmental and experiential factors encountered in earlier generations, although this is obviously controversial and highly speculative. For the most part, the epigenetic links have been studied among women and their offspring, but there is evidence that epigenetic changes in dad's sperm, brought about by phthalates, can also affect offspring (Pilsner, Parker, Sergeyev, & Suvorov, 2017).

Although there has been considerable attention that focused on the role of glucocorticoid receptors in relation to intergenerational effects of stressors and toxicants, there have been suggestions that estrogen receptors changes, as well as epigenetic effects on genes coding for BDNF also play a role in this regard. When assessed as adults, rats that had been raised by a stressed adult caretaker, which displayed abusive behaviors, expressed elevated methylation of BDNF-related genes in the prefrontal cortex. When these rats had their own litters, the very same BDNF epigenetic profile was apparent in the offspring, pointing to the transgenerational epigenetic effects of early life stressors (Roth, Lubin, Funk, & Sweatt, 2009). Although epigenetic effects can last throughout life, and be passed across generations, raising rats in an enriched environment, a manipulation that may increase BDNF, reversed the adverse transgenerational actions of a stressor (Gapp et al., 2016). Likewise, exposure to enrichment during the juvenile period could reverse the negative behavioral consequences ordinarily provoked by poor early life maternal care (Champagne & Meaney, 2007).

Transmission of stressor effects from a mother to her offspring could occur either through epigenetic effects, various actions related to the uterine environment, or actions directly attributable to the dam's behavior (e.g., high anxiety in the dam may be transmitted to pups through her behaviors), or even

behaviors of the male parent. There is also the question of whether transgenerational epigenetic effects are transmitted exclusively through the mother, or can the father's experiences also have such effects? It appeared that in male pups that experienced chronic maternal separation, depressive-like behaviors and altered behavioral responses to aversive stimuli were evident upon being stressed as an adult. In the offspring of these males, behavioral and neuroendocrine disturbances were also apparent even though these animals had not been subjected to any particular stressor (i.e., they were reared normally) (Franklin et al., 2010). It was likewise reported that among male mice that experienced social defeat, their anxiety- and depressive-like responses were transmitted to their offspring despite these males not being present after the dam was impregnated and hence had no direct effect on their postnatal environment (Dietz et al., 2011).

To determine whether stressed male mice might have caused females to become distressed, thereby affecting her behavior toward the pups, females were impregnated through IVF using sperm from a stressed mouse. Under these circumstances, however, the transmission of behavioral effects from parent to offspring was not present (Dietz & Nestler, 2012). These data suggest that paternal transmission of the stressor effects was, in fact, not linked to epigenetic changes. Yet, stressors experienced by the male could have influenced sperm quality, including sperm motility, and thus might have been least likely to fertilize an egg, hence precluding transgenerational effects that potentially might have occurred. Given these varied findings, there is still some question about the epigenetic involvement of transgenerational transmission of stressor effects.

Finally, it is important to underscore that even when an epigenetic effects appear, they often peter-out over several generations. It had typically been surmised that either this is not a genuine transgenerational effect or one that decays or dilutes with successive passages across generations. It has, however, been shown that genes exist (termed Modified Transgenerational Epigenetic Kinetics) that are able to switch epigenetic events on and off. Moreover, small RNAs are present that act to regulate these genes. What causes the RNAs to behave as they do is uncertain, and it isn't known whether these actions are modifiable (Houri-Ze'evi et al., 2016).

Transmission of Psychological and Physical Sequelae of Trauma in Humans

It has long been known that cultural differences exist in relation to many diseases and in the occurrence of various polymorphisms. As we've seen, these differences may stem from living conditions (poverty, pollution, smoking, and diet), but ethnic identity was also associated with hundreds of epigenetic differences, about 76% being inherited, whereas the remainder came from social and experiential events (Galanter et al., 2017).

Few studies examined transgenerational (or intergenerational) epigenetic effects in humans, although it was reported that among the children of Holocaust survivors the risk for PTSD (or subthreshold symptoms of this disorder) and abnormal HPA functioning was elevated (Yehuda & Biere, 2009), and epigenetic marks could be detected (Yehuda et al., 2016). Likewise, although second generation Holocaust survivors didn't exhibit an increased disposition toward schizophrenia, the severity of this illness, when it did occur, was more intense among individuals whose parents had been survivors of the Holocaust (Levine at al., 2016).

Some of the transgenerational effects that occur could be related to epigenetic actions, or it might be that being born in traumatic times,

could have affected children for any number of other reasons (e.g., poverty, food shortages, safety concerns, or the ability of parents to care for their children adequately). It is interesting that Holocaust survivors exhibited increased epigenetic marks on the *FKBP5* gene, which is considered to be a stress-related gene that predicts PTSD and depression. Curiously, however, in the children of Holocaust survivors, methylation of this gene occurred less frequently than among the children of individuals that hadn't been in the holocaust (Yehuda et al., 2016). Precisely how these polar opposite effects came about and what they imply isn't at all certain, but it is possible the children of traumatized individuals might inherit characteristics that favor resilience as well as those that favor vulnerability.

The intergenerational consequences attributable to traumatic experiences are not unique to the children of Holocaust survivors. For instance, the risk for hyperglycemia was elevated among individuals who had experienced a severe famine in China from 1959 through to 1961. This was also seen in the adult offspring whose mothers were pregnant with them during this period, such that 31% were hyperglycemic and 11% developed type 2 diabetes, twice that of children born after the famine (Li, Liu et al., 2017). What the consequences are during the next generation of offspring remains to be seen.

Epigenetic processes have been used to explain how environmental factors come to promote disease states, and the idea that trauma experiences may have intergenerational and transgenerational effects has important social ramifications. The notion that illnesses can be inherited in a nongenomic manner, being linked to the experiences of parents and grandparents is intriguing. At the same time, there is the question as to why some epigenetic changes are maintained across generations, whereas others are not, and there isn't certainty concerning why transgenerational effects occur in response to some experiences, but not others. As well, would altered expression occur in genes related to some neurobiological processes and not others, and how do we know that an epigenetic change is actually carried through from a given event to adulthood given that epigenetic effects are modifiable? Finally, are these epigenetic effects sufficiently potent to undermine or attenuate the influence of environmentally related neuroplasticity, especially as these may also vary across generations?

It ought to be underscored that although studies in rodents within well-controlled laboratory settings are amenable to analysis of transgenerational epigenetic actions, in humans this is far more difficult. Although early life trauma could set in motion gene suppression that provokes downstream neurobiological effects and pathological conditions, it could also create a behavioral or emotional milieu (e.g., detachment, social rejection sensitivity, or changes in family dynamics), independent of epigenetic factors, which would favor the later development of pathology (Bombay, Matheson, & Anisman, 2014). In fact, in some groups (e.g., Indigenous Peoples in North America), negative early experiences were later coupled with multiple toxic situations (poverty, abuse, and drug use), any of which might have epigenetic actions that favor pathology. It was suggested that society as a whole might serve as an environment, which through epigenetic actions could influence cognitive, emotional, and physical health (Branscombe & Reynolds, 2015). Selecting one event and then attempting to link this to epigenetic changes may not be all that meaningful when it is considered that a traumatic event can have multiple sequelae that could cumulatively affect health-related processes.

Collective, Historical Trauma

Too often, groups of individuals are not only affected by a single or multiple traumatic events (collective trauma), but this cumulative emotional and psychological wounding is experienced by groups over generations. According to social identity theory, sense of self, or identity, is derived from group memberships, and having a particular identity and affiliation with a group (irrespective of whether it is one that comprises race, religion, gender, or one that entails occupation, or being a breast cancer survivor) serves multiple adaptive functions. Social identities take on considerable importance, particularly when the group is challenged, and a variety of adaptive or counterproductive emotions can be elicited (e.g., collective shame or guilt vs collective anger). These threats can also come to provoke individual or group behaviors that can be either constructive (i.e., prosocial) or destructive (i.e., antisocial).

Members of some groups (e.g., a racial, religious, or cultural group) may accept the impermanence of their own lives, but believe that aspects of their group (e.g., values, morals, beliefs, traditions, and symbols representing the group) will be passed on from one generation to the next. The more they identify with the group, the more threatening they will perceive external threats, leading to efforts to push against the challenge. Thus, it is hardly surprising that a history of collective abuse and threats of extermination, would be associated with group members being especially vigilant, sensitive, and reactive to the perceived evil or malign intents of others. Outsiders might view the heightened vigilance of ingroup members as a neurotic obsession with past trauma (perhaps even resenting the constant referrals to the past collective abuses attributed to the outgroup). Yet, if history is any teacher, then the elevated vigilance might be perfectly reasonable and adaptive.

CARRYING THE LOAD OF COLLECTIVE, HISTORICAL TRAUMA

The psychological and physical consequences of "collective, historical trauma" experienced by a group can be passed down across generations through multiple processes. Epigenetic changes and parental influences might act in this capacity, but negative outcome can also occur through other processes. In the years following collective trauma, dislocation creates multiple hardships, living conditions may be exceptional disturbed, and poverty may be pervasive, so that the psychologically toxic environment promotes a wide range of physical and mental illnesses.

As with any stressor, wide individual differences exist in relation to the effects of collective trauma. Even within a circumscribed group, such as Holocaust survivors, considerable diversity exists regarding their experiences, and marked physical and behavioral variations occur among survivors and their children. While some survivors might not have seemed to carry the trauma of the Holocaust with them (or, might not have openly expressed their feelings), others frequently displayed one or more of several emotional disturbances, ranging from survivor's guilt, denial, agitation, mistrust, intrusive thoughts, nightmares, disorganized reasoning, difficulty expressing emotions, anxiety, and depression, although in most instances these were at sub-syndromal levels (Bar-On & Rottgardt, 1998). Whereas some survivors of collective trauma could not stop speaking of their trauma, others rarely spoke of their experiences (a "conspiracy of silence" that was frequently the norm; Danieli, 1998). Even if the trauma was not spoken about, it was nonetheless "silently present in the home." Despite not directly encountering the trauma, children of those who did might incorporate their parents or grandparents narratives and images into their own schema (Hirsch, 2001). These narratives might have been passed on through both verbal and nonverbal communication, and when blanks existed in the historical accounting, these may have been filled in by the fertile imaginations of children. As it turned out, the children who experienced this silence generally seemed to be more vulnerable to intergenerational transmission of trauma (Wiseman et al., 2002).

The consequences of collective, historical assaults among Indigenous People within the US and Canada (as well as many other countries) aimed at eliminating a group's identity ("taking the Indian out of the Indian") can reflect a "soul wound" that isn't readily eliminated (Duran, Duran, Heart, & Horse-Davis, 1998). These experiences can produce social and psychological consequences on family and communities, fostering disturbed social functioning, an erosion of leadership, basic trust, social norms, morals, and values, often persisting for generations (Bombay et al., 2011, 2014). The impact of collective trauma may diminish over successive generations, largely being recalled through rituals and symbols (e.g., holidays and remembrance days) or through story-telling. At the same time, "recovery" might never be complete as feelings concerning these traumatic events sit only slightly below the surface, reemerging with further reminders (e.g., discriminatory behaviors or other threats to the groups' well-being) of the indignities committed against their group. Once again, the transmission of trauma across generations of Indigenous people can also be influenced by the communications (either silence or continued recapitulation of experiences) between survivors and their children and grandchildren (Matheson, Bombay, Dixon, & Anisman, 2018).

It is difficult (and inappropriate) to make comparisons between the experiences and

---

(cont'd)

consequences associated with collective trauma across different groups. Each collectively traumatized group is unique, each involves its own series of experiences, and each has its own consequences. This said, the long-term impacts of collective, historical trauma are not "just" a result of the trauma per se, but might also be subject to the healing opportunities (or the lack of such opportunities) that came afterward, particularly if these groups continued to live in substandard conditions and experienced ongoing discrimination. As well, while ingroup and outgroup support and nurturance could potentially act against adverse effects emerging, in the end, traumatized groups may need to find their own ways of healing, without government agencies looking over their shoulders, pushing them one way or another. If nothing else, this will allow for the adoption of a strength-based approach to overcome current problems and may thus be instrumental in self-healing.

CONCLUDING COMMENTS

It might seem curious that stressful experiences in infancy or early childhood, of which we typically have no memory, can have profound and damaging long-term physical and psychological consequences. Early theorists struggled to understand this paradox, postulating that such memories were buried as unconscious thoughts that fought to emerge in some manner, thereby affecting behavior and cognitions. They hadn't considered the impact of prenatal stressors and likely would have had a hard time incorporating the action of these events into a formulation involving unconscious thoughts (Jung, perhaps being an exception, given his views on the collective unconscious). Recent explanations that are more parsimonious and testable include the possibility that the effects of early experiences, acting through neuroplasticity or epigenetic processes, influence the reactivity of neurons, thereby affecting developmental trajectories as well as reactivity in response to later stressors. In essence, the effects of early life stressor experiences may become "embedded" during childhood and then exacerbated by further stressor encounters (McGowan & Szyf, 2010).

These early life challenges can provoke a wide range of pathological conditions, paralleled by changes of reactivity within stress-relevant brain regions, which also predicted the efficacy of drug treatments in producing illness remission (Goldstein-Piekarski et al., 2016).

As we've seen, early life epigenetic programing can play a significant role in determining the appearance of particular abnormal behaviors, although adaptations in several forms can affect the developmental trajectory that might otherwise occur (Gershon & High, 2015). The brain's exquisite neuroplasticity is apparent throughout life, but is most notable during early life periods. Thus, both positive and negative experiences and environmental influences will have their greatest effects at this time. Because of the brains plasticity, genetic influences and early experiences are, in a sense, the seeds laid down, whose full blossom might not be realized for some time.

Beside stressors, we've seen that numerous other events experienced pre- and postnatally, ranging from bacteria and viruses, through to nutritional factors, can have lasting effects. It ought to come as no surprise that maternal malnutrition, and the loss of macro- and micronutrients disturb immune functioning in mom, and in fetal immune development, leaving them both vulnerable to opportunistic infections. In addition, stress responses will be elicited, including HPA activation, which will affect microbial and immune functioning, and thus may further undermine later health (Macpherson et al., 2017). For that matter, the development of adiposity, adult diabetes, and heart disease may have their roots in the uterine environment (Fall, 2011), possibly through epigenetic changes that occurred at that time. Not only are the repercussions apparent in relation to physical health and the development of inflammatory-related illnesses, but also with respect to mental health. Indeed, new interventions are being developed to modify microbial factors that could impinge on mental health. A strong case has been made that our conceptions of mental health need to be considered more broadly so that in addition neurotransmitter and growth factors, greater consideration needs to be given to the gut and its commensal microbiota (Jacka, 2017), including actions derived from prenatal and early postnatal experiences.

Just as early postnatal life may be a critical period during which numerous experiences can readily induce epigenetic changes, the prenatal environment may also be one that lends itself to such outcomes (Cao-Lei et al., 2017). Although this view, at least from the perspective of epigenetics, is fairly recent, the "fetal programming hypothesis" has been around for some time (Barker, 1990), and has been broadened and renamed as the Developmental Origins of Health and Disease (DOHaD) theory (Gluckman, Hanson, & Buklijas, 2010). The basic view expressed was that stressful experiences, through a variety of neurobiological processes, such as changes of fetal glucocorticoid levels, can program fundamental components of growth and metabolism, thereby leaving a lasting imprint that is manifested at later times. The increasing research pointing to the social environment as potentially having effects on epigenetic programing, especially if these occur during prenatal and early postnatal periods (Turecki & Meaney, 2016), attests to the importance of these periods to neurodevelopment and the emergence of health problems. Moreover, through epigenetic processes, the sins of one generation can be imposed on the next. This said, epigenetic processes are dynamic, and providing nurturing environments can alter previously programed epigenetic actions as well as psychosocial factors that affect well-being. We suggested earlier (Anisman, 2014) that in this case, it may be possible that the "bell can be unrung." This might well be a bit of exaggeration, but surely, its sound can be muffled.

Depressive Disorders

THE BURDEN OF DEPRESSION

Of the many mental disorders, depressive illnesses, are particularly common, affecting 10-−15% of individuals, but is much higher in some populations. Discouragingly, since 2005, depression has increased by 18%, rather than declining as one would have hoped (WHO, 2017). After being successfully treated, illness recurrence is evident in 50−70% of cases within a 5-year period, and thus the view has been offered that depression ought to be considered a life-long disorder.

Depressive disorders are pervasive among older individuals, with almost half of seniors in residential care homes being affected, and antidepressants have become the most widely used prescription medication in this population. Women also fair poorly in relation to depression, being twice (or three times) as likely to become depressed relative to men. Depressive disorders are exceptionally common and debilitating in young people, with 6% of males and 11% of females of the college-aged cohort experiencing clinically significant depression. Frighteningly, after automobile accidents, suicide related to depression is the leading cause of death among youth.

The financial burdens of medical care have become enormous across countries, and the cost of medical care related to mental illnesses is at the front of the pack. This is not only because depression reoccurs frequently, but also because depressed individuals utilize medical facilities (for issues unrelated to mental health) more frequently than do nondepressed individuals. As well, mental illnesses, and especially depressive disorders, make up a significant loss of potential labor supply. Indeed, in some countries, more than 50% of new disability benefit claims are due to mental health issues, far exceeding that of lung, colorectal, breast and prostate cancer combined, and the amalgam of all infectious diseases doesn't come close to the cost of treating depressive illnesses. What is especially disturbing is that the vast majority (more than 70%) of individuals who are depressed do not receive medical attention.

Given the burden of depression on health care systems, financial support for prevention and treatment has been exceptionally short-sighted, especially as every dollar spent on treatment for depression and anxiety yields a 4 dollar return on investment (WHO, 2016a). It has been said that resource allocation in relation to mental illness is the orphan of the health care system, and research concerning mental illness is the orphan of the orphan (Merali & Anisman, 2016).

The Immune System and Mental Health
DOI: https://doi.org/10.1016/B978-0-12-811351-6.00008-5

DEFINING DEPRESSION

From a precision medicine perspective, mental illness and its treatment should not be considered in terms of a syndrome, but instead ought to be broken down into specific symptoms, genes, and related factors, and on this basis determine treatment methods. This is especially germane for depressive disorders given the variability of symptoms that are common across individuals and the differential effectiveness of treatment strategies. Even though our inclination is to proselytize for a precision medicine approach, much like the Research Domain Criteria (RDoC) strategy (but with some modifications as described in Chapter 1: Multiple Pathways Linked to Mental Health and Illness), the Diagnostic and Statistical Manual of Mental Disorders (DSM-5) (American Psychiatric Association, 2013) is still the most widely adopted framework and thus we will first consider depressive disorders from this vantage.

Ordinarily, a diagnosis of "Typical" major depression based on the DSM criteria entails an individual presenting with a set of symptoms for at least a 2-week period. These comprise either depressed mood or anhedonia (i.e., no longer receiving pleasure from events or stimuli that had previously been rewarding), along with four additional symptoms from a prescribed list. These additional symptoms are composed of significant weight loss *or* weight gain, insomnia *or* hypersomnia almost every day, psychomotor agitation *or* retardation, fatigue *or* loss of energy, feelings of worthlessness or excessive, inappropriate guilt, diminished cognitive abilities (e.g., impaired concentration, difficulty making decisions), and recurrent thoughts of death or recurrent suicidal ideation. While not part of the diagnostic criteria, it is not unusual for auxiliary symptoms to be present, such as feelings of helplessness, hopelessness, social withdrawal, low-self-esteem and self-efficacy, high levels of rumination, and disturbed attention/memory.

At first blush, the characteristics of depression might seem unambiguous and a diagnosis might not seem particularly difficult. But, depressive disorders may share symptoms with other illnesses (e.g., the depressed phase of a bipolar disorder episode) and thus illnesses may be erroneously diagnosed (this, in part, accounts for the frequent delays in bipolar illness being appropriately diagnosed). It is curious that two individuals can present with widely different symptoms, many of which may even be opposites of one another, and yet they would still receive the same diagnosis, and often treated in precisely the same way. In light of this, is it any wonder that pharmacologically based treatments have not been overly impressive, and individualized treatment strategies might help in remedying the current diagnostic and treatment difficulties? Ideally, illness ought to be considered and treated based on individual symptoms, but several symptoms often appear together, and so it may be viable to consider symptom clusters rather than individual symptoms.

Subtypes of depression can be diagnosed on the basis of these clusters together with other features of an illness, such as the duration, severity, or specific features of the condition. Table 8.1 provides a description of depression "sub-types" based on DSM-5 features. These subtypes speak to the complexity of depressive illnesses. All too often research in depression has failed to distinguish between subtypes of depression, and throwing all patients into the same pot may have contributed to the mishmash of results that have so often been reported.

A depressive illness may begin fairly early in life (prior to the age of 16), and at one time early onset depression was considered to be a genetic disorder rather than one brought on by psychosocial stressor. However, it seems likely that environmental/social factors

TABLE 8.1 Subtypes of Depression

Typical depression: Characterized by sadness and/or anhedonia, coupled with symptoms such as reduced eating, weight loss, and sleep disturbances (e.g., early morning awakening), psychomotor agitation, fatigue, feelings of guilt or worthlessness, diminished cognitive functioning, including memory disturbances, and suicidal ideation

Atypical depression: This form of depression is reminiscent of typical depression, being characterized by poor mood or anhedonia, as well as many of the cognitive features of the illness. However, the atypical form of depression is associated with *reversed* neurovegetative features, which include increased eating, weight gain, and increased sleep, as well as a tendency towards persistent rejection sensitivity (often causing impairment of social functioning), feelings of heaviness in the limbs ("leaden paralysis") and mood reactivity even during well periods

Melancholic depression: Reflected by features of typical depression, but at a severe level. The intense depressed mood is accompanied by marked anhedonia, psychomotor agitation, early morning awakening, excessive weight loss, or excessive guilt accompanying grief or loss

Dysthymia: Symptoms are appreciably less intense than major depressive illness, and a personality disorder is often present. Dysthymia is diagnosed when the low grade depressive symptoms persist for at least 2 years, although symptoms typically wax and wane. If not treated, major depression may evolve, so that it is superimposed on a dysthymic background. This "double depression" is more difficult to treat

Seasonal affective disorder (SAD): This form of depression is tied to seasons (or duration of light over the course of a day), emerging in the autumn or winter, and then resolving in the spring. This diagnosis is made if episodes have occurred in colder months over a period of at least 2 years, without episodes occurring at other times

Treatment resistant depression: Typically diagnosed when antidepressant treatments fail to diminish symptoms appreciably, even after repeated (usually three) efforts were made using different compounds

Recurrent brief depression: Characterized by intermittent depressive episodes that occur, on average, about once a month over at least 1 year. The appearance of an episode, which lasts for 2–4 days, does not appear to be tied to any particular cycles (e.g., menstrual cycle). The diagnostic criteria for recurrent brief depression are much like those of major depressive disorder, but can be particularly severe, and may be accompanied by suicidal ideation and suicide attempts

Postpartum depression: Occurs after childbirth and affects about 5%–10% of women. Symptoms are much like those of major depression. The appearance of the illness may be linked to hormonal changes that accompany pregnancy or those that occur in association with pregnancy or childbirth; however, hormonal therapy has not been an effective treatment strategy

Minor Depression: Characterized by a mood disorder that does not fully meet the criteria for major depressive disorder, but persists for at least 2 weeks

(e.g., relationship problems, abuse, bullying, and cyber-bullying), in conjunction with genetic and biochemical disturbances are responsible for the development of the illness. A depressive disorder may begin with one or two disturbed thoughts, but insidiously evolves so that a range of symptoms develop that involve multiple cognitive, behavioral, and neurobiological processes. As in the case of many other illnesses, depressive disorders are more amenable to treatment if caught relatively early. However, individuals might not recognize their depression as being abnormal and thus fail to seek treatment. Moreover, as already indicated, the stigma associated with mental illness also undermines help-seeking, and so a good number of people, unfortunately, suffer in silence.

PREDICTING RECURRENCE OF DEPRESSION

Even when positive treatment effects are obtained, doesn't necessarily imply that things are back to normal, as residual symptoms may persist. What makes depressive disorders still more disturbing is the high rate of recurrence, and depressive-like features can be reinstated by specific contextual cues and experiences. Affected individuals might simply have a negative or ruminative personality style that makes them prone to depression. Alternatively, they might have experienced multiple stressors throughout their life, such that their biological systems are sensitized or primed, so that when stressors are again encountered, even if these are relatively modest, neurochemical responses will push them toward depression. By example, among currently depressed patients as well as those in remission, a sad mood induction procedure that entailed exposure to negative autobiographical memory scripts, resulted in a decline of cerebral blood flow within the medial orbitofrontal and anterior cingulate cortex. In line with these findings, when patients with recurring depression were assessed during a well period, having them view a sad film clip was sufficient to provoke changes in medial prefrontal cortex activity, which predicted the propensity toward rumination and elevated risk for relapse. However, like participants that had never been depressed, among patients who experienced sustained remission, the sad film did not markedly influence the prefrontal cortical response (Farb, Anderson, Bloch, & Segal, 2011). Essentially, in some patients in remission, certain brain regions remained fragile, so that a mood challenge effectively "unmasked" processes that could act as a marker for later depression given further stressor encounters. The neuronal response to a sad film, especially if it occurs together with rumination, might reflect the need for continued therapy (e.g., cognitive behavioral therapy) in order to thwart depressive relapse. Conducting an imaging procedure on each former patient to predict relapse may not be practical, but there are other approaches that could be feasible, including analyses of hormonal changes elicited by stressor (or hormone) challenges (Zobel et al., 2001).

In essence, even though patients may be free from depressive symptoms following treatment, this doesn't necessarily mean that they were "cured." Processes that governed their depressive symptoms might have been masked, but were ready to recur given the wrong circumstances.

THEORETICAL MODELS OF DEPRESSIVE ILLNESSES

Both cognitive and neurobiological views of depression have been offered, and within each framework several theoretical positions have been adopted. These perspectives aren't necessarily independent of one another, and both have implications regarding the treatment strategies that ought to be adopted.

Cognitive Models

Hopelessness

The "hopelessness" hypothesis advanced by Beck (1967) suggests that individuals develop schemas (or perspectives) based on their experiences, which come to influence the way events are interpreted (Beck, 2008). Negative events experienced early in life, such as abuse or neglect, might contribute to the development of dysfunctional self-referential schemas ("I'm unworthy" or "This is all my fault, and always will be"). Thus, when individuals encounter challenges at some later time, these schemas promote negative reactions, which become progressively more entrenched. Biased information processing will also develop so that memories of past events are viewed more negatively, and individuals may develop pervasive negative expectations of the future. The biased perspectives may become increasingly pronounced to the extent that individuals will selectively attend to stimuli that are consistent with their negative perspectives. These negative self-referential biases are not only apparent in relation to cognitive styles, but are also captured by EEG recordings (Dainer-Best, Trujillo, Schnyer, & Beevers, 2017). Concurrently, positive events and evidence inconsistent with their negative appraisals are filtered or ignored, and consequently the individual's warped cognitive views predominate (Beck & Dozois, 2011). Eventually, streams of negative "automatic thoughts" emerge spontaneously among depressed individuals, which reinforces their poor self-esteem and self-worth, and encourages feelings of hopelessness and negative rumination.

Framing Hopelessness in a Neurobiological Context

A neurobiological accounting was provided that mapped onto the cognitive aspects of a hopelessness view (Disner, Beevers, Haigh, & Beck, 2011). Given the array of characteristics that are associated with depressive illness, it was reasoned that multiple brain regions likely underlie such disorders, concurrently operating in both a bottom-up and a top-down fashion. Limbic brain regions were seen as regulating the emotional aspects of depressive disorders (bottom-up components), whereas

cognitive disturbances were attributed to top-down process that involved cortical regions. According to this model, depression evolves when inhibitory control (e.g., coming from the prefrontal cortex) is limited so that unrestrained neuronal activation occurs at particular brain sites that allow for the predominance of disturbed emotional processing (e.g., aspects of the amygdala). In essence, this disorder comprises the abdication of cortical control mechanisms that ordinarily limit negative automatic responses, together with a bias toward negative appraisals of experiences.

While disturbed medial prefrontal cortex functioning was seen as being fundamental for regulation and control of self-referential schemas, the dorsolateral prefrontal cortex was taken to be responsible for rumination and disturbed processing, and biased attention was determined through the ventrolateral prefrontal cortex. The development, progression, and maintenance of negative self-referential schemas in depression were viewed as being perpetuated by increases in the salience and self-referential elements of negative stimuli or events. These negative signals are generated through amygdala activity, which promotes increased anterior cingulate cortex neuronal activity. In the latter regard, different aspects of the anterior cingulate cortex serve diverse functions. The ventral portion contributes to the labeling of stimuli with emotional valence, whereas the rostral aspect is involved in labeling stimuli with self-reference values. The dorsal portion contributes to the relaying top-down cognitive inputs, but as it has limited functional connectivity with limbic regions, it contributes to attenuated cognitive control.

Aspects of the amygdala might also contribute to emotional memory, whereas disturbed hippocampal functioning may contribute to biased episodic memory (memory pertaining to autobiographical events, including times, places, and contextual knowledge). These memory disturbances were seen as fostering the negative biases as well as the negative processing of information, thereby perpetuating the memory biases characteristic of depression. Further to this, chronic stressor experiences may give rise to attentional control difficulties coupled with disturbed functional connectivity involving a frontoparietal network thus contributing to cognitive inflexibility and ineffective coping processes.

It was further surmised that nucleus accumbens responses would be blunted, so that appraisal and recognition of positive stimuli or events would be impaired, leading to anhedonia. Moreover, it was also proposed that reward processes involving the medial orbitofrontal cortex may be disturbed in depressed individuals. The lateral orbitofrontal cortex has been linked to a variety of behavioral processes, and may be fundamental for translating nonreward and punishing events, and might contribute to the development of a negative sense of self, and of poor self-esteem (Stalnaker, Cooch, & Schoenbaum, 2015). In essence, depression could be seen as a combination of diminished reactivity to rewarding stimuli and increased responsivity to a failure to receive reward or to reach desired end points. Further to this, in depressed individuals, reduced effective connectivity occurred in brain regions involved in reward processes, whereas increased connectivity occurred in brain regions associated with memory. In essence, depressed individuals may gain less reward from positive events and then are especially capable of retaining these memories of low reward (Rolls et al., 2017).

Cognitive Behavioral Therapy

The development of cognitive behavioural therapy (CBT) in the treatment of depression was developed from the hopelessness perspective, with the goal of using behavioral and cognitive methods to diminish dysfunctional behaviors, cognitions, and emotions. Through this form of therapy, individuals learn to challenge inappropriate and counterproductive patterns and beliefs, and to replace cognitive errors, such as overgeneralizing, magnification of negatives, minimization of positives and catastrophizing, with thoughts that are both realistic and effective. The therapy takes individuals through a series of steps that challenge an individual's way of thinking and their reactions to entrenched habits and behaviors. Ultimately, the benefits of cognitive behavioral therapy stem from patients coming to appraise stressors appropriately, contextualizing stressors and their depression, and coping effectively. While originally developed to diminish depression, CBT or variants of the basic procedure that are tailored for specific conditions, have found their way into the treatment of several other conditions, including anxiety disorders, PTSD, eating disorders, obsessive-compulsive disorder, substance abuse disorder, and psychotic illnesses.

It was consistently reported that CBT was as effective as pharmacotherapy in treating depression, and when used as an adjunct to standard pharmacotherapy, it had lasting effects on mood (Wiles et al., 2016). It has also contributed to the rise of other therapies, such as mindfulness meditation, which for a number of years has been gaining popularity (Kabat-Zinn, 1990; Segal, Williams, & Teasdale, 2002), and like CBT, mindfulness-based approaches have led to diminished depressive relapse. Alternative strategies, such as "Behavioral Activation" therapy, encourage patients to focus on activities that are personally meaningful to them based on their individual values. Although it hasn't been studied broadly, Behavioral Activation was reported to be as effective as CBT (Richards et al., 2016). A related approach, acceptance and commitment therapy (ACT), has also seen some success in treating depressive disorders. This procedure entails individuals becoming more aware and more focused on their therapeutic goals and more engaged in reaching these.

MINDFULNESS: FLAVOR OF THE DECADE

Think in the moment (as opposed to ruminating over the past or worrying about the future) and don't be judgemental are the mantras of mindfulness meditation. Mindfulness can reflect a dispositional (trait) characteristic wherein some individuals are generally more likely to engage in mindful behaviors. These individuals exhibit lower emotional reactivity in response to negative experiences, tend to be relatively non-judgemental, display lower neuroendocrine responses to a social evaluative threat (Brown, Weinstein, & Creswell, 2012), and thus ought to be less prone to stress-related illnesses.

A meta-analysis of the effectiveness of mindfulness training in clinical settings indicated that in 88% of 108 trials, positive outcomes were reported. Moreover, when administered to groups, mindfulness was as effective as individually administered CBT in diminishing psychiatric symptoms (Sundquist, Palmér, Johansson, & Sundquist, 2017). As encouraging as this seems, many of the published reports were under-powered, and it should be considered that negative findings are less likely to be published. It was suggested that research on this topic has lacked the rigor needed to make credible conclusions about its effectiveness, and few studies systematically evaluated the efficacy of the treatment (Van Dam et al., 2017). It was maintained that the field needs better data and less hype.

Is mindfulness for everybody? Certain types of individuals may be drawn to this type of therapy, whereas for others it may be a turn-off. This shouldn't be viewed as castigation of this procedure. It simply means that patients need to seek out the treatment with which they would be comfortable, and clinicians likewise need to determine whether their patient is suitably inclined for this therapeutic approach.

Helplessness

A popular cognitive perspective of depression is the "learned helplessness" hypothesis (Seligman & Maier, 1967). This perspective, initially developed on the basis of studies in animals, proposed that when uncontrollable experiences are encountered, people learn that their responses and subsequent outcomes are independent of one another (i.e., they have no control) and consequently adopt a cognitive perspective that they are helpless in controlling events. If individuals encounter uncontrollable negative experiences in significant situations, which include failures to reach important goals, the feelings of helplessness could eventually lead to depression.

The original helplessness hypothesis was not uniformly accepted, especially as the original conceptualization paid little attention to the obvious fact that only a small proportion of individuals who experience uncontrollable stressors actually become depressed. So, what were the essential features that resulted in some individuals succumbing to depression, whereas others were hardly affected, or could even benefit from life challenges? An elaboration of this hypothesis focused on the cognitive attributions individuals made in response to failure, arguing that specific attributional styles largely determined whether helplessness, and hence depression, would evolve (Abramson, Seligman, & Teasdale, 1978). Individuals were described as differing in three fundamental appraisal dimensions that were linked to "locus of control" (the extent to which individuals perceive themselves as having control over events that affect them). Upon encountering a failure experience, individuals make an appraisal and attribution as to why this occurred, which in particular combinations would be more or less likely to favor the emergence of helplessness. If individuals form internal, stable, global attributions ("It's my fault that I failed; that won't ever change, and I'll continue to fail; I'm no good at anything") regarding their inabilities, they may develop negative expectations of the future, broad feelings of inadequacy and poor self-esteem, culminating in cognitions and feelings related to helplessness, which then encourage the emergence of depressive disorders.

This explanation is in keeping with cognitive perspectives of depression which emphasizes that hopelessness and depression stem from negative appraisals and a general dysfunctional (maladaptive) pattern of thinking in relation to life events (Beck & Dozois, 2011). These dysfunctional attributional styles may have arisen through any number of processes, including negative early experiences, together with children learning (modeling) their parents behaviors.

There's little question that in clinically significant cases of depression, patients often describe themselves as feeling helpless and their situation as hopeless. However, are these feelings simply symptomatic of the illness rather than being a causal agent? Despite this caveat, the learned helplessness model was widely adopted in the psychological and psychiatric literature as being fundamental to the emergence of depression. It is possible that feelings of helplessness in response to uncontrollable experiences represent one of the early symptoms associated with depression, or that feelings of helplessness might be most readily induced by stressors among individuals with subclinical levels of depression. It is similarly possible that stressful events could provoke neurobiological outcomes (e.g., changes of monoamine functioning, neurotrophins, or inflammatory immune processes), thereby provoking feelings and cognitions that favored behavioral disturbances (Anisman et al., 2008)[1].

NEUROCHEMICAL PERSPECTIVES OF DEPRESSIVE DISORDERS

Analyses of several neurobiological correlates of depressive illness in humans can offer important clues regarding the processes responsible for depression. As sophisticated as these approaches have been, they typically don't provide a detailed understanding of the neurochemical changes that govern psychological disorders. As a result, much of what we know about these processes has come from studies using "animal models" of the disorder. A common critique of this approach is that depression may be a uniquely human disorder (although it seems likely that elephants grieve the loss of loved ones). Indeed, do rats and mice have the same feelings that humans do, and if present, are they governed by attributions like those associated with depression in humans? As indicated in Chapter 5, Stressor Processes and Effects on Neurobiological Functioning, the symptoms of depression in humans obviously can't be fully recapitulated in rodents, but several models are available that arguably comprise valid proxies of some aspects of the human disorder, and when used in parallel with human research, a better understanding of diseases may be possible.

For some time the effects of uncontrollable stressors on later escape performance in rodents was a favorite model of depression

---

[1] Loneliess can be distressing, agonizing and formulaic in promoting disease (Cacioppo, Cacioppo, Capitanio, & Cole, 2015). Although loneliness has frequently been tied to depressive disorders, it can have adverse effects beyond that of depression itself, having been linked to earlier mortality. These and other illnesses related to loneliness are discussed in Chapter 16, Comorbidities in Relation to Inflammatory Processes, in which comorbid illnesses are discussed.

(i.e., the learned helplessness paradigm). Several drug treatments that were effective in treating human depression (e.g., SSRIs) also attenuated the behavioral disturbances elicited by an uncontrollable stressor. Yet, several drugs that were ineffective as antidepressants also had positive effects in this paradigm (Anisman et al., 2008). Thus, the escape deficits introduced by uncontrollable stressors might not be a valid model of depression, but would be more comfortably used as a screen for anti-depressant agents[2].

As an alternative to assessing escape deficits engendered by uncontrollable stressors, increasing use of positively motivated behaviors have been adopted following exposure to diverse stressors. These have included responding for rewarding brain stimulation or intake of preferred substances, such as particular snacks (e.g., pieces of cookie or sweetened solutions, or lightly sweetened water), which might be useful for assessing the anhedonia associated with a stressor experience. Likewise, cognitive disturbances related to decision-making or memory, could potentially be used to mimic a portion of the depressive profile.

Recent efforts have focused on using natural stressors to induce a depressive-like profile in rodents. A social challenge that has proven useful in modeling depression comprised changes in housing conditions in order to create social stress (e.g., transferring mice from grouped to isolated housing or vice versa; Audet & Anisman, 2010). An increasingly popular approach to model depression and anxiety has been to expose a dominant, aggressive mouse to a subordinate on single or repeated occasions, and then assessing the neurobiological and behavioral changes that emerge (Takahashi et al., 2017)[3]. As we'll see, these paradigm have been instrumental in determining neurobiological effects of stressors, as well as the genetic and epigenetic interactions that operate in moderating stressor actions.

Monoamine Processes Associated With Depressive Disorders

Identifying the specific neurochemical mechanisms governing psychological illnesses, as we've repeatedly indicated, is exceptionally difficult given that they involve complex actions of several neurobiological processes within several brain regions. Moreover, the processes involved in one form of depression may be distinct from that of a different form of the illness (e.g., dysthymia vs major depression; typical vs atypical depression), and even within a particular subtype, vastly different symptoms might be present (Anisman, Ravindran, Griffiths, & Merali, 1999). As the efficacy of treatment strategies also vary between individuals, this potentially signals the involvement of different neurobiological processes. One gets the sense that we're actually dealing with several disorders that share common features comprising sadness or the feeling that life is not as pleasurable as it once was, but differ in many other respects.

To make the complexity still greater, males and females may differ considerably in relation to the biological substrates associated with this disorder. Specifically, an analysis of gene expression of postmortem tissues across several brain regions associated with depression

---

[2] Other behavioral tests, such as the passivity that occurs in a forced swim test (Porsolt forced swim test) are similarly useful as screens for antidepressant, but it is not unusual for this test to be inappropriately used to model depression.

[3] Exposing mice or rats to a predator (or predator scents) have been reported to be effective in some studies, but in other studies this manipulation did not produce appreciable effects. It may be that after generations of inbreeding and being raised in a laboratory context, these critters lost their innate tendency to react to certain threats.

indicated that there was limited overlap between depressed males and females, although many of the downstream pathways that were linked to depression were comparable in the two sexes. Paralleling the data from humans, among chronically stressed mice, the expression of numerous genes were altered relative to nonstressed mice (e.g., those linked to executive and reward processes), but these also occurred in a sex-specific fashion (Labonté et al., 2017). As we'll see in our discussion of pain processes in Chapter 11, Pain Processes, it is unfortunate that it has only been in recent years that studies attended to sex differences, particularly as this has important implications for the efficacy of treatments.

Many of the biological changes that have been linked to depressive disorders may stem from stressful experiences, particularly those that are uncontrollable and occur chronically. The effects of recent stressors may also be primed by negative early life experiences and may vary with several genetic factors. As in the case of many other illnesses, a multihit or sensitization hypothesis has been advanced wherein an early life trauma or specific genetic factors serve as a first hit (or two hits), and a stressor encountered subsequently represents the hit that takes the individual over the top (or perhaps, more appropriately, it takes the individual to the bottom) into a state of depression.

Norepinephrine

The attention devoted to norepinephrine in relation to depression is predictable given that stressful events, which most certainly contribute to depression, markedly affect this neurotransmitter's activity. Moreover, norepinephrine has been linked to anxiety that is so often comorbid with depressive disorders. Despite being displaced by serotonin as a primary player in depressive disorders, there has been a bit of resurgence with respect to norepinephrine, especially in regard to the role of specific receptors (Maletic, Eramo, Gwin, Offord, & Duffy, 2017). There is also evidence that stressful events that modify dopamine activity, and hence produce anhedonia, may be steered by norepinephrine variations (Isingrini et al., 2016). As the anhedonia present in depression is especially resistant to treatment, it is likely that "triple reuptake inhibitors," including each of the three key monoamines, may be among the next generation of antidepressants.

Dopamine

Studies in rodents had indicated that manipulations that elevated DA functioning reduced the impact of stressors on depressive-like behaviors, and stressors could engender epigenetic effects relevant to reward pathways (e.g., Maze et al., 2010). It might seem a bit puzzling that increased DA neuronal activity within the ventral tegmental—nucleus accumbens circuit is associated with reward processes, but also occurs in relation to stressful events. It is possible that DA functioning within the nucleus accumbens operates to make particular events more salient (Berridge & Robinson, 1998), and consequently, irrespective of whether events have a positive or negative valence, activation of the DA circuit will result in these events taking on more significance. Alternatively, positive and negative events can be distinguished by specific characteristics associated with DA neuronal activity (i.e., the nature of the neuronal firing patterns), possibly varying with the characteristics of the stressor experience (Tye et al., 2013). In effect, positive and negative events could send different messages from the nucleus accumbens to other brain regions, even though the messages may both involve dopamine functioning. For that matter, while anhedonia may stem from dysfunction of ventral tegmental-nucleus accumbens pathway, aspects of coping may be related to disturbances involving the ventral tegmental-prefrontal pathway (Bai et al., 2017). Thus, depressive disorders may involve more than a single dopamine pathway.

Consistent with other sources of evidence, genetic links involving dopamine and depression have been reported. It has been suggested that genes related to the dopamine transporter (DAT) were tied to depression, and in mice and in humans depression was associated with reduced expression of the gene *Slc6a15* on particular dopamine (D_2) neurons in brain regions that regulate reward processes. Furthermore, in mice that had been stressed the depressive-like state could be attenuated by enhancing *Slc6a1* expression (Chandra et al., 2017). Suicide was also accompanied by dysregulation of dopamine D_1 and D_2 receptors, and there were hints that these disturbances were linked to early life adversity (Fitzgerald et al., 2016).

As dopamine mediated anhedonia is a key symptom of depressive disorders, it had been expected that increasing DA levels through l-DOPA treatment would diminish depressive symptoms. However, this agent was ineffective in diminishing depression, which undercut interest in this transmitter's role in depression (a scientific version of "tossing the baby out with the bathwater"). It has since been demonstrated that several drugs (e.g., bupropion) that increase DA functioning were moderately effective antidepressants, but without side effects related to disturbed sexually related behaviors or weight gain (Patel, Allen et al., 2016).

Serotonin

Numerous stressors alter serotonin turnover and levels in rodents, and concurrently provoke depressive-like behavioral impairments. Consistent with the view that serotonin contributed to this behavioral profile, SSRIs administered repeatedly attenuated the behavioral deficits elicited by uncontrollable stressors. It had initially been assumed that the beneficial effects of SSRIs came about because of the inhibition of serotonin reuptake, resulting in increased availability of serotonin at the synapse. It is now understood that SSRIs have effects beyond that of inhibiting serotonin reuptake. By example, in mice genetically engineered so that functioning of the serotonin transporter [5-HT transporter (5-HTT)] was altered, 5-HT_{1A} receptors were also affected. It is possible that 5-HTT dysfunction might be responsible for some features of depression, whereas 5-HT_{1A} receptor variations may contribute to other symptoms. In addition to the involvement of 5-HT_{1A} receptors, it was proposed that 5-HT_{1B} receptors present on cholecystokinin inhibitory interneurons of the dentate gyrus are fundamental for the initiation of a cascade of changes that are responsible for the therapeutic benefits stemming from antidepressant treatments (Medrihan et al., 2017). As we'll see shortly, several other downstream effects of serotonin changes may also be responsible for the actions of antidepressant treatments.

Imaging and Postmortem Analyses

Imaging procedures have enormous potential for uncovering the processes leading to or associated with psychiatric and neurological disorders. Despite the call for precision medicine, and the acknowledgment that depression is a biochemically heterogeneous disorder (and that subtypes of depression exist that may involve different underlying processes), imaging and postmortem studies frequently failed to consider this, which might account for the variability observed in such studies. Indeed, when patient characteristics were considered, these mapped onto differences in brain connectivity. Taking this somewhat further, in a relatively large study of 1188 participants, of which about 400 were depressed, patients could be distinguished from one another based on symptoms and fMRI scans (Drysdale et al., 2017). Specifically, depressed patients clustered within several categories based on distinct patterns of dysfunctional connectivity that were apparent within the insula, orbitofrontal cortex, ventromedial prefrontal cortex, and several subcortical areas, which were tied to specific clinical symptoms (e.g., feelings of sadness, hopelessness, helplessness, anhedonia, and fatigue or anergia abnormal lack of energy). These profiles could also predict the effectiveness of repetitive transcranial magnetic stimulation (rTMS) directed at the dorsomedial prefrontal cortex.

In addition to potential disturbances of serotonin levels and turnover in depressive disorders, 5-HT_{1A} receptor reductions across multiple brain regions were associated with human depression. Of potential practical value, disturbed 5-HT_{1A} binding was predictive of the effectiveness of antidepressant treatment. A case has also been made for 5-HT_{2A} receptor involvement in depression, as the density of this receptor was elevated in the prefrontal cortex of depressed suicides. Moreover, symptom remission following antidepressant treatment was accompanied by receptor normalization. Once again, however, there were also reports that 5-HT_{2A} differences did not occur between controls and depressed individuals that died of suicide (see Stockmeier, 2003), and DNA microarray analyses failed to detect 5-HT_{2A} gene differences within the prefrontal cortex of those who died of suicide and controls (Sibille et al., 2004). It may turn out to especially significant, however, that among individuals with particularly elevated feelings of pessimism and hopelessness, 5-HT_{2A} binding was elevated within the dorsolateral prefrontal cortex (Meyer et al., 2003). Perhaps disturbances of this receptor are tied to specific symptoms or features of an illness, which might inform the choice of subsequent treatment strategies.

The binding of the 5-HTT transporter in relation to depression was likewise variable, but as we'll see, conclusions in this regard need to be considered in a broader context. Although early studies had revealed inconsistent outcomes in the link between 5-HTT functioning and depression, with more effective methods of assessing reuptake, 5-HTT alterations were more reliably observed within the prefrontal cortex and amygdala of depressed individuals.

Postmortem Analyses of Brain Neurochemicals

It is not unusual to be seduced by studies that evaluated neurochemical levels in human brain tissue among depressed individuals who died of suicide relative to that of people who died rapidly of causes other than suicide. Given the numerous problems associated with postmortem analyses, the variability that exists between studies isn't overly surprising. The many problems associated with such studies were reviewed in Anisman (2016), and hence only a few points will be mentioned here. Not all instances of suicide are accompanied by depressive illness, and in some instances it is difficult to know whether a particular neurochemical change is due to depression versus suicide. As well, there is no guarantee that the controls were actually not depressed, especially as depression may go undiagnosed or unreported. Furthermore, the nature of the depression is often not considered (e.g., depression subtypes), and whether the depression was comorbid with other illnesses that could have affected physiological processes. Importantly, individuals who die of suicide may be more impulsive than others and hence do not represent the population of depressed individuals.

Based on postmortem analyses, several reports pointed to depression being associated with variations of specific 5-HT receptors, but inconsistent findings were frequent regarding which 5-HT receptors were actually disturbed (see Stockmeier, 2003). A review of this issue indicated that of eight studies, four reported lower 5-HT_{1A} receptor density, two reported increased 5-HT_{1A} receptor density, and two reported no change (Shrestha et al., 2011). Once more, any number of factors related to problems associated with postmortem analyses could have been responsible for the discordant findings.

Genetic Links to Depression

It's been known for decades that a family history of depression comprised a risk factor for depression, and children of depressed parents are at increased risk of early onset depression and high incidence of recurrent depression (Weissman et al., 2016). It was hoped that once tools became available to identify the specific genes that operate in relation to depression, better treatment methods would follow based on the effective use of biomarkers. This is still some way off and it might actually be counterproductive to rely solely on genetic and epigenetic markers.

It needs to be said that many early studies often comprised relatively small sample sizes, and thus had limited power, and might thus have contributed to the frequent failures to replicate earlier findings. Of course, the inconsistent findings may have been due to symptom and biochemical heterogeneity that wasn't considered in many studies. It should also be added that in rodents the relation between particular genes and a depressive-like phenotype, were minimized if they were raised in a positive, stimulating environment (Mehta-Raghavan et al., 2016). It is possible that some nondepressed individuals may well have carried genes that were linked to illness, but their impact was limited by their social and developmental experiences.

These caveats notwithstanding, a polymorphism was identified that linked depression to a gene controlling the enzyme tryptophan hydroxylase-2, which is fundamental in 5-HT synthesis. This polymorphism was also predictive of diminished SSRI responsiveness, supporting the enzymes' involvement in affective disorders. A polymorphism was similarly detected on the promoter region of the gene regulating the 5-HT_{1A} receptor that was associated with depression (Albert, Vahid-Ansari, & Luckhart, 2014), and a polymorphism related to the norepinephrine uptake process and that of 5-HT_{1A} were linked to depression and the diminished hippocampal volume associated with this disorder (Phillips et al., 2015). It should be underscored that a polymorphism within the promoter region of the 5-HT_{1A} gene was linked to a range of psychiatric disturbances beyond that of depression, including suicidality and substance use (Donaldson et al., 2016). Given that most psychiatric illnesses involve multiple factors, each accounting for a modest portion of the variance, the 5-HT_{1A} polymorphism may simply be one among many neurobiological factors that contribute to illnesses.

Particular attention has been devoted to the finding that depression/suicide was associated with a gene promotor polymorphism for the 5-HT transporter (termed 5-HTTLPR) (Caspi et al., 2003). Several reports confirmed that individuals carrying a particular polymorphism on the alleles of the 5-HTT gene might be especially significant in predicting depression. These alleles can be either long (l) or short (s), based on the long allele having a segment that has 14 repeats, whereas the short allele has 12 repeats. As we'll see, individuals carrying the short form of the 5-HTT promoter on one or both alleles were at greater risk for the development of depression than individuals that were homozygous for the long allele, and as it turned out, individuals homozygous for the long allele also exhibited superior responses to SSRI treatment. What made these findings especially exciting was that depression in those carrying the short allele was most common if individuals also experienced major life stressors, early life trauma, or a stressful family environment (Caspi et al., 2003). It was likewise reported that ongoing distress associated with chronic pain was more likely to result in depression among individuals carrying the 5-HTT polymorphism (Hooten, Townsend, & Sletten, 2017). It seems that among individuals carrying the short allele, amygdala and hippocampal functioning was tonically (continuously) activated, as might be expected under conditions of hypervigilance, threat, and rumination, and the neuronal response to negative events was exacerbated. In essence, inheriting particular genes doesn't condemn individuals to depression, but instead creates the conditions wherein individuals will be more sensitive or reactive to stressors, thereby favoring the appearance of depression (Caspi, Hariri, Holmes, Uher, & Moffitt, 2010).

As exciting as these reports were, especially in light of their implications for personalized treatment, data inconsistent with the initial findings were reported, and several overviews of the relevant literature came to question the reliability of the findings given that most studies were underpowered. In fact, several meta-analyses actually came to very different conclusions on this topic, depending on the research papers that were included in these analyses. This included whether the 5-HTT gene was linked to chronic and recurrent depression, or that the replication studies were not entirely comparable to the original report. As well, some meta-analyses were tainted by inappropriate reliance on retrospective, self-report measures of life stressors, rather than being based on objective measures of stressor experiences. Furthermore, when participant samples were stratified on the basis of the specific stressor experienced, it seemed that early life maltreatment interacted with the 5-HTT gene in predicting depression,

whereas these effects were less pronounced in association with stressors encountered in adulthood (Karg, Burmeister, Shedden, & Sen, 2011). There were indications that the presence of the s allele, coupled with both childhood and adult stressors, modulated the occurrence of depression (Juhasz et al., 2015). This said, an analysis that included 31 independent data sets, meant to be the last word on the subject (even though there's rarely a last work on virtually any subject), did not support the contention that carrying the s allele conferred increased vulnerability to depression in the presence of stressors. This was apparent irrespective of whether current or life-time illness, childhood maltreatment, or broad stressor experiences during adulthood were considered (Culverhouse et al., 2018). Analyses of the mouse equivalent of 5-HTTLPR, namely mice carrying a knockout of the SERT gene similarly did not bring greater clarity to the issue (Houwing, Buwalda, van der Zee, de Boer, & Olivier, 2017).

Aside from the polymorphism related to 5-HTT, it was reported that DNA methylation of the transporter gene in peripheral cells was associated with childhood trauma, being male, and having a small hippocampal volume (Booij, Szyf et al., 2015). Further, it was maintained that methylation of the gene for 5-HTT could serve as an epigenetic marker for depression related to early life adversity. It has been maintained that simply assessing 5-HTTLPR was insufficient in this regard, and it may be necessary to concurrently consider potential epigenetic changes to this gene (Iurescia, Seripa, & Rinaldi, 2017).

It is tempting to argue that the story linking 5-HTTLPR and depression is actually a nonstory in the sense that the effect observed is a spurious one. It may be that a publication bias for studies with positive outcomes drove the view that the short form of the gene related to 5-HTT contributed to depressive disorders. Yet, there have been too many positive reports to dismiss these out of hand, particularly as they might be indicating that there is more unearthed[4].

depth to the findings, which still need to be unearthed[4].

A SLIGHTLY DIFFERENT PERSPECTIVE REGARDING THE 5-HTT X STRESSOR INTERACTIONS

The interaction between stressful events and the short allele for the 5-HTT promoter gene were not limited to predicting depression, but were also linked to Obsessive-compulsive disorder (OCD), anxiety, and PTSD, as well as suicidality among bipolar patients. Likewise, the 5-HTTLPR x Stress interaction was apparent in relation to the neuronal connectivity of executive control brain regions and activity within the default mode brain network, which had been linked to depression (van der Meer et al., 2017). Thus, the combination of the gene for 5-HTT and stressor experiences favor a general susceptibility to illness, possibly in the form of elevated stressor sensitivity or impaired coping, thereby favoring a range of stressor-related illnesses.

Yet another gene, p11, which functions to move particular serotonin receptors from inside the cell to the cell surface, has been implicated as a player in the development of depression (Anisman et al., 2008; Svenningsson et al., 2013). It seems that p11 is reduced in depressed humans, possibly acting through effects on BDNF. In mice with a p11 knockout, depressive-like symptoms were elevated, but were attenuated by delivering a functional gene for p11. It was surmised that the nucleus accumbens, which is involved in reward processes, might be important for the p11 effects.

Effectiveness of Antidepressant Medications

The early findings indicating that SSRIs could ameliorate depression were met with considerable enthusiasm. Unfortunately, the initial optimism concerning the effectiveness of SSRIs was largely wishful thinking, fueled by media hype. In fact, well-conducted human trials indicated that the treatment efficacy of SSRIs was modest, although side effects were less pronounced than those elicited by compounds that had been in use earlier. Current reports suggest that SSRIs are effective in about 50%−60%, but some studies suggested that drug efficacy was actually much lower, hardly being distinguishable from that associated with placebo treatment (Kirsch, 2014). An early meta-analysis of antidepressant effectiveness indicated that the effects observed were hardly noticeable among patients with moderate levels of depression, and had only mild positive effects among those with severe depression (Kirsch et al., 2008)[5].

---

[4] Some overlooked factors may also be at play, but what these might be is an open question. A possible starting place may be to consider the contribution of further variations on the gene coding for 5-HTT, rather than just assessing the contribution of the s and l alleles. For instance, the long allele may contain polymorphisms that make it more s-like, and thus might favor a depressive phenotype.

[5] It is of practical significance that among individuals experiencing depression secondary to a chronic illness, such as kidney disease or congestive heart failure, SSRIs are largely ineffective in modifying mood.

In children and adolescents, in whom depression occurs at rates of 3% and 6%, respectively, the success rate is still lower, and only Prozac seemed to have notable effects (Cipriani et al., 2016). The impact of SSRIs and SNRIs were slightly better than placebo in some trials in children, being somewhat more effective for anxiety disorders than for depression (Locher et al., 2017). Unfortunately, the relative effectiveness of drug and psychological treatments is uncertain in children and adolescents, and additional well controlled trials are needed.

The limited effects of SSRIs, in retrospect, shouldn't be all that surprising, given that the biological underpinnings of depressive illness involve multiple mechanisms and vary across individuals. In fact, this view prompted multi-targeted approaches in the treatment of depressive disorders, and pharmaceutical companies that at one time focused on increasingly more "specific" treatments, possibly to find "the magic bullet" to treat the illness, or hoping to limit side effects, reverted to the development of drug combinations that concurrently affected several neurotransmitters. There is, to be sure, considerable evidence attesting to the benefits derived from concurrently prescribing several drugs to treat the illness (Blier, 2016).

There are several further limitations of standard antidepressant therapies, aside from the fact that 2–3 weeks are needed for positive effects to emerge, and often several drugs are prescribed before one is found to be effective. Even when drug treatments are successful, the depressive symptoms may not be entirely ameliorated as residual symptoms frequently persist, sometimes presaging illness recurrence. Furthermore, individuals experiencing strong stressors in their lives will not necessarily experience positive outcomes so long as their stressful environment persists. In this regard, the effects of a repeated treatment with an SSRI (fluoxetine) in diminishing the adverse effects a chronic stressor regimen was most apparent if mice were also maintained in an enriched environment during the course of the antidepressant treatment. When the two treatments (medication plus enriched environment) were combined, the reduced depression-like behavior was accompanied by elevated BDNF, and HPA axis activity was normalized. As well, the reduction of neurogenesis and long-term potentiation that was otherwise elicited by the stressor was attenuated (Alboni et al., 2017).

Findings such as these suggest that the effects of SSRIs are not simply a result of biological alterations, but instead represent the effects of the drug being moderated by environmental contexts. This could occur for any number of reasons, but an intriguing possibility is that the drug treatment may come to make contextual factors more salient so that in a positive environment the beneficial effects will be realized more readily. This doesn't imply that positive environments are absolutely essential for beneficial effects of a treatment to be obtained. But, it does suggest that agents, such as SSRIs, might be most effective when accompanied by positive environments or cognitive types of therapy, or having the two therapies provided sequentially.

ANTIDEPRESSANT USE DURING PREGNANCY

Pregnancy may be accompanied by oxidative and nitrosative stress biomarkers, activated neuro-immune pathways, and diminished antioxidant defenses, all of which may predict the occurrence of perinatal depressive symptoms, as well as possible obstetric and behavioral complications (Roomruangwong et al., 2018). Left untreated, maternal depression may be associated with various negative consequences for the fetus (e.g., preterm birth, poor prenatal

(cont'd)

care, and depression that occurs years later) and diminished amygdala volume together with microstructural changes (Wen et al., 2017).

Given the frequency of depression, it can pretty well be expected that many women will be taking antidepressant medications while pregnant. It has been estimated that 2%—8% of pregnant women use antidepressants, with SSRIs being the most common (Malm et al., 2016). Thus, there had been concern that SSRIs and other antidepressant agents could affect the developing fetus. In fact, antidepressant use was associated with preterm delivery and low birth weight, as well as reduced Apgar scores in newborns, increased occurrence of seizures, respiratory distress, and persistent pulmonary hypertension (Tak et al., 2017). There were also indications that antidepressant use might lead to ADHD as well as autism in offspring (Boukhris, Sheehy, Mottron & Bérard, 2016; Sujan et al., 2017). These suppositions were not unreasonable given that SSRIs are able to cross the placental barrier and consequently influence fetal brain development. However, the data concerning antidepressant linkages to both disorders have been inconsistent (see Liu et al.,

2017), and a recent review concluded that the relationship is likely smaller than initially thought (Man et al., 2017).

Prenatal antidepressants use was associated with more frequent depression in offspring (Malm et al., 2016). This outcome, of course, might be related to the heritability of depression rather than effects attributable to the drug consumption during pregnancy. This said, a long-term analysis over more than 900,000 individuals indicated that antidepressant use among pregnant women revealed greater incidence of a range of psychiatric condition (Liu et al., 2017), leading to the suggestion that focusing simply on depression in offspring may be too conservative.

Despite the inconsistent findings, it is tempting to come down on the cautious side and recommend against the use of antidepressants. However, as women who are depressed are no doubt experiencing considerable strain, which could itself affect the offspring. Thus, not taking any steps to treat depression could be hazardous to the fetus. A reasonable approach would obviously be one of having the depression treated through psychotherapy or behavioral methods, before resorting to drug treatments.

GAMMA-AMINOBUYTRIC ACID AND GLUTAMATE PROCESSES ASSOCIATED WITH DEPRESSIVE DISORDERS

Interest in gamma-aminobuytric acid (GABA) activity as an essential player in depressive disorders and comorbid anxiety has been an off-and-on affair (Luscher, Shen, & Sahir, 2011). As we've seen, GABA is markedly affected by stressors as are changes in the composition of the $GABA_A$ receptor. Reinforcing the link to depression, GABA levels were

relatively low in the dorsolateral prefrontal and occipital cortex of depressed individuals, and the low GABA levels were especially notable among severely depressed (melancholic) patients, and were also tied to treatment resistance (Price et al., 2009). Likewise, levels of the enzyme involved in GABA synthesis, glutamic acid decarboxylase (GAD), were altered within the hippocampus and prefrontal cortex of depressed patients.

Several studies examined the link between depression and GABA receptor change. The $GABA_A$ receptor, it may be recalled, comprises five subunits that come from a much larger set of subunits. Whether a particular drug (e.g., alcohol, benzodiazepines) influences the receptor is determined by the subunit conformation, and it was similarly suggested that mood disorders are influenced by the presence of particular $GABA_A$ receptors. Within several limbic regions of depressed individuals that died of suicide, the mRNA expression of $GABA_A$ subunits may be up-regulated (Merali et al., 2004; Poulter et al., 2010), but more than this, $GABA_A$ subunit expression "patterns" were altered (Merali et al., 2004). Ordinarily, among nondepressed individuals that had died suddenly of causes unrelated to suicide, the mRNA expression of the various $GABA_A$ subunits were highly correlated (e.g., within the prefrontal cortex, hippocampus and amygdala). In contrast to this coordinated profile, among the depressed individuals that died of suicide the $GABA_A$, subunit interrelations (coordination) was far less notable. Coordination between these subunits may be needed for the integration of neural networks and for the neuronal rhythms that ordinarily occur (Poulter et al., 2010). The disintegration of the coordination between subunits could thus result in depression being present.

A broad-based genomic analysis confirmed that among individuals who died of suicide GABAergic-related gene expression was altered in several cortical and subcortical brain regions, including the prefrontal cortex and hippocampus. However, this was apparent irrespective of whether or not these individuals had been depressed, and hence the GABA changes were likely linked to suicide rather than depression itself (Sequeira et al., 2009).

There have been indications that GABA manipulations can be used to alter behavior, and diminish symptoms of depression, although these data have primarily been obtained through animal research. It was demonstrated, for instance, that GLO1, an enzyme that promotes reduced methylglyoxal (MG),

which acts as a competitive partial agonist at $GABA_A$ receptors, may be linked to depressive-like behaviors elicited by stressors. By inhibiting GLO1 for a limited time (5 days), antidepressant effects were achieved in several behavioral paradigms, coupled with hippocampal BDNF variations (McMurray et al., 2017). It is premature to know whether these data will turn out to have clinical benefits, but they point to GABA functioning as a reasonable target for antidepressant actions.

Glutamate

An impressive number of studies have pointed to depressive disorders being linked to disturbed glutamatergic functioning, either alone or in combination with other systems. Processes that regulate glutamate clearance and metabolism, and morphological changes within several brain areas have been identified that are associated with cognitive and emotional behaviors (Rubio-Casillas & Fernández-Guasti, 2016). In this respect, stressor-provoked glutamate changes involving the prefrontal cortex, hippocampus, and nucleus accumbens, were implicated in the production of many illness symptoms, presumably occurring in a subset of individuals who carry genetic vulnerabilities. The source for the glutamate changes in depression have not been fully defined, but they have been attributed to the reduction in mitochondrial energy production or a decline of neuronal input or synaptic strength (Abdallah et al., 2014). Likewise, GABA and glutamate activation patterns within specific aspects of the anterior cingulate could lead to disturbed engagement of prefrontal cortical regions that might lead to depression, as well as its treatment (Lener et al., 2016).

Aside from the impact of current stressor, early life negative experiences may increase glutamate dysfunction, which then causes microglial activation and increased production of inflammatory factors that promote several psychiatric disturbances (Mondelli, Vernon, Turkheimer, Dazzan, & Pariante, 2017). These include anxiety, depression and alcohol addiction, and might be a contributing factor in treatment-resistant depression.

THE FLEXIBLE BRAIN

Depressive symptoms may be determined by a shift in awareness related to the balance between external and internal mental focus, particularly increased self-focus and negative rumination. Disturbances of inhibitory GABA regulation manifested by an imbalance between the default-mode network (DMN; operative when an individual isn't focusing on the external world, and for all intents the brain is in a state of wakeful rest) and executive networks, may result in a shift in focus from external to internal mental content, thereby promoting features of depression (Northoff & Sibille, 2014). Hyperconnectivity within the DMN and the "salience network" could function to promote recurrent negative rumination, which leads to feelings of depression (Jacobs et al., 2014). It is significant that resting state connectivity within the DMN could be used to distinguish the effectiveness of antidepressant treatments (CBT and antidepressant drugs), and thus could serve as a marker to predict treatment efficacy (Dunlop et al., 2017).

From the symptoms expressed by depressed patients, one might assume that this illness is accompanied by diminished neuronal functioning or diminished connectivity. Contrary to this perspective, however, depressive disorders may reflect hyper-connectivity involving multiple brain regions, possibly because of a dysfunction related to several genes that direct cells through a molecular tag (ubiquitin), so that excessive synapses form. As much as having many connections might be considered to be a good thing, at some point connectivity may become excessive, resulting in disturbed selectivity concerning the interconnections that need to be activated, and hence miscommunication may occur within neuronal circuits. Alternatively having too many associations in memory circuits could lead individuals into negative ruminative loops wherein they are unable to turn off counterproductive thoughts—there may be many roads, but they all seem to lead to the same bad place. Resilience calls for being flexible so that individuals are capable of moving from one strategy to another as the situation demands. As GABA is fundamental in inhibiting messages between neurons, this transmitter may contribute to the inflexibility and selectivity of neuronal exchanges that are evident among depressed individuals.

Ketamine and Related Fast Acting Antidepressant Compounds

A dramatic shift in the focus of antidepressant treatments came about as a result of unexpected findings regarding ketamine, an NMDA receptor antagonist that is frequently used as a general anesthetic and analgesic in veterinary practice, and as a street drug where it is known as Special K. Formal analyses of the effects of this glutamate receptor antagonist confirmed anecdotal reports of ketamine's antidepressant effects, revealing that approximately 60% of patients who had been considered treatment resistant responded positively to ketamine (Blier, 2013; Murrough, Abdallah, & Mathew, 2017). In contrast to SSRIs and other antidepressants that do not have effects until patients have taken the drug for several weeks, the actions of ketamine can appear within hours.

The rapid actions of ketamine has made it an especially valuable tool in treating depression, especially as ketamine has potent actions in diminishing suicide risk (Murrough et al., 2015), independent of the antidepressant actions that are elicited (Ballard et al., 2017)[6].

The positive effects of ketamine may stem from the downstream actions of glutamate NMDA receptor antagonism. Ketamine provokes an increase of brain derived neurotrophic factor (BDNF), which may influence depressive disorders (Abdallah, Sanacora, Duman, & Krystal, 2015). As well, the treatment promotes synaptogenesis and spine formation in frontal cortical regions through stimulation of the "mammalian target of rapamycin" (mTOR) which regulates cell growth, cell proliferation, cell survival, protein synthesis, and transcription. Although these actions could be responsible for the positive impact of ketamine, there has been some debate as to whether mTOR is responsible for the effects observed (Popp et al., 2016).

The view was expressed that the antidepressant actions of ketamine stemmed from effects on inflammatory cytokines, which may influence glutamate metabolism, thereby affecting mood symptoms (Haroon & Miller, 2016). In fact, levels of serum IL-6 were useful as a marker in predicting ketamine's effectiveness (Yang et al., 2015). Later in this chapter, we'll be discussing the possibility that immune factors may operate on depression by a pathway that affects the production of neurodestructive factors. It seems that ketamine may have its effects by acting on this very same kynurenine pathway.

Owing to the many failures encountered, pharmaceutical companies have been reluctant to pursue new treatments for depressive disorders. Glutamate-acting agents may be one of the exceptions given the success seen with ketamine, including its rapid action. Given that the antidepressant actions of ketamine in humans are present for only a few days (but can be extended modestly with repeated treatment), one goal of novel compounds includes extending the duration of the antidepressant effects. Several new related formulations, such as traxoprodil (CP 101 606), GLYX-13, CGP3466B, and esketamine were developed that had rapid and potent effects without the negative side effects that other treatments might have elicited, although their positive effects dissipated relatively quickly over time (Singh, Fedgchin et al., 2016). Yet another agent, methoxetamine, also had a rapid onset of action and the effects lasted somewhat longer than those elicited by ketamine. Targets for the treatment of depression have also focused on metabotropic glutamate (mGlu) receptors that have fewer side effects than ketamine (Pałucha-Poniewiera & Pilc, 2016). Other agents that have their glutamatergic effects through varied routes are also being evaluated. For that matter, the anesthetic isofluorane, was found to have antidepressant effects in a subset of patients, possibly through actions like those of ketamine, including effects on BDNF-related processes (Antila et al., 2017). It hardly need to be said that although the prospects for ketamine-like agents seem promising, the hurdles may be appreciable, including analyses of safety and efficacy of long term use, and determination for whom the treatments will be most effective (e.g., influence of genetic factors, trauma history) (Murrough et al., 2017).

---

[6] In a clever way of assessing the mood changes associated with ketamine, an analysis was undertaken of 41,000 patients prescribed ketamine or other drugs to deal with chronic pain. The incidence of depression, which is frequently comorbid with chronic pain, was 50% lower among ketamine users relative to those taking other drugs (Cohen, Makunts et al., 2017).

THE BUZZ ON THE OTHER ANTIDEPRESSANT TREATMENTS

Ketamine is not the only drug with mind-bending actions that has found its way into the clinic. Magic mushrooms, whose active ingredient psilocybin, is thought to be useful as an antianxiety and antidepressant (Carhart-Harris et al., 2017), and was useful in attenuating these moods states among those with life-threatening cancer. Indeed, psilocybin even had effects in treatment resistant depressed patients that lasted for at least 5 weeks, accompanied by changes in amygdala and increased network stability (Carhart-Harris et al., 2017). Likewise, ayahuasca, which is also known as yagé, is consumed as a brew that has dimethyltryptamine present, and purportedly acts on MAO and BDNF, and thus has antidepressant qualities (Dos Santos, Balthazar, Bouso, & Hallak, 2016). Nitrous oxide (laughing gas), is known to have antianxiety effects, possibly through its actions on GABA$_A$ receptors, and it has been proposed as a potential remedy for treatment-resistant depression. The use of LSD has also been offered as a treatment for existential distress among individuals who confront diseases, such as cancer, as well as in control of addictions, such as alcoholism (Bogenschutz & Ross, 2017).

It isn't fully understood how these agents come to affect depression, but several distinct neurobiological processes have been suggested. These include changes of specific neurotransmitters and neurotrophins, as well as changes within key components of the default mode network, such as cingulate gyrus and the medial prefrontal cortex, so that connectivity within this network is reduced. It has also been maintained that these agents enhanced the interactions between the patient and the treating psychiatrist. It may be relevant that many psychedelic agents are potent inhibitors of inflammatory processes, including TNF-α binding to its receptor within the brain.

Corticotropin Releasing Hormone

The finding that major depressive disorder was accompanied by elevated cortisol levels (and altered responses to the synthetic corticoid dexamethasone), prompted greater attention on HPA functioning in relation to this illness. Several reports revealed that the presence of HPA-related polymorphisms, including diminished responsiveness to cortisol, were linked to individuals being at elevated risk of both developing major depressive disorder and diminished clinical responses to antidepressant treatments.

Given that CRH was fundamental in HPA functioning, increasing interest focused on this peptide in mediating depression. However, with the recognition that CRH and its receptors were present at sites other than the hypothalamus, such as the prefrontal cortex, amygdala and hippocampus, it was considered that depression might stem from CRH changes at these sites (Nemeroff & Vale, 2005). Consistent with this possibility, CRH was elevated in cerebrospinal fluid as well as in the prefrontal and dorsomedial prefrontal cortex of depressed individuals who had died of suicide, and mRNA expression of CRH$_1$ receptors was reduced (Merali et al., 2004). As expected, a SNP on the CRH$_1$ gene was associated with a diminished therapeutic response to SSRI treatment among anxious depressed patients (Liu et al., 2007). Although CRH$_1$ receptors were considered as playing a principle role in depression (Reul & Holsboer, 2002), it is premature to dismiss a role for CRH$_2$ receptors in mood disorders. Sex-dependent differences of stressors on CRH_2 receptor expression have been observed, and hence the involvement of these receptors in accounting for male-female differences in depression ought to be considered.

Despite impressive evidence supporting CRH involvement in anxiety and depression, the development of pharmacological treatments to deal with these conditions has been slow in coming. A CRH_1 antagonist reduced the symptoms of depression, but only in a subset of patients (Holsboer & Ising, 2010). If a treatment based on CRH variations is developed, it will be best to identify biomarkers that inform the nature of the biochemical disturbances (Spierling & Zorrilla, 2017).

Arginine Vasopressin

This hormone was elevated within the locus coeruleus and dorsomedial prefrontal cortex of depressed individuals that died of suicide (Merali, Kent et al., 2006). Furthermore, among depressed patients, treatment with desmopressin [a vasopressin analog that stimulates the V3 arginine vasopressin (AVP) receptor] elicited adrenocorticotropic hormone (ACTH) and cortisol release that exceeded that evident in nondepressed individuals, provisionally suggesting that AVP might be linked to depression through HPA changes.

An intriguing perspective has been advanced concerning the conjoint actions of AVP and CRH in the provocation of depression (Tilders & Schmidt, 1999). Specifically, as we saw in Chapter 5, Stressor Processes and Effects on Neurobiological Functioning, the external zone of the median eminence contains neuronal terminals that primarily contain CRH and a modest amount of AVP. In rodents that had been exposed to a stressor or received a cytokine treatment, storage of AVP within the CRH terminals increased with the passage of time (Schmidt, Binnekade, Janszen, & Tilders,

1996). When animals were subsequently stressed, both hormones were co-released, and synergistically stimulated pituitary ACTH release, and hence promoted greater adrenal corticosterone secretion (Schmidt et al., 1996). It seems that hypothalamic neurons have the capacity to change, and can thus have persistent behavioral ramifications. It is possible that early life stressors or inflammatory challenges might also have their long-term effects through such a mechanism, but this has not been widely assessed in relation to behavioral disturbances, nor is it known whether co-expression of these hormones normalizes after successful pharmacotherapy.

Neuropeptide Y

Neuropeptide Y (NPY) has been implicated as an important factor that could engender resilience in the face of stressors (see Chapter 10: Posttraumatic Stress Disorder, for its involvement in PTSD). In fact, NPY levels were lower in depressed patients than in controls, and increased in response to antidepressant medications (Ozsoy, Olguner Eker, & Abdulrezzak, 2016). The presence of an NPY (Y2) promoter polymorphism was also linked to the presence of depressive-like features (Treutlein et al., 2017). As it happens, subtypes of depression may exist with respect to cerebrospinal fluid (CSF) levels of NPY based upon whether the patients had experienced early life adversity, and consequently diverse effects can be expected with NPY treatment strategies.

There hasn't been a great deal of research reported concerning the links between NPY and immune activity in relation to depression. Nevertheless, immune cells can release NPY, and conversely, NPY influences immune functioning, and can attenuate some of the inflammatory effects provoked by stressors, and might thereby moderate depressive-like behaviors.

Oxytocin

Given that oxytocin favors the development of social interactions and might contribute to the salience of stimuli becoming more significant, it might be expected that treatments that increase oxytocin functioning would be useful in diminishing depressive disorders, most likely being used as an adjunct to standard medication (McQuaid, McInnis, Abizaid, & Anisman, 2014). As oxytocin variations can modulate dopamine functioning within brain regions supporting reward processes and hedonia, reinforces this perspective (Xiao, Priest, Nasenbeny, Lu, & Kozorovitskiy, 2017). The usefulness of oxytocin in attenuating depression would be predicated on oxytocin functioning being disturbed in patients. However, as discussed in Chapter 5, Stressor Processes and Effects on Neurobiological Functioning, if oxytocin functioning was not disturbed, increasing its levels through its exogenous administration might aggravate illness features by increasing sensitivity to or salience of negative stimuli. In fact, even neutral social stimuli might be more likely to be negatively misinterpreted.

NEUROTROPHINS

The volume of the prefrontal cortex may be reduced in association with chronic stressor experiences, frequently accompanied by cognitive and behavioral disturbances. Not unexpectedly, powerful stressors related to poverty have likewise been related to diminished brain volume, particularly in children (Yu et al., 2017). These brain changes are aligned with disturbed coping methods related to decision-making abilities, so that individuals generally resorted to previously established behavioral strategies (i.e., biases, habits), rather than framing and adopting new approaches. Like the brain changes associated with chronic stressors, major depression is accompanied by structural brain changes, such as reduced hippocampal volume, more so in patients who experienced repeated or lengthy periods of illness, and in nonremitted patients than among those in remission (Frodl et al., 2004). Paralleling these findings, early life stressful experiences, which favor the development of later depressive illness, were accompanied by reduced gray matter volume within cortical brain regions (Edmiston et al., 2011).

Once more, it isn't immediately evident whether depression provoked the reduced brain volume, or conversely, individuals with smaller hippocampal or cortical size were more prone to depression, and both scenarios are indeed possible. As well, small hippocampal volume was not limited to depression, having been observed in patients with PTSD and schizophrenia (e.g., O'Doherty, Chitty, Saddiqui, Bennett, & Lagopoulos, 2015). The essential point here is that diminished hippocampal volume may reflect a common risk factor for several disorders, which is a reasonable perspective in light of the role of this region for so many basic psychological and cognitive functions. The obvious question is which neurobiological factors act in favor or against neurogenesis that has been linked to psychopathology, and what gave rise to these actions. Several candidates have been implicated in this regard, including neurotrophins and downstream processes.

Brain Derived Neurotrophic Factor

Stressor experiences in rodents can influence hippocampal neuroplastic changes, such as dendritic remodeling, and reduce the number of synapses, which might contribute to the emergence of depression (Duman & Monteggia, 2006). The expression of BDNF protein and transcription was reduced in limbic brain regions following stressor experiences that elicited depressive-like behaviors in rodents. Similarly, a chronic stressor experienced over 31 days, increased BDNF gene methylation in the dorsal hippocampus, while reducing methylation in the ventral hippocampus. Once more, these findings are congruent with the proposition that stressors promote a depressive-like state, possibly being subserved by BDNF changes (Roth, Zoladz, Sweatt, & Diamond, 2011). It was similarly observed that BDNF signaling within the ventral tegmental area (VTA)-NAc circuit (which is fundamental to reward processes) was involved in the provocation of depressive-like outcomes elicited by chronic social defeat (Koo et al., 2016). Although dopamine functioning within this circuit has typically been implicated as the key player responsible for reward processes, these findings suggest that the BDNF receptor (tyrosine kinase B; TrkB) may contribute to altered reward-related processes stemming from a chronic stressor.

Understandably, the data in humans concerning BDNF variations in specific brain regions in relation to depression is limited. Among individuals that died of suicide, BDNF expression and protein levels and those of its TrkB receptor were reduced within the hippocampus and prefrontal cortex relative to that detected in age- and sex-matched controls (e.g., Dwivedi et al., 2003). As well, in a small set of depressed individuals who died of suicide, BDNF protein levels varied across brain regions in a gender-dependent manner, being reduced within the prefrontal cortex among females who had been depressed, whereas a reduction of BDNF was noted in the hippocampus of depressed males (Hayley et al., 2015). Reduced BDNF levels within the amygdala were also more prominent in females than in males (Guilloux et al., 2012). Within the anterior cingulate cortex of depressed individual who died of suicide, BDNF levels were comparable to that of controls, but the TrkB receptor for BDNF was lower in the depressed

individuals (Tripp et al., 2012). Overall, there is still insufficient information available to dissect the conditions in which BDNF is altered in depressed individuals, and which brain regions are most cogent in this regard.

Support for BDNF involvement in depression has also come from reports that the positive effects of electroconvulsive shock and that of several antidepressants were accompanied by elevated hippocampal neurogenesis in rodents, and these treatments prevented the down-regulation of hippocampal cell proliferation ordinarily provoked by stressors (Malberg & Duman, 2003). As well, the positive behavioral effects of antidepressants were diminished among mice with targeted deletion of genes for BDNF (Duman & Monteggia, 2006), indicating that functional BDNF activity is necessary for the positive effects of antidepressants to emerge. It was surmised that the beneficial effect of SSRIs might not simply be due to changes of 5-HT or its receptors, but might instead stem from BDNF variations that develop over days, thus accounting for the lag between treatment initiation and beneficial effects emerging.

As with BDNF, a case has been made for fibroblast growth factors (FGF)-2 involvement in depressive disorders. Stressors disrupt FGF-2 functioning, and this growth factor and its receptor were reduced in the prefrontal cortex of depressed individuals (Evans et al., 2004). The reduction of FGF transcripts was attenuated in animals that had been receiving antidepressant medication (Kajitani et al., 2015). As expected, the administration of FGF-2 directly into the brain largely eliminated stressor-provoked behavioral disturbances, and conversely, pretreatment with an FGF-2 antagonist prevented the positive effects of antidepressants in attenuating the effects of stressors (Elsayed et al., 2012). Other neurotrophic factors such as FGF9, have revealed interesting relations to depression. Specifically, the expression of this member of the FGF family was elevated in the hippocampus of individuals who had been diagnosed with depressive illness, and the presence of FGF9 was inversely related to FGF-2 (Aurbach et al., 2015).

Not to be left out of the mix, nerve growth factor (NGF), vascular endothelial growth factor (VEGF), glial cell line-derived neurotrophic factor (GDNF), erythropoietin (EPO), and insulin growth factor-1 (IGF-1) have also been implicated as contributing to depression. Some of these neurotrophin were, predictably, reduced with antidepressant treatment.

Neurotrophins Measured in Blood

Meta-analyses have generally confirmed that depression is accompanied by lowered serum BDNF levels (Molendijk et al., 2014). The extent of the BDNF reductions were more pronounced among patients with comorbid anxiety, greater suicidal ideation, as well as recurrent depressive episodes. With clinical improvement following antidepressant treatment, serum BDNF levels increased, but a similar rise did not occur among patients who did not show clinical improvement, supporting the link between the growth factor and the illness condition (Lee, Kim et al., 2007). This view was reinforced by the finding that serum BDNF and NGF were also accompanied by reduced amygdala size (e.g., Inal-Emiroglu et al., 2015).

Despite the impressive support, the data concerning serum BDNF levels in relation to depression have not been as uniform as one would like. Although serum BDNF was tied to depression, the aggregated data suggested that the association between serum BDNF concentrations was not reliably linked to the severity of depressive symptoms and considerable noise existed within the available findings (Molendijk et al., 2014). Of significance from a personalized medicine perspective, BDNF was not consistently associated with specific depressive symptoms. In fact, the positive effects of antidepressants on mood were not always accompanied by elevated serum BDNF. Nonetheless, among depressed patients in whom BDNF was relatively high, the response to antidepressants was more favorable than in patients with low BDNF levels (Wolkowitz et al., 2011). In essence, while levels of BDNF might not be directly linked to depressive symptoms in all patients, in those with a BDNF disturbance, the positive effects of antidepressants were most likely to emerge. This provided a hint that the effects of SSRIs might come about owing to changes of BDNF rather than effects on serotonin itself.

Consistent with so many other neurobiological factors, altered serum BDNF is promiscuous in the sense that it is not only related to depressive disorders, but is also related to schizophrenia, stroke occurrence, infarct size following ischemic stroke, lupus erythematosus, type 2 diabetes, and age-related cognitive impairments. Each of these conditions is highly comorbid with depression and has been linked to inflammatory processes. Thus, serum BDNF could be a common feature reflecting previous stressors or, perhaps, the distress created by serious illnesses or degenerative processes associated with these conditions.

Consistent with the view that BDNF is linked to depression, individuals with a polymorphism coding for BDNF expressed greater rumination following stressor experiences, which is predictive of a depressive propensity (Hosang, Shiles, Tansey, McGuffin, & Uher, 2014). Likewise, among at-risk adolescents carrying this single nucleotide polymorphism (SNP), and who also displayed increased morning cortisol levels, the incidence of depression was elevated (Goodyer, Croudace, Dudbridge, Ban, & Herbert, 2010). Whether this polymorphism (likely in combination with other factors) can be used to predict the later occurrence of depression isn't yet known. It will be recalled, however, in the context of early life stressors, this SNP may act against the development of depression, presumably by diminishing sensitivity to adverse events.

Epigenetic Processes and Depression

As we've seen, several studies pointed to epigenetic changes associated with depression in animal models. In contrast, only a limited number of reports are available concerning the epigenetic actions of stressful events in humans and how these map onto depressive disorders. However, among individuals who died of suicide, and who had a history of early childhood neglect/abuse, ribosomal RNA expression was hypermethylated throughout the promoter region of the gene for the glucocorticoid receptor within the hippocampus (McGowan et al., 2009). The gene coding for a microRNA, miR-124-3p, which is believed to be fundamental for brain plasticity, was also subject to epigenetic processes when comparisons were made between postmortem prefrontal cortex of depressed patients and controls, as well as in rodents that received chronic corticosterone to mimic the effects of a stressor (Roy, Dunbar, Shelton, & Dwivedi, 2016). As described earlier, epigenetic changes within the gene coding for the $GABA_A$ receptor was also found within the prefrontal cortex of depressed individuals who died by suicide (Poulter et al., 2008). These reports are in keeping with the position that glucocorticoid receptors, $GABA_A$ receptors, and neuronal plasticity, may all be linked to depression. Yet, as we've indicated repeatedly, many thousands of epigenetic marks are present on DNA, and thus studies showing that certain experiences or depression are associated with particular epigenetic changes does not mean that they are causally linked.

NEURONAL STIMULATION TO ATTENUATE DEPRESSION

Electroconvulsive treatment has long been known to have positive effects in ameliorating symptoms of depression, and the dangers of the procedure have diminished considerably over several decades. The positive effects of the treatment have been attributed to restorative and neurotrophic processes, moderating what otherwise would entail underactive cortical neuronal activity through overactive subcortical limbic functioning (Njau et al., 2017).

With the realization that stimulation of brain activity could diminish symptoms of depression, other forms of stimulation to elicit antidepressant responses were developed. In this regard, repetitive transcranial magnetic stimulation (rTMS) has received increasing attention, not just for depression, but for a large number of other brain disorders (e.g., PTSD, generalized anxiety disorder, pain relief, and substance abuse disorder). This entails an electrical coil being passed over the surface of the head in order to create a magnetic field that will excite neuronal activity, thereby diminishing depressed mood. A second method, transcranial direct-current stimulation (tDCS), applies a lower current (for a longer duration) to electrodes attached to the skull, which stimulates sites deeper in the brain.

A meta-analysis indicated that rTMS elicited positive effects comparable to that of electroconvulsive therapy (ECT) for nonpsychotic depression (Ren et al., 2014), and was even effective in some treatment-resistant patients. There is some question whether rTMS has beneficial effects on cognitive processes beyond promoting antidepressant actions, although tDCS influenced decision-making and problem solving. The brain processes by which TMS has its effects are not yet certain, but like ECT, it promotes increased glutamate functioning (Hone-Blanchet, Edden, & Fecteau, 2016), and it was suggested that its antidepressant actions stem from stimulation of

---

(cont'd)

the anterior cingulate cortex. Not all trials of TMS have elicited equally positive effects, and it was reported that an SSRI (escitalopram) yielded superior effects relative to tDCS (Brunoni et al., 2017).

Deep brain stimulation (as the name implies this involves direct stimulation of neurons deep in the brain) made its debut with the demonstration that it could be used to attenuate the symptoms of Parkinson's disease (Lang & Lozano, 1998). Subsequent trials demonstrated that this same procedure, targeting the subgenual cingulate cortex, could alleviate symptoms in patients with treatment resistant depression (Mayberg et al. 2005). Since then, other target sites have been proposed for deep brain stimulation, such as medial forebrain bundle, and it has also been used for other conditions, such as neuropathic pain, OCD, bipolar disorder, dementia, as well as several other neurological conditions. Once more, there are data suggesting that the beneficial effects of deep brain stimulation might come from prefrontal release of glutamate, which would activate α-amino-3-hydroxy-5-methyl-4-isoxazolepropionic acid (AMPA) receptors (Jiménez-Sánchez et al., 2016) or by indirectly producing inbition of neurons that would otherwise favor depressive features. Inflammatory processes have also been implicated in the actions of deep brain stimulation, as the positive effects were linked to regional inflammation, which was temporally correlated with elevations of glial-fibrillary-acidic-protein (GFAP) immunoreactivity (Perez-Caballero et al., 2014).

The down-side of deep brain stimulation, obviously, is that it is a surgical procedure, which has multiple drawbacks, even though as far as brain surgery goes, it's one of the less risky procedures. A new method has been developed that avoids surgery, but can activate specific deep brain regions using a set of individual high frequency fields that only activate neurons at sites where different electrical fields intersect (Grossman et al., 2017). It's too early to know how this procedure will fare in humans, but it would be ideal if it could serve as an alternative for deep brain stimulation.

INFLAMMATORY PROCESSES AND DEPRESSIVE DISORDERS

A link between immune functioning and depression is not intuitively obvious. In fact, as often recounted, there was a time when the immune system and the brain were seen as being independent of one another, and hence there was no reason to imagine that immune factors would affect psychological functioning. It is now certain that multidirectional communication occurs between the immune system, hormonal processes, growth factors, and gut bacteria, and could thereby influence a range of physical and psychological illnesses, including depression.

It is important to emphasize from the outset of this discussion that it is unlikely that inflammatory processes are involved in all instances of depression, but instead would only occur in a subset of individuals. Thus, treatments that do not influence inflammatory processes would be expected to be relatively ineffective in attenuating depression that is driven by the presence of inflammation. Conversely, manipulations of inflammatory processes would have little effect in cases where noninflammatory factors are responsible for the illness.

Inflammatory Markers Associated With Depression

Some of the early indications of an immune-depression link came from reports that depressive disorder, particularly melancholia, was accompanied by elevated levels of acute phase proteins, such as C-reactive protein (CRP) and alpha-1-acid glycoprotein, and proinflammatory cytokines and their soluble receptors (Maes, 1995). A detailed meta-analysis that included 58 studies concluded that IL-6 and CRP were elevated in major depressive disorder, and to lesser extent this was the case for TNF-α (Haapakoski, Mathieu, Ebmeier, Alenius, & Kivimäki, 2015). Such relations were especially notable among depressed patients with suicidal ideation, who displayed a combination of elevated inflammation, reduced neurotrophin concentrations, and increased stressor encounters (Priya, Rajappa, Kattimani, Mohanraj, & Revathy, 2016). As well, individuals who were already depressed were more sensitive to stressors, leading to still greater cytokine changes upon encountering further challenges.

From a personalized medicine vantage, it is of importance that elevated CRP levels were most closely aligned with vegetative features of depression (e.g., fatigue, restless sleep), and these relationships were no longer evident after antidepressant treatment (White, Kivimäki, Jokela, & Batty, 2017). Circulating IL-6 levels were also tied to the presence of impulsivity and the choice of a violent suicide attempt, and less likely to be associated with other characteristics of illness (Isung et al., 2014). To be sure, there have been questions as to whether the link between immune changes and depression is a direct one, particularly as the immune alterations could be secondary to life-style factors associated with the illness (Slavich & Irwin, 2014). At the same time, a longitudinal study in which individuals were followed for an average of 11.8 years indicated that the presence of elevated CRP and IL-6 was predictive of the later occurrence of the cognitive symptoms of depression (Gimeno et al., 2009). In addition to frank cytokine changes, depressive disorders were also associated with changes of Toll-like receptor 4 (TLR 4), which is fundamental in activating innate immune responses.

PERINATAL DEPRESSION AND INFLAMMATION

It is not unusual for depression to occur before and after birth of a child. Various neuroendocrine changes that occur at this time may increase vulnerability to mood changes, and the position was offered that prenatal placental CRH and HPA axis dysregulation represented risk factors for postpartum depression (O'Keane et al., 2013). This condition may also be accompanied by variations of the NLRP3 inflammasome, and a shift also occurs in the TH_1 and TH_2 balance, so that proinflammatory processes predominate (Leff-Gelman et al., 2016). This proinflammatory bias was also apparent in terms of the levels of omega-3, eicosapentaenoic acid, and docosahexaenoic acid and an increase of the omega-6/omega-3 ratio (Lin, Chang, Chong, Chen, & Su, 2017). In essence, at least some of the inflammatory changes associated with perinatal depression are similar to those that have been tied to unipolar depression in the absence of pregnancy.

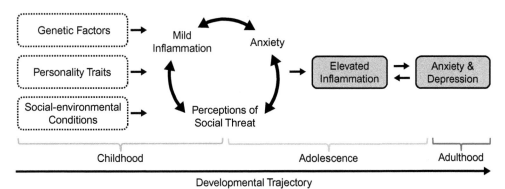

FIGURE 8.1 Threat perceptions and the impact of social challenges are influenced by a constellation of variables, comprising genetic factors related to inflammatory processes and serotonin, personality traits (e.g., neuroticism, interpersonal, and rejection sensitivity), and childhood and adolescence psychosocial stressors (e.g., social stress, abuse, and neglect). Elevated threat perception promotes mild inflammation and modest anxiety, but with further inflammation or protracted stressor experiences, anxiety may increase. Transcriptional changes related to inflammatory genes over the course of these experiences may promote a greater proinflammatory milieu, and the risk for anxiety is elevated accordingly. Depression may develop under these conditions, so that the features of anxiety are mixed with sad mood, anhedonia, fatigue, altered appetite, disturbed sleep, and social withdrawal. *Source: From Slavich and Irwin (2014).*

Further supporting inflammatory involvement in depressive illness, genome-wide association analyses and differential expression of genes in transcriptome analyses indicated that genes relevant to inflammatory processes are overrepresented among depressed individuals (Sharma, 2016). Similarly, chronic depression was associated with activation of several gene sets that were tied to inflammation (Elovainio et al., 2015), and expression of genes for IL-6 and natural killer cells were particularly elevated among currently depressed patients, relative to nondepressed controls or individuals who had previously been depressed. The findings of elevated IL-6 gene expression were further supported when depressed patients were assessed longitudinally over a 2-year period (Jansen et al., 2016).

Aside from the immediate consequences of stressors, the priming of neuroinflammatory processes governed by danger-associated molecular patterns and NLRP3 (inflammasome), can dispose individuals toward psychiatric disturbances (Frank, Weber, et al., 2015). As expected, agents that act as NLRP inflammasome antagonists, such as β-hydroxybutyrate, are able to diminish the behavioral and inflammatory actions of stressors (Yamanashi et al., 2017). In essence, when experiences that give rise to inflammation are followed by stressors, particularly those that were uncontrollable, and occurred chronically, neurobiological systems may be more likely to become overly taxed (allostatic overload), hence favoring the emergence of depression.

As described in Fig. 8.1, in addition to genetic contribution, several psychosocial and personality factors could influence stress-related inflammatory process that come to promote depression, and the emergence of these factors can vary over the course of development.

Neurochemical Effects of Immune Activation in Relation to Depressive Illnesses

There is currently little debate concerning the broad and pernicious effects of inflammatory immune activation, and various routes were offered as to how this came about in relation to affective illnesses. These processes are not independent of one another and may be relevant in different situations. One position was that immune factors could affect the vagal nerve, which then affected brain processes and might thus contribute to depressive illness. An alternative proposition was that proinflammatory cytokines released following immune activation could, despite their large size, gain access to the brain at sites where the blood brain barrier (BBB) is relatively permeable (i.e., at sites that surround the brain's ventricular system; circumventricular organs). As well, active transport mechanisms are present that could ferry cytokines from the periphery into the brain, thereby affecting neuronal functioning (Banks, 2016). There was also the suspicion that infection, traumatic insults, and strong stressors could undermine the integrity of the BBB, thus permitting greater cytokine passage into the brain, and hence greater psychological consequences would emerge.

While not dismissing the possibility that peripheral cytokines might be a link to psychological illness, as indicated in Chapter 1, Multiple Pathways Linked to Mental Health and Illness, a paradigm shift occurred with the finding that cytokines and their receptors are endogenously expressed in the brain, being released by activated microglia (Allen & Barres, 2009). Indeed, the concentrations of brain cytokines or their mRNA expression was increased appreciably in response to traumatic head injury, stroke, and seizure, as well as immune challenges, thereby promoting depressive disorders. Depression is also a fairly common postoperative problem, which has been linked to increased pain, delirium, delayed recovery, and recurrences of illnesses, potentially arising because of activation of inflammatory processes. It is especially significant that stressful events promote microglial activation and elevated cytokine expression (Tynan, Naicker, & Hinwood, 2010). These stressor effects, along with the anxiety-like behavior associated with microglial reactivity, could be attenuated by β-NE antagonism (Wohleb et al., 2011), pointing to the links between stress-microglia-norepinephrine linkages.

Cytokines that are released from brain microglia, both under pathological and non-pathological conditions, may have pertinent actions on neuronal processes that govern memory and executive functioning. In this regard, IL-6 within the orbitofrontal cortex may also contribute to cognitive flexibility, the individual's ability to make decisions, moderate responses as the context changes, and diminish cognitive deficit engendered by a chronic stressor. In essence, these data point to IL-6 within the orbitofrontal cortex as a potential therapeutic drug target for the treatment of stressor-produced illnesses that involve cognitive dysfunction (Donegan, Girotti, Weinberg, & Morilak, 2014).

MICROGLIA AS INDUCERS OF DEPRESSION

In response to stressors, circulating monocytes are increased, and find their way to the brain where along with cytokines they influence neuroplasticity. Some of the recruited monocytes remain in the brain, adopting characteristic of microglia (Hodes, Kana, Menard, Merad, & Russo, 2015). The importance of microglia comes from their ability to recruit monocytes in an effort to minimize damage that might have been detected, and in balancing pro- and anti-inflammatory functioning. However, excessive microglial activation and the ensuing cytokine release can disturb hippocampal neurogenesis and the incorporation of new neurons into existing neural circuits, thereby favoring the behavioral disturbances observed in inflammatory disorders (Chesnokova, Pechnick, & Wawrowsky, 2016). Moreover, although

---

(cont'd)

---

microglia are responsible for removing neurons that have died, in some instances (e.g., following seizure), microglia seem unable to detect these dead neurons, allowing for still greater damage to ensue, and hence the emergence of neuropsychiatric disorders.

Microglial senescence, which occurs with aging, neurodegenerative disorders, as well chronic stressor experiences, may instigate variations of neurotrophins or neurotransmitters that impair neuroplasticity and neurogenesis, thus encouraging depression (Yirmiya, Rimmerman, & Reshef, 2015). From this vantage, depression can be considered a disease generated by microglial disturbances that limit synaptic plasticity and neurogenesis. In fact, depression and anxiety associated with physical illnesses can be diminished by a TNF-α inhibitor, although the effects were relatively modest (Abbott et al., 2015). Interestingly, in response to stressors, several aspects of the microglial phenotype are preferentially influenced in females, and thus might contribute to sex differences in mood disorders (Bollinger, Bergeon Burns, & Wellman, 2016).

Stress, Inflammation, and Depression

Many of the neurobiological consequences engendered by immune activation were remarkably similar to those elicited by stressors. Specifically, proinflammatory cytokine or bacterial endotoxin [lipopolysaccharide (LPS)] treatment markedly increased HPA functioning and monoamine activity at hypothalamic and extra-hypothalamic sites, and affected BDNF functioning (Audet & Anisman, 2013). Thus, like stressors, inflammatory processes can instigate or exacerbate psychiatric or neurological conditions (Dantzer, O'Connor, Freund, Johnson, & Kelley, 2008; Raison, Capuron, & Miller, 2006).

Of course, bacterial or viral infections do not elicit appraisals comparable to those provoked by physical or psychological stressors, particularly as individuals might be unaware of inflammatory processes being activated. Nevertheless, inflammatory insults can elicit neurochemical effects that influence cognitive and affective processes. In this respect, psychogenic stressors stimulate HPA functioning through amygdala activation, whereas systemic stressors may do so through nonlimbic circuits, but in many respects, the end result is the same.

Ordinarily, inflammatory immune activation (e.g., by administering LPS or proinflammatory cytokines) increases the utilization of serotonin and norepinephrine in cortical and limbic regions (Anisman et al., 2008), and interferes with dopamine functioning within reward circuits (Felger et al., 2016). As we've seen, with persistent cytokine activation, protracted changes occurred with respect to CRH functioning, secretion of ACTH, beta-endorphin, and corticosterone, possibly leading to greater strain on physiological systems and hence greater risk for psychological disturbances.

Given that corticoids act to diminish cytokine functioning, it might seem puzzling that depression has been associated with both elevated cortisol and increased cytokine levels. Ordinarily, corticoids limit or prevent transcriptional activity of nuclear factor kappa beta (NF-kB), a primary inflammatory transcription factor, and thus suppress the production and secretion of proinflammatory cytokines. However, with continued HPA activation, which occurs with chronic stressors, glucocorticoid receptor sensitivity may be reduced, and as a result, the normal cortisol inhibition of cytokine functioning is diminished, thus allowing for cytokine secretion and levels to increase.

A second possibility exists concerning the elevated cortisol and cytokine functioning in depression. Specifically, under some circumstances, corticoids may promote proinflammatory effects, rather than cytokine suppression (Sorrells, Caso, Munhoz, & Sapolsky, 2009). It seems that the timing between cortisol changes and immune activation is critical in defining whether immune-enhancement or suppression will occur. Cortisol release that coincides or follows immune activation may lead to inhibition of inflammatory processes, whereas corticoid changes that occur in advance of immune promotion may be proinflammatory. When glucocorticoids are elevated, they provide a danger signal so that central and peripheral processes related to immune functioning are prepared to become reactive to potential threats stemming from microglia being primed to release cytokines. Factors that increase the presence of microglia, such as chronic stressors, serve to augment brain cytokine levels, possibly through activation of NMDA glutamate receptors, thus promoting depressive disorders (Tynan et al., 2010).

As we'll see in later chapters (see Chapter 16: Comorbidities in Relation to Inflammatory Processes), brain cytokine activation isn't only associated with depression, having been seen in schizophrenia and neurodegenerative disorders, as well as other conditions in which inflammatory immune functioning is elevated, such as rheumatoid arthritis, asthma, and inflammatory bowel disease (Raison & Miller, 2003). If nothing else, the accumulation of data has indicated that the impact of inflammatory factors varies across brain regions, and the development of a particular pathology is dependent on interactions with other neurochemical processes. Proinflammatory cytokines interact with CRH, monoamines (e.g., serotonin), neurotrophins (Patas et al., 2014), and enzymatic pathways that govern the production of oxidative species and other neurodegenerative processes (Bakunina, Pariante, & Zunszain, 2015). In essence, cytokines could have their damaging effects through different processes, thereby influencing several diseases in addition to depressive disorders.

Stressors and inflammatory immune activation, as we've seen, synergistically influence brain cytokine levels and brain neurotransmitter activity, and could promote cell loss. By example, application of a bacterial endotoxin (e.g., LPS) directly to the prefrontal cortex provoked an inflammatory response that could be markedly elevated in rats that had been stressed, thus leading to more pronounced loss of neurons and glial cells (de Pablos et al., 2006). Likewise, pronounced loss of hippocampal neurons was exacted by the combination of a chronic stressor plus LPS treatment. The combination of these treatments provoked depressive-like behaviors, which were attenuated in mice that had been genetically engineered to lack the IL-1 receptor (Goshen et al., 2008), attesting to the fundamental role of IL-1 in producing the observed outcomes.

While not speaking directly to depression, it might be significant that when female participants were put through a test in which they experienced social rejection, pronounced IL-6 elevations were observed (Eisenberger et al., 2010). This cytokine change, in turn, was associated with increased neural activity in the dorsal anterior cingulate cortex and anterior insula (brain regions implicated in decision-making and depression) as well as the ventral striatum (associated with reward processes). Moreover, neuronal activity within these brain regions mediated the relationship between IL-6 and depressed mood. Further to this issue, inflammation was associated with an increase in neuronal sensitivity ordinarily tied to threats, diminished reward processes, but augmented sensitivity to positive social experiences. As such, inflammation may both enhance avoidance of danger and the approach to positive social cues (Eisenberger, Moieni, Inagaki, Muscatell, & Irwin, 2017). To be sure, not all cytokines necessarily have such actions, but in addition to IL-6, it seems that IFNγ may also behave in this way, as blocking its action made mice less social. Presumably, the mix of cytokine variations that occur in the brain dictate the nature of the behavioral changes that emerge.

TRAUMATIC BRAIN INJURY AND DEPRESSIVE ILLNESS

Interest in head injury increased exponentially as it became clear that concussion could have lasting effects on behavioral and cognitive functioning. Indeed, of individuals who experienced traumatic head injury, about 50% developed depressive symptoms within one year, and these individuals often experienced a poor prognosis. The initial effect of head trauma may comprise necrotic death of brain cells, followed by a second set of damaging actions. These include excitotoxicity, oxidative stress, mitochondrial disturbances, compromised BBB integrity, and elevated inflammation, which together provoke persistent and progressive damage. It was considered that at least some of the downstream consequences of head injury arise owing to actions on damage-associated molecular patterns (DAMPs) that promote cytokine release from microglia, although peripheral immune changes may also be contributing factors (Corps, Roth, & McGavern, 2015). During an early phase following an insult, when various cells are involved in managing recovery from neural damage, the low cytokine levels may be advantageous (e.g., Russo & McGavern, 2016). However, neurodegenerative and depressive actions can emerge as inflammatory changes become more pronounced, and these outcomes can be limited by anti-inflammatory agents.

The altered levels of inflammatory factors within the brain can persist for years even after a single injury (Johnson et al., 2013), and depression and suicidal behavior associated with such injuries may persist for just as long (Fisher et al., 2016). In mice that experienced traumatic head injury in which behavioral symptoms dissipated after 1 week, a subset of microglia were still affected for as long as 30 days. When these mice were challenged with an immunological insult comprising LPS administration, the microglia exhibited a pronounced activation that corresponded with depressive-like symptoms (Fenn et al., 2014). Evidently, the trauma may have primed these cells so that depressive symptoms were more likely to emerge given a further challenge.

In view of the links between inflammatory immune activation following head trauma and neurodegenerative changes, it might be expected that anti-inflammatory agents, such as glucocorticoids, would have beneficial effects. But, this was not always the case and marked negative consequences may emerge. Given that the effects of head injury vary over time, the actions of treatments might similarly be time-dependent. Being able to identify the time following injury during which inflammatory promoting cells act to foster tissue repair versus damage will be important in the development of treatment strategies.

Different processes may be responsible for the psychological disturbances associated with head injury. Beyond excessive cytokine activity, the consequences of traumatic brain injury have been attributed to elevated reactive oxygen species coming from microglia. As well, as already mentioned, head injury could lead to excessive glutamate activation, which causes cell loss (Dorsett et al., 2017). Connectivity between brain regions associated with emotions, was also elevated between neuronal processes involving the thalamus, insula, and subgenual

---

(cont'd)

---

anterior cingulate cortex, which operate to provoke depressive symptomatology. The connectivity varied yet again with the nature of the symptoms expressed (i.e., cognitive vs affective symptoms) (Han, Chapman, & Krawczyk, 2015). It is of considerable practical and therapeutic importance that if SSRI treatment was initiated soon after head trauma, the development of later depression was appreciably curtailed (Jorge, Acion, Burin, & Robinson, 2016).

Given the potential for brain microglial involvement in depression, it might be expected that brain cytokine levels or mRNA expression would differ among depressed individuals who died of suicide relative to that of nondepressed individuals who died through causes other than suicide[7]. Understandably, however, the data available from postmortem tissue are scant, and to some extent inconclusive. Nonetheless, cytokine perturbances were associated mental health problems among teenaged individuals that died by suicide. Specifically, IL-1β, IL-6, and TNF-α mRNA as well as their protein expression were elevated in the prefrontal cortex, but these effects were more readily attributable to suicide rather than depression itself (Pandey et al., 2011). The mRNA expression of the anti-inflammatory cytokines IL-4 and IL-13 were also altered in the orbitofrontal cortex of women who died of suicide (Tonelli et al., 2008). Congruent with the cytokine variations, TLR3 and TLR4 protein overexpression in the dorsolateral prefrontal cortex (PFC) was linked to suicide (possibly reflecting greater impulsivity or other characteristics) rather than depression per se (Pandey, Rizavi, Ren, Bhaumik, & Dwivedi, 2014).

Suicide may occur for a variety of reasons, but depressive feelings and hopelessness are among the most frequent. Identifying who is at risk for suicide entails consideration of several behavioral and contextual factors, and there is reason to believe that biomarkers for suicidality might also exist. There has been the view, largely based on imaging studies, that depressive disorders and suicidal behaviors are tied to a deficiency in serotonin input to the anterior cingulate and ventromedial prefrontal cortex. Disturbances of these neural circuits and the impulsive/aggressive traits that accompany them may reflect trait characteristics that lend themselves to early onset mood disorders and elevated risk for suicidal behavior (Mann, 2013).

Suicidal ideation among moderately or severely depressed patients was associated with increased microglial inflammatory activity, particularly within the anterior cingulate cortex (Holmes et al., 2018). A meta-analysis indicated that most of the cytokines that had been assessed (TNF-α, IFNγ, transforming growth factor (TGF)-β, IL-4, and soluble IL-2 receptors) were not conclusively related to suicidal ideation, whereas IL-6 elevation showed a positive relationship (GananÇa et al., 2016).

---

[7] Although recent attention on microglia has focused on their actions related to inflammatory mechanisms, glial cells can influence depression through other processes. For instance, the NG2 form of glia elicits the release of FGF-2, which then causes astrocytes to regulate glutamate within the brain. Should NG2 decline owing to genetic factors or stressors, then regulatory processes may be disturbed, culminating in depression (Birey et al., 2015).

It was similarly concluded that IL-1β and IL-6 in blood, cerebrospinal fluid, and postmortem brain samples, were most consistently and robustly associated with suicidality (Black & Miller, 2015). In a telling study of 7.2 million individuals who were followed for up to 32 years, almost one quarter of those who died of suicide had been diagnosed with an infection, which far exceeds that expected. Moreover, risk of suicide increased with the more infections experienced (Lund-Sørensen, Benros, Madsen, Sørensen, & Eaton, 2016).

Several subtypes of suicidality were identified (depressed, anxious, combined anxiety and depression, nonaffective/psychotic), each with their own biomarkers. These included markers that had been linked to neurogenesis, mTOR signaling, the serotonin transporter, and the 5-HT$_{2A}$ receptor, as well as levels of IL-6 and APOE. Based on these biomarkers, together with a set of other genes, suicidal intent and future hospitalization could be identified with 90% and 77% accuracy, respectively (Niculescu et al., 2017).

Irrespective of whether cytokines are causally linked to depressive illnesses (or suicide), it is possible that elevated circulating cytokine levels might serve as a marker for these disorders and might also predict treatment efficacy. Indeed, while basal TNF-α levels were predictive of later depression, the levels of IL-6 was predictive of response to antidepressant treatment (Lanquillon, Krieg, Bening-Abu-Shach, & Vedder, 2000). It is similarly possible that having cytokines elevated even after symptom relief, could be a marker that a biological system remains perturbed, and could signal a disposition for depression recurrence.

Impact of Inflammatory Acting Agents in Rodents

The administration of immune activating agents and cytokines elicit a range of behavioral changes in animals. The most broadly studied of these is sickness behaviors, which comprises anorexia, fatigue, reduced motor activity, curled body posture, sleepiness, ruffled fur, and loose bowel functioning. These symptoms are much like those that accompany influenza infection, and in some respects they are also reminiscent of the atypical form of depression (Dantzer et al., 2008; Raison et al., 2006). As low doses of IL-1β or TNF-α administered directly into the brain can elicit the sickness profile, these symptoms are likely subserved by central processes.

Consistent with the suggestion that sickness features elicited by inflammatory factors reflect a depression-like state, antidepressant drugs diminish the behavioral symptoms (Yirmiya et al., 2001). If the effects of immune activation were limited to sickness behaviors, the case for depression might not be overly impressive. However, animals treated with LPS or IL-1β also display other depressive-like behaviors, including anhedonia and disrupted social interaction (Borowski, Kokkinidis, Merali, & Anisman, 1998). Furthermore, the sickness and depressive behaviors can be distinguished from one another, in that antidepressants preferentially influence the affective effects elicited by cytokines over those that involve sickness symptoms (Merali, Brennan, Brau, & Anisman, 2003). Moreover the sensitivity of the peripheral immune system (reflected by protracted elevations of IL-6) in response to a social stressor accounted for vulnerability to the depressive-like phenotype (Hodes et al., 2014).

Although IL-6 was correlated to depression in humans, in rodents the administration of IL-6 hardly produces any observable behavioral changes, such as sickness symptoms, possibly suggesting that sickness might not be an adequate facsimile of human depression, or that IL-6 affects depression processes that are not reflected by sickness behaviors. However, in behavioral testing paradigms assessing food motivation, which might be closely aligned with depression, IL-6 reduced the motivation to work for food, and dopamine release in the nucleus accumbens was concurrently reduced (Yohn et al., 2016). Thus, while IL-6 might be responsible for motivational and reward processes associated with depression-related inflammation, other cytokines may account for the neurovegetative features of depression.

In addition to the other neurovegetative features of depression, sleep disturbances, either in the form of early morning awakening or sleeping excessively, are hallmarks of depression, which may be indicative of the presence of neurobiological disturbances. Sleep disturbances can predict the occurrence of later depressive illness and may be one of the first signs of depression emerging. The suggestion was made that sleep disturbances, possibly owing to the distress this can create, might contribute to inflammatory processes that lead to depression, and consequently targeting both inflammation and sleep might be ideal to diminish depressive symptoms (Irwin & Opp, 2017).

Infection and Depression

When the immune system is activated by pathogen- or damage-associated molecular patterns (PAMPs or DAMPs), the emergence of sickness behavior is a common response that may be adaptive in the sense that energy expenditure will be limited, while immune mobilization and endocrine changes support the fight against invaders (Dantzer, 2017). However, viral infection (e.g., influenza viruses, varicella-zoster virus, herpes simplex virus, HIV/AIDS, and hepatitis C) and common parasites (e.g., Toxoplasma gondii) have been linked to the occurrence of depression and suicidality, especially among older women (e.g., Coughlin, 2012). Even seasonal allergic rhinitis, which is associated with inflammatory and endocrine responses, was related to depressive symptoms (Trikojat et al., 2017).

A causal connection between infection and depression was inferred from studies showing that administration of a low dose of an endotoxin that increased plasma TNF-α and IL-6 levels, elicited modestly elevated depressive symptoms and a feeling of "social disconnection" in otherwise healthy people (Reichenberg et al., 2001). The intensity of the depressive mood was correlated with the extent of the cytokine rise elicited (Yirmiya, 2000). Significantly, the mild depressive-like symptoms (e.g., lassitude, social anhedonia) brought on by an endotoxin were attenuated by pretreatment with the SSRI citalopram, but these behavioral changes were not accompanied by peripheral cytokine variations, suggesting that they were not direct modulators of mood state (Hannestad, DellaGioia, Ortiz, Pittman, & Bhagwagar, 2011). However, more recent reports indicated that SSRIs and SNRIs have anti-inflammatory actions that could contribute to their antidepressant actions (e.g., Gałecki, Mossakowska-Wójcik, & Talarowska, 2017). Consistent with these findings, elevated plasma IL-6 levels provoked by vaccination (e.g., typhoid vaccine), was accompanied by depressive-like mood, fatigue, confusion, and impaired concentration, which may have been related to altered neuronal activity within the anterior cingulate cortex (Harrison et al., 2009). Vaccination with live attenuated rubella virus similarly resulted in a long lasting (10 weeks) increase of depressed mood among high-risk (low socioeconomic status) teenage girls (Yirmiya, Pollak, Morag, Reichenberg, & Barak, 2000).

Antidepressant Treatment Effects

There has been the ongoing question of whether inflammatory processes might have influenced depression through serotonin related processes, and conversely whether antidepressants had their positive effects by diminishing inflammation. The data in this regard have been decidedly mixed. Using animal models of depression, several antidepressants caused the suppression of humoral and cell-mediated immunity, accompanied by diminished release of proinflammatory cytokines. The most prominent effects of antidepressants comprised a decline of IL-6 and CRP levels in blood (Hiles, Baker, De Malmanche, & Attia, 2012). Like SSRIs, the fast acting NMDA antagonist, ketamine, reduced IL-6 and TNF-α (De Kock, Loix, & Lavand'homme, 2013), and repeated ECT diminished immune and cytokine activity (Zincir, Öztürk, Bilgen, İzci, & Yükselir, 2016). The anti-inflammatory action of different antidepressant treatments could stem from their direct effects on macrophages as well as through changes of neurotransmitters sensed by these cells (Nazimek et al., 2016). Moreover, given that BDNF levels are linked to IL-6 concentrations among melancholic patients (Patas et al., 2014), and BDNF variations are increased by antidepressants, the potential involvement of the neurotrophin in the cytokine-depression link should be considered.

On the surface, it might seem that antidepressant might influence inflammatory processes, thereby diminishing depression, but it is also possible that the inflammatory changes stem from a reduction of depression. This can be addressed by evaluating the effects of nonpharmacological treatments of depression. In the main, these treatments also produced the normalization of cytokine levels (Dahl et al., 2016). Successful psychotherapy was accompanied by reduced levels of serum IL-6 and TNF-α, although levels of these cytokines did not seem to be correlated to the severity of depressive symptoms after the intervention (Del Grande da Silva et al., 2016). Mindfulness training was also accompanied by downregulated NF-κB-associated gene expression (Creswell et al., 2012), and among women with a history of interpersonal trauma experiences, mindfulness-based stress reduction was accompanied by plasma IL-6 declining

(Gallegos, Lytle, Moynihan, & Talbot, 2015). Clearly, nondrug interventions to treat depression can produce cytokine effects much like SSRIs, suggesting that the reduced depression was responsible for the inflammatory cytokine changes.

Modifying Depression Through Anti-inflammatory Treatments

If depression emerges as a result of inflammation, then anti-inflammatory treatments might also be effective in the treatment of this illness, or at least serve as a useful adjunctive treatment. Strictly speaking, this isn't necessarily correct, as the impetus for illness progression may involve factors that are secondary to the initial cause of the illness rather than inflammatory factors (e.g., life-style changes, disturbed sleep). Moreover, once an illness has been established through a particular biological process, essentially involving synaptic connections and neuronal circuits being altered, or neuronal loss occurring, it may not be possible to undo the damage created.

Among individuals being treated for rheumatoid arthritis or psoriasis, the TNF-α antagonist etanercept reduced comorbid depressive symptoms (Kekow et al., 2011), as did administration of ustekinumab, an antibody against interleukin 12/23, used to attenuate psoriasis (Langley et al., 2010). In line with these findings, treatment with infliximab, a monoclonal antibody against TNF-α, diminished depressed mood among patients with high basal levels of inflammation reflected by elevated circulating levels of CRP (Raison et al., 2013). These reports are encouraging, although the outcomes might have been due to reduced symptoms of the primary illness rather than being directly related to diminished immune activity.

A meta-analysis that included seven randomized controlled trials, indicated that drugs (e.g., adalimumab, etanercept, infliximab, and tocilizumab), which have been used as anti-inflammatory agents in the treatment of auto-immune conditions, had an antidepressant effect, and could be used effectively as an adjunctive treatment. The effects observed varied with baseline symptom severity, but were unrelated to alleviation of symptoms of the primary condition being treated. Once more, these findings are consistent with the view that cytokines are causally related to depression, and that cytokine manipulations can be used in the treatment of this disorder (Kappelmann, Lewis, Dantzer, Jones, & Khandaker, 2016). These immune acting agents are powerful and ought to be used cautiously, but their value in treatment resistant patients may outweigh their potential negative actions, provided that patients show indications of elevated inflammation prior to treatment.

The broad spectrum antibiotic minocycline seemed to have antidepressant actions in rodents, including modification of processes that have been linked to neuron loss. Likewise, in humans, minocycline enhanced the actions of standard medication among treatment-resistant patients (Husain et al., 2017). In both rodents and humans, depressive symptoms could similarly be reduced through the use of nonsteroidal anti-inflammatory agents (NSAIDs) and through cyclooxygenase-2 (COX-2) inhibitors (Eyre, Air, Proctor, Rositano, & Baune, 2015), and the effectiveness of standard antidepressant medication in treating depression in humans was augmented by NSAIDs (Abbasi, Hosseini, Modabbernia, Ashrafi, & Akhondzadeh, 2012). A meta-analysis confirmed these positive effects and indicated that celecoxib was particularly effective in this regard (Köhler et al., 2014). Yet, a modest number of studies that assessed the effects of NSAIDs (either as adjuvant treatments or as monotherapies) revealed that they had negligible effects on depression. There are obviously questions that need to be answered concerning the effects of anti-inflammatory agents in relation to depression. For instance, while the NSAID celecoxib seemed to have antidepressant actions, the data for other agents are not as striking. Moreover, in the main, tests of these agents have focused on acute depression, and it is less certain whether these agents can have prophylactic effects (or modify depression at the first signs of illness) or be useful in preventing illness recurrence (Baune, 2016).

Depression Associated With Immunotherapy: The Case of IFN-α

Interferon-α and IL-2 were among the earliest immunotherapies used to treat some forms of cancer (e.g., malignant melanoma), and owing to its antiviral properties, IFN-α had been used in the treatment of hepatitis C. The data from these studies provided appreciable support for cytokine involvement in depression, as a considerable portion of patients (30%−50%) developed depressive-like symptoms over the course of treatment with IFN-α. In some instances, the mood change was sufficiently severe to require treatment discontinuation (Raison et al., 2006).

The IFN-α immunotherapy effects are not limited to depressive symptoms, but also come with a constellation of other neurovegetative symptoms and cognitive disturbances. Patients sometimes report feeling "in a fog," and at the higher doses used in cancer patients, they may experience malaise, especially during the first few days of treatment. As well, concentration and memory may be disturbed, and nonspecific features such as a confusional state, disorientation, psychotic-like features, irritability, anxiety, disturbed vigilance and alertness may also occur. Thus, while the immunotherapy certainly elicits a depressive like state, it probably ought to be separated from the other actions elicited by the therapy. It is, after all, possible that IFN-α provokes a nonspecific state (e.g., general toxicity) that favors a depressive-like condition.

This caveat notwithstanding, the appearance of depression stemming from IFN-α treatment was also linked to factors that have been associated with the emergence of depression in other contexts. The presence of pretreatment depressive symptoms was associated with immunotherapy-elicited depression, and levels of IL-6, salivary cortisol, and the ratio between particular omega-6 and omega-3 fatty acids predicted the emergence of depressive symptoms (Machado et al., 2016). Moreover, depression was prominent among individuals with low levels of the 5-HT precursor tryptophan, the presence of high pretreatment levels of IL-6, and in patients who displayed relatively marked ACTH and cortisol elevations following IFN-α treatment (Capuron & Miller, 2011; Udina et al., 2013). Vulnerability to depression in response to IFN-α was also elevated among individuals carrying a variant of the 5HT-1$_A$ serotonin receptor (HTR1A-1019G) (Kraus et al., 2007) or a polymorphism on the gene coding for IL-6 (Udina et al., 2013). Consistent with the important role of psychosocial factors in relation to depression, a longitudinal study confirmed that social support was effective in

diminishing symptoms and in increasing compliance among melanoma patients treated with low dose IFN-α therapy (Kovács et al., 2015). Tellingly, many of the signs of depression were also attenuated by antidepressant treatment (Capuron & Miller, 2011; Raison et al., 2006), and could be used prophylactically when administered at the start of immunotherapy. As in the case of depression that appeared outside of IFN-α treatment, it is unfortunate that following alleviation of cytokine-elicited depressive symptoms, recurrence of illness was high (Chiu, Su, Su, & Chen, 2017).

A transcriptome analysis (which assesses the expression of mRNAs in a specific cell population) indicated that many genes were modulated in patients who became depressed after IFN-α therapy relative to patients who did not become depressed. The elevated gene expression was largely related to those linked to inflammation, neuroplasticity, and oxidative stress pathways (Hepgul et al., 2016). Thus, the inclination toward depression in relation to the cytokine therapy may reflect a greater sensitivity to IFN-α, which was captured by the broader gene expression changes, and also by several processes linked to inflammation and growth factors.

OF MEN, BUT NOT MICE

Sometimes, what happens in a petri dish stays in a petri dish, and doesn't generalize to the living mouse. Likewise, effects seen in mice and rats, as we've said repeatedly, don't always translate easily to humans. *In vitro* studies indicated that IFN-α provoked CRH release within the amygdala and hypothalamus, but *in vivo* administration of IFN-α to rodents had limited effects on circulating ACTH and corticosterone (Anisman, Poulter, Gandhi, Merali, & Hayley, 2007; De La Garza et al., 2005) monoamine levels and turnover in the brain (Anisman et al., 2007), and elicited moderate GABA and glutamate

changes within the hypothalamus and limbic regions. As expected, the modestly reduced serotonin turnover, and the depressive effects of the IFN-α treatment were attenuated in animals pretreated with an NSAID (De La Garza et al., 2005).

It might seem curious that IFN-α profoundly affects mood in humans treated for cancer or hepatitis C, yet has very modest behavioral effects in mice. It seems that the impact of cytokines on behavioral and neurochemical outcomes in mice is moderated by the background conditions upon which the treatments are aministered. When animals are exposed to a

---

(cont'd)

---

psychosocial stressor, and then treated with IFN-α (Gibb, Al-Yawer, & Anisman, 2013), the cortisol and central monoamine changes elicited are far greater than would be expected based on the additive effects of the two treatments. Under these conditions the behaviors taken to reflect depression were more pronounced than that elicited in the absence of a stressor (Anisman et al., 2007). The very same effects were also observed in response to a viral analog (poly I:C), or a bacterial endotoxin (LPS) (Gandhi, Hayley, Gibb, Merali, & Anisman, 2007; Gibb, Hayley, Gandhi, Poulter, & Anisman, 2008). The fact that

synergisms occur between stressors and these immune/cytokine challenges has implications for the development of illnesses and for their treatment. In particular, within clinical settings, cytokine therapy is administered to patients with hepatitis C or certain cancers, and the stress related to their illness may have interacted with the IFN-α treatment to produce the depressive symptoms. In essence, the impact of inflammatory agents in clinical situations likely reflects the conjoint actions attributable to the cytokine and the stress related to illnesss (or the expected rigours related to the treatment).

Despite the limited behavioral and monoamine changes seen in mice systemically treated with IFN-α, when administered directly into the brain, the effects of IFN-α were substantial, promoting increased brain cytokine mRNA expression, elevated hypothalamic neuronal firing, altered serotonin levels and turnover within the prefrontal cortex (De La Garza et al., 2005), and increased 5-HT_{2C} receptor mRNA editing. It was also observed that in a human progenitor cell line, IFN-α reduced neurogenesis and through activation of STAT1, apoptosis was increased (Borsini et al., 2018). Clearly, provided that the IFN-α is able to access the brain in sufficient concentrations, it is able to affect multiple neurobiological processes. The distress associated with the illness and the expected side effects of treatment may be sufficient compromise BBB integrity, thereby facilitating IFN-α entry into the brain.

IFN-α Linkages to Serotonin and Neurodestructive Outcomes

A view of the IFN-α depressive actions that has been especially well received is based on the reports that cytokines, such as IFN-α, stimulate indoleamine-2,3-dioxygenase (IDO) and GTP-cyclohydrolase activity, which then promote the degradation of tryptophan, the precursor of 5-HT, and thus reduced serotonin levels. Beyond these actions, as depicted in Fig. 8.2, by affecting IDO, IFN-α causes kynurenine to form the oxidative metabolites, 3-hydroxykynurenine and then quinolinic acid (an NMDA agonist), which can have neurotoxic actions (Maes, Galecki, Chang, & Berk, 2011; Wichers et al., 2005), thus promoting depressive disorders[8]. In rodents, systemic administration of LPS engenders a depressive-like behavioral profile that was accompanied

---

[8] This pathway is not unique to depression, as its activation may also promote the development of neurodegenerative diseases, such as Huntington's and Parkinson's disease, and AIDS dementia (Schwarcz & Stone, 2017), and as we'll see in Chapter 13, it has also been implicated in the emergence of schizophrenia.

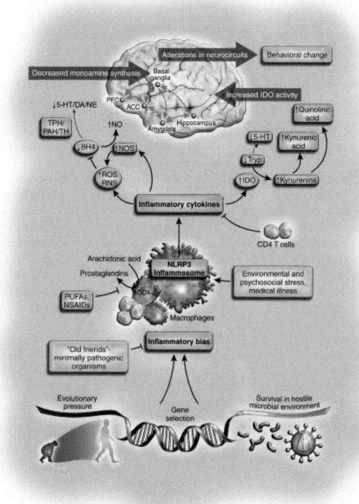

FIGURE 8.2   We are the way we are, in part, because our ancient ancestors were able to adapt to their environment. This entailed, among other things, being able to deal effectivity with multiple challenges, including a hostile microbial environment. These microbial processes may have provoked inflammatory changes that were enhanced over successive generations. Like so many other biological processes, inflammatory immune responses are modifiable by numerous psychosocial and experiential influences, thereby maintaining our ability to deal with varied challenges. These exceptional abilities involve a constellation of neurobiological systems, such as innate immune process that detect pathogenic microorganisms and sterile stressors system receptors. These receptors and sensors moderate caspase-1 (and related factors), which promotes inflammation upon exposure to infectious microbes. Having been activated, this inflammasome triggers the release of several cytokines, which activate enzyme pathways, such as IDO, together with the production of both reactive nitrogen and oxygen species (RNS and ROS).

As much as these changes ought to have been effective in protecting us, they are also able to provoke multiple neurobiological changes that can foster psychological disturbances. Among other things, they promote the release of neurotoxic metabolites of kynurenine, such as quinolinic acid, disrupt monoamine functioning by limiting the availability of their precursors, including tryptophan (Tryp) and tyrosine, as well as tetrahydrobiopterin (BH4), which is an essential co-factor for enzymes (TPH, PAH, and TH) needed for the formation of monoamines. The net result of these changes comprise variations of brain neural circuits, such as those that include the anterior cingulate cortex (ACC) and the prefrontal cortex (PFC), which are needed for survival in a stressful environment, particularly in the integration of immune and behavioral response to both pathogens and predators. When these circuits are overloaded as a result of chronic stressors these same responses may lend themselves to, as Raison and Miller (2013) indicated, "malaise, melancholy and madness, which are the inflammatory legacy of our evolutionary past."

NLRP3: NACHT domain-, leucine-rich repeat-, and pyrin domain-containing protein 3; NO-nitric oxide; NOS-nitric oxide synthase; NSAIDS-nonsteroidal anti-inflammatory drugs; PUFAs-polyunsaturated fatty acids. *Source: From Raison & Miller (2013).*

by elevated IDO activity. When IDO functioning was inhibited by a direct antagonist, or indirectly through minocycline treatment, the kynurenine/tryptophan ratio was normalized and the behavioral effects of LPS were reversed (O'Connor et al., 2009).

Consistent with this proposition, among hepatitis C patients treated with IFN-α over 12 week period, CSF kynurenine and quinolinic acid accumulation were associated with elevated depression, which was correlated to increased IFN-α within the cerebrospinal fluid, together with soluble TNF-α receptor 2 and monocyte chemoattractant protein-1 (Raison et al., 2010). Several reports in animals have confirmed this sequence of changes (e.g., Wang, Lawson, Dantzer, & Kelley, 2010), and indicated that a tricyclic antidepressant agent, such as desipramine, reduced the elevated IDO elicited by inflammatory agents, raising the possibility that the beneficial effects of common antidepressant agents may be operating through the kynurenine pathway (Brooks et al., 2017).

The consequences of kynurenine pathway activation are not restricted to responses related to immunotherapy. Levels of plasma kynurenine were also linked to depression (and suicidality) in the absence of IFN-α therapy (e.g., Sublette et al., 2011), and were also related to the ratios between kynurenine and tryptophan, and between glutamine and glutamate (Umehara et al., 2017). As well, attempted suicide was linked to enzymes in the kynurenine pathway that promoted glutamate receptor excitotoxicity and neuroinflammation (Brundin et al., 2016). These actions could come about because psychological stressors promote IDO activation, which could influence behavioral alterations (Kiank et al., 2010). Thus, individuals least able to cope or have limited coping resources available, will be most likely to develop depressed mood in response to stressors, and particularly vulnerable to the depressogenic actions of IFN-α immunotherapy.

An elaborated view of the kynurenine-depression hypothesis has been offered that might contribute to our understanding of the comorbidities that occur with depressive disorders (Cho et al., 2017). It was maintained that depressive illness develops through a combination of inflammatory-immune, oxidative, and nitrosative stress pathways (in which reactive nitrogen and reactive oxygen species cause damage to cells) together with somewhat lower levels of fundamental antioxidants (Maes et al., 2011). These neurobiological actions promote aggravation of inflammatory pathways that might contribute to disturbed synaptogenesis and neurodegeneration, hence promoting depression.

In addition to depressive disorders, inflammatory, oxidative and nitrosative stress pathways may also be a key feature leading to chronic fatigue syndrome (Lucas, Morris, Anderson, & Maes, 2015) and other fatigue-related disorders, and an assortment of autoimmune disorders, cancer, cardiovascular diseases, Parkinson's disease, stroke, obesity, and even schizophrenia (Kanchanatawan et al., 2018; Morris, Berk, Klein et al., 2016). As different as these disorders appear, they share common processes, such as intracellular inflammation, increased production of NF-kB, cyclooxygenase-2 (COX-2), inducible NO synthase (iNOS), and damage to membrane fatty acids (Maes, 2009). Moreover, Th17 cells and the secretion of IL-17, may be a key driver for these effects (Slyepchenko, Carvalho, Cha, Kasper, & McIntyre, 2016). Proposals such as these have been useful in consolidating wide ranging literatures that encompassed immune activation (and elevated cytokine levels), brain microglial release of cytokines, neuroplasticity, and how responses to stressors activate these processes, thereby eliciting a depressive like states (Wohleb, Franklin, Iwata, & Duman, 2016). This understanding may prove to be useful in depression associated with inflammatory activation, as well as the development of comorbid illnesses.

The data concerning links between IDO activation and depressive symptoms have been impressive, but there have been reports that were inconsistent with this position. Contrary to expectation, in the absence of a comorbid medical condition, postmortem analysis of the prefrontal cortex of depressed individuals indicated diminished conversion of tryptophan to kynurenine and reduced mRNA expression of IDO and tryptophan-2,3-dioxygenase, which correlated with levels of IFNγ and TNF-α, as well as lowered quinolinic acid levels (Clark et al., 2016). Why such unexpected outcomes emerged is uncertain, but many variables related to individual characteristics and symptom profiles need to be considered. As well, the

role of endocrine and inflammatory factors in depressive disorders may vary with typical versus atypical symptoms (Anisman et al., 1999; Dunjic-Kostic et al., 2013). Moreover, the response to IFN-α therapy also varies in relation to the presence or absence of polymorphisms associated with serotonin, dopamine, cortisol, and BDNF. Thus, as much as the kynurenine pathway is likely important for the development of depression, this condition is likely moderated by still other factors (Udina et al., 2016). The very fact that IFN-α induces a depressive-like state in only a subset (30%–50%) of patients again underscores the need to identify biomarkers that predict which patients will do best with certain treatments, and whether some would do well to receive antidepressant medications in advance of IFN-α therapy.

EXERCISE AS AN ANTIDEPRESSANT: EFFECTS THROUGH INFLAMMATORY PROCESSES

There has been considerable support for the position that physical exercise may diminish depressive characteristics, particularly the affective symptoms of depression (Booij, Bos, Jonge, & Oldehinkel, 2015). Exercise also diminishes feelings of sadness and suicide attempts in the context of an ongoing stressor, such as among bullied teens. Aside from potentially acting in a therapeutic capacity, moderate regular exercise (even 1 or 2 hours a week) can have prophylactic effects. A prospective study conducted over an 11-year period estimated that 12% of depressive occurrences could have been prevented through regular exercise (Harvey et al., 2017).

Engaging in exercise could reduce poor mood simply by acting as a distractor, but there's likely more to it than that. Engaging in exercise influences circulating cytokine levels, particularly those with anti-inflammatory actions (Svensson, Lexell, & Deierborg, 2015), and affects tryptophan availability, thereby influencing the kynurenine pathway (Badawy, 2015). Peripheral inflammation brought on by

nutritional or inflammatory signals, can instigate increased kynurenine accumulation in brain, which can be diminished by exercise through the activation of kynurenine clearance (Cervenka, Agudelo, & Ruas, 2017).

In instances where depressive disorders occur concurrently with elevated signs of inflammatory activity (e.g., owing to excessive abdominal fat), treatments such as exercise, which act to reduce inflammation, may promote better outcomes and reduce depression recurrence (Kiecolt-Glaser, Derry, & Fagundes, 2015). Exercise might also promote positive actions by increasing the activity erythropoietin (EPO), a cytokine that stimulates red blood cell production (made famous as way of blood doping to enhance performance in endurance sports). EPO increases neuronal functioning, promotes anti-inflammatory processes, serves as an antiapoptotic agent and antioxidant, as well as one that increases the synthesis and levels of BDNF (Osborn et al., 2013). Although it has not been widely assessed for its potential antidepressant actions, there is evidence that EPO can

(cont'd)

influence cognitive functioning and can diminish depressive symptoms in an animal model. Indeed, EPO administration elicited brain activity changes reminiscent of those produced by antidepressants, and may be useful as an adjunctive treatment of depression (Hayley, 2011).

The position was also advanced that through actions on microbiota, exercise and eating (the two influence different gut bacteria populations) can affect brain processes, thereby altering cognitive functioning and mood. Thus, the combination of exercise and consumption of probiotics might be considered as a way of enhancing mood (Grant & Baker, 2016).

MICROBIOME-IMMUNE INTERACTIONS

Aside from gut microbiota actions in relation to illnesses, such as autoimmune disorders (Jangi et al., 2016), there is also ample reason to believe that gut microbiota are linked to anxiety, depression, and PTSD (Dinan & Cryan, 2013; Leclercq, Forsythe, & Bienenstock, 2016). It has been about a decade since depression was first associated with dysfunction of intestinal mucosa, and the suggestion made that some cases of depression might be tied to leaky gut syndrome (Maes et al., 2008). The gut microbial involvement in illness conditions has, indeed, become part of our broader understanding of the holistic processes that are tied to a broad range of psychiatric and neurodegenerative disorders (Anderson &, Maes, 2016). With this in mind, novel and innovative treatment strategies are being developed in an effort to diminish mood disorders, or serve as an adjunct therapy (Rieder, Wisniewski, Alderman, & Campbell, 2017).

Support for the microbiota-depression link came from the finding that symptoms of depression could be elicited by microbiota depletion through antibiotic treatment in adult rodents (Hoban, Moloney et al., 2016), as well as in humans (Lurie, Yang, Haynes, Mamtani, & Boursi, 2015). Moreover, depressive symptoms associated with microbiota depletion could be attenuated by recolonization with commensal bacteria or by probiotic treatment (Slyepchenko, Carvalho, Cha, Kasper, & McIntyre, 2014). Concurrently, probiotic treatment diminished the corticosterone, serotonin, GABA, BDNF, and cytokine variations elicited by stressors (Ait-Belgnaoui et al., 2014).

Studies in rodents had made it clear that stressors markedly influenced the composition of intestinal bacterial communities, and that these microbial changes might contribute to the emergence or exacerbation of depressive- or anxiety-like conditions (Bravo et al., 2011). Consistent with the microbiota-depression linkage, the probiotic Bifidobacterium infantis attenuated immobility in a forced-swim test (a screen for antidepressant agents) among rats that had experienced early life distress, while concurrently diminishing circulating IL-6 levels (Desbonnet et al., 2010). Likewise, cortical, hippocampal, and amygdala functioning could be altered by chronic Lactobacillus rhamnosus (Bravo et al., 2011), and it was able to diminish stress-elicited corticosterone and behaviors that reflected anxiety and depression.

Among stressed rats the appearance of depressive-like symptoms coincided with gut bacterial changes, and when these microorganisms were transferred to naïve animals, a depressive profile was instigated, again speaking to a possible causal link between gut microbiota and mood disorders. In addition to behavioral signs of depression, when microbiota from stressed animals were transferred to germ-free mice, exaggerated corticosterone and ACTH elevations were promoted. Remarkably, transferring fecal microbiota from depressed patients to microbiota-depleted rats, instigated behavioral and neurobiological features of depression (Kelly, Borre et al., 2016).

Interestingly, microbiota may also contribute to myelination and myelin plasticity within the prefrontal cortex. Among germ free mice, axons were hypermyelinated, which normalized following bacterial colonization. These findings were taken to suggest that the microbiome may be fundamental for proper regulation of myelin-related processes, which could point to potential targets to deal with psychiatric disturbances (Hoban, Stilling et al., 2016). In line with preclinical studies, *B. Longum* administered to patients with irritable bowel syndrome altered brain responses to negative emotional stimuli and diminished signs of depression (Pinto-Sanchez et al., 2017). Findings such as these might be instrumental in accounting for the comorbidity often seen between depressive illness and both MS and irritable bowel syndrome.

As predicted, a probiotic combination comprising *L. helveticus* and *B. longum* reduced 24 hours urinary cortisol output in healthy volunteers, and a 4 week regimen of prebiotic intake was associated with a reduction of the cortisol awakening response, potentially reflecting diminished stress (Schmidt et al., 2015). Importantly, probiotic treatment was accompanied by reduced negative ruminative thinking and reduced self-reported depression and influenced functional brain activity in a cognitive task (Tillisch et al., 2013). Based on a systematic review, probiotics were taken to have positive effects on depressive symptoms, and their corresponding neurobiological processes. However, further double-blind longitudinal studies are still needed (Wallace &

Millev, 2017), especially as there have still been few randomized controlled trials showing that probiotic treatments have an ameliorative effect on mental disorders.

An imbalance among many bacteria (dysbiosis) are linked to stressor-elicited depression (Dinan & Cryan, 2013; Luna & Foster, 2015), and it was suggested that *Lactobacillus* may be particularly relevant in this regard (Marin et al., 2017). Lower *Bifidobacterium* and/or *Lactobacillus* were expressed more frequently in depressed than nondepressed individuals (Aizawa et al., 2016). In addition, the levels of several genera of the *Firmicutes* phylum were reduced in association with depression, as were *Faecalibacterium*, which varied in a dose-dependent fashion (Jiang et al., 2015). As this genus also has anti-inflammatory properties, its overrepresentation in healthy individuals might reflect a protective action. In this regard, early life treatment with *Bifidobacterium* pseudocatenulatum could attenuate the depressive-like effects in mice that experienced early maternal separation, possibly through its anti-inflammatory impact (Moya-Pérez et al., 2017). These are only a fraction of the studies supporting a link between stressor-provoked bacterial changes and the development of depression. Even if microbiota were instrumental in promoting clinical levels of depression, the essential questions that would remain are what processes are responsible for these effects, and could these be mined to alleviate the mood disorder?

Gut Bacteria and Immune Variations

Gut bacteria affect brain and immune functioning, thereby contributing to a slate of psychological and physical disturbances. Fig. 8.3 provides a theoretical model by which gut-related events could promote psychological disorders. This model offers multiple routes to get from the gut to the brain, including many

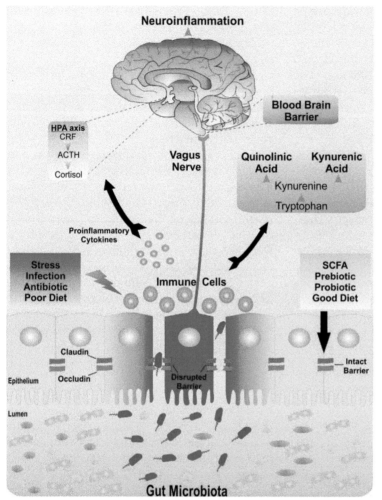

FIGURE 8.3 Postulated signaling pathways between the gut microbiota, the intestinal barrier, and the brain. Gut dysbiosis could produce a microbiota-driven proinflammatory state that has implications for inflammatory immune processes that lead to illness. *Source: From Kelly, Kennedy, Cryan, Dinan, Clarke, and Hyland (2015).*

of those that were discussed in earlier portions of this chapter, such as the actions of stressors, specific microbiota variations, and disruptions of the intestinal barrier (Dinan & Cryan, 2016). Gut microbiota, through effects on tryptophan availability, as well as several hormones, could affect kynurenine pathway metabolism and hence the development of depressive disorders (Kennedy, Cryan, Dinan, & Clarke, 2017). For instance, various short chain fatty acids (SCFAs), such as butyrate, through their actions on immune functioning and gut barrier

integrity, may come to affect depressive disorders (Morris, Berk, Carvalho et al., 2016). There has similarly been the view that by virtue of the systemic inflammatory response elicited, "leaky gut" syndrome may also contribute to inflammatory bowel disease, diabetes, allergies and depressive mood (Kelly et al., 2015). Likewise, both gut and brain processes affect autonomic functioning, which then influences visceral processes (Taché, Larauche, Yuan, & Million, 2017), thereby influencing the emergence of illness.

THE USEFULNESS OF PROBIOTICS?

Well over a hundred years ago, the Nobel Prize winner Élie Metchnikoff, who received the Nobel prize jointly with Paul Ehrlich, advanced the idea that consumption of fermented foods contributed to well-being and could promote longevity. Although this concept was long ignored, it has made a major comeback. In fact, todays "in" thing are probiotics or prebiotics, and the global market for these supplements exceeded $35B in 2015, and is increasing. However, do these pro- and prebiotics actually work? Like so many other foods and supplements, probiotics don't go through the approval process that drugs do, and it's not at all certain that what's stated on the label is actually in the capsule. In fact, in tests that have been conducted to assess the congruency between the label and the product, the ingredients have come up short! Yet, even if the label was accurate, how many consumers know what the ingredients signify—Faecalibacterium, Proteobacteria Actinobacteria, Alistipes, Bifidobacterium, Lactobacillus Firmicutes, Actinobacteria and Bacteroidetes aren't on most people's bucket list of things to know. In fact, even if an individual is deficient of particular bacteria, they likely wouldn't know it, and thus the bottle they picked up may or may not have the right bacteria in it. As it happens, although some bacteria are believed to have positive effects, it still isn't entirely certain which combination of bacteria might be beneficial in attenuating particular illnesses. In the end, the purchased probiotics may only be useful if individuals are experiencing a shortage of specific microbial species, or if they had received antibiotic treatments that diminished gut bacteria. Eventually, an individualized medicine approach may be possible, focusing on microbial dysbiosis, but this will not be simple given the complex interactions that exist between host genetics, gut microbiota, and diet (Ussar et al., 2015).

As foods consumed may affect microbiota, immune functioning, hormones and central neurotransmitters, it would be expected that dietary patterns would also be linked to depressive disorders. Diet-related data are notoriously unreliable, thus studies linking diet with depression ought to be taken with a grain of salt. Several studies have nonetheless pointed to the therapeutic benefits obtained through diet improvements as well as diet counseling in both cross-sectional and prospective analyses (Jacka et al., 2017).

Activation of the inflammasome promotes the maturation of caspase-1, an enzyme which comes to promote IL-1β and IL-18 activation, and may thus come to affect neuroinflammation, neurodegeneration, and psychiatric illnesses. Mice deficient of caspase-1 displayed lower levels of depressive- and anxiety-like behaviors under basal conditions and in response to a chronic stressor. Conversely, following treatment with minocycline, which diminished the response to the stressor, gut microbial changes were apparent that were consistent with the positive actions of diminished inflammation (Wong et al., 2016), although, as we saw earlier, minocyclin has many anti-inflammatory effects beyond the effects on the microbiome. Consistent with the effects of other microbial manipulations, L. helveticus NS8 treatment attenuated anxiety, depression and cognitive dysfunction associated with chronic restraint stress, being as effective as citalopram. The attenuated behavioral disturbances were also accompanied by a constellation of physiological changes expected with diminished depression, such as lower plasma corticosterone and ACTH levels, elevated plasma levels of the anti-inflammatory cytokine, IL-10, restored hippocampal 5-HT, and greater hippocampal BDNF mRNA expression (Liang et al., 2015).

Depression (or the accompanying distress), like chronic infection, could potentially promote changes of gut functioning, accompanied by disturbed feeding patterns, together with elevated proinflammatory cytokine activity and altered metabolism involving kynurenine/tryptophan pathways (Bercik & Collins, 2014; Kennedy et al., 2017), thereby further aggravating the existing depressive condition. As well, some gut bacteria, which comprise both commensal and pathogenic microorganisms can produce and deliver neuroactive substances, such as GABA and 5-HT, thereby contributing to the emergence of illness (Evrensel & Ceylan, 2015). It also appears that disturbing peripheral IL-6 receptors can normalize disturbed microbiota stemming from social defeat, thereby leading to antidepressant-like effects (Zhang, Yao et al., 2017). Given the multidirectional communication that occurs between gut microbiota and brain systems, hormonal, autonomic processes, and immune functioning, it is likely that microbial changes can have far-reaching consequences on psychological and gastrointestinal disorders (Bercik & Collins, 2014; De Palma, Collins, Bercik, & Verdu, 2014).

As described in Fig. 8.3, the balance between beneficial and harmful bacteria may be thrown off by stressor experiences or antibiotic intake, which can then affect immune, hormonal and brain processes. Changes of the gut microbiota may contribute to greater intestinal permeability, and as a result, various bacteria-derived neuroactive peptides and their metabolites may come to influence brain functioning and the corresponding behavioral outcomes. As well, SCFAs, particularly butyrate, that are produced may affect gene expression and inflammation within the CNS), thereby influencing brain functioning and behavioral disturbances (Stilling et al., 2016). Aside from potentially affecting mood states, butyrate produced in the gut by fermentation of dietary fiber, together with other fermentation-derived SCFAs (e.g., acetate, propionate) as well as ketone bodies (e.g., acetoacetate and d-β-hydroxybutyrate) may contribute to obesity, diabetes, inflammatory bowel disease, and colorectal cancer. It has been observed that SCFAs in the gut, primarily coming from fruits and vegetables, can produce epigenetic effects within cells of the gut lining, and can thus be instrumental in fighting infection (Fellows et al., 2018) and possibly affecting mood states. Long-chain omega-3 fatty acids, such as eicosapentaenic acid (EPA) and docosahexaenoic acid (DHA), can have inflammation-resolving properties, and could thereby diminish depressive symptoms (McNamara, 2015).

Despite the converging evidence implicating microbial processes in the development of psychological conditions, the bulk of the relevant data have come from studies in animals, and we know full well that translation to human diseases isn't always straight forward. To be sure, there have been many proponents of exploring the therapeutic effects of attenuating microbiota dysbiosis. Yet, at the moment there simply aren't sufficient data from humans to determine what microbial changes are optimal to treat psychological disorders. As such, it may still be premature to conclude that the gut microbiota is a simple target for the treatment of psychiatric conditions (Kelly, Clarke, Cryan, & Dinan, 2016). Once again, a personalized approach in this respect will likely be an ideal strategy.

BIPOLAR DISORDER

Several findings have pointed to the possibility that microbial factors may contribute to the evolution of bipolar features. Is it a coincidence that many individuals hospitalized with mania had been treated somewhat earlier with antibiotics to treat infection (Yolken et al., 2016)?

Acute episodes of mania have been linked to immune activation (Barbosa et al., 2014; Rosenblat & McIntyre, 2016), whereas the decline of mania was accompanied by reduced levels of inflammatory factors (Dickerson et al., 2013). The severity of bipolar symptoms was related to microbial alterations, with Faecalibacterium variations apparently being particularly significant (Evans et al., 2017). How immune system changes come to have such effects have yet to be determined, but it was suggested that NF-κB may contribute in this respect, and the kynurenine pathway was also linked to bipolar disorder (Anderson, Jacob, Bellivier, & Geoffroy, 2016).

CONCLUDING COMMENTS

Several positions have been advanced that attempted to incorporate monoamine functioning, CRH and corticoids, GABA and glutamate, neurotrophins, inflammatory and microbial processes (and multiple feedback loops involving these processes), in accounting for depressive disorders. The symptoms of depressive illnesses expressed differ markedly across individuals and vastly different neurobiological signatures may accompany these symptom variations. Thus, it is understandable that treatment efficacy will differ across individuals, and some patients will be resistant to treatments. For many patients the illness may stem from excessive inflammatory immune activation, which might also account for some of the comorbid conditions that are apparent. For these patients, treatments that influence inflammatory processes might be useful.

Even with successful treatment, patients often persist in presenting with residual symptoms, and should they stop medication, symptoms may reemerge. Even in the absence of these residual symptoms, the illness frequently returns within a few years. The treatments are clearly not providing a permanent 'cure'. If the depression stemmed solely from stressor-provoked neurochemical, immune, or microbial disturbances, then drug treatments might have beneficial effects, at least in the short-run, but for sustained benefits, the root of the problem obviously must be dealt with, irrespective of whether these comprise particular biological processes or cognitive disturbances.

Trial-and-error therapy is obviously not in the patient's best interests, and often can have harmful effects, possibly encouraging feelings of hopelessness as patients see one treatment after another fail to be effective. The solution might entail greater efforts need to be expended on precision medicine. This is easy to say, but exceptionally difficult to accomplish. There are so many genetic, biochemical, and behavioral variables that would need to go into understanding depression for each individual, the feasibility and sustainability of this approach might be questioned. On the other side, however, it is essential to ask what the consequences are of not developing an appropriate strategy, especially as depression tied to inflammatory processes has implications for so many other diseases.

In their thoughtful review, Raison and Miller (2013) tied numerous biological systems to one another, showing how these might be connected in the provocation of mental illness. It was suggested that evolution favored certain inflammatory responses, but today, this inflammatory bias is less well orchestrated and moderated (e.g., owing to life-styles involving poor eating, lack of exercise, disturbed sleep, as well as exposure to numerous toxicants) and consequently may favor diseases.

Anxiety Disorders

THE TWO SIDES OF ANXIETY

It's almost certain that all of us will, at some time, experience fear and anxiety. If the anxiety levels are not excessive, then this emotion might have considerable adaptive value, as it keeps individuals prepared and alert for potential dangers, and might instigate or contribute to high levels of performance in various milieus (stage performance, athletic competitions). When anxiety becomes excessive and persistent, behavioral and cognitive responses may be impaired, as might social functioning. Moreover, high levels of anxiety are often comorbid with other psychological disorders, such as depression, PTSD and substance use disorders, and may contribute to the emergence of physical illnesses, such as heart disease. For highly anxious individuals "the mind never rests," and some patients with comorbid anxiety and depression report that the anxiety is more debilitating than the depression itself.

Anxiety disorders are among the most common mental illness in Western countries, even appearing in about 12% of children, possibly being elicited by particularly distressing events, or they may be linked to a family history of anxiety. In many instances, adult anxiety may have first raised its threatening head in childhood, and if left untreated may persist throughout life.

It is important to distinguish between fear, which reflects a response to a real or imagined imminent threat, and anxiety that is more often seen as worry or uneasiness that is broad and unfocused. Anxiety also needs to be distinguished on the basis of it being a trait characteristic (a component of the individual's personality) or a state that is brought about by events being experienced. Several anxiety disorders have been identified (e.g., generalized anxiety disorder, phobias, social anxiety disorder). Anxiety can also come in the form of existential anxiety which may stem from individuals questioning whether their life has meaning or purpose (often termed existential crisis, but it also borders on nihilism). Existential anxiety can also encompass instances in which individuals face uncertainty and anxiety (e.g., regarding survival of their ingroup, e.g., a religion) owing to threats of annihilation or disappearance through other ways.

Aside from the common feelings of anxiety that are often experienced, various anxiety disorders exist, which can be exceptionally debilitating, often undermining individuals' social and work functioning. Given the frequency of anxiety disorders, it is somewhat remarkable that they haven't gained the notoriety of depression in the public mind. Like depression, the great majority of cases of anxiety disorders are not brought to the attention of physicians, and consequently go untreated (Craske et al., 2017). These disorders, like other mental health conditions are biochemically heterogeneous, and there has been a call for a personalized approach to treatment (Craske & Stein, 2016). Table 9.1 describes some of the features of the primary types of anxiety disorders.

In assessing anxiety related processes, it is essential to distinguish between the detection of threats, the response to threats in the form of conscious fear, and the anxiety that arises owing to unconscious processes. Aside from the symptoms related to conscious and unconscious processes differing from one another, the detection, appraisal, and response to threats can differ from those that elicit fear responses. Furthermore, they may be subject to distinct antecedent factors and predisposing influences, and attenuation of these feelings may require different treatment strategies (LeDoux, 2014; LeDoux & Pine, 2016).

Much like those people who can't handle ambiguity or uncertainty, some individuals generally seem to always see situations negatively (glass half-empty), which is accompanied by worry and rumination, as well as depression. When these worriers were shown a negative image and instructed to put a positive spin on it, they seemed incapable of doing so, and analyses of their brain activity suggested that their negativity actually worsened. It seems that negativity is an inherent characteristic that is difficult to change, although in slow steps, such as through cognitive behavioral therapy, this becomes possible.

Fear and anxiety are often considered together, although they are distinguishable from one another in several ways, even differing in the brain processes that underlie these emotions. Studies in both animals and in humans have pointed to the amygdala and the bed nucleus of the stria terminalis (the extended amygdala) being differentially involved in fear and anxiety, respectively (Davis, Walker, Miles, & Grillon, 2010). Recent formulations, however, suggested that these regions should be viewed as an integrated unit, particularly since they respond similarly to an assortment of threats and challenges, even those that are uncertain and are not imminent (LeDoux & Pine, 2016; see Fig. 9.1), as well as persistent experiences within nonspecific threatening contexts (Shackman & Fox, 2016).

NEUROBIOLOGICAL FACTORS AND TREATMENT OF ANXIETY DISORDERS

There are clearly many differences between the various anxiety disorders, but given that feelings of anxiety are common to each, there ought to be overlapping neurobiological features associated with them. Multiple neurochemical processes have been linked to anxiety and anxiety disorders, and many brain regions may contribute in this regard, including those tied to appraisal processes and memory of aversive or threatening events. For instance, diminished volume of the inferior frontal cortex was accompanied by relatively high levels of anxiety together with a negativity bias (Hu & Dolcos, 2017), which could potentially graduate to a clinical level of pathology. Moreover, dispositional negativity, hypervigilance, and attentional biases serve to promote the development and maintenance of anxiety, likely involving amygdala, prefrontal cortical, and

TABLE 9.1 Anxiety Disorders

Generalized anxiety disorder (GAD): Persistent (at least 6 months) anxiety or worry that is not focused on any single subject, object, or situation. It is accompanied by at least three of the following: restlessness (or feeling on edge), easily fatigued, difficulty concentrating, irritability, muscle tension, sleep disturbance comprising either difficulty falling asleep, or staying asleep (or restless sleep). The life-time prevalence of GAD is about 5%, occurring twice as often in females relative to males, frequently first appearing when individuals are in their 20's or 30's, but can initially appear early in life.

Phobias: The presence of significant anxiety (fear) elicited by certain situations (e.g., heights, open spaces, snakes, spiders), activities, things, or people. It becomes a problem when individuals display excessive and unreasonable desire to avoid or escape from the feared object or situation. Phobias are common, with the estimated prevalence being approximately 8%—9%, and are more frequent in women than in men.

Social phobia and social anxiety disorder: These conditions are characterized by intense fear of public scrutiny, embarrassment, or humiliation (negative public scrutiny) often being evident across multiple situations, but may be restricted to specific venues. The anxiety is especially pronounced when elevated scrutiny is likely, such as when an individual is required to speak or perform publicly, or when they must interact with others. Social anxiety is common, occurring in 6.8% of people, equally distributed among females and males. It often begins during childhood, persisting into adolescence and adulthood. In children it may be incapacitating, to the extent that they are fearful of playing with others or speaking to teachers.

Separation anxiety disorder: Characterized by fear or anxiety in relation to separation from an attachment figure, which is inappropriate for the person's development age. Persistent worries and anxiety about harm being experienced by the attachment figures could be responsible for the development of the disorder. Children may experience physical symptoms characteristic of distress and may have nightmares. Although symptoms occur most often during childhood, separation anxiety can be diagnosed in adults.

Panic disorder: Characterized by discrete periods (usually 1—20 minute) of sudden intense apprehension or terror, with episodes occurring for a period longer than 1 month. Panic episodes are accompanied by additional symptoms, including shortness of breath, chest pain, palpitations, feelings of choking or smothering, nausea/abdominal distress, sweating, trembling, feeling dizzy, unsteady or lightheaded, depersonalization (person feels detached from themselves) or derealization (a feeling of unreality), fear of dying, hot flashes, and a fear of losing control. Panic disorder occurs within 2.7% of the population, being twice as likely to appear in women as in men.

Obsessive-compulsive disorder (OCD): This disorder has been separated from anxiety disorders in the DSM-5, but is nevertheless characterized by the presence of repetitive obsessions that involve distressing, persistent, and intrusive thoughts that provoke anxiety and compulsions. The obsessions may involve a preoccupation with particular thoughts (e.g., with sexual or religious behaviors) that provoke anxiety. The anxiety may be alleviated, for a time, by behaviors being acted out (which eventually become compulsive behaviors), although obsessions and compulsions can be independent of one another. Individuals may display ritualistic behaviors, such as checking and rechecking (e.g., whether the door is locked, or the stove is off) or engaging in specific behavioral sequences. Moreover, obsessions can become more formed so that individuals will interpret certain objects as being significant or "meaningful," or the obsession can take the form of delusional behaviors (e.g., presence of conspiracies). This disorder has a life-time prevalence of 1%—2%, being equally common in females and males, often first appearing in childhood or adolescence.

locus coeruleus functioning (e.g., Shackman et al., 2016).

Determining the neurobiological mechanisms associated with anxiety disorders is difficult for several reasons, including that anxiety can be elicited by many stressful stimuli, which as we've seen, can have different neurobiological consequences. This is made still more complicated by the distress that individuals may experience as a result of their psychological illness (this also applies to depression, PTSD, and other mental illnesses), and consequently

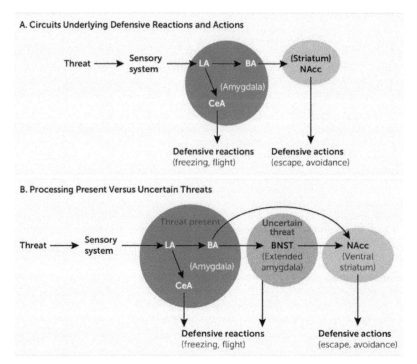

FIGURE 9.1   The amygdala is the central hub of circuits that control reactions and actions elicited by an immediately present threat (Panel A). The lateral amygdala (LA) receives sensory inputs about the threat. Connections from LA to the central nucleus of the amygdala (CeA) control defensive reactions, whereas connections from LA to the basal nucleus (BA), and from there to the ventral striatum (nucleus accumbens, NAcc), control the performance of actions, such as escape and avoidance. When the threat is uncertain (Panel B), connections from the amygdala and hippocampus (not shown) to the extended amygdala (the bed nucleus of the stria terminalis, BNST) are engaged, which give rise to more complex (possibly planned) defensive behaviors. *Source: Figure and figure caption from LeDoux & Pine (2016).*

it is uncertain whether brain neurochemical alterations are linked to the anxiety disorder or to the distress that occurs. Moreover, as LeDoux (2014) indicated, some forms of anxiety may be linked to conscious experiences, whereas others may be tied to nonconscious events. These difficulties notwithstanding, imaging studies revealed that although anxiety conditions can be distinguished from one another, they also share common features. In this regard, PTSD, GAD, social anxiety disorder, OCD, and panic disorder may be accompanied by altered network activity, but default mode network connectivity and the salience network may be comparable in only some of

these conditions (Peterson, Thome, Frewen, & Lanius, 2014).

In diagnosing and treating anxiety disorders, as in any other psychiatric illness, it would be ideal to be able to count on biomarkers or behavioral indices that could direct clinicians to particular treatments. Efforts have been made to use neuroimaging and genetic markers to this end, as well as various hormonal substrates (CRH, ACTH, cortisol, cholecystokinin, neurokinins, atrial natriuretic peptide, and oxytocin), neurotrophic factors (NGF, BDNF), inflammatory factors (cytokines), and neurophysiological measures (EEG, heart rate variability). Several markers proved to be tempting candidates in this regard, but a consensus statement from a broad group of researchers suggested that currently none are sufficient and specific to the extent that they can be used as diagnostic tools (Bandelow et al., 2016, 2017).

Neurobiology of Anxiety

To a significant extent, data relevant to human anxiety have come from animal models with a particular focus on the impact of stressors. The anxiety elicited in many of these models should not be misconstrued as necessarily reflecting anxiety disorders, especially as many of these manipulations also elicit depressive-like behaviors and PTSD. For that matter, studies of the behavioral effects of stressors often lump anxiety and depression together given that they may be difficult to disentangle in some animal models, and even many of the treatments used to treat anxiety disorders are precisely those that are used to diminish depressive illnesses and PTSD (i.e., SSRIs).

As observed in other psychiatric illnesses, anxiety disorders are accompanied by variations of brain morphology and connectivity. Some of the structural variations seen in adults were also evident in pediatric anxiety disorders in which ventromedial prefrontal cortex and left precentral gyrus thickness was increased, possibly being related to disturbed emotional processing. The enlarged size of the striatum seen in many anxiety patients is linked to an inability to deal with uncertainty and ambiguity in relation to future threats (Kim, Shin et al., 2017), again speaking to specific symptoms being predicted by particular brain changes. It was also thought that stressful events could lead to anxiety and PTSD through neurogenesis and neuroplasticity that occurred within the hippocampus and the amygdala. Indeed, it seems that specialized cells within the hippocampus may only fire in response to

anxiety-provoking stimuli or situations, and do so without input from cortical brain regions (Jimenez et al., 2018). Once activated, these cells trigger hypothalamic neurons involved in controlling several stress hormones and affect cardiovascular responses.

The symptom variations observed between patients may be linked to specific genetic differences. For instance, the right amygdala and orbitofrontal regions were enlarged among patients with anxiety who carried a polymorphism related to FKBP5, which influences glucocorticoid receptor sensitivity (Hirakawa et al., 2016), and may contribute to emotional processing by affecting amygdala connections and interactions with the orbital frontal cortex (Holz et al., 2015). A polymorphism for FK506-binding protein (FKBP51), which serves to modulate glucocorticoid and androgen sensitivity, has been linked to anxiety, and an antagonist of FKBP51 was effective in reducing anxiety in mice (Hartmann et al., 2015).

Corticotrophin Releasing Hormone

As described in Chapter 5, Stressor Processes and Effects on Neurobiological Functioning, stressors rapidly provoke secretion of CRH from the paraventricular nucleus of the hypothalamus, which instigates HPA activation (Zhang et al., 2016), and is clearly essential for allostasis. However, as we've seen, the involvement of CRH in stress responsivity goes beyond HPA functioning, as this hormone's role in fear and anxiety was attributed to changes that occurred within specific aspects of the amygdala. Stressors also influence CRH activity within the prefrontal cortex and the hippocampus, contributing to appraisals, decision making, and memory, all of which are involved in the development of anxiety. Moreover, CRH activation by stressors may influence locus coeruleus activity, possibly promoting vigilance.

Consistent with the involvement of CRH in anxiety, the provocation of CRH overexpression in the dorsal amygdala of monkeys promoted anxious temperament, and connectivity within components of anxiety circuits was altered (Kalin et al., 2016). Predictably, treatments that increase CRH in rodents typically promoted anxiety, whereas CRH receptor antagonists had the opposite effect (Slater, Yarur, & Gysling, 2016). These actions, however, were dependent on the type of CRH receptor that was activated. Anxiety stemming from the overproduction of CRH was primarily attenuated by pharmacologically antagonizing the CRH_1 receptor subtype, and was likewise diminished by genetic deletion of CRH_1 receptors (Muller et al., 2003).

Although the functions of CRH_1 and CRH_2 receptors in relation to anxiety have yet to be fully deduced, the view was offered that CRH_1 receptors primarily contribute to emotional responses, whereas CRH_2 receptors are more important in regulating coping responses (Liebsch, Landgraf, Engelmann, Lörscher, & Holsboer, 1999). Still another view has it that CRH_1 receptors mediate emotional, as well as executive functions, attention, and learning about emotions. The activation of CRH_2 receptors, in contrast, contributes to stress-related changes of basic functions necessary for survival, such as feeding, reproduction, and defense. From an applied perspective, patients presenting with anxiety and depression might benefit most from treatments that modify CRH_1 receptors, patients with eating disorders would benefit more from treatments that affect CRH_2 receptors. Unfortunately, for a variety of reasons, limited headway has been realized with respect to the development of CRH_1 antagonists for clinical purposes.

Norepinephrine

Peripheral norepinephrine produces signs of anxiety (e.g., elevated heart rate), which may serve as a signal to the individual (feedback) that they are anxious, thus giving rise to emotional responses. By virtue of effects on HPA functioning as well as neuronal activity within other brain regions, norepinephrine may affect anxiety responses. Specifically, stressor-provoked activation of the prefrontal cortex resulted in stimulation of the locus coeruleus, from which norepinephrine neurons originate, leading to the heightened vigilance and anxiety (e.g., Borodovitsyna, Flamini, & Chandler, 2018). The view that norepinephrine played a role in anxiety was reinforced by the anti-anxiety effects provoked by β-norepinephrine antagonists (e.g., propranolol).

Serotonin

There has long been the view that serotonin functioning may contribute to the development of anxiety. As SSRIs are effective as anxiolytics points to serotonin involvement in anxiety (e.g., Curtiss, Andrews, Davis, Smits, & Hofmann, 2017), as is the finding that the serotonin transporter gene may be predictive of the efficacy of drug treatments (Lueken et al., 2016). Specific serotonin receptors (e.g., 5-HT_{1A}), which might affect other neuronal processes, have also been implicated in anxiety (Nikolaus, Antke, Beu, & Müller, 2010). Among mice in which the 5-HT_{1A} was knocked out, anxiety-related behaviors were more pronounced (e.g., diminished open field activity and elevated fear in threatening environments), although these mice also exhibited disturbed swim performance that is usually used as a marker (screen) of antidepressant treatments (Akimova, Lanzenberger, and Kasper, 2009). In addition to 5-HT_{1A} receptor involvement, there is reason to suppose that 5-HT_{2A} and 5-HT_{2C} receptors also contribute to anxiety and depression (Millan, 2005). Serotonin certainly doesn't act alone in generating anxiety, and potent anti-anxiety effects could be achieved by a combination treatment that affected both norepinephrine and serotonin (Goddard et al., 2010).

GABA and Glutamate

One of the more prominent consequences of stressors is the change of GABA activity and that of the subunits that make up $GABA_A$ receptors (Poulter, Du, Zhurov, Merali & Ansiman, 2010), and as indicated in our discussion of depression, threats effective in eliciting anxiety also altered GABA activity relative to that evident during a safe period. In line with a role for GABA, anxiogenic outcomes were provoked by negative modulators of GABA receptors (Kalueff & Nutt, 2007; Nuss, 2015), and the effectiveness of benzodiazepines in reducing anxiety has been attributed to their actions on $GABA_A$ functioning. Conversely, agents that augment GABAergic tone, such as valproate, vigabatrin, and tiagabine, diminish anxiety. In should be said that although, most studies focused on the role of $GABA_A$ in relation to anxiety, elements of $GABA_B$ functioning may also contribute to aspects of anxiety. Behavioral studies also indicated that $GABA_B$ modulators influence anxiety (Pizzo, O'Leary, & Cryan, 2017), and may contribute to the long-term consolidation of contextual memories that influence generalized contextual fear (Lynch, Winiecki, Gilman, Adkins, & Jasnow, 2017) (Fig. 9.2).

As discussed in the context of depressive disorders, GABA and glutamate acted in a coordinated fashion, and thus glutamate can be expected to play a prominent role in anxiety. In fact, diminished anxiety has been associated with glutamate activation and kappa opioid receptor functioning. Upon being activated, kappa opioid receptors cause glutamate release from the basolateral amygdala (BLA) inputs to the BNST, and hence anxiety levels increase. Conversely, deletion of kappa receptors within the amygdala produces an anxiolytic phenotype (Crowley et al., 2016). It was similarly observed that glucocorticoid functioning within several glutamate circuits, notably in the forebrain and BLA, are fundamental to the regulation of fear and anxiety (Hartmann et al., 2017). In line with this, the synthetic corticoid dexamethasone administered chronically can undermine coordination of GABA and glutamate functioning, thereby promoting stress-related anxiety (Wang, Zhu, Cui, & Wang, 2016). Given that glutamate and particular glutamate receptors (e.g., mGluR5), are involved in stress responses (e.g., Peterlik et al., 2017) and may contribute to depressive disorders, it's likely that the attention to glutamate as a target in treating anxiety disorders will continue. Aside from these interactions, GABA modulates the links between serotonin and CRH, which might also contribute to anxiety symptoms, and it appears that $GABA_A$ receptor expression may be affected by ovarian hormones, such as progesterone, which could account for sex differences in anxiety-related conditions.

Cannabinoids

Cannabinoid activity has been tied to changes of anxiety and depression, although in a subset of individuals, it can elicit excitation and elevated reactivity. Aside from generalized anxiety, cannabinoids have been implicated as potential treatments for panic disorder, social anxiety disorder, OCD as well as PTSD. For the most part, the actions of cannabis have been assessed acutely, and much less information is available concerning the influence of its long-term use (Blessing, Steenkamp, Manzanares, & Marmar, 2015; Soares & Campos, 2017). As discussed earlier, this largely is a result of government prohibitions related to cannabis use, including its clinical testing.

It will be recalled that cannabis' psychoactive component, Δ9-tetrahydrocannabinol (THC), acts through endogenous molecules

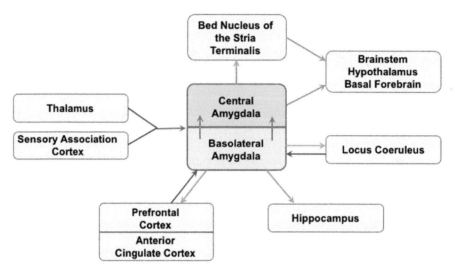

FIGURE 9.2   Disturbed GABA functioning was associated with several anxiety disorders (Kalueff & Nutt, 2007), which likely reflect the fine ability of GABA to inhibit amygdala neuronal functioning (Nuss, 2015). The basolateral amygdala (BLA) and central nucleus of the amygdala (CeA) are particularly germane to anxiety processes. Upon receiving threat-relevant information from the thalamus and the sensory association cortex, the BLA stimulates the CeA through an excitatory glutamatergic pathway. Additionally, inhibitory GABAergic interneurons—the intercalated neurons—serve to inhibit CeA neuronal functioning. The inhibitory GABAergic neurons from the CeA project to the hypothalamus and brainstem, promoting physical (somatic) signs of anxiety. Dysphoria and vigilance are elicited by projections to basal fore-brain nuclei and the locus coeruleus, respectively. The projections from the BLA also act on neurons within the bed nucleus of the stria terminalis, which acts like the central amygdala, although the amygdala and bed nucleus might be differentially involved in fear vs. anxiety. The medial prefrontal cortex and anterior cingulate cortex also receive and send glutamatergic projections to and from the basolateral amygdala, serving to influence amygdala activity in response to threatening stimuli. Essentially, by influencing neuronal activity in the basolateral amygdala, the medial prefrontal cortex regulates the perception and expression of anxiety. The more dorsal aspects of the cortex are thought to govern conscious, voluntary control of anxiety, whereas ventral regions regulate subconscious aspects of anxiety. In this way, the functioning of the amygdala is determined through top-down processes (Source: see details in *Nuss (2015)*, from which this figure is adapted).
Blue arrows denote principal inputs to the BLA; green arrows represent principal outputs of the BLA; orange arrows show principal outputs of the CeA and BNST.

(endocannabinoids; eCBs) that bind to particular cannabinoid receptors (e.g., CB_1 and CB_2) that are located within numerous brain regions, many of which have been tied to anxiety (Hill et al., 2010). It may be of particular relevance that among its other effects, as described in Fig. 9.3, cannabinoids can inhibit microglial activation, thereby diminishing inflammatory responses within the brain (Lisboa, Gomes, Guimaraes, & Campos, 2016). Similarly, cannabinoids can inhibit cyclooxygenase-2 (COX-2) and thus might act like other COX-2 inhibitors in diminishing inflammation (Patel, Hill, Cheer, Wotjak, &

Holmes, 2017). In addition to these anti-inflammatory effects, eCBs affect hippocampal cannabinoid receptors and modulate synaptic plasticity, thereby altering stress responses (Scarante et al., 2017).

As described in Chapter 5, Stressor Processes and Effects on Neurobiological Functioning, basal and threat-elicited anxiety may be governed, at least in part, by the functioning of two brain endocannabinoids, anandanine (AEA), and 2-arachodonoylglycerol (2-AG), which serve as the gatekeepers of the stress response (Hill & Tasker, 2012). A calm disposition can be sustained because AEA

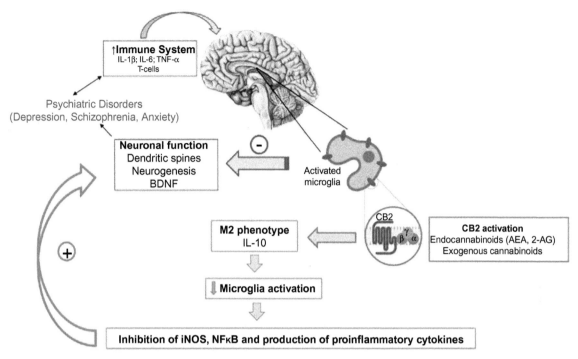

FIGURE 9.3 Disorders such as depression, schizophrenia, and anxiety have been associated with an increase of activated microglia as well as neuronal and synaptic damage. Activation of cannabinoids receptors (primarily CB2 on activated microglia) may drive M2 microglia, creating a state in which the proinflammatory factors would be diminished, whereas anti-inflammatories increased. As a result, neurons would be protected from damage and psychopathology would thus be reversed or prevented. *Source: From Lisboa et al. (2016).*

tonically affects the CB_1 receptor; however, in response to stressors, CRH is released at the amygdala, which ultimately reduces the signaling by AEA, thereby contributing to anxiety. In essence, AEA's role is that of maintaining individuals in a relaxed state during non-stress periods, and contributes to the HPA stress response and anxiety that stems from stressor experiences. Like AEA, 2-AG activation is provoked by stressors, largely acting to keep the stress response in check, and may be involved in the adaptation associated with chronic stressors.

Activation of eCBs and CB_1 receptors regulate the release of several neurotransmitters and hormones that influence stress reactions, and application of a CB_1 agonist into the basolateral amygdala diminished stressor-induced HPA activation, and might contribute to the consolidation of stressor-provoked emotional memories (Hill et al., 2010). As expected, a stressor in the form of social defeat in mice resulted in eCB promotion of synaptic plasticity within the nucleus accumbens that might contribute to anxiety- and depressive symptoms (Bosch-Bouju, Larrieu, Linders, Manzoni, & Layé, 2016). Thus, manipulating the endocannabinoid system may provide a way of diminishing anxiety and anxiety-related disorders.

ANXIETY DISORDERS

Generalized Anxiety Disorder

It is generally accepted that inappropriate threat appraisals or a failure to inhibit responses to non-existent threats might contribute to the emergence of GAD. Relatedly, this disorder may reflect overgeneralization so that individuals see non-threatening events as being a threat (Dunsmoor & Paz, 2015). As communication between the anterior cingulate cortex and the basolateral portion of the amygdala (BLA) ordinarily serves in the appraisal of threat and safety, dysfunction of these connections could be responsible for persistent threat appraisals. Specifically, aspects of the prefrontal cortex, notably the anterior cingulate cortex, ordinarily inhibit amygdala activity, thereby limiting anxiety. A disturbance of the white matter tract connecting the anterior cingulate cortex and amygdala (the uncinate fasciculus) would result in the inhibition of the amygdala being lost, hence leading to persistent anxiety (Tromp do et al., 2012). Based on a systematic review, it was concluded that generalized anxiety was more complex, being accompanied by deficient prefrontal cortex and anterior cingulate cortex functioning together with disturbed top-down control operations during emotion regulation tasks (Mochcovitch, da Rocha Freire, Garcia, & Nardi, 2014). This pathway was also viewed as important for the extinction of fear responses, as well as their reinstatement by specific environmental triggers, and thus a disturbance in this regard would result in fear and anxiety being persistent as observed in GAD (Likhtik & Paz, 2015), as well as in PTSD.

In GAD patients, amygdala and prefrontal cortex activation were elevated, while reduced functional connectivity existed between these regions, possibly accounting for the excessive worry and autonomic dysregulation that is so often apparent in this disorder (Makovac et al., 2016). The links between the prefrontal cortex and the amygdala likely involve multiple neurochemical alterations. The involvement of norepinephrine and cortisol have long been suspected, and there is good reason to suppose that cortical GABA, together with midbrain serotonin, 5-HT$_{1A}$ receptors, and 5-HT reuptake processes influence dopamine functioning, which then affects several anxiety disorders, including GAD (Nikolaus et al., 2010).

Treatments of GAD

Benzodiazepines can be effective for acute anxiety, as well as for several anxiety related disorders, including GAD, panic disorder, and social anxiety disorder, and when given together with SSRIs or psychological treatments, still better outcomes may be realized (Starcevic, 2014). However, benzodiazepines are not recommended for long-term use owing to the development of tolerance, physical dependence, and withdrawal symptoms. There has also been concern that the positive effects of benzodiazepines stemmed from their sedative/hypnotic properties rather than effects on anxiety (Chagraoui, Skiba, Thuillez, & Thibaut, 2016).

Owing to these concerns, other treatments, such as SSRIs and SNRIs, are more commonly used for GAD, especially when depressive symptoms are also present. One of the problems that can be encountered in a subset of patients is that early in treatment with SSRIs, feelings of anxiety may increase, possibly owing to CRH$_1$ receptor activation within the BNST (Marcinkiewcz et al., 2016). Several other 5-HT acting drugs have also been used in the treatment of GAD, such as buspirone (BuSpar), a 5-HT$_{1A}$ receptor partial agonist that also acts as a dopamine D$_2$ and α-adrenergic antagonist. The GABA acting agent, pregabalin (Lyrica), which is known for its effects on neuropathic pain, has also been used in treating GAD (Frampton, 2014). Although it has anti-anxiety actions commensurate with those of benzodiazepines, the risk for dependence is low.

In addition to pharmacological treatments, cognitive behavioral therapy (CBT) has been used to diminish GAD symptoms. It seems that CBT may have more sustained effects, whereas the pharmacological route appears to promote a more immediate fix, although even in the case of pregabalin, about 1 week of treatment is needed for ideal effects to become evident. Mindfulness training may also be effective in ameliorating symptoms of GAD and diminish changes of ACTH and proinflammatory cytokines ordinarily elicited by a laboratory social evaluative stressor in the form of public speaking (Hoge et al., 2013). None of these treatments are effective for everybody, as they only produce appreciable remission in 50−60% of individuals.

Panic Disorder

Several cognitive explanations were advanced to account for the development and maintenance of panic disorder (Schmidt & Keough, 2010). An emotion-based perspective suggests that some individuals, owing to genetic factors or life experiences, are disposed to overreacting to stressors. These reactions are persistent, possibly owing to Pavlovian conditioning, so that internal sensations may be "catastrophically" misinterpreted as being especially threatening, possibly owing to disturbances within the prefrontal cortex, which then promotes psychological and physical arousal and elevated perceived threat.

Several potential neurochemical mechanisms were again offered to account for panic disorder. These included alterations of neuropeptide factors, such as CRH, AVP, and cholecystokinin (the latter is a gut peptide better known for its role in digestion and satiety). Recent genetic analyses have also pointed to several variants of the glycine receptor B gene (GLRB) as playing a role in panic disorder by affecting a fear network (Deckert et al., 2017).

This disorder was also attributed to lower concentrations of GABA in the anterior cingulate cortex and basal ganglia, and serotonin was similarly offered as being a primary contributor to panic disorder. Among individuals with panic disorder, 5-HT$_{1A}$ receptor binding was altered (Nash et al., 2008), and indirect support for serotonin involvement in this condition has come from reports that SSRIs can be used to diminish panic disorder (Stein, Steckler, Lightfoot, Hay, & Goddard, 2010).

Patients who were exposed to emotionally salient cues that comprised anxiety-provoking visual stimuli or threatening words, exhibited markedly elevated neuronal activity in brain regions involved in appraisal and executive processes, such as the anterior cingulate cortex, posterior cingulate cortex, orbital frontal cortex, and hippocampus (Beutel, Stark, Pan, Silbersweig, & Dietrich, 2010). It is especially interesting from a clinical perspective that these altered brain processes normalized following successful psychotherapy. Curiously, however, when successfully treated patients were assessed in an emotional conflict paradigm (patients were exposed to emotional faces and words that were either congruent or incongruent with one another), exaggerated neuronal activity was still evident in the anterior cingulate cortex, dorsal medial prefrontal cortex, and amygdala (Chechko et al., 2009). These findings suggest that either these brain regions do not underlie panic disorder, or that persistent dysfunction of this network is responsible for a high probability of relapse.

Treatment of Panic Disorder

Treatment with SSRIs and benzodiazepines have been the most common methods of dealing with panic disorder. There have been studies that compared the efficacy of these treatments, but typically they were underpowered and firm conclusions couldn't be made based on the available information (Bighelli et al., 2016). Several psychotherapeutic approaches were used in an effort to control panic disorder, with the greatest focus being on CBT (Kaczkurkin & Foa, 2015). In addition, several alternative behavioral approaches were adopted to treat panic disorder, including psychoeducation, supportive psychotherapy, and psychodynamic therapies. Of the various treatment approaches, CBT was somewhat superior to other methods (Pompoli, Furukawa, Imai, Tajika, & Efthimiou, 2016), being effective in about 60% of patients. This method was particularly effective when it explicitly dealt with the perceived likelihood of panic occurring, the perceived consequences of panic, and an individual's ability to cope with panic. However, CBT was not especially effective when panic disorder was comorbid with PTSD, which is not all that uncommon.

Panic disorder has been treated with combined psychotherapy and drug treatments, but there were indications that aside from some short-term benefits of this approach, enhanced effects were not apparent over the longer run (Hofmann, Sawyer, Korte, & Smits, 2009; Schmidt & Keough, 2010). Similarly, although benzodiazepine treatment may be effective in panic, added value was not obtained by combining psychotherapy and benzodiazepine treatment (Schmidt & Keough, 2010).

Obsessive-Compulsive Disorder

The underpinnings of OCD have been difficult to define given that the specific obsessions and compulsions come in so many varieties, and obsessions and compulsions can also appear independently of each other. Still, the nature of the disorder, which involves repetitive behaviors has encouraged the position that OCD reflects a disturbance within a complex loop involving cortical brain regions associated with executive functioning and decision making (i.e., the anterior cingulate cortex, ventromedial, dorsolateral, and lateral-orbital cortex)

together with processes related to reward (nucleus accumbens, caudate). In turn, these regions send trajectories that activate the thalamus and the basal ganglia, which then transmit messages back to cortical regions (Milad & Rauch, 2011). Since its initial inception, this view was reformulated so that the lateral and medial orbitofrontal cortex (lOFC and mOFC) have been given key roles in OCD, being responsible for processing of information with a negative or a positive valence, respectively (Milad & Rauch, 2011). These regions act together to determine appraisals of threatening and non-threating situations, and then controlling (inhibiting) the functioning of other cortical regions that contribute to the emergence of anxiety.

The anterior cingulate and orbitofrontal cortex have frequently been discussed in the context of depressive disorders, but some of its presumed functions, such as identifying cognitive conflict, error monitoring, and decision-making, may also contribute to the development of OCD (Ball, Stein, & Paulus, 2014). Ordinarily, when dual and inconsistent messages are received, or when noise comes from external sources, the anterior cingulate cortex is especially necessary for decision-making. Among individuals with OCD, this brain region is hyperactive in decision making situations, possibly indicating the difficulty in making appropriate appraisals and decisions, or it may reflect a disturbance involving improper feedback, thus leading to repeated behavioral responses. In this regard, it was demonstrated that GABA concentrations in the orbital frontal cortex of OCD patients were reduced, and there were indications of such an action in the anterior cingulate cortex (Zhang et al., 2016). An interesting view of some subtypes of OCD is that it reflects the failure of a security motivation system in which individuals need a "feeling of knowing" (e.g., Did I turn off the stove?), before they can move on to other tasks. Should feedback systems not operate as they should, then OCD symptoms, such as repeated checking, would persist (Woody & Szechtman, 2011).

Several animal models of OCD have been developed, although these were complicated owing to the diversity of OCD typologies that exist (Does hoarding involve the same processes as persistent checking, and do these involve the same processes as obsessive hair pulling or handwashing?) In a timely critique, it was suggested that the current animal models of OCD probably should not be viewed as "models" in the traditional sense, but instead should be considered as reflecting processes linked to several psychopathological conditions. These include compulsivity, stereotypy, or perseverance, which are seen in OCD, but are also common in several other psychiatric or neurological disorders (Alonso, López-Solà, Real, Segalàs, & Menchón, 2015).

Despite the limitations, there have been clues concerning the primary neurobiological processes that govern OCD. For instance, knocking out kainate receptor subunits within the dorsal striatum (iontophoretic receptors that are stimulated by glutamate) resulted in mice showing signs of OCD (Xu et al., 2017). As well, largely based on pharmacological studies and those that involve genetically engineered mice, several genes were identified that were tied to OCD. Deletion of the *Sapap3* gene, which is important in communication between neurons, was accompanied by prodigious self-grooming (Chen et al., 2011), which could be attenuated by treatment with a MGluR5 antagonist. Moreover, these effects occurred very quickly, rather than the many days needed for SSRIs to exert a positive effect (Ade et al., 2016).

Treatment of OCD

It is common for SSRIs to be used in the treatment of OCD, usually at high doses (Fineberg et al., 2015), which may be especially valuable as OCD and depression are frequently comorbid conditions. It likewise appears that manipulations of glutamate functioning may attenuate OCD symptoms (Marazziti, Carlini, & Dell'Osso, 2012), and there have been indications based on animal studies that dopamine manipulations might be useful in diminishing the characteristics of OCD (Dvorkin et al., 2010). At the moment, SSRI treatment continues to be a first line treatment of OCD, but adjunctive treatments that influence glutamatergic process or inflammatory factors (e.g., riluzole, ketamine, memantine, N-acetylcysteine, lamotrigine, celecoxib, and ondansetron), may enhance the effects observed (Hirschtritt, Bloch, & Mathews, 2017).

As an alternative to drug treatments, especially when these fail to be effective, behavioral and cognitive treatment approaches may be adopted. Appreciable success was achieved through CBT (Lewin, Wu, McGuire, & Storch, 2014) and with "exposure and response prevention" (ERP) in which patients learn, typically in graded steps, to tolerate the anxiety that comes about when they are unable to engage in the compulsive behavior (e.g., Wheaton et al., 2016). Especially positive outcomes were realized with the combination of CBT and antidepressant agents (Bandelow et al., 2012). An alternative method that has yielded some success has been a learning based approach in which an attempt is made to alter the associations normally present in response to obsessive thoughts. For instance, an individual obsessed with not touching anything "contaminated by germs" might be "taught" to appraise external objects less negatively by having them pair these objects with neutral thoughts or emotions. In cases of treatment resistant OCD, deep brain stimulation directed at the bed nucleus of the stria terminalis, which has been associated with anxiety, promoted a rapid and sustained positive effect on symptoms (Raymaekers et al., 2017).

Phobias

Phobias may develop through classical conditioning processes wherein events or stimuli associated with a negative feeling or emotion may come to elicit fear or anxiety upon later encounters with these stimuli. It wouldn't be unreasonable to assume that many of the processes that are associated with other anxiety-related disorders also contribute to phobic responses, and involve brain regions that are associated with appraisals, decision making, and fear responses, and their extinction. A phobic reaction is accompanied by clear neurophysiological responses comprising elevated amygdala and anterior cingulate cortex activity, which normalized with alleviation of a phobia through behavioral therapy (Fredrikson & Faria, 2014).

Exposure therapy similarly promotes extinction of the fear/anxiety response. Gradual desensitization treatment (achieved by diminishing distress in small steps or by having the feared situation or object come progressively closer or more real) have a high success rate. Therapies for phobia frequently involve behavioral/emotional methods to extinguish this conditioned response. To this end, imagery and virtual reality treatments have been used to desensitize individuals by promoting emotional and behavioral change in incremental steps. Likewise, CBT has been offered so that individuals will begin to understand their negative thought patterns, and then to take steps to modify their behavioral and emotional responses.

Social Anxiety

As in the case of other anxiety conditions, social anxiety is accompanied by exaggerated neuronal activity in the amygdala and in cortical regions linked to attention and processing of social threats (Miskovic & Schmidt, 2011). Moreover, among social anxiety patients without comorbid depression, cortical thickness was increased (e.g., within the left insula, right anterior cingulate), which was consistent with greater attentional and executive control functioning (Brühl et al., 2014). Moreover, the elevated left amygdala volume associated with social anxiety disorder was diminished 1 year after successful CBT treatment (Månsson et al., 2017).

As in the case of other anxiety disorders, serotonin, norepinephrine, glutamate and GABA, have been implicated in social anxiety disorder (Marazziti et al., 2015). In line with this, a SNP related to the 5-HT transporter was linked to social anxiety disorder (Forstner et al., 2017). Significantly, in response to fearful faces, activation of the insula was greater among individuals carrying the ss genotype for the 5-HT transporter than among individuals carrying the ll genotype (Klumpp et al., 2014). As with other anxiety conditions, social anxiety is often comorbid with depressive illness and thus it may be difficult to uncouple the link between 5-HTT and social anxiety versus that of depression. Moreover, subtypes (or dimensional differences) exist with respect to social anxiety disorders, and thus heterogeneity appears regarding the neurobiological substrates for this condition (Binelli et al., 2015).

There has been increasing evidence consistent with the proposition that oxytocin may be related to anxiety. When oxytocin was administered directly to the prelimbic portion of the prefrontal cortex, activation of GABA neurons was provoked in this region, and amygdala neuronal functioning was reduced (Sabihi, Dong, Maurer, Post, & Leuner, 2017). Oxytocin may be particularly relevant to social anxiety (van Honk, Bos, Terburg, Heany, & Stein, 2015) so that when an oxytocin antagonist was administered directly into the BNST the effects of a stressor on social behavior in female rodents was attenuated (Duque-Wilckens et al., 2018).

In humans, social anxiety was associated with a less secure attachment style, which varied with the presence of an OXTR polymorphism (Notzon et al., 2016). A prospective study of 400 adolescents revealed that the combination of chronic interpersonal stressors and the presence of the CD38 polymorphism (which predicts lower oxytocin levels rather than changes of the oxytocin receptor), predicted trait social anxiety and symptoms of depression over a 6 year period (Tabak, Vrshek-Schallhorn et al., 2016). Further, a polymorphism for the oxytocin receptor was related to greater amygdala volume and exaggerated amygdala fMRI activity in response to socially-relevant face stimuli (Marusak et al., 2015). Similarly, epigenetic changes related to the oxytocin receptor were tied to social anxiety (Kumsta, Hummel, Chen, & Heinrichs, 2015) as well as increased social phobia and anxiety stemming from social interactions (Ziegler et al., 2015). Congruent with these findings, epigenetic variations related to the OXTR receptor were accompanied by increased amygdala responsiveness in a test that involved social phobia-related word processing (Ziegler et al., 2015). Furthermore, although oxytocin treatment did not affect amygdala neuronal activity in response to emotional faces in healthy control participants, in generalized social anxiety patients the hyperactive amygdala response ordinarily observed in response to emotional faces was eliminated by oxytocin administration (Labuschagne et al., 2010).

Treatment of Social Anxiety

Oxytocin treatment can be used to attenuate stressor effects on several brain processes as well as in the promotion of emotional and social behaviors (e.g., Sánchez-Vidaña et al., 2016). Thus, manipulations of oxytocin (and arginine vasopressin) may serve as targets for social anxiety disorder (Neumann & Slattery, 2016). Yet, the possibility has been raised that under some circumstances oxytocin treatments can have deleterious effects by increasing the salience of negative social cues. Indeed, among individuals with elevated social anxiety, social working memory was disturbed by oxytocin, supporting the contention that among individuals with social anxiety, exogenous oxytocin administration may impair social cognitive functioning (Tabak, Meyer et al., 2016).

Many reports have attested to the effectiveness of SSRIs (e.g., escitalopram) in the treatment of social anxiety disorder, just as these agents have been effective in treating other anxiety conditions. Indeed, SSRIs and SNRIs were considered the treatment of choice, although only small to medium effect sizes were attributable to the treatment (Curtiss et al., 2017). When SSRIs and SNRIs were ineffective or not well tolerated, the GABA-like agent pregabalin was effective in diminishing social anxiety disorder (Kawalec, Cierniak, Pilc, & Nowak, 2015). Agents such as SSRIs and SNRIs have also yielded positive findings in children and adolescents, as have psychotherapeutic interventions (Weisz et al., 2017).

A systematic review confirmed that both behavioral and pharmacological treatment strategies may be effective in the treatment of social anxiety, although in many cases drug treatments promoted greater effect sizes relative to behavioral approaches (Bandelow et al., 2015). Of the behavioral approaches, CBT was deemed to be the most effective treatment (Mayo-Wilson et al., 2014), and could be enhanced by concurrent escitalopram administration. In line with the treatment efficacy, successful treatment with CBT corresponded to structural changes within brain regions linked to emotional regulation (Steiger et al., 2016). Likewise, the amygdala-frontal cortical co-activation profiles could distinguish between effective and ineffective treatments (Faria et al., 2014). In essence, when the amygdala is not operating as it should, particular treatments could be used to attenuate anxiety, but these same treatments may be ineffective if the source of the anxiety entails a non-amygdala process.

INFLAMMATORY FACTORS IN ANXIETY AND ANXIETY DISORDERS

Inflammation and Anxiety Responses in Animals

The data linking inflammatory processes and anxiety have largely come from studies in rodents. Individual differences of anxiety in rodents (or differences across strains of mice exhibiting varied phenotypes) have been linked to inflammatory factors, which can vary still further in the presence of stressors. Brain inflammatory responses also tended to be more pronounced among rats bred for high anxiety, and in response to neonatal LPS treatment hippocampal neuroinflammation, microglia morphology. and anxiety-like behaviors were more pronounced in anxious rats (Claypoole, Zimmerberg, & Williamson, 2017).

As we've seen, stressors trigger peripheral inflammatory responses, which might influence brain functions that contribute to anxiety. As well, peripheral norepinephrine release provoked by stressors may promote the release of T cells from secondary immune organs, such as the spleen, which could trigger brain neurochemical changes that elicit emotional responses. Stressor-elicited anxiety may, alternatively, have stemmed from the microglial recruitment of monocytes to the brain endothelium and the subsequent IL-1β activation of IL-1R1 (type 1 receptors) (McKim et al., 2017). Brain microglial cell changes varied systematically with levels of anxiety across strains of mice, supporting the view that these inflammatory factors account for individual differences of anxiety (Li, Ma, Kulesskaya, Võikar, & Tian, 2014).

Treating mice with an agent that activated the immune system (e.g., LPS), so that cytokines were released into circulation or were released by microglia within the brain, elicited dose-dependent signs of anxiety in several behavioral paradigms. Treatment with IL-1β likewise elicited anxiety, reinforcing the possibility that the actions of LPS might largely be due to this cytokine's release (Sulakhiya et al., 2016), although it could also involve the synergistic action of IL-1β and other cytokine variations that are provoked by the endotoxin (Brebner, Hayley, Zacharko, Merali, & Anisman, 2000). Congruent with the finding that depressive-like behaviors were linked to IDO-kynurenine and their downstream actions, activation of the kynurenine pathway also elicited anxiety and anhedonia, whereas behavioral outcomes elicited by LPS could be diminished by pretreating animals with an IDO inhibitor, 1-methyltryptophan. As anxiety and depression frequently appear together, the relative contribution of IDO-kynurenine functioning in mediating these behaviors are not readily dissociated (Salazar, Gonzalez-Rivera, Redus, Parrott, & O'Connor, 2012).

The anxiety elicited by LPS was accompanied by increased expression of mRNA for IL-1β, IL-6 and TNF-α within the central amygdala, temporally corresponding with measures of anxiety (Engler et al., 2011). Predictably, the LPS-elicited anxiety was attenuated by treatment with an antioxidant (esculetin) that also had anti-inflammatory and neuroprotective actions (Sulakhiya et al., 2016). These findings are consistent with the view that systemic insults can engender anxiety through amygdala stimulation, but other brain regions are also affected by inflammatory factors (Anisman & Merali, 2003). For instance, inflammation can elicit activation of the locus coeruleus, which may contribute to one element of anxiety, that of vigilance.

A social stressor that reliably induces an anxious/depressed profile was most prominent in mice that exhibited the greatest IL-6 changes, and the anxiety could be modified by manipulations that altered IL-6 production (Hodes et al., 2014). There have also been indications that repeated treatment with the anti-inflammatory agent ibuprofen could reduce anxiety in several behavioral tests, while concurrently reducing the expression of proinflammatory mediators, such as TNF-α and IL-1β, as well as hippocampal BDNF (Lee, Schnitzlein, et al., 2016). Tellingly, the stressor-provoked anxiety/depressive profile and elevated IL-6 levels normalized with repeated antidepressant (imipramine) treatment (Ramirez & Sheridan, 2016).

Although LPS and staphylococcal enterotoxin B (SEB), are both bacterial products that elicit distinct immune responses and provoke elevated amygdala neuronal activity and anxiety responses, only LPS elicited proinflammatory changes within the amygdala. These data suggest that the brain may be sensitive to varying immune challenges, but the anxiety provoked likely involves factors other than or in addition to cytokines (Prager et al., 2013). Indeed, LPS that elicited anxiety were accompanied by variations of norepinephrine and serotonin in brain regions that have been linked to anxiety (Lacosta, Merali, & Anisman, 1999), although the experiments to verify causal connections to anxiety remain to be conducted. It also appeared that the anxiety-inducing effects of IL-1β could be diminished by treatment with a compound that attenuated the effects of α-melanocyte-stimulating hormone, supporting the view that the cytokine acts on anxiety through indirect actions on specific melanocyte receptors (Cragnolini, Schiöth, & Scimonelli, 2006).

There have been indications that super-antigens (which elicit non-specific T-cell activation and marked cytokine release), such as SEB and SEA, might not necessarily elicit anxiety in several behavioral tests, but are apparent if the test situation is one that creates a degree of distress (Rossi-George, LeBlanc, Kaneta, Urbach, & Kusnecov, 2004). Consistent with such findings, among mice with the gene for the anti-inflammatory cytokine IL-4 knocked out, so that proinflammatory processes predominated, anxiety was generally not provoked, unless the test situation involved a mild challenge, again suggesting that inflammatory changes produced state anxiety (Moon et al., 2015).

As in the case of a bacterial endotoxin, anxiety induced among virally-infected mice (e.g., by murine cytomegalovirus) was accompanied by IL-6 and TNF-α elevations, which could be restrained by the presence of corticosterone (Silverman et al., 2007). Anxiety was also accompanied by elevation of several proinflammatory cytokines within the brain of mice infected with the protozoan parasite *Plasmodium berghei* (de Miranda et al., 2011). Congruent with such findings, the administration of the viral mimic poly I:C elicited an anxious-depressed behavioral profile, coupled with reduced BDNF expression, and activation of the kynurenine pathway (Gibney, McGuinness, Prendergast, Harkin, & Connor, 2013). In essence, anxiety could be provoked through several routes that all provoked an inflammatory immune response.

Further support for inflammatory factors in relation to anxiety has come from the demonstration that benzodiazepines, which diminish short-term anxiety, also inhibited stressor-provoked IL-6 elevations, the accumulation of macrophages within the CNS, and inhibited trafficking of monocytes and granulocytes in circulation (Ramirez, Niraula, & Sheridan, 2016). As we saw earlier, antidepressant agents, such as SSRIs may also diminish anxiety by affecting immune functioning. Of course, each of these agents have anti-anxiety effects that could occur through other pathways (e.g., by affecting GABA or serotonin functioning), making it premature to conclude that the observed effects stem from immune changes elicited by the drug. It is certainly more likely that the immune changes are actually secondary to the diminished anxiety created by the drug treatments.

Inflammation and Anxiety in Humans

Anxiety introduced by a stressor in a laboratory context could promote increase circulating IFNγ and IL-1β levels (Moons & Shield, 2015), which were associated with elevated feelings of anxiety. As well, greater circulating CRP levels in men (but not women) were associated with increased threat-related amygdala activation (Swartz, Prather, & Hariri, 2017). Similarly, administration of an endotoxin in humans elicited feelings of anxiety (and depressive-like mood) that were accompanied by elevated TNF-α and its soluble receptor, IL-6, and cortisol levels (Reichenberg et al., 2001). Congruent with this report, administration of low doses of LPS elicited sub-clinical anxiety symptoms together with elevated IL-6 and TNF-α as well as the anti-inflammatory IL-10 (Lasselin et al., 2016).

Anxiety Responses Associated With Chronic Illness

The links between anxiety and inflammatory processes has been inferred from reports indicating that anxiety was associated with a substantial number of illnesses, particularly those related to immune disturbances (e.g., inflammatory bowel disease, autoimmune disorders, such as rheumatoid arthritis and multiple sclerosis), and even food allergies were associated with childhood anxiety, particularly social anxiety (Goodwin et al., 2017). In the main, these studies were correlational, precluding causal conclusions. Indeed, anxiety associated with illnesses may be related to the distress created, rather than immune changes. Nonetheless, the data from these studies have been informative and bear reviewing however briefly.

Both anxiety and depression were seen among patients with multiple sclerosis, which were tied to increased CSF levels of inflammatory factors. These outcomes were diminished among patients whose symptoms were remitting, and it appeared that IL-2 was aligned with anxiety, whereas IL-1β and TNF-α were more closely linked to depressive symptoms (Rossi et al., 2017). Increased inflammation and the presence of psychiatric symptoms were not unique to multiple sclerosis, having been apparent in other autoimmune conditions, including rheumatoid arthritis and inflammatory bowel disease. The levels of IL-17 were elevated in human rheumatoid arthritis patients, being most notable in those patients that exhibited greatest anxiety (Liu, Ho, & Mak, 2012). In systemic lupus erythematosus patients, variations of IFNγ were also closely aligned with anxiety symptoms (Figueiredo-Braga et al., 2009), and anxiety levels were elevated in a mouse model of this autoimmune disorder (Schrott & Crnic, 1996). Several studies likewise attested to the frequent anxiety associated with inflammatory bowel diseases (Abautret-Daly, Dempsey, Parra-Blanco, Medina, & Harkin, 2017; Bannaga & Selinger, 2015). As in the case of autoimmune disorders, state anxiety associated with the presence of several types of cancer were accompanied by variations of genes coding for IL-1R2 and TNF-α, and trait anxiety was associated with alterations in the expression of genes for IL-1β, IL-1R2, and Nf-KB (Miaskowski et al., 2015).

Reiterating our earlier point, although anxiety and depression are frequent in immune-mediated inflammatory diseases (Marrie et al., 2017), these outcomes might result from the distress related to these disorders, quite apart from whether or not they involve inflammation. However, studies in animals that allow for better experimental control, suggested that the inflammation itself was a prime mediator of the mood changes. Specifically, using an animal model of inflammatory bowel disease in which *Citrobacter rodentium* was administered, an increase of anxiety was apparent within 8 hr, well before inflammation or elevated cytokine levels were apparent (Lyte, Li, Opitz, Gaykema, & Goehler, 2006). Congruent with such illness-related findings, in an animal model of multiple sclerosis, the co-occurring anxiety was mediated by circulating TNF-α (Haji et al., 2012), and anxiety elicited by inflammatory pain in mice was accompanied by amygdala TNF-α elevations. In the latter instance, the anxiety could be attenuated by a TNF-α neutralizing antibody, suggesting causal involvement of this cytokine in producing the anxiety response (Chen et al., 2013). Likewise, sepsis in mice was accompanied by anxiety together with elevated plasma TNF-α and that of brain TNF-α, IFNγ, IL-1β and IL-6 measured 10 days after the surgical procedure that induced sepsis (Calsavara et al., 2013). In a review of this literature, largely based on animal studies (given how few studies were available in humans), it was suggested that chronic illnesses alter resident brain microglia and disturbed hippocampal neurogenesis. This, in turn, may instigate cognitive and mood disturbances, such as anxiety and depression (Chesnokova, Pechnick, & Wawrowsky, 2016).

Despite the sensitivity of both IL-6 and TNF-α in response to stressors, it might be unrealistic to believe that anxiety would exclusively be mediated by these cytokines (Vogelzangs, Beekman, de Jonge, & Penninx, 2013). Given that anxiety is biochemically heterogeneous, these cytokines would only be expected to be altered in a subset of individuals. Further to this, although anxiety was accompanied by dysregulation of several blood cytokines, many of these relations disappeared when controlling for life-style factors, pointing to the indirect relationship that might exist between anxiety and immune factors. However, other cytokines, such as IL-8 (also referred to as CXCL8, a chemokine produced by macrophages, epithelial, and endothelial cells) were elevated in both current and remitted patients, even after controlling for life-style (Vogelzangs, de Jonge, Smit, Bahn, & Penninx, 2016). In addition to frank changes of cytokine levels, anxiety and depressive disorders have been associated with changes of Toll-like receptor 4 (TLR 4) necessary for activation of innate immune responses. Consistent with these findings, among major depressive patients, the appearance of anxiety and weight loss were accompanied by altered TLR 4 mRNA levels (Wu, Huang, Huang, Huang, & Hung, 2015).

While not discounting the diverse routes by which anxiety may be linked to inflammatory processes and various illness conditions, it has been maintained that psychological interventions may be exceptionally beneficial for patients dealing with disorders, such as inflammatory bowel disease (Bannaga & Selinger, 2015; Filipovic & Filipovic, 2014), in which inter-relations exist between neural, hormonal and inflammatory factors (Abautret-Daly et al., 2017). For that matter, among some individuals with inflammatory bowel disease, coping through religion was effective in diminishing psychological distress and enhanced treatment adherence (Freitas et al., 2015).

Genetic Correlates of Anxiety

Genetic analyses linking anxiety and inflammatory processes have come from linkage studies, genome-wide association studies, and candidate gene studies. These reports have implicated the usual suspects in relation to anxiety disorders (Gottschalk & Domschke, 2017). Specifically, generalized anxiety was related to several candidate genes associated with monoamines (5-HTT, 5-HT$_{1A}$, MAOA) and BDNF, and their interactions with stressful experiences (e.g., early adversity). Together with psychosocial determinants of anxiety, these factors could potentially serve as markers to predict treatment responses to therapeutic drugs.

Cytokine links to anxiety have also been observed based on such studies. For example, anxiety was linked to a SNP within the gene associated with IL-8 (Janelidze et al., 2015), and in the gene coding for IL-1β and in the NF-κβ p100 subunit gene. As observed in other contexts, the influence of a polymorphism on anxiety interacted with stressor experiences. In particular, a polymorphism related to the gene coding for IL-1β was associated with anxiety and depression, provided that individuals had encountered a significant life-stressor (Kovacs et al., 2016).

The presence of a haplotype (referring to several genes that are inherited as a group) related to the proinflammatory IL-18 was associated with increased central and medial amygdala responsivity. Moreover, in women the IL-18 haplotype was also linked to symptoms of anxiety and depression following a stressor encounter (Swartz, Prather, Di Iorio, Bogdan, & Hariri, 2016). Evidently, the presence of a gene haplotype related to inflammatory processes can be predictive of amygdaloid reactivity as well as anxiety in response to a threatening context, but as tempting as it might be to link the brain, immune, and anxiety changes, the directionality of any such relations can't be deduced from these data.

Consistent with the epigenetic effects provoked by stressful experiences, global DNA methylation levels were elevated among anxious relative to non-anxious individuals. Moreover, among anxious patients, symptom severity was associated with epigenetic processes related to IL-6 gene expression (Murphy et al., 2015), as well as to several other immune-relevant genes, such as the TLR-2 promoter and the inducible nitric oxide synthase promoter (Kim, Kubzansky, et al., 2016). As mentioned earlier, there are likely many epigenetic changes that might similarly have been associated with anxiety, including on other genes associated with

inflammatory processes, but at this time the data relevant to this are limited.

ANXIETY DISORDERS IN RELATION TO INFLAMMATORY PROCESSES

As expected, many anxiety disorders share several common features related to immune functioning, just as common neurotransmitter processes have been identified, but they are also distinct from one another in several respects (Furtado & Katzman., 2015). Most studies that assessed these linkages involved only a modest number of participants, and different inflammatory immune factors were assessed across studies, thus limiting the conclusions that can be drawn. We emphasize once again that many anxiety disorders are comorbid with depression, and they are accompanied by eating and sleep disturbances, as well as poor life-styles (e.g., elevated smoking), any of which can engender elevated levels of C-reactive protein and IL-6 (Slavich & Irwin, 2014). These inflammatory alterations may be secondary to these symptoms rather than being directly linked to anxiety-related illnesses.

Generalized Anxiety Disorder

Elevated levels of CRP accompanied GAD (Khandaker, Zammit, Lewis, & Jones, 2016), and a longitudinal study of children who were followed for more than a decade, confirmed these findings (Copeland, Shanahan, Worthman, Angold, & Costello, 2012). Both current and remitted anxiety disorders were also accompanied by elevated levels of the inflammatory marker CRP, suggesting that CRP is a trait feature of GAD. Several studies indicated that IL-2 and IL-6 were elevated in anxious adolescents who were depressed relative to healthy controls (Pallavi et al., 2015), and in clinically anxious individuals the elevated IL-6 was evident even after controlling for age, sex, presence of depressive symptoms, and neuroticism (O'Donovan et al., 2010). The ratio between pro- and anti-inflammatory cytokines also favored proinflammatory factors in GAD patients (Hou et al., 2017), and serum levels of a constellation of inflammatory cytokines was elevated (Tang et al., 2018). Following a psychosocial challenge in the form of public speaking, cytokine levels and ACTH were increased further among patients with GAD, but these actions were limited among patients who had received Mindfulness-Based Stress Reduction (Hoge et al., 2013).

Analysis of inflammatory factors that were stimulated *ex vivo* indicated that the production of IL-8 was directly related to severity of anxiety and depressive symptoms (Vogelzangs et al., 2016). As well, the concentrations of IL-2 and IFNγ, as well as the anti-inflammatory cytokine IL-4, were lower in cell cultures of GAD patients than in control participants, whereas TNF-α and IL-17 were elevated (Vieira et al., 2010).

Panic Disorder

Like other anxiety conditions, panic disorder was associated with elevated levels of IL-1β (Brambilla et al., 1994) and IL-18 (Kokai, Kashiwamura, Okamura, Ohara, & Morita, 2002). Other studies indicated a much greater range of cytokine differences in panic disorder, including elevated levels of IL-2, IL-1β, IL-6, IL-8, and the anti-inflammatory IL-4 (Hoge et al., 2009). In contrast to these reports, it was reported that panic disorder was accompanied by reduced levels of IL-12 and IFNγ, whereas the levels of IL-1β, TNF-α, IL-6 and IL-2 did not vary between patients and controls (Tükel et al., 2012). Clearly, it is premature to draw conclusions regarding the specific cytokines linked to this condition or particular symptoms of the disorder.

Obsessive-Compulsive Disorder

As in the case of other anxiety disorders, limited data are available concerning the link between inflammation and OCD. A population-based cohort study revealed that streptococcal infection was associated with elevated risk for developing OCD (Orlovska et al., 2017), although these findings do not speak to the processes responsible for this association. However, it is of both theoretical and clinical significance that minocycline treatment as an adjunct to the SSRI fluvoxamine, was accompanied by an enhanced therapeutic response in diminishing OCD symptoms (Esalatmanesh et al., 2016).

Studies that examined peripheral indices of inflammation in relation to OCD provided mixed results. In some reports (e.g., Maes, Meltzer, & Bosmans, 1994; Weizman et al., 1996), cytokine levels among OCD patients did not differ from controls, although the severity of OCD symptoms was positively correlated with circulating IL-6 and that of the soluble IL-6 receptor (Maes et al., 1994). Reduced plasma TNF-α also accompanied OCD, whereas IL-1β and IL-6 were not affected (Monteleone, Catapano, Fabrazzo, Tortorella, & Maj, 1998). In contrast to these reports, elevated basal levels of IL-2, IL-6, TNF-α as well as IL-4 and IL-10, were present in drug naïve OCD patients (Rao, Venkatasubramanian, Ravi, Kalmady, & Cherian, 2015). It was also observed that OCD was accompanied by elevated levels of several circulating chemokines (e.g., CCKL3 and CXCL8) and soluble TNF receptors, and these relations were dependent on the symptom dimensions expressed (e.g., washing, hoarding) (Fontenelle et al., 2012).

Reduced ex vivo TNF-α production was similarly observed in blood of OCD patients, whereas changes were not apparent with respect to IL-6 and IFNγ, or the anti-inflammatory cytokines IL-4 and IL-10 (Denys, Fluitman, Kavelaars, Heijnen, & Westenberg, 2004). Lower levels of IL-6 were also present in blood cells of OCD patients that had been stimulated with LPS, but this outcome was not evident in patients with social anxiety disorder (Fluitman et al., 2010), pointing to specificity regarding the relationships.

A systematic review of the few published studies available concluded that although OCD was accompanied by reduced IL-1β (even though plasma IL-1β in humans is difficult to detect), the concentrations of IL-6 and TNF-α were not different between those with OCD and control participants. There were indications based on a stratified analysis that cytokine differences might be present in relation to age and the use of psychotropic medications (e.g., reduced IL-6 and elevated TNF-α when OCD was accompanied by depression), but the available data were not sufficiently compelling for firm conclusions to be offered (Gray & Bloch, 2012). Given the limited analyses of blood cytokines related to OCD, it's hardly surprising that few data are available concerning brain changes related to inflammation. This said, the occurrence of OCD was accompanied by inflammation within the orbital frontal cortex (Attwells et al., 2017).

The occurrence of OCD (and several other anxiety conditions) was associated with a polymorphism for the TNF-α gene (Hounie et al., 2008), as well as a SNP within the gene for FKBP5 (the glucocorticoid modulator) (Minelli et al., 2013). There have also been reports indicating that epigenetic factors on inflammatory-related genes were associated with OCD (Cappi et al., 2016), although this doesn't rule out similar correlations with other epigenetic changes.

Even if inflammatory factors are linked to OCD, this relationship is likely moderated by a constellation of factors. Moreover, as OCD is so often comorbid with depressive disorder, it can be difficult to dissociate the relative contributions of these illnesses on inflammatory processes. In addition, it warrants mentioning that OCD is characteristically different from GAD and social anxiety disorder, and it probably should be distinguished from these other conditions (as it is in the DSM-5), and hence comparable inflammatory responses would not be expected.

LINKING MICROBIOTA AND CYTOKINES IN THE PROMOTION OF ANXIETY

Gut bacteria, as we've seen, can influence peripheral and brain neural functioning, and in this way may promote anxiety (Foster & McVey Neufeld, 2013). Messaging between the microbiota and the brain may occur through vagal processes, or through other neural, endocrine, and metabolic mechanisms, as well as through immune homeostasis, thereby influencing psychological disorders. In this regard, diet-provoked alterations of gut bacteria might also predict later anxiety (Luna & Foster, 2015; Phillips, Shivappa, Hébert, & Perry, 2017), but given the difficulties in having individuals comply and report on their food consumption, the data linking diet and anxiety in humans are often questionable. The alternative approach is that of assessing microbiota among patients with specific anxiety disorders. However, even here problems of interpretation can arise. For instance, pediatric autoimmune neuropsychiatric disorders (PANDAS) leading to OCD was associated with streptococcal infections, but this outcome might not have been due to the infection itself, but instead may have stemmed from the antibiotics used to treat the infection (Rees, 2014).

During the early stages of intestinal (enteric) infection, rodents display anxiety-like behaviors, possibly mediated by inflammatory immune processes that influence CNS functioning via the vagal nerve (Bercik & Collins, 2014). In mice deficient of Caspase-1 [also referred to as interleukin-1 converting enzyme (ICE)], which ordinarily cleaves proteins, such as precursors of IL-1β and IL-18, anxiety and depression were reduced, as would be expected with reduced levels of proinflammatory cytokines. In addition, following treatment with a Caspase-1 antagonist, the usual depression and anxiety associated with a chronic stressor were absent. It is especially interesting that treatments that diminished anxiety also provoked a rebalancing of several gut bacteria, potentially implicating gut microbiota in the protective effects ordinarily associated with caspase-1 (Wong et al., 2016).

PARASITIC INFECTION

Intestinal parasites come in two primary forms, protozoa (single celled organisms, such as giardia, cryptosporidium, microsporidia, and isospora) and helminths (e.g., tapeworms, hookworms, pinworms, roundworms, whipworms, blood flukes). Carrying these parasites are hardly rare conditions as they affect hundreds of millions people world-wide. Parasites come in so many variations, it's not practical at this point to list the varied symptoms that they promote. Some parasites consume food that is ingested, so that individuals feel perpetually hungry, whereas other parasites feed off red blood cells, causing anemia. Parasites can also disturb digestive processes, but it may be less obvious that parasites may contribute to autoimmune disorders (e.g., Hanevik, Dizdar, Langeland, & Hausken, 2009). These parasites can also influence brain functioning and thus affect cognitive and affective processes. Following the administration of the noninvasive parasite Trichuris muris to produce gut inflammation, anxiety behaviors were seen in rodents, which could be attenuated by anti-inflammatory treatments, such as a TNF-α antagonist. A probiotic treatment also attenuated the anxiety brought about by Trichuris muris, but did so without affecting cytokine or kynurenine levels (Bercik et al., 2010), suggesting that brain functioning might be affected through some other route (e.g., through the vagal nerve).

Modifying Anxiety Through Microbial Processes

Consistent with the view that microbial factors affect brain functioning and hence mood states, when the presence of gut microbiota were reduced from weaning onward through an antibiotic cocktail, anxiety-like behavior was reduced and cognitive disturbances were provoked. These behavioral changes were accompanied by changes within the tryptophan metabolic pathway, together with diminished BDNF, as well as oxytocin and vasopressin expression (Desbonnet et al., 2015). It was similarly observed that an antibiotic cocktail administered to BALB/c mice (a highly anxious strain) for several days through their drinking water produced a change of the microbiota, although for a relatively brief time, and an increase of exploratory behaviors, possibly reflecting diminished anxiety. These behavioral changes were accompanied by elevated hippocampal BDNF, whereas little change was apparent with respect to autonomic nervous system activity, gastrointestinal neurotransmitter presence, or inflammation. Interestingly, behavior was not altered among these mice following systemic injection of antimicrobials nor was behavior altered among germ-free Swiss-Webster mice. However, among the BALB/c that were colonized with microbiota from the Swiss-Webster mice exploratory behavior and hippocampal BDNF was elevated. Conversely, when the germ-free Swiss-Webster mice were colonized with microbiota from the BALB/c, exploratory behavior was diminished (Bercik et al., 2011). Thus, microbial dysbiosis could affect later anxiety-like behaviors, independent of inflammation or gut neurotransmitter changes.

Although the anxiogenic actions of stressors are usually attributed to brain neurochemical changes, it has been suggested that disruption of intestinal microbiota elicited by stressor exposure may promote overproduction of inflammatory mediators, hence contributing to anxiety (Bailey, 2014). As this effect was precluded among germ-free mice, attests to the critical role of commensal bacteria in the provocation of stressor-provoked inflammatory and anxiety responses. Commensurate with this finding, diminished anxiety was apparent in germ-free mice (Clarke et al., 2013; Neufeld et al., 2011a), which was accompanied by elevated hippocampal serotonin utilization in males. Although anxiety and circulating tryptophan could be reestablished through postweaning bacterial colonization, the hippocampal serotonin changes remained altered (Clarke et al., 2013). It was similarly observed that among germ-free mice, lasting morphological alterations were detected. Specifically, dendritic hypertrophy was present within pyramidal neurons and spiny interneurons within the basolateral amygdala, while in the ventral hippocampus, neurons were both shorter and less well branched (Luczynski et al., 2016). It was also observed that in germ-free mice as well as those that had gut bacteria diminished by an antibiotic cocktail, marked dysregulation of microRNA occurred within the amygdala and prefrontal cortex, which partially normalized following bacterial recolonization (Hoban,

Stilling, Moloney, Moloney et al., 2017). Together, these studies indicate that early-life microbiota depletion may have lasting effects on neuronal morphology and neurotransmitter functioning, possibly rendering these mice vulnerable to adult psychopathology.

Relevant to the microbiota—anxiety linkage, diminished recall and emotional responses in response to cues that had been paired with a stressor was seen in germ-free mice. These mice also displayed a neural transcription profile that could be differentiated from conventional mice, particularly in relation to genes thought to be involved in synaptic transmission and neuronal activity, raising the possibility that these factors mediate the link between microbiota and fear/anxiety. Unlike the persistent hippocampal changes seen in germ-free mice, upon recolonization with microbiota the fear reactions normalized (Hoban, Stilling, Moloney, Shanahan et al., 2017). In line with such findings, reduced anxiety in germ-free mice, corresponded with a decrease in central amygdala NMDA receptor expression, together with increased BDNF expression and decreased $5HT_{1A}$ mRNA in the dentate granule layer of the hippocampus (Neufeld et al., 2011b). However, when these germ-free mice received bacteria from control animals during adulthood, the anxious phenotype persisted, which differed from the findings of previously mentioned studies. It may be that microbiota may contribute to anxiety states, and although this could be altered by prebiotics, there might be a narrow window during which this could be achieved (Neufeld et al., 2011a).

As expected, 2 weeks of treatment with a prebiotic, such as 3′Sialyllactose (3′SL) or 6′Sialyllactose (6′SL), limited the microbiota changes otherwise elicited by a stressor and similarly attenuated the anxiety evident in stressed mice (Tarr et al., 2015). A prebiotic mix administered over 3 weeks likewise attenuated the anxiety provoked by LPS, possibly through modulation of cortical IL-1β and

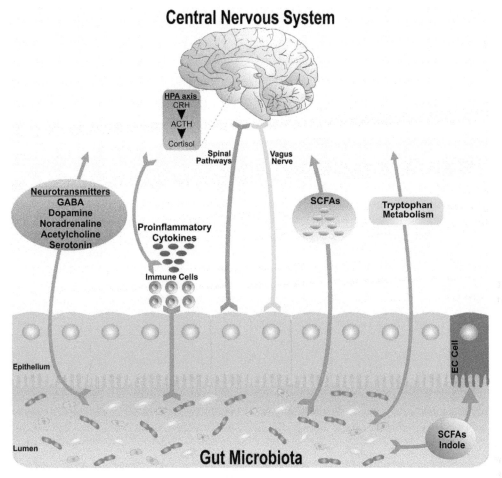

FIGURE 9.4 Microbiota -gut-brain axis. Gut microbiota communication with the brain can occur through several pathways. Microbiota can influence immune activity and the production of proinflammatory cytokines that can stimulate the HPA axis activity. Alternatively, microbiota can impact brain functioning through the production of short chain fatty acids (SCFAs), such as propionate, butyrate, and acetate. Microbial produced SCFAs also influence enterochromaffin (EC) cells of the enteric nervous system. As well, brain functioning can be affected by microbiota through modulation of tryptophan metabolism and downstream variations of serotonin, kynurenic acid, and quinolinic acid. Neuronal and spinal pathways, including afferent vagal fibers, could also mediate the effects of the gut microbiota on brain function and behavior. Abbreviations: *ACTH*, adrenocorticotropin hormone; *CRH*, corticotropin-releasing hormone; *EC*, enterochromaffin cells; *GABA*, gamma-aminobutyric acid; *HPA*, hypothalamic-pituitary-adrenal; *SFCAs*, short-chain fatty acids. *Source: From Kennedy et al. (2017).*

5-HT_{2A} receptor expression within the frontal cortex (Savignac et al., 2016). In line with such findings, chronic treatment with the L. rhamnosus (JB-1) altered GABA(B1b) and GABA (Aα2) mRNA in brain regions associated with both depression and anxiety, and reduced stressor-provoked corticosterone and behaviors that reflected anxiety and depression. These actions were not evident in vagotomized mice, pointing to the contribution of the vagus nerve in mediating the behavioral actions associated with microbial changes (Bravo et al., 2011).

On the basis of such data, it was considered that manipulation of the gut bacteria through prebiotics may turn out to be an effective therapeutic approach to deal with mood disorders (Cryan & Dinan, 2012). The precise processes that govern the microbiota-behavior linkage remain to be fully worked out, although several candidates have been offered. Given that 90% of peripheral serotonin is present within the gut, it was a good bet that it might serve in communicating with the brain. It is conceivable that the kynurenine pathway may contribute to the development of anxiety (Kennedy, Cryan, Dinan & Clarke, 2017; see Fig. 9.4) just as it might play a role in the development of depression. As well, epigenetic changes might contribute to the microbial dysbiosis thereby influencing mood states (Stilling, Dinan, & Cryan, 2014). This is a reasonable view, but supportive data are still scant.

CONCLUSIONS

There's little question that anxiety can be more than just an uncomfortable feeling, and may become exceptionally debilitating. It is especially disconcerting that some anxiety disorders (e.g., social anxiety) begin during childhood, and carried forward to affect adult anxiety. The processes responsible for anxiety and related disorders have yet to be fully defined, although CRH, GABA/glutamate, monoamine, and cannabinoid functioning have all been implicated as possible mediators of anxiety. Considerable evidence has also supportted a role for microbiota and inflammatory immune dysregulation in the provocation of anxiety. These actions could occur owing to the effects of microbial and cytokine effects on brain hormonal, neurotransmitter, and neurotrophin changes. It is unlikely that any one of these processes acts in isolation of others in promoting and maintaining anxiety, and

although several contribute to more than a single anxiety disorder, the neurobiology of these disorders are distinguishable from one another.

The risk factors for anxiety disorders comprise genetic components together with psychosocial stressors, and specific hormonal alterations, such as the elevated afternoon cortisol levels common in social anxiety disorder. These disorders have been treated by a variety of pharmacological agents, the most common being SSRIs, benzodiazepines, β-norepinephrine blockers, as well as by psychotherapeutic treatments (e.g., CBT, exposure therapy, stress management). But, here again, the symptoms presented vary remarkably across the different anxiety disorders, and even within any disorder, marked individual differences exist with respect to the symptoms presented. This is particularly notable in the case of OCD in which some patients may exhibit very different behaviors, but share the common feature of anxiety declining when the obsessive behavior is expressed.

Even if it was accepted that inflammatory immune processes might contribute to anxiety, it is entirely uncertain why inflammatory processes would give rise to one or another form of anxiety disorder. It is likely that inflammatory factors serve to either aggravate an already existing condition, or alternatively serve as a general challenge to mental health, much as psychogenic stressors act this way, with the specific condition that emerges varying with other factors that occur concurrently or are already present. Despite these uncertainties, there is the view that microbial changes or anti-inflammatory manipulations could be used as adjunctive treatments for anxiety-related conditions. As more is learned about the specific immune or microbial factors that accompany different forms and symptoms of anxiety, the prospects for adjunctive treatments will be improved.

10

Posttraumatic Stress Disorder

IT AIN'T OVER EVEN AFTER IT'S OVER

It isn't unusual for people to mistakenly believe that once a traumatic event is over, so is the distress, and consequently individuals should recover. We even have expressions that speak to this, such as "time heals all wounds," and there are occasions when trauma survivors are told "just get over it." Time might allow some wounds to heal in some individuals. For others, time moves ever so slowly, and the trauma memories linger and so do the psychological and physical consequences of the trauma. Parents who have lost a child, individuals that experienced the horrors of warfare, or being a survivor of genocidal efforts, don't simply forget. Some might never speak about their experiences (a conspiracy of silence), whereas others can't stop speaking about it. Although the majority of individuals go on with life, the traumatic memories are often just below the surface, and their effects can reemerge.

There's little question that some of Freud's views were off the wall. Still, many of his thoughts on defense mechanisms and other processes are still considered to be useful. He had postulated, among many other things, that early experiences might mark individuals for life, although he didn't understand how this came about from a biological perspective. After all, he didn't have the luxury of knowing about the workings of neurochemical systems, and neuroplasticity in relation to memory was still far off. He did understand that traumatic memories, especially those from childhood, represented a causative agent that continued to undermine an individual's mental health, often throughout life. This could occur through the unconscious, but could also have their damaging effects through persistent negative rumination and incidents being replayed.

A PERSPECTIVE ON TRAUMA-RELATED DISORDERS

For some years, posttraumatic stress disorder (PTSD) has been a top of mind issue among those working in mental health fields, and thanks to media attention, public awareness increased, and thus the stigma of PTSD has fallen away. Still, some affected individuals continue to suffer in silence or attempt to diminish their distress through self-medication (e.g., use of alcohol). With greater understanding of PTSD came the realization that the disorder was common among people who encountered a wide range of traumatic events, and it has been estimated that this disorder is present in about 10% of the US population.

Within the DSM-5, two separate, but related syndromes are described that have been linked to severe or chronic stressor experiences. These comprise acute stress disorder (ASD) and PTSD, which differ from one another in several respect, but still share several features.

ACUTE STRESS DISORDER

Soon after a catastrophic event, intense emotional reactions may appear, typically diminishing over time, although the emotional upheaval can persist or even become more pronounced. A diagnosis of ASD is applied if symptoms appear within 1 month of a trauma experience and persist for at least 3 days. In addition to severe anxiety, a diagnosis of ASD also requires the presence of three or more peritraumatic (at the time of the trauma) "dissociative" symptoms. These may comprise (1) a sense of numbing, detachment, or absence of emotional responses, (2) diminished awareness of surroundings (feeling as if in a daze), (3) derealization, in which perceptions or experiences of the external world seem unreal, (4) depersonalization, wherein an individual has the feeling of watching themselves act, but feel that they have no control, and (5) dissociative amnesia, which is characterized by memory gaps, such that individuals are unable to recall information concerning events of a traumatic or stressful nature.

Aside from these symptoms, ASD may also be accompanied by "re-experiencing" the traumatic event by way of either recurrent images, thoughts, dreams, illusions, and flashbacks. Individuals may also feel as if they're reliving the traumatic experience. As these features can

be promoted by reminder stimuli, individuals often exhibit strong avoidance responses. Features of anxiety and arousal are typically present, including hypervigilance, hyperreactivity, irritability, impaired concentration, disturbed sleep, and motor restlessness.

There had been the belief that ASD symptoms might be predictive of the subsequent development of PTSD, but this turned out not to be the case. Although PTSD sometimes followed ASD, most people who developed PTSD did so without having displayed ASD (Bryant, Creamer, O'Donnell, Silove, & McFarlane, 2012). Thus, there is some question as to the usefulness of ASD as an independent category of illness. Although dissociative symptoms weren't useful in predicting later PTSD, these characteristics had predictive value in other contexts. A 15-year prospective study indicated that among patients who experienced a heart attack, the presence of in-hospital depression-dissociative symptoms predicted all-cause mortality (Ginzburg et al., 2016).

POSTTRAUMATIC STRESS DISORDER

A wide range of stressful events can promote PTSD, including unique or unusual experiences (being held hostage), but also traumatic experiences that are common (car accident, medical complications, being told about a severe medical condition, assault, rape, and repeated exposure to images of traumatic events). In addition, chronic stressors, such as bullying and racial discrimination, can instill PTSD symptoms, as can distal stressors (i.e., when individuals were not directly confronted with the trauma) as seen among US residents following the 9/11 terrorist attacks.

A diagnosis of PTSD, based on the DSM-5, is considered within the context of five broad categories. (1) Exposure to actual or threatened death, serious injury, sexual violation, or witnessing others experience these traumatic events; (2) presence of one or more intrusive symptoms (e.g., involuntary distressing memories of the traumatic event(s), or intense psychological distress or physiological reactions triggered by reminders of the trauma); (3) persistent avoidance of trauma associated stimuli; (5) disturbed cognitions and mood (e.g., impaired recall of an important aspect of the event, persistent and distorted self- or other blame regarding the cause or consequences of the traumatic event(s) (feelings of detachment or estrangement from others); (5) marked arousal and reactivity associated with the traumatic event(s), reflected as hypervigilance, hyperarousal, or exaggerated startle response.

PTSD Vulnerability and Resilience

At some time in their life, most individuals (> 60%) encounter stressors that are sufficiently chronic or traumatic to produce PTSD. Yet, only a modest proportion of these individuals actually develop this disorder. Numerous risk factors have been identified in relation to the development of PTSD including being female, childhood trauma, experiencing multiple previous traumatic events, preexisting mental disorders, a history of interpersonal violence, and relatively few years of education. A detailed accounting of the demographic and epidemiological variables linked to PTSD, together with the incidence in relation to varied types of trauma, the biological factors associated with PTSD, as well as its treatments were provided by Shalev, Liberzon, and Marmar (2017).

Fundamental to the impact of stressors is how they are appraised (i.e., perceived level of threat) and what sorts of responses emerge following a trauma experience. Peritraumatic experiences might contribute to the emergence of PTSD, as can a constellation of psychosocial factors, including the individual's coping styles and strategies, and perceived social support. Using emotional-focused coping, particularly reliance upon emotional avoidance and adopting limited emotional expression, predicted elevated PTSD symptoms. Thus, it was maintained that in order to diminish PTSD symptoms, individuals ought to separate their emotional responses from the cognitive representation of the event (cognitive-emotional distinctiveness). As in so many other stress-related conditions, the emergence and continuation of PTSD symptoms were particularly notable among those with low social support, diminished adaptive cognitive coping methods (i.e., worry, self-punishment) and the inappropriate use of avoidant coping strategies (e.g., social and nonsocial avoidance coping). In the latter regard, in some instances avoidant strategies can be useful, as this might allow for compartmentalization of the stressor experience, thereby limiting memories of the trauma, as well as the cues that encourage these memories. Thus, treatments aimed at diminishing maladaptive coping strategies (worry, self-punishment, and social avoidance), and encouraging support and understanding from a person's social network, might limit PTSD symptoms.

Pretrauma Experiences

Stressful experiences, as we've seen, may result in the sensitization of neurobiological systems so that they are more responsive to later challenge, essentially creating a disposition toward PTSD. Aside from greater reactivity of neuronal processes, earlier trauma experiences, including abuse or neglect in childhood, can have lasting effects on the individual's ability to cope with subsequent stressors, thereby influencing the development of PTSD in response to further challenges. In fact, one of the most prominent predictors of PTSD is having a history of multiple traumas. It similarly appeared that having mental health problems prior to the trauma determined whether or not PTSD would emerge. Individuals who score highly on a neuroticism scale (characterized by constant worrying, chronic anxiety, and over reactivity to daily negative experiences) were most likely to develop PTSD. Not surprisingly, childhood trauma, interpersonal violence, severe symptoms of an anxiety/affective disorder, predicted the development of PTSD and poor remission from the disorder.

PHYSICAL ILLNESS IN RELATION TO PTSD

The progression of cancer and the response to cancer treatment can be affected by life-style factors and to stressor experiences. Conversely, cancer and cancer-related experiences can promote psychological disturbances, including PTSD and depression (e.g., Caruso et al., 2015). As such, it would be propitious for cancer treatment teams to determine an individual's psychiatric history prior to the initiation of therapy, particularly as the development of a psychiatric condition can undermine the efficacy of the cancer therapy (Cordova, Riba, & Spiegel, 2017).

In addition to being comorbid with depression and poor quality of life, PTSD is often comorbid with other pathological conditions tied to immune, HPA, or autonomic disturbances, including diabetes, heart disease, autoimmune disorders, and may be linked to premature cellular aging. Given that the primary symptoms of PTSD are relatively diverse, it is likely that many neuronal mechanisms are linked to this condition and contribute to a wide range of comorbid illnesses. Inflammatory genes are prime culprits that have been implicated as acting in this capacity (Zass, Hart, Seedat, Hemmings, & Malan-Müller, 2017).

PTSD AND STRUCTURAL BRAIN ALTERATION

Although few studies have been able to assess PTSD-related neurochemical changes in brain tissues of human participants, interesting findings have come from imaging studies that have pointed to variations of connectivity within neuronal networks, and changes of brain architecture. Many reports focused on community participants who encountered some form of trauma that led to PTSD, whereas others assessed PTSD among war veterans who had also sustained traumatic brain injuries of varying degrees. It is important to assess the mechanisms operating improperly among these individuals, but in the analysis of the mechanisms supporting PTSD, the presence of physical injuries, particularly head injury, is a confounding factor, just as the presence of comorbid illnesses makes it difficult to identify the mechanisms responsible for PTSD. Further to this, it is important to distinguish between the brain changes seen among individuals that experienced trauma, and those that experienced trauma and developed PTSD, rather than just assessing the effects of trauma and PTSD relative to no-trauma controls.

An impressive number of reports indicated that PTSD was accompanied by diminished size of the hippocampus and prefrontal cortex, and that symptom severity was closely aligned with reduced left hippocampal volume (e.g., Nelson & Tumpap, 2016). It was concluded, based on a meta-analysis across a number of different trauma-induced incidents, that PTSD involved a common neural network that included the cingulate cortex, insula, and the parietal, frontal and limbic regions. At the same time, there were also differences of network involvement in relation to the type of trauma experienced (Boccia et al., 2016). This is not altogether unexpected given that diverse trauma experiences (e.g., abuse vs accident) may have different cognitive effects, and varied stressor networks can be engaged in response to diverse stressors (as described in Chapter 4: Life-Style Factors Affecting Biological Processes and Health).

Unlike the fairly consistent hippocampal findings, the data concerning amygdala volume in PTSD have not been entirely consistent. Nevertheless, it seems that adult PTSD related to childhood maltreatment was accompanied by reduced amygdala volume, leading to the provisional suggestion that early trauma, such as sexual abuse, could disturb normal amygdala development, thus increasing vulnerability to adult PTSD (Veer et al., 2015). This does not belie the inconsistencies regarding PTSD-related amygdala volumetric differences. There is reason to suppose, however, that the variability was related to subregional differences of connectivity and gray matter volume within the amygdaloid complex. Indeed, PTSD was accompanied by lower right basolateral amygdala connectivity and volume, but elevated centro-medial amygdala connectivity, the latter being most notable with severe PTSD (Aghajani et al., 2016)[1].

Although bilateral amygdala volume was reduced relative to that seen in healthy controls, a meta-analysis indicated that amygdala volume did not differ between individuals with trauma-related PTSD and trauma-exposed individuals without PTSD. Thus, the trauma rather than PTSD itself seemed to be the key factor linked to amygdala volume (O'Doherty et al., 2015). Yet, it also appeared that relative to individuals who experienced trauma without developing PTSD, individuals with PTSD exhibited more pronounced reductions of gray matter volume within the hippocampus and amygdala, as well the anterior cingulate cortex, frontal medial cortex, middle frontal gyrus, superior frontal gyrus, paracingulate gyrus, and precuneus cortex (O'Doherty et al., 2017). Moreover, with treatment that diminished PTSD symptoms, left amygdala volume increased, whereas volumetric changes were not apparent within the hippocampus (Laugharne et al., 2016).

White matter tracts have also been shown to vary with trauma and PTSD. Regardless of whether PTSD was present, participants who experienced trauma exhibited disturbances of white matter integrity involving the uncinate fasciculus, cingulate gyrus, and corpus callosum, relative to nontraumatized individuals. Moreover, PTSD severity was inversely correlated with variations in several of these tracts, suggesting that trauma was linked to compromised white matter integrity in pathways linking the amygdala to frontal brain regions, such as the anterior cingulate cortex (O'Doherty et al., 2018). As indicated earlier, disturbances of this pathway permit the emergence (disinhibition) of anxiety and fear, and thus might underlie the anxiety and fear that form part of the PTSD profile. In fact, greater amygdala neuronal responsivity was present among individuals with PTSD, and amygdala reactivity 1 month after the trauma predicted the presence of PTSD 12 months later (Stevens et al., 2017).

---

[1] Individuals who experienced trauma might exhibit heightened fear conditioning, impaired fear extinction, suppressed cortisol functioning, and altered amygdala volume, much as these features are seen in PTSD. However, it may be important to distinguish the characteristics stemming from the trauma experience versus those that are more closely aligned with the actual emergence of PTSD.

Based on morphological changes that were related to particular symptoms among individuals with relatively severe signs of PTSD, it was maintained that hippocampal abnormalities were most closely aligned with arousal symptoms, whereas the amygdala abnormality was most closely linked to reexperiencing symptoms (Akiki et al., 2017). Other symptoms seemed to involve diminished cortical thickness that was apparent among veterans with PTSD, as well as among those who scored highly on the Combat Exposure Scale even in the absence of frank PTSD (Wrocklage et al., 2017). Bilaterally, anterior cingulate cortex volume reductions among individuals with PTSD, were linked to attentional disturbances and impairments of emotional regulation (O'Doherty et al., 2015). It may be of particular significance that when PTSD was accompanied by depressive illness, which is fairly common, symptom severity was associated with reduced gray matter volume in the left dorsomedial prefrontal cortex, and shape variations within aspects of the amygdala seemed to be related to both emotional reactivity and emotional components of memory (Knight, Naaz, Stoica, Depue, & Alzheimer's Disease Neuroimaging Initiative, 2017).

Functionally, PTSD was accompanied by hyperactivity within the mid- and dorsal anterior cingulate cortex and the amygdala, whereas hypoactivity was present within the ventromedial prefrontal cortex, which may have been linked to the elevated amygdala activity (Hayes, Hayes, & Mikedis, 2012). As expected, with successful psychotherapy to deal with PTSD, these activity changes were largely normalized (Malejko, Abler, Plener, & Straub, 2017). However, those patients that exhibited particularly high levels of amygdala activity prior to treatment were less likely to exhibit positive effects of treatment, suggesting that they may have been most affected by the trauma and hence less amenable to treatment.

Worsening of PTSD severity and disturbed neuropsychological performance were related to lower medial prefrontal cortex and rostral anterior cingulate cortex activity, and disturbed connectivity between these regions and the inferior frontal gyrus that were detected during performance of a cognitive interference task. It was posited that PTSD may stem from disturbed regulation of medial prefrontal cortical regions that are part of the default mode network (Clausen et al., 2017). Further to this, among individuals with PTSD, resting-state functional connectivity was greater between the salience network and the default mode network, as well as the dorsal attention network and the ventral attention network, which may have accounted for disturbed disengagement seen in a test of spatial attention (Russman Block et al., 2017).

Vulnerability to PTSD and to the structural brain changes was also linked to genetic influences. For instance, altered amygdala functioning in individuals with PTSD were tied to the presence of a specific polymorphism on genes within the dorsolateral prefrontal cortex (Bharadwaj et al., 2018). Likewise, PTSD symptom severity and the most pronounced reductions of left hippocampal volume were particularly evident among individuals who displayed a polymorphism for the gene expressing COMT (the enzyme involved in norepinephrine degradation) (Hayes et al., 2017). These findings mapped on to a report indicating that a polymorphism related to the gene coding for COMT also interacted with trauma experiences in determining increased risk for PTSD symptoms (van Rooij et al., 2016). As well, epigenetic changes related to glucocorticoid receptors within the hippocampus were also related to PTSD (McNerney et al., 2018). These are only a few of the many genes and epigenetic changes that have been detected in relation to PTSD. In ensuing sections, as we discuss specific mechanisms that may underlie PTSD, we will introduce still other genes that have been linked to the disorder. As interesting as such findings are, analyses are necessary linking these to specific behavioral characteristics associated with trauma and PTSD related to trauma, and in the efficacy of various treatments.

IMAGING THE CHILD'S BRAIN

With brain connections not yet mature, children may be particularly vulnerable to long-lasting changes of brain structure and functioning, thereby influencing later vulnerability to PTSD. Functional brain abnormalities were apparent among children and adolescents with PTSD, but these were not fully congruent with those seen in adults (Milani, Hoffmann, Fossaluza, Jackowski, & Mello, 2017). A meta-analysis of cross sectional reports suggested that maltreatment may not engender immediate hippocampal volume changes, but these may emerge over the courses of development, appearing fully in adulthood (Woon & Hedges, 2008). A review of studies that assessed white matter volume in relation to pediatric PTSD and the presence of PTSD stemming from pediatric trauma, indicated that clusters of white matter alterations were present in these conditions. Although differences were observed across studies, there was congruity regarding changes within the cingulum (Daniels, Lamke, Gaebler, Walter, & Scheel, 2013).

The damaging effects of trauma and PTSD in children went well beyond structural changes. For instance, among children that had experienced an earthquake in China in 2008, the map of neural connections within the brain (connectome) varied between individuals who experienced PTSD and those that experienced the trauma but did not develop PTSD. This amounted to diminished local and global network efficiency stemming from disturbances or disconnections between brain regions, and neuronal networks tended to be relatively localized so that connections between diverse brain areas required more synapses or junctions for messages to traverse between networks (Lei et al., 2015).

THEORETICAL PERSPECTIVES ON PTSD

PTSD as a Failure of Recovery Systems

Ideally, when stressors are imminent or present, neuronal activity should be elevated, and then decline when the stressor is no longer present. However, owing to the sensitization of neurochemical systems, altered brain structure and functioning related to anxiety, fear, and cognitive alterations may persist, and the opportunity for rest and safety is diminished. Indeed, even cues that were related to a traumatic experience were effective in modifying amygdala functioning (Liberzon & Abelson, 2016).

Some of the brain alterations associated with PTSD normalized with treatment, but this was not apparent across all areas. More importantly, as we know from studies in animals, even when "extinction" has occurred, so that anxiety and fear are not apparent, this doesn't mean that things are back to normal. Instead, the behaviors, cognitions, and emotions that reflect psychopathology are kept in abeyance, but can readily reemerge given the presence of potential threats.

PTSD as a Reflection of Failure to Distinguish Danger (Threat Detection) From Safety Cues

Related to the view that PTSD reflects an inability to turn off danger (e.g., fear) cues, it could be argued that this condition reflects the inability to distinguish between safety and danger cues. Contextual memories, which comprise those associated with specific experiences, are essential for the proper responses to cues that signal danger. It turns out that functioning of the hippocampus is modulated by inhibitory projections from the entorhinal cortex, which could play a significant role in mice being able to distinguish between danger and safety (Basu et al., 2016). A similar model related to context processing has been offered in which it was suggested that dysregulation related to the hippocampal-prefrontal-thalamic circuitry results in impairments of fear learning, salience, threat detection, emotional and executive regulation, and the application of appropriate meaning to various situations (Liberzon & Abelson, 2016; Shalev et al., 2017).

One of the most fundamental components that determine stress responses is being able to appraise stressors appropriately. Often, appraisals depend upon the context in which an event occurs. Encountering danger while walking in a dark alley leads to very different responses than walking into our front yard at night. Contextual processing essentially means being able to distinguish between safe and dangerous places, which is thought to involve the medial prefrontal cortex, amygdala, and hippocampus. It turns out that these regions may contribute differently to PTSD. A subpopulation of hippocampal neurons that concurrently project to the medial prefrontal cortex and amygdala might play an organizational or coordinating role in fear learning, but in PTSD this role might be compromised and hence extinction of the fear response would occur less readily (Lang et al., 2009).

PTSD as a Disturbance of Memory Processes

As interesting as reports of diminished hippocampal size in PTSD might be, especially in relation to memory processes, they typically don't provide any indication of whether these outcomes are causally linked, the direction of this action, or even if both outcomes are related to some other common factor. Thus, it was of particular significance that reduced hippocampal size was not only present in a person with combat-related PTSD, but also in their co-twin who had not been traumatized or suffered PTSD (Pitman et al., 2006). In effect, having a relatively small hippocampus might increase vulnerability to PTSD in response to severe stressor experiences, rather than stemming from the trauma or from the illness. At the same time, reduced hippocampal volume was less common among individuals that had shown symptom remission relative to that evident among those with continued PTSD, implicating hippocampal volume in sustaining PTSD symptoms (Apfel, Ross et al., 2011).

Given that neuroplasticity is as pronounced as it is within the hippocampus, it has been a prime suspect in the development of the adverse effects of trauma, and points to memory processes being involved in PTSD. Having been traumatized by a particular event, it would be reasonable to expect that strong behavioral and neurobiological reactions would emerge in response to stimuli relevant to the trauma. While these reactions would be highly adaptive to maintain safety and readiness to respond to similar threats, it would be maladaptive if the hyper-responsivity generalized to other situations that bore limited resemblance to the initial trauma event.

Ordinarily, when a memory is well entrenched, strong responses are elicited by cues that are reminiscent of the primary aversive stimulus, and moderate generalization should be expected. Among those with PTSD, strong reactions are elicited by cues highly reminiscent of the original trauma, pointing to a well-entrenched memory. However, stress responses also occur in response to vague cues, as would be evident if the memory was not well entrenched or entirely accurate. The fact that memory processes may be activated even when cues are distinct from a previously encountered traumatic experience, may occur because these cues engage the same neural circuits that had been activated by a previously encountered stressor. From this perspective, PTSD might not stem from disturbed memory, but instead reflects sensitized biological responses to stressors. However, as we saw earlier, these findings can also be taken to imply that PTSD is not a disorder of strong memories being instigated, but instead reflects a disorder in which the normal extinction of fear responses does not occur readily.

Memory Reconsolidation

When memories are first being established and are present in short-term storage, they are labile and easily disturbed or altered. Once these memories are consolidated and in long-term storage, they are more resistant to being altered. However, upon the events being recalled, the memories are, in a sense, back in a labile state, and once again they can readily be altered. Accordingly, when animals are provided with reminders of a previous stressor experience, thereby bringing the memories back to short-term storage, treatment with a protein synthesis inhibitor can alter or disturb the memory (Nader, Schafe, & LeDoux, 2000). It was supposed that by taking advantage of the memory lability when it is being recalled, it might be possible to eliminate memories relevant to a PTSD-like state.

As we'll see later, norepinephrine and cortisol have been implicated as playing an integral role in memory processes and in the reconsolidation of memories. In fact, PTSD symptoms could be diminished by treatment with a β-blocker administered soon after a trauma or after exposure to trauma-related cues (Giustino & Maren, 2015). This would occur because disruption of norepinephrine signaling interferes with pathologic emotional memories, but without affecting declarative memory. The notion that treatments of PTSD might be particularly effective when tied to reconsolidation has become a favorite strategy.

Attention Processes

Subtle impairments related to response inhibition and regulation of attention mechanisms involving the anterior cingulate cortex were implicated as risk factors for PTSD. As we've seen repeatedly, this brain region plays a pivotal role in decision-making processes, and functioning of neurons at this site is exceptionally sensitive to stressor experiences. Much like the hippocampus, the volume of this brain region is reduced among patients with PTSD, but unlike the hippocampus, data from a twin study suggested that this was not related to genetic factors, but stemmed primarily from PTSD being present (Kasai et al., 2008). It is especially instructive that neuronal activity within the anterior cingulate cortex and dorsal cingulum predicted PTSD scores over both short- and long-term follow-up analyses (Kennis, van Rooij, Reijnen, & Geuze, 2017). Moreover, the severity of PTSD symptoms prior to and following prolonged exposure treatment was associated with reduced subgenual anterior cingulate cortex and parahippocampal activation during recall of fear extinction (Helpman et al., 2016).

Sensitized Responses

Simply because an individual has seemingly dealt effectively with a traumatic experience does not mean that everything is fine, as PTSD could develop some time later. The development of PTSD could involve different neural circuits (e.g., Shalev et al., 2017), and could theoretically emerge at different times following a traumatic experience. From a psychodynamic perspective, the distress associated with trauma might provoke persistent rumination and the replaying of events, which can ultimately have a negative cumulative impact, just as chronic stressors are able to do so. As well, the traumatic experience can influence neuronal responsivity so that subsequent stressor events or reminders of negative experiences more readily trigger these responses. The sensitization of these neurons may grow with the passage of time, peaking several weeks after the initial trauma (Anisman, Hayley, & Merali, 2003), which would account for the delayed emergence of PTSD. With further stressor experiences or reminders of the initial trauma, the neuronal network associated with the emotional memories stemming from trauma could be strengthened, thereby diminishing the dissipation of emotional trauma memories.

There is high probability that the emergence of PTSD may reflect the cumulative effects of multiple trauma experiences, which can also reflect sensitization. When a rodent that had been moderately stressed, reexperiencing another stressor gives rise to enhanced memory of the earlier trauma, enhanced anxiety, threat generalization, and resistance to extinction, all of which are characteristic of PTSD. Paralleling these effects, corticosterone treatment soon after the initial trauma could promote PTSD symptoms. These findings are in line with the view that corticosterone may be a key player for the multihit effects that contribute to PTSD. While not denying a key role for glucocorticoids, it needs to be kept in mind that sensitization effects also occur with respect to NE, 5-HT, gamma-aminobutyric acid (GABA), BDNF, and cytokine activity within several brain regions, making it difficult to say which of

these (or their combinations) might be relevant to the emergence of PTSD (Fig. 10.1).

BIOCHEMICAL CORRELATES OF PTSD

Defining the specific neurochemical and hormonal mechanisms that underlie PTSD is difficult as individuals might not only differ in the specific symptoms expressed, but the neural circuitry activated may vary as a function of the specific stressor encountered as well as synaptic connections laid down based on earlier experiences. The efficacy of drug treatments in attenuating an illness may provide clues as to the mechanisms supporting the disturbance. Unfortunately, treatments of PTSD have not been impressive, achieving a success rate of about 30%, and the drugs used are so varied that they tell us little concerning the disorder's neurochemical underpinnings. The drugs assessed have included β2 norepinephrine antagonists, dopamine-β-hydroxylase inhibitors, dopamine antagonists, partial dopamine D2 receptor agonists, glucocorticoid receptor agonists, tropomyosin receptor kinase B agonists, selective serotonin reuptake inhibitors, COMT inhibitors, GABA receptor agonists, glutamate receptor inhibitors, MAO B inhibitors, N-methyl-D-aspartate (NMDA) receptor antagonists, and fatty acid amide hydrolase antagonists (e.g., Naß & Efferth, 2017). There was a good rationale for each of these agents being assessed, but in retrospect, it seems like a fishing expedition that went awry.

Consideration of biomarkers could lead to better accuracy in selecting appropriate treatments from the array of drugs that can potentially be used. However, to date, none of the available biomarkers has been sufficiently specific to serve as a diagnostic tool or one that predicted treatment efficacy (Bandelow et al., 2016, 2017). As PTSD is a biochemically and genetically heterogeneous disorder with many

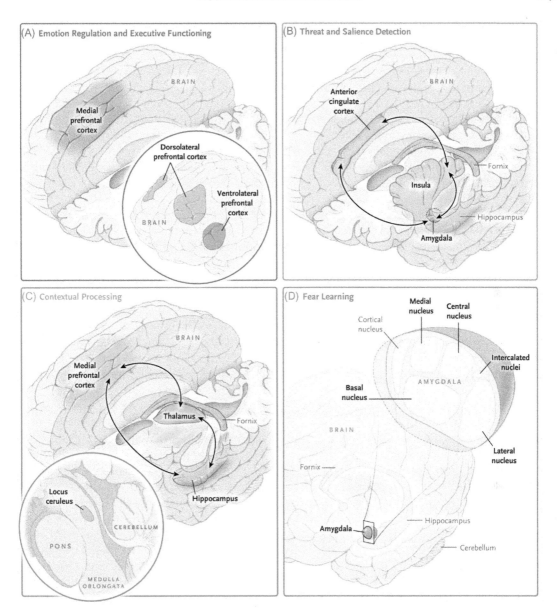

FIGURE 10.1 Brain regions implicated in the varied behavioral and emotional components of PTSD. Emotional regulation and executive functioning (A), threat and salience detection (B), contextual processing (C), and fear learning (D). Emotional regulation and executive functioning are determined by functioning of aspects of the prefrontal cortex, primarily the dorsal and ventrolateral prefrontal region, whereas the threat and salience detection are principally determined through bidirectional communication between the anterior cingulate cortex, insula, and the amygdala. Contextual processing and memory consolidation comes about with bilateral communication between the prefrontal cortex, thalamus, and the hippocampus, as well as inputs from the locus coeruleus that plays a role in vigilance. Fear expression and fear learning are largely the domain of various amygdala nuclei, each of which has a somewhat different function in relation to fear and anxiety processes. *Source: From Shalev et al. (2017).*

comorbidities, defining best treatment strategies will necessarily require some sort of culling to determine the single best set of methods from so many that are possible. As in other conditions, it is difficult to decipher whether treatments are actually affecting mechanisms responsible for the illness or are simply masking symptoms. However, even if the treatments were only masking symptoms, for an affected patient, this might be sufficient for the moment.

Once more, in an effort to determine causal connections between neuronal processes and PTSD, there has been considerable reliance on animal models of disorders, but it is questionable, as we discussed earlier, whether complex human pathologies can be accurately simulated in rodents. Some characteristics of PTSD, such as hyperarousal, can readily be assessed in animal models (e.g., by evaluating startle responses to sudden noise), but it is obviously difficult to determine whether rodents are "re-experiencing" the trauma.

Several behavioral paradigms in animals have been developed that might be useful in mimicking PTSD[2]. One of these capitalizes on natural responses to predators (e.g., rats' responses to a cat), although as mentioned earlier, there have been questions as to the suitability and reliability of such predator models. Another approach to model PTSD has entailed prolonged exposure to a stressor, followed by stressor reexposure sessions over several weeks, leading to symptoms such as an exaggerated startle response (Olson et al., 2011). Mice treated this way also exhibited diminished social interaction, exaggerated aggressiveness toward an intruder, and elevated resistance to extinction of the fear response, much as humans with PTSD are resistant to the elimination of their fear reactions. The behaviors in this model were also accompanied by disturbed corticoid functioning, and the behavioral disturbances were attenuated by D-cycloserine, which is effective in some instances of human PTSD (Yamamoto et al., 2009). Still another model involves a single prolonged exposure to a compound stressor that varies over the course of a session. This model remains to be widely used, but it does seem that symptoms in this paradigm can be reduced by treatments that are somewhat effective in human patients.

Norepinephrine

Considerable evidence from both animal and human studies has implicated disturbed norepinephrine functioning in the development and treatment of PTSD. Elevated norepinephrine activity may be accompanied by hyperarousal, nightmares, and sleep disturbances in patients with PTSD, possibly owing to stimulation of α_1-NE receptors within the prefrontal cortex, and norepinephrine activity within the amygdala might contribute to behavioral hyperarousal (Ronzoni, Del Arco, Mora, & Segovia, 2016). As well, baseline unit activity of norepinephrine locus coeruleus neurons was diminished in rodents that developed PTSD-like symptoms, but was enhanced relative to controls upon subsequent exposure to a

---

[2] Comorbid features of PTSD often include depression and general anxiety, which creates difficulties in simulating PTSD in animal models. In fact, in many studies, the behavioral paradigms used to model depression or anxiety are the same as those used to model PTSD. Thus, it is unclear whether PTSD, depression, anxiety-related features, or some combination of these syndromes was actually being measured. If this weren't sufficient to undermine some animal models, their predictive validity was questionable as the treatments used to attenuate features of PTSD (e.g., SSRIs) are also used to diminish anxiety and depression.

novel stressor, pointing to persistent potential reactivity of norepinephrine neurons within this site (George et al., 2013).

Among individuals with PTSD, peripheral norepinephrine accumulation was increased, and elevated norepinephrine receptor sensitivity was evident in some brain regions (Krystal & Neumeister, 2009). It also appeared that negative emotional stimuli, which were accompanied by intrusive memories, were linked to the interaction of peripheral norepinephrine and cortisol (Nicholson, Bryant, & Felmingham, 2014). A prospective analysis conducted among police officers (deemed to be a high-risk population) indicated that increased peripheral norepinephrine utilization predicted later development of PTSD symptoms following a critical incident (Apfel, Otte et al., 2011). Beyond these reports, positron emission tomography (PET) analyses indicated that the availability of the norepinephrine transporter within the locus coeruleus was reduced among individuals with PTSD relative to healthy controls or individuals who experienced trauma but did not develop PTSD (Pietrzak et al., 2013). Thus, norepinephrine would remain in the synaptic cleft longer and have the opportunity to cause neuronal excitation. In line with this, patients with PTSD exhibited greater indices of autonomic reactivity in response to somewhat aversive stimuli, which was accompanied by elevated locus coeruleus neuronal activity. These actions might account for the hypervigilance and exaggerated startle responses that are characteristic of PTSD (Naegeli et al., 2017).

Norepinephrine's role in the development of PTSD might comprise the induction of high levels of vigilance, together with behavioral and cognitive flexibility, or might be involved in facilitating the consolidation of fear memories. In the latter regard, stressor-provoked activation of glucocorticoid receptors (GR) and norepinephrine release within the amygdala can promote the mobilization of glutamate (AMPA) receptors, which may be involved in "firming-up" fear memories (Aubry, Serrano, & Burghardt, 2016). Furthermore, free cortisol within the prefrontal cortex and hippocampus was linked to the immediate behavioral and emotional responses to stressors, whereas the release of norepinephrine in the prefrontal cortex predicted PTSD symptoms (arousal) that appeared 1 month later (Kao, Stalla, Stalla, Wotjak, & Anderzhanova, 2015).

In rodent models, manipulations that attenuated norepinephrine functioning, such as activation of inhibitory autoreceptors or blockade of postsynaptic α1-NE receptors acted against the PTSD-like symptoms (Olson et al., 2011). Moreover, in humans, the α1-antagonist prazosin reduced some of the symptoms of PTSD, such as having distressing dreams (De Berardis et al., 2015), and doxazosin similarly diminished PTSD symptoms (Rodgman et al., 2016). It has also been suggested that like α1-antagonists, treatments with an α2-agonist, such as clonidine or guanfacine, which ultimately diminished postsynaptic NE receptor activation, could reduce symptoms of PTSD (Arnsten, Raskind, Taylor, & Connor, 2015). In fact, the β-blocker propranolol diminished the retention of fear memory, possibly by altering glutamate receptors (particularly GluA1 subunits) within the lateral aspect of the amygdala (Zhou, Luo et al., 2015). The potential involvement of norepinephrine in mediating PTSD was also supported by reports that propranolol could diminish the stressor-provoked disruption of fear extinction if it was administered soon after fear conditioning (Fitzgerald, Giustino, Seemann, & Maren, 2015). The extinction of fear memories in an animal model of PTSD was not only linked to elevated norepinephrine, but was reportedly achieved through diminished dopamine functioning within the prefrontal cortex and amygdala (Lin, Tung, Lin, Huang, & Liu, 2016).

Serotonin

Stressors influence serotonin functioning in several brain regions, and may contribute to anxiety and depression. Ample data are available that provisionally suggest a role for serotonin in the development of PTSD. As in the case of depression, individuals carrying the short allele of the 5-HTT gene were at increased risk of developing PTSD (Liu et al., 2015), especially in the context of having experienced both a childhood and adult trauma (Xie et al., 2009). As depression and PTSD are often comorbid, it is not surprising that PTSD would be associated with a gene mutation related to serotonin reuptake. It is uncertain, however, whether this gene mutation is present among individuals with PTSD who do not present with comorbid depression.

Previous stressor experiences, as described earlier, influence the appearance of PTSD symptoms upon later exposure to a strong challenge, and likewise affected amygdala 5-HT_{2C} receptor presence. When the functioning of these receptors was blocked, the signs of PTSD were prevented, raising the possibility that targeting these receptors, may be a viable strategy to diminish PTSD (Baratta et al., 2016). In addition to these particular 5-HT receptors, PTSD characteristics in rodents were accompanied by elevated expression of the 5-HT_{1A} receptor at the dorsal raphe nucleus, and conversely, knocking down the 5-HTT transporter (SERT) gene within the dorsal raphe nucleus diminished the PTSD symptoms (Wu et al., 2016).

The selective serotonin reuptake inhibitors (SSRIs) are among the most common treatments used for PTSD, and as a result it has often been assumed that the positive effects observed (as limited as they are) stem from actions on serotonin processes. However, as reported in relation to depressive illnesses, it is possible that the beneficial effects could be due to downstream changes that occur (e.g., on growth factors, such as BDNF). It is also possible that SSRIs affect anxiety and depressive symptoms, making it appear as if PTSD has diminished, even though fear memory is unaltered (Lin, Tung, & Liu, 2016). Indeed, among mice exposed to a PTSD-eliciting stressor, some of the amygdala disturbances persisted even after SSRI treatment (Han, Xiao et al., 2015).

Corticotropin Releasing Hormone

As CRH activity and CRH receptors within aspects of the amygdala and prefrontal cortex are intimately linked to anxiety and fear, it isn't a great leap to suggest that CRH would also be related to PTSD. At the same time, bidirectional signaling also occurs between CRH and norepinephrine processes, and the combination of these biochemical variations may contribute to PTSD symptoms (Hendrickson & Raskind, 2016). Increasing evidence has indeed accumulated implicating this peptide's involvement in PTSD. For instance, plasma CRH levels were higher in combat veterans with PTSD than in combat veterans without these symptoms (de Kloet, Vermetten, Geuze et al., 2008a), although these peripheral changes might have little bearing on brain CRH processes.

Studies using animal models confirmed that CRH manipulations had effects as strong as those associated with SSRIs and that of NMDA receptor antagonist D-cycloserine (Philbert, Beeské, Belzung, & Griebel, 2015), which has seen increased use for PTSD. It was also reported that a CRH_1 receptor antagonist administered prior to or 30 minutes after stressor exposure (during the period when memory was being consolidated), diminished stress-related behavioral disturbances measured days later, and was also effective in altering CRH activity within regions of the hippocampus. These outcomes were accompanied by up-regulation of BDNF and pERK1/2 protein, possibly indicating their involvement in the actions of the CRH receptor manipulations (Kozlovsky, Zohar, Kaplan, & Cohen, 2012). Somewhat curiously, CRH was also tied to remote fear memories, but not recent fear memories, as they could be disrupted by a CRH_1 receptor antagonist, possibly by enhancing AMPA receptor (GluR1) signaling within the dentate gyrus (Thoeringer et al., 2012).

The contribution of CRH_2 receptors in PTSD is less clear. This said, in mice genetically engineered to exhibit CRH overexpression early in life, later trauma-induced PTSD symptoms were accompanied by CRH_2 receptor elevations, varying with sex (Toth et al., 2016). Consistent with such findings, among mice that displayed a symptom profile reminiscent of PTSD, long lasting upregulation of CRH_2 mRNA expression occurred within the bed nucleus of the stria terminalis, whereas knocking down this receptor's expression culminated in diminished susceptibility to PTSD-like characteristics (Henckens et al., 2017).

Aside from these actions, variations of CRH functioning during early life may have long-term ramifications on responses to trauma (as we saw in Chapter 7: Prenatal and Early Postnatal Influences on Health). Among mice that overexpressed forebrain CRH during early life, vulnerability to an adult PTSD-like state was elevated (Toth et al., 2016), and a strong stressor experienced during adolescence elicited behavioral signs of stress reactivity, accompanied by HPA dysregulation and elevated CRH_1 receptor expression (Li, Liu et al., 2015).

Glucocorticoids

Given the sensitivity of glucocorticoids in response to stressors, considerable attention was devoted to assessing cortisol involvement in the provocation and maintenance of PTSD.

It might have been thought that cortisol, acting in an adaptive capacity, would be particularly elevated in association with PTSD. To the contrary, there were many reports of reduced cortisol levels among individuals with PTSD (e.g., Yehuda, 2002). As well, conditions that might lead to PTSD, such as chronic stressors and early life abuse, were not uniformly associated with elevated cortisol, and in many instances, these experiences were accompanied by diminished levels, including attenuation of the early morning cortisol rise (Michaud, Matheson, Kelly, & Anisman, 2008). The diminished cortisol was also apparent among individuals who experienced early life emotional abuse, parental desertion, and particularly low levels of care (Carpenter et al., 2009).

Aside from the finding that PTSD was associated with reduced cortisol levels, among soldiers that were about to be deployed (i.e., pretrauma), blunted stressor-elicited levels of cortisol were predictive of later occurrence of PTSD provided that testosterone levels were also reduced (Josephs, Cobb, Lancaster, Lee, & Telch, 2017). Evidently, the link to predict PTSD involved more than a single hormone, which might speak to some of the inconsistent results previously observed, especially as testosterone ordinarily acts to suppress cortisol. Furthermore, although basal levels of cortisol might be important in predicting who will be most vulnerable to illness, hormone functioning might be best assessed under a modest challenge in order to understand how the system will operate later, when more severe stressors are encountered.

Simply because HPA functioning is downregulated among individuals with PTSD doesn't necessarily imply that this system is incapable of responding. As described earlier, among previously abused women who displayed diminished cortisol levels, a novel challenge or reminders of their abuse, generated particularly elevated ACTH or cortisol levels (Elzinga, Schmahl, Vermetten, van Dyck, &

Bremner, 2003; Heim & Nemeroff, 2002; Matheson & Anisman, 2012). It was surmised that the down-regulated HPA functioning among women who had previously been traumatized might have been instrumental in preventing hippocampal cell loss that would otherwise occur, but reduced corticoid functioning could be counterproductive if the individual had to deal with further challenges. Thus, in the face of meaningful stressors, relevant brain regions would be engaged, overriding the process that produces the down-regulated HPA system.

Among its many other functions, glucocorticoid stimulation of receptors in the amygdala and hippocampus may be involved in emotional memory formation, and might be instrumental in reconsolidation of such memories (Wolf, Atsak, de Quervain, Roozendaal, & Wingenfeld, 2016). In fact, studies in animals supported the view that corticoid levels soon after a trauma may predict the later development of PTSD, and that cortisol treatment could augment memory of negative or arousing events. When corticosterone is administered to rodents shortly after initial fear conditioning, cue-specific memory consolidation and hippocampal long-term potentiation may be augmented (Abrari, Rashidy-Pour, Semnanian, & Fathollahi, 2009), possibly through activation of norepinephrine within the basolateral amygdala. Moreover, inhibiting corticosterone synthesis during the formation of fear memories may exacerbate cued fear extinction memory deficits (Keller, Schreiber, Stanfield, & Knox, 2015). Such effects vary with the strain of mouse tested, possibly reflecting differences in trait anxiety that exist within these strains (Brinks, De Kloet, & Oitzl, 2009).

Norepinephrine and glucocorticoids may act collaboratively, both peripherally and within the brain, in supporting the development and maintenance of PTSD symptoms. As we saw earlier, concurrent glucocorticoid and norepinephrine activation may promote glutamate functioning, thereby affecting memory consolidation (Aubry et al., 2016). As well, repeated exposure to a predator provoked persistent sensitization of α_1 norepinephrine receptors within the basolateral amygdala through a mechanism that involved CRH_1 (Rajbhandari, Baldo, & Bakshi, 2015). The variations of HPA functioning and hypothalamic-sympathetic processes may feedback to promote several brain changes, yielding the cognitive and emotional features that are part of the PTSD syndrome.

There have been few studies in humans concerning the effects of cortisol on reconsolidation, although in patients with PTSD, treatment with a low-dose of cortisol over a 1-month period diminished the signs of traumatic memories (de Quervain & Margraf, 2008). Yet, this hormone had no effect on reconsolidation of a fear memory (Meir Drexler, Merz, Hamacher-Dang, & Wolf, 2016). In contrast to these negative findings, a whole genome analysis in postmortem tissue, revealed that PTSD was accompanied by low expression of a particular gene, SGK1, within the prefrontal cortex. This gene encodes serum and glucocorticoid-regulated kinase 1, and seems to contribute to marked behavioral disturbances indicative of a depressive-like state coupled with high levels of fear (Licznerski et al., 2015). It also contributes to the regulation of multiple stress-related processes, including neuronal excitability, hormone release and functioning, inflammatory processes, cell proliferation, and apoptosis (Lang, Strutz-Seebohm, Seebohm, & Lang, 2010), and may thus be involved in other stressor-related conditions, such as the development of hypertension, diabetic neuropathy, ischemia, and neurodegenerative diseases.

Gamma-Aminobutyric Acid

A prospective analysis over a 1-month period following deployment among military personnel revealed that the development of PTSD was accompanied by a rise of plasma GABA levels (Schür et al., 2016). Among combat veterans presenting with PTSD, the presence of benzodiazepine receptors was reduced, again implicating GABA functioning in this disorder. As sensitivity to substances that stimulate $GABA_A$ receptors were diminished, benzodiazepines and other $GABA_A$ receptor compounds became less effective among PTSD patients (Möller, Bäckström, Nyberg, Söndergaard, & Helström, 2016).

Although benzodiazepine treatment administered after the trauma did not deter the pathology from evolving, the GABA analog pregabalin effectively reduced the severity of PTSD symptoms, and augmented the effects of antidepressants (Baniasadi, Hosseini, Bordbar, Ardani, & Toroghi, 2014). The case for GABA involvement in PTSD was strengthened by the finding that treatment with GABAergic neuroactive steroids (as well as agents that enhance their synthesis) facilitated extinction of fear memories (Rasmusson et al., 2017). In humans, baclofen which acts as a $GABA_B$ agonist, was also an effective add-on agent in SSRI treatment of PTSD (Manteghi et al., 2014).

HIDDEN MEMORIES

The concept of state-dependent learning, which has been around for more than 50 years, asserts that if individuals learn particular materials in one state (e.g., after having consumed a particular drug), then later recall will be better if they return to that same state. In the same way, the memories of traumatic memories can remain buried unless individuals are in the same state as they had been shortly after the trauma (e.g., high norepinephrine or cortisol levels). It was suggested that extra-synaptic GABA receptors, which are involved in arousal processes, also contribute to memory encoding. If these receptors were activated during the initial trauma, but are not activated at a subsequent time, then the memory will be difficult to recall. When these receptors are again activated (e.g., by relevant environmental cues or internal milieu), thereby recreating the "trauma state," memory retrieval may be facilitated. Indeed, in mice treated with gaboxadol, which stimulated extra-synaptic GABA receptors, prior to being exposed to a stressor, later recall was impaired. If, however, mice were again treated with this agent prior to the test session, then recall was adequate. It might be that different memory circuits exist, one of which is difficult to access and is reserved for traumatic memories (Jovasevic et al., 2015). Thus, even though particular memories aren't readily accessible doesn't mean that they don't have pernicious effects.

Glutamate

As glutamate is involved in anxiety, it might be reasonable to expect that this excitatory transmitter would also contribute to PTSD or changes in anxiety associated with PTSD, and perhaps influence memory and consolidation processes associated with this condition (Averill et al., 2017). Supporting this view, a single prolonged stressor session, which elicits PTSD-like characteristics in rodents, reduced glutamate and creatine levels within the frontal cortex (Lim, Song, Yoo, Woo, & Choe, 2017). As variations of glutamate and glutamate receptors were associated with fear extinction, pharmacological treatments to alter glutamate might be a useful therapeutic approach.

Imaging studies indicated that among recently traumatized individuals, the ratio of glutamate to glutamine was predictive of current and subsequent development of PTSD (Harnett et al., 2017). It was similarly observed that PTSD symptoms, particularly upon reexperiencing the trauma, were accompanied by indications of hippocampal glutamate excess together with compromised neuron integrity (Rosso et al., 2017), although signs of PTSD were accompanied by reduced glutamate and its metabolites within medial prefrontal cortex (Knox, Perrine, George, Galloway, & Liberzon, 2010).

As discussed earlier, at low concentrations, glutamate stimulates neural functioning, but at high concentrations, it can be neurotoxic. Indeed, at high levels, glutamate might encourage the development of PTSD, but its action on reconsolidation processes can be very different, particularly as different types of NMDA receptors might be aligned with different types of fear memories (e.g., cue-specific vs contextual memories). Moreover, they could also act differently in varied portions of the cortex (e.g., prelimbic and infralimbic), and could even act in opposition to one another (Giustino & Maren, 2015).

It has been considered that fear memories associated with PTSD were tied to an imbalance between GABA and glutamate within the hippocampus, but not in other regions such as the prefrontal cortex (Gao et al., 2014). In this regard, the control of stress-related epigenetic and gene transcriptional responses involving hippocampal functioning may stem from distinct glutamatergic and glucocorticoid-driven processes, which are regulated by both GABAergic interneuron functioning and by

limbic inputs (Reul, 2014). The position was also advanced that neuronal activity stimulated by glutamate and norepinephrine might cooperate in affecting attention, and memory consolidation, thereby influencing the development of PTSD (Abdallah, Averill, Krystal, Southwick, & Arnsten, 2017). Similarly, interactions between glutamate and cortisol might contribute to PTSD, and interventions based on alterations of these systems interfere with stress-related memories (Reul & Nutt, 2008).

Neuropeptide Y

Data are available pointing to the possibility that neuropeptide Y (NPY) confers resilience so that PTSD is less likely to develop in reaction to trauma (Kautz, Charney, & Murrough, 2017). In rodents, the anxiolytic actions of NPY are mediated by neuronal changes within the CA1 region of hippocampus. Ordinarily, stressors inhibit NPY functioning in this region, and hippocampal NPY injection diminished stressor-elicited anxiety (Li, Liu et al., 2017). Likewise, manipulations that affect PTSD symptoms in an animal model (e.g., environmental enrichment and exposure) influenced NPY receptors, pointing to NPY-Y1 receptors within the basolateral amygdala as a possible target in the treatment of this disorder (Hendriksen et al., 2014).

A chronic variable stressor that ordinarily leads to behavioral disturbances was accompanied by diminished amygdala NPY levels (McGuire, Larke, Sallee, Herman, & Sah, 2011). However, in animals with seemingly better methods of dealing with a stressor, the levels of NPY in the hippocampus, amygdala, and BNST were elevated, and behavioral resilience in a model of PTSD was predictive of NPY levels (Cohen, Liu et al., 2012). Consistent with the involvement of NPY in resilience, anxiety levels were exacerbated among mice engineered to lack NPY receptors.

As expected, the administration of NPY limited the occurrence of later stressor-provoked PTSD-like outcomes, and could attenuate the effects of a previously administered PTSD-inducing stressor (Serova, Mulhall, & Sabban, 2017). Consistent with these findings, the anxiety and hyper-reactivity elicited by the odor of a predator were accompanied by reduced hippocampal and amygdala NPY expression, and administration of NPY directly into the brain could attenuate the behavioral disturbances that were otherwise apparent (Cohen, Liu et al., 2012). In line with the therapeutic potential of NPY, when administered soon after a traumatic stressor, it prevented the later development of PTSD features (Laukova, Alaluf, Serova, Arango, & Sabban, 2014), and when administered intranasally, an NPY agonist prevented the development of depressive-like behaviors and dysregulation of the CRF/HPA system seen in models of PTSD (Serova et al., 2017). Early intervention using intranasal NPY attenuated the CRH and behavioral changes that were otherwise apparent (Sabban, Laukova, Alaluf, Olsson, & Serova, 2015) and prevented stressor-elicited dysregulation of HPA functioning, possibly by reinstating effective negative feedback inhibition through glucocorticoid receptor alterations (Laukova et al., 2014).

Studies in humans supported the contention that NPY serves to enhance resilience and acts against the development of PTSD. Among soldiers with high levels of NPY, symptoms of PTSD were unlikely to develop following combat (Sah, Ekhator, Jefferson-Wilson, Horn, & Geracioti, 2014). Predictably, NPY levels among Special Forces soldiers were higher than in non-Special Forces soldiers, and increased NPY was elicited during the course of mock interrogation, which was negatively associated with the development of dissociative symptoms (Morgan et al., 2000). Conversely, NPY levels were directly related to symptom improvement and positive coping in

veterans with a history of PTSD (Yehuda, Brand, & Yang, 2006). In general, across numerous situations, individuals with high NPY levels tended not to develop stress-related disturbances as readily as those with low levels of this peptide (Schmeltzer, Herman, & Sah, 2016). Moreover, baseline NPY levels and that stimulated by an α2 norepinephrine autoreceptor antagonist were lower among individuals experiencing PTSD than in healthy volunteers that had not experienced trauma (Rasmusson et al., 2000).

It has been suggested that resilience in relation to PTSD can be augmented by targeting NPY and glutamate (NMDA) functioning (Horn, Charney, & Feder, 2016; Schmeltzer et al., 2016), and that NPY expression could be used as a biomarker to identify individuals at risk for this illness. How NPY comes to have the positive effects that it does is uncertain. However, an interesting proposition in this regard is that it may be tied to memory extinction, and the possibility was raised that this involved the infralimbic cortex in which neuronal activity is diminished among PTSD patients (Vollmer et al., 2016).

INFLAMMATORY PROCESSES ASSOCIATED WITH PTSD

Given the pronounced effects of stressors on immune functioning, circulating cytokines, and cytokine expression, it is predictable that PTSD, like major depressive disorder, would also be associated with immune and cytokine variations. PTSD-like symptoms in animal models were associated with increased levels of inflammatory factors both within systemic circulation and in the brain (Wilson et al., 2013). As indicated earlier, altered microglial functioning instigated by chronic or severe stressors can promote excessive production of proinflammatory and neurotoxic factors (e.g., free radicals, nitric oxide, and superoxide) that give rise to neuronal injury and cell death (de Pablos et al., 2014), which could potentially favor the development of PTSD.

The presence of PTSD was accompanied by the presence of low-grade inflammation (Spitzer et al., 2010). These signs of inflammation comprised increased circulating inflammatory markers (e.g., CRP), a reduction in regulatory T cells, expansion of activated T cells, changes in peripheral leukocyte sensitivity in response to glucocorticoids, as well as increased reactivity to antigen skin tests (Pace & Heim, 2011). Elevated inflammatory markers, such as CRP, were also detected with different types of stressors that promoted PTSD, including interpersonal violence and terrorism experiences.

Given the comorbidity between depression and PTSD, it wasn't altogether surprising that when controlling for the presence of depression the association between PTSD and CRP was no longer evident (Plantinga et al., 2013). Yet, in view of the frequency of the depression being comorbid with PTSD, it might be inadvisable to actually attempt to separate their actions. In this regard, inflammation related to CNS processes was linked to PTSD symptoms, but was most prominent when comorbid depression was present (Baker, Nievergelt, & O'Connor, 2012). Moreover, among military personnel with PTSD, the proinflammatory cytokine elevations were present even after controlling for early life stressors and adult depression (Lindqvist et al., 2017), and normalization of immune and cytokine activity occurred with the decline of PTSD symptoms associated with trauma (Morath et al., 2014). It should be added that in monozygotic twins discordant for PTSD, both CRP and Intercellular Adhesion Molecule-1 (ICAM-1) were present in low concentrations in membranes of endothelial cells and leukocytes in the twin without PTSD, whereas proinflammatory cytokines were elevated in the twin with

PTSD (Plantinga et al., 2013). Thus, these inflammatory reactions were not antecedent condition for the development of PTSD, but were specifically associated with PTSD developing.

Among patients with primary PTSD, the levels of 18 of 20 peripheral cytokines were elevated relative to age- and gender-matched healthy controls (Hoge et al., 2009), and spontaneous production of interleukin IL-1β, IL-6, and TNF-α in isolated peripheral blood mononuclear cells was greater among those with PTSD, and correlated with symptom severity (Gola et al., 2013). It similarly appeared that PTSD was accompanied by elevated levels of circulating IL-6 and its soluble receptor (Newton, Fernandez-Botran, Miller, & Burns, 2014) as well as diminished levels of anti-inflammatory cytokines (Cohen, Meir et al., 2011). Furthermore, resilience to trauma stemming from urban violence was associated with elevated levels of the anti-inflammatory IL-10 (Teche et al., 2017). Although such effects could be related to depression that may accompany PTSD, the elevated levels of proinflammatory cytokines among combat-related PTSD patients were present even after controlling for depression and early life trauma (Lindqvist et al., 2014).

Several immune-related genetic factors have been linked to PTSD in genome-wide association studies, as has dysregulation of particular gene networks related to innate immunity (Breen et al., 2015). The latter study, for instance, revealed that before and following deployment among soldiers, a set of co-regulated genes were tied to overexpression of interferon. In addition to the cytokine variations, PTSD has been associated with increased NF-κB, a transcription factor for genes involved in inflammation, as well as cell survival and differentiation. Genes encoding proteins of the NF-κB family (e.g., p65 and c-Rel) were elevated among individuals with PTSD (O'Donovan et al., 2011), and among women who had experienced childhood abuse and later presented with PTSD characteristics. Not only was NF-κB pathway activity elevated compared to controls, but this transcription factor was positively correlated with the severity of symptoms, possibly being attributable to reduced glucocorticoid sensitivity of immune cells (Pace et al., 2012). In rodents, "extreme" responders to a stressor exhibited elevated levels of hippocampal gene expression of several factors that would increase NF-kB. However, if rats received postexposure treatment with the combination of a high-dose of corticosterone and an NF-kB inhibitor, the prevalence of the extreme responders was diminished, which was accompanied by the normalization of gene expression otherwise altered by the traumatic event (Cohen, Kozlovsky, Matar, Zohar, & Kaplan, 2011).

Consistent with the linkage between PTSD and the inflammatory marker CRP, a polymorphism within the CRP gene was associated with increased PTSD symptoms, particularly in relation to individuals "being overly alert," which might reflect a precursor to PTSD-related hyperarousal (Michopoulos et al., 2015). A polymorphism on the gene coding for TNF-α was also associated with later PTSD (Bruenig et al., 2017), once again raising the possibility that inflammatory reactivity may be a precondition that increases vulnerability to PTSD. The expression of genes for other proinflammatory cytokines that have not been well studied in relation to stressors, namely IL-16 and IL-18, were also elevated in blood of individuals suffering PTSD as a result of witnessing a traumatic event that could have injured them (Zieker et al., 2007).

As basal levels of CRP among military personnel predicted postdeployment PTSD, this inflammatory factor may be a useful biomarker for vulnerability to later PTSD (Eraly et al., 2014). Likewise, elevated IL-6 and diminished levels of the regulatory cytokine TGF-β also predicted later PTSD symptoms among individuals who had experienced various orthopedic injuries (Cohen, Meir et al., 2011) pointing to their potential value as biomarkers for later illness. A substantial number of other immune factors and hormones have been linked to PTSD, many of which could potentially serve as biomarkers for this condition (Michopoulos, Vester, & Neigh, 2016), although none of these is specific to this disorder. Most have been linked to other mental conditions (e.g., depression, anxiety, and schizophrenia) as well as physical illnesses (chronic heart disease, type 2 diabetes). Still, it is possible that a constellation of biomarkers and life events (e.g., stressor history) might be useful in predicting which individuals will be most vulnerable to PTSD.

As microbiota changes are readily promoted by stressors, finding that traumatic events would also have such effects would be expected. Persistent microbiota changes might likewise be apparent, given the distress associated with this condition. Among PTSD patients, the levels of several microbiota phyla were lower than in individuals who experienced trauma, but were not diagnosed with PTSD (Hemmings et al., 2017). Although these data do not speak to a causal connection between the two, it has nevertheless been considered that microbial processes may come to affect the emergence and maintenance of PTSD (Leclercq, Forsythe, & Bienenstoc, 2016). If this were correct, then probiotic or prebiotic treatments, alone or in combination with more usual treatments, should have positive effects in alleviating PTSD symptoms. There are scant data available that are relevant to this. A review of this literature indicated that there was only a single study relevant to this, and three others that assessed traumatic brain injury. There were hints of ameliorative effects of such treatments, but in the main there was insufficient evidence to arrive at definitive conclusions (Brenner et al., 2017).

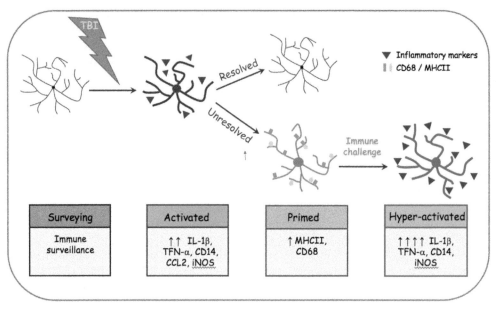

FIGURE 10.2 Traumatic Brain Injury (*TBI*) results in priming of microglia, as well as elevated reactivity to immune challenges. Activated microglia secrete cytokines and chemokines that can operate in a protective capacity, but at high levels can be destructive. Following trauma, the immediate cytokine response resolves, but within some microglia a primed or proinflammatory response persists, which is characterized by elevated MHCII and CD68 expression. With a second hit, such as a peripheral immune challenge, primed microglia become especially reactive, culminating in amplified and prolonged proinflammatory responses, causing depressive illness and disturbed memory recall. Likewise, repeated TBI can provoke exaggerated neuroinflammation, which can promote white matter reductions and cognitive decline. Aβ deposition can also occur causing accelerated plaque formation, impaired phagocytosis, and hence cognitive decline. *Source: Modified from Witcher et al. (2015).*

Anxiety and PTSD Related to Head Injury

Traumatic brain injury has frequently been associated with anxiety and with PTSD (Mallya, Sutherland, Pongracic, Mainland, & Ornstein, 2015). This could develop through any number of processes, including the possibility that traumatic brain injury may prime microglia, so that a subsequent second hit will result in neuropsychiatric and neurodegenerative conditions (Witcher, Eiferman, & Godbout, 2015). As head injury is often accompanied by elevated cytokine levels, likely of microglial origin, it might be expected that PTSD would emerge, as described in Fig. 10.2.

EPIGENETIC CONTRIBUTIONS

As we've seen, strong stressors are able to engender epigenetic changes, which can contribute to both psychological and physical pathologies. Epigenetic changes can presumably occur at any time in life, but early life events are particularly adept in creating such outcomes (Szyf, 2009), and might increase vulnerability to PTSD upon subsequent trauma experiences (Yehuda et al., 2010). In fact, the epigenetic profiles of those with PTSD who encountered adult versus childhood trauma experiences were distinguishable from one another (Mehta et al., 2013). As discussed earlier, individual differences in response to traumatic events have been linked to preexisting risks for PTSD, which may result in differential DNA methylation of genes that govern endocrine functioning.

In view of the influence of glucocorticoids on fear memory consolidation, and its potential role in the development of PTSD, epigenetic changes related to glucocorticoid receptors may be relevant in relation to PTSD (Maddox, Schafe, & Ressler, 2015). Traumatic events were, as predicted, linked to DNA methylation in promoters of glucocorticoid receptors, which were accompanied by hypoactive HPA functioning in individuals with PTSD (Labonte, Azoulay, Yerko, Turecki, & Brunet, 2014). For instance, among male survivors (but not females) of the Rwandan genocide, increased DNA methylation relevant to the binding site of the glucocorticoid receptor gene *NR3C1*, was linked to alterations of intrusive memories of the trauma and PTSD risk (Vukojevic et al., 2014). Moreover, methylation of this gene promoter predicted the efficacy of subsequent treatments for PTSD. Methylation of the *FKBP5* gene (which functions to reduce glucocorticoid binding to the receptor), in contrast, did not predict treatment response, but was associated with diminished recovery from symptoms (Yehuda et al., 2013). Of importance, epigenetic processes have been implicated in the transgenerational effects of traumatic experiences. For instance, Holocaust experiences were tied to *FKBP5* methylation, an outcome that was also apparent in the children of survivors, which might render them at increased risk for pathology (Yehuda et al., 2016).

Once more, it isn't certain which systems modified through epigenetic processes are responsible for the emergence of PTSD. Glucocorticoid receptor changes associated with early life trauma might be relevant to later pathology, as might BDNF gene methylation within the hippocampus, which has been linked to fear-related memory processes (Roth, Zoladz, Sweatt, & Diamond, 2011). In a review of this literature, the case was made that epigenetic processes relevant to hippocampal and amygdala functioning are responsible for the formation and stabilization of fear-based memories and may be tied to PTSD (Zovkic & Sweatt, 2013).

Given the many epigenetic changes that occur following stressor experiences, it comes as little surprise that among individuals with PTSD, epigenetic changes were present in genes that had downstream effects on immune functioning (Uddin et al., 2010). As well, methylation was increased in relation to several genes that have been associated with inflammation, and plasma levels of IL-2 and TNF-α, and the anti-inflammatory cytokine IL-4, were elevated in relation earlier childhood abuse and overall life stress (Smith et al., 2011). Similarly, among armed forces personnel, DNA methylation related to yet another cytokine, IL-18, was elevated among individuals who developed PTSD after deployment (Rusiecki et al., 2013).

Within peripheral blood mononuclear cells of patients with PTSD, several miRNA variations relevant to proinflammatory cytokines were detected, and Th1 and Th17 cells were elevated, whereas regulatory T cells (Tregs) were diminished (Zhou et al., 2016). It was similarly reported that increased methylation occurred in the promoters of several genes related to inflammatory processes (e.g., IFNγ and IL-12), and the expressions of these proinflammatory cytokines were regulated by miRNAs (Bam, Yang, Zhou et al., 2016). A subsequent analysis in war vets with PTSD indicated that relative to controls there were expression differences of 326 genes and 190 miRNAs measured in peripheral blood mononuclear cells, and there were indications, albeit modest, of epigenetic changes of genes linked to inflammatory processes (Bam Yang, Zumbrun et al., 2016).

As a whole, the data point to a relationship between epigenetic factors and the presence of

PTSD. The epigenetic variations associated with PTSD primarily involved glucocorticoid, neurotrophin, and inflammatory processes, but it remains to be established whether other epigenetic changes are associated with trauma, and how these play out over time in relation to PTSD. Studies in animals have made it clear that although epigenetic effects may be persistent and even be transmitted over generations, they are subject to modification with time and experience, thus the potential adverse consequences of trauma, at least with respect to the suppression of gene expression, could potentially be undone.

TREATING PTSD

Cognitive and Behavioral Therapies

Several behavioral, cognitive, and psychotherapeutic methods [e.g., cognitive behavioral therapy (CBT)], family or interpersonal therapy, and trauma management therapy have been used in an effort to treat PTSD. It generally seemed that CBT, trauma-focused psychotherapies, and prolonged exposure therapy, were most effective when administered soon after the trauma (Lee, Schnitzlein et al., 2016; Shalev et al., 2012). Some of these treatments may also be effective in children and adolescents, and are preferable to drug treatments in these populations (Gillies, Taylor, Gray, O'Brien, & D'Abrew, 2013). Only a cursory overview of these approaches is provided here, but more detailed descriptions are available in numerous reports (e.g., Cusack et al., 2016).

Cognitive Behavioral Therapy

Individualized CBT has been among the most effective of the behavioral approaches to treat PTSD (Roberts, Roberts, Jones, & Bisson, 2015). This treatment has even been effective under conditions in which threats persisted, such as ongoing terrorism. As in the case of depression, CBT isn't equally effective for all individuals, being least useful among individuals carrying the short allele for 5-HTT (Bryant et al., 2010), which could potentially be a useful marker in selecting the treatment for PTSD. Other biomarkers that could be used to predict psychotherapy outcomes included genotypes linked to serotonin and glucocorticoid functioning, glucocorticoid sensitivity and metabolism, heart rate responses that might be indicative of fear habituation, and greater baseline heart rate (e.g., Colvonen et al., 2017; O'Donovan et al., 2017). The case was also made that pretreatment activation and density of brain regions involved in appraisal, memory, emotional regulation, and decision-making (amygdala, hippocampus, as well as anterior cingulate cortex and insula) were associated with treatment response (Colvonen et al., 2017) and thus, volumetric measures of particular brain regions could be used to predict treatment efficacy.

Exposure Therapy

Prolonged exposure therapy has also seen some success in the treatment of PTSD. This approach, which follows from desensitization methods, involves the individual reexperiencing the traumatic event through remembering and engaging with it, rather than avoiding the memories. The goal is to extinguish emotional and cognitive responses to danger cues that had persisted, which had allowed individuals to continue in their excessive responses to potential danger signals. In progressive steps, the memory and cues that had elicited the powerful negative cognitive, emotional, and physiological responses might no longer engender the same degree of action. Significantly, among individuals who responded positively to prolonged exposure therapy, baseline hippocampal volume was greater than it was among treatment nonresponders (Rubin et al., 2016). Thus, based on pretreatment markers, some patients may be better suited for certain treatments.

DYNAMIC NEUROBIOLOGICAL CHANGES IN PTSD AND THE TIMING OF TREATMENTS

In addition to the diversity of neurochemical processes that could underlie PTSD, the characteristics of the disorder also evolve over time following the trauma, and it may be that dynamic variations of neurobiological processes also occur with the passage of time. The neurochemical alterations apparent soon after a trauma, presumably acting in an adaptive capacity to protect behavioral and physical integrity, may be replaced (or overwhelmed) by less beneficial neurobiological changes (e.g., receptor dysregulation or cell loss). Thus, it ought to be considered that the effectiveness of treatment strategies would also vary over the different phases of the disorder. It is conceivable that a given treatment administered during a window of opportunity soon after the traumatic experience (or during reconsolidation), such as β-blockers and corticoid manipulations, can diminish the signs of PTSD (e.g., Dębiec & LeDoux, 2006), but might be less useful at other times. Likewise, as chronically stressed mice show a transient increase of hippocampal mGlu2 receptors or epigenetic regulation of these receptors shortly after the last stressor session, it is possible that treatments that affect glutamate functioning might be useful at this time. In contrast, treatments or treatment combinations (e.g., cognitive behavioral therapy in combination with an SSRI or D-cycloserine), might be efficacious in diminishing symptoms once PTSD is established. In the same way, specific treatments might be most useful in attempts to diffuse PTSD by administering them following presentation of reminder cues (i.e., during reconsolidation). This suggestion is consistent with the view that illnesses, such as PTSD, require a multitargeted approach, but we are suggesting here that the treatment or treatment combinations effective at a particular phase of the illness might be less efficacious at a second phase in the evolution of the condition.

The nature of the stressor (familiar vs unfamiliar) and the timing of treatments relative to trauma were also linked to the behavioral and glutamate changes that occurred (Nasca et al., 2015). These data were taken to suggest that a window might also exist wherein modifications of epigenetically driven hippocampal responses can be modified and hence might be useful in treating PTSD. It likewise seemed that both prolonged exposure therapy and CBT undertaken a short time after the trauma yielded diminution of symptoms more readily than when treatment was delayed (Shalev et al., 2016).

Eye Movement Desensitization and Reprocessing

Another form of therapy, that of eye movement desensitization and reprocessing (EMDR), consists of a series of sessions during which patients are asked to focus on a vivid image of the traumatic event, while at the same time engaging in particular eye movements (e.g., tracking a finger that moves across their visual field), although stimulation of other senses have also been explored. It is thought that while the memory of the trauma is in short-term storage (e.g., participants are asked to recall the traumatic experience) EMDR allows for association to be made with nonthreatening stimuli or generally allowing for the dissociation of the traumatic memory and the emotions ordinarily elicited by these memories.

This may sound a bit flaky and might not readily be accepted as a legitimate method of treatment. Yet, there is evidence supporting its usefulness (Cusack et al., 2016), and it was effective when administered to children soon after a traumatic experience (de Roos et al., 2017)[3]. A small meta-analysis suggested that EMDR is just as effective as CBT, and it was actually superior in eliminating some symptoms (Chen, Zhang, Hu, & Liang, 2015). Furthermore, EMDR-like manipulations were associated with a reduction of neuronal activity in brain regions associated with emotional processing (Thomaes, Engelhard, Sijbrandij, Cath, & Van den Heuvel, 2016). It also appeared that the volume of the left amygdala increased following EMDR treatment, an outcome that was not seen in the hippocampus, and was not provoked by prolonged exposure therapy (Laugharne et al., 2016).

Pharmacological Therapies

Selective Serotonin Reuptake Inhibitors

Pharmacological interventions, including SSRIs and SNRIs, have been a primary treatment for PTSD, even though the success achieved by most agents has been modest (e.g., Lee, Schnitzlein et al., 2016), and the effect sizes obtained were typically relatively small (Hoskins, Pearce, Bethell, Dankova, & Barbui, 2015). Unlike the moderate effects of SSRIs in reducing PTSD symptoms that are already present, treatment with escitalopram soon after the occurrence of a traumatic event, did not prevent the development of PTSD (Shalev et al., 2012; Zohar et al., 2017). Despite the limited success achieved, in a subset of patients a reasonably good response was obtained through escitalopram treatment (Qi, Gevonden, & Shalev, 2017). It also appears

that some antidepressants may be useful for specific PTSD symptoms, such as nightmares and difficulty sleeping.

Even though SSRIs and CBT both offer some relief from symptoms, the combination of the two did not engender a better outcome than either treatment alone, although superior long-term benefits could potentially be derived from the combination therapy. Specifically, whereas psychotherapy may be effective in the treatment of PTSD, relapse frequently occurs following cessation of treatment (Lancaster, Teeters, Gros, & Back, 2016), but with combination therapy, the positive effects may be more sustained.

Norepinephrine Receptor Antagonists

Several studies assessed the impact of α- and β-adrenergic antagonists in ameliorating PTSD symptoms. As indicated earlier, treatments that modified norepinephrine activity during the early posttrauma period could prevent the development of PTSD (Giustino, Fitzgerald, & Maren, 2016), but were less effective in modifying these features once PTSD was well entrenched. Compounds that diminished norepinephrine release, such as α_2 agonist (clonidine), or blocked α_1 and β receptors (such as prazosin and propranolol, respectively) were able to reduce symptoms of PTSD (Krystal & Neumeister, 2009; Pitman, Rasmusson, Koenen, Shin, & Orr, 2012) and enhance cognitive performance (Mahabir, Ashbaugh, Saumier, & Tremblay, 2016). However, the specific symptoms abolished varied with different treatments. By example, the α_1 adrenergic antagonist prazosin reduced

---

[3] Among patients waiting in hospital after a motor vehicle accident who were provided with reminders of their experience and asked to play the computer game Tetris, exhibited diminished PTSD development. Engaging in this task, which entails high visuospatial demands, presumably disturbs the consolidation of trauma memories and hence PTSD symptoms would be prevented. Indeed, intrusive trauma memories could be precluded, although other signs of PTSD were not affected (Iyadurai et al., 2018).

trauma-related nightmares and insomnia, whereas the β-blocker propranolol was more useful in attenuating the emotional disturbances associated with traumatic memories. Accordingly, it was maintained that PTSD would be especially amenable to a treatment that affected both receptors concurrently.

As we discussed earlier, it was maintained that when memories were recalled, essentially being dredged from long-term to short-term stores, they might be more vulnerable to being changed. During these times, the traumatic memories and the emotional responses associated with them could be dissociated from one another, and then reconsolidated in an altered form (Nader, 2015; Nader et al., 2000). Thus, when a fear memory in rodents was reactivated by reminder cues, and hence present in short-term storage, norepinephrine stimulation of the amygdala enhanced reconsolidation of fear memories, thus making them resistant to extinction, possibly contributing to the persistence and severity of traumatic memories. In contrast, treatment with a norepinephrine antagonist (e.g., propranolol) during this period, or disrupting lateral amygdala norepinephrine activity, had the effect of impairing subsequent fear recall (Zhou, Luo et al., 2015). This said, there have been reports that propranolol was ineffective in limiting the development of PTSD in either animal models or in humans (e.g., Cohen, Liu et al., 2012). A meta-analysis indicated that the data generally don't support a role for propranolol in disrupting reconsolidation of memories and thereby affecting PTSD (Steenen et al., 2016). Clearly, despite some positive findings, firm conclusions regarding the effectiveness of propranolol and similar drugs in attenuating fear reconsolidation and hence PTSD will have to await further data becoming available. Nonetheless, the possibility of targeting memory and memory reconsolidation to deal with PTSD likely is a viable option (Nader, Hardt, & Lanius, 2013).

Glutamate Manipulations

Support for glutamate as an important player in PTSD comes from the finding that the partial NMDA glutamate receptor agonist D-cycloserine (Seromycin), facilitated fear extinction in animals (Ledgerwood, Richardson, & Cranney, 2005). Importantly, D-cycloserine also had promising effects in human PTSD patients, especially if it was combined with CBT or exposure (extinction) therapy. The NMDA antagonist, ketamine, which was so effective in treating recalcitrant depression, was also posited as potentially being effective in attenuating PTSD symptoms (Horn et al., 2016), possibly by acting through mTOR signaling (Girgenti, Ghosal, LoPresto, Taylor, & Duman, 2017).

Cannabinoids

In an informative review of this literature, it was indicated that in mice engineered to lack the CB_1 receptor, or following treatment with a CB_1 antagonist, impairments occurred in the promotion of cued fear conditioning, contextual fear conditioning, fear potentiated startle, consolidation of fear memories and fear extinction was facilitated (Papini, Sullivan, Hien, Shvil, & Neria, 2015), all of which have been used to model PTSD decline. Studies in humans have also indicated that cannabinoid treatments might be useful in managing PTSD, but the data have been inconsistent, possibly because of the diversity of approaches and methods used to assess the effects of the drug. It should be added that the use of marijuana and other cannabinoids in the treatment of PTSD may have several downsides. It has been associated with negative psychiatric outcomes, including psychosis and illicit substance use, and to some extent may promote anxiety and depression.

ECSTASY

There was the notion in some quarters, dating back at least four decades, that 3,4-methylene-dioxy-methamphetamine (MDMA), known on the street as "molly" or "ecstasy," could be used clinically. A damper was placed on this because MDMA was considered a Schedule I illicit drug, and some researchers believed that the drug simply wasn't ready for prime time (Parrott, 2014). Since then, there has been the impression that MDMA can be helpful in alleviating PTSD (Sessa, 2015). This may occur through effects on serotonin, norepinephrine, and dopamine, thereby influencing fear- and appraisal-related processes involving the amygdala, hippocampus, and anterior cingulate cortex (Amoroso, 2015). It is believed that MDMA treatment diminishes the distress elicited when trauma events are recalled, and the drug is effective in limiting the emotional responses related to the trauma event, entirely overriding negative cognitive appraisals. When combined with psychotherapy, it is still more useful, possibly because the ecstasy allows for better connections with the clinician. It appears that the effectiveness of MDMA is far more impressive than that of other treatments, and the FDA has declared it a "breakthrough therapy," so that it is being fast-tracked for possible clinical use.

Anti-inflammatory Agents

A line of research that provided support for inflammatory involvement in PTSD came from studies that assessed the effects of immune-acting agents in altering symptomatology. Specifically, repeated treatment with anti-inflammatory agents (ibuprofen, minocycline) reduced the anxiety symptoms in rodents as well as the expression of hippocampal IL-1β, TNF-α, and BDNF (Lee et al., 2016). The actions of these agents in rodents were particularly effective when administered soon after the trauma (Levkovitz, Fenchel, Kaplan, Zohar, & Cohen, 2015). In humans who had been treated with the antibiotic doxycycline prior to fear conditioning, diminished reactivity was evident in response to a startle stimulus presented in the presence of the conditioned stimulus (Bach, Tzovara, & Vunder, 2017). While these findings do not deal with PTSD directly, they are relevant to the role of inflammatory processes in fear conditioning that may be fundamental for this disorder.

We've said a great deal concerning biomarkers linked to PTSD, and in predicting the response to treatment. Markers reflecting glucocorticoid sensitivity and metabolism, and particular genes and gene methylation patterns associated with serotonin, glucocorticoids, and glucocorticoid receptor sensitivity, as well as inflammatory factors, predicted improved response to PTSD treatment (Daskalakis et al., 2016). Inflammation within the brain has also been implicated as a potent marker of PTSD, particularly if depression was also present (Baker et al., 2012). Particular gene expression markers might, in theory, also be used to predict the development of PTSD. Once more, however, the vast array of genes that could act in this capacity could make this difficult. Yet, sets of blood-based gene markers were found to achieve accuracy of 90% among deployed soldiers (Tylee et al., 2015).

Transcranial Magnetic Stimulation

A fair number of newer therapies have been attempted with modest success (Metcalf et al., 2016), of these transcranial magnetic stimulation that was effective in treating depression, was also somewhat effective in mitigating

PTSD symptoms (Yan et al., 2017), and was effective when administered during consolidation of fear extinction (Van't Wout et al., 2017).

COMORBID ILLNESSES LINKED TO PTSD

Comorbid illnesses, as we will discuss in Chapter 16, Comorbidities in Relation to Inflammatory Processes, are frequent between various psychiatric illness, as well as between mental illnesses and a range of physical conditions. This not only complicates analysis of the mechanisms responsible for a specific illness, but is obviously a major problem for a patient's health and treatment. In the case of drug addiction, which is a frequent comorbid condition of depression and PTSD, treatments are often not administered for the primary condition (e.g., PTSD) until the addicted individual is clean. It is of both theoretical and clinical value to determine whether addiction is secondary to PTSD in any given individual, or whether specific triggers affect several illnesses concurrently, but independently of one another. In this regard, it may also be important to determine whether comorbid conditions appear owing to a pleiotropic gene effect, such that a particular gene (or set of core genes) influences two or more phenotypes that are seemingly independent of one another.

Depression is exceptionally common among individuals suffering PTSD, as is anxiety and drug use, and PTSD is also associated with an increase in the frequency of suicide, although this may vary between civilian and noncivilian populations and may be tied to other extenuating factors (Shalev et al., 2017). It appeared that PTSD patients with and without comorbid depression could be distinguished from one another based on their ACTH levels following a challenge with the synthetic corticoid dexamethasone (de Kloet, Vermetten, Lentjes et al., 2008b). In contrast, both illnesses share several common inflammatory disturbances (e.g., elevated proinflammatory cytokines), and these cytokine alterations, or factors that modulate them (e.g., glucocorticoid functioning), might be responsible for this comorbidity. Inflammatory factors have also been associated with anxiety disorders, and could potentially underlie the comorbidity that is common between PTSD and these conditions (e.g., Furtado & Katzman, 2015).

The occurrence of PTSD is accompanied by increased vulnerability to immune-based illnesses, such as autoimmune disorders, ranging from psoriasis to rheumatoid arthritis (e.g., Gupta, Jarosz, & Gupta, 2017), as well as to inflammatory-related conditions, such as heart disease and cancer (Brudey et al., 2015). Indirectly related to these findings, whole genome analyses among soldiers returning from Iraq or Afghanistan conflicts revealed that PTSD was linked to a gene, *ANKRD55*, which has been associated with autoimmune and inflammatory disorders, such as type 2 diabetes, multiple sclerosis, rheumatoid arthritis, and celiac disease (Stein et al., 2016). Likewise, PTSD was associated with low levels of adiponectin, a hormone linked to inflammatory illnesses, such as obesity, diabetes, and heart disease, possibly through its actions on proinflammatory cytokines, particularly TNF-α (Zhang, Wang et al., 2016). In view of these many comorbities, it is no wonder that PTSD is also associated with earlier mortality (Schlenger et al., 2015).

It is not uncommon for PTSD to be comorbid with both schizophrenia and bipolar disorder. Paralleling the inflammatory profile seen in relation to PTSD, several inflammatory genes were also upregulated in association with these disorders (Pandey, Ren, Rizavi, & Zhang, 2015). There is also evidence that a single nucleotide polymorphisms coding for TNF-α was linked to schizophrenia, and polymorphisms on the IL-6 gene predicted the age of onset of bipolar disorder (Clerici et al., 2009). There have been indications that glucocorticoids play into the links between inflammatory processes and the PTSD-bipolar connection, but the available data have not been altogether uniform (see Zass et al., 2017). There is obviously reason to suspect that the inflammatory factors might underlie such comorbidity, but it is uncertain whether these reflect causal connections, pleiotropic actions, or simply that the inflammatory changes emerge with stress or general illness.

PAIN AND PTSD

Aside from the comorbidities that exist between PTSD and other mental disorders, chronic pain may be a companion of PTSD, possibly owing to excessive release of proinflammatory cytokines (Tursich et al., 2014). It may be particularly significant that among those with PTSD, painful stimulation caused by injection of capsaicin (an irritant that comes from chili peppers), was accompanied by elevated cerebrospinal fluid levels of IL-1β coupled with a delayed increase of the anti-inflammatory IL-10 (Lerman et al., 2016). These findings point to central inflammatory processes being related in some fashion to the elevated pain that may accompany PTSD.

A good deal of data is consistent with inflammatory factors being a common denominator for a variety of illnesses that have been linked to PTSD, but long-term prospective studies will be needed to determine whether the relationships are bidirectional and what factors operate along with inflammatory factors in predicting these conditions. The changes of immune and cytokine functioning associated with PTSD (and depression) may be related to norepinephrine changes which promote release of immune cell from secondary immune organs. As well, serotonin also modulates immune activity and SSRIs can, in fact, be used to modulate immune activity (Gobin, Van Steendam, Denys, & Deforce, 2014).

CONCLUSION

The development of PTSD has come to the attention of the public owing to the many wounded soldiers that have returned home from various conflicts with severe problems. However, PTSD is a much broader problem, appearing in civilian populations in great number. At one time, PTSD was considered as a condition that developed owing trauma that was "outside the norm of human experiences." This notion was abandoned with reports that PTSD was widespread in response to a variety of traumas that were clearly within the normal course of human experiences (giving birth certainly fits within this category).

Understanding the mechanisms that operate in the development of PTSD have been hampered by numerous factors, including the lack of appropriate animal models for the disorder (although this has changed in recent years), and differentiating the processes underlying PTSD from the many comorbidities that accompany this disorder. It is equally clear that although many treatment options have been offered for PTSD, including behavioral and pharmacological approaches, none have turned out to be particularly impressive, although in most instances considerable variability was apparent. Like so many other conditions, PTSD is a multidimensional disorder, and hence it's unlikely that one treatment suits all, nor that any single treatment is sufficient to deal with this condition.

It seems that PTSD is a problem linked to memory and overgeneralization, difficulty distinguishing safe from unsafe stimuli, and difficulties letting go or extinguishing fear reactions. Moreover, dynamic processes operate during the course of an illness, changing over time, and the effectiveness of treatments could potentially vary in this regard. There have been indications that some treatments administered soon after a trauma can prevent the later development of PTSD, and although there were suggestions that treatments (e.g., SSRIs, β-blockers, corticoids) administered during the 1st month following a traumatic experience, a meta-analysis concluded that this wasn't the case (Sijbrandij, Kleiboer, Bisson, Barbui, & Cuijpers, 2015).

The involvement of inflammatory factors in relation to PTSD is a relative newcomer to this field. To some extent, it's difficult to understand how immune functioning could in any way be germane to this mental condition, unless it is considered that stressors, especially traumatic events, can have actions on glial processes and cytokine release that occurs. Many peripheral immune correlates have been identified in relation to PTSD, but it is questionable which, if any, plays a causal role in the illness or which might sustain this condition once it has developed. Given the paucity of data, it is even difficult to appreciate which of the inflammatory changes could serve as markers of PTSD development and the response to treatments.

Before closing off this chapter, an additional thought should be added at the risk of sounding a bit too preachy. In Chapter 1, Multiple Pathways Linked to Mental Health and Illness, we mentioned some of the historical views that guided (or misguided) research and treatment of mental illnesses. One perspective that has been expressed concerning mental illness is that it is a normal response to an abnormal world (situation). The original statement in this regard, provided by the great theorist and Holocaust survivor Viktor Frankl in his 1959 book *Man's Search for Meaning* (initially published under the title *From Death-Camp to Existentialism*) was that "an abnormal reaction to an abnormal situation is normal behavior." Those with PTSD suffer greatly, to be sure, but given the trauma experienced, it's neither unusual nor a reflection of weakness. As the frequency of trauma related to diseases, wars, forced immigration, natural disasters, and those brought on secondary to climate change, we can probably count on many more cases of PTSD and related pathologies occurring. Failing our ability to limit traumatic events, we need to be ready to deal with the normal mental health issues that will evolve.

Pain Processes

PAIN AS AN ADAPTIVE RESPONSE?

An important element of pain responses is that they inform others of an ongoing problem and that their help might be needed, and indeed, empathy and social support can act as a pain buffer. As well, shared pain may serve as a trigger that brings people together so that they form a tighter and more cooperative group (Bastian, Jetten, & Ferris, 2014). It also seems that as people make greater use of social resources in response to painful stimulation, pain-relevant areas of the brain are activated to a lesser degree, possibly reflecting the adaptive nature of pain communication, group membership, and social support (Ferris, Jetten, Molenberghs, Bastian, & Karnadewi, 2016).

Studies in mice indicated that they actually do share their pain; when one animal watches another in distress, their own pain sensitivity increases (Smith, Hostetler, Heinricher, & Ryabinin, 2016), and to some extent this also occurs in humans (López-Solà, Koban, Krishnan, & Wager, 2017). Watching others endure trauma or pain can affect us emotionally, as most of us know, and may also affect the same brain circuits that are excited when pain is actually felt. Based on responses to conditioned stimuli it was suggested that such reactions stem from the release of opioids within the amygdala and periaqueductal gray (Haaker, Yi, Petrovic, & Olsson, 2017).

It is also assumed that pain serves in an adaptive capacity, providing feedback concerning the occurrence of potential tissue damage so that appropriate defensive actions can be taken to preclude further harm and to allow for healing to occur. Yet, one needs to wonder whether the pain that acts as a warning needed to be as intense and chronic as it sometimes is. For that matter, was intractable pain (severe, constant, and incurable) really necessary for the realization that something was amiss? Nature could have been much, much kinder.

PAIN AS THE UBIQUITOUS MALADY

Chronic illness and chronic pain are experiences that nobody wants to endure, yet too often these chronic conditions are the fate of many. There's hardly anybody who hasn't experienced different forms of acute pain, but unfortunately, in Western countries about 20%–30% of people experience chronic pain. The impact of chronic pain (defined as lasting longer than 6 months, or longer than it ought to take for normal healing to occur) on well-being is enormous, particularly as it gives rise to other illnesses, such as depression (50%–80% of individuals with chronic pain experience significant depression), and can reduce life span. Chronic pain sometimes recedes over time, but frequently it remains stable even in the absence of signs of tissue or nerve damage. The severity of pain may worsen with time (chronic progressive pain), as in the case of rheumatoid arthritis or some types of cancer, and in some instances the pain arises from unknown causes (idiopathic pain).

Classification systems have been devised in an effort to have a uniform perspective on pain. The International Association for the Study of Pain (IASP) has offered the position that pain ought to be considered on the basis of the body region affected, as well as the intensity, duration, and pattern of pain occurrence,. In addition, it is important to consider the conditions that led to the pain, and the possible mechanisms that operate to produce the pain (Merskey & Bogduk, 1994). Pain sensations come in various flavors that can be clinically meaningful, as they influence treatment and the recovery process. Neuropathic pain arises owing to damage to nerve endings, and appears as numbness along the course of an affected nerve, or by a burning or heavy sensation. This form of pain can occur because of infections, toxins, particular diseases, as well as by poor nutrition and excessive alcohol use. Nociceptive pain, which entails damage to tissues, can be detected by specialized sensory nerves. This type of pain may be felt as sharp, aching, or throbbing sensations, possibly having developed secondary to some other condition, such as the growth and spread of tumors that crowd other body organs or create blood vessels blockage. It may also appear in response to thermal or chemical stimulation (intense heat or cold, or to noxious substances that stimulate particularly sensitive body regions) or to mechanically elicited damage (cutting, crushing). Pain has also been classified as being somatic (coming from bones, muscles, and other soft tissues) or visceral pain (coming from internal organs). In both instances, painful stimuli activate nociceptors that have specialized nerve endings, which are able to detect temperature, pressure, and stretching within and around affected tissue.

A chronic pain condition that develops following a limb injury, complex regional pain syndrome (CRPS), can occur owing to peripheral or central neuronal damage. In this syndrome, a burning sensation or the feelings of pins and needles are often present, reflecting *allodynia* (elevated sensitivity to stimuli that ordinarily do not produce pain) that can spread beyond the site of injury (e.g., from a part of the hand to the entire arm). These actions seem to be mediated by brain neuroplasticity, as well as interactions with inflammatory processes, and autonomic nervous system functioning.

PSYCHOLOGICAL IMPACT OF CHRONIC PAIN

Experiencing pain day-after-day exacts an enormous toll on psychological health and encourages other comorbid physical

conditions, and addiction may develop because of opioid medications. Chronic pain forces behavioral changes as an individual's activities might revolve around accommodating their physical discomfort. Even sleeping can be difficult, and the resulting changes in alertness may increase pain sensitivity (Alexandre et al., 2017). Predictably, chronic pain can engender various psychological disorders (anxiety, depression, PTSD), which can also morph into catastrophizing that further exacerbate chronic pain (Fig. 11.1).

Chronic Pain Leading to Depression and PTSD

Not unexpectedly, in response to chronic pain, some individuals develop a pattern of negative thinking as well as diminished coping flexibility to deal with their pain and other stressors. By promoting counterproductive coping methods (e.g., rumination, wishful-thinking, social withdrawal, and drug abuse) and feelings of helplessness, chronic pain can create a milieu that fosters psychological disturbances. In some instances, social interactions necessarily need to

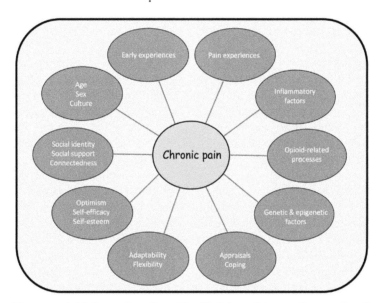

FIGURE 11.1 A wide range of risk factors have been identified that might contribute to chronic pain. These comprise innate genetic influences, specific polymorphisms, or epigenetic changes that developed as a result of earlier pain experiences or other environmental influences. Genetic factors could act additively or interactively with acquired processes in affecting pain perception. Differences in pain perception can also be related to the presence of particular neurotrophins. As well, early-life stressor or pain experiences, sensitization of neuronal processes, or memory imprints of pain, which affect multiple neurochemical systems, may alone or in combination, influence the response to painful stimuli (Burke Finn, et al., 2017). Furthermore, characteristics of the individual (age, sex), including a large constellation of cognitive changes and personality factors (optimism, neuroticism, self-esteem, and self-efficacy) can affect pain processing. As well, gender, cultural, social, and developmental factors contribute to pain perceptions, which could be related to genetic influences, but also to the marked individual differences in the ability to cope with pain. They might also be driven by the engagement of neuronal circuits involved in emotional appraisals, including circuits involving the anterior cingulate cortex (Ashar, Chang, & Wager, 2017). It's also a good bet that genetic and epigenetic factors related to inflammatory processes contribute to differences in pain perception (Wang, Stefano, & Kream, 2014). In this regard, inflammatory pain experienced early in life can also alter hippocampal-dependent memory processes, and this outcome can be worsened by other chronic stressor experiences.

be curtailed, so that individuals may find themselves with fewer social supports that might ordinarily help them cope with the pain challenge. With physical health deteriorating, the occurrence of anger, anxiety, depression, suicidal ideation, or suicide attempts increase.

The development of PTSD occurs in 10%—20% of individuals experiencing chronic pain, being most common among those with a history of depressive mood, and among individuals who believe that they will be living in pain for the rest of their lives. It is understandable that chronic pain would instigate depression, but not all individuals react this way and it seems that genetic factors might also contribute to the evolution of depression associated with chronic pain (McIntosh et al., 2016). Understanding the determinants of depression associated with differing forms of chronic pain might inform clinicians regarding the diagnosis and effective treatments of these comorbid conditions. For the moment, it appears that the sensory component of chronic pain can be dissociated from its affective components, and that the latter are determined by functioning of the anterior cingulate cortex (Barthas, Sellmeijer, Hugel, Waltisperger, & Barrot, 2015), just as this region is associated with depression brought about through other stressors. In addition, variations of neuronal activity within the mesolimbic dopamine circuit, which has been associated with reward salience, motivation, and mood state, may be accompanied by altered pain perception. As well, this circuit might contribute to the responsiveness to opioids and to the pain reducing effects of antidepressants (Mitsui & Zachariou, 2016).

Psychological and Contextual Factors Influence Pain Perception

It's been known since the seminal work of Melzack and Wall (1965) that psychological factors moderate pain perception, and pain perception may vary with the context in which it is assessed. For instance, individuals who had been in a motor vehicle accident or who experienced rape can develop a chronic pain condition that is exacerbated by further negative psychological factors. Such effects were especially marked among individuals carrying the FKBP5 gene that is fundamental in regulating HPA axis functioning (Bortsov et al., 2013), suggesting an interaction between physical and psychological factors in determining long-term pain perception. It was similarly observed that negative or positive experiences in response to an analgesic medication influenced the behavioral responses to subsequent treatment, which were accompanied by changes of neuronal activity within brain regions that encode pain and analgesia (Kessner et al., 2014). In this regard, chronic pain was also accompanied by epigenetic changes within the prefrontal cortex that could serve as a "genomic memory" to influence later responses to painful stimuli (Alvarado et al., 2015).

Given the multiple influences and mechanisms responsible for some forms of pain, a *neurological pain signature* was sought based on brain imaging procedures. Developing such a signature could be exceptionally valuable in identifying the source for allodynia, which could then be incorporated into personalized treatment strategies. As worthwhile as this might be, owing to the limits of technology, it might not always be possible to identify subtle characteristics that comprise the neural signature for certain types of pain. False negative will appear on some occasions (in which a person in pain is believed not to actually have these sensations), and thus these patients might not receive appropriate and necessary treatments, fail to obtain time off from work to recover, or might even be denied insurance benefits. The issue had become sufficiently serious to have clinicians, brain imaging experts, and those practicing neuroethics and law, to indicate that the methodologies are not foolproof, and they have discouraged their use in the diagnosis of chronic pain (Davis et al., 2017).

SEX DIFFERENCES IN PAIN PERCEPTION

It seems that females are more sensitive and less tolerant of painful stimuli (Mogil, 2016), and women are more often affected by chronic pain than are men, and tend to be less responsive to analgesic treatments (Jazin & Cahill, 2010). Multiple factors contribute to pain sensitivity, including sex hormones (Mogil, 2016) that also vary with menstrual cycle. As well, in rodents, males and females differ in the way the immune system is incorporated in dealing with pain: among males, microglia within the spinal cord are involved in pain perception to some degree, whereas T cells are more involved in this capacity among females (Sorge & Totsch, 2017). In view of the sex differences in pain perception, it is unfortunate that females had been vastly under-represented in pain research.

Placebo Responses

Placebo responses comprise positive responses to treatments (e.g., analgesics or antidepressants) that don't have direct organic effects on physiological processes. Typically, placebo effects are discussed in the context of drug responses, but they should also be considered in relation to nonpharmacologic treatments, including mechanical or electrical devices to reduce pain perception, acupuncture needles (e.g., when inserted into inappropriate locations), as well as faith healing. Placebo effects can be exceptionally powerful, to the extent that in dealing with mild or moderate levels of pain, the placebo response can often match the effectiveness of low morphine doses. Furthermore, a portion of the remedial effects of "genuine" treatments, such as those of antidepressants, is attributable to placebo effects. Although placebo outcomes are commonly discussed in relation to mood changes and pain perception, they have also been reported in relation to agents that purportedly act as muscle relaxants, or drugs that influence blood pressure. Remarkably, symptoms of neurological conditions, such as Parkinson's disease, can be abated, possibly owing to an increased dopamine release within the striatum (e.g., Lidstone et al., 2010).

From a research perspective, a placebo effect is problematic as it confounds the actions of a genuine treatment. For that matter, simply recruiting participants for a drug trial, or interactions that occur among participants in such a trial is sufficient to alter outcomes, which has considerable bearing on how data are interpreted in relation to the actual efficacy of treatments (Benedetti, Carlino, & Piedimonte, 2016). As most researchers and clinicians know, individuals frequent modify their behavior when they are aware that they are being observed (Hawthorne effect), which can influence the effects of drug treatments and placebo responses. Yet, an impressive case can be made that understanding the basis of placebo and nocebo responses (the latter referring to occasions in which ordinarily effective drugs have no positive effects) may facilitate their use in clinical practice (Belcher, Ferré, Martinez, & Colloca, 2017; Benedetti, 2014). This is particularly the case as a placebo run-in phase of a clinical trial may inform the effects of subsequent drug treatment.

As much as placebos may have an important place in treatment, their usefulness shouldn't be overstated as only a minority of patients (25%–50%) display analgesic responses, and the placebo response for antidepressants is still lower. Many factors play into the effectiveness of placebos, including optimism and altruism, previous experiences with medications, and the presence of certain genes.

The expectancy of positive outcomes, even if these are not explicit (individuals might not consciously be aware of these expectations), are especially important in creating a placebo response (Schwarz, Pfister, & Büchel, 2016). For instance, if patients experienced side effects that were provoked by the treatments, they were more likely to believe they were in the drug arm of a placebo controlled study, and thus more likely to show a positive outcome (Berna, Kirsch, Zion, Lee, & Jensen, 2017).

The entire social milieu that goes along with treatments (the presence of a doctor or a nurse, the hospital environment) may contribute to a placebo response emerging. For that matter, having a trusted doctor (or with whom the patient identified) was associated with greater pain relief (Losin, Anderson, & Wager, 2017). Just as contextual and environmental stimuli are potent in affecting the re-occurrence (reinstatement) of drug addictions as well as the reestablishment of depressive and PTSD symptoms, these same features may contribute to placebo effects (Carlino, Piedimonte, & Benedetti, 2016).

The specific environmental factors that instigate a placebo response can be subtle, so that patients may be unaware of them. For example, pain relief was greater among patients informed when a morphine drip began than in patients who were unaware of when this occurred. Simply seeing another patient obtain pain relief following treatment resulted in their own response to this agent being enhanced, supporting the suggestion that social comparisons, social interactions, and social learning can promote a placebo response (Benedetti, 2014). Oddly, pain reduction obtained through a placebo treatment for several days, persisted even after patients had been informed that they had been receiving a placebo, possibly reflecting a conditioned analgesia. Likewise, even when it was made clear to participants that they were in a placebo condition (in a test of analgesia), they responded as if they had received an active pain relieving treatment (Rosén et al., 2017). As well, patients who were aware that they were taking a placebo in conjunction with a standard treatment reported a sizable reduction of lower back pain, again pointing to the importance of the general environment and the patient-doctor connection in promoting pain relief.

Neurobiological Correlates of Placebo Responses

Given the multimodal attributes of pain relief, it was of interest to determine whether particular brain changes accompanied placebo responses with the view that this could inform the development of novel targets to reduce pain. As it turned out, placebo analgesia responses were accompanied by a rich network of brain changes, including activation of pain and stress-related sub-cortical neuronal processes, such as the connections present between the midbrain periaqueductal gray and limbic brain regions. Predictably, marked neuronal changes were also detected at brain sites that govern executive functioning, such as the anterior cingulate, prefrontal, orbitofrontal, and insular cortices (Ashar et al., 2017). Supporting the importance of emotional processes, analgesia elicited by placebo treatment was paralleled by activation of brain regions fundamental for emotional and anxiety responses as well as reward processes (hypothalamus, amygdala, and nucleus accumbens) (Atlas & Wager, 2014). It is especially significant that a clinical placebo response could be predicted on the basis of resting-state brain connectivity determined by fMRI, which can be important in relation to clinical applications. By example, in patients with chronic knee osteoarthritis, connectivity involving the right parahippocampal gyrus predicted placebo responders (95% correct) as well as the magnitude of the response (Tétreault et al., 2016).

It had been thought that pain-related placebo responses were mediated by endorphins, as the opioid antagonist, naloxone, attenuated the placebo response on postoperative dental pain. However, there may be several different types of placebo responses that can be invoked by different treatments and involve diverse mechanisms, such as cholecystokinin, dopamine, as well as endocannabinoids and prostaglandins (Carlino et al., 2016). Some of these neurochemical changes might contribute to feelings of relief or reward, whereas others might be more closely aligned with expectancies, or to the actual nature of the disturbance being treated (e.g., Denk, McMahon, & Tracey, 2014). As expected in light of the emotional components of pain induction and pain relief, hyperalgesia stemming from traumatic stressors seemed to be mediated by CRH_1 signaling within the central amygdala (Itoga et al., 2016).

NO, YOU DEFINITELY DON'T FEEL MY PAIN

Most people are able to commiserate with others who are experiencing pain, but there are instances in which the pain of others is dismissed, and may even be met with unsupportive responses. Even though pain can emerge as a result of neuronal sensitization so that benign stimulation may elicit pain (allodynia), reports of chronic, unexplainable pain, may be met with skepticism. To an extent, the same holds for psychosomatic (psychogenic) pain that is diagnosed when other causes of pain have "seemingly" been eliminated. Unfortunately, this may exacerbate poor mood in pain sufferers, and stigmatization is often encountered by individuals experiencing psychogenic pain.

For some time, chronic fatigue syndrome and myalgic encephalomyelitis (CFS/ME) had fit within this category. This illness is characterized by severe fatigue, widespread pain, depression, and a constellation of other symptoms that weren't attributable to a specific cause. The diagnosis of this condition was hampered because its etiology was uncertain, and an objective diagnostic test didn't exist. Patients with CFS were said to have their stress-response systems continuously on alert, which ultimately led to physical and psychological disturbances. Understandably, many patients became increasingly embittered; especially as they had to weave their way between attitudes (even blunt accusations) that implied that they were either malingering or that the illness was a reflection of psychological instability (McInnis Bombay, Matheson & Anisman, et al., 2015). Not infrequently, patients were stigmatized by their own GPs and some psychiatrists seemed to be reluctant to deal with patients with "undiagnosable" conditions that might be secondary to personality disorders (e.g., borderline, histrionic), and hence exceptionally difficult to treat.

Skip ahead just few years. It became clear that the occurrence of CFS symptoms were not solely due to psychological disturbances, but came about owing to inflammatory factors, and could even be an auto-immune disorder (Morris, Berk, Galecki, & Maes, 2014). A meta-analysis suggested that other than transforming growth factor-β (TGF-β), there was little evidence of other cytokines being linked to CFS (Blundell, Ray, Buckland, & White, 2015). Since then, increasing evidence indicated that a constellation of cytokines, including IL-1β, IL-6, TNF-α, IFN-α, and TGF-β contributed to the emergence CFS symptoms (Montoya et al., 2017). Indeed, systemic inflammation provoked by administration of a low dose of LPS increased pain sensitivity relative to that evident among placebo treated individuals. The elevated pain sensitivity was accompanied by diminished neuronal activity within the rostral anterior cingulate and prefrontal cortex, but elevated within the anterior insular cortex (IC), which is thought to determine affective and interoceptive pain (Karshikoff et al., 2016). As

(cont'd)

many of these inflammatory cytokines were similarly associated with autoimmune disorders (e.g., rheumatoid arthritis, systemic lupus erythematosus, and Sjögren's disease), and other illnesses in which fatigue was a prominent feature, it is conceivable that they were responsible for that aspect of the illnesses.

Chronic fatigue syndrome was also associated with inflammatory immune activation that stimulated the basal ganglia, which is involved in reward processes, cognitive functioning, and motor acts. As well, this condition was associated with bilateral white matter atrophy and disturbances within the arcuate fasciculus (fibers that connect the temporal and inferior parietal cortex to the frontal cortex), which could potentially be used as a biomarker of CFS (Zeineh et al., 2014).

Aside from cytokine variations, CFS was accompanied by limited diversity of gut bacteria, and it was possible to distinguish individuals who were affected with CFS and those who were not based on their gut bacteria composition. Several studies suggested that CFS

was triggered by gut dysbiosis, and that common gut bacteria could account for the frequent comorbidity that exists between CFS and intestinal bowel disease, although the two conditions could also be differentiated from one another on the basis of gut bacterial composition (Giloteaux et al., 2016).

So, we're now at the point that CFS is widely believed to be a genuine illness, with specific neurological and immunological markers, even if its etiology is still not well understood. A sure cure for CFS isn't available, but both CBT and graded exercise therapy have been helpful for some patients (Sharpe et al., 2015). The findings that CFS seemed to be a genuine illness didn't initially lead to a decline in the stigma that so often followed these individuals. In an effort to deal with the stigma, and to legitimize it as a physical illness, in 2015 CFS had its name changed to "systemic exertion intolerance disease." In essence, people who had been considered neurotic or had personality disorders could take solace in their physical illness being legitimized.

NEUROPHYSIOLOGICAL CHANGES THAT ACCOMPANY PAIN PERCEPTION

For several decades, the *gate control theory* was the leading perspective regarding pain processes (Melzack & Wall, 1965), which held that different types of sensory fibers largely determined pain perception. It was proposed that information is transmitted along fibers from the site of injury to the dorsal horn of the spinal cord (where sensory fibers synapse onto spinal neurons), which then send pain signals to the projection neurons, from which they are transmitted to the brain. Fibers were hypothesized to exist that could produce either fast or

slow signals (the Aβ fibers and C fibers, respectively) and nerve fibers could have either activating or inhibitory effects, as well as actions that comprised the inhibition of inhibitory effects. These varied neuronal inputs would result in the figurative opening or closing of a "gate" within the spinal cord, which would determine pain sensations. Painful stimuli influence non-myelinated C fibers, and respond to strong stimuli, resulting in deep pain that is spread over a wide area. The myelinated Aβ fibers, in contrast, are responsible for rapid, shallow pain, and they are also stimulated by touch, pressure or vibration, thereby acting to dampen painful sensations. Of particular significance, at least from the perspective of the present discussion, is that a central state comprising emotional and cognitive processes moderates the perceptions of pain.

Attributes of the central state that could affect pain perception were subsequently elaborated to include sensory-discriminative (intensity, location, quality, and duration), affective-motivational (unpleasantness), and cognitive-evaluative processes (appraisal, cultural values, context, and cognitive state) (Melzack & Casey, 1968). The incorporation of these processes and their underlying nonpain fibers, together with inhibitory neuronal processes, allowed for an accounting of pain-related phenomena (e.g., placebo effects, phantom limb pain, and the influence of distraction) that wouldn't fit comfortably into an analysis that was limited to physiological processes.

This theorizing occurred well before a good understanding was available concerning diverse neurotransmitter processes, and knowledge of different fiber types was also limited. The initial perspectives have since been updated and reformulated. Nonetheless, the Gate Control theory was pivotal to the understanding that multiple processes, particularly those of a psychological nature, could influence pain sensations. This gave credibility to the development of cognitive treatments in modifying pain perception (Moayedi & Davis, 2013).

Several more recent formulations have advanced the earlier positions, incorporating several new elements to views of pain perception, but maintaining central state as a core feature, as described in Fig. 11.2 (Grace, Hutchinson, Maier, & Watkins, 2014). Later in this chapter, we'll also see that inflammatory factors are a major contributor to this formulation.

Opioids and Opioid Receptors

The powerful analgesic effects of opioids come from their action on the endogenous peptides enkephalins, dynorphins, and endorphins stimulating three receptor subtypes, mu (μ), delta (δ), and kappa (κ). In addition to these brain receptors being triggered, opioid receptors have also been identified on immune cells, which could influence pain perception (Celik et al., 2016). As we all know, opioids are effective in alleviating many types of pain, but their addictive potential ought to have tempered their use. Sadly, this hasn't been the case, and thus continued attention has been directed toward the goal of developing better targets for pain reduction, but without the addiction potential ordinarily generated by opioid acting agents, or the uncomfortable side effects that often emerge (e.g., nausea, constipation, drowsiness, and even respiratory arrest). A relatively new treatment without the addiction potential has been able to provoke morphine-like actions within inflamed tissues while leaving healthy tissue unaffected. This amounted to treating rats with a compound (NFEPP) modeled after fentanyl, which selectively activated μ-opioid receptors at low pH (Spahn et al., 2017).

Studies in which genes involved in formation of opioid receptors were deleted, revealed that the μ receptors mediated the analgesic and addictive properties of morphine, as well as the rewarding properties of several other compounds (Matthes et al., 1996). In this regard, the actions of cannabinoids, nicotine, and alcohol, were diminished in μ receptor knock-out mice, even though each of these agents operates by stimulating different processes (e.g., Contet, Kieffer, & Befort, 2004). The μ- and δ-opioid receptors seemed to have different effects on emotional responses, as mice with deletions of δ-receptors exhibited increased levels of anxiety, and a depressive-like behavioral profile (Filliol et al., 2000). Thus, while μ receptors might be responsible for features related to pain and addiction, δ-receptors are more aligned with the regulation of emotional responses. These findings also raised the possibility that pain, emotions, and addictions,

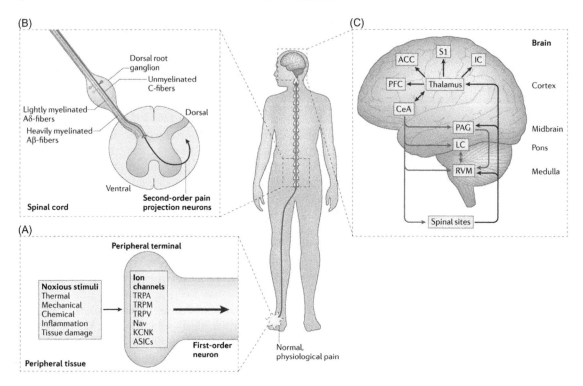

Nature Reviews | Immunology

FIGURE 11.2   Processes related to pain perception.

A. Peripheral nociceptive stimuli of various sorts will activate sensory neurons (first-order primary afferent neurons). The peripheral terminals of these neurons contain ion channels that include several "transient receptor potential" channel subtypes (TRPA, TRPM, and TRPV), sodium channel isoforms (Nav), potassium channel subtypes, (KCNK) and acid-sensing ion channels (ASICs), which can cause membrane depolarization.

B. The resulting activation of nociceptive Aβ- and C-fibers trigger glutamate release at central synapses located within the spinal dorsal horn. The axons of these nociceptive neurons project to the brainstem and thalamic nuclei through the anterolateral system. The firing rate and the specific fibers activated code for the quality, intensity and duration of the painful stimulation. Nociception may be influenced by several mechanisms, including glutamate and neuropeptides, such as substance P and calcitonin gene-related peptide (CGRP). Glutamate activates AMPA (α-amino-3-hydroxy-5-methyl-4-isoxazole proprionic acid), kainate, and NMDA receptors, which differentially contribute to different types of pain. Nociceptive signaling at the level of the spinal cord can be altered by activation of local GABA and glycine inhibitory interneurons, excitatory glutamatergic interneurons, as well as descending serotonergic and noradrenergic efferents from the brain. The second-order spinal projection neurons connect to supra-spinal sites in the brainstem and thalamus, which then project to cortical and subcortical regions, which ultimately encode the perception of the multidimensional pain experience.

C. Several brain regions may contribute to pain perception, with each having somewhat different functions. As depicted in Panel C of Fig. 11.2, neurons from the thalamus project to a variety of cortical and subcortical regions that govern sensory-discriminative [e.g., somatosensory cortex (S1)], emotional responses [anterior cingulate cortex (ACC), amygdala (CeA) and insular cortex (IC)], as well as cognitive processes [prefrontal cortex (PFC)]. As well, pain modulation occurs through brainstem sites (gray arrows), including the periaqueductal gray (PAG), locus coeruleus (LC), and rostral ventromedial medulla (RVM).

Source: Figure and somewhat modified caption from Grace et al. (2014).

could eventually be modified by manipulating activity of specific genes that influence these opioid-related processes.

Cannabinoids

Cannabinoids act on processes involving the periaqueductal gray matter and the rostral ventromedial portion of the medulla that comprise part of the descending pain control system, in which reducing the inhibition attributable to GABA, serves to reduce pain (Vanegas, Vazquez, & Tortorici, 2010). As pain perception is dependent on contextual factors and mood, it is possible that the analgesic effects of cannabinoids are also related to their actions on anxiety (Morena, Leitl et al., 2016a) or its buffering effects in relation to stressors (Morena, Patel, Bains, & Hill, 2016b). This said, it is important to emphasize that while low doses of THC have an antistress action, the opposite effects are associated with intake of higher doses (Childs, Lutz, & de Wit, 2017), and the optimal dose likely varies across individuals.

It was postulated that by affecting CB_1 receptors within the paraventricular nucleus of the hypothalamus, hippocampus, prefrontal cortex, and amygdala, the cannabinoid (eCB) system can regulate stress responses (Hill et al., 2010), thereby affecting pain perception. The psychoactive component of cannabis, THC, binds to receptors present at these sites, and upon being activated the eCB system influences the release of GABA and glutamate, thus inducing a variety of behavioral outcomes, including stress responses. Endocannabinoids may also interact with nerve growth factor and BDNF, thereby affecting pain sensitivity (Luongo, Maione, & Di Marzo, 2014). In addition, the link between amygdala and hypothalamic functioning involved in stress responses is modifiable by variations of eCB functioning. Specifically, application of a CB_1 receptor agonist into the basolateral amygdala diminished

stressor-provoked activation of the HPA axis, whereas pharmacological antagonism of CB_1 receptors increased basal HPA activity and increased the response to acute psychogenic stressors.

Despite the wide-spread belief that THC is effective in diminishing neuropathic pain, a systematic review of this literature indicated that this outcome was not uniformly evident. A positive effect of the treatment was observed in 7 of 11 studies (Lynch & Ware, 2015), but the long-term consequences of this treatment are still not fully understood (Deshpande, Mailis-Gagnon, Zoheiry, & Lakha, 2015). It seemed that only a minority of patients responded positively to inhaled cannabis (Andreae et al., 2015), and sleep quality was modestly improved, whereas mood disturbances were not attenuated (Ware et al., 2010). Moreover, the effectiveness of cannabis might not only vary with individual difference factors, but also with different types of pain and the processes presumably responsible for the perceived pain.

Although it was initially thought that its positive actions were limited to reducing neuropathic pain, cannabis can diminish adverse actions brought about by inflammation. There has indeed been a growing movement toward the use of medicinal marijuana to reduce pain associated with illnesses that involve inflammatory factors. These have included pain associated with multiple sclerosis (Fitzpatrick & Downer, 2016; American Association of Neurology, 2016), symptoms of Crohn's disease, and the pain stemming from some types of cancer. As well, cannabinoids influence brain processes that could affect brain-gut axis functioning, thereby modifying gastrointestinal symptoms, including pain (Sharkey & Wiley, 2016).

Effects like these could come about through activation of toll-like receptors expressed on sentinel cells, such as dendritic cells and macrophages, and a decline of the proinflammatory IL-17 producing T cells or the chemokine CCL-2 (also known as monocyte chemotactic protein1) (Fitzpatrick & Downer, 2016). The analgesic effects of cannabis may also occur through activation of CB_1 receptors at brain sites (e.g., amygdala, hippocampus) that govern fear and pain processing, and may be linked to neuro-immune interactions that stem from inflammatory changes (Huang, Chen, & Zhang, 2016). It ought to be said that protracted use of THC in adolescent mice has lasting effects on both proinflammatory and anti-inflammatory cytokines related to macrophage functioning and has effects on brain cytokines, notably within the hypothalamus and hippocampus, and could thus have potential effects on adult immune and behavioral responses (Moretti et al., 2015) and could also affect pain processes.

A degree of tolerance develops to the pain altering actions of CB_1 receptor stimulation, and preclinical studies suggested that treatments that target CB_2 receptors could diminish pain, but without the development of tolerance (Woodhams, Sagar, Burston, & Chapman, 2015). In fact, in the presence of inflammation, CB_2 expression within the brain was elevated and CB_2 manipulations promoted anti-inflammatory outcomes. In this regard, treatments that activate CB_2 receptors can suppress the proliferation and activation of microglia, and it is reasonable to believe that this approach could be adopted to diminish pain resulting from elevated inflammatory microglial activity (Huang, Chen et al., 2016).

Aside from potential effects on pain processes, cannabinoids can also affect symptoms of some illnesses and the adverse actions of certain treatments, including nausea related to cancer therapy, the consequences of traumatic brain injury, seizure among children who have multiple daily episodes or Drevet syndrome in which childhood seizures are largely drug-resistant. It also seems that low doses of cannabinoids can diminish cognitive disturbances that were brought about by inflammation.

Given the presumed benefits of cannabis, the move for the legalization of marijuana gained considerable traction over the last decade, although there has been concern that the alleged benefits of marijuana might have been overblown. However, a meta-analysis indicated that while there is only low quality evidence for a reduction of nausea and vomiting, there is moderate quality support attesting to cannabinoids having pain reducing qualities, and diminishing spasticity associated with some illnesses (Whiting et al., 2015).

Inflammatory Processes

There is no question that inflammatory factors are associated with chronic pain conditions, and could serve as a link to comorbid depression that is so often apparent. While peripheral inflammatory factors could affect pain processes, there is ample reason to believe that centrally acting cytokines, including both IL-1β and TNF-α released from glial cells, could promote hyperalgesia (Grace et al., 2014). As expected, intrathecal administration of an IL-1 receptor antagonist (IL-1Ra) effectively diminished pain, and the broad-spectrum tetracycline antibiotic, minocycline, reduced the expression of proinflammatory factors and concurrently diminished allodynia (Ledeboer et al., 2005).

Insults to peripheral tissue gives rise to immune processes being activated, and may promote migration of specific cells to the spinal dorsal horn, but the nature of these cells may differ with the type of insult or injury sustained. Preclinical studies indicated that the characteristics of the immune cells activated could predict changes of nociception (Grace et al., 2012). It is especially important, as well, that neurons can be activated by factors released from immune cells, and conversely, immune cell functioning can be affected by neurotransmitters and hormones. Thus, immune-related functioning could regulate pain signals, just as other centrally mediated processes can be affected in this way. There may be a significant down-side to the effects of opioids on inflammatory processes. Specifically, rats that had received a weeklong course of morphine treatment to diminish pain stemming from traumatic injury, subsequently exhibited a more sustained chronic pain condition, possibly owing to elevations of TNF-α, IL-1β, TLR4, and NLRP3 at the site of spinal cord injury (Ellis et al., 2016). In essence, the short-term gain attributable to opioids was accompanied, literally, by long-term pain.

Glial Functioning

Glial cells play a fundamental role in pain processes, operating at the site of tissue damage in the periphery and also at the dorsal horn within the spinal cord where microglia are responsive to cytokines and chemokines (Old, Clark, & Malcangio, 2015). Once peripheral neuronal damage occurs, reactive gliosis is engendered, which comprises a nonspecific reactive change of astrocytes, microglia, and oligodendrocytes. As a result of cytokine release, immune cell infiltration and astrogliosis occur together with several other alterations, including inflammatory mediators being released from reactive microglia, and activation of some toll-like receptors (TLR-2 and TLR-4) (Grace, Rolan, & Hutchinson, 2011). Aside from actions within the periphery and spinal cord, painful stimulation provokes an increase of microglia within several aspects of the brain, such as the thalamus, sensory cortex, and amygdala. Cytokines released by brain microglia contribute to gliosis, and the gliopathy that occurs in the spinal cord and peripheral nervous system, might contribute to chronic pain, including that seen in some disease conditions, such as autoimmune disorders (Mifflin & Kerr, 2017). The brain changes did not seem to simply result from general distress, as the microglial variations occurred primarily on the side contralateral to painful injury (Taylor, Mehrabani, Liu, Taylor, & Cahill, 2017). As chronic pain was accompanied by brain inflammation, it was considered that related processes could be a useful target for novel pain medications.

Peripheral damage can induce allodynia, but glial cells within the brain were likely responsible for this action (Kim, Hayashi et al., 2016) given that pain perception was modifiable by blocking the activity of glial cells, which are ordinarily activated by painful stimuli. As well, neuropathic pain stemming from spinal cord injury, which is often intractable, has been linked to inflammatory processes, and can be somewhat abated by anti-inflammatory agents (Walters, 2014). It also appeared that peripheral monocytes and microglia acted synergistically in modulating chronic pain, and chemotherapeutic agents that prevent microglia from proliferating soon after injury could reduce pain (Peng et al., 2016).

It may be particularly significant that following an immune challenge, transcriptional activity, or epigenetic modifications within microglia may be elevated so that the response to later challenges was augmented (Romero-Sandoval & Sweitzer, 2015). For instance, priming (or sensitization) of pain responses, which can be elicited by a variety of factors, such as stressor occurrences, could instigate chronic pain. Paralleling such findings, allodynia elicited by an endotoxin (LPS), could be augmented by a stressor, such as abdominal surgery, and once more this could be attenuated by minocycline (Hains et al., 2010). Like several neurotransmitters and hormones, cytokine levels and functioning is subject to a sensitization effect (Anisman, Hayley, & Merali, 2003), making it possible that these inflammatory factors are related to the priming effects that have been reported in relation to pain processes.

Given the apparent involvement of glia and cytokines in relation to pain processes, it was considered that they could potentially represent an ideal target for pain management (Old et al., 2013). Unfortunately, studies examining glial reactivity in moderating pain perception have not been encouraging, at least within animal studies (Romero-Sandoval & Sweitzer, 2015). This said, modifying glial-neurotransmitter interactions could potentially initiate beneficial changes through the altered release of anti-inflammatory factors, thereby precluding or limiting neurotoxicity and may diminish sensitized pain responses so that "normal" pain is experienced (Tiwari, Guan, & Raja, 2014). As we'll see shortly, manipulating inflammatory processes has become a focus for pain management, with signs of success being apparent in some conditions.

Summarizing, in response to tissue damage, a constellation of brain cytokine changes are apparent, including both proinflammatory and anti-inflammatory factors. If a persistent imbalance occurs in favor of proinflammatory factors, chronic pain may emerge, whereas treatments that increase anti-inflammatory factors (i.e., IL-1Ra, IL-4, and IL-10) may diminish hypersensitivity (Loram et al., 2013). The link between glial cells and cytokines in relation to pain processes certainly ought to be considered within a broader context, and there is still information that needs to be uncovered concerning glia and pain processes that are unrelated to inflammation (Graeber & Christie, 2012). Further information concerning individual difference factors would be welcome as these may be useful in directing treatment strategies. For instance, age is accompanied by longer lasting inflammation, and might thus contribute to differences of pain perception. Other factors, ranging from sex, culture, and experiences, may similarly operate in modulating pain-related inflammatory processes.

MICROBIOTA AND PAIN

Given the influence of microbiota on neuroendocrine, neurotransmitter, and inflammatory processes, it is hardly unexpected that gut processes also seem to be related to pain perception, particularly visceral pain (Mayer & Tillisch, 2011). The visceral hypersensitivity evident in germ-free mice was accompanied by elevated gene expression of cytokines and with Toll-like receptors within the spinal cord, which normalized following recolonization with microbiota from ordinary mice. It is especially significant that the visceral sensitivity was associated with a greater volume of periaqueductal gray, and diminished volume of the anterior cingulate cortex, both of which have been linked to pain perception (Luczynski et al., 2017).

Gut dysbiosis early in life can have proactive effects that influence visceral pain in adulthood (O'Mahony et al., 2014), and in female rodents, early-life stressors could have effects on adult visceral pain that varied with the estrus phase (Moloney et al., 2016). It is believed that visceral pain in such instances occurs through the effects of specific gut bacteria on TRPV receptors, together with inflammatory factors, protease release, polyunsaturated fatty acids, and short chain fatty acid (SCFA) production. As well, lipids, such as N-acylethanolamine (NAE) and SCFAs, such as butyrate, are fundamental in altering inflammatory processes, thereby influencing pain mechanisms (Russo et al., 2017). It is still a bit early to be certain, but other treatments that act against pain, such as opioid and endocannabinoid variations, could potentially involve gut microbiota changes (O'Mahony, Clarke, Dinan, & Cryan, 2017a, b; Rea, O'Mahony, Dinan, & Cryan, 2017).

Emotional Processing Affects Pain Perception

Pain processing, as described earlier (e.g., Fig. 11.2) is influenced by cognitive appraisal processes, and as we'll see shortly, the anterior insula along with the anterior cingulate cortex appear to serve as part of a *"salience network"* in the interpretation of impending pain. Expectation of pain may promote greater perceived unpleasantness of a stimulus, coupled with increased activity within the anterior cingulate cortex and other frontal regions thought to underlie appraisal and decision-making processes.

The possibility that psychological factors could moderate pain perception has led to the view that individuals can "think pain away." To some extent this is a possibility, although it should be said that the *neurological pain signature,* which is regarded as sensitive to varied pain intensities was not affected when individuals were able to think pain away. What was also observed, however, was that connections to the nucleus accumbens and other aspects of the prefrontal cortex were activated, thus supporting the view that pain perceptions can be modified by cognitive processes that are linked to neurological pain profiles (Woo, Roy, Buhle, & Wager, 2015).

Although somewhat controversial, it was suggested that physical and emotional pain (e.g., elicited by pictures of a former romantic partner) shared neural circuitry involving the anterior cingulate cortex (Lieberman & Eisenberger, 2015). When participants received a nasal spray (placebo) that would allegedly diminish their emotional pain, this was also associated with a change of their neuronal activation profile (Koban, Kross, Woo, Ruzic, & Wager, 2017). Further to this, just as physical pain influences neuronal activity within the anterior cingulate cortex, it seems that feelings of social exclusion (rejection) are accompanied by increased neuronal activity within this region (Lieberman & Eisenberger, 2015). For that matter, simply witnessing rejection experienced by another person, which presumably elicits empathy in the witness, also activates aspects of the cingulate cortex, as does viewing pictures of a romantic ex-partner and thinking about rejection by this partner (Kross, Berman, Mischel, Smith, & Wager, 2011). Such findings have been consistent with the view that the anterior cingulate cortex is part of a "neural alarm system" that is sensitive to both physical and social threats (Lieberman & Eisenberger, 2015). The fact that transcranial direct stimulation applied to the right ventrolateral prefrontal cortex reduced the pain of social rejection (Riva, Romero Lauro, Vergallito, DeWall, & Bushman, 2015) was consistent with the belief that emotional pain is modifiable by activation of neuronal processes, even if these are somewhat diffuse.

Consistent with the view that psychological and physical pain are linked, treatment with a mild anti-inflammatory (acetaminophen) that reduces physical pain, diminished the psychological pain of social rejection while concurrently reducing the neural responses within the dorsal anterior cingulate cortex and anterior insula (DeWall et al., 2010). It similarly appeared that acetaminophen could reduce empathy associated with viewing another person's pain (Mischkowski, Crocker, & Way, 2016). It's somewhat ironic that acetaminophen had these effects, given that this compound (see following text box) has limited effects in diminishing various forms of acute pain, and still less in affecting chronic pain perception. Obviously, this raises questions about the link between social rejection and physical pain, at least in relation to the involvement of the anterior cingulate cortex.

The neurophysiological concordance between physical and psychological pain could potentially have clinical implications. However, the view concerning overlapping prefrontal emotional and physical pain circuits has not gone unchallenged. Although physical and psychological pain elicited similar brain activation profiles, detailed analyses revealed that they were not identical, just as differences in connectivity were apparent in other brain regions (Woo, Roy, Buhle, & Wager, 2015). Even though the dorsal anterior cingulate cortex may be relevant to "survival-relevant goals," activation of this brain region is also apparent in response to a variety of nonstressful conditions, including attention, emotion, reward expectancy, skeletomotor, and visceromotor activity (Wager, Atlas, & Botvinick, 2016). Thus, activity within the dorsal anterior cingulate cortex should not be considered to uniquely reflect either physical or emotional pain, especially as different forms of pain might engage somewhat different neural circuits. This said, even if different challenges have distinct pain profiles, they might have overlapping features and it is possible that emotional pain can modify different physical pain sensations through actions at the anterior cingulate cortex and insula.

In addition to the anterior and dorsal cingulate, the mid-cingulate also appears to be involved in nociceptive processes. The position was taken that this region does not mediate pain sensations, but by acting through a broad cortical and subcortical network, particularly serotonergic projections to the spinal cord, the mid-cingulate may contribute to sensory hypersensitivity (Tan et al., 2017). It is interesting that the progressive changes that can occur in serotonergic functioning within this descending pathway might contribute to the transition from acute pain to a more troubling chronic pain condition.

NEXT STEPS IN PAIN MANAGEMENT METHODS: ADVANCES AND CAVEATS

The management of chronic pain hasn't been nearly as effective as one would have hoped, and disparities have been reported in the under-treatment of pain based on race and gender (Hoffman, Trawalter, Axt, & Oliver, 2016).

Drug Treatments to Diminish Pain

The efficacy of many drugs varies with the nature of the pain being experienced. For instance, pain stemming from nerve damage might be best treated by compounds that affect GABA neuronal functioning, such as pregabalin (Lyrica), whereas corticoid injections at the site of musculoskeletal injuries can be used to reduce pain stemming from inflammation. The anticonvulsant medication gabapentin also turned out to be effective in many cases of chronic neuropathic pain and chronic fatigue syndrome/fibromyalgia (Moore, Wiffen, Derry, Toelle, & Rice, 2014). Unfortunately, the drug can promote experiences like those of opioids and benzodiazepines, and has become increasingly misused.

ACUTE PAIN EXPERIENCES

The most widely used pain management treatments comprise nonopioid drugs containing acetaminophen, NSAIDs, or COX-2 inhibitors (a form of NSAID that influences inflammatory processes), even though their effectiveness is often limited. In fact, several detailed reviews of the literature indicated that acetaminophen actually had only a modest effect on acute lower back pain (Saragiotto et al., 2016) and tension-type headaches (Stephens, Derry, & Moore, 2016), only transiently reduced postoperative pain and did so in only in a third of patients (McNicol et al., 2016). To make matters worse, many people inadvertently misuse

---

(cont'd)

---

acetaminophen, sometimes leading to serious health consequences. Some agents may pose risks for cardiovascular health, and overuse may provoke gastrointestinal damage and bleeding, disturbed kidney health, (fluid retention), and may have serious effects on children in whom liver damage and death could occur (Reye's syndrome) when taken while being infected with a viral infection (e.g., influenza or chicken pox).

Inflammation produces hyperalgesia, together with poor mood. Thus, ibuprofen, a nonsteroidal anti-inflammatory drug (NSAID; Advil, Motrin), fairs relatively well in diminishing some cases of acute pain, particularly those linked to inflammation (e.g., following dental surgery). However, it only helps a minority of people in diminishing tension headaches, and is about as (in)effective as acetaminophen in alleviating acute back pain. The combination of acetaminophen and ibuprofen provided superior postoperative (dental) pain relief than either treatment alone (Derry, Derry, & Moore, 2015). Acetylsalicylic acid (aspirin) is also used to manage pain, fever, and inflammation, although its effectiveness as a pain reliever is thought to be less than that of ibuprofen. Its ability to control fever, owing to its actions on prostaglandins is well established, and it is used to reduce acute and chronic inflammation, although in the latter instance only short-term use is recommended given its potential for inducing intestinal bleeding.

Opioids

As just about everyone is aware, the most popular prescription medications for severe pain relief have come from the opioid family of drugs (e.g., morphine, diamorphin, buprenorphine, oxymorphone, oxycodone, hydromorphone, and fentanyl). However, increasing evidence has indicated opioids may only be modestly effective in the treatment of chronic noncancer pain. Coupled with the development of tolerance and opioid addiction, their use has become problematic (Volkow, Benveniste, & McLellan, 2017). Ironically, just a few days of opioid treatments can put glial cells into a heightened state of activation, provoking inflammatory responses within the spinal cord. The combination of the initial pain signal and the later opioid treatment give rise to a signaling cascade, including activation of IL-1β release, which promotes increased responsiveness of neural processes, leading to increase pain perception that persists for months. So, the very treatments meant to diminish pain intensity, may actually have the effect of increasing pain duration (Grace et al., 2016).

Because of the addiction potential, the CDC has prepared guidelines for physicians concerning opioids being prescribed for adult patients experiencing chronic pain (apart from that associated with cancer treatment, palliative, and end-of-life care) with an eye toward consideration of when the benefits of opioid use outweighs risks, and whether alternatives to opioids are a reasonable option (Dowell, Haegerich, & Chou, 2016). Unfortunately, in some countries where long delays in treatment are often encountered, patients may be kept on opioid acting agents for extended periods, thereby promoting the development of addictions. How inexcusable is it for patients to develop chronic illnesses, such as devastating addictions, while simply awaiting treatment in an overloaded medical system?

Not every person who is treated with an opioid will develop an addiction and there are certainly many variables that determine who will or will not be a victim. Personality characteristics and previous experiences with drugs have been tied to addiction, as have the drug dose used, the route of administration, how quickly the compound's rewarding effects appear, what other drugs may be on board, the context in which the drug is taken, and the expectations concerning the effects that will be obtained (Volkow et al., 2017). In the latter regard, the rewarding effects attributable to the drug as well as its addiction potential may be less notable when taken for a medical condition relative to that apparent when a drug is taken in order to obtain pleasurable feelings.

The opioid epidemic and the large number of deaths (increasingly attributable to fentanyl, which is orders of magnitude more powerful than other agents) is hardly news. A report from the National Center for Health Statistics (US) indicated that opioid use within the US doubled (some say tripled) in a 20 year period, and drug-related deaths in the US reached 64,000 (this is likely an underestimate), increasing from 52,400 in 2015. Overdoses are thus the top cause of death among younger individuals, outstripping car accidents that reached 35,100 that year. The media has done its work in bringing the problems in to public focus, but unfortunately, it may have the unintended consequences of denying opioid-based medications to those who are in absolutely great need. Obviously, it would be ideal to have a set of biomarkers that could inform pain specialists who would or would not be most as risk for developing additions, as well as in determining the best strategies to deal with addiction.

Dealing With Opioid-Based Treatments

Efforts to diminish the addiction potential associated with opioid-based pain medications have included the modification of oxycodone tablets, making them more difficult to crush or dissolve and thus less likely to be snorted or injected. But, those with addictions frequently outmaneuvered pharmaceutical companies by switching to other drugs, rather than ceasing drug use. It has become clear that modest manipulations of the drugs cannot be expected to suddenly alter the behavior of dedicated drug users.

The preferred alternative is obviously to develop new and better treatments to diminish pain, yet limit the odds of addiction occurring. Compounds were developed based on the transformation of opioid agents so that they carried reduced tolerance liabilities, as well as compounds that affect nerve signals that carry pain. As indicated earlier, drugs have also been developed to antagonize the μ opioid receptor so that the analgesic effects of opioids would be maintained, but the euphoria and reward induced by these opioids would not be present (e.g., Sutton et al., 2016). Still another method has been that of combining Nociceptin/orphanin FQ peptide receptor (NOP) agonists, which have pain-reducing actions, with that of a μ opioid receptor peptide (MOP). Thus, pain relieving effects would be obtained with lower doses of the MOP, but with diminished risk of addiction (Ding et al., 2016). Other agents have been developed that block calcium channels fundamental for electrical nerve conduction, or act on both particular calcium channels and endocannabinoid receptors, thereby potently diminishing inflammatory pain and tactile allodynia. Still other approaches that are being evaluated has involved manipulations of TRPV4 and TRPA1 receptors, which are thought to sense particular types of painful stimuli involving inflammation (e.g., joint pain and abdominal pain, respectively) (e.g., Kanju et al., 2016).

ITCH AND SCRATCH

Typically, an itch appears transiently, leading to a scratch reflex. The cause for an itch might be due to a mild skin irritation, or a moderate allergic reaction following contact with substances that are released from plants, such as poison ivy or poison oak. Aside from body, head, or pubic lice, and insect (mosquito) bites, itch can be provoked by viral illnesses (e.g., chicken pox, herpes), skin conditions (eczema) or severe allergic reactions (e.g., in response to bee or hornet stings). Burns and sites of surgical wounds may also be accompanied by a hard to relieve itch, and in some chronic conditions, such as kidney or liver failure, diabetes, and cancer, incessant itchiness may appear. At times, scratching diminishes or relieves an itch, but this typically provides only temporary relief.

The neural circuitry associated with itch comprise some of those involved in pain perception, so that altered pain sensitivity is present around an itchy area, and individuals who are congenitally insensitive to pain tend not to feel itch. Moreover, stimuli that promote pain and itch activate spinal neurons, and both sensations are transmitted through the spinothalamic tract, and involve activation of many of the same brain regions (Akiyama & Carstens, 2013). At the same time, the processes related to itch can be dissociated from those related to pain, as treatments that reduce itch do not necessarily affect pain perception (Pitake, Debrecht, & Mishra, 2017).

Receptors exist that are sensitive to pain, itch or both, and while antihistamines reduce itch stemming from mild allergic reactions, they hardly affect itch stemming from diabetes or kidney failure. Likewise, opioids, which reduce pain, may cause worsening of itch, again suggesting that the two involve independent processes. Nonetheless, there has been interest in capitalizing on the links between itch and pain in the development of new targets to treat chronic pain, and medications for pain relief might also be used to diminish itch.

Itch is elicited by activation of somatosensory neurons that express a particular ion channel TRPV1 (transient receptor potential cation channel subfamily V member 1). It seems that a neurotransmitter, natriuretic polypeptide b (Nppb), is present in a subset of TRPV1 or TRPV4 neurons. Upon being activated, vigorous scratching responses are elicited (Chen Fang, et al., 2016), but even in the absence of Nppb, the hormone gastrin releasing peptide (GRP) could elicit marked scratch responses, pointing to its involvement in itch/scratch processes (Mishra & Hoon, 2013). A newly developed compound, triazole 1.1, a κ opioid receptor agonist, is able to attenuate pain and itch, and does so without producing euphoria, hence limiting the addiction potential (Brust et al., 2016). Such a compound would obviously be ideal for pain prevention given the current opioid epidemic.

Glutamate likely plays a role in pain perception, and thus treatments that diminish glutamate receptor functioning (notably mGluR1) could be useful to diminish chronic pain. As the inhibitory neurotransmitter, GABA, has also been implicated in pain perception (Lau & Vaughan, 2014), it may also serve as a target for pain management. The possibility has even been raised that norepinephrine processes involving locus coeruleus activation can be commandeered to alleviate pain perception (Taylor & Westlund, 2017). Manipulations of locus coeruleus norepinephrine functioning may indirectly diminish pain by altering arousal or vigilance.

Although antidepressants might have direct pain reducing actions over and above effects secondary to reduced depression or distress, the available data in this regard have been equivocal (Hearn, Moore, Derry, Wiffen, & Phillips, 2014). In rodents, a chronic stressor regimen increased stressor sensitivity, and these effects could be ameliorated by treatment with the SSRI fluoxetine (Lian et al., 2017). Likewise, among patients in whom depressive symptoms diminished, a degree of pain reduction also occurred, supporting the importance of mood states in modifying pain perception. As it happens, reductions of chronic pain and depression are regulated by a specific gene (RGS9), and RGS9-2, the protein for which it codes, and antidepressants that regulate chronic pain and depression can influence neuronal functioning within the nucleus accumbens, possibly by affecting RGS9-2 processes (Mitsi et al., 2015).

Manipulating Inflammatory Process to Modify Pain

With the realization that inflammatory factors could affect brain or spinal processes, thereby influencing pain perception, new drug targets became available in an effort to diminish pain. These comprised agents that affected local proinflammatory signaling, stimulation of anti-inflammatory processes, inhibition of specific immune mediators, and blocking particular cytokine or chemokine receptors (e.g., using agents to inhibit TNF-α or antagonize IL-1β receptors). These methods aren't without problems, as inhibiting cytokine functioning may leave individuals vulnerable to some illnesses, and immune inhibiting agents can interfere with neuroprotective processes. Still, as Grace et al. (2014) indicated, these approaches are, at least for the moment, at the forefront in the search for chronic pain treatments.

Inhibition of Proinflammatory Signaling

Several anti-inflammatory drugs (minocycline, propentofylline, ibudilast, and methotrexate) diminished allodynia in preclinical models of neuropathic pain (e.g., Ellis et al., 2014). It was thought that agents such as ibudilast, which acts as a toll-like receptor 4 antagonist, may have positive effects for pain related to diabetic neuropathy, CRPS, and neuropathic pain stemming from multiple sclerosis (Hutchinson, Loram et al., 2010; Hutchinson, Zhang et al., 2010). This compound also diminished symptoms of relapsing multiple sclerosis, although it did not actually seem to reduce focal inflammatory activity (Goodman, Gyang, & Smith, 2016).

Purinergic factors, which have numerous cellular functions, such as the proliferation and migration of neural stem cells, also regulate immune functioning (Cekic & Linden, 2016) and are effective in exciting microglia and cytokine release (Tsuda, 2017). Following peripheral nerve damage, pain hypersensitivity may develop owing to activation of purinergic P2X4 receptors located on microglia. The chemokine CCL21, which is part of a family of cytokines, contributes to pain associated with upregulation of P2X4 receptors (Biber et al., 2011). As well, the 5-HT and NE reuptake inhibitor duloxetine has an inhibitory effect on P2X4 receptors and might prove to be a viable antiallodynic agent (Stokes, Layhadi, Bibic, Dhuna, & Fountain, 2017).

The purine nucleoside adenosine, which is involved in energy processes, may also affect pain processes. Altering the activity of adenosine A3 receptors, which stimulates GABA, also turned off the pain signal (Salvemini & Jacobson, 2017). As well, pharmacologically targeting A2A and A2B receptors on immune cells and glia, which provoked elevated appearance of the anti-inflammatory IL-10 and decreased the levels of the proinflammatory TNF-α, attenuated peripheral nerve injury-nociceptive hypersensitivity for an extended period (Loram et al., 2013). In addition to these targets, novel TLR4 antagonists were effective in preclinical neuropathic pain models (Wang et al., 2013) and administration of a TLR4 antagonist precluded morphine-induced amplification of neuropathic pain.

REPETITIVE TRANSCRANIAL MAGNETIC STIMULATION AND DEEP BRAIN STIMULATION

Better known for its effects in ameliorating depression, repetitive transcranial magnetic stimulation (rTMS) has also been used to alleviates lower back pain, neuropathic pain associated with spinal cord injury, and phantom limb pain among land mine victims (Malavera, Silva, Fregni, Carrillo, & Garcia, 2016). Pain related to mild traumatic brain injury could likewise be diminished by rTMS, although the studies of this sort usually involved a small number of participants. Several reports also attested to the efficacy of rTMS for both episodic and chronic migraine, but this has been somewhat controversial, and a meta-analysis indicated that little benefit was obtained through magnetic brain stimulation (Shirahige, Melo, Nogueira, Rocha, & Monte-Silva, 2016).

There have been limitations concerning the efficacy of rTMS, often stemming from procedural factors. The precision of the coil's placement may be imperfect, and in the early studies, the current penetrated only a small distance into the human brain, and in some instances portions of the cortex could not be adequately stimulated. However, improved methods have been developed that may provide better outcomes.

Deep brain stimulation (DBS) was effective in ameliorating neuropathic pain and phantom limb pain, as well as intractable chronic headache, pelvic, and perineal pain (Nguyen, Nizard, Keravel, & Lefaucheur, 2011). It also appeared that DBS that targeted the ventral striatum/anterior limb of the internal capsule had positive actions in relation to the affective sphere of pain among poststroke pain syndrome patients (Lempka et al., 2017). Given that chronic pain is often linked to depressive mood, which can certainly exacerbate pain sensitivity and reactivity, it may be profitable to modify the affective component of pain syndromes in an effort to diminish severe discomfort.

Behavioral and Cognitive Approaches to Reduction of Chronic Pain

As pain perception is influenced by psychological and experiential factors (e.g., previous pain experiences, distraction), there has been a push for therapies based on psychological and cognitive treatments, particularly methods to modify perspectives and expectancies of pain relief. The effectiveness of these strategies are, to a considerable extent, dependent on patients having an action oriented attitude, including being ready for change and taking charge of pain management. Adherence to treatment is also fundamental in determining whether positive effects will be obtained, and pain-related beliefs, particularly self-efficacy, was particularly important. Some of these approaches have been moderately useful, and have been beneficial when combined with pharmacological attempts at pain reduction.

Relaxation therapy has been used to diminish acute pain, with varying results, and distraction has been useful to some extent in attenuating acute pain. Hypnosis may promote pain relief, even among those who were not fully hypnotizable (Jensen, Day, & Miró, 2014), and self-hypnosis was also effective as a pain reducing procedure. These actions of hypnosis could not entirely be attributed to placebo-like effects, but were nonetheless related to expectancies created in highly suggestible people. Not surprisingly, the positive impact of hypnosis varies with the nature of the pain experienced. Few patients with spinal cord injuries obtain pain relief, somewhat better results are promoted in patients with multiple sclerosis, and relatively good outcomes are seen in patients with phantom limb pain (Jensen et al., 2014). Hypnosis was also effective in relation to pain associated with irritable bowel syndrome as well as pain secondary to breast cancer and procedures that were used in children being treated for cancer. Several brain changes accompany the pain reduction associated with hypnosis, including diminished "connectivity" within circuits related to pain and pain perceptions, but it is unlikely that this would account for the differences observed in relation to different pain conditions.

Guided imagery, which entails individuals visualizing and focusing intensely on a place or event that promotes positive thoughts and feelings as well as feeling relaxed and at peace has been used to reduce pain, at least transiently, possibly owing to the relaxation it elicits or because it acts as a distractor. The usefulness of this procedure may be especially notable as a component of integrative therapies that include music, art, massage, therapeutic play, and distraction. In combination with other approaches, guided imagery has been used with some success in children and adolescents dealing with cancer, and in diminishing symptoms among patients with fibromyalgia (Thrane, 2013).

Biofeedback training typically comprises feedback based on autonomic nervous system responses (heart rate, blood pressure, and galvanic skin response) or brain activity (EEG) in an effort to gradually promote particular physiological or behavioral outcomes. This procedure has also been used to diminish anxiety and to promote calmness, which may be linked to altered heart functioning, and reduced migraine and musculoskeletal pain, as well as rehabilitation following injury or surgery. Feedback based on changes of activity within the anterior cingulate cortex (using imaging methods) likewise achieved a measure of control in patients experiencing chronic pain (Chapin, Bagarinao, & Mackey, 2012).

Physical exercise was also found to reduce neuropathic pain introduced by mechanical allodynia and partial sciatic nerve destruction. Multiple processes were implicated in producing such effects (Kami, Tajima, & Senba, 2017), including actions operating at peripheral nerves, the spinal dorsal horn and dorsal root ganglion, brainstem, and higher brain centers. In animal models, exercise in the form of running or swimming, altered the levels of inflammatory cytokines, neurotrophins, several neurotransmitters, and endogenous opioids, acting at peripheral nerves, dorsal root ganglia, and the spinal dorsal horn. At least some of these actions (e.g., suppression of GAD65 the enzyme that converts glutamate to GABA) could stem from epigenetic modifications secondary to inflammation or pain perception (Zhang, Cai, Zou, Bie, & Pan, 2011). In addition, variations within the mesolimbic reward system may contribute to changes in pain perception and modulation (Mitsui & Zachariou 2016). However, it was cautioned that it is important to distinguish between voluntary exercise that may be rewarding from the effects of forced exercise that may be aversive (Kami et al., 2017). It is conceivable that the analgesic effects of exercise in humans also vary based on the individual's appraisal of exercise being a chore versus a pleasure.

Cognitive Behavior Therapy, Acceptance and Commitment Therapy, and Mindful Meditation

Catastrophizing, negative appraisal, depressive affect, and the tendency to inhibit the expression of anger have all been associated with elevated pain perception among individuals with illnesses, such as cancer. Conversely, diminished pain among cancer patients could be promoted by psychosocial interventions and empowerment, and by augmenting feelings of self-efficacy, as well as through the availability of adequate social support. Each of these factors, individually, might have modest effects on pain perception, but multimodal treatment strategies that include specific medications as well as cognitive therapies, might provide still better pain attenuating actions.

In view of the widespread positive effects of CBT, several studies examined whether this treatment would also be effective in diminishing pain perception. Unfortunately, the beneficial effects observed were small, and only infrequently superior to other psychologically-based treatments. Yet, being faced with chronic pain may promote anxiety and depression, which can aggravate pain perception and a degree of pain relief could be achieved by treatments to reduce anxiety and depression. Because pain perception and psychological disturbances may be intertwined, perceived pain reduction by means of pharmacological treatments might benefit from patients initially obtaining psychological therapies to diminish disturbed emotional states.

As we saw in discussing placebo effects, the effectiveness of analgesic agents is dependent on patient beliefs about their pain, and their perceived self-efficacy in relation to pain management. Negative attitudes at the beginning of treatment predict relatively unsuccessful outcomes using CBT. In contrast, pain relief could more readily be obtained among individuals who reported high level of social support and who were actively involved in treatment (e.g., self-evaluation, writing about their experience, and exchanges with others concerning their condition). Likewise, CBT was effective when offered as part of a chronic pain management program. In essence, maximal positive effects could be attained if (1) patients maintained the belief that their pain could be managed, and (2) they actively engaged in achieving pain relief, rather than being passive bystanders waiting for relief to come. At the same time, patients must avoid the negative cognitive perspectives that can develop owing to a track record of failed treatments. Moreover, when pain relief is obtained, it is important that patients attribute this to the cognitive therapy.

As described in discussing depression, an important element in dealing with chronic pain is to remain flexible in the use of specific methods that may or may not be effective. Acceptance and Commitment therapy (ACT), has been used to regulate pain. As well as individuals becoming more aware and more focused on their therapeutic goals, patients are trained to relentlessly develop proactive behavioral patterns (e.g., thinking or *acting* based on the presence of pain) instead of simply focusing on reduction of their pain symptoms. This approach has been useful in attenuating features secondary to chronic pain (e.g., sleep disturbances) or actual pain perception, as well as pain catastrophizing, acceptance, and endurance (e.g., in relation to multiple sclerosis) (Baranoff, Hanrahan, Burke, & Connor, 2016).

The combination of meditation and CBT, offered in the form of mindful meditation to deal with depression and anxiety, has also been offered as a treatment strategy to diminish chronic pain perception. The efficacy of mindfulness has been demonstrated in several pain conditions, and in some instances (e.g., treating chronic back pain), it was as effective as CBT (Cherkin et al., 2016). If nothing else, mindfulness may be an effective strategy to limit pain catastrophizing, pain acceptance, and self-efficacy, as well as affective changes that come about with chronic pain (Jensen et al., 2014). Mindful meditation might also act on the depression and anxiety that occurs with chronic pain, and thus could indirectly affect pain perception, or at least dampen the exacerbated pain that accompanies negative thinking.

Changes of neuronal activity within insula and anterior cingulate cortex, which are thought to govern attention, emotional responses and interoceptive awareness were instigated by mindful meditation (Jensen et al., 2014), thereby affecting pain perception. As with so many other actions of this method, the outcomes observed varied between novice and expert practitioners; however, even limited mindfulness training could affect variations of brain activity and hippocampal and parietal cortical volume, which mapped on to a corresponding decline of pain perception (Zeidan et al., 2015). In fact, just a single 15-minute session of mindfulness training gave rise to pain reduction like that provided by an opioid (Garland et al., 2017).

Despite reports pointing to the benefits of mindful meditation, a systematic review and meta-analysis that evaluated relevant studies indicated that many of the published reports were of mixed methodological quality, and that there was, in fact, limited support for the position that mindfulness interventions were useful for chronic pain (Bawa et al., 2015). This conclusion notwithstanding, for some individuals, mindfulness meditation could represent a viable alternative to opioid-based therapy for chronic pain (Jacob, 2016), and may be useful in helping patients taper their use of high doses of opioid-acting agents.

PAINFUL MEMORIES AND MEMORIES OF PAIN

Memories of pain might be created and maintained in much the same way as other memories, although the specific networks related to pain may be differ from that of other forms of memory. This raises the question as to whether memories of pain perceptions in humans are modifiable much as other types of memory can be altered (Price & Inyang, 2016), such as those related to PTSD. As memories of pain might influence the maintenance of chronic pain, modifying these memories (just as memories related to PTSD can be altered) could also have remedial actions.

Alternative, Complementary, and Integrative Medicine

For those who experience chronic pain that isn't ameliorated by standard pain treatments, it isn't uncommon for alternative medicines to be sought, not only in desperation, but also to feel a semblance of control over one's own destiny. In fact, alternative treatments have been used in an effort to assuage pain more than for any other illness or ailment. Despite many reports debunking alternative treatments, alternative medicines and complementary medicine have maintained their popularity in some circles.

Aside from acupuncture, pain relief was also sought through spinal manipulation, mobilization, and massage techniques, and approaches that focused on thoracic pain management. Whatever bit of relief these treatments provide is typically very modest. However, when approaches such as mindfulness are combined with other treatments, often as part of CAM therapies within hospital settings, the likelihood of perceived pain reduction is increased (e.g., Thrane, 2013). Should patients choose alternative approaches, the downside isn't particularly bad so long as they aren't adopted as a long-term alternative to those approaches that could have genuine positive effects. Given that chronic pain is often untreatable with the current methods available, it's likely that alternative therapies will continue to be used.

CONCLUDING COMMENTS

Chronic pain is unquestionably an exceptionally debilitating condition, and patients often have little recourse in efforts to modify the pain experienced. The downstream toll taken by chronic pain is considerable, affecting quality of life, and contributes to the development of other illness conditions (e.g., depression, addiction). Inflammatory processes have been implicated as being fundamental in the emergence of some forms of pain, and these same processes, as we've seen repeatedly, may contribute to comorbid psychological disorders as well as the emergence of other physical illnesses. Chronic pain processes have become better understood, and although some forms of chronic pain cannot easily be thwarted, other forms of pain have been manageable through behavioral and pharmacological treatments.

The most common means of limiting pain are through agents with high addiction potential. It is doubly unfortunate that patients with chronic pain that is treatable (e.g., surgery for disc problems), often encounter remarkably long wait lists, all the while consuming opioids-based agents, thereby increasing the risk for addiction. Socialized medicine is the norm within Western countries (the US being only part-way there), which certainly has improved the health condition for many people, although inequities persist as a function of wealth, race, culture, and social settings. When treatment becomes unavailable (or severely delayed) for serious, but treatable pain conditions, per force leading to addiction, this signals the existence of problems within the government directed medical system that need to be resolved, sooner rather than later.

12

Autism

THE CHANGING FACE OF AUTISM OVER A CENTURY

Autism is the most prevalent developmental disorder of behavior in the current biomedical era. This condition has been recognized for many years, and only recently became a separate diagnostic entity in major psychiatry classification systems (e.g., International Classification of Diseases (ICD), Diagnostic and Statistical Manual (DSM) of the American Psychiatric Association). Since its initial use, the term "autism" has experienced a variety of dramatic transformations. It was first used in 1911 by the German psychiatrist, Eugene Bleuler, who applied the term to schizophrenia. In his view, "autism" was a mental strategy or defense, in which unwelcome thoughts and circumstances were supplanted by fantasy and hallucinations (Evans, 2013). In this conceptualization, "autism" denoted a rich and vibrant inner mental life, if perhaps marked by the turbulence and psychopathology that is often witnessed in schizophrenia. Later, in the 1960s, the term took on new meanings, describing mental life that was bereft of fantasy and imagination. This turnaround in what was meant by "autism," created the notion of poverty or absence of normal mental life. Indeed, this appeared to be the manner in which the term was used when applied to a particular syndrome observed in children by Kanner (1946). He was struck by a unique constellation of behavioral features in 23 children who presented with marked social withdrawal and an inability to form conventional relations with people, beginning very early in life. Additional features, such as isolation from others, a proclivity for sameness and repetitious behavior (including echolalia), cases of mutism, and a lack of reactivity to surrounding events (which he noted, led parents to suspect deafness), all converged on Kanner coining what he saw as "early infantile autism." Interestingly, it was noted that symptoms overlapped with what was also observed in schizophrenia, and may have initially eclipsed the emergence of autism as a unique, developmental disorder, which did not receive its own separate entry until the third version of the DSM in 1980. Even today, clinicians may still grapple with defining whether someone has autism or early onset schizophrenia.

The Immune System and Mental Health
DOI: https://doi.org/10.1016/B978-0-12-811351-6.00012-7

FEATURES OF AUTISM SPECTRUM DISORDER

A diagnosis of "autism" or "autism spectrum disorder" (ASD) now encompasses a far less restrictive conceptualization than it had in earlier years, referring to a range of social and emotional behaviors that are incompletely developed and/or inaccurately or inappropriately expressed[1]. Intelligence, cognition, and motor function are similarly incorporated into this terminology, widening the range of potential problems that might be identified in children eventually diagnosed with ASD (see Fig. 12.1). Within the revisions to the DSM-IV, where autism was "autistic disorder," the latest version (DSM-5) applies the broad term ASD, and incorporates conditions previously considered as distinct entities: Aspergers's disorder, childhood disintegrative disorder, Rett's disorder, and pervasive developmental disorder. What unifies these conditions are the core symptoms of (1) deficient social communication and interaction, and (2) repetitive patterns of behavior, interests, and activities (see Fig. 12.1).

The prevalence of autism appears to have increased over the past several decades. This was prominent from the 1970s through to the 1990s, and was driven in part by increased awareness of childhood developmental delays, including those involving language and social behavior. Clinicians swept up by this interest, expanded the range of potentially relevant symptoms thought to constitute autism, thus casting a wider diagnostic net, which likely accounted for the dramatic increase in the perceived prevalence of autism. For example, in 1992, for every 10,000 children examined, 19 cases of autism were recorded in the United States, a number that spiked to 90 (per 10,0000

children) in 2006, and more recently stands at around 1 per 68 children (or 146 per 10,000 children) (Christensen et al., 2016). This sharp rise in numbers has given way to speculation that we are in the midst of an "autism epidemic," and explanations for the spike, vacillate from overly inclusive diagnostic criteria to the contribution of environmental and genetic risk factors. All are likely to be relevant to varying degrees. But, whatever the reason, autism has a substantially increased profile and an intensive research agenda. Much of this focuses on identifying its etiology, and in particular identifying mutations in candidate genes (Yuen, Merico, Bookman, L Howe, & Thiruvahindrapuram, 2017; Vorstman et al., 2017). This approach has a solid foundation, given that autism and autism-like traits have strong familial associations. In fact, while individual gene variants provide little risk, combinations of many different gene variants contributes to ASD within a range of 15%–50% (Vorstman et al., 2017). Moreover, frank neurodevelopmental disorders linked to genetic mutations (e.g., fragile X syndrome; tuberous sclerosis, and Rett's disorder) present with behavioral symptoms that overlap with those observed in autism (Betancur, 2011). This supports the notion that ASD is a disorder driven by a strong genetic contribution. However, this position must be tempered by the likelihood that nongenetic factors play an important role, since any given ASD individual can display a unique set of genetic mutations (Yuen et al., 2017).

Aside from genetics, autism etiology may also encompass multiple environmental factors, including chemical toxins, environmental pollutants, diet, and infectious disease (Bilbo, Block, Bolton, Hanamsagar, & Tran, 2018), and paternal age and reproductive factors have

---

[1] The term is derived from the Greek "auto" for self. The suffix "—ism" implies a state of self-directed actions, and this very much captures the isolated nature of the autistic individual, who fails to engage with others, and in extreme cases, directs harmful attention to himself.

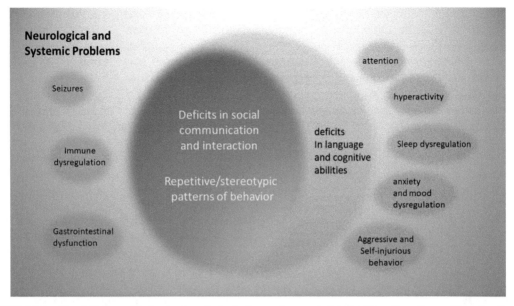

FIGURE 12.1 **Constellation of behavioral and physical symptoms associated with autism.** Recognition of autism as a diagnosis emerges from the presentation of core defining symptoms (central dark blue sphere) which relate to impaired social communication and repetitive and/or stereotypic behavior. Matching onto these cores may be ancillary intellectual, emotional and attentional problems, some of which are more often present (light blue crescent), than not (gray ovals) in the most commonly seen cases of autism. Importantly, along with a wide range of other factors that orbit the core features of autism, the immune system and GI disorders (closely related to immunity, as we shall discuss) are now associated with autism. Given the peripheral (as opposed to central) presence of some of these symptoms, it is not surprising that a multiplicity of "autisms", each with their unique profile, likely exist in the clinical population. *Source: Figure inspired by https:// www.autismspeaks.org/what-autism/symptoms.*

been investigated. Not surprisingly, at present, there is no one single factor that has emerged as a game-changing explanation for the development of autism. This is likely the case because of the heterogeneity of autism described in Fig. 12.1. Indeed, a survey of the many reviews on specific aspects of autism can be dizzying in the range of factors that may or may not be important in contributing to this developmental disorder. In this chapter, we will address mainly those factors relevant to the immune system. We will do so, however,

after briefly reviewing some of the history and major positions maintained with regard to the etiology and neuropathology of autism.

The immune system is now considered a potential variable that is either dysregulated in autism, or contributes in unknown ways to the etiology of the disorder[2]. In this chapter, we will also highlight the many different approaches taken to examine the relationship between the immune system and autism. Since autism is a distinctly human disorder, it is critical to identify what is immunologically awry in the

[2] At times we will refer to "autism," and at other times ASD. Since we are summarizing material from different sources, and spanning several decades, it can be difficult to know what criteria were used to achieve a given diagnosis. Was it for ASD or simply autism? In either case, there can be considerable overlap, and therefore, we will at times use these terms interchangeably.

autistic individual. However, animal research has proven an ally in generating conceivable hypotheses for how neurodevelopment may either be facilitated or impaired by immune alterations that occur around the perinatal period. This research approach will be addressed again in Chapter 13, Schizophrenia, as behavioral animal models of both disorders are quite similar.

WHAT IS AUTISM: TODAY!

One feature of autism that distinguishes it from childhood onset schizophrenia is that it is usually evident by the 2nd or 3rd year of life, and thereafter shows a range of developmental delays in different compartments of behavior, such as language and motor function. In about 35% of cases, however, developmental progress and acquisition of important communication and motor skills appears to achieve expected and normal milestones, but then, usually between 18 and 36 months, shows marked and persistent regression to poor functioning (Xue, Brimacombe, Chaaban, Zimmerman-Bier, & Wagner, 2008). A prominent feature of autism is impaired social behavior, characterized by aloofness, disinterest, or odd and unexpected styles of interaction. Children also display poor recognition or perception of the intentions and feelings of others, which together may reflect disturbed social cognition. This represents one of the three core features of autism, or "triad of impairments," identified, which comprise social interaction, communication, and repetitive/stereotypic behavior (Wing & Gould, 1979). This understanding of autism has endured to the present day, as shown in Fig. 12.1[3].

Social interaction has been a particularly strong research and clinical focus, given its importance in learning, affective experience, and communication. As such, it influences the creation of various representations of the self across different social contexts. It is here that deficits or deviations from normative aspects of behavior can be particularly distressing to parents. Moreover, research on the genetics and neurobiology of social cognition and/or social behavior has considered the identification of factors important in exercising trust (and conversely distrust), promoting affiliative behavior, perceiving faces, and differentiating between biological organisms and inanimate objects. The human brain is essentially "wired" for initiating and maintaining social behavior, and in autism, this aspect of brain function appears to be altered. Consequently, many animal studies focused on modeling autism-like problems, by focusing on social behavior. There is more to say about this later, but suffice for the moment that several studies observed changes in animal social behavior that were linked to immunogenetics, as well as to key cytokines, including IFNγ and IL-4 (Filiano, Gadani, & Kipnis, 2017).

BRAIN DEVELOPMENT AND AUTISM: BEHAVIORAL DEFICITS LINKED TO SPECIFIC CIRCUITS

The onset of autism typically occurs before the age of 2 years, when development of the brain is still dynamic and the full assembly of circuits and synaptic connections is years away from its final state. Indeed, children can vary in their rate of development, and therefore, appearances can be deceiving: The preternaturally gifted eventually fall back to the norm, while the seemingly delayed or immature may

---

[3] In the DSM-5, social interaction and communication are combined as one core feature. However, given the heterogeneity of behavioral problems presented by ASD children, it is simply a starting point to an inevitable parsing process.

work their way to the head of the pack. Consequently, determining whether a 2-year old child has autism can at times be difficult, and requires diagnostic restraint. Experienced clinicians presented with a full display of the core symptoms (delays in language and social interaction), can confidently determine suspected autism by ages 2–3 years. Recent brain scan studies have raised the possibility that diagnosis can be more confidently determined within the first 2 years (Hazlett et al., 2017). However, this study involved participants from ASD-prone families, and thus predictions were made on the likelihood of a pending appearance of autism in the scanned infants. In essence, false positives were less likely to be observed in this study. Nonetheless, the use of brain scans is dependent on the long-standing belief that autism is a neurodevelopmental disorder in which there is abnormal brain growth and brain size.

Early support for this view came from postmortem and electroencephalographic (EEG) brain analyses that revealed alterations in cell numbers, cell morphology, impaired neural circuits, regulatory function, neurochemical stability/imbalance, and regional and functional interconnectivity (DiCicco-Bloom et al., 2006; Moldin, Rubenstein, & Hyman, 2006). Results from functional connectivity magnetic resonance imaging (fcMRI) studies have generated considerable excitement, but have encountered problems of replication and consistency (Mash, Reiter, Linke, Townsend, & Muller, 2017; Muller et al., 2011). The physical substrates for variations in functional connections within and between brain regions is the quantity of synaptic input and dendritic spine numbers (i.e., spine density). Functional connectivity MRI cannot actually address this level of refinement or resolution, but can measure the frequency and intensity of dynamic changes—or oscillations—that occur across distinct cortical and subcortical neuronal groups. This can generate time-stamped correlations that provide

some notion of the functional connections operating within or between various brains sites during a specific behavioral event, be it cognitive, emotional, or sensorimotor. For autism, it was thought that connections between brain regions were overly busy or synaptically dense (i.e., "over-connectivity"), whereas at another level, there was insufficient or weak connectivity (i.e., "under-connectivity"). Evidence has been offered for both scenarios, with under-connectivity evident for more distant connections (e.g., anterior-posterior connections within the brain), whereas over-connectivity appears to be more prominent within localized areas. While this understanding has become the norm, confusion and various inconsistencies were noted, since intrinsic functional over-connectivity can vary according to intraindividual characteristics during data collection. For instance, whether participants close or open their eyes during resting state data collection in fcMRI, can reveal or dissolve any intrinsic connectivity variations. That is, if subjects rest with their eyes open, connectivity within the posterior visual regions of the cortex appears to be greater for ASD participants than among typically developing individuals (Nair et al., 2017). This difference is not evident when participants rest with their eyes closed. Moreover, differences in connectivity patterns between ASD and control participants vary across the cingulate gyrus, based on whether their eyes are closed or open (Nair et al., 2017). These procedural details may have a significant impact on how to interpret functional connections within and between brain regions in ASD subjects. Nonetheless, there appears to be little dispute that coordinated activity within the brain is different for ASD individuals. The challenge is in knowing why this happens.

The neural eccentricities noted in ASD underlie the behavioral deficits that occur at specific developmental stages. For example, language development progresses along a well-recognized trajectory that involves vocabulary building, sentence construction, grammatical rule development, and conceptual expression and understanding—all initiated during infancy by active listening and extraction of word sounds and their meaning. These appear at well-defined steps that emerge from the initial rudimentary babbling and single syllable word formation. In addition to other key deficits, children with autism may show significant delays in language development and expression (Baird & Norbury, 2016), as well as delays and abnormalities in motor development, such as alterations in gait, and performance of repetitive or perseverative movements. The motor abnormalities are likely linked to known disturbances in the striatum and the cerebellum. The latter is responsible for fine motor coordination and temporal resolution of movements. However, the cerebellum has important connections to other regions of the forebrain, including the striatum (or basal ganglia) and the frontal cortex. Postmortem studies have revealed marked cellular abnormalities in the cerebellum. For this reason, and given some of the motor deficits seen in autism, cerebellar dysfunction is considered a prominent neuroanatomical feature of the disorder (D'Mello & Stoodley, 2015).

Finally, the core feature of autism comprising a deficit in social behavior, has received particular attention and was instrumental in encouraging the development of the relatively young field of social cognition. This line of inquiry has been approached from multiple perspectives. One obvious approach, when considered in the context of autism, is to understand the neurobiology of social behavior. Deficits of social behavior among ASD children are characterized by a poverty of eye contact and a lack of directed attention and/or awareness related to mutual interactivity. This represents a failure of engagement and cooperative behavior. Children with autism may display this in various forms, such as maintaining isolation from others, or at the other extreme,

exhibiting overt displays of aggression, which can take the form of self-injurious behavior. The affectless state or overt angry outbursts belie problems with emotion-regulation, and are consistent with studies showing abnormalities in the development of the limbic system, which includes the amygdala and which has been related to social judgment and social perception. In neurological patients who sustained damage to the amygdala, more trusting interactions and/or perceptions of strangers is common. Thus, it was suggested that in ASD, dysregulation of the amygdala may account for deficits in social interaction (Rutishauser, Mamelak, & Adolphs, 2015).

The amygdala receives multiple inputs from brain areas that process sensory information, as well as those involved in assigning meaning to information (e.g., the hippocampus and temporal cortex), and executing decisions based on this meaning (e.g., inputs from the prefrontal cortex). Since the amygdala interacts in reciprocal fashion with these brain regions, decisions and actions are dependent on the quality of these interactions. The application of this knowledge to social cognition and autism has minimized regional bias and single locus attribution when reaching for neuroanatomical and neurobiological explanations for autism. Nonetheless, the well-documented abnormalities already noted in the cerebellum, as well as recent evidence of abnormal cell numbers and synaptic connections in specific cell layers of the frontal cortex, highlight the fact that the behavioral problems in autism very likely arise from impairments between the interregional connections or pathways, and from within the nodes that make up these networks.

Additional neural circuits that subserve motor behavior and reward may also be impaired in autism. The cortico-striatal circuit regulates voluntary motor behaviors and habit formation. Disturbances in this circuit, may result in stereotypic and repetitive behaviors, and may arise from over-stimulation of the nigrostrial dopaminergic pathway, which is known to promote stereotypic behaviors. Maintenance of these behaviors may require some form of reward or reinforcement, which can be provided by dopamine pathways from the midbrain to the nucleus accumbens (i.e., the mesocorticolimbic dopamine system). What we already understand about this system in driving reward, and its interactions with the nigrostriatal dopamine system (in shaping habits), may explain the persistence of stereotypic behavior. Alternatively, while these circuits may be important in explaining stereotypy, it may also be possible to address social deficits as a failure of experiencing social reward. The role of this pathway in ASD and schizophrenia has recently been reviewed (Bissonette & Roesch, 2016), but it is important to recognize that these processes are also associated with drug addiction and depression. Thus, while this dopamine pathway may contribute to social reward, it is likely just one of many contributing factors.

The circuits that are involved in attention and social perception include the parieto-frontal and temporo-frontal systems. The former is particularly relevant in relation to attention to spatially distributed stimulus information and the accurate determination of stimulus location. Visual attention is modulated by collicular nuclei in the midbrain that guide eye movements. This ensures cognitive capture of visual information so that their meaning and significance can be determined. Detection of eye gaze, for example, and attending to this to discern meaning and reciprocal or joint attention, is a primary aspect of social interaction. Children with autism display poor eye contact and perception of gaze. Visual scanning of faces by ASD children is seldom directed at the eyes, with attention given more to the lower regions of the face (Chita-Tegmark, 2016). That is, children with autism show aberrant processing of gaze, fail to recognize emotional expressions, and perform poorly in recall of previously seen faces. Moreover, when presented with different gaze conditions, there is poor attribution of intention (e.g., seeing another gaze at candy does not imply desire for a sweet, when most children would conclude otherwise). Examples such as these represent instances of theory of mind, and the failure of such mental processes to operate normally in ASD. These types of problems may reside in poor neural processing in the superior temporal sulcus, as well as amygdala disturbances, ostensibly because gaze processing possesses strong emotional valence (Rutishauser et al., 2015).

The point to take away from all this is that the behavioral problems in autism align with many of the key circuits driving cognitive-emotional and motor behavior. This is consistent with what is already known about brain development in autism: physical changes in the development of the brain, and related alterations in cell numbers and neuronal connectivity. Since this can range in severity, distribution, and frequency, it has naturally revealed marked heterogeneity in the display of symptoms, as we discussed earlier. However, it does raise problems with regard to etiology, which has also plagued other areas of psychiatry (e.g., schizophrenia, mood disorders). When disorders are syndromal in nature, the symptoms rarely coalesce in a consistent enough fashion to make the search for causes simple and easy. This problem is compounded by the symptoms of ASD overlapping with numerous concurrent behavioral problems, including epilepsy, sleep disorder, depression and other mood disorders, obsessive compulsive disorder, aggression, and self-injurious behavior (Xue et al., 2008). As we indicated earlier, a precision medicine approach might prove useful in defining the links between symptoms and particular neurobiological processes, which ultimately may be relevant to treatment methods.

ETIOLOGY OF AUTISM: THE IMMUNE HYPOTHESIS

The foregoing section characterized the brain of an ASD individual as possessing different levels of connectivity and absolute variations in synaptic and spine density. In the current section, we will consider what an examination of the immune system can tell us about why this happens. Several reviews have focused purely on the immunological changes in ASD individuals, or have addressed immunological factors that might drive the neurodevelopmental aberrations observed in ASD (Ashwood, Wills, & Van de Water, 2006; Careaga et al., 2017; Edmiston, Ashwood, & Van de Water, 2017; Estes & McAllister, 2015; Meltzer & Van de Water, 2017).

A prominent hypothesis of ASD is that antecedent autoimmune processes interfere with neuronal development and impair behavior. This is distinct from autoimmune effects operating throughout childhood, and which undermine the normal quality of functional activity in the brain. We will consider autoimmune phenomena, and the strength of this evidence, after first addressing immunogenetics and autism, a topic that typically underlies explanations of autoimmune dysregulation. Much of the research regarding immunological factors that might affect neuronal development is aligned with the same themes present in schizophrenia research. To the extent that it is largely indistinguishable from investigations that examine early life immune events and how they may affect alterations in social, cognitive, and emotional behavior in animal models, this aspect of autism-related research will be discussed in Chapter 13, Schizophrenia.

Research on autoimmune antecedents highlights the possibility that among individuals with ASD immune system functioning may already be compromised. The degree to which this approach is fruitful in explaining the onset or development of ASD is uncertain. Dysregulation in the immune system is likely to affect neurodevelopment at different stages, prenatally and postnatally, and may do so in a highly individualized manner. Critical periods of influence, and what types of aberrant immune processes might be of greatest impact, is not entirely clear, although cytokines are often considered as potential primary mediators of disruption. Furthermore, immune dysregulation may derail neurodevelopment in only a subset of people, while many others who suffer from altered immunity, remain resilient or perhaps succumb to other problems unrelated to ASD. Confronting these types of questions, or caveats, is now starting to become more prominent, especially in the face of the growing number of genetic risk factors that are being identified. It does not necessarily generate treatment or preventative strategies, but does highlight the possibility that the etiology of ASD may be unique and in need of personalized approaches to treatment (Hodson, 2016). This problem of indeterminate etiology due to a heterogeneity of gene candidates has arisen in the case of schizophrenia, as well as mood disorders, where confidence in heritable risk factors is moderate or high. It is possible that certain inherited immune-related genes might contribute to ASD.

Immunogenetics

The main genetic approach in the immunological domain has focused on MHC genes, which have been linked to several diseases, and in particular, autoimmune disorders. Recall from Chapter 2, The Immune System: An Overview, that unlike the mouse system, nomenclature for human major histocompatibility complex (MHC) is based on the human leukocyte antigen (HLA) system of classification. The HLA region is found on chromosome 6 and encodes over 200 different proteins or antigens that serve important functions in the immune system, including cytokines, receptors, transcription factors, and factors that allow the immune system to recognize host tissue and foreign antigens (e.g., on microbes, tissue transplants). Interest in how the HLA figured in autism, was piqued by early studies (Stubbs, Ritvo, & Mason-Brothers, 1985) that noted greater similarity in HLA antigen profiles between parents of children with autism relative to parental pairs with non-ASD children. This was considered important, since less sharing of HLA antigens between parents was thought to provide greater protection for the fetus, resulting in reduced chances of rejection and/or other complications. However, it is uncertain how such disparity in HLA confers such protection.

Moving away from parents to the actual child, reports pointed to HLA antigen expression being different in ASD children. In this regard, ASD individuals are more likely to carry the 44.1 HLA haplotype (also known as B44-SC30-DR4), which contains alleles for A2 and B44, DR1*04 (DR4), as well as a deletion of complement C4B (Torres et al., 2016). These alleles are found in class I, II, and III regions of the HLA. Moreover, in determining which Class I alleles are carried more frequently in ASD, it was found that HLA-A2 is more common. The functional significance of these associations is not known, but it should be noted that HLA-A2 is associated with autoimmune and neurodegenerative diseases (Torres et al., 2016). Furthermore, a reduction of naïve CD4 T cells, and an increase in CD4 T cells bearing a memory phenotype was reported in ASD children, and this was linked to ASD children being more likely to carry HLA-A2 and HLA-DR11 alleles (Ferrante et al., 2003).

Other immune genes associated with HLA Class I molecules have been linked to ASD (reviewed by Torres et al., 2016). For instance, the killer-cell immunoglobulin-like receptors (KIRs) are found on NK cells, and certain HLA

products of the Class I region serve as ligands for KIRs. The presence of HLA and KIR genes in the placenta suggests that HLA-KIR interactions may occur prenatally to influence the development of the fetus. However, as KIR genes appear at increased frequency in ASD individuals, suggests that HLA-KIR interactions are more likely to be important in the developing organism. Indeed, individuals with ASD have a higher frequency of activating cB01/tA01 KIR gene haplotype, and similarly had a higher frequency of the cognate ligand HLA-C1$_k$, which activates the KIR haplotype. Thus, carriers of these two haplotypes are more likely to show increased NK cell activation. How this contributes to the search for an etiology to the neurodevelopmental basis of ASD is not known, but it does offer some genetic support for possible immune dysregulation in autism. Finally, complement proteins have received considerable attention in relation to neurological and psychiatric disorders, and in particular neurodegenerative conditions (Morgan, 2015), and was linked to autism. We noted in Chapter 2, The Immune System: An Overview, that a region of the MHC codes for complement proteins (in the HLA this is the Class III region). Interestingly, increased risk for autism has been linked to the complement system, which consists of multiple blood-borne proteins that are activated in sequence to eventually implement cytotoxic effects on pathogenic cells. One of these complement proteins, C4B, is coded by the eponymous gene, *C4B*, the absence of which increases the risk for autoimmune disease, and was also found in over a third of ASD individuals, compared to less than 10% of healthy controls (Mostafa & Shehab, 2010). The argument was made that when ASD individuals show a deficit in the C4B allele, and reduced amounts of circulating C4B, it may reflect increased risk for autoimmune disease (Estes & McAllister, 2015). The challenge is to extend these associations and show direct evidence that an autoimmune process is directed at specific targets that would derail brain development in children.

In fact, there is considerable evidence that children who develop ASD are more likely to have mothers with circulating anti-fetal brain autoantibodies (Edmiston et al., 2017). In animal studies, maternal antibrain autoantibodies altered species-typical social interactions with unfamiliar peers, as well as increase frontal lobe size and white matter density (Bauman et al., 2013). Similarly, in mice exposed prenatally to antibrain maternal autoantibodies, brain overgrowth and enlargement of cortical neurons occurred, which pointed to an autism-like neuroanatomical phenotype (Martinez-Cerdeno et al., 2016). Among ASD children, there is evidence for generation of antineuronal autoantibodies, including antibodies that bind cerebellar antigens (Wills et al., 2009.) It is thus possible that genetic deficits, such as those in the MHC, and which increased risk for autoimmune disease, may also be the basis of anti-neuronal antibody formation, and possibly impaired brain development.

Autoimmune Disease and Autism From a Broader Perspective

As we've seen, it has been a long-standing hypothesis that the development of autism is driven by autoimmune processes (Estes & McAllister, 2015; Ornoy, Weinstein-Fudim, & Ergaz, 2016). In one of the earliest studies, the average number of autoimmune diseases was greatest in families that had children with ASD (Comi, Zimmerman, Frye, Law, & Peeden, 1999). Indeed, the odds ratio for having a child with ASD moved from 1.9 to 5.5 as the number of autoimmune diseases multiplied within a family. Moreover, approximately one-in-five mothers with an ASD child reported having an autoimmune disorder. A review of the relevant literature concluded that ASD risk increased with the presence of familial autoimmune

disease (Wu, Ding et al., 2015), and was particularly notable in the presence of maternal autoimmune disease (Chen, Zhong et al., 2016).

A meta-analysis that included 10 studies, 9 of which were case-control studies (9775 pregnant autoimmune cases and 952,211 pregnant healthy controls), examined the link between maternal autoimmune disease and ASD diagnoses in the offspring. The studies were conducted globally, with most coming from the United States, Canada, and Europe, with two reports from Asia. The overall conclusion of the review was that the presence of autoimmune disease *during pregnancy* increased the risk for an ASD diagnosis in the offspring by 30%, with the main contribution coming from autoimmune thyroiditis (Chen, Zhang, Hu, & Liang, 2015). Significantly, as four studies reported no significant association between maternal autoimmune disease and ASD, it seems that the maternal autoimmune disease link is not universal. Moreover, a connection to ASD did not appear in cases where systemic lupus erythematosus (SLE), inflammatory bowel disease (IBD), psoriasis, and rheumatoid arthritis (RA) occurred during pregnancy. Since these diseases were associated with the familial connection to ASD, it seems that alternative mechanisms—such as genetics—may contribute to ASD.

Registry Studies of Autoimmune Disease

Large epidemiological studies that relied on well-organized nationwide registers have served to determine whether familial history of autoimmune disease represents a risk factor for development of ASD. An analysis was conducted within 3 Swedish registries and compared each of 1227 ASD children with 25 controls matched for sex, birth date, and birth hospital (total number of controls used: 30,295) (Keil et al., 2010). Information regarding whether either or both parents of each child had an autoimmune disorder was obtained *a priori* through the Swedish Hospital Discharge Register, with diagnoses of up to 19 different autoimmune diseases being identified. An odds ratio analysis revealed that children with ASD had a marginally higher likelihood of having parents with autoimmune disease, and of these, type 1 diabetes and ulcerative colitis/Crohn's disease were most strongly associated with ASD. However, a different set of autoimmune diseases were associated with ASD in a Danish study. Specifically, of 3325 ASD children (1089 with Infantile Autism diagnoses), and 26 autoimmune disorders examined, children with ASD were most likely to have mothers with RA or celiac disease, and in the case of an infantile autism, a family history of type 1 diabetes (Atladottir et al., 2009). It was interesting that the strongest associations were for non-CNS (central nervous system) autoimmune diseases, suggesting that familial history of auto-aggressive immune responses directed at the CNS is not as much of a risk factor as other conditions.

It is somewhat paradoxical that RA is an ASD risk, since pregnancy is typically associated with amelioration of RA symptoms in up to 75% of pregnant women with RA prior to pregnancy (Adams Waldorf & Nelson, 2008). Given that autoantibodies are known to transfer to the fetus during gestation, failure to inhibit key autoimmune processes in preexisting autoimmune disorders may influence neural development in the fetus. Not enough is known at present to determine whether mothers who gave birth to children that developed ASD were more symptomatic or had flare ups during pregnancy, and whether this predicts births ultimately diagnosed with ASD.

With regard to other autoimmune diseases, amelioration during pregnancy is not as common as for RA, but certainly has been observed. SLE in particular represents a potential source of complication for pregnancy, influencing outcomes like preterm labor or fetal death (Adams Waldorf & Nelson, 2008). A Canadian study, examined the frequency of ASD births in 509 women diagnosed with SLE prior to pregnancy and 5824 normal, healthy control women (Vinet et al., 2015). In the SLE group, 1.4% of the births resulted in a diagnosis of ASD, as opposed to an incidence of 0.6% in the healthy controls. This is a relatively small association, and is in keeping with the absence of an association between SLE during pregnancy and ASD outcomes for mothers with SLE (Chen, Zhang et al., 2015). The latter point was reinforced in another study involving pregnant women with antiphospholipid syndrome (APS) or SLE (Abisror et al., 2013). The results were modest and inconclusive, with no ASD diagnoses among the offspring of SLE mothers, while 0.08% of the offspring of the APS mothers were diagnosed with ASD.

Given the severity of SLE, and the relationship to symptoms during pregnancy, these results are somewhat surprising. For example, a recent analysis of the literature (Bundhun, Soogund, & Huang, 2017) revealed that pregnancy in women who have SLE is strongly associated with increased risk of requiring caesarean delivery, episodes of preeclampsia and hypertension, as well as increased risk of spontaneous abortion or premature birth, and classification of infants as "small for gestational age." Lupus flares occur during pregnancy, with an incidence rate of 25%−65% (Lateef & Petri, 2013), and may pose a particular health and pregnancy risk if associated with nephritis (Moroni & Ponticelli, 2016). Given these multiple effects of SLE, it is predictable that behavioral risks for the offspring are significant, and several prominent neurobehavioral deficits were observed, in particular learning disorders and dyslexia (Yousef Yengej, van Royen-Kerkhof, Derksen, & Fritsch-Stork, 2017).

Other maternal autoimmune diseases linked to ASD included autoimmune thyroiditis (Brown et al., 2015). Serum samples obtained from mothers who had children with ASD and those with healthy offspring, were assayed for antibody against thyroid peroxidase (TPO-Ab), an indicator of autoimmune thyroiditis. For pregnancies that resulted in ASD cases, 6.15% of mothers were positive for TPO-Ab, whereas in those mothers who did not give birth to children diagnosed with ASD, 3.54% had detectable TPO-Ab. Further, relative to TPO-Ab negative pregnancies, the chance of obtaining an ASD birth was increased by 80% in the presence of TPO-Ab positive pregnancies. Once again, while these results are statistically encouraging, they show only a modest association between maternal autoantibody and a possible influence on autism. In fact, the autoantibody data are not necessarily sufficient to explain the presence of autism in offspring, since TPO-Ab was present in pregnancies that did not result in an ASD diagnosis. Therefore, as with other studies, it is not the case that maternal autoimmune disease is driving the development of an ASD phenotype in the developing fetus[4].

Immune Dysregulation in ASD Individuals

In a review of the literature more than a decade ago it was highlighted that immune function in ASD individuals shows an altered phenotype (Ashwood et al., 2006). Many of the studies reviewed focused on lymphocyte function, the capacity for cytokine production, and antibody (or immunoglobulin) production. Abnormalities in cytokine production, as well as altered basal levels of circulating cytokines, may have several implications. Cytokines are necessary to regulate the various activities of lymphocytes, dendritic cells and

macrophages/monocytes, and may influence neural and behavioral functions. This can operate in the adult nervous system to produce behavioral disturbances, but in the nascent nervous system of a young child, the impact of cytokines is likely to be more profound.

The organization and patterning of the nervous system continues well into adolescence, and immune dysregulation has the capacity to disrupt this process. However, to determine the relevance of immune dysregulation to ASD itself, the information gathered needs to demonstrate an immunological effect on relevant brain and behavioral functions. With certain types of immune measures, this is not always clear. For example, suppression of T cell mitogenic function does not provide evidence of a proinflammatory response that can potentially impact CNS function. To be sure, it is possible that reduced cell division may underlie insufficient T cell involvement in a given inflammatory process, which could potentially influence the brain. However, this is pure speculation, and is based on an ill-defined or characterized premise. In short, data like these (viz, T cell proliferative function) is important, but limited. At best, one can simply identify an aberrant state of immunophysiology in ASD. Much of the information on immune function in ASD is of this nature—associative and lacking in explanatory power with regard to ASD etiology. Nonetheless, it has been suggested that eruptions of immunological activity may exacerbate certain ASD conditions, and may even be relevant to regression of autism-like symptoms (Xue et al., 2008). At the same time, certain forms of immune activity might have benefits. Febrile conditions typically arise from inflammatory responses, and children with ASD, who

---

[4] A recent report added 18 more genes to the hundreds now being investigated (Yuen et al., 2017), and has suggested that each ASD individual may have a unique etiology. While this is consistent with the principles of precision medicine, it makes the search for critical elements still more difficult to identify. The involvement of autoimmune factors is just one piece of the puzzle, and although relevant, it does seem to have a relatively modest effect in human studies.

present with fever show reductions in a variety of behavioral symptoms, including communication (Grzadzinski, Lord, Sanders, Werling, & Bal, 2017). In studies of this nature there is typically no information concerning potential immune mechanisms, or even whether fever was linked to immunological factors. However, it is an intriguing phenomenon, and one which deserves further exploration, as it may reveal important clues regarding which brain mechanisms are seemingly corrected (albeit temporarily), thereby opening up some avenues for therapeutic development.

Cytokines and Autism

Brain Measures of Cytokine and/or Microglial Cells

Cytokines hold an important place in the pantheon of immunological factors that might disturb or impair CNS function, and we will look closely at some attempts to link cytokines to the autistic brain. In a relatively early study, cerebrospinal fluid (CSF) and serum samples were collected from individuals diagnosed with autism, as well as age-matched control subjects with non-ASD diagnoses (Zimmerman et al., 2005). None of the cytokines pursued (IL-1β, IL-2, IFNγ, and TGFβ) were detected in the CSF of those with autism, although trace amounts of soluble IL-1 receptor antagonist and IL-6 were detected in a few participants in both the autism and control groups, which was also the case for measures of soluble tumor necrosis factor receptors I and II (TNFRI and TNFRII)[5]. In contrast, a small increase in soluble TNFRII was present in the serum of the

autism group. Overall, this study was notable for its access to proximate material from the brain (viz, CSF), although the lack of any cytokine elevations, or even detectable amounts of these cytokines—failed to support an activated inflammatory cytokine network in the CNS of children with autism.

The same group of researchers (Vargas, Nascimbene, Krishnan, Zimmerman, & Pardo, 2005) assessed astrocytes and microglia for levels of glial fibrillary acid protein (GFAP) and HLA-DR (i.e., MHC II expression), which identify astrocytes and microglia, respectively, in postmortem brains from autistic and control brains. The numbers of GFAP and HLA-DR expressing cells in the cerebellum was elevated in the autism brain sections, and this was associated with neuronal cell loss. Moreover, the anterior cingulate gyrus, an area heavily involved in bidirectional communication between the prefrontal cortex and other deeper structures that drive motivational and emotional behaviors, showed elevated IL-6, IL-10, and TGFβ1. A postmortem analysis of the orbitofrontal cortex of deceased ASD individuals revealed elevated levels of the transcription factor NF-κB[6], which was associated with activated microglia (Young, Campbell, Lynch, Suckling, & Powis, 2011). It thus appears that there is reason to think that the activity of innate immune cells in the brain may be more vigorous in the autistic brain, driving increased cytokine production.

Why cytokines such as TGFß, IL-6, and IL-10 are elevated in the autistic brain, and what types of neurobiological functions they mediate, is open to speculation. However, it is worth noting that TGFβ1 has assumed a

---

[5] Soluble receptors for different cytokines, such as IL-1 and TNF, represent a regulatory system that "mops" up and inactivates excess amounts of cytokine. An increase in such receptors may reflect an increased level of the cytokine to which they bind.

[6] This transcription factor, it will be recalled, is induced by proinflammatory cytokines. Once activated, it translocates to the cell nucleus, where it binds to relevant regions of DNA and triggers further transcription of these and other cytokines. As such, it is a major mechanism for proinflammatory cytokine generation.

neuroprotective, anti-inflammatory role in the brain (as it does in the immune system), while IL-10 is the canonical anti-inflammatory cytokine that regulates levels of proinflammatory activity in the immune system—and may do so in the brain. With regard to IL-6, we should note that this cytokine generally receives a bad rap, and typically gets lumped with the proinflammatory gang of cytokines (like IL-1ß and TNFα), when in fact, it can function in an anti-inflammatory capacity and mediates neuroprotective influences (Peng Qiu, Lu, & Wang, 2005; Westberg et al., 2007; Wolf, Rose-John, & Garbers, 2014). Therefore, an elevation of these cytokines in postmortem tissue both implies that neuroinflammatory activity may be elevated in the brain of an autistic individual, and also suggests the generation of anti-inflammatory mechanisms[7].

Additional studies of brain tissue have reinforced this view, in that increased concentrations of TNFα and IL-6, as well as GM-CSF, a growth promoting cytokine with chemokine-like actions, were present in the frontal cortex of autistic samples (Li et al., 2009). In addition, elevations were observed in IL-8 (which functions as a chemokine), and IFNγ. Finally, measures of cytokine mRNA in the frontal gyrus of the autistic brain, revealed modest elevations in IL-1β and IFNγ, with a more robust increase for IL-6 (Patel et al., 2016), which is consistent with other evidence for increased IL-6 protein in the autistic brain (Li et al., 2009).

Studies of this type suggest that some form of inflammation-related activity is present in postmortem brain tissue of individuals with autism. However, we should note that measures were taken in the brains of individuals who died traumatic deaths (e.g., drowning), and who did so, many years after the onset of autism. Consequently, while offering important clues, it is not obvious from these studies what

is learned regarding the etiology of autism. Of some importance, it provides investigators with target molecules to investigate in very young children, although sampling in this age group is obviously highly restrictive.

Cytokine Measures in the Periphery

We have focused in some detail on research using postmortem brain and CSF samples, since these studies offer relatively direct measures of intra-CNS inflammatory profiles. Measures of immunological factors, such as cytokines and chemokines in serum or plasma, provide the next level of investigation that may be relevant to cytokine impact on brain function. There are, however, several caveats to be made here. In the absence of infection or other immune activators, typical concentrations of some cytokine in serum or plasma, tend to be very low, and in some cases may be undetectable. Statistical tools and data transformation are applied in some studies to enable some form of meaningful interpretation to be taken from the work. This is often done in the psychosocial literature, but rarely is of much concern in immunological studies, where frank increases in a given cytokine or chemokine measure are of greater value, than minimal or trace levels that introduce ambiguity and uncertainty regarding the effect of a particular variable. Still, mechanisms exist for entering brain parenchyma in the case of a number of different proinflammatory cytokines (Banks, 2015). Thus, it is conceivable that a slight elevation in a given cytokine may increase its potential interaction with neurons and glial cells once the cytokine leaves the blood. Studies that report on circulating cytokine levels typically allude to this possibility, but is highly speculative.

Early reports indicated that IL-12 (a macrophage-derived cytokine) and IFNγ were elevated in serum of individuals diagnosed

---

[7] Later we will point out that immune assessments in ASD subjects partly lean toward a more anti-inflammatory cytokine profile.

with autism (Ashwood et al., 2006). It was proposed that this may involve enhanced activation of Th1 cells, although no *in vitro* analyses of T cell function were performed to test this hypothesis. More recently, the serum levels of IL-1, IL-6, IL-12, IL-23, and TNFα were elevated in autistic individuals (Ricci et al., 2013), although this was largely a statistical conclusion, as many ASD subjects had undetectable cytokine levels, or were at the same level as control sera. The authors provided the individual concentrations for all ASD and control serum samples, revealing that for IL-6, 48% of samples in the ASD group were negative (undetectable for IL-6), while 41% of controls were undetectable. Better odds were obtained for TNFα, since 20% of the ASD samples were undetectable, while 60% of the controls failed to provide a cytokine signal. Furthermore, specific cytokine detectability varied within each subject. For instance, 6 ASD subjects that measured over 200 pg/mL IL-1β, were negative for IL-6; and 10 subjects that tested negative for IL-12, measured with positive values for IL-23 (Ricci et al., 2013). Consequently, when the cytokine data are expressed as a summary of group levels, there is a failure to consider that testing across a panel of cytokines results in variability within the ASD group. Group summaries, therefore, convey the erroneous impression that all ASD individuals respond with a given cytokine, when the opposite actually applies (e.g., an individual ASD subject is positive for IL-6, but negative for TNF, which is not what is conveyed when the ASD group as a whole is said to be higher for IL-6 and TNF). When this information is taken into consideration (as done by Ricci et al., 2013), it is evident that hypotheses regarding cytokine influences and/or changes in ASD may not be specific to a given cytokine. This complicates interpretations that would favor a given role for a proinflammatory cytokine in autism generally.

Finally, autism was linked to IL-18, a cytokine with a growing reputation (Businaro et al., 2016). Interleukin-18 is part of the IL-1 family of cytokines, and its principal immune sources are monocytes, macrophages, and dendritic cells. It is a prominent IFNγ inducer, and has been associated with a variety of pathological conditions. Elevations of IL-18, or its intrinsic production in the CNS (ostensibly by microglial cells), may contribute to the accumulation of amyloid precursor protein (APP), which has been found in the brains of autistic individuals. Moreover, IL-18 reactivity was present in the brains of individuals with viral encephalitis and tuberous sclerosis, in which ASD behavioral symptoms are common (Businaro et al., 2016). This was less directly linked to the ASD samples measured, but it was noted that ASD serum samples showed significantly lower IL-18 concentrations relative to control samples. Those younger than 10 years old, tended to show higher values of IL-18, which dropped by 33% in the older ASD samples. Apparently, although IL-18 is more highly expressed by activated glial cells, the peripheral levels of IL-18 are inversely related to brain elevations.

Discrepancies in the measurement of cytokines limit the ability to form conclusions regarding whether peripheral cytokine elevations occur in ASD. Many laboratories tend to run cytokine arrays, in the hope of observing one or more cytokines that might be systematically high in ASD. When reductions in circulating cytokines are present, as has been reported, it can be difficult to conceptualize the formation and/or persistence of autism-like behavior in terms of a cytokine impact on CNS function. This is because the main premise for assessing cytokines is predicated on the notion that they modify neural and behavioral functions. Consequently, when cytokines are reduced or undetectable, this fails to support the hypothesis that a proinflammatory cytokine mechanism contributes to autism-like symptoms.

Therefore, while low levels, in fact, may be reflective of the general status of blood cytokine levels in ASD individuals, as a mechanistic explanation for behaviors typical of ASD, the lower values hold little explanatory power. In conclusion, there do not appear to be consistent patterns regarding the relationship between circulating cytokines and ASD.

Ex Vivo *Measures of Immune Function and Autism*

The emphasis on measuring cytokines in serum/plasma and CSF is based on the expectation of observing spontaneous production and release of cytokines in ASD individuals. But the source of the stimulus and cells involved is not known when cytokine elevations are reported. Moreover, it is never clear whether the measures would be consistent over multiple sampling periods, nor whether actual behavior of the ASD individual accounts for the elevation. As discussed in Chapter 6, Stress and Immunity, psychological stressors can elevate circulating cytokine levels, and given that ASD individuals may be emotionally labile, it is likely that under some circumstances this might account for changes in cytokine concentrations in the blood.

These points will be relevant for our next discussion of the relationship between immunity and ASD. Much of this concerns measurement of lymphocyte response capacity using *in vitro* stimulation procedures. Consequently, caution should be exercised in how to interpret these studies, since it is not always clear that the response of immune cells *ex vivo* is predictive of *in vivo* functional capacity (which was a point made clear in Chapter 6: Stress and Immunity). In these cases, blood is collected from ASD individuals and leukocytes processed for analysis. One common procedure is to run a mitogen assay to determine the proliferation of immune cells that are challenged with an antigen. In earlier studies, blood leukocytes from autistic individuals were shown to have depressed proliferative responses to the T cell mitogens PHA and Con A, and the mixed T and B cell stimulus, Pokeweed Mitogen (PWM) (Ashwood et al., 2006), indicating that T cell proliferative capacity can be depressed in autism.

Other approaches to assessing the functional status of the immune system include quantitation or relative amounts of lymphocytes and their subtypes. Various surface markers on lymphocytes can provide clues as to their prior experience. Cells can be naïve, not having been activated by antigen, or they can exhibit signs of prior activation. Indeed, blood leukocytes from autistic individuals have more pronounced reductions in $CD4^+$ T cells, as well as in the percentage of naïve CD4 T cells (designated $CD4^+/CD45RA^+$) (Ashwood et al., 2006). A percent reduction in naïve T cells might reflect the presence of more activated, or experienced T cell populations. However, autistic individuals present with higher surface levels of HLA-DR on T cells, which are expressed de novo, once activated by antigen. Therefore, while T lymphocytes from autistic individuals can appear to be functionally normal (i.e., proliferation is neither augmented nor depressed), their DR+ status might suggest prior antigenic experience. In some cases, this activational status can appear incomplete, and more akin to that seen in autoimmune diseases, since the DR+ T cells of autistic individuals may also lack IL-2 receptor (CD25) expression. Because various autoimmune diseases are associated with a greater number of DR^+ T cells that are negative for IL-2 receptor expression, it is possible that in the autistic individual, T cells exhibit more autoimmune-like behavior. This notion is reinforced by earlier reports that autism is accompanied by higher numbers of $CD4^+$ T cell memory cells (designated $CD4^+/CD45R0^+$) (Ferrante et al., 2003).

As with any evolving hypothesis, data may emerge that punch holes in what would seem to be a nice and neat story. Evidence was gathered in ASD individuals indicating that the percentage of $CD4^+$ and $CD8^+$ memory T cells was similar to that determined in healthy sibling controls (Saresella et al., 2009). While this suggests that more activated T cell phenotypes are not universal in ASD, the study itself does not outright dismiss the notion that propensity toward greater reactivity or sensitivity to activation resides among T cells in ASD individuals. In both ASD individuals and their healthy siblings, there was an additional interesting observation. The distribution of activated T cell phenotypes (CD4 and CD8), when compared to nonfamilial, but healthy and age-matched control subjects, was quite different. Specifically, both those affected with ASD and their healthy siblings showed more CD8 naïve T cells, but fewer terminally differentiated (i.e., effector) and memory CD4 and CD8 T cells (Saresella et al., 2009).

How can these data be interpreted? The presence of naïve cells is an indication of possible overcompensation of thymic CD8 T cell output, or simply a lack of antigenic opportunity among this T cell subset population, which is supported by the reduction of effector memory T cells. Moreover, terminally differentiated CD4 T cells are considered effector cells with little proliferative capacity, and reductions in this population further support reduced levels of antigenic encounters. Nonetheless, these data are interesting, and together with other evidence discussed earlier, point to the idea that T cells in autistic individuals present with varying degrees of experiential history with antigenic encounters. Whether these antigens are linked to the self or to prior infectious encounters, is not known. With regard to the latter, infectious encounters in the first 2 years of life were not found to differ between those diagnosed with autism and nonautistic populations (Rosen, Yoshida, & Croen, 2007).

Consequently, greater antigenic load due to increased incidence of combatting infections may not be the answer.

The immunological similarity between ASD individuals and their siblings implies a hereditary link to an ASD diagnosis that may influence the development of a more reactive immune system. The outcome of this reactivity, however, does not have to result in autoantibodies to CNS antigens, since Saresella et al. did not find significant concentrations of antibodies directed against a small selection of up to eight different CNS antigens in the plasma of ASD subjects. Of course, this does not exclude the presence of other antibodies that might be directed against the CNS. Indeed, older research had already alluded to the possibility of higher autoantibody levels that were directed against brain antigens (Ashwood et al., 2006). Although it does appear that B cell activity is generally restricted and not substantially greater among ASD individuals that span ages from middle childhood through late adolescence.

Effector Immune Function and Autism

To this point, we have been considering the distribution of leukocyte subtypes in the blood of ASD individuals. This is useful, but reveals very little about function. Earlier, we talked about mitogen-induced proliferation in ASD subjects, and noted that deficiencies had been observed. However, T and B lymphocyte expansion is only one aspect of effector function, which evolves to produce and amplify several other critical effector functions. This includes cytokine production, cytotoxic activity, T cell regulatory function (e.g., helper functions, or suppressor functions), and antibody production. These are specific to the adaptive arm of the immune system, which has been the major focus with regard to immunological assessments in ASD individuals.

Leukocyte Production of Cytokines

Earlier, we discussed studies that measured *in vivo* cytokine levels in ASD serum, CSF, and postmortem brain tissue samples. However, *in vitro* cytokine production by T lymphocytes and cells of the innate immune system, such as monocytes and polymorphonuclear leukocytes, has also received some attention. Normally, in nonactivated T cells, the level of most cytokines in the cytoplasm is very low. However, after activation, these intracellular cytokine concentrations can increase substantially. In an early study (Gupta, Aggarwal, Rashanravan, & Lee, 1998), leukocytes from ASD children were stimulated *in vitro* to increase T cell cytokine expression. Relative to healthy controls, the percentage of T cells (either CD4 or CD8) positive for IL-2 and IFNγ was decreased, whereas IL-4 secreting T cells were increased. This was interpreted as evidence that children with autism have a greater Th2 cell functional capacity, and lower Th1 cell function. It was suggested that the more pronounced Th2 cell cytokine production (IL-4) supports the hypotheses that there is a greater likelihood for autoimmune reactivity in autism.

In another study, involving a cohort of Italian children with autism, T cells were stimulated with a staphylococcal enterotoxin (SEB), and their responses compared to those of unaffected siblings and unrelated healthy children (Saresella et al., 2009). Relative to stimulated cells from their sibling controls, cells from the autism group had greater numbers of CD4 T cells that made TNFα and IFNγ, and CD8 T cells that produced IFNγ. Interestingly, this stood in contrast to the earlier findings discussed that had found fewer IFNγ T cells (Gupta et al., 1998). The latter finding, which implies a less inflammatory state, is consistent with another finding that anti-inflammatory Th2 cytokines, such as IL-4, IL-5, and IL-13 were produced in higher amounts in cultured blood leukocytes obtained from ASD

individuals (Molloy et al., 2006). More recently (Krakowiak et al., 2017), using neonatal heel-derived dried blood spots that were archived and then assayed several years later, those derived from neonates subsequently diagnosed with ASD had higher IL-4 levels.

To some degree, these observations imply that ASD individuals may present with an anti-inflammatory propensity, although this is by no means universal. Indeed, a recent review of the immunological literature, including cytokine analyses, has noted that there is presently no consensus on whether the pattern of cytokine production in ASD is skewed toward an inflammatory or anti-inflammatory profile (Meltzer & Van de Water, 2017). However, this may be an important aspect of immunity to explore further, as recent evidence gathered in the mouse has suggested that the presence of IFNγ-secreting T cells in the meninges may influence prosocial behaviors (Filiano et al., 2016, 2017). Consequently, immunologic profiles that suppress IFNγ output, may contribute to the development of an autism-like behavioral profile.

In summary, we have pointed to several lines of evidence linking the immune system to autism, which are presented in Fig. 12.2. At present, much of the data that have been gathered reveals little about the etiology of autism, other than perhaps the influence of antineuronal antibodies, whether of maternal origin or arising from the individual with autism. However, it is obvious that some form of immune dysregulation is operating in ASD, although there is little coherence and uniformity to the information that has been collected. This is a sentiment expressed by others in the field (Careaga et al., 2017), who have pointed to the possibility that there may be an immune endophenotype that is present in a subset of ASD individuals. However, this needs to be isolated from the general ASD population to allow for a more focused investigation of

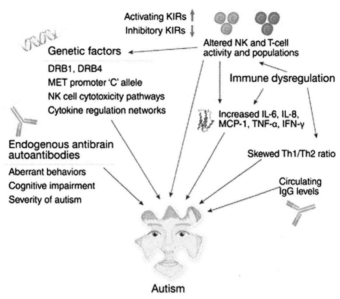

Autism

FIGURE 12.2 Ongoing immune dysregulation persists in autism. After birth and at least throughout childhood, an individual with ASD may have endogenous antibrain autoantibodies, separate from any maternal IgG, which correlate with aberrant behaviors and impaired development. There also exists a broad picture of ASD-related immune dysregulation, including an increased inflammatory cytokine milieu (e.g., IL-6, IL-8, and MCP-1), thus leading to an increased, proinflammatory Th1/Th2 ratio. T-cell and NK-cell populations may also be skewed, displaying a shift in cell subpopulations. NK cells in particular show an increased baseline activity but a decreased response to activation, rendering the cells unable to properly respond to stimuli. NK cells interact with activating and inhibitory KIRs (killer-cell immunoglobulin-like receptors), many of which are genetically linked to ASD. Other genetic factors include the oncogene MET and members of the diverse family of HLA genes. The broader background of immunogenetic factors of ASD includes multiple networks of the immune system, such as pathways that regulate cytokines and NK cells, which together constitute a broad, endogenous environment of atypical immune regulation and response. *Source: Figure and caption from Meltzer & Van de Water (2017)*

immunologic contributions to the disorder. This view is likely to gain some favor, given the heterogeneity of symptoms and states of severity that exist within the population of diagnosed autistic individuals.

Gastrointestinal Immunity and Autism

Increased attention has focused on gastrointestinal functions in autism and whether this is related to the development and/or exacerbation of autism-like symptoms. This interest has intensified based on findings that the gut microbiome is capable of influencing neural functions (see Chapter 3: Bacteria, Viruses, and the Microbiome). In fact, there have been numerous reviews of this literature, although mechanistic studies that examine brain and behavioral changes through a functional link between bacterial entities in the gut and the immunological apparatus in the small intestine are still being explored. Most certainly, there are many other ways that the microbiome might affect brain function, and some of the data are relevant to autism. For example, germ-free mice exhibit deficits in social interactions, anxiety, and motor function, which can be corrected by replacement with microbial-rich fecal material (Bruce-Keller, Salbaum, & Berthoud, 2018). The degree to which immune factors are involved in providing recovery of behavior, is not clear. As mentioned earlier,

T cell-derived IFNγ may facilitate social behavior (Filiano et al., 2016). This study involved a thorough dissection of T cell factors and adhesion molecules to establish that a loss of social exploration and/or interaction was dependent on the production of IFNγ by T cells that resided in the meninges. Similar approaches are needed to investigate whether intestinal bacteria are shaping the immunological profile of gut immune processes to maintain normal behavioral functions.

The gut immune system is perhaps the most abundant of the regionally distributed components of the common mucosal immune system (which also incorporates the lungs and upper respiratory tract and urogenital regions). Altered microbiota may either impact or be a result of immunological activity in the duodenum and small intestine. Research into this possibility has not been overly extensive, but it is known that GI distress is comorbid with autism. This may influence the behavioral symptoms of the autistic individual, and as such may also drive further changes in the gut and associated immune system. To some extent, dietary manipulations have been reported to improve symptoms, but many of the observations are anecdotal and based on parental observations. For example, a dairy or wheat-free diet (referred to as a casein-free and gluten-free [cf/gf] diet) may have benefits, and might suggest food sensitivity or the presence of food allergies.

There are generally three different hypotheses regarding the importance of the gut in autism and potential causes of GI symptoms. These include the *leaky gut hypothesis, dysbiosis,* and *autism colitis* (Jyonouchi, 2009). The leaky gut hypothesis, as we saw in Chapter 3, Bacteria, Viruses, and the Microbiome, refers to a possible deficit in intestinal permeability that compromises physical and immunological defenses against possible toxins and pathogens encountered through the diet. Alternatively, macromolecules (e.g., milk proteins) may be able to diffuse through the intestinal wall and sensitize resident immune cells, which then results in chronic allergic-like reactions and physical discomfort. Aside from the intraepithelial and lamina propria lymphocytes, macrophages and dendritic cells of the gut wall, there are also more organized groups of immune cells present in the peyer's patches, the quasi lymph nodes of the small intestine. These represent the immune defense of the gut and will drain into the mesenteric lymph nodes (MLN) where further processing and general immune reactivity takes place. Other more nonspecific defense mechanisms operating in the gut include antimicrobial enzymes produced by specialized local cells (e.g., Paneth cells found in the intestinal crypts), which degrade gram-positive and gram-negative bacteria. These types of defenses limit the entry of luminal bacteria through the gut epithelium (Wells et al., 2017). Further, the colonic epithelium can maintain gut homeostasis and an anti-inflammatory state.

Evidence for a breach of these defenses or at the very least, signs of a "leaky" gut has been pursued using the Lactulose: Mannitol ratio measure. Mannitol is normally absorbed by the gut, but entry of lactulose, a larger molecule, is less permissive. This approach is used clinically for testing gut permeability, and there is some support for autism individuals having increased gut permeability, but this is not a well-replicated finding (Samsam, Ahangari, & Naser, 2014). Evidently, it is not fully certain that a leaky gut is present in all ASD subjects.

Given the hundreds of different species of bacteria in the gut, a balance needs to be maintained between these bacteria and the immune system. Furthermore, commensal bacteria are also responsible for ensuring that pathogenic bacteria do not exert toxic effects on host tissue. Hence, a further delicate balance exists within the gut (and other mucosal surfaces) to ensure that host physiological functions are maintained, while the bacterial flora, in turn, also benefit from this situation. Dietary consistency is likely to maintain some sort of status quo in the makeup of the intestinal flora. Indeed, in children with regressive autism, treatment with peroral vancomycin resulted in temporary attenuation of their behavioral symptoms (reviewed by Jyonouchi, 2009). Because antibiotics may shift the makeup of commensal bacteria, the implication of this finding is that vancomycin had corrected a dysbiotic state in the ASD children. Yet, antibiotics destroy both "good" and "bad" bacteria, making it uncertain what actually may be occurring in children with ASD.

Assessment of colonic content of regressive ASD individuals has revealed newer emergent clostridial bacterial species, and a preponderance of nonspore forming anaerobes and microaerophilic bacteria. Similarly, there is a greater incidence of *clostridium histolyticum* in the intestinal flora (Ding, Taur, & Walkup, 2017). This type of bacteria is a recognized producer of illness-inducing toxins, and potentially can account for GI distress and other symptoms in ASD individuals. In fact, such distress can impose a burden on the immune system in ASD, resulting in impaired production of innate immune cytokines, including IL-1β, IL-6 and IL-12 (Jyonouchi, Geng, Streck, & Toruner, 2011). Body compartments like the gastrointestinal system rely on innate immune responses to aid adaptive immune responses. Therefore, the GI distress experienced in ASD may perpetuate problems that are both physiological and behavioral.

The presence of potentially toxic bacteria in the gut of ASD children focuses attention on the protective tools present in the gastrointestinal system. Aseptic or sterile conditions need to be maintained outside the enteric system in order to prevent infectious disease. Immunological surveillance within the intestinal wall ensures that any bacterial or parasitic infiltrate is eliminated, and in recent years, more information has been gained concerning just how the immune system monitors the state of the intestinal flora (Powell, Walker, & Talley, 2017). This prepares and ultimately engages the immune system when potentially pathogenic dysbiosis takes place in the gut. Dendritic cells, the professional antigen presenting cells of the immune system, are a particularly important source of immune surveillance in the gut. These cells widen the normally tight junctions between intestinal epithelial cells and extend their dendritic branches into the gut lumen, where they can sample for the presence of novel changes in microbial antigenicity and microbial by-products. The small intestine is also rich in another type of antigen-processing cell, namely, M (for *microfold*) cells. These cells are abundant in the follicle-associated epithelium, where they present or transport antigen to lymphocytes in the peyer's patches. This presentation may induce immune responses as well as promote oral tolerance to food antigens.

Oral tolerance is a case of immunological tolerance that is best demonstrated experimentally. Animals that receive foreign proteins for the first time via the gastrointestinal tract (i.e., by feeding), will display reduced immune responsiveness to the antigen when it is injected systemically. In contrast, animals injected with the antigen without prior exposure to the protein through feeding, display enhanced T and B cell immune responses to the antigen. Evidently, initial encounters with foreign proteins through oral ingestion induce tolerogenic mechanisms that suppress the adaptive immune response to food antigens. The precise cellular interactions involved in producing tolerance to enteric antigens are currently being determined. Dendritic cells of the lamina propria (nonpeyer's patch region of the small intestine) can be transported from the intestine to the mesenteric lymph nodes via draining lymphatics (see Chapter 2: The Immune System: An Overview, for a discussion of the lymphatic system). Retinoic acid, a vitamin A metabolite, is released by DCs in the mesenteric lymph nodes (MLN) and local stromal cells (connective tissue cells of the MLN). This induces expression of gut-homing receptors (e.g., alpha4/beta7 integrin and CCR9) on activated T cells that then influence, in conjunction with TGFß and FoxP3 expression, the development of Tregs. These cells then migrate back to the lamina propria, where IL-10 derived from macrophages coaxes Treg expansion and subsequent egress from the gut to extra-enteric lymphoid compartments (e.g., spleen and lymph nodes) and induction of systemic antigen-specific tolerance (Pabst & Mowat, 2012). Indeed, given that autoimmune and inflammatory responses involve the generation of Th17 cells, it is pertinent that induction of retinoic acid by DCs, may serve to inhibit the generation of Th17 cells (Mucida et al., 2007), and thereby establish an anti-inflammatory state. Interestingly, excess IL-17 production is found in ASD, along with enhanced expression of the IL-17 receptor on monocytes (Nadeem et al., 2018). This is thought to create a proinflammatory state in ASD. Moreover, restoration of vitamin A deficiencies in ASD children improves behavioral symptoms (Guo et al., 2017). However, in spite of this increased understanding of gut immunoregulatory systems, it is not known whether intestinal tolerogenic and anti-inflammatory mechanisms are deficient in autism patients. Nonetheless, given the presence of food hypersensitivity in subsets of ASD patients (Jyonouchi, 2009), and the high prevalence of gastrointestinal complaints (Kang, Wagner, & Ming, 2014), this represents an important area of immunological research to pursue in future studies.

Very little research has been conducted in which gastrointestinal cells have been examined for immunological properties in ASD patients. However, the opportunity for such research has been provided by clinical examinations of patients with colitis and Crohn's disease, which are highly associated with autism

(Lee et al., 2017). In fact, up to 50% of ASD patients show inflammation of the gut, which was characterized by lymphoid nodule hyperplasia, increased infiltrating eosinophils (which are cells of the myeloid lineage), and lymphoid aggregates (Kang et al., 2014). For a more specific examination of the properties of gut-related immunopathology, one can turn to a report by Ashwood et al. (2006), who obtained tissue samples from ASD patients with ileocolonic lymphoid nodular hyperplasia (LNH). They were able to stimulate isolated CD3 + T cells to induce cytokine production, which revealed enhanced concentrations of TNFα and IL-12 production, but lower IL-10 levels. It was further reported that gut biopsies obtained from autistic individuals had increased numbers of CD3 + T cells that were positive for TNFα and IFNγ. Whether these immunological characteristics are in any way etiologically linked to autism is not known. However, they may contribute to an exacerbation of symptoms, and in cases of remission, might influence autistic regression (Xue et al., 2008).

Microbiota in Relation to ASD

A thorough review of the relevant literature suggested that in the majority of studies ASD was accompanied by microbiota alterations (Kelly et al., 2017). It was nevertheless cautioned, based on a systemic review (Cao, Lin, Jiang, & Li, 2013), that in early studies, the data were limited and the findings variable. Still, there were reports showing that this disorder was associated with elevated levels of short chain fatty acids as well as particular microbiota species, notably *Bacteroides vulgatus* and *Clostridium Bolteae* (Finegold et al., 2010; Wang et al., 2012), as well as diminished diversity of the gut microbiota community. Subsequent studies also pointed to gut microbiota variations at the genus level among children with ASD (De Angelis et al., 2013; Strati et al., 2017). It is of practical importance that bacterial diversity in saliva and dental samples were similarly lower in children with ASD relative to controls, as well as lower levels of Prevotella, Selenomonas, Actinomyces, Porphyromonas, and Fusobacterium (Qiao et al., 2018). At this time there have been too few studies that assessed microbiota in saliva or dental samples, but it's simple accessibility (relative to gut microbiota), might turn out to be a good method of assessing microbiota in relation to illness. Of course, it needs to be determined that oral and gut microbiota are correlated, and what their implications are relative to various disease conditions, especially those that operate through inflammatory changes provoked by microbiota.

At this point, it ought to be clear that microbiota could potentially affect autism symptoms through different routes, including through their well-established effects on immune processes that come to affect brain functioning (e.g., Doenyas, 2018; Needham, Tang, & Wu, 2018). As we've seen, most studies have been of a correlational nature, but in a study with a limited number of participants (N = 18) transfer of fecal microbiota resulted in a diminution of autism symptoms together with attenuation of GI symptoms (Kang et al., 2017), still being apparent 8 weeks after the transfer. Admittedly, the number of participants in this study was small, and it is uncertain how long the positive action of the treatment lasts. The importance of microbiota in relation to autisms was underscored by a report that the antibiotic vancomycin, which acts only in the gut, provided a degree of symptom alleviation, although these actions were only short-lasting (Sandler, Finegold, Bolte, Buchanan, & Maxwell, 2000). It is still too early to begin broader analyses of the effects of fecal transplants (even in oral form) on ASD symptoms, and even the Kang et al. (2017) study only received approval for a small number of individuals (not in young children). Nonetheless, based on the available data, it was suggested that the gut microbiome might be a

viable target in the treatment of ASD (e.g., Yang, Tian, & Yang, 2018). However, identifying the specific bacteria that should be manipulated is still uncertain, making fecal transplants, which influence many types of microbiota, a better bet than probiotic treatments that are more restricted in their actions. There is also the issue of whether optimal treatments would have to be applied early in life, before autism symptoms are present (e.g., Watkins, Stanton, Ryan, & Ross, 2017).

ASD and Food Allergies

A final word needs to be said regarding food allergies. These have been categorized as IgE-mediated or non-IgE-mediated. Atopic conditions (i.e., those that involve allergic reactions) are typically immune mediated, with IgE responses serving as the trigger that liberates symptoms such as redness, swelling, itching, sneezing, and runny nasal and lacrimal (i.e., tear ducts) discharge. Mast cells are a major target for IgE, and release of histamine and other molecules by these cells produces the foregoing allergy symptoms. Other types of allergic reactions, such as asthma, can involve T cells that have become hypersensitized to their respective allergenic molecules. Atopic conditions are quite common in ASD individuals, and although food allergies are frequently reported, ASD is also associated with asthma, eczema and psoriasis (Theoharides, Tsilioni, Patel, & Doyle, 2016). Interestingly, examination of atopic symptoms in those ASD individuals who also have comorbid GI difficulties has not revealed uniquely higher incidence of IgE-mediated food allergies (Jyonouchi, 2009).

With regard to non-IgE-mediated food allergies, peripheral blood leukocytes from ASD individuals do react to common dietary proteins, such as whole cow's milk protein (and its major derivatives, including casein and beta-lactoglobulin), and the wheat protein, gliadin. Specifically, leukocytes obtained from those

ASD subjects who have pronounced GI symptoms, responded with higher TNFα and IL-12 production when challenged with milk protein antigens and gliadin (Jyonouchi et al., 2011). This was not uniquely due to the presence of GI problems, as those ASD individuals who were negative for GI symptoms also responded with significantly greater TNF and IL-12; however, unlike the GI-positive ASD subjects, they were significantly less responsive to gliadin and casein and beta-lactoglobulin. In effect, GI symptoms in ASD individuals are associated with a broader panel of food antigen sensitivity.

Overall, despite a paucity of close immunological assessment of ASD patients with GI disturbances, there are sufficient data to consider seriously the hypothesis that gut food processing, as well as gut-associated immune surveillance, may be involved in autism. Whether these conditions are antecedent to the development of autism, or evolve concurrently with or following the onset of the disorder, remain to be determined. In fact, such problems can emerge from 18 months to 3 years of life (Bresnahan et al., 2015), but can become less severe as children with ASD grow older (> 6 years) and display less food hypersensitivity (Jyonouchi, 2009). This may be due to maturation of oral tolerance mechanisms, which would be in keeping with other states of developmental lag present in autism.

Perinatal Immune Activation

For many years, there has been a stirring and at times emotional debate regarding whether vaccinations contribute to the development of autism. The pediatric and neurological medical communities have vociferously rejected this notion, citing a lack of convincing scientific evidence. Nonetheless, the debate continues, as do investigations into the impact of vaccinations on psychiatric conditions (Leslie, Kobre, Richmand, Aktan Guloksuz, & Leckman, 2017). In addition to childhood vaccinations, there has been intense interest in the immune response of the mother during pregnancy. The autoimmune disease literature, already discussed earlier, is but one part of the probe into the impact of a mother's immune system on the developing embryo or fetus. Another approach is to consider the presence of maternal infection and the immune system's response to this on intrauterine neurodevelopment and subsequent postnatal behavioral development. Epidemiological observations precipitated experimental investigations into this question, and for the most part, much of the current literature is represented by animal studies. Because much of the animal literature pertains to neurodevelopmental alterations and behaviors that include not only those relevant to autism, but also schizophrenia and many other psychiatric disorders, discussion of this literature is presented in the chapter on prenatal and postnatal manipulations and their impact on behavior (as well as Chapter 13: Schizophrenia). Suffice to say that recent studies that examined influenza vaccines taken by women during pregnancy, have not found a significant relationship between vaccination and risk for autism among their offspring (Zerbo et al., 2017). Vaccinations, however, are not equivalent to actual infection, which has a different time course, impacts behavior, and can vary in the level of immune cell recruitment and responsiveness. Nonetheless, if studies of prenatal infection in animals warn against receiving vaccinations during pregnancy, there does not appear to be strong evidence, at present, that this is a neurological risk for the developing fetus. Consequently, the CDC continues to support and urge vaccinations in pregnant women to ensure prevention of infection, which, in being a different animal altogether, could be a greater threat than receiving a vaccine.

CONCLUDING COMMENTS

In general, as with many other investigations, the relationship of the immune system to autism has proven to be complex. It is difficult to determine cause and effect, when individuals with autism are sampled well after neurodevelopmental delays have already been initiated. Moreover, it is virtually impossible to separate the immune alterations in autism with neurohormonal influences on the immune system as a function of their aberrant behavior and changes in neural tone.

One of the challenges of human studies is to assess the immune status of newly born infants and form predictions regarding the likelihood of whether a particular immune profile is indicative of future development of autism. However, the immune system is developing prenatally, and at birth contains a full complement of adaptive and innate immune components ready to receive antigen stimulation, which will further shape the immune system. How prenatal immune development is shaping neurodevelopment is not well known, and of course, human studies are not in a position to address this question with as much detail as animal studies. The animal studies, however, are limited with regard to this question. To be sure, there is a growing movement that identifies microglial cells as shapers of synaptic density, although this is something that very likely is most important during the postnatal period. At present, the research data are strongly suggestive of immune dysregulation in autism. However, to what extent this is another index of aberrant physiological behavior in the ASD individual is still a matter of speculation. Longitudinal studies of immune function in ASD will help resolve the issue of whether the various immunological deviations are permanent or temporary states.

Schizophrenia

THE EXTRA BURDEN OF SCHIZOPHRENIA

Schizophrenia is a highly debilitating psychiatric disease defined by the presence of psychosis (a loss of contact with reality). Understanding this disorder and identifying its etiology has long engaged thinkers, philosophers, scientists and clinicians. Among the general public, and especially throughout the 20th century, it has been misidentified as multiple personality disorder or what many termed "split" personality. Even before this, it was thought of as a malady of the blood or the result of meddling by the devil and other possessive spirits. Among other confusions, schizophrenia is wrongly thought to be an aggression disorder with patients having a predominantly violent predisposition, an impression that leads to the perception that schizophrenia causes criminal behavior. Not uncommonly, a large proportion of the mentally ill, who fit the diagnosis of schizophrenia, are incarcerated, which is more a problem of poor public health policy, rather than the defining trait of criminals. Additional misunderstandings—mostly propagated by popular films—is that some people with schizophrenia are intellectually brilliant, bordering on genius, or at least display a high level of intelligence. As with many mental illnesses, there are high functioning individuals, although for schizophrenia, the harsh statistic is that close to 20% of people diagnosed with the disorder are unemployed, and in some cases with little prospect of gaining employment. This is due to significant deficits in cognitive functions such as impaired attention and working memory, which has been recognized for many years, and diagnostically, has become a core symptom domain of schizophrenia. For all these reasons and more, schizophrenia remains the most debilitating and disturbing mental illness as judged by the magnitude of functional disruption and its relatively poor prognosis. Of course, it takes an exceptional toll on family members, and has an enormous societal cost that in the United States, exceeds 150 billion dollars a year.

OVERVIEW OF SCHIZOPHRENIA

There are presently no adequate explanations for the etiology of schizophrenia, and the disease continues to remain elusive in relation to preemptive intervention and/or eradication. This is not to say that new clues for what causes schizophrenia are not forthcoming. In contrast to the views of clinicians and psychiatrists prior to the 1950s, who approached etiology from a psychodynamic perspective inspired by Freudian psychoanalytic thinking, the disease overwhelmingly is now considered to have a biological origin. This was inspired by early familial studies of heredity, and in the past decade, large-scale genetic studies. However, the presence of genetic mutations—many of which are inconsistent across different individuals with schizophrenia, and which makes the task of blaming specific genes more difficult—is not a sufficient condition for the appearance of the disease. Most certainly, it may be a potential necessary requirement, but complete emergence of the disease appears to occur in conjunction with particular environmental conditions. These additional conditions are considered under the general rubric of "stress," and as such, are championed by the diathesis-stress—or the related "two-hit"—hypotheses of schizophrenia. Into this theoretical framework, one can integrate the immune system, and the relationship of immune-related cells and molecules (e.g., microglia and certain cytokines) to neurodevelopment and neural modulation.

The legitimacy of an immune approach to schizophrenia has steadily increased over the past few decades. Many studies had attempted to link the immune system to psychosis, but for a variety of reasons, the bulk of these studies lacked well-controlled longitudinal assessments that spanned the first 20–30 years of life. Thus, mainstream views had not fully embraced the role of the immune system in the etiology of schizophrenia. But this all changed with a series of landmark studies that examined a large array of genes and located a significant number of variations within the MHC region (i.e., the HLA genes) of chromosome 6. There is now a concerted effort to learn more about how the major histocompatibility complex (MHC) region figures in the etiology of schizophrenia. One such role appears to involve the gene for the C4 component of complement[1], and which may be involved in synaptic pruning (Sekar et al., 2016). Therefore, there is good reason to take seriously the hypothesis that schizophrenia is an immune-related disorder.

In this chapter, we will consider some of this evidence and consider new avenues that might be worth exploring. Following a brief overview of the nature of schizophrenia and efforts to determine etiology and optimal therapy, we will turn our attention squarely on how the immune system fits into this complex and seemingly intractable disease. To a degree, we will face questions similar to those addressed in assessing autism. Are autoimmune elements directed at the central nervous system (CNS) responsible for schizophrenia? Is the immune system in schizophrenia different from that of healthy individuals? Do early life prenatal and postnatal immune perturbations account for the appearance of schizophrenia? And finally, can we target the immune system (e.g., with anti-inflammatory drugs) to attenuate clinical symptoms of schizophrenia?

---

[1] Recall from Chapter 2, The Immune System: An Overview, that the various genes that line up on chromosome 6 to form the MHC, have as their neighbors genes for key components of the complement system. Evidently, mutations in these genes may also affect how the nervous system is sculpted, and if not done right, might facilitate the development of schizophrenia.

DEFINING SCHIZOPHRENIA

The term schizophrenia did not exist until early in the 20th century, when Eugen Bleuler, a German psychiatrist, coined the term as a combination of the Greek words for "split" (viz, *schizo-*) and "mind" (viz, *phrene*). Bleuler was urged by the need to differentiate the condition from Kraeplin's earlier formulation of "dementia praecox," which involved early (childhood) loss of mental faculties due to a suspected degenerative process; this placed Kraeplin firmly in the biological camp with regard to the etiology of psychological disorders (Moskowitz & Heim, 2011). It was Bleuler's contention that the disorder actually emerged later than childhood or early adolescence, and was not necessarily unremitting and degenerative. There may have been merit to this general understanding of mental illness, and the fact that it subscribed to individual differences in its presentation. Moreover, the conceptualization that Bleuler formulated, appealed more to psychoanalytic forms of thinking, since greater emphasis was placed on psychological anomalies, as opposed to organic explanations that sought causality in unique biological or constitutional states of the individual. Nonetheless, throughout the 20th century, there was a gradual movement to restore Kraeplin's more biocentric conceptualizations, and over time schizophrenia gained prominence as a neurodevelopmental disorder. This view is in keeping with contemporary clinical and research findings, and is a restoration of Kraeplin's original conception of the disease as having origins early in childhood.

Schizophrenia has multiple domains of dysfunction. Individual's displays aberrant forms of thought and behavior, as well as distorted sensory and perceptual experiences, which converge on a diagnosis of psychosis. The key elements of psychosis are hallucinations, delusional and/or disorganized thinking and speech, grossly disorganized (or catatonic) behavior, and paranoid ideation. These are often referred to as *positive* symptoms, not because they are highly valued, but because they stand out as bizarre and unusual in the context of normal cognitive and social experience. A constellation of additional factors may then gravitate about the core presence of psychosis, and which are also emblematic of schizophrenia. This typically includes a set of "negative" symptoms, which are so-called for the general failure to express various normal cognitive and emotional behaviors. This includes absent or reduced verbal communication (i.e., poverty of speech or alogia) and loss of affect. Negative symptoms, therefore, commonly comprise blank, featureless facial expressions, as well as anhedonic and avolitional states. This plethora of "a"- based prefixes served as a useful mnemonic, since schizophrenia appeared to subscribe neatly to the four A's rule, comprising altered a̲ssociations, impaired a̲ffect (flat or inappropriate), a̲mbivalence, and a̲utistic isolation (Insel, 2010). These features reflect a loosening of *associations* (often in thought) considered critical in a diagnosis of schizophrenia (Moskowitz & Heim, 2011)[2]. In

---

[2] Note the incorporation of "autistic isolation" into this general conceptualization. Indeed, historically it was not uncommon to think of the child with autism as having a schizophreniform disorder (i.e., a seemingly less full-blown form of schizophrenia). However, the differentiating factor is that schizophrenia appears in its frank, clinical form much later than the core features of autism, which are diagnosed after 3 years. Still, recent evidence has indicated that schizophrenia, autism and bipolar disorder shared common patterns of gene expression, although distinct differences were also detected (Gandal et al., 2018).

addition to these features, the illness may be accompanied by disturbed sleep patterns, dysphoria, hostility, derealization (feeling that surroundings are not real), depersonalization (feeling detached or disconnect from the self), lack of insight regarding one's illness, and deficits of social cognition. As with many psychiatric conditions, all these symptoms need to be differentiated from possible organic causes, such as trauma due to injury or stroke. However, once an organic basis is dismissed, the presence of negative and positive symptoms, as well as psychosis, can point the clinician squarely toward a diagnosis of schizophrenia, or some variant of this disorder. Indeed, schizophrenia can appear in multiple forms, and core symptoms—such as psychosis—can be seen in other disorders. This can make a pure diagnosis difficult. For example, "affective psychosis," is a subcategory of schizophrenia, but might occur in mood disorders, such as bipolar disorder[3]. Therefore, in recognizing that the syndromal nature of schizophrenia may result in there being many "schizophrenias," as based on the preponderance of a given symptom(s), we nonetheless will follow suit with much of the literature, and use the term "schizophrenia."

THE PATHOPHYSIOLOGY OF SCHIZOPHRENIA

Although we focus largely on the immunological hypothesis of schizophrenia, immunity is essentially a distal variable, since mental functions are driven by neurons. Proximal influences on neuronal and synaptic functions typically represent factors within and between neurons. To this end, much of the effort to try and understand the cause of schizophrenia has focused on neurochemistry, and in particular, the monoamine neurotransmitters.

The Dopamine Hypothesis

Of the monoamines, altered functional properties of dopamine and its receptors has figured prominently as an explanatory tool for the symptoms of schizophrenia. According to the "dopamine hypothesis," schizophrenia is due to dopamine over-activity in widespread areas of the brain. It was believed to account largely for the positive symptoms of the disorder, and it emerged as a result of the biological revolution in psychiatry triggered by the introduction of the drug chlorpromazine, a derivative of the antihistamine, promethazine (Baumeister, 2013).

THE ORIGIN OF NEUROLEPTIC AGENTS

The drug, chlorpromazine, was used to treat bipolar patients in Europe during the early 1950s, as it reduced manic symptoms, such as agitation and excitement. It caught the attention of American psychiatrists who administered the drug to psychotic patients, with reports of reduced positive symptoms and increased clarity of thought and emotional stability. Chlorpromazine soon achieved celebrity status, and ushered in several decades of

pharmaceutical development that saw the persistent use of neuroleptic drugs. Because there was a reduction in motivational or volitional deficits in behavior, along with improved control of emotional lability (without obvious sedation), the term "neuroleptic" was used to describe chlorpromazine and other drugs that were soon developed. This was reinforced by animal studies, in which these drugs suppressed spontaneous movements, reduced initiative and interest in the

---

[3] It is instructive that the common variants among the genes of the MHC, were observed in populations that suffered either schizophrenia or bipolar disorder.

(cont'd)

---

environment, suppressed emotional responses, and lowered aggressive and impulsive behavior. With the advent of chlorpromazine, therefore, it was evident that behavior could be controlled, and it also implied that psychotic and mood disorders emerged from unspecified biological dysfunction. The nature of this dysfunction soon gave birth in 1963 to the dopamine hypothesis, which has endured to this day (Howes, McCutcheon, Owen, & Murray, 2017).

The success of chlorpromazine in attenuating the more bizarre and disruptive symptoms of psychosis-related disorders resulted in the development of other drugs, which came to be referred to simply as *antipsychotics*. Parallel with these pharmacotherapeutic developments was a better understanding of the monoamine neurotransmitters, which eventually led to the proposal that antipsychotics worked by blocking monoamine actions, and in particular, they antagonized the dopamine D_2 receptor. Together with other evidence that dopamine agonists could produce psychotic-like states, and that dopamine was important in brain pathways that regulated cognitive, emotional and motor functions (viz, the mesocorticolimbic and nigrostriatal pathways), it became evident that the positive symptoms of schizophrenia, could be due to excessive release of dopamine and/or heightened dopamine receptor stimulation. This version of the dopamine hypothesis was eventually modified to incorporate certain nagging problems regarding the precise efficacy of antipsychotics, and the disconnect between cognitive deficiencies (e.g., hypofrontality)[4] and negative symptoms with more active dopamine release and D_2 receptor transmission. Therefore, as opposed to a predominant *hyper*-dopamine activity, which was felt to be exclusively in subcortical regions, such as the striatum (which regulates voluntary motor functions, as well as habitual, repetitive, and stereotypic movements), room was made for evidence that suggested *hypo*-dopamine activity in the frontal cortex. To date, much of the molecular imaging work in patients has tended to point in this direction: too much dopamine below the cortex; and not enough in the cortex (Howes & Kapur, 2009). How this inverse dopamine involvement in different regions of the brain developed continues to be a driving question, and may be linked to neurodevelopmental disruptions.

Enter Glutamate…

The dopamine hypothesis has enjoyed a long run, and despite having certain problems, it persisted and expanded into a more general neurochemical theory of schizophrenia that now includes glutamate, and to a lesser degree, gamma-aminobutyric acid (GABA) (Howes et al., 2017). The reason for this expansion resides in the failure of the dopamine hypothesis to explain fully the cognitive and negative symptoms of schizophrenia, the general failure of antipsychotics to affect negative symptoms, and the fact that not all patients respond to

---

[4] Hypofrontality is a term that refers to reduced or deficient function in the frontal lobe of the brain. Although it originally was derived from observations of a physiological deficiency, such as reduced blood flow, it can sometimes be applied to the associated deficits in psychological functions deemed unique to the frontal lobe. For instance, executive functions (attention, decision-making, emotional regulation, and working memory) in the prefrontal areas of the frontal lobe, are less well implemented, and therefore, behavioral indicators of *hypofrontality*.

antipsychotic treatment. The glutamate hypothesis was designed to remedy these problems. The chief premise is that the pathophysiology of schizophrenia involves insufficient glutamate activity, or NMDA receptor *hypofunctioning*. Support for this has come from studies that showed NMDA antagonists could produce negative, positive and cognitive symptoms that are aligned with the notion of a schizophrenia-like phenotype.

From this perspective, a disturbed excitatory system in the brain creates imbalances in the integration of neuronal firing patterns, such that there is a bias in the flow of excitatory and inhibitory influences on key behavioral actions, such as response-inhibition, maintenance of working memory, regulation of emotion, and control over perceptual functions. These actions may have a neurodevelopmental origin, given that during the preadolescent years there is a gradual balancing of inhibitory and excitatory synaptic functions in the prefrontal cortex (Insel, 2010). In fact, a thorough analysis of genome wide studies that identified well over 100 genetic abnormalities, pointed to a general loss of balance between excitation and inhibition in the brains of schizophrenic patients (Devor et al., 2017.) In any case, we will talk more about the glutamate hypothesis later when discussing evidence regarding antibodies against the NMDA glutamate receptor.

To sum up, if there is a major pathophysiological profile of schizophrenia, this is dominated by a general appreciation that dopaminergic transmission is dysregulated, and may, in fact, be tied to altered functions of regulatory neurotransmitters that are essentially responsible for traffic control in the brain. To this end, glutamate and GABA, our "green light" and "stop sign," respectively, take part in converging simultaneously on most cells in the brain, and by doing so, dictate whether the target cell will generate a response (glutamate wins) or not (GABA wins). This is the general scenario in virtually all areas of the brain, and

an important part of the neural circuits driving many of our behaviors. There is no mystery that schizophrenia is a condition characterized by competing neural commands and scrambled bits of neurochemical information. In essence, genetic abnormalities that control many of the slow and fast forms of transmission in the central nervous system, offer a good explanation for the origins of the discordant and disorganized nature of the behaviors displayed in schizophrenia (Devor et al., 2017). It is knowing precisely how we get from genetic variability to neurochemical dysregulation that is the over-arching problem—the solution to which, remains elusive.

THE IMMUNE SYSTEM AND THE TWO-HIT MODEL OF SCHIZOPHRENIA

Immunological factors have long been considered a viable influence in the etiology and/or precipitation of psychosis and diagnosis of schizophrenia. Viral and bacterial infections, including syphilis, have been linked to psychosis and dementia, while other findings have highlighted increased inflammation in the brains of schizophrenic patients using neuroimaging approaches. As previously reviewed (Feigenson, Kusnecov, & Silverstein, 2014), the role of inflammation in schizophrenia is very likely just one element of a range of other potential influences—some from the environment, some from constitutional anomalies—that converge on the individual and skew development toward an abnormal phenotype. One particular focus is the perinatal period, when heightened sensitivity to immunological stressors might derail normal brain development.

Much of this research is animal-based, and involves a restricted set of immune activation models. Moreover, it is concerned mainly with immune responses induced by the maternal immune system, and how this may propagate a sequence of physiological events that impact the developing fetus. The outcome of this work has met with mixed results, but the immune hypothesis remains well-respected, and a legitimate piece of the overall puzzle in relation to the origins of schizophrenia. New pieces are being introduced regularly, but the cause and treatment of schizophrenia remains enigmatic. There is certainly more known about the disorder, but this is largely a better characterization of the problem, and less a clear path to prevention and better treatment. Most certainly, efforts to repurpose certain anti-inflammatory drugs (such as acetylsalicylic acid, celecoxib, and minocycline) for the treatment of schizophrenia have begun, and some modest, but inconclusive, signs of promise have been reported (Bumb, Enning, & Leweke, 2015).

With the data gathered from genetic studies, and increased understanding of the predictive power of prodromal behavioral factors (Fusar-Poli et al., 2013), it is clear that schizophrenia arises from a convergence of multiple influences that favor the development of disorganized and/or anomalous neural functions. For example, the current prominence of the glutamate hypothesis has raised the notion that glutamate-induced neurotoxicity may involve some interaction with inflammatory processes. Moreover, immune factors—cytokines and proinflammatory stimuli—modulate monoamine neurotransmitter systems, including dopamine. Consequently, inappropriate or prolonged release of cytokines in the brain and/or periphery may very well provide a destabilizing influence on neural functions operating in the context of a diseased state.

We will explore more closely some of the areas just mentioned, but again raise a note of caution: the immune system alone may not fully explain schizophrenia. More likely, the immunological theory of schizophrenia helps to bridge the contributions of genetics and the environment, since neither one alone is sufficient to explain how the disorder develops. It is indisputable that schizophrenia is a genetically based disorder. This has been a driving hypothesis for more than three decades, influenced by studies that revealed schizophrenia to have strong family ties, a high rate of concordance among monozygotic twins, and occurrence in adopted children that had been born to (biological) parents with schizophrenia (Insel, 2010). Unfortunately, the specific gene or genes that trigger the disorder have not been identified to a degree of confidence that serves to predict accurately whether an individual will develop schizophrenia. Important discoveries have been made linking mutations in the MHC region, as well as other more prominent mutations, such as in the genes for neuregulin and transcription factor 4, which are important in neuronal development and differentiation. In addition, genes for neurotrophins, serotoninergic neurotransmission, cell adhesion molecules, sodium channels, and genes related to the dopamine system have been linked to schizophrenia (Zai, Robbins, Sahakian, & Kennedy, 2017). As impressive as this is, no single gene appears to account for all cases of schizophrenia. That is, their individual effect sizes are small, suggesting that constellations of gene variants may drive susceptibility and/or increased risk for schizophrenia.

Aside from influencing gene expression, immunological influences may also impact delicate postnatal neurodevelopmental stages in preadolescence. Neuroimaging data have revealed that there is excessive cortical thinning in childhood onset schizophrenia (Thompson et al., 2001), and that cortical degeneration can persist into adulthood (Cobia, Smith, Wang, & Csernansky, 2012). In fact, it is not unusual to observe fluctuating changes in cortical volume during early life. The development of synapses, particularly within the cortex, is highly dynamic during preadolescence. Gray matter volume expands dramatically during these early years, followed by a synaptic pruning phase that creates a seemingly optimal level of functional connections. The frontal cortex, in particular, does not fully attain a stable state of synaptic connections until the early 20's (Insel, 2010). Until then, the cortex is sufficiently plastic, but vulnerable to toxic damage and/or modifying environmental experiences. In schizophrenia, the early cortical changes appear to be unregulated, and therefore sustain more pronounced reductions in neurons and synapses. Whether an aberrant immune system contributes to this is an open question. Animal studies have suggested that microglial cells influence synaptic pruning, which appears to be part of a normal set of operations in the developing brain (Paolicelli et al., 2011). Although we will return to microglial cells, we most certainly can acknowledge that their contribution to neuropsychiatric disorders is conceivable. Moreover, whatever other mechanisms are involved in tending to the number and quality of synapses that are formed in the brain, immunological factors—whether through cytokines or other means—represent a potential source of disruption to this process.

The Two-Hit Hypothesis

And so we come to the two-hit hypothesis. Two prominent features of schizophrenia that might interact with the immune system have been emphasized: genetics and development. Both are in keeping with what is likely to be the most promising approach to unraveling the cause of schizophrenia (Feigenson et al., 2014; Howes et al., 2017). We note that as in the case of other illnesses that are thought to be regulated by "two hits," this perspective is conceptual in nature, since "two" (hits) could be three, four, or more factors that all interact in some way to generate disease. As we have seen, in relation to other psychiatric and neurodegenerative disorders, this hypothesis adopts an interactive approach to the origins and emergence of schizophrenia. To the extent that genetic abnormalities may be necessary for schizophrenia to develop, they do not appear to be sufficient. Moreover, more than one gene is involved, and the culprit genes may exert their influence at critical stages of development, both prenatally and after birth. Moreover, development is a dynamic process, during which nothing is complete until a particular age-related milestone has been reached; any additional changes are layered into a mosaic of biological mechanisms that drive growth toward functional maturity. Throughout this process, periodic fluctuations in specific gene function may fully express a given abnormal phenotype[5], or weaken the ability of specific biological systems to adapt to discrete environmental events. This in turn may lead to pathology.

What we know about schizophrenia fits neatly into the two-hit hypothesis. Genetic risk is one factor; developmental critical periods sensitive to perturbation serve as the second—and stressors (or environmental events) serve as the third. In the latter category, we can include perturbations of the immune system. No one element seems to be sufficient to explain the etiology and progression of the disorder; but in combination, something might give. In what follows, we will attend to some of the features of immunity that have been related to schizophrenia.

---

[5] A good example of this is transient perinatal disruption of the DISC1 gene in the prefrontal cortex of mice. This interruption in DISC1 function affected postnatal dopamine maturation, and in adulthood, impaired sensorimotor gating as measured by the prepulse inhibition procedure (and which is impaired in schizophrenia) (Niwa et al., 2010). Therefore, interference in the function of important genes, if only briefly, can have dramatic consequences on the evolution of neurochemical and behavioral development.

AUTOIMMUNE DISEASE AND SCHIZOPHRENIA

In the same vein as the link between autism and the immune system received attention through a focus on autoimmune disease (see Chapter 12: Autism), schizophrenia has strong ties to an autoimmune hypothesis. This idea has been circulating for some time, and was dusted off, so to speak, in the early 1990s by Ganguli, Rabin and colleagues, who revived earlier thinking about the possible importance of an autoimmune process facilitating, or at the very least, maintaining schizophrenia symptoms. Their studies supported the notion that autoimmune abnormalities are more evident in schizophrenia, and characterized by reduced IL-2 production in first episode, unmedicated schizophrenia patients (Ganguli et al., 1995). Since IL-2 may be an important inhibitor of autoimmune responses (Dooms & Abbas, 2010), this was in keeping with an autoimmune hypothesis. However, others reported increased IL-2 production by peripheral blood lymphocytes in unmedicated patients (O'Donnell et al., 1996), a contradiction that might be related to the heterogeneity of schizophrenia phenotypes, as well as differences in age of onset, but this is only speculation.

Based on relatively more observations, there has been a resurgence of interest in the autoimmune hypothesis. First, autoantibodies made against the NMDA glutamate receptor are prominent in schizophrenia patients, while registry studies have linked autoimmune phenomena to schizophrenia. Second, genome-wide association studies have pointed to the MHC locus, where allelic variations in the MHC have consistently been linked to autoimmune disease (Mokhtari & Lachman, 2016). Clearly, there are compelling reasons to think that schizophrenia has close affiliations to autoimmune disease.

Epidemiological research on data from tens of thousands of individuals, has shown that a high proportion of people with schizophrenia have strong associations with a number of different autoimmune diseases. Danish registry studies revealed that a history of autoimmune disease increased the likelihood of schizophrenia by 29%–45%, and pushed this to 60% with the inclusion of infection-related hospitalizations (Benros et al., 2011; Eaton et al., 2006). Moreover, the incidence ratio for schizophrenia was virtually tripled (from 1.29 to 3.4) if cases of combined multiple infection and autoimmune disease were considered (Benros et al., 2011). In addition, while the autoimmune hypothesis is more compelling when conceptualizing autoimmunity as antecedent to the onset of schizophrenia, in-patients can display a range of concurrent autoimmune diseases, including Grave's disease, psoriasis, pernicious anemia, celiac disease, and hypersensitivity vasculitis (Chen et al., 2012). Notably, the odds ratio for a link to schizophrenia vacillates between 1.32 and 5.0, suggesting differential associations with specific autoimmune diseases (Chen et al., 2012). Unfortunately, registry studies offer little clue to why schizophrenia is associated with autoimmune disease. But at best, it is consistent with the notion of immunological dysregulation and possible impairment of tolerance to self-antigens.

Autoimmune diseases, in and of themselves, are notorious in eluding a well-defined etiology. The presence of autoantibodies explains the cause of organ or tissue damage in autoimmune disease, but what prompted B cells to make antibodies directed at self-antigens, and to relinquish the various control mechanisms that delete such B cells, is not known. Genetics plays some role in autoimmune disease, and this relationship was explored over two decades ago for schizophrenia. For example, first-degree relatives of patients with psychosis had a higher incidence of thyrotoxicosis and insulin-dependent diabetes mellitus (Type 1 diabetes), and thyrotoxicosis was five times greater in the mothers of psychotic patients (Gilvarry et al., 1996). Furthermore, patient registries in Denmark confirmed that the parents of schizophrenic individuals presented with a higher incidence of autoimmune disease than did the parents of healthy individuals (Eaton et al., 2006). This supports the contention that associations between autoimmune disease and schizophrenia have a heritable component, although the precise genes involved remain unknown. The HLA genes of the MHC are currently strong candidates.

Given these seemingly strong associations between autoimmunity and schizophrenia, there are some observations that should give pause to any convictions regarding the autoimmune hypothesis. In particular, CNS-directed autoimmune diseases, such as multiple sclerosis (MS) and SLE, are not preferentially more comorbid in schizophrenia than in the general population. This poses a challenge to the concept of CNS-reactive autoimmune diseases being a necessary precursor for psychosis. It is true that comorbidity of mental illness can be high in MS, with close to one-half of MS patients reporting some level of anxiety, depression, bipolar disorder, or schizophrenia (Marrie et al., 2017). But much of this is accounted for by clinical depression, and not schizophrenia. This appears to suggest that if schizophrenia involves autoimmune processes, prominent CNS-directed events are not sufficient to cause psychotic conversion. This is quite surprising, given that a hallmark feature of MS, that of demyelination, is observed in schizophrenia, while whole brain atrophy and ventricular enlargement—classical neuropathological indices of schizophrenia—are also observed in MS (Zivadinov et al., 2016). To be sure, these observations underlie different causal processes and follow different timelines[6], but it must be considered that if schizophrenia involves antineural autoantibody production, other co-regulatory elements are required to induce psychosis.

Rheumatoid Arthritis and Schizophrenia

It is the general approach of most investigations to seek a positive association between autoimmune disease and schizophrenia. Remarkably, rheumatoid arthritis (RA), is one autoimmune disease that seems to share a strong *negative*, or mutually exclusive, relationship with schizophrenia. Rheumatoid arthritis, as we discussed in Chapter 6, Stress and Immunity, is an inflammatory disease with a worldwide prevalence rate of 0.5% in the adult population, and a heritability component that has a 15%–30% genetic link, as determined by monozygotic twin concordance studies (Boissier, Semerano, Challal, Saidenberg-Kermanac'h, & Falgarone, 2012). A chief characteristic of the disease is chronic inflammation that aggravates and destroys the joints of the body, and mortality is accelerated by 3–10 years due to the impact of inflammation on cardiovascular function, as well as increased susceptibility to infection, perhaps the primary cause of mortality. Autoantibodies have been identified, such as rheumatoid factor and anticitrullinated peptide antibody, as well as multiple proinflammatory cytokines (e.g., IL-1 and TNF) that are released due to a coordinated convergence of adaptive and innate immune components in the affected joints. This is not necessarily a disease of adolescence, like schizophrenia, since most cases of RA tend to emerge after the age of 30. Therefore, this would appear to be a disease with a tenuous link to schizophrenia. In fact, it was noted years ago that with regard to various comorbidities present in RA, there is actually a

[6] The onset of MS is usually after age 30, and most prominent in women. Schizophrenia has a much earlier onset, and may even present with prodromal symptoms well before the initial psychotic episode. This is typically in late adolescence and early twenties. Interestingly, females tend to have a delayed onset of schizophrenia, most commonly in their mid-twenties; this is the commencement of the risk period for MS, which extends into the early 50's.

markedly low proportion that present with schizophrenia (Chen et al., 2012; Eaton et al., 2006). In fact, the risk for developing schizophrenia among those individuals who develop RA is half that of the general population who do not develop RA.

These findings beg the question of whether something about the propensity for RA protects against the development of psychosis, and whether this differs from other autoimmune diseases with a stronger link to schizophrenia. Confirmation for this idea was suggested in a study that prospectively followed 220 patients with RA and 196 control patients with medical conditions other than RA (Gorwood et al., 2004). Interestingly, patients with RA were less likely (by 25%) to display signs of psychosis (such as paranoid ideation). To some extent this was somewhat surprising given that RA can be an extremely debilitating disease, and has a high prevalence of depression and anxiety. It is clearly premature to conclude that the immunopathology present in RA somehow retards or inhibits the development of cognitive and emotional disruptions inherent in schizophrenia (or vice versa), and the nature of the particular influence remains a mystery.

Investigations looking for clues among susceptibility genes for RA in the MHC, have proposed only tenuous links to schizophrenia. For example, single nucleotide polymorphism (SNP) analyses in TNF-associated and other regions of the MHC showed only weak associations to psychosis, while a more direct selection of five RA-associated genes, including those for the cytokine-regulating transcription factor, NF-κB, did not find significant overlap with schizophrenia (Watanabe et al., 2009). The import of such findings was that the *absence* of certain allelic variants normally observed in RA, confers a risk to the development of schizophrenia. How this would operate is uncertain. An alternative view suggests that schizophrenia and RA have different forms (alleles) of similar genes, as well as SNPs in up to eight different genes (including two in the HLA region) (Wang, Lopez et al., 2015). These allelic variations might exert pleiotropic effects (i.e., individual genes give rise to more than one particular phenotype), so that genes related to RA preclude the appearance of other phenotypes, including schizophrenia. This is an intriguing idea, and one wonders whether the variations among the HLA genes, direct different forms of immunological influence, some of which facilitate the development of psychotic symptoms.

Autoimmunity, Inflammation, and the Glutamate Hypothesis of Schizophrenia

Psychopharmacological treatment of schizophrenia has unmasked a number of hypotheses based on potential alterations in neurotransmitter signaling. Historically, the dopamine hypothesis dominated this perspective, but as we noted earlier, an idea gaining prominence is focused on glutamate—namely, the *glutamate hypofunction* hypothesis of schizophrenia. It is believed that low levels of glutamate signaling can explain the symptomatology associated with schizophrenia, since glutamate antagonists, can cause negative symptoms, and developmental abnormalities involve poor glutamate activity. This has been supported by several *in vitro* and *in vivo* studies that showed dysregulation of the glutamate signaling systems in humans and in animal models of schizophrenia, and many genes implicated in schizophrenia are involved in glutamate signaling (Lin, Lane, & Tsai, 2012). The dopamine system is still relevant here, but it is a question of how it is controlled (or for that matter, not controlled), by aberrant glutamate signaling. To understand how this scenario may come into play as part of an autoimmune process that affects the glutamate system, we can briefly inspect the known neural circuitry in the dopamine pathway, and how it is controlled by the activities of glutamate and GABA neurons (see Fig. 13.1 and text box).

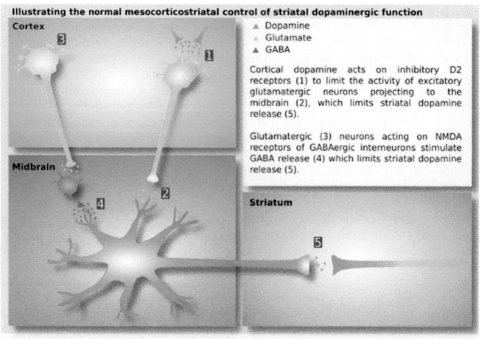

Illustrating the normal mesocorticostriatal control of striatal dopaminergic function

Cortex

▲ Dopamine
▴ Glutamate
▲ GABA

Cortical dopamine acts on inhibitory D2 receptors (1) to limit the activity of excitatory glutamatergic neurons projecting to the midbrain (2), which limits striatal dopamine release (5).

Glutamatergic (3) neurons acting on NMDA receptors of GABAergic interneurons stimulate GABA release (4) which limits striatal dopamine release (5).

Midbrain

Striatum

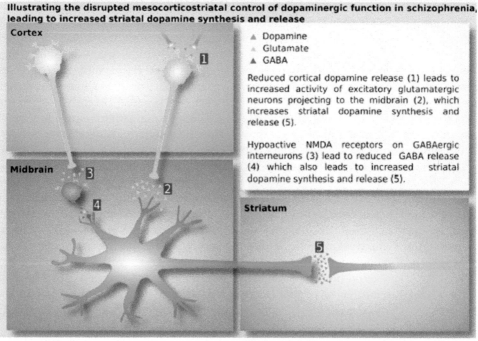

Illustrating the disrupted mesocorticostriatal control of dopaminergic function in schizophrenia, leading to increased striatal dopamine synthesis and release

Cortex

▲ Dopamine
▴ Glutamate
▲ GABA

Reduced cortical dopamine release (1) leads to increased activity of excitatory glutamatergic neurons projecting to the midbrain (2), which increases striatal dopamine synthesis and release (5).

Hypoactive NMDA receptors on GABAergic interneurons (3) lead to reduced GABA release (4) which also leads to increased striatal dopamine synthesis and release (5).

Midbrain

Striatum

FIGURE 13.1   Illustration of glutamatergic and GABAergic interactions with dopamine neurons in the mesocorticolimbic and nigrostriatal brain regions. Afferent glutamate signaling to inhibitory GABA-ergic interneurons in the descending cortico-brainstem pathway are responsible for maintaining normal functioning, thereby creating strong levels of arousal, attention and information processing. When this direct glutamate stimulation is impaired, there is less dopamine released in the cortex, and therefore, a depressed state of cortical functioning predominates, as corroborated by neuroimaging studies (Howes & Kapur, 2009). Overall, less effective glutamate signaling can explain both positive, negative, and cognitive symptoms of schizophrenia. *Source: From Howes et al. (2017).*

The inflammation argument for glutamate dysregulation is based on the disruptive influence of cytokines and glial activity. One of the better studied signaling cascades that regulates this interaction is the kynurenine system (Schafer & Stevens, 2010), which was briefly described in relation to depressive illnesses (Chapter 8: Depressive Disorders), but will be covered here in a bit more detail. It will be recalled that in this pathway, kynurenine (KYN) is converted to tryptophan (TRY) with the help of indoleamine 2, 3-dioxygenase (IDO), expressed predominantly by CNS astrocytes, or tryptophan 2,3-dioxygenase (TDO), predominantly expressed by microglia. Through a signaling cascade, various metabolites are formed, including kynurenic acid (KYNA), a naturally occurring NMDA antagonist, and quinolinic acid (QUIN), a natural NMDA agonist. Tellingly, the levels of KYNA in the CSF or tissue of schizophrenia patients were higher than in those of healthy controls (Linderholm et al., 2012), thereby indirectly supporting the glutamate hypofunction hypothesis. In addition, impaired tryptophan metabolism, and associated generation of kynurenic acid, is associated with deterioration of cognitive symptoms in schizophrenia (Kanchanatawan et al., 2017).

The kynurenine signaling system is tied to the immune system. Its two major catalysts, TDO and IDO, are both produced by glial cells, which become upregulated in response to inflammation and stress. In addition, levels of IDO and KYNA are increased in response to TNF-α and IFNγ, inflammatory cytokines known to be upregulated in schizophrenia. An inflammation-induced elevation in KYNA may then alter glutamatergic signaling through noncompetitive antagonism, and also via downstream signaling through α7 nicotinic acetylcholine receptors and GABA-ergic interneurons (Feigenson et al., 2014). This can result in extensive neurotransmitter dysregulation, altering levels of glutamate, dopamine, acetylcholine, and GABA, that could be relevant in causing or exacerbating symptoms in schizophrenia.

Finally, in animal studies, more direct manipulations have provided support for the position that the kynurenine pathway may be involved in neurodevelopmental deficits. For example, animals exposed to KYNA during early development, subsequently have higher levels as adults, coupled with lower glutamate levels, and increased appearance of abnormal behaviors in animal models of schizophrenia (Alexander, Wu, Schwarcz, & Bruno, 2012; Pocivavsek, Wu, Elmer, Bruno, & Schwarcz, 2012). Thus, it is possible that glutamatergic hypofunctioning emerges early, facilitated by inflammatory regulation of KYNA activity.

Anti-NMDA Receptor Antibodies and Schizophrenia

An alternative perspective on glutamatergic hypofunctioning involves a B cell mediated autoimmune response that interferes with signaling at the NMDA receptor. There is no better direct form of immunological *inactivation* of a given molecular signal than the blocking or antagonistic actions of antibody. With regard to schizophrenia, the detection of anti-NMDA-R antibodies has varied across different studies. In a cohort of 121 first-episode patients, 10% showed multiple types of NMDA-R antibodies (Steiner et al., 2013), while only 3% were detected in another large cohort of first-episode patients (Lennox et al., 2017). Finally, a complete lack of detection was noted in a screening of plasma among Taiwanese patients (Chen, Cheng et al., 2017). It appears, then, that detecting antibody for NMDA-R is a fickle business, but seems to be more reliable than looking for other antineuronal antibodies, which are equally present in schizophrenic and healthy subjects (Chen, Cheng et al., 2017;

Lennox et al., 2017). An instructive piece of information is that toxoplasmosis may be a co-requisite condition for the appearance of anti-NMDA-R antibodies in both animal models and in first-episode patients with schizophrenia (Kannan et al., 2017). This means that better confirmation of anti-NMDA-R antibody status may be enhanced by a consideration of infection history. Overall, however, the ability of anti-NMDA-R antibodies to help buttress an immunological hypothesis for glutamate hypofunctioning is still a work in progress.

AUTOANTIBODIES AND THE CASE OF MYASTHENIA GRAVIS

Myasthenia gravis is an autoimmune disease in which antibodies are made against the acetylcholine receptor expressed on skeletal muscles. This results in muscle weakness that not only affects movement, but also speech, breathing and swallowing. Such autoantibody production is not uncommon, but when directed against neural components, the lesson of myasthenia gravis is instructive. In the last decade, there has been greater attention on autoantibody related encephalitis disorders, most prominently anti-NMDA-R encephalitis. This progressive illness first manifests with psychotic features, memory deficits, and seizures, subsequently followed by dyskinesias, hypoventilation, loss of consciousness, and death. The mechanism of action is a reduction in NMDA receptors caused by excessive antibodies to the NR1 subunit of the NMDA receptor. Many patients eventually diagnosed with anti-NMDA-R encephalitis initially present with positive, disorganized, and negative symptoms that might mistakenly result in a diagnosis of schizophrenia. However, when given a correct diagnosis, many patients respond with complete or partial recovery with immunosuppressive therapy (Dalmau, Lancaster, Martinez-Hernandez, Rosenfeld, & Balice-Gordon, 2011).

PERINATAL INFECTION AND SCHIZOPHRENIA

Infection and other environmental factors have been raised as viable contributions to the etiology of schizophrenia. Epidemiological literature noted what came to be known as the "seasonal effect" of schizophrenia, the observation that offspring whose gestational development overlapped with the winter months were more at risk of developing schizophrenia. This came to be questioned on the basis of modest effects, which were often statistically nonsignificant (Fouskakis et al., 2004). Another proposed hypothesis was famine-related starvation stress. Here it was reported that pregnancy during periods of famine, such as the Dutch famine in the latter stages of World War II, resulted in a higher incidence of eventual schizophrenia diagnoses (Brown, 2011). These reports, and the hypotheses that emerged from them, focused attention on the prenatal effects of environmental stressors. In particular, the role of prenatal infection—and the immune response of the pregnant mother—has become a major focus, such that there are now hundreds of studies in human subjects and animal models that argue for the potential influence of the maternal immune response on neural and behavioral development.

Prenatal Infections

In discussing the effects of various prenatal and early postnatal insults on later pathology in Chapter 7, Prenatal and Early Postnatal

Influences on Health, it was indicated that a variety of bacterial and viral infections have been linked to psychological disturbances in adulthood. In this regard, the link between viruses and schizophrenia was fairly impressive, varying with the specific virus encountered. Here, we will go into a bit more detail regarding some of these reports.

Of the various infections that have been addressed in humans (e.g., rubella, herpes simplex) influenza has received the most scrutiny. Following on from epidemiologic reports that cited an increased incidence of schizophrenia among offspring of pregnancies that coincided with major influenza epidemics, Brown et al. (2004) accessed archived maternal serological data from births recorded between 1959 and 1966. Antibody to influenza was measured in the serum to confirm exposure to influenza virus during pregnancy. The appearance of antibody during the first half of gestation suggested that exposure to influenza at this time was associated with a three-fold increase of risk for later development of schizophrenia among the offspring. This risk increased to seven-fold among those who showed appearance of antibody during the first trimester (Brown et al., 2004). This was the first empirical evidence using a nested case-control design that an immune response to influenza during pregnancy increased the risk for development of schizophrenia among the offspring. Other infectious agents presenting during pregnancy, such as *Toxoplasma Gondii* (*T. Gondii*) and Herpes Simplex Virus Type 2 (HSV-2), were also linked to subsequent diagnosis for schizophrenia in the offspring (Brown & Derkits, 2010). Furthermore, while rubella, mumps, respiratory infection, measles, and polio have also been associated with infection during pregnancy, the odds ratio for development of schizophrenia from pregnancies that involved these infections was variable (ranging from 0.62 to 3.58) (Scharko, 2011). Since many of these studies were retrospective or based on registry data, prospective studies are needed to address this issue.

Of course, if maternal infection poses a risk to the offspring, what is the mechanism? Is it the maternal immune response, or direct infection of the fetus? Much of the attention on maternal immune effects on postnatal psychiatric disorders is focused on viral infections. In general, it is rare for the fetus to be infected during a maternal viral infection, given the physical and immunological blockade set up by the placenta. However, on occasion, certain viruses [e.g., cytomegalovirus (CMV), herpes simplex virus-2 (HSV-2), or rubella] are known to breach the placenta and infect the fetus, causing impaired development of neurosensory systems, learning deficits, and in the case of the zika virus, microcephaly (Racicot & Mor, 2017). Needless to say, such events are rare, and in the US, congenital rubella is virtually nonexistent, the prevalence of fetal exposure to CMV is 0.05%, and 95% of the very low 0.06% rate of neonatal HSV infections are developed postnatally. This likely is due to the lack of adequate receptors for different viruses in the placenta, an evolutionary benefit that provides an effective impediment to transmission of viruses across the placenta. In contrast, maternal susceptibility to viral infection is quite high, ranging anywhere from 20% to 90%, depending on the type of virus involved (e.g., maternal influenza virus infections are around 40%, HSV-2 infections range from 18% to 22%, and CMV infections can range from 60% to 90%) (Racicot & Mor, 2017).

It appears that if there were an infection-related insult to the developing fetus, it would more likely come from the maternal side of the placental interface between mother and fetus. Moreover, given the rich armamentarium of innate, as well as adaptive immune cells in the decidual tissue of the uterus, the immune system seems the most likely threat to the developing fetus, although, one should be prepared to question the virtue of allowing an immune apparatus to engage an infectious microbial antigen without also ensuring protection of the fetus. The cytokine and antibody cascade that would ensue as a direct assault on an invading virus, should hardly serve to compromise safe and healthy fetal development, and progression toward timely parturition. Intuitively, it makes little sense to accuse the immune system of interfering with intrauterine development. Nonetheless, as argued by others with regard to obstetric complications (Forsyth et al., 2013), given the high degree of genetic abnormalities in schizophrenia, it is likely that in the spirit of the two-hit hypothesis, special cases of vulnerability to the maternal immune response might affect the developing fetus.

One possibility in this respect is the influence of soluble factors, such as cytokines and antibody. Indeed, elevated levels of inflammatory cytokines, such as TNF-α and IL-8 (a chemotactic cytokine), are significantly elevated in the serum of mothers of whom offspring later develop schizophrenia (Brown & Patterson, 2011; Ellman et al., 2010). In addition, high levels of maternal antibody to dietary antigens were associated with an increased risk for psychosis in offspring (Karlsson et al., 2012). These observations support the notion that during gestation maternal immune factors can be a vulnerability factor for developing schizophrenia. This is reinforced by animal studies discussed later, which provide more direct evidence of immune activation and cytokine analysis.

Postnatal Infections

Newborns are particularly susceptible to all manner of insults, including nutritional imbalances, and of course, infection. This may be superimposed on a genetic constitution that may be present (with significant anomalies or variations) some of which are genomic, and others that are epigenetic (Weber-Stadlbauer, 2017). Whereas prenatal immune events might alter postnatal behavioral propensities, additional immunologic events during the early (and even late) postnatal period may continue to disrupt development, and tip things in favor of abnormal behavior. But how does infection after birth affect the developing nervous system and promote schizophrenia-like phenotypes? Surprisingly, little is known about this, compared to prenatal immune events. Most studies involve animals, and these often use postnatal nonimmunologic stressors (e.g., maternal separation) that have been linked to epigenetic and behavioral alterations. In contrast, the analysis of the impact of postnatal immune changes on schizophrenia-relevant neurobiological changes, such as reduced hippocampal neurogenesis, are rare relative to maternal immune activation studies (Musaelyan et al., 2014). This is certainly an important gap in knowledge that needs to be addressed. This is especially important in view of an increased interest in determining how prodromal predictors of psychosis might be driven by biological factors.

In response to infection, postnatal and childhood immune responses are superimposed on a sensitive and incompletely developed CNS. Normal and pathological MRI data show brain development to be a highly dynamic process right into adulthood, with changes in white and gray matter proceeding through regionally determined phases of cellular expansion followed by pruning and refinement of synaptic connections. Cortical gray matter increases during early childhood, then enters an elimination stage during which the cortex thins, and seems to reach maturity sooner in fundamental areas of sensory and motor processing, being delayed in association cortices and in the prefrontal cortex (Fields, 2008; Gogtay et al., 2004). Upon moving past the neonatal period and into childhood, excessive exposure to stressors, such as maltreatment and abuse, can promote later mental health problems, including psychosis and schizophrenia (Schenkel, Spaulding,

DiLillo, & Silverstein, 2005). Relatedly, stressors can alter cortical development (Howes et al., 2017), providing reasonable support for the notion that periods' of dynamic neuronal sculpting can be derailed by traumatic and disturbing life events.

We learned in previous chapters that infection can modify cognitive functioning, motor performance, motivational behavior, social investigation, and other sickness-related behaviors. These types of behavioral deficits are among many of the premorbid signs for the development of schizophrenia (Schenkel et al., 2005). Delirium and mental confusion are not uncommon in serious infections, but are commonly thought to be a by-product of a medical condition. Nonetheless, given certain predisposing mental conditions and potentially malleable dynamic maturation of the brain in young individuals, infection operating as a stressor, may introduce enduring changes in brain and behavioral development that sides with the appearance of a psychiatric disorder. In the case of schizophrenia, proof of concept has been provided by clinical studies.

A review of the literature revealed a significant link between viral infections of the CNS in childhood and future development of adult schizophrenia (Khandaker, Zimbron, Dalman, Lewis, & Jones, 2012). Infections that were most strongly correlated with later development of schizophrenia were cytomegalovirus and mumps. Interestingly, an epidemiological analysis of almost 2 million children in a Swedish registry revealed that infections during pregnancy, combined with psychiatric complications, increased the likelihood of both psychotic development in the offspring, and higher rates of childhood infection (Blomstrom et al., 2016). Notably, this study controlled for confounding variables previously linked to development of psychosis, such as urban births, socioeconomic status, winter births, and being small for gestational age (i.e., birth weight). It appeared that although maternal infection alone did not account for development of schizophrenia, interactions with maternal (but not paternal) psychiatric complications predicted later life psychosis in offspring. Again, this is in keeping with the two-hit hypothesis, emphasizing that children destined for a diagnosis of schizophrenia, may reach this point as a function of converging influences of prenatal and postnatal experiences.

As discussed in Chapter 5, Stressor Processes and Effects on Neurobiological Functioning, repeated exposure to various stressors can modify immune function and increase susceptibility to infection. Stress and emotional lability is common in schizophrenia, and while we have noted increased parameters of inflammation in schizophrenia, this may actually be secondary to the behavioral disorder. That is, the burden of having schizophrenia alters immune function. Among children and adults with diagnoses of schizophrenia or nonaffective psychosis, relapse during or after remission is often associated with a recent infection, and in particular, concurrent urinary tract infections appear to be prominent (Carson, Phillip, & Miller, 2017). Under these circumstances, infection may be viewed as a precipitant of psychiatric symptoms, such as hypermania, as well as being a consequence of stress-reactions experienced by the patient. For the former condition, we know that infection and associated changes in proinflammatory cytokines can impact stress pathways in the brain. As such, it is not unlikely that infections could precipitate relapse, or even initial conversion to psychosis during the prodromal phase. With regard to incurring infection, this can arise from poor hygiene, especially during a psychotic period, or the presence of certain negative symptoms (e.g., avolition, depressed mood) that compromise self-care. Alternatively, a preexisting immunological dysregulation may already be present, emerging during development or altered by the various prodromal traits observed in people who eventually experience a psychotic episode.

Toxoplasmosis: Toxoplasma Gondii Infection

There has been attention given to the possibility that Toxoplasmosis, an infection due to the coccidian protozoan, *T. Gondii*, causes schizophrenia in a subset of individuals. Particularly relevant to neuropsychiatric concerns is the fact that this parasite has neurotrophic properties and can replicate in the brain (Wohlfert, Blader, & Wilson, 2017. The preferential host for *T. Gondii* is the cat, which becomes infected by virtue of exposure to mice, birds and other infected animal species. The parasite replicates in cats and is present in oocysts shed through feces, such that household litter or outdoor regions can become areas of *T. Gondii* exposure for humans and other animals (Tenter, 2000). It has been urged that pregnant women, young children, and any immunosuppressed individuals avoid exposure to specific areas (such as cat litter) or undercooked meat that might contain *T.Gondii*-containing oocysts. According to the Center for Disease Control (CDC) in the United States, at least 10% of the US population over the age of 6 years, has been exposed to *T. Gondii*, while in parts of the world with a warm, humid climate, over 95% of the population is positive for Toxoplasmosis. Given the high rate of exposure, most people exposed to *T. Gondii* are asymptomatic due to effective immune surveillance. However, in immunocompromised individuals, infection results in flu-like symptoms (fever, weakness, muscle aches, and pains), and in more severe cases, toxoplasmosis leads to neurologic problems (e.g., seizures and encephalitis).

Reviews of various studies that examined the relationship of serological levels of *T. Gondii* antibodies and schizophrenia revealed a significant link between infection and schizophrenia (Torrey, Bartko, & Yolken, 2012). Diagnosed individuals were significantly more likely to have higher levels of *T. Gondii* antibody, which conferred risk levels comparable to using cannabis and having minor physical anomalies. Similarly, register-based studies in Denmark revealed that higher serum levels of antibodies from pregnant mothers and infants were predictive of future development of schizophrenia in both mothers and their offspring (Pedersen, Stevens, Pedersen, Norgaard-Pedersen, & Mortensen, 2011). In fact, for individuals already diagnosed with schizophrenia, testing positive for *T. Gondii* antibodies is predictive of more severe psychopathology. In particular, previously infected patients have more positive symptoms, low education status, and familial history of mental illness, as well as reduced gray matter density (Holub et al., 2013). There is also evidence that *T. Gondii* in pregnant women is a risk factor for the development of schizophrenia in offspring (Blomstrom et al., 2012). This implies a possible sensitivity of the developing nervous system to Toxoplasmosis, and it is important to consider that postnatal exposure to *T. Gondii* may similarly alter neural development either directly or via the immune response. Indeed, infection with *T. Gondii* is negatively correlated with intelligence, memory, novelty seeking, and reaction time, as well as being positively linked to alterations in personality and increased suicide attempts (Feigenson et al., 2014). The impairment in cognitive skills, may account for the reported increase in accidents during driving or in work-related environments, where attention and reaction time are important human factors. These cognitive changes are supported by animal studies, which have more directly established that infection with *T. Gondii* impairs learning and memory, eliminates fear, and produces hyperlocomotor activity (Kannan & Pletnikov, 2012).

The mechanisms promoting these cognitive and behavioral effects are not fully understood. One potential mechanism is that acute *T. Gondii* infection can directly alter neurotransmitter release and signaling. For example, infection may alter dopamine metabolism, kynurenic acid activity, and glutamate signaling (Haroon et al., 2012; Prandovszky et al., 2011). It is also known that *T. Gondii* can effectively colonize both glial and neuronal cells; consequently, astrocyte and microglial cell function may be modified by *T. Gondii* infection. In addition, infection can alter cellular migration, cytokine production and release, and regulation of neurotransmitter activity. Indeed, reactive astrocytes and microglia initiate a series of complex signaling cascades in response to *T. Gondii*, generating greater levels of prostaglandins and members of the transforming growth factor, interleukin, and interferon cytokine families (Wohlfert et al., 2017).

The effect of *T. Gondii* on the peripheral immune response may also play a role in the development of mental illness. Given the prominence of schizophrenia-associated gene variants occurring in the MHC region, it is notable that the response to *T. Gondii* infection appears to be influenced by specific genes in the MHC region (Mack et al., 1999). Asymptomatic hosts infected with *T. Gondii* typically have elevated immune responses to counter the infection, and this creates a dynamic environment in which parasite and host alter their respective environments to seek mutual advantages. To escape immune surveillance, *T. Gondii* exerts an inhibitory effect on proinflammatory cytokines, while simultaneously augmenting anti-inflammatory and/or regulatory cytokines. Additionally, the production of nitric oxide (NO), which can promote cytotoxic effects, is reduced, thereby allowing the parasite to propagate within the host cells. In the brain, control of parasitic replication and protection of neurons is supported by infiltrating T cells, which can then limit infection (Wohlfert et al., 2017). As an act of self-preservation, *T. Gondii* infection reduces the number of cytotoxic CD8 + T cells, an effect that may underlie promotion of cognitive disturbances in schizophrenia (Bhadra, Cobb, Weiss, & Khan, 2013).

In sum, there is strong evidence that *T. Gondii* infection—as a neurotrophic parasite—can interfere significantly with neural and behavioral functions. In the brain, it can potentially contribute to increased glutamate neurotransmission, and contribute to excitotoxic effects and interfere with dopaminergic functions (Prandovszky et al., 2011). Given the presence of high antibody titers to *T. Gondii* in schizophrenia, it is not unreasonable to consider that genetically and environmentally produced vulnerabilities interact with toxoplasmosis and precipitate psychosis. At present, we are only at the stage of hypothesizing such possibilities, and while the evidence does seem to be pointing in this direction, there are some data that mitigate this view. For example, an idea proposed by Torrey and colleagues—known as the "cat ownership" hypothesis—argued that children raised in families containing cats were more likely to develop psychosis-related mental illness. This is a controversial idea, for which there was some support. However, a recent study in the UK found no evidence for an increased risk for schizophrenia among children raised in cat-containing households (Solmi, Hayes, Lewis, & Kirkbride, 2017). It seems we will have to wait a little longer, before turning our feline friends out onto the street!

Endogenous Retroviruses

A potential bridge between environmental stressors and later development of schizophrenia is the human endogenous retrovirus type-W. Approximately 8% of the human genome is composed of retroviruses; HERVs are a family of retroviruses that are heritable, can code for proteins, are found in the CNS, and are associated with disorders, such as MS and schizophrenia (Slokar & Hasler, 2015). The expression of HERV-W, however, is variable and typically dormant, but can be activated by specific environmental triggers, which can lead to an immune response against HERV-W associated retroviral envelope proteins; these can induce a proinflammatory response consisting of cytokines such as IL-1, IL-6, and TNF-α, which are often associated with schizophrenia (Rolland et al., 2006). Increased HERV-W element expression in schizophrenia has been obtained through measures of serum, CSF, and tissue analyses, and is prominent in both first episode and chronic forms of psychosis (Feigenson et al., 2014). During embryonic development, there is a reduction of epigenetic modifications, such as DNA-methylation, which creates an inductive environment for the reactivation of HERV-W elements (Perron & Lang, 2010). The potentially elevated expression induced by environmental stressors aligns with the multifactorial model embodied by the two-hit hypothesis. Responses to HERV-W can have both developmental effects, as well as acute inflammatory consequences once development is complete. Influenza infection during pregnancy may also help activate a second proinflammatory cascade in the embryo through induced expression of HERV-W elements, and subsequently alter neuronal development (Perron et al., 2012). This may also sensitize individuals to respond to immune stressors at later time points, creating a cyclic reaction, whereby future stressors and infections initiate additional HERV-W expression and subsequent inflammatory events. Indeed, TNF-α, BDNF, and the dopamine receptor D_3 may be elevated during the immune response to HERV-W, all of which may play roles in altering neuronal communication during the later phases of schizophrenia (Huang et al., 2011; Rolland et al., 2006). Finally, interactions have been noted with toxoplasmosis, in that greater expression of HERV-W can occur in some cases of schizophrenia with evidence of prior *T. Gondii* infection (Perron et al., 2012). Once again, it seems that the convergence of different infectious and immune elements may be complicit in precipitating and/or maintaining schizophrenia.

Cytokines and Schizophrenia

Several studies that assessed c-reactive protein (CRP) as a predictor of inflammation and psychiatric disease indicated that this inflammatory marker was highly correlated with severity of psychiatric illness. In particular, circulating levels of CRP are higher in patients presenting with catatonia, negative symptoms, and aggression (Orsolini et al., 2018). In a comprehensive review designed to determine whether schizophrenia presents with a unique immunophenotype (Miller & Goldsmith, 2017), it was noted that elevated circulating IL-1β, IL-6, TGFβ, IL-12, TNF-α, and the soluble form of the IL-2 receptor (sIL-2R) were frequent findings in both first episode and chronic forms of schizophrenia, whereas changes in IFNγ were more variable. In the far fewer studies that assessed CSF samples, IL-1β, IL-6 and IL-8 were elevated, while sIL-2R was reduced (Wang & Miller, 2018). Interestingly, postmortem analyses of protein or mRNA levels in the cortex also showed upregulation of IL-1β, IL-6, and IL-8 (Fillman et al., 2013). This type of consistency is encouraging and provides some confidence regarding which cytokines might be useful targets of therapeutic intervention.

From a mechanistic perspective, however, these findings are a little uncertain, as similar types of changes in circulating cytokines have been noted for nonpsychotic psychiatric disorders, such as depression (see Chapter 8: Depressive Disorders). Furthermore, these changes may be reflections of the stress typically experienced by patients, rather than antecedent biomarkers of disease onset. This problem is being addressed in prodrome studies that can determine whether circulating cytokines can be used as biomarkers of pending psychotic breakdowns. For example, a consortium group, the North American Prodrome Longitudinal Study (NAPLS 2), has been formed to identify predictors and mechanisms of conversion to psychosis. One of the first reports regarding blood biomarkers, was short on details regarding specific cytokines that might represent key therapeutic targets, but did note that running blood samples of individual prepsychotic patients against a data base of known inflammatory factors linked to psychosis, might prove to be a useful predictive tool for schizophrenia onset (Perkins et al., 2015). While this is preliminary, there is reason to think that increased activation of the immune system might influence the onset of psychotic episodes. Alternatively, as discussed in relation to other psychiatric disorders, inflammatory indices may reflect a general health risk, and still other factors are necessary in order to accurately predict whether a specific disorder will emerge.

ANIMAL MODELS OF IMMUNITY AND SCHIZOPHRENIA

Earlier, we discussed human investigations and epidemiological analyses that suggested a link between maternal infection and schizophrenia. Here we consider experimental approaches using animal models of immune activation. By way of context, as described in Chapter 7, Prenatal and Early Postnatal Influences on Health, immunobiological and psychosocial manipulations in the young animal have a significant impact on later-life endocrine, immune, and behavioral functions. With regard to the etiology of schizophrenia-like symptoms, the strongest arguments have emerged from studies that examined the postnatal development of mice, rats and, in some cases, nonhuman primates that received prenatal immunologic and/or infectious exposure (Estes & McAllister, 2015). This has come to be referred to as maternal immune activation

(MIA), and capitalizes on the notion that the immune response influences the developing nervous system of the embryo and/or fetus. The number of studies documenting significant postnatal effects has grown dramatically over the past 15 years, and much of this has been thoroughly reviewed (Meyer & Feldon, 2012). A variety of outcomes have been assessed, including dopaminergic, GABA-ergic, and glutamatergic changes, as well as several tests of altered cognitive and emotional behavior. It is understood that the full spectrum of schizophrenia symptoms is difficult to replicate in animals, and to some extent the behaviors that are assessed may apply to other conditions, such as autism and depression. For example, changes in social exploration, hedonic capacity, anxiety, learning and memory, and prepulse inhibition (PPI) can be found in a range of psychiatric conditions. Thus, while the effects of maternal (or prenatal) infection or immune activation are relevant to schizophrenia, it is probably more helpful to view these findings as indicators of the contribution that prenatal infection makes toward specific endophenotypes that cluster with other symptoms relevant to a diagnosis of schizophrenia.

An important perspective here is that an animal model is useful for understanding schizophrenia if it can reproduce one or more aspects of the condition, even if that aspect is not unique to the disorder. For example, social deficits are observed in both schizophrenia and autism. Therefore, as pointed out in Chapter 12, Austim, our discussion of MIA will be as relevant to autism, as it may be to schizophrenia, depression, and any other relevant psychiatric disorder with overlapping behavioral deficits. As described in our discussion of animal models, it is unlikely that any single test will reproduce all aspects of schizophrenia, and it is recognized that this is an impossible goal given some of its distinctly human aspects (e.g., formal thought disorder and hallucinations). Furthermore, given that significant postnatal and early adolescent environmental events are likely to be superimposed on intrauterine neurodevelopmental dysregulation due to prenatal infection, it is likely that additive or synergistic interactions may trigger truly abnormal behavioral changes that warrant psychiatric intervention. To this end, some studies have tested the two-hit hypothesis, in which postnatal exposure to psychogenic stressors or cannabinoid receptor agonists among offspring from infected mothers showed more dramatic behavioral deficits than those seen due to maternal infection alone (Dalton, Verdurand, Walker, Hodgson, & Zavitsanou, 2012; Giovanoli et al., 2013).

Immunologic Promotion of Altered Neurobehavioral Development: Preclinical Studies

The form of immune stimulation used in most MIA experiments has largely relied on the use of molecular agents that activate the innate immune system (see Chapter 2: The Immune System: An Overview). Most studies have utilized either the endotoxin LPS or poly I:C, a double-stranded RNA that mimics viral infections. Both these immunogenic molecules stimulate macrophages, monocytes and neutrophils through TLR 3 and TLR4[7].

As noted in earlier, LPS stimulates monocytes and macrophages to produce a range of proinflammatory cytokines (e.g., IL-1, IL-6, and TNF-α) that can have a variety of neurobehavioral effects. Poly I:C similarly induces proinflammatory cytokine production and exerts

neural and behavioral effects, including increased turnover of monoamine neurotransmitter in the brain and the induction of modest sickness behavior (Gibb, Hayley, Poulter, & Anisman, 2011; Meyer & Feldon, 2012). It should be noted, however, that in normal adult mice, differences have been observed in the profile of neurochemical changes after challenge with either LPS or Poly I:C. For example, LPS can have pronounced effects on brain monoamine neurotransmitter alterations, although dopaminergic changes are minimal in the prefrontal cortex. Similarly, the effects of poly I:C appear to be relatively modest or absent in the prefrontal and limbic brain regions (Gandhi, Hayley, Gibb, Merali, & Anisman, 2007). Whether these differences are important with regard to impact on the developing embryo and fetus is not known. Postnatal behaviors assessed in animals from mothers that have been subjected to MIA have included exploratory behavior, social interaction, cognitive function, and sensorimotor gating, many of which were affected (Meyer & Feldon, 2012). In studies that used LPS or Poly I:C, a single injection is given at some point during gestation (typically, at around 12–13 days after conception). Interestingly, when immune challenge with Poly I:C was varied relative to gestational age, injections on gestational day 9 (GD9), but not GD17, resulted in impaired sensorimotor gating, as measured by the acoustic PPI procedure described in our discussion of PTSD, while injection of Poly I:C on GD17 impaired working memory[8] (Meyer & Feldon, 2012). This demonstrates that

---

[7] As we noted in Chapter 2, The Immune System: An Overview, these are known as Toll-like receptors, first identified in flies, and found to be important for immune defense. The TLR3 is preferentially stimulated by Poly I:C, whereas LPS tends to stimulate TLR4.

[8] Working memory is the term most often used for what used to be short-term memory. In animal studies, working memory can be tested by letting animals know that a behavioral option is always available to them, but the opportunity (such as location of an escape platform in the Morris water maze) may vary from one trial to another. Attentional and working memory problems are part of the range of cognitive deficits seen in schizophrenia, and fall into the "hypofrontality" domain attributed to impaired prefrontal cortex functions.

immune activation can cause differential phenotypic changes in the nervous system, depending on critical stages of embryonic development.

Of particular relevance to any preclinical animal study purporting to identify antecedents to schizophrenia-like abnormalities is evidence of neuroanatomical and neurotransmitter and receptor alterations. As previously stated the dominant hypothesis for the pathophysiology of schizophrenia is dysregulation of the dopamine system, and impaired functioning of the inhibitory GABAergic and excitatory glutamatergic systems. Accordingly, it is noteworthy that in rat offspring from mothers challenged with LPS on GD15/16, dopamine D_2 receptor expression in the medial PFC, and numbers of D_2 receptor expressing cells, were reduced on postnatal days between 35 and 60 (Baharnoori, Bhardwaj, & Srivastava, 2013). This was consistent with murine postnatal dopaminergic changes in the PFC and hippocampus pursuant to maternal Poly I:C challenge (Bitanihirwe, Peleg-Raibstein, Mouttet, Feldon, & Meyer, 2010; Meyer & Feldon, 2012). Changes in the brain GABAergic system of offspring have also been reported after MIA. This includes a reduction in the concentration of GABA and glutamic acid decarboxylase (GAD), the synthetic enzyme needed to make GABA, in the prefrontal cortex and hippocampus. These changes correlated with impaired working memory performance (Richetto, Calabrese, Riva, & Meyer, 2014), and as such, support various neurotransmitter models of schizophrenia, which may have their origins in the prenatal period.

Extending these neurochemical analyses are electrophysiological approaches used to determine neurodevelopmental alterations in synaptic activity in offspring from LPS-challenged pregnant rats. A common method for studying synaptic plasticity involves high frequency electrical stimulation of presynaptic neurons to generate enhanced electrophysiological effects

in postsynaptic neurons [i.e., long-term potentiation (LTP)]. This typically involves glutamate signaling, and likely reflects the physical basis of memory formation. Conversely, downregulation of the postsynaptic glutamate receptor can attenuate postsynaptic electrophysiological responses after high rates of presynaptic stimulation, an effect referred to as long-term depression (LTD). Interestingly, glutamatergic signaling is altered in offspring from mothers given LPS. Specifically, LTD in the hippocampus was impaired as a result of disturbed glutamate NMDA receptor signaling (Burt, Tse, Boksa, & Wong, 2013), which is in agreement with other evidence that prenatal immune activation alters the glutamate system (see Meyer & Feldon, 2012). Importantly, it provides support for the glutamate hypofunctioning hypothesis of schizophrenia that was discussed earlier.

Finally, the normal operation of the brain is dependent on efficient communication between different regions that specialize in cognitive, emotional, and motoric functions. This is mediated through well-defined neuroanatomical circuits that integrate cortical activity with subcortical information processing (e.g., sensory events). Prenatal challenge of pregnant rats with Poly I:C can produce asynchronous EEG activity in the prefrontal cortex and hippocampus in offspring rats (Dickerson, Wolff, & Bilkey, 2010). Given that synchronous activity (i.e., activity that is temporally coincidental) in these two areas underlies successful working memory performance, the asynchronous activity may reflect altered cognitive capacity. Recent evidence supports the hypothesis that following prenatal immune challenge impaired communication occurs between important areas of the brain that subserve reasoning and memory skills, potentially arising from glutamatergic hypofunction.

Since epidemiological infection data has energized the immunological hypothesis for developmental origins of schizophrenia, it is relevant to determine in animal models whether actual infection of the mother alters the neurobehavioral development of the off-spring. It was indeed demonstrated that intra-nasal infection of pregnant mice with a specific strain of influenza A virus resulted in deposition of the virus in the fetal brain, and the viral RNA persisted in the brains of offspring up to the age of 90 days (Aronsson et al., 2002). This suggests that aside from known immunologic changes (e.g., cytokine expression) that can occur in the fetal brain, the infectious agent in and of itself can traverse the placenta and gain access to the developing organism. However, this interpretation is not altogether conclusive, as others have found no evidence for cross-placental migration of influenza virus into the fetal brain (Shi, Tu, & Patterson, 2005). Although this issue is largely unresolved, and may depend on a variety of different factors, the general consensus is that altered neurobiological functioning following maternal viral infection is the result of maternal immune reactivity. Indeed, as with the noninfectious models (e.g., LPS, Poly I:C), prenatal infection with human influenza produces deficits in PPI of the acoustic startle response, as well as reduced exploration of novel environments and objects (Shi, Fatemi, Sidwell, & Patterson, 2003). Moreover, this can result in reduced numbers of Purkinje cells in specific layers of the cerebellum (Shi et al., 2009). Finally, maternal infection of rhesus monkeys with influenza virus reduced the amount of white matter in the cerebellum and the gray matter density in the prefrontal cortex, frontal cortex (extending throughout the precentral and secondary motor areas), cingulate, insula, parietal cortex, and the superior region of the temporal cortex (Short et al., 2010). No major behavioral deficits were observed in the infants, although accelerated autonomy from the mother and reduced orienting or attentional behavior were noted in primates from infected mothers.

Overall, there is sufficient evidence to suggest that maternal immune activation interferes with the normal development of the fetal and neonatal brain. The particular mechanisms involved in exerting this effect are presently unknown, although the role of proinflammatory factors related to microglial cells in the CNS has received increased attention. Moreover, the possibility exists that prenatal immunologic influences are more profound in genetically vulnerable individuals (Lipina, Zai, Hlousek, Roder, & Wong, 2013). Finally, there is also evidence for altered sensorimotor gating and cognitive deficits in animals that were infected or immunologically challenged as neonates (Rothschild, O'Grady, & Wecker, 1999; Williamson, Sholar, Mistry, Smith, & Bilbo, 2011). Combined, the immune response and/or inflammation can be considered as significant factors in the multifactorial conceptualization of the two-hit hypothesis and schizophrenia development.

MICROGLIAL ACTIVITY AND SCHIZOPHRENIA

The role of microglia during psychiatric disease is beginning to be better understood in relation to schizophrenia, and as we've discussed at length, they are also relevant to other psychiatric disorders. Microglia produce, and respond to, many different cytokines, and a number of antipsychotic drugs directly act by limiting microglial activity and inflammatory cytokine production (Monji, Kato, & Kanba, 2009), all of which implicate microglia in the etiology of schizophrenia. In adults, postmortem, serological, and imaging studies have shown increased microglia activity in individuals with schizophrenia. A review of postmortem analyses on brains from individuals with schizophrenia revealed consistent evidence for greater microglial density (van Kesteren et al., 2017). This suggests greater microglial proliferation, whether *in situ* or through increased recruitment of myeloid precursor cells. It is uncertain, however, whether gray matter reductions in schizophrenia are related to increased microglial engagement in phagocytic activity, and removal of synapses and neurons (Selvaraj et al., 2017). It is conceivable that the increased microglia in schizophrenia might be inherent to the neurobiology of the disorder or a by-product of the stress of living with the disorder. Since stress can increase microglial cell activation (Howes & McCutcheon, 2017), this is not an unreasonable explanation. Animal studies of both prenatal and early-life infections or stress can prime microglia to be more responsive to later life stressors and infections (van Kesteren et al., 2017), and once activated, the composition of glutamate receptors on microglial cells can vary, resulting in augmented proinflammatory cytokine release (Beppu et al., 2013).

Microglia can also extensively regulate the glutamatergic signaling pathway. They express and respond to glutamate, have dynamic interactions with NMDA receptors, and alter glutamatergic synaptic function when activated (Roumier et al., 2008). Furthermore, microglia regulate synapses through a highly regulated pruning process, which can include chemokine signaling and complement (Paolicelli et al., 2011; Schafer & Stevens, 2010). Synapses are normally culled in an activity dependent manner from early life into late adolescence, essentially preserving high load synaptic connections at the expense of extraneous, low use synaptic terminals. Several reports indicated that individuals with schizophrenia have a disproportionate decrease in the amount of dendritic spines beginning at around adolescence (Bennett, 2011; Glantz & Lewis, 2000; Glausier & Lewis, 2012). This may represent aberrant synaptic pruning, which could lead to disruptions in a number of neurotransmitter systems. The overlap of critical periods of change in the density of spines and active synapses with the first onset of schizophrenia symptoms (or during the prodromal period) implicate pruning

as a potentially altered process in schizophrenia (Bennett, 2011). Whether aberrant microglial activity is at the heart of this process remains to be determined.

MICROBIOTA AND SCHIZOPHRENIA

Although the microbiota has received considerable attention in relation to stress-related disorders, such as anxiety and depression, and gut-related inflammatory disorders, limited research has been undertaken to assess the link to schizophrenia. However, if it is accepted that immune factors contribute to schizophrenia (and to autism), then it's essential to at least introduce the potential relationship between these illnesses and microbial factors given that microbiota affect immune functioning. We will cover this literature fairly briefly, but an excellent review on this topic, as well as the relation between microbiota and autism is available in Kelly, Minuto, Cryan, Clarke, & Dinan (2017).

Patients diagnosed with schizophrenia exhibit microbial differences from controls at the family, genus, and species level, and specific bacteria varied with the illness characteristics expressed. Specifically, the severity of symptoms was directly related to the amount of gut *Lactobacillus* bacteria (Kelly, Minuto et al., 2017), and *Lactobacillus*, *Bifidobacterium*, and *Ascomycota* were elevated in the mouth and pharynx of schizophrenic patients (Castro-Nallar et al., 2015). There have also been indications that in male patients the appearance of *C. albicans*, a fungal factor, was elevated in schizophrenia (Severance et al., 2016), and the appearance of *C. albicans* antibodies declined in association with probiotic treatment, as did schizophrenia symptoms, although only to a modest extent (Severance et al., 2017). As described in Chapter 7, Prenatal and Early Postnatal Influences on Health, prenatal maternal diet as well as microbial changes secondary to maternal stressors and immune-related illnesses have been linked to schizophrenia, and indeed, some of the effects of altered microbiota disturbances in rodents may not appear until adolescence, much like schizophrenic symptoms appear at this time.

As we indicated earlier, it is uncertain how microbial factors are related to psychological disturbances. Microbiota has been associated with various brain neurochemical, hormonal, neurotrophin, microglial, and immune changes, any of which could affect psychological state. Indeed, Kelly et al. (2017) have made an impressive case for one or more of these processes being mediators between microbiota and mental illnesses. As tempting as it might be to experimentally follow-up on these possibilities, it needs to be kept in mind that microbiota are sensitive to stressful events, exercise, and diet, all of which are likely altered secondary to schizophrenia. Likewise, antipsychotic agents that most patients were receiving might influence microbiota. Thus, it is probably premature to make any firm conclusions concerning the connection between microbiota and schizophrenia, but in view of the broad actions of microbial factors, it is also premature not to pay much more attention to their involvement in mental illnesses.

CONCLUDING COMMENTS

This chapter has provided an overview of some of the major areas that have sought answers for the etiology of schizophrenia in the inflammatory or immunologic domain. Fig. 13.2 provides an overall summary of the various topics that we have discussed, and which are pertinent to the evolution of a schizophrenic disorder. To some extent, to include the immune system as a pathophysiologic index is reasonable enough. But then this would place the immune system in a category with diabetes and metabolic disturbances, as well as cardiovascular diseases, all of which may be sequelae of a discordant and stressful life style and/or poor neural regulation of systemic biological processes needed for good health. These are important variables to control and evaluate, and perhaps even investigate as potentially linked to the etiology of schizophrenia.

However, it seems that the immune system may be more than just an accessory biological entity altered and mismanaged in the context of an aberrant set of neural functions. It may actually be an important impetus for change in the CNS that is already weakened and susceptible as a result of inherited gene variations. This is based on research regarding the impact of prenatal immune activation, as well as studies that linked infections to exacerbation and possibly the triggering of psychotic episodes. In the meantime, as researchers try to nail down the precise immunological effects on the development and appearance of psychotic disorders, there are opportunities to try modulating inflammatory processes in the hope of improving symptoms and the quality of life for patients and family members. The use of nonsteroidal anti-inflammatory drugs (NSAIDs) as an adjunct to ongoing antipsychotic treatment is slowly increasing, with preliminary trials suggesting they might be helpful (Bumb et al., 2015; De Picker, Morrens, Chance, & Boche, 2017). Perhaps the strongest argument was based on an analysis of a small number of double-blind, placebo controlled studies, which suggested that NSAIDs significantly reduced the severity of positive and negative symptoms (Sommer, de Witte, Begemann, & Kahn, 2012). However, other reports suggested that NSAIDs were less effective than targeting serotonin receptors (Andrade, 2014), while in a large study of over 16,000 patients, the use of NSAIDs increased the risk of relapse (rehospitalization for schizophrenia) in incident-free patients (Köhler, Petersen, Benros, Mors, & Gasse, 2016). It is likely given these findings,

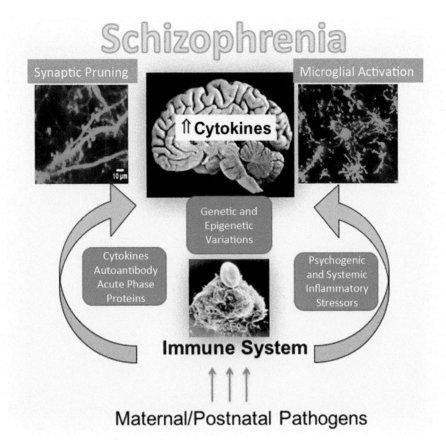

FIGURE 13.2 Summary of various precipitating factors for the onset of schizophrenia. Immune activation during pregnancy and/or during the postnatal and childhood period may generate aberrations in cytokine production, self-directed auto-antibody production, as well as other inflammatory factors. As well, inflammatory responses and psychological stressors, that interact with various genetic and epigenetic variants identified in large populations of schizophrenic patients, may also favor the emergence of pathology. Eventual neuropathology may involve excess microglial activation, which might contribute to synaptic pruning during critical phases of neuronal development. In addition, neurochemical disturbances might ensue from many of these interactions (e.g., glutamate hypofunctioning and poor tryptophan metabolism).

that there is a need for greater precision and control over the use of different types of NSAIDs (e.g., celecoxib, acetylsalicylic acid, ibuprofen), as well as an assessment of the inflammatory profile of individual patients. The latter may be important, since the efficacy of NSAID treatment may rely on predisposing states of inflammation, not all of which will apply to different patients. Moreover, such treatments may be provided earlier, if prodromal research identifies high-risk groups in need of anti-inflammatory treatment.

14

Inflammatory Roads to Parkinson's Disease

A MULTI-HIT FRAMEWORK FOR PARKINSON'S DISEASE

A growing consensus has revolved around the view of a "multi-hit" hypothesis for Parkinson's disease (PD). Environmental and inflammatory insults experienced over the lifetime collectively contribute to the genesis of PD (Fig. 14.1). In the case of idiopathic PD (which accounts for greater than 90% of cases of the disease), heavy metals, organic pollutants, several pesticides, and even viral/bacterial infections have been linked to the disorder in humans, and can induce PD-like pathology in animals. The first "hit" could be of a genetic nature, and gene polymorphisms, such as LRRK2, PINK, and DJ-1, have been implicated in the disease. Additive or even synergistic interactions between genetic and stressors could shape the evolution of pathology. For instance, increased incidence of PD has been correlated to exposure to the herbicide, paraquat, in combination with risk alleles for PD.

Several lines of evidence suggested that microglial-mediated neuroinflammatory processes play a prominent role in PD and that environmental stressors can impact neurons through inflammatory mechanisms. In the broadest sense, environmental insults can prime neuroinflammatory cascades, resulting in augmented responses that over time (and in conjunction with genetic vulnerabilities) can damage the susceptible midbrain dopamine neurons that normally are lost in PD. Of course, many processes independent of inflammation are also likely to be involved in PD.

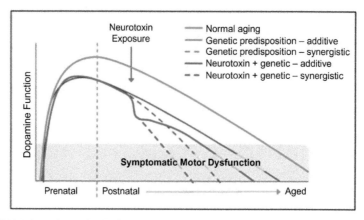

FIGURE 14.1 Multi-hit hypothesis for Parkinson's disease. Normal aging results in diminished dopaminergic cell levels and function, however, this process is exacerbated by certain genetic polymorphisms, which can have either additive or synergistic effects. These actions can be exacerbated by environmental toxins, which again can have additive or synergistic actions. If enough environmental "hits" occur over the lifespan then a threshold of neuronal loss/pathology may be crossed and Parkinson's symptoms begin to emerge. Of course, environmental hits at developmentally sensitive times may be especially deleterious and may augment the impact of subsequent hits. *Source: Modified from Boger, Granholm, McGinty, & Middaugh (2010).*

GENERAL INTRODUCTION TO PARKINSON'S DISEASE

This chapter focuses on the various environmental and genetic elements that contribute to Parkinson's disease (PD), emphasizing the role of the inflammatory immune system in the development, progression, and possibly the treatment of PD. Indeed, putative environmental triggers, such as pesticides and infectious agents, in concert with genetic vulnerabilities, most notably LRRK2 mutations, have well documented proinflammatory consequences. Animal models of PD have revealed that the administration of chemical or other toxicants that provoke robust neuroinflammatory effects, including activation of the brain immunocompetent microglial cells and the production of cytokines that serve as inflammatory signaling molecules, can engender effects reminiscent of PD.

The brains of many PD patient display similar inflammatory signatures and appreciable effort is being devoted to devising

pharmacological treatments that tame these destructive inflammatory cascades, while still preserving beneficial immune processes. Such treatments could include specific anti-inflammatory drugs or immunomodulatory agents that shift microglia and/or other immune cell states toward a protective or recovery mode. Throughout this chapter, we take a broad view of PD, focusing not only on the obvious primary motor pathology, but also consider its common comorbid features, such as depression, as well as the many life-style factors that could contribute to the many facets of the disease.

PD the second most common age-related neurodegenerative disorder, occurring in up to 0.2% of the population by 60 years of age and sharply rising to 4% in later years (Hirsch, Jette, Frolkis, Steeves, & Pringsheim, 2016). The disease is characterized primarily as a movement disorder, in which patients display the four cardinal motor symptoms; resting tremor, rigidity, slowness of movement, (bradykinesia) and postural instability. These motor disturbances emerge as a result of the progressive degeneration of dopamine (DA) neurons in substantia nigra compacta (SNc) region of the midbrain (Grünblatt, Mandel, & Youdim, 2000).

The SNc DA neurons project to the caudate and putamen regions of the brain, forming the nigrostriatal pathway, which regulates many aspects of motor functioning. In rodents, the caudate and putamen brain regions are fused and collectively referred to as the striatum. This striatal (or in humans caudate-putamen) brain area is considered a master regulator of many aspects of motor functioning and is a fundamental relay and integrative center for many other brain regions, which at a systems level are collectively referred to as the basal ganglia. With relevance to PD, other key basal ganglia regions that communicate with the striatum include the internal and external globus pallidus and subthalamic and ventrolateral thalamus. Signals originating from these areas ultimately project to the motor cortex to facilitate voluntary movement.

Another defining characteristic of PD comprises the presence of Lewy body inclusions (Del Tredici & Braak, 2012). Lewy bodies (named after Fritz Heinrich Lewy who first characterized them over a 100 years ago) are essentially abnormal intracellular clumps of protein that are somewhat analogous to the senile plaques that characterize Alzheimer's disease (AD). These inclusions are present throughout the brain, being found in both the soma and dendrites of neurons. Lewy bodies are primarily composed of accumulated misfolded protein aggregates that principally encompass α-synuclein, as well as the limited expression of other proteins including parkin and ubiquitin. The α-synuclein is important for sending synaptic signals between neurons, but in the case of PD, this protein becomes abnormally activated (through the addition of a phosphate group; i.e., they become hyper-phosphorylated) and tend to clump together.

Although considerable effort has focused on the role of α-synuclein rich Lewy bodies in PD, it remains unclear whether these actually contribute to the initial development of disease or are secondarily produced and are more important for disease progression. In fact, there is the view that Lewy bodies might not represent a pathological process, but instead reflect compensatory "clean up" mechanisms that are aimed at repair or disposal of damaged neurons. That said, support for this hypothesis is sparse and a body of data has amassed consistent with the view that Lewy bodies are part of the deleterious degenerative process. In this regard, most studies have suggested that abnormally activated α-synuclein somehow causes the misfolding of proteins, resulting in their pathological clumping, or alternatively, interferes with the normal breakdown of cellular debris or metabolic products (Fujiwara et al., 2002). In either case, the abnormal protein clusters interfere with basic neuronal functions required for signaling and even survival.

One of the most intriguing aspects of α-synuclein associated Lewy body pathology is that these inclusions appear to be transmissible in a manner similar to viral or prion agents (Kordower, Chu, Hauser, Freeman, & Olanow, 2008). Indeed, monomeric α-synuclein species are present in PD plasma and cerebrospinal fluid, and there is evidence that they may spread from transplanted embryonic PD tissue (Brundin, Dave, & Kordower, 2017). Moreover, animal models of PD have involved "seeding" the brains of rodents with mutant α-synuclein fibrils (linear rod-like form), which resulted in a spread of α-synuclein pathology akin to other PD models (Thakur et al., 2017). The ability for α-synuclein spread is consistent with the progression of Lewy body inclusions in PD, which are believed to follow an anatomical spread from the posterior to frontal parts of the brain (Rietdijk, Perez-Pardo, Garssen, van Wezel, & Kraneveld, 2017). In fact, it was proposed that

Lewy body pathology in PD first begins in the gut and then later spreads to the brain (Lionnet et al., 2017). Thereafter, Lewy pathology is thought to occur first in the brainstem of the CNS and only later moves to the nigrostriatal pathway. Eventually, in later stages of disease, widespread distribution of the Lewy bodies occurs throughout the cortex, thalamus, and hypothalamus.

The spread of PD pathology through distinct stages first occurring outside the basal ganglia, and later spread to motor regions was first described by Braak et al., (1996). Essentially, the crux of the hypothesis is that PD may result from some pathogen entering through nasal membranes and/or the gut, causing the formation of Lewy bodies in peripheral tissues, which then spreads into the brainstem and then to basal ganglia and forebrain regions (Hawkes, Del Tredici, & Braak, 2009). This is consistent with the early comorbid symptoms, such as olfactory disturbances, depression, and autonomic difficulties that precede the motor pathology and eventual PD diagnosis.

PARKINSON'S MIGHT START OUTSIDE THE BRAIN

Lewy bodies are present in the periphery and brainstem before they spread to the midbrain (Fig. 14.2), which is consistent with the possibility that peripheral immune processes might be involved in the early stages of disease initiation. It is also consistent with viral or bacterial routes to disease, as these typically are first encountered through muscoal membranes within nasal, intestinal, or respiratory tracts.

These microbes could then incubate in peripheral tissue, or in some cases, directly invade the brain. In fact, PD is known to have many comorbidities, such as gastrointestinal and olfactory disturbances, as well as depressive and anxious symptoms. These clinical manifestations are influenced by inflammatory cytokines and related factors. In effect, multisystem disturbances may be at the heart of the disease.

PARKINSON'S COMORBIDITY AND STRESS CONNECTIONS

Although motor function deficits are the most recognizable clinical features of PD, as already indicated, a large proportion of PD patients also experience comorbid symptoms, including anxiety, depression, along with autonomic, gastrointestinal dysfunction and olfactory problems (Litteljohn et al., 2017). These comorbid symptoms often precede the onset of motor decline or essential tremor (Gustafsson, Nordström, & Nordström, 2015), suggesting that pathological processes are simultaneously operative across multiple brain and body sites. As degeneration progresses, symptoms begin to include cognitive deficits, most notably problems with executive functioning, and frank dementia occurs in a subgroup of PD patients

in the later stages. These symptoms are major determinants of poor quality of life among PD patients and collectively they might even influence the progression of disease.

Unlike AD, the majority of PD patients do not display dementia or obvious memory loss, at least not until the very end stages of disease (Coelho & Ferreira, 2012). What the patients do show is very specific executive deficits, involving difficulty planning and organizing for future events, as well as set shifting problems (Maetzler, Liepelt, & Berg, 2009). Set shifting is basically the ability to exercise flexibility in switching between one task and another, or shifting strategies in problem solving situations (e.g., within the Wisconsin Card Sorting Task). Essentially, PD seems to be associated with impaired cognitive flexibility, which has been predominantly ascribed to prefrontal cortex

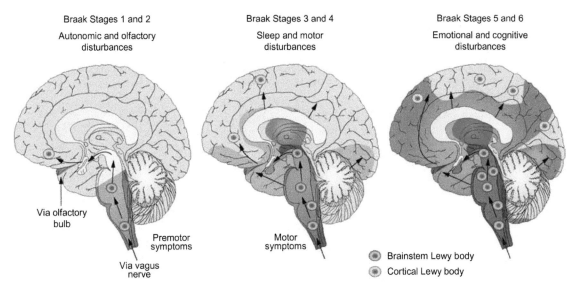

Braak Stages 1 and 2

Autonomic and olfactory
disturbances

Braak Stages 3 and 4

Sleep and motor
disturbances

Braak Stages 5 and 6

Emotional and cognitive
disturbances

Via olfactory
bulb

Premotor
symptoms

Motor
symptoms

Via vagus
nerve

⦿ Brainstem Lewy body

⦿ Cortical Lewy body

FIGURE 14.2 **Spread of Parkinson's pathology.** Pathology (e.g., pathological Lewy body aggregates) may begin outside the brain and then spread to the brainstem via vagal or olfactory routes. Within the brain, pathology then spreads forward toward the midbrain and finally cortical layers. As depicted by the different Braak stages, differing symptoms emerge coincident with this pattern of spread of pathology from the brainstem (Stages 1–2) to the motor regulatory basal ganglia brain regions (Stages 2–4) to finally the higher cortical areas that promote cognitive and other symptoms (Stages 5–6). *Source: From Doty (2012).*

functioning. In fact, several animal and human studies have confirmed dysfunctional frontal cortex functioning in PD, which appears to be related to a loss of dopaminergic and noradrenergic input to this region, together with the local accumulation of Lewy bodies and inflammatory species (Caspell-Garcia et al., 2017; Santpere et al., 2017). A small subgroup of PD patients that eventually develop dementia are characterized in the separate classification of Parkinson's disease dementia (PDD). These individuals experience the "worst of both worlds," clinically sharing symptoms from both PD and AD.

The two primary psychiatric comorbid symptoms, depression and anxiety, are not unique to PD, being the most common comorbidites across many neurological conditions, including AD, cerebral stroke, and multiple sclerosis. As mentioned in earlier chapters (see Chapter 3: Bacteria, Viruses, and the Microbiome and Chaper 4: Life-Style Factors Affecting Biological Processes and Health), they are also commonly associated with cardiovascular and metabolic conditions, such as heart disease, diabetes, and general illness. Because of their evolutionary importance for threat detection and social bonding, the emotional brain circuits linked to depressive illness might be especially sensitive to any perturbations in brain functioning. In effect, for evolutionary reasons, depression might be the "canary in the coal mine" across several different pathological states. Brain processes that govern emotional state are easily disturbed in many different threatening situations/states because they specifically evolved to be highly plastic and highly responsive to anything that challenges biological homeostasis.

Depression in Relation to Parkinson's Disease

Approximately 40%–60% of PD patients present with clinical depression, just as this sometimes occurs with other neurological disturbances. In some of these conditions, most notably stroke, the depression most often arises after the specific primary disease event and is likely caused by the inflammatory and other neurobiological processes associated with the traumatic event. In other instances, the distress associated with the illness clearly promotes depression.

At first blush, it would seem that depression would be an obvious response to a PD diagnosis and the major changes in quality of life that come with such a diagnosis. Certainly, in some cases, this distress is exactly what precipitates the depression. Yet, considerable evidence points to alternate causes in most cases, especially as depression is often evident long before the first motor symptom presentation and subsequent PD diagnosis (Ishihara & Brayne, 2006). Imaging studies revealed that comorbid depression in PD was related to widespread differences in functional connectivity between the cortex, hippocampus, caudate, and thalamus. In particular, diminished caudate to thalamus and frontal cortex to hippocampus activity was present and when anxiety symptoms were evident, further disturbances were also present (Dan et al., 2017). Further to this, besides the degeneration of midbrain DA neurons, other neurotransmitter systems are affected in PD. Most notably, the primary norepinephrine (NE) producing neurons in the locus coeruleus, also degenerate in PD, perhaps developing early in the disease processes (Buddhala et al., 2015). In fact, animal studies suggested that the loss of these NE neurons might contribute to the subsequent loss of midbrain DA neurons, possibly by reducing the availability of essential growth factors that are required for proper neuronal survival. Whatever the case, the loss of brainstem NE have cascading effects throughout the brain, which could adversely influence the functioning of emotional brain circuits. The monoaminergic dysfunctions within multiple interconnected cortical, limbic and brainstem regions likely play an important role in comorbid depressive pathology. However, PD comorbidity cannot simply be reduced to neurotransmitter changes, and the role of the inflammatory and associated systems is of particular importance.

Stress in Relation to Parkinson's Disease

No discussion of PD would be complete without mentioning the many aspects of stressor exposure relevant for depression, and possibly for some of primary motor deficits. It is also important to reemphasize the position that stressors across the entire spectrum, ranging from psychological to chemical to immune processes, collectively contribute to PD and as such, it is difficult to parse out the effects of any single type of stressor. That said, some stressors (e.g., chemical pesticides) might have more direct damaging effects on motor brain regions, whereas psychological stressors might be more aligned with neurochemically related emotional symptoms. Importantly, the impact of a given stressor experience may depend upon its temporal proximity to other stressors, its severity and duration, as well as the actual disease stage and state of the brain's different microenvironments at the time of exposure.

We know that when certain lag periods are imposed between stressors, such as immune (cytokines, LPS), drugs of abuse (amphetamine), and laboratory stressors (restraint, footshock), sensitized (augmented) or de-sensitized (diminished) brain and behavioral responses can be realized (Anisman et al., 2008). This is illustrated by the example of ischemic tolerance, in which a very small stroke induced in rats, limits the damage induced by a second, larger stroke that follows within a discrete period of time (usually 1–2 days), whereas such effects are not apparent after a longer delay (Li, Hafeez et al., 2017). In fact, damage is actually increased at the later time. Likewise, LPS pretreatment 4 hours prior to an ischemic event increased damage, whereas 24 hours following treatment with the endotoxin, the extent of damage was reduced (Ikeda, Yang, Ikenoue, Mallard, & Hagberg, 2006).

Consistent with these findings, immune priming (with bacterial or viral challenges) enhanced the impact of later pesticide exposure on midbrain dopamine neurons, but this sensitizing effect was only evident with certain time intervals (Mangano & Hayley, 2009). Pretreatment with LPS or the proinflammatory cytokine, TNF-α, likewise time-dependently sensitized rodents to the effects of later reexposure to the same challenge, but also to a different (restraint) stressor (Hayley, Staines, Merali, & Anisman, 2001). Evidently, even different categories of stressor might reinforce each other actions, depending upon their temporal sequence.

Various stressors could differentially influence PD symptomatology, depending upon the state of the brain at the time of exposure. By state of the brain, we are referring to the current level of functioning of neurons and extent of prodeath processes that might be engaged, together with the activational state of accompanying cells, particularly microglia and astrocytes. Importantly, given the varied microenvironments across different brain regions, it follows that the stressor effects will also vary across regions. A further layer of complexity is that these different brain regions are connected to one another, and thus a change in one region could reverberate through these circuits.

Virtually all stressors stimulate the HPA axis, resulting in glucocorticoid release, which has already seen in earlier chapters, easily penetrates the brain parenchyma and at high enough levels, is neurotoxic (Vyas et al., 2016). Of interest is that the corticoid inducing hypothalamic peptide, CRH, may alter the activity of brain mast cells and astrocytes, which

collectively can influence blood-brain-barrier (BBB) permeability and can thus increase neuronal pathology, as observed with experimental autoimmune encephalitis. As well, as shown in Fig. 14.3, psychological stressors activate resident microglia, leading to the enhanced secretion of proinflammatory factors (Delpech et al., 2015). These proinflammatory cytokines can themselves increase BBB permeability (e.g., via upregulating endothelial cell adhesion molecules), which has the dual effect of further enhancing cytokine production at vascular sites (i.e., a positive feedback loop) and augmenting peripheral immune cell trafficking across the BBB.

Given that psychological, immunological, and chemical stressors were all able to induce BBB disruption, these challenges would place the already vulnerable brain in contact with a host of potentially dangerous substances. As will be discussed in upcoming sections, a variety of environmental factors, including pesticides, heavy metals, and microbial agents, could all enter the brain through a compromised BBB to promote local neuroinflammatory or other consequences. Thus, stressors could directly or indirectly affect brain functioning, which in turn, could create a microenvironment in the brain that (1) further contributes to prodeath processes operative in PD, thereby increasing motor impairment, and (2) contributes to comorbid symptoms, such as depression/anxiety, that might exacerbate primary motor symptoms.

Chronic psychological distress may also influence symptom presentation owing to the loss of neurotrophic factors which normally provide neuronal support (Anisman & Hayley, 2012a,b). Certain acute and chronic stressors might also influence PD by promoting proinflammatory effects through the activation of brain microglial cells or by fostering the mobilization of peripheral immune cells (Blandino et al., 2009). Interestingly, extended chronic restraint (8 hours of restraint 5 days a week for 2–16 weeks) caused the loss of SNc dopamine

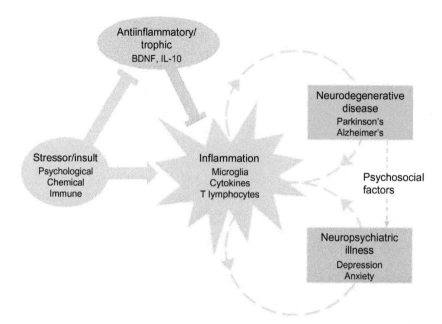

FIGURE 14.3   Inflammation as a common mechanism of neurodegeneration and comorbid depression. Environmental insults ranging from psychological stressors to chemical or immune insults can activate inflammatory immune elements. Most notably, microglia and the cytokines they release (and possibly in conjunction with T lymphocytes) are known to cause neurochemical changes that favor emotional symptoms, as well as potentially induce prodeath factors aligned with neurodegeneration. These toxic processes may be buffered by factors that can promote anti-inflammatory or trophic support. *Source: From Litteljohn & Hayley (2012).*

neurons with progressively more neuronal loss over time (Sugama et al., 2016). These same investigators subsequently demonstrated that dopamine neurons of the ventral tegmental area (VTA) were also vulnerable to cell death with a repeated restraint stressor, leading to the suggestion that damage to the VTA reward pathway might be involved in the comorbid depression often observed in PD (Sugama & Kakinuma, 2016).

Psychological stressors might also influence the responses provoked by subsequent immune challenges. In this regard, chronically stressed animals displayed augmented microglia reactivity and proinflammatory factors upon a subsequent immune insult (LPS). Indeed, augmented IL-1β, IL-6, and TNF-α responses, along with sickness behaviors, were evident in response to LPS in mice that were

exposed to a strong stressor 4 days earlier (Johnson, O'Connor, Hansen, Watkins, & Maier, 2003). It was subsequently shown that the stressor-induced sensitization of the LPS inflammatory response was related to modulation of the microglial phenotype. Specifically, the stressor primed microglia so that they upregulated NLRP3 inflammasome expression and the release of endogenous "danger" signals, such as HMGB-1 (Weber, Frank, Tracey, Watkins, & Maier, 2015). Hence, the brain appears to be sensing the psychological stress in a manner akin to the way it responds to pathogenic challenges. These examples illustrate that stressful events predispose biological systems to be more vulnerable to future insults.

Several studies have also indicated that psychological stressors are able to induce bacterial translocation from the gastrointestinal tract

(Zareie et al., 2006). This is particularly significant in light of the microbiome research that has implicated variations in microbial species with virtually every disease one can think of!

LIFE-STYLE AND AGING

The one unequivocal risk factor for PD is age and not surprisingly, age may act to augment the deleterious effects of virtually all environmental insults. However, for now, the focus will be on how age, in combination with other typical life-style factors (and not environmental insults per se) can collectively influence the risk of developing PD. The life-style factors of interest include diet, smoking, alcohol, drug use, and exercise, and how these influence the onset and/or severity of PD progression, as well as its comorbidities.

As already mentioned, the typical age of onset of PD is around 60–65 and generally the incidence of new cases drops sharply by 80 years of age (Tysnes & Storstein, 2017). This narrow window regarding age of onset is consistent with the time-dependent unfolding of genetic-driven processes and/or the slow accumulation of multiple environmental hits throughout one's lifetime. However, there are cases of PD that can occur as early as the 30s. As we'll see shortly, early onset PD is more often associated with specific genetic mutations (Schormair et al., 2017), than the typical "idiopathic" or "sporadic" cases, in which the origins of disease are not known. The absence of a clear genetic link in the majority of PD cases has, in fact, been one of the driving factors that have spurred the search for environmental causes. Of course, thinking that PD is caused by a single environmental factor would be short-sighted. Rather, a more realistic view is that genetics sets the stage for disease, but the "wear and tear" produced by multiple environmental challenges ultimately contribute to the development of PD (Rudyk, Litteljohn,

Syed, Dwyer, & Hayley, 2015). This is the crux of the multi-hit hypothesis which will be discussed in upcoming sections.

Most studies have generally reported the incidence of PD to be higher in men than women, at a ratio of about 1:5. Although this has been taken to suggest a protective effects of estrogen in PD (Bourque, Dluzen, & Di Paolo, 2009), the alternative is that men are more likely to be exposed to toxins or physical insults in the workplace (Wirdefeldt, Adami, Cole, Trichopoulos, & Mandel, 2011). There is also evidence concerning geographic distribution of PD, occurring less frequently in Africa and South America than in Europe or North America (Wirdefeldt et al., 2011). Yet, PD prevalence varies across African subregions, being lowest in eastern countries, which have a relatively low life expectancy and higher in the more affluent north African countries (Callixte et al., 2015). Similarly, in the more industrialized regions of South America, including Brazil, PD rates were similar to that in Europe and North America. Thus, life expectancy might account for difference in PD distribution or alternatively, it might be related to life-styles in the more Westernized countries.

The incidence of PD has generally increased over the past half a century, which could simply be a manifestation of people now living longer. Yet, life-style changes could be fueling the increased PD incidence, including changes in diet, a more sedentary life-style, greater reliance on chemical additives, and the presence of an increasing number of synthetic pesticides and other pollutants (Chin-Chan, Navarro-Yepes, & Quintanilla-Vega, 2015), as well as antibiotics increasingly appearing in foods.

Less obvious and indeed, somewhat surprisingly, PD risk factors include reductions in the smoking rate, certain occupations, such as farming, as well as certain high impact sports, most notably boxing. Of course, with these types of studies, it is difficult to determine causal connections and is further compounded by the long prodromal state that exists for PD. It is thought that the PD prodeath disease processes might actually begin decades before the first motor symptoms appear.

The long incubation time for PD is quite scary in that certain people could literally be "ticking time bombs," with a gradual decay in brain tissue occurring over time. But viewed from another perspective, such a long prodromal phase could potentially allow time for interventions to be undertaken. Of course, this rests on finding the best treatment(s) at the best time, which requires a better understanding of the early disease processes. This necessitates identification of biomarkers to identify who is to risk of developing the disease. It would also be necessary to have biomarkers that speak to the type of PD that could develop, as knowing the nature of potential comorbidities would also be exceptionally valuable, as the different flavors of primary and comorbid disease might call for different intervention or treatment targets.

Few studies have examined the impact of exercise on PD, and those that have generally revealed modest, albeit significant, links between the extent of physical activity and likelihood of developing PD. For instance, self-reports of regular high levels of physical activity before age 65 or participation in competitive sports before age 25 cut the risk of developing PD in half (Shih, Liew, Krause, & Ritz, 2016). Similarly, daily vigorous exercise modestly reduced risk in men, but not women (Chen, Zhang, Schwarzschild, Hernán, & Ascherio, 2005), but other reports pointed to increased risk with heavy exercise as we saw earlier in relation to other diseases. Unfortunately, many of the reported studies did not control for confounding variables, including smoking and diet.

Paralleling the link to physical activity, a clear relationship has not emerged with regard to PD

and dietary factors. Nonetheless, there were indications that specific foods might have some bearing in relation to PD. An association was not observed between PD and consumption of fruits and vegetables, but a modestly reduced risk was associated with consumption of meat and fish. A strong positive relationship was apparent between milk consumption and PD risk, whereas an inverse relationship was evident for polyunsaturated fat (Kyrozis et al., 2013).

Some evidence, however scant, suggested that particular variations in vitamins and minerals could play a role in PD development. Much of this has focused on the dietary metal, iron, since its levels are increased in the PD brain and iron has well known oxidative stress effects (Jiang, Wang, Rogers, & Xie, 2017). While some researchers failed to find a link between iron and PD, several others reported increased disease risk with high dietary iron levels (Powers et al., 2003). Several studies also reported significant, but modest positive associations between dietary zinc and copper levels and the occurrence of PD (Stelmashook et al., 2014). Reduced PD was also linked to high dietary intake of vitamin E, possibly being related to its antioxidant potential.

Perhaps the most surprising relationship with life-style factors was the robust inverse association between smoking and PD. More than 20 case-controlled studies, as well as prospective studies, reported that smoking (particularly current smoking) was associated with reduced risk of PD. For instance, a 26 year follow up study of over 8,000 men in Honolulu found a 60% reduced risk of developing PD and this effect was dose-dependent, being inversely correlated with the amount smoked (Grandinetti, Morens, Reed, & MacEachern, 1994). A meta-analysis that examined 48 studies found 40% or 60% reduced PD risk for past or current smokers, respectively, compared to those who never smoked (Hernán, Takkouche, Caamaño-Isorna, & Gestal-Otero, 2002).

Understandably, there has been a push to study the potential protective effects of nicotine or possibly other active compounds found in cigarette smoke. Several animal studies showed that certain nicotine dosing regimens lessened the impact of PD-relevant toxins, and that chronic nicotine treatment was neuroprotective in rodent and monkey PD models that involved lesioning with the dopaminergic toxicants (Huang, Parameswaran, Bordia, Michael McIntosh, & Quik, 2009). One study reported that chronic oral nicotine administration in monkeys prevented striatal terminal loss, but did not influence the loss of dopamine cell bodies in the SNc (Quik et al., 2006). Thus, nicotine might act on terminals to preserve their functioning and possibly promote compensatory sprouting. This is also consistent with findings showing that the cell bodies and terminals of nigrostriatal dopamine neurons degenerate at different rates in PD, and that the mechanisms responsible for pathology in these two biological compartments may be autonomous. Moreover, even with a 95% loss of striatal terminals, preservation of motor behaviors was exhibited in response to dopaminergic acting drugs (Golden et al., 2013), suggesting marked compensatory processes even in the face of severe degeneration.

In an attempt, to translate preclinical findings to actual patients, clinical trials using nicotine are ongoing (Villafane et al., 2017). Discouragingly, a short clinical report found that nicotine did not provide symptomatic relief in 40 PD patients that received either transdermal nicotine therapy or placebo over 39 weeks (Wood, 2017). Likewise, no benefit was obtained by PD patients treated with nicotine for a total of 45 weeks (with dose escalation up to 90 mg/day and then de-escalating before clinical testing following drug wash out) (Villafane et al., 2017). Of course, these PD patients may be past the critical period where nicotine might actually be beneficial, or alternatively, heterogeneity of disease manifestation and mechanisms of cell death across the patients might dilute the magnitude or ability to detect any clinical effects. Besides, the data available had indicated that nicotine might be useful in a prophylactic capacity, but not diminishing already existent PD. Thus, biomarkers to identify potential PD candidates during the prodromal disease phase might help triage the different subtypes of disease that might be required for efficacious nicotine treatment.

Much like smoking, the majority of studies that has assessed a link between coffee and tea consumption and PD have reported an inverse relationship, and a meta-analysis indicated that regular coffee consumption reduced the relative risk of PD by about 30% (Hernán, Checkoway, et al., 2002). It is believed that the potential protective effects of these beverages stem from the well-known adenosine antagonistic properties of caffeine (Chen et al., 2001). The brain adenosine pathway controls wakefulness, and blocking this pathway accounts for the stimulating effects of caffeine; an added benefit might be the protection against insults relevant for PD (Schwarzschild, Chen, & Ascherio, 2002). Interestingly, caffeine was reported to be neuroprotective against 6-hydroxydopamine (6-OHDA) lesion in rodents, which was related to an anti-inflammatory effect, characterized by reduced IL-1β and TNF-α levels (Machado-Filho et al., 2014). Further, direct application of caffeine on cultured macrophages promoted an antiinflammatory phenotype in these cells (Shushtari & Abtahi Froushani, 2017) and caffeine similarly reduced the inflammatory and oxidative stress effects of hypoxia in neonatal rats (Endesfelder et al., 2017).

There have been a few studies that failed to find a link between caffeine consumption and PD (Wirdefeldt et al., 2011), possibly owing to differential dose effects of caffeine. A U-shaped dose relationship was reported with regard to caffeine and PD in women, such that the most protective effects were observed with intermediate coffee (one to three cups/day) consumption rates (Ascherio et al., 2001). Thus, gender effects may also be at play, with estrogen possibly serving a mediating role in the biological implications of adenosine receptor blockade by caffeine.

In contrast to smoking and caffeine consumption, alcohol has not been consistently linked to PD incidence. In fact, a meta-analysis of 16 high quality reports revealed a weak protective effect in 7 studies, but 2 showed a negative effect, whereas the remaining 7 showed no effect (Bettiol, Rose, Hughes, & Smith, 2015). Such a range of outcomes underscores the complex nature of the human research studies, particularly in isolating alcohol consumption from other factors. A general summary of the environmental and genetic factors positively and negatively related to PD are shown in Fig. 14.4.

EXPERIMENTAL ANIMAL MODELS OF PARKINSON'S

This section will introduce the reader to the most common and best validated animal models for PD. The majority of studies were conducted in rodents, although several involved nonhuman primates, typically in late-stage preclinical testing of new drug treatments. A growing number of studies used lower sentient organisms, most notably roundworms and fruit flies, but these are generally restricted to experiments utilizing very specific genetic manipulation that require more basic neuronal systems, and thus will not be focused upon here.

Classic Toxicant Based Models: 6-OHDA and MPTP

The most commonly used animal models of PD are those that involve toxicant exposure, most notably, 6-OHDA and 1-methyl-4-phenyl-

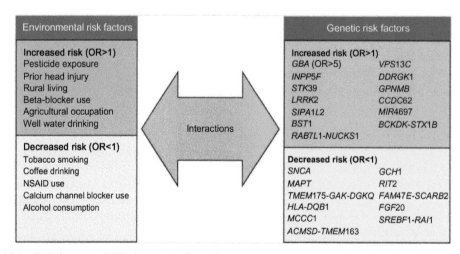

FIGURE 14.4 Risk factors and PD. Summarized are the environmental and genetic factors that either increase (Odds Ratio > 1) or decrease (Odds Ratio < 1) in the risk of developing Parkinson's disease. Pesticide exposure and head injury are most strongly linked to increased risk, while smoking and coffee drinking are most robustly related to decreased risk. A number of different genes are correlated with increased or decreased risk. Many of these genes are very rare, but the most common, LRRK2, also is a robust inflammatory immune system regulator. *Source: From Kalia & Lang, (2015).*

1,2,3,6-tetrahydropyridine (MPTP) (Cannon & Greenamyre, 2010). These well validated models and have provided a wealth of data regarding the degenerative process in PD and are ideal for testing new therapeutic regimens. The former (6-OHDA) is a dopamine analog that is taken up by the dopamine transporter and acts as a potent inducer of oxidative stress. In fact, molecules akin to 6-OHDA are naturally produced as part of the autooxidative processes that characterize normal dopamine metabolism.

THE DARK SIDE OF DOPAMINE

The high oxidative nature of dopamine metabolism suggests that high concentrations of dopamine may be neurotoxic, which raises a potential dark side to treatments, such as l-DOPA, that artificially raise dopamine levels[1]. So, efficient metabolic processes are essential for breakdown of such products and detoxifying any oxidative or nitrogen radicals. There is evidence that these processes might become somewhat weaker with age. Similarly, intracellular protein handling and detoxification (e.g., via the proteasome) are also thought to be compromised with advanced age. It also seems that the dopamine neurons in the SNc are appreciably more vulnerable than others, such as those in the ventral tegmental neurons. As we've seen, this is not unique to this neurotransmitter, as glutamate causes excitotoxicity at high concentrations. Thus, as the toxicologist Paracleus said "The dose makes the poison."

For about 50 years, 6-OHDA has been used to induce highly reproducible Parkinsonism (Blandini & Armentero, 2012). The administration of 6-OHDA through infusion into either the SNc, striatum or the medial forebrain bundle induces degeneration of the nigrostriatal system. However, because 6-OHDA does not cross the BBB, it must be infused directly into the brain to produce its toxic consequences. Though there are several variations in the infusion paradigms, generally the dopaminergic neurons and terminals are lost within 5 days after 6-OHDA infusion. When injected into the SNc, cell death is thought to begin somewhat earlier than when it is delivered to striatal terminals.

Of the animal PD models, 6-OHDA induces the most dramatic motor outcomes, including bradykinesia, disturbed coordination, and gait imbalances (Baldwin, Koivula, Necarsulmer, Whitaker, & Harvey, 2017). One of its most noticeable and reliable effects is that animals lesioned on only one side of the brain exhibit abnormal unilateral turning, rapidly circling toward the noninjected contralateral side. Unilaterally compromised DA motor regulatory apparatus also occurs in human PD patients who typically display unilateral hand tremors, and autopsy has confirmed that SNc DA loss tends to be greater on one side, thus giving rise to the asymmetry of pathology (Riederer & Sian-Hülsmann, 2012).

In contrast to 6-OHDA, MPTP readily crosses the BBB and hence, is administered

---

[1] Studies that showed cellular toxicity of l-DOPA were generally done *in vitro*, and little *in vivo* evidence exists to support toxicity. Indeed, a recent study that profiled potential toxicity of 20 day continuous *in vivo* infusion of l-DOPA and carbidopa formulations (the latter allows for greater dopamine entry into the brain rather than being broken down in the periphery) found no adverse effects, with the exception of some modest inflammatory changes that resolved after a few days (Ramot et al., 2017).

systemically (typically intraperitoneally) with mice and monkeys being the species of choice[2]. This synthetic agent is converted into its active toxic form, MPP+, by the MAO-B enzyme within astrocytes. The MPP+ is then extracellularly released and taken up, much like 6-OHDA, into DA neurons via the dopamine transporter (Richardson, Quan, Sherer, Greenamyre, & Miller, 2005). Importantly, MPTP induces nigral DA neuronal loss in humans virtually identical to that observed in the typical idiopathic cases of PD. We know this because of an unfortunate event that occurred in California in 1983, wherein a small group of individuals' self-administered a drug that was contaminated with MPTP. Rapidly, these people developed acute Parkinsonian symptoms that were responsive to l-DOPA (just like idiopathic PD) and upon their deaths years later, it was apparent that they had lost a substantial number of SNc DA neurons (Langston, Ballard, Tetrud, & Irwin, 1983). This is a case of a "naturally occurring experiment" in humans that resulted in the subsequent development of an animal model, essentially reflecting the "bedside to bench" translation of knowledge[3].

One drawback of the MPTP model is that the type and severity of motor pathology varies across species and even across rodent strains (Dauer & Przedborski, 2003). Also, while some PD features are typically observed following MPTP, most notably the reduced basal movement (bradykinesia), others features of PD, particularly the tremors, are not seen. Another weakness of the MPTP model (like that of 6-OHDA) is that it does not cause Lewy body inclusions. Indeed, no such inclusions are observed in rodents, and monkeys display only slight intraneuronal α-synuclein inclusions (even after 10 years of treatment) (Halliday et al., 2009).

Despite these shortcomings, MPTP is an extremely useful tool for assessing cellular and molecular mechanisms of DA neuronal death. It has highly reproducible neurodegenerative effects upon the nigrostriatal system, which have been ascribed to its ability to induce energetic cell failure and oxidative radical generation owing to blockage of electron transport by targeting mitochondrial complex I (Jackson-Lewis & Przedborski, 2007). However, as will be discussed shortly, MPTP, like pretty well all of the animal models of PD, also provokes substantial neuroinflammatory consequences, centered around enhanced microglial proinflammatory and prooxidative activities (Pisanu et al., 2014). As we'll see, it is unclear whether such neuroinflammatory processes are a cause or consequence of the MPTP induced prodeath processes; however, at the very least, inflammation influences the evolution of the disease state.

One final point that bears mention is that besides their individual effects, joint actions of multiple toxicants or even other stressors and toxicant challenges could influence the development of PD. The nigrostriatal system may be particularly sensitive, even to psychological stressor challenges (Finlay & Zigmond, 1997). In fact, animal models of PD using 6-OHDA have shown that chronic stressors both prior to and after 6-OHDA lesioning, accelerated the rate and enhanced the magnitude of damage to the nigrostriatal system, and concurrently exacerbated motor behavioral impairments (Hemmerle, Dickerson, Herman, & Seroogy, 2014). These findings are similar to reports of enhanced dopamine neuronal loss in MPTP

---

[2] When administered to young animals, 6-OHDA will enter the brain to produce destruction of dopamine neurons.

[3] Many instances of antidepressant and antipsychotic agents were discovered by happenstance, and have found their way to the laboratory where more detailed analyses could be performed both to understand how these actions came about, and in the hope of better agents being developed.

treated mice that were also exposed to chronic mild stress. Moreover, anhedonia was only observed in MPTP treated mice that were also exposed to a stressor (Janakiraman et al., 2016). Together, these findings raise the possibility that psychologically relevant stressors experienced prior and during the course of PD could affect both neurodegenerative processes and the primary motor symptoms, as well as certain nonmotor or comorbid neuropsychiatric manifestations.

ENVIRONMENTAL STRESSES AND PARKINSON'S DISEASE

Mutations in genes, including *SNCA* (PARK1/4), *LRRK2* (PARK8), *parkin* (PARK2), *PINK1*(PARK6), have been implicated in the manifestation and progression of PD (Klein & Schlossmacher, 2006). Yet, familial-related PD accounts for only a small percentage of cases, with the vast majority being idiopathic and sporadic in nature (Deleidi & Gasser, 2013). With respect to sporadic cases, several factors have been associated with higher risk of PD development, including age, genetic polymorphisms, stress, and cumulative exposure to environmental factors (Breckenridge, Berry, Chang, Sielken, & Mandel, 2016). In addition to heavy metals (lead and manganese), air pollutants, head trauma, viral infections (Breckenridge et al., 2016; Lee, Raaschou-Nielsen et al., 2016), and exposure to pesticides have repeatedly been implicated in PD (Ritz, Paul, & Bronstein, 2016).

Pesticides and Parkinson's Disease

Support for pesticide exposure in disease provocation has come from epidemiological studies primarily conducted in agricultural communities (Ritz et al., 2016). In fact, several compelling lines of evidence suggest a role for specific pesticides, including the organic insecticide rotenone, and the nonselective herbicide paraquat as major risk factors for disease development (Baltazar et al., 2014). These epidemiological findings are supported by animal studies (both primate and rodent) demonstrating that the administration of pesticides, such as paraquat and rotenone, induce many of the neuropathological and behavioral features characteristic of PD (Rudyk et al., 2015). These included motor behavioral disturbances, reduction in striatal fiber density, microglia activation, oxidative stress, and aggregated protein inclusions (Bobyn et al., 2012). It was likewise demonstrated that paraquat dose-dependently induced the loss of dopamine neurons in the SNc, reminiscent of what occurs in PD (Mangano & Hayley, 2009). Interestingly, paraquat has chemical structure that is strikingly similar to the active MPTP metabolite, MPP+ (Dauer & Przedborski, 2003), and is able to gain entry into the CNS via a neutral amino acid transporter present at the tight junction BBB. Upon entry into the brain, the toxin is distributed throughout the prefrontal cortex, hippocampus, olfactory bulbs, striatum, and SNc (Peng, Peng, Stevenson, Doctrow, & Andersen, 2007).

In the CNS, paraquat can promote a central inflammatory cascade through microglia activation, and subsequently can gain entry to neurons to directly perturb intracellular functions (Mangano et al., 2012; Rappold et al., 2011). Once in the neuron, paraquat can disrupt calcium homeostasis and mitochondrial electron transport chain complexes (i.e., complex I and IV), as well as increase mitochondrial membrane permeability (Huang, Chao et al., 2016). These effects collectively (1) reduce ATP energy production, thereby starving the neuron, (2) increase excitotoxicy, and (3) promote apoptotic factor release. Paraquat can also cause vesicular damage, induce endoplasmic reticulum stress, and even react with α-synuclein, resulting in further neuronal dysfunction and damage (Cochemé & Murphy, 2008). Together, these actions can endanger neuronal survival, but caution should be exercised with respect to the potential hazards of paraquat. It is unlikely that a substantial number of PD cases stem simply from exposure to paraquat. Rather, it is far more likely that paraquat is simply acting as one hit among many that are encountered over time.

Rotenone is an organic insecticide derived from a South American plant root, and like paraquat, has been associated with increased incidence of PD (Tanner et al., 2011). It is highly lipophilic and can cross the BBB, and acts as a potent inhibitor of the mitochondrial respiratory processes (Greenamyre, Cannon, Drolet, & Mastroberardino, 2010). This compound causes the loss of SNc dopamine neurons, as well as a loss of noradrenergic locus coeruleus neurons, as occurs in PD (Duty & Jenner, 2011), and also induces α-synuclein rich aggregations reminiscent of Lewy bodies (Cannon & Greenamyre, 2010). One major problem of the rotenone model is that there have been difficulties replicating the size of lesion induced. Of course, all the toxicant models encounter this issue to some degree; nevertheless, the rotenone model may be particularly vulnerable to this issue. As well, the route and timing of administration have a profound impact on the nature of the lesion produced (Fig. 14.5).

Infectious Agents and PD

The earliest evidence linking infectious agents to PD dates back to the highly virulent influenza epidemic of 1918 that killed millions of individuals. Notably, among the survivors that did not succumb to the disease, there were a sizeable number developed PD-like symptoms. These individuals displayed shuffling gait and difficulty with coordination, and later autopsy revealed marked inflammation within

the brain, coupled with dopamine neuronal loss (Ravenholt & Foege, 1982). In effect, it appeared that these people had severe encephalitis, which contributed to the PD syndrome (Casals, Elizan, &Yahr, 1998). Several further studies reported cases of PD that were associated with viral infections and it was suggested that intrauterine influenza infection could lead to PD, but this has never been directly demonstrated. However, direct testing of viral infection using the Japanese encephalitis virus, Herpes C, and H5N1 influenza virus in rodents clearly provoked the loss of SNc DA neurons (Jang et al., 2009).

Animal studies have provided evidence that the gut microbiota can influence the development of AD pathology. For instance, α-synuclein overexpressing transgenic mice that were raised in a germ-free environment or were treated with antibiotics to deplete their microbiota, displayed diminished microglial activation and motor deficits (Sampson et al., 2016). This effect could be reversed by administration of short chain fatty acids that the typical microbiota secretes. Even more astonishingly, transfer of microbiota from PD patients resulted in enhanced pathology in α-synuclein mice. These findings seem to suggest that metabolites from the microorganisms that colonize the intestinal tract carry a message that modulates the vulnerability of dopamine neurons in PD.

The involvement of gut processes in PD is all the more intriguing given that gastrointestinal disturbances are common in the disease. Indeed, gastrointestinal symptoms evident in PD include constipation, abdominal bloating, and nausea, which could be secondary to CNS pathology, although it is also possible that peripheral processes, such as inflammation, are primary driving factors. This is particularly the case considering that the microorganisms that colonize the gut play an important role in shaping immunity, which could then impact CNS processes.

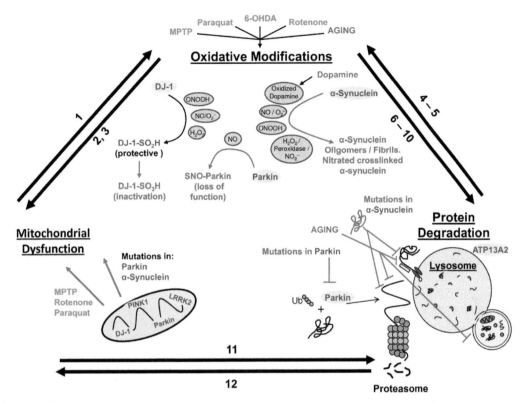

FIGURE 14.5 Oxidative stress, mitochondrial pathology, and impaired protein handling together form a "Bermuda triangle" of related processes that can give rise to PD. Multiple toxicants can impact these processes; most notably, MPTP, paraquat, rotenone, and 6-OHDA, along with the general wear and tear of aging. The numbers shown on the arrows depict the following interrelated processes: (1) mitochondrial respiration failure that generates oxidative radicals. (2) Antioxidants like SOD protect against mitochondrial toxins. (3) The protein DJ-1 might also protect against mitochondrial failure. (4–5) Impaired protein degradation elevates sensitivity to α-synuclein and oxidative stress. (6–10) Oxidative stresses and α-synuclein impair protein folding and degradation leading to faulty protein aggregates. (11–12) Autophagy and proteasome inhibition further contributes to protein aggregates and metabolic derangement. *Source: From Malkus, Tsika, and Ischiropoulos (2009).*

GUT FEELINGS AND PARKINSON'S DISEASE

Being the most prominent reservoir of microbes in our body, it is not surprising that the gut has received so much attention and has been implicated in numerous brain-related health issues. Certainly, PD is no different, and there is reason to think that the gut-brain axis is important for this disorder. Besides the obvious gastrointestinal symptoms evident in PD, the gut microbes are known to be critical for the proper development of the BBB, as well as for "training" immune cells. With regard to the former, the composition of the gut microbiome dictates how well the BBB forms, which is presumably a result of evolutionary pressure. As a result of a "leaky" BBB, environmental toxicants would have greater access to the brain, and consequently, greater likelihood of producing neurodegenerative actions. With regard to the

(cont'd)

second issue, that of immune cell training, it is known that endogenous microbes play a key role in modulating immune cell maturity and possibly phenotype. The ramifications are that environmental stressors that alter the microbiome in favor of pathological microbes can tip the scales in favor of an inflammatory environment. Further, stressors that limit the complexity of species present in the gut might result in impaired BBB functioning, as the normal evolutionary-derived pressures from microbes are diminished.

INFLAMMATORY MECHANISMS OF PARKINSON'S DISEASE

A major shift in thinking about brain-immune interactions and the development PD has evolved, with numerous studies emerging over the past 15 years linking various inflammatory immune processes to PD. This view was initially met with considerable resistance. It is no secret that the dogma regarding the immune privileged brain dominated much of neuroscience research until only two decades ago. In some ways the brain certainly is a privileged organ, but this does not mean that the brain is entirely off-limits to the immune system. As we've seen, cytokines and immune cells gain entry, albeit in limited concentration, to the colony stimulating factor (CSF) and the brain parenchyma (Filiano, Gadani, & Kipnis, 2017). The acceptance that immune processes greatly influence brain functioning has led to the current flourishing of neuroimmunology and related fields. Although the PD field was slow to catch up, the pendulum has swung sharply in favor of neuroinflammatory mechanisms being a mainstream player in PD pathology, as well as in virtually all neurological and neuropsychiatric disorders.

As with so many other diseases, multiple environmental stressors (ranging from chemical to microbial to even psychological challenges) collectively contribute to PD and its comorbidities by affecting neuroimmune processes. Genetic vulnerabilities set the stage for the impact of these insults, but multiple roads may converge at the neuroinflammatory interface to modulate PD pathology. We do not discount other nonimmune processes, such as apoptotic or other intracellular prodeath pathways, but these too can be influenced by an inflammatory milieu.

Adaptive Immune Processes in Parkinson's Disease

As indicated in Chapter 2, The Immune System: An Overview, the immune system can be roughly divided into the innate and adaptive immune branches, both of which can orchestrate inflammatory processes. The innate system has received particular attention in the development of PD, although several studies implicated adaptive immune processes in the disease.

The core of the adaptive immune response, as described in earlier chapters (see Chapter 1: Multiple Pathways Linked to Mental Health and Illness, Chapter 2: The Immune System: An Overview, and Chapter 3: Bacteria, Viruses, and the Microbiome), is largely made up of the T and B lymphocytes interacting with antigen presenting cells (e.g., macrophages, dendritic cells, or microglia) to provide immunological specificity and memory. Any pathogen or other exogenous threat detected by innate antigen presenting cells is processed and presented via the HLA (or MHC in rodents) to the adaptive immune lymphocyte, thereby priming it specifically against this one insult and secondarily, provoking clonal expansion and memory against future insults.

In the case of autoimmune diseases, such as MS, this process mistakenly identifies a naturally occurring part of the myelin sheath as being a nonself threat. Hence, T lymphocytes are primed against the myelin protein and orchestrate an attack that results in neuroinflammation and the breakdown of myelin. This leads to inefficient neuronal signaling and eventually neuronal demise, and the accumulation of cellular "junk" referred to as sclerotic plaques hence the name, multiple sclerosis.

Although PD is not an autoimmune disease in the strictest sense, it does have certain immune elements in common with such conditions. Most notably, the up-regulation of CD4+ and CD8+ positive T cells in the nigrostriatal system was demonstrated in PD brains, as well as in animal models of the disease (Alberio et al., 2012; Cebrián et al., 2014). It is thought that peripheral immune T cells infiltrate the brain and interact with microglia, placing them in a highly active proinflammatory state that subsequently is toxic to DA neurons (Brochard et al., 2009). Accordingly, in animal models, treatments that reduced neuronal damage (i.e., were neuroprotective) reduced T cell infiltration into the midbrain (Gendelman & Appel, 2011).

It was observed that PD pathology was linked to the levels of the different T cell subsets. In particular, PD patients had higher Th1 levels and reduced Th2 levels, compared to age matched controls (Chen, Qi et al., 2015). Further, Th1 lymphocytes were particularly toxic to DA neurons in MPTP treated mice, whereas the Th2 cells had the opposite effect, being neuroprotective (Olson et al., 2015). This is in keeping with the view of the Th1-Th2 dichotomy, which considers Th1 cells as driving proinflammatory and enzymatic cascades that are highly damaging to DA neurons, whereas Th2 lymphocytes favor the release of anti-inflammatory cytokines, such as IL-4 and IL-10, along with beneficial growth factors.

The fact that Th cells may be involved in PD suggests that humoral antibody dependent toxicity is occurring. In this regard, elevated brain specific and viral (herpes simplex) specific antibody concentrations have been observed in the blood and CSF of PD patients. There was the interesting observation that patients with familial inherited PD, but not the more common idiopathic form of the disease, displayed somewhat higher peripheral serum levels of α-synuclein specific autoantibodies compared to controls (Papachroni et al., 2007). Curiously, however, others reported elevated autoantibody levels in both plasma and CSF of mild and newly diagnosed cases of idiopathic PD, but these levels were somewhat lower in the more severe cases of PD (Horvath, Iashchishyn, Forsgren, & Morozova-Roche, 2017). Such autoantibodies normally are involved in clearing pathological proteins, but in the case of PD, this process breaks down, resulting in accumulation of α-synuclein rich Lewy body inclusions.

Besides α-synuclein, there have been indications that other autoantibodies are involved in PD, including those reactive against heat shock proteins and myelin associated glycoprotein (Papuć, Kurys-Denis, Krupski, & Rejdak, 2015). In particular, elevated serum levels of IgG and IgM antibodies against the β-crystalline heat shock protein and against myelin basic protein were observed in PD patients. Caution should be exercised when considering these data as others failed to find a relationship between autoantibody levels and PD (Maetzler et al., 2009). It has been proposed, as already indicated with respect to other PD-related biological processes, that many of these autoantibodies may have evolved as a protective "clean up" mechanism to prevent toxic protein aggregates (Nagele et al., 2013), and in this case, the elevations present in PD might actually represent a failed attempt to rid the brain of potentially toxic aggregates. The case could be made, however, that although antibody dependent toxicity could remove faulty/damaged proteins, the associated inflammatory reaction might produce collateral damage and harm adjacent healthy tissue.

Moving beyond CD4 + Th mediated antibody dependent humoral immunity, we now turn to the evidence for cellular T lymphocyte toxicity in PD. Indeed, CD8 + cytotoxic T cells have also been implicated in PD, suggesting that these cells have access to dopamine neurons for direct cell-to-cell toxicity. Evidence in favor of this comes from the finding that SNc, but not adjacent VTA dopamine neurons, normally express the MHC I molecule (Cebrián et al., 2014). This provides a mechanism through which CD8 + T cells can directly lyse dopamine neurons and helps explain why the SNc displays elevated sensitivity relative to the VTA and other brain regions. Predictably, PD animal models involving MPTP or 6-OHDA treatment, as well as α-synuclein overexpression, are associated with increased infiltration of CD8 + cells into the brain (Thakur et al., 2017). Although the data concerning the actual CD8 + T cell levels in the PD brain are scant, some studies have reported alterations within the periphery and curiously, these amounted to both increases and decreases, depending upon the particular study (Baba, Kuroiwa, Uitti, Wszolek, & Yamada, 2005; Jiang, Gao, Luo, Wang, & Yang, 2017). As well, it was reported an increased ratio of CD8 + T cells over the CD4 + subtype, and these cells were producing increased IFNγ, consistent with an inflammatory phenotype (Baba et al., 2005).

Consistent with the importance of CD4 + and CD8 + cells in PD, an immunization strategy that boosted T cell infiltration was neuroprotective in a MPTP model of PD (Benner et al., 2004). In this report mice were immunized with copolymer 1 (which is a random mix of polypeptides that carry the amino acids found in myelin and is used to treat MS) and then the primed T cell were given to a second cohort of MPTP treated mice (adoptive transfer), which conferred protection against dopamine neuron loss. Moreover, depleting these mice of T cells reversed the protective effects (Benner et al., 2004). Further studies that isolated the CD4 + CD25 + T cells (i.e., T reg cells), suggested that they might be responsible for neuroprotective consequences. In fact, direct application of CD4 + CD25 + T lymphocytes themselves conferred protection against the neurotoxic effects of MPTP (Kosloski, Kosmacek, Olson, Mosley, & Gendelman, 2013).

Intriguingly, adoptive transfer of T reg cells that were induced by GM-CSF (a potent stimulator of myeloid progenitor cells), downregulated the microglial inflammatory response elicited by MPTP (Kosloski et al., 2013). Conversely, systemic administration of GM-CSF alone was sufficient to induce neuroprotection in a paraquat-based toxicant model of PD, which was related to downregulation of the microglial response (Mangano et al., 2011). It seems that GM-CSF imparts protective consequences owing to modulation of neuroinflammatory processes. Yet, GM-CSF also has trophic and antiapoptotic effects and promotes increased BDNF levels, but had no effect on the antiapoptotic protein, Bcl-2 (Mangano et al., 2011). Thus, trophic and immunomodulatory effects could be at play in determining the beneficial effects of GM-CSF.

A novel idea encompassing adaptive immunity in PD is that neoepitopes are created in the brain that could fuel the infiltration of peripheral immune cells and subsequent degradation of DA neurons. For instance, metabolic products might damage DA neurons creating novel cell fragments that act as neoepitopes, or newly created epitopes that are immunologically recognized as being foreign. This could then result in microglia taking up the epitopes and transmitting a signal for the mobilization of peripheral immune cells, which then invade the brain to encounter the neoepitope being presented by the glial cell (Fig. 14.6).

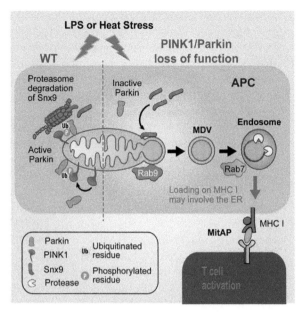

FIGURE 14.6 Mitochondrial antigen induced autoimmunity and PD. Environmental stressors (in this case LPS or heat stress) can impact mitochondrial functioning. Normally, the Parkinson's-linked genes *Parkin* and *PINK1* act to regulate mitochondrial functioning. These proteins ensure that any damaged mitochondria are removed through the process of being tagged by ubiquitin (ubiquitinated) for removal by the proteasome. However, when *Parkin* or *PINK1* are altered through genetic mutation, this process breaks down. The end result is that antigens bud off the damaged mitochondria and are placed in an endosome and then presented on the cell surface via an MHC I molecule. This attracts T cells which can initiate a destructive autoimmune attack that kills the cell. *Source: From Matheoud et al. (2016).*

IS PD AN AUTOIMMUNE DISEASE?

In considering the involvement of immunity in PD, it seems the pendulum has swung sharply from one side (i.e., in thinking there is no immune involvement) to the other (i.e., to thinking that PD might even be an autoimmune disease). While the bulk of evidence has favored a role for the innate branch of immunity in PD, several convincing studies implicated the adaptive immune response, suggesting that T-cell dependent autoimmunity might even be at play. One account had it that the presentation of mitochondrial antigens might trigger an autoimmune response against SNc dopamine neurons (Matheoud et al., 2016). With mitochondria being thought to have essentially come from ancient bacteria, it is important that their antigens be hidden from the immune system, as any breach could lead to autoimmunity.

This is consistent with the considerable data showing mitochondrial dysfunction in PD. Furthermore, it was demonstrated that PINK and PARKIN genes regulate mitochondrial antigen presentation, and mutations in these proteins might confer PD risk by removing the normally protective ability of these proteins to limit antigen processing and presentation (Matheoud et al., 2016). There have been other reports indicating that damage to dopamine neurons might produce neoepitopes that could fuel autoimmunity. Likewise, the concept of molecular mimicry, wherein certain endogenous neural motifs might be falsely recognized as foreign, has been applied to account for PD pathology. Whatever the case, the microglia are undoubtedly at center stage with regard to neuroimmune mechanisms.

Microglia and Parkinson's Disease: The Master Innate Neuroinflammatory Orchestrator

Microglia normally provide neuronal support through the release of trophic factors, including BDNF, as well as by clearing any cellular debris or metabolic by-products through phagocytosis (Trang, Beggs, & Salter, 2011). During this relatively quiescent resting state, microglia have an immobile ramified morphology characterized by a small cell body with long, thin processes that extend out into the extracellular milieu. These processes possess pathogen-associated molecular pattern (PAMP) receptors or damage-associated molecular pattern (DAMP) receptors, allowing them to detect and deal with microbial or other threats (Venegas & Heneka, 2017). Damaged or stressed cells release ATP into the extracellular space, which microglia can detect and place them in a "clean up" state for debris removal (Davalos et al., 2005). While removing cellular debris, microglia also extend their processes to shield healthy neurons from dying tissue, and also release attractant cytokines (chemokines), to recruit neighboring microglia (Gao & Hong, 2008).

Although there is substantial evidence that microglial-driven chronic inflammation is important for the genesis and the progression of PD (Ramirez et al., 2017), it is not clear what precisely drives microglial reactivity. One strong possibility is that it is a reaction to the accumulation of α-synuclein rich Lewy bodies. Misfolded α-synuclein that comprises a core feature of these pathological inclusions induces strong microglial activation, initiating inflammatory cytokine expression (Mrak & Griffin, 2007). In essence, microglia likely recognize these pathological features as DAMPs and respond accordingly. Thus, one can conceive different precipitating stressors (e.g., pesticides, heavy metals, microbial agents, and psychological stressors) as being interpreted as

either DAMPs or PAMPs. Even in the case of psychological stressors, the metabolic, energetic, and neurochemical changes provoked might be interpreted by microglia or other immune invaders as DAMPs. If such responding becomes chronic and/or excessive, then pathology may ensue.

Whatever the case, once activated, microglia may be responsible for phagocytic processes and release cytotoxic factors in order to ingest and eliminate noxious compounds. Microglial release of proinflammatory cytokines (i.e., IFNγ, TNF-α, IL-6, and IL-1β) and oxidative radicals (superoxide, nitric oxide) comprises their core reaction to PAMPs and DAMPs. Some of these same inflammatory factors were observed in postmortem PD patients, or in their CSF (Shi et al., 2011). In fact, in comparison to inflammation observed in arthritic joints, the levels of some of these factors appeared to be even greater in PD (McGeer & McGeer, 2004).

Microglia are particularly adept at producing inflammatory factors, and represent a typical reaction designed to neutralize threats. The problem comes when microglia are in a chronic or permanently active state. In such a case, these normally defensive molecules begin to damage over wise healthy tissue, particularly DA neurons within the SNc. As indicated earlier, these specific DA neurons are far more vulnerable to microglial-inflammation than other neuronal subtypes found throughout the brain.

It was observed that LPS infusion into the SNc had far more damaging effects on local DA neurons relative neurons in the frontal cortex, hippocampus, thalamus, or hindbrain. It was speculated that this enhanced sensitivity stemmed from the higher density of microglia in the SNc. Interestingly, the differences in the sensitivity to MPTP evident between mouse strains were correlated with SNc glial density (Smeyne, Jiao, Shepherd, & Smeyne, 2005), and the vulnerabilities to MPTP could be reversed by manipulating the number of glial cells present *in vitro* (Smeyne et al., 2005). Yet, many other factors could also contribute to nigral dopamine neuronal vulnerability. Among other things, enhanced iron or melanin content, as well as the high oxidative potential of these neurons, could render them vulnerable to various threats. Whatever the case, it is clear that microglia are somehow involved. Ultimately, the SNc neurons themselves may have an intrinsic vulnerability, and the SNc microglia are phenotypically different or more reactive than microglia from other brain regions.

Dopamine neurons of the SNc have a very high basal oxidative potential and are exceptionally metabolically active (Reale, Pesce, Priyadarshini, Kamal, & Patruno, 2012). These features make them prone to energetic disturbances and also to the generation of high levels of potentially destructive oxidative radicals as by-products of normal cellular reactions. These neurons also contain high levels of iron in PD patients, and this high iron load may provide a substrate for inflammatory molecules with which they can react (Zucca et al., 2017). This is consistent with the epidemiological evidence showing that exposure to heavy metals is accompanied by increased risk of PD. As well, exposing mice to iron or manganese augmented the neurodegenerative effects of paraquat or MPTP (Peng et al., 2007).

The SNc microglia express more of a "hair pin trigger" compared to those of adjacent brain regions. The extent of microgliosis (or essentially the activation state and mobilization) in response to many different challenges is often more robust in the SNc than in other brain regions. Beside the magnitude of the microglial response, the phenotype or characteristics of the factors expressed by these cells also appear to differ in PD. Specifically, microglia are generally polarized toward an M1 state, wherein they produce more inflammatory cytokines and oxidative factors, and they may be locked into this state for

extended periods of time (Tang & Le, 2016) The unfortunate humans that self-administered MPTP (described earlier) displayed signs of massive microglial activation in the SNc when their brains were assessed 3–16 years after an MPTP injection (Langston et al., 1999). As well, time course studies indicated that microglia activation occurs prior to dopaminergic degeneration in the face of 6-OHDA administration, and that microglia displayed a phagocytosis phenotype that could engulf compromised dopamine neurons (Marinova-Mutafchieva et al., 2009).

In keeping with the idea that stressors might be interpreted as DAMPs or PAMPs and reinforce microglial inflammatory activity, systemic infection exaggerated symptoms in patients, and interacted with environmental or even psychological insults to increase neurodegeneration in animals (Ascherio & Schwarzschild, 2016). Mechanistically, exposure to pathogens or cytokines concomitant with psychological stressors, traumatic brain injury, or toxin insults, can augment BBB permeability, in turn, promoting the central infiltration of immune factors and pathogens that can have vast and widespread effects on CNS functioning (Coureuil, Lécuyer, Bourdoulous, & Nassif, 2017). As described earlier, exposure to the bacterial endotoxin, LPS, caused exaggerated dopamine neuron and terminal loss in the SNc and striatum, respectively, upon subsequent exposure to the pesticide paraquat (Mangano & Hayley, 2009). Thus, inflammatory stimuli may prime or sensitize microglia, so that later challenges provoke exaggerated responses that have the potential to exacerbate the neurodegenerative process.

An interesting route to PD pathology involves the intersection between innate inflammatory and α-synuclein processes. This is supported by the finding that α-synuclein was associated with HLA + microglia in postmortem PD brains (Orr, Rowe, Mizuno, Mori, & Halliday, 2005). Moreover, microglial activation using inflammatory stimuli promoted the spread of α-synuclein pathology and neuronal degeneration (Shavali, Combs, & Ebadi, 2006), and it seems that the PAMPs used to detect pathogens might fuel the development of synucleinopathy. In fact, TLR2 was localized to α-synuclein rich Lewy bodies and its activation promoted α-synuclein accumulation, whereas its inhibition countered this effect (Dzamko et al., 2017). As well, priming isolated microglial cells with α-synuclein greatly influenced the inflammatory responses to subsequent immune insults, promoting greater cytokine release and a prominent shift in their morphological state (Roodveldt et al., 2013). Not surprisingly, direct application of α-synuclein into the SNc markedly activated microglia and induced dopaminergic neuronal death. Together, these findings point to a reciprocal relationship between microglial inflammatory and α-synuclein processes, such that one likely feeds into the other to fuel the development of pathology (Fig. 14.7).

Cytokines in Parkinson's Disease

Cytokines have been implicated in virtually all neurodegenerative and neuropsychiatric diseases. As we've seen, cytokines and their receptors are expressed endogenously in the CNS and mounting evidence suggests this is largely due to *de novo* synthesis by microglia. However, beyond glial cells, it was suggested that neurons might also be capable of synthesizing small amounts of proinflammatory and anti-inflammatory cytokines during disease or distress.

The general picture that has emerged is one in which proinflammatory cytokines are viewed as deleterious factors within the brain. This is certainly supported by the strong link between chronically and/or very high levels of TNF-α, IL-1β, and IFNγ and the incidence of brain pathology (and reactive microgliosis). Conversely, anti-inflammatory agents (e.g., minocycline, nonsteroidal and anti-inflammatory drugs) have generally been associated with neuroprotective consequences in a range of preclinical animal models (Subramaniam & Federoff, 2017). However, as we've seen in relation to other pathologies, unlike high levels of proinflammatory cytokines that can have deleterious effects on brain function, low physiological levels of cytokines may induce neuroprotection and adaptive neuroplasticity (e.g., via the release of free radical scavengers and trophic factors) (Litteljohn et al., 2014).

Complicating the picture, cytokines typically display a high degree of redundancy, pleiotropy, and synergy (but also antagonism) (Brebner, Hayley, Zacharko, Merali, & Anisman, 2000). The complex nature of cytokine physiology has led to the view that cytokines are best considered as a network of biologically active mediators whose collective output determines physiological and pathological consequences.

Cytokines have been implicated in numerous oxidative and excitotoxic pathways that produce neuronal damage and demise. The induction of the prooxidant/inflammatory enzymes COX-2 and inducible nitric oxide synthase (iNOS) result in the production of free radicals, such as superoxide and peroxynitrate, which promote oxidative damage to DNA, proteins, and lipids (Grünblatt et al., 2000; Mangano et al., 2012). Beyond this, certain proinflammatory cytokines, such as IL-1β and TNF-α, can also trigger the activation of caspase-dependent apoptotic pathways and excitatory glutamatergic signaling (Bakunina, Pariante, & Zunszain, 2015). Thus, such inflammatory cytokines are well placed to cause neuronal damage through multiple pathways, and consequently designing novel treatment options is highly complex.

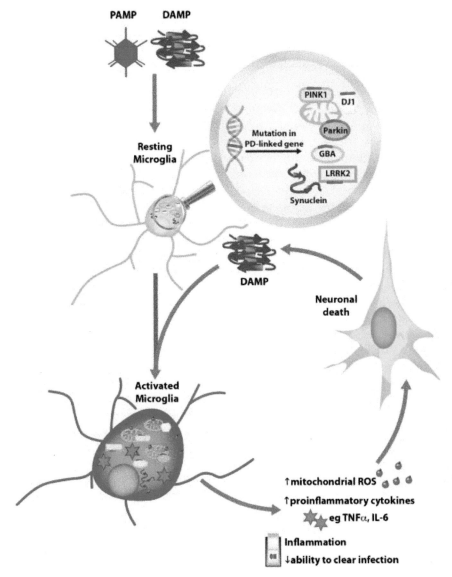

FIGURE 14.7   Microglial activation in PD. LRRK2, α-synuclein, Parkin, Pink1, and glucocerebrosidase (GBA) mutations can prime microglia to be more reactive to microbial agents (PAMPs) or even the sterile inflammation provoked by endogenous danger signals (DAMPs). This can lead to enhanced oxidative radicals and proinflammatory cytokines. *Source: From Dzamko, Geczy, & Halliday (2015).*

IFNγ and PD: A Specific Inflammatory Link

Among the proinflmmatory cytokines linked to PD, particularly strong evidence suggests

the importance of the interferons (IFNs). These are broadly divided into either type I IFNs, including the IFN-α and IFN-β isoforms, or the type II IFNγ. The latter is biologically active as a homodimer and upon binding to its receptor, induces various intracellular signaling factors (JAKs and STATs). Although it was initially believed that IFNγ is secreted exclusively from NK cells and Th1 lymphocytes, it appears that antigen-presenting cells, such as dendritic cells and macrophages, also produce IFNγ (Schroder, Hertzog, Ravasi, & Hume, 2004). Additionally, low levels of the cytokine are synthesized *de novo* within the brain by activated microglia (Kawanokuchi et al., 2006) and astrocytes (Lau & Yu, 2001).

The IFNs are fundamental mediators of both early innate and adaptive immune responses to microbial infection, including antigen processing, activation of macrophages, NK cell effector functions, and stimulation of antigen-specific T cells. In addition to their antiviral actions, IFNγ plays an important role in host defense against certain bacterial (e.g., mycobacteria), fungal, and parasitic pathogens, as well having antitumor properties. Like other cytokines, IFNγ mainly signals through JAK and STAT intracellular proteins. Ligand binding induces a conformational change in the IFNγ receptor-1/2 chains, which leads to the sequential activation through phosphorylation of JAK1 and JAK2, which then promotes the recruitment, phosphorylation, and homodimerization of STAT1 (Schroder et al., 2004). These STAT1 dimers translocate to the nucleus and bind to IFNγ-activation sites in the promoter region of IFNγ-responsive genes. IFNγ-STAT1 signaling can then initiate or suppress the transcription of a host of immunologically relevant target genes. Additionally, IFNγ signaling can produce STAT1 heterodimers, which can bind to IFN-stimulated response element promoter regions to regulate further rounds of transcription.

Given its potent inflammatory actions, several endogenous mechanisms exist to naturally inhibit IFNγ levels. These include (1) degradation and/or recycling of the IFNγ:IFNγ receptor-1 complex, (2) disruption of JAK phospho-activity and/or targeting of JAKs for proteasomal degradation by suppressors of cytokine signaling-1 (SOCS-1) and SOCS-3, and (3) dephosphorylation of IFNγ receptors and STAT1 by the protein tyrosine phosphatases (Majoros et al., 2017). Similarly, glucocorticoids and the anti-inflammatory cytokines IL-4, IL-10, and TGFβ, inhibit IFNγ activity/production (Majoros et al., 2017).

IFNγ induces the activation of proinflammatory cytokine cascades and various inducible enzyme systems, in particular iNOS, COX-2, and nicotinamide adenine dinucleotide phosphate (NADPH) oxidase (Majoros et al., 2017). Although these inflammatory enzymes confer protection against pathogenic invasion, chronic activation of these enzyme systems has been linked to behavioral deficits and impaired structural and functional neuroplasticity (Litteljohn, Mangano, & Hayley, 2008). Of note, IFNγ signaling may also drive the downregulation of neuroprotective species in glial cells, which could increase progenitor cell and neuronal vulnerability to oxidative and inflammatory damage (Litteljohn et al., 2017). For example, IFNγ might contribute to paraquat-induced neurodegeneration, in part by mediating an early occurring reduction in central BDNF (Mangano et al., 2012). Moreover, IFNγ directly damages cultured neurons through induced glutamatergic neurotoxicity (Mizuno et al., 2008), and IFNγ deficiency protected against the toxic effects of MPTP treatment, which was tied to microglial inflammatory processes (Mount et al., 2007). Not surprisingly, STAT1 mediated the effects of IFNγ and likely controlled the production of many different trophic and inflammatory factors relevant for pathology.

GENETIC VULNERABILITY

As we've seen, it is thought that factors responsible for PD likely exist on a spectrum with familial linked forms (e.g., SNCA, LRRK2, DJ-1, Parkin, and PINK1) that result in early provocation at one end, and purely environmental impact at the other, leaving the bulk to the interactive effects of genetic vulnerability and environmental influence (Klein & Schlossmacher, 2006; Ritz et al., 2016). The fact that not all genes associated with PD give rise to Parkinsonian symptoms, together with reports that many polymorphisms have been linked to PD, implicates the influence of environmental factors (Chan et al., 1998). Evidence from animal research and epidemiological studies over the last decade has supported gene-environment interactions in PD pathophysiology (Lee, Raaschou-Nielsen et al., 2016; Tysnes & Storstein, 2017).

Several mechanisms have been suggested by which genetic vulnerability enhances susceptibility to environmental insults (i.e., paraquat exposure), hence resulting in PD. For example, an epidemiological revealed increased risk of developing PD in individuals who possessed either the rs1045642 or rs2032582 polymorphisms in the ABCB1 gene in frequent pesticide sprayers (Narayan et al., 2015). Importantly, the ABCB1 gene encodes for P-glycoprotein, which acts as a cellular efflux transporter of lipophilic compounds across the BBB. Thus, variations in P-glycoprotein would be expected to facilitate the transport of toxicants to contact nigral dopaminergic neurons. Indeed, rs1045642 and rs2032582 polymorphisms altered P-glycoprotein expression on epithelial cells lining the BBB, which resulted in elevated concentrations of neurotoxic substances (i.e., xenobiotics, such as pesticides) penetrating the brain (Narayan et al., 2015). Moreover, among paraqaut sprayers, an 11-fold increase in risk for PD was noted, but only in individuals who also lacked the glutathione S transferase T1 (GSST1) gene, which is primarily responsible for detoxifying xenobiotic compounds (Goldman et al., 2010).

Other polymorphisms have similar PD relevant effects. For instance, individuals possessing polymorphisms in the DAT/SLC6A3 are at increased risk of developing PD when also exposed to pesticides (Kelada et al., 2006). It was hypothesized that the DAT/SLC6A3 polymorphisms enhance pesticide binding to DAT and entry into dopamine neurons (Kelada et al., 2006). Carriers of the DAT SNPs, rs2652511, and rs2937639, exhibit an increase in DAT expression in striatal regions (van de Giessen et al., 2009) and variations in the transporter can enhance susceptibility to paraquat entry into neurons and subsequent toxicity (Ritz et al., 2016). Thus, again, it seems that these polymorphisms alone do not cause disease, but increase susceptibility by modifying the impact of environmental challenges.

LRRK2 in Relation to PD and Other Inflammatory Disorders

The LRRK2 mutations are inherited in an autosomal dominant manner, but usually display low penetrance and variable expressivity; not all carriers of a given LRRK2 mutation will go on to develop PD and among those who do, considerable clinical and neuropathological variability exists (Zimprich et al., 2004). These findings led to the suggestion that environmental events, such as exposure to toxins and chronic stressors, as well as genetic factors (e.g., other PD-associated genes or loci), may be triggers or modifiers of LRRK2-related PD (Karuppagounder et al., 2016; Litteljohn et al., 2017). Consistent with this view, the LRRK2 gene has been implicated in both familial and sporadic forms of PD (Cookson, 2017; Klein & Schlossmacher, 2006). Indeed, while LRRK2 mutations are the most frequent known genetic cause of the familial form of PD, several LRRK2 polymorphisms are also associated with heightened risk of developing the more common sporadic form of the disease (Wallings, Manzoni, & Bandopadhyay, 2015).

It is likely that the LRRK2 genetic mutations can be a dominant genetic factor that causes a circumscribed early onset form of disease, or alternatively, can act as a vulnerability factor in which some other insult produces the later age-dependent form of the disease. The finding that individuals that possess the G2019S mutation, but do not display PD, showed alterations in brain metabolism, is consistent with the position that the gene might be a vulnerability factor. Imaging work confirmed that nonmanifesting carriers of the LRRK2 G2019S mutation displayed altered brain regional activity and interregional functional connectivity (van Nuenen et al., 2012).

Although the LRRK2-related genetic form of Parkinsonism has been considered to closely resemble sporadic PD, there have been indications that LRRK2 mutations might actually cause an altered neuropsychiatric phenotype (Hayley & Litteljohn, 2013). In particular, several clinical studies indicated that PD patients and asymptomatic carriers harboring the LRRK2 G2019S point mutation (glycine-to-serine substitution at position 2019, which is the most commonly encountered LRRK2 mutation) have higher rates of depressive and cognitive symptoms, which also tend to be more severe. Yet, a study with over 500 G2019S PD patients found slightly lower depressive scores, compared to idiopathic non-G2019S patients (Marras et al., 2016). These differences might stem from the fundamentally different populations used in the studies that are known to differ with respect to these mutations.

In addition to PD, LRRK2 has been linked to Crohn's disease, leprosy, and diabetic neuropathy as well as HIV-1 associated neurocognitive disorders (Lewis & Manzoni, 2012; Puccini et al., 2015). An in-depth discussion of the proposed role of LRRK2 in these conditions exceeds the scope of the present work (but see Greggio, Civiero, Bisaglia, & Bubacco, 2012). However, it bears mentioning that all of these LRRK2-associated conditions have a prominent immuno-inflammatory component in common with PD.

LRRK2: A Novel Regulator of CNS-Immune System Interactions

LRRK2 is a large multimeric protein (286 kDa) with several distinct domains and is unique in that it has both guanine triphosphate hydrolysis (GTPase) and MAP kinase functions (Cookson, 2015; Wallings et al., 2015). Substantial evidence suggests that enhanced kinase activity of the gene is a major contribution to the pathogenesis of LRRK2-related PD (e.g., the LRRK2 G2019S mutation increases kinase activity $\sim$ twofold to threefold) (Greggio et al., 2012; West, 2017). However, not all LRRK2 variants are associated with increased kinase activity (Rudenko, Chia, & Cookson, 2012). Moreover, given the complexity of the gene, other activities including GTPase activity, and its ability to act as "structural scaffolding" for protein-protein interactions, might also be involved in its links to PD.

Despite the clear importance of LRRK2 in PD, the normal role(s) of LRRK2 remains elusive. Nonetheless, LRRK2 has provisionally been implicated in a diverse range of cellular functions, including autophagy, cytoskeletal dynamics, intracellular membrane trafficking, synaptic vesicle cycling/neurotransmission, and the inflammatory response (Cookson, 2015; Wallings et al., 2015). As it relates to LRRK2 kinase functions, the growing list of suspected or validated LRRK2 kinase substrates includes LRRK2 itself (auto phosphorylation), MAP kinases (which are immediately upstream of JNK and p38 MAP kinase), as well as NF-κB (Gloeckner, Schumacher, Boldt, & Ueffing, 2009). Similarly, NF-κB transcriptional activity and vascular cell adhesion molecule expression were exaggerated by LRRK2 G2019S over-expression. This same mutation also potentiated leukocyte chemotaxis (Moehle et al., 2015) and enhanced the activity of several major intracellular inflammatory signaling pathways, namely NF-κB, and the MAP kinases, p38 and JNK (Kim, Yang et al., 2012). Thus, LRRK2 may promote PD pathology by activating these pathways through kinase-dependent phosphorylation.

In addition to being expressed in a variety of peripheral organs (e.g., kidney, lung, liver, heart, and spleen), LRRK2 is readily detectable in neurons over a wide range of brain regions. Included here are the SNc and striatum (surprisingly low in the former, but high in the latter), as well as the hippocampus, PFC, locus coeruleus, and various hypothalamic and amygdaloid nuclei. This led to the suggestion that beyond its role in nigrostriatal motor function, LRRK2 may contribute to the regulation of emotional and cognitive processes (Litteljohn et al., 2017). In fact, the G2019S mutation was associated with altered dopaminergic activity (Litteljohn et al., 2017) and nonmotor behavioral impairments (Karuppagounder et al., 2016).

Subcellular localization studies indicated that LRRK2 associates with membranous and vesicular structures, including mitochondria, lysosomes, endosomes, lipid rafts, and transport/synaptic vesicles. Moreover, the regulation of synaptic vesicle cycling/neurotransmission might factor importantly in this regard (Migheli et al., 2013). Interestingly, however, the cell types that most highly express LRRK2 are immune cells, mostly monocytes, macrophages, B lymphocytes, dendritic cells, and microglia (Gardet et al., 2010; Thévenet, Pescini Gobert, Hooft van Huijsduijnen, Wiessner, & Sagot, 2011). Perhaps not surprisingly, emerging evidence has highlighted a prominent role for LRRK2 and its mutations in the inflammatory response (Dzamko et al., 2017). For instance, LRRK2 silencing in macrophages and/or microglia attenuated LPS-induced NF-κB transcriptional activity, iNOS, and COX-2

expression, along with TNF-α, IL-1β, and IL-6 release (Kim, Yang et al., 2012; Moehle et al., 2012). In contrast, overexpression of the PD-linked LRRK2 R1441G mutation exacerbated many of these endotoxin-induced effects (Gillardon, Schmid, & Draheim, 2012).

Among the proinflammatory cytokines, IFNγ stands out as a preferential inducer of LRRK2. This cytokine markedly increases LRRK2 expression in human peripheral blood mononuclear cells (Gardet et al., 2010), particularly CD14 + CD16 + monocytes (typical of inflammatory conditions) (Thévenet et al., 2011), as well as in circulating macrophages and brain-resident microglia (Gillardon et al., 2012). Thus, inflammatory driven stimulation of IFNγ can mobilize LRRK2 expression and may promote pathology. This may explain how IFNγ knockout was protective against MPTP or paraquat exposure (Mangano et al., 2011; Mount et al., 2007).

Finally, LRRK2 plays an important role in autophagy (breakdown of cellular waste) and hence, could be important for Lewy bodies or other pathological aggregates. An excess of undigested proteins are found in kidneys of LRRK2 KO mice, consistent with a role in autophagy (Gómez-Suaga et al., 2012). Moreover, mutant versions of LRRK2 can bind to the lysosome and interfere with the degradation of proteins, most notably α-synuclein (Orenstein et al., 2013). This can lead to the pathological accumulation of α-synuclein in Lewy body inclusions (Orenstein et al., 2013). Additionally, LRRK2 interacts with rab5b and rab7, preventing proper fusion of autophagosomes to lysosomes, resulting in decreased protein degradation (Roosen & Cookson, 2016).

Taken together, these studies suggest that LRRK2 acts through multiple intracellular substrates to interfere with protein recycling and processing, which could ultimately produce inclusions as those seen in PD. Although this could be how LRRK2 directly affects dopamine neurons, as just mentioned, the fact that

LRRK2 is found at much higher levels in microglia and immune cells and is involved in a variety of immune processes, indicates that LRRK2 can also secondarily impact neurons through either resident microglia or infiltrating leukocytes.

FUTURE IMMUNOMODULATORY TREATMENTS AND PARKINSON'S DISEASE

Preclinical work has pointed to different treatments that are neuroprotective in PD. Unfortunately, as we've seen with so many other illnesses, none of these treatments has successfully translated into clinical applications that stop or even appreciably slow the disease. Part of the problem may reside in the differences in complexity of the human brain compared to experimental animals. Similarly, the degree of epigenetic changes between species is substantial (Lin et al., 2014). When comparing transcriptional profiles in 15 different tissues between humans and rodents, there were more similarities between the differing tissues within each species than between the species (Lin et al., 2014).

These problems are difficult to overcome; yet, a major issue that could be more easily addressed experimentally is the better modeling of different stages of PD by adopting specialized treatments that suit the particular disease state. As we've indicated in relation to other diseases, animal preparations that model early, middle, and late disease states, as well as representative comorbid symptoms, such as depression and anxiety, might be of particular use. In this way the efficacy of different treatments might be tested, allowing for determination of whether differential effects appear depending upon disease stage and type of comorbid features present. Furthermore, it might also be that multiple concomitant treatments that target different disease processes

are required to produce clinically meaningful outcomes. Such combined treatment strategies are now commonplace for psychiatric conditions, such as depression (Oluboka et al., 2017), and such a comprehensive approach may also be warranted for PD.

In the following sections, we will describe potential emerging treatments that act upon (at least in part) immunological processes to influence PD pathology. These treatment strategies will be interpreted within the framework of their efficacy at different stages of PD pathology and with different symptom presentation. We will also emphasize how some treatments might be used together or might interact with existing standard therapies, such as l-DOPA, that simply aim to manage motor symptomatology without actually affecting underlying disease mechanisms. Finally, it might be fruitful to combine immunomodulatory treatments with emerging genetic cell-based strategies.

The three main types of immunomodulatory treatments that we will focus upon are anti-inflammatory drugs, trophic cytokines, and LRRK2 inhibitors. With each of these strategies, the objectives are to (1) stabilize or prevent further degeneration, and (2) promote some degree of functional recovery by facilitating neuronal plasticity and compensation within neural circuitry.

Anti-inflammatory PD Treatments

The curious finding that rheumatoid arthritis patients that were on chronic high NSAID doses, displayed a particularly low incidence of AD (Auriel, Regev, & Korczyn, 2014), kickstarted the notion that anti-inflammatory drugs might treat neurodegeneration. Indeed, this spurred many preclinical and clinical trials assessing the utility of NSAIDs as neuroprotective agents, quickly spreading to the PD area, given the substantial neuroinflammatory component implicated in PD.

Unfortunately, the majority of clinical trials using NSAIDs for either Alzheimer's or PD have provided disappointing results (Miguel-Álvarez et al., 2015; Samii, Etminan, Wiens, & Jafari, 2009). A few studies indicated that ibuprofen, but not aspirin or acetaminophen, modestly reduced PD risk, and a meta-analysis supported a modest reduction of symptoms with ibuprofen, but not with NSAIDs (Gagne & Power, 2010). At least one study also report that regular aspirin use reduced PD risk, but only among women (Wahner, Bronstein, Bordelon, & Ritz, 2007). Despite the disappointing findings, the combination of high levels of NSAIDs, together with smoking and high coffee intake, was associated with the greatest reduction of PD risk (Powers et al., 2008). It seems likely that NSAIDs or other general anti-inflammatories would not be useful for treatment in existing PD patients, but might have value as a prophylactic treatment in vulnerable populations. Limiting inflammation during the early incubation stages of disease might allow for some degree of endogenous plasticity. However, once the PD pathology has taken hold, the point of no return may have passed, and recovery would be exceedingly difficult to achieve simply through anti-inflammatories.

Another caveat concerning the use of broad spectrum anti-inflammatory agents is that they not only limit harmful inflammatory processes, but also other inflammatory cascades that might be critical for recovery or reparative processes. Cytokines, for example, while harmful in high doses are required in low concentrations for normal brain homeostasis (Subramaniam & Federoff, 2017). Instead of simply applying broad spectrum anti-inflammatory treatment to shut down microglia, it might be more efficacious to target the modulation of the microglial phenotype. Thus, an immunomodulatory rather than anti-inflammatory approach might be fruitful. This would represent a departure in the

conceptualization of the disease and instead of demonizing all inflammatory processes, accepting that certain microglial states could promote CNS repair. This is most apparent when considering the M1 and M2 dichotomy and all the intermediate states. In particular, more M2-like states (which are characterized by the release of anti-inflammatory cytokines and trophic factors) might release protective trophic factors, as well as help with the clearance of potentially destructive plaques. This is in opposition to the more proinflammatory M1 phenotype. Thus, in a way, this could be viewed as immunomodulation of microglia toward a neuroprotective phenotype.

It is important to underscore once again that M1 and M2 are not discrete phenotypes (states), but that microglia exist on a continuum between these two extremes (Zhou et al., 2017). Moreover, different microglia expressing these varied states can co-exist in the same brain region. Consequently, at any one time, there may be a complex tug of war between the different phenotypic states. Clearly, in the later stages of disease, the M1-like inflammatory microglia are the winners and predominate lesions sites. So, a potential goal of therapy could be to encourage microglia to adopt a more M2-like state, rather than attempting to blunt their overall activity. Indeed, the M2-like state is not simply resting or blunted, but rather acts toward processes more aligned with cell survival, repair, and removal (via phagocytosis) of damaged tissue.

When in an M2 state, microglia are able to engulf and remove debris or damaged cells that display phosphatidylserine, which essentially acts as an "eat me" signal (Xia, Zhang, Gao, Wang, & Chen, 2015). Removal of compromised cells and synapses would presumably help strengthen the functioning of healthy cells. At the same time, M2 microglia upregulate antioxidants, such as glutathione and heme oxygenase-1, which help detoxify potentially harmful reactive oxygen species that typically characterize PD. These microglia further assist neurons by releasing trophic growth factors, including GNDF, NGF, and TGFβ, particularly in the presence of the anti-inflammatory cytokines, IL-10 or IL-4 (de Bilbao et al., 2009). Finally, other transmembrane receptors up-regulated on M2 microglia, including Ym-1 and Arg-1 (which has been implicated in wound healing), also foster recovery from injury by positively modulating the extracellular compartment (Cherry, Olschowka, & O'Banion, 2014).

The M1 and M2 perspective came from *in vitro* studies involving cultured microglia in which IFNγ and IL-4 promoted these respective states. An analogous situation may not be reflected *in vivo*. Deficiencies in the M1-M2 dichotomy have been highlighted, and it was argued that *in vivo*, microglia cannot be neatly grouped into these two states (Ransohoff, 2016). In fact, microglia may co-express both M1 and M2 markers simultaneously, but the levels and precise roles of these factors vary with the stimulus/insult and state of the microenvironment (Perego, Fumagalli, & De Simoni, 2013). Irrespective of how the range of microglia states are described, the important and unequivocal point is that microglia sometimes help and sometimes hinder whatever pathological state exists. Thus, even if the M1-M2-like dichotomy was accepted, the basic premise of immunomodulatory regulation of microglia for neuroprotection is still valid. But, the critical issue is when and how microglia can be modulated to produce positive effects. This brings us to the notion of specific target proteins on or related to microglia that can be pharmacologically or genetically manipulated, and once more, LRRK2 may be an ideal target for this.

LRRK2 Inhibition and Recovery

There are four distinct processes through which LRRK2 can act, and hence could serve as targets for pharmacological inhibitors. These comprise (1) kinase mediated phosphorylation, (2) GTPase mediated control of G protein cascades, (3) scaffold proteins, and (4) modulation of protein translation. Although numerous small molecule LRRK2 inhibitors have been developed, the majority focused upon interfering with kinase functioning (Hatcher, Choi, Alessi, & Gray, 2017). For instance, in preclinical trials, the LRRK2 kinase inhibitor, PF-06447475, halted α-synuclein induced neurodgeneration and neuroinflammation (Daher et al., 2015). Other early broad spectrum LRRK2 kinase inhibitors, LRRK2-INI and GNE-7915, were shown to have good BBB penetrably and *in vivo* potency (Qin et al., 2017).

In addition to kinase inhibitors, there has also been a focus on developing GTPase inhibitors. The compound, FX2149, acts by inhibiting the LRRK2 GTP binding site (Li, He et al., 2015), thereby preventing GTPase activity (which involves signaling by switching on/off a pathway through the shuttling between GTP and GDP). Ordinarily, GTPase activity is critical for facilitating G protein signaling and inhibiting GTP hydrolysis, which would essentially prohibit protein shuttling between a resting and active state, thereby locking normal LRRK2 mediated signal transduction. Certain LRRK2 mutations (specifically those that affect its ROC-COR domain) impair GTPase hydrolysis and G protein signaling (Gilsbach & Kortholt, 2014), and thus could influence PD symptomatology.

A study that screened 640 LRRK2 inhibitory compounds, found that only three actually reversed motor disability and dopamine neuron loss in LRRK2-G2019S transgenic flies (that normally have a degeneration of their dopamine neurons). Of the three, lovastatin had the greatest efficacy and was further found to induce antiapoptotic Akt/Nrf signaling, while inhibiting caspase 3 and GSK3β activity (Lin, Lin et al., 2016). Lovastatin also had anti-inflammatory effects, blocking the release of the proinflammatory cytokines following 6-OHDA administration (Yan et al., 2015). Similarly, a statin, simvastatin, promoted an anti-inflammatory M2-like microglia state in a model of cerebral stroke, and had *in vivo* and *in vitro* neuroprotective and anti-inflammatory effects in an MPTP and high dose LPS models of PD (Ghosh et al., 2009). Thus, besides their beneficial anticholesterol effects, statins might also unexpectedly convey protective properties with regard to PD, and at least some of these actions could be linked to LRRK2 processes. To be sure, many of its effects may be independent of LRRK2, and by limiting cholesterol, statins may prevent the accumulation of pathological Lewy bodies. Whatever the mechanism, it is still too early to ascertain whether regular statin use might diminish PD risk.

Trophic Cytokines

Aside from their impact on immune processes, many cytokines are potent inducers of the growth factors GDNF and BDNF, and could potentially be harnessed for therapeutic purposes. Major problems with the administration of GDNF or BDNF themselves stems from their poor ability to cross the BBB, as well as the many deleterious side effects (including pain sensitivity and potential for tumor growth) that limit their widespread clinical application (Allen, Watson, Shoemark, Barua, & Patel, 2013). In contrast, CSF cytokines, such as granulocyte macrophage-CSF (GM-CSF), as well as some hematopoietic cytokines, most notably erythropoietin (EPO), readily cross the BBB and have fewer complicating effects. Importantly, these cytokines all are potent trophic factor inducers, and preclinical studies established their neuroprotective potential (Maurer, Schäbitz, & Schneider, 2008). Interestingly, as well, it was argued that the correlation between the low incidence of PD in NSAID treated arthritic patients is not due to the drug treatment, but rather that the

autoimmune disease itself increases endogenous GM-CSF levels in an effort to cope with the injury. This would also help explain why the NSAID clinical trials in PD were not successful.

It appeared that GM-CSF may have neuroprotective actions for PD-like pathology. In particular, systemic GM-CSF administration prevented the loss of DA neurons associated with MPTP (Kim et al., 2009) and acted additively with another cytokine, IL-3, to block the neurodegenerative effects of 6-OHDA (Choudhury et al., 2011). Moreover, GM-CSF had neuroprotective consequences against paraquat exposure alone and in mice that were also pretreated with LPS prior to receiving the paraquat (Mangano et al., 2011).

Strikingly, some of GM-CSF's neuroprotective effects might be related to the mobilization of peripheral immune cells, as it promoted protection of SNc dopaminergic neurons by stimulating adaptive immunity through the release of Treg cells (Kosloski et al., 2013). The Treg cells normally act as suppressors of other inflammatory cells and are thought to be critical for the induction of immunological tolerance, thereby warding off autoimmunity. In fact, adoptive transfer of GM-CSF primed Treg cells was enough to protect against MPTP induced cell death (Kosloski et al., 2013). This is consistent with studies showing that adoptive transfer of GM-CSF primed splenic cells had anti-inflammatory consequences and promoted tissue healing in an animal model of inflammatory colitis (Bernasconi et al., 2010).

Besides preventing frank neuronal loss, GM-CSF may have particularly marked effects on neuronal recovery when administered following CNS lesion. Delayed administration of GM-CSF following spinal cord injury promoted axonal regeneration and increased BDNF expression, which could facilitate neuroplastic processes, reflected by dendritic sprouting and neurogenesis (Bouhy et al., 2006). Similarly, GM-CSF induced the differentiation of neural stem cells into mature neurons (Krüger, Laage, Pitzer, Schäbitz, & Schneider, 2007), and could promote the re-innervation of the striatum after 6-OHDA had already caused a lesion, potentially making this a clinically important finding (Farmer, Rudyk, Prowse, & Hayley, 2015). This raises the possibility that trophic cytokines, such as GM-CSF, could promote a degree of neural recovery after the disease has already taken hold by bolstering endogenous neuroplastic processes.

It seems that GM-CSF acts through the JAK-STAT pathway to influence cellular proliferation and inhibit the actions of proinflammatory cytokines. Phosphorylation (activation) of STAT3 and STAT5 can result in neuroprotective effects upon on local DA neurons, either through the production of trophic or antiapoptotic factors or by buffering the impact of extracellular excitotoxic/oxidative species. After nuclear translocation, STAT3 and STAT5 can regulate transcription of anti-inflammatory/growth factors, as well as the antiapoptotic factor, bcl-2, together with the antioxidant, manganese superoxide dismutase.

Turning to the hemapoietic cytokine, EPO, it is clear that this factor might also hold promise in the treatment of PD by promoting the production of red blood cells, thereby increasing oxygen carrying capacity. Importantly from a clinical perspective, EPO readily crosses the BBB and is clinically used to treat anemia safely. In addition to EPO itself, several synthetic analogs and receptor agonists that have antioxidant and anti-inflammatory effects (Punnonen, Miller, Collier, & Spencer, 2015) may also have clinical benefits for PD.

Most of the studies that assessed the brain effects of EPO have focused on either its ability to stimulate cognitive process, as well as its potential use as a protective agent in the context of stroke. Similarly, EPO treatment stimulated hippocampal neurogenesis and improved memory and spatial learning in rodents (Wang, Yan et al., 2017). This growth factor also protected hippocampal neurons from stressor-induced apoptosis, and through such actions might have cognitive enhancing effects, as indicated by improvements in several neuropsychological indices of cognitive performance (Peng et al., 2014). Preclinical data also suggested its possible use as an agent to promote neuronal recovery, having neuroprotective consequences in models of stroke and traumatic brain injury (Yao et al., 2016).

The potential utility of EPO has also been demonstrated in 6-OHDA and MPTP animal models of PD in which EPO limited the dopamine loss otherwise observed (Zhang et al., 2006). Likewise, it reduced 6-OHDA-induced apoptosis and dysregulation of Bcl-2, Bax, and Caspase-3 in striatal neurons. Interestingly, EPO also enhanced the survival of transplanted neural precursor cells into the SNc (McLeod et al., 2006), suggesting its potential utility as an adjunctive agent administered together with stem cells. Some evidence also suggested that EPO might have clinical value for the treatment comorbid nonmotor PD symptoms, such as autonomic functioning, fatigue, mood, and attention.

The homodimerization of two EPO molecules and their subsequent binding to their receptor, results in the activation (phosphorylation) of several intracellular adaptor proteins (e.g., JAK2 and STAT5). These signaling cascades promote antiapoptotic factors, cell differentiation, cellular growth, and modulation of plasticity and can occur on either neurons or glial cells. In parallel with STAT5 activation, EPO can signal through the mTOR pathway. Ordinarily, mTOR induces protein synthesis by inactivation of translation repressor 4E-binding protein 1, while concomitantly stimulating p70S6 kinase, which influences protein translation at ribosomes. Within the brain, mTOR has been implicated in several aspects of synaptic plasticity, such as BDNF production, dendritic remodeling, neurogenesis, and synaptogenesis (Russo, Citraro, Constanti, & De Sarro, 2012).

Besides its trophic actions, EPO could have beneficial effects for PD by down-regulating inflammatory pathways, through reduced microglial activation and proinflammatory cytokines. It might also exert protective effects by modulation of autophagy, which is dysregulated in PD and likely contribute to toxic Lewy body formation. In fact, EPO reversed the ability of rotenone to down-regulate the autophagy markers, Beclin-1, AMPK, and ULK-1 in cultured SH-SY5Y cells (Jang et al., 2016). In parallel, EPO also reversed the rotenone induced α-synuclein expression (Jang et al., 2016), suggesting that EPO may prevent toxic protein aggregation.

CONCLUDING COMMENTS

There is little question that multiple players contribute to the development of PD, including disturbances of inflammatory processes. As in the case of so many other diseases, heterogeneity exists regarding the processes leading to PD and its comorbid symptomatology. This underscores the importance of personalized approaches to treatment that consider the multiple mechanisms that ought to be targeted, which could vary over the course of the illness. In the final section of this chapter, we considered several novel emerging therapies that revolve around the theme of immunomodulation. In the main, these strategies target microglia and/or peripheral immune cells in an attempt to foster anti-inflammatory and/or trophic phenotypes.

A primary problem that will require some change in the way we approach PD is that virtually none of the preclinical success have been translated into meaningful clinical treatments. Some of the reasons for the lack of translation of neuroprotective treatments from animal to humans have already been mentioned. In addition to these, a primary mechanism for neuronal death in human tissue was absent in the

rodent. In humans, dopamine contributes to its own demise by accumulating an oxidative form of dopamine that was toxic to the cells mitochondria and lysosomes; however, this toxic dopamine metabolic pathway was not evident in animal models of PD (Burbulla et al., 2017). Clearly, caution needs to be exercised with an eye toward an optimal balance between mechanistic animal experiments and clinically relevant human studies. Hopefully, their combination can inform us about the best possible way forward in understanding and treating PD.

These shortcomings notwithstanding, a strong aspect of PD animal models is that there are many reliable means of dose-dependently destroying SNc dopamine neurons in both rodents and nonhuman primates. Moreover, the basal ganglia circuitry is very well delineated in both the human and animal models. There is also reason for optimism given the emerging technological advances and the many new drugs in development. An exciting advance was the use of deep brain stimulation (DBS), which revolutionized the treatment of the tremor and bradykinesia in PD. To be sure, this is an invasive procedure and its efficacy tends to wane over time. What is fundamentally needed is a treatment that can stabilize existing neurons, thereby protecting them from further degeneration and some means of enhancing compensatory mechanisms to overcome the dramatic cell loss that has already occurred. We have touched upon the use of trophic cytokines or immunodulatory agents that could be used in this capacity. But these factors have pleiotropic actions and could lead to unpredictable outcomes, plus it is very difficult to regulate their optimal levels, so as to minimize unwanted effects.

Ultimately, it would be desirable to have ways of preventing the disease from taking hold in the first place. This obviously requires excellent biomarkers to aid in the identification of who is at risk for the disease. This is major problem as the majority of PD cases are idiopathic and, at best, we can only predict elevated risk based on exposure patterns (e.g., pesticide applier, welder) and carrying certain polymorphisms (e.g., G2019S LRRK2 mutation). Recent advances in genetic and proteomic multiplex screens might hold the key to finding a profile or collection of markers that together determine disease risk. Likewise, it might also be possible to determine which profiles indicate the severity of disease progression, as well as the type of comorbidities that might appear. It should also be kept in mind that the disease processes involve elements outside the brain. As described earlier, in addition to cytokines, microbiota might fill this bill. Finally, the LRRK2 gene is key for a number of immune and metabolic processes. So, this should be included in a systems level view of the disease and the intricate connections between the various nodes within the overall system.

A Neuroinflammatory View of Alzheimer's Disease

ARE OUR BRAINS BEING INFLAMED AS WE AGE?

The human lifespan is finite because the body eventually wears out, possibly being exacerbated by multiple challenges encountered together with genetic determinants. The brain may be especially vulnerable to the ravages of age, as it is the most energetically expensive organ in the body. A large proportion of energy is shuttled to the brain and the mitochondria present in neurons are especially active, and this high level of activity can produce free radicals and other potentially harsh byproducts. Neurons are also exquisitely sensitive to a wide range of environmental stressors. At the same time, the brain's glial cells, primarily microglia and astrocytes, respond rapidly to stressful challenges, and over time, they may adopt long "active" states, wherein they can chronically release factors that foster an inflammatory environment within the brain. These inflammatory processes could prematurely age the brain by placing excessive demands on various neuronal circuits. At the same, pathological proteins aggregates, amyloid plaques, may fuel inflammatory pathology still further (see Fig. 15.1). Considerable debate currently exists as to whether these plaques should be the main targets for treating disease. Whatever the case, the possibility exists that anti-inflammatory agents, and even vaccines, could be used to combat the neurobiological processes that contribute to Alzheimer's disease (AD).

The past decade has seen a virtual exponential increase in published papers supporting a link between immune factors and neurodegeneration. Just 15−20 years ago the role inflammatory immune factors in AD or Parkinson's disease (PD) was considered controversial, whereas now it's pretty much considered mainstream. In fact, there has been a growing consensus that immunocompetent microglial cells and the inflammatory and oxidative factors they release are involved in some sort of cell to cell combat in the diseased brain. What is not clear is whether such processes are involved in the genesis of the disease, or secondarily turned on by dying neurons. Understanding these processes and placing them in the context of potential novel treatments strategies, is a key focus of research to curtail or prevent the progression of AD.

The Immune System and Mental Health
DOI: https://doi.org/10.1016/B978-0-12-811351-6.00015-2

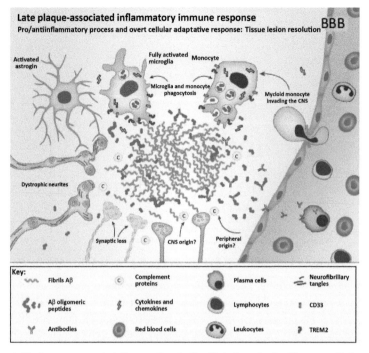

FIGURE 15.1   β-amyloid plaques promote inflammation in the Alzheimer brain. The centrally located β-amyloid plaque (containing pathological fibrils and oligomeric peptides, which are depicted as the orange and red "squiggles" in the middle of the figure) creates a microenvironment that activates microglia and to an extent astroglia (upper left in figure), which can help break down plaques, but also have the ability to damage neurons or their projections (e.g., causing pathological dystrophic neurites; lower portion of figure). Plaques themselves can also damage neurons and cause synaptic loss. Concomitant with these brain changes, monocytes/macrophage immune cells can be recruited from the periphery (shown on the right). These cells can also interact with plaques to help with their disposal but again, like glia, they also have the capacity to contribute to synaptic loss and the creation of abnormal (dystrophic) neurites and eventually neuronal degeneration. *Source: From Cuello (2017). © Massouh BioMEDia 2017.*

ALZHEIMER'S DISEASE BACKGROUND

Alzheimer's disease (AD) is the most common neurodegenerative disease throughout the world. It strikes at the very core of what makes us who we are, namely our memories and sense of self! The disease ravages the elements of the brain that subserve cognitive functions and particularly the encoding of new memories, and eventually the recall of older ones. The disease is currently incurable, and at best, symptoms can be slowed among some individuals, but only by mere months! As in the case of PD, a wealth of evidence has accumulated implicating inflammatory immune processes in AD. In fact, Alois Alzheimer first remarked over a century ago on the large number and distinct morphology of microglial cells present in the AD brain. This chapter will first outline the disease characteristics and current theories of AD and then delve into specifics regarding peripheral and central immune processes and their relevance to brain pathology. We will end on a discussion of potential new immunotherapeutic that might hold some clinical utility.

Typically, AD patients first reach the clinic after some aspects of their memory begins to fail.

This is often comprises forgetfulness related to common daily events, along with some problems with autobiographical memory. The individual may forget names and significant life events, or may even become temporarily lost on their way home from a friend's house. Of course, we all show some slight memory lapses from time to time, but in the AD patient the *progressive* memory loss eventually extends to all facets of life. Often, within only a few years the person requires total supervision and essentially has completely lost their memory faculties, sense of self, and the ability to interact meaningfully with the world. At this stage, other secondary health issues also arise, including infections, blood clots, cardiovascular problems, and pneumonia that eventually results in death.

Not all individuals showing dementia necessarily are afflicted by AD, as dementia can arise from cerebrovascular insults, certain drugs or toxicants, or may temporarily arise from neuropsychiatric states. The main difference is that the insidious progressive nature that characterizes AD may not be present in other forms of dementia. By example, another condition involving dementia, that of Korsakoff's syndrome, arises from chronic alcoholism (and is related the vitamin B deficiency this causes), but again, the pathology of this disease is far more circumscribed and its course far less progressive that AD. In fact, about 25% of individuals may show some degree of recovery. So, the lesion and pathology may be more static, whereas in AD the entire brain eventually becomes compromised.

A separate functional diagnostic category related to AD, that of mild cognitive impairment (MCI), is reserved for cases too "mild" to receive an AD diagnosis and in which obvious dementia is not apparent. Instead, diminished cognitive functioning is present to the extent that interferes with daily tasks, but is not of sufficient severity as to be totally disabling. Although some individuals remain in this condition without showing appreciable progression of pathology, another subgroup of individuals go on to eventually develop full-blown AD. Thus, there may be a protracted prodromal state reflected by MCI, raising the possibility that early therapeutic interventions might hold promise for MCI patients. Although many drugs are in the pipeline, none have yet shown appreciable clinical promise.

Because of the widespread pathology in AD, late stage patients essentially become unresponsive, with symptoms touching virtually all aspects of functioning. As well, more than a quarter of AD patients typically experience a range of psychiatric symptoms, ranging from anxiety and depression to psychosis and agitation. Similarly, a variety of nonpsychiatric features are also associated with AD, including sleep and hormonal disturbances, along with motor and gastrointestinal problems.

The earliest memory deficits evident in AD are thought to come about because of the loss of neurons and their connections within the parahippocampal gyrus located within the temporal region of the brain (Ferreira et al., 2017). The portion of the brain that links the hippocampus to the cerebral cortex (entorhinal cortex) seems to be hit hard early in the disease process. This brain area contains the perforant path, a dense fiber bundle comprising axons connecting the tri-synaptic hippocampal circuit with the cortex. This pathway is thought to be absolutely critical for the construction (or consolidation) or new memories, as well as acting as a pathway for the retrieval of older memories.

Hippocampal Processes and AD

Both postmortem AD human tissue and that obtained through animal models of the disease show a profound loss of neurons and axons within the hippocampus and the adjacent entorhinal cortex. Although this region is probably not the initial source of disease pathology, it is believed to occur early and to represent a tipping point in disease progression, giving rise to the first signs of cognitive impairment (Yakushev et al., 2011). As such, a large proportion of human and animal AD studies have focused on this area, with fewer, but still a large number of studies researching forebrain cortical regions, which also have an important role in the disease.

The hippocampus proper is comprised of the so called tri-synaptic circuit, which involves the relay of information between the CA1 (*Cornu Ammonis*), CA3 and dentate gyrus subregions. Briefly, cortical information arrives from the entorhinal cortex via the performant path to first synapse on CA1 pyramidal neurons. These cells send dense bushy fiber projections (mossy fibers) to CA3 neurons, which project via their Schaffer collaterals, to the dentate gyrus. The small rounded granule neurons of the dentate gyrus ultimately feedback onto CA1 and CA3, as well as feeding forward to the subiculum, and then out through the performant path to the cortex. The reverberation of signals within this circuit is critical for memory consolidation and the appropriate processing of input from multiple sensory modalities. Hence, it is not hard to see how disturbances in this circuit could give rise to many of the dementia features that characterize AD.

In addition to the parahippocampal gyrus, forebrain cortical and subcortical regions are greatly affected in AD. For instance, cholinergic neurons of the nucleus basalis of mynert seem to be susceptible to degeneration in AD. These nuclei typically project diffusely to the numerous cortical region, which also show substantial neuronal loss with disease progression. Together, these regions are important for learning and memory, as well as emotional regulation and processing, so their loss would be expected to take many cognitive functions "offline" and contribute to the vast AD symptomatology.

The hippocampus and medial temporal cortical neurons are known to be particularly plastic, such that they readily exhibit long-term potentiation (LTP), which is thought to represent learning at the most basic, electrophysiological level. LTP is characterized as the increased neuronal excitability (both electrically and chemically) following high frequency stimulation (Kandel, Dudai, & Mayford, 2014). Importantly, LTP processes are impaired in numerous AD animal models. As an example, β-amyloid impaired LTP by disrupting intracellular calcium handling or through the α-amino-3-hydroxy-5-methyl-4-isoxazolepropionic acid (AMPA) glutamate signaling pathways. As well, mice deficient of the AD-linked gene, ApoE, exhibited disturbed LTP (Valastro, Ghribi, Poirier, Krzywkowski, & Massicotte, 2001). Several different insults, including chronic stressors or immune insults [e.g., lipopolysaccharide (LPS) or IL-1β], also interfere with LTP processes and subsequently cause learning or memory problems (Abareshi et al., 2016). Thus, while a fundamental cellular correlate for memory processes, LTP is not uniquely important for AD and likely reflects one of many processes that are disturbed in the disease.

Adult hippocampal neurogenesis is a fundamental plastic processes that is disrupted in AD. While neurogenesis (birth of new neurons) is critical for the initial development of the brain, there are a couple of neurogenic niches that maintain low levels of neurogenesis throughout adulthood. These are primarily limited to the subventricular zone (around the lateral ventricles) and the dentate gyrus region of the hippocampus. Interestingly, postmortem tissue from early AD patient displayed reduced hippocampal neuronal maturation, but an actual increase in stem cell precursors (Boekhoorn, Joels, & Lucassen, 2006). This may represent an attempt by the hippocampus to compensate for the loss of neurons or a last-ditch effort to preserve tri-synaptic functioning. In fact, it has been argued, along the lines of the cognitive reserve hypothesis, that enhanced synaptic or neurogenic activity might endow individuals with the luxury of being able to lose appreciable neuronal tissue before showing clinical symptoms. In effect, the continuous replacement of neurons resulting from hippocampal neurogenesis might be one way of maintaining hippocampal neurogenic reserve (Valero, Mastrella, Neiva, Sánchez, & Malva, 2014). Further evidence suggests that the medial septum brain region (which has many projections to the hippocampus) greatly influences adult hippocampal neurogenesis through gamma-aminobutyric acid (GABA) functioning, and that targeting this inhibitory pathway might hold benefit for AD (Bao et al., 2017).

Genetic-based animal models of AD have been associated with disturbances of adult hippocampal neurogenesis (Radad et al., 2017). For instance, transgenic expression of the AD linked Presenilin-1 gene disturbed basal hippocampal neurogenesis, as well as the normal elevation observed with environmental enrichment (Hollands et al., 2016). Similarly, overexpression of a pathological form of the tau protein (which forms a critical component of the AD neurofibrillary tangles) reduced adult neurogenesis. This is important given, as we'll see shortly, a hyper-phosphorylated form of the tau protein contributes to the characteristic neurofibrillary tangles that are evident in the AD brain.

Besides genetic factors, chronic inflammation observed with AD might contribute to the suppression of hippocampal neurogenesis (Valero et al., 2014). In this regard, both central and systemic injection of LPS reduced adult hippocampal neurogenesis. Moreover, this antineurogenic effect was related to elevated hippocampal microglial reactivity, and could be reversed by anti-inflammatory nonsteroidal anti-inflammatory drug (NSAID) drugs (Ormerod et al., 2013).

The most likely mechanism through which LPS disturbs neurogenesis is through the production of proinflammatory cytokines. Consistent with this suggestion, transgenic overexpression of IL-6 diminished adult hippocampal neurogenesis (Vallières, Campbell, Gage, & Sawchenko, 2002), as well as reducing neurite length in surviving neurons (Oh et al., 2010). Similarly, TNF-α administration reduced the number of neurogenic neurons, whereas knockout of the TNF type 1 receptor increased proliferation of hippocampal cells (Iosif et al., 2006). Accumulating evidence has also implicated IL-1 as a negative regulator of adult neurogenesis, as central IL-1β administration reduced adult hippocampal neurogenesis, whereas the infusion of IL-1 inhibitors had the opposite effect (Koo & Duman, 2008). It has, indeed, been maintained that IL-1 might underlie the antineurogenic effects of acute stressors, since an IL-1 receptor antagonist (IL-1Ra) administered before the stressor was applied, prevented its impact on neurogenesis (Koo & Duman, 2008). Similarly, chronic mild stressor exposure elevated hippocampal IL-1b, and these effects were attenuated in mice that over-expressed IL-1Ra (Goshen et al., 2008).

Besides the obvious loss of neurons and their projections, the two hallmark pathological features of AD are the presence of senile plaques and neurofibrillary tangles. The so called "senile" plaques largely consist of a truncated form of β-amyloid as a core component that is known to have toxic effects on neurons and their synapses. It has even be reported that along with the early memory deficits, β-amyloid induces alterations in neuronal firing, leading to synaptic disturbances (Pozueta, Lefort, & Shelanski, 2013). Of particular significance in relation to inflammation, β-amyloid itself has dramatic activating effects on microglia within both *in vitro* and *in vivo* studies, showing microglial-driven release of inflammatory and oxidative factors (McGeer & McGeer, 2010).

The amyloid precursor protein (APP) is normally cleaved by the β- and γ-secretase enzymes resulting in β-amyloid peptides. In the case of AD, cleavage is altered, leading to an abnormal accumulation of the 42-amino acid form of β-amyloid, which has a high rate of fibrillization and insolubility. This results in the formation of diffuse and eventually mature dense-core plaques. The dense-core plaques also contain abnormal neuronal processes (dystrophic neurites), as well as various inflammatory and acute-phase proteins, such as α2-macroglobulin, IL-1β, IL-6, and intercellular adhesion molecules (Serrano-Pozo et al., 2011). The amyloid plaques aggregate within multiple cortical areas and to a lesser extent within the hippocampus, brainstem, and basal ganglia. Importantly, whereas diffuse amyloid plaques are commonly present in the brains of cognitively intact elderly people, dense-core plaques, particularly those with neuritic dystrophies, are often found in patients with AD dementia.

A three-stage model for β-amyloid plaque progression in AD has been proposed (Braak & Braak, 1995). It was suggested that aberrant β-amyloid proteins accumulate in frontal, temporal, and occipital cortices, which then progress to cortical association areas and hippocampus, before finally spreading to all subcortical areas, most notably the cerebellum striatum, hypothalamus, subthalamic nucleus, and thalamus. Intriguingly, a recent report indicated that the earliest signs of β-amyloid fibril accumulation occurred in the precuneus, medial orbitofrontal, and posterior cingulate cortices (Palmqvist et al., 2017). The affected individuals were in the earliest preclinical AD stages, thus amyloid deposition and the associated mechanisms could serve as biomarkers of disease. These brain regions comprise much of the default mode network (DMN), which is thought to underlie basic processes that give rise to the general experience of consciousness and self. These specific brain region disturbances are consistent with reports indicating that DMN activity can be used to differentiate AD patients from healthy agers (Greicius, Srivastava, Reiss, & Menon, 2004).

The neurofibrillary tangles are composed of a hyper-phosphorylated and misfolded form of the microtubule associated protein, tau, along with various other microfilament proteins. Also, dendritic and axonal degradation gives rise to neuropil threads which accompany the tangles. The neurofibrillary tangles impair axonal transport of vital proteins and lead to abnormal accumulation and/or degradation of proteins. Tangle accumulations are thought to begin at the medial temporal lobe (entorhinal cortex), spreading to limbic regions (hippocampus, amygdala, and thalamus), and then to sensory cortical areas and possibly the nigrostriatal system (Braak & Braak, 1995).

In considering the most common neurodegenerative states, namely, AD, Parkinson's, and multiple sclerosis, it seems that they share the common histopathological feature of having some aberrant plaque-like brain deposits. As just discussed, the AD brain possesses senile plaques, and as we learned in Chapter 14, Inflammatory Roads to Parkinson's Disease, the Parkinson's brain is littered with Lewy bodies (which are essentially α-synuclein rich plaques) and multiple sclerotic plaques characterize multiple sclerosis.

Different mechanisms give rise to the plaque formations and it is unclear whether the plaques play a primary or secondary role in each of the separate pathological states, and the possibility exists that the plaques might arise from failed protective processes. It is tempting to speculate that common fundamental biological processes are involved including the inflammatory immune system. As outlined in Chapter 14, Inflammatory Roads to Parkinson's Disease, substantial evidence points to innate (and to a certain degree adaptive) immune cells and their soluble factors in the genesis and progression of neurodegeneration. Before we dive into these processes further, we will first turn to the lifestyle and genetic factors that have been implicated in AD. As we discussed in relation to PD, we favor a multihit hypothesis for AD, wherein multiple "hits" over the lifetime interact with genetic constitution to shape the evolution of disease. We posit that the inflammatory immune system is the common thread "weaving" these various "hits" together into what is manifested as the primary disease and its comorbidities.

β-AMYLOID: TO TARGET OR NOT TO TARGET

It is clear that β-amyloid has become the pivotal, and yet controversial target for treatment options. Although amyloid is a key feature of senile plaques and that its' soluble form can have toxic effects, some investigators believe that it is not a viable target. Indeed, all methods that have been used to remove plaques, ranging from vaccines to drugs, have met with clinical failure. This has been the case despite the many preclinical studies have shown benefits to using these treatments. Efforts have thus been made to develop a variety of non-amyloid directed strategies to limit microglial activation in the AD brain. These include targeting cannabinoid receptors on microglia, or targeting the complement C1q receptor that induces inflammatory effects. However, it would be inappropriate to completely "turn off" microglia, as their normal brain functions are critical for many housekeeping duties. For that matter, it might be the case that microglia help clear senile plaques. This may or may not be key for clinical benefit. In fact, together with peripheral antibodies and immune cells, microglia might be a mechanism that could be used to clear β-amyloid plaques.

LIFE STYLE FACTORS AND AD

The majority of AD cases occur after 65 years of age, and becoming highly prevalent (almost 40%) into the late 80s. Again, early onset cases occur among individuals in their 30s or 40s, but these are rare. When considering the impact of life style factors, it is important to bear in mind that these are acting within the context of an aging individual and their effects may vary across the lifespan. Indeed, we know that many different stressors can have especially negative effects on brain health when they occur early in life. However, this is also true for stressor experiences during later stages of life when the degree of plasticity and resiliency is already taxed and when compensatory processes are diminished by the aging process itself. So, it is likely a complex scenario plays out, much like PD, in which the multiple hits faced over the entire lifespan collectively act to determine AD risk.

Many aspects of aging may dispose individuals for the development of AD, including a general reduction of blood flow to the brain (Dong, Maniar, Manole, & Sun, 2017). Indeed, the brain hypoperfusion that occurs in the elderly begins during early to middle adulthood and progresses with age. The deterioration of cerebral blood flow limits vital nutrient supplies for the high energetic demands of neurons. Accordingly, high blood pressure and diabetes, both of which strongly impact vascular functioning, greatly increase AD risk, possibly through their links with inflammatory processes (McKenzie et al., 2017). Even healthy aging is associated with the accumulation of β-amyloid plaques, though clearly not to the extent observed in AD. Also, aging itself is associated with a degree of elevated inflammatory tone throughout the brain, including increased reactive microglia (Lourbopoulos, Ertürk, & Hellal, 2015).

A strong link exists between cerebrovascular disease and dementia in general, and AD in particular. It was reported that 7% incidence of dementia follows a first cerebral stroke (Pendlebury & Rothwell, 2009). Predictably, hypertension and diabetes (which are major risk factors for stroke) were also associated with AD. Accordingly, when hypertension is evident at middle age, the risk of dementia is much higher than when it first occurs later in life (Whitmer, Gunderson, Barrett-Connor, Quesenberry, & Yaffe, 2005). This could stem from the greater severity and chronicity of high blood pressure or it could be a sign of some other processes at work, such as inflammation. It is known that hypertension is associated with the increased production of proinflammatory cytokines, C-reactive protein (CRP), and prostaglandins. Hypertension can occur, owing to blockages or constrictions in blood vessels and elevated platelets and leukocytes contribute in this regard. These inflammatory factors can have long-term central nervous system (CNS) effects and might even confer sensitization, so that they augment neuronal and behavioral responses to variety of subsequently encountered psychogenic, neurogenic, or systemic stressors (Anisman & Hayley, 2012a, b). In effect, early or midlife inflammatory processes might proactively increase the risk for later pathology given that sufficient challenges are encountered.

Type 2 diabetes is associated with a twofold increased risk of AD, possibly being related to the disruption of enzymes that is critical for the clearance of extracellular β-amyloid, which contributes to pathological plaque buildup. Clearance of AD senile plaques will be covered in detail in an upcoming section that deals with the importance of the primary immunocompetent CNS cells, namely, the microglia. Suffice it for the moment that transgenic mice that expressed diabetic pathology showed increased microglial density around senile plaque. This was associated with a reduction of insoluble β-amyloid, but with elevated levels of the toxic soluble species (Infante-Garcia, Ramos-Rodriguez, Galindo-Gonzalez, & Garcia-Alloza, 2016). Thus, the microglial phenotype is likely altered in the AD and possibly the diabetic brain.

Intriguingly, insulin treatment helped improve scores on cognitive tests in early stage, mild AD patients and pioglitazone (an antihyperglycemic diabetes medication) administration had small cognitive benefits, and reduced microglial activation and β-amyloid oligomer deposition (Gad, Zaitone, & Moustafa, 2016). These findings are consistent with animal studies showing that intranasal insulin administration reduced hippocampal lesion size following traumatic injury (Brabazon et al., 2017), and also had anti-inflammatory properties that were linked to cognitive improvement in an animal model of HIV (Mamik et al., 2016). Finally, it should be mentioned that insulin growth factor-1 (IGF-1) is neuroprotective in several animal models related to AD, possibly stemming from both its' trophic and anti-inflammatory effects (George et al., 2017).

Perhaps not surprisingly given its obvious ties to diabetes, hypertension, and cerebral stroke, obesity has been linked to AD (Alford, Patel, Perakakis, & Mantzoros, 2017). Yet, low body weight has also been linked to elevated AD risk. Thus, dramatic changes away from an "ideal" weight might contribute to the elevated AD risk. Adipose tissue, as we've seen earlier, is a rich reservoir for proinflammatory cytokines, such as TNF-α and IL-6 (Engin, 2017), and the accumulation of these factors could elevate inflammatory status. Conversely, the rapid loss of adipose tissue could remove this reservoir, but could also liberate these inflammatory cytokines into circulation. Any spike in circulating cytokines could have negative effects throughout the body and CNS, particularly if this occurs on chronic basis. Another interesting element to consider is that many environmental toxicants, including pesticides, polychlorinated biphenyls (PCBs), and methyl-mercury, are all highly lipophilic compounds that readily accumulate in adipose tissue (Artacho-Cordón et al., 2016). With aging, there is often a loss of some long-standing fat cells, possibly resulting in high levels of toxicants entering circulation, thereby heightening inflammatory and oxidative stress on the brain and body.

Diet, along with physical and mental activity levels, is high on the list of lifestyle factors important for AD. Lower AD risk was associated with consumption of fish, nuts, tea, and turmeric (Jaroudi et al., 2017). The omega-3 polyunsaturated fatty acids and eicosapentaenoic acid (EPA) found in fish might be responsible for its beneficial effects given their potent antioxidant and anti-inflammatory actions (Devassy, Leng, Gabbs, Monirujjaman, & Aukema, 2016). Indeed, EPA can inhibit microglial activation and reduce antigen presentation and T cell proliferation. Another prominent omega-3, docosahexaenoic acid (DHA), is a critical phospholipid in brain membranes. Although these reports would lead to the expectations that omega-3s would have positive effects in the course of AD, but the results regarding the clinical utility of omega-3s have been discouraging. Placebo controlled trials, in fact, revealed no significant effect on cognition (Shinto et al., 2014). This is not altogether surprising, as the beneficial effects of fatty acids would be expected to be modest, and if anything, would be more suited to a preventative or prophylactic role, rather than as a frank clinical treatment. It should be added that unlike polyunsaturated fats, it was indicated that saturated and trans fats might increase AD risk.

Considerable research has supported a neuroprotective role for curcumin, which is the source of the turmeric spice found in many Indian curries. This was supported by the substantially lower rates of AD in India, coupled with epidemiological evidence showing that Asian people who regularly consumed turmeric-rich curry, performed better on the Mini Mental State Examination (MMSE) cognitive test, compared to those who did not or very rarely consumed curry. Once again, these are only correlational data and no causation can be inferred, but, experimental data indicated that even low doses of curcumin modulated microglial proliferation and differentiation and produced neuroprotective properties (Sharma, Sharma, & Nehru, 2017). As it turns out, curcumin readily crosses the blood brain barrier (BBB) and reduced β-amyloid plaque burden in AD transgenic mice (Sundaram et al., 2017). Although it is unclear exactly how curcumin might have such benefits, its anti-inflammatory and antioxidant actions are well-documented. Curcumin is a potent inhibitor of the prostaglandin synthesizing enzyme, COX-2 (much like NSAIDs), and reduces the inflammatory transcription factor, NF-κB (Deng et al., 2014). These effects have downstream consequences that result in diminished microglial activation as well as the release or proinflammatory cytokines and reactive oxygen species.

Gastrointestinal problems are frequent in AD patients and increasing evidence has indicated that the gut-brain axis influences processes that could be important for neurodegeneration. Moreover, fragments from gram-negative bacteria that are normally found in high concentrations within the gut (*Bacteroides fragilis* and *Escherichia coli*) were evident in the AD brain. Fungal infections and accumulated herpes simplex virus DNA were likewise found in the hippocampus and cortex of AD patients. Thus, some aspect of the disease process might be related to the translocation of gut bacteria into circulation and subsequently, the brain parenchyma. Such bacterial product movement would be associated with obvious stimulation of various elements of the immune system. For instance, within a drosophila fly model, immune hemocytes (phagocytic innate immune cells found in invertebrates) were critical for gut-brain communication and mediated AD-like pathology. Specifically, enterobacteria infection mobilized hemocytes that infiltrated the brain and promoted TNF-α dependent neurodegeneration, and conversely, hemocyte depletion attenuated inflammation and neurodegeneration (Wu, Cao, Chang, & Juang, 2017).

As repeatedly mentioned, the potential role of probiotics in affecting brain health has gained substantial momentum. In the case of AD, SLAB51 (a mixture of lactic acid bacteria and bifidobacteria) had beneficial effects in a transgenic AD mouse model. In particular, SLAB51 increased gut levels of the anti-inflammatory bacteria, *Bifidobacterium*, while reducing the more proinflammatory species, *Campylobacterales*. This was associated with behavioral improvements in conjunction with reductions of cerebral β-amyloid levels and hence, diminished plaque pathology (Bonfili et al., 2017). However, 12 weeks of probiotic administration in AD patients revealed that while the supplements reduced serum CRP levels, it had no effect on any other inflammatory or oxidative stress markers, and provoked only modest improvement on several measures of cognition (Akbari et al., 2016).

Several recent reviews have drawn attention to the gut-brain axis in relation to virtually all neurological conditions, reaching the consensus that bacterial gut species can affect CNS functioning. Yet, the evidence pertaining to microbiota actually modulating neuronal degeneration is still scant and it is likely that any connection would be complex and indirect. One potential indirect route would be through the inflammatory immune cells and cytokines.

Whatever the case, this emerging gut-brain story is fascinating and might foster novel perspectives, although it is likely that these will be relevant to potential prophylactic actions in relation to AD, but it is difficult to envisage therapeutic outcomes.

With regard to physical and mental activity, the evidence is pretty simple, "use it or lose it." The more physically and mentally active individuals have the lowest risk of developing dementia. Obviously, there are exceptions and an active lifestyle holds no absolute guarantee of any sort. It might be that initiating exercise earlier in life has protracted benefits. For instance, starting regular cardio fitness training at middle age reduced the likelihood of late life dementia by 50% (Liu et al., 2010). Interestingly modest cognitive benefits was apparent in patients that initiated exercise regimens after AD diagnosis. This is in keeping with the animal literature that suggested that aerobic exercise in rodents limited β-amyloid plaque development and improved performance on spatial memory tasks (Tapia-Rojas, Aranguiz, Varela-Nallar, & Inestrosa, 2016).

Both mental and physical activity promote processes that could be protective and stave off dementia, notably, the increased blood flow to the brain, and the up-regulation of trophic and anti-inflammatory factors. Regular exercise, as we saw in Chapter 4, Life-Style Factors Affecting Biological Processes and Health, augmented hippocampal-dependent spatial learning in rodents and such effects were related to increased brain derived neurotrophic factor (BDNF) levels, along with elevated adult neurogenesis and dendritic branching. Similarly, regular, voluntary aerobic exercise reduced proinflammatory CRP and interleukin-6 levels (Tyml et al., 2017).

In line with the potential protective effects of increased mental activity, low education was associated with increased AD risk. This has fueled the so called "cognitive reserve" hypothesis, that holds that the more cognitively advanced and correspondingly more complex synaptic projections present, the greater the protection against dementia symptoms developing. A meta-analysis confirmed a negative association (46% reduced risk) between dementia risk and education, IQ, and mental activities (Valenzuela & Sachdev, 2006), and 25 of 33 studies demonstrated a significant protective effect. However, there have been failures to replicate such findings. Most troubling, some prospective AD studies found that higher education or occupational achievement resulted in AD patients dying more quickly and experienced a more rapid cognitive decline. This said, it was posited that this finding is still consistent with the "cognitive reserve" hypothesis, in that by the time the "intellectually stimulated" individuals first show clinical symptoms, they might already have significantly more pathology than other AD patients (i.e., their extra cognitive reserve managed to keep cognitive processes together despite pathological changes). So, intellectual activity doesn't necessarily slow the biological processes operative in AD, but rather the intellectually stimulated brain has more alternative and rich neural connections that can, for a time, keep memory disturbances from being manifested.

As a final lifestyle issue, one that is especially significant, are reports that fragmented sleep was associated with poor cognition and increased AD risk. Recent work suggests that sleep might play an important role in the homeostasis and metabolism of β-amyloid in the AD brain (Kastanenka et al., 2017). Similarly, sleep disturbances were associated with widespread pathology in positron emission tomography (PET) scans of AD patients (Liguori et al., 2017). Interestingly as well, blood and cerebrospinal fluid (CSF) levels of β-amyloid itself show a circadian pattern of distribution (Cicognola, Chiasserini, & Parnetti, 2015). The duration and quality of sleep may be the two most important characteristics to consider as these were most closely correlated with the buildup of β-amyloid in the prefrontal cortex and even the density of cortical neurofibrillary tangles in AD patients (Fjell et al., 2017).

It is uncertain how sleep affects β-amyloid levels, but, animal studies raised the possibility that sleep disruptions might impair glial ability to metabolize β-amyloid. Among other things, sleep deprivation impaired the ability of astrocytes to interact with ApoE and help clear β-amyloid (Yulug, Hanoglu, & Kilic, 2017). Likewise, sleep deprivation altered microglia and inflammatory processes and increased β-amyloid levels (Ju et al., 2017). As described in Chapter 4, Life-Style Factors Affecting Biological Processes and Health, sleep disturbances are associated with multiple changes of immune and inflammatory functioning, and might thus contribute to cell loss. The impact of sleep deprivation on cognition and β-amyloid is not restricted to AD, having been reported among non-AD individuals. In fact, poor quality of sleep or sleep deprivation was accompanied by increased brain β-amyloid aggregation and elevated β-amyloid in cognitively healthy adults (Yulug et al., 2017). Thus, the elevations of β-amyloid might be considered as an additional "hit" on the road to AD.

GENETICS OF AD

Although AD has a higher familial prevalence than PD, the majority of cases are not strictly genetic, but vulnerability genes may still play a role. The early onset form of AD, which accounts for less than 1% of AD cases, has been associated autosomal dominant inheritance, whereas the more common late form generally has a much more complicated genetic pattern of inheritance involving many low penetrance genes (Lanoiselée et al., 2017). Finally, the e4 APOE allele has also been linked to both forms of AD and as will be discussed shortly, two "inflammatory genes" TREM2 and CD33, have also been implicated in AD.

Advanced age allows for the accumulation of the damage accumulated from multiple hits encountered through life, hence leading to the age-related formation of AD. It is equally possible that some genetic program(s) unfold in a time-dependent manner. Essentially a person might be born with a "ticking time bomb" genetic program that only becomes activated at advanced ages. Another twist on this idea is that such a genetic program(s) might only become activated after certain environmental exposures are encountered. It might be that some threshold of basic damage has to be incurred before symptomatology arises; but, once certain biological processes are in motion, they take on a life of their own and may be extremely difficult to slow down.

Studies examining genetic loading have often assessed genes in terms of early onset versus late onset AD. In general, there is much more evidence for multiple genes, most notably APP, PSENs 1 and 2, being operative in the less common early onset form of the disease (Olson, Goddard, & Dudek, 2001). Indeed, genetic linkage studies have identified mutations in APP on chromosome 21q, PSEN1 on 14q, and PSEN2 on 1q in early onset AD (Cruchaga et al., 2017). This said, several large scale genetic linkage studies have implicated the e4 allele of the ApoE gene in both early and late onset AD (Di Battista, Heinsinger, & Rebeck, 2016). As well, APP and the PSENs have also been implicated in familial forms of late onset AD and might even be vulnerability factors for the more commonly observed cases.

We will first describe how APP processing can give rise to amyloid pathology that has been linked to AD. The APP gene found on chromosome 21 is a transmembrane protein that is processed by three different enzymes, α-, β-, and γ-secretases. Three forms of the protein result from alternate splicing: APP695, APP751, and APP770. The first isoform is found mainly in neurons, whereas the other two are located throughout the body. It is the neuronal form that is thought to be dysregulated in AD. This mutation can cause faulty enzymatic APP processing by the β-secretase and γ-secretases, producing truncated pathological Aβ fragments that can form the core of amyloid plaques (Wang et al., 2016).

Two b-amyloid species cleaved from APP have received particular attention in relation to AD pathology; these comprise β-amyloid 40 and β-amyloid 42. The latter is thought to be more toxic owing to its hydrophobic nature and propensity to form fibrils. β-amyloid can be secreted and act on extracellular targets or may be produced and act at intracellular sites. Pathological β-amyloid can accumulate in extracellular senile plaques, or can accumulate within the neuron to form intraneuronal inclusions in AD. Such inclusions have been found in the AD hippocampus early in the disease, prior to major senile plaque deposition (Kadokura, Yamazaki, Lemere, Takatama, & Okamoto, 2009). The fact that similar effects were observed in genetic animal models, lends support for intra-neuronal β-amyloid accumulation as an early pathological event, with extracellular plaques increasing later, possibly fueling disease progression.

Most of the more than 40 APP mutations that are found in over a hundred different families are dominantly inherited. The majority of these mutations occur in the vicinity of β- and γ-secretase sites on the gene. The most studied APP mutation, the so called Swedish APP Mutation (KM670/671NL), results in dramatically increased β-amyloid levels together with widespread cortical atrophy and ventricular enlargement (Balakrishnan et al., 2015). Indeed, a double APP mutation that occurs in a Swedish family increased β-amyloid production by modifying beta-secretase enzymatic activity. Similarly, London and Flemish mutations in APP lead to deficits in γ-secretase activity, culminating in accumulation of β-amyloid rich senile plaques (Acx et al., 2017). Finally, Down syndrome patients who have triplicate version of APP, develop AD-like pathology at a relatively early age, again, suggesting a link between APP levels and/or processing and cognitive pathology.

A specific role for APP and its cleaved β-amyloid proteins has yet to be determined. At one time, it was believed that these were simply abnormally created pathological proteins with little "normal" physiological functions. This was subsequently shown to be incorrect, with some studies pointing to a role of APP proteins in the modulation of synaptogenesis, cell adhesion and transport, as well intracellular signaling (Copenhaver & Kögel, 2017). Given its heavy processing, the different intracellular fragments likely have markedly different roles. Such effects are most likely dose-dependent given that high levels of intracellular β-amyloid are extremely toxic, but lower more physiologically relevant levels enhance neuroplasticity, facilitate neurochemical signaling, and buffer against neuronal toxicity stemming from excessive heavy metals, such as iron or copper. Hence, pathological plaque evolution might reflect normally protective endogenous processes that have gone awry, or alternatively, processes that have simply been overtaken by some other independent disease mechanisms.

Besides APP, mutations of PSEN1 are associated with familial AD (Kelleher & Shen, 2017). The PSEN1 and PSEN2 genes are vital components of the γ-secretase complex and are localized on the endoplasmic reticulum. PSEN1 acts as the catalytic subunit of gamma-secretase with over 200 pathogenic mutations being reported for the gene (Szaruga et al., 2017). These mutations have been associated with increased levels of pathological β-amyloid 42 that forms the core of senile plaques. Indeed, astrocytes derived from AD patients that had a PSEN1 mutation produced increased β-amyloid (Oksanen et al., 2017). Hippocampal pathology and

accumulation of β-amyloid peptides was confirmed in rodents bearing the mutant form of the gene. The PSEN2 gene mutations are somewhat more rare, with 13 pathological mutations reported that were associated with altered β-secretase activity (Lanoiselée et al., 2017). Like PSEN1, these PSEN2 mutations were associated with accumulation of senile plaques formation and AD-like neurodegeneration.

In addition to its impact on β-amyloid levels, PSEN1 mutations might promote disease by reducing trophic support to neurons. In fact, endogenous presenilin is thought to be involved in learning and may be important for neuronal health during aging. Curiously, a study that analyzed over a hundred different PSEN1 mutations found that 90% *decreased* production of β-amyloid 40 and 42 (Sun et al., 2017). This obviously contrasts with the findings that were just mentioned, and are inconsistent with the general β-amyloid hypothesis for AD. However, Sun et al. (2017) found that PSEN1 reduced β-amyloid 40 to a greater extent than β-amyloid 42, thereby altering the ratio between the two. Thus, it is possible that the ratio between the two is the fundamental component driving pathology.

The strongest genetic risk factor for typical nonfamilial late onset AD is possession of the ε4 allele of the APOE gene, of which there are three alleles; ε2, ε3, and ε4. The primary function of APOE is the shuttling of cholesterol around the body and brain, but is also thought to contribute to regulation of inflammatory processes and synaptic plasticity (Belinson & Michaelson, 2009). Strong evidence has linked the ε4 allele to increased AD risk, which was gene dose-dependent, being more than 10 times higher in homozygous individuals that possess two ε4 alleles. The ε4 allele has also been linked to less efficient Aβ metabolism and increased inflammation in the brain (Dong, Zhou, Ji, Pan, & Zheng, 2017). Indeed, APOE ε4 homozygotic AD patients displayed increased hippocampal atrophy and senile plaque accumulation that was associated with signs of enhanced microglial reactivity (Schreiber et al., 2017).

APOE ε4 has also been implicated in increasing risk for the early onset form of the disease. A gene dosing effect was once again evident, with homozygous e4 carriers showing increased risk, but heterozygous individuals only displaying enhanced risk in the context of a family history of AD (Sando et al., 2008). Thus, the e4 allele might interact with other genetic variables to modulate risk. In contrast, the e2 APOE allele can reduce disease risk (Chung et al., 2016), although the reason behind this is uncertain.

Whatever the case, APOE functioning appears to be important for the formation of β-amyloid plaques and the associated histopathology. There may be several mechanistic reasons for this, but the two that immediately stand out are the cholesterol handling and inflammatory role of APOE. With regard to the former, accumulating data have indicated that cholesterol metabolism is altered in AD, which might contribute to vascular blockages or other blood vessel problems that could influence the development of AD (Shahbazi et al., 2017). As mentioned earlier, stroke, cardiovascular illness, and the chronic hypoperfusion of the brain that occurs with aging are all major risk factors for dementia. The involvement of inflammatory processes has also come from the finding that genetic overexpression of APOE e4 (in knockin mice) increased the inflammatory response to LPS administration, and APOE e4 was associated with a greater *in vivo* inflammatory response in humans (Gale et al., 2014). APOE e4 mice also basally exhibited spatial memory deficits, along with a profound loss of synapses and dendritic spines, indicating that even in an unchallenged state, e4 expression can promote deficits in neuronal plasticity and AD-like memory problems.

In addition to the genes that are secondarily linked to inflammatory processes, AD genetic studies have begun to identify specific single nucleotide polymorphisms in genes that directly control microglial functioning, including TREM2 and CD33. The TREM2 R47H mutation is particularly associated with increased AD risk (Ghani et al., 2016). Interestingly, TREM2 can have inhibitory as well as excitatory effects on macrophage and microglial functioning that might be important for AD (Wang, Ulland et al., 2016). In this regard, TREM2 was reported to control microglial responsivity to β-amyloid and it was posited that TREM2 alterations might underlie the inability of microglia to properly clear senile plaques (Colonna & Wang, 2016). As shown in Fig. 15.3, impaired TREM2 functioning stemming from genetic alterations can diminish the capability of microglia to remove plaques and fuel the development of neuronal distress. This makes sense given the well-known role for TREM2 in mediating phagocytosis, as well as microglial proliferation and the release of cytokines. Indeed, TREM2 is a major regulator of actin remodeling in microglia that is required for their ability to adopt an active "M1-like" inflammatory state (Sasaki, 2017) (Fig. 15.2).

In vitro assessment of microglia revealed that when these cells were induced to overexpress TREM2 they were far better at phagocytizing damaged neurons and β-amyloid (Krasemann et al., 2017). Similarly, TREM2 expression was critical for macrophage engulfment of bacteria. Importantly, the enhanced phagocytic potential induced by viral expression of TREM2 benefitted cognitive outcomes in AD mouse models. Conversely, AD mice that carried multiple pathogenic mutations, APP and 5XFAD, which normally display marked microglial activation and β-amyloid deposition, failed to do so in TREM2-null mice (Golde, Streit, & Chakrabarty, 2013). Microglia from TREM2 mutants were also deficient in their ability to bind to phospholipids and expressing proinflammatory cytokines.

In addition to microglia, TREM2 might also be important for phagocytic potential and

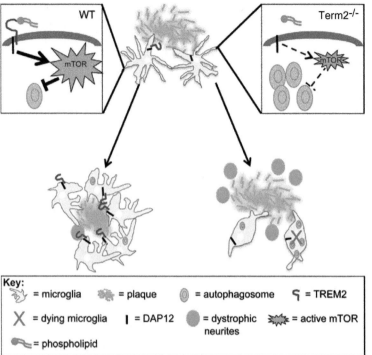

FIGURE 15.2 TREM2 deficiency fuels β-amyloid plaque development. Mice lacking TREM2 had faulty phagocytic ability in the context of β-amyloid plaques. Many of these glial cells were also vulnerable to autophagic death. It is thought that these microglial deficiencies stem from TREM's ability to activate the mTOR pathway, which is critical for basic protein synthesis. Thus, TREM2-mTOR regulated protein synthesis may be required for proper microglial clearance of plaques. *Source: From Ulland et al. (2017).*

recruitment of peripheral immune cells. Increased expression of TREM2 was found on macrophages that congregated around senile plaques in APP mutant mice (Raha et al., 2017). Hence, it is possible that TREM2 contributes to the recruitment of peripheral phagocytes (mostly macrophages, but possibility also neutrophils and dendritic cells) into the brain in an attempt to clear β-amyloid plaques.

The second microglial related gene, cluster of differentiation 33 (CD33), has been linked to both early and late onset AD in several genome-wide association study (GWAS) studies (Dos Santos et al., 2017). Importantly, CD33, along with APOE, are the only two genes confirmed to be involved in late onset AD using GWAS and family-based analyses (Siddiqui et al., 2017; Wang, Liu et al., 2017). Moreover, CD33 expression was elevated in microglial cells found in the brains of AD patients and the levels were reported to be related to degree of clinical decline (Jiang, Yu et al., 2014). It was also reported that individuals with probable AD or MCI and who also possessed CD33 single nucleotide polymorphisms (SNPs), which has been implicated in AD, displayed reduced cortical thickness and diminished hippocampal volume (Wang, Liu et al., 2017). This finding is consistent with the possibility that CD33 might be involved in the initial pathology underlying development of cognitive impairment and the possible evolution to full blown AD. Of course, future longitudinal studies would be required to confirm this hypothesis.

CD33 is a transmembrane receptor found on microglia and certain peripheral immune cells, notably macrophages, neutrophils, and mast cells. It is a member of the sialic acid−binding immunoglobulin family, and like antibodies is critical for the recognition and removal of biological threats. In the case of AD, CD33 alterations were involved in the recognition and clearance of β-amyloid plaques (Malik et al., 2013). As expected, dysregulation of CD33 can alter phagocytosis and endocytosis, and over-expression of CD33 impairs β-amyloid phago-cytosis (Griciuc et al., 2013). This can result in a failure to clear extracellular toxic aggregates and hence, accumulation of senile plaque.

There is evidence that different CD33 gene variants might convey either increased or reduced risk, and that these might further interact with other AD-associated genes. In fact, CD33 processing can result in two splice variants that differentially influence late onset AD risk, with a full-length form that elevates risk, whereas a shorter version lacking the sialic acid-binding domain, was reported to hold protective effects (Siddiqui et al., 2017). It seems that the several different processed ver-sion of the CD33 gene may convey either increased or decreased AD risk by modulating microglial state (Malik et al., 2013). The protec-tive isoform is associated with reduced β-amyloid levels, which were posited to be related to shifts in the cellular location of CD33. Specifically, a shift from being trans-membrane bound to associating with intracel-lular proteins (Siddiqui et al., 2017), would result in less inhibitory control over microglia, and hence greater potential for plaque clear-ance. This proposition clearly contrasts with the transmembrane location of the other CD33 isoforms.

Yet another interesting facet of CD33 func-tioning is that it may interact with other AD vulnerability genes. Specifically, CD33 modu-lated the impact of TREM2 on AD risk, sug-gesting a convergence of genetic vulnerability genes (Chan et al., 2015). This is particularly interesting since both of these genes are media-tors of innate immunity in general, and micro-glial functioning in particular. In fact, CD33 influences the relative overall TREM1/TREM2 expression ratio, as well as altering specific TREM2 transmembrane levels. Moreover, the increased TREM2 expression observed with the CD33 AD risk allele, enhanced β-amyloid plaques and infiltration of peripheral immune cells (Chan et al., 2015). Conversely, inhibition of TREM2 functioning in an AD mouse model diminished microglia pathology and infiltra-tion of inflammatory immune cells (Jay et al., 2015). Yet, at least one study reported a benefi-cial role for TREM2, with the protein reducing inflammatory processes and augmenting phagocytosis of void (Raha et al., 2017).

So once again, the jury is out on whether TREM2, CD33, or other mutations of immune factors are ultimately beneficial, deleterious, or benign in relation to AD. As emphasized here and in Chapter 14, Inflammatory Roads to Parkinson's Disease, it is unwise to simply brand inflammatory processes as either "good or bad," and it is much more useful to consider the specific aspects of inflammatory responses within the context of the particular stage and comorbid features of AD. This approach should emphasize, for example, the important beneficial consequences of microglia in scav-enging β-amyloid species and plaques, while at the same time noting the potential deleterious effects of proinflammatory microglia that could be releasing large concentrations of unstable oxidative species. Exactly what aspects of the disease processes that favor cer-tain polarized microglial activation states over another, is still being worked out. But it is in these phenotypic specificities and their relation to the microenvironment in which they are embedded that may determine pathological and hence, clinical outcomes.

ENVIRONMENTAL FACTORS AND ALZHEIMER'S DISEASE

In contrast to PD, there is a relatively weak link between environmental toxicant exposure and AD. That said, it was reported that long-term exposure to small particulate matter (2.5 μm or less) was associated with a modestly increased risk of AD (Calderón-Garcidueñas et al., 2016). Similarly, smoking, which results in exposure to hundreds of fine toxic particles, was related to elevated AD risk (Wallin et al., 2017). This is particularly interesting in light of the fact, as described in Chapter 14, Inflammatory Roads to Parkinson's Disease, that smoking is strongly linked to a *reduced* risk of PD. However, the inverse link in PD is much stronger and supported by multiple studies, whereas the evidence for the direct relation to AD is weak and data are sparse.

Much like PD, epidemiological data exist linking pesticide exposure to AD. Elevated blood levels of pesticide metabolites [e.g., dichlorodiphenyldichloroethylene (DDE)] correlated with a greater likelihood of an AD diagnosis (Bible, 2014), and acute toxic doses of organophosphate pesticides were associated with the appearance of dementia (Zaganas et al., 2013). Interestingly, the PON1 gene for Paraoxonase 1 (which is an enzyme involved in the metabolism of organophosphates) was associated with increased risk of AD (Erlich et al., 2012). As well, occupational pesticide exposure was correlated with a twofold elevated risk of developing vascular dementia (Hebert et al., 2000).

Organophosphate pesticides disrupt acetylcholine receptor binding within the brain, thereby producing cognitive problems (Sánchez-Santed, Colomina, & Herrero Hernández, 2016). The cholinergic system running from the basal telencephalon to the cerebral cortex and hippocampus, has long been known to modulate learning and memory, and these same neural circuits are ravaged in AD. Hence, although high dose organophosphate exposure can cause severe disability or even death, of more relevance to AD are the changes that might occur with long-term low dose exposure. A meta-analysis showed that low dose pesticide exposure that included organophosphates and other common pesticides was associated with modest, but significantly

increased risk of AD (Yan, Zhang, Liu, & Yan, 2016).

Although clearly linked to PD, scant data are available for the insecticide, rotenone, with regards to AD, although rotenone was directly toxic to cholinergic neurons in brain slices (Ullrich & Humpel, 2009). Further studies revealed an increased risk of AD among those occupationally exposed to pesticides and fertilizers, but the association was no longer significant when education level was considered (Lindsay, Sykes, McDowell, Verreault, & Laurin, 2004). Another such study in rural living individuals (> 70 years old) did not find a significant risk of AD related to exposure to pesticides (Gauthier et al., 2001). Thus, the available data are equivocal.

Heavy metals, including iron, zinc, copper, and aluminum have been anecdotally and experimental linked to AD to some degree. A meta-analysis of 901 relevant studies of AD and aluminum exposure found that of 8 studies that met their basic inclusion criteria, 71% revealed increased AD risk among individuals with chronic exposure to the heavy metal (Wang, Wei et al., 2016). Of course, as only 8/901 studies met the strict inclusion criteria, raises questions regarding the generalization of these data to the AD community. Yet, several other studies revealed correlations between the incidence of dementia and aluminum exposure that was presumed to come from drinking water or dietary sources. An older study conducted in England, indicated a 1.5-fold increase in AD in an area where aluminum water levels were high (0.11 mg/L compared to 0.01 mg/L in other areas). Another early study revealed a clear dose-response relationship with risk of AD increasing linearly in relation to aluminum exposure concentration, with a peak of 1.46-fold elevated risk in those estimated to have been exposed to > 0.2 mg/L. Yet, we would be remiss if we did not point out that several other studies failed to find a relationship between aluminum exposure and

AD risk (Wettstein, Aeppli, Gautschi, & Peters, 1991). Likewise, other studies examining dietary sources of aluminum, most notably tea and herbs (both of which accumulate the metal) found no significant relationship with AD risk (Molloy et al., 2007).

Experimental data certainly indicate that dietary exposure to aluminum results in a buildup of the metal at brain concentrations that can potentially be neurotoxic. In fact, chronic exposure to aluminum through drinking water led to the accumulation of the metal throughout the brain of rodents, with highest concentrations occurring in the hippocampus and cortex (Gómez, Esparza, Cabré, García, & Domingo, 2008). Aluminum can enter the cell and accumulate in the nucleus by binding phosphate groups on the DNA, thus potentially influencing genome integrity and apoptotic processes.

Aluminum, iron, copper, and zinc were associated with β-amyloid plaques in brain tissue from AD patients and at a mechanistic level, metals can interact with β-amyloid to influence their aggregative behavior and potential toxicity (Singh et al., 2013). For instance, excessively aged mice showed copper deposits in neural capillaries and associated with elevated β-amyloid levels (Singh et al., 2013). Furthermore, AD transgenic mice (APPsw/0) also displayed elevated β-amyloid and copper both in the capillaries, as well as in the brain parenchyma. Such metal ions might influence b-sheet formation and assemblage of fibrils in β-amyloid plaques (Innocenti et al., 2010).

The fact that environmental toxicants might contribute to AD has informed the development of specific toxicant-based animal models that aim to recapitulate some basic AD symptoms. These have often focused on how lesions of the hippocampus or frontal cortex give rise to deficits in learning and memory. In this respect, intrahippocampal infusion of the potent excitotoxin kainic acid caused the loss of glutamate producing CA1 and CA3 pyramidal cells, along with many cholinergic neurons. The neurodegeneration was associated with marked cognitive deficits, most notably a profound inability to navigate a Morris water maze, which is indicative of a spatial learning disability (Park et al., 2012), much as this occurs in humans.

The impact of the excitotoxins appeared to be moderated by genotype. Mice that bore vulnerability genes for AD, most notably APOe e4, were more sensitive to kainic acid injections (Duan et al., 2006). In contrast, the e3 APOE allele actually conferred protection against kainic acid-induced neurodegeneration (Buttini et al., 1999). It is of interest that although overexpression of mutant presenilin-1 augmented kainic acid-induced hippocampal neuronal loss, it had no impact on pathology induced by cerebral stroke (MCAO) (Grilli et al., 2000), indicating specificity regarding the gene-toxicant interaction.

The final portion of this section will delve into the evidence that infectious pathogens or other immunological agents might be linked to AD. The idea that infection could lead to AD dates back to the early 1900s when Alois Alzheimer first noticed the similarity between the symptoms exhibited by what came to become known as AD and those that had dementia that arose from syphilis infection. More recent evidence has pointed to the possibility that infectious agents that spread to the brain might be involved in the origins of AD. A meta-analysis reported a 4–10-fold increase in the occurrence of AD in those with spirochetal or *Chlamydophila pneumoniae* infection (Maheshwari & Eslick, 2015). Remnants of the bacteria that cause Lyme disease and pneumonia were also found in the brains of deceased AD patients. Similarly, DNA sequencing and polymerase chain reaction (PCR) analyses revealing the presence of a variety of fungal species in the AD brain (Alonso et al., 2014). It is likely that any infectious agents that breach the brain might interact with other factors to collectively augment risk. This was the case for individuals with the ApoE4 allele that showed a 12 times higher AD risk when they were also exposed to herpes simplex virus (HSV) infection (Bu et al., 2015). Thus, as in the case for PD, infections might be a vector, which in conjunction with other hits (including genetic vulnerability), increases AD risk.

Quite surprisingly, given the mainstream resistance, a large body of evidence has amassed that supports a link between microbes and AD. In fact, several postmortem studies reported signs of pathogen exposure (e.g., HSV1 DNA) in conjunction with amyloid plaques and neurofibrillary tangles (Itzhaki et al., 2016). Likewise, HSV infection provoked AD-like pathology in mice bearing a mutation in the APOE gene, which has been strongly implicated in AD (Burgos, Ramirez, Sastre, & Valdivieso, 2007). Perhaps most remarkably, AD brain tissue can actually confer pathology when transferred to healthy mice or nonhuman primates (Clavaguera et al., 2013), indicating that the disease could spread in a manner akin to a typical infectious or prion-like disease.

The possibility was entertained that a latent virus that is reactivated by environmental stressors could give rise to AD. It is also possible that certain genetic polymorphisms might be common vulnerability factors that influence the impact of infectious insults on AD. For instance, the APOE e4 allele is known to modulate susceptibility to infection, and genes for viral receptors have been found in the AD brain (Fujioka et al., 2013).

Exposure of mice to HSV provoked β-amyloid deposition and tau pathology reminiscent of AD. Several studies implicated HSV1 in AD, indicating that the impact of the virus is especially great in APOE e4 carriers (Itzhaki, 2017). Likewise, increased risk of AD (fourfold increase of odds ratio) was reported in patients who carried an increased bacterial or viral burden comprising infection with Borrelia burgdorferi, C. pneumoniae, and Helicobacter pylori (Bu et al., 2015). As well, a relationship existed between peripheral proinflammatory cytokine and β-amyloid levels. Not surprisingly, AD patients exposed to infections declined more precipitously, possibly being due to priming of innate macrophages or central microglia (Perry, Nicoll, & Holmes, 2010).

Growth hormones derived from postmortem tissue were previously used to treat individuals with developmental growth deficits. Unfortunately, a number of the recipients developed Creutzfeldt–Jakob disease (CJD) as a result of receiving tissue contaminated with human prions. Dramatically, some of the individuals that subsequently died from CJD displayed marked accumulation of β-amyloid plaques in the brain that mirrored that observed in AD (Jaunmuktane et al., 2015). It may be telling that CJD and AD share a prolonged prodromal phase, with the diseases only manifesting decades after exposure. Thus, the iatrogenic CJD cases just described and AD could conceivably be related to some latent infectious agent or alternatively, some genetic program that unfolds over time.

Animal studies have supported an "infection-like" spread of β-amyloid pathology that occurs with long incubation periods. Dense β-amyloid plaques were evident throughout the brain of marmoset monkeys that were centrally infused with cerebral tissue from AD patients 6–7 years previously (Baker, Ridley, Duchen, Crow, & Bruton, 1994). Similarly, while oral, intravenous, intraocular, and intranasal administration of β-amyloid containing brain extracts had no effect, the placement of β-amyloid contaminated steel wires in the brain yielded β-amyloid pathology in mice (Eisele et al., 2009). This finding suggests that direct contact with the brain is required for the spread of plaques and that the process is less effective than typical infections that can be propagated through other routes. Interestingly, central inoculation with brain tissue from transgenic AD mice could transfer β-amyloid pathology to another mouse and this effect was very much strain-dependent (Meyer-Luehmann et al., 2006). Curiously, there was no tau pathology in these animal studies indicating that although β-amyloid might be transferrable between individuals and spread throughout the brain, this alone is not sufficient to produce full blown AD pathology. Of course, the possibility exists that with further aging the pathology might emerge.

A radical idea that is beginning to gain traction is that β-amyloid acts in an antimicrobial capacity. Indeed, β-amyloid is induced by the same pathogens that were linked to AD, including HSV, HIV, spirochetes and chlamydia, and genetic knockout of β-amyloid reduced survival rate against these infections, suggesting its importance in fighting infections. Conversely, infecting the brain with pathogens accelerates β-amyloid deposition in AD transgenic mice and in C. elegans (Kumar, Choi, Washicosky, Eimer, & Tucker, 2016). The β-amyloid fibrils fight off pathogens by increasing their clumping together (agglutination) and entrapment, much in the same way that immune cells engulf or neutralize invading pathogens. It was posited that Aβ oligomers bind to pathogen cell wall carbohydrates and promote agglutination and engulfment of pathogens. In this respect, β-amyloid might act as an innate immune protective antimicrobial peptide, which sharply contrasts with previous notions that β-amyloid was strictly a pathological peptide in the context of AD.

In conjunction with its other actions, β-amyloid properties might also convey AD pathology by acting akin to the cathelicidian and definsin families of antimicrobial peptides. These are found in many immune cells, most notably macrophages and neutrophils, and serve as critical mediators of innate immune defense. These antimicrobials can kill pathogens by, among other things, "punching" holes in microbial membranes causing the leakage of essential ions and nutrients. It is known that antimicrobial peptides, such as LL-37 (which is a member of the cathelicidin peptide family), can exhibit both protective and detrimental actions, depending upon the particular situation (Kahlenberg & Kaplan, 2013). Thus, while LL-37 normally acts as an important innate immune defender against infection, its elevated levels can be toxic. It is possible that β-amyloid might have effects like LL-37, in that both molecules bind microbial cell walls and are structurally very similar (Kumar et al., 2016). If it turns out that AD is related, at least in part, to pathogen infection and that β-amyloid deposition represents a reaction to that primary infection, then the entire β-amyloid cascade theory will need to be revisited. As a consequence, the push toward treatments that target β-amyloid plaques (to be covered in an upcoming section) will likewise need to be reconsidered. That said, even if β-amyloid is protective at certain stages of disease (such as early on by fighting off infections), there are overwhelming data suggesting that it is toxic at later disease stages. Perhaps there is some tipping point at which β-amyloid levels and/or processes become dysregulated so that their antimicrobial and/or other functions become harmful to delicate brain tissues.

B-AMYLOID AND INFECTION LINK

It has been considered, as already mentioned, that b-amyloid might spread through the brain in a manner analogous to how pathogens or prions spread. It seems possible that viral and bacterial agents might enhance such a spread. This is a scary proposition indeed, especially, as β-amyloid fibrils are also associated with HIV and are thought to boost the infectivity of the virus. Remarkably, it seems that amyloid fibrils are present in semen and could

(cont'd)

increase the spread of infection. It has even been posited that amyloid fibrils might normally play a role in fertility, but are exploited in HIV. Moreover, their spread into the brain of HIV infected individuals may give rise to HIV dementia, which is strikingly similar to AD.

INFLAMMATORY MECHANISMS OF ALZHEIMER'S DISEASE

Although inflammation likely plays a role AD, the question remains whether it plays a primary provocative role or is secondarily involved in modulating the course of illness, or perhaps both. In this context, it is significant that increased circulating inflammatory cytokines, such as IL-6, were reported to occur in AD patients about 5 years prior to the onset of disease (Engelhart et al., 2004). The general sickness symptoms, such as malaise, fever, aches, and pains that often begin to occur prior to the onset of dementia might similarly be explained by variations in cytokine and associated inflammatory factors (Holmes, Cunningham, Zotova, Culliford, & Perry, 2011). In a cohort of AD patients that had been followed for over 20 years, those that displayed elevated signs of inflammation (e.g., increased systemic leukocyte counts) at middle age had a significantly lower hippocampal volume (~5% reduction) later in life (Walker et al., 2017). Further reports using genetic and proteomic approaches to profile inflammatory markers, including systemic leukocytes, allowed for an impressive (>90%) ability to predict future cognitive pathology and eventual AD diagnosis (Delvaux et al., 2017). Thus, peripheral inflammatory processes are likely related to disease progression in some manner, and their detection could help foster early treatments.

Accumulating evidence has supported the possibility that peripheral inflammation may come to promote inflammatory processes within the brain, which can give rise to neurodegeneration and cognitive deficits. In animal, repeated systemic LPS injections caused spatial memory deficits, coupled with increased β-amyloid levels in the hippocampus (Kahn et al., 2012). As peripheral poly I:C (a viral mimic) administration essentially had the same effect (Weintraub et al., 2014), indicates that viral and bacterial agents are equally capable of producing AD-like pathology.

Specific cytokines induced by viral or bacterial agents might underlie their impact on the brain in AD. For instance, systemic TNF-α administration produced cognitive dysfunction and when administered in the context of a neurodegenerative model of prion disease, it elicited exaggerated behavioral and inflammatory responses (Hennessy et al., 2017). Similarly, IL-1β provoked cognitive deficits, and at the cellular level could suppress LTP (Lynch, 2015), suggesting that this cytokine could underlie some of its effects on memory. Moreover, IL-1β also altered APP processing and promoted tau phosphorylation (Mrak and Griffin, 2007), which has been linked to senile plaque and fibrillary tangle formation. Conversely, IL-1 receptor blocking antibody was able to inhibit tau pathology and diminish cognitive deficits evident in AD mice (Kitazawa et al., 2011). Ultimately, cytokine changes could influence AD relevant pathology through their interactions with microglia, which may be the actual source of the brain cytokines. In this regard, damaged cells typically release ATP which can be sensed by the microglial purineric receptors (Davalos et al., 2005).

When they encounter damage or a threat, microglia can extend their processes and release soluble factors, as well as physically interact with the threat in an attempt to neutralize it or remediate damage.

Microglial overactivation could be damaging to neurons via inflammatory processes, but as already mentioned their underactivation might be equally destructive (in the context of TREM2 and CD33 mutations) (Leyns et al., 2017; Siddiqui et al., 2017) owing to the failure to properly clear toxic β-amyloid species. Hence, in considering inflammatory microglial involvement in AD, it might be productive to be mindful of the remarkable plasticity of microglia, as well as how specific phenotypic states may be beneficial under certain conditions, but deleterious in others.

With advanced age, microglia tend to show impaired or at least altered functioning (Block, Zecca, & Hong, 2007), and display evidence of cellular senescence, characterized by abnormal morphology and shorter telomere length (Miller & Streit, 2007). Aged microglia also exhibit diminished reactivity, reduced phagocytosis, as well as impaired overall motility and migration to sites of damage (Matt & Johnson, 2016). Such microglia senescence could occur because these cells are highly active, with proliferation and maturation occurring throughout the life cycle, and consequently are prone to DNA damage (Raj et al., 2014). Alternatively, the increased myelin fragmentation that can occur with age was postulated to give rise to microglia dysfunction by interfering with their intracellular lysosomal functioning (Safaiyan et al., 2016). Besides phagocytic deficits, aged microglia displayed sensitized reactions to their usual trigger stimuli, such as ATP (Lai, Dibal, Armitage, Winship, & Todd, 2013). Whatever the case, their diminished state could contribute to AD pathology owing to a lack of protection from microbial insults or an impaired ability to clear toxic aggregates.

In culture, exposing old microglial cells to the culture medium obtained from new microglia, resulted in the older cells proliferating coupled with enhancement of their ability to clear β-amyloid plaques that were added to the culture (Daria et al., 2017). The addition of the trophic cytokine, GM-CSF, also provoked such effects, raising the possibility that diminished microglial functioning that may occur with age can be reversed by trophic cytokines released by younger glial cells.

New in vivo PET imaging methods using the marker, PK11195, have allowed for the identification of an increased number of highly activated microglia in the brains of living AD patients (Parbo et al., 2017). Correspondingly, autopsied brains from either AD patients or animal models of the disease, unequivocally demonstrated signs of robust neuroinflammation (Bachstetter et al., 2015). In the human and animal brain, this invariably involves marked microglial activation (as determined by morphological stains) along with proinflammatory cytokines and/or other inflammatory enzymes (Bachstetter et al., 2015). Of particular note is that most activated microglia tended to be especially dense around the β-amyloid senile plaques. This could reflect that inflammatory microglia were being provoked by the plaques, or conversely, that inflammatory cells fueled the development of the plaques. The combination of plaques and immune cells could represent an aberrant deleterious state or alternatively, they could conceivably represent a protective clean up response. It will be recalled from Chapter 14, Inflammatory Roads to Parkinson's Disease, that a similar position applies to the Lewy bodies observed with PD.

Although inflammatory insults alone might cause AD-like pathology, when it was present within an existing genetic animal model, an especially convincing phenotypic state was apparent. Likewise, systemic LPS treatment markedly augmented the pathological consequences observed in a triple transgenic model of AD (Valero et al., 2014). It seems that superimposing inflammation on a genetic vulnerability more fully promotes the expression of an AD-like syndrome. Puzzlingly, other investigators have found that LPS priming increased the clearance of β-amyloid plaques and improved AD-like behavioral pathology (Qin et al., 2016), indicating that the effects of LPS treatment are not straightforward, likely being influenced by many variables. In effect, inflammation can be a doubled edged sword, being either deleterious or beneficial depending upon the circumstances.

It appears that priming certain innate immune elements might combat AD pathology. Priming phagocytic cells, notably peripheral macrophages and/or brain microglia with specific immune agents can place the cells in a state that has them migrate to and then decorate the surface of β-amyloid senile plaques (Fig. 15.3), thereby promoting their breakdown through phagocytosis and the release of lytic enzymes (Michaud et al., 2013). An interesting set of studies using mice expressing a green fluorescent protein in newly created cells, indicated that LPS administration caused the production of new macrophages (that appeared green) derived from the bone marrow, and then migrated to the brain to attack β-amyloid plaques in affected mice (Michaud, Richard, & Rivest, 2012). Further, it appeared that infiltrating immune cells and local microglia act as beneficial "clean up" cells, but become overwhelmed at later stages of disease owing to excessive plaque buildup and a constant onslaught of inflammatory cytokines (El Ali & Rivest, 2016).

These findings are particularly fascinating in light of the revelation that, in contrast to previous dogma, the resident microglia are of distinct embryonic origin, different from macrophages. Indeed, microglia are thought to arise from the yolk sac earlier in development than the macrophages, which emerge from bone marrow stem cells (Elmore et al., 2014).

After microglial colonization of the brain during development, the majority of newly generated microglia during the lifetime of the organism come from resident brain stem cells, with only a small minority coming from bone marrow (Bruttger et al., 2015). Accordingly, microglia and peripheral macrophages, although sharing many functions, may have different roles in sculpting brain processes. In the case of AD plaques, it could be envisioned that resident microglia have a more direct and precise impact on the microenvironment, likely acting to shape rapid synaptic processes and deal with metabolic by products. In contrast, peripheral macrophages may only be recruited after some threshold of pathology has been reached or when microglia become overwhelmed, and in this case, the invading macrophages might be tasked with the removal of plaques.

While supporting the notion of increased microglial activity in AD, postmortem studies on AD brains generally pointed to a complex interplay between genotype, microglial phenotype, and clinical state. Microglia express APOE, and genetic variants of this gene could influence microglia functioning and the protein is, indeed, increased in microglia within genetic AD models (Holtman et al., 2015). Although it is unclear how APOE acts in microglia, it is suspected to be a contributor to innate immune responses, including control of inflammatory cytokines (TNF-α, IL-6) and production of reactive oxygen species (Vitek, Brown, & Colton, 2009). Similarly, as described earlier, the TREM2 gene, which has been implicated in AD is an important regulator of microglial functioning. TREM2 normally diminishes microglial release of inflammatory cytokines, but enhances their phagocytosis of pathogens or damaged cells (Kleinberger et al., 2014). In fact, the R47H TREM2 mutation that increases AD risk, also happens to reduce TREM2's phagocytic ability (Wang, Lopez et al., 2015), which is consistent with the evidence for a role

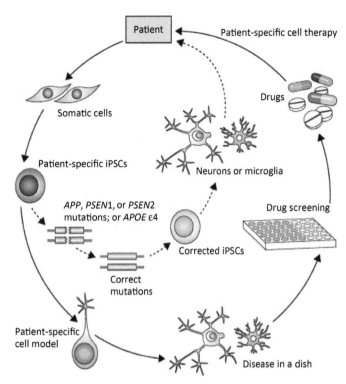

FIGURE 15.3 Emerging genetic technology in the personalized treatment of Alzheimer's disease. Stem cell reprograming techniques allows for Alzheimer patient somatic (e.g., skin) cells to be reprogramed into neuronal cells, which can then be used to access mechanisms of degeneration and potentially to determine the efficacy of new drugs. This approach also allows for genetic engineering to correct any mutations (e.g., APP, PSEN, or APOE ε4) and then transplant cells back into the same patient. *Source: From Hunsberger et al. (2016).*

for microglia in cleaning up β-amyloid aggregates and plaques.

Somewhat troubling are the findings concerning the *in vitro* and *in vivo* effects of TREM that appear to be at odds with one another. For instance, TREM2 knockout reduced levels of inflammatory genes in a genetic AD mouse model (Wang, Lopez et al., 2015). Conversely, endotoxin stimulated primary microglia that lacked TREM2 displayed increased proinflammatory cytokine levels (Neumann & Takahashi, 2007). These discordant findings might stem from differences in the animal models used. While TREM2 knockout diminished β-amyloid plaques in APPPS1-21 mice (Jay et al., 2015), the opposite was seen in 5xFAD mice (Wang, Lopez et al., 2015). Once again, these findings point to the complexity of the microenvironment and degree of ongoing pathology provoked by the particular model,

which can have marked consequences on the role of TREM2 in microglia.

To complicate matters further, the correlation between microglial and clinical state also varies with the particular microglial marker being assessed. This is not surprising considering that these markers subserve differing microglial functions. For instance, while CD68 is more aligned with phagocytic functions, HLA-DR is typically associated with antigen presentation. Further, the commonly used marker, IBa1, is thought to be required for actin bundling and hence, microglial motility (Ohsawa, Imai, Sasaki, & Kohsaka, 2004). In line with this, brain samples from a large United Kingdom-based aging study showed that *reductions* of Iba1, but *increases* of CD68 and HLA-DR, were related to the presence of dementia prior to death (Minett et al., 2016). The presence of APOE alleles further impact the microglial phenotype, with the e4 risk factor allele being associated with increased CD68 and HLA-DR levels together with dementia, whereas the e2 allele was tied to high levels of Iba1 and protection against dementia (Kim, Seo et al., 2017). It was posited that the Iba1 reduction reflects a failure of microglia to respond and mobilize to fight off potentially harmful plaques, whereas the CD68/HLA-DR elevations might reflect overzealous responding to plaques that could damage adjacent neurons (Zotova et al., 2013).

There has been substantial evidence going back several decades, showing the presence of complement proteins in AD senile plaques (Aiyaz, Lupton, Proitsi, Powell, & Lovestone, 2012). The complement system, it will be recalled, comprises a series of more than 30 proteins that sequentially activate one another (usually following antibody binding), causing the lysis or phagocytosis of some microbial or other foreign invader. Direct *in vitro* evidence revealed that β-amyloid aggregates can directly trigger the complement cascade inducing the production of numerous anaphylatoxins including, C3a and C3b, along with C5, and the ultimate production of the attack complex. Furthermore, *in vivo* studies indicated that C1q was responsible for the toxic effects of soluble β-amyloid oligomers (Hong et al., 2016), and C1q mediated amyloid-dependent reductions in hippocampal LTP. Similarly, the C3 complement protein was increased in the AD brain and was reported to contribute to cognitive deficits (Shi et al., 2017). Consistent with these findings, C3 deficiency was neuroprotective and reversed functional impairments, although curiously, resulted in elevated cerebral plaque load (Shi et al., 2017). This again fits with the "double edged sword" view regarding immune processes. In this case, C3 may be an important element in the clearance of β-amyloid plaques; yet, it clearly has other inflammatory actions that are destructive in the AD brain. It was suggested that the actual plaque load is less important than the microglial reactions to the plaques, and that this microglial reaction mediates a critical loss of hippocampal synapses (Shi et al., 2017).

The fact that elevated levels of the cleaved form of C1 occurred in the AD brain in conjunction with increased levels of its endogenous inhibitor, might represent a natural defense mechanism being engaged in response to pathology. As well, since BBB pericyte cells (that are embedded in the BBB in proximity to the tight junctions) and cerebral blood vessels can also secrete C1q upon interaction with β-amyloid, these cells might also contribute to the mix of toxic factors in AD.

The mechanisms through which β-amyloid plaques interact with complement is not entirely known, but may initially involve interactions between C1q and the first few residues of the β-amyloid molecule (Velazquez, Cribbs, Poulos, & Tenner, 1997) and C3 might similarly bind to the aggregated β-amyloid fragments. Given these interactions, efforts were made to assess the impact of inhibiting complement activation in AD. *In vitro* C1r and C1s protease inhibitors have been considered in this respect, as have small molecules aimed at interfering with the β-amyloid binding site on C1.

In addition to microglia, there is evidence that astrocytes might contribute to AD through their impact on inflammatory processes, particularly IL-1β. Indeed, triple transgenic AD mice were protected from β-amyloid pathology and cognitive deficits by using an IL-1 receptor neutralizing antibody, which was attributed to astrocytic and not microglial functioning (Lim et al., 2015). As well, astrocytes secrete C3, which then binds to the C3a receptor on microglia, resulting in augmented inflammatory drive and β-amyloid pathology in APP transgenic mice (Lian et al., 2016). Hence astrocytes might act in parallel with microglia to modulate complement signaling in the context of β-amyloid pathology.

IMMUNOMODULATORY TREATMENTS FOR AD

Anti-inflammatory Treatments

The first indication that anti-inflammatories might be useful for AD treatment, as described earlier, came from the clinical observation that rheumatoid arthritis patients prescribed NSAIDs show very low rates of developing AD (Miguel-Álvarez et al., 2015). Animal studies further indicated that ibuprofen decreased the activation of the inflammatory kynurenine pathway, and this effect was associated with cognitive improvement in APPSwe-PS1 mice. Moreover, administration of prostaglandin PGI2, which acts as a powerful inhibitor of platelet aggregation and vasodilator, inhibited the production of proinflammatory cytokines, in an APP/PS1 transgenic AD mouse (Wang, Zhao et al., 2016). Likewise, the β-amyloid 1-42 oligomer-induced IFNγ production, which in turn, facilitated β-amyloid plaque accumulation (Wang, Zhao et al., 2016). Administration of the COX-1 inhibitor, SC-560, to transgenic AD (3 × Tg-AD) mice diminished β-amyloid plaques and tau hyperphosphorylation, along with improving cognition (Choi et al., 2013). The drug also reduced microglial activation, but facilitated their phagocytic potential. Similarly, the COX-2 inhibitor, NS-398, improved cognition and reduced microglial activation and β-amyloid pathology. This same COX-2 drug also rescued TNF-α production in brain slices obtained from TgAPPsw mice (Quadros et al., 2003).

Despite the promising actions on brain inflammatory processes, relatively recent studies assessing the clinical efficacy of general anti-inflammatory drugs have failed to show any benefit in AD patients or in elderly individuals with a family history of the disease. For instance, treatment with naproxen or celecoxib for 1−3 years had no significant influence on cognitive functioning in elderly individuals with a family history of AD (FADAPT-FS Research Group, 2015) or in recently diagnosed AD patients. Two further research trials confirmed the absence of clinical benefit for cognition after 1 year of daily treatment with the NSAIDs, naproxen, indomethacin, or rofecoxib, in mild—moderate AD patients (Aisen et al., 2003; de Jong et al., 2008). Similarly, in a large multicenter double blind placebo trial the NSAID derivative, Tarenflurbil, also had no beneficial effect on measures of cognition or daily living in mild AD patients (Green et al., 2009). The lack of clinical efficacy could be related to the fact that the individuals already had AD, whereas NSAIDs might only be useful in a prophylactic capacity (Policicchio, Ahmad, Powell, & Proitsi, 2017).

Several clinical studies addressed potential beneficial effects of certain nutritional agents, most notably curcumin and resveratrol. Given its proven anti-inflammatory and antioxidant effects, along with neuroprotective consequences in AD animal models, recent clinical trials focused on the potential beneficial effects of curcumin. A 24-week clinical trial with curcumin revealed no positive effects in a cohort of mild-moderate AD patients (Ringman et al., 2012). However, resveratrol, which has potent anti-inflammatory and antioxidant actions, reduced inflammatory markers in the CSF of AD patients and modestly attenuated cognitive decline, but this only occurred in a subset of subjects (Sawada et al., 2012).

Cytokines and Trophic Factors

Whatever strategy is adopted, it must promote the distribution of "protective" or "recovery" agents broadly across the brain. This necessarily requires a clinical agent that can access large portions of the brain and probably modulate several different pathways relevant for pathology. In the case of AD, this would likely mean targeting inflammatory and other pro-death processes, in conjunction with those aligned with the accumulation of faulty amyloid aggregates. We will briefly cover two relevant cytokines in this section, GM-CSF and IL-1β, with the rationale being that they reflect two of the most relevant cytokine categories with regards to AD pathology.

The unfortunate lack of clinical efficacy of NSAIDs in AD has resulted in further analyses concerning why there is such a low prevalence of AD in patients with rheumatoid arthritis. One possible explanation that has been offered is that elevated levels of endogenous trophic factors in these patients might buffer them against AD. In addition to elevated levels of several proinflammatory cytokines in their synovial fluid, arthritic patients also have increased concentrations of the trophic cytokine, GM-CSF. While GM-CSF is believed to contribute to arthritic damage (Avci, Feist, & Burmester, 2016), it may confer protection against neuronal and synaptic loss.

It will be recalled from the section on inflammatory mechanisms that several studies suggested that GM-CSF might have neuroprotective effects in AD. Systemic GM-CSF reduced β-amyloid deposition and cognitive deficits (Sanchez-Ramos, Song, Sava, Catlow, & Lin, 2009), and it was suggested that GM-CSF can improve cognition in cancer patients at risk of cognitive dysfunction owing to their chemotherapeutic regimens (Jim, Boyd, Booth-Jones, Pidala, & Potter, 2012). Of course, caution should be exercised given that GM-CSF can increase BBB permeability (by reducing claudin-5 and ZO-1 BBB tight junction proteins) and facilitate peripheral monocyte infiltration (Shang et al., 2016). Finally, inhibiting GM-CSF through specific antibodies also reduced β-amyloid deposition (Volmar, Ait-Ghezala, Frieling, Paris, & Mullan, 2008), suggesting that GM-CSF can contribute to plaque formation. Using a trophic cytokine, such as GM-CSF, is currently a contentious issue and further data will be required before this can be resolved. It might be that GM-CSF is effective in only a subgroup of patients, but might be deleterious in others. Thus, a precision medicine approach will be required for clinical trials.

Targeting the proinflammatory cytokine IL-1β to modulate inflammation in AD has received some interest, particularly given that IL-1β polymorphisms were associated with AD (Payão et al., 2012). Moreover, several studies reported elevated IL-1β or its intermediate signaling factors within the brain and CSF of AD patients. Importantly, pathologically elevated levels of IL-1β levels have been associated with plaque evolution and the rate of cognitive decline among clinically diagnosed MCI patients. Similarly, increased brain levels of IL-1β activity are thought to occur early in the AD brain and might represent a novel biomarker of early detection of AD (Dursun et al., 2015).

In animal studies, several transgenic AD models demonstrated the age-dependent emergence of increased IL-1β expression that correlated with β-amyloid plaque deposition and spatial memory deficits (Stampanoni Bassi et al., 2017). Mechanistic studies provided evidence that IL-1β may fuel the development of senile plaques and neurofibrillary tangles, and conversely, that the endogenous antagonist, IL-1Ra, may limit such processes (Craft, Watterson, Hirsch, & Van Eldik, 2005). In fact, IL-1β promoted faulty APP cleavage, resulting in the accumulation of pathological β-amyloid deposits, whereas IL-1β inhibition using selective antibodies prevented tau pathology and improved cognition in an AD animal model (Kitazawa et al., 2011).

There have been efforts to produce compounds that target IL-1β (Mitroulis, Skendros, & Ritis, 2010), including rilonacept, a long-acting IL-1β receptor fusion protein, and canakinumab, a fully humanized anti-IL-1β monoclonal antibody have been marketed. As well, the IL-1Ra, anakinra, which has been used to treat rheumatoid arthritis, was also reported to reduce the extent of brain damage in stroke patients. Clinical use in AD might also come from IL-1Ra since it acts within the hippocampus to reverse β-amyloid-induced disruption of LTP and cognitive dysfunction (Schmid, Lynch, & Herron, 2009), and following hippocampal infusion of IL-Ra in an animal model, the development of AD-like pathology was moderately diminished. It was particularly interesting that the transplantation of neural precursor cells that genetically overexpressed IL-1Ra produced even more impressive results, reversing the cognitive deficits in a Tg2576 genetic AD mouse model (Ben-Menachem-Zidon, Ben Menachem-Zidon, Ben-Menahem, Ben-Hur, & Yirmiya, 2014). It could be that positive effects occurred with the transplanted cells themselves and the IL-1Ra further augmented such outcomes.

Curiously, chronic overexpression of IL-1β in transgenic mice resulted in enhanced plaque clearance by microglia (Ghosh et al., 2013), once again indicating both beneficial as well as deleterious actions of IL-1β. Beyond the toxic oxidative factors they release, microglia can also help in the scavenging of senile plaques and associated pathological tissue, and might thus be protective. In this regard, the chemokine, fractalkine (CXCL1), signals through its CX3CR1 receptor on microglia to inhibit phagocytosis, and hence could be another important mechanism involved in plaque clearance. In line with this suggestion, CX3CR1 knockout exacerbated pathology in transgenic hAPP-J20 AD mice (Cho et al., 2011), and viral-induced fractalkine overexpression reduced tau pathology and prevented neurodegeneration in a Tg4510 mice that normally show tau pathology (Merino, Muñetón-Gómez, Alvárez, & Toledano-Díaz, 2016). Thus, CX3CR1 might be a viable new target for treating AD. However, as discussed earlier, cytokines often have complex biphasic actions in regulating CNS processes, and hence caution needs to be exercised in considering cytokine

manipulations. Certainly, it is not our intent to recommend throwing the baby out with the bathwater or promote remedies that can worsen the disease or cause unexpected comorbidity. Altering cytokine levels not only influences neuronal survival, but could also impact the growth rates of many different cell types, including those that could potentially become cancerous.

Vaccine Therapy

From an immune perspective, perhaps the most intriguing potential therapeutic option is that of vaccination against AD pathology. Given the substantial evidence indicating a deleterious role for β-amyloid in AD, considerable work focused on using either active or passive immunization strategies to fight the disease. In contrast to conventional drug treatments, immunotherapy utilizes the individuals own immune system to produce lasting elevated concentrations of polyclonal antibodies that can attack the β-amyloid plaques. To this end, heavy investments have been in developing vaccines against β-amyloid fragments that result in the clearance of amyloid plaques. Indeed, both active (direct injection of the amyloid or hyper-phosphorylated tau antigen) or passive (administration of specific amyloid or tau primed antibodies) immunization strategies have been used in Alzheimer animal models and with limited success in human clinical trials. In both cases, the β-amyloid or tau specific antibodies were shown to infiltrate the brain and bind to monomers or oligomers of the respective antigens, as well as to decorate the surface of plaques and/or neurofibrillary tangles (NFTs). Microglia and macrophages can then interact more efficiently with antibody coated aggregates to promote their clearance.

It was first reported in 1999 that inoculating transgenic AD mice with β-amyloid resulted in a dramatic clearance of plaques, coupled with improved cognitive functioning

(Schenk, Barbour, Dunn, Gordon, & Grajeda, 1999). Immediately, thereafter, animal studies revealed that vaccine-provoked plaque clearance was associated with improved cognitive functioning. Of particular interest from an inflammatory perspective, was that the vaccination also modulated microglia, which were posited to be critical for the plaque clearance and subsequent cognitive improvement (Wilcock et al., 2004).

These findings prompted clinical trials, the first of which used active immunization against the full toxic β-amyloid 42 peptide fragment (AN1792). Unfortunately, significant cognitive changes were not observed (Bayer et al., 2005). Follow up analyses revealed a significant reduction of β-amyloid plaques in the treated AD patients, but once again, this was not accompanied by clinical benefits. Even patients with virtually complete plaque removal still displayed severe dementia with no evidence of improved survival (Holmes, Boche, Wilkinson, Yadegarfar, & Hopkins, 2008), thus casting doubt on whether existing plaques contribute to cognitive decline. Finally, the immunized patients show a down-regulation of microglial activity, with variations in the different subpopulations of microglia; CD68, CD64, and CD32 positive microglia were reduced, while Iba-1 positive cells were unaffected.

Of considerable concern was that serious side effects were noted. Meningeal encephalitis and vascular hemorrhages were noted in several inoculated patients (Boche et al., 2008), and several patients died during the trials. As well, a substantial number of proinflammatory Th1 lymphocytes had infiltrated the AD brains following vaccination, which may have been responsible for the harmful inflammatory side effects. Hence, more recent efforts sought to develop "cleaner" or more specific vaccines that lacked these side effects.

To this end, improvements of vaccines have aimed to increase plaque directed antibodies, while diminishing general off-target T lymphocyte-mediated inflammation. This entailed synthesizing highly specific antigens that represent very confined immunogenic portions of the β-amyloid molecule. The neuroinflammation caused by clinical vaccination using synthetic β-amyloid (AN1792) was thought to result from recruited proinflammatory Th1 lymphocytes. A more truncated form of β-amyloid, together with an adenovirus vector encoding GM-CSF, provoked an anti-inflammatory Th2 response, along with antibodies that coated the amyloid plaques (Kim et al., 2005). Other vaccines included vanutide cridificar (ACC-001), together with a specific (QS-21) adjuvant, which was assessed for safety and efficacy in Phase 2 clinical trials. While ACC-001 was found to be safe and well tolerated, it did not affect any cognitive or volumetric brain measures.

It is likely that microglia are a major part of the process through which vaccination could be clearing β-amyloid plaques. The presence of β-amyloid is apparent in microglial cells from immunized mice and in the microglia from patients immunized with AN1792 (Nicoll et al., 2003). It was posited that the Fc receptors found on the surface of microglial cells trigger the phagocytosis of β-amyloid. All microglia express multiple classes of Fc receptors, and the nature of the reaction provoked depends upon the ratio of Fc receptors activated, as well as the therapeutic antibody isotype (Vidarsson, Dekkers, & Rispens, 2014). Specifically, IgG types found in vaccinations decorate the β-amyloid plaques, which are then recognized by the Fc microglial receptor, resulting in and degradation of the β-amyloid species. Finally, the finding that Fc receptor expression normally increased with advanced age (Okun, Mattson, & Arumugam, 2010) is consistent with an enhanced inflammatory milieu being manifest with the repeated challenges associated with aging.

In addition to phagocytosis, activating Fc microglial receptors also provokes proinflammatory cytokine release, as well as activation of the complement cascade. The proinflammatory cytokines could help in plaque breakdown through the production of lytic enzymes or other chemical inflammatory enzymes. The role of complement in AD pathology is not entirely understood, as described earlier, but has nonetheless been suggested to be a useful therapeutic target (Shen, Yang, & Li, 2013). Indeed, complement activation has been reported in the AD brain and in associated with β-amyloid in animal models of the disease. Interestingly, the same antibody isotypes that bind to Fc, also showed a high affinity for C1q, raising the tantalizing possibility of interactions between these systems. As described in earlier chapters, C1q triggers the activation of proteins that eventually engage the membrane attack complex, rupturing the membrane and degrading cellular integrity. Thus, complement activation observed in the AD brain could reflect a failed attempt to clean up damage, and thus bolstering this system could aid this process.

Interesting evidence has pointed to a role for complement in the so called astrocyte-microglia-neuron tripartite synapse in AD. It was posited that β-amyloid can activate NF-κB within astrocytes, resulting in C3 release, which can then activate microglial phagocytosis, while also modulating neuronal dendritic morphology (Lian et al., 2016). Astrocytes express C5a and C3a, and microglia similarly express complement receptors CR1, CR3, and CR4 that can act as phagocytic receptors for complement-attached complexes. There has even been the suggestion that neurons, under certain conditions, can likewise express complement receptors.

There are many reasons that clinical immunotherapy trials have failed, including problems with side effects, low immunogenicity, and poor pharmacokinetics of the antibody. A novel vaccine has been developed that takes into account these past problems. Specifically, an active particle-based vaccine that targets the β-amyloid epitope, PP-3copy-Aβ1-6-loop123, produced very high antibody titers that were specific for this PP region (Fu et al., 2017). The vaccine cleared β-amyloid plaques and enhanced cognitive performance in the Morris water maze test of spatial learning and memory, and importantly, no toxicity was observed and the vaccine had no effect on proinflammatory T cells. These data are particularly important in light of the fact that previous vaccines tended to utilize a small B-cell β-amyloid epitope that had low immunogenicity. It was also of note that immunization earlier in the course of AD-like pathology had more profound effects than when treatment commenced at a time when pathology was severe. Hence, immunotherapy may be more effective early in disease or possibly during a prodromal stage, or when patients might be suffering from MCI.

The mechanisms through which vaccination might clear β-amyloid were proposed to be due to either a direct effect of antibodies reaching plaques within the brain, or alternatively, through the so called "peripheral sink" pathway (Fu et al., 2017). In the first instance, antibodies bound to their β-amyloid target interact with the Fc receptors found on microglia or infiltrating phagocytes, which would elicit the phagocytosis of the antigen-antibody complex. According to the peripheral sink route, the anti-amyloid antibodies generated in the periphery that interact with blood or lymph β-amyloid would result in a concentration gradient effect, favoring the movement of brain β-amyloid out into the periphery. This could then lead to the easier clearance of plaques and diminished burden within the brain.

It was long held that β-amyloid senile plaques are primary players in the genesis of AD pathology, but as indicated earlier, many researchers do not accept this view. Indeed, some studies reported poor correlations between the extent of plaques and degree of cognitive disability in AD. Further, the soluble β-amyloid oligomers are more closely aligned with AD symptomatology. As senile plaques might reflect a compensatory response, recent views have included the notion that disequilibrium between β-amyloid fibrils-oligomers-monomers is the key to AD pathology. It should also be considered that β-amyloid normally has a role in synaptic functioning (Musardo & Marcello, 2017) and hence, indiscriminately blocking its function could have unwanted effects.

As a final point on AD immunotherapy, efforts recently involved targeting the pathological tau proteins found in neurofibrillary tangles, rather than the β-amyloid plaques. In the first such Phase 1 clinical trial, the vaccine was safely tolerated by most individuals with only 2 of 30 withdrawing because of serious side effects (Novak et al., 2017). Importantly, 29 of the 30 patients developed significant antibody titers over the 12-week trial, suggesting that further Phases II and III clinical trials assessing efficacy were warranted.

COMBINED THERAPY APPROACH FOR AD

As in the case for many brain conditions, a combination approach that is tailored for each patient will likely be the future of treatment methods. In the case of AD, this might require vaccines against both tau-tangles and amyloid-plaques and possibly anti-inflammatory agents that specifically modulate the microglial phenotype. Of course, it will also be necessary to identify and treat the disease at a relatively early stage, when the brain still has sufficient neuroplasticity, making it capable of rebounding. This will require well-validated biomarkers, probably comprising a panel that provides an overall disease signature. The final piece of the puzzle might be the identification of specific combinations of environmental insults that certain people need to avoid, based on the specific polymorphisms they possess. This said, in many individuals there might be certain genetic alterations that give rise to disease regardless of exposure history. In such cases, future gene editing strategies might be required. This almost "science fiction" approach could be feasible at some point in the future, especially given the marked advances in gene technology, such as CRISPR-Cas9, that allow easy deletion or modification of certain genes (especially given the recent advances in diminishing off-target effects). Similarly, technologies are available that enable the over or under expression of certain genes in both neurons and glial cells in rodents. It may even be possible to harvest the patient's own somatic cells and reprogram them to healthy neurons that can be transplanted back into the patient's own brain (Fig. 15.3). Thus, one could conceivably have their microglia or neurons express a factor for which they might be deficient, owing to a specific polymorphism. At this point, it's hard to imagine such strategies will not eventually be attempted in humans.

CONCLUDING COMMENTS

We have attempted to cover basic well-established evidence, as well as emerging novel findings, regarding the factors that contribute to the onset and evolution of AD. Animal models, with their many strengths and weakness, have been important in our understanding of how particular processes map on to the wide range of neurobiological disturbances evident in AD patients. At the same time, human studies help pinpoint potential environmental triggers of disease and possible useful biomarkers, but in the main, these studies are only correlational.

At this time, pretty much all clinical trials for AD have been disappointing. There are many variables that could account for such discouraging results, including the use of poor animal models of AD, the failure to target useful mechanisms, the heterogeneity of the disease and problems with drug bioactivity/bioavailability/distribution.

The animal models have primarily involved overexpression of some combination of genes implicated in AD. This has resulted in rodents showing cognitive deficits and the accumulation of β-amyloid deposits, and these effects appeared to be age-dependent, usually emerging after 6 months. There have also been models in which specific toxicants were injected into the hippocampus, causing cognitive deficits, possibly reflecting those that occur early in AD when the parahippocampal gyrus is first compromised. While these models have recapitulated some aspects of the disease, they have failed to reflect other aspects, such as the disintegration of higher level cortical processes that are unique to humans. In fact, a critical element that profoundly differs between the human and rodent models is the extent of cortical involvement. The rodent cortex is profoundly underdeveloped (being lissencephalic) compared to the highly convoluted, richly developed human cortex. Consequently, a substantial proportion of the higher cognitive functions ascribed to the cortex, and which are disturbed in AD, are absent in the rodent. Furthermore, most of the neuronal loss in the rodent is subcortical, compared to the greater cortical involvement in humans. The simple fact is that mice aren't humans, and as in other illnesses, positive effects in rodents frequently do not translate into meaningful effects in humans. It has been said that "if mice were humans, cancer would have been cured thousands of times." The same most certainly holds for AD.

Determining which mechanisms to focus on in AD research is no easy task and judging by the complexity of processes covered in this chapter, it is likely that multiple processes, which vary over the course of disease progression, should be targeted. Determining which and when is exceedingly difficult and will surely draw upon emerging technological advances made in fields of genetics, neuroimmune pharmacology, proteomics, and areas related to microbiome functioning. It would be short sighted to simply focus on one aspect of the disease and neglect the complexity of glial-neuron and immune-neuron interactions. It should be clear that microglial and immune cells may be involved in processes related to AD and targeting these may yield fruitful therapeutic options. Yet, as discussed earlier, this treatment route is complicated by the fact that there are beneficial and deleterious elements to the inflammatory immune response in AD. As we've seen, microglial cells can release toxic factors that are damaging to neurons, but at the same time, these cells also perform valuable phagocytic responses that may help rid the brain of β-amyloid plaques. Similarly, invading peripheral immune cells (and local microglia and astrocytes) can release trophic factors that foster recovery, while at the same time engaging in prodeath intracellular signaling pathways. In essence, it may be necessary to target discrete subpopulations of cells, considering that their importance to AD may change over the course of disease.

The heterogeneity of AD is yet another issue that needs to be addressed before large-scale clinical translation can be successful. The complexity of the genetics and environmental stresses involved in the provocation of the disease surely give rise to the differing symptom profiles and progression of the disease. Thus, finding biomarkers that allow the streamlining of specialized treatments for differing patient subgroups could be essential. These biomarkers, however, might be most useful if they could also indicate when treatments ought to be adjusted or changed at different stages of disease progression. Of course, determining what mechanisms to target to prevent the disease from taking hold in the first place will be the greatest challenge.

A final basic consideration is that we need very specific acting pharmacological or genetic tools that allow manipulation of the various neural and glial circuits involved in AD. This will be driven by technological advances that allow us to delve deeper and more specifically into the various brain circuits that are disturbed in AD. Undoubtedly, this will fuel clinical trials with safer and more precise modes of action that can be tailored to suit the particular patient profile. Finding the means to harness the reparative potential of immune cells and their soluble factors is a potentially important route for future therapies. That said, considerable basic research is still needed to better understand not only the mechanisms of disease, but also yet uncovered mechanisms of recovery and how basic immune factors interface with both the healthy and diseased CNS.

Comorbidities in Relation to Inflammatory Processes

ILLNESS COMORBIDITY AND WHAT THIS IMPLIES

While cross-country skiing at the edge of a very old graveyard (don't ask why I chose that route), I came across a burial stone that was overly descriptive. This was in about 1971 or 72, obviously before cameras were built into cell phones (or even before cell phones, for that matter), so I'll describe the stone's inscription taking a few liberties given my inexact memory. But the main message will be clear enough.

John Smith
1857–1896
A hard life. Orphaned. Scrabbled for a living.
Lost the kids and wife to the influenza.
Pneumonia, came and went and came again.
Then tuberculosis, and then his heart gave out.
Died of sadness.

What first struck me was how unlucky this guy was. Then again, this was before antibiotics had been developed, and vaccines wouldn't be mass produced for many years. Life span at that time was, on average, less than 40 years, although once a person reached 30, having passed over many threats, they might count on living about another 34 years. So, depending on how it's calculated, this guy lived to an average age, or he died relatively young. Being an orphan in the mid-1800s, he likely had a tough early life, and he subsequently encountered repeated nasty life experiences. Given the immune, microbial, and other biological processes that might have been disrupted, he was rendered vulnerable to physical and psychological diseases.

Common sayings sometimes seem to have little basis in reality, but at other times there may be something to them. The expression "bad luck comes in threes" is probably inaccurate with respect to the frequency of events, but likely is telling in relation to negative events possibly leading to still other distressing experiences (stress generation). In relation to illness, it is common for some conditions to be comorbid with one another. Depression often predicts the later occurrence of cardiovascular disease and stroke, and depression and heart disease have also been linked to Alzheimer's disease.

(cont'd)

Moreover, the greater the number of cardiac risk factors present, the greater the presence of brain myeloid deposition, and those affected by one type of cancer may subsequently be victimized by a second type of cancer.

Mood disorders have been associated with a constellation of other chronic illnesses. More than 50% of depressed patients reported chronic pain, 33% experienced respiratory illnesses, and diabetes and arthritis occurred in almost 10% of depressed individuals, and cardiovascular illnesses were elevated by as much as 50%.

Likewise, individuals with schizophrenia experience increased occurrence of OCD, panic disorder, depression, and substance abuse, as well as viral illness (e.g., hepatitis). In the case of multiple sclerosis (MS), patients frequently suffer hypertension, type 2 diabetes, epilepsy, inflammatory bowel disease (IBD), chronic lung disease, fibromyalgia, and especially high rates of depression and anxiety. Given that each of these conditions may have further adverse ramifications, it's fair to say that "bad luck may come in multiples of threes."

SOMETHING ABOUT COMORBIDITY

As we've seen, many illnesses don't come about owing to a single factor, but instead evolve as a result of many genes, various biological processes, sex, early negative experiences or challenges encountered during later periods, cultural influences, as well as a constellation of other psychosocial and environmental factors. Some of these contributing agents are independent of one another, but may come together in pathology-inducing combinations. One illness or negative experience may serve as a "first hit" that primes biological systems so that a second hit has a greater negative impact.

Although comorbid illnesses might stem from common underlying features, such as particular genes that are linked to hormones or inflammatory factors, they may also emerge as a result of many life-style influences that are common among diseases. Depression and schizophrenia may be accompanied by poor diet, smoking, sedentary life-styles, and failure to receive medical attention for these illnesses. As a result, still other illnesses may emerge

that create problems for treatment, and make it difficult to evaluate the neurobiological processes that are responsible for the primary illness. Epidemiological studies, especially those that involve a longitudinal component, may be necessary to uncover the processes that lead to illness and illness comorbidities. As described in Fig. 16.1, there are several general routes by which comorbid illnesses may develop, which have implications for illness prediction and treatment.

INFLAMMATORY FACTORS IN RELATION TO COMORBID ILLNESSES

The Case of Autoimmune Disorders

Stressors, depression, and anxiety have been linked to numerous immune-mediated inflammatory diseases. These include illnesses stemming from immune functioning being compromised, or those in which the immune system turns on the self, causing some form of autoimmune disorder. For instance, depression was evident in about half the patients diagnosed with an

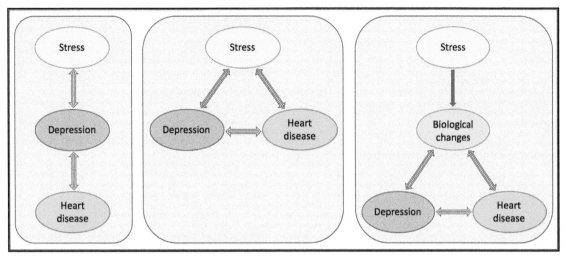

FIGURE 16.1   Comorbidity doesn't necessarily imply that one illness caused another, although it certainly might. By example an environmental event, such as a stressor, could give rise to depression, which then promotes heart disease. The heart disease may in turn, exacerbate depression and distress, so that the disease cycle is perpetuated (left panel). It is equally likely that the two illnesses might be provoked by common environmental triggers (e.g., stressors), and then act upon one another (middle panel). Finally, in a third scenario, a particular environmental trigger may instigate biological processes, say activation of inflammatory or microbiotic factors, so that other neurobiological disturbances evolved (e.g., hormones, neurotransmitters, and growth factors) that could have favored the development of the two illnesses, but these conditions are independent of one another. The two illnesses could affect one another. But, even if they don't, the biological mediating factors could serve as a predictor (biomarker) of subsequent illness. In some cases they might be useful in predicting specific illnesses emerging, but in other instances they might reflect general risk factors, and still other markers would need to be identified that predict specific disease outcomes. It is of particular significance that if one illness appears, it may itself be a marker that the individual is at increased risk for other types of illnesses, perhaps signaling that preventive measures ought to be taken to diminish further illnesses developing. *Source: From Anisman (2016).*

autoimmune disorder, possibly being tied to immune-related alterations of clock genes that affect neuronal timing mechanisms (Pryce & Fontana, 2016). Stressful events and depression may promote flares among patients with MS, and may interfere with coping abilities. Individuals with rheumatoid arthritis were likewise more likely to experience depression and attempt suicide than healthy individuals. In fact, the odds of a suicide attempt was three-times greater among individuals with arthritis who had also experienced early life stressors (parental domestic violence or sexual abuse) compared to those that only experienced arthritis, pointing to the interactive effects of inflammatory disorders and other stressor experiences (Fuller-Thomson, Ramzan, & Baird, 2016).

Crohn's disease was similarly linked to depressive symptoms, poor sleep quality, and impaired cognitive functioning (van Langenberg, Yelland, Robinson, & Gibson, 2016), and inflammatory factors were prominent in patients with comorbid depression. As expected, the level of depression in patients with Chrohn's disease was reduced by anti-TNF-α treatment (Guloksuz et al., 2013), supporting the position that the link between the two illnesses was mediated by inflammatory factors. Depression and asthma symptoms also appear in tandem, accompanied by greater IL-1β, IL-6, and TNF-α levels, as well as that of the anti-inflammatory cytokine, IL-4 (Jiang M et al., 2014). Of particular importance was that in some instances, beneficial effects were achieved through treatment with an IL-1β antagonist (Mendiola & Cardona, 2017).

For those experiencing another immune-related disorder, psoriasis, it was similarly not uncommon for anxiety and depression to be present, typically being attributed to the distress created by individuals experiencing the damaging skin condition. To an extent this is likely accurate, but it also seems that the anxiety and depression aggravate this autoimmune disorder so that the skin condition worsens. Aside from the distress created by psoriasis, some psychological symptoms that appear may be related to activation of inflammatory processes (Connor, Liu, & Fiedorowicz, 2015), as might other immune-related illnesses. In fact, psoriasis was accompanied by a substantial increase in the subsequent appearance of type 2 diabetes, which was also linked to inflammatory processes (Wan et al., 2017).

INFLAMMATORY BOWEL DISEASE (IBD): THE PROTOTYPICAL BRAIN-GUT AXIS DISORDER

IBD (Crohn's disease and ulcerative colitis being the two main forms of this condition) appears to be an autoimmune condition, but also involves other immune-related factors. The tie between the microbiome and IBD has become increasingly more apparent. Signals generated within the brain may come to influence microbial communities within the gut, which in turn, can affect brain structure and functioning (Labus et al., 2017). In a set of individuals, psychological disturbances, such as anxiety and depression, precede the gut disturbances, but in others the gut symptoms appear first. Aside from pointing to the interrelations that exist between gut disturbances and psychological factors, the order of illness appearance may be significant in treatment decisions.

The appearance of IBD was accompanied by several brain changes, including structural variations and changes of HPA activity, as well as psychological disturbances (Kennedy, Cryan, Dinan, & Clarke, 2014). The induction of intestinal inflammation to model IBD in mice, led to arrested hippocampal neurogenesis, and may thereby influence behavioral outcomes, such as depression (Zonis et al., 2015). In fact, the link between IBD and depression has been attributed to inflammatory processes, oxidative and nitrosative stress factors, as well as tryptophan catabolites (Martin-Subero, Anderson, Kanchanatawan, Berk, & Maes, 2016).

In view of the gut disturbances characteristic of IBD, it was not altogether unexpected to find microbiota alterations comprising diminished bacterial diversity, diminished stability of the microbiota, as well as their metabolic activity (Jeffery et al., 2012), which can have long lasting repercussions. In rodents, perturbing the gut microbiota in early life through antibiotic treatment can induce visceral hypersensitivity during adulthood (O'Mahony et al., 2014), whereas increasing probiotic strains such as Bifidobacteria infantis can alleviate hypersensitivity and diminish symptoms in IBD patients (Clarke, Cryan, Dinan, & Quigley, 2012). When fecal microbiota were transplanted from patients with IBD to germ-free mice, the latter exhibited increased immune activation coupled with elevated anxiety, again pointing to the causal link between gut microbiota and anxiety (De Palma et al., 2017).

Despite the increasing appreciation of the microbiome's potential actions, it is still uncertain what specific gut bacteria do in relation to illness, although it is clear that some are protective and others destructive. In the case of IBD, firmicutes are increased whereas Bacteroidetes are reduced (Jeffery et al., 2012; Kennedy et al., 2014), and it also appeared that this illness was accompanied by increased Bifidobacterium and the Lactobacillus groups, whereas butyrate producing bacteria were reduced (Wang, Chen

(cont'd)

et al., 2014), which it may be recalled were also associated with depression. There is also appreciable evidence that gut microbiota and changes of immune functioning acts as mediators linking IBD, colorectal cancer, and type 2 diabetes (Jurjus et al., 2015). In addition to these comorbidities, the immune disturbances associated with IBD may also result in spondyloarthritis (inflammatory disorders that involve joint as well as sites at which ligaments and tendons attach to bones), raising the possibility that microbial manipulations might be useful in treating both Crohn's disease and comorbid spondyloarthritis, (Viladomiu et al., 2017).

In relation to some disorders, positive effects of treatments emerge in the presence of other factors. For example, the influence of gut bacteria on IBD may only emerge in the presence of particular genes related to regulatory T cell function (Chu et al., 2016). Moreover, the symptoms that are present, which may come and go, correspond with frequent changes of inflammation and gut microbiota variations (Halfvarson et al., 2017). Indeed, expression of genes associated with immune processes were linked to the clinical manifestations of the disorder among patients at different stages of IBD (Peters et al., 2017), indicating that the underlying processes may vary over the course of the illnesses' progression.

In addition to the linkages to IBD, there is ample reason to suggest that the immune processes involved in autoimmune disorders may be tied to microbial factors. By example, gut bacteria dysbiosis predicted the development of rheumatoid arthritis, and elevated levels of Collinsella were tied to arthritis, possibly through activity of the cytokine IL-17. Conversely, the commensal bacterium Prevotella histicola given to mice as either a prophylactic or therapeutic treatment diminished both arthritis symptoms and inflammation (Marietta et al., 2016). It is unfortunate that insufficient information was available that assessed whether these treatments also diminished the signs of comorbid depression.

It was suggested some years ago that environmental events, including stressors, either promote or exacerbate several illnesses because they involved some common elements, such as excessive cytokine functioning (Anisman et al., 2008). As indicated in Fig. 16.2, these illnesses may come about through different triggers, yet they ultimately give rise to overlapping inflammatory mediators, and might thus not only be comorbid, but also serve to reinforce one another's presence. This depiction is certainly simplistic and hardly addresses all of the factors that contribute to these comorbidities. At the same time, it may be useful in defining the intersections between disease conditions.

Inverse Relationships Between Disease Occurrences

Just as some illnesses predict the occurrence of other diseases, there are instances in which the appearance of one illness is inversely related to the occurrence of some other condition. As we saw in Chapter 14, Inflammatory Roads to Parkinson's Disease, the risk for Parkinson's disease was reduced in association with smoking, caffeine consumption, physical activity, and the use of ibuprofen. It likewise appears that rheumatoid arthritis and cancer occurrence was lower among people with schizophrenia than in the general population,

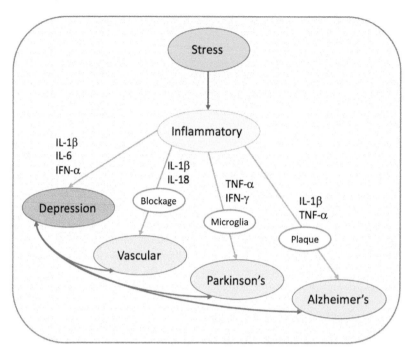

FIGURE 16.2   Stressors induce depressive illness, as well as several comorbid conditions, such as vascular illness (e.g., stroke and myocardial infarction), and neurodegenerative conditions, including Parkinson's (PD) and Alzheimer's (AD) disease. Inflammatory processes activated in response to stressors may be a common denominator for these conditions, even though the development and progression of these illnesses may involve many very different processes. Specifically, these cytokines, among others, might contribute to comorbid pathologies through several divergent routes, including activation of microglia (release of oxidative and inflammatory factors), the promotion of plaque/protein aggregate formation, and by altering vascular factors that contribute to arterial blockage (e.g., clot formation). *Source: From Anisman, Merali & Hayley (2008).*

despite their poorer life-styles (e.g., smoking) (Barak, Achiron, Mandel, Mirecki, & Aizenberg, 2005). Does this mean that people affected by schizophrenia also have some sort of protective feature (e.g., the presence of a particular gene) that diminished the occurrence of a second illness? This is possible given that cancer incidence in siblings and parents of a schizophrenic individual was also lower than that observed in the general population (Lichtermann, Ekelund, Pukkala, Tanskanen, & Lönnqvist, 2001), but it might also reflect common environments or experiences that operated in a protective capacity.

Let's examine yet another negative relationship with schizophrenia. Cardiovascular diseases were diminished among individuals diagnosed with schizophrenia, which might appear as good news for these individuals. But, in this case the inverse relationship might not be a genuine one, but instead reflects under-diagnosis of cardiovascular problems in mentally ill patients. This view is, in fact, supported by the finding that the proportion of schizophrenic individuals who die owing to ischemic heart disease is higher than in the general population (Crump, Winkleby, Sundquist, & Sundquist, 2013).

Other curious relationships have also been uncovered. For instance, within Tanzania and Uganda, HIV patients carrying a gene variant that codes for IL-12, were at much lower risk of developing tuberculosis (Sobota et al., 2016). It similarly appears that carrying the gene for cystic fibrosis may imbue individuals with immunity to tuberculosis. There are other instances in which there is a selective advantage to carrying a particular characteristic on only one allele, but the advantages presented may be unique to particular places (Africa in the case of sickle cell anemia promoting resistance to malaria).

Assessing the comorbidities between illnesses may be useful in identifying mechanisms responsible for diseases, and there may likewise be benefits in determining why the occurrence of two (or more) illnesses are negatively related, provided that the contribution of confounding factors are considered.

TWO FOR THE PRICE OF ONE

Just as illnesses are often comorbid with one another, often sharing underlying biological processes, there are many instances in which treatments that were initially developed for one purpose are found to be effective for a second condition. In some instances, the connection between the two can be predicted because the illnesses share obvious common processes. In other instances, the links between the two aren't immediately obvious, but this dual drug effect may offer clues regarding the processes responsible for each of the illnesses.

There are a huge number of such occurrences, and we'll only highlight a few of these because of their value or their surprise factor. Treatments that are used for their psychological or neurological impact, have found their way into multiple other conditions. It is well known that some SSRIs initially developed to treat depression are especially effective as antianxiety treatments. Surprisingly, several drugs that have been used as either antidepressants (e.g., SSRIs) or antipsychotics, were also helpful in eliminating C. difficile and salmonella. Gabapentin and its cousin pregabalin, which are used as antiepileptic agents, are commonly used to diminish anxiety disorders and neuropathic pain. Ropinirole, which was developed to diminish signs of Parkinson's disease, found its way to the treatment of Restless Legs Syndrome and SSRI-induced sexual dysfunction, while both the antibiotic doxycycline and MSDC-0160 developed to treat diabetes have the potential to delay the progression of Parkinson's disease.

In light of the breadth of illnesses that have been linked to elevated inflammation, it can be expected that an anti-inflammatory agent created for one immune-related condition might have a positive effect on a second disorder that involves immune actions. Thus, Gleevec used to treat some cancers, may turn out to be useful in treating type 2 diabetes; cancer acting agents that target the protein BRD4 might act against heart disease; histone deacetylase inhibitors that are used in treating some cancers, could be used for psoriasis treatments. Likewise, the β-blocker propranolol has been offered as a way of limiting soft tissue sarcomas, and pan β-blockers (a specific type of receptor antagonist) markedly increased the effectiveness immunotherapy in the treatment of melanoma.

In some instances, the source for a drug having multiple beneficial actions isn't all that clear. One of the most effective treatments for type 2 diabetes, metformin, which primarily acts by reducing glucose production by the liver, may have beneficial effects in slowing down heart disease and seemed to be useful in the treatment of drug-resistant breast cancer. The combination of metformin and the antihyptertensive syrosingopine could also prevent cancer cells from receiving the nutrients needed for them to survive. Still more surprising has been the report that nitazoxanide, a treatment for parasites such as Giardia and Tapeworms, acts against prostate and colon cancer by diminishing Beta-catenin that ordinarily encourages cancer cell multiplication.

(cont'd)

Obviously, repurposing drugs can have enormous benefits in the treatment of "off-site" disorders, and in some instances, their effectiveness may be more important or more effective than for the condition they were initially developed. Repurposing drugs has the benefits of having previously been tested extensively in relation to side effects and contraindications. Of course, the cost of bringing a drug to market for treating a second illness is appreciably reduced (although this doesn't necessarily mean that the patient will see these savings). The repurposing of drugs is hardly new, and many drugs are prescribed as off-label medications because they had been found to have strong alternate effects. In fact, in some cases these agents are prescribed because the usual meds haven't been working or because a tolerance had developed. This is all fine and good, but in some instances the drugs may not have gone through the usual evidence based analyses that would be needed to confirm their effectiveness for other illnesses. In fact, some agents may actually be contraindicated (e.g., using antipsychotic drugs to treat agitated dementia patients). Similarly, as most drugs are initially evaluated in adults, their off-label use in children could be especially problematic.

These are only a few of the very many examples of direct or inverse relationships between illnesses. We'll now explore how various comorbidities related to psychiatric conditions come about, and consider the theoretical and practical implications of these connections. To a significant extent, we'll deal primarily with the linkages between depressive/anxiety disorders in relation to heart disease, stroke, and diabetes, which are some of the more serious illness that individuals ordinarily encounter. Obviously, comorbid illnesses go well beyond those described here, but many of the principles outlined hold for these conditions as well.

DEPRESSION AND HEART DISEASE

A recent Lancet (2017) editorial that accompanied several articles concerning the Global Burden of Disease indicated that 72.3% (39.5 million) of all deaths in 2016 stemmed from noncommunicable diseases. The greatest killer, at least between 2006 and 2016, was ischemic heart disease, which increased by 19% globally. During this same period, diabetes was also a prime killer and in 2016 it was responsible for 1.43 million deaths, which represents an increase of 31.1% since 2006. Mental health problems have also been a major source of disability, and since 1990, a period of almost 30 years, there has been no decline of these conditions.

Although heart disease has been declining in most developed countries, it is still very high among some ethnic groups, such as Native Americans. In some places, including the Russian Federation, Belarus, Ukraine, and Central Asian republics, coronary artery disease (CAD) is at record levels, largely owing to poor life-style choices (smoking, alcohol consumption, and poor diet). It has likewise been increasing in developing or transitional countries, again stemming from life-style changes that accompanied industrial development. Of the various forms of heart disease that can occur, the most pernicious are hypertension and CAD. In many instances in which comorbidity occurs with depression, several common etiological factors are present, including activation of inflammatory processes[1].

PHYSIOLOGICAL STRESS RESPONSES AND HEART DISEASE

Central and Sympathetic Nervous System Processes

Stressful experiences, through their actions on sympathetic nervous system (SNS) activity, have been implicated in hypertension and eventual atherosclerosis, left ventricular hypertrophy (thickening of the left ventricle), arrhythmia, and plaques rupturing, thus leading to heart attack. "Hot reactors" in whom decidedly elevated sympathetic hyper-reactivity is commonly observed in response to challenges, may be especially likely to developing atherosclerosis. Depressed individuals may similarly be more reactive to certain challenges and the exaggerated sympathetic activity and circulating norepinephrine levels might favor cardiac disturbances.

Stimulation of β-norepinephrine receptors increases blood flow and pressure, and thus administration of drugs that block these receptors ought to diminish blood pressure, although questions have now arisen as to the effectiveness of these agents. Aside from peripheral actions, blood pressure is influenced by hypothalamic mechanism, as well as top down regulation of blood pressure. Specifically, brain regions that are involved in cognition and mood states, such as cortical and limbic brain areas, also contribute to cardiac functioning. Likewise, neuronal activity within the cingulate cortex and insula, which are involved in decision-making and information processing relevant to stressors, as well as the locus coeruleus, which has been associated with vigilance, have also been linked to cardiovascular activity (Wang, Chen et al., 2014). Not surprisingly, sympathetic reactivity in response to stressors is also moderated by a constellation of psychological factors related to coping styles, personality traits, previous stressor experiences, and lifestyle factors.

Hypertension

Hypertensive heart disease, characterized by elevated blood pressure affects about 20% of people, differing with age, sex, ethnicity, and a constellation of life-style factors, such as exercise, diet, weight, job strain, and sleep profile. Often, however, the cause of hypertension is unknown (essential hypertension or idiopathic hypertension) or it may be secondary to ongoing medical conditions, such as thyroid or kidney disease (secondary hypertension). Hypertension is often not accompanied by obvious symptoms, and left untreated, damage can occur to the arterial walls, and plaque accumulation increases so that arteries narrow and harden. Hypertension has been linked to immunological changes, including the accumulation of macrophages and T cells in blood vessel walls, especially in perivascular fat as well as in the kidney (Harrison, Vinh, Lob, & Madhur, 2010).

Many genes contribute to the emergence of hypertension, with each gene accounting for a

---

[1] The occurrence of herpes zoster (shingles), which reflects reactivation of varicella zoster virus (chicken pox), occurs in about 1% of individuals above the age of 60. However, it occurs at unusually high levels among depressed individuals, especially those at middle age and/or experiencing renal disease, hyperlipidemia, rheumatic illnesses, hypertension, anxiety, or sleep disorders (Liao, Chang, Muo, & Kao, 2015). Herpes zoster was similarly associated with increased risk for both heart attack and stroke (Kim et al., 2017), which aligns with the perspective that inflammatory factors act as a common mediator for these different conditions. Fortunately, a vaccine against shingles is available for those who choose to avoid this condition, although this doesn't necessarily imply that the vaccine will act against the comorbid illnesses.

small portion of the variance. A large scale, multisite study identified 31 new genes that were linked to hypertension, bringing the total number of hypertension-related genes to almost 100. Some of the same genes were also associated with CAD, potentially allowing for a more detailed analysis of the processes by which hypertension might culminate in CAD (Warren et al., 2017). It is relevant in this regard that ethnicity has also been linked to hypertension, and while this may be related to genetic influences, the cultural differences in hypertension are also attributable to other variables, such as poverty and life-style factors.

Emotional Contributions and Personality

It was widely believed that emotions, such as anger and hostility, were tied to hypertension. In fact, having a disposition of expressing anger was associated with elevated risk of hypertension, whereas having a forgiving personality was associated with reduced risk for hypertension (May et al., 2014). It is notable, as well, that while social competence was accompanied by reduced risk for cardiovascular illness, hostility was associated with elevated levels of the cardiac hormone atrial natriuretic peptide (ANP), which has been used as a potential biomarker of later heart disease (Smith et al., 2013). Aside from the physiological changes associated with strong emotional responses, hostile individuals may lose their social coping resources, thereby increasing risk for stress-related hypertension. In contrast, social competence was accompanied by reduced risk for cardiovascular illness.

CORONARY ARTERY DISEASE (CAD)

In response to damage to the endothelium (the thin layer of cells that lines the inner portion of blood vessels), immune factors, including monocytes, macrophages, and T cells, infiltrate the site of damage. If dead and dying cells are present, macrophages will ingest them, but there are some cells that carry a message, a CD47 protein, which essentially tells the macrophage "don't eat me". The accumulation of these dead cells favors plaque formation, perhaps being driven by TNF-α that is involved in programed cell removal. Importantly as well, CD47 can also promote angiogenesis (new blood vessel sprouting), which might have significance with regards to fueling aberrant vascular development. As expected, antibodies against CD47 facilitated the clearance of diseased vascular tissue, and in mouse models atherosclerosis was diminished by this treatment (Kojima, Volkmer, McKenna, Civelek, & Lusis, 2016).

Immune cells that gather at damaged areas release cytokines that encourage inflammation and facilitate plaque formation made up of cholesterol, fat, calcium, and fibrin. With recurrent damage, and with the passage of time, plaque appearance becomes progressively greater, eventually restricting blood flow to the heart (i.e., atherosclerosis). However, symptoms may not be noticeable until blood flow is appreciably restricted (more than 75%), whereupon individuals might feel chest pain (angina pectoris) owing to insufficient oxygenated blood reaching the heart.

It is not uncommon for myocardial ischemia and angina to be transient, resolving with discontinuation of behaviors that placed a load on the heart (stable angina) or through medications that increase blood supply (e.g., nitroglycerine placed under the tongue). With progression of the illness, symptoms may occur even with minimal energy output or when individuals are at rest, and ischemic periods become more persistent. This unstable angina may predict arrhythmias (disturbed heart rhythm) and myocardial infarction.

Factors That Promote Heart Disease

Numerous risk factors for CAD have been identified, including negative early adverse experiences, chronic stressors encountered during adulthood, lower socioeconomic class, the occurrence of depressive illness, as well as anxiety, anger, and hostility coupled with low social support. Risk for the development and progression of CAD increases further when several of these factors are present concurrently (Steptoe & Kivimäki, 2013), especially in the presence of gene mutations involved in neurochemical regulation or in the presence of type 2 diabetes. It also appeared that among individual that exhibited minor symptoms of depression, possibly reflecting a first hit (or an existing vulnerability), the occurrence of trauma was accompanied by a 75% increase of C-reactive protein (CRP), as well as the development of CAD (Murdock, Stowe, Peek, Lawrence, & Fagundes, 2017).

ADIPOSITY, CYTOKINES AND HEART DISEASE

The consequences of obesity may be exceptionally long-lasting given that obese children and adolescents will continue to exhibit biological risk factors relevant to later heart disease (Kit et al., 2015). Obesity is accompanied by the presence of excess cholesterol and fat (dyslipidemia), disturbed glucose tolerance, metabolic syndrome, and type 2 diabetes, all of which incrementally contribute to CAD. Fat located around the midsection, as we saw earlier, is notoriously rich in inflammatory cytokines, which may be among the most essential factors determining the link between obesity and heart disease as well as other illnesses mediated by inflammatory factors. For that matter, even among normal weight individuals, the presence of elevated abdominal fat is associated with increased risk for all-cause mortality.

Although bad cholesterol [low-density lipoprotein (LDL)] was long considered as being responsible for the link to heart disease, only half of the people who had heart attacks displayed elevated cholesterol, leading to the view that some other factor, either alone or in combination with cholesterol, was fundamental in the promotion of heart problems. While it was generally believed that good cholesterol (HDL) didn't fully make up for excessive bad cholesterol, it now appears that the gene coding for HDL may contain a mutation (a SCARB1 variant) that undermines the functioning of the receptor for HDL, and might thereby increase the risk for heart disease (Zanoni et al., 2016).

Genetic Factors

A genome-wide association study identified approximately 240 genetic signals tied to heart disease, including genes that might contribute to cholesterol level and plaque formation (Roberts, 2014). Given the very large number of genes related to CAD, most of these account for a small portion of the variance, although a core set of genes likely has more of an influence. A variant of *PCSK9* (which was associated with reduced LDL) may be particularly relevant, and this realization led to the development of a monoclonal antibody that seemed to have promising effects (Natarajan & Kathiresan, 2016). A mutation at the gene *ANGPTL3* has also been associated with very low triglyceride and LDL levels, and thus efforts are being made to develop treatments that target this gene in order to diminish heart disease. It is particularly relevant that genes related to inflammatory processes may contribute to heart functioning, and it might turn out that treatments that target both cholesterol and inflammatory processes will yield still better outcomes (Musunuru & Kathiresan, 2017).

As powerful as the effects of genetic factors might be, once again their presence doesn't necessarily guarantee that individuals are bound to develop heart disease. Many genetic factors may interact with particular foods consumed, stressors, or other life-style factors, thereby encouraging or discouraging CAD. Heart disease is known to vary with cultural and ethnic influences, potentially being related to genetic differences that contribute to glucose intolerance, obesity, elevated triglycerides, and insulin levels; once again, however, many of these variables could reflect life-style factors. Indeed, even if individuals were at a high genetic risk for heart disease (based on the contribution of an array of genes), a favorable life-style diminished risk of cardiac illness by about 50% (Khera et al., 2016). Of course, as depicted in Fig. 16.3, with age we can count on meeting many illnesses in addition to heart disease, which can also promote the evolution of other conditions.

Given the powerful effects of stressors and early adverse events in promoting epigenetic changes, it might be expected that such changes could also contribute to heart disease (Loche & Ozanne, 2016). As expected, among individuals living in disadvantaged neighborhoods, epigenetic changes may develop that are tied to stress reactivity and inflammatory processes, and might thus affect atherosclerosis (Smith et al., 2017). Moreover, having a heart attack was associated with particular epigenetic changes (Rask-Andersen et al., 2016), which could potentially serve as biomarkers for further events or perhaps in predicting response to treatments.

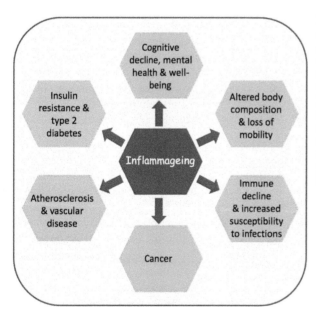

FIGURE 16.3 Central role of "inflammageing" in chronic conditions of ageing. There's little question that ageing is among the variables that are most likely to be associated with illnesses. Low grade inflammation appears with ageing (hence the term inflammageing), which then influences multiple disease states. Microbiota and increased inflammation may be linked to several life-style factors (e.g., nutrition) that may affect and health conditions. With age and experience the occurrence of epigenetic changes increase, and thus epigenetic marks across multiple tissues can be used to predict cellular ageing (Stubbs et al., 2017). *Source: Modified from Calder et al. (2017).*

PREDICTING LONGEVITY

Humans have sought every imaginable method of extending life. The search for the fountain of youth dates back at least five decades BCE, although it was made famous by the Spanish explorer Juan Ponce de León in the 16th century and the famous painting of the fountain by Lucas Cranach the Elder.

Longevity may be determined by multiple mutations that act to suppress the development of senescence and hence favor longer life (Moorad & Walling, 2017). In a sense, longevity might be viewed as being a reflection of not dying from a wide number of diseases, owing to the protective effects of particular genes (or particular mutations and epigenetic changes). Based on data obtained from over 116,000 people, a set of 16 genetic markers were identified that predicted longevity, whereas certain combinations predicted shortened lifespan (McDaid et al., 2017). A cross-species analysis revealed common genes that might somehow contribute to longevity. One gene in particular, b-cat1, seemed to be especially notable in this regard, since blocking its action increased the life-span of C. elegans (a type of worm frequently used in research) by 25%, possibly because this increased the availability of important amino acids (Mansfeld et al., 2015). Working on the assumption that a single gene is unlikely to account for considerable variance in relation to longevity, an alternative was developed in which patterns or "signatures" of biomarkers were identified that are predictive of aging or disease conditions that affect longevity (Sebastiani et al., 2017).

One view of extending life or at least well-being, entails removing senescent cells that have little positive value, but still use up resources. It was observed that through the deletion of a particular set of bacterial genes the life-span of C. elegans was extended. This same treatment also acted against tumor growth, and also protected these worms from excessive β-amyloid accumulation, which has been implicated in human Alzheimer's disease (Han et al., 2017). Admittedly, fruit flies and worms are some distance from humans, but these data nonetheless provide interesting hints regarding ageing processes and disease.

Balances between molecular processes related to growth factors and stress response pathways (e.g., mTOR and the p53 DNA damage response pathway) may also play a fundamental role in determining cellular senescence and aging, and thus manipulating these balances can affect biological ageing (Hasty & Christy, 2013). As well, members of the sirtuin deacetylase/ADP-ribosyltransferase/deacylase family (typically referred to simply as sirtuins) have been linked to longevity, presumably through their actions on stem cell regulation or epigenetic processes (Yu & Dang, 2017). Sirtuins may also have positive effects by affecting glucose and fat oxidation, increasing resveratrol, regulating macrophage renewal (Imperatore et al., 2017), or by inhibiting inflammation that otherwise takes a toll on various organ systems. Among the benefits of sirtuins is their capacity to diminish lung inflammation and limit the development of COPD, as well as the occurrence of illnesses, such as cancer (Guarente, 2014).

It has also been demonstrated that cardiosphere-derived cells from young rats (cardiac stem cells) injected into old rats improved heart functioning and exercise capability, and their cardiac cell telomeres were lengthened (Grigorian-Shamagian et al., 2017). The suggestion also arose that transfer of "young blood" to older individuals might have beneficial effects (vampire therapy). Young blood, it was suggested, may provide ingredients that enhance cognitive performance in old mice, even being able to reverse the cognitive effects attributable to ageing (Katsimpardi et al., 2014; Villeda et al., 2014). These outcomes were

---

(cont'd)

---

accompanied by transcriptional changes tied to enhanced hippocampal synaptic plasticity and increased dendritic spine density. These findings still need to be repeated, but it was observed that transfer of old blood into younger mice reduced hippocampal neurogenesis and impaired performance reflecting neuromuscular ability (Rebo et al., 2016).

Personality Factors

The potential contribution of psychological factors to heart disease, even if these were vastly overstated, came to the fore with reports that heart disease was especially common among individuals with a Type A personality, characterized by being ambitious, competitive, impatient, hostile, and rushed (Friedman & Rosenman, 1971). This view, as exciting as it initially was, did not fully stand the test of further replication, although a component of this personality style that comprised hostility and anger was linked to coronary problems. This was most evident among individuals who directed their anger inwardly (Denollet & Kupper, 2007).

As an alternative to the Type A personality, heart disease was proposed as being more tightly linked to the distressed (Type D) personality (Denollet, Gidron, Vrints, & Conraads, 2010). This personality type was characterized as being particularly attentive to negative stimuli, and tending toward the expression of negative emotions, such as depressed mood, anxiety, anger, hostile feelings, and worrying, and highly stressor reactive in relation to cardiac functioning. Type D individuals are socially immature in the sense that they were uncomfortable with strangers, tense and insecure with other people, and typically inhibited self-expression in social interactions. Thus, they may not be equipped with the social coping mechanisms needed to deal with many life stressors. This view, at least at first blush, seems interesting, but a strong relationship between Type D personality and heart disturbances was not always apparent. Questions also arose as to whether this concept provides information beyond that associated with a depression-CAD link, whether the Type D personality is tied to neurobiological processes that affect CAD, or whether this personality type predicts CAD through the adoption of unhealthy behaviors, or poorer adherence to medication. Ultimately, detailed prospective analyses will be needed to determine whether the Type D personality acts as a causal agent in relation to CAD or can serve as a behavioral marker that predicts the illness.

Stressor Effects

In response to stressors (or physical exertion), circulating inflammatory factors are increased and together with lipids bound to the cell wall, increase plaque formation and the development of atherosclerosis. Further, blood pressure increases produced by stressors can engender myocardial ischemia (reduced oxygen) if the coronary arteries are already narrowed, which predicts later cardiac events. Myocardial infarction can be provoked by acute trauma or catastrophic events, being most evident among individuals with preexisting cardiac problems (Kop & Mommersteeg, 2014).

Persistent psychological distress (e.g., chronic caregiving, bereavement, and job strain) was associated with new diagnoses of CAD, and was predictive of greater mortality among individuals with stable CAD (Stewart et al., 2017). As in the case of other illnesses, adverse events early in life may influence the developmental trajectory related to cardiovascular functioning so that adult atherosclerosis was more likely to occur. Of the many chronic stressors that can affect heart disease, social rank within the workplace appeared to be especially significant, and was predictive of a constellation of other illnesses, such as type 2 diabetes, certain forms of cancer, gastrointestinal illnesses, chronic lung disease, and not surprisingly, depression and suicide, and individuals often suffered more than a single condition. A particularly germane feature related to increased occurrence of CAD, and earlier mortality was "job strain" experienced. This amounts to high job demands together with low decision latitude, as well as high work efforts and low rewards and a perceived a lack of justice (unfairness). The impacts of these factors were exacerbated if they occurred on a backdrop of other ongoing stressors that promoted anxiety, irritability, and marital problems, as well as the adoption of poor lifestyles (e.g., smoking, alcohol consumption, and poor sleep), which might actually have been secondary to job-related distress (Marmot, Rose, Shipley, & Hamilton, 1978).

POSITIVE PSYCHOLOGY IN RELATION TO HEART HEALTH

Just as stressful events and depression favor the development of cardiovascular illness, experiences and events that diminish distress, such as having supportive relationships, may have cardiovascular health benefits. The availability of social support can diminish the stress that might otherwise instigate depressive illness, which in turn, may protect against heart disease. It is hardly unexpected that having support or other effective means of coping with stressors will diminish their impact on health. But, is it also the case that positivity and optimism are accompanied by enhanced general well-being and improved heart health?

A meta-analysis that included both retrospective and prospective studies indicated that the relationship between positive mood and heart disease was not simply a result of diminished emotional distress or blunting of stressor actions (Chida & Steptoe, 2008). Among other things, having a general sense of well-being was accompanied by the adoption of better health behaviors, which in turn, were associated with enhanced physiological indices of heart health. In essence, the presence of positives in relation to heart health reflects more than simply the absence of negative events. As we've seen so often, most reports of linkages between positivity and well-being comprised correlational studies within the general population (this was also the case for studies that assessed the links to specific illnesses, including cancer, heart disease, and type 2 diabetes). There has been a scarcity of studies that used positive psychology interventions in the treatment of these disorders, and the few that were conducted (usually amounting to 1–3 studies for any given disorder) haven't always been high quality studies, and thus have had limited applied value (Macaskill, 2016).

Depressive Illness and Heart Disease

Mental illness of virtually any sort (schizophrenia, bipolar disorder, anxiety disorder, and major depression) is associated with a dramatic increase in the risk for cardiovascular problems, and if the psychiatric illness is sufficiently pronounced to require medications, the risk of heart attack and stroke was particularly elevated. This relationship was particularly prominent among individuals with chronic depression, and indeed childhood and adolescent depression were predictive of the later development of heart disease (Rottenberg et al., 2014). The link was especially evident in relation to chronic depression, which predicted 6-month survival following myocardial infarction (Frasure-Smith, Lesperance, & Talajic, 1993). The occurrence of PTSD stemming from chronic stress among soldiers at war, which is often comorbid with depression, was likewise related to an increase of later heart disease (Vaccarino et al., 2013).

Among individuals with atypical depression, heart disease is more likely to occur relative to that seen with typical depression (Penninx, 2017), possibly owing to the life-style changes this involves. However, the doubling of CAD among depressed individuals was evident even after controlling for life-style factors (Baune et al., 2012). Thus, other characteristics of atypical depression, including circulating cytokine levels, may account for the link to heart disease. It also appeared that the presence of hopelessness and pessimism was accompanied by slower recovery following cardiac events, and was predictive of future cardiac problems (Kop & Mommersteeg, 2014). The more frequently depressive symptoms appeared among adults (on successive assessments over a 10 year period), the greater the risk for heart disease and stroke (Péquignot et al., 2016). Even though the connection between depression and heart disease is well known, and depression was even proposed as

a cogent marker for later heart disease, all too often patients do not receive treatment to diminish feelings of depression, and even less frequently are the long-term consequences considered.

The comorbidity between depression and CAD may stem from any number of factors. These conditions share many features, including altered sympathetic and parasympathetic functioning, and a bias in favor of proinflammatory over anti-inflammatory processes (Halaris, 2013, 2017). Likewise, depression, schizophrenia, and bipolar disorder share similar patterns of cytokine alterations (Goldsmith, Rapaport, & Miller, 2016). These comorbid conditions may be related to HPA dysregulation that affects sympathoadrenal hyperactivity or altered immune functioning, which may lead to increased vasoconstrictive tone, heart rate, and platelet activation. Once corticoid dysregulation occurs, the inhibition on proinflammatory cytokines is diminished, which may produce damage to the heart. As we've seen in earlier chapters (Chapter 8: Depressive Disorders), depressive disorders are accompanied by several inflammatory changes, and it's fairly certain that the elevated inflammation contributes to the development of CAD[2].

Given that depression exacerbates heart disease and limits recovery, it might be expected that heart problems would resolve following successful treatment of depression. However, once cardiac problems are present, probably after years in the making, it may be far too late to institute a cure simply by diminishing depressive symptoms, although reducing depression might limit further damage from occurring.

---

[2] Cardiovascular illness has been related to "vital exhaustion", comprising feeling excessive fatigue and lack of energy, progressively greater irritability, and demoralization. These features are reminiscent of the somatic characteristic of depressive illnesses (sleep problems, appetite changes, and psychomotor alterations). Inflammatory processes have been linked to vital exhaustion, depressive illness, and to heart disease. While vital exhaustion may be brought about owing to a failing heart, it may also stem from the release of inflammatory factors that engender most of the symptoms described.

Sex Differences in Heart Disease

Relative to men of the same age, premenopausal women are less apt to develop heart disease, but they catch up after menopause. Apparently, estrogen provides "female protection" against heart disease; however, when estrogen levels decline following menopause, or in association with menstrual irregularities, the risk for CAD increases. As in men, CAD was less common among women employed in administrative positions than among those in the lower ranks. Job satisfaction could act as a buffer to diminish life stressors that might otherwise be detrimental to heart health, but when women faced the additional load of home work and taking care of a family, CAD risk was elevated. Women, as we've seen, are more stress reactive than men, and the greater neuroendocrine and inflammatory changes that occur, may favor the occurrence of heart disease.

Knowing the symptoms of heart attack is obviously important to survive such an event. Many of these symptoms are the same in men and women, but there are also some very marked differences. Among men, more than in women, heart attack is accompanied by discomfort on the right-side (opposite the heart), dull ache, and a feeling much like that of indigestion. Among women, throat discomfort, pressing on the chest, and vomiting, are more common than in men. But, these symptoms are more readily misidentified, leading to women taking longer to get to hospital, making for poorer outcomes.

Sleep Disturbances, Inflammation and Cardiovascular Disease

As we've seen, sleep disruption has been associated with immune disturbances and related health risks (Irwin, 2015), including hypertension, CAD, and all-cause mortality (Azevedo Da Silva et al., 2014). The negative consequences of sleep disturbances are long lasting in that sleep problems during adolescence were predictive of heart disease in adulthood, although it should be said that sleep problems were often accompanied by poor life-style choices.

Meta-analyses that included both cross-sectional and prospective studies confirmed that short sleep durations were frequently linked to hypertension, CAD, and stroke (Azevedo Da Silva et al., 2014). Cardiovascular problems were also apparent among individuals who slept long hours, which it will be recalled, are also associated with elevated inflammation. To no surprise, obstructive sleep apnea has also been linked to cardiovascular disease, and inflammatory elevation was a key element in producing this outcome (Bouloukaki et al., 2017).

Inflammatory Processes in Heart Disease

It ought to be clear by this point that systemic inflammation contributes to plaque formation and is thus involved in the promotion of atherosclerosis. Among individuals diagnosed with CAD, the presence of antibodies reflecting earlier infection (e.g., for Chlamydia pneumoniae and cytomegalovirus) were especially common (Kop & Mommersteeg, 2014). Thus, the promotion of inflammation (e.g., through infectious agents, chemical agents, or tissue damage) that increases circulating proinflammatory cytokines may contribute to CAD. Further to this, total pathogen burden (total number of infections experienced) together with socioeconomic status predicted heart disease (Steptoe et al., 2008). Chronic stressors, as we've seen, promote circulating cytokine elevations, and their cumulative actions over years can instigate heart disease (O'Donovan, Neylan, Metzler, & Cohen, 2012), and may contribute to complications related to atherosclerosis.

Although depressive illness has been associated with several immune-related changes, including elevated levels of proinflammatory cytokines, and reductions of anti-inflammatory cytokines, a review of available data suggested that some forms of depression (e.g., dysthymia; chronic depression) may be more closely aligned with heart disease and inflammation (e.g., Baune et al., 2012). The cytokine links to depression and heart disease have largely focused on elevated IL-1β, IL-6, TNF-α, and reduced levels of the anti-inflammatories IL-4 and IL-10. However, IL-18 released from cardiac muscle cells (cardiomyocytes), may also contribute to the development of heart disease (Reddy et al., 2010). In response to challenges, such as chronic hemodynamic stress and hormonal variations, IL-18 may promote "vascular remodeling" wherein cell production and growth is altered, perhaps in an effort to deal with the ongoing strain. Over the course of these changes occurring, lesions can develop, which then engender structural alterations that can lead to narrowing of blood vessels and the development of atherosclerosis. It is still not known whether IL-18 varies with different forms of depressive illness.

Cytokine variations may directly or indirectly promote a series of other physiological changes (macrophage and T cell actions, oxidative stress, and altered clotting factors) that favor the development of CAD. Concurrently, depression may be associated with metabolic syndrome and diabetes, which also have powerful direct and indirect effects on CAD. In essence, depressive illnesses and CAD involve mutually generating processes.

LONELINESS AND HEART DISEASE

Feelings of loneliness can become pervasive and all-encompassing, provoking distress and a sense of being disconnected and apart from others. Although the appearance of loneliness is most certainly a result of social and environmental factors, there also seems to be a heritable component (Gao et al., 2016). Loneliness has been linked to elevated inflammation (Jaremka et al., 2013; Pressman et al., 2005), and thus it is not overly surprising that loneliness is associated with physical and psychological disturbances (Cacioppo, Cacioppo, Capitanio, & Cole, 2015).

The link between loneliness and heart disease was especially notable among older people, as well as in those with disabilities (e.g., Holwerda, van Tilburg, Deeg, Schutter, & Van, 2016). In older individuals, many of whom had lost their social connections (friends may have died, moved away, or had simply been out of touch) and often suffer multiple indignities, including age discrimination, loneliness was not only identified as being instrumental in promoting heart disease, but also vascular disease, stroke, and a variety of immune related disorders (Holt-Lunstad, Smith, Baker, Harris, & Stephenson, 2015). The relationship between loneliness and heart disease was particularly evident among women, and it has been suggested that in relation to heart disease, loneliness is more dangerous than is smoking or obesity. A systematic review and meta-analysis that included 16 longitudinal data sets revealed that loneliness and poor social relationships were linked to a 29% increase in risk of incident CAD and a slightly greater (32%) increase in risk of stroke (Valtorta, Kanaan, Gilbody, Ronzi, & Hanratty, 2016). We can pretty well expect that in populations at-risk for other illnesses, the distress and provocation of inflammatory alterations will exacerbate their occurrence and may disturb recovery from illness. For instance, the impact of loneliness has been witnessed among breast cancer survivors assessed 2–3 months following treatment. Women who were relatively lonely experienced elevated depression, fatigue and pain, as well greater antibody titers for cytomegalovirus than more socially connected individuals. As our social groups and networks can have important health benefits (Steffens, Cruwys, Haslam, Jetten, & Haslam, 2016), it's hardly surprising that loneliness can have the adverse actions that it does.

Just as activation of the kynurenine pathway was offered as a mechanism by which cytokines may come to promote depression, this pathway could also affect heart disease as well as other illness conditions, and consequently might account for the comorbidity between depression and heart disease (Halaris, 2017). Specifically, kynurenine metabolites synergistically generate free radicals that can promote endothelial dysfunction that eventually engenders cardiovascular problems (Halaris, 2017; Nagy et al., 2017). Further, both depression and fatigue that often occur following stroke may similarly involve the very same pathway.

A related view is that the NLRP3 inflammasome is fundamental in linking psychological stress to depression (as described in Chapter 8: Depressive Disorders), as well as other immune-related conditions. The NLRP3 inflammasome might serve as a sensor for systemic changes that could pose a threat to the organism's well-being (Schroder & Tschopp, 2010). It will be recalled that toll-like receptors (TLRs) recognize pathogen-associated molecular patterns (PAMPs) as well as damage-associated molecular patterns (DAMPs), leading to the release of "alarm" cytokines, such as IL-1β, TNF-α, and IL-6 (Schroder & Tschopp, 2010). These and other inflammatory cytokines are potentially responsible for comorbidities between depression and a constellation of illnesses, such as heart disease, diabetes, stroke, chronic pain, obesity, diabetes, rheumatoid arthritis, MS, COPD, asthma, and Alzheimer's disease (Benatti et al., 2016; Iwata, Ota, & Duman, 2013). These systemic diseases have all been associated with NLRP3, as well as the different types of danger signals that lead to IL-1β release (or activate caspase-1), which have been linked to elevated depressive illness relative to the general population.

Predicting Heart Disease Through Inflammatory Markers

In line with the view that prevention is more desirable than treatment (after heart disease has developed), there has been a search for biomarkers that are indicative of impending problems. Consistent with a role for inflammation in heart disease, severe infection (pneumonia and sepsis) predicted heart disease over the ensuing 5 years (Bergh et al., 2017). The presence of particular genes could be used to predict heart disease, and there has been attention to using inflammatory markers and heart hormones in this capacity. Elevated CRP released from the liver in response to inflammation has been used for some time as a marker for later heart disease. Likewise, elevated levels of a protein, troponin, occurred in approximately 40% of individuals with type 2 diabetes and stable ischemic heart disease. These individuals were at twice the risk of dying of heart disease or stroke over the ensuing 5 years (Everett et al., 2015), making this protein a potentially valuable marker. As well, cardiac problems have reportedly been accompanied by the increased appearance of two heart hormones, ANP and brain natriuretic peptide (BNP), whose release is stimulated by cytokines. Among older individuals, these factors, together with elevated IL-6 and fibrinogen predicted increased risk for fatal heart attack relative to nonfatal events (Sattar et al., 2009) (Fig. 16.4).

Cardiovascular disturbances, including heart failure, are accompanied by disturbed hippocampal functioning and reduced volume, which might influence memory loss and the occurrence of depressive disorders. Not surprisingly, central processes are also likely to contribute to CAD. The very fact that stressors affect corticolimbic functioning, which determines appraisal, cognitive, and emotional responses can affect heart functioning, and ultimately heart disease (Myers, 2017). Moreover, individuals who displayed higher levels of

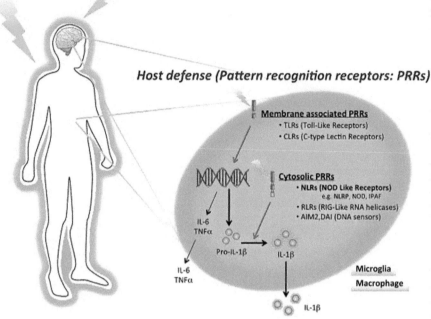

FIGURE 16.4   Danger signals and host defense mechanisms may contribute to the development of depression and several systemic illnesses. Pattern recognition receptors (*PRRs*) sense the presence of danger signals, leading to biological defensive responses being initiated. Membrane associated PRRs, through TLRs stimulate the production of IL-6 and TNF-α. Cytosolic (water-soluble components of the cytoplasm) PRRs, notably NOD-like receptors (*NLRs*), detect diverse ligands, such as ATP, fatty acids, and β-amyloid. The best characterized NLR, that of NLRP3, forms inflammasomes that comprise large multiprotein complexes. Once danger signals are detected, synthesis of pro-IL-1β is promoted, and the NLRP3 complex causes the secretion of IL-1β. The resulting inflammation favors the development of depression as well as physical illnesses, such as heart disease. Abbreviations: AIM, absent in melanoma; DAI, DNA-dependent activator of interferon (IFN)-regulatory factors; IPAF, ICE-protease activating factor; NLRP, NLR family, pyrin domain-containing; NOD, nucleotide-binding oligomerization domain. *Source: From Iwata et al. (2013).*

stress-like amygdala activity were more likely to develop heart disease, pointing to a connection between the brain's ability to regulate emotions and heart disease (Tawakol et al., 2017).

Microbiota and Heart Disease

Given the links between microbial factors and inflammatory processes, it is tempting to suggest that the microbiome also contributes to the emergence of hypertension and CAD (e.g., Emoto et al., 2016). This perspective is bolstered by the knowledge that dietary factors can influence gut microbiota, as well as the occurrence of diabetes that is so often linked to CAD. In fact, the presence of excess body fat and microbial dysbiosis, are risk factors for cardiometabolic diseases (Aron-Wisnewsky & Clement, 2016), possibly involving neuroendocrine processes and their effects on immune functioning.

Consistent evidence has emerged in line with this view, particularly in the relation between disease condition and the presence of trimethylamine N-oxide (TMAO), a metabolite derived from gut microbes and dietary phosphatidylcholine. Among patients with stable CAD, elevated TMAO levels predicted greater mortality risk (Tang et al., 2014). As heart disease was associated with obesity, and both conditions are accompanied by elevated TMAO, it was considered that TMAO and related metabolic pathways were responsible for the comorbidity (Tang & Hazen, 2017). By generating TMAO, gut microbiota may directly contribute to both platelet hyperreactivity and enhanced potential for thrombosis (Zhu et al., 2016). Furthermore, high TMAO was also accompanied by more frequent type 2 diabetes, raising the possibility that TMAO might also be the link between obesity and diabetes (Dambrova et al., 2016; Tang et al., 2014), although the increased mortality risk appeared to be independent of glycemic control (Tang et al., 2017)

It was especially noteworthy that when mice lacked a gene for a particular enzyme, flavin-containing monooxygenase 3 (FMO3), which is responsible for the conversion of TMAO to its active form, obesity was not present, and could thus limit the occurrence of CAD. It was also observed that among germ-free mice and following antibiotic-induced suppression of gut microbiota, TMAO production was reduced and atherosclerosis was less likely to occur (Wang et al., 2011). Likewise, reducing TMAO levels through diet was able to diminish the development of plaque in mice susceptible to atherosclerosis (Wang, Roberts et al., 2015).

Among mice maintained on a Western diet, body weight increased, accompanied by dyslipidemia and elevated TMAO levels. In these mice, blood pressure was tied to several proinflammatory cytokines being elevated, and anti-inflammatory cytokines reduced. Many of these effects were diminished if mice received a TMAO inhibitor, pointing to the causal connection between TMAO and heart problems (Chen, Zheng, Feng, Li, & Zhang, 2017). Conversely, among mice maintained on a diet rich in choline, plasma TMAO levels increased, culminating in the development of atherosclerotic plaque, which could be attenuated by antibiotic treatment.

GUM DISEASE IN RELATION TO HEART DISEASE

Occasionally, correlations between diseases appear that, on the surface, hardly seem to make much sense. There had long been indications that the incidence of heart disease and the severity of heart attack was elevated among individuals with periodontal (inflammatory gum) disease (Marfil-Álvarez et al., 2014). As well, gum disease was linked to the occurrence of ischemic stroke (Leira et al., 2016), often being associated with especially elevated levels of particular bacteria. It seems likely that the presence of inflammation stemming from gum disease and possibly dysbiosis of bacteria within the mouth, including changes of Streptococcus, could potentially be involved in these comorbid conditions (Tonomura et al., 2016).

Meta-analyses of prospective cohort studies confirmed that periodontal disease was independently related to elevated risk of CAD (Leng, Zeng, Kwong, & Hua, 2015). As it happens, periodontal disease was also associated with the development of type 2 diabetes (Branchereau et al., 2016), and hence the development of heart disease. It has been maintained that cytokines released from diseased gums may travel through the blood, eventually coming to affect pancreatic beta cells, and promoting insulin resistance and diabetes. Indeed, when patients received periodontal treatments,

---

(cont'd)

A1c levels declined, reflecting improved glucose levels over the preceding 3 months. Despite these associations, it ought to be underscored that the American Heart Association had previously indicated that conclusions linking gum disease and heart conditions were tenuous given that the disorders share other features, including age, and smoking habits, which alone might account for the comorbidity (Lockhart, Bolger, Papapanou, Osinbowale, & Trevisan, 2012).

Preventing and Treating Heart Disease

Prevention of hypertension could be aided through maintaining proper life-styles, but at times alternative strategies are needed. Varied relaxation techniques and methods of reducing or coping with stressors (e.g., CBT, mindful meditation, yoga, relaxation training, and bio-feedback) can be used in an effort to control blood pressure, and to diminish the drug dose being used to control this condition. In the main, however, once hypertension developed, individuals typically rely on medications to control blood pressure. For decades, norepinephrine β-receptor blockers were the most common treatment used to this end, but there has been some question as to whether they actually reduced the risk of heart attacks, deaths from heart attacks, or stroke. In fact, β-blockers in the treatment of hypertension could actually increase the risk of cardiovascular events and death, and increase adverse events after noncardiac surgery. Similarly, the use of β-blockers following acute myocardial infarction (without heart failure) was not accompanied by reduced death (Dondo et al., 2017).

As an alternative to β-blockers, angiotensin-converting-enzyme (ACE) inhibitors that control blood pressure by regulating fluid volumes have been widely used. The ACE inhibitors prevent the conversion of the hormone angiotensin I to the vasoconstrictor angiotensin II, and the attenuated blood vessel constriction allows for better flow of blood to the heart, thus reducing blood pressure, and diminishing cardiovascular death stroke. A related drug class, angiotensin II receptor blockers (ARBs), which limit angiotensin II from stimulating blood vessel receptors, have also been useful in the treatment of hypertension.

In view of the influence of inflammatory factors in the development of heart disease, anti-inflammatory manipulations have been used in a treatment capacity. Unfortunately, TNF-α inhibiting agents, such as etanercept or infliximab that were effective in treatment of other conditions, were ineffective for CAD and in some instances could provoke further problems (Cacciapaglia et al., 2011). The anti-inflammatory treatments were likely administered when problems were already well advanced, and it may have been too late to gain benefits from the treatment, although it might have been expected that the progression of the illness would be slowed.

Because of their anti-inflammatory effects, omega-3 fatty acids have received considerable attention in relation to heart health. Although high doses of omega-3 from fish oils was reported to lower blood pressure and triglyceride levels, and enhance heart functioning and diminish heart scarring among heart attack survivors, not all reports were so positive. In fact, a meta-analysis of 20 clinical trials indicated that omega-3 didn't lower the risk of heart attack, stroke, or death from heart disease (Rizos, Ntzani, Bika, Kostapanos, & Elisaf, 2012). A more recent meta-analysis of 19 studies indicated that biomarker concentrations of seafood and plant-derived omega-3 fatty acids were linked to modestly lower incidence of fatal CAD (Del Gobbo et al., 2016). These reports notwithstanding, it is not uncommon for omega-3s to be used for the purpose of "cardiovascular protection", and as a supplement for patients that experienced a heart attack.

As certain types of cholesterol contribute to the build-up of plaque, statins have often been prescribed to reduce lipids and LDL cholesterol levels, particularly among older individuals and in those with diabetes. Among individuals with high LDL levels, but without any apparent heart problems, the use of a statin reduced the occurrence of heart disease by about 25% when reassessed in a 20-year follow-up (Vallejo-Vaz et al., 2017). A systematic review indicated that heart disease and death among individuals with high bad cholesterol levels could be reduced by statins (Taylor et al., 2013). Indeed, the Cholesterol Treatment Trialists' Collaboration indicated that for every 39-point drop in LDL cholesterol, the likelihood of a stroke or a heart attack decline dropped by 21 percent, and that mortality linked to CAD diminished by 19%. The 2008 JUPITER trials also indicated that statin use was accompanied by a marked decrease of surgery for blocked arteries, and diminished risk for heart attack and stroke (Ridker, Lonn, Paynter, Glynn, & Yusuf, 2017).

Despite this enthusiasm, there has been considerable controversy regarding the usefulness of statins. Aside from the possibility that statins might favor the development of Parkinson's disease, and increase the risk for type 2 diabetes, their effectiveness in limiting heart disease was questioned. Specifically, the usefulness of statins in limiting heart disease among women was uncertain, and in individuals over 65 years of age, no benefits were accrued (Han et al., 2017). The debate concerning statin use hasn't been fully resolved, and in

some respects has escalated (see Collins et al., 2016; Redberg & Katz, 2017). Yet, it should be acknowledged that the lack of data concerning the comparative efficacy and safety of specific statins across individuals, has pointed to the need for a personalized treatment strategy, particularly as an unacceptable number of individuals experience adverse effects in response to standard statin treatment (Collins et al., 2016). Despite the very different views that have been offered, both the American College of Cardiology and the American Heart Association, perhaps to stay on the safe side, continued to recommend the use of statins to limit or prevent cardiovascular disease among individuals with high levels of bad (LDL) cholesterol.

Several alternative treatments are being developed in an effort to diminish cholesterol. One of the more interesting methods of dealing with heart disease has involved delivery of nanoparticles that course through the blood stream breaking apart plaques. These nanoparticles need a target so that they can act most expeditiously, and as macrophages and other immune cells typically gather around the site of plaque formation, nanoparticles have been developed that bind to molecules present on the surface of macrophages. In this way, plaques could be markedly reduced in mouse models (Lewis et al., 2015). Related approaches are being developed, but for the moment, they are only being considered for use in very ill patients.

STROKE

Stroke comes in several basic forms. Transient ischemic attack (TIA) reflects a brief interruption of blood to some aspect of the brain, and may promote symptoms much like those that accompany more serious stroke. However, a TIA is often thought not to have permanent functional effects, but may be a risk factor for more serious stroke. A TIA should be distinguished from a silent cerebral infarct, more commonly referred to as a silent stroke, which may have lasting cognitive and mood altering effects. Multiple events of this nature may create considerable damage to cells, and among individuals with cerebrovascular problems, pronounced and lasting cognitive disturbances may occur (Summers et al., 2017).

Hemorrhagic and ischemic are the two most serious forms of major stroke. The hemorrhagic type, which makes up about 10%–15% of cases, comprises a blood vessel rupture so that bleeding occurs directly into the brain or into the subarachnoid space surrounding brain tissue. This form of stroke, which has been associated with hypertension, is generally not treatable, but it was observed that IL-27 may have the effect of shifting neutrophils from causing damage to brain functioning to that of promoting recovery. Thus, by harnessing neutrophils appropriately, it may be possible to treat hemorrhagic stroke patients (Zhao, Ma et al., 2017), although this doesn't imply anything like a return to normal functioning.

Ischemic stroke, which is the most common, occurs as a result of blood vessel blockage owing to a thrombosis or arterial embolism (clot, fat globule, or gas bubble), or as a result of cerebral hypoperfusion (general reduction of blood supply). The damaging effects that follow stroke can go on for some time owing to the excessive release of the excitatory neurotransmitter glutamate causing cell death (Kostandy, 2012). If "clot busters", such as tPA, are administered within the "golden hours", usually considered within 3 hour following a stroke, then damage to the brain might be limited (some positive effects can be obtained up to 4.5 hour since stroke occurrence). Because of its anti-inflammatory properties, aspirin use soon after a minor stroke can reduce the risk of a major stroke later on (Hankey, 2016). Predictably, in rodents, agents that limit the inflammatory response, such as an IL-1 receptor antagonist (IL-1Ra) facilitated recovery from stroke, and encouraged neurogenesis (Pradillo et al., 2017).

The symptoms expressed may vary as a function of the site and the extent of damage left in the wake of a stroke (the infarct). The infarct that encroaches upon the basal ganglia and cortical regions gives rise to the well-known motor and cognitive problems. Speech areas can also be affected, as can limbic brain regions that lead to anxious and depressive symptoms. The infarct itself can be further subdivided into the internal core and outer penumbra subregions. The core comprises very rapid and massive inflammatory and oxidative neuronal damage that is exceptionally difficult to prevent. In contrast, the outer penumbral region involves slower cell death that can occur over several days and involves marked "programmed" apoptotic cell death. The delayed cell death has been the focus of efforts to mitigate the infarct size following stroke. Indeed, treatments that modify inflamamotry cell infiltration, limit excessive glutamateric signaling, and buffer mitochondrial energetic functioning, are all potential strategies of achieving his goal.

Stroke and Depression

Depression following stroke is not unusual, and was more common among individuals who had previously experienced a major depressive episode. It will be recalled that depression is accompanied by changes in the volume of several brain regions, and microstructural changes are also present within the midbrain. Such microstructural changes have been associated with ageing and might contribute to stroke occurrence, and it seems that the occurrence of poststroke depression may predict later stroke occurrence (Sun J, Ma et al., 2016).

Typically, depression that follows ischemic stroke may occur relatively soon after the event, but among individuals 18–50 years of age, depression and anxiety features occurred as long as a decade later (Maaijwee et al., 2016). Like more usual instances of depression, poststroke depression was accompanied by lower serum BDNF levels, diminished amygdala volume, and disturbed HPA feedback processes (Noonan, Carey, & Crewther, 2013). It is especially significant that aside from the psychic drain created by depression in stroke victims, the occurrence of depression signals a poor prognosis for functional recovery and future global functioning, as well as being predictive of further strokes and increased mortality over the ensuing few years (Lichtman et al., 2014). Several meta-analyses revealed that antidepressants reduced the incidence of poststroke depression, and when administered early, both physical and cognitive recovery from stroke was enhanced as was survival (Robinson & Jorge, 2016). Anxiety following stroke has received somewhat less attention than depression, but the occurrence of GAD, phobic disorder, and OCD are three to four times higher following stroke relative to that apparent in the general population (Cumming, Blomstrand, Skoog, & Linden, 2016).

As in the case of damage to neurons elicited by other traumatic events, microglial release of cytokines increases following stroke, as do glutamate levels. The cytokines may help in the clearance of debris and diminishing infection, and may even promote cell growth. However, within a brief poststroke period the build-up of cytokines may act in a neurodestructive capacity (Shichita, Ito, & Yoshimura, 2014), and

excessive glutamate release may similarly be neurodestructive (Kostandy, 2012), thereby promoting depressive symptoms (Feng, Fang, & Liu, 2014). In this regard, following stroke the M1 microglial subtype that is proinflammatory predominates over the M2 that secretes anti-inflammatory cytokines (Zhao, Rasheed et al., 2017). The position was advanced that these cytokine changes promote IDO metabolism, which then engenders a reduction of serotonin in limbic brain regions, and promotes neuronal loss stemming from neurotoxic conditions, thus leading to depression.

Although several cytokines may be related to poststroke depression, IL-6, and IL-18 elevations were most common among individuals who developed depression over a 1-year period, being especially apparent in patients who did not use a statin (Kang et al., 2016). Consistent with a stroke-cytokine perspective, in animals subjected to middle cerebral artery occlusion, which mimics ischemic stroke, a depressive-like profile emerged, which could be limited by antagonizing the actions of IL-1β (Craft & DeVries, 2006). Particular inflammatory cytokines could be fundamental in accounting for disturbed functional recovery among individuals who experience poststroke depression, and could serve as predictors of later strokes and mortality, or might signify later functional recovery. Knowing which cytokines perform particular functions or predict future outcomes may contribute to the development of treatment strategies and targets to treat or limit some of the negative consequences associated with stroke.

TAKING ADVANTAGE OF SYNAPTIC PLASTICITY IN REVERSING THE EFFECTS OF STROKE

Neuroplasticity is an important element in fostering functional recovery following stroke and, indeed, activity-dependent rewiring and strengthening of synapses can be encouraged by various behavioral manipulations (Murphy

& Corbett, 2009). Soon after a stroke has occurred, various gene expression and protein changes occur, including increased production of growth factors (such as BDNF, insulin-like growth factor, and nerve growth factor, which

(cont'd)

augment synaptogenesis) and cytokine activity, which could be acting in a reparative capacity.

Stroke rehabilitation treatment during the first few months following stroke might be helpful by encouraging synaptic plasticity. Through the engagement of particular behaviors during this time, synapses controlling relearning of various skills are augmented and recovery may progress more readily. If partial neuronal functioning is present, appreciable restoration of neural circuit activity can be achieved, likely owing to the establishment and strengthening of compensatory rewiring and remapping of the neural circuitry by activation of growth factors. Aerobic exercise and environmental enrichment may also have significant rehabilitative effects through the neuroplastic changes provoked (Livingston-Thomas et al., 2016). While aerobic exercise facilitates motor recovery, acts against depression, and limits

degradation that occurs with normal ageing, more can be done to encourage recovery from stroke, especially in relation to the disturbed cognitive impairments. New multimodal approaches have been used to attenuate the consequences of stroke, including transcranial direct current brain stimulation (Allman et al., 2016), exercise, maintaining a Mediterranean diet, and environmental enrichment, including video games and virtual reality-based exercise. The latter have the add-on value of facilitating therapy conducted at home, and favor social interaction that serve to diminish distress and encourage the therapy to be undertaken. Although optimal effects of rehabilitation procedures are realized if these procedures occur within 2–3 months of stroke, it was suggested that these effects can be enhanced using broader therapy that encourages synaptic plasticity, and some recuperation can still occur for some time following stroke (Corbett et al., 2017).

It is widely accepted that following stroke, astrocytes and neural stem/progenitor cells (NSPCs) could be fundamental for healing and recovery by modulating inflammation, and by facilitating a new blood supply (angiogenesis) (Lindvall & Kokaia, 2015). It seemed that a particular type of astrocyte, Olig2PC-Astros, protected neurons from oxidative stress and was thus fundamental for stroke recovery (Jiang et al., 2013). As well, monocyte-derived macrophages that are recruited to the site of ischemic injury contribute to recovery by diminishing the damaging inflammation that is present following stroke. This notion led to the view that incomplete functional recovery among many patients may stem from insufficient recruitment of M2-like anti-inflammatory macrophages (or

alternatively an excessive of M1-like proinflammatory macrophages) to injured brain sites, and that developing anti-inflammatory treatments that home-in on injured sites could promote improvement after stroke (Wattananit et al., 2016).

This brings us to potential alternative forms of treatment that have been in the cards for some time. In animals with brain damage provoked by stroke, administration of transplanted human stem cells together with a protein, 3K3A-APC, that spurs the transplanted cells to become functional neurons, had beneficial actions (Wang, Zhao et al., 2016). Administration of a naturally occurring molecule C3a peptide may similarly be useful in stroke recovery as it has the ability to increase nerve growth and synapse formation, although to this point the research has been restricted to mice (Stokowska et al., 2016). It will certainly be some time before this or other such approaches are used in humans, but it may occur, and together with multimodal therapeutic approaches very positive outcomes may develop.

As microbial changes are associated with diet, stress, anxiety, and depression (Kleiman et al., 2015), it would reasonably be expected that stroke would likewise be reflected in such changes. Gut microbiota dysbiosis and diminished levels of TMAO, elevated numbers of opportunistic pathogens, and a lower number of commensal bacteria accompanied stroke and TIA (Yin et al., 2015). As gut microbiota alterations can affect inflammatory processes, there is reason to expect that treatments that influence microbial functioning (e.g., diet) might have positive effects in attenuating the damage associated with stroke (Benakis et al., 2016). This is not to say that gut bacteria are in some fashion related to the primary effect of stroke, but only that the poststroke period involves dynamic inflammatory changes, and that microbial processes might impact these events, thereby influencing outcomes.

DIABETES

Diabetes comes in several forms, including type 1, type 2, and gestational diabetes. The most obvious symptoms of type 1 and type 2 diabetes typically comprise elevated thirst and hunger, frequent urination, weight loss, and elevated blood sugar levels. As glucose levels vary over the day, as well as from day to day, diagnosis of diabetes relies more heavily on glycated hemoglobin (A1c) levels, which reflects the average blood sugar level over approximately the preceding 3 months.

Gestational Diabetes

The symptoms of gestational diabetes aren't as readily apparent as those of type 1 and type 2 forms, and are typically identified in the course of health screening during pregnancy, and thus may go undiagnosed among women without an obstetrician. This form of diabetes occurs in about 9% of pregnant women, and can ultimately lead to preeclampsia (high blood pressure and high levels of protein in the urine) if the condition is not treated. Women who developed gestational diabetes were also more likely to display symptoms of postpartum depression. The risk of developing this form of diabetes was elevated among women who were depressed during the initial two trimesters of pregnancy. Moreover, weight gain, even as little as 1.5%–2.5%, doubled the risk for gestational diabetes, and a gain of 2.5% over each of several years, tripled the risk (Adane et al., 2017). Related to this, gut microbiome composition during weeks 24–28 of pregnancy could predict gestational diabetes (Guo et al., 2017).

Type 1 Diabetes

Type 1 diabetes, which typically appears in the young (peaking at about 14 years of age) is an autoimmune disorder, possibly brought about by genetic factors, chemical agents, and environmental challenges, including viruses or bacteria. Unlike other autoimmune disorders, type 1 diabetes occurs equally often in males and females. This condition develops owing to the dysfunction of pancreatic beta cells so that insulin is not produced, which is necessary for body cells to take up sugars from the blood. In the absence of insulin, cells will not receive adequate nutrition, and elevated sugars in the blood will cause damage to cells of various organs. The onset of type 1 diabetes may be preceded by modification of particular immune cells, namely MAIT lymphocytes, which are capable of recognizing and responding to microbiota. The MAIT lymphocytes, which are ordinarily involved in mucosal homeostasis, may be undermined so that the bacterial entry into (and out of) the gut is increased, leading to the autoimmune response (Rouxel et al., 2017).

In an effort to preserve the health of normal cells, the immune system increases the presence of molecules, termed immune checkpoints, which either turn an immune signal up or down, thereby preventing T cells from attacking normal tissues. Checkpoints have become increasingly important in relation to cancer, and checkpoint inhibitors have become one of the new cadres of immunotherapies to treat certain cancers. It seems that checkpoints are diminished in patients with type 1 diabetes, and enhancing levels of the immunosuppressive protein PD-L1 can diminish type 1 diabetes in a mouse model (Ben Nasr, Tezza, D'Addio, Mameli, & Usuelli, 2017).

The mainstay for treatment of type 1 diabetes is insulin administration, which allows for glucose to be taken up into cells. However, insulin doesn't affect the inflammation and oxidative stress that may be present and still better outcomes could potentially be obtained by concurrently targeting these processes. It is possible that a cure will eventually be found, rather than a treatment that comprises exogenous insulin administration. For instance, an individual's own stem cells could potentially be modified so that they begin to make insulin (Millman et al., 2016). Alternatively, stem cell-derived beta cells encapsulated in a polymer can be transplanted, allowing for insulin release, without these cells being attacked by the immune system, which would occur without the polymer coating (Vegas et al., 2016). The development of artificial beta cells is also a possibility, and a smart patch has been developed that is replete with beta cells that can secrete insulin through tiny needles.

Type 2 Diabetes

Being a frequent precursor of type 2 diabetes, it is curious that metabolic syndrome seems to be taken less seriously than it should. This condition is diagnosed when insulin resistance (i.e., cells have become less responsive to insulin) is accompanied by at least three additional features, including abdominal obesity, high blood pressure, elevated fasting blood glucose, decreased HDL cholesterol, and high triglycerides. This syndrome may also be accompanied by low grade inflammation, and upregulation of proinflammatory cytokines (Esser, Legrand-Poels, Piette, Scheen, & Paquot, 2014), stemming from the presence of elevated visceral fat.

Type 2 diabetes develops as individuals become insulin resistant (i.e., cells fail to respond to insulin), or when release of insulin is dialed down so that insufficient levels of this hormone appear in the blood. Prediabetes and early diabetes might be managed by diet, exercise, and maintaining low weight. However, as the illness continues, a variety of medications are used to keep blood sugars within a normal range, but diabetes is a hard beast to conquer, especially if patients are resistant to advice.

Diabetes and Its Comorbidities

Diabetes is a wicked illness that engenders a constellation of secondary disturbances, such as circulatory problems, neuropathy, blindness, and impaired wound healing leading to lower limb amputations. It is also highly comorbid with a broad range of other illnesses and conditions, most notably heart disease, stroke, depression, kidney disease, fatty liver disease, immune-related disorders, and some types of cancer. Many of these conditions have also been linked to dyslipidemia (elevated plasma cholesterol and triglycerides), as well as increased inflammatory activity, oxidative stress, endothelial disturbances, and hypercoagulability, which are all correlated to one another to some extent. These factors may come together to produce the development of microvascular and macrovascular complications, leading to nephrological and cardiovascular disturbances. Of particular significance is that these conditions can be predicted by endothelial, inflammatory, and procoagulant biomarkers (Domingueti et al., 2016).

That's hardly the end of it, as individuals with diabetes may also experience reduced volume of temporal, prefrontoparietal, motor, and occipital cortices (Yoon et al., 2017). The brain atrophy and the accompanying cognitive decline were identified as a risk factor for the development of dementia (Ohara et al., 2011). In fact, the vascular disturbances that may stem from diabetes can affect blood flow to the brain, and thus result in impaired cognitive skills (Chung, Pimentel et al., 2015). Among older individuals, diabetes has also been associated with cognitive deficits, which may be related to vascular disturbances, as well as to oxidative stress and inflammation stemming from gut microbial disturbances (Xu, Zhou, & Zhu, 2017).

Fat cells, as we have already discussed, are a reservoir for proinflammatory cytokines, and thus may contribute to the comorbidity that has been observed between type 2 diabetes and depression, rheumatoid arthritis, heart disease, and other vascular illnesses (Paragh, Seres, Harangi, & Fülöp, 2014). In mice, insulin resistance in vascular endothelial cells was accompanied by increased cell-adhesion proteins and neutrophils, and could potentially favor intestinal tumor production (Wang, Häring et al., 2017).

When more than a single comorbid illness is present, they additively or synergistically promote further health disturbances. For instance, when depression accompanies diabetes, the risk of dying as a result of cardiovascular illness was appreciably increased (Park, Katon, & Wolf, 2013). This may occur owing to the conjoint biological actions of these different diseases, as well as the distress they create. It is

also possible that treatment adherence is relatively likely to be disrupted among depressed individuals, leading to poorer outcomes. The presence of the stress-reactive cytokine growth differentiation factor-15 (GDF-15), which contributes to regulation of inflammatory responses, is elevated in a wide number of diseases that involve cardiovascular functioning. Thus, its increase among individuals with type 2 diabetes could serve as a useful biomarker in predicting the development of diabetes-related heart disease (Berezin, 2016).

Given the clear dangers associated with diabetes, beyond that of heart disease and stroke, it is unsettling that both type 1 and type 2 diabetes in youth have been increasing progressively since 2002. Type 1 diabetes increased yearly by about 2% and type 2 diabetes that was associated with obesity increased by 5% annually. These rates were still greater among Blacks, Asian Americans and Hispanic people within the US (Mayer-Davis et al., 2017).

FACTORS THAT FAVOR THE DEVELOPMENT OF TYPE 2 DIABETES

Multiple interlinking factors contribute to the development of type 2 diabetes and also influence the response to treatments (Fig. 16.5). These influences can act additively or synergistically in promoting illness, and the actions of life style factors (e.g., obesity, stressors) are most prominent if they occur on a chronic basis.

Genetic Contributions

Diabetes is more common among individuals with a family history of this condition, supporting the involvement of genetic factors in its development. Many genes have been identified that are linked to dysfunction of pancreatic beta cells (Dooley et al., 2016), and

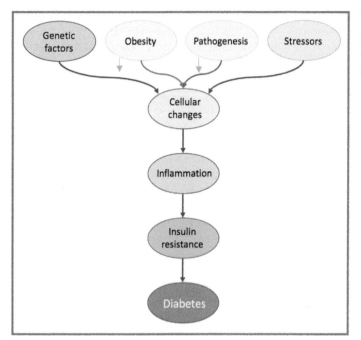

FIGURE 16.5 Genetic factors, diet, obesity, pathogenesis (immune and microbial factors) factors, and stressors, come together to favor inflammatory changes that promote insulin resistance and diabetes.

numerous genes (possibly exceeding 150) have been associated with type 2 diabetes, each of which accounts for only a small fraction of the variance related to this disorder. Other factors linked to diabetes, such as obesity, may similarly be regulated by genetic factors, and may thus indirectly contribute to the development of this disorder. Based on an analysis of 3,000 genes that distinguished healthy controls from those with type 2 diabetes, 168 major hubs (key genes with multiple links) seemed to be particularly germane to this disease, although certain core genes (e.g., SOX5) appear to be primary contributing factors (Axelsson et al., 2017).

Although it is often thought that heart disease is a consequence of diabetes, the possibility was explored that these two illnesses might share common genetic predictors. In line with this supposition, a genome-wide analysis involving more than 260,000 participants revealed 16 previously unreported genetic links between these illnesses. Of these, eight genes were tied to elevated diabetes, which also conferred elevated risk for heart disease, and pointed to at least two of these genes as potential targets for treating both diseases (Zhao, Rasheed et al., 2017). Many genetic association analyses have pointed to multiple genes coding for a wide variety of endocrine and immune factors in relation to diabetes as well as its many comorbid conditions, making it clear how complex analyses of these diseases can be.

Most of the gene variants identified have been associated with increased risk for type 2 diabetes, but in some instances, as in the case of carriers of a mutation in a gene, SLC30A8, a protective effect was evident (Flannick et al., 2014). Specifically, among older individuals the risk for diabetes was diminished by 65% in those carrying this mutation. If this gene's expression could be manipulated or the protein that it produces altered, this could potentially serve as a way of managing or treating diabetes.

Epigenetic changes may also occur that could influence metabolic syndrome and diabetes. Such actions were detected in a gene coding for "G-protein pathway suppressor 2" (GPS2) that influences macrophage functioning within fat cells, and which may be linked to the development of type 2 diabetes. Consistent with this suggestion, genetically engineered mice lacking GPS2 in macrophages, displayed inflammation of adipose tissue and developed insulin resistance upon being fed a high fat diet (Fan et al., 2016). Yet another epigenetic change has been identified on the TXNIP gene, being hypomethylated in individuals who were not controlling their glucose levels well (Soriano-Tárraga et al., 2016). Once again, these findings don't allow for causal conclusions to be drawn between the gene methylation and diabetes, but the findings may turn out to be particularly relevant.

Impact of Stressors

Although it is unlikely that stressors cause type 1 diabetes, serious stressful events in childhood were accompanied by a tripling in the risk for this condition (Nygren, Carstensen, Koch, Ludvigsson, & Frostell, 2015). To be sure, the data in this domain are sparce, but it does seem that chronic stressful experiences can exacerbate symptoms once diabetes is already present. The case for stressors in the development of metabolic disorder and type 2 diabetes is much stronger. Chronic life stressors can be instrumental in promoting these conditions, particularly if individuals indulged in high-fat, high-sugar diets. In a 1-year prospective study, the levels of A1c were related to the occurrence of stressful experiences (Lloyd et al., 1999), and both stressor experiences and the presence of anxiety similarly predicted the development of type 2 diabetes, which was linked to circulating IL-6 levels (Murdock et al., 2016).

As type 2 diabetes takes years to develop, it is difficult to determine whether stressful life experiences during particular periods were critical for its appearance. In fact, it seems that early life stressors and even those experienced in utero, could potentially contribute to the subsequent emergence of type 2 diabetes (Eriksson et al., 2014), even after adjusting for socioeconomic status and obesity in adulthood. Psychosocial stressors encountered prenatally were likewise associated with insulin resistance in young adults, and diabetes incidence was elevated among the children of women who lost a loved one during pregnancy. As expected, diabetes was also tied to stressors encountered during adulthood. Burnout, occupational class, and chronic distress at work or home, was accompanied by increased incidence of type 2 diabetes (Xu et al., 2018). As in the case of heart disease, type 2 diabetes was relatively common in association with a perceived mismatch between efforts and rewards received and in the presence of high job strain (Heraclides, Chandola, Witte, & Brunner, 2012).

Type 2 diabetes that develops as a result of stressors could occur through any of several processes. As we've seen repeatedly, stressors can favor increased eating among some individuals (particularly foods high in carbs) and can undermine efforts to engage in regular exercise, leading to weight gain, which then favors the development of metabolic syndrome and diabetes. As we saw earlier, stressors also affect hormones associated with food and energy regulation (e.g., ghrelin, leptin and cortisol), which influence dopamine processes associated with reward perception, thereby further encouraging consumption of tasty snacks rich in carbs (Abizaid et al., 2006). At the same time, the elevated cortisol may interfere with the action of insulin, hence contributing to the development of diabetes. Parenthetically, insulin insufficiency or insulin resistance can reduce the rewarding feelings derived from comfort foods, and as in the case of drug addictions, individuals might compensate for this by indulging in these foods to a still greater extent in order to regain the pleasure they previously obtained from these snacks (Khanh, Choi, Moh, Kinyua, & Kim, 2014). The net result of these stressor actions, among other things, is the development of visceral fat, which increases the availability of proinflammatory cytokines and instigation of type 2 diabetes.

Relation to Depression

Prospective studies revealed that type 2 diabetes was linked to depressive disorders, particularly when symptoms of metabolic syndrome were present (Schmitz et al., 2016). Consistent with the involvement of stressors and depression in the provocation and aggravation of type 2 diabetes, the availability of effective coping strategies, including social support, could act against the development of diabetes, and could augment treatment adherence and diabetes control. Likewise, the combination of psychological interventions and diabetes education facilitated glycemic control, as did a combination of stress management training or mindfulness training together with standard care (Surwit et al., 2002). In concert with these findings, the use of effective coping strategies together with high self-efficacy, self-esteem and optimism, were accompanied by better glucose control. It is of obvious practical significance that among diabetic individuals over 60 years of age who had been diagnosed with depression, treating the mood condition was accompanied by reduced symptoms related to other illnesses (Bogner et al., 2016).

Inflammatory Immune Factors and Type 2 Diabetes

Ordinarily, when food is consumed, macrophage activity is elevated in the intestine and the resulting secretion of IL-1β stimulates insulin production and release from pancreatic beta cells. However, a chronic increase of elevated cytokine presence secondary to stressors and increased abdominal fat, can disturb pancreatic beta cell functioning, leading to insulin resistance and thus metabolic syndrome and diabetes (Strissel, Denis, & Nikolajczyk, 2014; Stuart & Baune, 2012). To an extent, treatments that inhibit IL-1β, TNF-α, and NF-κB, can enhance the secretion of insulin from isolated islet cells (Nordmann et al., 2017).

As we saw in the case of heart disease (and other illnesses), there has been the view that omega-3 polyunsaturated fats may have positive effects in relation to type 2 diabetes, possibly through actions on inflammatory processes (Forouhi et al., 2016). Indeed, omega-3 treatment is accompanied by a decrease of CRP (Lin, Shi et al., 2016), and attenuates experimentally induced diabetogenic actions of tunicamycin-, streptozotocin-, or high fat diet-induced β-cell disturbances[3].

Obesity and Diabetes

Abdominal obesity has been linked to several genetic variants, which may favor the development of type 2 diabetes and heart disease (Emdin et al., 2017). Adiposity, as we've seen, is accompanied by elevated inflammatory stores, and when obese individuals encounter stressors, exaggerated inflammatory responses

---

[3] Diabetes is accompanied by elevated cholesterol (and triglycerides) that drive monocytes to form into macrophages. The macrophages engulf lipids (cholesterol and triglycerides) that have adhered to artery walls, and the build up of this complex contributes to atherosclerosis. These actions are helped along the way by a form of protein kinase C, which contribute to both the macrovascular and microvascular complications. However, one form of protein kinase C, that of protein kinase Cδ, may defend against inflammation. Thus, by selectively manipulating this form of PKC it may be possible to modify atherosclerosis (Li, Park et al., 2017).

occur, potentially leading to type 2 diabetes and depression, rheumatoid arthritis, heart disease, and other vascular illnesses (Paragh et al., 2014). In mice, insulin resistance in vascular endothelial cells was accompanied by increased cell-adhesion proteins and neutrophils, and as already mentioned, this could potentially favor intestinal tumor production, which may account for the elevated tumor risk among diabetic patients (Wang, Häring et al., 2017). Given these linkages, prevention or treatment of diabetes could be expected by acting on these processes. By example, the generation of healthy blood vessels within adipose tissues could be achieved by vascular endothelial growth factor B (VEGFB) and blocking the VEGF receptor-1, which would result in reduced inflammation and enhanced insulin functioning (Robciuc et al., 2016). In addition, through actions on hypothalamic neural stem cells and synaptic plasticity, FGF1 might ultimately be developed as a method of controlling glucose levels (Gasser, Moutos, Downes, & Evans, 2017).

In addition to being a storehouse for inflammatory cytokines that can promote diabetes, the continued release of the inflammatory substance leukotriene B_4 (LTB4) from visceral fat can promote metabolic disturbances. Significantly, improved metabolic health could be achieved among obese mice that were treated with an LTB4 receptor antagonist, which was initially developed to treat inflammatory disease (Li, Bandyopadhyay et al., 2015). Further, a form of white blood cell, eosinophils, could potentially contribute to diabetes and may hold prospects for its treatment. Eosinophils are ordinarily present in perivascular adipose tissue that surrounds blood vessels, operating to regulate artery contraction, but the presence of eosinophils is diminished in diabetes and hypertension (Withers et al., 2017). The addition of eosinophils could enhance perivascular adipose tissue functioning, leading to a rapid change of vascular functioning. These findings are still some way from being clinically practical, but for patients who don't respond to existing therapies, there is hope for alternative treatments coming down the pipeline based on immune manipulations.

ENVIRONMENTAL TOXICANTS AS A COMMON DENOMINATOR FOR ILLNESSES

One hardly needs to mention that various environmental agents have been linked to the development of mutations that lead to cancer. There is also evidence that pesticides are associated with increased incidence of type 2 diabetes (Evangelou et al., 2016), and endocrine disrupting chemicals (e.g., bisphenol A and phthalates) increased the risk for diabetes, particularly in children (Diamanti-Kandarakis et al., 2009). The endocrine disrupting agents also affect the microbiome, causing disrupted glucose homeostasis, thereby provoking hyperglycemia (Velmurugan, Ramprasath, Gilles, Swaminathan, & Ramasamy, 2017).

Gut Bacteria Influence the Emergence of Diabetes

Microbiota might also contribute to type 1 diabetes, possibly though their actions on inflammatory processes (Pellegrini et al., 2017) and in a set of genetically susceptible individuals, gut bacteria may trigger an autoimmune response against beta cells, thereby producing type 1 diabetes (Dunne et al., 2014). As expected, antibiotic treatment to young mice provoked the equivalent of type 1 diabetes, apparently owing to altered gut microbiota

(Livanos et al., 2016). Further to this, in female mice genetically at risk for type 1 diabetes, treatment with normal gut bacteria obtained from adult male mice, reduced the incidence of diabetes by 85%. These actions were accompanied by elevated testosterone, reduced islet inflammation and diminished autoantibody production, raising the possibility that sex differences in autoimmune disorders may be influenced by microbial factors (Markle et al., 2013).

Considerable data have supported a causal link between microbiota and type 2 diabetes (Pedersen, Gudmundsdottir, Nielsen, Hyotylainen, & Nielsen, 2016). In humans, type 2 diabetes was linked to elevated intake of antibiotics in the years preceding onset of this condition (Mikkelsen, Knop, Frost, Hallas, & Pottegård, 2015). Similarly, mice maintained on a high-fat diet were at elevated risk for the development of type 2 diabetes, but germ-free mice were less likely to become glucose intolerant and insulin resistant (Bäckhed, Manchester, Semenkovich, & Gordon, 2007). However, when microbiota transplants were initiated, their vulnerability to metabolic illnesses was elevated.

A high-fat diet may promote gut dysbiosis, which causes mucosal inflammation, and chronic systemic inflammation, which then favors the occurrence of type 2 diabetes (Bleau, Karelis, St-Pierre, & Lamontagne, 2015). Consistent with this view, administration of an anti-inflammatory agent, 5-aminosalicyclic acid (5-ASA), often used in the treatment of IBD, acted against insulin resistance that developed among mice that had been maintained on a high fat diet (Luck et al., 2015). Furthermore, a genetically engineered form of the gut bacteria lactobacillus, secretes glucagon-like peptide-1 (GLP-1), which reduces glucose levels among individuals with type 2 diabetes (Duan, Liu, & March, 2015) again pointing to the gut microbiota linkage to this condition.

Especially interesting findings came after reports that gut microbiota were altered among patients with diabetes, and that treatment with the common diabetic medication metformin was associated with an increase of several bacterial species that seemed to be tied to improved metabolism. In fact, when microbiota were taken from patients before and after metformin treatment, and then transplanted to mice, the microbiota that had been transformed by metformin led to positive outcomes, suggesting that these benefits of the drug came through microbiotic processes (Wu et al., 2017).

Fig. 16.6 depicts some of the pathways linking diet, microbiota, inflammatory processes, endocrine functioning, and brain processes (Calder et al., 2017). These same links may be fundamental in the provocation of type 2 diabetes. Increasingly, treatments to deal with metabolic syndrome and type 2 diabetes have involved approaches that targeted the microbiome (e.g. probiotics, prebiotics, and fecal transfer), with the aim of defining specific bacteria that should be altered to treat metabolic disorders (e.g., Delzenne, Cani, Everard, Neyrinck, & Bindels, 2015). In addition to beneficial effects that could be gained from prebiotics or probiotics, "post-biotic factors" that are derived from bacteria, can also have a positive impact on health, and muramyl dipeptide (MDP), a component of some types of bacteria, lowered adipose inflammation and diminished glucose intolerance in obese mice (Cavallari, Fullerton, Duggan, Foley, & Denou, 2017). As described earlier, dietary supplementation of short chain fatty acids butyrate, acetate, propionate, may limit weight gain, enhance energy expenditure, and thus augment insulin sensitivity, thereby acting against diabetes occurring (Herrema, IJzerman, & Nieuwdorp, 2017). Based on a small clinical trial, it also appeared that fecal transplants obtained from a lean donor, diminished insulin resistance in about half of obese men involved in this trial, and the clusters of bacteria that were present in the recipient could predict whether the fecal transplant would be effective (Kootte et al., 2017).

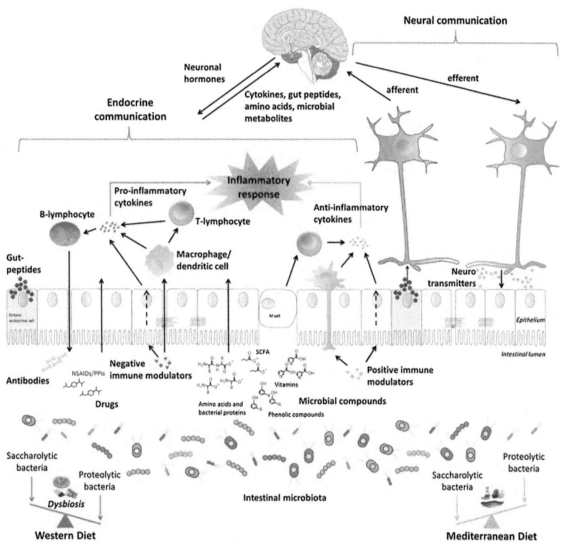

FIGURE 16.6 Overview of general mechanisms by which the gut microbiota affects host intestinal epithelium, immune-inflammatory response, and brain functioning. The epithelial layer consists of a single layer of epithelial cells that are sealed by tight junction proteins preventing paracellular passage. The connective tissue close to the epithelial cells (lamina propria) contains a large number of immune cells, both of the innate immune system (e.g., macrophages, dendritic cells, and mast cells) and the adaptive immune system (e.g., T cells, antibody-producing B cell derived plasma cells). In addition, cells of the central and enteric nervous system are innervated in the lamina propria. Factors affecting intestinal barrier function include food-derived allergens, (pathogenic and commensal) bacteria, and microbial compounds [lipopolysaccharides, metabolites such as short chain fatty acids (SCFA), tryptophan-related metabolites, neurotransmitters and peptides] as well as nonsteroidal anti-inflammatory drugs (NSAIDs) and proton pump inhibitors (PPIs). When activated by immune modulators, lymphocytes release anti-inflammatory and/or proinflammatory cytokines which trigger or regulate an inflammatory response. In addition, the released cytokines signal the brain to activate immunomodulatory mechanisms like the cholinergic anti-inflammatory pathway, the HPA axis as well as the SNS. Furthermore, intestinal neurotransmitters or their precursors can modulate functions of the central nervous systems. The neuronal efferent activation may also impact directly the epithelium and the gut microbiota composition. *Source: From Calder et al. (2017).*

TREATMENTS FOR TYPE 2 DIABETES

Within the US, 12%–14% of individuals have been diagnosed as being type 2 diabetic, and three times that number are prediabetic. It would be a reasonable guess that a great number of individuals who are prediabetic (blood glucose levels are near the cusp of being a problem, but A1c levels are still within the normal range) will say that they can ward off the illness through exercise and a proper diet. In theory that's likely correct, but the reality is another issue entirely. This is probably why they are where they are at the moment (the expression that "The road to hell is paved with good intentions" didn't develop for no reason).

Drugs such as metformin, have been remarkably effective, reducing the amount of sugar formed by the liver, and enhancing the sensitivity of insulin receptors. If metformin is not fully effective, or its effectiveness diminishes over time, additional treatments may be prescribed. Numerous options are available in this regard, each operating through somewhat different routes. One effective treatment comprises a sodium-glucose cotransporter-2 (SGLT2) inhibitor that increases glucose excretion in urine. As it turns out, an SGLT2 inhibitor (e.g., empagliflozi) was associated with a large ($\sim$35%) reduction of heart failure and death from heart disease (Zinman et al., 2015), although some agents in this class have been linked to increased risk of toe or foot amputation. Sulfonylureas drugs may be prescribed to increase insulin release from beta cells, or alpha-glucosidase inhibitors can be used to prevent digestion of carbohydrates, and injection of GLP-1 can be used to increase insulin secretion from the pancreas and increase insulin-sensitivity. Incidentally, variants of the GLP-1 receptor gene might also be linked to heart disease, and targeting GLP-1 receptors could thus have beneficial cardiac

effects (Scott, Freitag et al., 2016). Thiazolidinediones can fight insulin resistance and diminish certain cytokines, such as IL-6, but there have been indications that over the longer run pancreatic beta cells may lose their ability to respond. It ought to be noted that although different treatments can be used to reduce A1c levels to the same extent, the results on other measures may differ, as in the case of kidney disease (Lipska & Krumholz, 2017). As with so many illnesses, each of the available treatments has benefits that vary across individuals, but they can also engender uncomfortable side effects that also vary across individuals. Understandably, a precision medicine approach would be ideal in the selection of the appropriate treatments.

CONCLUDING COMMENTS

Many illnesses, such as type 2 diabetes and heart disease, are among a number of readily preventable illness that evolve with poor lifestyle choices, particularly in the presence of specific gene influences. The likelihood of type 2 diabetes and heart disease evolving can be diminished by favoring foods low in carbs, exercising regularly, getting enough sleep, and being able to cope effectively with distressing events. However, once these illnesses develop, they're difficult dragons to slay, particularly as many people simply continue with some of the same behaviors that brought them to the ill state. Thus, treatment of diabetes ought to involve diabetes education, self-management training, social support, and coaching to diminish those factors that interfere with self-control.

We focused on the comorbidities involving only a small number of illnesses, but a very great number of comorbidities exist, as we've seen with respect to psychiatric and neurodegenerative disorders. Many illnesses share common elements, including specific genetic factors, as well as microbial, inflammatory, and hormonal disturbances, raising the possibility that the shared features may be a common denominator responsible for the emergence of comorbid illnesses. Inflammatory factors may be a particularly important common element in this regard, and it has been suggested that NF-κB, which is involved in cellular responses to a variety of stimuli (e.g., stressors, bacterial and viral challenges, and cytokines) may play a prominent role in the development of disturbed immune functioning, inflammatory illnesses, autoimmune disorders, cancer, and responses to viral illnesses. Many chronic illnesses ought to be subject to modification by treatments that diminish inflammation (Laveti et al., 2013), and it may be productive to develop a slate of biomarkers to predict the later development of illnesses before they become intractable. In this regard, the presence of some illnesses may be effective markers for the development of other illnesses, and it was suggested that physicians attend to these possibilities in the prevention of future possible disorders (Anisman & Hayley, 2012a,b).

References

Abadier, M., Pramod, A. B., McArdle, S., Marki, A., Fan, Z., et al. (2017). Effector and regulatory T cells roll at high shear stress by inducible tether and sling formation. *Cell Reports, 21*(13), 3885–3899.

Abareshi, A., Anaeigoudari, A., Norouzi, F., Shafei, M. N., Boskabady, M. H., et al. (2016). Lipopolysaccharide-induced spatial memory and synaptic plasticity impairment is preventable by Captopril. *Advances in Medicine, 2016*, 7676512.

Abautret-Daly, Á., Dempsey, E., Parra-Blanco, A., Medina, C., & Harkin, A. (2017). Gut–Brain actions underlying comorbid anxiety and depression associated with inflammatory bowel disease. *Acta Neuropsychiatrica, 8*, 1–22.

Abbas, A. K., Lichtman, A. H., & Pillai, S. (2007). *Cellular and molecular immunology* (6th ed.). Saunders.

Abbas, A. K., Lichtman, A. H., & Pillai, S. (2015). *Cellular and molecular immunology* (8th ed.). Elsevier.

Abbasi, S. H., Hosseini, F., Modabbernia, A., Ashrafi, M., & Akhondzadeh, S. (2012). Effect of celecoxib add-on treatment on symptoms and serum IL-6 concentrations in patients with major depressive disorder: Randomized double-blind placebo-controlled study. *Journal of Affective Disorders, 141*, 308–314.

Abbink, M. R., Naninck, E. F. G., Lucassen, P. J., & Korosi, A. (2017). Early-life stress diminishes the increase in neurogenesis after exercise in adult female mice. *Hippocampus, 27*(8), 839–844.

Abbott, R., Whear, R., Nikolaou, V., Bethel, A., Coon, J. T., et al. (2015). Tumour necrosis factor-α inhibitor therapy in chronic physical illness: A systematic review and meta-analysis of the effect on depression and anxiety. *The Journal of Psychosomatic Research, 79*(3), 175–184.

Abdallah, C. G., Averill, L. A., Krystal, J. H., Southwick, S. M., & Arnsten, A. F. (2017). Glutamate and norepinephrine interaction: Relevance to higher cognitive operations and psychopathology. *Behavioral and Brain Sciences, 39*, e201.

Abdallah, C. G., Jiang, L., De Feyter, H. M., Fasula, M., Krystal, J. H., et al. (2014). Glutamate metabolism in major depressive disorder. *American Journal of Psychiatry, 171*(12), 1320–1327.

Abdallah, C. G., Sanacora, G., Duman, R. S., & Krystal, J. H. (2015). Ketamine and rapid-acting antidepressants:

A window into a new neurobiology for mood disorder therapeutics. *Annual Review of Medicine, 66*, 509–523.

Abeles, S. R., Jones, M. B., Santiago-Rodriguez, T. M., Ly, M., Klitgord, N., et al. (2016). Microbial diversity in individuals and their household contacts following typical antibiotic courses. *Microbiome, 4*(1), 39.

Aberle, J., Flitsch, J., Beck, N. A., Mann, O., Busch, P., et al. (2008). Genetic variation may influence obesity only under conditions of diet: Analysis of three candidate genes. *Molecular Genetics and Metabolism, 95*, 188–191.

Abildgaard, A., Elfving, B., Hokland, M., Lund, S., & Wegener, G. (2017). Probiotic treatment protects against the pro-depressant-like effect of high-fat diet in Flinders Sensitive Line rats. *Brain, Behavior, & Immunity, 65*, 33–42.

Abisror, N., Mekinian, A., Lachassinne, E., Nicaise-Roland, P., De Pontual, L., Chollet-Martin, S., & Fain, O. (2013). Autism spectrum disorders in babies born to mothers with antiphospholipid syndrome. *Seminars in Arthritis and Rheumatism, 43*(3), 348–351. Available from https://doi.org/10.1016/j.semarthrit.2013.07.001.

Abizaid, A. (2009). Ghrelin and dopamine: New insights on the peripheral regulation of appetite. *Journal of Neuroendocrinology, 21*(9), 787–793.

Abizaid, A., & Horvath, T. L. (2008). Brain circuits regulating energy homeostasis. *Regulatory Peptides, 149*(1–3), 3–10.

Abizaid, A., Liu, Z. W., Andrews, Z. B., Shanabrough, M., Borok, E., et al. (2006). Ghrelin modulates the activity and synaptic input organization of midbrain dopamine neurons while promoting appetite. *Journal of Clinical Investigation, 116*(12), 3229–3239.

Abizaid, A., Luheshi, G., & Woodside, B. C. (2014). Interaction between immune and energy-balance signals in the regulation of feeding and metabolism. In A. V. Kusnecov, & H. Anisman (Eds.), *The Wiley Blackwell handbook of psychoneuroimmunology* (pp. 488–503). Chichester, UK: John Wiley & Sons Ltd.

Abramson, L. Y., Seligman, M. E., & Teasdale, J. D. (1978). Learned helplessness in humans: Critique and reformulation. *Journal of Abnormal Psychology, 87*(1), 49–74.

Abrari, K., Rashidy-Pour, A., Semnanian, S., & Fathollahi, Y. (2009). Post-training administration of corticosterone enhances consolidation of contextual fear memory and hippocampal long-term potentiation in rats. *Neurobiology of Learning and Memory, 91*(3), 260−265.

Abul-Husn, N. S., Manickam, K., Jones, L. K., Wright, E. A., Hartzel, D. N., et al. (2016). Genetic identification of familial hypercholesterolemia within a single US health care system. *Science, 354*(6319), pii: aaf7000.

Ackerman, K. D., Heyman, R., Rabin, B. S., Anderson, B. P., Houck, P. R., et al. (2002). Stressful life events precede exacerbations of multiple sclerosis. *Psychosomatic Medicine, 64*(6), 916−920.

Ackerman, K. D., Martino, M., Heyman, R., Moyna, N. M., & Rabin, B. S. (1998). Stressor-induced alteration of cytokine production in multiple sclerosis patients and controls. *Psychosomatic Medicine, 60*(4), 484−491.

Acx, H., Serneels, L., Radaelli, E., Muyldermans, S., Vincke, C., et al. (2017). Inactivation of γ-secretases leads to accumulation of substrates and non-Alzheimer neurodegeneration. *EMBO Molecular Medicine, 9*(8), 1088−1099.

Adam, E. K., Heissel, J. A., Zeiders, K. H., Richeson, J. A., Ross, E. C., et al. (2015). Developmental histories of perceived racial discrimination and diurnal cortisol profiles in adulthood: A 20-year prospective study. *Psychoneuroendocrinology, 62*, 279−291.

Adams, E. F., Lee, A. J., Pritchard, C. W., & White, R. J. (2010). What stops us from healing the healers? A survey of help-seeking behaviour, stigmatization and depression within the medical profession. *International Journal of Social Psychiatry, 56*, 359−370.

Adams Waldorf, K. M., & Nelson, J. L. (2008). Autoimmune disease during pregnancy and the microchimerism legacy of pregnancy. *Immunological investigations, 37*(5), 631−644. Available from https://doi.org/10.1080/08820130802205886.

Adane, A. A., Tooth, L. R., & Mishra, G. D. (2017). Prepregnancy weight change and incidence of gestational diabetes mellitus: A finding from a prospective cohort study. *Diabetes Management, 124*, 72−80.

Ade, K. K., Wan, Y., Hamann, H. C., O'Hare, J. K., Guo, W., et al. (2016). Increased mGluR5 signaling underlies OCD-like behavioral and striatal circuit abnormalities in mice. *Biological Psychiatry, 80*(7), 522−533.

Ader, R., Felten, D., & Cohen, N. (1990). Interactions between the brain and the immune system. *Annual Review of Pharmacology and Toxicology, 30*, 561−602.

Adlan, A. M., Lip, G. Y., Paton, J. F., Kitas, G. D., & Fisher, J. P. (2014). Autonomic function and rheumatoid arthritis: A systematic review. *Seminars in Arthritis and Rheumatism, 44*(3), 283−304.

Afzal, S., Tybjærg-Hansen, A., Jensen, G. B., & Nordestgaard, B. G. (2016). Change in body mass index associated with lowest mortality in Denmark. *The Journal of the American Medical Association, 315*(18), 1989−1996.

Aghajani, M., Veer, I. M., van Hoof, M. J., Rombouts, S. A., van der Wee, N. J., et al. (2016). Abnormal functional architecture of amygdala-centered networks in adolescent posttraumatic stress disorder. *Human Brain Mapping, 37*(3), 1120−1135.

Aguilera, M., Arias, B., Wichers, M., Barrantes-Vidal, N., Moya, J., et al. (2009). Early adversity and 5-HTT/BDNF genes: New evidence of gene environment interactions on depressive symptoms in a general population. *Psychosomatic Medicine, 39*(9), 1425−1432.

Ahmadizar, F., Vijverberg, S. J., Arets, H. G., de Boer, A., Lang, J. E., et al. (2016). Early life antibiotic exposure is associated with an increased risk of allergy. *European Respiratory Society, 48*, PA3639.

Ahmadizar, F., Vijverberg, S. J. H., Arets, H. G. M., de Boer, A., Garssen, J., et al. (2017). Breastfeeding is associated with a decreased risk of childhood asthma exacerbations later in life. *Pediatric Allergy and Immunology, 28*(7), 649−654.

Aisen, P. S., Schafer, K. A., Grundman, M., Pfeiffer, E., Sano, M., et al. (2003). Effects of rofecoxib or naproxen vs placebo on Alzheimer disease progression: A randomized controlled trial. *The Journal of the American Medical Association, 289*(21), 2819−2826.

Ait-Belgnaoui, A., Colom, A., Braniste, V., Ramalho, L., Marrot, A., et al. (2014). Probiotic gut effect prevents the chronic psychological stress-induced brain activity abnormality in mice. *Neurogastroenterology and Motility, 26*(4), 510−520.

Aiyaz, M., Lupton, M. K., Proitsi, P., Powell, J. F., & Lovestone, S. (2012). Complement activation as a biomarker for Alzheimer's disease. *Immunobiology, 217*(2), 204−215.

Aizawa, E., Tsuji, H., Asahara, T., Takahashi, T., Teraishi, T., et al. (2016). Possible association of Bifidobacterium and Lactobacillus in the gut microbiota of patients with major depressive disorder. *Journal of Affective Disorders, 202*, 254−547.

Akbari, E., Asemi, Z., Daneshvar Kakhaki, R., Bahmani, F., Kouchaki, E., et al. (2016). Effect of probiotic supplementation on cognitive function and metabolic status in Alzheimer's disease: A randomized, double-blind and controlled trial. *Frontiers in Aging Neuroscience, 8*, 256.

Akdeniz, C., Tost, H., Streit, F., Haddad, L., Wüst, S., et al. (2014). Neuroimaging evidence for a role of neural social stress processing in ethnic minority-associated environmental risk. *JAMA Psychiatry, 71*(6), 672−680.

Akiki, T. J., Averill, C. L., Wrocklage, K. M., Schweinsburg, B., Scott, J. C., et al. (2017). The association of PTSD symptom severity with localized hippocampus and amygdala abnormalities. *Chronic Stress*. Available from https://doi.org/10.1177/2470547017724069.

Akimova, E., Lanzenberger, R., & Kasper, S. (2009). The serotonin-1A receptor in anxiety disorders. *Biological Psychiatry, 66*(7), 627–635.

Akiyama, T., & Carstens, E. (2013). Neural processing of itch. *Neuroscience, 250,* 697–714.

Alberio, T., Pippione, A. C., Zibetti, M., Olgiati, S., Cecconi, D., et al. (2012). Discovery and verification of panels of T-lymphocyte proteins as biomarkers of Parkinson's disease. *Scientific Reports, 2,* 953.

Albert, P. R., Vahid-Ansari, F., & Luckhart, C. (2014). Serotonin-prefrontal cortical circuitry in anxiety and depression phenotypes: Pivotal role of pre- and post-synaptic 5-HT1A receptor expression. *Frontiers in Behavioral Neuroscience, 8,* 199.

Alboni, R. M., van Dijk, S., Poggini, G., Milior, M., Perrotta, T., et al. (2017). Fluoxetine effects on molecular, cellular and behavioral endophenotypes of depression are driven by the living environment. *Molecular Psychiatry, 22*(4), 552–561.

Alda, M., Puebla-Guedea, M., Rodero, B., Demarzo, M., Montero-Marin, J., et al. (2016). Zen meditation, length of telomeres, and the role of experiential avoidance and compassion. *Mindfulness, 7,* 651–659.

Alexander, K. S., Wu, H. Q., Schwarcz, R., & Bruno, J. P. (2012). Acute elevations of brain kynurenic acid impair cognitive flexibility: Normalization by the alpha7 positive modulator galantamine. *Psychopharmacology (Berlin, Germany), 220*(3), 627–637.

Alexandre, C., Latremoliere, A., Ferreira, A., Miracca, G., Yamamoto, M., et al. (2017). Decreased alertness due to sleep loss increases pain sensitivity in mice. *Nature Medicine (New York, NY, United States), 23*(6), 768–774.

Alford, S., Patel, D., Perakakis, N., & Mantzoros, C. S. (2017). Obesity as a risk factor for Alzheimer's disease: Weighing the evidence. *Obesity Reviews, 19*(2), 269–280.

Alhabbab, R., Blair, P., Elgueta, R., Stolarczyk, E., Marks, E., et al. (2015). Diversity of gut microflora is required for the generation of B cell with regulatory properties in a skin graft model. *Scientific Reports, 5,* 11554.

Allen, J. M., Mailing, L. J., Cohrs, J., Salmonson, C., Fryer, J. D., et al. (2017). Exercise training-induced modification of the gut microbiota persists after microbiota colonization and attenuates the response to chemically-induced colitis in gnotobiotic mice. *Gut Microbes, 1,* 1–16.

Allen, N. J., & Barres, B. A. (2009). Neuroscience: Glia—More than just brain glue. *Nature, 457*(7230), 675–677.

Allen, S. J., Watson, J. J., Shoemark, D. K., Barua, N. U., & Patel, N. K. (2013). GDNF, NGF and BDNF as therapeutic options for neurodegeneration. *Pharmacology & Therapeutics, 138*(2), 155–175.

Allman, C., Amadi, U., Winkley, A. M., Wilkins, I., Filippini, N., et al. (2016). Ipsilesional anodal tDCS enhances the functional benefits of rehabilitation in patients after stroke. *Science Translational Medicine, 8* (330), 330re1.

Alonso, P., López-Solà, C., Real, E., Segalàs, C., & Menchón, J. M. (2015). Animal models of obsessive-compulsive disorder: Utility and limitations. *Neuropsychiatric Disease and Treatment, 11,* 1939–1955.

Alonso, R., Pisa, D., Marina, A. I., Morato, E., Rábano, A., et al. (2014). Fungal infection in patients with Alzheimer's disease. *Journal of Alzheimer's Disease, 41*(1), 301–311.

Al-Safadi, S., Al-Safadi, A., Branchaud, M., Rutherford, S., Dayanandan, A., et al. (2014). Stress-induced changes in the expression of the clock protein PERIOD1 in the rat limbic forebrain and hypothalamus: Role of stress type, time of day, and predictability. *PLoS One, 9*(10), e111166.

Alvarado, S., Tajerian, M., Suderman, M., Machnes, Z., Pierfelice, S., et al. (2015). An epigenetic hypothesis for the genomic memory of pain. *Frontiers in Cellular Neuroscience, 9,* 88.

Amat, J., Christianson, J. P., Aleksejev, R. M., Kim, J., Richeson, K. R., et al. (2014). Control over a stressor involves the posterior dorsal striatum and the act/outcome circuit. *The European Journal of Neuroscience, 40*(2), 2352–2358.

Ambeskovic, M., Roseboom, T. J., & Metz, G. A. S. (2017). Transgenerational effects of early environmental insults on aging and disease incidence. *Neuroscience & Biobehavioral Reviews,* pii: S0149-7634(16)30714-X.

Amenyogbe, N., Kollmann, T. R., & Ben-Othman, R. (2017). Early-life host—Microbiome interphase: The key frontier for immune development. *Front in Pediatrics, 5,* 111.

American Association of Neurology (2016). *Efficacy and safety of the therapeutic use of medical marijuana (cannabis) in selected neurologic disorders.* Retrieved from https://www.aan.com/Guidelines/home/GetGuidelineContent/651. Accessed May 2017.

American Psychiatric Association. (2013). *Diagnostic and statistical manual of mental disorders* (5th ed.). Arlington, WA: American Psychiatric Publishing.

Amoroso, T. (2015). The psychopharmacology of ±3,4 methylenedioxymethamphetamine and its role in the treatment of posttraumatic stress disorder. *Journal of Psychoactive Drugs, 47*(5), 337–344.

Anacker, C., & Hen, R. (2017). Adult hippocampal neurogenesis and cognitive flexibility—Linking memory and mood. *Nature Reviews Neuroscience, 18*(6), 335–346.

Ancelin, M. L., Scali, J., Norton, J., Ritchie, K., Dupuy, A. M., et al. (2017). The effect of an adverse psychological environment on salivary cortisol levels in the elderly differs by 5-HTTLPR genotype. *Neurobiology of Stress, 7,* 38–46.

Anderson, G., Jacob, A., Bellivier, F., & Geoffroy, P. A. (2016). Bipolar disorder: The role of the kynurenine and melatonergic pathways. *Current Pharmaceutical Design*, 22(8), 987–1012.

Anderson, G., & Maes, M. (2016). How immune-inflammatory processes link CNS and psychiatric disorders: Classification and treatment implications. *CNS & Neurological Disorders Drug Targets, 16*(3), 266–278.

Andrade, C. (2014). Nonsteroidal anti-inflammatory drugs and 5-HT(3) serotonin receptor antagonists as innovative antipsychotic augmentation treatments for schizophrenia. *Journal of Clinical Psychiatry, 75*, e707–709.

Andrade, C. (2016). Cannabis and neuropsychiatry, 1: Benefits and risks. *Journal of Clinical Psychiatry, 77*, e551–e554.

Andreae, M. H., Carter, G. M., Shaparin, N., Suslov, K., Ellis, R. J., et al. (2015). Inhaled cannabis for chronic neuropathic pain: A meta-analysis of individual patient data. *The Journal of Pain, 16*(12), 1221–1232.

Andrews, Z. B., & Abizaid, A. (2014). Neuroendocrine mechanisms that connect feeding behavior and stress. *Frontiers in Neuroscience, 8*, 312–322.

Angelier, F., & Chastel, O. (2009). Stress, prolactin and parental investment in birds: A review. *General and Comparative Endocrinology, 163*(1–2), 142–148.

Anisman, H. (2009). Cascading effects of stressors and inflammatory immune system activation: Implications for major depressive disorder. *Journal of Psychiatry & Neuroscience, 34*(1), 4–20.

Anisman, H. (2014). *An introduction to stress and health.* London: Sage.

Anisman, H. (2016). *Health psychology.* London: SAGE Publications.

Anisman, H., Du, L., Palkovits, M., Faludi, G., Kovacs, G. G., et al. (2008). Serotonin receptor subtype and p11 mRNA expression in stress-relevant brain regions of suicide and control subjects. *Journal of Psychiatry & Neuroscience, 33*(2), 131–141.

Anisman, H., & Hayley, S. (2012a). Illness comorbidity as a biomarker? *Journal of Psychiatry & Neuroscience, 37*(4), 221–223.

Anisman, H., & Hayley, S. (2012b). Inflammatory factors contribute to depression and its comorbid conditions. *Science Signaling, 5*(244), pe45.

Anisman, H., Hayley, S., & Merali, Z. (2003). Cytokines and stress: Sensitization and cross-sensitization. *Brain, Behavior, and Immunity, 17*, 86–93.

Anisman, H., Lu, Z. W., Song, C., Kent, P., McIntyre, D. C., et al. (1997). Influence of psychogenic and neurogenic stressors on endocrine and immune activity: Differential effects in fast and slow seizing rat strains. *Brain, Behavior, and Immunity, 11*(1), 63–74.

Anisman, H., & Matheson, K. (2005). Stress, anhedonia and depression: Caveats concerning animal models. *Neuroscience & Biobehavioral Reviews, 29*(4–5), 525–546.

Anisman, H., & Merali, Z. (1999). Anhedonic and anxiogenic effects of cytokine exposure. *Advances in Experimental Medicine and Biology, 461*, 199–233.

Anisman, H., & Merali, Z. (2002). Cytokines, stress, and depressive illness. *Brain, Behavior, and Immunity, 16*(5), 513–524.

Anisman, H., & Merali, Z. (2003). Cytokines, stress and depressive illness: Brain-immune interactions. *Annals of Medicine, 35*(1), 2–11.

Anisman, H., Merali, Z., & Hayley, S. (2008). Neurotransmitter, peptide and cytokine processes in relation to depressive disorder: Comorbidity of depression with neurodegenerative disorders. *Progress in Neurobiology, 85*(1), 1–74.

Anisman, H., Poulter, M. O., Gandhi, R., Merali, Z., & Hayley, S. (2007). Interferon-alpha effects are exaggerated when administered on a psychosocial stressor backdrop: Cytokine, corticosterone and brain monoamine variations. *Journal of Neuroimmunology, 186*(1–2), 45–53.

Anisman, H., Ravindran, A., Griffiths, J., & Merali, Z. (1999). Endocrine and cytokine correlates of major depression and dysthymia with typical or atypical features. *Molecular Psychiatry, 4*(2), 182–188.

Anisman, H., Zaharia, M. D., Meaney, M. J., & Merali, Z. (1998). Do early-life events permanently alter behavioral and hormonal responses to stressors? *International Journal of Developmental Neuroscience, 16*(3–4), 149–164.

Antila, H., Ryazantseva, M., Popova, D., Sipilä, P., Guirado, R., et al. (2017). Isoflurane produces antidepressant effects and induces TrkB signaling in rodents. *Scientific Reports, 7*(1), 7811.

Apfel, B. A., Otte, C., Inslicht, S. S., McCaslin, S. E., Henn-Haase, C., et al. (2011). Pretraumatic prolonged elevation of salivary MHPG predicts peritraumatic distress and symptoms of post-traumatic stress disorder. *Journal of Psychiatric Research, 45*(6), 735–741.

Apfel, B. A., Ross, J., Hlavin, J., Meyerhoff, D. J., Metzler, T. J., et al. (2011). Hippocampal volume differences in Gulf War veterans with current versus lifetime posttraumatic stress disorder symptoms. *Biological Psychiatry, 69*(6), 541–548.

Arambula, S. E., Jima, D., & Patisaul, H. B. (2017). Prenatal bisphenol A (BPA) exposure alters the transcriptome of the neonate rat amygdala in a sex-specific manner: A CLARITY-BPA consortium study. *Neurotoxicology, 65*, 207–220, pii: S0161-813X(17)30209-7.

Arnsten, A. F., Raskind, M. A., Taylor, F. B., & Connor, D. F. (2015). The effects of stress exposure on prefrontal cortex: Translating basic research into successful treatments for post-traumatic stress disorder. *Neurobiology of Stress, 1*, 89–99.

Aronsson, F., Lannebo, C., Paucar, M., Brask, J., Kristensson, K., et al. (2002). Persistence of viral RNA in the brain of offspring to mice infected with influenza A/WSN/33 virus during pregnancy. *Journal of Neuroinflammation, 8*(4), 353−357.

Aron-Wisnewsky, J., & Clement, K. (2016). The gut microbiome, diet, and links to cardiometabolic and chronic disorders. *Nature Reviews Nephrology, 12*(3), 169−181.

Arrieta, M. C., Stiemsma, L. T., Dimitriu, P. A., Thorson, L., Russell, S., et al. (2015). Early infancy microbial and metabolic alterations affect risk of childhood asthma. *Science Translational Medicine, 7*(307), 307ra152.

Artacho-Cordón, F., León, J., Sáenz, J.M., Fernández, M.F., Martin-Olmedo, P., et al. (2016). Contribution of persistent organic pollutant exposure to the adipose tissue oxidative microenvironment in an adult cohort: A multipollutant approach. *Environmental Science and Technology, 50*, 13529–13538.

Ascherio, A., & Schwarzschild, M. A. (2016). The epidemiology of Parkinson's disease: Risk factors and prevention. *Lancet Neurology, 15*(12), 1257−1272.

Ascherio, A., Zhang, S. M., Hernán, M. A., Kawachi, I., Colditz, G. A., et al. (2001). Prospective study of caffeine consumption and risk of Parkinson's disease in men and women. *Annals of Neurology, 50*(1), 56−63.

Ashar, Y. K., Chang, L. J., & Wager, T. D. (2017). Brain mechanisms of the placebo effect: An affective appraisal account. *Annual Review of Clinical Psychology, 13*, 73−98.

Ashhurst, T. M., van Vreden, C., Niewold, P., & King, N. J. (2014). The plasticity of inflammatory monocyte responses to the inflamed central nervous system. *Cellular Immunology, 291*(1−2), 49−57.

Ashwood, P., Wills, S., & Van de Water, J. (2006). The immune response in autism: A new frontier for autism research. *Journal of Leukocyte Biology, 80*(1), 1−15. Available from https://doi.org/10.1189/jlb.1205707.

Askew, K., Li, K., Olmos-Alonso, A., Garcia-Moreno, F., Liang, Y., et al. (2017). Coupled proliferation and apoptosis maintain the rapid turnover of microglia in the adult brain. *Cell Reports, 18*(2), 391−405.

Asnis, G. M., De La Garza, R., Kohn, S. R., Reinus, J. F., Henderson, M., et al. (2003). IFN-induced depression: A role for NSAIDs. *Psychopharmacology Bulletin, 37*(3), 29−50.

Atladottir, H. O., Pedersen, M. G., Thorsen, P., Mortensen, P. B., Deleuran, B., Eaton, W. W., & Parner, E. T. (2009). Association of family history of autoimmune diseases and autism spectrum disorders. *Pediatrics, 124*(2), 687−694. Available from https://doi.org/10.1542/peds.2008-2445.

Atlas, L. Y., & Wager, T. D. (2014). A meta-analysis of brain mechanisms of placebo analgesia: Consistent findings and unanswered questions. *Handbook of Experimental Pharmacology, 225*, 37−69.

Attwells, S., Setiawan, E., Wilson, A. A., Rusjan, P. M., Mizrahi, R., Miler, L., et al. (2017). Inflammation in the neurocircuitry of obsessive-compulsive disorder. *JAMA Psychiatry, 74*(8), 833−840.

Aubry, A. V., Serrano, P. A., & Burghardt, N. S. (2016). Molecular mechanisms of stress-induced increases in fear memory consolidation within the amygdala. *Frontiers in Behavioral Neuroscience, 10*, 191.

Audet, M. C., & Anisman, H. (2010). Neuroendocrine and neurochemical impact of aggressive social interactions in submissive and dominant mice: Implications for stress-related disorders. *International Journal of Neuropsychopharmacology, 13*(3), 361−372.

Audet, M. C., & Anisman, H. (2013). Interplay between pro-inflammatory cytokines and growth factors in depressive illnesses. *Frontiers in Cellular Neuroscience, 7*, 68.

Audet, M. C., Mangano, E. N., & Anisman, H. (2010). Behavior and pro-inflammatory cytokine variations among submissive and dominant mice engaged in aggressive encounters: Moderation by corticosterone reactivity. *Frontiers in Behavioral Neuroscience, 4*, 156.

Audet, M. C., McQuaid, R. J., Merali, Z., & Anisman, H. (2014). Cytokine variations and mood disorders: Influence of social stressors and social support. *Frontiers in Neuroscience, 8*, 416.

Aune, D., Keum, N., Giovannucci, E., Fadnes, L. T., Boffetta, P., et al. (2016). Whole grain consumption and the risk of cardiovascular disease, cancer, and all-cause and cause-specific mortality: A systematic review and dose-response meta-analysis of prospective studies. *British Medical Journal, 353*, i2716.

Aurbach, E. L., Inui, E. G., Turner, C. A., Hagenauer, M. H., Prater, K. E., et al. (2015). Fibroblast growth factor 9 is a novel modulator of negative affect. *Proceedings of the National Academy of Sciences of the United States of America, 112*(38), 11953−11958.

Auriel, E., Regev, K., & Korczyn, A. D. (2014). Nonsteroidal anti-inflammatory drugs exposure and the central nervous system. *Handbook of Clinical Neurology, 119*, 577−584.

Avci, A. B., Feist, E., & Burmester, G. R. (2016). Targeting GM-CSF in rheumatoid arthritis. *Clinical and Experimental Rheumatology, 34*(4 Suppl 98), 39−44.

Avella-Garcia, C. B., Julvez, J., Fortuny, J., Rebordosa, C., García-Esteban, R., et al. (2016). Acetaminophen use in pregnancy and neurodevelopment: Attention function and autism spectrum symptoms. *International Journal of Epidemiology, 45*(6), 1987−1996.

Averill, L. A., Purohit, P., Averill, C. L., Boesl, M. A., Krystal, J. H., et al. (2017). Glutamate dysregulation and glutamatergic therapeutics for PTSD: Evidence from human studies. *Neuroscience Letters, 649*, 147−155.

Aviles, H., & Monroy, F. P. (2001). Immunomodulatory effects of cold stress on mice infected intraperitoneally with a 50% lethal dose of Toxoplasma gondii. *Neuroimmunomodulation, 9*(1), 6–12.

Avitsur, R., Levy, S., Goren, N., & Grinshpahet, R. (2015). Early adversity, immunity and infectious disease. *Stress, 18*(3), 289–296.

Avitsur, R., Stark, J. L., Dhabhar, F. S., Padgett, D. A., & Sheridan, J. F. (2002). Social disruption-induced glucocorticoid resistance: Kinetics and site specificity. *Journal of Neuroimmunology, 124*(1–2), 54–61.

Axelsson, A. S., Mahdi, T., Nenonen, H. A., Singh, T., Hänzelmann, S., et al. (2017). Sox5 regulates beta-cell phenotype and is reduced in type 2 diabetes. *Nature Communications, 8*, 15652.

Azad, M. B., Abou-Setta, A. M., Chauhan, B. F., Rabbani, R., Lys, J., et al. (2017). Nonnutritive sweeteners and cardiometabolic health: A systematic review and meta-analysis of randomized controlled trials and prospective cohort studies. *Canadian Medical Association Journal, 189*(28), E929–E939.

Azad, M. B., Konya, T., Guttman, D. S., Field, C. J., Sears, M. R., et al. (2015). Infant gut microbiota and food sensitization: Associations in the first year of life. *Clinical and Experimental Allergy, 45*(3), 632–643.

Azevedo Da Silva, M., Singh-Manoux, A., Shipley, M. J., Vahtera, J., Brunner, E. J., et al. (2014). Sleep duration and sleep disturbances partly explain the association between depressive symptoms and cardiovascular mortality: The Whitehall II cohort study. *Journal of Sleep Rresearch, 23*(1), 94–97.

Baba, Y., Kuroiwa, A., Uitti, R. J., Wszolek, Z. K., & Yamada, T. (2005). Alterations of T-lymphocyte populations in Parkinson disease. *Parkinsonism & Related Disorders, 11*(8), 493–498.

Bach, D. R., Seymour, B., & Dolan, R. J. (2009). Neural activity associated with the passive prediction of ambiguity and risk for aversive events. *Journal of Neuroscience, 29*(6), 1648–1656.

Bach, D. R., Tzovara, A., & Vunder, J. (2017). Blocking human fear memory with the matrix metalloproteinase inhibitor doxycycline. *Molecular Psychiatry*. Available from https://doi.org/10.1038/mp.2017.65.

Bachstetter, A. D., Van Eldik, L. J., Schmitt, F. A., Neltner, J. H., Ighodaro, E. T., et al. (2015). Disease-related microglia heterogeneity in the hippocampus of Alzheimer's disease, dementia with Lewy bodies, and hippocampal sclerosis of aging. *Acta Neuropathologica Communications, 3*, 32.

Bäckhed, F., Ding, H., Wang, T., Hooper, L. V., Koh, G. Y., et al. (2004). The gut microbiota as an environmental factor that regulates fat storage. *Proceedings of the National Academy of Sciences of the United States of America, 101*(44), 15718–15723.

Bäckhed, F., Manchester, J. K., Semenkovich, C. F., & Gordon, J. I. (2007). Mechanisms underlying the resistance to diet-induced obesity in germ-free mice. *Proceedings of the National Academy of Sciences of the United States of America, 104*(3), 979–984.

Bäckhed, F., Roswall, J., Peng, Y., Feng, Q., Jia, H., et al. (2015). Dynamics and stabilization of the human gut microbiome during the first year of life. *Cell Host & Microbe, 17*(5), 690–703.

Badawy, A. A. (2015). Tryptophan availability for kynurenine pathway metabolism across the life span: Control mechanisms and focus on aging, exercise, diet and nutritional supplements. *Neuropharmacology, 112*(Pt B), 248–263.

Baeza-Raja, B., Sachs, B. D., Li, P., Christian, F., Vagena, E., et al. (2016). p75 neurotrophin receptor regulates energy balance in obesity. *Cell Reports, 14*, 255–268.

Baharnoori, M., Bhardwaj, S. K., & Srivastava, L. K. (2013). Effect of maternal lipopolysaccharide administration on the development of dopaminergic receptors and transporter in the rat offspring. *PLoS One, 8*(1), e54439.

Bai, M., Zhu, X., Zhang, L., Zhang, Y., Xue, L., et al. (2017). Divergent anomaly in mesocorticolimbic dopaminergic circuits might be associated with different depressive behaviors, an animal study. *Brain and Behavior, 7*(10), e00808.

Bailey, M. T. (2014). Influence of stressor-induced nervous system activation on the intestinal microbiota and the importance for immunomodulation. *Advances in Experimental Medicine and Biology, 817*, 255–276.

Baily, M. T. (2016). Psychological stress, immunity, and the effects on indigenous microflora. *Advances in Experimental Medicine and Biology, 874*, 225–246.

Bailey, M. T., Engler, H., Powell, N. D., Padgett, D. A., & Sheridan, J. F. (2007). Repeated social defeat increases the bactericidal activity of splenic macrophages through a Toll-like receptor-dependent pathway. *American Journal of Physiology Regulatory Integrative and Comparative Physiology, 293*(3), R1180–R1190.

Bailey, S. M., Udoh, U. S., & Young, M. E. (2014). Circadian regulation of metabolism. *Journal of Endocrinology, 222* (2), R75–R96.

Baird, G., & Norbury, C. F. (2016). Social (pragmatic) communication disorders and autism spectrum disorder. *Archives of Disease in Childhood, 101*(8), 745–751. Available from https://doi.org/10.1136/archdischild-2014-306944.

Baker, D. G., Nievergelt, C. M., & O'Connor, D. T. (2012). Biomarkers of PTSD: Neuropeptides and immune signaling. *Neuropharmacology, 62*(2), 663–673.

Baker, H. F., Ridley, R. M., Duchen, L. W., Crow, T. J., & Bruton, C. J. (1994). Induction of beta (A4)-amyloid in primates by injection of Alzheimer's disease brain homogenate. Comparison with transmission of spongiform encephalopathy. *Molecular Neurobiology, 8*(1), 25–39.

Baker, S. R., & Stephenson, D. (2000). Prediction and control as determinants of behavioural uncertainty: Effects on task performance and heart rate reactivity. *Integrative Physiological and Behavioral Science, 35*(4), 235–250.

Bakunina, N., Pariante, C. M., & Zunszain, P. A. (2015). Immune mechanisms linked to depression via oxidative stress and neuroprogression. *Immunology, 144*(3), 365–373.

Balakrishnan, K., Rijal Upadhaya, A., Steinmetz, J., Reichwald, J., Abramowski, D., et al. (2015). Impact of amyloid β aggregate maturation on antibody treatment in APP23 mice. *Acta Neuropathologica Communications, 3*, 41.

Baldwin, H. A., Koivula, P. P., Necarsulmer, J. C., Whitaker, K. W., & Harvey, B. K. (2017). Step sequence is a critical gait parameter of unilateral 6-OHDA Parkinson's rat models. *Cell Transplantation, 26*(4), 659–667.

Ball, T. M., Stein, M. B., & Paulus, M. P. (2014). Toward the application of functional neuroimaging to individualized treatment for anxiety and depression. *Depression and Anxiety, 31*(11), 920–933.

Ballard, E. D., Wills, K., Lally, N., Richards, E. M., Luckenbaugh, D. A., et al. (2017). Anhedonia as a clinical correlate of suicidal thoughts in clinical ketamine trials. *Journal of Affective Disorders, 218*, 195–200.

Baltazar, M. T., Dinis-Oliveira, R. J., de Lourdes Bastos, M., Tsatsakis, A. M., Duarte, J. A., et al. (2014). Pesticides exposure as etiological factors of Parkinson's disease and other neurodegenerative diseases--A mechanistic approach. *Toxicology Letters, 230*(2), 85–103.

Bam, M., Yang, X., Zhou, J., Ginsberg, J. P., Leyden, Q., et al. (2016). Evidence for epigenetic regulation of proinflammatory cytokines, interleukin-12 and interferon gamma, in peripheral blood mononuclear cells from PTSD patients. *Journal of Neuroimmune Pharmacology, 11*(1), 168–181.

Bam, M., Yang, X., Zumbrun, E. E., Zhong, Y., Zhou, J., et al. (2016). Dysregulated immune system networks in war veterans with PTSD is an outcome of altered miRNA expression and DNA methylation. *Scientific Reports, 6*, 31209.

Bandelow, B., Baldwin, D., Abelli, M., Altamura, C., Dell'Osso, B., et al. (2016). Biological markers for anxiety disorders, OCD and PTSD: A consensus statement. Part I: Neuroimaging and genetics. *The World Journal of Biological Psychiatry, 17*(5), 321–365.

Bandelow, B., Baldwin, D., Abelli, M., Bolea-Alamanac, B., Bourin, M., et al. (2017). Biological markers for anxiety disorders, OCD and PTSD: A consensus statement. Part II: Neurochemistry, neurophysiology and neurocognition. *The World Journal of Biological Psychiatry, 18*(3), 162–214.

Bandelow, B., Reitt, M., Röver, C., Michaelis, S., Görlich, Y., et al. (2015). Efficacy of treatments for anxiety disorders: A meta-analysis. *International Clinical Psychopharmacology, 30*(4), 183–192.

Bandelow, B., Sher, L., Bunevicius, R., Hollander, E., Kasper, S., et al. (2012). Guidelines for the pharmacological treatment of anxiety disorders, obsessive-compulsive disorder and posttraumatic stress disorder in primary care. *International Journal of Psychiatry in Clinical Practice, 16*(2), 77–84.

Bangasser, D. A., Wiersielis, K. R., & Khantsis, S. (2016). Sex differences in the locus coeruleus-norepinephrine system and its regulation by stress. *Brain Research, 1641* (Pt B), 177–188.

Baniasadi, M., Hosseini, G., Fayyazi Bordbar, M. R., Rezaei Ardani, A., & Mostafavi Toroghi, H. (2014). Effect of pregabalin augmentation in treatment of patients with combat-related chronic posttraumatic stress disorder: A randomized controlled trial. *Journal of Psychiatric Practice, 20*(6), 419–427.

Banks, W. A. (2015). The blood-brain barrier in neuroimmunology: Tales of separation and assimilation. *Brain, Behavior, and Immunity, 44*, 1–8. Available from https://doi.org/10.1016/j.bbi.2014.08.007.

Banks, W. A. (2016). From blood-brain barrier to blood-brain interface: New opportunities for CNS drug delivery. *Nature Reviews Drug Discovery, 15*(4), 275–292.

Bannaga, A. S., & Selinger, C. P. (2015). Inflammatory bowel disease and anxiety: Links, risks, and challenges faced. *Clinical and Experimental Gastroenterology, 8*, 111–117.

Bao, H., Asrican, B., Li, W., Gu, B., Wen, Z., et al. (2017). Long-range GABAergic inputs regulate neural stem cell quiescence and control adult hippocampal neurogenesis. *Cell Stem Cell, 21*(5), 604–617.e5.

Barak, Y. (2006). The immune system and happiness. *Autoimmunity Reviews, 5*(8), 523–527.

Barak, Y., Achiron, A., Mandel, M., Mirecki, I., & Aizenberg, D. (2005). Reduced cancer incidence among patients with schizophrenia. *Cancer, 104*(12), 2817–2821.

Baranoff, J. A., Hanrahan, S. J., Burke, A. L., & Connor, J. P. (2016). Changes in acceptance in a low-intensity, group-based acceptance and commitment therapy (ACT) chronic pain intervention. *International Journal of Behavioral Medicine, 23*(1), 30–38.

Baratta, M. V., Kodandaramaiah, S. B., Monahan, P. E., Yao, J., Weber, M. D., et al. (2016). Stress enables reinforcement-elicited serotonergic consolidation of fear memory. *Biological Psychiatry, 79*(10), 814–822.

Barbosa, I. G., Rocha, N. P., Assis, F., Vieira, É. L., Soares, J. C., et al. (2014). Monocyte and lymphocyte activation in bipolar disorder: A new piece in the puzzle of immune dysfunction in mood disorders. *International Journal of Neuropsychopharmacology, 18*, 1–7.

Barker, D. J. (1990). The fetal and infant origins of adult disease. *British Medical Journal, 301*(6761), 1111.

Bar-On, D., & Rottgardt, E. (1998). Reconstructing silenced biographical issues through feeling-facts. *Psychiatry, 61*(1), 61–83.

Bar-On, Y., Seidel, E., Tsukerman, P., Mandelboim, M., & Mandelboim, O. (2014). Influenza virus uses its neuraminidase protein to evade the recognition of two activating NK cell receptors. *Journal of Infectious Diseases, 210*(3), 410–418.

Barrès, R., Yan, J., Egan, B., Treebak, J. T., Rasmussen, M., et al. (2012). Acute exercise remodels promoter methylation in human skeletal muscle. *Cell Metabolism, 15*(3), 405–411.

Barrès, R., & Zierath, J. R. (2016). The role of diet and exercise in the transgenerational epigenetic landscape of T2DM. *Nature Reviews Endocrinology, 12*(8), 441–451.

Barroso-Batista, J., Demengeot, J., & Gordo, I. (2015). Adaptive immunity increases the pace and predictability of evolutionary change in commensal gut bacteria. *Nature Communications, 6*, 8945.

Barthas, F., Sellmeijer, J., Hugel, S., Waltisperger, E., Barrot, M., et al. (2015). The anterior cingulate cortex is a critical hub for pain-induced depression. *Biological Psychiatry, 77*(3), 236–245.

Barthels, C., Ogrinc, A., Steyer, V., Meier, S., Simon, F., et al. (2017). CD40-signalling abrogates induction of RORγt + Treg cells by intestinal CD103 + DCs and causes fatal colitis. *Nature Communications, 8*, 14715.

Baruch, K., & Schwartz, M. (2016). Circulating monocytes in between the gut and the mind. *Cell Stem Cell, 18*(6), 689–691.

Bastian, B., Jetten, J., & Ferris, L. J. (2014). Pain as social glue: Shared pain increases cooperation. *Psychological Science, 25*(11), 2079–2085.

Basu, J., Zaremba, J. D., Cheung, S. K., Hitti, F. L., Zemelman, B. V., et al. (2016). Gating of hippocampal activity, plasticity, and memory by entorhinal cortex long-range inhibition. *Science, 351*(6269), aaa5694.

Bauer, M. E., Vedhara, K., Perks, P., Wilcock, G. K., Lightman, S. L., et al. (2000). Chronic stress in caregivers of dementia patients is associated with reduced lymphocyte sensitivity to glucocorticoids. *Journal of Neuroimmunology, 103*(1), 84–92.

Bauman, M. D., Iosif, A. M., Ashwood, P., Braunschweig, D., Lee, A., Schumann, C. M., ... Amaral, D. G. (2013). Maternal antibodies from mothers of children with autism alter brain growth and social behavior development in the rhesus monkey. *Translational Psychiatry, 3*, e278. Available from https://doi.org/10.1038/tp.2013.47.

Baumeister, A. A. (2013). The chlorpromazine enigma. *Journal of the History of the Neurosciences, 22*(1), 14–29.

Baune, B. T. (2016). Are non-steroidal anti-inflammatory drugs clinically suitable for the treatment of symptoms in depression-associated inflammation? *Current Topics in Behavioral Neurosciences, 31*, 303–319.

Baune, B. T., Stuart, M., Gilmour, A., Wersching, H., Arolt, V., et al. (2012). The relationship between subtypes of depression and cardiovascular disease: A systematic review of biological models. *Translational Psychiatry, 2*, e92.

Bawa, F. L. M., Mercer, S. W., Atherton, R. J., Clague, F., Keen, A., et al. (2015). Does mindfulness improve outcomes in patients with chronic pain? Systematic review and meta-analysis. *British Journal of General Practice, 65* (635), e387–e400.

Bayer, A. J., Bullock, R., Jones, R. W., Wilkinson, D., Paterson, K. R., et al. (2005). Evaluation of the safety and immunogenicity of synthetic Abeta42 (AN1792) in patients with AD. *Neurology, 64*(1), 94–101.

Beaver, K. M., & Belsky, J. (2012). Gene-environment interaction and the intergenerational transmission of parenting: Testing the differential-susceptibility hypothesis. *The Psychiatric Quarterly, 83*(1), 29–40.

Beck, A. T. (1967). *Depression: Clinical, experimental, and theoretical aspects.* New York: Harper & Row.

Beck, A. T. (2008). The evolution of the cognitive model of depression and its neurobiological correlates. *The American Journal of Psychiatry, 165*(8), 969–977.

Beck, A. T., & Dozois, D. J. (2011). Cognitive therapy: Current status and future directions. *Annual Review of Medicine, 62*, 397–409.

Belcher, A. M., Ferré, S., Martinez, P. E., & Colloca, L. (2017). Role of placebo effects in pain and neuropsychiatric disorders. *Progress in Neuro-Psychopharmacology & Biological Psychiatry*, pii: S0278-5846(17)30164-1.

Belinson, H., & Michaelson, D. M. (2009). ApoE4-dependent Abeta-mediated neurodegeneration is associated with inflammatory activation in the hippocampus but not the septum. *Journal of Neural Transmission, 116*(11), 1427–1434.

Belkaid, Y., & Hand, T. W. (2014). Role of the microbiota in immunity and inflammation. *Cell, 157*(1), 121–141.

Bell, J. A., Hamer, M., Sabia, S., Singh-Manoux, A., Batty, G. D., et al. (2015). The natural course of healthy obesity over 20 years. *Journal of the American College of Cardiology, 65*(1), 101–102.

Bellesi, M., Pfister-Genskow, M., Maret, S., Keles, S., Tononi, G., et al. (2013). Effects of sleep and wake on oligodendrocytes and their precursors. *Journal of Neuroscience, 33*(36), 14288–14300.

Bellinger, D. L., Nance, D. M., & Lorton, D. (2013). Innervation of the immune system. In A. V. Kusnecov, & H. Anisman (Eds.), *The Wiley Blackwell handbook of psychoneuroimmunology*. Chichester, UK: John Wiley & Sons Ltd.

Bellono, N. W., Bayrer, J. R., Leitch, D. B., Castro, J., Zhang, C., et al. (2017). Enterochromaffin cells are gut chemosensors that couple to sensory neural pathways. *Cell, 170*, 185−198.e16.

Benakis, C., Brea, D., Caballero, S., Faraco, G., Moore, I., et al. (2016). Commensal microbiota affects ischemic stroke outcome by regulating intestinal γδ T cells. *Nature Medicine (New York, NY, United States), 22*(5), 516−523.

Benatti, C., Blom, J. M., Rigillo, G., Alboni, S., Zizzi, F., et al. (2016). Disease-induced neuroinflammation and depression. *CNS & Neurological Disorders Drug Targets, 15*(4), 414−433.

Benedetti, F. (2014). Drugs and placebos: What's the difference? Understanding the molecular basis of the placebo effect could help clinicians to better use it in clinical practice. *EMBO Reports, 15*(4), 329−332.

Benedetti, F., Carlino, E., & Piedimonte, A. (2016). Increasing uncertainty in CNS clinical trials: The role of placebo, nocebo, and Hawthorne effects. *Lancet Neurology, 15*(7), 736−747.

Ben-Menachem-Zidon, O., Ben-Menahem, Y., Ben-Hur, T., & Yirmiya, R. (2014). Intra-hippocampal transplantation of neural precursor cells with transgenic overexpression of IL-1 receptor antagonist rescues memory and neurogenesis impairments in an Alzheimer's disease model. *Neuropsychopharmacol, 39*(2), 401−414.

Ben Nasr, M., Tezza, S., D'Addio, F., Mameli, C., Usuelli, V., et al. (2017). PD-L1 genetic overexpression or pharmacological restoration in hemotopoietic stem and progenitor cells reverses autoimmune diabetes. *Science Translational Medicine, 9*(416), eaam7543.

Benner, E. J., Mosley, R. L., Destache, C. J., Lewis, T. B., Jackson-Lewis, V., et al. (2004). Therapeutic immunization protects dopaminergic neurons in a mouse model of Parkinson's disease. *Proceedings of the National Academy of Sciences of the United States of America, 101* (25), 9435−9440.

Bennett, M. R. (2011). Schizophrenia: Susceptibility genes, dendritic-spine pathology and gray matter loss. *Progress in Neurobiology, 95*(3), 275−300.

Benros, M. E., Nielsen, P. R., Nordentoft, M., Eaton, W. W., Dalton, S. O., & Mortensen, P. B. (2011). Autoimmune diseases and severe infections as risk factors for schizophrenia: a 30-year population-based register study. *The American Journal of Psychiatry, 168*(12), 1303−1310. Available from https://doi.org/10.1176/appi.ajp.2011.11030516.

Ben-Shaanan, T. L., Azulay-Debby, H., Dubovik, T., Starosvetsky, E., Korin, B., et al. (2016). Activation of the reward system boosts innate and adaptive immunity. *Nature Medicine, 22*(8), 940−944.

Beppu, K., Kosai, Y., Kido, M. A., Akimoto, N., Mori, Y., et al. (2013). Expression, subunit composition, and function of AMPA-type glutamate receptors are changed in activated microglia; possible contribution of GluA2 (GluR-B)-deficiency under pathological conditions. *Glia, 61*(6), 881−891.

Berardi, A., Schelling, G., & Campolongo, P. (2016). The endocannabinoid system and post traumatic stress disorder (PTSD): From preclinical findings to innovative therapeutic approaches in clinical settings. *Pharmacological Research, 111*, 668−678.

Bercik, P., & Collins, S. M. (2014). The effects of inflammation, infection and antibiotics on the microbiota-gut-brain axis. *Advances in Experimental Medicine and Biology, 817*, 279−289.

Bercik, P., Denou, E., Collins, J., Jackson, W., Lu, J., et al. (2011). The intestinal microbiota affect central levels of brain-derived neurotropic factor and behavior in mice. *Gastroenterology, 141*(2), 599−609.

Bercik, P., Verdu, E. F., Foster, J. A., Macri, J., Potter, M., et al. (2010). Chronic gastrointestinal inflammation induces anxiety-like behavior and alters central nervous system biochemistry in mice. *Gastroenterology, 139*(6), 2102−2112.

Berer, K., Gerdes, L. A., Cekanaviciute, E., Jia, X., Xiao, L., et al. (2017). Gut microbiota from multiple sclerosis patients enables spontaneous autoimmune encephalomyelitis in mice. *Proceedings of the National Academy of Sciences of the United States of America, 114*(40), 10719−10724.

Berer, K., Mues, M., Koutrolos, M., Rasbi, Z. A., Boziki, M., et al. (2011). Commensal microbiota and myelin autoantigen cooperate to trigger autoimmune demyelination. *Nature, 479*(7374), 538−541.

Berezin, A. E. (2016). Diabetes mellitus related biomarker: The predictive role of growth-differentiation factor-15. *Diabetes & Metabolic Syndrome, 10*(1 Suppl 1), S154−S157.

Berger, M., & Sarnyai, Z. (2015). "More than skin deep": Stress neurobiology and mental health consequences of racial discrimination. *Stress, 18*(1), 1−10.

Bergh, C., Fall, K., Udumyan, R., Sjöqvist, H., Fröbert, O., et al. (2017). Severe infections and subsequent delayed cardiovascular disease. *European Journal of Preventive Cardiology, 24*(18), 1958−1966.

Berna, C., Kirsch, I., Zion, S. R., Lee, Y. C., Jensen, K. B., et al. (2017). Side effects can enhance treatment response through expectancy effects: An experimental analgesic randomized controlled trial. *Pain, 158*(6), 1014−1020.

Bernasconi, E., Favre, L., Maillard, M. H., Bachmann, D., Pythoud, C., et al. (2010). Granulocyte-macrophage colony-stimulating factor elicits bone marrow-derived cells that promote efficient colonic mucosal healing. *Inflammatory Bowel Disease, 16*(3), 428−441.

Berridge, K. C., & Robinson, T. E. (1998). What is the role of dopamine in reward: Hedonic impact, reward learning, or incentive salience? *Brain Research Reviews, 28*(3), 309−369.

Berridge, K. C., & Robinson, T. E. (2016). Liking, wanting, and the incentive-sensitization theory of addiction. *American Psychologist, 71*(8), 670−679.

Bertone-Johnson, E. R., Whitcomb, B. W., Missmer, S. A., Karlson, E. W., & Rich-Edwards, J. W. (2012). Inflammation and early-life abuse in women. *Am J Prev Med, 43*(6), 611−620.

Besedovsky, L., Dimitrov, S., Born, J., & Lange, T. (2016). Nocturnal sleep uniformly reduces numbers of different T-cell subsets in the blood of healthy men. *American Journal of Physiology Regulatory Integrative and Comparative Physiology, 311*(4), R637−R642.

Besedovsky, L., Lange, T., & Born, J. (2012). Sleep and immune function. *Pflugers Archives, 463*(1), 121−137.

Betancur, C. (2011). Etiological heterogeneity in autism spectrum disorders: More than 100 genetic and genomic disorders and still counting. *Brain Research, 1380,* 42−77. Available from https://doi.org/10.1016/j.brainres.2010.11.078.

Bettiol, S. S., Rose, T. C., Hughes, C. J., & Smith, L. A. (2015). Alcohol consumption and Parkinson's disease risk: A review of recent findings. *Journal of Parkinson's Disease, 5*(3), 425−442.

Beura, L. K., Hamilton, S. E., Bi, K., Schenkel, J. M., Odumade, O. A., et al. (2016). Normalizing the environment recapitulates adult human immune traits in laboratory mice. *Nature, 532*(7600), 512−516.

Beutel, M. E., Stark, R., Pan, H., Silbersweig, D., & Dietrich, S. (2010). Changes of brain activation pre- post short-term psychodynamic inpatient psychotherapy: An fMRI study of panic disorder patients. *Psychiatry Research, 184*(2), 96−104.

Bhadra, R., Cobb, D. A., Weiss, L. M., & Khan, I. A. (2013). Psychiatric disorders in toxoplasma seropositive patients--The CD8 connection. *Schizophrenia Bulletin, 39*(3), 485−489.

Bharadwaj, R. A., Jaffe, A. E., Chen, Q., Deep-Soboslay, A., Goldman, A. L., et al. (2018). Genetic risk mechanisms of posttraumatic stress disorder in the human brain. *Journal of Neuroscience Research, 96*(1), 21−30.

Bharwani, A., Mian, M. F., Foster, J. A., Surette, M. G., Bienenstock, J., et al. (2016). Structural & functional consequences of chronic psychosocial stress on the microbiome & host. *Psychoneuroendocrinology, 63,* 217−227.

Bharwani, A., Mian, M. F., Surette, M. G., Bienenstock, J., & Forsythe, P. (2017). Oral treatment with Lactobacillus rhamnosus attenuates behavioural deficits and immune changes in chronic social stress. *BMC Medicine, 15*(1), 7.

Bhatnagar, S., Shanks, N., & Meaney, M. J. (1996). Plaque-forming cell responses and antibody titers following injection of sheep red blood cells in nonstressed, acute, and/or chronically stressed handled and nonhandled animals. *Dev Psychobiol, 29*(2), 171−181.

Bian, G., Gloor, G. B., Gong, A., Jia, C., Zhang, W., et al. (2017). The gut microbiota of healthy aged chinese is similar to that of the healthy young. *mSphere, 2*(5), pii: e00327-17.

Biber, K., Tsuda, M., Tozaki-Saitoh, H., Tsukamoto, K., Toyomitsu, E., et al. (2011). Neuronal CCL21 up-regulates microglia P2X4 expression and initiates neuropathic pain development. *EMBO Journal, 30*(9), 1864−1873.

Bible, E. (2014). Alzheimer disease: High serum levels of the pesticide metabolite DDE--A potential environmental risk factor for Alzheimer disease. *Nature Reviews Neurology, 10*(3), 125.

Bighelli, I., Trespidi, C., Castellazzi, M., Cipriani, A., Furukawa, T. A., et al. (2016). Antidepressants and benzodiazepines for panic disorder in adults. *Cochrane Database of Systematic Reviews, 9,* CD011567.

Bigley, A. B., & Simpson, R. J. (2015). NK cells and exercise: Implications for cancer immunotherapy and survivorship. *Discovery Medicine, 19*(107), 433−443.

Bilbo, S. D., Block, C. L., Bolton, J. L., Hanamsagar, R., & Tran, P. K. (2018). Beyond infection − Maternal immune activation by environmental factors, microglial development, and relevance for autism spectrum disorders. *Experimental Neurology, 299*(Pt A), 241−251. Available from https://doi.org/10.1016/j.expneurol.2017.07.002.

Bilbo, S. D., Dhabhar, F. S., Viswanathan, K., Saul, A., Yellon, S. M., et al. (2002). Short day lengths augment stress-induced leukocyte trafficking and stress-induced enhancement of skin immune function. *Proceedings of the National Academy of Sciences of the United States of America, 99,* 4067−4072.

Bilbo, S. D., & Schwarz, J. M. (2012). The immune system and developmental programming of brain and behavior. *Frontiers in Neuroendocrinology, 33*(3), 267−286.

Binelli, C., Muñiz, A., Sanches, S., Ortiz, A., Navines, R., et al. (2015). New evidence of heterogeneity in social anxiety disorder: Defining two qualitatively different personality profiles taking into account clinical, environmental and genetic factors. *European Psychiatry, 30*(1), 160−165.

Bird, A. (2007). Perceptions of epigenetics. *Nature, 447*(7143), 396−398.

Birey, F., Kloc, M., Chavali, M., Hussein, I., Wilson, M., et al. (2015). Genetic and stress-induced loss of NG2 glia triggers emergence of depressive-like behaviors through reduced secretion of FGF2. *Neuron, 88*(5), 941–956.

Birney, E., Smith, G. D., & Greally, J. M. (2016). Epigenome-wide association studies and the interpretation of disease-omics. *PLoS Genetics, 12*(6), e1006105.

Bissonette, G. B., & Roesch, M. R. (2016). Development and function of the midbrain dopamine system: What we know and what we need to. *Genes, Brain and Behavior, 15*(1), 62–73. Available from https://doi.org/10.1111/gbb.12257.

Biswas, A., Oh, P. I., Faulkner, G. E., Bajaj, R. R., Bajaj, R. R., et al. (2015). Sedentary time and its association with risk for disease incidence, mortality, and hospitalization in adults: A systematic review and meta-analysis. *Annals of Internal Medicine, 162*(2), 123–132.

Bitanihirwe, B. K., Peleg-Raibstein, D., Mouttet, F., Feldon, J., & Meyer, U. (2010). Late prenatal immune activation in mice leads to behavioral and neurochemical abnormalities relevant to the negative symptoms of schizophrenia. *Neuropsychopharmacology, 35*(12), 2462–2478.

Bjelakovic, G., Nikolova, D., Gluud, L. L., Simonetti, R. G., & Gluud, C. (2012). Antioxidant supplements for prevention of mortality in healthy participants and patients with various diseases. *Cochrane Database of Systematic Reviews, 14*(3), CD007176.

Black, C., & Miller, B. J. (2015). Meta-analysis of cytokines and chemokines in suicidality: Distinguishing suicidal versus nonsuicidal patients. *Biological Psychiatry, 78*(1), 28–37.

Blalock, J. E. (2005). The immune system as the sixth sense. *Journal of Internal Medicine, 257*(2), 126–138.

Bland, S. T., Tamlyn, J. P., Barrientos, R. M., Greenwood, B. N., Watkins, L. R., et al. (2007). Expression of fibroblast growth factor-2 and brain-derived neurotrophic factor mRNA in the medial prefrontal cortex and hippocampus after uncontrollable or controllable stress. *Neuroscience, 144*(4), 1219–1228.

Bland, S. T., Twining, C., Watkins, L. R., & Maier, S. F. (2003). Stressor controllability modulates stress-induced serotonin but not dopamine efflux in the nucleus accumbens shell. *Synapses, 49*(3), 206–208.

Blandini, F., & Armentero, M. T. (2012). Animal models of Parkinson's disease. *FEBS Journal, 279*(7), 1156–1166.

Blandino, P., Barnum, C. J., Solomon, L. G., Larish, Y., Lankow, B. S., et al. (2009). Gene expression changes in the hypothalamus provide evidence for regionally-selective changes in IL-1 and microglial markers after acute stress. *Brain, Behavior, and Immunity, 23*(7), 958–968.

Bleau, C., Karelis, A. D., St-Pierre, D. H., & Lamontagne, L. (2015). Crosstalk between intestinal microbiota, adipose tissue and skeletal muscle as an early event in systemic low-grade inflammation and the development of obesity and diabetes. *Diabetes/Metabolism Research and Reviews, 31*, 545–561.

Blessing, E. M., Steenkamp, M. M., Manzanares, J., & Marmar, C. R. (2015). Cannabidiol as a potential treatment for anxiety disorders. *Neurotherapeutics, 12*(4), 825–836.

Blevins, C. L., Sagui, S. J., & Bennett, J. M. (2017). Inflammation and positive affect: Examining the stress-buffering hypothesis with data from the National Longitudinal Study of Adolescent to Adult Health. *Brain, Behavior, and Immunity, 61*, 21–26.

Blier, P. (2013). Exploiting N-methyl-d-aspartate channel blockade for a rapid antidepressant response in major depressive disorder. *Biological Psychiatry, 74*, 238–239.

Blier, P. (2016). Neurotransmitter targeting in the treatment of depression. *Journal of Clinical Psychiatry, 74*(Suppl 2), 19–24.

Block, M. L., Zecca, L., & Hong, J. S. (2007). Microglia-mediated neurotoxicity: Uncovering the molecular mechanisms. *Nature Reviews Neuroscience, 8*(1), 57–69.

Blomstrom, A., Karlsson, H., Gardner, R., Jorgensen, L., Magnusson, C., et al. (2016). Associations between maternal infection during pregnancy, childhood infections, and the risk of subsequent psychotic disorder--A Swedish cohort study of nearly 2 million individuals. *Schizophrenia Bulletin, 42*(1), 125–133.

Blomstrom, A., Karlsson, H., Wicks, S., Yang, S., Yolken, R. H., et al. (2012). Maternal antibodies to infectious agents and risk for non-affective psychoses in the offspring--A matched case-control study. *Schizophrenia Research, 140*(1–3), 25–30.

Bloomfield, S. F., Rook, G. A., Scott, E. A., Shanahan, F., Stanwell-Smith, R., et al. (2016). Time to abandon the hygiene hypothesis: New perspectives on allergic disease, the human microbiome, infectious disease prevention and the role of targeted hygiene. *Perspectives in Public Health, 136*(4), 213–224.

Bluett, R. J., Báldi, R., Haymer, A., Gaulden, A. D., Hartley, N. D., et al. (2017). Endocannabinoid signalling modulates susceptibility to traumatic stress exposure. *Nature Communications, 8*, 14782.

Blundell, S., Ray, K. K., Buckland, M., & White, P. D. (2015). Chronic fatigue syndrome and circulating cytokines: A systematic review. *Brain, Behavior, and Immunity, 50*, 186–195.

Bobyn, J., Mangano, E. N., Gandhi, A., Nelson, E., Moloney, K., et al. (2012). Viral-toxin interactions and Parkinson's disease: Poly I:C priming enhanced the neurodegenerative effects of paraquat. *Journal of Neuroinflammation, 9*, 86.

Boccia, M., D'Amico, S., Bianchini, F., Marano, A., Giannini, A. M., et al. (2016). Different neural modifications underpin PTSD after different traumatic events: An fMRI meta-analytic study. *Brain, Behavior, and Immunity, 10*, 226–237.

Boche, D., Zotova, E., Weller, R. O., Love, S., Neal, J. W., et al. (2008). Consequence of Abeta immunization on the vasculature of human Alzheimer's disease. *brain*. *Brain: A Journal of Neurology*, *131*(Pt 12), 3299–3310.

Bode, L., Kunz, C., Muhly-Reinholz, M., Mayer, K., Seeger, W., et al. (2004). Inhibition of monocyte, lymphocyte, and neutrophil adhesion to endothelial cells by human milk oligosaccharides. *Journal of Thrombosis and Haemostasis*, *92*(6), 1402–1410.

Boekhoorn, K., Joels, M., & Lucassen, P. J. (2006). Increased proliferation reflects glial and vascular-associated changes, but not neurogenesis in the presenile Alzheimer hippocampus. *Neurobiology of Disease*, *24*(1), 1–14.

Bogenschutz, M. P., & Ross, S. (2017). Therapeutic applications of classic hallucinogens. *Current Topics in Behavioral Neurosciences*, *36*, 361–391. Available from https://doi.org/10.1007/7854_2016_464.

Boger, H. A., Granholm, A. C., McGinty, J. F., & Middaugh, L. D. (2010). A dual-hit animal model for age-related parkinsonism. *Progress in Neurobiology*, *90*(2), 217–229.

Bogner, H. R., Joo, J. H., Hwang, S., Morales, K. H., Bruce, M. L., et al. (2016). Does a depression management program decrease mortality in older adults with specific medical conditions in primary care? An exploratory analysis. *Journal of the American Geriatrics Society*, *64*(1), 126–131.

Boissier, M. C., Semerano, L., Challal, S., Saidenberg-Kermanac'h, N., & Falgarone, G. (2012). Rheumatoid arthritis: From autoimmunity to synovitis and joint destruction. *Journal of Autoimmunity*. Available from https://doi.org/10.1016/j.jaut.2012.05.021.

Bollinger, J. L., Bergeon Burns, C. M., & Wellman, C. L. (2016). Differential effects of stress on microglial cell activation in male and female medial prefrontal cortex. *Brain, Behavior, and Immunity*, *52*, 88–97.

Bombay, A., Matheson, K., & Anisman, H. (2011). The impact of stressors on second generation Indian Residential School survivors. *Transcultural Psychiatry*, *48*(4), 367–391.

Bombay, A., Matheson, K., & Anisman, H. (2014). The intergenerational effects of Indian Residential Schools: Implications for the concept of historical trauma. *Transcultural Psychiatry*, *51*(3), 320–338.

Bondar, N. P., & Merkulova, T. I. (2016). Brain-derived neurotrophic factor and early-life stress: Multifaceted interplay. *Journal of Bioscience*, *41*(4), 751–758.

Bonfili, L., Cecarini, V., Berardi, S., Scarpona, S., Suchodolski, J. S., et al. (2017). Microbiota modulation counteracts Alzheimer's disease progression influencing neuronal proteolysis and gut hormones plasma levels. *Scientific Reports*, *7*(1), 2426.

Bonneau, R. H. (1996). Stress-induced effects on integral immune components involved in herpes simplex virus (HSV)-specific memory cytotoxic T lymphocyte activation. *Brain, Behavior, and Immunity*, *10*(1996), 139–163.

Bonneau, R. H., Zimmerman, K. M., Ikeda, S. C., & Jones, B. C. (1998). Differential effects of stress-induced adrenal function on components of the herpes simplex virus-specific memory cytotoxic T-lymphocyte response. *Journal of Neuroimmunology*, *82*(2), 191–199.

Booij, L., Szyf, M., Carballedo, A., Frey, E. M., Morris, D., et al. (2015). DNA methylation of the serotonin transporter gene in peripheral cells and stress-related changes in hippocampal volume: A study in depressed patients and healthy controls. *PLoS One*, *10*(3), e0119061.

Booij, S. H., Bos, E. H., Jonge, P., & Oldehinkel, A. J. (2015). Markers of stress and inflammation as potential mediators of the relationship between exercise and depressive symptoms: Findings from the TRAILS study. *Psychophysiology*, *52*(3), 352–358.

Borgland, S. L., Chang, S. J., Bowers, M. S., Thompson, J. L., Vittoz, N., et al. (2009). Orexin A/hypocretin-1 selectively promotes motivation for positive reinforcers. *Journal of Neuroscience*, *29*(36), 11215–11225.

Borodovitsyna, O., Flamini, M. D., & Chandler, D. J. (2018). Acute stress persistently alters locus coeruleus function and anxiety-like behavior in adolescent rats. *Neuroscience*, *373*, 7–19.

Borovikova, L. V., Ivanova, S., Zhang, M., Yang, H., Botchkina, G. I., et al. (2000). Vagus nerve stimulation attenuates the systemic inflammatory response to endotoxin. *Nature*, *2000*(405), 458.

Borowski, T., Kokkinidis, L., Merali, Z., & Anisman, H. (1998). Lipopolysaccharide, central in vivo biogenic amine variations, and anhedonia. *NeuroReport*, *9*, 3797–3801.

Borsini, A., Cattaneo, A., Malpighi, C., Thuret, S., Harrison, N. A., et al. (2018). Interferon-alpha reduces human hippocampal neurogenesis and increases apoptosis via activation of distinct STAT1-dependent mechanisms. *International Journal of Neuropsychopharmacology*, *21*(2), 187–200.

Bortsov, A. V., Smith, J. E., Diatchenko, L., Soward, A. C., Ulirsch, J. C., et al. (2013). Polymorphisms in the glucocorticoid receptor co-chaperone FKBP5 predict persistent musculoskeletal pain after traumatic stress exposure. *Pain*, *154*(8), 1419–1426.

Bosch-Bouju, C., Larrieu, T., Linders, L., Manzoni, O. J., & Layé, S. (2016). Endocannabinoid-mediated plasticity in nucleus accumbens controls vulnerability to anxiety after social defeat stress. *Cell Reports*, *16*, 1237–1242.

Bottasso, O., Bay, M. L., Besedovsky, H., & del Rey, A. (2007). The immuno-endocrine component in the pathogenesis of tuberculosis. *Scandinavian Journal of Immunology*, *66*(4), 166–175.

Bouhy, D., Malgrange, B., Multon, S., Poirrier, A. L., Scholtes, F., et al. (2006). Delayed GM-CSF treatment stimulates axonal regeneration and functional recovery in paraplegic rats via an increased BDNF expression by endogenous macrophages. *FASEB Journal, 20*(8), 1239−1241.

Boukhris, T., Sheehy, O., Mottron, L., & Bérard, A. (2016). Antidepressant use during pregnancy and the risk of autism spectrum disorder in children. *JAMA Pediatrics, 170,* 117−120.

Bouloukaki, I., Mermigkis, C., Tzanakis, N., Kallergis, E., Moniaki, V., et al. (2017). Evaluation of inflammatory markers in a large sample of obstructive sleep apnea patients without comorbidities. *Mediators of Inflammation, 2017,* 1−13.

Bourassa, K., & Sbarra, D. A. (2016). Body mass and cognitive decline are indirectly associated via inflammation among ageing adults. *Brain, Behavior, and Immunity, 60,* 63−70.

Bourque, M., Dluzen, D. E., & Di Paolo, T. (2009). Neuroprotective actions of sex steroids in Parkinson's disease. *Frontiers in Neuroendocrinology, 30*(2), 142−157.

Boyce, R., Glasgow, S. D., Williams, S., & Adamantidis, A. (2016). Causal evidence for the role of REM sleep theta rhythm in contextual memory consolidation. *Science, 352*(6287), 812−816.

Boyle, E. A., Li, Y. I., & Pritchard, J. K. (2017). An expanded view of complex traits: From polygenic to omnigenic. *Cell, 169*(7), 1177−1186.

Braak, H., & Braak, E. (1995). Staging of Alzheimer's disease-related neurofibrillary changes. *Neurobiology of Aging, 16*(3), 271−278.

Braak, H., Braak, E., Yilmazer, D., de Vos, R. A., Jansen, E. N., & Bohl, J. (1996). Pattern of brain destruction in Parkinson's and Alzheimer's diseases. *J Neural Transm (Vienna), 103*(4), 455−490.

Brabazon, F., Wilson, C. M., Jaiswal, S., Reed, J., Frey, W. H., et al. (2017). Intranasal insulin treatment of an experimental model of moderate traumatic brain injury. *Journal of Cerebral Blood Flow & Metabolism, 37*(9), 3203−3218.

Braithwaite, E. C., Kundakovic, M., Ramchandani, P. G., Murphy, S. E., & Champagne, F. A. (2015). Maternal prenatal depressive symptoms predict infant NR3C1 1F and BDNF IV DNA methylation. *Epigenetics, 10*(5), 408−417.

Brambilla, F., Bellodi, L., Perna, G., Bertani, A., Panerai, A., et al. (1994). Plasma interleukin-1 beta concentrations in panic disorder. *Psychiatry Research, 54*(2), 135−142.

Brancato, A., Bregman, D., Ahn, H. F., Pfau, M. L., Menard, C., et al. (2017). Sub-chronic variable stress induces sex-specific effects on glutamatergic synapses in the nucleus accumbens. *Neuroscience, 350,* 180−189.

Branchereau, M., Reichardt, F., Loubieres, P., Marck, P., Waget, A., et al. (2016). Periodontal dysbiosis linked to periodontitis is associated with cardio-metabolic adaptation to high-fat diet in mice. *American Journal of Physiology Gastrointestinal and Liver Physiology, 310*(11), G1091−G1101.

Braniste, V., Al-Asmakh, M., Kowal, C., Anuar, F., Abbaspour, A., et al. (2014). The gut microbiota influences blood-brain barrier permeability in mice. *Science Translational Medicine, 6*(263), 263ra158.

Branscombe, N. R., & Reynolds, K. J. (2015). Toward person plasticity: Individual and collective approaches. In K. J. Reynolds, & N. R. Branscombe (Eds.), *Psychology of Change: Life contexts, experiences, and identities* (pp. 3−22). New York: Psychology Press.

Bravo, J. A., Forsythe, P., Chew, M. V., Escaravage, E., Savignac, H. M., Dinan, T. G., et al. (2011). Ingestion of Lactobacillus strain regulates emotional behavior and central GABA receptor expression in a mouse via the vagus nerve. *Proceedings of the National Academy of Sciences of the United States of America, 108*(38), 16050−16055.

Bravo, J. A., Julio-Pieper, M., Forsythe, P., Kunze, W., Dinan, T. G., et al. (2012). Communication between gastrointestinal bacteria and the nervous system. *Current Opinion in Pharmacology, 12*(6), 667−672.

Bravo-Blas, A., Wessel, H., & Milling, S. (2016). Microbiota and arthritis: Correlations or cause? *Current Opinion in Rheumatology, 28*(2), 161−167.

Brebner, K., Hayley, S., Zacharko, R., Merali, Z., & Anisman, H. (2000). Synergistic effects of interleukin-1beta, interleukin-6, and tumor necrosis factor-alpha: Central monoamine, corticosterone, and behavioral variations. *Neuropsychopharmacology, 22*(6), 566−580.

Breckenridge, C. B., Berry, C., Chang, E. T., Sielken, R. L., Jr., & Mandel, J. S. (2016). Association between Parkinson's disease and cigarette smoking, rural living, well-water consumption, farming and pesticide use: Systematic review and meta-analysis. *PLoS One, 11*(4), e0151841.

Breen, M. S., Maihofer, A. X., Glatt, S. J., Tylee, D. S., Chandler, S. D., et al. (2015). Gene networks specific for innate immunity define post-traumatic stress disorder. *Molecular Psychiatry, 20*(12), 1538−1545.

Brenner, L. A., Stearns-Yoder, K. A., Hoffberg, A. S., Penzenik, M. E., Starosta, A. J., et al. (2017). Growing literature but limited evidence: A systematic review regarding prebiotic and probiotic interventions for those with traumatic brain injury and/or posttraumatic stress disorder. *Brain, Behavior, and Immunity, 65,* 57−67.

Bresnahan, M., Hornig, M., Schultz, A. F., Gunnes, N., Hirtz, D., Lie, K. K., & Lipkin, W. I. (2015). Association of maternal report of infant and toddler gastrointestinal symptoms with autism: Evidence from a prospective birth cohort. *JAMA Psychiatry, 72*(5), 466−474. Available from https://doi.org/10.1001/jamapsychiatry.2014.3034.

Breton, J., Tennoune, N., Lucas, N., Francois, M., Legrand, R., et al. (2015). Gut commensal E. coli proteins activate host satiety pathways following nutrient-induced bacterial growth. *Cell Metabolism, 23*(2), 324−334.

Bricou, O., Taïeb, O., Baubet, T., Gal, B., Guillevin, L., et al. (2006). Stress and coping strategies in systemic lupus erythematosus: A review. *Neuroimmunomodulation, 13* (5−6), 283−293.

Bridgewater, L. C., Zhang, C., Wu, Y., Hu, W., Zhang, Q., et al. (2017). Gender-based differences in host behavior and gut microbiota composition in response to high fat diet and stress in a mouse model. *Scientific Reports, 7*(1), 10776.

Brinks, V., De Kloet, E. R., & Oitzl, M. S. (2009). Corticosterone facilitates extinction of fear memory in BALB/c mice but strengthens cue related fear in C57BL/6 mice. *Experimental Neurology, 216*(2), 375−382.

Brochard, V., Combadière, B., Prigent, A., Laouar, Y., Perrin, A., et al. (2009). Infiltration of CD4 + lymphocytes into the brain contributes to neurodegeneration in a mouse model of Parkinson disease. *Journal of Clinical Investigation, 119*(1), 182−192.

Brody, G. H., Yu, T., Miller, G. E., & Chen, E. (2015). Discrimination, racial identity, and cytokine levels among African-American adolescents. *The Journal of Adolescent Health, 56*(5), 496−501.

Brooks, A. K., Janda, T. M., Lawson, M. A., Rytych, J. L., Smith, R. A., et al. (2017). Desipramine decreases expression of human and murine indoleamine-2,3-dioxygenases. *Brain, Behavior, and Immunity, 62*, 219−229.

Broug-Holub, E., Persoons, J. H., Schornagel, K., Mastbergen, S. C., & Kraal, G. (1998). Effects of stress on alveolar macrophages: A role for the sympathetic nervous system. *American Journal of Respiratory Cell and Molecular Biology, 19*(5), 842−848.

Brown, A. S. (2011). The environment and susceptibility to schizophrenia. *Progress in Neurobiology, 93*(1), 23−58.

Brown, A. S., Begg, M. D., Gravenstein, S., Schaefer, C. A., Wyatt, R. J., et al. (2004). Serologic evidence of prenatal influenza in the etiology of schizophrenia. *Archives of General Psychiatry, 61*(8), 774−780.

Brown, A. S., & Derkits, E. J. (2010). Prenatal infection and schizophrenia: A review of epidemiologic and translational studies. *American Journal of Psychiatry, 167*(3), 261−280.

Brown, A. S., & Patterson, P. H. (2011). Maternal infection and schizophrenia: Implications for prevention. *Schizophrenia Bulletin, 37*(2), 284−290.

Brown, A. S., Surcel, H. M., Hinkka-Yli-Salomaki, S., Cheslack-Postava, K., Bao, Y., & Sourander, A. (2015). Maternal thyroid autoantibody and elevated risk of autism in a national birth cohort. *Progress in Neuro-Psychopharmacology & Biological Psychiatry, 57*, 86−92. Available from https://doi.org/10.1016/j.pnpbp.2014.10.010.

Brown, K. W., Weinstein, N., & Creswell, J. D. (2012). Trait mindfulness modulates neuroendocrine and affective responses to social evaluative threat. *Psychoneuroendocrinology, 37*(12), 2037−2041.

Brown, R. F., Tennant, C. C., Dunn, S. M., & Pollard, J. D. (2005). A review of stress-relapse interactions in multiple sclerosis: Important features and stress-mediating and -moderating variables. *Multiple Sclerosis, 11*(4), 477−484.

Browne, H. P., Forster, S. C., Anonye, B. O., Kumar, N., Neville, B. A., et al. (2016). Culturing of 'unculturable' human microbiota reveals novel taxa and extensive sporulation. *Nature, 533*(7604), 543−546.

Bruce-Keller, A. J., Salbaum, J. M., & Berthoud, H. R. (2018). Harnessing gut microbes for mental health: Getting from here to there. *Biological Psychiatry, 83*(3), 214−223. Available from https://doi.org/10.1016/j.biopsych.2017.08.014.

Brudey, C., Park, J., Wiaderkiewicz, J., Kobayashi, I., Mellman, T. A., et al. (2015). Autonomic and inflammatory consequences of posttraumatic stress disorder and the link to cardiovascular disease. *American Journal of Physiology Regulatory Integrative and Comparative Physiology, 309*(4), R315−R321.

Bruenig, D., Mehta, D., Morris, C. P., Harvey, W., Lawford, B., et al. (2017). Genetic and serum biomarker evidence for a relationship between TNFα and PTSD in Vietnam war combat veterans. *Comprehensive Psychiatry, 74*, 125−133.

Brugiroux, S., Beutler, M., Pfann, C., Garzetti, D., Ruscheweyh, H. J., et al. (2016). Genome-guided design of a defined mouse microbiota that confers colonization resistance against Salmonella enterica serovar Typhimurium. *Nature Microbiology, 2*, 16215.

Brühl, A. B., Hänggi, J., Baur, V., Rufer, M., Delsignore, A., et al. (2014). Increased cortical thickness in a frontoparietal network in social anxiety disorder. *Human Brain Mapping, 35*(7), 2966−2977.

Brundin, L., Sellgren, C. M., Lim, C. K., Grit, J., Pålsson, E., et al. (2016). An enzyme in the kynurenine pathway that governs vulnerability to suicidal behavior by regulating excitotoxicity and neuroinflammation. *Translational Psychiatry, 6*(8), e865.

Brundin, P., Dave, K. D., & Kordower, J. H. (2017). Therapeutic approaches to target alpha-synuclein pathology. *Experimental Neurology, 298*(Pt B), 225−235.

Brunoni, A. R., Moffa, A. H., Sampaio-Junior, B., Borrione, L., Moreno, M. L., et al. (2017). Trial of electrical direct-current therapy versus escitalopram for depression. *New England Journal of Medicine, 376* (26), 2523−2533.

Brusca, S. B., Abramson, S. B., & Scher, J. U. (2014). Microbiome and mucosal inflammation as extraarticular triggers for rheumatoid arthritis and autoimmunity. *Current Opinion in Rheumatology, 26*(1), 101−107.

Brust, T. F., Morgenweck, J., Kim, S. A., Rose, J. H., Locke, J. L., et al. (2016). Biased agonists of the kappa opioid receptor suppress pain and itch without causing sedation or dysphoria. *Science Signaling, 9*(456), ra117.

Bruttger, J., Karram, K., Wörtge, S., Regen, T., Marini, F., et al. (2015). Genetic cell ablation reveals clusters of local self-renewing microglia in the mammalian central nervous system. *Immunity, 43*(1), 92–106.

Bryant, R. A., Creamer, M., O'Donnell, M., Silove, D., & McFarlane, A. C. (2012). The capacity of acute stress disorder to predict posttraumatic psychiatric disorders. *Journal of Psychiatric Research, 46*(2), 168–173.

Bryant, R. A., Felmingham, K. L., Falconer, E. M., Pe Benito, L., Dobson-Stone, C., et al. (2010). Preliminary evidence of the short allele of the serotonin transporter gene predicting poor response to cognitive behavior therapy in posttraumatic stress disorder. *Biological Psychiatry, 67*, 1217–1219.

Brydon, L., Walker, C., Wawrzyniak, A. J., Chart, H., & Steptoe, A. (2009). Dispositional optimism and stress-induced changes in immunity and negative mood. *Brain, Behavior, and Immunity, 23*(6), 810–816.

Bu, X. L., Yao, X. Q., Jiao, S. S., Zeng, F., Liu, Y. H., et al. (2015). A study on the association between infectious burden and Alzheimer's disease. *European Journal of Neurology, 22*(12), 1519–1525.

Bucci, V., Tzen, B., Li, N., Simmons, M., Tanoue, T., et al. (2016). MDSINE: Microbial dynamical systems inference engine for microbiome time-series analyses. *Genome Biology, 17*(1), 121.

Buddhala, C., Loftin, S. K., Kuley, B. M., Cairns, N. J., Campbell, M. C., et al. (2015). Dopaminergic, serotonergic, and noradrenergic deficits in Parkinson disease. *Annals of Clinical and Translational Neurology, 2*(10), 949–959.

Buffie, C. G., Bucci, V., Stein, R. R., McKenney, P. T., Ling, L., et al. (2015). Precision microbiome reconstitution restores bile acid mediated resistance to clostridium difficile. *Nature, 517*(7533), 205–208.

Buffington, S. A., Di Prisco, G. V., Auchtung, T. A., Ajami, N. J., Petrosino, J. F., et al. (2016). Microbial reconstitution reverses maternal diet-induced social and synaptic deficits in offspring. *Cell, 165*(7), 1762–1775.

Bumb, J. M., Enning, F., & Leweke, F. M. (2015). Drug repurposing and emerging adjunctive treatments for schizophrenia. *Expert Opinion on Pharmacotherapy, 16*(7), 1049–1067.

Bundhun, P. K., Soogund, M. Z., & Huang, F. (2017). Impact of systemic lupus erythematosus on maternal and fetal outcomes following pregnancy: A meta-analysis of studies published between years 2001-2016. *Journal of Autoimmunity, 79*, 17–27. Available from https://doi.org/10.1016/j.jaut.2017.02.009.

Burbulla, L. F., Song, P., Mazzulli, J. R., Zampese, E., Wong, Y. C., et al. (2017). Dopamine oxidation mediates mitochondrial and lysosomal dysfunction in Parkinson's disease. *Science, 357*(6357), 1255–1261.

Burgos, J. S., Ramirez, C., Sastre, I., & Valdivieso, F. (2007). Apolipoprotein E genotype influences vertical transmission of herpes simplex virus type 1 in a gender specific manner. *Aging Cell, 6*(6), 841–842.

Burgos-Robles, A., Kimchi, E. Y., Izadmehr, E. M., Porzenheim, M. J., Ramos-Guasp, W. A., et al. (2017). Amygdala inputs to prefrontal cortex guide behavior amid conflicting cues of reward and punishment. *Nature Neuroscience, 20*(6), 824–835.

Burke, L. K., Darwish, T., Cavanaugh, A. R., Virtue, S., Roth, E., et al. (2017). mTORC1 in AGRP neurons integrates exteroceptive and interoceptive food-related cues in the modulation of adaptive energy expenditure in mice. *eLife, 6*, e22848.

Burke, N. N., Finn, D. P., McGuire, B. E., & Roche, M. (2017). Psychological stress in early life as a predisposing factor for the development of chronic pain: Clinical and preclinical evidence and neurobiological mechanisms. *Journal of Neuroscience Research, 95*(6), 1257–1270.

Burt, M. A., Tse, Y. C., Boksa, P., & Wong, T. P. (2013). Prenatal immune activation interacts with stress and corticosterone exposure later in life to modulate N-methyl-D-aspartate receptor synaptic function and plasticity. *International Journal of Neuropsychopharmacology, 16*(8), 1835–1848.

Businaro, R., Corsi, M., Azzara, G., Di Raimo, T., Laviola, G., Romano, E., & Ricci, S. (2016). Interleukin-18 modulation in autism spectrum disorders. *Journal of Neuroinflammation, 13*, 2. Available from https://doi.org/10.1186/s12974-015-0466-6.

Buske-Kirschbaum, A., Gierens, A., Hollig, H., & Hellhammer, D. H. (2002). Stress-induced immunomodulation is altered in patients with atopic dermatitis. *Journal of Neuroimmunology, 129*(1–2), 161–167.

Buske-Kirschbaum, A., Kern, S., Ebrecht, M., & Hellhammer, D. H. (2007). Altered distribution of leukocyte subsets and cytokine production in response to acute psychosocial stress in patients with psoriasis vulgaris. *Brain, Behavior, and Immunity, 21*(1), 92–99.

Buttini, M., Orth, M., Bellosta, S., Akeefe, H., Pitas, R. E., et al. (1999). Expression of human apolipoprotein E3 or E4 in the brains of Apoe-/- mice: Isoform-specific effects on neurodegeneration. *Journal of Neuroscience, 19*(12), 4867–4880.

Byrne, C. S., Chambers, E. S., Alhabeeb, H., Chhina, N., Morrison, D. J., et al. (2016). Increased colonic propionate reduces anticipatory reward responses in the human striatum to high energy foods. *American Journal of Clinical Nutrition, 104*(1), 5–14.

Caballero, S., Carter, R., Ke, X., Sušac, B., Leiner, I. M., et al. (2015). Distinct but spatially overlapping intestinal niches for vancomycin-resistant enterococcus faecium and carbapenem-resistant klebsiella pneumoniae. *PLoS Pathogens*, 11(9), e1005132.

Cacciapaglia, F., Navarini, L., Menna, P., Salvatorelli, E., Minotti, G., et al. (2011). Cardiovascular safety of anti-TNF-alpha therapies: Facts and unsettled issues. *Autoimmunity Reviews*, 10(10), 631–635.

Cacioppo, J. T., Cacioppo, S., Capitanio, J. P., & Cole, S. W. (2015). The neuroendocrinology of social isolation. *Annual Review of Psychology*, 66, 733–767.

Cain, D. W., & Cidlowski, J. A. (2017). Immune regulation by glucocorticoids. *Nature Reviews Immunology*, 17(4), 233–247.

Calder, P. C., Bosco, N., Bourdet-Sicard, R., Capuron, L., Delzenne, N., et al. (2017). Health relevance of the modification of low grade inflammation in ageing (inflammageing) and the role of nutrition. *Ageing Research Reviews*, 40, 95–119.

Calderón-Garcidueñas, L., Avila-Ramírez, J., Calderón-Garcidueñas, A., González-Heredia, T., Acuña-Ayala, H., et al. (2016). Cerebrospinal fluid biomarkers in highly exposed PM2.5 urbanites: The risk of Alzheimer's and Parkinson's diseases in young Mexico City residents. *Journal of Alzheimer's Disease*, 54(2), 597–613.

Caldwell, T. (2015). *Is gwyneth paltrow wrong about everything?* Toronto: Penguin.

Caldwell, W., McInnis, O. A., McQuaid, R. J., Liu, G., Stead, J. D., et al. (2013). The role of the Val66Met polymorphism of the brain derived neurotrophic factor gene in coping strategies relevant to depressive symptoms. *PLoS One*, 8(6), e65547.

Caleyachetty, R., Thomas, G. N., Toulis, K. A., Mohammed, N., Gokhale, K. M., et al. (2017). Metabolically healthy obese and incident cardiovascular disease events among 3.5 Million men and women. *Journal of the American College of Cardiology*, 70(12), 1429–1437.

Callixte, K. T., Clet, T. B., Jacques, D., Faustin, Y., François, D. J., et al. (2015). The pattern of neurological diseases in elderly people in outpatient consultations in Sub-Saharan Africa. *BMC Research Notes*, 8, 159.

Calsavara, A. C., Rodrigues, D. H., Miranda, A. S., Costa, P. A., Lima, C. X., et al. (2013). Late anxiety-like behavior and neuroinflammation in mice subjected to sublethal polymicrobial sepsis. *Neurotoxicity Research*, 24(2), 103–108.

Cameron, E. A., & Sperandio, V. (2015). Frenemies: Signaling and nutritional integration in pathogen-microbiota-host interactions. *Cell Host & Microbe*, 18(3), 275–284.

Campbell, S. C., Wisniewski, P. J., Noji, M., McGuinness, L. R., Häggblom, M. M., et al. (2016). The effect of diet and exercise on intestinal integrity and microbial diversity in mice. *PLoS One*, 11(3), e0150502.

Cani, P. D., & Everard, A. (2016). Talking microbes: When gut bacteria interact with diet and host organs. *Molecular Nutrition & Food Research*, 60(1), 58–66.

Cannon, J. R., & Greenamyre, J. T. (2010). Neurotoxic in vivo models of Parkinson's disease recent advances. *Progress in Brain Research*, 184, 17–33.

Cao, J., Min, L., Lansing, B., Foxman, B., & Mody, L. (2016). Multidrug-resistant organisms on patients' hands: A missed opportunity. *JAMA Internal Medicine*, 176(5), 705–706.

Cao, X., Lin, P., Jiang, P., & Li, C. (2013). Characteristics of the gastrointestinal microbiome in children with autism spectrum disorder: A systematic review. *Shanghai Archives of Psychiatry journal*, 25(6), 342–353.

Cao-Lei, L., De Rooij, S. R., King, S., Matthews, S. G., Metz, G. A., et al. (2017). Prenatal stress and epigenetics. *Neuroscience & Biobehavioral Reviews*, pii: S0149-7634(16)30726-6.

Cappi, C., Diniz, J. B., Requena, G. L., Lourenço, T., Lisboa, B. C., et al. (2016). Epigenetic evidence for involvement of the oxytocin receptor gene in obsessive-compulsive disorder. *BMC Neuroscience*, 17(1), 79.

Cappuccio, F. P., Cooper, D., D'Elia, L., Strazzullo, P., & Miller, M. A. (2011). Sleep duration predicts cardiovascular outcomes: A systematic review and meta-analysis of prospective studies. *European Heart Journal*, 32(12), 1484–1492.

Capuron, L., & Miller, A. H. (2011). Immune system to brain signaling: neuropsychopharmacological implications. *Pharmacology & Therapeutics*, 130, 226–238.

Cardoso, C., Kingdon, D., & Ellenbogen, M. A. (2014). A meta-analytic review of the impact of intranasal oxytocin administration on cortisol concentrations during laboratory tasks: Moderation by method and mental health. *Psychoneuroendocrinology*, 49, 161–170.

Careaga, M., Rogers, S., Hansen, R. L., Amaral, D. G., Van de Water, J., & Ashwood, P. (2017). Immune endophenotypes in children with autism spectrum disorder. *Biological Psychiatry*, 81(5), 434–441. Available from https://doi.org/10.1016/j.biopsych.2015.08.036.

Carhart-Harris, R. L., Roseman, L., Bolstridge, M., Demetriou, L., Pannekoek, J. N., et al. (2017). Psilocybin for treatment-resistant depression: fMRI-measured brain mechanisms. *Science Reports*, 7, 13187.

Carlino, E., Piedimonte, A., & Benedetti, F. (2016). Nature of the placebo and nocebo effect in relation to functional neurologic disorders. *Handbook of Clinical Neurology*, 139, 597–606.

Carlson, A. L., Xia, K., Azcarate-Peril, M. A., Goldman, B. D., Ahn, M., et al. (2018). Infant gut microbiome associated with cognitive development. *Biological Psychiatry*, pii: S0006-3223(17)31720-1.

Carney, M. C., Tarasiuk, A., DiAngelo, S. L., Silveyra, P., Podany, A., et al. (2017). Metabolism-related microRNAs in maternal breast milk are influenced by premature delivery. *Pediatric Research*, 82(2), 226–236.

Carpenter, L. L., Tyrka, A. R., Ross, N. S., Khoury, L., Anderson, G. M., et al. (2009). Effect of childhood emotional abuse and age on cortisol responsivity in adulthood. *Biological Psychiatry, 66*(1), 69–75.

Carroll, J. E., Esquivel, S., Goldberg, A., Seeman, T. E., Effros, R. B., et al. (2016). Insomnia and telomere length in older adults. *Sleep, 39*(3), 559–564.

Carson, C. M., Phillip, N., & Miller, B. J. (2017). Urinary tract infections in children and adolescents with acute psychosis. *Schizophrenia Research, 183*, 36–40.

Caruso, R., Nanni, M. G., Riba, M., Sabato, S., Mitchell, A. J., et al. (2015). Depressive spectrum disorders in cancer: Prevalence, risk factors and screening for depression: A critical review. *Acta Oncologica, 56*(2), 146–155.

Carver, C. S., & Connor-Smith, J. (2010). Personality and coping. *Annual Review of Psychology, 61*, 679–704.

Carver, C. S., Johnson, S. L., Joormann, J., Lemoult, J., & Cuccaro, M. L. (2011). Childhood adversity interacts separately with 5-HTTLPR and BDNF to predict lifetime depression diagnosis. *Journal of Affective Disorders, 132*(1–2), 89–93.

Carver, C. S., Smith, R. G., Antoni, M. H., Petronis, V. M., Weiss, S., et al. (2005). Optimistic personality and psychosocial well-being during treatment predict psychosocial well-being among long-term survivors of breast cancer. *Health Psychology, 24*(5), 508–516.

Casals, J., Elizan, T. S., & Yahr, M. D. (1998). Postencephalitic parkinsonism--A review. *Journal of Neural Transmission, 105*(6–7), 645–676.

Caspell-Garcia, C., Simuni, T., Tosun-Turgut, D., Wu, I.-W., Zhang, Y., et al. (2017). Multiple modality biomarker prediction of cognitive impairment in prospectively followed de novo Parkinson disease. *PLoS One, 12*(5), e0175674.

Caspi, A., Hariri, A. R., Holmes, A., Uher, R., & Moffitt, T. E. (2010). Genetic sensitivity to the environment: The case of the serotonin transporter gene and its implications for studying complex diseases and traits. *The American Journal of Psychiatry, 167*(5), 509–527.

Caspi, A., Sugden, K., Moffitt, T. E., Taylor, A., Craig, I. W., et al. (2003). Influence of life stress on depression: Moderation by a polymorphism in the 5-HTT gene. *Science, 301*(5631), 386–389.

Castro-Nallar, E., Bendall, M. L., Perez-Losada, M., Sabuncyan, S., Severance, E. G., et al. (2015). Composition, taxonomy and functional diversity of the oropharynx microbiome in individuals with schizophrenia and controls. *PeerJ, 3*, e1140.

Cattaneo, A., Macchi, F., Plazzotta, G., Veronica, B., Bocchio-Chiavetto, L., et al. (2015). Inflammation and neuronal plasticity: A link between childhood trauma and depression pathogenesis. *Frontiers in Cellular Neuroscience, 9*, 40.

Cavallari, J. F., Fullerton, M. D., Duggan, B. M., Foley, K. P., & Denou, E. (2017). Muramyl diapeptide-based postbiotics mitigate obesity-induced insulin resistance via IRF4. *Cell Metabolism, 25*(5), 1063–1074.

Cebrián, C., Zucca, F. A., Mauri, P., Steinbeck, J. A., Studer, L., et al. (2014). MHC-I expression renders catecholaminergic neurons susceptible to T-cell-mediated degeneration. *Nature Communications, 5*, 3633.

Cekanaviciute, E., Yoo, B. B., Runia, T. F., Debelius, J. W., Singh, S., et al. (2017). Gut bacteria from multiple sclerosis patients modulate human T cells and exacerbate symptoms in mouse models. *Proceedings of the National Academy of Sciences of the United States of America, 114*(40), 10713–10718.

Cekic, C., & Linden, J. (2016). Purinergic regulation of the immune system. *Nature Reviews Immunology, 16*(3), 177–192.

Celik, M. Ö., Labuz, D., Henning, K., Busch-Dienstfertig, M., Gaveriaux-Ruff, C., et al. (2016). Leukocyte opioid receptors mediate analgesia via Ca2 + -regulated release of opioid peptides. *Brain, Behav, Immun, 57*, 227–242.

Centre for Disease Control. (2016, September). *Antibiotic resistance threats in the United States, 2013*. Retrieved from http://www.cdc.gov/drugresistance/pdf/ar-threats-2013-508.pdf. Accessed March 2017.

Cerdá, B., Pérez, M., Pérez-Santiago, J. D., Tornero-Aguilera, J. F., González-Soltero, R., et al. (2016). Gut microbiota modification: Another piece in the puzzle of the benefits of physical exercise in health? *Frontiers in Physiology, 7*, 51.

Cervenka, I., Agudelo, L. Z., & Ruas, J. L. (2017). Kynurenines: Tryptophan's metabolites in exercise, inflammation, and mental health. *Science, 357*(6349), eaaf9794.

Cervera, R., Khamashta, M. A., & Hughes, G. R. V. (2009). The Euro-lupus project: Epidemiology of systemic lupus erythematosus in Europe. *Lupus, 18*(10), 869–874.

Chae, D. H., Nuru-Jeter, A. M., Adler, N. E., Brody, G. H., Lin, J., et al. (2014). Discrimination, racial bias, and telomere length in African-American men. *American Journal of Preventive Medicine, 46*(2), 103–111.

Chagraoui, A., Skiba, M., Thuillez, C., & Thibaut, F. (2016). To what extent is it possible to dissociate the anxiolytic and sedative/hypnotic properties of GABAA receptors modulators? *Progress in Neuro-Psychopharmacology & Biological Psychiatry, 71*, 189–202.

Champagne, F. A. (2010). Epigenetic influence of social experiences across the lifespan. *Developmental Psychobiology, 52*(4), 299–311.

Champagne, F. A., & Meaney, M. J. (2007). Transgenerational effects of social environment on variations in maternal care and behavioral response to novelty. *Behavioral Neuroscience, 121*(6), 1353–1363.

Chan, D. K., Woo, J., Ho, S. C., Pang, C. P., Law, L. K., et al. (1998). Genetic and environmental risk factors for Parkinson's disease in a Chinese population. *Journal of Neurology, Neurosurgery and Psychiatry, 65*, 781–784.

Chan, G., White, C. C., Winn, P. A., Cimpean, M., Replogle, J. M., et al. (2015). CD33 modulates TREM2: Convergence of Alzheimer loci. *Nature Neuroscience, 18*(11), 1556–1558.

Chandra, R., Francis, T. C., Nam, H., Riggs, L. M., Engeln, M., et al. (2017). Reduced Slc6a15 in nucleus accumbens D2-neurons underlies stress susceptibility. *Journal of Neuroscience, 37*(27), 6527–6538, 3250-16.

Chao, A. M., Jastreboff, A. M., White, M. A., Grilo, C. M., & Sinha, R. (2017). Stress, cortisol, and other appetite-related hormones: Prospective prediction of 6-month changes in food cravings and weight. *Obesity, 25*(4), 713–720.

Chapin, H., Bagarinao, E., & Mackey, S. (2012). Real-time fMRI applied to pain management. *Neuroscience Letters, 520*(2), 174–181.

Charney, D. S. (2004). Psychobiological mechanisms of resilience and vulnerability: Implications for successful adaptation to extreme stress. *The American Journal of Psychiatr, 161*(2), 195–216.

Chassaing, B., Koren, O., Goodrich, J. K., Poole, A. C., Srinivasan, S., et al. (2015). Dietary emulsifiers impact the mouse gut microbiota promoting colitis and metabolic syndrome. *Nature, 519*(7541), 92–96.

Chechko, N., Wehrle, R., Erhardt, A., Holsboer, F., Czisch, M., et al. (2009). Unstable prefrontal response to emotional conflict and activation of lower limbic structures and brainstem in remitted panic disorder. *PLoS One, 4*(5), e5537.

Chen, C. H., Cheng, M. C., Liu, C. M., Liu, C. C., Lin, K. H., et al. (2017). Seroprevalence survey of selective anti-neuronal autoantibodies in patients with first-episode schizophrenia and chronic schizophrenia. *Schizophrenia Researchs, 190*, 28–31.

Chen, C. J., Wu, G. H., Kuo, R. L., & Shih, S. R. (2017). Role of the intestinal microbiota in the immunomodulation of influenza virus infection. *Microbes and Infection, 19*(12), 570–579.

Chen, G. Y., & Nunez, G. (2010). Sterile inflammation: Sensing and reacting to damage. *Nature Reviews Immunology, 10*(12), 826–837.

Chen, H., Zhang, S. M., Schwarzschild, M. A., Hernán, M. A., et al. (2005). Physical activity and the risk of Parkinson disease. *Neurology, 64*, 664–669.

Chen, J., Song, Y., Yang, J., Zhang, Y., Zhao, P., et al. (2013). The contribution of TNF-α in the amygdala to anxiety in mice with persistent inflammatory pain. *Neuroscience Letters, 541*, 275–280.

Chen, J., Wright, K., Davis, J. M., Jeraldo, P., Marietta, E. V., et al. (2016). An expansion of rare lineage intestinal microbes characterizes rheumatoid arthritis. *Genome Medicine, 8*(1), 1.

Chen, J. F., Xu, K., Petzer, J. P., Staal, R., Xu, Y. H., et al. (2001). Neuroprotection by caffeine and A(2A) adenosine receptor inactivation in a model of Parkinson's disease. *Journal of Neuroscience, 21*(10), RC143.

Chen, K., Zheng, X., Feng, M., Li, D., & Zhang, H. (2017). Gut microbiota-dependent metabolite trimethylamine N-oxide contributes to cardiac dysfunction in western diet-induced obese mice. *Frontiers in Physiology, 8*, 1–9.

Chen, L., Wilson, J. E., Koenigsknecht, M. J., Chou, W. C., Montgomery, S. A., et al. (2017). NLRP12 attenuates colon inflammation by maintaining colonic microbial diversity and promoting protective commensal bacterial growth. *Nature Immunology, 18*(5), 541–551.

Chen, L., Zhang, G., Hu, M., & Liang, X. (2015). Eye movement desensitization and reprocessing versus cognitive-behavioral therapy for adult posttraumatic stress disorder: Systematic review and meta-analysis. *Journal of Nervous and Mental Disease, 203*(6), 443–451.

Chen, M., Wan, Y., Ade, K., Ting, J., Feng, G., et al. (2011). Sapap3 deletion anomalously activates short-term endocannabinoid-mediated synaptic plasticity. *Journal of Neuroscience, 31*(26), 9563–9573.

Chen, R., Shi, L., Hakenberg, J., Naughton, B., Sklar, P., et al. (2016). Analysis of 589,306 genomes identifies individuals resilient to severe Mendelian childhood diseases. *Nature Biotechnology, 34*(5), 531–538.

Chen, S. G., Stribinskis, V., Rane, M. J., Demuth, D. R., Gozal, E., et al. (2016). Exposure to the functional bacterial amyloid protein curli enhances alpha-synuclein aggregation in aged Fischer 344 rats and Caenorhabditis elegans. *Scientific Reports, 6*, 34477.

Chen, S. J., Chao, Y. L., Chen, C. Y., Chang, C. M., Wu, E. C., et al. (2012). Prevalence of autoimmune diseases in inpatients with schizophrenia: Nationwide population-based study. *British Journal of Psychiatry, 200*(5), 374–380.

Chen, S. W., Zhong, X. S., Jiang, L. N., Zheng, X. Y., Xiong, Y. Q., Ma, S. J., & Chen, Q. (2016). Maternal autoimmune diseases and the risk of autism spectrum disorders in offspring: A systematic review and meta-analysis. *Behavioural Brain Research, 296*, 61–69. Available from https://doi.org/10.1016/j.bbr.2015.08.035.

Chen, Y., Fang, Q., Wang, Z., Zhang, J. Y., MacLeod, A. S., et al. (2016). Transient receptor potential vanilloid 4 ion channel functions as a pruriceptor in epidermal keratinocytes to evoke histaminergic itch. *Journal of Biological Chemistry, 291*(19), 10252–10262.

Chen, Y., Qi, B., Xu, W., Ma, B., Li, L., et al. (2015). Clinical correlation of peripheral CD4 + -cell sub-sets, their imbalance and Parkinson's disease. *Molecular Medicine Reports, 12*(4), 6105–6111.

Chen, Z., Guo, L., Zhang, Y., Walzem, R. L., et al. (2014). Incorporation of therapeutically modified bacteria into gut microbiota inhibits obesity. *Journal of Clinical Investigation, 124*(8), 3391–3406.

Cheng, C., Lau, H. P. B., & Chan, M. P. S. (2014). Coping flexibility and psychological adjustment to stressful life changes: A meta-analytic review. *Psychological Bulletin, 140*, 1582–1607.

Cheng, C. W., Adams, G. B., Perin, L., Wei, M., Zhou, X., et al. (2014). Prolonged fasting reduces IGF-1/PKA to promote hematopoietic-stem-cell-based regeneration and reverse immunosuppression. *Cell Stem Cell, 14*(6), 810–823.

Cheng, Y., Jope, R. S., & Beurel, E. (2015). A preconditioning stress accelerates increases in mouse plasma inflammatory cytokines induced by stress. *BMC Neuroscience, 16*, 31.

Cherkin, D. C., Sherman, K. J., Balderson, B. H., Cook, A. J., Anderson, M. L., et al. (2016). Effect of mindfulness-based stress reduction vs cognitive behavioral therapy or usual care on back pain and functional limitations in adults with chronic low back pain: A randomized clinical trial. *The Journal of the American Medical Association, 315*, 1240–1249.

Cherry, J. D., Olschowka, J. A., & O'Banion, M. K. (2014). Neuroinflammation and M2 microglia: The good, the bad, and the inflamed. *Journal of Neuroinflammation, 11*, 98.

Chesnokova, V., Pechnick, R. N., & Wawrowsky, K. (2016). Chronic peripheral inflammation, hippocampal neurogenesis, and behavior. *Brain, Behavior, and Immunity, 58*, 1–8.

Chetty, R., Stepner, M., Abraham, S., Lin, S., Scuderi, B., et al. (2016). The association between income and life expectancy in the United States, 2001-2014. *The Journal of the American Medical Association, 315*, 1750–1766.

Chew, E. Y., Clemons, T. E., Agrón, E., Launer, L. J., Grodstein, F., et al. (2015). Effect of omega-3 fatty acids, lutein/zeaxanthin, or other nutrient supplementation on cognitive function: The AREDS2 randomized clinical trial. *The Journal of the American Medical Association, 314* (8), 791–801.

Chida, Y., & Steptoe, A. (2008). Positive psychological well-being and mortality: A quantitative review of prospective observational studies. *Psychosomatic Medicine, 70*(7), 741–756.

Childs, E., Lutz, J. A., & de Wit, H. (2017). Dose-related effects of delta-9-THC on emotional responses to acute psychosocial stress. *Drug and Alcohol Dependence, 177*, 136–144.

Chin-Chan, M., Navarro-Yepes, J., & Quintanilla-Vega, B. (2015). Environmental pollutants as risk factors for neurodegenerative disorders: Alzheimer and Parkinson diseases. *Frontiers in Cellular Neuroscience, 9*, 124.

Chita-Tegmark, M. (2016). Social attention in ASD: A review and meta-analysis of eye-tracking studies. *Research in Developmental Disabilities, 48*, 79–93. Available from https://doi.org/10.1016/j.ridd.2015.10.011.

Chiu, W. C., Su, Y. P., Su, K. P., & Chen, P. C. (2017). Recurrence of depressive disorders after interferon-induced depression. *Translational Psychiatry, 7*(2), e1026.

Cho, H. J., Savitz, J., Dantzer, R., Teague, T. K., Drevets, W. C., et al. (2017). Sleep disturbance and kynurenine metabolism in depression. *The Journal of Psychosomatic Research, 99*, 1–7.

Cho, S. H., Sun, B., Zhou, Y., Kauppinen, T. M., Halabisky, B., et al. (2011). CX3CR1 protein signaling modulates microglial activation and protects against plaque-independent cognitive deficits in a mouse model of Alzheimer disease. *Journal of Biological Chemistry, 286* (37), 32713–32722.

Choi, I. Y., Piccio, L., Childress, P., Bollman, B., Ghosh, A., et al. (2016). A diet mimicking fasting promotes regeneration and reduces autoimmunity and multiple sclerosis symptoms. *Cell Reports, 15*(10), 2136–2146.

Choi, S. H., Aid, S., Caracciolo, L., Minami, S. S., et al. (2013). Cyclooxygenase-1 inhibition reduces amyloid pathology and improves memory deficits in a mouse model of Alzheimer's disease. *Journal of Neurochemistry, 124*(1), 59–68.

Choudary, P. V., Molnar, M., Evans, S. J., Tomita, H., Li, J. Z., et al. (2005). Altered cortical glutamatergic and GABAergic signal transmission with glial involvement in depression. *Proceedings of the National Academy of Sciences of the United States of America, 102*(43), 15653–15658.

Choudhury, M. E., Sugimoto, K., Kubo, M., Nagai, M., Nomoto, M., et al. (2011). A cytokine mixture of GM-CSF and IL-3 that induces a neuroprotective phenotype of microglia leading to amelioration of (6-OHDA)-induced Parkinsonism of rats. *Brain and Behavior, 1*(1), 26–43.

Chowdhury, R., Warnakula, S., Kunutsor, S., Crowe, F., Ward, H. A., et al. (2014). Association of dietary, circulating, and supplement fatty acids with coronary risk: A systematic review and meta-analysis. *Annals of Internal Medicine, 160*(6), 398–406.

Christensen, D. L., Baio, J., Van Naarden Braun, K., Bilder, D., Charles, J., Constantino, J. N., … Yeargin-Allsopp, M. (2016). Prevalence and Characteristics of Autism Spectrum Disorder Among Children Aged 8 Years– Autism and Developmental Disabilities Monitoring Network, 11 Sites, United States, 2012. *MMWR. Surveillance Summaries : Morbidity and Mortality Weekly Report. Surveillance Summaries / CDC, 65*(3), 1–23.

Available from https://doi.org/10.15585/mmwr.ss6503a1.

Christopherson, K. S., Ullian, E. M., Stokes, C. C., Mullowney, C. E., Hell, J. W., et al. (2005). Thrombospondins are astrocyte-secreted proteins that promote CNS synaptogenesis. *Cell, 120*(3), 421−433.

Chu, H., Khosravi, A., Kusumawardhani, I. P., Kwon, A. H., Vasconcelos, A. C., et al. (2016). Gene-microbiota interactions contribute to the pathogenesis of inflammatory bowel disease. *Science, 352*(6289), 1116−1120.

Chu, H., & Mazmanian, S. K. (2013). Innate immune recognition of the microbiota promotes host-microbial symbiosis. *Nature Immunology, 14*(7), 668−675.

Chua, M. C., Ben-Amor, K., Lay, C., Neo, A. G. E., Chiang, W. C., et al. (2017). Effect of synbiotic on the gut microbiota of cesarean delivered infants: A randomized, double-blind, multicenter study. *Journal of Pediatric Gastroenterology and Nutrition, 65*(1), 102−106.

Chuang, J. C., & Zigman, J. M. (2010). Ghrelin's roles in stress, mood, and anxiety regulation. International. *The Journal of Peptide*, pii: 460549.

Chung, C. C., Pimentel, D., Jor'dan, A. J., Hao, Y., Milberg, W., et al. (2015). Inflammation-associated declines in cerebral vasoreactivity and cognition in type 2 diabetes. *Neurology, 10*, 1212.

Chung, W. S., Verghese, P. B., Chakraborty, C., Joung, J., Hyman, B. T., et al. (2016). Novel allele-dependent role for APOE in controlling the rate of synapse pruning by astrocytes. *Proceedings of the National Academy of Sciences of the United States of America, 113*(36), 10186−10191.

Chung, W. S., Welsh, C. A., Barres, B. A., & Stevens, B. (2015). Do glia drive synaptic and cognitive impairment in disease? *Nature Neuroscience, 18*(11), 1539−1545.

Cicognola, C., Chiasserini, D., & Parnetti, L. (2015). Preanalytical confounding factors in the analysis of cerebrospinal fluid biomarkers for Alzheimer's disease: The issue of diurnal variation. *Frontiers in Neurology, 6*, 143.

Cipriani, A., Zhou, X., Del Giovane, C., Hetrick, S. E., Qin, B., et al. (2016). Comparative efficacy and tolerability of antidepressants for major depressive disorder in children and adolescents: A network meta-analysis. *Lancet, 388*(10047), 881−890.

Cirulli, F., Francia, N., Branchi, I., Antonucci, M. T., Aloe, L., et al. (2009). Changes in plasma levels of BDNF and NGF reveal a gender-selective vulnerability to early adversity in rhesus macaques. *Psychoneuroendocrinology, 34*(2), 172−180.

Civelek, M., Wu, Y., Pan, C., Raulerson, C. K., Ko, A., et al. (2017). Genetic regulation of adipose gene expression and cardio-metabolic traits. *American Journal of Human Genetics, 100*(3), 428−443.

Claassen, D. O., Josephs, K. A., Ahlskog, J. E., Silber, M. H., Tippmann-Peikert, M., et al. (2010). REM sleep behavior disorder preceding other aspects of synucleinopathies by up to half a century. *Neurology, 75*(6), 494−499.

Clark, A., & Mach, N. (2017). The crosstalk between the gut microbiota and mitochondria during exercise. *Frontiers in Physiology, 8*, 319.

Clark, P. J., Amat, J., McConnell, S. O., Ghasem, P. R., Greenwood, B. N., et al. (2015). Running reduces uncontrollable stress-evoked serotonin and potentiates stress-evoked dopamine concentrations in the rat dorsal striatum. *PLoS One, 10*(11), e0141898.

Clark, S. M., Pocivavsek, A., Nicholson, J. D., Notarangelo, F. M., Langenberg, P., et al. (2016). Reduced kynurenine pathway metabolism and cytokine expression in the prefrontal cortex of depressed individuals. *Journal of Psychiatry & Neuroscience, 41*(6), 386−394.

Clark, S. M., Sand, J., Francis, T. C., Nagaraju, A., Michael, K. C., et al. (2014). Immune status influences fear and anxiety responses in mice after acute stress exposure. *Brain, Behavior, and Immunity, 38*, 192−201.

Clarke, G., Cryan, J. F., Dinan, T. G., & Quigley, E. M. (2012). Review article: Probiotics for the treatment of irritable bowel syndrome—Focus on lactic acid bacteria. *Alimentary Pharmacology & Therapeutics., 35*(4), 403−413.

Clarke, G., Grenham, S., Scully, P., Fitzgerald, P., Moloney, R. D., et al. (2013). The microbiome-gut-brain axis during early life regulates the hippocampal serotonergic system in a sex-dependent manner. *Molecular Psychiatry, 18*(6), 666−673.

Clausen, A. N., Francisco, A. J., Thelen, J., Bruce, J., Martin, L. E., et al. (2017). PTSD and cognitive symptoms relate to inhibition-related prefrontal activation and functional connectivity. *Depression and Anxiety, 24*(5), 427−436.

Claussnitzer, M., Dankel, S. N., Kim, K., Quon, G., Meuleman, W., et al. (2015). FTO obesity variant circuitry and adipocyte browning in humans. *New England Journal of Medicine, 373*(10), 895−907.

Clavaguera, F., Akatsu, H., Fraser, G., Crowther, R. A., Frank, S., et al. (2013). Brain homogenates from human tauopathies induce tau inclusions in mouse brain. *Proceedings of the National Academy of Sciences of the United States of America, 110*(23), 9535−9540.

Claypoole, L. D., Zimmerberg, B., & Williamson, L. L. (2017). Neonatal lipopolysaccharide treatment alters hippocampal neuroinflammation, microglia morphology and anxiety-like behavior in rats selectively bred for an infantile trait. *Brain, Behavior, and Immunity, 59*, 135−146.

Clerici, M., Arosio, B., Mundo, E., Cattaneo, E., Pozzoli, S., et al. (2009). Cytokine polymorphisms in the pathophysiology of mood disorders. *CNS Spectrums, 14*(8), 419−425.

Cobia, D. J., Smith, M. J., Wang, L., & Csernansky, J. G. (2012). Longitudinal progression of frontal and temporal lobe changes in schizophrenia. *Schizophrenia Research, 139*(1−3), 1−6.

Cochemé, H. M., & Murphy, M. P. (2008). Complex I is the major site of mitochondrial superoxide production by paraquat. *Journal of Biological Chemistry, 283*(4), 1786−1798.

Coelho, M., & Ferreira, J. J. (2012). Late-stage Parkinson disease. *Nature Reviews Neurology, 8*(8), 435−442.

Coelho, R., Viola, T. W., Walss-Bass, C., Brietzke, E., & Grassi-Oliveira, R. (2014). Childhood maltreatment and inflammatory markers: A systematic review. *Acta Psychiatrica Scandinavica, 129*(3), 180−192.

Cohen, S., & Williamson, G. M. (1991). Stress and infectious disease in humans. *Psychol Bull, 109*(1), 5−24.

Cohen, H., Kozlovsky, N., Matar, M. A., Zohar, J., & Kaplan, Z. (2011). The characteristic long-term upregulation of hippocampal NF-κB complex in PTSD-like behavioral stress response is normalized by high-dose corticosterone and pyrrolidine dithiocarbamate administered immediately after exposure. *Neuropsychopharmacology, 36*(11), 2286−2302.

Cohen, H., Liu, T., Kozlovsky, N., Kaplan, Z., Zohar, J., et al. (2012). The neuropeptide Y (NPY)-ergic system is associated with behavioral resilience to stress exposure in an animal model of post-traumatic stress disorder. *Neuropsychopharmacology, 37*(2), 350−363.

Cohen, I. V., Makunts, T., Atayee, R., & Abagyan, R. (2017). Population scale data reveals the antidepressant effects of ketamine and other therapeutics approved for non-psychiatric indications. *Scientific Reports, 7*(1), 1450.

Cohen, L. J., Esterhazy, D., Kim, S. H., Lemetre, C., Aguilar, R. R., et al. (2017). Commensal bacteria make GPCR ligands that mimic human signalling molecules. *Nature, 549*(7670), 48−53.

Cohen, M., Meir, T., Klein, E., Volpin, G., Assaf, M., et al. (2011). Cytokine levels as potential biomarkers for predicting the development of posttraumatic stress symptoms in casualties of accidents. *International Journal of Ppsychiatry in Medicine., 42*(2), 117−131.

Cohen, S., Doyle, W. J., Alper, C. M., Janicki-Deverts, D., & Turner, R. B. (2009). Sleep habits and susceptibility to the common cold. *Archives of Internal Medicine, 169*(1), 62−67.

Cohen, S., Frank, E., Doyle, W. J., Skoner, D. P., Rabin, B. S., et al. (1998). Types of stressors that increase susceptibility to the common cold in healthy adults. *Health Psychology, 17*(3), 214−223.

Cohen, S., Janicki-Deverts, D., & Doyle, W. J. (2015). Self-rated health in healthy adults and susceptibility to the common cold. *Psychosomatic Medicine, 77*(9), 959−968.

Cohen, S., Janicki-Deverts, D., Doyle, W. J., Miller, G. E., Frank, E., et al. (2012). Chronic stress, glucocorticoid receptor resistance, inflammation, and disease risk. *Proceedings of the National Academy of Sciences of the United States of America, 109*(16), 5995−5999.

Cohen, S., Janicki-Deverts, D., Turner, R. B., Marsland, A. L., Casselbrant, M. L., et al. (2013). Childhood socioeconomic status, telomere length, and susceptibility to upper respiratory infection. *Brain, Behavior, and Immunity, 34*, 31−38.

Cohen, S., Miller, G. E., & Rabin, B. S. (2001). Psychological stress and antibody response to immunization: A critical review of the human literature. *Psychosomatic Medicine, 63*(1), 7−18.

Cohen, S., Tyrrell, D. A., & Smith, A. P. (1991). Psychological stress and susceptibility to the common cold. *New England Journal of Medicine, 325*(9), 606−612.

Colditz, G. A., & Sutcliffe, S. (2015). The preventability of cancer stacking the deck. *Science, 347*, 78−81.

Collins, R., Reith, C., Emberson, J., Armitage, J., Baigent, C., et al. (2016). Interpretation of the evidence for the efficacy and safety of statin therapy. *Lancet, 388*(10059), 2532−2561.

Collins, S. M., Surette, M., & Bercik, P. (2012). The interplay between the intestinal microbiota and the brain. *Nature Reviews Microbiology, 10*(11), 735−742.

Colonna, M., & Wang, Y. (2016). TREM2 variants: New keys to decipher Alzheimer disease pathogenesis. *Nature Reviews Neuroscience, 17*(4), 201−207.

Colvonen, P. J., Glassman, L. H., Crocker, L. D., Buttner, M. M., Orff, H., et al. (2017). Pretreatment biomarkers predicting PTSD psychotherapy outcomes: A systematic review. *Neuroscience & Biobehavioral Reviews, 75*, 140−156.

Comi, A. M., Zimmerman, A. W., Frye, V. H., Law, P. A., & Peeden, J. N. (1999). Familial clustering of autoimmune disorders and evaluation of medical risk factors in autism. *J Child Neurol, 14*(6), 388−394. Available from https://doi.org/10.1177/088307389901400608.

Connor, C. J., Liu, V., & Fiedorowicz, J. G. (2015). Exploring the physiological link between psoriasis and mood disorders. *Dermatology Research and Practice, 2015*, 409637.

Connor, T. J., Brewer, C., Kelly, J. P., & Harkin, A. (2005). Acute stress suppresses pro-inflammatory cytokines TNF-alpha and IL-1 beta independent of a catecholamine-driven increase in IL-10 production. *Journal of Neuroimmunology, 159*(1−2), 119−128.

Contet, C. S., Kieffer, B. L., & Befort, K. (2004). Mu opioid receptor: A gateway to drug addiction. *Current Opinion in Neurobiology, 14*(3), 370−379.

Cook, M. D., Allen, J. M., Pence, B. D., Wallig, M. A., Gaskins, H. R., et al. (2016). Exercise and gut immune function: Evidence of alterations in colon immune cell homeostasis and microbiome characteristics with exercising training. *Immunology & Cell Biology, 94*(2), 158−163.

Cookson, M. R. (2017). Mechanisms of Mutant LRRK2Neurodegeneration. *Advances in Neurobiology, 14,* 227–239.

Cooper, D.G. (1970). *Psychiatry and anti-psychiatry* (Ed.), London: Paladin.

Copeland, W. E., Shanahan, L., Worthman, C., Angold, A., & Costello, E. J. (2012). Generalized anxiety and C-reactive protein levels: A prospective, longitudinal analysis. *Psychological Medicine, 42*(12), 2641–2650.

Copenhaver, P. F., & Kögel, D. (2017). Role of APP Interactions with heterotrimeric G Proteins: Physiological functions and pathological consequences. *Frontiers in Molecular Neuroscience, 10,* 3.

Corbett, D., Carmichael, S. T., Murphy, T. H., Jones, T. A., Schwab, M. E., et al. (2017). Enhancing the alignment of the preclinical and clinical stroke recovery research pipeline: Consensus-based core recommendations from the stroke recovery and rehabilitation roundtable translational working group. *Neurorehabilitation and Neural Repair, 31*(8), 699–707.

Cordova, M. J., Riba, M. B., & Spiegel, D. (2017). Post-traumatic stress disorder and cancer. *The Lancet Psychiatry, 4*(4), 330–338.

Corps, K. N., Roth, T. L., & McGavern, D. B. (2015). Inflammation and neuroprotection in traumatic brain injury. *JAMA Neurolology, 72*(3), 355–362.

Corrigan, P. W., Rafacz, J., & Rüsch, N. (2011). Examining a progressive model of self-stigma and its impact on people with serious mental illness. *Psychiatry Research, 189* (3), 339–343.

Costa, R. J. S., Snipe, R. M. J., Kitic, C. M., & Gibson, P. R. (2017). Systematic review: Exercise-induced gastrointestinal syndrome-implications for health and intestinal disease. *Alimentary Pharmacology & Therapeutics, 46*(3), 246–265.

Cotman, C. W., Berchtold, N. C., & Christie, L. A. (2007). Exercise builds brain health: Key roles of growth factor cascades and inflammation. *Trends in Neurosciences, 30* (9), 464–472.

Coughlin, S. S. (2012). Anxiety and depression: Linkages with viral diseases. *Public Health Reviews, 34,* 92.

Coureuil, M., Lécuyer, H., Bourdoulous, S., & Nassif, X. (2017). A journey into the brain: Insight into how bacterial pathogens cross blood-brain barriers. *Nature Reviews Microbiology, 15*(3), 149–159.

Courties, G., Moskowitz, M. A., & Nahrendorf, M. (2014). The innate immune system after ischemic injury: Lessons to be learned from the heart and brain. *JAMA Neurolology, 71*(2), 233–236.

Courtney, C. M., Goodman, S. M., Nagy, T. A., Levy, M., Bhusal, P., et al. (2017). Potentiating antibiotics in drug-resistant clinical isolates via stimuli-activated superoxide generation. *Science Advances, 3*(10), e1701776.

Cowardin, C. A., Buonomo, E. L., & Saleh, M. M. (2016). The binary toxin CDT enhances Clostridium difficile virulence by suppressing protective colonic eosinophilia. *Nature Microbiology, 1*(8), 16108.

Cox, L. M., & Blaser, M. J. (2015). Antibiotics in early life and obesity. *Nature Reviews Endocrinology, 11*(3), 182–190.

Crabbe, J. C., Wahlsten, D., & Dudek, B. C. (1999). Genetics of mouse behavior: Interactions with laboratory environment. *Science, 284*(5420), 1670–1672.

Craft, J. M., Watterson, D. M., Hirsch, E., & Van Eldik, L. J. (2005). Interleukin 1 receptor antagonist knockout mice show enhanced microglial activation and neuronal damage induced by intracerebroventricular infusion of human beta-amyloid. *Journal of Neuroinflammation, 2,* 15.

Craft, T. K., & DeVries, A. C. (2006). Role of IL-1 in post-stroke depressive-like behavior in mice. *Biological Psychiatry, 60*(8), 812–818.

Cragnolini, A. B., Schiöth, H. B., & Scimonelli, T. N. (2006). Anxiety-like behavior induced by IL-1β is modulated by α-MSH through central melanocortin-4 receptors. *Peptides (New York, 1979-1987), 27,* 1451–1456.

Crane, J. D., Palanivel, R., Mottillo, E. P., Bujak, A. L., Wang, H., et al. (2015). Inhibiting peripheral serotonin synthesis reduces obesity and metabolic dysfunction by promoting brown adipose tissue thermogenesis. *Nature Medicine (New York, NY, United States), 21*(2), 166–172.

Craske, M. G., & Stein, M. B. (2016). Anxiety. *Lancet, 388* (10063), 3048–3059.

Craske, M. G., Stein, M. B., Eley, T. C., Milad, M. R., Holmes, A., et al. (2017). Anxiety disorders. *Nature Reviews Disease Primers, 3,* 17024.

Craveiro, M., Cretenet, G., Mongellaz, C., Matias, M. I., Caron, O., et al. (2017). Resveratrol stimulates the metabolic reprogramming of human CD4 + T cells to enhance effector function. *Science Signaling, 10*(501), pii: eaal3024.

Creswell, J. D., Irwin, M. R., Burklund, L. J., Lieberman, M. D., Arevalo, J. M., et al. (2012). Mindfulness-based stress reduction training reduces loneliness and pro-inflammatory gene expression in older adults: A small randomized controlled trial. *Brain, Behavior, and Immunity, 26,* 1095–1101.

Crotty, S. (2015). A brief history of T cell help to B cells. *Nature Reviews Immunology, 15*(3), 185–189.

Crowley, N. A., Bloodgood, D. W., Hardaway, J. A., Kendra, A. M., McCall, J. G., et al. (2016). Dynorphin controls the gain of an amygdalar anxiety circuit. *Cell Reports, 14,* 2774–2783.

Cruchaga, C., Del-Aguila, J. L., Saef, B., Black, K., Fernandez, M. V., et al. (2018). Polygenic risk score of sporadic late-onset Alzheimer's disease reveals a shared architecture with the familial and early-onset forms.

Alzheimer's & Dementia, *14*, 205–214, pii: S1552-5260(17)33708-1.

Crumeyrolle-Arias, M., Jaglin, M., Bruneau, A., Vancassel, S., Cardona, A., et al. (2014). Absence of the gut microbiota enhances anxiety-like behavior and neuroendocrine response to acute stress in rats. *Psychoneuroendocrinology*, *42*, 207–217.

Crump, C., Winkleby, M. A., Sundquist, K., & Sundquist, J. (2013). Comorbidities and mortality in persons with schizophrenia: A Swedish national cohort study. *American Journal of Psychiatry*, *170*(3), 324–333.

Cruwys, T., Haslam, S. A., Dingle, G. A., Haslam, C., & Jetten, J. (2014). Depression and social identity: An integrative review. *Personality and Social Psychology Review*, *18*(3), 215–238.

Cryan, J. F., & Dinan, T. G. (2012). Mind-altering microorganisms: The impact of the gut microbiota on brain and behavior. *Nature Reviews Neuroscience*, *13*, 701–712.

Cuello, A. C. (2017). Early and late CNS inflammation in Alzheimer's disease: Two extremes of a continuum? *Trends in Pharmacological Sciences*, *38*(11), 956–966.

Culverhouse, R. C., Saccone, N. L., Horton, A. C., Ma, Y., Anstey, K. J., et al. (2018). Collaborative meta-analysis finds no evidence of a strong interaction between stress and 5-HTTLPR genotype contributing to the development of depression. *Molecular Psychiatry*, *23*(1), 133–142.

Cumming, T. B., Blomstrand, C., Skoog, I., & Linden, T. (2016). The high prevalence of anxiety disorders after stroke. *The American Journal of Geriatric Psychiatry*, *24*(2), 154–160.

Curtin, N. M., Boyle, N. T., Mills, K. H., & Connor, T. J. (2009). Psychological stress suppresses innate IFN-gamma production via glucocorticoid receptor activation: Reversal by the anxiolytic chlordiazepoxide. *Brain, Behavior, and Immunity*, *23*(4), 535–547.

Curtin, N. M., Mills, K. H., & Connor, T. J. (2009). Psychological stress increases expression of IL-10 and its homolog IL-19 via beta-adrenoceptor activation: Reversal by the anxiolytic chlordiazepoxide. *Brain, Behavior, and Immunity*, *23*(3), 371–379.

Curtiss, J., Andrews, L., Davis, M., Smits, J., & Hofmann, S. G. (2017). A meta-analysis of pharmacotherapy for social anxiety disorder: An examination of efficacy, moderators, and mediators. *Expert Opinion on Pharmacotherapy*, *18*(3), 243–251.

Cusack, K., Jonas, D. E., Forneris, C. A., Wines, C., Sonis, J., et al. (2016). Psychological treatments for adults with posttraumatic stress disorder: A systematic review and meta-analysis. *Clinical Psychology Review*, *43*, 128–141.

Cuthbert, B. N., & Insel, T. R. (2013). Toward the future of psychiatric diagnosis: The seven pillars of RDoC. *BMC Medicine*, *11*, 126.

Daher, J. P. L., Abdelmotilib, H. A., Hu, X., Volpicelli-Daley, L. A., Moehle, M. S., et al. (2015). Leucine-rich Repeat Kinase 2 (LRRK2) pharmacological inhibition abates α-Synuclein gene-induced neurodegeneration. *Journal of Biological Chemistry*, *290*, 19433–19444.

Dahl, J., Ormstad, H., Aass, H. C., Sandvik, L., Malt, U. F., et al. (2016). Recovery from major depressive disorder episode after non-pharmacological treatment is associated with normalized cytokine levels. *Acta Psychiatrica Scandinavica*, *134*(1), 40–47.

Dainer-Best, J., Trujillo, L. T., Schnyer, D. M., & Beevers, C. (2017). Sustained engagement of attention is associated with increased negative self-referent processing in major depressive disorder. *Biological Psychology*, *129*, 231–241.

Dalgaard, K., Landgraf, K., Heyne, S., Lempradl, A., Longinotto, J., et al. (2016). Trim28 haploinsufficiency triggers bi-stable epigenetic obesity. *Cell*, *164*(3), 353–364.

Dallman, M. F. (2010). Stress-induced obesity and the emotional nervous system. *Trends in Endocrinology and Metabolism*, *21*(3), 159–165.

Dalmau, J., Lancaster, E., Martinez-Hernandez, E., Rosenfeld, M. R., & Balice-Gordon, R. (2011). Clinical experience and laboratory investigations in patients with anti-NMDAR encephalitis. *Lancet Neurology*, *10*(1), 63–74.

Dalton, V. S., Verdurand, M., Walker, A., Hodgson, D. M., & Zavitsanou, K. (2012). Synergistic effect between maternal infection and adolescent cannabinoid exposure on serotonin 5HT1A receptor binding in the hippocampus: Testing the "two hit" hypothesis for the development of schizophrenia. *ISRN Psychiatry*, *2012*, 451865.

Dambrova, M., Latkovskis, G., Kuka, J., Strele, I., Konrade, I., et al. (2016). Diabetes is associated with higher trimethylamine N-oxide plasma levels. *Experimental and Clinical Endocrinology & Diabetes*, *124*(4), 251–256.

Dan, R., Růžička, F., Bezdicek, O., Růžička, E., Roth, J., et al. (2017). Separate neural representations of depression, anxiety and apathy in Parkinson's disease. *Scientific Reports*, *7*(1), 12164.

Danese, A., & J Lewis, S. (2017). Psychoneuroimmunology of early-life stress: The hidden wounds of childhood trauma? *Neuropsychopharmacology*, *42*(1), 99–114.

Danieli, Y. (1998). *International handbook of multigenerational legacies of trauma*. New York: Plenum.

Daniels, J. K., Lamke, J. P., Gaebler, M., Walter, H., & Scheel, M. (2013). White matter integrity and its relationship to PTSD and childhood trauma--A systematic review and meta-analysis. *Depression and Anxiety*, *30*(3), 207–216.

Danielson, A. M., Matheson, K., & Anisman, H. (2011). Cytokine levels at a single time point following a reminder stimulus among women in abusive dating relationships: Relationship to emotional states. *Psychoneuroendocrinology, 36*(1), 40–50.

Dantzer, R. (2017). Role of the kynurenine metabolism pathway in inflammation-induced depression: Preclinical approaches. *Current Topics in Behavioral Neurosciences, 31*, 117–138.

Dantzer, R., O'Connor, J. C., Freund, G. G., Johnson, R. W., & Kelley, K. W. (2008). From inflammation to sickness and depression: When the immune system subjugates the brain. *Nature Reviews Neuroscience, 9*(1), 46.

Daria, A., Colombo, A., Llovera, G., Hampel, H., Willem, M., et al. (2017). Young microglia restore amyloid plaque clearance of aged microglia. *EMBO Journal, 36* (5), 583–603.

Daskalakis, N. P., Cohen, H., Nievergelt, C. M., Baker, D. G., Buxbaum, J. D., et al. (2016). New translational perspectives for blood-based biomarkers of PTSD: From glucocorticoid to immune mediators of stress susceptibility. *Experimental Neurology, 284*(Pt B), 133–140.

Dauer, W., & Przedborski, S. (2003). Parkinson's disease: Mechanisms and models. *Neuron, 39*(6), 889–909.

Davalos, D., Grutzendler, J., Yang, G., Kim, J. V., Zuo, Y., et al. (2005). ATP mediates rapid microglial response to local brain injury in vivo. *Nature Neuroscience, 8*, 752–758.

David, L. A., Maurice, C. F., Carmody, R. N., Gootenberg, D. B., Button, J. E., et al. (2014). Diet rapidly and reproducibly alters the human gut microbiome. *Nature, 505* (7484), 559–563.

Davis, M., Walker, D. L., Miles, L., & Grillon, C. (2010). Phasic vs sustained fear in rats and humans: role of the extended amygdala in fear vs anxiety. *Neuropsychopharmacology, 35*, 105–135.

Davis, E. P., Waffarn, F., & Sandman, C. A. (2011). Prenatal treatment with glucocorticoids sensitizes the HPA axis response to stress among full-term infants. *Developmental Psychobiology, 53*(2), 175–183.

Davis, K. D., Flor, H., Greely, H. T., Iannetti, G. D., Mackey, S., et al. (2017). Brain imaging tests for chronic pain: Medical, legal and ethical issues and recommendations. *Nature Reviews Neurology, 13*(10), 624–638.

De Angelis, M., Piccolo, M., Vannini, L., Siragusa, S., De Giacomo, A., et al. (2013). Fecal microbiota and metabolome of children with autism and pervasive developmental disorder not otherwise specified. *PLoS One, 8* (10), e76993.

De Berardis, D., Marini, S., Serroni, N., Iasevoli, F., Tomasetti, C., et al. (2015). Targeting the noradrenergic system in posttraumatic stress disorder: A systematic review and meta-analysis of prazosin trials. *Current Drug Targets, 16*(10), 1094–1106.

de Berker, A. O., Rutledge, R. B., Mathys, C., Marshall, L., Cross, G. F., et al. (2016). Computations of uncertainty mediate acute stress responses in humans. *Nature Communications, 7*, 10996.

Dębiec, J., Bush, D. E., & LeDoux, J. E. (2011). Noradrenergic enhancement of reconsolidation in the amygdala impairs extinction of conditioned fear in rats—A possible mechanism for the persistence of traumatic memories in PTSD. *Depression and Anxiety, 28*(3), 186–193.

Dębiec, J., & LeDoux, J. E. (2006). Noradrenergic signaling in the amygdala contributes to the reconsolidation of fear memory: Treatment implications for PTSD. *Annals of the New York Academy of Sciences, 1071*, 521–524.

de Bilbao, F., Arsenijevic, D., Moll, T., Garcia-Gabay, I., Vallet, P., et al. (2009). In vivo over-expression of interleukin-10 increases resistance to focal brain ischemia in mice. *Journal of Neurochemistry, 110*(1), 12–22.

Deckert, J., Weber, H., Villmann, C., Lonsdorf, T. B., Richter, J., et al. (2017). GLRB allelic variation associated with agoraphobic cognitions, increased startle response and fear network activation: A potential neurogenetic pathway to panic disorder. *Molecular Psychiatry, 22*(10), 1431–1439.

de Cossío, L. F., Fourrier, C., Sauvant, J., Everard, A., Capuron, L., et al. (2017). Impact of prebiotics on metabolic and behavioral alterations in a mouse model of metabolic syndrome. *Brain, Behavior, and Immunity, 64*, 33–49.

De Dreu, C. K. (2012). Oxytocin modulates cooperation within and competition between groups: An integrative review and research agenda. *Hormones and Behavior, 61* (3), 419–428.

de Groot, J., Boersma, W. J., Scholten, J. W., & Koolhaas, J. M. (2002). Social stress in male mice impairs long-term antiviral immunity selectively in wounded subjects. *Physiology & Behavior, 75*, 277–285.

de Groot, J., van Milligen, F. J., Moonen-Leusen, B. W., Thomas, G., & Koolhaas, J. M. (1999). A single social defeat transiently suppresses the anti-viral immune response in mice. *Journal of Neuroimmunology, 95*(1–2), 143–151.

de Groot, P. F., Frissen, M. N., de Clercq, N. C., & Nieuwdorp, M. (2017). Fecal microbiota transplantation in metabolic syndrome: History, present and future. *Gut Microbes, 8*, 253–267.

Dehghan, M., Mente, A., Zhang, X., Swaminathan, S., Li, W., et al. (2017). Associations of fats and carbohydrate intake with cardiovascular disease and mortality in 18 countries from five continents (PURE): A prospective cohort study. *Lancet, 390*(10107), 2050–2062. Available from https://doi.org/10.1016/S0140-6736(17)32252-3.

de Jong, D., Jansen, R., Hoefnagels, W., Jellesma-Eggenkamp, M., Verbeek, M., et al. (2008). No effect of one-year treatment with indomethacin on Alzheimer's disease progression: A randomized controlled trial. *PLoS One, 3*(1), e1475.

de Kloet, C., Vermetten, E., Lentjes, E., Geuze, E., van Pelt, J., et al. (2008b). Differences in the response to the combined DEX-CRH test between PTSD patients with and without co-morbid depressive disorder. *Psychoneuroendocrinology, 33*(3), 313–320.

de Kloet, C. S., Vermetten, E., Geuze, E., Lentjes, E. G., Heijnen, C. J., et al. (2008a). Elevated plasma corticotrophin-releasing hormone levels in veterans with posttraumatic stress disorder. *Progress in Brain Research, 167*, 287–291.

de Kloet, E. R. (2014). From receptor balance to rational glucocorticoid therapy. *Endocrinology, 155*(8), 2754–2769.

De Kock, M., Loix, S., & Lavand'homme, P. (2013). Ketamine and peripheral inflammation. *CNS Neuroscience & Therapeutics, 19*(6), 403–410.

De La Garza, R., 2nd, Asnis, G. M., Pedrosa, E., Stearns, C., Migdal, A. L., et al. (2005). Recombinant human interferon-alpha does not alter reward behavior, or neuroimmune and neuroendocrine activation in rats. *Progress in Neuro-Psychopharmacology & Biological Psychiatry, 29*, 781–792.

Deleidi, M., & Gasser, T. (2013). The role of inflammation in sporadic and familial Parkinson's disease. *Cellular and Molecular Life Sciences, 70*(22), 4259–4273.

Del Gobbo, L. C., Imamura, F., Aslibekyan, S., Marklund, M., Virtanen, J. K., et al. (2016). ω-3 polyunsaturated fatty acid biomarkers and coronary heart disease: Pooling project of 19 cohort studies. *JAMA Internal Medicine, 176*(8), 1155–1166.

Del Grande da Silva, G., Wiener, C. D., Barbosa, L. P., Gonçalves Araujo, J. M., Molina, M. L., et al. (2016). Pro-inflammatory cytokines and psychotherapy in depression: Results from a randomized clinical trial. *Journal of Psychiatric Research, 75*, 57–64.

Delpech, J. C., Madore, C., Nadjar, A., Joffre, C., Wohleb, E. S., et al. (2015). Microglia in neuronal plasticity: Influence of stress. *Neuropharmacology, 96*(Pt A), 19–28.

del Rey, A., & Besedovsky, H. O. (2000). The cytokine-HPA axis circuit contributes to prevent or moderate autoimmune processes. *Zeitschrift fuer Rheumatologie, 59* (Suppl 2), Ii/31–35.

Del Tredici, K., & Braak, H. (2012). Lewy pathology and neurodegeneration in premotor Parkinson's disease. *Movement Disorders, 27*(5), 597–607.

Delvaux, E., Mastroeni, D., Nolz, J., Chow, N., Sabbagh, M., et al. (2017). Multivariate analyses of peripheral blood leukocyte transcripts distinguish Alzheimer's, Parkinson's, control, and those at risk for developing Alzheimer's. *Neurobiology of Aging, 58*, 225–237.

Delzenne, N. M., Cani, P. D., Everard, A., Neyrinck, A. M., & Bindels, L. B. (2015). Gut microorganisms as promising targets for the management of type 2 diabetes. *Diabetologia, 58*(10), 2206–2217.

De Miguel, Z., Haditsch, U., Palmer, T. D., Azpiroz, A., & Sapolsky, R. M. (2018). Adult-generated neurons born during chronic social stress are uniquely adapted to respond to subsequent chronic social stress. *Molecular Psychiatry*. Available from https://doi.org/10.1038/s41380-017-0013-1.

de Miranda, A. S., Lacerda-Queiroz, N., de Carvalho Vilela, M., Rodrigues, D. H., Rachid, M. A., et al. (2011). Anxiety-like behavior and proinflammatory cytokine levels in the brain of C57BL/6 mice infected with Plasmodium berghei (strain ANKA). *Neuroscience Letters, 491*(3), 202–206.

Demoruelle, M. K., Deane, K. D., & Holers, V. M. (2014). When and where does inflammation begin in rheumatoid arthritis? *Current Opinion in Rheumatology, 26*(1), 64–71.

Deng, Y., Lu, X., Wang, L., Li, T., Ding, Y., et al. (2014). Curcumin inhibits the AKT/NF-κB signaling via CpG demethylation of the promoter and restoration of NEP in the N2a cell line. *AAPS Journal, 16*(4), 649–657.

Dengler, V., Westphalen, K., & Koeppen, M. (2015). Disruption of circadian rhythms and sleep in critical illness and its impact on innate immunity. *Current Pharmaceutical Design, 21*(24), 3469–3476.

Denk, F., McMahon, S. B., & Tracey, I. (2014). Pain vulnerability: A neurobiological perspective. *Nature Neuroscience, 17*(2), 192–200.

Denollet, J., Gidron, Y., Vrints, C. J., & Conraads, V. M. (2010). Anger, suppressed anger, and risk of adverse events in patients with coronary artery disease. *American Journal of Cardiology, 105*(11), 1555–1560.

Denollet, J., & Kupper, N. (2007). Type-D personality, depression, and cardiac prognosis: Cortisol dysregulation as a mediating mechanism. *Journal of Psychosomatic Research, 62*(6), 607–609.

Denys, D., Fluitman, S., Kavelaars, A., Heijnen, C., & Westenberg, H. (2004). Decreased TNF-α and NK activity in obsessive-compulsive disorder. *Psychoneuroendocrinology, 29*(7), 945–952.

de Pablos, R. M., Herrera, A. J., Espinosa-Oliva, A. M., Sarmiento, M., Muñoz, M. F., et al. (2014). Chronic stress enhances microglia activation and exacerbates death of nigral dopaminergic neurons under conditions of inflammation. *Journal of Neuroinflammation, 11*, 34.

De Pablos, R. M., Villarán, R. F., Argüelles, S., Herrera, A. J., Venero, J. L., et al. (2006). Stress increases vulnerability to inflammation in the rat prefrontal cortex. *Journal of Neuroscience, 26*(21), 5709–5719.

De Palma, G., Blennerhassett, P., Lu, J., Deng, Y., Park, A. J., et al. (2015). Microbiota and host determinants of behavioural phenotype in maternally separated mice. *Nature Communications, 6, 7735.*

De Palma, G., Collins, S. M., Bercik, P., & Verdu, E. F. (2014). The microbiota-gut-brain axis in gastrointestinal disorders: Stressed bugs, stressed brain or both? *Journal of Physiology (Cambridge, United Kingdom), 592*(14), 2989–2997.

De Palma, G., Lynch, M. D., Lu, J., Dang, V. T., Deng, Y., et al. (2017). Transplantation of fecal microbiota from patients with irritable bowel syndrome alters gut function and behaviour in recipient mice. *Science Translational Medicine, 9, 379.*

De Picker, L. J., Morrens, M., Chance, S. A., & Boche, D. (2017). Microglia and brain plasticity in acute psychosis and schizophrenia illness course: A meta-review. *Frontiers in Psychiatry, 8, 238.*

Depino, A. M. (2015). Early prenatal exposure to LPS results in anxiety-and depression-related behaviors in adulthood. *Neuroscience, 299,* 56–65.

Depino, A. M. (2017). Perinatal inflammation and adult psychopathology: From preclinical models to humans. *Seminars in Cell and Developmental Biology, 77,* 104–114, pii: S1084-9521(17)30309-9.

de Quervain, D. J., & Margraf, J. (2008). Glucocorticoids for the treatment of post-traumatic stress disorder and phobias: A novel therapeutic approach. *European Journal of Pharmacology, 583*(2–3), 365–371.

Derks, N. A., Krugers, H. J., Hoogenraad, C. C., Joëls, M., & Sarabdjitsingh, R. A. (2016). Effects of early life stress on synaptic plasticity in the developing hippocampus of male and female rats. *PLoS One, 11*(10), e0164551.

de Roos, C., van der Oord, S., Zijlstra, B., Lucassen, S., Perrin, S., et al. (2017). Comparison of eye movement desensitization and reprocessing therapy, cognitive behavioral writing therapy, and wait-list in pediatric posttraumatic stress disorder following single-incident trauma: A multicenter randomized clinical trial. *Journal of Child Psychology and Psychiatry, and Allied Disciplines., 58*(11), 1219–1228.

Derry, C. J., Derry, S., & Moore, R. A. (2015). Single dose oral analgesics for acute postoperative pain in adults — An overview of Cochrane reviews. *Cochrane Database of Systematic Reviews,* (9), CD008659.

Desbonnet, L., Clarke, G., Traplin, A., O'Sullivan, O., Crispie, F., et al. (2015). Gut microbiota depletion from early adolescence in mice: Implications for brain and behaviour. *Brain, Behavior, and Immunity, 48,* 165–173.

Desbonnet, L., Garrett, L., Clarke, G., Kiely, B., Cryan, J. F., et al. (2010). Effects of the probiotic Bifidobacterium infantis in the maternal separation model of depression. *Neuroscience, 170*(4), 1179–1188.

Deshpande, A., Mailis-Gagnon, A., Zoheiry, N., & Lakha, S. F. (2015). Efficacy and adverse effects of medical marijuana for chronic noncancer pain: Systematic review of randomized controlled trials. *Canadian Family Physician Medecin de famille Canadien, 61*(8), e372–e381.

Devassy, J. G., Leng, S., Gabbs, M., Monirujjaman, M., & Aukema, H. M. (2016). Omega-3 polyunsaturated fatty acids and oxylipins in neuroinflammation and management of alzheimer disease. *Advances in Nutrition, 7*(5), 905–916.

Devor, A., Andreassen, O. A., Wang, Y., Maki-Marttunen, T., Smeland, O. B., et al. (2017). Genetic evidence for role of integration of fast and slow neurotransmission in schizophrenia. *Molecular Psychiatry, 22,* 792–801.

DeWall, C. N., MacDonald, G., Webster, G. D., Masten, C. L., Baumeister, R. F., et al. (2010). Acetaminophen reduces social pain: Behavioral and neural evidence. *Psychological Science, 2,* 931–937.

Dewey, F. E., Murray, M. F., Overton, J. D., Habegger, L., Leader, J. B., et al. (2016). Distribution and clinical impact of functional variants in 50,726 whole-exome sequences from the DiscovEHR study. *Science, 354,* aaf6814.

Dhabhar, F. S. (2009). Enhancing versus suppressive effects of stress on immune function: Implications for immunoprotection and immunopathology. *Neuroimmunomodulation, 16*(5), 300–317.

Dhabhar, F. S. (2014). Effects of stress on immune function: The good, the bad, and the beautiful. *Immunologic Research, 58*(2–3), 193–210.

Dhabhar, F. S., & McEwen, B. S. (1996). Stress-induced enhancement of antigen-specific cell-mediated immunity. *Journal of Immunology, 156*(7), 2608–2615.

Dhabhar, F. S., & McEwen, B. S. (1999). Enhancing versus suppressive effects of stress hormones on skin immune function. *Proceedings of the National Academy of Sciences of the United States of America, 96,* 1059–1064.

Dhabhar, F. S., Saul, A. N., Daugherty, C., Holmes, T. H., Bouley, D. M., et al. (2010). Short-term stress enhances cellular immunity and increases early resistance to squamous cell carcinoma. *Brain, Behavior, and Immunity, 24,* 127–137.

Dhami, G. K., & Ferguson, S. S. (2006). Regulation of metabotropic glutamate receptor signaling, desensitization and endocytosis. *Pharmacology & Therapeutics, 111,* 260–271.

Diamanti-Kandarakis, E., Bourguignon, J. P., Giudice, L. C., Hauser, R., Prins, G. S., et al. (2009). Endocrine-disrupting chemicals: An Endocrine Society scientific statement. *Endocrine Reviews, 30*(4), 293–342.

Di Battista, A. M., Heinsinger, N. M., & Rebeck, G. W. (2016). Alzheimer's disease genetic risk factor APOE-ε4 also affects normal brain function. *Current Alzheimer Research, 13,* 1200–1207.

DiCicco-Bloom, E., Lord, C., Zwaigenbaum, L., Courchesne, E., Dager, S. R., Schmitz, C., & Young, L. J. (2006). The developmental neurobiology of autism spectrum disorder. *Journal of Neuroscience, 26*(26), 6897–6906. Available from https://doi.org/10.1523/jneurosci.1712-06.2006.

Dickerson, D. D., & Bilkey, D. K. (2013). Aberrant neural synchrony in the maternal immune activation model: Using translatable measures to explore targeted interventions. *Frontiers in Behavioral Neuroscience, 7,* 217.

Dickerson, D. D., Wolff, A. R., & Bilkey, D. K. (2010). Abnormal long-range neural synchrony in a maternal immune activation animal model of schizophrenia. *Journal of Neuroscience, 30*(37), 12424–12431.

Dickerson, F., Stallings, C., Origoni, A., Vaughan, C., Katsafanas, E., et al. (2013). A combined marker of inflammation in individuals with mania. *PLoS One, 8,* e73520.

Dickerson, S., & Kemeny, M. (2004). Acute stressors and cortisol responses: A theoretical integration and synthesis of laboratory research. *Psychological Bulletin, 130*(9), 355–391.

Dietz, D. M., Laplant, Q., Watts, E. L., Hodes, G. E., Russo, S. J., et al. (2011). Paternal transmission of stress-induced pathologies. *Biological Psychiatry, 70*(5), 408–414.

Dietz, D. M., & Nestler, E. J. (2012). From father to offspring: Paternal transmission of depressive-like behaviors. *Neuropsychopharmacology, 37,* 311–312.

Dimatelis, J. J., Pillay, N. S., Mutyaba, A. K., Russell, V. A., Daniels, W. M. U., et al. (2012). Early maternal separation leads to down-regulation of cytokine gene expression. *Metabolic Brain Disease, 27*(3), 393–397.

Dimitrov, S., Hulteng, E., & Hong, S. (2017). Inflammation and exercise: Inhibition of monocytic intracellular TNF production by acute exercise via β 2-adrenergic activation. *Brain, Behavior, and Immunity, 61,* 60–68.

Dinan, T. G., & Cryan, J. F. (2013). Melancholic microbes: A link between gut microbiota and depression? *Neurogastroenterology and Motility, 25*(9), 713–719.

Dinan, T. G., & Cryan, J. F. (2016). Microbes, immunity, and behavior: Psychoneuroimmunology meets the microbiome. *Neuropsychopharmacology, 42,* 178–192.

Dinan, T. G., & Cryan, J. F. (2017). Gut-brain axis in 2016: Brain-gut-microbiota axis – Mood, metabolism and behaviour. *Nature Reviews Gastroenterology & Hepatology, 14,* 69–70.

Dinel, A. L., Joffre, C., Trifilieff, P., Aubert, A., Foury, A., et al. (2014). Inflammation early in life is a vulnerability factor for emotional behavior at adolescence and for lipopolysaccharide-induced spatial memory and

neurogenesis alteration at adulthood. *Journal of Neuroinflammation, 11,* 155.

Ding, H., Czoty, P. W., Kiguchi, N., Cami-Kobeci, G., Sukhtankar, D. D., et al. (2016). A novel orvinol analog, BU08028, as a safe opioid analgesic without abuse liability in primates. *Proceedings of the National Academy of Sciences of the United States of America, 113,* E5511–E5518.

Ding, H. T., Taur, Y., & Walkup, J. T. (2017). Gut Microbiota and Autism: Key Concepts and Findings. *Journal of Autism and Developmental Disorders, 47*(2), 480–489. Available from https://doi.org/10.1007/s10803-016-2960-9.

Disner, S. G., Beevers, C. G., Haigh, E. A., & Beck, A. T. (2011). Neural mechanisms of the cognitive model of depression. *Nature Reviews Neuroscience, 12,* 467–477.

D'Mello, A. M., & Stoodley, C. J. (2015). Cerebro-cerebellar circuits in autism spectrum disorder. *Frontiers in Neuroscience, 9,* 408. Available from https://doi.org/10.1016/j.semarthrit.2013.07.001.

D'Mello, C., Ronaghan, N., Zaheer, R., Dicay, M., Le, T., et al. (2015). Probiotics improve inflammation-associated sickness behavior by altering communication between the peripheral immune system and the brain. *Journal of Neuroscience, 35,* 10821–10830.

Dobbs, C. M., Feng, N., Beck, F. M., & Sheridan, J. F. (1996). Neuroendocrine regulation of cytokine production during experimental influenza viral infection: Effects of restraint stress-induced elevation in endogenous corticosterone. *Journal of Immunology, 157,* 1870–1877.

Dodd, G. T., Andrews, Z. B., Simonds, S. E., Michael, N. J., DeVeer, M., et al. (2017). A hypothalamic phosphatase switch coordinates energy expenditure with feeding. *Cell Metabolism, 26*(2), 375–393.

Doenni, V. M., Song, C. M., Hill, M. N., & Pittman, Q. J. (2017). Early-life inflammation with LPS delays fear extinction in adult rodents. *Brain, Behavior, and Immunity, 63,* 176–185.

Doenyas, C. (2018). Gut Microbiota, Inflammation, and Probiotics on Neural Development in Autism Spectrum Disorder. *Neuroscience, 374,* 271–286, pii: S0306-4522(18) 30099-X.

Dolzani, S. D., Baratta, M. V., Amat, J., Agster, K. L., Saddoris, M. P., et al. (2016). Activation of a habenulo-raphe circuit is critical for the behavioral and neurochemical consequences of uncontrollable stress in the male rat. *eNeuro, 3*(5), ENEURO.0229-16.2016.

Domingueti, C. P., Dusse, L. M., Carvalho, M. D., de Sousa, L. P., Gomes, K. B., et al. (2016). Diabetes mellitus: The linkage between oxidative stress, inflammation, hypercoaguailily and vascular complications. *Journal of Diabetes and Its Complications, 30*(4), 738–745.

Dominguez-Bello, M. G., De Jesus-Laboy, K. M., Shen, N., Cox, L. M., Amir, A., et al. (2016). Partial restoration of the microbiota of caesarean-born infants via vaginal microbial transfer. *Nature Medicine (New York, NY, United States), 22,* 250–253.

Donaldson, Z. R., le Francois, B., Santos, T. L., Almli, L. M., Boldrini, M., et al. (2016). The functional serotonin 1a receptor promoter polymorphism, rs6295, is associated with psychiatric illness and differences in transcription. *Translational Psychiatry, 6,* e746.

Dondo, T. B., Hall, M., West, R. M., Jernberg, T., Lindahl, B., et al. (2017). β-blockers and mortality after acute myocardial infarction in patients without heart failure or ventricular dysfunction. *Journal of the American College of Cardiology, 69*(22), 2710–2720.

Donegan, J. J., Girotti, M., Weinberg, M. S., & Morilak, D. A. (2014). A novel role for brain interleukin-6: Facilitation of cognitive flexibility in rat orbitofrontal cortex. *Journal of Neuroscience, 34,* 953–962.

Dong, M., Zhou, C., Ji, L., Pan, B., & Zheng, L. (2017). AG1296 enhances plaque stability via inhibiting inflammatory responses and decreasing MMP-2 and MMP-9 expression in ApoE-/- mice. *Biochemical and Biophysical Research Communications, 489,* 426–431.

Dong, S., Maniar, S., Manole, M. D., & Sun, D. (2017). Cerebral hypoperfusion and other shared brain pathologies in ischemic stroke and Alzheimer's disease. *Translational Stroke Research.*

Dooley, J., Tian, L., Schonefeldt, S., Delghingaro-Augusto, V., Garcia-Perez, J. E., et al. (2016). Genetic predisposition for beta cell fragility underlies type 1 and type 2 diabetes. *Nature Genetics, 48,* 519–527.

Dooms, H., & Abbas, A. K. (2010). Revisiting the role of IL-2 in autoimmunity. *European Journal of Immunology, 40* (6), 1538–1540. Available from https://doi.org/10.1002/eji.201040617.

Dorsett, C. R., McGuire, J. L., DePasquale, E. A., Gardner, A. E., Floyd, C. L., et al. (2017). Glutamate neurotransmission in rodent models of traumatic brain injury. *Journal of Neurotrauma, 34*(2), 263–272.

Dos Santos, L. R., Pimassoni, L. H. S., Sena, G. G. S., Camporez, D., Belcavello, L., et al. (2017). Validating GWAS variants from microglial genes implicated in Alzheimer's disease. *Journal of Molecular Neuroscience, 62,* 215–221.

Dos Santos, R. G., Balthazar, F. M., Bouso, J. C., & Hallak, J. E. (2016). The current state of research on ayahuasca: A systematic review of human studies assessing psychiatric symptoms, neuropsychological functioning, and neuroimaging. *Journal of Psychopharmacology, 30*(12), 1230–1247.

Doty, R. L. (2012). Olfactory dysfunction in Parkinson disease. *Nature Reviews Neurology, 8,* 329–339.

Dou, H., Feher, A., Davila, A. C., Romero, M. J., Patel, V. S., et al. (2017). Role of adipose tissue endothelial ADAM17 in age-related coronary microvascular dysfunction. *Arteriosclerosis, Thrombosis, and Vascular Biology, 37,* 1180–1193.

Douglass, A. M., Kucukdereli, H., Ponserre, M., Markovic, M., Gründemann, J., et al. (2017). Central amygdala circuits modulate food consumption through a positive-valence mechanism. *Nature Neuroscience, 20,* 1384–1394.

Dowell, D., Haegerich, T. M., & Chou, R. (2016). CDC guideline for prescribing opioids for chronic pain—United States, 2016. *The Journal of the American Medical Association, 315,* 1624–1645.

Doyle, D. M., & Molix, L. (2014). Perceived discrimination and well-being in gay men: The protective role of behavioural identification. *Psychology and Sexuality, 5,* 117–130.

Drummond, E. M., & Gibney, E. R. (2013). Epigenetic regulation in obesity. *Current Opinion in Clinical Nutrition and Metabolic Care, 16,* 392–397.

Drysdale, A. T., Grosenick, L., Downar, J., Dunlop, K., Mansouri, F., et al. (2017). Resting-state connectivity biomarkers define neurophysiological subtypes of depression. *Nature Medicine (New York, NY, United States), 23,* 28–38.

Duan, F. F., Liu, J. H., & March, J. C. (2015). Engineered commensal bacteria reprogram intestinal cells into glucose-responsive insulin-secreting cells for the treatment of diabetes. *Diabetes, 64,* 1794–1803.

Duan, R.-S., Chen, Z., Dou, Y. C., Concha Quezada, H., Nennesmo, I., et al. (2006). Apolipoprotein E deficiency increased microglial activation/CCR3 expression and hippocampal damage in kainic acid exposed mice. *Experimental Neurology, 202*(2), 373–380.

Duda-Chodak, A., Tarko, T., Satora, P., & Sroka, P. (2015). Interaction of dietary compounds, especially polyphenols, with the intestinal microbiota: A review. *European Journal of Nutrition, 54,* 325–341.

Duman, R. S. (2014). Pathophysiology of depression and innovative treatments: Remodeling glutamatergic synaptic connections. *Dialogues in Clinical Neuroscience, 16* (1), 11–27.

Duman, R. S., & Li, N. (2012). A neurotrophic hypothesis of depression: Role of synaptogenesis in the actions of NMDA receptor antagonists. *Philosophical Transactions of the Royal Society of London. Series B, Biological Sciences., 367*(1601), 2475–2484.

Duman, R. S., & Monteggia, L. M. (2006). A neurotrophic model for stress-related mood disorders. *Biological Psychiatry, 59*(12), 1116–1127.

Dunjic-Kostic, B., Ivkovic, M., Radonjic, N. V., Petronijevic, N. D., Pantovic, M., et al. (2013). Melancholic and atypical major depression--Connection between cytokines, psychopathology and treatment. *Progress in Neuro-Psychopharmacology & Biological Psychiatry, 43,* 1–6.

Dunlop, B. W., Rajendra, J. K., Craighead, W. E., Kelley, M. E., McGrath, C. L., et al. (2017). Functional connectivity of the subcallosal cingulate cortex and differential outcomes to treatment with cognitive-behavioral therapy or antidepressant medication for major depressive disorder. *American Journal of Psychiatry, 174*, 533−545.

Dunne, J. L., Triplett, E. W., Gevers, D., Xavier, R., Insel, R., et al. (2014). The intestinal microbiome in type 1 diabetes. *Clinical and Experimental Immunology, 177*, 30−37.

Dunsmoor, J. E., & Paz, R. (2015). Fear generalization and anxiety: Behavioral and neural mechanisms. *Biological Psychiatry, 78*, 336−343.

Duque-Wilckens, N., Steinman, M. Q., Busnelli, M., Chini, B., Yokoyama, S., et al. (2018). Oxytocin receptors in the anteromedial bed nucleus of the stria terminalis promote stress-induced social avoidance in females. *Biological Psychiatry, 83*(3), 203−213.

Duran, E., Duran, B., Heart, M. Y. H. B., & Horse-Davis, S. Y. (1998). *Healing the American Indian soul wound. International handbook of multigenerational legacies of trauma* (pp. 341−354). US: Springer:.

Duranti, S., Ferrario, C., van Sinderen, D., Ventura, M., & Turroni, F. (2017). Obesity and microbiota: An example of an intricate relationship. *Genes and Nutrition, 12*, 18.

Durgan, D. J. (2017). Obstructive sleep apnea-induced hypertension: Role of the gut microbiota. *Current Hypertension Reports, 19*, 35.

Dursun, E., Gezen-Ak, D., Hanağası, H., Bilgiç, B., Lohmann, E., et al. (2015). The interleukin 1 alpha, interleukin 1 beta, interleukin 6 and alpha-2-macroglobulin serum levels in patients with early or late onset Alzheimer's disease, mild cognitive impairment or Parkinson's disease. *Journal of Neuroimmunology, 283*, 50−57.

Duty, S., & Jenner, P. (2011). Animal models of Parkinson's disease: A source of novel treatments and clues to the cause of the disease. *British Journal of Pharmacology, 164*, 1357−1391.

Dvorkin, A., Silva, C., McMurran, T., Bisnaire, L., Foster, J., et al. (2010). Features of compulsive checking behavior mediated by nucleus accumbens and orbital frontal cortex. *The European Journal of Neuroscience, 32*(9), 1552−1563.

Dwivedi, Y., Rizavi, H. S., Conley, R. R., Roberts, R. C., Tamminga, C. A., et al. (2003). Altered gene expression of brain-derived neurotrophic factor and receptor tyrosine kinase B in postmortem brain of suicide subjects. *Archives of General Psychiatry, 60*, 804−815.

Dzamko, N., Geczy, C. L., & Halliday, G. M. (2015). Inflammation is genetically implicated in Parkinson's disease. *Neuroscience, 302*, 89−102.

Dzamko, N., Gysbers, A., Perera, G., Bahar, A., Shankar, A., et al. (2017). Toll-like receptor 2 is increased in neurons in Parkinson's disease brain and may contribute to alpha-synuclein pathology. *Acta Neuropathologica, 133*, 303−319.

Earnshaw, V. A., Elliott, M. N., Reisner, S. L., Mrug, S., Windle, M., et al. (2017). Peer victimization, depressive symptoms, and substance use: A longitudinal analysis. *Pediatrics, 139*(6), e20163426.

Eaton, W. W., Byrne, M., Ewald, H., Mors, O., Chen, C. Y., Agerbo, E., & Mortensen, P. B. (2006). Association of schizophrenia and autoimmune diseases: linkage of Danish national registers. *The American Journal of Psychiatry, 163*(3), 521−528. Available from https://doi.org/10.1176/appi.ajp.163.3.521.

Eban-Rothschild, A., Rothschild, G., Giardino, W. J., Jones, J. R., & de Lecea, L. (2016). VTA dopaminergic neurons regulate ethologically relevant sleep-wake behaviors. *Nature Neuroscience, 19*(10), 1356−1366.

Edgar, R. S., Stangherlin, A., Nagy, A. D., Nicoll, M. P., Efstathiou, S. M., et al. (2016). Cell autonomous regulation of herpes and influenza virus infection by the circadian clock. *Proceedings of the National Academy of Sciences of the United States of America, 113*(36), 10085−10090.

Edmiston, E., Ashwood, P., & Van de Water, J. (2017). Autoimmunity, Autoantibodies, and Autism Spectrum Disorder. *Biological Psychiatry, 81*(5), 383−390. Available from https://doi.org/10.1016/j.biopsych.2016.08.031.

Edmiston, E. E., Wang, F., Mazure, C. M., Guiney, J., Sinha, R., et al. (2011). Corticostriatal-limbic gray matter morphology in adolescents with self-reported exposure to childhood maltreatment. *Archives of Pediatrics & Adolescent Medicine, 165*, 1069−1077.

Ehrlich, K. B., Ross, K. M., Chen, E., & Miller, G. E. (2016). Testing the biological embedding hypothesis: Is early life adversity associated with a later proinflammatory phenotype? *Development and Psychopathology, 28*, 1273−1283.

Eidelman, S., & Biernat, M. (2003). Derogating black sheep: Individual or group protection? *Journal of Experimental Social Psychology, 39*, 602−609.

Eisele, Y. S., Bolmont, T., Heikenwalder, M., Langer, F., Jacobson, L. H., et al. (2009). Induction of cerebral beta-amyloidosis: Intracerebral versus systemic Abeta inoculation. *Proceedings of the National Academy of Sciences of the United States of America, 106*, 12926−12931.

Eisenberg, D. T., Borja, J. B., Hayes, M. G., & Kuzawa, C. W. (2017). Early life infection, but not breastfeeding, predicts adult blood telomere lengths in the Philippines. *American Journal of Human Biology, 29*.

Eisenberger, N. I. (2012). The pain of social disconnection: Examining the shared neural underpinnings of physical and social pain. *Nature Reviews Neuroscience, 13*, 421−434.

Eisenberger, N. I., Berkman, E. T., Inagaki, T. K., Rameson, L. T., Mashal, N. M., et al. (2010). Inflammation-induced anhedonia: Endotoxin reduces ventral striatum responses to reward. *Biological Psychiatry*, *68*(8), 748−754.

Eisenberger, N. I., Moieni, M., Inagaki, T. K., Muscatell, K. A., & Irwin, M. R. (2017). In sickness and in health: The co-regulation of inflammation and social behavior. *Neuropsychopharmacology*, *42*, 242−253.

Eisenstein, M. (2016). Microbiome: Bacterial broadband. *Nature*, *533*(7603), S104−S106.

Ekdahl, C. T., Kokaia, Z., & Lindvall, O. (2009). Brain inflammation and adult neurogenesis: The dual role of microglia. *Neuroscience*, *158*, 1021−1029.

El Ali, A., & Rivest, S. (2016). Microglia in Alzheimer's disease: A multifaceted relationship. *Brain, Behavior, and Immunity*, *55*, 138−150.

Ellenbogen, M. A., Linnen, A. M., Cardoso, C., & Joober, R. (2013). Intranasal oxytocin impedes the ability to ignore task-irrelevant facial expressions of sadness in students with depressive symptoms. *Psychoneuroendocrinology*, *38*, 387−398.

Ellis, A., Grace, P. M., Wieseler, J., Favret, J., Springer, K., et al. (2016). Morphine amplifies mechanical allodynia via TLR4 in a rat model of spinal cord injury. *Brain, Behavior, and Immunity*, *58*, 348−356.

Ellis, A., Wieseler, J., Favret, J., Johnson, K. W., Rice, K. C., et al. (2014). Systemic administration of propentofylline, ibudilast, and (+)-naltrexone each reverses mechanical allodynia in a novel rat model of central neuropathic pain. *The Journal of Pain*, *15*, 407−421.

Ellman, L. M., Deicken, R. F., Vinogradov, S., Kremen, W. S., Poole, J. H., et al. (2010). Structural brain alterations in schizophrenia following fetal exposure to the inflammatory cytokine interleukin-8. *Schizophrenia Research*, *121*(1−3), 46−54.

Elmore, M. R. P., Najafi, A. R., Koike, M. A., Dagher, N. N., Spangenberg, E. E., et al. (2014). Colony-stimulating factor 1 receptor signaling is necessary for microglia viability, unmasking a microglia progenitor cell in the adult brain. *Neuron*, *82*(2), 380−397.

Elovainio, M., Taipale, T., Seppälä, I., Mononen, N., Raitoharju, E., et al. (2015). Activated immune-inflammatory pathways are associated with long-standing depressive symptoms: Evidence from gene-set enrichment analyses in the Young Finns Study. *Journal of Psychiatric Research*, *71*, 120−125.

Elsayed, M., Banasr, M., Duric, V., Fournier, N. M., Licznerski, P., et al. (2012). Antidepressant effects of fibroblast growth factor-2 in behavioral and cellular models of depression. *Biological Psychiatry*, *72*, 258−265.

Elwood, P., Galante, J., Pickering, J., Palmer, S., Bayer, A., et al. (2013). Healthy lifestyles reduce the incidence of chronic diseases and dementia: Evidence from the Caerphilly cohort study. *PLoS One*, *8*(12), e81877.

Elzinga, B. M., Schmahl, C. G., Vermetten, E., van Dyck, R., & Bremner, J. D. (2003). Higher cortisol levels following exposure to traumatic reminders in abuse-related PTSD. *Neuropsychopharmacology*, *28*(9), 1656.

Emdin, C. A., Khera, A. V., Natarajan, P., Klarin, D., Zekavat, S. M., et al. (2017). Genetic association of waist-to-hip ratio with cardiometabolic traits, type 2 diabetes, and coronary heart disease. *The Journal of the American Medical Association*, *317*(6), 626−634.

Emoto, T., Yamashita, T., Sasaki, N., Hirota, Y., Hayashi, T., et al. (2016). Analysis of gut microbiota in coronary artery disease patients: A possible link between gut microbiota and coronary artery diease. *Journal of Atherosclerosis and Thrombosis*, *23*, 908−921.

Endesfelder, S., Weichelt, U., Strauß, E., Schlör, A., Sifringer, M., et al. (2017). Neuroprotection by caffeine in hyperoxia-induced neonatal brain injury. *International Journal of Molecular Sciences*, *18*, pii: E187.

Engelhart, M. J., Geerlings, M. I., Meijer, J., Kiliaan, A., Ruitenberg, A., et al. (2004). Inflammatory proteins in plasma and the risk of dementia: The rotterdam study. *Archives of Neurology*, *61*, 668−672.

Engin, A. (2017). The pathogenesis of obesity-associated adipose tissue inflammation. *Advances in Experimental Medicine and Biology*, *960*, 221−245.

Engler, H., Bailey, M. T., Engler, A., Stiner-Jones, L. M., Quan, N., et al. (2008). Interleukin-1 receptor type 1-deficient mice fail to develop social stress-associated glucocorticoid resistance in the spleen. *Psychoneuroendocrinology*, *33*(1), 108−117.

Engler, H., Doenlen, R., Engler, A., Riether, C., Prager, G., et al. (2011). Acute amygdaloid response to systemic inflammation. *Brain, Behavior, and Immunity*, *25*, 1384−1392.

Entringer, S., Epel, E. S., Kumsta, R., Lin, J., Hellhammer, D. H., et al. (2011). Stress exposure in intrauterine life is associated with shorter telomere length in young adulthood. *Proceedings of the National Academy of Sciences of the United States of America*, *108*, E513−E518.

Entringer, S., Kumsta, R., Hellhammer, D. H., Wadhwa, P. D., & Wüst, S. (2009). Prenatal exposure to maternal psychosocial stress and HPA axis regulation in young adults. *Hormones and Behavior*, *55*(2), 292−298.

Entringer, S., Kumsta, R., Nelson, E. L., Hellhammer, D. H., Wadhwa, P. D., et al. (2008). Influence of prenatal psychosocial stress on cytokine production in adult women. *Developmental Psychobiology*, *50*(6), 579−587.

Eraly, S. A., Nievergelt, C. M., Maihofer, A. X., Barkauskas, D. A., Biswas, N., et al. (2014). Assessment of plasma C-reactive protein as a biomarker of posttraumatic stress disorder risk. *JAMA Psychiatry*, *71*, 423−431.

Eriksen, K. G., Radford, E. J., Silver, M. J., Fulford, A. J., Wegmüller, R., et al. (2017). Influence of intergenerational in utero parental energy and nutrient restriction on offspring growth in rural Gambia. *FASEB Journal, 31*, 4928–4934.

Eriksson, M., Räikkönen, K., & Eriksson, J. G. (2014). Early life stress and later health outcomes: Findings from the Helsinki Birth Cohort Study. *American Journal of Human Biology, 26*(2), 111–116.

Erlich, P. M., Lunetta, K. L., Cupples, L. A., Abraham, C. R., Green, R. C., et al. (2012). Serum paraoxonase activity is associated with variants in the PON gene cluster and risk of Alzheimer disease. *Neurobiology of Aging, 33*, 1015.e7–1015.e23.

Erny, D., Hrabě de Angelis, A. L., Jaitin, D., Wieghofer, P., Staszewski, O., et al. (2015). Host microbiota constantly control maturation and function of microglia in the CNS. *Nature Neuroscience, 18*(7), 965–977.

Esalatmanesh, S., Abrishami, Z., Zeinoddini, A., Rahiminejad, F., Sadeghi, M., et al. (2016). Minocycline combination therapy with fluvoxamine in moderate to severe obsessive-compulsive disorder: A placebo-controlled, double-blind, randomized trial. *Psychiatry and Clinical Neurosciences, 70*(11), 517–526.

Espelage, D. L., Hong, J. S., & Mebane, S. (2016). Recollections of childhood bullying and multiple forms of victimization: Correlates with psychological functioning among college students. *Social Psychology of Education, 19*(4), 715–728.

Esser, N., Legrand-Poels, S., Piette, J., Scheen, A. J., & Paquot, N. (2014). Inflammation as a link between obesity, metabolic syndrome and type 2 diabetes. *Diabetes Management, 105*, 141–150.

Estes, M. L., & McAllister, A. K. (2015). Immune mediators in the brain and peripheral tissues in autism spectrum disorder. *Nature Reviews Neuroscience, 16*(8), 469–486. Available from https://doi.org/10.1038/nrn3978.

Evangelou, F., Ntritsos, G., Chondrogiorgi, M., Kvvoura, F. K., Kernández, A. F., et al. (2016). Exposure to pesticides and diabetes: A systematic review and meta-analysis. *Environment International, 91*, 60–68.

Evans, B. (2013). How autism became autism: The radical transformation of a central concept of child development in Britain. *History of the Human Sciences, 26*(3), 3–31.

Evans, S. J., Bassis, C. M., Hein, R., Assari, S., Flowers, S. A., et al. (2017). The gut microbiome composition associates with bipolar disorder and illness severity. *Journal of Psychiatric Research, 87*, 23–29.

Evans, S. J., Choudary, P. V., Neal, C. R., Li, J. Z., Vawter, M. P., et al. (2004). Dysregulation of the fibroblast growth factor system in major depression. *Proceedings of the National Academy of Sciences of the United States of America, 101*, 15506–15511.

Everett, B. M., Brooks, M. M., Vlachos, H. E., Chaitman, B. R., Frye, R. L., et al. (2015). Troponin and cardiac events in stable ischemic heart disease and diabetes. *New England Journal of Medicine, 373*(7), 610–620.

Evrensel, A., & Ceylan, M. E. (2015). The gut-brain axis: The missing link in depression. *Clinical Psychopharmacology and Neuroscience, 13*(3), 239–244.

Eyre, H. A., Air, T., Proctor, S., Rositano, S., & Baune, B. T. (2015). A critical review of the efficacy of non-steroidal anti-inflammatory drugs in depression. *Progress in Neuro-Psychopharmacology & Biological Psychiatry, 57*, 11–16.

Fagundes, C. P., Glaser, R., & Kiecolt-Glaser, J. K. (2013). Stressful early life experiences and immune dysregulation across the lifespan. *Brain, Behavior, and Immunity, 27*, 8–12.

Fali, T., Vallet, H., & Sauce, D. (2018). Impact of stress on aged immune system compartments: Overview from fundamental to clinical data. *Experimental Gerontology, 105*, 19–26, pii: S0531-5565(17)30782-9.

Fall, C. H. (2011). Evidence for the intra-uterine programming of adiposity in later life. *Annals of Human Biology, 38*(4), 410–428.

Fan, R., Toubal, A., Goñi, S., Drareni, K., Huang, Z., et al. (2016). Loss of the co-repressor GPS2 sensitizes macrophage activation upon metabolic stress induced by obesity and type 2 diabets. *Nature Medicine (New York, NY, United States), 22*, 780–791.

Fan, W., Waizenegger, W., Lin, C. S., Sorrentino, V., He, M. X., et al. (2017). PPARδ promotes running endurance by preserving glucose. *Cell Metabolism, 25*, 1186–1193.

Fang, S., Suh, J. M., Reilly, S. M., Yu, E., Osborn, O., et al. (2015). Intestinal FXR agonism promotes adipose tissue browning and reduces obesity and insulin resistance. *Nature Medicine (New York, NY, United States), 21*, 159–165.

Farb, N. A., Anderson, A. K., Bloch, R. T., & Segal, Z. V. (2011). Mood-linked responses in medial prefrontal cortex predict relapse in patients with recurrent unipolar depression. *Biological Psychiatry, 70*(4), 366–372.

Faria, V., Ahs, F., Appel, L., Linnman, C., Bani, M., et al. (2014). Amygdala-frontal couplings characterizing SSRI and placebo response in social anxiety disorder. *International Journal of Neuropsychopharmacology, 17*, 1149–1157.

Farmer, K., Rudyk, C., Prowse, N. A., & Hayley, S. (2015). Hematopoietic cytokines as therapeutic players in early stages Parkinson's disease. *Frontiers in Aging Neuroscience, 7*, 126.

Faust, T. W., Chang, E. H., Kowal, C., Berlin, R., Gazaryan, I. G., et al. (2010). Neurotoxic lupus autoantibodies alter brain function through two distinct mechanisms. *Proceedings of the National Academy of Sciences of the United States of America, 107*, 18569–18574.

Feigenson, K. A., Kusnecov, A. W., & Silverstein, S. M. (2014). Inflammation and the two-hit hypothesis of schizophrenia. *Neuroscience & Biobehavioral Reviews, 38*, 72–93.

Felger, J. C., Li, Z., Haroon, E., Woolwine, B. J., Jung, M. Y., et al. (2016). Inflammation is associated with decreased functional connectivity within corticostriatal reward circuitry in depression. *Molecular Psychiatry, 21*, 1358–1365.

Fellows, R., Denizot, J., Stellato, C., Cuomo, A., Jain, P., et al. (2018). Microbiota derived short chain fatty acids promote histone crotonylation in the colon through histone deacetylases. *Nature Communications, 9*, 105.

Feng, C., Fang, M., & Liu, X. Y. (2014). The neurobiological pathogenesis of poststroke depression. *The Scientific World Journal, 2014*, 521349.

Fenn, A. M., Gensel, J. C., Huang, Y., Popovich, P. G., Lifshitz, J., et al. (2014). Immune activation promotes depression 1 month after diffuse brain injury: A role for primed microglia. *Biological Psychiatry, 76*(7), 575–584.

Ferrante, P., Saresella, M., Guerini, F. R., Marzorati, M., Musetti, M. C., & Cazzullo, A. G. (2003). Significant association of HLA A2-DR11 with CD4 naive decrease in autistic children. *Biomedicine & Pharmacotherapy, 57*(8), 372–374.

Ferreira, D., Verhagen, C., Hernández-Cabrera, J. A., Cavallin, L., Guo, C. J., et al. (2017). Distinct subtypes of Alzheimer's disease based on patterns of brain atrophy: Longitudinal trajectories and clinical applications. *Scientific Reports, 7*, 46263.

Ferris, L. J., Jetten, J., Molenberghs, P., Bastian, B., & Karnadewi, F. (2016). Increased pain communication following multiple group memberships salience leads to a relative reduction in pain-related brain activity. *PLoS One, 11*, e0163117.

Fields, R. D. (2008). White matter in learning, cognition and psychiatric disorders. *Trends in Neurosciences, 31*, 361–370.

Figueiredo-Braga, M., Mota-Garcia, F., O'Connor, J. E., Garcia, J. R., Mota-Cardoso, R., et al. (2009). Cytokines and anxiety in systemic lupus erythematosus (SLE) patients not receiving antidepressant medication. *Annals of the New York Academy of Sciences, 1173*, 286–291.

Filaire, E., Larue, J., Portier, H., Abed, A., Graziella, P. D., et al. (2011). Lecturing to 200 students and its effects on cytokine concentration and salivary markers of adrenal activation. *Stress and Health, 27*(2), e25–e35.

Filiano, A. J., Gadani, S. P., & Kipnis, J. (2017). How and why do T cells and their derived cytokines affect the injured and healthy brain? *Nature Reviews Neuroscience, 18*(6), 375–384. Available from https://doi.org/10.1038/nrn.2017.39.

Filiano, A. J., Xu, Y., Tustison, N. J., Marsh, R. L., Baker, W., Smirnov, I., … Kipnis, J. (2016). Unexpected role of interferon-gamma in regulating neuronal connectivity and social behaviour. *Nature, 535*(7612), 425–429. Available from https://doi.org/10.1038/nature18626.

Filipovic, B. R., & Filipovic, B. F. (2014). Psychiatric comorbidity in the treatment of patients with inflammatory bowel disease. *World Journal of Gastroenterology, 20*, 3552–3563.

Filliol, D., Ghozland, S., Chluba, J., Martin, M., Matthes, H. W., et al. (2000). Mice deficient for delta- and mu-opioid receptors exhibit opposing alterations of emotional responses. *Nature Genetics, 25*, 195–200.

Fillman, S. G., Cloonan, N., Catts, V. S., Miller, L. C., Wong, J., et al. (2013). Increased inflammatory markers identified in the dorsolateral prefrontal cortex of individuals with schizophrenia. *Molecular Psychiatry, 18*(2), 206–214.

Fineberg, N. A., Reghunandanan, S., Simpson, H. B., Phillips, K. A., Richter, M. A., et al. (2015). Obsessive-compulsive disorder (OCD): Practical strategies for pharmacological and somatic treatment in adults. *Psychiatry Research, 227*, 114–125.

Finegold, S. M., Dowd, S. E., Gontcharova, V., Liu, C., Henley, K. E., et al. (2010). Pyrosequencing study of fecal microflora of autistic and control children. *Anaerobe, 16*, 444–453.

Finger, B. C., Dinan, T. G., & Cryan, J. F. (2012). The temporal impact of chronic intermittent psychosocial stress on high-fat diet-induced alterations in body weight. *Psychoneuroendocrinology, 37*, 729–741.

Finlay, J. M., & Zigmond, M. J. (1997). The effects of stress on central dopaminergic neurons: Possible clinical implications. *Neurochemical Research, 22*, 1387–1394.

Fisher, C. K., Mora, T., & Walczak, A. M. (2017). Variable habitat conditions drive species covariation in the human microbiota. *PLoS Computational Biology, 13*, e1005435.

Fisher, L. B., Pedrelli, P., Iverson, G. L., Bergquist, T. F., Bombardier, C. H., et al. (2016). Prevalence of suicidal behaviour following traumatic brain injury: Longitudinal follow-up data from the NIDRR Traumatic Brain Injury Model Systems. *Brain Injury, 30*(11), 1311–1318.

Fitzgerald, M. L., Kassir, S. A., Underwood, M. D., Bakalian, M. J., Mann, J. J., et al. (2016). Dysregulation of striatal dopamine receptor binding in suicide. *Neuropsychopharmacology, 42*, 974–982.

Fitzgerald, P. J., Giustino, T. F., Seemann, J. R., & Maren, S. (2015). Noradrenergic blockade stabilizes prefrontal activity and enables fear extinction under stress. *Proceedings of the National Academy of Sciences of the United States of America, 112*, E3729–E3737.

Fitzpatrick, J. K., & Downer, E. J. (2016). Toll-like receptor signalling as a cannabinoid target in Multiple Sclerosis. *Neuropharmacology, 113*, 618−626.

Fjell, A. M., Idland, A. V., Sala-Llonch, R., Watne, L. O., Borza, T., et al. (2017). Neuroinflammation and Tau interact with amyloid in predicting sleep problems in aging independently of atrophy. *Cerebral Cortex*, 1−11.

Flannick, J., Thorleifsson, G., Beer, N. L., Jacobs, S. B., Grarup, N., et al. (2014). Loss-of-function mutations in SLC30A8 protect against type 2 diabetes. *Nature Genetics, 46*, 357−363.

Flegal, K. M., Kit, B. K., Orpana, H., & Graubard, B. I. (2013). Association of all-cause mortality with overweight and obesity using standard body mass index categories: A systematic review and meta-analysis. *The Journal of the American Medical Association, 309*(1), 71−82.

Fleshner, M. (2013). Stress-evoked sterile inflammation, danger associated molecular patterns (DAMPs), microbial associated molecular patterns (MAMPs) and the inflammasome. *Brain, Behavior, and Immunity, 27*, 1−7.

Flint, H. J., Scott, K. P., Louis, P., & Duncan, S. (2012). The role of the gut microbiota in nutrition and health. *Nature Reviews Gastroenterology & Hepatology, 9*, 577−589.

Flint, M. S., Valosen, J. M., Johnson, E. A., Miller, D. B., & Tinkle, S. S. (2001). Restraint stress applied prior to chemical sensitization modulates the development of allergic contact dermatitis differently than restraint prior to challenge. *Journal of Neuroimmunology, 113*, 72−80.

Flores, C., & Stewart, J. (2000). Basic fibroblast growth factor as a mediator of the effects of glutamate in the development of long-lasting sensitization to stimulant drugs: Studies in the rat. *Psychopharmacology (Berlin, Germany), 151*, 152−165.

Fluitman, S., Denys, D., Vulink, N., Schutters, S., Heijnen, C., et al. (2010). Lipopolysaccharide-induced cytokine production in obsessive-compulsive disorder and generalized social anxiety disorder. *Psychiatry Research, 178*(2), 313−316.

Fontenelle, L. F., Barbosa, I. G., Luna, J. V., de Sousa, L. P., Abreu, M. N., et al. (2012). A cytokine study of adult patients with obsessive-compulsive disorder. *Comprehensive Psychiatry, 53*(6), 797−804.

Forouhi, N. G., Imamura, F., Sharp, S. J., Koulman, A., Schulze, M. B., et al. (2016). Association of plasma phospholipid n-3 and n-6 polyunsaturated fatty acids with type 2 diabetes: The EPIC-InterAct Case-Cohort Study. *PLoS Medicine, 13*, e1002094.

Forslund, K., Hildebrand, F., Nielsen, T., Falony, G., Le Chatelier, E., et al. (2015). Disentangling type 2 diabetes and metformin treatment signatures in the human gut microbiota. *Nature, 528*, 262−266.

Forstner, A. J., Rambau, S., Friedrich, N., Ludwig, K. U., Böhmer, A. C., et al. (2017). Further evidence for genetic variation at the serotonin transporter gene SLC6A4 contributing toward anxiety. *Psychiatric Genetics, 27*(3), 96−102.

Forsyth, J. K., Ellman, L. M., Tanskanen, A., Mustonen, U., Huttunen, M. O., et al. (2013). Genetic risk for schizophrenia, obstetric complications, and adolescent school outcome: Evidence for gene-environment interaction. *Schizophrenia Bulletin, 39*, 1067−1076.

Foster, J. A., & McVey Neufeld, K. A. (2013). Gut−brain axis: How the microbiome influences anxiety and depression. *Trends in Neurosciences, 36*, 305−312.

Fouskakis, D., Gunnell, D., Rasmussen, F., Tynelius, P., Sipos, A., & Harrison, G. (2004). Is the season of birth association with psychosis due to seasonal variations in foetal growth or other related exposures? A cohort study. *Acta Psychiatrica Scandinavica, 109*(4), 259−263.

Frampton, J. E. (2014). Pregabalin: A review of its use in adults with generalized anxiety disorder. *CNS Drugs, 28*, 835−854.

Francis, D. D., Young, L. J., Meaney, M. J., & Insel, T. R. (2002). Naturally occurring differences in maternal care are associated with the expression of oxytocin and vasopressin (V1a) receptors: Gender differences. *Journal of Neuroendocrinology, 14*, 349−353.

Frank, M. G., & Cantera, R. (2014). Sleep, clocks, and synaptic plasticity. *Trends in Neurosciences, 37*, 491−501.

Frank, M. G., Watkins, L. R., & Maier, S. F. (2015). The permissive role of glucocorticoids in neuroinflammatory priming: Mechanisms and insights. *Current Opinion in Endocrinology, Diabetes and Obesity, 22*, 300.

Frank, M. G., Weber, M. D., Watkins, L. R., & Maier, S. F. (2015). Stress-induced neuroinflammatory priming: A liability factor in the etiology of psychiatric disorders. *Neurobiology of Stress, 4*, 62−70.

Frankl, V. E. (1959). *Man's search for meaning*. Boston, MA: Beacon Press.

Franklin, T. B., Russig, H., Weiss, I. C., Gräff, J., Linder, N., et al. (2010). Epigenetic transmission of the impact of early stress across generations. *Biological Psychiatry, 68*, 408−415.

Fransen, F., van Beek, A. A., Borghuis, T., El Aidy, S., Hugenholtz, F., et al. (2017). Aged gut Microbiota contributes to systemical inflammaging after Transfer to germ-Free Mice. *Frontiers in immunology, 8*.

Frasure-Smith, N., Lesperance, F., & Talajic, M. (1993). Depression following myocardial infraction: impact on 6-month survival. *The Journal of the American Medical Association, 270*, 1819−1825.

Fredrikson, M., & Faria, V. (2014). Neuroimaging in anxiety disorders. In D. S. Baldwin, & B. E. Leonard (Eds.), *Anxiety disorders. Mod tends pharmacopsychiatry* (Vol. 29, pp. 47−66). Basel: Karger Publishers.

Freedberg, D. E., Salmasian, H., Cohen, B., Abrams, J. A., & Larson, E. L. (2016). Receipt of antibiotics in hospitalized patients and risk for Clostridium difficile infection in subsequent patients who occupy the same bed. *JAMA Internal Medicine, 176*(12), 1801–1808.

Freitas, T. H., Hyphantis, T. N., Andreoulakis, E., Quevedo, J., Miranda, H. L., et al. (2015). Religious coping and its influence on psychological distress, medication adherence, and quality of life in inflammatory bowel disease. *Revista Brasileira de Psiquiatria, 37*, 219–227.

Fridman, O., Goldberg, A., Ronin, I., Shoresh, N., & Balaban, N. Q. (2014). Optimization of lag time underlies antibiotic tolerance in evolved bacterial populations. *Nature, 513*, 418–421.

Friedman, M., & Rosenman, R. H. (1971). Type A behavior pattern: Its association with coronary heart disease. *Annals of Clinical Research, 3*, 300–312.

Fritschy, J. M., & Brünig, I. (2003). Formation and plasticity of GABAergic synapses: Physiological mechanisms and pathophysiological implications. *Pharmacology & Therapeutics, 98*, 299–323.

Frodl, T., Meisenzahl, E. M., Zetzsche, T., Höhne, T., Banac, S., et al. (2004). Hippocampal and amygdala changes in patients with major depressive disorder and healthy controls during a 1-year follow-up. *The Journal of Clinical Psychiatry, 65*, 492–499.

Frodl, T., & O'Keane, V. (2013). How does the brain deal with cumulative stress? A review with focus on developmental stress, HPA axis function and hippocampal structure in humans. *Neurobiology of Disease, 52*, 24–37.

Fu, L., Li, Y., Hu, Y., Zheng, Y., Yu, B., et al. (2017). Norovirus P particle-based active Aβ immunotherapy elicits sufficient immunogenicity and improves cognitive capacity in a mouse model of Alzheimer's disease. *Scientific Reports, 7*, 41041.

Fujioka, H., Phelix, C. F., Friedland, R. P., Zhu, X., Perry, E. A., et al. (2013). Apolipoprotein E4 prevents growth of malaria at the intraerythrocyte stage: Implications for differences in racial susceptibility to Alzheimer's disease. *Journal of Health Care for the Poor and Underserved, 24*(4 Suppl), 70–78.

Fujimura, K. E., Sitarik, A. R., Havstad, S., Lin, D. L., Levan, S., et al. (2016). Neonatal gut microbiota associates with childhood multisensitized atopy and T cell differentiation. *Nature Medicine, 22*, 1187–1191.

Fujiwara, H., Hasegawa, M., Dohmae, N., Kawashima, A., Masliah, E., et al. (2002). alpha-Synuclein is phosphorylated in synucleinopathy lesions. *Nature Cell Biology, 4*, 160–164.

Fuller-Thomson, E., Ramzan, N., & Baird, S. L. (2016). Arthritis and suicide attempts: Findings from a large nationally representative Canadian survey. *Rheumatology International, 36*, 1237–1248.

Fulton, S., Pissios, P., Manchon, R. P., Stiles, L., Frank, L., et al. (2006). Leptin regulation of the mesoaccumbens dopamine pathway. *Neuron, 51*, 811–822.

Fumagalli, F., Bedogni, F., Perez, J., Racagni, G., & Riva, M. A. (2004). Corticostriatal brain derived neurotrophic factor dysregulation in adult rats following prenatal stress. *The European Journal of Neuroscience, 20*(5), 1348–1354.

Fumagalli, F., Bedogni, F., Slotkin, T. A., Racagni, G., & Riva, M. A. (2005). Prenatal stress elicits regionally selective changes in basal FGF-2 gene expression in adulthood and alters the adult response to acute or chronic stress. *Neurobiology of Disease, 20*, 731–737.

Fumagalli, M., Moltke, I., Grarup, N., Racimo, F., Bjerregaard, P., et al. (2015). Greenlandic Inuit show genetic signatures of diet and climate adaptation. *Science, 349*, 1343–1347.

Furtado, M., & Katzman, M. A. (2015). Neuroinflammatory pathways in anxiety, posttraumatic stress, and obsessive-compulsive disorders. *Psychiatry Research, 229*, 37–48.

Fusar-Poli, P., Borgwardt, S., Bechdolf, A., Addington, J., Riecher-Rossler, A., et al. (2013). The psychosis high-risk state: A comprehensive state-of-the-art review. *JAMA Psychiatry, 70*, 107–120.

Gacias, M., Gaspari, S., Santos, P. M., Tamburini, S., Andrade, M., et al. (2016). Microbiota-driven transcriptional changes in prefrontal cortex override genetic differences in social behavior. *eLife, 5*, e13442.

Gad, E. S., Zaitone, S. A., & Moustafa, Y. M. (2016). Pioglitazone and exenatide enhance cognition and downregulate hippocampal beta amyloid oligomer and microglia expression in insulin-resistant rats. *Canadian Journal of Physiology and Pharmacology, 94*, 819–828.

Gagne, J. J., & Power, M. C. (2010). Anti-inflammatory drugs and risk of Parkinson disease: A meta-analysis. *Neurology, 74*, 995–1002.

Galanter, J. M., Gignoux, C. R., Oh, S. S., Torgerson, D., Pino-Yanes, M., et al. (2017). Differential methylation between ethnic sub-groups reflects the effect of genetic ancestry and environmental exposures. *eLife, 6*, e20532.

Gale, S. C., Gao, L., Mikacenic, C., Coyle, S. M., Rafaels, N., et al. (2014). APOε4 is associated with enhanced in vivo innate immune responses in human subjects. *Journal of Allergy and Clinical Immunology, 134*, 127–134.

Gałecki, P., Mossakowska-Wójcik, J., & Talarowska, M. (2017). The kynurenine pathway and the brain: Challenges, controversies and promises. *Neuropharmacology, 112*, 237–247.

Galland, L. (2014). The gut microbiome and the brain. *Journal of Medicinal Food, 17*, 1261–1272.

Gallegos, A. M., Lytle, M. C., Moynihan, J. A., & Talbot, N. L. (2015). Mindfulness-based stress reduction to enhance psychological functioning and improve inflammatory biomarkers in trauma-exposed women: A pilot study. *Psychological Trauma: Theory, Research, Practice and Policy*, 7(6), 525–532.

Ganança, L., Oquendo, M. A., Tyrka, A. R., Cisneros-Trujillo, S., Mann, J. J., et al. (2016). The role of cytokines in the pathophysiology of suicidal behavior. *Psychoneuroendocrinology*, 63, 296–310.

Gandal, M. J., Haney, J. R., Parikshak, N. N., Leppa, V., Ramaswami, G., et al. (2018). Shared molecular neuropathology across major psychiatric disorders parallels polygenic overlap. *Science*, 359, 693–697.

Gandhi, R., Hayley, S., Gibb, J., Merali, Z., & Anisman, H. (2007). Influence of poly I:C on sickness behaviors, plasma cytokines, corticosterone and central monoamine activity: Moderation by social stressors. *Brain, Behavior, and Immunity*, 21, 477–489.

Ganea, D., Hooper, K. M., & Kong, W. (2015). The neuropeptide vasoactive intestinal peptide: Direct effects on immune cells and involvement in inflammatory and autoimmune diseases. *Acta Physiologica*, 213(2), 442–452.

Ganguli, R., Brar, J. S., Chengappa, K. R., DeLeo, M., Yang, Z. W., Shurin, G., & Rabin, B. S. (1995). Mitogen-stimulated interleukin-2 production in never-medicated, first-episode schizophrenic patients. The influence of age at onset and negative symptoms. *Archives of General Psychiatry*, 52(8), 668–672.

Ganusov, V. V., & Auerbach, J. (2014). Mathematical modeling reveals kinetics of lymphocyte recirculation in the whole organism. *PLoS Computational Biology*, 10, e1003586.

Gao, H. M., & Hong, J. S. (2008). Why neurodegenerative diseases are progressive: Uncontrolled inflammation drives disease progression. *Trends in Immunology*, 29, 357–365.

Gao, J., Davis, L. K., Hart, A. B., Sanchez-Roige, S., Han, L., et al. (2016). Genome-wide association study of loneliness demonstrates a role for common variation. *Neuropsychopharmacology*, 42, 811–821.

Gao, J., Wang, H., Liu, Y., Li, Y. Y., Chen, C., et al. (2014). Glutamate and GABA imbalance promotes neuronal apoptosis in hippocampus after stress. *Medical Science Monitor*, 20, 499–512.

Gapp, K., Bohacek, J., Grossmann, J., Brunner, A. M., Manuella, F., et al. (2016). Potential of environmental enrichment to prevent transgenerational effects of paternal trauma. *Neuropsychopharmacology*, 41, 2749–2758.

Gardet, A., Benita, Y., Li, C., Sands, B. E., Ballester, I., Stevens, C., ... Podolsky, D. K. (2010). LRRK2 is involved in the IFN-gamma response and host response to pathogens. *Journal of Immunology*, 185(9), 5577–5585.

Gardner, K. L., Hale, M. W., Lightman, S. L., Plotsky, P. M., & Lowry, C. A. (2009). Adverse early life experience and social stress during adulthood interact to increase serotonin transporter mRNA expression. *Brain Research*, 1305, 47–63.

Garland, E. L., Baker, A. K., Larsen, P., Riquino, M. R., Priddy, S. E., et al. (2017). Randomized controlled trial of brief mindfulness training and hypnotic suggestion for acute pain relief in the hospital setting. *Journal of General Internal Medicine*, 32(10), 1106–1113.

Gasser, E., Moutos, C. P., Downes, M., & Evans, R. M. (2017). FGF1 – A new weapon to control type 2 diabetes mellitus. *Nature Reviews Endocrinology*, 13, 599–609.

Gauthier, E., Fortier, I., Courchesne, F., Pepin, P., Mortimer, J., et al. (2001). Environmental pesticide exposure as a risk factor for Alzheimer's disease: A case-control study. *Environmental Research*, 86, 37–45.

Gazda, L. S., Smith, T., Watkins, L. R., Maier, S. F., & Fleshner, M. (2003). Stressor exposure produces long-term reductions in antigen-specific T and B cell responses. *Stress*, 6, 259–267.

Gendelman, H. E., & Appel, S. H. (2011). Neuroprotective activities of regulatory T cells. *Trends in Molecular Medicine*, 17, 687–688.

Gensollen, T., Iyer, S. S., Kasper, D. L., & Blumberg, R. S. (2016). How colonization by microbiota in early life shapes the immune system. *Science*, 352, 539–544.

George, C., Gontier, G., Lacube, P., François, J. C., Holzenberger, M., et al. (2017). The Alzheimer's disease transcriptome mimics the neuroprotective signature of IGF-1 receptor-deficient neurons. *Brain: A Journal of Neurology*, 140(7), 2012–2027.

George, S. A., Knox, D., Curtis, A. L., Aldridge, J. W., Valentino, R. J., et al. (2013). Altered locus coeruleus–norepinephrine function following single prolonged stress. *The European Journal of Neuroscience*, 37(6), 901–909.

Geraghty, A. C., Muroy, S. E., Zhao, S., Bentley, G. E., Kriegsfeld, L. J., et al. (2015). Knockdown of hypothalamic RFRP3 prevents chronic stress-induced infertility and embryo resorption. *eLife*, 4, e04316.

Gerngroß, C., Schretter, J., Klingenspor, M., Schwaiger, M., & Fromme, T. (2017). Active brown fat during 18FDG-PET/CT imageing defines a patient group with characteristic traits and an increased probability of brown fat redetection. *Journal of Nuclear Medicine*, 58, 1104–1110.

Gershon, N. B., & High, P. C. (2015). Epigenetics and child abuse: Modern-day Darwinism--The miraculous ability of the human genome to adapt, and then adapt again. *American Journal of Medical Genetics. Part C, Seminars in Medical Genetics*, 169(4), 353–360.

Ghaemmaghami, P., Dainese, S. M., La Marca, R., Zimmermann, R., & Ehlert, U. (2014). The association between the acute psychobiological stress response in second trimester pregnant women, amniotic fluid glucocorticoids, and neonatal birth outcome. *Developmental Psychobiology*, 56(4), 734–747.

Ghani, M., Sato, C., Kakhki, E. G., Gibbs, J. R., Traynor, B., et al. (2016). Mutation analysis of the MS4A and TREM gene clusters in a case-control Alzheimer's disease data set. *Neurobiology of Aging*, 42, 217.e7–217.e13.

Ghosal, S., Hare, B., & Duman, R. S. (2017). Prefrontal cortex GABAergic deficits and circuit dysfunction in the pathophysiology and treatment of chronic stress and depression. *Current Opinion in Behavioral Sciences*, *14*, 1−8.

Ghosh, A., Roy, A., Matras, J., Brahmachari, S., Gendelman, H. E., et al. (2009). Simvastatin inhibits the activation of p21ras and prevents the loss of dopaminergic neurons in a mouse model of Parkinson's disease. *Journal of Neuroscience*, *29*, 13543−13556.

Ghosh, S., Wu, M. D., Shaftel, S. S., Kyrkanides, S., LaFerla, F. M., et al. (2013). Sustained interleukin-1β overexpression exacerbates tau pathology despite reduced amyloid burden in an Alzheimer's mouse model. *Journal of Neuroscience*, *33*, 5053−5064.

Gibb, J., Al-Yawer, F., & Anisman, H. (2013). Synergistic and antagonistic actions of acute or chronic social stressors and an endotoxin challenge vary over time following the challenge. *Brain, Behavior, and Immunity*, *28*, 149−158.

Gibb, J., Hayley, S., Gandhi, R., Poulter, M. O., & Anisman, H. (2008). Synergistic and additive actions of a psychosocial stressor and endotoxin challenge: Circulating and brain cytokines, plasma corticosterone and behavioral changes in mice. *Brain, Behavior, and Immunity*, *22*, 573−589.

Gibb, J., Hayley, S., Poulter, M. O., & Anisman, H. (2011). Effects of stressors and immune activating agents on peripheral and central cytokines in mouse strains that differ in stressor responsivity. *Brain, Behavior, and Immunity*, *25*, 468−482.

Gibney, S. M., McGuinness, B., Prendergast, C., Harkin, A., & Connor, T. J. (2013). Poly I: C-induced activation of the immune response is accompanied by depression and anxiety-like behaviours, kynurenine pathway activation and reduced BDNF expression. *Brain, Behavior, and Immunity*, *28*, 170−181.

Gibson, M. K., Wang, B., Ahmadi, S., Burnham, C. A. D., Tar, P. I., et al. (2016). Developmental dynamics of the preterm infant gut microbiota and antibiotic resistome. *Nature Microbiology*, *1*, 16024.

Gillardon, F., Schmid, R., & Draheim, H. (2012). Parkinson's disease-linked leucine-rich repeat kinase 2 (R1441G) mutation increases proinflammatory cytokine release from activated primary microglial cells and resultant neurotoxicity. *Neuroscience*, *208*, 41−48.

Gillies, D., Taylor, F., Gray, C., O'Brien, L., & D'Abrew, N. (2013). Psychological therapies for the treatment of post-traumatic stress disorder in children and adolescents (Review). *Evidence-based Child Health*, *8*, 1004−1116.

Gilman, S. E., Cherkerzian, S., Buka, S. L., Hahn, J., Hornig, M., et al. (2016). Prenatal immune programming of the sex-dependent risk for major depression. *Translational Psychiatry*, *6*, e822.

Giloteaux, L., Goodrich, J. K., Walters, W. A., Levine, S. M., Ley, R. E., et al. (2016). Reduced diversity and altered composition of the gut microbiome in individuals with myalgic encephalomyelitis/chronic fatigue syndrome. *Microbiome*, *4*, 30.

Gilsbach, B. K., & Kortholt, A. (2014). Structural biology of the LRRK2 GTPase and kinase domains: Implications for regulation. *Frontiers in Molecular Neuroscience*, *7*, 32.

Gilvarry, C. M., Sham, P. C., Jones, P. B., Cannon, M., Wright, P., Lewis, S. W., ... Murray, R. M. (1996). Family history of autoimmune diseases in psychosis. *Schizophrenia Research*, *19*(1), 33−40.

Gimblet, C., Meisel, J. S., Loesche, M. A., Cole, S. D., Horwinski, J., et al. (2017). Cutaneous leishmaniasis induces a transmissible dysbiotic skin microbiota that promotes skin inflammation. *Cell Host & Microbe*, *22*, 13−24.

Gimeno, D., Kivimäki, M., Brunner, E. J., Elovainio, M., De Vogli, R., et al. (2009). Associations of C-reactive protein and interleukin-6 with cognitive symptoms of depression: 12-year follow-up of the Whitehall II study. *Psychological Medicine*, *39*, 413−423.

Ginzburg, K., Kutz, I., Koifman, B., Roth, A., Kriwisky, M., et al. (2016). Acute stress disorder symptoms predict all-cause mortality among myocardial infarction patients: A 15-year longitudinal study. *Annals of Behavioral Medicine*, *50*, 177−186.

Giovanoli, S., Engler, H., Engler, A., Richetto, J., Voget, M., et al. (2013). Stress in puberty unmasks latent neuropathological consequences of prenatal immune activation in mice. *Science*, *339*, 1095−1099.

Girgenti, M. J., Ghosal, S., LoPresto, D., Taylor, J. R., & Duman, R. S. (2017). Ketamine accelerates fear extinction via mTORC1 signaling. *Neurobiology of Disease*, *100*, 1−8.

Girotti, M., Pace, T. W., Gaylord, R. I., Rubin, B. A., Herman, J. P., et al. (2006). Habituation to repeated restraint stress is associated with lack of stress-induced c-fos expression in primary sensory processing areas of the rat brain. *Neuroscience*, *138*, 1067−1081.

Giustino, T. F., Fitzgerald, P. J., & Maren, S. (2016). Revisiting propranolol and PTSD: Memory erasure or extinction enhancement? *Neurobiology of Learning and Memory*, *130*, 26−33.

Giustino, T. F., & Maren, S. (2015). The role of the medial prefrontal cortex in the conditioning and extinction of fear. *Frontiers in Behavioral Neuroscience*, *9*, 298.

Givens, J. L., & Tjia, J. (2002). Depressed medical students' use of mental health services and barriers to use. *Academic Medicine*, *77*(9), 918−921.

Glantz, L. A., & Lewis, D. A. (2000). Decreased dendritic spine density on prefrontal cortical pyramidal neurons in schizophrenia. *Archives of General Psychiatry*, *57*, 65−73.

Glaser, R., & Kiecolt-Glaser, J. K. (2005). Stress-induced immune dysfunction: Implications for health. *Nature Reviews Immunology, 5,* 243–251.

Glaser, R., MacCallum, R. C., Laskowski, B. F., Malarkey, W. B., Sheridan, J. F., et al. (2001). Evidence for a shift in the Th-1 to Th-2 cytokine response associated with chronic stress and aging. *The Journals of Gerontology. Series A, Biological sciences and Medical Sciences, 56*(8), M477–M482.

Glaser, R., Rabin, B., Chesney, M., Cohen, S., & Natelson, B. (1999). Stress-induced immunomodulation: Implications for infectious diseases? *The Journal of the American Medical Association, 281,* 2268–2270.

Glausier, J. R., & Lewis, D. A. (2012). Dendritic spine pathology in schizophrenia. *Neuroscience, 251,* 90–107.

Gloeckner, C. J., Schumacher, A., Boldt, K., & Ueffing, M. (2009). The Parkinson disease-associated protein kinase LRRK2 exhibits MAPKKK activity and phosphorylates MKK3/6 and MKK4/7, in vitro. *Journal of Neurochemistry, 109,* 959–968.

Glover, V. (2011). Annual research review: Prenatal stress and the origins of psychopathology: An evolutionary perspective. *Journal of Child Psychology and Psychiatry, and Allied Disciplines, 52*(4), 356–367.

Gluckman, P. D., Hanson, M. A., & Buklijas, T. (2010). A conceptual framework for the developmental origins of health and disease. *Journal of Developmental Origins of Health and Disease, 1*(1), 6–18.

Gobin, V., Van Steendam, K., Denys, D., & Deforce, D. (2014). Selective serotonin reuptake inhibitors as a novel class of immunosuppressants. *International Immunopharmacology, 20,* 148–156.

Goddard, A. W., Ball, S. G., Martinez, J., Robinson, M. J., Yang, C. R., et al. (2010). Current perspectives of the roles of the central norepinephrine system in anxiety and depression. *Depression and Anxiety, 27*(4), 339–350.

Goebel, M. U., Mills, P. J., Irwin, M. R., & Ziegler, M. G. (2000). Interleukin-6 and tumor necrosis factor-alpha production after acute psychological stress, exercise, and infused isoproterenol: Differential effects and pathways. *Psychosomatic Medicine, 62*(4), 591–598.

Gogtay, N., Giedd, J. N., Lusk, L., Hayashi, K. M., Greenstein, D., Vaituzis, A. C., et al. (2004). Dynamic mapping of human cortical development during childhood through early adulthood. *Proceedings of the National Academy of Sciences of the United States of America, 101*(21), 8174–8179.

Gola, H., Engler, H., Sommershof, A., Adenauer, H., Kolassa, S., et al. (2013). Posttraumatic stress disorder is associated with an enhanced spontaneous production of proinflammatory cytokines by peripheral blood mononuclear cells. *BMC Psychiatry, 13,* 40.

Golde, T. E., Streit, W. J., & Chakrabarty, P. (2013). Alzheimer's disease risk alleles in TREM2 illuminate innate immunity in Alzheimer's disease. *Alzheimer's Research & Therapy, 5*(3), 24.

Golden, J. P., Demaro, J. A., Knoten, A., Hoshi, M., Pehek, E., et al. (2013). Dopamine-dependent compensation maintains motor behavior in mice with developmental ablation of dopaminergic neurons. *Journal of Neuroscience, 33,* 17095–17107.

Goldman, N., Chen, M., Fujita, T., Xu, Q., Peng, W., et al. (2010). Adenosine A1 receptors mediate local antinociceptive effects of acupuncture. *Nature Neuroscience, 13,* 883–888.

Goldsmith, D. R., Rapaport, M. H., & Miller, B. J. (2016). A meta-analysis of blood cytokine network alterations in psychiatric patients: Comparisons between schizophrenia, bipolar disorder and depression. *Molecular Psychiatry, 21,* 1696–1709.

Goldstein-Piekarski, A. N., Korgaonkar, M. S., Green, E., Suppes, T., Schatzberg, A. F., et al. (2016). Human amygdala engagement moderated by early life stress exposure is a biobehavioral target for predicting recovery on antidepressants. *Proceedings of the National Academy of Sciences of the United States of America, 113,* 11955–11960.

Golubeva, A. V., Crampton, S., Desbonnet, L., Edge, D., O'Sullivan, O., et al. (2015). Prenatal stress-induced alterations in major physiological systems correlate with gut microbiota composition in adulthood. *Psychoneuroendocrinology, 60,* 58–74.

Gómez, M., Esparza, J. L., Cabré, M., García, T., & Domingo, J. L. (2008). Aluminum exposure through the diet: Metal levels in AbetaPP transgenic mice, a model for Alzheimer's disease. *Toxicology, 249,* 214–219.

Gómez-Suaga, P., Luzón-Toro, B., Churamani, D., Zhang, L., Bloor-Young, D., et al. (2012). Leucine-rich repeat kinase 2 regulates autophagy through a calcium-dependent pathway involving NAADP. *Human Molecular Genetics, 21,* 511–525.

Goodman, A. D., Gyang, T., & Smith, A. D., 3rd (2016). Ibudilast for the treatment of multiple sclerosis. *Expert Opinion on Investigational Drugs, 25*(10), 1231–1237.

Goodrich, J. K., Davenport, E. R., Beaumont, M., Jackson, M. A., Knight, R., et al. (2016). Genetic determinants of the gut microbiome in UK Twins. *Cell Host & Microbe, 19,* 731–743.

Goodwin, R. D., Rodgin, S., Goldman, R., Rodriguez, J., DeVos, G., et al. (2017). Food allergy and anxiety and depression among ethnic minority children and their caregivers. *Journal of Pediatrics, 187,* 258–264.e1.

Goodyer, I. M., Croudace, T., Dudbridge, F., Ban, M., & Herbert, J. (2010). Polymorphisms in BDNF (Val66Met) and 5-HTTLPR, morning cortisol and subsequent depression in at-risk adolescents. *The British Journal of Psychiatry, 197*(5), 365–371.

Gordon, L. B., Knopf, P. M., & Cserr, H. F. (1992). Ovalbumin is more immunogenic when introduced into brain or cerebrospinal fluid than into extracerebral sites. *Journal of Neuroimmunology, 40*, 81−87.

Gorski, D. H., & Novella, S. P. (2014). Clinical trials of integrative medicine: Testing whether magic works? *Trends in Molecular Medicine, 20*, 473−476.

Gorwood, P., Pouchot, J., Vinceneux, P., Puechal, X., Flipo, R. M., De Bandt, M., & Ades, J. (2004). Rheumatoid arthritis and schizophrenia: a negative association at a dimensional level. *Schizophrenia Research, 66*(1), 21−29.

Goshen, I., Kreisel, T., Ben-Menachem-Zidon, O., Licht, T., Weidenfeld, J., et al. (2008). Brain interleukin-1 mediates chronic stress-induced depression in mice via adrenocortical activation and hippocampal neurogenesis suppression. *Molecular Psychiatry, 13*(7), 717−728.

Gottesman, I. I., & Gould, T. D. (2003). The endophenotype concept in psychiatry: Etymology and strategic intentions. *American Journal of Psychiatry, 160*, 636−645.

Gottschalk, M. G., & Domschke, K. (2017). Genetics of generalized anxiety disorder and related traits. *Dialogues in Clinical Neuroscience, 19*(2), 159−168.

Goujon, E., Parnet, P., Laye, S., Combe, C., Kelley, K. W., et al. (1995). Stress downregulates lipopolysaccharide-induced expression of proinflammatory cytokines in the spleen, pituitary, and brain of mice. *Brain, Behavior, and Immunity, 9*, 292−303.

Grabe, H. J., Wittfeld, K., Van der Auwera, S., Janowitz, D., Hegenscheid, K., et al. (2016). Effect of the interaction between childhood abuse and rs1360780 of the FKBP5 gene on gray matter volume in a general population sample. *Human Brain Mapping, 37*, 1602−1613.

Grace, P. M., Hurley, D., Barratt, D. T., Tsykin, A., Watkins, L. R., et al. (2012). Harnessing pain heterogeneity and RNA transcriptome to identify blood-based pain biomarkers: A novel correlational study design and bioinformatics approach in a graded chronic constriction injury model. *Journal of Neurochemistry, 122*, 976−994.

Grace, P. M., Hutchinson, M. R., Maier, S. F., & Watkins, L. R. (2014). Pathological pain and the neuroimmune interface. *Nature Reviews Immunology, 14*, 217−231.

Grace, P. M., Rolan, P. E., & Hutchinson, M. R. (2011). Peripheral immune contributions to the maintenance of central glial activation underlying neuropathic pain. *Brain, Behavior, and Immunity, 25*, 1322−1332.

Grace, P. M., Strand, K. A., Galer, E. L., Urban, D. J., Wang, X., et al. (2016). Morphine paradoxically prolongs neuropathic pain in rats by amplifying spinal NLRP3 inflammasome activation. *Proceedings of the National Academy of Sciences of the United States of America, 113*, E3441−E5340.

Graeber, M. B., & Christie, M. J. (2012). Multiple mechanisms of microglia: A gatekeeper's contribution to pain states. *Experimental Neurology, 234*, 255−261.

Graham, B. S., & Ambrosino, D. M. (2015). History of passive antibody administration for prevention and treatment of infectious diseases. *Current Opinion in HIV and AIDS, 10*(3), 129−134.

Graham, L. C., Harder, J. M., Soto, I., de Vries, W. N., John, S. W., et al. (2016). Chronic consumption of a western diet induces robust glial activation in ageing mice and in a mouse model of Alzheimer's disease. *Scientific Reports, 6*, 21568.

Grandinetti, A., Morens, D. M., Reed, D., & MacEachern, D. (1994). Prospective study of cigarette smoking and the risk of developing idiopathic Parkinson's disease. *American Journal of Epidemiology, 139*, 1129−1138.

Grant, M. C., & Baker, J. S. (2016). An overview of the effect of probiotics and exercise on mood and associated health conditions. *Critical Reviews in Food Science and Nutrition, 57*, 3887−3893.

Gray, S. M., & Bloch, M. H. (2012). Systematic review of proinflammatory cytokines in obsessive-compulsive disorder. *Current Psychiatry Reports, 14*(3), 220−228.

Green, E., Goldstein-Piekarski, A. N., Schatzberg, A. F., Rush, A. J., Ma, J., et al. (2017). Personalizing antidepressant choice by sex, body mass index, and symptom profile: An iSPOT-D report. *Personalized Medicine in Psychiatry, 1*, 65−73.

Green, R. C., Schneider, L. S., Amato, D. A., Beelen, A. P., Wilcock, G., et al. (2009). Effect of tarenflurbil on cognitive decline and activities of daily living in patients with mild Alzheimer disease: A randomized controlled trial. *The Journal of the American Medical Association, 302*, 2557−2564.

Green, S. I., Kaelber, J. T., Ma, L., Trautner, B. W., Ramig, R. F., et al. (2017). Bacteriophages from ExPEC reservoirs kill pandemic multidrug-resistant strains of clonal group ST131 in animal models of bacteremia. *Scientific Reports, 7*, 46151.

Greenamyre, J. T., Cannon, J. R., Drolet, R., & Mastroberardino, P. G. (2010). Lessons from the rotenone model of Parkinson's disease. *Trends in Pharmacological Sciences, 31*(4), 141−142, author reply142−143.

Greenberg, M. V., Glaser, J., Borsos, M., El Marjou, F., Walter, M., et al. (2017). Transient transcription in the early embryo sets an epigenetic state that programs postnatal growth. *Nature Genetics, 49*, 110−118.

Greenhill, C. (2017). Obesity: Fermentable carbohydrates increase satiety signals. *Nature Reviews Endocrinology, 13*, 3.

Greer, R. L., Dong, X., Moraes, A. C., Zielke, R. A., Fernandes, G. R., et al. (2016). Akkermansia muciniphila mediates negative effects of IFN-γ on glucose metabolism. *Nature Communications, 7*, 13329.

Greggio, E., Civiero, L., Bisaglia, M., & Bubacco, L. (2012). Parkinson's disease and immune system: Is the culprit

LRRKing in the periphery? *Journal of Neuroinflammation*, *9*, 94.

Greicius, M. D., Srivastava, G., Reiss, A. L., & Menon, V. (2004). Default-mode network activity distinguishes Alzheimer's disease from healthy aging: Evidence from functional MRI. *Proceedings of the National Academy of Sciences of the United States of America*, *101*, 4637–4642.

Griciuc, A., Serrano-Pozo, A., Parrado, A. R., Lesinski, A. N., Asselin, C. N., et al. (2013). Alzheimer's disease risk gene CD33 inhibits microglial uptake of amyloid beta. *Neuron*, *78*, 631–643.

Grigorian-Shamagian, L., Liu, W., Fereydooni, S., Middleton, R. C., Valle, J., et al. (2017). Cardiac and systemic rejuvenation after cardiosphere-derived cell therapy in senescent rats. *European Heart Journal*, *38*, 2957–2967.

Grilli, M., Diodato, E., Lozza, G., Brusa, R., Casarini, M., et al. (2000). Presenilin-1 regulates the neuronal threshold to excitotoxicity both physiologically and pathologically. *Proceedings of the National Academy of Sciences of the United States of America*, *97*, 12822–12827.

Grollman, A. P., & Marcus, D. M. (2016). Global hazards of herbal remedies: Lessons from Aristolochia: The lesson from the health hazards of Aristolochia should lead to more research into the safety and efficacy of medicinal plants. *EMBO Reports*, *17*(5), 619–625.

Grønli, J., Bramham, C., Murison, R., Kanhema, T., Fiske, E., et al. (2006). Chronic mild stress inhibits BDNF protein expression and CREB activation in the dentate gyrus but not in the hippocampus proper. *Pharmacology, Biochemistry, and Behavior*, *85*, 842–849.

Grossman, N., Bono, D., Dedic, N., Kodandaramaiah, S. B., Rudenko, A., et al. (2017). Noninvasive deep brain stimulation via temporally interfering electric fields. *Cell*, *169*, 1029–1041.

Grünblatt, E., Mandel, S., & Youdim, M. B. (2000). MPTP and 6-hydroxydopamine-induced neurodegeneration as models for Parkinson's disease: Neuroprotective strategies. *Journal of Neurology*, *247* (Suppl 2), II95–II102.

Grundwald, N. J., & Brunton, P. J. (2015). Prenatal stress programs neuroendocrine stress responses and affective behaviors in second generation rats in a sex-dependent manner. *Psychoneuroendocrinology*, *62*, 204–216.

Grzadzinski, R., Lord, C., Sanders, S. J., Werling, D., & Bal, V. H. (2017). Children with autism spectrum disorder who improve with fever: Insights from the Simons Simplex Collection. *Autism Research*, *11*(1), 175–184. Available from https://doi.org/10.1002/aur.1856.

Gu, D., Dong, N., Zheng, Z., Lin, D., Huang, M., et al. (2018). A fatal outbreak of ST11 carbapenem-resistant hypervirulent Klebsiella pneumoniae in a Chinese hospital: A molecular epidemiological study. *The Lancet Infectious Diseases*, *18*(1), 37–46.

Guarente, L. (2014). The many faces of sirtuins: Sirtuins and the Warburg effect. *Nature Medicine (New York, NY, United States)*, *20*, 24–25.

Guida, F., Turco, F., Iannotta, M., De Gregorio, D., Palumbo, I., et al. (2017). Antibiotic-induced microbiota perturbation causes gut endocannabinoidome changes, hippocampal neuroglial reorganization and depression in mice. *Brain, Behavior, and Immunity*, *67*, 230–245.

Guilloux, J. P., Douillard-Guilloux, G., Kota, R., Wang, X., Gardier, A. M., et al. (2012). Molecular evidence for BDNF- and GABA-related dysfunctions in the amygdala of female subjects with major depression. *Molecular Psychiatry*, *17*, 1130–1142.

Guloksuz, S., Wichers, M., Kenis, G., Russel, M. G., Wauters, A., et al. (2013). Depressive symptoms in Crohn's disease: Relationship with immune activation and tryptophan availability. *PLoS One*, *8*, e60435.

Gumusoglu, S. B., Fine, R. S., Murray, S. J., Bittle, J. L., & Stevens, H. E. (2017). The role of IL-6 in neurodevelopment after prenatal stress. *Brain, Behavior, and Immunity*, *65*, 274–283.

Guo, Y., Kuang, Y. S., Li, S. H., Yuan, M. Y., He, J. R., et al. (2017). Connections between the gut microbiome and gestational diabetes mellitus. *American Journal of Obstetrics and Gynecology*, *216*, S293–S294.

Gupta, S., Aggarwal, S., Rashanravan, B., & Lee, T. (1998). Th1- and Th2-like cytokines in CD4 + and CD8 + T cells in autism. *Journal of Neuroimmunology*, *85*(1), 106–109.

Gupta, M. A., Jarosz, P., & Gupta, A. K. (2017). Posttraumatic stress disorder (PTSD) and the dermatology patient. *Clinical Dermatology*, *35*, 260–266.

Guo, M., Zhu, J., Yang, T., Lai, X., Liu, X., Liu, J., & Li, T. (2017). Vitamin A improves the symptoms of autism spectrum disorders and decreases 5-hydroxytryptamine (5-HT): A pilot study. *Brain Research Bulletin*, *137*, 35–40. Available from https://doi.org/10.1016/j.brainresbull.2017.11.001.

Gur, T. L., & Bailey, M. T. (2016). Effects of stress on commensal microbes and immune system activity. *Advances in Experimental Medicine and Biology*, *874*, 289–300.

Gur, T. L., Shay, L., Palkar, A. V., Fisher, S., Varaljay, V. A., et al. (2017). Prenatal stress affects placental cytokines and neurotrophins, commensal microbes, and anxiety-like behavior in adult female offspring. *Brain, Behavior, and Immunity*, *64*, 50–58.

Gustafsson, H., Nordström, A., & Nordström, P. (2015). Depression and subsequent risk of Parkinson disease: A nationwide cohort study. *Neurology*, *84*, 2422–2429.

Guyon, A., Balbo, M., Morselli, L. L., Tasali, E., Leproult, R., et al. (2014). Adverse effects of two nights of sleep restriction on the hypothalamic-pituitary-adrenal axis in healthy men. *Journal of Clinical Endocrinology and Metabolism, 99*, 2861–2868.

Haack, M., Sanchez, E., & Mullington, J. M. (2007). Elevated inflammatory markers in response to prolonged sleep restriction are associated with increased pain experience in healthy volunteers. *Sleep, 30*(9), 1145–1152.

Haaker, J., Yi, J., Petrovic, P., & Olsson, A. (2017). Endogenous opioids regulate social threat learning in humans. *Nature Communications, 8*, 15495.

Haapakoski, R., Mathieu, J., Ebmeier, K. P., Alenius, H., & Kivimäki, M. (2015). Cumulative meta-analysis of interleukins 6 and 1β, tumour necrosis factor α and C-reactive protein in patients with major depressive disorder. *Brain, Behavior, and Immunity, 49*, 206–215.

Hadad-Ophir, O., Ardi, Z., Brande-Eilat, N., Kehat, O., Anunu, R., et al. (2017). Exposure to prolonged controllable or uncontrollable stress affects GABAergic function in sub-regions of the hippocampus and the amygdala. *Neurobiology of Learning and Memory, 138*, 271–280.

Hains, L. E., Loram, L. C., Weiseler, J. L., Frank, M. G., Bloss, E. B., et al. (2010). Pain intensity and duration can be enhanced by prior challenge: Initial evidence suggestive of a role of microglial priming. *The Journal of Pain, 11*(10), 1004–1014.

Haji, J., Hamilton, J. K., Ye, C., Swaminathan, B., Hanley, A. J., et al. (2014). Delivery by Caesarean section and infant cardiometabolic status at one year of age. *Journal of Obstetrics and Gynaecology Canada, 36*(10), 864–869.

Haji, N., Mandolesi, G., Gentile, A., Sacchetti, L., Fresegna, D., et al. (2012). TNF-α-mediated anxiety in a mouse model of multiple sclerosis. *Experimental Neurology, 237*, 296–303.

Halaris, A. (2013). Co-morbidity between cardiovascular pathology and depression: Role of inflammation. *Modern Trends in Pharmacopsychiatry, 28*, 144–161.

Halaris, A. (2017). Inflammation-associated co-morbidity between depression and cardiovascular disease. *Current Topics in Behavioral Neurosciences, 31*, 45–70.

Hale, K. D., Weigent, D. A., Gauthier, D. K., Hiramoto, R. N., & Ghanta, V. K. (2003). Cytokine and hormone profiles in mice subjected to handling combined with rectal temperature measurement stress and handling only stress. *Life Sciences, 72*, 1495–1508.

Halfvarson, J., Brislawn, C. J., Lamendella, R., Vázquez-Baeza, Y., Walters, W. A., et al. (2017). Dynamics of the human gut microbiome in inflammatory bowel disease. *Nature Microbiology, 2*, 17004.

Hall, A. B., Tolonen, A. C., & Xavier, R. J. (2017). Human genetic variation and the gut microbiome in disease. *Nature Reviews Genetics, 1*, 690–699.

Hall, H. (2013). Uncertainty in medicine. *Skeptic Magazine, 18.4*. Retrieved from http://www.skeptic.com/reading_room/uncertainty-in-medicine/. Accessed February 2017.

Halliday, G., Herrero, M. T., Murphy, K., McCann, H., Ros-Bernal, F., et al. (2009). No Lewy pathology in monkeys with over 10 years of severe MPTP Parkinsonism. *Movement Disorders, 24*, 1519–1523.

Hamers, L. (2016). Big biological datasets map life's networks: Multi-omics offers a new way of doing biology. *Science News*. Retrieved from https://www.sciencenews.org/article/big-biological-datasets-map-lifes-networks. Accessed April 2017.

Hammerschlag, A. R., Stringer, S., de Leeuw, C. A., Sniekers, S., Taskesen, E., et al. (2017). Genome-wide association analysis of insomnia complaints identifies risk genes and genetic overlap with psychiatric and metabolic traits. *Nature Genetics, 49*, 1584–1592.

Han, B., Sivaramakrishnan, P., Lin, C. J., Neve, I. A. A., He, J., et al. (2017). Microbial genetic composition tunes host longevity. *Cell, 169*, 1249–1262.

Han, B. H., Sutin, D., Williamson, J. D., Davis, B. R., Piller, L. B., et al. (2017). Effect of statin treatment vs usual care on primary cardiovascular prevention among older adults: The ALLHAT-LLT randomized clinical trail. *JAMA Internal Medicine, 177*, 955–965.

Han, F., Xiao, B., Wen, L., & Shi, Y. (2015). Effects of fluoxetine on the amygdala and the hippocampus after administration of a single prolonged stress to male Wistar rates: In vivo proton magnetic resonance spectroscopy findings. *Psychiatry Research, 232*, 154–161.

Han, K., Chapman, S. B., & Krawczyk, D. C. (2015). Altered amygdala connectivity in individuals with chronic traumatic brain injury and comorbid depressive symptoms. *Frontiers in Neurology, 6*, 231.

Hanevik, K., Dizdar, V., Langeland, N., & Hausken, T. (2009). Development of functional gastrointestinal disorders after Giardia lamblia infection. *BMC Gastroenterology, 9*, 27.

Hankey, G. J. (2016). The benefits of aspirin in early secondary stroke prevention. *Lancet, 388*, 312–314.

Hannestad, J., DellaGioia, N., Ortiz, N., Pittman, B., & Bhagwagar, Z. (2011). Citalopram reduces endotoxin-induced fatigue. *Brain, Behavior, and Immunity, 25*, 256–259.

Hao, S., Dey, A., Yu, X., & Stranahan, A. M. (2016). Dietary obesity reversibly induces synaptic stripping by microglia and impairs hippocampal plasticity. *Brain, Behavior, and Immunity, 51*, 230–239.

Harnett, N. G., Wood, K. H., Ference, E. W., 3rd, Reid, M. A., Lahti, A. C., et al. (2017). Glutamate/glutamine concentrations in the dorsal anterior cingulate vary with post-traumatic stress disorder symptoms. *Journal of Psychiatric Research, 91*, 169–176.

Haroon, E., & Miller, A. H. (2016). Inflammation effects on brain glutamate in depression: Mechanistic considerations and treatment implications. *Current Topics in Behavioral Neurosciences, 31*, 173–198.

Haroon, F., Handel, U., Angenstein, F., Goldschmidt, J., Kreutzmann, P., et al. (2012). Toxoplasma gondii actively inhibits neuronal function in chronically infected mice. *PLoS One, 7*, e35516.

Harrington, W. E., Kanaan, S. B., Muehlenbachs, A., Morrison, R., Stevenson, P., et al. (2017). Maternal microchimerism predicts increased infection but decreased disease due to plasmodium falciparum during early childhood. *Journal of Infectious Diseases, 215*, 1445–1451.

Harrison, D. G., Vinh, A., Lob, H., & Madhur, M. S. (2010). Role of the adaptive immune system in hypertension. *Current Opinion in Pharmacology, 10*, 203–207.

Harrison, N. A., Brydon, L., Walker, C., Gray, M. A., Steptoe, A., et al. (2009). Inflammation causes mood changes through alterations in subgenual cingulated activity and mesolimbic connectivity. *Biological Psychiatry, 66*, 407–414.

Hartmann, J., Dedic, N., Pöhlmann, M. L., Häusl, A., Karst, H., et al. (2017). Forebrain glutamatergic, but not GABAergic, neurons mediate anxiogenic effects of the glucocorticoid receptor. *Molecular Psychiatry, 22*, 466–475.

Hartmann, J., Wagner, K. V., Gaali, S., Kirschner, A., Kozany, C., et al. (2015). Pharmacological inhibition of the psychiatric risk factor FKBP51 has anxiolytic properties. *Journal of Neuroscience, 35*, 9007–9016.

Harvey, L., & Boksa, P. (2012). Prenatal and postnatal animal models of immune activation: Relevance to a range of neurodevelopmental disorders. *Developmental Neurobiology, 72*, 1335–1348.

Harvey, S. B., Øverland, S., Hatch, S. L., Wessely, S., Mykletun, A., et al. (2017). Exercise and the prevention of depression: Results of the HUNT Cohort study. *American Journal of Psychiatry, 175*, 28–36.

Haslam, C., Cruwys, T., & Haslam, S. A. (2014). 'The we's have it': evidence for the distinctive benefits of group engagement in enhancing cognitive health in aging. *Social Science & Medicine, 120*, 57–66.

Hasty, P., & Christy, B. A. (2013). p53 as an intervention target for cancer and aging. *Pathobiology of Aging & Age Related Diseases, 3*. Available from https://doi.org/10.3402/pba.v3i0.22702.

Hatcher, J. M., Choi, H. G., Alessi, D. R., & Gray, N. S. (2017). Small-molecule inhibitors of LRRK2. *Advances in Neurobiology, 14*, 241–264.

Havekes, R., Park, A. J., Tudor, J. C., Luczak, V. G., Hansen, R. T., et al. (2016). Sleep deprivation causes memory deficits by negatively impacting neuronal connectivity in hippocampal area CA1. *eLife, 5*, e13424.

Hawkes, C. H., Del Tredici, K., & Braak, H. (2009). Parkinson's disease: The dual hit theory revisited. *Annals of the New York Academy of Sciences, 1170*, 615–622.

Hayes, J. P., Hayes, S. M., & Mikedis, A. M. (2012). Quantitative meta-analysis of neural activity in post-traumatic stress disorder. *Biology of Mood & Anxiety Disorders, 2*, 9.

Hayes, J. P., Logue, M. W., Reagan, A., Salat, D., Wolf, E. J., et al. (2017). COMT Val158Met polymorphism moderates the association between PTSD symptom severity and hippocampal volume. *Journal of Psychiatry& Neuroscience, 42*(2), 95–102.

Hayley, S. (2011). Toward an anti-inflammatory strategy for depression. *Frontiers in Behavioral Neuroscience, 5*, 19.

Hayley, S., Du, L., Litteljohn, D., Palkovits, M., Faludi, G., et al. (2015). Gender and brain regions specific differences in brain derived neurotrophic factor protein levels of depressed individuals who died through suicide. *Neuroscience Letters, 600*, 12–16.

Hayley, S., Lacosta, S., Merali, Z., van Rooijen, N., & Anisman, H. (2014). Central monoamine and plasma corticosterone changes induced by a bacterial endotoxin: Sensitization and cross-sensitization effects. *The European Journal of Neuroscience, 13*(6), 1155–1165.

Hayley, S., & Litteljohn, D. (2013). Neuroplasticity and the next wave of antidepressant strategies. *Frontiers in Cellular Neuroscience, 7*, 218.

Hayley, S., Staines, W., Merali, Z., & Anisman, H. (2001). Time-dependent sensitization of corticotropin-releasing hormone, arginine vasopressin and c-fos immunoreactivity within the mouse brain in response to tumor necrosis factor-alpha. *Neuroscience, 106*, 137–148.

Hayward, A. C., Fragaszy, E. B., Bermingham, A., Wang, L., Copas, A., et al. (2014). Comparative community burden and severity of seasonal and pandemic influenza: Results of the Flu Watch cohort study. *The Lancet, Respiratory Medicine, 2*, 445–454.

Hazlett, H. C., Gu, H., Munsell, B. C., Kim, S. H., Styner, M., Wolff, J. J., & Piven, J. (2017). Early brain development in infants at high risk for autism spectrum disorder. *Nature, 542*(7641), 348–351. Available from https://doi.org/10.1038/nature21369.

He, Q., Zhang, P., Li, G., Dai, H., & Shi, J. (2017). The association between insomnia symptoms and risk of cardio-cerebral vascular events: A meta-analysis of prospective cohort studies. *European Journal of Preventive Cardiology, 24*(10), 1071–1082.

He, Y., Gao, H., Li, X., & Zhao, Y. (2014). Psychological stress exerts effects on pathogenesis of hepatitis B via type-1/type-2 cytokines shift toward type-2 cytokine response. *PLoS One, 9*, e105530.

Hearn, L., Moore, R. A., Derry, S., Wiffen, P. J., & Phillips, T. (2014). Desipramine for neuropathic pain in adults. *Cochrane Database of Systematic Reviews*, CD011003.

Hebert, L. ,E., Wilson, R. S., Gilley, D. W., Beckett, L. A., Scherr, P. A., et al. (2000). Decline of language among women and men with Alzheimer's disease. *The Journals of Gerontology. Series B, Psychological Sciences and Social Sciences, 55*, 354–360.

Heim, C., & Nemeroff, C. B. (2002). Neurobiology of early life stress: clinical studies. *Sem Clin Neuropsychiat, 7*, 147–159.

Heim, C., Newport, D. J., Mletzko, T., Miller, A. H., & Nemeroff, C. B. (2008). The link between childhood trauma and depression: Insights from HPA axis studies in humans. *Psychoneuroendocrinology, 33*(6), 693–710.

Held, B. S. (2002). The tyranny of the positive attitude in America: Observation and speculation. *Journal of Clinical Psychology, 58*, 965–991.

Helpman, L., Marin, M. F., Papini, S., Zhu, X., Sullivan, G. M., et al. (2016). Neural changes in extinction recall following prolonged exposure treatment for PTSD: A longitudinal fMRI study. *NeuroImage, Clinical, 12*, 715–723.

Hemmerle, A. M., Dickerson, J. W., Herman, J. P., & Seroogy, K. B. (2014). Stress exacerbates experimental Parkinson's disease. *Molecular Psychiatry, 19*, 638–640.

Hemmings, S. M. J., Malan-Muller, S., van den Heuvel, L. L., Demmitt, B. A., Stanislawski, M. A., et al. (2017). The microbiome in posttraumatic stress disorder and trauma-exposed controls: An exploratory study. *Psychosomatic Medicine, 79*(8), 936–946.

Hemmingsson, E., Johansson, K., & Reynisdottir, S. (2014). Effects of childhood abuse on adult obesity: A systematic review and meta-analysis. *Obesity Reviews, 15*, 882–893.

Henckens, M. J., Printz, Y., Shamgar, U., Dine, J., Lebow, M., et al. (2017). CRF receptor type 2 neurons in the posterior bed nucleus of the stria terminalis critically contribute to stress recovery. *Molecular Psychiatry, 22*, 1691–1700.

Hendrickson, R. C., & Raskind, M. A. (2016). Noradrenergic dysregulation in the pathophysiology of PTSD. *Experimental Neurology, 284*, 181–195.

Hendriksen, H., Bink, D. I., Daniels, E. G., Pandit, R., Piriou, C., et al. (2014). Re-exposure and environmental enrichment reveal NPY-Y1 as a possible target for posttraumatic stress disorder. *Neuropharmacology, 63*, 733–742.

Hennessy, E., Gormley, S., Lopez-Rodriguez, A. B., Murray, C., Murray, C., et al. (2017). Systemic TNF-α produces acute cognitive dysfunction and exaggerated sickness behavior when superimposed upon progressive neurodegeneration. *Brain, Behavior, and Immunity, 59*, 233–244.

Hepgul, N., Cattaneo, A., Agarwal, K., Baraldi, S., Borsini, A., et al. (2016). Transcriptomics in interferon-α-treated patients identifies inflammation-, neuroplasticity-and

oxidative stress-related signatures as predictors and correlates of depression. *Neuropsychopharmacology, 41*, 2502–2511.

Heppner, F. L., Ransohoff, R. M., & Becher, B. (2015). Immune attack: The role of inflammation in Alzheimer disease. *Nature Reviews Neuroscience, 16*, 358–372.

Heraclides, A. M., Chandola, T., Witte, D. R., & Brunner, E. J. (2012). Work stress, obesity and the risk of type 2 diabetes: Gender-specific bidirectional effect in the Whitehall II study. *Obesity, 20*, 428–433.

Herman, J. P., & Tasker, J. G. (2016). Paraventricular hypothalamic mechanisms of chronic stress adaptation. *Frontiers in Endocrinology, 7*, 137.

Hernán, M. A., Checkoway, H., O'Brien, R., Costa-Mallen, P., De Vivo, I., Colditz, G. A., ... Ascherio, A. (2002). MAOB intron 13 and COMT codon 158 polymorphisms, cigarette smoking, and the risk of PD. *Neurology, 58*(9), 1381–1387.

Hernán, M. A., Takkouche, B., Caamaño-Isorna, F., & Gestal-Otero, J. J. (2002). A meta-analysis of coffee drinking, cigarette smoking, and the risk of Parkinson's disease. *Annals of Neurology, 52*, 276–284.

Herrema, H., IJzerman, R. G., & Nieuwdorp, M. (2017). Emerging role of intestinal microbiota and microbial metabolites in metabolic control. *Diabetologia, 60*(4), 613–617.

Higgins, G. A., Sellers, E. M., & Fletcher, P. J. (2013). From obesity to substance abuse: Therapeutic opportunities for 5-HT 2C receptor agonists. *Trends in Pharmacological Sciences, 34*, 560–570.

Hiles, S. A., Baker, A. L., De Malmanche, T., & Attia, J. (2012). Interleukin-6, C-reactive protein and interleukin-10 after antidepressant treatment in people with depression: A meta-analysis. *Psychological Medicine, 42*(10), 2015–2026.

Hill, M. N., Patel, S., Campolongo, P., Tasker, J. G., Wotjak, C. T., et al. (2010). Functional interactions between stress and the endocannabinoid system: From synaptic signaling to behavioral output. *The Journal of Neuroscience, 30*, 14980–14986.

Hill, M. N., & Tasker, J. G. (2012). Endocannabinoid signaling, glucocorticoid-mediated negative feedback, and regulation of the hypothalamic-pituitary-adrenal axis. *Neuroscience, 204*, 5–16.

Hirakawa, H., Akiyoshi, J., Muronaga, M., Tanaka, Y., Ishitobi, Y., et al. (2016). FKBP5 is associated with amygdala volume in the human brain and mood state: A voxel-based morphometry (VBM) study. *International Journal of Psychiatry in Clinical Practice, 20*(2), 106–115.

Hirsch, L., Jette, N., Frolkis, A., Steeves, T., & Pringsheim, T. (2016). The incidence of Parkinson's disease: A systematic review and meta-analysis. *Neuroepidemiology, 46*, 292–300.

Hirsch, M. (2001). Surviving images: Holocaust photographs and the work of postmemory. *Yale Journal of Criticism, 14*, 5–37.

Hirschtritt, M. E., Bloch, M. H., & Mathews, C. A. (2017). Obsessive-compulsive disorder: Advances in diagnosis and treatment. *The Journal of the American Medical Association, 317*, 1358–1367.

Hjorth, M. F., Roager, H. M., Larsen, T. M., Poulsen, S. K., Licht, T. R., et al. (2018). Pre-treatment microbial Prevotella-to-Bacteroides ratio, determines body fat loss success during a 6-month randomized controlled diet intervention. *International Journal of Obesity, 42*, 824.

Ho, C. S., Lopez, J. A., Vuckovic, S., Pyke, C. M., Hockey, R. L., et al. (2001). Surgical and physical stress increases circulating blood dendritic cell counts independently of monocyte counts. *Blood, 98*, 140–145.

Hoban, A. E., Moloney, R. D., Golubeva, A. V., Neufeld, K. M., O'Sullivan, O., et al. (2016). Behavioural and neurochemical consequences of chronic gut microbiota depletion during adulthood in the rat. *Neuroscience, 339*, 463–477.

Hoban, A. E., Stilling, R. M., Moloney, G. M., Moloney, R. D., Shanahan, F., et al. (2017). Microbial regulation of microRNA expression in the amygdala and prefrontal cortex. *Microbiome, 5*, 102.

Hoban, A. E., Stilling, R. M., Moloney, G., Shanahan, F., Dinan, T. G., et al. (2018). The microbiome regulates amygdala-dependent fear recall. *Molecular Psychiatry, 23*, 1134–1144.

Hoban, A. E., Stilling, R. M., Ryan, F. J., Shanahan, F., Dinan, T. G., et al. (2016). Regulation of prefrontal cortex myelination by the microbiota. *Translational Psychiatry, 6*, e774.

Hodes, G. E., Kana, V., Menard, C., Merad, M., & Russo, S. J. (2015). Neuroimmune mechanisms of depression. *Nature Neuroscience, 18*, 1386–1393.

Hodes, G. E., Pfau, M. L., Leboeuf, M., Golden, S. A., Christoffel, D. J., et al. (2014). Individual differences in the peripheral immune system promote resilience versus susceptibility to social stress. *Proceedings of the National Academy of Sciences of the United States of America, 111*, 16136–16141.

Hodgkin, P. D., Heath, W. R., & Baxter, A. G. (2007). The clonal selection theory: 50 years since the revolution. *Nature Immunology, 8*, 1019–1026.

Hodson, R. (2016). Precision medicine*Nature, 537*(7619), S49-S49. Available from https://doi.org/10.1038/537S49a.

Hoeksema, M. A., & de Winther, M. P. (2016). Epigenetic regulation of monocyte and macrophage function. *Antioxidant and Redox Signaling, 25(14)*, 758–774.

Hoffman, K. M., Trawalter, S., Axt, J. R., & Oliver, M. N. (2016). Racial bias in pain assessment and treatment recommendations, and false beliefs about biological differences between blacks and whites. *Proceedings of the National Academy of Sciences of the United States of America, 113*, 4296–4301.

Hoffman-Goetz, L., & Quadrilatero, J. (2003). Treadmill exercise in mice increases intestinal lymphocyte loss via apoptosis. *Acta Physiologica Scandinavica, 179*, 289–297.

Hofmann, S. G., Sawyer, A. T., Korte, K. J., & Smits, J. A. (2009). Is it beneficial to add pharmacotherapy to cognitive-behavioral therapy when treating anxiety disorders? A meta-analytic review. *International Journal of Cognitive Therapy, 2(2)*, 162–178.

Hoge, E. A., Brandstetter, K., Moshier, S., Pollack, M. H., Wong, K. K., et al. (2009). Broad spectrum of cytokine abnormalities in panic disorder and posttraumatic stress disorder. *Depression and Anxiety, 26(5)*, 447–455.

Hoge, E. A., Bui, E., Marques, L., Metcalf, C. A., Morris, L. K., et al. (2013). Randomized controlled trial of mindfulness meditation for generalized anxiety disorder: Effects on anxiety and stress reactivity. *Journal of Clinical Psychiatry, 74*, 786–792.

Hoge, E. A., Bui, E., Palitz, S. A., Schwarz, N. R., Owens, M. E., et al. (2017). The effect of mindfulness meditation training on biological acute stress responses in generalized anxiety disorder. *Psychiatry Research, 262*, 328–332, pii: S0165-1781(16)30847-2.

Högström, G., Nordström, A., & Nordström, P. (2014). High aerobic fitness in late adolescence is associated with a reduced risk of myocardial infarction later in life: A nationwide cohort study in men. *European Heart Journal, 35*, 3133–3140.

Hollands, G. J., French, D. P., Griffin, S. J., Prevost, A. T., Sutton, S., et al. (2016). The impact of communicating genetic risks of disease on risk-reducing health behaviour: Systematic review with meta-analysis. *British Medical Journal, 352*, i1102.

Holm, C. K., Rahbek, S. H., Gad, H. H., Bak, R. O., Jakobsen, M. R., et al. (2016). Influenza A virus targets a cGAS-independent STING pathway that controls enveloped RNA viruses. *Nature Communications, 7*, 10680.

Holmes, C., Boche, D., Wilkinson, D., Yadegarfar, G., & Hopkins, V. (2008). Long-term effects of Abeta42 immunisation in Alzheimer's disease: follow-up of a randomised, placebo-controlled phase I trial. *Lancet, 372*, 216–223.

Holmes, C., Cunningham, C., Zotova, E., Culliford, D., & Perry, V. H. (2011). Proinflammatory cytokines, sickness behavior, and Alzheimer disease. *Neurology, 77*, 212–218.

Holmes, S. E., Hinz, R., Conen, S., Gregory, C. J., Matthews, J. C., et al. (2018). Elevated translocator protein in anterior cingulate in major depression and a role for inflammation in suicidal thinking: A positron emission tomography study. *Biological Psychiatry, 83*, 61–69.

Holsboer, F., & Ising, M. (2010). Stress hormone regulation: Biological role and translation into therapy. *Annual Review of Psychology*, 61, 81–109.

Holt-Lunstad, J., Smith, T. B., Baker, M., Harris, T., & Stephenson, D. (2015). Loneliness and social isolation as risk factors for mortality: A meta-analytic review. *Perspectives on Psychological Science*, 10, 7–237.

Holtman, I. R., Raj, D. D., Miller, J. A., Schaafsma, W., Yin, Z., et al. (2015). Induction of a common microglia gene expression signature by aging and neurodegenerative conditions: A co-expression meta-analysis. *Acta Neuropathological Communication*, 3, 31.

Holub, D., Flegr, J., Dragomirecka, E., Rodriguez, M., Preiss, M., et al. (2013). Differences in onset of disease and severity of psychopathology between toxoplasmosis-related and toxoplasmosis-unrelated schizophrenia. *Acta Psychiatrica Scandinavica*, 127(3), 227–238.

Holwerda, T. J., van Tilburg, T. G., Deeg, D. J., Schutter, N., & Van, R. (2016). Impact of loneliness and depression on mortality: Results from the Longitudinal Ageing Study Amsterdam. *British Journal of Psychiatry*, 209, 127–134.

Holz, N. E., Buchmann, A. F., Boecker, R., Blomeyer, D., Baumeister, S., et al. (2015). Role of FKBP5 in emotion processing: Results on amygdala activity, connectivity and volume. *Brain Structure & Function*, 220, 1355–1368.

Honda, K., & Littman, D. R. (2016). The microbiota in adaptive immune homeostasis and disease. *Nature*, 535, 75–84.

Hone-Blanchet, A., Edden, R. A., & Fecteau, S. (2016). Online effects of transcranial direct current stimulation in real time on human prefrontal and striatal metabolites. *Biological Psychiatry*, 80, 432–438.

Hong, S., Beja-Glasser, V. F., Nfonoyim, B. M., Frouin, A., Li, S., et al. (2016). Complement and microglia mediate early synapse loss in Alzheimer mouse models. *Science*, 352, 712–716.

Hood, L., & Auffray, C. (2013). Participatory medicine: A driving force for revolutionizing healthcare. *Genome Medicine*, 5, 110.

Hood, S., & Amir, S. (2017). Neurodegeneration and the circadian clock. *Frontiers Ageing Neurosci*, 9, 170.

Hooper, L. V. (2012). Interactions between the microbiota and the immune system. *Science*, 336, 1268–1273.

Hooten, W. M., Townsend, C. O., & Sletten, C. D. (2017). The triallelic serotonin transporter gene polymorphism is associated with depressive symptoms in adults with chronic pain. *Journal of Pain Research*, 10, 1071–1078.

Horn, S. R., Charney, D. S., & Feder, A. (2016). Understanding resilience: New approaches for preventing and treating PTSD. *Experimental Neurology*, 284, 119–132.

Horn, T., & Klein, J. (2013). Neuroprotective effects of lactate in brain ischemia: Dependence on anesthetic drugs. *Neurochemistry International*, 62, 251–257.

Horvath, I., Iashchishyn, I. A., Forsgren, L., & Morozova-Roche, L. A. (2017). Immunochemical detection of α-synuclein autoantibodies in Parkinson's disease: Correlation between plasma and cerebrospinal fluid levels. *ACS Chemical Neuroscience*, 8, 1170–1176.

Hosang, G. M., Shiles, C., Tansey, K. E., McGuffin, P., & Uher, R. (2014). Interaction between stress and the BDNF Val66Met polymorphism in depression: A systematic review and meta-analysis. *BMC Medicine*, 12, 7.

Hoskins, M., Pearce, J., Bethell, A., Dankova, L., & Barbui, C. (2015). Pharmacotherapy for post-traumatic stress disorder: Systematic review and meta-analysis. *British Journal of Psychiatry*, 206, 93–100.

Hostinar, C. E., Davidson, R. J., Graham, E. K., Mroczek, D. K., Lachman, M. E., et al. (2017). Frontal brain asymmetry, childhood maltreatment, and low-grade inflammation at midlife. *Psychoneuroendocrinology*, 75, 152–163.

Hou, R., Garner, M., Holmes, C., Osmond, C., Teeling, J., et al. (2017). Peripheral inflammatory cytokines and immune balance in generalised anxiety disorder: Case-controlled study. *Brain, Behavior, and Immunity*, 62, 212–218.

Hounie, A. G., Cappi, C., Cordeiro, Q., Sampaio, A. S., Moraes, I., et al. (2008). TNF-alpha polymorphisms are associated with obsessive-compulsive disorder. *Neuroscience Letters*, 442, 86–90.

Houri-Ze'evi, L., Korem, Y., Sheftel, H., Faigenbloom, L., Toker, I. A., et al. (2016). A tunable mechanism determines the duration of the transgenerational small RNA inheritance in C. elegans. *Cell*, 165, 88–99.

Houwing, D. J., Buwalda, B., van der Zee, E. A., de Boer, S. F., & Olivier, J. D. (2017). The serotonin transporter and early life stress: Translational perspectives. *Frontiers in Cellular Neuroscience*, 11, 117.

Howard, C. D., Li, H., Geddes, C. E., & Jin, X. (2017). Dynamic nigrostriatal dopamine biases action selection. *Neuron*, 93(6), 1436–1450.

Howes, O. D., & Kapur, S. (2009). The dopamine hypothesis of schizophrenia: Version III--the final common pathway. *Schizophrenia Bulletin*, 35, 549–562.

Howes, O. D., & McCutcheon, R. (2017). Inflammation and the neural diathesis-stress hypothesis of schizophrenia: A reconceptualization. *Translational Psychiatry*, 7, e1024.

Howes, O. D., McCutcheon, R., Owen, M. J., & Murray, R. M. (2017). The role of genes, stress, and dopamine in the development of schizophrenia. *Biological Psychiatry*, 81, 9–20.

Hryhorczuk, C., Florea, M., Rodaros, D., Poirier, I., Daneault, C., et al. (2015). Dampened mesolimbic dopamine function and signaling by saturated but not monounsaturated dietary lipids. *Neuropsychopharmacology, 41*, 811−821.

Hu, Y., & Dolcos, S. (2017). Trait anxiety mediates the link between inferior frontal cortex volume and negative affective bias in healthy adults. *Social Cognitive and Affective Neuroscience, 12*(5), 775−782.

Hua, Y., Yang, Y., Sun, S., Iwanowycz, S., Westwater, C., et al. (2017). Gut homeostasis and regulatory T cell induction depend on molecular chaperone gp96 in CD11c + cells. *Scientific Reports, 7*(1), 2171.

Huang, C. L., Chao, C. C., Lee, Y. C., Lu, M. K., Cheng, J. J., et al. (2016). Paraquat induces cell death through impairing mitochondrial membrane permeability. *Molecular Neurobiology, 53*, 2169−2188.

Huang, L. Z., Parameswaran, N., Bordia, T., Michael McIntosh, J., & Quik, M. (2009). Nicotine is neuroprotective when administered before but not after nigrostriatal damage in rats and monkeys. *Journal of Neurochemistry, 109*, 826−837.

Huang, W., Li, S., Hu, Y., Yu, H., Luo, F., et al. (2011). Implication of the env gene of the human endogenous retrovirus W family in the expression of BDNF and DRD3 and development of recent-onset schizophrenia. *Schizophrenia Bulletin, 37*(5), 988−1000.

Huang, W. J., Chen, W. W., & Zhang, X. (2016). Endocannabinoid system: Role in depression, reward and pain control (review). *Molecular Medicine Reports, 14*, 2899−2903.

Huang, Y., Shen, Z., Hu, L., Xia, F., Li, Y., et al. (2016). Exposure of mother rats to chronic unpredictable stress before pregnancy alters the metabolism of gamma-aminobutyric acid and glutamate in the right hippocampus of offspring in early adolescence in a sexually dimorphic manner. *Psychiatry Research, 246*, 236−245.

Hughes, G. C., & Choubey, D. (2014). Modulation of autoimmune rheumatic diseases by oestrogen and progesterone. *Nature Reviews Rheumatology, 10*, 740−751.

Huizink, A. C., Mulder, E. J. H., & Buitelaar, J. K. (2004). Prenatal stress and risk for psychopathology: Specific effects or induction of general susceptibility? *Psychological Bulletin, 130*(1), 115−142.

Hunsberger, J. G., Rao, M., Kurtzberg, J., Bulte, J. W., Atala, A., et al. (2016). Accelerating stem cell trials for Alzheimer's disease. *Lancet Neurology, 15*, 219−230.

Husain, M. I., Chaudhry, I. B., Husain, N., Khoso, A. B., Rahman, R. R., et al. (2017). Minocycline as an adjunct for treatment-resistant depressive symptoms: A pilot randomised placebo-controlled trial. *Journal of Psychopharmacology, 31*, 1166−1175.

Hussey, S., Purves, J., Allcock, N., Fernandes, V. E., Monks, P. S., et al. (2017). Air pollution alters Staphylococcus aureus and Streptococcus pneumoniae biofilms, antibiotic tolerance and colonisation. *Environmental Microbiology, 19*, 1868−1880.

Hutchinson, M. R., Loram, L. C., Zhang, Y., Shridhar, M., Rezvani, N., et al. (2010). Evidence that tricyclic small molecules may possess toll-like receptor and myeloid differentiation protein 2 activity. *Neuroscience, 168*, 551−563.

Hutchinson, M. R., Zhang, Y., Shridhar, M., Evans, J. H., Buchanan, M. M., et al. (2010). Evidence that opioids may have toll-like receptor 4 and MD-2 effects. *Brain, Behavior, and Immunity, 24*, 83−95.

Hutson, M. R., Keyte, A. L., Hernández-Morales, M., Gibbs, E., Kupchinsky, Z. A., et al. (2017). Temperature-activated ion channels in neural crest cells confer maternal fever-associated birth defects. *Science Signaling, 10* (500), eaal4055.

Huypens, P., Sass, S., Wu, W., Dyckhoff, D., Tschöp, M., et al. (2016). Epigenetic germline inheritance of diet-induced obesity and insulin resistance. *Nature Genetics, 48*, 497−499.

Hyde, C. L., Nagle, M. W., Tian, C., Chen, X., Paciga, S. A., et al. (2016). Identification of 15 genetic loci associated with risk of major depression in individuals of European descent. *Nature Genetics, 48*, 1031−1036.

Hyland, N. P., & Cryan, J. F. (2016). Microbe-host interactions: Influence of the gut microbiota on the enteric nervous system. *Developmental Biology (Amsterdam Netherlands), 417*, 182−187.

Hypericum Depression Trial Study Group. (2002). Effect of hypericum perforatum (St John's wort) in major depressive disorder: A randomized controlled trial. *The Journal of the American Medical Association, 287*, 1807−1814.

Hyphantis, T., Palieraki, K., Voulgari, P. V., Tsifetaki, N., & Drosos, A. A. (2011). Coping with health-stressors and defense styles associated with health-related quality of life in patients with systemic lupus erythematosus. *Lupus, 20*, 893−903.

Iggman, D., Ärnlöv, J., Cederholm, T., & Risérus, U. (2016). Association of adipose tissue fatty acids with cardiovascular and all-cause mortality in elderly men. *JAMA Cardiology, 1*(7), 745−753.

Ikeda, T., Yang, L., Ikenoue, T., Mallard, C., & Hagberg, H. (2006). Endotoxin-induced hypoxic-ischemic tolerance is mediated by up-regulation of corticosterone in neonatal rat. *Pediatric Research, 59*, 56−60.

Imeri, L., & Opp, M. R. (2009). How (and why) the immune system makes us sleep. *Nature Reviews Neuroscience, 10*, 199−210.

Imperatore, F., Maurizio, J., Vargas Aguilar, S., Busch, C. J., Favret, J., et al. (2017). SIRT1 regulates macrophage self-renewal. *The EMBO Journal, 36*(16), 2353−2372.

Inal-Emiroglu, F. N., Karabay, N., Resmi, H., Guleryuz, H., Baykara, B., et al. (2015). Correlations between amygdala volumes and serum levels of BDNF and NGF as a neurobiological markerin adolescents with bipolar disorder. *Journal of Affective Disorders, 182,* 50–56.

Infante-Garcia, C., Ramos-Rodriguez, J. J., Galindo-Gonzalez, L., & Garcia-Alloza, M. (2016). Long-term central pathology and cognitive impairment are exacerbated in a mixed model of Alzheimer's disease and type 2 diabetes. *Psychoneuroendocrinology, 65,* 15–25.

Ingram, K. M., Betz, N. E., Mindes, E. J., Schmitt, M. M., & Smith, N. G. (2001). Unsupportive responses from others concerning a stressful life event: Development of the unsupportive social interactions inventory. *Journal of Social and Clinical Psychology, 20*(2), 173–207.

Innocenti, M., Salvietti, E., Guidotti, M., Casini, A., Bellandi, S., et al. (2010). Trace copper(II) or zinc(II) ions drastically modify the aggregation behavior of amyloid-beta1-42: An AFM study. *Journal of Alzheimers Disease, 19*(4), 1323–1329.

Innominato, P. F., Roche, V. P., Palesh, O. G., Ulusakarya, A., Spiegel, D., et al. (2014). The circadian timing system in clinical oncology. *Annals of Medicine, 46,* 191–207.

Insel, T. R. (2010). Rethinking schizophrenia. *Nature, 468,* 187–193.

Insel, T. R. (2014). The NIMH research domain criteria (RDoC) project: Precision medicine for psychiatry. *American Journal of Psychiatry, 171,* 395–397.

Insel, T., Cuthbert, B., Garvey, M., Heinssen, R., Pine, D. S., et al. (2010). Research domain criteria (RDoC): Toward a new classification framework for research on mental disorders. *American Journal of Psychiatry, 167,* 748–751.

Insel, T. R., & Hulihan, T. J. (1995). A gender-specific mechanism for pair bonding: Oxytocin and partner preference formation in monogamous voles. *Behavioral Neuroscience, 109,* 782–789.

Iosif, R. E., Ekdahl, C. T., Ahlenius, H., Pronk, C. J. H., Bonde, S., et al. (2006). Tumor necrosis factor receptor 1 is a negative regulator of progenitor proliferation in adult hippocampal neurogenesis. *Journal of Neuroscience, 26,* 9703–9712.

Irwin, M., Mascovich, A., Gillin, J. C., Willoughby, R., Pike, J., et al. (1994). Partial sleep deprivation reduces natural killer cell activity in humans. *Psychosometic Medicine, 56* (6), 493–498.

Irwin, M. R. (2015). Why sleep is important for health: A psychoneuroimmunology perspective. *Annual Review of Psychology, 66,* 143–172.

Irwin, M. R., Olmstead, R., Carrillo, C., Sadeghi, N., Breen, E. C., et al. (2014). Cognitive behavioral therapy versus tai chi for late life insomnia and inflammation: A randomized controlled comparative efficacy trial. *Sleep, 37* (9), 1543–1552.

Irwin, M. R., Olmstead, R., & Carroll, J. E. (2016). Sleep disturbance, sleep duration, and inflammation: A systematic review and meta-analysis of cohort studies and experimental sleep deprivation. *Biological Psychiatry, 80* (1), 40–52.

Irwin, M. R., & Opp, M. R. (2017). Sleep health: Reciprocal regulation of sleep and innate immunity. *Neuropsychopharmacology, 42,* 129–155.

Irwin, M. R., Wang, M., Ribeiro, D., Cho, H. J., Olmstead, R., et al. (2008). Sleep loss activates cellular inflammatory signaling. *Biological Psychiatry, 64*(6), 538–540.

Ishihara, L., & Brayne, C. (2006). A systematic review of depression and mental illness preceding Parkinson's disease. *Acta Neurologica Scandinavica, 113,* 211–220.

Isingrini, E., Perret, L., Rainer, Q., Amilhon, B., Guma, E., et al. (2016). Resilience to chronic stress is mediated by noradrenergic regulation of dopamine neurons. *Nature Neuroscience, 19,* 560–563.

Ismail, N., Garas, P., & Blaustein, J. D. (2011). Long-term effects of pubertal stressors on female sexual receptivity and estrogen receptor-α expression in CD-1 female mice. *Hormones and Behavior, 59*(4), 565–571.

Isung, J., Aeinehband, S., Mobarrez, F., Nordström, P., Runeson, B., et al. (2014). High interleukin-6 and impulsivity: Determining the role of endophenotypes in attempted suicide. *Translational Psychiatry, 4,* e470.

Itani, O., Jike, M., Watanabe, N., & Kaneita, Y. (2017). Short sleep duration and health outcomes: A systematic review, meta-analysis, and meta-regression. *Sleep Medicine, 32,* 246–256.

Itoga, C. A., Hellard, E. A. R., Whitaker, A. M., Lu, Y. L., Schreiber, A. L., et al. (2016). Traumatic stress promotes hyperalgesia via corticotropin-releasing factor-1 receptor (CRFR1) signaling in central amygdala. *Neuropsychopharmacology, 41,* 2463–2472.

Itzhaki, R. F. (2017). Herpes simplex virus type 1 and Alzheimer's disease: possible mechanisms and signposts. *The FASEB Journal : Official Publication of the Federation of American Societies for Experimental Biology, 31,* 3216–3226.

Itzhaki, R. F., Lathe, R., Balin, B. J., Ball, M. J., Bearer, E. L., et al. (2016). Microbes and Alzheimer's disease. *Journal of Alzheimers Disease, 51*(4), 979–984.

Iurescia, S., Seripa, D., & Rinaldi, M. (2017). Looking beyond the 5-HTTLPR polymorphism: Genetic and epigenetic layers of regulation affecting the serotonin transporter gene expression. *Molecular Neurobiology, 54,* 8386–8403.

Iwakabe, K., Shimada, M., Ohta, A., Yahata, T., Ohmi, Y., et al. (1998). The restraint stress drives a shift in Th1/Th2 balance toward Th2-dominant immunity in mice. *Immunology Letters, 62,* 39–43.

Iwata, M., Ota, K. T., & Duman, R. S. (2013). The inflammasome: Pathways linking psychological stress, depression, and systemic illnesses. *Brain, Behavior, and Immunity, 31,* 105–114.

Iyadurai, L., Blackwell, S. E., Meiser-Stedman, R., Watson, P. C., Bonsall, M. B., et al. (2018). Preventing intrusive memories after trauma via a brief intervention involving Tetris computer game play in the emergency department: A proof-of-concept randomized controlled trial. *Molecular Psychiatry, 23,* 674–679.

Jacka, F. N. (2017). Nutritional psychiatry: Where to next? *EBioMedicine, 17,* 24–29.

Jacka, F. N., O'Neil, A., Opie, R., Itsiopoulos, C., Cotton, S., et al. (2017). A randomised controlled trial of dietary improvement for adults with major depression (the "SMILES" trial). *BMC Medicine, 15,* 23.

Jackson, S. E., Kirschbaum, C., & Steptoe, A. (2017). Hair cortisol and adiposity in a population based sample of 2,527 men and women aged 54 to 87 years. *Obesity, 25,* 539–544.

Jackson-Lewis, V., & Przedborski, S. (2007). Protocol for the MPTP mouse model of Parkinson's disease. *Nature Protocols, 2,* 141–151.

Jacob, J. A. (2016). As opioid prescribing guidelines tighten, mindfulness meditation holds promise for pain relief. *The Journal of the American Medical Association, 315*(22), 2385–2387.

Jacobs, M. C., Haak, B. W., Hugenholtz, F., & Wiersinga, W. J. (2017). Gut microbiota and host defense in critical illness. *Current Opinion in Critical Care, 23*(4), 257–263.

Jacobs, R., Pawlak, C. R., Mikeska, E., Meyer-Olson, D., Martin, M., et al. (2001). Systemic lupus erythematosus and rheumatoid arthritis patients differ from healthy controls in their cytokine pattern after stress exposure. *Rheumatology, 40,* 868–875.

Jacobs, R. H., Jenkins, L. M., Gabriel, L. B., Barba, A., Ryan, K. A., et al. (2014). Increased coupling of intrinsic networks in remitted depressed youth predicts rumination and cognitive control. *PLoS One, 9*(8), e104366.

Jacobson-Pick, S., Audet, M. C., McQuaid, R. J., Kalvapalle, R., & Anisman, H. (2012). Stressor exposure of male and female juvenile mice influences later responses to stressors: Modulation of GABAA receptor subunit mRNA expression. *Neuroscience, 215,* 114–126.

Jacobson-Pick, S., Elkobi, A., Vander, S., Rosenblum, K., & Richter-Levin, G. (2008). Juvenile stress-induced alteration of maturation of the GABAA receptor alpha subunit in the rat. *International Journal of Neuropsychopharmacology, 11,* 891–903.

Janakiraman, U., Manivasagam, T., Thenmozhi, A. J., Essa, M. M., Barathidasan, R., et al. (2016). Influences of chronic mild stress exposure on motor, non-motor impairments and neurochemical variables in specific brain areas of MPTP/probenecid induced neurotoxicity in mice. *PLoS One, 11*(1), e0146671.

Janelidze, S., Suchankova, P., Ekman, A., Erhardt, S., Sellgren, C., et al. (2015). Low IL-8 is associated with anxiety in suicidal patients: Genetic variation and decreased protein levels. *Acta Psychiatrica Scandinavica, 131,* 269–278.

Jang, W., Kim, H. J., Li, H., Jo, K. D., Lee, M. K., & Yang, H. O. (2016). The neuroprotective effect of erythropoietin on rotenone-induced neurotoxicity in SH-SY5Y cells through the induction of autophagy. *Molecular Neurobiology, 53*(6), 3812–3821.

Jang, H., Boltz, D., Sturm-Ramirez, K., Shepherd, K. R., Jiao, Y., et al. (2009). Highly pathogenic H5N1 influenza virus can enter the central nervous system and induce neuroinflammation and neurodegeneration. *Proceedings of the National Academy of Sciences of the United States of America, 106,* 14063–14068.

Jangi, S., Gandhi, R., Cox, L. M., Li, N., Von Glehn, F., et al. (2016). Alterations of the human gut microbiome in multiple sclerosis. *Nature Communications, 7,* 12015.

Jansen, R., Penninx, B. W., Madar, V., Xia, K., Milaneschi, Y., et al. (2016). Gene expression in major depressive disorder. *Molecular Psychiatry, 21,* 339–347.

Janusek, L. W., Tell, D., Albuquerque, K., & Mathews, H. L. (2013). Childhood adversity increases vulnerability for behavioral symptoms and immune dysregulation in women with breast cancer. *Brain, Behavior, and Immunity, 30,* S149–S162.

Jaremka, L. M., Fagundes, C. P., Peng, J., Bennett, J. M., Glaser, R., et al. (2013). Loneliness promotes inflammation during acute stress. *Psychological Science, 24*(7), 1089–1097.

Jaroudi, W., Garami, J., Garrido, S., Hornberger, M., Keri, S., et al. (2017). Factors underlying cognitive decline in old age and Alzheimer's disease: The role of the hippocampus. *Reviews of Neuroscience, 28,* 705–714.

Jaunmuktane, Z., Mead, S., Ellis, M., Wadsworth, J. D. F., Nicoll, A. J., et al. (2015). Evidence for human transmission of amyloid-β pathology and cerebral amyloid angiopathy. *Nature, 525,* 247–250.

Jay, T. R., Miller, C. M., Cheng, P. J., Graham, L. C., Bemiller, S., et al. (2015). TREM2 deficiency eliminates TREM2 + inflammatory macrophages and ameliorates pathology in Alzheimer's disease mouse models. *Journal of Experimental Medicine, 212,* 287–295.

Jazin, E., & Cahill, L. (2010). Sex differences in molecular neuroscience: From fruit flies to humans. *Nature Reviews. Neuroscience, 11*(1), 9–17.

Jeffery, I. B., O'Toole, P. W., Öhman, L., Claesson, M. J., Deane, J., et al. (2012). An irritable bowel syndrome subtype defined by species-specific alterations in faecal microbiota. *Gut, 61,* 997–1006.

Jembrek, M. J., Auteri, M., Serio, R., & Vlainić, J. (2017). GABAergic system in action: Connection to gastrointestinal stress-related disorders. *Current Pharmaceutical Design, 23*, 4003–4011.

Jensen, M. P., Day, M. A., & Miró, J. (2014). Neuromodulatory treatments for chronic pain: efficacy and mechanisms. *Nat Rev Neurol, 10*, 167–178.

Jensen, P. S., Zhu, Z., & van Opijnen, T. (2017). Antibiotics disrupt coordination between transcriptional and phenotypic stress responses in pathogenic bacteria. *Cell Reports, 20*, 1705–1716.

Jiang, D. G., Jin, S. L., Li, G. Y., Li, Q. Q., Li, Z. R., et al. (2017). Serotonin regulates brain-derived neurotrophic factor expression in select brain regions during acute psychological stress. *Neural Regeneration Research, 11*(9), 1471–1479.

Jiang, H., Ling, Z., Zhang, Y., Mao, H., Ma, Z., et al. (2015). Altered fecal microbiota composition in patients with major depressive disorder. *Brain Behavior, and Immunity, 48*, 186–194.

Jiang, H., Wang, J., Rogers, J., & Xie, J. (2017). Brain iron metabolism dysfunction in Parkinson's disease. *Molecular Neurobiology, 54*, 3078–3101.

Jiang, M., Qin, P., & Yang, X. (2014). Comorbidity between depression and asthma via immune-inflammatory pathways: A meta-analysis. *Journal of Affective Disorders, 166*, 22–29.

Jiang, P., Chen, C., Wang, R., Chechneva, O. V., Chung, S. H., et al. (2013). hESC-derived Olig2 + progenitors generate a subtype of astroglia with protective effects against ischaemic brain injury. *Nature Communications, 4*, 2196.

Jiang, S., Gao, H., Luo, Q., Wang, P., & Yang, X. (2017). The correlation of lymphocyte subsets, natural killer cell, and Parkinson's disease: A meta-analysis. *Neurological Sciences, 38*(8), 1373–1380.

Jiang, T., Yu, J. T., Hu, N., Tan, M. S., Zhu, X. C., et al. (2014). CD33 in Alzheimer's disease. *Molecular Neurobiology, 49*, 529–535.

Jim, H. S., Boyd, T. D., Booth-Jones, M., Pidala, J., & Potter, H. (2012). Granulocyte macrophage colony stimulating factor treatment is associated with improved cognition in cancer patients. *Brain Disorder & Therapy, 1*(1), pii: 1000101.

Jimenez, J. C., Su, K., Goldberg, A. R., Luna, V. M., Biane, J. S., et al. (2018). Anxiety cells in a hippocampal-hypothalamic circuit. *Neuron, 97*, 670–683.

Jiménez-Castellanos, J. C., Wan Nur Ismah, W. A. K., Takebayashi, Y., Findlay, J., Schneiders, T., et al. (2018). Envelope proteome changes driven by RamA overproduction in Klebsiella pneumoniae that enhance acquired β-lactam resistance. *Journal of Antimicrobial Chemotherapy, 73*, 88–94.

Jiménez-Sánchez, L., Castañé, A., Pérez-Caballero, L., Grifoll-Escoda, M., López-Gil, X., et al. (2016). Activation of AMPA receptors mediates the antidepressant action of

deep brain stimulation of the infralimbic prefrontal cortex. *Cerebral Cortex, 26*(6), 2778–2789.

Johnson, J. D., O'Connor, K. A., Deak, T., Stark, M., Watkins, L. R., et al. (2002). Prior stressor exposure sensitizes LPS-induced cytokine production. *Brain, Behavior, and Immunity, 16*(4), 461–476.

Johnson, J. D., O'Connor, K. A., Hansen, M. K., Watkins, L. R., & Maier, S. F. (2003). Effects of prior stress on LPS-induced cytokine and sickness responses. *American Journal of Physiology Regulatory Integrative and Comparative Physiology, 284*, R422–R432.

Johnson, T. A., Stedtfeld, R. D., Wang, Q., Cole, J. R., Hashsham, S. A., et al. (2016). Clusters of antibiotic resistance genes enriched together stay together in swine agriculture. *mBio, 7*, e02214–e02215.

Johnson, V. E., Stewart, J. E., Begbie, F. D., Trojanowski, J. Q., Smith, D. H., et al. (2013). Inflammation and white matter degeneration persist for years after a single traumatic brain injury. *Brain, 136*, 28–42.

Johnston, B. C., Kanters, S., Bandayrel, K., Wu, P., Naji, F., et al. (2014). Comparison of weight loss among named diet programs in overweight and obese adults: A meta-analysis. *The Journal of the American Medical Association, 312*(9), 923–933.

Jonas, W., & Woodside, B. (2016). Physiological mechanisms and behavioral and psychological factors influencing the transfer of milk from mothers to their young. *Hormones and Behavior, 77*, 167–181.

Jones, M. E., Lebonville, C. L., Paniccia, J. E., Balentine, M. E., Reissner, K. J., et al. (2018). Hippocampal interleukin-1 mediates stress-enhanced fear learning: A potential role for astrocyte-derived interleukin-1β. *Brain, Behavior, and Immunity, 67*, 355–363.

Jonker, I., Rosmalen, J. G. M., & Schoevers, R. A. (2017). Childhood life events, immune activation and the development of mood and anxiety disorders: The TRAILS study. *Translational Psychiatry, 7*, e1112.

Jorge, R. E., Acion, L., Burin, D. I., & Robinson, R. G. (2016). Sertraline for preventing mood disorders following traumatic brain injury. *JAMA Psychiatry, 73*(10), 1041–1047.

Jorgensen, E. M., Alderman, M. H., & Taylor, H. S. (2016). Preferential epigenetic programming of estrogen response after in utero xenoestrogen (bisphenol-A) exposure. *FASEB Journal, 30*(9), 3194–3201.

Josephs, R. A., Cobb, A. R., Lancaster, C. L., Lee, H. J., & Telch, M. J. (2017). Dual-hormone stress reactivity predicts downstream war-zone stress-evoked PTSD. *Psychoneuroendocrinology, 78*, 76–84.

Jovasevic, V., Corcoran, K. A., Leaderbrand, K., Yamawaki, N., Guedea, A. L., et al. (2015). GABAergic mechanisms regulated by miR-33 encode state-dependent fear. *Nature Neuroscience, 18*, 1265–1271.

Joyner, M. J. (2016). Precision medicine, cardiovascular disease and hunting elephants. *Progress in Cardiovascular Diseases*, 58(6), 651–660.

Ju, Y. E. S., Ooms, S. J., Sutphen, C., Macauley, S. L., Zangrilli, M. A., et al. (2017). Slow wave sleep disruption increases cerebrospinal fluid amyloid-β levels. *Brain: A Journal of Neurology*, 140(8), 2104–2111.

Juhasz, G., Gonda, X., Hullam, G., Eszlari, N., Kovacs, D., et al. (2015). Variability in the effect of 5-HTTLPR on depression in a large European population: The role of age, symptom profile, type and intensity of life stressors. *PLoS One*, 10, 1–15.

Jurjus, A., Eid, A., Al Kattar, S., Zeenny, M. N., Gerges-Geagea, A., et al. (2015). Inflammatory bowel disease, colorectal cancer and type 2 diabetes mellitus: The links. *BBA Clinical*, 5, 16–24.

Jyonouchi, H. (2009). Food allergy and autism spectrum disorders: Is there a link? *Current Allergy and Asthma Reports*, 9(3), 194–201.

Jyonouchi, H., Geng, L., Streck, D. L., & Toruner, G. A. (2011). Children with autism spectrum disorders (ASD) who exhibit chronic gastrointestinal (GI) symptoms and marked fluctuation of behavioral symptoms exhibit distinct innate immune abnormalities and transcriptional profiles of peripheral blood (PB) monocytes. *Journal of Neuroimmunology*, 238(1–2), 73–80. Available from https://doi.org/10.1016/j.jneuroim.2011.07.001.

Kabat-Zinn, J. (1990). *Full catastrophe living: Using the wisdom of your body and mind to face stress, pain, and illness.* New York, NY: Delacourt.

Kaczkurkin, A. N., & Foa, E. B. (2015). Cognitive-behavioral therapy for anxiety disorders: An update on the empirical evidence. *Dialogues Clinical Neuroscience*, 17(3), 337–346.

Kaczorowski, K. J., Shekhar, K., Nkulikiyimfura, D., Dekker, C. L., Maecker, H., et al. (2017). Continuous immunotypes describe human immune variation and predict diverse responses. *Proceedings of the National Academy of Sciences of the United States of America*, 114 (30), E6097–E6106.

Kadam, R. U., Juraszek, J., Brandenburg, B., Buyck, C., Schepens, W. B. G., et al. (2017). Potent peptidic fusion inhibitors of influenza virus. *Science*, 358(6362), 496–502.

Kadokura, A., Yamazaki, T., Lemere, C. A., Takatama, M., & Okamoto, K. (2009). Regional distribution of TDP-43 inclusions in Alzheimer disease (AD) brains: Their relation to AD common pathology. *Neuropathology*, 29(5), 566–573.

Kahlenberg, J. M., & Kaplan, M. J. (2013). Little peptide, big effects: The role of LL-37 in inflammation and autoimmune disease. *Journal of Immunology*, 191, 4895–4901.

Kahn, M. S., Kranjac, D., Alonzo, C. A., Haase, J. H., Cedillos, R. O., et al. (2012). Prolonged elevation in hippocampal Aβ and cognitive deficits following repeated endotoxin exposure in the mouse. *Behavioural Brain Research*, 229, 176–184.

Kahneman, D. (2011). *Thinking, fast and slow.* New York: Farrar, Straus and Giroux.

Kahneman, D., & Tversky, A. (1996). On the reality of cognitive illusions. *Psychological Review*, 103(3), 582–591.

Kajitani, N., Hisaoka-Nakashima, K., Okada-Tsuchioka, M., Hosoi, M., Yokoe, T., et al. (2015). Fibroblast growth factor 2 mRNA expression evoked by amitriptyline involves extracellular signal-regulated kinase-dependent early growth response 1 production in rat primary cultured astrocytes. *Journal of Neurochemistry*, 135, 27–37.

Kalia, L. V., & Lang, A. E. (2015). Parkinson's disease. *Lancet*, 386, 896–912.

Kalin, N. H., Fox, A. S., Kovner, R., Riedel, M. K., Fekete, E. M., et al. (2016). Overexpressing corticotropin-releasing factor in the primate amygdala increases anxious temperament and alters its neural circuit. *Biological Psychiatry*, 80, 345–355.

Kalueff, A. V., & Nutt, D. J. (2007). Role of GABA in anxiety and depression. *Depression and Anxiety*, 24(7), 495–517.

Kami, K., Tajima, F., & Senba, E. (2017). Exercise-induced hypoalgesia: Potential mechanisms in animal models of neuropathic pain. *Anatomical Science International*, 92(1), 79–90.

Kammel, A., Saussenthaler, A., Jähnert, M., Jonas, W., Stirm, L., et al. (2016). Early hypermethylation of hepatic Igfbp2 results in its reduced expression preceding fatty liver in mice. *Human Molecular Genetics*, 25, 2588–2599.

Kanchanatawan, B., Hemrungrojn, S., Thika, S., Sirivichayakul, S., Ruxrungtham, K., et al. (2017). Changes in tryptophan catabolite (TRYCAT) pathway patterning are associated with mild impairments in declarative memory in schizophrenia and deficits in semantic and episodic memory coupled with increased false-memory creation in deficit schizophrenia. *Molecular Neurobiology*. Available from https://doi.org/10.1007/s12035-017-0751-8.

Kanchanatawan, B., Sirivichayakul, S., Carvalho, A. F., Anderson, G., Galecki, P., et al. (2018). Depressive, anxiety and hypomanic symptoms in schizophrenia may be driven by tryptophan catabolite (TRYCAT) patterning of IgA and IgM responses directed to TRYCATs. *Progress in Neuro-Psychopharmacology & Biological Psychiatry*, 80(Pt C), 205–216.

Kandel, E. R., Dudai, Y., & Mayford, M. R. (2014). The molecular and systems biology of memory. *Cell*, 157, 163–186.

Kang, D. W., Adams, J. B., Gregory, A. C., Borody, T., Chittick, L., et al. (2017). Microbiota transfer therapy alters gut ecosystem and improves gastrointestinal and autism symptoms: An open-label study. *Microbiome, 5,* 10.

Kang, H. J., Bae, K. Y., Kim, S. W., Kim, J. T., Park, M. S., et al. (2016). Effects of interleukin-6, interleukin-18, and statin use, evaluated at acute stroke, on post-stroke depression during 1-year follow-up. *Psychoneuroendocrinology, 72,* 156–160.

Kang, V., Wagner, G. C., & Ming, X. (2014). Gastrointestinal dysfunction in children with autism spectrum disorders. *Autism Research, 7*(4), 501–506. Available from https://doi.org/10.1002/aur.1386.

Kanju, P., Chen, Y., Lee, W., Yeo, M., Lee, S. H., et al. (2016). Small molecule dual-inhibitors of TRPV4 and TRPA1 for attenuation of inflammation and pain. *Scientific Reports, 6,* 26894.

Kannan, G., Gressitt, K. L., Yang, S., Stallings, C. R., Katsafanas, E., et al. (2017). Pathogen-mediated NMDA receptor autoimmunity and cellular barrier dysfunction in schizophrenia. *Translational Psychiatry, 7,* e1186.

Kannan, G., & Pletnikov, M. V. (2012). Toxoplasma gondii and cognitive deficits in schizophrenia: An animal model perspective. *Schizophrenia Bulletin, 38*(6), 1155–1161.

Kao, C. Y., Stalla, G., Stalla, J., Wotjak, C. T., & Anderzhanova, E. (2015). Norepinephrine and corticosterone in the medial prefrontal cortex and hippocampus predict PTSD-like symptoms in mice. *European Journal of Neuroscience, 41,* 1139–1148.

Kappelmann, N., Lewis, G., Dantzer, R., Jones, P. B., & Khandaker, G. M. (2016). Antidepressant activity of anti-cytokine treatment: A systematic review and meta-analysis of clinical trials of chronic inflammatory conditions. *Molecular Psychiatry, 23,* 335–343.

Karg, K., Burmeister, M., Shedden, K., & Sen, S. (2011). The serotonin transporter promoter variant (5-HTTLPR), stress, and depression meta-analysis revisited: Evidence of genetic moderation. *Archives of General Psychiatry, 68* (5), 444–454.

Karkhanis, A. N., Rose, J. H., Weiner, J. L., & Jones, S. R. (2016). Early-life social isolation stress increases kappa opioid receptor responsiveness and downregulates the dopamine system. *Neuropsychopharmacology, 41,* 2263–2274.

Karlsson, H., Blomstrom, A., Wicks, S., Yang, S., Yolken, R. H., et al. (2012). Maternal antibodies to dietary antigens and risk for nonaffective psychosis in offspring. *American Journal of Psychiatry, 169,* 625–632.

Karlsson, H. K., Tuominen, L., Tuulari, J. J., Hirvonen, J., Parkkola, R., et al. (2015). Obesity is associated with decreased μ-opioid but unaltered dopamine D2 receptor availability in the brain. *Journal of Neuroscience, 35,* 3959–3965.

Karnam, G., Rygiel, T. P., Raaben, M., Grinwis, G. C., Coenjaerts, F. E., et al. (2012). CD200 receptor controls sex-specific TLR7 responses to viral infection. *PLoS Pathogens, 8,* e1002710.

Kanner, L. (1946). Irrelevant and metaphorical language in early infantile autism. *The American Journal of Psychiatry, 103*(2), 242–246. Available from https://doi.org/10.1176/ajp.103.2.242.

Karsch-Völk, M., Barrett, B., Kiefer, D., Bauer, R., Ardjomand-Woelkart, K., et al. (2014). Echinacea for preventing and treating the common cold. *The Cochrane Database of Systematic Reviews, 2,* CD000530.

Karshikoff, B., Jensen, K. B., Kosek, E., Kalpouzos, G., Soop, A., et al. (2016). Why sickness hurts: A central mechanism for pain induced by peripheral inflammation. *Brain, Behavior, and Immunity, 57,* 38–46.

Karuppagounder, S. S., Xiong, Y., Lee, Y., Lawless, M. C., Kim, D., et al. (2016). LRRK2 G2019S transgenic mice display increased susceptibility to 1-methyl-4-phenyl-1,2,3,6-tetrahydropyridine (MPTP)-mediated neurotoxicity. *Journal of Chemical Neuroanatomy, 76*(Pt B), 90–97.

Kasai, K., Yamasue, H., Gilbertson, M. W., Shenton, M. E., Rauch, S. L., et al. (2008). Evidence for acquired pregenual anterior cingulate gray matter loss from a twin study of combat-related posttraumatic stress disorder. *Biological Psychiatry, 63*(6), 550–556.

Kastanenka, K. V., Hou, S. S., Shakerdge, N., Logan, R., Feng, D., et al. (2017). Optogenetic restoration of disrupted slow oscillations halts amyloid deposition and restores calcium homeostasis in an animal model of Alzheimer's disease. *PLoS One, 12*(1), e0170275.

Katsimpardi, L., Litterman, N. K., Schein, P. A., Miller, C. M., Loffredo, F. S., et al. (2014). Vascular and neurogenic rejuvenation of the aging mouse brain by young systemic factors. *Science, 344,* 630–634.

Katz, L. Y., Kozyrskyj, A. L., Prior, H. J., Enns, M. W., Cox, B. J., et al. (2008). Effect of regulatory warnings on antidepressant prescription rates, use of health services and outcomes among children, adolescents and young adults. *The Canadian Medical Association Journal, 178,* 1005–1011.

Kaufmann, T., Elvsåshagen, T., Alnæs, D., Zak, N., Pedersen, P. Ø., et al. (2016). The brain functional connectome is robustly altered by lack of sleep. *NeuroImage, 127,* 324–332.

Kautz, M., Charney, D. S., & Murrough, J. W. (2017). Neuropeptide Y, resilience, and PTSD therapeutics. *Neuroscience Letters, 649,* 164–169.

Kawaguchi, Y., Okada, T., Konishi, H., Fujino, M., Asai, J., et al. (1997). Reduction of the DTH response is related to morphological changes of Langerhans cells in mice exposed to acute immobilization stress. *Clinical and Experimental Immunology, 109*(2), 397–401.

Kawalec, P., Cierniak, A., Pilc, A., & Nowak, G. (2015). Pregabalin for the treatment of social anxiety disorder. *International Clinical Psychopharmacology, 30*(4), 183–192.

Kawanokuchi, J., Mizuno, T., Takeuchi, H., Kato, H., Wang, J., et al. (2006). Production of interferon-gamma by microglia. *Multiple Sclerosis, 12*(5), 558–564.

Kazi, A. I., & Oommen, A. (2014). Chronic noise stress-induced alterations of glutamate and gamma-aminobutyric acid and their metabolism in the rat brain. *Noise and Health, 16*(73), 343–349.

Keil, A., Daniels, J. L., Forssen, U., Hultman, C., Cnattingius, S., et al. (2010). Parental autoimmune diseases associated with autism spectrum disorders in offspring. *Epidemiology, 21*(6), 805–808.

Kekow, J., Moots, R., Khandker, R., Melin, J., Freundlich, B., & Singh, A. (2011). Improvements in patient-reported outcomes, symptoms of depression and anxiety, and their association with clinical remission among patients with moderate-to-severe active early rheumatoid arthritis. *Rheumatology, 50*(2), 401–409.

Kelada, S. N. P., Checkoway, H., Kardia, S. L. R., Carlson, C. S., Costa-Mallen, P., et al. (2006). 5′ and 3′ region variability in the dopamine transporter gene (SLC6A3), pesticide exposure and Parkinson's disease risk: A hypothesis-generating study. *Human Molecular Genetics, 15*(20), 3055–3062.

Kelleher, R. J., & Shen, J. (2017). Presenilin-1 mutations and Alzheimer's disease. *Proceedings of the National Academy of Sciences of the United States of America, 114*(4), 629–631.

Keller, S. M., Schreiber, W. B., Stanfield, B. R., & Knox, D. (2015). Inhibiting corticosterone synthesis during fear memory formation exacerbates cued fear extinction memory deficits within the single prolonged stress model. *Behavioural Brain Research, 287*, 182–186.

Kelly, J. R., Allen, A. P., Temko, A., Hutch, W., Kennedy, P. J., et al. (2017). Lost in translation? The potential psychobiotic Lactobacillus rhamnosus (JB-1) fails to modulate stress or cognitive performance in healthy male subjects. *Brain, Behavior, and Immunity, 61*, 50–59.

Kelly, J. R., Borre, Y., O'Brien, C., Patterson, E., El Aidy, S., et al. (2016). Transferring the blues: Depression-associated gut microbiota induces neurobehavioural changes in the rat. *Journal of Psychiatric Research*, 109–118.

Kelly, J. R., Clarke, G., Cryan, J. F., & Dinan, T. G. (2016). Brain-gut-microbiota axis: Challenges for translation in psychiatry. *Annals of Epidemiology, 26*(5), 366–372.

Kelly, J. R., Kennedy, P. J., Cryan, J. F., Dinan, T. G., Clarke, G., et al. (2015). Breaking down the barriers: The gut microbiome, intestinal permeability and stress-related psychiatric disorders. *Frontiers in Cellular Neuroscience, 9*, 392.

Kelly, J. R., Minuto, C., Cryan, J. F., Clarke, G., & Dinan, T. G. (2017). Cross talk: The microbiota and neurodevelopmental disorders. *Frontiers in Neuroscience, 11*, 490.

Kennedy, P. J., Cryan, J. F., Dinan, T. G., & Clarke, G. (2014). Irritable bowel syndrome: A microbiome-gut-brain axis disorder? *World Journal of Gastroenterology, 20* (39), 14105–14125.

Kennedy, P. J., Cryan, J. F., Dinan, T. G., & Clarke, G. (2017). Kynurenine pathway metabolism and the microbiota-gut-brain axis. *Neuropharmacology, 112*(Pt B), 399–412.

Kennis, M., van Rooij, S. J., Reijnen, A., & Geuze, E. (2017). The predictive value of dorsal cingulate activity and fractional anisotropy on long-term PTSD symptom severity. *Depression and Anxiety, 34*(5), 410–418.

Kertes, D. A., Bhatt, S. S., Kamin, H. S., Hughes, D. A., Rodney, N. C., et al. (2017). BNDF methylation in mothers and newborns is associated with maternal exposure to war trauma. *Clinical Epigenetics, 9*, 68.

Kessner, S., Forkmann, K., Ritter, C., Wiech, K., Ploner, M., et al. (2014). The effect of treatment history on therapeutic outcome: Psychological and neurobiological underpinnings. *PLoS One, 9*(10), e109014.

Khan, D., Fernando, P., Cicvaric, A., Berger, A., Pollak, A., et al. (2014). Long-term effects of maternal immune activation on depression-like behavior in the mouse. *Translational Psychiatry, 4*, e363.

Khandaker, G. M., Zammit, S., Lewis, G., & Jones, P. B. (2016). Association between serum C-reactive protein and DSM-IV generalized anxiety disorder in adolescence: Findings from the ALSPAC cohort. *Neurobiology of Stress, 4*, 55–61.

Khandaker, G. M., Zimbron, J., Dalman, C., Lewis, G., & Jones, P. B. (2012). Childhood infection and adult schizophrenia: A meta-analysis of population-based studies. *Schizophrenia Research, 139*(1–3), 161–168.

Khanh, D. V., Choi, Y. H., Moh, S. H., Kinyua, A. W., & Kim, K. W. (2014). Leptin and insulin signaling in dopaminergic neurons: relationship between energy balance and reward system. *Frontiers in Psychology, 5*, 846.

Khera, A. V., Emdin, C. A., Drake, I., Natarajan, P., Bick, A. G., et al. (2016). Genetic risk, adherence to a healthy life-style, and coronary disease. *New England Journal of Medicine, 375*(24), 2349–2358.

Khoury, M. J., & Galea, S. (2016). Will precision medicine improve population health? *The Journal of the American Medical Association, 316*(13), 1357–1358.

Khoury, M. J., Iademarco, M. F., & Riley, W. T. (2016). Precision public health for the era of precision medicine. *American Journal of Preventive Medicine, 50*(3), 398–401.

Kiank, C., Zeden, J. P., Drude, S., Domanska, G., Fusch, G., et al. (2010). Psychological stress-induced, IDO1-dependent tryptophan catabolism: Implications on immunosuppression in mice and humans. *PLoS One, 5* (7), e11825.

Kiecolt-Glaser, J. K., Derry, H. M., & Fagundes, C. P. (2015). Inflammation: Depression fans the flames and feasts on the heat. *American Journal of Psychiatry, 172* (11), 1075–1091.

Kiecolt-Glaser, J. K., Glaser, R., Gravenstein, S., Malarkey, W. B., & Sheridan, J. (1996). Chronic stress alters the immune response to influenza virus vaccine in older adults. *Proceedings of the National Academy of Sciences of the United States of America, 93*(7), 3043–3047.

Kiecolt-Glaser, J. K., Gouin, J. P., Weng, N. P., Malarkey, W. B., Beversdorf, D. Q., et al. (2011). Childhood adversity heightens the impact of later-life caregiving stress on telomere length and inflammation. *Psychosomatic Medicine, 73*(1), 16–22.

Kierkegaard, M., Lundberg, I. E., Olsson, T., Johansson, S., Ygberg, S., et al. (2016). High-intensity resistance training in multiple sclerosis – An exploratory study of effects on immune markers in blood and cerebrospinal fluid, and on mood, fatigue, health-related quality of life, muscle strength, walking and cognition. *Journal of the Neurological Sciences, 362,* 251–257.

Kim, B., Yang, M. S., Choi, D., Kim, J. H., Kim, H. S., et al. (2012). Impaired inflammatory responses in murine Lrrk2-knockdown brain microglia. *PLoS One, 7*(4), e34693.

Kim, D., Kubzansky, L. D., Baccarelli, A., Sparrow, D., Spiro, A., et al. (2016). Psychological factors and DNA methylation of genes related to immune/inflammatory system markers: The VA Normative Aging Study. *British Medical Journal Open, 6*(1), e009790.

Kim, D. Y., Hao, J., Liu, R., Turner, G., Shi, F. D., et al. (2012). Inflammation-mediated memory dysfunction and effects of a ketogenic diet in a murine model of multiple sclerosis. *PLoS One, 7*(5), e35476.

Kim, E. J., & Dimsdale, J. E. (2007). The effect of psychosocial stress on sleep: A review of polysomnographic evidence. *Behavioral Sleep Medical, 5*(4), 256–278.

Kim, H. D., Cao, Y., Kong, F. K., Van Kampen, K. R., Lewis, T. L., et al. (2005). Induction of a Th2 immune response by co-administration of recombinant adenovirus vectors encoding amyloid beta-protein and GM-CSF. *Vaccine, 23*(23), 2977–2986.

Kim, H. S., Sherman, D. K., Mojaverian, T., Sasaki, J. Y., Park, J., et al. (2011). Gene-culture interaction: Oxytocin receptor polymorphism (OXTR) and emotion regulation. *Social Psychological Personality Science, 2*(6), 665–672.

Kim, J. S., Lee, Y. H., Kim, J. C., Ko, Y. H., Yoon, C. S., et al. (2014). Effect of exercise training of different intensities on anti-inflammatory reaction in streptozotocin-induced diabetic rats. *Biology of Sport, 31*(1), 73–79.

Kim, K. H., Kim, Y. H., Son, J. E., Lee, J. H., Kim, S., et al. (2017). Intermittent fasting promotes adipose thermogenesis and metabolic homeostasis via VEGF-mediated alternative activation of macrophage. *Cell Research, 27* (11), 1309–1326.

Kim, K. S., Hong, S. W., Han, D., Yi, J., Jung, J., et al. (2016). Dietary antigens limit mucosal immunity by inducing regulatory T cells in the small intestine. *Science, 351* (6275), 858–863.

Kim, M. C., Yun, S. C., Lee, H. B., Lee, P. H., Lee, S. W., et al. (2017). Herpes zoster increases the risk of stroke and myocardial infraction. *Journal of the American College of Cardiology, 70*(2), 295–296.

Kim, M. J., Shin, J., Taylor, J. M., Mattek, A. M., Chavez, S. J., et al. (2017). Intolerance of uncertainty predicts increased striatal volume. *Emotion, 17*(6), 895–899.

Kim, N. K., Choi, B. H., Huang, X., Snyder, B. J., Bukhari, S., et al. (2009). Granulocyte-macrophage colony-stimulating factor promotes survival of dopaminergic neurons in the 1-methyl-4-phenyl-1,2,3,6-tetrahydropyridine-induced murine Parkinson's disease model. *European Journal of Neuroscience, 29*(5), 891–900.

Kim, S. K., Hayashi, H., Ishikawa, T., Shibata, K., Shigetomi, E., et al. (2016). Cortical astrocytes rewire somatosensory cortical circuits for peripheral neuropathic pain. *Journal of Clinical Investigation, 126*(5), 1983–1997.

Kim, Y. J., Seo, S. W., Park, S. B., Yang, J. J., Lee, J. S., et al. (2017). Protective effects of APOE e2 against disease progression in subcortical vascular mild cognitive impairment patients: A three-year longitudinal study. *Scientific Reports, 7,* 1910.

Kinnally, E. L., Capitanio, J. P., Leibel, R., Deng, L., LeDuc, C., et al. (2010). Epigenetic regulation of serotonin transporter expression and behavior in infant rhesus macaques. *Genes, Brain and Behavior, 9*(6), 575–582.

Kiraly, D. D., Walker, D. M., Calipari, E. S., Labonte, B., Issler, O., et al. (2016). Alterations of the host microbiome affect behavioral responses to cocaine. *Scientific Reports, 6,* 35455.

Kirby, E. D., Muroy, S. E., Sun, W. G., Covarrubias, D., Leong, M. J., et al. (2013). Acute stress enhances adult rat hippocampal neurogenesis and activation of newborn neurons via secreted astrocytic FGF2. *eLife, 2,* e00362.

Kirk, J. A., Gebhart, D., Buckley, A. M., Lok, S., Scholl, D., et al. (2017). New class of precision antimicrobials redefines role of clostridium difficile S-layer in virulence and viability. *Science Translational Medicine, 9*(406), eaah6813.

Kirmayer, L. J., & Crafa, D. (2014). What kind of science for psychiatry? *Frontiers in Human Neuroscience, 8,* 435.

Kirsch, I. (2014). The emperor's new drugs: Medication and placebo in the treatment of depression. *Handbook of Experimental Pharmacology, 225*, 291−303.

Kirsch, I., Deacon, B. J., Huedo-Medina, T. B., Scoboria, A., Moore, T. J., et al. (2008). Initial severity and antidepressant benefits: A meta-analysis of data submitted to the Food and Drug Administration. *PLoS Medicine, 5*(2), e45.

Kit, B. K., Kuklina, E., Carroll, M. D., Ostchega, Y., Freedman, D. S., et al. (2015). Prevalence of and trends in dyslipidemia and blood pressure among US children and adolescents, 1999-2012. *The Journal of the American Medical Association Pediatrics, 169*(3), 272−279.

Kitamura, H., Konno, A., Morimatsu, M., Jung, B. D., Kimura, K., et al. (1997). Immobilization stress increases hepatic IL-6 expression in mice. *Biochemical and Biophysical Research Communications, 238*(3), 707−711.

Kitazawa, M., Cheng, D., Tsukamoto, M. R., Koike, M. A., Wes, P. D., et al. (2011). Blocking IL-1 signaling rescues cognition, attenuates tau pathology, and restores neuronal β-catenin pathway function in an Alzheimer's disease model. *Journal of Immunology, 187*(12), 6539−6549.

Kleiman, S. C., Watson, H. J., Bulik-Sullivan, E. C., Huh, E. Y., Tarantino, L. M., et al. (2015). The intestinal microbiota in acute anorexia nervosa and during renourishment: Relationship to depression, anxiety, and eating disorder psychopathology. *Psychosomatic Medicine, 77* (9), 969−981.

Klein, C., & Schlossmacher, M. G. (2006). The genetics of Parkinson disease: Implications for neurological care. *Nature Clinical Practice Neurology, 2*(3), 136−146.

Kleinberger, G., Yamanishi, Y., Suárez-Calvet, M., Czirr, E., Lohmann, E., et al. (2014). TREM2 mutations implicated in neurodegeneration impair cell surface transport and phagocytosis. *Science Translational Medicine, 6*(243), 243ra86.

Kleyn, C. E., Schneider, L., Saraceno, R., Mantovani, C., Richards, H. L., et al. (2008). The effects of acute social stress on epidermal Langerhans' cell frequency and expression of cutaneous neuropeptides. *Journal of Investigative Dermatology, 128*(5), 1273−1279.

Kliewer, K. L., Ke, J. Y., Lee, H. Y., et al. (2015). Short-term food restriction followed by controlled refeeding promotes gorging behavior, enhances fat deposition, and diminishes insulin sensitivity in mice. *Journal of Nutritional Biochemistry, 26*, 721−728.

Klumpers, F., Kroes, M. C. W., Baas, J. M. P., & Fernández, G. (2017). How human amygdala and bed nucleus of the stria terminalis may drive distinct defensive responses. *Journal of Neuroscience, 37*(40), 9645−9656.

Klumpp, H., Fitzgerald, D. A., Cook, E., Shankman, S. A., Angstadt, M., et al. (2014). Serotonin transporter gene alters insula activity to threat in social anxiety disorder. *NeuroReport, 25*(12), 926−931.

Klumpp, H., Roberts, J., Kapella, M. C., Kennedy, A. E., Kumar, A., et al. (2017). Subjective and objective sleep quality modulate emotion regulatory brain function in anxiety and depression. *Depression and Anxiety, 34*(7), 651−660.

Knight, L. K., Naaz, F., Stoica, T., Depue, B. E., & Alzheimer's Disease Neuroimaging Initiative. (2017). Lifetime PTSD and geriatric depression symptomatology relate to altered dorsomedial frontal and amygdala morphometry. *Psychiatry Research, 267*, 59−68.

Knox, D., Perrine, S. A., George, S. A., Galloway, M. P., & Liberzon, I. (2010). Single prolonged stress decreases glutamate, glutamine, and creatine concentrations in the rat medial prefrontal cortex. *Neuroscience Letters, 480* (1), 16−20.

Knutie, S. A., Wilkinson, C. L., Kohl, K. D., & Rohr, J. R. (2017). Early-life disruption of amphibian microbiota decreases later-life resistance to parasites. *Nature Communications, 8*, 86.

Koban, L., Kross, E., Woo, C. W., Ruzic, L., & Wager, T. D. (2017). Frontal-brainstem pathways mediating placebo effects on social rejection. *Journal of Neuroscience, 37*(13), 3621−3631.

Köhler, O., Benros, M. E., Nordentoft, M., Farkouh, M. E., Iyengar, R. L., et al. (2014). Effect of anti-inflammatory treatment on depression, depressive symptoms, and adverse effects: A systematic review and meta-analysis of randomized clinical trials. *JAMA Psychiatry, 71*(12), 1381−1391.

Köhler, O., Petersen, L., Benros, M. E., Mors, O., & Gasse, C. (2016). Concomitant NSAID use during antipsychotic treatment and risk of 2-year relapse − A population-based study of 16,253 incident patients with schizophrenia. *Expert Opinion on Pharmacotherapy, 17*(8), 1055−1062.

Kohman, R., & Kusnecov, A. W. (2009). Stress, immunity and dendritic cells in cancer. In R. Salter, & M. Shurin (Eds.), *Dendritic cells in cancer*. New York, NY: Springer.

Kohut, M. L., Cooper, M. M., Nickolaus, M. S., Russell, D. R., & Cunnick, J. E. (2002). Exercise and psychosocial factors modulate immunity to influenza vaccine in elderly individuals. *The journals of gerontology Series A, Biological sciences and medical sciences, 57*(9), M557−562.

Kojima, Y., Volkmer, J. P., McKenna, K., Civelek, M., Lusis, A. J., et al. (2016). CD47-blocking antibodies restore phagocytosis and prevent antherosclerosis. *Nature, 536* (7614), 86−90.

Kokai, M., Kashiwamura, S. I., Okamura, H., Ohara, K., & Morita, Y. (2002). Plasma interleukin-18 levels in patients with psychiatric disorders. *Journal of Immunotherapy, 25*(Supply 1), S68−S71.

Kokaia, Z., & Lindvall, O. (2012). Stem cell repair of striatal ischemia. *Progress in Brain Research*, *201*, 35–53.

Koo, J. W., & Duman, R. S. (2008). IL-1beta is an essential mediator of the antineurogenic and anhedonic effects of stress. *Proceedings of the National Academy of Sciences of the United States of America*, *105*(2), 751–756.

Koo, J. W., Labonté, B., Engmann, O., Calipari, E. S., Juarez, B., et al. (2016). Essential role of mesolimbic brain-derived neurotrophic factor in chronic social stress-induced depressive behaviors. *Biol Psychiat*, *80*(6), 469–478.

Kootte, R. S., Levin, E., Salojärvi, J., Smits, L. P., Hartstra, A. V., et al. (2017). Improvement of insulin sensitivity after lean donor feces in metabolic syndrome is driven by baseline intestinal microbiota composition. *Cell Metabolism*, *26*(8), 611–619.

Kop, W. J., & Mommersteeg, P. M. C. (2014). Psychoneuroimmunological processes in coronary artery disease and heart failure. In A. Kusnecov, & H. Anisman (Eds.), *The Wiley-Blackwell handbook of psychoneuroimmunology*. Chichester, UK: John Wiley & Sons Ltd.

Kordower, J. H., Chu, Y., Hauser, R. A., Freeman, T. B., & Olanow, C. W. (2008). Lewy body-like pathology in long-term embryonic nigral transplants in Parkinson's disease. *Nature Medicine (New York, NY, United States)*, *14*(5), 504–506.

Kosloski, L. M., Kosmacek, E. A., Olson, K. E., Mosley, R. L., & Gendelman, H. E. (2013). GM-CSF induces neuroprotective and anti-inflammatory responses in 1-methyl-4-phenyl-1,2,3,6-tetrahydropyridine intoxicated mice. *Journal of Neuroimmunology*, *265*(1–2), 1–10.

Kostandy, B. B. (2012). The role of glutamate in neuronal ischemic injury: The role of spark in fire. *Neurological Sciences*, *33*, 223–237.

Kovacs, D., Eszlari, N., Petschner, P., Pap, D., Vas, S., et al. (2016). Effects of IL1B single nucleotide polymorphisms on depressive and anxiety symptoms are determined by severity and type of life stress. *Brain, Behavior, and Immunity*, *56*, 96–104.

Kovács, P., Pánczél, G., Balatoni, T., Liszkay, G., Gonda, X., et al. (2015). Social support decreases depressogenic effect of low-dose interferon alpha treatment in melanoma patients. *Journal of Psychosomatic Research*, *78*, 579–584.

Kowal, C., Degiorgio, L. A., Lee, J. Y., Edgar, M. A., Huerta, P. T., et al. (2006). Human lupus autoantibodies against NMDA receptors mediate cognitive impairment. *Proceedings of the National Academy of Sciences of the United States of America*, *103*(52), 19854–19859.

Koziol, M. J., Bradshaw, C. R., Allen, G. E., Costa, A. S. H., Frezza, C., et al. (2016). Identification of methylated deoxyadenosines in vertebrates reveals diversity in DNA modifications. *Nature Structural & Molecular Biology*, *23*(1), 24–30.

Kozlovsky, N., Zohar, J., Kaplan, Z., & Cohen, H. (2012). Microinfusion of a corticotrophin-releasing hormone receptor 1 antisense oligodeoxynucleotide into the dorsal hippocampus attenuates stress responses at specific times after stress exposure. *Journal of Neuroendocrinology*, *24*(3), 489–503.

Krakowiak, P., Goines, P. E., Tancredi, D. J., Ashwood, P., Hansen, R. L., Hertz-Picciotto, I., & Van de Water, J. (2017). Neonatal cytokine profiles associated with autism spectrum disorder. *Biological Psychiatry*, *81*(5), 442–451. Available from https://doi.org/10.1016/j.biopsych.2015.08.007.

Krasemann, S., Madore, C., Cialic, R., Baufeld, C., Calcagno, N., et al. (2017). The TREM2-APOE pathway drives the transcriptional phenotype of dysfunctional microglia in neurodegenerative diseases. *Immunity*, *47*(3), 566–581.e9.

Kraus, M. R., Al-Taie, O., Schäfer, A., Pfersdorff, M., Lesch, K. P., et al. (2007). Serotonin-1A receptor gene HTR1A variation predicts interferon-induced depression in chronic hepatitis C. *Gastroenterology*, *132*(4), 1279–1286.

Krishnan, V., & Nestler, E. J. (2011). Animal models of depression: Molecular perspectives. *Current Topics in Behavioral Neurosciences*, *7*, 121–147.

Kronfol, Z., Nair, M., Zhang, Q., Hill, E. E., & Brown, M. B. (1997). Circadian immune measures in healthy volunteers: Relationship to hypothalamic-pituitary-adrenal axis hormones and sympathetic neurotransmitters. *Psychosomatic Medicine*, *59*(1), 42–50.

Kross, E., Berman, M. G., Mischel, W., Smith, E. E., & Wager, T. D. (2011). Social rejection shares somatosensory representations with physical pain. *Proc Nat Acad Sci*, *108*, 6270–6275.

Krueger, J. M., & Opp, M. R. (2016). Sleep and microbes. *International Review of Neurobiology*, *131*, 207–225.

Krüger, C., Laage, R., Pitzer, C., Schäbitz, W. R., & Schneider, A. (2007). The hematopoietic factor GM-CSF (granulocyte-macrophage colony-stimulating factor) promotes neuronal differentiation of adult neural stem cells in vitro. *BioMed Central Neuroscience*, *8*, 88.

Krystal, J. H., & Neumeister, A. (2009). Noradrenergic and serotonergic mechanisms in the neurobiology of posttraumatic stress disorder and resilience. *Brain Research*, *1293*, 13–23.

Kudielka, B. M., & Kirschbaum, C. (2005). Sex differences in HPA axis responses to stress: A review. *Biological Psychology*, *69*(1), 113–132.

Kumar, D. K., Choi, S. H., Washicosky, K. J., Eimer, W. A., & Tucker, S. (2016). Amyloid-β peptide protects against microbial infection in mouse and worm models of Alzheimer's disease. *Sci Transl Med.*, *8*, 340ra72.

Kumsta, R., Hummel, E., Chen, F. S., & Heinrichs, M. (2015). Epigenetic regulation of the oxytocin receptor gene: Implications for behavioral neuroscience. *Frontiers in Neuroscience, 7*, 83.

Kundaje, A., Meuleman, W., Ernst, J., Bilenky, M., Yen, A., et al. (2015). Integrative analysis of 111 reference human epigenomes. *Nature, 518*(539), 317–330.

Kundakovic, M., Gudsnuk, K., Herbstman, J. B., Tang, D., Perera, F. P., et al. (2015). DNA methylation of BDNF as a biomarker of early-life adversity. *Proceedings of the National Academy of Sciences of the United States of America, 112*(22), 6807–6813.

Kusnecov, A. W., & Rabin, B. S. (1993). Inescapable footshock exposure differentially alters antigen- and mitogen-stimulated spleen cell proliferation in rats. *Journal of Neuroimmunology, 44*(1), 33–42.

Kusnecov, A. W., & Rossi-George, A. (2001). Potentiation of interleukin-1beta adjuvant effects on the humoral immune response to antigen in adrenalectomized mice. *Neuroimmunomodulation, 9*(2), 109–118.

Kusnecov, A. W., & Rossi-George, A. (2002). Stressor-induced modulation of immune function: A review of acute, chronic effects in animals. *Acta Neuropsychiatrica, 14*(6), 279–291.

Kyrozis, A., Ghika, A., Stathopoulos, P., Vassilopoulos, D., Trichopoulos, D., et al. (2013). Dietary and life-style variables in relation to incidence of Parkinson's disease in Greece. *European Journal of Epidemiology, 28*(1), 67–77.

Labonte, B., Azoulay, N., Yerko, V., Turecki, G., & Brunet, A. (2014). Epigenetic modulation of glucocorticoid receptors in posttraumatic stress disorder. *Translational Psychiatry, 4*, e368.

Labonté, B., Engmann, O., Purushothaman, I., Menard, C., Wang, J., et al. (2017). Sex-specific transcriptional signatures in human depression. *Nature Medicine (New York, NY, United States), 23*(9), 1102–1111.

Labonté, B., Suderman, M., Maussion, G., Lopez, J. P., Navarro-Sánchez, L., et al. (2013). Genome-wide methylation changes in the brains of suicide completers. *American Journal of Psychiatry, 170*, 511–520.

Labrecque, N., Whitfield, L. S., Obst, R., Waltzinger, C., Benoist, C., et al. (2001). How much TCR does a T cell need? *Immunity, 15*(1), 71–82.

Labus, J. S., Hollister, E. B., Jacobs, J., Kirbach, K., Oezguen, N., et al. (2017). Differences in gut microbial composition correlate with regional brain volumes in irritable bowel syndrome. *Microbiome, 5*(1), 49.

Labuschagne, I., Phan, K. L., Wood, A., Angstadt, M., Chua, P., et al. (2010). Oxytocin attenuates amygdala reactivity to fear in generalized social anxiety disorder. *Neuropsychopharmacology, 35*(12), 2403–2413.

Lacosta, S., Merali, Z., & Anisman, H. (1999). Behavioral and neurochemical consequences of lipopolysaccharide in mice: Anxiogenic-like effects. *Brain Research, 818*(2), 291–303.

Lafuse, W. P., Gearinger, R., Fisher, S., Nealer, C., Mackos, A. R., et al. (2017). Exposure to a social stressor induces translocation of commensal lactobacilli to the spleen and priming of the innate immune system. *Journal of Immunology, 198*(6), 2383–2393.

Lai, A. Y., Dibal, C. D., Armitage, G. A., Winship, I. R., & Todd, K. G. (2013). Distinct activation profiles in microglia of different ages: A systematic study in isolated embryonic to aged microglial cultures. *Neuroscience, 254*, 185–195.

Laine, M. A., Sokolowska, E., Dudek, M., Callan, S. A., Hyytia, P., et al. (2017). Brain activation induced by chronic psychosocial stress in mice. *Scientific Reports, 7* (1), 15061.

Laing, R. D. (1960). *The divided self: An existential study in sanity and madness.* Harmondsworth: Penguin.

Lancaster, C. L., Teeters, J. B., Gros, D. F., & Back, S. E. (2016). Posttraumatic stress disorder: Overview of evidence-based assessment and treatment. *Journal of Clinical Medicine, 5*(11), pii: E105.

Lancet. (2017). Life, death, and disability in 2016. *Lancet, 390*(10100), 1083.

Lang, A. E., & Lozano, A. M. (1998). Parkinson's disease: First of two parts. *The New England Journal of Medicine, 339*(15), 1044–1053.

Lang, F., Strutz-Seebohm, N., Seebohm, G., & Lang, U. E. (2010). Significance of SGK1 in the regulation of neuronal function. *Journal of Physiology (Cambridge, United Kingdom), 588*, 3349–3354.

Lang, S., Kroll, A., Lipinski, S. J., Wessa, M., Ridder, S., et al. (2009). Context conditioning and extinction in humans: Differential contribution of the hippocampus, amygdala and prefrontal cortex. *The European Journal of Neuroscience, 29*(4), 823–832.

Lange, T., Dimitrov, S., & Born, J. (2010). Effects of sleep and circadian rhythm on the human immune system. *Annals of the New York Academy of Sciences, 1193*, 48–59.

Langley, R. G., Feldman, S. R., Han, C., Schenkel, B., Szapary, P., et al. (2010). Ustekinumab significantly improves symptoms of anxiety, depression, and skin-related quality of life in patients with moderate-to-severe psoriasis: Results from a randomized, double-blind, placebo-controlled phase III trial. *Journal of the American Academy of Dermatology, 63*(3), 457–465.

Langston, J. W., Ballard, P., Tetrud, J. W., & Irwin, I. (1983). Chronic Parkinsonism in humans due to a product of meperidine-analog synthesis. *Science, 219*(4587), 979–980.

Langston, J. W., Forno, L. S., Tetrud, J., Reeves, A. G., Kaplan, J. A., et al. (1999). Evidence of active nerve cell degeneration in the substantia nigra of humans years after 1-methyl-4-phenyl-1,2,3,6-tetrahydropyridine exposure. *Annals of Neurology, 46*(4), 598–605.

Lanoiselée, H.-M., Nicolas, G., Wallon, D., Rovelet-Lecrux, A., Lacour, M., et al. (2017). APP, PSEN1, and PSEN2 mutations in early-onset Alzheimer disease: A genetic screening study of familial and sporadic cases. *PLoS Medicine, 14*(3), e1002270.

Lanquillon, S., Krieg, J. C., Bening-Abu-Shach, U., & Vedder, H. (2000). Cytokine production and treatment response in major depressive disorder. *Neuropsychopharmacology, 22*(4), 370–379.

Lashinger, L. M., Rossi, E. L., & Hursting, S. D. (2014). Obesity and resistance to cancer chemotherapy: Interacting roles of inflammation and metabolic dysregulation. *Clinical Pharmacology and Therapeutics, 96*(4), 458–463.

Lasselin, J., Elsenbruch, S., Lekander, M., Axelsson, J., Karshikoff, B., et al. (2016). Mood disturbance during experimental endotoxemia: Predictors of state anxiety as a psychological component of sickness behavior. *Brain, Behavior, and Immunity, 57*, 30–37.

Lateef, A., & Petri, M. (2013). Managing lupus patients during pregnancy. *Best Practice and Research. Clinical Rheumatology, 27*(3), 435–447. Available from https://doi.org/10.1016/j.berh.2013.07.005.

Lau, B. K., & Vaughan, C. W. (2014). Descending modulation of pain: The GABA disinhibition hypothesis of analgesia. *Current Opinion in Neurobiology, 29*, 159–164.

Lau, L. T., & Yu, A. C. (2001). Astrocytes produce and release interleukin-1, interleukin-6, tumor necrosis factor alpha and interferon-gamma following traumatic and metabolic injury. *Journal of Neurotrauma, 18*(3), 351–359.

Laugharne, J., Kullack, C., Lee, C. W., McGuire, T., Brockman, S., et al. (2016). Amygdala volumetric change following psychotherapy for posttraumatic stress disorder. *Journal of Neuropsychiatry and Clinical Neurosciences*, appineuropsych16010006.

Laukova, M., Alaluf, L. G., Serova, L. I., Arango, V., & Sabban, E. L. (2014). Early intervention with intranasal NPY prevents single prolonged stress-triggered impairments in hypothalamus and ventral hippocampus in male rats. *Endocrinology, 155*(10), 3920–3933.

Laveti, D., Kumar, M., Hemalatha, R., Sistla, R., Naidu, V. G., et al. (2013). Anti-inflammatory treatments for chronic diseases: A review. *Inflamm & Allergy Drug Targets, 12*(5), 349–361.

Lazarus, R. S., & Folkman, S. (1984). *Stress, appraisal, and coping.* New York: Springer.

Le Bastard, Q., Al-Ghalith, G. A., Grégoire, M., Chapalet, G., Javaudin, F., et al. (2018). Systematic review: Human gut dysbiosis induced by non-antibiotic prescription medications. *Alimentary Pharmacology & Therapeutics, 47*(3), 332–345.

Leclercq, S., Forsythe, P., & Bienenstock, J. (2016). Posttraumatic stress disorder: Does the gut microbiome hold the key? *Canadian Journal of Psychiatry, 61*(4), 204–213.

Ledeboer, A., Sloane, E. M., Milligan, E. D., Frank, M. G., Mahony, J. H., et al. (2005). Minocycline attenuates mechanical allodynia and proinflammatory cytokine expression in rat models of pain facilitation. *Pain, 115* (1–2), 71–83.

Ledgerwood, L., Richardson, R., & Cranney, J. (2005). D-cycloserine facilitates extinction of learned fear: Effects on reacquisition and generalized extinction. *Biological Psychiatry, 57*(8), 841–847.

LeDoux, J. E. (2000). Emotion circuits in the brain. *Annual Review of Neuroscience, 23*, 155–184.

LeDoux, J. E. (2014). Coming to terms with fear. *Proceedings of the National Academy of Sciences of the United States of America, 111*(8), 2871–2878.

LeDoux, J. E., & Pine, D. S. (2016). Using neuroscience to help understand fear and anxiety: A two-system framework. *American Journal of Psychiatry, 173*(11), 1083–1093.

Lee, B., Sur, B., Yeom, M., Shim, I., Lee, H., et al. (2016). Effects of systemic administration of ibuprofen on stress response in a rat model of post-traumatic stress disorder. *The Korean Journal of Physiology & Pharmacology, 20*, 357–366.

Lee, B. H., Kim, H., Park, S. H., & Kim, Y. K. (2007). Decreased plasma BDNF level in depressive patients. *Journal of Affective Disorders, 101*(1–3), 239–244.

Lee, D. J., Schnitzlein, C. W., Wolf, J. P., Vythilingam, M., Rasmusson, A. M., et al. (2016). Psychotherapy versus pharmacotherapy for posttraumatic stress disorder: Systemic review and meta-analyses to determine first-line treatments. *Depression and Anxiety, 33*(9), 792–806.

Lee, M., Krishnamurthy, J., Susi, A., Sullivan, C., Gorman, G. H., Hisle-Gorman, E., & Nylund, C. M. (2017). Association of autism spectrum disorders and inflammatory bowel disease. *Journal of Autism and Developmental Disorders, 48*(5), 1523–1529. Available from https://doi.org/10.1007/s10803-017-3409-5, Epub ahead of print.

Lee, P. C., Raaschou-Nielsen, O., Lill, C. M., Bertram, L., Sinsheimer, J. S., et al. (2016). Gene-environment interactions linking air pollution and inflammation in Parkinson's disease. *Environmental Research, 151*, 713–720.

Lee, P. R., Brady, D. L., Shapiro, R. A., Dorsa, D. M., & Koenig, J. I. (2007). Prenatal stress generates deficits in rat social behavior: Reversal by oxytocin. *Brain Research, 1156*, 152–167.

Leff-Gelman, P., Mancilla-Herrera, I., Flores-Ramos, M., Cruz-Fuentes, C., Reyes-Grajeda, J. P., et al. (2016). The immune system and the role of inflammation in perinatal depression. *Neuroscience Bulletin, 32*(4), 398–420.

Lei, D., Li, L., Li, L., Suo, X., Huang, X., et al. (2015). Microstructural abnormalities in children with posttraumatic stress disorder: A diffusion tensor imaging study at 3.0T. *Scientific Reports, 5*, 8933.

Lei, Y., Chen, C. J., Yan, X. X., Li, Z., & Deng, X. H. (2017). Early-life lipopolysaccharide exposure potentiates forebrain expression of NLRP3 inflammasome proteins and anxiety-like behavior in adolescent rats. *Brain Research, 1671*, 43—54, pii: S0006-8993(17)30262-7.

Lei, Y. M., Chen, L., Wang, Y., Stefka, A. T., Molinero, L. L., et al. (2016). The composition of the microbiota modulates allograft rejection. *Journal of Clinical Investigation, 126*(7), 2736—2744.

Leira, Y., López-Dequidt, I., Arias, S., Rodríguez-Yáñez, M., Leira, R., et al. (2016). Chronic periodontitis is associated with lacunar infarct: A case-control study. *European Journal of Neurology, 23*(10), 1572—1579.

LeMay, L. G., Vander, A. J., & Kluger, M. J. (1990). The effects of psychological stress on plasma interleukin-6 activity in rats. *Physiology & Behavior, 47*(5), 957—961.

Lemos, J. C., Wanat, M. J., Smith, J. S., Reyes, B. A., Hollon, N. G., et al. (2012). Severe stress switches CRF action in the nucleus accumbens from appetitive to aversive. *Nature, 490*(7420), 402—406.

Lempka, S. F., Malone, D. A., Hu, B., Baker, K. B., Wyant, A., et al. (2017). Randomized clinical trial of deep brain stimulation for poststroke pain. *Annals of Neurology, 81*(5), 653—663.

Lener, M. S., Niciu, M. J., Ballard, E. D., Park, M., Park, L. T., et al. (2016). Glutamate and gamma-aminobutyric acid systems in the pathophysiology of major depression and antidepressant response to ketamine. *Biological Psychiatry, 81*(10), 886—897.

Leng, W. D., Zeng, X. T., Kwong, J. S., & Hua, X. P. (2015). Periodontal disease and risk of coronary heart disease: An updated meta-analysis of prospective cohort studies. *International Journal of Cardiology, 201*, 469—472.

Lennox, B. R., Palmer-Cooper, E. C., Pollak, T., Hainsworth, J., Marks, J., et al. (2017). Prevalence and clinical characteristics of serum neuronal cell surface antibodies in first-episode psychosis: A case-control study. *The Lancet Psychiatry, 4*(1), 42—48.

Lenz, T. L., Spirin, V., Jordan, D. M., & Sunyaev, S. R. (2016). Excess of deleterious mutations around HLA genes reveals evolutionary cost of balancing selection. *Molecular Biology and Evolution, 33*(10), 2555—2564.

Lerman, I., Davis, B. A., Bertram, T. M., Proudfoot, J., Hauger, R. L., et al. (2016). Posttraumatic stress disorder influences the nociceptive and intrathecal cytokine response to a painful stimulus in combat veterans. *Psychoneuroendocrinology, 73*, 99—108.

Lerner, M. J., & Montada, L. (1998). An overview: Advances in belief in a just world theory and methods. In L. Montada, & M. J. Lerner (Eds.), *Responses to victimizations and belief in a just world* (pp. 1—7). New York: Plenum Press.

Leslie, D. L., Kobre, R. A., Richmand, B. J., Aktan Guloksuz, S., & Leckman, J. F. (2017). Temporal association of certain neuropsychiatric disorders following vaccination of children and adolescents: A pilot case-control study. *Frontiers in Psychiatry, 8*, 3. Available from https://doi.org/10.3389/fpsyt.2017.00003.

Letenneur, L., Proust-Lima, C., Le Gouge, A., Dartigues, J. F., & Barberger-Gateau, P. (2007). Flavonoid intake and cognitive decline over a 10-year period. *American Journal of Epidemiology, 165*(12), 1364—1371.

Levine, S. Z., Kodesh, A., Viktorin, A., Smith, L., Uher, R., et al. (2018). Association of maternal use of folic acid and multivitamin supplements in the periods before and during pregnancy with the risk of autism spectrum disorder in offspring. *Journal of the American Medical Association Psychiatry, 75*(2), 176—184.

Levine, S. Z., Levav, I., Goldberg, Y., Pugachova, I., Becher, Y., et al. (2016). Exposure to genocide and the risk of schizophrenia: A population-based study. *Psychological Medicine, 46*(4), 855—863.

Levkovitz, Y., Fenchel, D., Kaplan, Z., Zohar, J., & Cohen, H. (2015). Early post-stressor intervention with minocycline, a second-generation tetracycline, attenuates posttraumatic stress response in an animal model of PTSD. *European Neuropsychopharmacology, 25*(1), 124—132.

Lewin, A. B., Wu, M. S., McGuire, J. F., & Storch, E. A. (2014). Cognitive behavior therapy for obsessive-compulsive and related disorders. *The Psychiatric Clinics of North America, 37*(3), 415—445.

Lewis, D. R., Petersen, L. K., York, A. W., Zablocki, K. R., Joseph, L. B., et al. (2015). Sugar-based amphiphilic nanoparticles arrest atherosclerosis in vivo. *Proceedings of the National Academy of Sciences of the United States of America, 112*(9), 2693—2698.

Lewis, P. A., & Manzoni, C. (2012). LRRK2 and human disease: A complicated question or a question of complexes? *Science Signaling, 5*(207), pe2.

Ley, R. E., Bäckhed, F., Turnbaugh, P., Lozupone, C. A., Knight, R. D., et al. (2005). Obesity alters gut microbial ecology. *Proceedings of the National Academy of Sciences of the United States of America, 11070—11075.

Leyns, C. E. G., Ulrich, J. D., Finn, M. B., Stewart, F. R., Koscal, L. J., et al. (2017). TREM2 deficiency attenuates neuroinflammation and protects against neurodegeneration in a mouse model of tauopathy. *Proceedings of the National Academy of Sciences of the United States of America, 114*(43), 11524—11529.

Li, C., Liu, Y., Yin, S., Lu, C., Liu, D., et al. (2015). Long-term effects of early adolescent stress: Dysregulation of hypothalamic-pituitary-adrenal axis and central corticotropin releasing factor receptor 1 expression in adult male rats. *Behavioural Brain Research*, *288*, 39–49.

Li, H., Achour, I., Bastarache, L., Berghout, J., Gardeux, V., et al. (2016). Integrative genomics analyses unveil downstream biological effectors of disease-specific polymorphisms buried in intergenic regions. *NPJ Genomic Medicine*, *1*, 16006.

Li, J., Liu, S., Li, S., Feng, R., Na, L., et al. (2017). Prenatal exposure to famine and the development of hyperglycemia and type 2 diabetes in adulthood across consecutive generations: A population-based cohort study of families in Suihua, China. *American Journal of Clinical Nutrition*, *105*(1), 221–227.

Li, J., Olsen, J., Vestergaard, M., Kristensen, J. K., & Olsen, J. (2012). Prenatal exposure to bereavement and type-2 diabetes: A Danish longitudinal population based study. *PLoS One*, *7*(8), e43508.

Li, P., Bandyopadhyay, G., Lagakos, W. S., Lagakos, W. S., Talukdar, S., et al. (2015). LTB4 promotes insulin resistance in obese mice by acting on macrophages, hepatocytes and myocytes. *Nature Medicine (New York, NY, United States)*, *21*(3), 239–247.

Li, Q., Park, K., Xia, Y., Matsumoto, M., Qi, W., et al. (2017). Regulation of macrophage apoptosis and atherosclerosis by lipid-induced PKCδ isoform activation. *Circulation Research*, *121*(10), 1153–1167.

Li, T., Harada, M., Tamada, K., Abe, K., & Nomoto, K. (1997). Repeated restraint stress impairs the antitumor T cell response through its suppressive effect on Th1-type CD4 + T cells. *Anticancer Research*, *17*(6D), 4259–4268.

Li, T., He, X., Thomas, J. M., Yang, D., Zhong, S., et al. (2015). A novel GTP-binding inhibitor, FX2149, attenuates LRRK2 toxicity in Parkinson's disease models. *PLoS One*, *10*(3), e0122461.

Li, X., Chauhan, A., Sheikh, A. M., Patil, S., Chauhan, V., Li, X. M., & Malik, M. (2009). Elevated immune response in the brain of autistic patients. *Journal of Neuroimmunology*, *207*(1–2), 111–116. Available from https://doi.org/10.1016/j.jneuroim.2008.12.002.

Li, Z., Ma, L., Kulesskaya, N., Võikar, V., & Tian, L. (2014). Microglia are polarized to M1 type in high-anxiety inbred mice in response to lipopolysaccharide challenge. *Brain, Behavior, and Immunity*, *38*, 237–248.

Lian, H., Litvinchuk, A., Chiang, A. C. A., Aithmitti, N., Jankowsky, J. L., et al. (2016). Astrocyte-microglia cross talk through complement activation modulates amyloid pathology in mouse models of Alzheimer's disease. *Journal of Neuroscience*, *36*(2), 577–589.

Lian, Y. N., Chang, J. L., Lu, Q., Wang, Y., Zhang, Y., et al. (2017). Effects of fluoxetine on changes of pain sensitivity in chronic stress model rats. *Neuroscience Letters*, *651*, 16–20.

Liang, S., Wang, T., Hu, X., Luo, J., Li, W., et al. (2015). Administration of Lactobacillus helveticus NS8 improves behavioral, cognitive, and biochemical aberrations caused by chronic restraint stress. *Neuroscience*, *310*, 561–577.

Liao, C. H., Chang, C. S., Muo, C. H., & Kao, C. H. (2015). High prevalence of herpes zoster in patients with depression. *Journal of Clinical Psychiatry*, *76*(9), e1099–e1104.

Liao, S., & von der Weid, P. Y. (2015). Lymphatic system: An active pathway for immune protection. *Seminars in Cell and Developmental Biology*, *38*, 83–89.

Liberzon, I., & Abelson, J. L. (2016). Context processing and the neurobiology of post-traumatic stress disorder. *Neuron*, *92*(5), 14–30.

Lichtermann, D., Ekelund, J., Pukkala, E., Tanskanen, A., & Lönnqvist, J. (2001). Incidence of cancer among persons with schizophrenia and their relatives. *Archives of General Psychiatry*, *58*(6), 573–578.

Lichtman, J. H., Froelicher, E. S., Blumenthal, J. A., Carney, R. M., Doering, L. V., et al. (2014). Depression as a risk factor for poor prognosis among patients with acute coronary syndrome: Systematic review and recommendations. A scientific statement from the American Heart Association. *Circulation*, *129*(12), 1350–1369.

Licznerski, P., Duric, V., Banasr, M., Alavian, K. N., Ota, K. T., et al. (2015). Decreased SGK1 expression and function contributes to behavioral deficits induced by traumatic stress. *PLoS Biology*, *13*(10), e1002282.

Liddelow, S. A., Guttenplan, K. A., Clarke, L. E., Bennett, F. C., Bohlen, C. J., et al. (2017). Neurotoxic reactive astrocytes are induced by activated microglia. *Nature*, *541*(7638), 481–487.

Lidstone, S. C., Schulzer, M., Dinelle, K., Mak, E., Sossi, V., et al. (2010). Effects of expectation on placebo-induced dopamine release in Parkinson disease. *Archives of General Psychiatry*, *67*(8), 857–865.

Lieberman, M. D., & Eisenberger, N. I. (2015). The dorsal anterior cingulate cortex is selective for pain: Results from large-scale reverse inference. *Proceedings of the National Academy of Sciences of the United States of America*, *112*(49), 15250–15255.

Liebsch, G., Landgraf, R., Engelmann, M., Lörscher, P., & Holsboer, F. (1999). Differential behavioural effects of chronic infusion of CRH 1 and CRH 2 receptor antisense oligonucleotides into the rat brain. *Journal of Psychiatric Research*, *33*(2), 153–163.

Liguori, C., Chiaravalloti, A., Nuccetelli, M., Izzi, F., Sancesario, G., et al. (2017). Hypothalamic dysfunction is related to sleep impairment and CSF biomarkers in Alzheimer's disease. *Journal of Neurology*, *264*(11), 2215–2223.

Likhtik, E., & Paz, R. (2015). Amygdala-prefrontal interactions in (mal)adaptive learning. *Trends in Neurosciences, 38*(3), 158–166.

Lim, J., Altman, M. D., Baker, J., Brubaker, J. D., Chen, H., et al. (2015). Identification of N-(1H-pyrazol-4-yl)carboxamide inhibitors of interleukin-1 receptor associated kinase 4: Bicyclic core modifications. *Bioorganic & Medicinal Chemistry Letters, 25*(22), 5384–5388.

Lim, S. I., Song, K. H., Yoo, C. H., Woo, D. C., & Choe, B. Y. (2017). Decreased glutamatergic activity in the frontal cortex of single prolonged stress model: In vivo and ex vivo proton MR spectroscopy. *Neurochemical Research, 42*(8), 2218–2229.

Lin, C. C., Tung, C. S., Lin, P. H., Huang, C. L., & Liu, Y. P. (2016). Traumatic stress causes distinctive effects on fear circuit catecholamines and the fear extinction profile in a rodent model of posttraumatic stress disorder. *European Neuropsychopharmacology, 26*(9), 1484–1495.

Lin, C. C., Tung, C. S., & Liu, Y. P. (2016). Escitalopram reversed the traumatic stress-induced depressed and anxiety-like symptoms but not the deficits of fear memory. *Psychopharmacology (Berl), 233*(7), 1135–1146.

Lin, C. H., Lane, H. Y., & Tsai, G. E. (2012). Glutamate signaling in the pathophysiology and therapy of schizophrenia. *Pharmacology, Biochemistry and Behavior, 100*(4), 665–677.

Lin, C. H., Lin, H. I., Chen, M. L., Lai, T. T., Cao, L. P., et al. (2016). Lovastatin protects neurite degeneration in LRRK2-G2019S parkinsonism through activating the Akt/Nrf pathway and inhibiting GSK3β activity. *Human Molecular Genetics, 25*(10), 1965–1978.

Lin, N., Shi, J. J., Li, Y. M., Zhang, X. Y., Chen, Y., et al. (2016). What is the impact of n-3 PUFAs on inflammation markers in Type 2 diabetic mellitus populations? A systematic review and meta-analysis of randomized controlled trials. *Lipids in Health and Disease., 15*, 1–8.

Lin, P. Y., Chang, C. H., Chong, M. F., Chen, H., & Su, K. P. (2017). Polyunsaturated fatty acids in perinatal depression: A systematic review and meta-analysis. *Biological Psychiatry, 82*(2), 560–569.

Lin, S., Lin, Y., Nery, J. R., Urich, M. A., Breschi, A., et al. (2014). Comparison of the transcriptional landscapes between human and mouse tissues. *Proceedings of the National Academy of Sciences of the United States of America, 111*(48), 17224–17229.

Linderholm, K. R., Skogh, E., Olsson, S. K., Dahl, M. L., Holtze, M., et al. (2012). Increased levels of kynurenine and kynurenic acid in the CSF of patients with schizophrenia. *Schizophr Bull, 38*(3), 426–432.

Lindqvist, D., Dhabhar, F. S., Mellon, S. H., Yehuda, R., Grenon, S. M., et al. (2017). Increased proinflammatory milieu in combat related PTSD — A new cohort replication study. *Brain, Behavior, and Immunity, 59*, 260–264.

Lindqvist, D., Wolkowitz, O. M., Mellon, S., Yehuda, R., Flory, J. D., et al. (2014). Proinflammatory milieu in combat-related PTSD is independent of depression and early life stress. *Brain, Behavior, and Immunity, 42*, 81–88.

Lindsay, J., Sykes, E., McDowell, I., Verreault, R., & Laurin, D. (2004). More than the epidemiology of Alzheimer's disease: Contributions of the Canadian study of health and aging. *The Canadian Journal of Psychiatry, 49*(2), 83–91.

Lindvall, O., & Kokaia, Z. (2015). Neurogenesis following stroke affecting the adult brain. *Cold Spring Harbor Perspectives in Biology, 7*(11), 1–19.

Lionnet, A., Leclair-Visonneau, L., Neunlist, M., Murayama, S., Takao, M., et al. (2017). Does Parkinson's disease start in the gut? *Acta Neuropathologica, 135*(1), 1–12.

Lipina, T. V., Zai, C., Hlousek, D., Roder, J. C., & Wong, A. H. (2013). Maternal immune activation during gestation interacts with disc1 point mutation to exacerbate schizophrenia-related behaviors in mice. *Journal of Neuroscience, 33*(18), 7654–7666.

Lipska, K. J., & Krumholz, H. M. (2017). Is haemoglobin A1c the right outcome for studies of diabetes? *The Journal of the American Medical Association, 317*(10), 1017–1018.

Lisboa, S. F., Gomes, F. V., Guimaraes, F. S., & Campos, A. C. (2016). Microglial cells as a link between cannabinoids and the immune hypothesis of psychiatric disorders. *Frontiers in neurology, 7*, 5.

Lisnevskaia, L., Murphy, G., & Isenberg, D. (2014). Systemic lupus erythematosus. *Lancet, 384*(9957), 1878–1888.

Li-Tempel, T., Larra, M. F., Winnikes, U., Tempel, T., DeRijk, R. H., et al. (2016). Polymorphisms of genes related to the hypothalamic-pituitary-adrenal axis influence the cortisol awakening response as well as self-perceived stress. *Biological Psychology, 119*, 112–121.

Litteljohn, D., & Hayley, S. (2012). Cytokines as potential biomarkers for Parkinson's disease: A multiplex approach. *Methods in Molecular Biology, 934*, 121–144.

Litteljohn, D., Mangano, E. N., & Hayley, S. (2008). Cyclooxygenase-2 deficiency modifies the neurochemical effects, motor impairment and co-morbid anxiety provoked by paraquat administration in mice. *The European Journal of Neuroscience, 28*(4), 707–716.

Litteljohn, D., Nelson, E., & Hayley, S. (2014). IFN-γ differentially modulates memory-related processes under basal and chronic stressor conditions. *Frontiers in Cellular Neuroscience, 8*, 391.

Litteljohn, D., Rudyk, C., Dwyer, Z., Farmer, K., Fortin, T., et al. (2017). The impact of murine LRRK2 G2019S transgene overexpression on acute responses to inflammatory challenge. *Brain, Behavior, and Immunity, 67*, 246–256.

Littlefield, A. M., Setti, S. E., Prister, C., & Kohman, R. A. (2015). Voluntary exercise attenuates LPS-induced reductions in neurogenesis and increases microglia expression of proneurogenic phenotype in aged mice. *Journal of Neuroinflammation, 12,* 138.

Liu, F., Pardo, L. M., Schuur, M., Sanchez-Juan, P., Isaacs, A., et al. (2010). The apolipoprotein E gene and its age-specific effects on cognitive function. *Neurobiology of Aging, 31*(10), 1831–1833.

Liu, L. Y., Coe, C. L., Swenson, C. A., Kelly, E. A., Kita, H., et al. (2002). School examinations enhance airway inflammation to antigen challenge. *American Journal of Respiratory and Critical Care Medicine, 165*(8), 1062–1067.

Liu, R. T., & Alloy, L. B. (2010). Stress generation in depression: A systematic review of the empirical literature and recommendations for future study. *Clinical Psychology Review, 30*(5), 582–593.

Liu, X., Agerbo, E., Ingstrup, K. G., Musliner, K., Meltzer-Brody, S., et al. (2017). Antidepressant use during pregnancy and psychiatric disorders in offspring: Danish nationwide register based cohort study. *British Medical Journal, 358,* j3668.

Liu, Y., Garrett, M. E., Dennis, M. F., Green, K. T., et al. (2015). An examination of the association between 5-HTTLPR, combat exposure, and PTSD diagnosis among U.S. veterans. *PLoS One, 10*(3), e0119998.

Liu, Y., Ho, R. C., & Mak, A. (2012). The role of interleukin (IL)-17 in anxiety and depression of patients with rheumatoid arthritis. *International Journal of Rheumatic Diseases, 15*(2), 183–187.

Liu, Z., Zhu, F., Wang, G., Xiao, Z., Tang, J., et al. (2007). Association study of corticotropin-releasing hormone receptor1 gene polymorphisms and antidepressant response in major depressive disorders. *Neuroscience Letters, 414*(2), 155–158.

Livanos, A. E., Greiner, T. U., Vangay, P., Pathmasiri, W., Stewart, D., et al. (2016). Antibiotic-mediated gut microbiome perturbation accelerates development of type 1 diabetes in mice. *Nature Microbiology, 1*(11), 16140.

Livingston-Thomas, J., Nelson, P., Karthikeyan, S., Antonescu, S., Jeffers, M. S., et al. (2016). Exercise and environment enrichment as enablers of task-specific neuroplasticity and stroke recovery. *Neurotherapeutics, 13*(2), 395–402.

Livneh, Y., Ramesh, R. N., Burgess, C. R., Levandowski, K. M., Madara, J. C., et al. (2017). Homeostatic circuits selectively gate food cue responses in insular cortex. *Nature, 546*(7660), 611–616.

Llewelyn, M. J., Fitzpatrick, J. M., Darwin, E., Tonkin-Crine, S., Gorton, C., et al. (2017). The antibiotic course has had its day. *British Medical Journal, 358,* j3418.

Lloyd, C. E., Dyer, P. H., Lancashire, R. J., Harris, T., Daniels, J. E., et al. (1999). Association between stress and glycemic control in adults with type 1 (insulin-dependent) diabetes. *Diabetes Care, 22*(8), 1278–1283.

Loche, E., & Ozanne, S. E. (2016). Early nutrition, epigenetics, and cardiovascular disease. *Current Opinion in Lipidology, 27*(5), 449–458.

Locher, C., Koechlin, H., Zion, S. R., Werner, C., Pine, D. S., et al. (2017). Efficacy and safety of selective serotonin reuptake inhibitors, serotonin-norepinephrine reuptake inhibitors, and placebo for common psychiatric disorders among children and adolescents: A systematic review and meta-analysis. *JAMA Psychiatry, 74*(10), 1011–1020.

Lockhart, P. B., Bolger, A. F., Papapanou, P. N., Osinbowale, O., & Trevisan, M. (2012). Periodontal disease and atherosclerotic vascular disease: does the evidence support an independent association?: A scientific statement from the American Heart Association. *Circulation, 125,* 2520–2544.

Lodato, M. A., Rodin, R. E., Bohrson, C. L., Coulter, M. E., Barton, A. R., et al. (2018). Aging and neurodegeneration are associated with increased mutations in single human neurons. *Science, 359*(6375), 555–559.

Lopatkin, J. A., Meredith, H. R., Srimani, J. K., Pfeiffer, C., Durrett, R., et al. (2017). Persistence and reversal of plasmid-mediated antibiotic resistance. *Nature Communications, 8*(1), 1689.

López, P., Rodríguez-Carrio, J., Caminal-Montero, L., Mozo, L., & Suárez, A. (2016). A pathogenic IFNα, BLyS and IL-17 axis in Systemic Lupus Erythematosus patients. *Scientific Reports, 6,* 20651.

López-Solà, M., Koban, L., Krishnan, A., & Wager, T. D. (2017). When pain really matters: A vicarious-pain brain marker tracks empathy for pain in the romantic partner. *Neuropsychologia,* pii: S0028-3932(17)30265-8.

Loram, L. C., Taylor, F. R., Strand, K. A., Harrison, J. A., Rzasalynn, R., et al. (2013). Intrathecal injection of adenosine 2A receptor agonists reversed neuropathic allodynia through protein kinase (PK)A/PKC signaling. *Brain, Behavior, and Immunity, 33,* 112–122.

Losin, E. A. R., Anderson, S. R., & Wager, T. D. (2017). Feelings of clinician-patient similarity and trust influence pain: Evidence from simulated clinical interactions. *The Journal of Pain, 18*(7), 787–799.

Lourbopoulos, A., Ertürk, A., & Hellal, F. (2015). Microglia in action: How aging and injury can change the brain's guardians. *Frontiers in Cellular Neuroscience, 9,* 54.

Louveau, A., Plog, B. A., Antila, S., Alitalo, K., Nedergaard, M., et al. (2017). Understanding the functions and relationships of the glymphatic system and meningeal lymphatics. *Journal of Clinical Investigation, 127*(9), 3210–3219.

Lowry, C. A., Smith, D. G., Siebler, P. H., Schmidt, D., Stamper, C. E., et al. (2016). The microbiota, immunoregulation, and mental health: Implications for public health. *Current Environmental Health Reports, 3*(3), 270–286.

Lu, Z. W., Song, C., Ravindran, A. V., Merali, Z., & Anisman, H. (1998). Influence of a psychogenic and a neurogenic stressor on several indices of immune functioning in different strains of mice. *Brain, Behavior, and Immunity*, 12(1), 7−22.

Luby, J. L., Belden, A., Harms, M. P., Tillman, R., & Barch, D. M. (2016). Preschool is a sensitive period for the influence of maternal support on the trajectory of hippocampal development. *Proceedings of the National Academy of Sciences of the United States of America*, 113 (20), 5742−5747.

Lucas, K., Morris, G., Anderson, G., & Maes, M. (2015). The Toll-like receptor radical cycle pathway: A new drug target in immune-related chronic fatigue. *CNS & Neurological Disorders Drug Targets*, 14(7), 838−854.

Luck, H., Tsai, S., Chung, J., Clemente-Casares, X., Ghazarian, M., et al. (2015). Regulation of obesity-related insulin resistance with gut anti-inflammatory agents. *Cell Metabolism*, 21, 527−542.

Luczynski, P., Tramullas, M., Viola, M., Shanahan, F., Clarke, G., et al. (2017). Microbiota regulates visceral pain in the mouse. *eLife*, pii: e25887.

Luczynski, P., Whelan, S. O., O'Sullivan, C., Clarke, G., Shanahan, F., et al. (2016). Adult microbiota-deficient mice have distinct dendritic morphological changes: differential effects in the amygdala and hippocampus. *The European Journal of Neuroscience*, 44(9), 2654−2666.

Lueken, U., Zierhut, K. C., Hahn, T., Straube, B., Kircher, T., et al. (2016). Neurobiological markers predicting treatment response in anxiety disorders: A systematic review and implications for clinical application. *Neuroscience & Biobehavioral Reviews*, 66, 143−162.

Lukkes, J. L., Meda, S., Thompson, B. S., Freund, N., & Andersen, S. L. (2017). Early life stress and later peer distress on depressive behavior in adolescent female rats: Effects of a novel intervention on GABA and D2 receptors. *Behavioural Brain Research*, 330, 37−45.

Luna, R. A., & Foster, J. A. (2015). Gut brain axis: Diet microbiota interactions and implications for modulation of anxiety and depression. *Current Opinion in Biotechnology*, 32, 35−41.

Lund, F. E. (2008). Cytokine-producing B lymphocytes-key regulators of immunity. *Current Opinion in Immunology*, 20(3), 332−338.

Lund-Sørensen, H., Benros, M. E., Madsen, T., Sørensen, H. J., & Eaton, W. W. (2016). A nationwide cohort study of the association between hospitalization with infection and risk of death by suicide. *JAMA Psychiatry*, 73(9), 912−919.

Luongo, L., Maione, S., & Di Marzo, V. (2014). Endocannabinoids and neuropathic pain: Focus on neuron-glia and endocannabinoid-neurotrophin interactions. *The European Journal of NeuroscienceEur*, 39(3), 401−408.

Lupinsky, D., Moquin, L., & Gratton, A. (2017). Interhemispheric regulation of the rat medial prefrontal cortical glutamate stress response: role of local GABA- and dopamine-sensitive mechanisms. *Psychopharmacology (Berlin, Germany)*, 234(3), 353−363.

Lurie, I., Yang, Y. X., Haynes, K., Mamtani, R., & Boursi, B. (2015). Antibiotic exposure and the risk for depression, anxiety, or psychosis: A nested case-control study. *Journal of Clinical Psychiatry*, 76(11), 1522−1528.

Luscher, B., Shen, Q., & Sahir, N. (2011). The GABAergic deficit hypothesis of major depressive disorder. *Molecular Psychiatry*, 16(4), 383−406.

Lutz, P. E., Tanti, A., Gasecka, A., Barnett-Burns, S., Kim, J. J., et al. (2017). Association of a history of child abuse with impaired myelination in the anterior cingulate cortex: Convergent epigenetic, transcriptional, and morphological evidence. *The American Journal of Psychiatry*, 174(12), 1185−1194.

Lynch, J. F., Winiecki, P., Gilman, T. L., Adkins, J. M., & Jasnow, A. M. (2017). Hippocampal GABAB(1a) receptors constrain generalized contextual fear. *Neuropsychopharmacology*, 42(4), 914−924.

Lynch, M. A. (2015). Neuroinflammatory changes negatively impact on LTP: A focus on IL-1β. *Brain Research*, 1621, 197−204.

Lynch, M. E., & Ware, M. A. (2015). Cannabinoids for the treatment of chronic non-cancer pain: An updated systematic review of randomized controlled trials. *Journal of Neuroimmune Pharmacology*, 10(2), 293−301.

Lynch, S. V., & Pedersen, O. (2016). The human intestinal microbiome in health and disease. *New England Journal of Medicine*, 375(24), 2369−2379.

Lynn, D. J., & Pulendran, B. (2017). The potential of the microbiota to influence vaccine responses. *Journal of Leukocyte Biology*, 103(2), 225−231, pii: jlb. 5MR0617-216R.

Lyte, M., Li, W., Opitz, N., Gaykema, R. P., & Goehler, L. E. (2006). Induction of anxiety-like behavior in mice during the initial stages of infection with the agent of murine colonic hyperplasia Citrobacter rodentium. *Physiology & Behavior*, 89(3), 350−357.

Lyte, M., Nelson, S. G., & Thompson, M. L. (1990). Innate and adaptive immune responses in a social conflict paradigm. *Clinical Immunology and Immunopathology*, 57(1), 137−147.

Ma, K., Xu, A., Cui, S., Sun, M. R., Xue, Y. C., et al. (2016). Impaired GABA synthesis, uptake and release are associated with depression-like behaviors induced by chronic mild stress. *Translational Psychiatry*, 6(10), e910.

Maaijwee, N. A., Tendolkar, I., Rutten-Jacobs, L. C., Arntz, R. M., Schaapsmeerders, P., et al. (2016). Long-term depressive symptoms and anxiety after transient ischaemic attack or ischaemic stroke in young adults. *European Journal of Neurology*, 23(8), 1262−1268.

Macaskill, A. (2016). Review of positive psychology applications in clinical medical populations. *Healthcare, 4*(3), E66.

Machado, M. O., Oriolo, G., Bortolato, B., Köhler, C. A., Maes, M., et al. (2016). Biological mechanisms of depression following treatment with interferon for chronic hepatitis C: A critical systematic review. *Journal of Affective Disorders, 209,* 235–245.

Machado-Filho, J. A., Correia, A. O., Montenegro, A. B. A., Nobre, M. E. P., Cerqueira, G. S., et al. (2014). Caffeine neuroprotective effects on 6-OHDA-lesioned rats are mediated by several factors, including pro-inflammatory cytokines and histone deacetylase inhibitions. *Behavioural Brain Research, 264,* 116–125.

Mack, D. G., Johnson, J. J., Roberts, F., Roberts, C. W., Estes, R. G., et al. (1999). HLA-class II genes modify outcome of Toxoplasma gondii infection. *International Journal for Parasitology, 29*(9), 1351–1358.

Macpherson, A. J., de Agüero, M. G., & Ganal-Vonarburg, S. C. (2017). How nutrition and the maternal microbiota shape the neonatal immune system. *Nature Reviews Immunology, 17*(8), 508–517.

Maddox, S. A., Schafe, G. E., & Ressler, K. J. (2015). Exploring epigenetic regulation of fear memory and biomarkers associated with post-traumatic stress disorder. *Frontiers in Psychiatry, 4,* 62.

Madigan, S., Wade, M., Plamondon, A., Maguire, J. L., & Jenkins, J. M. (2017). Maternal adverse childhood experience and infant health: Biomedical and psychosocial risks as intermediary mechanisms. *Journal of Pediatrics, 187,* 282–289.e1.

Maekawa, M., Watanabe, A., Iwayama, Y., Kimura, T., Hamazaki, K., et al. (2017). Polyunsaturated fatty acid deficiency during neurodevelopment in mice models the prodromal state of schizophrenia through epigenetic changes in nuclear receptor genes. *Translational Psychiatry, 7*(9), e1229.

Maes, M. (1995). Evidence for an immune response in major depression: A review and hypothesis. *Progress in Neuro-Psychopharmacology & Biological Psychiatry, 19*(1), 11–38.

Maes, M. (2009). Inflammatory and oxidative and nitrosative stress pathways underpinning chronic fatigue, somatization and psychosomatic symptoms. *Current Opinion in Psychiatry, 22*(1), 75–83.

Maes, M., Christophe, A., Bosmans, E., Lin, A., & Neels, H. (2000). In humans, serum polyunsaturated fatty acid levels predict the response of proinflammatory cytokines to psychologic stress. *Biological Psychiatry, 47,* 910–920.

Maes, M., Galecki, P., Chang, Y. S., & Berk, M. (2011). A review on the oxidative and nitrosative stress (O&NS) pathways in major depression and their possible contribution to the (neuro)degenerative processes in that illness. *Progress in Neuro-Psychopharmacology & Biological Psychiatry, 35*(3), 676–692.

Maes, M., Kubera, M., & Leunis, J. C. (2008). The gut-brain barrier in major depression: intestinal mucosal dysfunction with an increased translocation of LPS from gram negative enterobacteria (leaky gut) plays a role in the inflammatory pathophysiology of depression. *Neuro Endocrinology Letters, 29*(1), 117–124.

Maes, M., Meltzer, H. Y., & Bosmans, E. (1994). Psychoimmune investigation in obsessive-compulsive disorder: Assays of plasma transferrin, IL-2 and IL-6 receptor, and IL-1β and IL-6 concentrations. *Neuropsychobiology, 30*(2–3), 57–60.

Maes, M., Song, C., Lin, A., De Jongh, R., Van Gastel, A., et al. (1998). The effects of psychological stress on humans: Increased production of pro-inflammatory cytokines and a Th1-like response in stress-induced anxiety. *Cytokine, 10*(Pt B), 313–318.

Maetzler, W., Liepelt, I., & Berg, D. (2009). Progression of Parkinson's disease in the clinical phase: Potential markers. *Lancet Neurology, 8*(12), 1158–1171.

Magri, G., Comerma, L., Pybus, M., Sintes, J., Lligé, D., et al. (2017). Human secretory IgM emerges from plasma cells clonally related to gut memory B cells and targets highly diverse commensals. *Immunity, 47*(1), 118–134.

Mahabir, M., Ashbaugh, A. R., Saumier, D., & Tremblay, J. (2016). Propranolol's impact on cognitive performance in post-traumatic stress disorder. *Journal of Affective Disorders, 192,* 98–103.

Maheshwari, P., & Eslick, G. D. (2015). Bacterial infection and Alzheimer's disease: A meta-analysis. *Journal of Alzheimer's Disease, 43*(3), 957–966.

Mahmoud, R., Wainwright, S. R., & Galea, L. A. (2016). Sex hormones and adult hippocampal neurogenesis: Regulation, implications, and potential mechanisms. *Front Neuroendocrinology, 41,* 129–152.

Maier, S. F., & Seligman, M. E. P. (1976). Learned helplessness: Theory and evidence. *Journal of Experimental Psychology General, 105,* 3–46.

Majidi, J., Kosari-Nasab, M., & Salari, A. A. (2016). Developmental minocycline treatment reverses the effects of neonatal immune activation on anxiety- and depression-like behaviors, hippocampal inflammation, and HPA axis activity in adult mice. *Brain Research Bulletin, 120,* 1–13.

Majoros, A., Platanitis, E., Kernbauer-Hölzl, E., Rosebrock, F., Müller, M., et al. (2017). Canonical and non-canonical aspects of JAK-STAT signaling: Lessons from interferons for cytokine responses. *Frontiers in Immunology, 8,* 29.

Makovac, E., Meeten, F., Watson, D. R., Herman, A., Garfinkel, S. N., et al. (2016). Alterations in amygdala-prefrontal functional connectivity account for excessive worry and autonomic dysregulation in generalized anxiety disorder. *Biological Psychiatry, 80*(10), 786–795.

Malavera, A., Silva, F. A., Fregni, F., Carrillo, S., & Garcia, R. G. (2016). Repetitive transcranial magnetic stimulation for phantom limb pain in land mine victims: A double-blinded, randomized, sham-controlled trial. *The Journal of Pain, 17*(8), 911−918.

Malberg, J. E., & Duman, R. S. (2003). Cell proliferation in adult hippocampus is decreased by inescapable stress: Reversal by fluoxetine treatment. *Neuropsychopharmacology, 28*, 1562−1671.

Male, D. K., Brostoff, J., Roth, D., & Roitt, I. (2006). *Immunology* (7th ed.). Mosby.

Malejko, K., Abler, B., Plener, P. L., & Straub, J. (2017). Neural correlates of psychotherapeutic treatment of post-traumatic stress disorder: A systematic literature review. *Frontiers in Psychiatry, 8*, 85.

Maletic, V., Eramo, A., Gwin, K., Offord, S. J., & Duffy, R. A. (2017). The role of norepinephrine and its α-Adrenergic receptors in the pathophysiology and treatment of major depressive disorder and schizophrenia: A systematic review. *Frontiers in Psychiatry, 8*, 42.

Malik, M., Simpson, J. F., Parikh, I., Wilfred, B. R., Fardo, D. W., et al. (2013). CD33 Alzheimer's risk-altering polymorphism, CD33 expression, and exon 2 splicing. *Journal of Neuroscience, 33*(33), 13320−13325.

Malkus, K. A., Tsika, E., & Ischiropoulos, H. (2009). Oxidative modifications, mitochondrial dysfunction, and impaired protein degradation in Parkinson's disease: How neurons are lost in the Bermuda triangle. *Molecular Neurodegeneration, 4*, 24.

Mallya, S., Sutherland, J., Pongracic, S., Mainland, B., & Ornstein, T. J. (2015). The manifestation of anxiety disorders after traumatic brain injury: A review. *J Neurotrauma, 32*, 411−421.

Malm, H., Brown, A. S., Gissler, M., Gyllenberg, D., Hinkka-Yli-Salomäki, S., et al. (2016). Gestational exposure to selective serotonin reuptake inhibitors and offspring psychiatric disorders: A national register-based study. *The Journal of the American Academy of Child & Adolescent Psychiatry, 55*(5), 359−366.

Mamik, M. K., Asahchop, E. L., Chan, W. F., Zhu, Y., Branton, W. G., et al. (2016). Insulin treatment prevents neuroinflammation and neuronal injury with restored neurobehavioral function in models of HIV/AIDS neurodegeneration. *Journal of Neuroscience, 36*(41), 10683−10695.

Man, K. K. C., Chan, E. W., Ip, P., Coghill, D., Simonoff, E., et al. (2017). Prenatal antidepressant use and risk of attention-deficit/hyperactivity disorder in offspring: Population based cohort study. *British Medical Journal, 357*, j2350.

Mander, B. A., Winer, J. R., Jagust, W. J., & Walker, M. P. (2016). Sleep: A novel mechanistic pathway, biomarker, and treatment target in the pathology of Alzheimer's disease? *Trends in Neurosciences, 39*(8), 552−566.

Manderino, L., Carroll, I., Azcarate-Peril, M. A., Rochette, A., Heinberg, L., et al. (2017). Preliminary evidence for an association between the composition of the gut microbiome and cognitive function in neurologically healthy older adults. *Journal of the International Neuropsychological Society, 23*(8), 700−705.

Mangalam, A., Shahi, S. K., Luckey, D., Karau, M., Marietta, E., et al. (2017). Human gut-derived commensal bacteria suppress CNS inflammatory and demyelinating disease. *Cell Reports, 20*(6), 1269−1277.

Mangano, E. N., & Hayley, S. (2009). Inflammatory priming of the substantia nigra influences the impact of later paraquat exposure: Neuroimmune sensitization of neurodegeneration. *Neurobiology of Aging, 30*(9), 1361−1378.

Mangano, E. N., Litteljohn, D., So, R., Nelson, E., Peters, S., et al. (2012). Interferon-γ plays a role in paraquat-induced neurodegeneration involving oxidative and proinflammatory pathways. *Neurobiology of Aging, 33*(7), 1411−1426.

Mangano, E. N., Peters, S., Litteljohn, D., So, R., Bethune, C., et al. (2011). Granulocyte macrophage-colony stimulating factor protects against substantia nigra dopaminergic cell loss in an environmental toxin model of Parkinson's disease. *Neurobiology of Disease, 43*(1), 99−112.

Mangino, M., Roederer, M., Beddall, M. H., Nestle, F. O., & Spector, T. D. (2017). Innate and adaptive immune traits are differentially affected by genetic and environmental factors. *Nature Communications, 8*, 13850.

Manikkam, M., Haque, M. M., Guerrero-Bosagna, C., Nilsson, E. E., & Skinner, M. K. (2014). Pesticide methoxychlor promotes the epigenetic transgenerational inheritance of adult-onset disease through the female germline. *PLoS One, 9*, e102091.

Mann, J. J. (2013). The serotonergic system in mood disorders and suicidal behaviour. *Philosophical Transactions of the Royal Society of London. Series B, Biological sciences, 368*(1615), 20120537.

Manousaki, D., Paternoster, L., Standl, M., Moffatt, M. F., Farrall, M., et al. (2017). Vitamin D levels and susceptibility to asthma, elevated immunoglobulin E levels, and atopic dermatitis: A Mendelian randomization study. *PLoS Medicine, 14*, e1002294.

Mansfeld, J., Urban, N., Priebe, S., Groth, M., Frahm, C., et al. (2015). Branched-chain amino acid catabolism is a conserved regulator of physiological ageing. *Nature Communications, 6*, 10043.

Månsson, K. N., Salami, A., Carlbring, P., Boraxbekk, C. J., Andersson, G., et al. (2017). Structural but not functional neuroplasticity one year after effective cognitive behaviour therapy for social anxiety disorder. *Behavioural Brain Research, 318*, 45−51.

Manteghi, A. A., Hebrani, P., Mortezania, M., Haghighi, M. B., & Javanbakht, A. (2014). Baclofen add-on to citalopram in treatment of posttraumatic stress disorder. *Journal of Clinical Psychopharmacology, 34*(2), 240−243.

Marazziti, D., Abelli, M., Baroni, S., Carpita, B., Ramacciotti, C. E., et al. (2015). Neurobiological correlates of social anxiety disorder: An update. *CNS Spectrums, 20,* 100–111.

Marazziti, D., Carlini, M., & Dell'Osso, L. (2012). Treatment strategies of obsessive-compulsive disorder and panic disorder/agoraphobia. *Current Topics in Medicinal Chemistry, 12,* 238–253.

Marcinkiewcz, C. A., Mazzone, C. M., D'Agostino, G., Halladay, L. R., Hardaway, J. A., et al. (2016). Serotonin engages an anxiety and fear-promoting circuit in the extended amygdala. *Nature, 537,* 97–101.

Marfil-Álvarez, R., Mesam, F., Arrebola-Morenom, A., Magán-Fernández, A., O'Valle, F., et al. (2014). Acute myocardial infarct size is related to periodontitis extent and severity. *Journal of Dental Research, 93,* 993–998.

Margolis, A. E., Herbstman, J. B., Davis, K. S., Thomas, V. K., Tang, D., et al. (2016). Longitudinal effects of prenatal exposure to air pollutants on self-regulatory capacities and social competence. *Journal of Child Psychology and Psychiatry, and Allied Disciplines, 57*(7), 851–860.

Marietta, E. V., Murray, J. A., Luckey, D. H., Jeraldo, P. R., Lamba, A., et al. (2016). Suppression of inflammatory arthritis by human gut-derived prevotella histicola in humanized mice. *Arthritis & Rheumatology, 68*(12), 2878–2888.

Marin, I. A., Goertz, J. E., Ren, T., Rich, S. S., Onengut-Gumuscu, S., et al. (2017). Microbiota alteration is associated with the development of stress-induced despair behavior. *Scientific Reports, 7,* 43859.

Marinova-Mutafchieva, L., Sadeghian, M., Broom, L., Davis, J. B., Medhurst, A. D., et al. (2009). Relationship between microglial activation and dopaminergic neuronal loss in the substantia nigra: A time course study in a 6-hydroxydopamine model of Parkinson's disease. *Journal of Neurochemistry, 110,* 966–975.

Markle, J. G., Frank, D. N., Mortin-Toth, S., Robertson, C. E., Feazel, L. M., et al. (2013). Sex differences in the gut microbiome drive hormone-dependent regulation of autoimmunity. *Science, 339,* 1084–1088.

Marmot, M. G., Rose, G., Shipley, M., & Hamilton, P. J. (1978). Employment grade and coronary heart disease in British civil servants. *Journal of Epidemiology and Community Health, 32*(4), 244–249.

Marras, C., Alcalay, R. N., Caspell-Garcia, C., Coffey, C., Chan, P., et al. (2016). Motor and nonmotor heterogeneity of LRRK2-related and idiopathic Parkinson's disease. *Movement Disorders, 31,* 1192–1202.

Marrie, R. A., Walld, R., Bolton, J. M., Sareen, J., Walker, J. R., et al. (2017). Increased incidence of psychiatric disorders in immune-mediated inflammatory disease. *Journal of Psychosomatic Research, 101,* 17–23.

Marrocco, J., Reynaert, M. L., Gatta, E., Gabriel, C., Mocaër, E., et al. (2014). The effects of antidepressant treatment in prenatally stressed rats support the glutamatergic hypothesis of stress-related disorders. *Journal of Neuroscience, 34,* 2015–2024.

Marshall, G. D., Jr., Agarwal, S. K., Lloyd, C., Cohen, L., Henninger, E. M., et al. (1998). Cytokine dysregulation associated with exam stress in healthy medical students. *Brain, Behavior, and Immunity, 12*(4), 297–307.

Marsland, A. L., Walsh, C., Lockwood, K., & John-Henderson, N. A. (2017). The effects of acute psychological stress on circulating and stimulated inflammatory markers: A systematic review and meta-analysis. *Brain, Behavior, and Immunity, 64,* 208–219.

Martel, J., Ojcius, D. M., Chang, C. J., Lin, C. S., Lu, C. C., et al. (2016). Anti-obesogenic and antidiabetic effects of plants and mushrooms. *Nature Reviews Endocrinology, 13,* 149–160.

Martin, J. B. (2002). The integration of neurology, psychiatry, and neuroscience in the 21st century. *American Journal of Psychiatry, 159*(5), 695–704.

Martineau, A. R., Jolliffe, D. A., Hooper, R. L., Greenberg, L., Aloia, J. F., et al. (2017). Vitamin D supplementation to prevent acute respiratory tract infections: Systematic review and meta-analysis of individual participant data. *British Medical Journal, 356,* i6583.

Martinez, M., Phillips, P. J., & Herbert, J. (1998). Adaptation in patterns of c-fos expression in the brain associated with exposure to either single or repeated social stress in male rats. *Eur J Neurosci, 10*(1), 20–33.

Martinez-Cerdeno, V., Camacho, J., Fox, E., Miller, E., Ariza, J., Kienzle, D., ... de Water, J. (2016). Prenatal Exposure to Autism-Specific Maternal Autoantibodies Alters Proliferation of Cortical Neural Precursor Cells, Enlarges Brain, and Increases Neuronal Size in Adult Animals. *Cerebral Cortex (New York, NY), 26*(1), 374–383. Available from https://doi.org/10.1093/cercor/bhu291.

Martin-Subero, M., Anderson, G., Kanchanatawan, B., Berk, M., & Maes, M. (2016). Comorbidity between depression and inflammatory bowel disease explained by immune-inflammatory, oxidative, and nitrosative stress; tryptophan catabolite; and gut-brain pathways. *CNS Spectrums, 21,* 184–198.

Marusak, H. A., Furman, D. J., Kuruvadi, N., Shattuck, D. W., Joshi, S. H., et al. (2015). Amygdala responses to salient social cues vary with oxytocin receptor genotype in youth. *Neuropsychologia, 79*(Pt A), 1–9.

Mash, L. E., Reiter, M. A., Linke, A. C., Townsend, J., & Muller, R. A. (2017). Multimodal approaches to functional connectivity in autism spectrum disorders: An integrative perspective. *Developmental Neurobiology.* Available from https://doi.org/10.1002/dneu.22570.

Maslanik, T., Bernstein-Hanley, I., Helwig, B., & Fleshner, M. (2012). The impact of acute-stressor exposure on splenic innate immunity: A gene expression analysis. *Brain, Behavior, and Immunity, 26*, 142–149.

Masri, S., Kinouchi, K., & Sassone-Corsi, P. (2015). Circadian clocks, epigenetics, and cancer. *Current Opinion in Oncology, 27*(1), 50–56.

Matheoud, D., Sugiura, A., Bellemare-Pelletier, A., Laplante, A., & Rondeau, C. (2016). Parkinson's disease-related proteins PINK1 and parkin repress mitochondrial antigen presentation. *Cell, 166*(2), 314–327.

Matheson, K., & Anisman, H. (2003). Systems of coping associated with dysphoria, anxiety and depressive illness: A multivariate profile perspective. *Stress, 6*(3), 223–234.

Matheson, K., & Anisman, H. (2012). Biological and psychosocial responses to discrimination. In J. Jetten, C. Haslam, & S. A. Haslam (Eds.), *The social cure* (pp. 133–154). New York: Psychology Press.

Matheson, K., Bombay, A., & Anisman, H. (2018). Culture as an ingredient of personalized medicine. *Journal of Psychiatry & Neuroscience, 43*(1), 3–6.

Matheson, K., Bombay, A., Dixon, K., & Anisman, H. (2018). Intergenerational communication regarding Indian Residential Schools: Implications for cultural identity, perceived discrimination, and depressive symptoms. *Transcultural Psychiatry*, in press.

Matheson, K., Skomorovsky, A., Fiocco, A., & Anisman, H. (2007). The limits of "adaptive" coping: Well-being and affective reactions to stressors among women in abusive dating relationships. *Stresss, 10*(1), 75–92.

Matt, S. M., & Johnson, R. W. (2016). Neuro-immune dysfunction during brain aging: New insights in microglial cell regulation. *Current Opinion in Pharmacology, 26*, 96–101.

Matthes, H. W., Maldonado, R., Simonin, F., Valverde, O., Slowe, S., et al. (1996). Loss of morphine-induced analgesia, reward effect and withdrawal symptoms in mice lacking the mu-opioid-receptor gene. *Nature, 383*, 819–823.

Maurer, M. H., Schäbitz, W. R., & Schneider, A. (2008). Old friends in new constellations--the hematopoetic growth factors G-CSF, GM-CSF, and EPO for the treatment of neurological diseases. *Current Medicinal Chemistry, 15*, 1407–1411.

May, R. W., Sanchez-Gonzalez, M. A., Hawkins, K. A., Batchelor, W. B., Fincham, F. D., et al. (2014). Effect of anger and trait forgiveness on cardiovascular risk in young adult females. *American Journal of Cardiology, 114*, 47–52.

Mayberg, H. S., Lozano, A. M., Voon, V., McNeely, H. E., Seminowicz, D., et al. (2005). Deep brain stimulation for depression. *Neuron, 45*, 651–660.

Mayer, E. A., Knight, R., Mazmanian, S. K., Cryan, J. F., & Tillisch, K. (2014). Gut microbes and the brain: Paradigm shift in neuroscience. *Journal of Neuroscience, 34*(46), 15490–15496.

Mayer, E. A., & Tillisch, K. (2011). The brain-gut axis in abdominal pain syndromes. *Annual Review of Medicine, 62*, 381–396.

Mayer-Davis, E. J., Lawrence, J. M., Dabelea, D., Divers, J., Isom, S., et al. (2017). Incidence trends of Type 1 and Type 2 diabetes among youths, 2002–2012. *New England Journal of Medicine, 376*, 1419–1429.

Mayo-Wilson, E., Dias, S., Mavranezouli, I., Kew, K., Clark, D. M., et al. (2014). Psychological and pharmacological interventions for social anxiety disorder in adults: A systematic review and network meta-analysis. *The Lancet Psychiatry, 1*(5), 368–376.

Mays, J. W., Bailey, M. T., Hunzeker, J. T., Powell, N. D., Papenfuss, T., et al. (2010). Influenza virus-specific immunological memory is enhanced by repeated social defeat. *Journal of Immunology, 184*, 2014–2025.

Maze, I., Covington, H. E., 3rd, Dietz, D. M., LaPlant, Q., Renthal, W., et al. (2010). Essential role of the histone methyltransferase G9a in cocaine-induced plasticity. *Science, 327*, 213–216.

McAuley, P. A., Artero, E. G., Sui, X., Lee, D., Church, T. S., et al. (2012). The obesity paradox, cardiorespiratory fitness, and coronary heart disease. *Mayo Clinic Proceedings, 87*, 443–451.

McCrory, E. J., De Brito, S. A., Sebastian, C. L., Mechelli, A., Bird, G., et al. (2011). Heightened neural reactivity to threat in child victims of family violence. *Current Biology, 21*, R947–R948.

McDade, T. W., Ryan, C., Jones, M. J., MacIsaac, J. L., Morin, A. M., et al. (2017). Social and physical environments early in development predict DNA methylation of inflammatory genes in young adulthood. *Proceedings of the National Academy of Sciences of the United States of America, 114*, 7611–7616.

McDaid, A. F., Joshi, P. K., Porcu, E., Komljenovic, A., Li, H., et al. (2017). Bayesian association scan reveals loci associated with human lifespan and linked biomarkers. *Nature Communications, 8*, 15842.

McDonald, D., Ackermann, G., Khailova, L., Baird, C., Heyland, D., et al. (2016). Extreme dysbiosis of the microbiome in critical illness. *mSphere, 1*(4), e00199–e00216.

McDougle, D. R., Watson, J. E., Adbeen, A. A., Abdeen, A. A., Adili, R., et al. (2017). Anti-inflammatory ω-3 endocannabinoid epoxides. *Proceedings of the National Academy of Sciences of the United States of America., 114* (30), E6034–E6043.

McEwen, B. S. (2000). Allostasis and allostatic load: Implications for neuropsychopharmacology. *Neuropsychopharmacology, 22*, 108–124.

McEwen, B. S., Bowles, N. P., Gray, J. D., Hill, M. N., Hunter, R. G., et al. (2015). Mechanisms of stress in the brain. *Nature Neuroscience, 18*(10), 1353–1363.

McEwen, B. S., & Gianaros, P. J. (2011). Stress- and allostasis-induced brain plasticity. *Annual Review of Medicine, 62,* 431–445.

McEwen, B. S., & Wingfield, J. C. (2003). The concept of allostasis in biology and biomedicine. *Hormones and Behavior, 43*(1), 2–15.

McGeer, E. G., & McGeer, P. L. (2010). Neuroinflammation in Alzheimer's disease and mild cognitive impairment: A field in its infancy. *Journal of Alzheimer's Disease, 19*(1), 355–361.

McGeer, P. L., & McGeer, E. G. (2004). Inflammation and the degenerative diseases of aging. *Annals of the New York Academy of Sciences, 1035,* 104–116.

McGovern, N., Shin, A., Low, G., Low, D., Duan, K., et al. (2017). Human fetal dendritic cells promote prenatal T-cell immune suppression through arginase-2. *Nature, 546,* 662–666.

McGowan, P. O., Sasaki, A., D'Alessio, A. C., Dymov, S., Labonté, B., et al. (2009). Epigenetic regulation of the glucocorticoid receptor in human brain associates with childhood abuse. *Nature Neuroscience, 12,* 342–348.

McGowan, P. O., & Szyf, M. (2010). The epigenetics of social adversity in early life: Implications for mental health outcomes. *Neurobiology of Disease, 39,* 66–72.

McGrath, J. J. (2017). Vitamin D and mental health—the scrutiny of science delivers a sober message. *Acta Psychiatrica Scandinavica, 135,* 183–184.

McGuire, J. L., Larke, L. E., Sallee, F. R., Herman, J. P., & Sah, R. (2011). Differential regulation of neuropeptide Y in the amygdala and prefrontal cortex during recovery from chronic variable stress. *Frontiers in Behavioral Neuroscience, 5,* 54.

McInnes, I. B., & Schett, G. (2007). Cytokines in the pathogenesis of rheumatoid arthritis. *Nature Reviews Immunology, 7,* 429–442.

McInnis, O. A., McQuaid, R. J., Bombay, A., Matheson, K., & Anisman, H. (2015). Finding benefit in stressful uncertain circumstances: Relations to social support and stigma among women with unexplained illnesses. *Stress, 18*(2), 169–177.

McIntosh, A. M., Hall, L. S., Zeng, Y., Adams, M. J., Gibson, J., et al. (2016). Genetic and environmental risk for chronic pain and the contribution of risk variants for major depressive disorder: A family-based mixed-model analysis. *PLoS Medicine, 13,* e1002090.

McIntosh, J., Anisman, H., & Merali, Z. (1999). Short-and long-periods of neonatal maternal separation differentially affect anxiety and feeding in adult rats: Gender-dependent effects. *Developmental Brain Research, 113* (1–2), 97–106.

McKenzie, J. A., Spielman, L. J., Pointer, C. B., Lowry, J. R., Bajwa, E., et al. (2017). Neuroinflammation as a common mechanism associated with the modifiable risk factors for Alzheimer's and Parkinson's diseases. *Current Aging Science, 10*(3), 158–176.

McKim, D. B., Weber, M. D., Niraula, A., Sawicki, C. M., Liu, X., et al. (2017). Microglial recruitment of IL-1β-producing monocytes to brain endothelium causes stress-induced anxiety. *Molecular Psychiatry.* Available from https://doi.org/10.1038/mp.2017.64.

McKlveen, J. M., Morano, R. L., Fitzgerald, M., Zoubovsky, S., Cassella, S. N., et al. (2016). Chronic stress increases prefrontal inhibition: A mechanism for stress-induced prefrontal dysfunction. *Biological Psychiatry, 80*(10), 754–764.

McLeod, M., Hong, M., Mukhida, K., Sadi, D., Ulalia, R., & Mendez, I. (2006). Erythropoietin and GDNF enhance ventral mesencephalic fiber outgrowth and capillary proliferation following neural transplantation in a rodent model of Parkinson's disease. *The European Journal of Neuroscience, 24*(2), 361–370.

McMurray, K. M., Ramaker, M. J., Barkley-Levenson, A. M., Sidhu, P. S., Elkin, P. K., et al. (2017). Identification of a novel, fast-acting GABAergic antidepressant. *Molecular Psychiatry, 23,* 384–391.

McNamara, R. K. (2015). Mitigation of inflammation-induced mood dysregulation by long-chain omega-3 fatty acids. *Journal of the American College of Nutrition, 34,* 48–55.

McNerney, M. W., Sheng, T., Nechvatal, J. M., Lee, A. G., Lyons, D. M., et al. (2018). Integration of neural and epigenetic contributions to posttraumatic stress symptoms: The role of hippocampal volume and glucocorticoid receptor gene methylation. *PLoS One, 13,* e0192222.

McNicol, E. D., Ferguson, M. C., Haroutounian, S., Carr, D. B., Schumann, R., et al. (2016). Single dose intravenous paracetamol or intravenous propacetamol for postoperative pain. *Cochrane Database of Systematic Reviews,* CD007126.

McQuaid, R. J., McInnis, O. A., Abizaid, A., & Anisman, H. (2014). Making room for oxytocin in understanding depression. *Neuroscience & Biobehavioral Reviews, 45,* 305–322.

McQuaid, R. J., McInnis, O. A., Matheson, K., & Anisman, H. (2015). Distress of ostracism: Oxytocin receptor gene polymorphism confers sensitivity to social exclusion. *Social Cognitive and Affective Neuroscience, 10*(8), 1153–1159.

McQuaid, R. J., McInnis, O. A., Paric, A., Al-Yawer, F., Matheson, K., et al. (2016). Relations between plasma oxytocin and cortisol: The stress buffering role of social support. *Neurobiology of Stress, 3,* 52–60.

Meagher, M. W., Johnson, R. R., Young, E. E., Vichaya, E. G., Lunt, S., et al. (2007). Interleukin-6 as a mechanism for the adverse effects of social stress on acute Theiler's virus infection. *Brain, Behavior, and Immunity, 21,* 1083–1095.

Medrihan, L., Sagi, Y., Inde, Z., Krupa, O., Daniels, C., et al. (2017). Initiation of behavioral response to antidepressants by cholecystokinin neurons of the dentate gyrus. *Neuron, 95*, 564–576.

Medzhitov, R. (2009). Approaching the asymptote: 20 years later. *Immunity, 30*, 766–775.

Mehta, D., Klengel, T., Conneely, K. N., Smith, A. K., Altmann, A., et al. (2013). Childhood maltreatment is associated with distinct genomic and epigenetic profiles in post-traumatic stress disorder. *Proceedings of the National Academy of Sciences of the United States of America, 110*(20), 8302–8307.

Mehta-Raghavan, N. S., Wert, S. L., Morley, C., Graf, E. N., & Redei, E. E. (2016). Nature and nurture: Environmental influences on a genetic rat model of depression. *Translational Psychiatry, 6*, e770.

Meir Drexler, S., Merz, C. J., Hamacher-Dang, T. C., & Wolf, O. T. (2016). Cortisol effects on fear memory reconsolidation in women. *Psychopharmacology (Berlin, Germany), 233*, 2687–2697.

Meltzer, A., & Van de Water, J. (2017). The Role of the Immune System in Autism Spectrum Disorder. *Neuropsychopharmacology, 42*(1), 284–298. Available from https://doi.org/10.1038/npp.2016.158.

Melzack, R., & Casey, K. L. (1968). Sensory, motivational, and central control determinants of pain. In D. R. Kenshalo (Ed.), *The skin senses* (pp. 423–439). Springfield, IL: Charles C. Thomas.

Melzack, R., & Wall, P. D. (1965). Pain mechanisms: A new theory. *Science, 150*, 971–979.

Menard, C., Pfau, M. L., Hodes, G. E., Kana, V., Wang, V. X., et al. (2017). Social stress induces neurovascular pathology promoting depression. *Nature Neuroscience, 20*, 1752–1760.

Mendelsohn, A. R., & Larrick, J. W. (2013). Sleep facilitates clearance of metabolites from the brain: Glymphatic function in ageing and neurodegenerative diseases. *Rejuvenation Research, 16*(6), 518–523.

Mendiola, A. S., & Cardona, A. E. (2017). The IL-1β phenomena in neuroinflammatory diseases. *Journal of Neural Transmission, 125*, 781–795. Available from https://doi.org/10.1007/s00702-017-1732-9.

Méquinion, M., Langlet, F., Zgheib, S., Dickson, S., Dehouck, B., et al. (2013). Ghrelin: Central and peripheral implications in anorexia nervosa. *Frontiers in Endocrinology, 4*, 15.

Merali, Z., & Anisman, H. (2016). Deconstructing the mental health crisis: 5 uneasy pieces. *Journal of Psychiatry & Neuroscience, 41*(4), 219–221.

Merali, Z., Bédard, T., Andrews, N., Davis, B., McKnight, A. T., et al. (2006). Bombesin receptors as novel anti-anxiety therapeutic target: Non-peptide antagonist PD 176252 reduces anxiety and 5-HT release through BB1 receptor. *Journal of Neuroscience, 26*, 10387–10396.

Merali, Z., Brennan, K., Brau, P., & Anisman, H. (2003). Dissociating anorexia and anhedonia elicited by interleukin-1beta: Antidepressant and gender effects on responding for "free chow" and "earned" sucrose intake. *Psychopharmacology (Berlin, Germany), 165*, 413–418.

Merali, Z., Du, L., Hrdina, P., Palkovits, M., Faludi, G., et al. (2004). Dysregulation in the suicide brain: mRNA expression of corticotropin releasing hormone receptors and GABA_A receptor subunits in frontal cortical brain region. *Journal of Neuroscience, 24*, 1478–1485.

Merali, Z., Graitson, S., Mackay, J. C., & Kent, P. (2013). Stress and eating: A dual role for bombesin like peptides. *Frontiers in Neuroscience, 7*, 193.

Merali, Z., Kent, P., Du, L., Hrdina, P., Palkovits, M., et al. (2006). Corticotropin-releasing hormone, arginine vasopressin, gastrin-releasing peptide, and neuromedin B alterations in stress-relevant brain regions of suicides and control subjects. *Biological Psychiatry, 59*(7), 594–602.

Merali, Z., McIntosh, J., Kent, P., Michaud, D., & Anisman, H. (1998). Aversive and appetitive events evoke the release of corticotropin-releasing hormone and bombesin-like peptides at the central nucleus of the amygdala. *The Journal of Neuroscience: The Official Journal of the Society for Neuroscience, 18*(12), 4758–4766.

Merino, J. J., Muñetón-Gómez, V., Alvárez, M. I., & Toledano-Díaz, A. (2016). Effects of CX3CR1 and fractalkine chemokines in amyloid beta clearance and p-Tau accumulation in Alzheimer's disease (AD) rodent models: Is fractalkine a systemic biomarker for AD? *Current Alzheimer Research, 13*, 403–412.

Merlot, E., Moze, E., Dantzer, R., & Neveu, P. J. (2004). Cytokine production by spleen cells after social defeat in mice: activation of T cells and reduced inhibition by glucocorticoids. *Stress, 7*(1), 55–61.

Merskey, H., & Bogduk, N. (1994). *Classification of chronic pain* (2nd ed.). Seattle, WA: International Association for the Study of Pain.

Metcalf, O., Varker, T., Forbes, D., Phelps, A., Dell, L., et al. (2016). Efficacy of fifteen emerging interventions for the treatment of posttraumatic stress disorder: A systematic review. *Journal of Traumatic Stress, 29*(1), 88–92.

Meyer, J. H., McMain, S., Kennedy, S. H., Korman, L., Brown, G. M., et al. (2003). Dysfunctional attitudes and 5-HT2 receptors during depression and self-harm. *American Journal of Psychiatry, 160*, 90–99.

Meyer, U., & Feldon, J. (2009). Prenatal exposure to infection: A primary mechanism for abnormal dopaminergic development in schizophrenia. *Psychopharmacology (Berlin, Germany), 206*, 587–602.

Meyer, U., & Feldon, J. (2012). To poly(I:C) or not to poly(I: C): Advancing preclinical schizophrenia research through the use of prenatal immune activation models. *Neuropharmacology, 62*, 1308–1321.

Meyer, U., Nyffeler, M., Yee, B. K., Knuesel, I., & Feldon, J. (2008). Adult brain and behavioral pathological markers of prenatal immune challenge during early/ middle and late fetal development in mice. *Brain, Behavior, and Immunity, 22*, 469−486.

Meyer-Luehmann, M., Coomaraswamy, J., Bolmont, T., Kaeser, S., Schaefer, C., et al. (2006). Exogenous induction of cerebral beta-amyloidogenesis is governed by agent and host. *Science, 313*, 1781−1784.

Michailidou, K., Lindström, S., Dennis, J., Beesley, J., Hui, S., et al. (2017). Association analysis identifies 65 new breast cancer risk loci. *Nature, 551*, 92−94.

Michaud, J. P., Hallé, M., Lampron, A., Thériault, P., Préfontaine, P., et al. (2013). Toll-like receptor 4 stimulation with the detoxified ligand monophosphoryl lipid A improves Alzheimer's disease-related pathology. *Proceedings of the National Academy of Sciences of the United States of America, 110*, 1941−1946.

Michaud, J. P., Richard, K. L., & Rivest, S. (2012). Hematopoietic MyD88-adaptor protein acts as a natural defense mechanism for cognitive deficits in Alzheimer's disease. *Stem cell Reviews, 8*, 898−904.

Michaud, K., Matheson, K., Kelly, O., & Anisman, H. (2008). Impact of stressors in a natural context on release of cortisol in healthy adult humans: A meta-analysis. *Stress, 11*(3), 177−197.

Michopoulos, V., Rothbaum, A. O., Jovanovic, T., Almli, L. M., Bradley, B., et al. (2015). Association of CRP genetic variation and CRP level with elevated PTSD symptoms and physiological responses in a civilian population with high levels of trauma. *American Journal of Psychiatry, 172*, 353−362.

Michopoulos, V., Vester, A., & Neigh, G. (2016). Posttraumatic stress disorder: A metabolic disorder in disguise? *Experimental Neurology, 284*, 220−229.

Mifflin, K. A., & Kerr, B. J. (2017). Pain in autoimmune disorders. *Journal of Neuroscience Research, 95*, 1282−1294.

Migheli, R., Del Giudice, M. G., Spissu, Y., Sanna, G., Xiong, Y., et al. (2013). LRRK2 affects vesicle trafficking, neurotransmitter extracellular level and membrane receptor localization. *PLoS One, 8*, e77198.

Miguel-Álvarez, M., Santos-Lozano, A., Sanchis-Gomar, F., Fiuza-Luces, C., Pareja-Galeano, H., et al. (2015). Non-steroidal anti-inflammatory drugs as a treatment for Alzheimer's disease: A systematic review and meta-analysis of treatment effect. *Drugs & Aging, 32*(2), 139−147.

Mika, A., Day, H. E., Martinez, A., Rumian, N. L., Greenwood, B. N., et al. (2017). Early life diets with prebiotics and bioactive milk fractions attenuate the impact of stress on learned helplessness behaviours and alter gene expression within neural circuits important for stress resistance. *The European Journal of Neuroscience, 45*(3), 342−357.

Mika, A., & Fleshner, M. (2016). Early-life exercise may promote lasting brain and metabolic health through gut

bacterial metabolites. *Immunology & Cell Biology, 94*, 151−157.

Mika, A., Rumian, N., Loughridge, A. B., & Fleshner, M. (2016). Exercise and prebiotics produce stress resistance: Converging impacts on stress-protective and butyrate-producing gut bacteria. *International Review of Neurobiology, 131*, 165−191.

Mikkelsen, K. H., Knop, F. K., Frost, M., Hallas, J., & Pottegård, A. (2015). Use of antibiotics and risk of type 2 diabetes: a population-based case-control study. *The Journal of Clinical Endocrinology & Metabolism, 100*, 3633−3640.

Milad, M. R., & Rauch, S. L. (2011). Obsessive-compulsive disorder: Beyond segregated corticostriatal pathways. *Trends in Cognitive Sciences, 16*(1), 43−51.

Milani, A. C., Hoffmann, E. V., Fossaluza, V., Jackowski, A. P., & Mello, M. F. (2017). Does pediatric post-traumatic stress disorder alter the brain? Systematic review and meta-analysis of structural and functional magnetic resonance imaging studies. *Psychiatry and Clinical neurosciences, 71*(3), 154−169.

Mildenberger, J., Johansson, I., Sergin, I., Kjøbli, E., Damås, J. K., et al. (2017). N-3 PUFAs induce inflammatory tolerance by formation of KEAP1-containing SQSTM1/ p62-bodies and activation of NFE2L2. *Autophagy, 13*, 1664−1678.

Millan, M. J. (2005). Serotonin 5-HT 2C receptors as a target for the treatment of depressive and anxious states: Focus on novel therapeutic strategies. *Therapie, 60*(5), 441−460.

Millan, S., Gonzalez-Quijano, M. I., Giordano, M., Soto, L., Martin, A. I., et al. (1996). Short and long restraint differentially affect humoral and cellular immune functions. *Life Sciences, 59*, 1431−1442.

Miller, B. J., & Goldsmith, D. R. (2017). Towards an immunophenotype of schizophrenia: Progress, potential mechanisms, and future directions. *Neuropsychopharmacology, 42*, 299−317.

Miller, G. E., Chen, E., Fok, A. K., Walker, H., Lim, A., et al. (2009). Low early-life social class leaves a biological residue manifested by decreased glucocorticoid and increased proinflammatory signaling. *Proc Natl Acad Sci U S A, 106*, 14716−14721.

Miller, G. E., Chen, E., & Parker, K. J. (2011). Psychological stress in childhood and susceptibility to the chronic diseases of aging: Moving toward a model of behavioral and biological mechanisms. *Psychological Bulletin, 137* (6), 959−997.

Miller, G. E., Cohen, S., Pressman, S., Barkin, A., Rabin, B. S., et al. (2004). Psychological stress and antibody response to influenza vaccination: When is the critical period for stress, and how does it get inside the body? *Psychosomatic Medicine, 66*(2), 215−223.

Miller, G. E., Cohen, S., & Ritchey, A. K. (2002). Chronic psychological stress and the regulation of pro-inflammatory cytokines: A glucocorticoid-resistance model. *Health Psychology, 21*(6), 531–541.

Miller, G. E., Lachman, M. E., Chen, E., Gruenewald, T. L., Karlamangla, A. S., et al. (2011). Pathways to resilience: Maternal nurturance as a buffer against the effects of childhood poverty on metabolic syndrome at midlife. *Psychological Science, 22*(12), 1591–1599.

Miller, K. R., & Streit, W. J. (2007). The effects of aging, injury and disease on microglial function: A case for cellular senescence. *Neuron Glia Biology, 3*(3), 245–253.

Miller, L., Bansal, R., Wickramaratne, P., Hao, X., Tenke, C. E., et al. (2014). Neuroanatomical correlates of religiosity and spirituality: A study in adults at high and low familial risk for depression. *JAMA Psychiat, 71*, 128–135.

Millman, J. R., Xie, C., Van Dervort, A., Gürtler, M., Pagliuca, F. W., et al. (2016). Generation of stem cell-derived β-cells from patients with type 1 diabetes. *Nature Communications, 7*, 11463.

Mina, M. J., Metcalf, C. J., de Swart, R. L., Osterhaus, A. D., & Grenfell, B. T. (2015). Vaccines: Long-term measles-induced immunomodulation increases overall childhood infectious disease mortality. *Science, 348*, 694–699.

Minelli, A., Maffioletti, E., Cloninger, C. R., Magri, C., Sartori, R., et al. (2013). Role of allelic variants of FK506-binding protein 51 (FKBP5) gene in the development of anxiety disorders. *Depression and Anxiety, 30*(12), 1170–1176.

Minett, T., Classey, J., Matthews, F. E., Fahrenhold, M., Taga, M., et al. (2016). Microglial immunophenotype in dementia with Alzheimer's pathology. *Journal of Neuroinflammation, 13*, 135.

Mischkowski, D., Crocker, J., & Way, B. M. (2016). From painkiller to empathy killer: Acetaminophen (paracetamol) reduces empathy for pain. *Social Cognitive and Affective Neuroscience, 11*(9), 1345–1353.

Mishra, S. K., & Hoon, M. A. (2013). The cells and circuitry for itch responses in mice. *Science, 340*(6135), 968–971.

Miskovic, V., & Schmidt, L. A. (2011). Social fearfulness in the human brain. *Neuroscience & Biobehavioral Reviews, 36*, 459–478.

Mitroulis, I., Skendros, P., & Ritis, K. (2010). Targeting IL-1beta in disease; the expanding role of NLRP3 inflammasome. *European Journal of Internal Medicine, 21*(3), 157–163.

Mitsi, V., Terzi, D., Purushothaman, I., Manouras, L., Gaspari, S., et al. (2015). RGS9-2--controlled adaptations in the striatum determine the onset of action and efficacy of antidepressants in neuropathic pain states. *Proceedings of the National Academy of Sciences of the United States of America, 112*, E5088–E5097.

Mitsui, V., & Zachariou, V. (2016). Modulation of pain, nociception, and analgesia by the brain reward center. *Neuroscience, 338*, 81–92.

Miyake, S., Kim, S., Suda, W., Oshima, K., Nakamura, M., et al. (2015). Dysbiosis in the gut microbiota of patients with multiple sclerosis, with a striking depletion of species belonging to clostridia XIVa and IV clusters. *PLoS One, 10*, e0137429.

Miyasaka, M., & Tanaka, T. (2004). Lymphocyte trafficking across high endothelial venules: Dogmas and enigmas. *Nature Reviews Immunology, 4*, 360–370.

Mizuno, T., Zhang, G., Takeuchi, H., Kawanokuchi, J., Wang, J., et al. (2008). Interferon-gamma directly induces neurotoxicity through a neuron specific, calcium-permeable complex of IFN-gamma receptor and AMPA GluR1 receptor. *FASEB Journal, 22*, 1797–1806.

Moayedi, M., & Davis, K. D. (2013). Theories of pain: From specificity to gate control. *Journal of Neurophysiology, 109*, 5–12.

Mochcovitch, M. D., da Rocha Freire, R. C., Garcia, R. F., & Nardi, A. E. (2014). A systematic review of fMRI studies in generalized anxiety disorder: Evaluating its neural and cognitive basis. *Journal of Affective Disorders, 167*, 336–342.

Moehle, M. S., Daher, J. P., Hull, T. D., Boddu, R., Abdelmotilib, H. A., Mobley, J., ... West, A. B. (2015). The G2019S LRRK2 mutation increases myeloid cell chemotactic responses and enhances LRRK2 binding to actin-regulatory proteins. *Human Molecular Genetics, 24*(15), 4250–4267.

Moehle, M. S., Webber, P. J., Tse, T., Sukar, N., Standaert, D. G., et al. (2012). LRRK2 inhibition attenuates microglial inflammatory responses. *Journal of Neuroscience, 32*, 1602–1611.

Mogil, J. S. (2016). Perspective: Equality need not be painful. *Nature, 535*, S7.

Mohr, D. C., Goodkin, D. E., Nelson, S., Cox, D., & Weiner, M. (2002). Moderating effects of coping on the relationship between stress and the development of new brain lesions in multiple sclerosis. *Psychosomatic Medicine, 64*(5), 803–809.

Mohr, D. C., Hart, S. L., Julian, L., Cox, D., & Pelletier, D. (2004). Association between stressful life events and exacerbation in multiple sclerosis: A meta-analysis. *British Medical Journal, 328*, 731.

Moisan, A., Lee, Y. K., Zhang, J. D., Hudak, C. S., Meyer, C. A., et al. (2015). White-to-brown metabolic conversion of human adipocytes by JAK inhibition. *Nature Cell Biology, 17*, 57–67.

Mokhtari, R., & Lachman, H. M. (2016). The major histocompatibility complex (MHC) in schizophrenia: A Review. *Journal of Clinical & Cellular Immunology, 7*.

Moldin, S. O., Rubenstein, J. L., & Hyman, S. E. (2006). Can autism speak to neuroscience? *Journal of Neuroscience, 26* (26), 6893–6896. Available from https://doi.org/10.1523/jneurosci.1944-06.2006.

Molendijk, M. L., Spinhoven, P., Polak, M., Bus, B. A., Penninx, B. W., et al. (2014). Serum bdnf concentrations as peripheral manifestations of depression: Evidence from a systematic review and meta-analyses on 179 associations (n = 9484). *Molecular Psychiatry, 19,* 791–800.

Möller, A. T., Bäckström, T., Nyberg, S., Söndergaard, H. P., & Helström, L. (2016). Women with PTSD have a changed sensitivity to GABA-A receptor active substances. *Psychopharmacology (Berlin, Germany), 233,* 2025–2033.

Molloy, C. A., Morrow, A. L., Meinzen-Derr, J., Schleifer, K., Dienger, K., Manning-Courtney, P., . . . Wills-Karp, M. (2006). Elevated cytokine levels in children with autism spectrum disorder. *Journal of Neuroimmunology, 172*(1-2), 198–205. Available from https://doi.org/10.1016/j.jneuroim.2005.11.007.

Molloy, D. W., Standish, T. I., Nieboer, E., Turnbull, J. D., Smith, S. D., et al. (2007). Effects of acute exposure to aluminum on cognition in humans. *Journal of Toxicology and Environmental Health. Part A, 70*(23), 2011–2019.

Moloney, R. D., Sajjad, J., Foley, T., Felice, V. D., Dinan, T. G., et al. (2016). Estrous cycle influences excitatory amino acid transport and visceral pain sensitivity in the rat: Effects of early-life stress. *Biology of Sex Differences, 7* (1), 33.

Mondelli, V., Vernon, A. C., Turkheimer, F., Dazzan, P., & Pariante, C. M. (2017). Brain microglia in psychiatric disorders. *The Lancet Psychiatry, 4*(7), 563–572.

Monji, A., Kato, T., & Kanba, S. (2009). Cytokines and schizophrenia: Microglia hypothesis of schizophrenia. *Psychiatry and Clinical Neurosciences, 63*(3), 257–265.

Monteleone, P., Catapano, F., Fabrazzo, M., Tortorella, A., & Maj, M. (1998). Decreased blood levels of tumor necrosis factor-alpha in patients with obsessive-compulsive disorder. *Neuropsychobiology, 37,* 182–185.

Montoya, J. G., Holmes, T. H., Anderson, J. N., Maecker, H. T., Rosenberg-Hasson, Y., et al. (2017). Cytokine signature associated with disease severity in chronic fatigue syndrome patients. *Proceedings of the National Academy of Sciences of the United States of America, 114,* E7150–E7158.

Moody, L., Chen, H., & Pan, Y. X. (2017). Postnatal diet remodels hepatic DNA methylation in metabolic pathways established by a maternal high-fat diet. *Epigenomics, 9,* 1387–1402.

Moon, M. L., Joesting, J. J., Blevins, N. A., Lawson, M. A., Gainey, S. J., et al. (2015). IL-4 knock out mice display anxiety-like behavior. *Behavior Genetics, 45,* 451–460.

Moons, W. G., Eisenberger, N. I., & Taylor, S. E. (2010). Anger and fear responses to stress have different biological profiles. *Brain, Behavior, and Immunity, 24,* 215–219.

Moons, W. G., & Shields, G. S. (2015). Anxiety, not anger, induces inflammatory activity: An avoidance/approach model of immune system activation. *Emotion, 15,* 463–476.

Moorad, J. A., & Walling, C. A. (2017). Measuring selection for genes that promote long life in a historical human population. *Nature Ecology & Evolution, 1*(11), 1773–1781.

Moore, R. A., Wiffen, P. J., Derry, S., Toelle, T., & Rice, A. S. (2014). Gabapentin for chronic neuropathic pain and fibromyalgia in adults. *Cochrane Database of Systematic Reviews,* CD007938.

Morath, J., Gola, H., Sommershof, A., Hamuni, G., Kolassa, S., et al. (2014). The effect of trauma-focused therapy on the altered T cell distribution in individuals with PTSD: Evidence from a randomized controlled trial. *Journal of Psychiatric Research, 54,* 1–10.

Morena, M., Leitl, K. D., Vecchiarelli, H. A., Gray, J. M., Campolongo, P., et al. (2016a). Emotional arousal state influences the ability of amygdalar endocannabinoid signaling to modulate anxiety. *Neuropharmacology, 111,* 59–69.

Morena, M., Patel, S., Bains, J. S., & Hill, M. N. (2016b). Neurobiological interactions between stress and the endocannabinoid system. *Neuropsychopharmacology, 41,* 80–102.

Moretti, S., Franchi, S., Castelli, M., Amodeo, G., Somaini, L., et al. (2015). Exposure of adolescent mice to delta-9-tetrahydrocannabinol induces long-lasting modulation of pro- and anti-inflammatory cytokines in hypothalamus and hippocampus similar to that observed for peripheral macrophages. *Journal of Neuroimmune Pharmacology, 10,* 371–379.

Morgan, B. P. (2015). The role of complement in neurological and neuropsychiatric diseases. *Expert Review of Clinical Immunology, 11*(10), 1109–1119. Available from https://doi.org/10.1586/1744666x.2015.1074039.

Morgan, C. A., 3rd, Wang, S., Southwick, S. M., Rasmusson, A., Hazlett, G., et al. (2000). Plasma neuropeptide-Y concentrations in humans exposed to military survival training. *Biological Psychiatry, 47,* 902–909.

Morgan, C. P., & Bale, T. L. (2011). Early prenatal stress epigenetically programs dysmasculinization in second-generation offspring via the paternal lineage. *Journal of Neuroscience, 31,* 11748–11755.

Moroni, G., & Ponticelli, C. (2016). Pregnancy in women with systemic lupus erythematosus (SLE). *European Journal of Internal Medicine, 32,* 7–12. Available from https://doi.org/10.1016/j.ejim.2016.04.005.

Morreall, J., Kim, A., Liu, Y., Degtyareva, N., Weiss, B., et al. (2015). Evidence for retromutagenesis as a mechanism for adaptive mutation in Escherichia coli. *PLoS Genetics, 11*, e1005477.

Morris, G., Berk, M., Carvalho, A., Caso, J. R., Sanz, Y., et al. (2016). The role of the microbial metabolites including tryptophan catabolites and short chain fatty acids in the pathophysiology of immune-inflammatory and neuroimmune disease. *Molecular Neurobiology, 54*, 4432–4451.

Morris, G., Berk, M., Galecki, P., & Maes, M. (2014). The emerging role of autoimmunity in myalgic encephalomyelitis/chronic fatigue syndrome (ME/ cfs). *Molecular Neurobiology, 49*, 741–756.

Morris, G., Berk, M., Klein, H., Walder, K., Galecki, P., et al. (2016). Nitrosative stress, hypernitrosylation, and autoimmune responses to nitrosylated proteins: New pathways in neuroprogressive disorders including depression and chronic fatigue syndrome. *Molecular Neurobiology, 54*, 4271–4291.

Morris, M. J., Beilharz, J. E., Maniam, J., Reichelt, A. C., Westbrook, R. F., et al. (2016). Why is obesity such a problem in the 21st century? The intersection of palatable food, cues and reward pathways, stress, and cognition. *Neuroscience & Biobehavioral Reviews, 58*, 36–45.

Moskowitz, A., & Heim, G. (2011). Eugen Bleuler's Dementia praecox or the group of schizophrenias (1911): A centenary appreciation and reconsideration. *Schizophrenia Bulletin, 37*(3), 471–479.

Mostafa, G. A., & Shehab, A. A. (2010). The link of C4B null allele to autism and to a family history of autoimmunity in Egyptian autistic children. *Journal of Neuroimmunology, 223*(1-2), 115–119. Available from https://doi.org/10.1016/j.jneuroim.2010.03.025.

Mostafalou, S., & Abdollahi, M. (2017). Pesticides: An update of human exposure and toxicity. *Archives of Toxicology, 91*, 549–599.

Motivala, S., & Irwin, M. R. (2007). Sleep and immunity: Cytokine pathways linking sleep and health outcomes. *Current Directions in Psychological Science, 16*, 21–25.

Mou, Z., Hyde, T. M., Lipska, B. K., Martinowich, K., Wei, P., et al. (2015). Human obesity associated with an intronic SNP in the brain-derived neurotrophic factor locus. *Cell Reports, 13*, 1073–1080.

Mouihate, A. (2013). Long-lasting impact of early life immune stress on neuroimmune functions. *Medical Principles and Practice, 22*(Suppl. 1), 3–7.

Mount, M. P., Lira, A., Grimes, D., Smith, P. D., Faucher, S., et al. (2007). Involvement of interferon-gamma in microglial-mediated loss of dopaminergic neurons. *Journal of Neuroscience, 27*, 3328–3337.

Moussaoui, N., Jacobs, J. P., Larauche, M., Biraud, M., Million, M., et al. (2017). Chronic early-life stress in rat pups alters basal corticosterone, intestinal permeability, and fecal microbiota at weaning: Influence of sex. *Journal of Neurogastroenterology and Motility, 23*, 135–143.

Moya-Pérez, A., Perez-Villalba, A., Benítez-Páez, A., Campillo, I., Sanz, Y., et al. (2017). Bifidobacterium CECT 7765 modulates early stress-induced immune, neuroendocrine and behavioral alterations in mice. *Brain, Behavior, and Immunity, 65*, 43–56.

Mrak, R. E., & Griffin, W. S. T. (2007). Common inflammatory mechanisms in Lewy body disease and Alzheimer disease. *Journal of Neuropathology & Experimental Neurology, 66*, 683–686.

Mucida, D., Park, Y., Kim, G., Turovskaya, O., Scott, I., et al. (2007). Reciprocal TH17 and regulatory T cell differentiation mediated by retinoic acid. *Science, 317*, 256–260.

Mugambi, M. N., Musekiwa, A., Lombard, M., Young, T., & Blaauw, R. (2012). Synbiotics, probiotics or prebiotics in infant formula for full term infants: A systematic review. *Nutrition Journal, 11*, 81.

Muller, M. B., Zimmermann, S., Sillaber, I., Hagemeyer, T. P., Deussing, J. M., et al. (2003). Limbic corticotropin-releasing hormone receptor 1 mediates anxiety-related behavior and hormonal adaptation to stress. *Nature Neuroscience, 6*, 1100–1107.

Muller, R. A., Shih, P., Keehn, B., Deyoe, J. R., Leyden, K. M., & Shukla, D. K. (2011). Underconnected, but how? A survey of functional connectivity MRI studies in autism spectrum disorders. *Cerebral Cortex, 21*(10), 2233–2243. Available from https://doi.org/10.1093/cercor/bhq296.

Mullins, N., & Lewis, C. M. (2017). Genetics of depression: Progress at last. *Current Psychiatry Reports, 19*(8), 43.

Murdock, K. W., LeRoy, A. S., Lacourt, T. E., Duke, D. C., Heijnen, C. J., et al. (2016). Executive functioning and diabetes: The role of anxious arousal and inflammation. *Psychoneuroendocrinology, 71*, 102–109.

Murdock, K. W., Stowe, R. P., Peek, M. K., Lawrence, S. L., & Fagundes, C. P. (2017). An elevation of perceived health risk and depressive symptoms prior to a disaster in predicting post-disaster inflammation. *Psychosomatic Medicine, 80*(1), 49–54.

Murphy, T. H., & Corbett, D. (2009). Plasticity during stroke recovery: From synapse to behaviour. *Nature Reviews Neuroscience, 10*, 861–872.

Murphy, T. M., O'Donovan, A., Mullins, N., O'Farrelly, C., McCann, A., et al. (2015). Anxiety is associated with higher levels of global DNA methylation and altered expression of epigenetic and interleukin-6 genes. *Psychiatric Genetics, 25*, 71–78.

Murrough, J. W., Abdallah, C. G., & Mathew, S. J. (2017). Targeting glutamate signalling in depression: Progress and prospects. *Nature Reviews Drug Discovery, 16*, 472–486.

Murrough, J. W., Soleimani, L., DeWilde, K. E., Collins, K. A., Lapidus, K. A., et al. (2015). Ketamine for rapid reduction of suicidal ideation: A randomized controlled trial. *Psychological Medicine, 45*, 3571–3580.

Musaelyan, K., Egeland, M., Fernandes, C., Pariante, C. M., Zunszain, P. A., et al. (2014). Modulation of adult hippocampal neurogenesis by early-life environmental challenges triggering immune activation. *Neural Plasticity, 2014*, 194396.

Musardo, S., & Marcello, E. (2017). Synaptic dysfunction in Alzheimer's disease: From the role of amyloid β-peptide to the α-secretase ADAM10. *European Journal of Pharmacology, 817*, 30–37.

Musazzi, L., Tornese, P., Sala, N., & Popoli, M. (2014). Acute stress is not acute: Sustained enhancement of glutamate release after acute stress involves readily releasable pool size and synapsin I activation. *Molecular Psychiatry, 22*, 1226–1227.

Musazzi, L., Treccani, G., Mallei, A., & Popoli, M. (2013). The action of antidepressants on the glutamate system: Regulation of glutamate release and glutamate receptors. *Biological Psychiatry, 73*, 1180–1188.

Musunuru, K., & Kathiresan, S. (2017). Cardiovascular endocrinology: Is ANGPTL3 the next PCSK9? *Nature Reviews Endocrinology, 13*, 503–504.

Muto, V., Jaspar, M., Meyer, C., Kussé, C., Chellappa, S. L., et al. (2016). Local modulation of human brain responses by circadian rhythmicity and sleep debt. *Science, 353*, 687–690.

Mychasiuk, R., Schmold, N., Ilnytskyy, S., Kovalchuk, O., Kolb, B., et al. (2011). Prenatal bystander stress alters brain, behavior, and the epigenome of developing rat offspring. *Developmental Neuroscience, 33*, 159–169.

Myers, B. (2017). Corticolimbic regulation of cardiovascular responses to stress. *Physiology & Behavior, 172*, 49–59.

Naß, J., & Efferth, T. (2017). Pharmacogenetics and pharmacotherapy of military personnel suffering from posttraumatic stress disorder. *Current Neuropharmacology, 15*, 831–860.

Nadeem, A., Ahmad, S. F., Attia, S. M., Bakheet, S. A., Al-Harbi, N. O., & Al-Ayadhi, L. Y. (2018). Activation of IL-17 receptor leads to increased oxidative inflammation in peripheral monocytes of autistic children. *Brain, Behavior, and Immunity, 67*, 335–344. Available from https://doi.org/10.1016/j.bbi.2017.09.010.

Nader, K. (2015). Reconsolidation and the dynamic nature of memory. *Cold Spring Harbor Perspectives in Biology, 7* (10), a021782.

Nader, K., Hardt, O., & Lanius, R. (2013). Memory as a new therapeutic target. *Dialogues in Clinical Neuroscience, 15*, 475–486.

Nader, K., Schafe, G. E., & LeDoux, J. E. (2000). The labile nature of consolidation theory. *Nature, 1*, 216–219.

Naegeli, C., Zeffiro, T., Piccirelli, M., Jaillard, A., Weilenmann, A., et al. (2017). Locus coeruleus activity mediates hyper-responsiveness in posttraumatic stress disorder. *Biological Psychiatry, 83*, 254–262.

Nagele, E. P., Han, M., Acharya, N. K., DeMarshall, C., Kosciuk, M. C., et al. (2013). Natural IgG autoantibodies are abundant and ubiquitous in human sera, and their number is influenced by age, gender, and disease. *PLoS One, 8*, e60726.

Nagy, B. M., Nagaraj, C., Meinitzer, A., Sharma, N., Papp, R., et al. (2017). Importance of kynurenine in pulmonary hypertension. *American Journal of Physiology: Lung Cellular and Molecular Physiology, 313*, L741–L751.

Nair, S., Jao Keehn, R. J., Berkebile, M. M., Maximo, J. O., Witkowska, N., & Muller, R. A. (2017). Local resting state functional connectivity in autism: Site and cohort variability and the effect of eye status. *Brain Imaging and Behavior, 12*(1), 168–179. Available from https://doi.org/10.1007/s11682-017-9678-y.

Nallu, A., Sharma, S., Ramezani, A., Muralidharan, J., & Raj, D. (2017). Gut microbiome in chronic kidney disease: challenges and opportunities. *Translational Research, 179*, 24–37.

Naneix, F., Tantot, F., Glangetas, C., Kaufling, J., Janthakhin, Y., et al. (2017). Impact of early consumption of high-fat diet on the mesolimbic dopaminergic system. *eNeuro, 4*, pii: ENEURO.0120-17.2017.

Naninck, E. F., Oosterink, J. E., Yam, K. Y., de Vries, L. P., Schierbeek, H., et al. (2017). Early micronutrient supplementation protects against early stress-induced cognitive impairments. *FASEB Journal: Official Publication of the Federation of American Societies for Experimental Biology, 31*, 505–518.

Narayan, S., Sinsheimer, J. S., Paul, K. C., Liew, Z., Cockburn, M., et al. (2015). Genetic variability in ABCB1, occupational pesticide exposure, and Parkinson's disease. *Environmental Research, 143*, 98–106.

Nasca, C., Zelli, D., Bigio, B., Piccinin, S., Scaccianoce, S., et al. (2015). Stress dynamically regulates behavior and glutamatergic gene expression in hippocampus by opening a window of epigenetic plasticity. *Proceedings of the National Academy of Sciences of the United States of America, 112*, 14960–14965.

Nash, J. R., Sargent, P. A., Rabiner, E. A., Hood, S. D., Argyropoulos, S. V., et al. (2008). Serotonin 5-HT1A receptor binding in people with panic disorder: Positron emission tomography study. *The British Journal of Psychiatry, 193*, 229–234.

Natarajan, P., & Kathiresan, S. (2016). PCSK9. *Cell, 165*, 1037.

Natarajan, R., Forrester, L., Chiaia, N. L., & Yamamoto, B. K. (2017). Chronic-stress-induced behavioral changes associated with subregion-selective serotonin cell death in the dorsal raphe. *Journal of Neuroscience, 37,* 6214–6223.

Natrajan, R., Sailem, H., Mardakheh, F. K., Arias Garcia, M., Tape, C. J., et al. (2016). Microenvironmental heterogeneity parallels breast cancer progression: A histology-genomic integration analysis. *PLoS Medicine, 13,* e1001961.

Nazimek, K., Strobel, S., Bryniarski, P., Kozlowski, M., Filipczak-Bryniarska, I., et al. (2016). The role of macrophages in anti-inflammatory activity of antidepressant drugs. *Immunobiology, 222,* 823–830.

Needham, B. D., Tang, W., & Wu, W. L. (2018). Searching for the gut microbial contributing factors to social behavior in rodent models of autism spectrum disorder. *Developmental Neurobiology, 78*(5), 474–499. Available from https://doi.org/10.1002/dneu.22581, [Epub ahead of print].

Nelson, M. D., & Tumpap, A. M. (2016). Posttraumatic stress disorder symptom severity is associated with left hippocampal volume reduction: A meta-analytic study. *CNS Spectrums, 22,* 363–372.

Nemeroff, C. B., & Vale, W. (2005). The neurobiology of depression: Inroads to treatment and new drug discovery. *The Journal of Clinical Psychiatry, 66,* 5–13.

Netea, M. G., Latz, E., Mills, K. H., & O'Neill, L. A. (2015). Innate immune memory: A paradigm shift in understanding host defense. *Nature Immunology, 16,* 675–679.

Neufeld, K. M., Kang, N., Bienenstock, J., & Foster, J. A. (2011a). Reduced anxiety-like behavior and central neurochemical change in germ-free mice. *Neurogastroenterol Motil, 23,* 255–264, e119.

Neufeld, K. A. M., Kang, N., Bienenstock, J., & Foster, J. A. (2011b). Effects of intestinal microbiota on anxiety-like behavior. *Communicative & Integrat Biol, 4,* 492–494.

Neumann, H., & Takahashi, K. (2007). Essential role of the microglial triggering receptor expressed on myeloid cells-2 (TREM2) for central nervous tissue immune homeostasis. *Journal of Neuroimmunology, 184,* 92–99.

Neumann, I. D., & Slattery, D. A. (2016). Oxytocin in general anxiety and social fear: A translational approach. *Biological Psychiatry, 79,* 213–221.

Newton, T. L., Fernandez-Botran, R., Miller, J. J., & Burns, V. E. (2014). Interleukin-6 and soluble interleukin-6 receptor levels in posttraumatic stress disorder: Associations with lifetime diagnostic status and psychological context. *Biological Psychology, 99,* 150–159.

Ng, W. L., & Bassler, B. L. (2009). Bacterial quorum-sensing network architectures. *Annual Review of Genetics, 43,* 197–222.

Nguyen, J. P., Nizard, J., Keravel, Y., & Lefaucheur, J. P. (2011). Invasive brain stimulation for the treatment of neuropathic pain. *Nature Reviews Neurology, 7,* 699–709.

Nicholson, E. L., Bryant, R. A., & Felmingham, K. L. (2014). Interaction of noradrenaline and cortisol predicts negative intrusive memories in posttraumatic stress disorder. *Neurobiology of Learning and Memory, 112,* 204–211.

Nicoll, J. A. R., Wilkinson, D., Holmes, C., Steart, P., Markham, H., et al. (2003). Neuropathology of human Alzheimer disease after immunization with amyloid-beta peptide: A case report. *Nature Medicine (New York, NY, United States), 9,* 448–452.

Nicolucci, A. C., Hume, M. P., Martínez, I., Mayengbam, S., Walter, J., et al. (2017). Prebiotic reduces body fat and alters intestinal microbiota in children with overweight or obesity. *Gastroenterology, 153,* 711–722.

Niculescu, A. B., Le-Niculescu, H., Levey, D. F., Phalen, P. L., Dainton, H. L., et al. (2017). Precision medicine for suicidality: From universality to subtypes and personalization. *Molecular Psychiatry, 22,* 1250–1273.

Nie, B., Nie, T., Hui, X., Gu, P., Mao, L., et al. (2017). Brown adipogenic reprogramming induced by a small molecule. *Cell Reports, 18,* 624–635.

Nikolaus, S., Antke, C., Beu, M., & Müller, H. W. (2010). Cortical GABA, striatal dopamine and midbrain serotonin as the key players in compulsive and anxiety disorders--Results from in vivo imaging studies. *Reviews of Neuroscience, 21,* 119–139.

Niwa, M., Kamiya, A., Murai, R., Kubo, K., Gruber, A. J., et al. (2010). Knockdown of DISC1 by in utero gene transfer disturbs postnatal dopaminergic maturation in the frontal cortex and leads to adult behavioral deficits. *Neuron, 65,* 480–489.

Njau, S., Joshi, S. H., Espinoza, R., Leaver, A. M., Vasavada, M., et al. (2017). Neurochemical correlates of rapid treatment response to electroconvulsive therapy in patients with major depression. *Journal of Psychiatry & Neuroscience, 42,* 6–16.

Nolen-Hoeksema, S. (1998). Ruminative coping with depression. In J. Heckhausen, & C. S. Dweck (Eds.), *Motivation and self-regulation across the life span* (pp. 237–256). Cambridge, UK: Cambridge University Press.

Noonan, K., Carey, L. M., & Crewther, S. G. (2013). Meta-analyses indicate associations between neuroendocrine activation, deactivation in neurotrophic and neuroimaging markers in depression after stroke. *Journal of Stroke & Cerebrovascular Diseases, 22,* e124–e135.

Nordmann, T. M., Dror, E., Schulze, F., Traub, S., Berishvili, E., et al. (2017). The role of inflammation in β-cell dedifferentiation. *Scientific Reports, 7,* 6285.

Northoff, G. (2013). *Unlocking the brain: Volume 2.* New York: Oxford University Press.

Northoff, G., & Sibille, E. (2014). Why are cortical GABA neurons relevant to internal focus in depression? A cross-level model linking cellular, biochemical and neural network findings. *Molecular Psychiatry, 19,* 966–977.

Notzon, S., Domschke, K., Holitschke, K., Ziegler, C., Arolt, V., et al. (2016). Attachment style and oxytocin receptor gene variation interact in influencing social anxiety. *The World Journal of Biological Psychiatry, 17*, 76–83.

Novak, P., Schmidt, R., Kontsekova, E., Zilka, N., Kovacech, B., et al. (2017). Safety and immunogenicity of the tau vaccine AADvac1 in patients with Alzheimer's disease: A randomised, double-blind, placebo-controlled, phase 1 trial. *Lancet Neurology, 16*, 123–134.

Nowacka, M. M., Paul-Samojedny, M., Bielecka, A. M., & Obuchowicz, E. (2014). Chronic social instability stress enhances vulnerability of BDNF response to LPS in the limbic structures of female rats: A protective role of antidepressants. *Neurosciences Research, 88*, 74–88.

Nowak, G. J., Sheedy, K., Bursey, K., Smith, T. M., & Basket, M. (2015). Promoting influenza vaccination: Insights from a qualitative meta-analysis of 14 years of influenza-related communications research by US Centers for Disease Control and Prevention (CDC). *Vaccine, 33*, 2741–2756.

Nuss, P. (2015). Anxiety disorders and GABA neurotransmission: A disturbance of modulation. *Neuropsychiatric Disease and Treatment, 11*, 165–175.

Nygren, M., Carstensen, J., Koch, F., Ludvigsson, J., & Frostell, A. (2015). Experience of a serious life event increases the risk for childhood type 1 diabetes: The ABIS population-based prospective cohort study. *Diabetologia, 58*, 1188–1197.

O'Connor, J. C., Lawson, M. A., Andre, C., Moreau, M., Lestage, J., et al. (2009). Lipopolysaccharide-induced depressive-like behavior is mediated by indoleamine 2, 3-dioxygenase activation in mice. *Molecular Psychiatry, 14*, 511–522.

Odio, M., Brodish, A., & Ricardo, M. J., Jr. (1987). Effects on immune responses by chronic stress are modulated by aging. *Brain, Behavior, and Immunity, 1*, 204–215.

O'Doherty, D. C., Chitty, K. M., Saddiqui, S., Bennett, M. R., & Lagopoulos, J. (2015). A systematic review and meta-analysis of magnetic resonance imaging measurement of structural volumes in posttraumatic stress disorder. *Psychiatry Research, 232*, 1–33.

O'Doherty, D. C. M., Ryder, W., Paquola, C., Tickell, A., Chan, C., et al. (2018). White matter integrity alterations in post-traumatic stress disorder. *Human Brain Mapping, 39*(3), 1327–1338.

O'Doherty, D. C. M., Tickell, A., Ryder, W., Chan, C., Hermens, D. F., et al. (2017). Frontal and subcortical grey matter reductions in PTSD. *Psychiatry Research, 266*, 1–9.

O'Donnell, M. C., Catts, S. V., Ward, P. B., Liebert, B., Lloyd, A., Wakefield, D., & McConaghy, N. (1996). Increased production of interleukin-2 (IL-2) but not soluble interleukin-2 receptors (sIL-2R) in unmedicated patients with schizophrenia and schizophreniform disorder. *Psychiatry Research, 65*(3), 171–178.

O'Donovan, A., Ahmadian, A. J., Neylan, T. C., Pacult, M. A., Edmondson, D., et al. (2017). Current posttraumatic stress disorder and exaggerated threat sensitivity associated with elevated inflammation in the your heart study. *Brain, Behavior, and Immunity, 60*, 198–205.

O'Donovan, A., Hughes, B. M., Slavich, G. M., Lynch, L., Cronin, M. T., et al. (2010). Clinical anxiety, cortisol and interleukin-6: Evidence for specificity in emotion-biology relationships. *Brain, Behavior, and Immunity, 24*, 1074–1077.

O'Donovan, A., Neylan, T. C., Metzler, T., & Cohen, B. E. (2012). Lifetime exposure to traumatic psychological stress is associated with elevated inflammation in the heart and soul study. *Brain, Behavior, and Immunity, 26*, 642–649.

O'Donovan, A., Sun, B., Cole, S., Rempel, H., Lenoci, M., et al. (2011). Transcriptional control of monocyte gene expression in post-traumatic stress disorder. *Disease Markers, 30*, 123–132.

Oftedal, B. E., Hellesen, A., Erichsen, M. M., Bratland, E., Vardi, A., et al. (2015). Dominant mutations in the autoimmune regulator AIRE are associated with common organ-specific autoimmune diseases. *Immunity, 42*, 1185–1196.

Oh, J., McCloskey, M. A., Blong, C. C., Bendickson, L., Nilsen-Hamilton, M., et al. (2010). Astrocyte-derived interleukin-6 promotes specific neuronal differentiation of neural progenitor cells from adult hippocampus. *Journal of Neuroscience Research, 88*, 2798–2809.

Ohara, T., Doi, Y., Ninomiya, T., Hirakawa, Y., Hata, J., Iwaki, T., et al. (2011). Glucose tolerance status and risk of dementia in the community: The Hisayama Study. *Neurology, 77*, 1126–1134.

Ohsawa, K., Imai, Y., Sasaki, Y., & Kohsaka, S. (2004). Microglia/macrophage-specific protein Iba1 binds to fimbrin and enhances its actin-bundling activity. *Journal of Neurochemistry, 88*, 844–856.

Okada, H., Kuhn, C., Feillet, H., & Bach, J. F. (2010). The "hygiene hypothesis" for autoimmune and allergic diseases: An update. *Clinical and Experimental Immunology, 160*, 1–9.

O'Keane, V., Lightman, S., Patrick, K., Marsh, M., Papadopoulos, A. S., et al. (2013). Changes in the maternal hypothalamic-pituitary-adrenal axis during the early puerperium may be related to the postpartum "blues". *Journal of Neuroendocrinology, 23*, 1149–1155.

Oksanen, M., Petersen, A. J., Naumenko, N., Puttonen, K., Lehtonen, Š., et al. (2017). PSEN1 mutant iPSC-derived model reveals severe astrocyte pathology in Alzheimer's disease. *Stem Cell Reports, 9*(6), 1885–1897.

Okun, E., Mattson, M. P., & Arumugam, T. V. (2010). Involvement of Fc receptors in disorders of the central nervous system. *Neuromolecular Medicine, 12*(2), 164–178.

Old, E. A., Clark, A. K., & Malcangio, M. (2015). The role of glia in the spinal cord in neuropathic and inflammatory pain. *Handbook of Experimental Pharmacology, 227*, 145–170.

Olofsson, P. S., Rosas-Ballina, M., Levine, Y. A., & Tracey, K. J. (2012). Rethinking inflammation: Neural circuits in the regulation of immunity. *Immunological Reviews, 248*, 188–204.

Olson, J. M., Goddard, K. A., & Dudek, D. M. (2001). The amyloid precursor protein locus and very-late-onset Alzheimer disease. *American Journal of Human Genetics, 69*, 895–899.

Olson, K. E., Kosloski-Bilek, L. M., Anderson, K. M., Diggs, B. J., Clark, B. E., et al. (2015). Selective VIP receptor agonists facilitate immune transformation for dopaminergic neuroprotection in MPTP-intoxicated mice. *Journal of Neuroscience, 35*, 16463–16478.

Olson, V. G., Rockett, H. R., Reh, R. K., Redila, V. A., Tran, P. M., et al. (2011). The role of norepinephrine in differential response to stress in an animal model of posttraumatic stress disorder. *Biological Psychiatry, 70*(5), 441–448.

Oluboka, O. J., Katzman, M. A., Habert, J., McIntosh, D., MacQueen, G. M., et al. (2017). Functional recovery in major depressive disorder: Providing early optimal treatment for the individual patient. *International Journal of Neuropsychopharmacology, 21*, 128–144.

O'Mahony, S. M., Clarke, G., Dinan, T. G., & Cryan, J. F. (2017a). Early-life adversity and brain development: Is the microbiome a missing piece of the puzzle? *Neuroscience, 342*, 37–54.

O'Mahony, S. M., Clarke, G., Dinan, T. G., & Cryan, J. F. (2017b). Irritable bowel syndrome and stress-related psychiatric co-morbidities: Focus on early life stress. *Handbook of Experimental Pharmacology, 239*, 219–246.

O'Mahony, S. M., Dinan, T. G., & Cryan, J. F. (2017). The gut microbiota as a key regulator of visceral pain. *Pain, 158*, S19–S28.

O'Mahony, S. M., Felice, V. D., Nally, K., Savignac, H. M., Claesson, M. J., et al. (2014). Disturbance of the gut microbiota in early-life selectively affects visceral pain in adulthood without impacting cognitive or anxiety-related behaviors in male rats. *Neuroscience, 277*, 885–901.

O'Mahony, S. M., Felice, V. D., Nally, K., Savignac, H. M., Claesson, M. J., Scully, P., et al. (2014). Disturbance of the gut microbiota in early-life selectively affects visceral pain in adulthood without impacting cognitive or anxiety-related behaviors. *Neuroscience, 277*, 885–901.

O'Mahony, S. M., Hyland, N. P., Dinan, T. G., & Cryan, J. F. (2011). Maternal separation as a model of brain-gut axis dysfunction. *Psychopharmacology (Berlin, Germany), 214*, 71–88.

Oosterhof, C. A., El Mansari, M., Merali, Z., & Blier, P. (2016). Altered monoamine system activities after prenatal and adult stress: A role for stress resilience? *Brain Research, 1642*, 409–418.

Orenstein, S. J., Kuo, S. H., Tasset, I., Arias, E., Koga, H., et al. (2013). Interplay of LRRK2 with chaperone-mediated autophagy. *Nature Neuroscience, 16*, 394–406.

Orlovska, S., Vestergaard, C. H., Bech, B. H., Nordentoft, M., Vestergaard, M., et al. (2017). Association of streptococcal throat infection with mental disorders: Testing key aspects of the PANDAS hypothesis in a nationwide study. *JAMA Psychiatry, 74*, 740–746.

Ormerod, B. K., Hanft, S. J., Asokan, A., Haditsch, U., Lee, S. W., et al. (2013). PPARγ activation prevents impairments in spatial memory and neurogenesis following transient illness. *Brain, Behavior, and Immunity, 29*, 28–38.

Ornoy, A., Weinstein-Fudim, L., & Ergaz, Z. (2016). Genetic syndromes, maternal diseases and antenatal factors associated with autism spectrum disorders (ASD). *Frontiers in Neuroscience, 10*, 316. Available from https://doi.org/10.3389/fnins.2016.00316.

Orr, C. F., Rowe, D. B., Mizuno, Y., Mori, H., & Halliday, G. M. (2005). A possible role for humoral immunity in the pathogenesis of Parkinson's disease. *Brain: A Journal of Neurology, 128*, 2665–2674.

Orsolini, L., Sarchione, F., Vellante, F., Fornaro, M., Matarazzo, I., et al. (2018). Protein-C reactive as biomarker predictor of schizophrenia phases of illness? A systematic review. *Current Neuropharmacology*.

Osborn, M., Rustom, N., Clarke, M., Litteljohn, D., Rudyk, C., et al. (2013). Antidepressant-like effects of erythropoietin: A focus on behavioural and hippocampal processes. *PLoS One, 8*, e72813.

Ott, B., Skurk, T., Hastreiter, L., Lagkouvardos, I., & Fischer, S. (2017). Effect of caloric restriction on gut permeability, inflammation markers, and fecal microbiota in obese women. *Scientific Reports, 7*, 11955.

Ozsoy, S., Olguner Eker, O., & Abdulrezzak, U. (2016). The effects of antidepressants on neuropeptide Y in patients with depression and anxiety. *Pharmacopsychiatry, 49*, 26–31.

Pabst, O., & Mowat, A. M. (2012). Oral tolerance to food protein. *Mucosal Immunology, 5*(3), 232–239. Available from https://doi.org/10.1038/mi.2012.4.

Pace, T. W., & Heim, C. M. (2011). A short review on the psychoneuroimmunology of posttraumatic stress disorder: From risk factors to medical comorbidities. *Brain, Behavior, and Immunity, 25*(1), 6–13.

Pace, T. W., Hu, F., & Miller, A. H. (2007). Cytokine-effects on glucocorticoid receptor function: Relevance to glucocorticoid resistance and the pathophysiology and treatment of major depression. *Brain, Behavior, and Immunity, 21*, 9−19.

Pace, T. W., Wingenfeld, K., Schmidt, I., Meinlschmidt, G., Hellhammer, D. H., et al. (2012). Increased peripheral NF-κB pathway activity in women with childhood abuse-related posttraumatic stress disorder. *Brain, Behavior, and Immunity, 26*, 13−17.

Padro, C. J., McAlees, J. W., & Sanders, V. M. (2013). Regulation of immune cell activity by norepinephrine and beta2-adrenergic receptor engagement. In A. W. Kusnecov, & H. Anisman (Eds.), *The Wiley-Blackwell handbook of psychoneuroimmunology*. Chichester, UK: John Wiley & Sons Ltd.

Pallavi, P., Sagar, R., Mehta, M., Sharma, S., Subramanium, A., et al. (2015). Serum cytokines and anxiety in adolescent depression patients: Gender effect. *Psychiatry Research, 229*, 374−380.

Palmqvist, S., Schöll, M., Strandberg, O., Mattsson, N., Stomrud, E., et al. (2017). Earliest accumulation of β-amyloid occurs within the default-mode network and concurrently affects brain connectivity. *Nature Communications, 8*, 1214.

Pałucha-Poniewiera, A., & Pilc, A. (2016). Glutamate-based drug discovery for novel antidepressants. *Expert Opinion on Drug Discovery, 11*, 873−883.

Pandey, G. N., Ren, X., Rizavi, H. S., & Zhang, H. (2015). Abnormal gene expression of proinflammatory cytokines and their receptors in the lymphocytes of patients with bipolar disorder. *Bipolar Disorders, 17*, 636−644.

Pandey, G. N., Rizavi, H. S., Ren, X., Bhaumik, R., & Dwivedi, Y. (2014). Toll-like receptors in the depressed and suicide brain. *Journal of Psychiatric Research, 53*, 62−68.

Pandey, G. N., Rizavi, H. S., Ren, X., Fareed, J., Hoppensteadt, D. A., et al. (2011). Proinflammatory cytokines in the prefrontal cortex of teenage suicide victims. *Journal of Psychiatric Research, 46*(1), 57−63.

Paolicelli, R. C., Bolasco, G., Pagani, F., Maggi, L., Scianni, M., et al. (2011). Synaptic pruning by microglia is necessary for normal brain development. *Science, 333*(6048), 1456−1458.

Papachroni, K. K., Ninkina, N., Papapanagiotou, A., Hadjigeorgiou, G. M., Xiromerisiou, G., et al. (2007). Autoantibodies to alpha-synuclein in inherited Parkinson's disease. *Journal of Neurochemistry, 101*(3), 749−756.

Papini, S., Sullivan, G. M., Hien, D. A., Shvil, E., & Neria, Y. (2015). Toward a translational approach to targeting the endocannabinoid system in posttraumatic stress disorder: A critical review of preclinical research. *Biological Psychology, 104*, 8−18.

Papuć, E., Kurys-Denis, E., Krupski, W., & Rejdak, K. (2015). Humoral response against small heat shock proteins in Parkinson's disease. *PLoS One, 10*(1), e0115480.

Paradis, J., Boureau, P., Moyon, T., Nicklaus, S., Parnet, P., et al. (2017). Perinatal western diet consumption leads to profound plasticity and GABAergic phenotype changes within hypothalamus and reward pathway from birth to sexual maturity in rat. *Frontiers in Endocrinology, 8*, 216.

Paragh, G., Seres, I., Harangi, M., & Fülöp, P. (2014). Dynamic interplay between metabolic syndrome and immunity. *Advances in Experimental Medicine and Biology, 824*, 171−190.

Parbo, P., Ismail, R., Hansen, K. V., Amidi, A., Mårup, F. H., et al. (2017). Brain inflammation accompanies amyloid in the majority of mild cognitive impairment cases due to Alzheimer's disease. *Brain: A Journal of Neurology, 140*(7), 2002−2011.

Park, C. L. (2010). Making sense of the meaning literature: An integrative review of meaning making and its effects on adjustment to stressful life events. *Psychological Bulletin, 136*(2), 257−301.

Park, D., Joo, S. S., Kim, T. K., Lee, S. H., Kang, H., et al. (2012). Human neural stem cells overexpressing choline acetyltransferase restore cognitive function of kainic acid-induced learning and memory deficit animals. *Cell Transplantation, 21*(1), 365−371.

Park, M., Katon, W. J., & Wolf, F. M. (2013). Depression and risk of mortality in individuals with diabetes: A meta-analysis and systematic review. *General Hospital Psychiatry, 35*(3), 217−225.

Parker, K. E., McCabe, M. P., Johns, H. W., Lund, D. K., Odu, F., et al. (2015). Neural activation patterns underlying basolateral amygdala influence on intra-accumbens opioid-driven consummatory versus appetitive high-fat feeding behaviors in the rat. *Behav NeurosciBehavioralNeuroscience, 129*(6), 812−821.

Parrott, A. C. (2014). The potential dangers of using MDMA for psychotherapy. *Journal of Psychoactive Drugs, 46*(1), 37−43.

Partridge, J. G., Forcelli, P. A., Luo, R., Cashdan, J. M., Schulkin, J., et al. (2016). Stress increases GABAergic neurotransmission in CRF neurons of the central amygdala and bed nucleus stria terminalis. *Neuropharmacology, 107*, 239−250.

Pase, M. P., Himali, J. J., Beiser, A. S., Aparicio, H. J., Satizabal, C. L., et al. (2017). Sugar- and artificially sweetened beverages and the risks of incident stroke and dementia: A prospective cohort study. *Stroke, 48*(5), 1139−1146.

Patas, K., Penninx, B. W., Bus, B. A., Vogelzangs, N., Molendijk, M. L., et al. (2014). Association between serum brain-derived neurotrophic factor and plasma interleukin-6 in major depressive disorder with melancholic features. *Brain, Behavior, and Immunity, 36*, 71−79.

Patel, K., Allen, S., Haque, M. N., Angelescu, I., Baumeister, D., et al. (2016). Bupropion: A systematic review and meta-analysis of effectiveness as an antidepressant. *Therapeutic Advances in Psychopharmacology, 6* (2), 99−144.

Patel, N., Crider, A., Pandya, C. D., Ahmed, A. O., & Pillai, A. (2016). Altered mRNA levels of glucocorticoid receptor, mineralocorticoid receptor, and co-chaperones (FKBP5 and PTGES3) in the middle frontal gyrus of autism spectrum disorder subjects. *Molecular Neurobiology, 53*(4), 2090−2099.

Patel, S., Hill, M. N., Cheer, J. F., Wotjak, C. T., & Holmes, A. (2017). The endocannabinoid system as a target for novel anxiolytic drugs. *Neuroscience & Biobehavioral Reviews, 76*(Pt A), 56−66.

Patterson, Z. R., & Abizaid, A. (2013). Stress induced obesity: Lessons from rodent models of stress. *Frontiers in Neuroscience, 7*, 130.

Patterson, Z. R., Khazall, R., Mackay, H., Anisman, H., & Abizaid, A. (2013). Central ghrelin signaling mediates the metabolic response of C57BL/6 male mice to chronic social defeat stress. *Endocrinology, 154*(3), 1080−1091.

Paules, C. I., Marston, H. D., Eisinger, R. W., Baltimore, D., & Fauci, A. S. (2017). The pathway to a universal influenza vaccine. *Immunity, 47*(4), 59−603.

Paulose, J. K., Wright, J. M., Patel, A. G., & Cassone, V. M. (2016). Human gut bacteria are sensitive to melatonin and express endogenous circadian rhythmicity. *PLoS One, 11*(1), e0146643.

Payão, S. L. M., Gonçalves, G. M., de Labio, R. W., Horiguchi, L., Mizumoto, I., et al. (2012). Association of interleukin 1β polymorphisms and haplotypes with Alzheimer's disease. *Journal of Neuroimmunology, 247* (1−2), 59−62.

Paz Levy, D., Sheiner, E., Wainstock, T., Sergienko, R., Landau, D., et al. (2017). Evidence that children born at early term (37-38 6/7 weeks) are at increased risk for diabetes and obesity-related disorders. *American Journal of Obstetrics and Gynecology, 217*(5), 588.e1−588.e11.

Pedersen, A. F., Zachariae, R., & Bovbjerg, D. H. (2009). Psychological stress and antibody response to influenza vaccination: A meta-analysis. *Brain, Behavior, and Immunity, 23*(4), 427−433.

Pedersen, H. K., Gudmundsdottir, V., Nielsen, H. B., Hyotylainen, T., Nielsen, T., et al. (2016). Human gut microbes impact host serum metabolome and insulin sensitivity. *Nature, 535*(7621), 376−381.

Pedersen, L., Idorn, M., Olofsson, G. H., Lauenborg, B., Nookaew, I., et al. (2016). Voluntary running suppresses tumor growth through epinephrine-and IL-6-dependent NK cell mobilization and redistribution. *Cell Metabolism, 23*(3), 554−562.

Pedersen, M. G., Stevens, H., Pedersen, C. B., Norgaard-Pedersen, B., & Mortensen, P. B. (2011). Toxoplasma infection and later development of schizophrenia in mothers. *American Journal of Psychiatry, 168*(8), 814−821.

Peeters, A., Barendregt, J. J., Willekens, F., Mackenbach, J. P., Al Mamun, A., et al. (2003). Obesity in adulthood and its consequences for life expectancy: A life-table analysis. *Annals of Internal Medicines, 138*(1), 24−32.

Pellegrini, S., Sordi, V., Bolla, A. M., Saita, D., Ferrarese, R., et al. (2017). Duodenal mucosa of patients with type 1 diabetes shows distinctive inflammatory profile and microbiota. *Journal of Clinical Endocrinology and Metabolism, 102*(5), 1468−1477.

Peña, C. J., Kronman, H. G., Walker, D. M., Cates, H. M., Bagot, R. C., et al. (2017). Early life stress confers lifelong stress susceptibility in mice via ventral tegmental area OTX2. *Science, 356*(6343), 1158−1188.

Pendlebury, S. T., & Rothwell, P. M. (2009). Prevalence, incidence, and factors associated with pre-stroke and post-stroke dementia: A systematic review and meta-analysis. *The Lancet Neurology, 8*(11), 1006−1018.

Peng, J., Gu, N., Zhou, L., Eyo, U. B., Murugan, M., et al. (2016). Microglia and monocytes synergistically promote the transition from acute to chronic pain after nerve injury. *Nature Communications, 7*, 12029.

Peng, J., Peng, L., Stevenson, F. F., Doctrow, S. R., & Andersen, J. K. (2007). Iron and paraquat as synergistic environmental risk factors in sporadic Parkinson's disease accelerate age-related neurodegeneration. *The Journal of Neuroscience, 27*(26), 6914−6922.

Peng, Y. P., Qiu, Y. H., Lu, J. H., & Wang, J. J. (2005). Interleukin-6 protects cultured cerebellar granule neurons against glutamate-induced neurotoxicity. *Neuroscience Letters, 374*(3), 192−196. Available from https://doi.org/10.1016/j.neulet.2004.10.069.

Peng, W., Xing, Z., Yang, J., Wang, Y., Wang, W., et al. (2014). The efficacy of erythropoietin in treating experimental traumatic brain injury: A systematic review of controlled trials in animal models. *Journal of Neurosurgery, 121*(3), 653−664.

Penninx, B. W. (2017). Depression and cardiovascular disease: Epidemiological evidence on their linking mechanisms. *Neuroscience & Biobehavioral Reviews, 74*(Pt B), 277−286.

Penteado, S. H., Teodorov, E., Kirsten, T. B., Eluf, B. P., Reis-Silva, T. M., et al. (2014). Prenatal lipopolysaccharide disrupts maternal behavior, reduces nest odor preference in pups, and induces anxiety: Studies of F1 and F2 generations. *European Journal of Pharmacology, 738*, 342−351.

Pepino, M. Y., Eisenstein, S. A., Bischoff, A. N., Klein, S., Moerlein, S. M., et al. (2016). Sweet dopamine: Sucrose preferences relate differentially to striatal D2 receptor binding and age in obesity. *Diabetes*, 65(9), 2618–2623.

Péquignot, R., Dufouil, C., Prugger, C., Pérès, K., Artero, S., et al. (2016). High level of depressive symptoms at repeated study visits and risk of coronary heart disease and stroke over 10 years in older adults: The three-city study. *Journal of the American Geriatrics Society*, 64(1), 118–125.

Peralta-Ramírez, M. I., Jiménez-Alonso, J., Godoy-García, J. F., & Pérez-García, M. (2004). The effects of daily stress and stressful life events on the clinical symptomatology of patients with lupus erythematosus. *Psychosamatic Medicine*, 66(5), 788–794.

Perandini, L. A., Sales-de-Oliveira, D., Almeida, D. C., Azevedo, H., Moreira-Filho, C. A., et al. (2016). Effects of acute aerobic exercise on leukocyte inflammatory gene expression in systemic lupus erythematosus. *Excercise Immunology Review*, 22, 64–81.

Perego, C., Fumagalli, S., & De Simoni, M. G. (2013). Three-dimensional confocal analysis of microglia/macrophage markers of polarization in experimental brain injury. *Journal of Visualized Experiments*, 79, 50605.

Pereira, O. C., Bernardi, M. M., & Gerardin, D. C. (2006). Could neonatal testosterone replacement prevent alterations induced by prenatal stress in male rats? *Life Sciences*, 78(24), 2767–2771.

Perez-Caballero, L., Pérez-Egea, R., Romero-Grimaldi, C., Puigdemont, D., Molet, J., et al. (2014). Early responses to deep brain stimulation in depression are modulated by anti-inflammatory drugs. *Molecular Psychiatry*, 19(5), 607–614.

Perissinotto, C. M., Cenzer, I. S., & Covinsky, K. E. (2012). Loneliness in older persons: A predictor of functional decline and death. *Archives of Internal Medicine*, 172(14), 1078–1083.

Perkins, D. O., Jeffries, C. D., Addington, J., Bearden, C. E., Cadenhead, K. S., et al. (2015). Towards a psychosis risk blood diagnostic for persons experiencing high-risk symptoms: Preliminary results from the NAPLS project. *Schizophrenia Bulletin*, 41(2), 419–428.

Perron, H., Hamdani, N., Faucard, R., Lajnef, M., Jamain, S., Daban-Huard, C., et al. (2012). Molecular characteristics of Human Endogenous Retrovirus type-W in schizophrenia and bipolar disorder. *Translational Psychiatry*, 2, e201.

Perron, H., & Lang, A. (2010). The human endogenous retrovirus link between genes and environment in multiple sclerosis and in multifactorial diseases associating neuroinflammation. *Clinical Reviews in Allergy & Immunology*, 39(1), 51–61.

Perry, V. H., Nicoll, J. A. R., & Holmes, C. (2010). Microglia in neurodegenerative disease. *Nature Reviews Neurology*, 6(4), 193–201.

Peterlik, D., Stangl, C., Bauer, A., Bludau, A., Keller, J., Grabski, D., et al. (2017). Blocking metabotropic glutamate receptor subtype 5 relieves maladaptive chronic stress consequences. *Brain, Behavior, and Immunity*, 59, 79–92.

Peters, A., & McEwen, B. S. (2015). Stress habituation, body shape and cardiovascular mortality. *Neuroscience & Biobehavioral Reviews*, 56, 139–150.

Peters, L. A., Perrigoue, J., Mortha, A., Iuga, A., Song, W., et al. (2017). A functional genomics predictive network model identifies regulators of inflammatory bowel disease. *Nature Genetics*, 49(10), 1437–1449.

Peterson, A., Thome, J., Frewen, P., & Lanius, R. A. (2014). Resting-state neuroimaging studies: A new way of identifying differences and similarities among the anxiety disorders? *Canadian Journal of Psychiatry. Revue Canadienne de Psychiatrie*, 59(6), 294–300.

Philbert, J., Beeské, S., Belzung, C., & Griebel, G. (2015). The CRF receptor antagonist SSR125543 prevents stress-induced long-lasting sleep disturbances in a mouse model of PTSD: Comparison with paroxetine and d-cycloserine. *Behavioural Brain Research*, 279, 41–46.

Phillips, C. M., Shivappa, N., Hébert, J. R., & Perry, I. J. (2017). Dietary inflammatory index and mental health: A cross-sectional analysis of the relationship with depressive symptoms, anxiety and well-being in adults. *Clinical Nutrition*, pii: S0261-5614(17)30312-6.

Phillips, J. L., Batten, L. A., Tremblay, P., Aldosary, F., Du, L., et al. (2015). Impact of monoamine-related gene polymorphisms on hippocampal volume in treatment-resistant depression. *Acta Neuropsychiatrica*, 27(6), 353–361.

Piantadosi, S. C., French, B. J., Poe, M. M., Timić, T., Marković, B. D., et al. (2016). Sex-dependent anti-stress effect of an α5 subunit containing GABAA receptor positive allosteric modulator. *Frontiers in Pharmacology*, 7, 446.

Pierrehumbert, B., Torrisi, R., Laufer, D., Halfon, O., Ansermet, F., et al. (2010). Oxytocin response to an experimental psychosocial challenge in adults exposed to traumatic experiences during childhood or adolescence. *Neuroscience*, 166(1), 168–177.

Pietrzak, R. H., Gallezot, J. D., Ding, Y. S., Henry, S., Potenza, M. N., et al. (2013). Association of posttraumatic stress disorder with reduced in vivo norepinephrine transporter availability in the locus coeruleus. *JAMA Psychiatry*, 70(11), 1199–1205.

Pigneur, B., & Sokol, H. (2016). Fecal microbiota transplantation in inflammatory bowel disease: The quest for the holy grail. *Mucosal Immunology*, 9(6), 1360–1365.

Piirainen, S., Youssef, A., Song, C., Kalueff, A. V., Landreth, G. E., et al. (2017). Psychosocial stress on neuroinflammation and cognitive dysfunctions in

Alzheimer's disease: The emerging role for microglia? *Neuroscience & Biobehavioral Reviews, 77*, 148–164.

Pilsner, J. R., Parker, M., Sergeyev, O., & Suvorov, A. (2017). Spermatogenesis disruption by dioxins: Epigenetic reprograming and windows of susceptibility. *Reproductive Toxicology, 69*, 221–229.

Pinto-Sanchez, M. I., Hall, G. B., Ghajar, K., Nardelli, A., Bolino, C., et al. (2017). Probiotic bifidobacterium longum NCC3001 reduces depression scores and alters brain activity: A pilot study in patients with irritable bowel syndrome. *Gastroenterology, 153*(2), 448–459.

Pisanu, A., Lecca, D., Mulas, G., Wardas, J., Simbula, G., et al. (2014). Dynamic changes in pro- and anti-inflammatory cytokines in microglia after PPAR-γ agonist neuroprotective treatment in the MPTPp mouse model of progressive Parkinson's disease. *Neurobiology of Disease, 71*, 280–291.

Pitake, S., Debrecht, J., & Mishra, S. K. (2017). Brain natriuretic peptide (BNP) expressing sensory neurons are not involved in acute, inflammatory or neuropathic pain. *Molecular Pain, 13*, 1744806917736993.

Pitman, R. K., Gilbertson, M. W., Gurvits, T. V., May, F. S., Lasko, N. B., & Orr, S. P. (2006). Harvard/VA PTSD twin study investigators. *Annals of the New York Academy of Science, 1071*, 242–254.

Pitman, R. K., Rasmusson, A. M., Koenen, K. C., Shin, L. M., & Orr, S. P. (2012). Biological studies of posttraumatic stress disorder. *Nature Reviews Neuroscience, 13*(11), 769–787.

Pizzo, R., O'Leary, O. F., & Cryan, J. F. (2017). Elucidation of the neural circuits activated by a GABA_B receptor positive modulator: Relevance to anxiety. *Neuropharmacology*, pii: S0028-3908(17)30352-0.

Plantinga, L., Bremner, J. D., Miller, A. H., Jones, D. P., Veledar, E., et al. (2013). Association between posttraumatic stress disorder and inflammation: A twin study. *Brain, Behavior, and Immunity, 30*, 125–132.

Playfair, J. H. L., & Chain, B. M. (2013). *Immunology at a Glance* (10th ed.). West Sussex, UK: Wiley-Blackwell.

Plog, B. A., & Nedergaard, M. (2017). The glymphatic system in central nervous system health and disease: Past, present, and future. *Annual Review of Pathology, 13*, 379–394.

Plotsky, P. M., & Meaney, M. J. (1993). Early, postnatal experience alters hypothalamic corticotropin-releasing factor (CRF) mRNA, median eminence CRF content and stress-induced release in adult rats. *Brain ResearchM molecular BrainResearch, 18*(3), 195–200.

Plovier, H., Everard, A., Druart, C., Depommier, C., Van Hul, M., et al. (2017). A purified membrane protein from akkermansia muciniphila or the pasteurized bacterium improves metabolism in obese and diabetic mice. *Nature Medicine (New York, NY, United States), 23*(1), 107–113.

Pocivavsek, A., Wu, H. Q., Elmer, G. I., Bruno, J. P., & Schwarcz, R. (2012). Pre- and postnatal exposure to kynurenine causes cognitive deficits in adulthood. *The European Journal of Neuroscience, 35*(10), 1605–1612.

Policicchio, S., Ahmad, A. N., Powell, J. F., & Proitsi, P. (2017). Rheumatoid arthritis and risk for Alzheimer's disease: A systematic review and meta-analysis and a Mendelian Randomization study. *Scientific Reports, 7*(1), 12861.

Pompoli, A., Furukawa, T. A., Imai, H., Tajika, A., Efthimiou, O., et al. (2016). Psychological therapies for panic disorder with or without agoraphobia in adults: A network meta-analysis. *The Cochrane Database of Systematic Reviews, 4*, CD011004.

Popp, S., Behl, B., Joshi, J. J., Lanz, T. A., Spedding, M., et al. (2016). In search of the mechanisms of ketamine's antidepressant effects: How robust is the evidence behind the mTor activation hypothesis. *F1000 Research, 5*, 634.

Posillico, C. K., & Schwarz, J. M. (2016). An investigation into the effects of antenatal stressors on the postpartum neuroimmune profile and depressive-like behaviors. *Behavioural Brain Research, 298*(Pt B), 218–228.

Poulter, M. O., Du, L., Weaver, I. C., Palkovits, M., Faludi, G., et al. (2008). GABAA receptor promoter hypermethylation in suicide brain: implications for the involvement of epigenetic processes. *Biological Psychiatry, 64*, 645–652.

Poulter, M. O., Du, L., Zhurov, V., Merali, Z., & Anisman, H. (2010). Plasticity of the GABA(A) receptor subunit cassette in response to stressors in reactive versus resilient mice. *Neuroscience, 165*(4), 1039–1051.

Pournajafi-Nazarloo, H., Kenkel, W., Mohsenpour, S. R., Sanzenbacher, L., Saadat, H., et al. (2013). Exposure to chronic isolation modulates receptors mRNAs for oxytocin and vasopressin in the hypothalamus and heart. *Peptides (New York, 1979-1987), 43*, 20–26.

Powell, N., Walker, M. M., & Talley, N. J. (2017). The mucosal immune system: Master regulator of bidirectional gut-brain communications. *Nature Reviews Gastroenterology & Hepatology, 14*(3), 143–159. Available from https://doi.org/10.1038/nrgastro.2016.191.

Power, S. E., O'Toole, P. W., Stanton, C., Ross, R. P., & Fitzgerald, G. F. (2014). Intestinal microbiota, diet, and health. *The British Journal of Nutrition, 111*(3), 387–402.

Powers, K. M., Kay, D. M., Factor, S. A., Zabetian, C. P., Higgins, D. S., et al. (2008). Combined effects of smoking, coffee, and NSAIDs on Parkinson's disease risk. *Movement Disorders, 23*(1), 88–95.

Powers, K. M., Smith-Weller, T., Franklin, G. M., Longstreth, W. T., Swanson, P. D., et al. (2003). Parkinson's disease risks associated with dietary iron, manganese, and other nutrient intakes. *Neurology, 60*(11), 1761–1766.

Pozueta, J., Lefort, R., & Shelanski, M. L. (2013). Synaptic changes in Alzheimer's disease and its models. *Neuroscience, 251*, 51−65.

Pradillo, J. M., Murray, K. N., Coutts, G. A., Moraga, A., Oroz-Gonjar, F., et al. (2017). Reparative effects of interleukin-1 receptor antagonist in young and aged/co-morbid rodents after cerebral ischemia. *Brain, Behavior, and Immunity, 61*, 117−126.

Prager, G., Hadamitzky, M., Engler, A., Doenlen, R., Wirth, T., et al. (2013). Amygdaloid signature of peripheral immune activation by bacterial lipopolysaccharide or staphylococcal enterotoxin B. *Journal of Neuroimmune Pharmacology, 8*(1), 42−50.

Prandovszky, E., Gaskell, E., Martin, H., Dubey, J. P., Webster, J. P., et al. (2011). The neurotropic parasite Toxoplasma gondii increases dopamine metabolism. *PLoS One, 6*(9), e23866.

Pratchett, L. C., & Yehuda, R. (2011). Foundations of post-traumatic stress disorder: Does early life trauma lead to adult posttraumatic stress disorder? *Development and Psychopathology, 23*(2), 477−491.

Pressman, S. D., & Cohen, S. (2005). Does positive affect influence health? *Pschological Bulletin, 131*(6), 925−971.

Pressman, S. D., Cohen, S., Miller, G. E., Barkin, A., Rabin, B. S., et al. (2005). Loneliness, social network size, and immune response to influenza vaccination in college freshmen. *Health Psychology: Official Journal of the Division of Health Psychology, American Psychological Association, 24*(3), 297−306.

Price, L. H., Kao, H. T., Burgers, D. E., Carpenter, L. L., & Tyrka, A. R. (2013). Telomeres and early-life stress: An overview. *Biological Psychiatry, 73*(1), 15−23.

Price, R. B., Shungu, D. C., Mao, X., Nestadt, P., Kelly, C., et al. (2009). Amino acid neurotransmitters assessed by proton magnetic resonance spectroscopy: Relationship to treatment resistance in major depressive disorder. *Biological Psychiatry, 65*(9), 792−800.

Price, T. J., & Inyang, K. E. (2016). Commonalities between pain and memory mechanisms and their meaning for understanding chronic pain. *Progress in Molecular Biology and Translational Science, 131*, 409−434.

Priya, P. K., Rajappa, M., Kattimani, S., Mohanraj, P. S., & Revathy, G. (2016). Association of neurotrophins, inflammation and stress with suicide risk in young adults. *Clinica Chimica Acta;International Journal of Clinical Chemistry, 457*, 41−45, See comment in PubMed Commons below.

Pryce, C. R., & Fontana, A. (2016). Depression in autoimmune diseases. *Current Topics in Behavioral Neurosciences, 31*, 139−154.

Puccini, J. M., Marker, D. F., Fitzgerald, T., Barbieri, J., Kim, C. S., et al. (2015). Leucine-rich repeat kinase 2 modulates neuroinflammation and neurotoxicity in models of human immunodeficiency virus 1-associated neurocognitive disorders. *Journal of Neuroscience, 35*(13), 5271−5283.

Punnonen, J., Miller, J. L., Collier, T. J., & Spencer, J. R. (2015). Agonists of the tissue-protective erythropoietin receptor in the treatment of Parkinson's disease. *Current Topics in Medicinal Chemistry, 15*(10), 955−969.

Pusceddu, M. M., Kelly, P., Ariffin, N., Cryan, J. F., Clarke, G., et al. (2015). n-3 PUFAs have beneficial effects on anxiety and cognition in female rats: Effects of early life stress. *Psychoneuroendocrinology, 58*, 79−90.

Qi, W., Gevonden, M., & Shalev, A. (2017). Efficacy and tolerability of high-dose escitalopram in posttraumatic stress disorder. *Journal of Clinical Psychopharmacology, 37*(1), 89−93.

Qiao, Y., Wu, M., Feng, Y., Zhou, Z., Chen, L., & Chen, F. (2018). Alterations of oral microbiota distinguish children with autism spectrum disorders from healthy controls. *Scientific Reports, 8*(1), 1597.

Qin, Q., Zhi, L. T., Li, X. T., Yue, Z. Y., Li, G. Z., et al. (2017). Effects of LRRK2 inhibitors on nigrostriatal dopaminergic neurotransmission. *CNS Neuroscience & Therapeutics, 23*(2), 162−173.

Qin, Y., Liu, Y., Hao, W., Decker, Y., Tomic, I., et al. (2016). Stimulation of TLR4 attenuates Alzheimer's disease-related symptoms and pathology in Tau-transgenic mice. *Journal of Immunology, 197*(8), 3281−3292.

Quadros, A., Patel, N., Crescentini, R., Crawford, F., Paris, D., et al. (2003). Increased TNFalpha production and Cox-2 activity in organotypic brain slice cultures from APPsw transgenic mice. *Neuroscience Letters, 353*(1), 66−68.

Quan, N., Avitsur, R., Stark, J. L., He, L., Shah, M., et al. (2001). Social stress increases the susceptibility to endotoxic shock. *Journal of Neuroimmunology, 115*(1−2), 36−45.

Quik, M., Chen, L., Parameswaran, N., Xie, X., Langston, J. W., & McCallum, S. E. (2006). Chronic oral nicotine normalizes dopaminergic function and synaptic plasticity in 1-methyl-4-phenyl-1,2,3,6-tetrahydropyridine-lesioned primates. *The Journal of Neuroscience, 26*(17), 4681−4689.

Rabl, U., Meyer, B. M., Diers, K., Bartova, L., Berger, A., et al. (2014). Additive gene-environment effects on hippocampal structure in healthy humans. *Journal of Neuroscience, 34*(30), 9917−9926.

Racicot, K., & Mor, G. (2017). Risks associated with viral infections during pregnancy. *Journal of Clinical Investigation, 127*(5), 1591−1599.

Radad, K., Moldzio, R., Al-Shraim, M., Kranner, B., Krewenka, C., et al. (2017). Recent Advances on the role of neurogenesis in the adult brain: Therapeutic potential in Parkinson's and Alzheimer's diseases. *CNS & Neurological Disorders Drug Targets, 16*(7), 740−748.

Radulovic, J., Jovasevic, V., & Meyer, M. A. (2017). Neurobiological mechanisms of state-dependent learning. *Current Opinion in Neurobiology, 45*, 92–98.

Raha, A. A., Henderson, J. W., Stott, S. R. W., Vuono, R., Foscarin, S., et al. (2017). Neuroprotective effect of TREM-2 in aging and Alzheimer's disease model. *Journal of Alzheimer's Disease, 55*(1), 199–217.

Raison, C. L., Capuron, L., & Miller, A. H. (2006). Cytokines sing the blues: Inflammation and the pathogenesis of depression. *Trends in Immunology, 27*(1), 24–31.

Raison, C. L., Dantzer, R., Kelley, K. W., Lawson, M. A., Woolwine, B. J., et al. (2010). CSF concentrations of brain tryptophan and kynurenines during immune stimulation with IFN-alpha: Relationship to CNS immune responses and depression. *Molecular Psychiatry, 15*(4), 393–403.

Raison, C. L., & Miller, A. H. (2003). When not enough is too much: The role of insufficient glucocorticoid signaling in the pathophysiology of stress-related disorders. *American Journal of Psychiatry, 160*(9), 1554–1565.

Raison, C. L., & Miller, A. H. (2013). Malaise, melancholia and madness: The evolutionary legacy of an inflammatory bias. *Brain, Behavior, and Immunity, 31*, 1–8.

Raison, C. L., Rutherford, R. E., Woolwine, B. J., Shuo, C., Schettler, P., et al. (2013). A randomized controlled trial of the tumor necrosis factor antagonist infliximab for treatment-resistant depression: The role of baseline inflammatory biomarkers. *JAMA Psychiatry, 70*(1), 31–41.

Raj, D. D. A., Jaarsma, D., Holtman, I. R., Olah, M., Ferreira, F. M., et al. (2014). Priming of microglia in a DNA-repair deficient model of accelerated aging. *Neurobiology of Aging, 35*(9), 2147–2160.

Rajbhandari, A. K., Baldo, B. A., & Bakshi, V. P. (2015). Predator stress-induced CRF release causes enduring sensitization of basolateral amygdala norepinephrine systems that promote PTSD-like startle abnormalities. *Journal of Neuroscience, 35*(42), 14270–14285.

Rakers, F., Rupprecht, S., Dreiling, M., Bergmeier, C., Witte, O. W., et al. (2017). Transfer of maternal psychosocial stress to the fetus. *Neuroscience & Biobehavioral Reviews*, pii: S0149-7634(16)30719-9.

Rakoff-Nahoum, S., Foster, K. R., & Comstock, L. E. (2016). The evolution of cooperation within the gut microbiota. *Nature, 533*(7602), 255–259.

Ramikie, T. S., & Patel, S. (2012). Endocannabinoid signaling in the amygdala: Anatomy, synaptic signaling, behavior, and adaptations to stress. *Neuroscience, 204*, 38–52.

Ramirez, A. I., de Hoz, R., Salobrar-Garcia, E., Salazar, J. J., Rojas, B., et al. (2017). The role of microglia in retinal neurodegeneration: Alzheimer's disease, Parkinson, and Glaucoma. *Frontiers in Aging Neuroscience, 9*, 214.

Ramirez, K., Niraula, A., & Sheridan, J. F. (2016). GABAergic modulation with classical benzodiazepines prevent stress-induced neuro-immune dysregulation and behavioral alterations. *Brain, Behavior, and Immunity, 51*, 154–168.

Ramirez, K., & Sheridan, J. F. (2016). Antidepressant imipramine diminishes stress-induced inflammation in the periphery and central nervous system and related anxiety-and depressive-like behaviors. *Brain, Behavior, and Immunity, 57*, 293–303.

Ramot, Y., Nyska, A., Maronpot, R. R., Shaltiel-Karyo, R., Tsarfati, Y., et al. (2017). Ninety-day local tolerability and toxicity study of ND0612, a novel formulation of levodopa/carbidopa, administered by subcutaneous continuous infusion in minipigs. *Toxicologic Pathology, 45*(6), 764–773.

Ransohoff, R. M. (2016). A polarizing question: Do M1 and M2 microglia exist? *Nature Neuroscience, 19*(8), 987–991.

Rao, M., & Gershon, M. D. (2016). The bowel and beyond: The enteric nervous system in neurological disorders. *Nature Reviews Gastroenterology & Hepatology, 13*(9), 517–528.

Rao, M., & Gershon, M. D. (2017). Neurogastroenterology: The dynamic cycle of life in the enteric nervous system. *Nature Reviews Gastroenterology & Hepatology, 14*(8), 453–454.

Rao, N. P., Venkatasubramanian, G., Ravi, V., Kalmady, S., & Cherian, A. (2015). Plasma cytokine abnormalities in drug-naïve, comorbidity-free obsessive-compulsive disorder. *Psychiatry Research, 229*(3), 949–952.

Raphael, I., Nalawade, S., Eagar, T. N., & Forsthuber, T. G. (2015). T cell subsets and their signature cytokines in autoimmune and inflammatory diseases. *Cytokine, 74*(1), 5–17.

Rappeneau, V., Blaker, A., Petro, J. R., Yamamoto, B. K., & Shimamoto, A. (2016). Disruption of the glutamate-glutamine cycle involving astrocytes in an animal model of depression for males and females. *Frontiers in Behavioral Neuroscience, 10*, 231.

Rappold, P. M., Cui, M., Chesser, A. S., Tibbett, J., Grima, J. C., et al. (2011). Paraquat neurotoxicity is mediated by the dopamine transporter and organic cation transporter-3. *Proceedings of the National Academy of Sciences of the United States of America, 108*(51), 20766–20771.

Rask-Andersen, M., Martinsson, D., Ahsan, M., Enroth, S., Ek, W. E., et al. (2016). Epigenome-wide association study reveals differential DNA methylation in individuals with a history of myocardial infraction. *Human Molecular Genetics, 25*(21), 4739–4748.

Rasmusson, A. M., Hauger, R. L., Morgan, C. A., Bremner, J. D., Charney, D. S., et al. (2000). Low baseline and yohimbine-stimulated plasma neuropeptide Y (NPY) levels in combat-related PTSD. *Biological Psychiatry, 47* (6), 526–539.

Rasmusson, A. M., Marx, C. E., Pineles, S. L., Locci, A., Scioli-Salter, E. R., et al. (2017). Neuroactive steroids and PTSD treatment. *Neuroscience Letters, 649*, 156–163.

Raspopow, K., Abizaid, A., Matheson, K., & Anisman, H. (2014). Anticipation of a psychosocial stressor differentially influences ghrelin, cortisol and food intake among emotional and non-emotional eaters. *Appetite, 74*, 35–43.

Raspopow, K., Matheson, K., Abizaid, A., & Anisman, H. (2013). Unsupportive social interactions influence emotional eating behaviors. The role of coping styles as mediators. *Appetite, 62*, 143–149.

Ravenholt, R. T., & Foege, W. H. (1982). 1918 influenza, encephalitis lethargica, parkinsonism. *Lancet, 2*(8303), 860–864.

Ravindran, A. V., Balneaves, L. G., Faulkner, G., Ortiz, A., & McIntosh, D. (2016). Canadian Network for Mood and Anxiety Treatments (CANMAT) 2016 clinical guidelines for the management of adults with major depressive disorder: section 5. complementary and alternative medicine treatments. *The Canadian Journal of Psychiatry, 61*, 576–587.

Raymaekers, S., Vansteelandt, K., Luyten, L., Bervoets, C., Demyttenaere, K., et al. (2017). Long-term electrical stimulation of bed nucleus of stria terminalis for obsessive-compulsive disorder. *Molecular Psychiatry, 22* (6), 931–934.

Rea, K., O'Mahony, S. M., Dinan, T. G., & Cryan, J. F. (2017). Visceral pain: Role of the microbiome-gut-brain axis. *The Microbiome, 39*, 6–9.

Reale, M., Pesce, M., Priyadarshini, M., Kamal, M. A., & Patruno, A. (2012). Mitochondria as an easy target to oxidative stress events in Parkinson's disease. *CNS & Neurological Disorders Drug Targets, 11*(4), 430–438.

Reber, S. O., Siebler, P. H., Donner, N. C., Morton, J. T., Smith, D. G., et al. (2016). Immunization with a heat-killed preparation of the environmental bacterium Mycobacterium vaccae promotes stress resilience in mice. *Proceedings of the National Academy of Sciences of the United States of America, 113*(22), E3130–3139.

Rebo, J., Mehdipour, M., Gathwala, R., Causey, K., Liu, Y., et al. (2016). A single heterochronic blood exchange reveals rapid inhibition of multiple tissues by old blood. *Nature Communications, 7*, 13363.

Recker, M., Laabei, M., Toleman, M. S., Reuter, S., Saunderson, R. B., et al. (2017). Clonal differences in staphylococcus aureus bacteraemia-associated mortality. *Nature Microbiology, 2*(10), 1381–1388.

Redberg, R. F., & Katz, M. H. (2017). Statins for primary prevention: The debate is intense but the data are weak. *JAMA Internal Medicine, 177*(1), 21–23.

Reddy, V. S., Prabhu, S. D., Mummidi, S., Valente, A. J., Venkatesan, B., et al. (2010). Interleukin-18 induces EMMPRIN expression in primary cardiomyocytes via JNK/Sp1 signaling and MMP-9 in part via EMMPRIN and through AP-1 and NF-kappaB activation. *American journal of physiology – Heart and circulatory physiology, 299*(4), H1242–H1254.

Redwine, L., Hauger, R. L., Gillin, J. C., & Irwin, M. (2000). Effects of sleep and sleep deprivation on interleukin-6, growth hormone, cortisol, and melatonin levels in humans. *Journal of Clinical Endocrinology and Metabolism, 85*(10), 3597–3603.

Rees, J. C. (2014). Obsessive-compulsive disorder and gut microbiota dysregulation. *Medical Hypotheses, 82*(2), 163–166.

Reeve, S. M., Scocchera, E. W., G-Dayanadan, N., Keshipeddy, S., Krucinska, J., et al. (2016). MRSA isolates from United States hospitals carry dfrG and dfrK resistance genes and succumb to propargyl-linked antifolates. *Cell Chemical Biology, 23*(12), 1458–1467.

Reiche, E. M., Nunes, S. O., & Morimoto, H. K. (2004). Stress, depression, the immune system, and cancer. *Lancet Oncology, 5*(10), 617–625.

Reichenberg, A., Yirmiya, R., Schuld, A., Kraus, T., Haack, M., et al. (2001). Cytokine-associated emotional and cognitive disturbances in humans. *Archives of General Psychiatry, 58*(5), 445–452.

Reisinger, S. N., Kong, E., Khan, D., Schulz, S., Ronovsky, M., et al. (2016). Maternal immune activation epigenetically regulates hippocampal serotonin transporter levels. *Neurobiology of Stress, 4*, 34–43.

Remely, M., Hippe, B., Zanner, J., Aumueller, E., Brath, H., et al. (2016). Gut microbiota of obese, type 2 diabetic individuals is enriched in Faecalibacterium prausnitzii, Akkermansia muciniphila and Peptostreptococcus anaerobius after weight loss. *Endocrine, Metabolic & Immune Disorders Drug Targets, 16*(2), 99–106.

Ren, J., Li, H., Palaniyappan, L., Liu, H., Wang, J., et al. (2014). Repetitive transcranial magnetic stimulation versus electroconvulsive therapy for major depression: A systematic review and meta-analysis. *Progress in Neuro-psychopharmacology & Biological Psychiatry, 51*, 181–189.

Rethorst, C. D., Greer, T. L., Toups, M. S., Bernstein, I., Carmody, T. J., et al. (2015). IL-1β and BDNF are associated with improvement in hypersomnia but not insomnia following exercise in major depressive disorder. *Translational Psychiatry, 5*, e611.

Reul, J. M. (2014). Making memories of stressful events: A journey along epigenetic, gene transcription, and signaling pathways. *Frontiers in Psychiatry, 5*, 5.

Reul, J. M., & Nutt, D. J. (2008). Glutamate and cortisol--a critical confluence in PTSD? *Journal of Psychopharmacology, 22*(5), 469–472.

Reul, M. H. M., & Holsboer, F. (2002). Corticotropin-releasing factor receptors 1 and 2 in anxiety and depression. *Current Opinion in Pharmacology*, 2(1), 23–33.

Ricci, S., Businaro, R., Ippoliti, F., Lo Vasco, V. R., Massoni, F., Onofri, E., & Archer, T. (2013). Altered cytokine and BDNF levels in autism spectrum disorder. *Neurotoxicity Research*, 24(4), 491–501. Available from https://doi.org/10.1007/s12640-013-9393-4.

Rice, F., Harold, G. T., Boivin, J., van den Bree, M., Hay, D. F., & Thapar, A. (2010). The links between prenatal stress and offspring development and psychopathology: Disentangling environmental and inherited influences. *Psychological Medicine*, 40(2), 335–345.

Richards, D. A., Ekers, D., McMillan, D., Taylor, R. S., Byford, S., et al. (2016). Cost and outcome of behavioural activation versus cognitive behavioural therapy for depression (COBRA): A randomised, controlled, non-inferiority trial. *Lancet*, 388(10047), 871–880.

Richardson, J. R., Quan, Y., Sherer, T. B., Greenamyre, J. T., & Miller, G. W. (2005). Paraquat neurotoxicity is distinct from that of MPTP and rotenone. *Toxicological Sciences*, 88(1), 193–201.

Richetto, J., Calabrese, F., Riva, M. A., & Meyer, U. (2014). Prenatal immune activation induces maturation-dependent alterations in the prefrontal GABAergic transcriptome. *Schizophrenia Bulletin*, 40(2), 351–361.

Richter, C., Woods, I. G., & Schier, A. F. (2014). Neuropeptidergic control of sleep and wakefulness. *Annual Review of Neuroscience*, 37, 503–531.

Ridaura, V. K., Faith, J. J., Rey, F. E., Cheng, J., Duncan, A. E., et al. (2013). Gut microbiota from twins discordant for obesity modulate metabolism in mice. *Science*, 341, 1241214.

Ridker, P. M., Lonn, E., Paynter, N. P., Glynn, R., & Yusuf, S. (2017). Primary prevention with Statin therapy in the elderly: New meta-analyses from the contemporary JUPITER and HOPE-3 randomized trials. *Circulation*, 135(20), 1979–1981.

Rieder, R., Wisniewski, P. J., Alderman, B. L., & Campbell, S. C. (2017). Microbes and mental health: A review. *Brain, Behavior, and Immunity*, 66, 9–17.

Riederer, P., & Sian-Hülsmann, J. (2012). The significance of neuronal lateralisation in Parkinson's disease. *Journal of Neural Transmission*, 119(8), 953–962.

Riera Romo, M., Perez-Martinez, D., & Castillo Ferrer, C. (2016). Innate immunity in vertebrates: an overview. *Immunology*, 148, 125–139.

Rietdijk, C. D., Perez-Pardo, P., Garssen, J., van Wezel, R. J. A., & Kraneveld, A. D. (2017). Exploring Braak's hypothesis of Parkinson's disease. *Frontiers in Nuerology*, 8, 37.

Rijlaarsdam, J., Cecil, C. A., Walton, E., Mesirow, M. S., Relton, C. L., et al. (2017). Prenatal unhealthy diet, insulin-like growth factor 2 gene (IGF2) methylation, and attention deficit hyperactivity disorder symptoms in youth with early-onset conduct problems. *Journal of Child Psychology and Psychiatry, and Allied Disciplines*, 58(1), 19–27.

Ringman, J. M., Frautschy, S. A., Teng, E., Begum, A. N., Bardens, J., et al. (2012). Oral curcumin for Alzheimer's disease: Tolerability and efficacy in a 24-week randomized, double blind, placebo-controlled study. *Alzheimer's Research & Theory*, 4(5), 43.

Ritz, B. R., Paul, K. C., & Bronstein, J. M. (2016). Of pesticides and men: A California story of genes and environment in Parkinson's disease. *Current Environment Health Reports*, 3(1), 40–52.

Riva, P., Romero Lauro, L. J., Vergallito, A., DeWall, C. N., & Bushman, B. J. (2015). Electrified emotions: Modulatory effects of transcranial direct stimulation on negative emotional reactions to social exclusion. *Social Neuroscience*, 10, 46–54.

Rivest, S. (2009). Regulation of innate immune responses in the brain. *Nature Reviews Immunology*, 9(6), 429–439.

Rizos, E. C., Ntzani, E. E., Bika, E., Kostapanos, M. S., & Elisaf, M. S. (2012). Association between omega-3 fatty acid supplementation and risk of major cardiovascular disease events: A systematic review and meta-analysis. *JAMA*, 308(10), 1024–1033.

RK, C. Y., Merico, D., Bookman, M., J, L. H., Thiruvahindrapuram, B., Patel, R. V., & Scherer, S. W. (2017). Whole genome sequencing resource identifies 18 new candidate genes for autism spectrum disorder. *Nature Neuroscience*, 20(4), 602–611. Available from https://doi.org/10.1038/nn.4524.

Robciuc, M. R., Kivelä, R., Williams, I. M., de Boer, J. F., van Dijk, T. H., et al. (2016). VEGFB/VEGFR1-induced expansion of adipose vasculature counteracts obesity and related metabolic complications. *Cell Metabolism*, 23(4), 712–724.

Roberts, N. P., Roberts, P. A., Jones, N., & Bisson, J. I. (2015). Psychological interventions for post-traumatic stress disorder and comorbid substance use disorder: A systematic review and meta-analysis. *Clinical Psychology Review*, 38, 25–38.

Roberts, R. (2014). A genetic basis for coronary artery disease. *Trends in Cardiovascular Medicine*, 25(3), 171–178, pii: S1050-1738(14)00183-2.

Robinson, M., Mattes, E., Oddy, W. H., Pennell, C. E., van Eekelen, A., et al. (2011). Prenatal stress and risk of behavioral morbidity from age 2 to 14 years: The influence of the number, type, and timing of stressful life events. *Development and Psychopathology*, 23(2), 507–520.

Robinson, R. G., & Jorge, R. E. (2016). Post-stroke depression: A review. *American Journal of Psychiatry*, 173(3), 221–231.

Rodgman, C., Verrico, C. D., Holst, M., Thompson-Lake, D., Haile, C. N., et al. (2016). Doxazosin XL reduces symptoms of posttraumatic stress disorder in veterans with PTSD: A pilot clinical trial. *Journal of Clinical Psychiatry, 77*(5), e561−e565.

Rojas, J. M., Avia, M., Martin, V., & Sevilla, N. (2017). IL-10: A multifunctional cytokine in viral infections. *Journal of Immunology Research, 2017*, 6104054.

Rolland, A., Jouvin-Marche, E., Viret, C., Faure, M., Perron, H., et al. (2006). The envelope protein of a human endogenous retrovirus-W family activates innate immunity through CD14/TLR4 and promotes Th1-like responses. *Journal of Immunology, 176*(12), 7636−7644.

Rolls, E. T., Cheng, W., Gilson, M., Qiu, J., Hu, Z., et al. (2017). Effective connectivity in depression. *Biological Psychiatry: Cognitive Neuroscience and Neuroimaging, 3*, 187−197.

Romero-Sandoval, E.A., & Sweitzer, S. (2015). Nonneuronal central mechanisms of pain: Glia and immune response. *Progress in Molecular Biology and Translational Science, 131*, 325−358.

Ronin, I., Katsowich, N., Rosenshine, I., & Balaban, N. Q. (2017). A long-term epigenetic memory switch controls bacterial virulence bimodality. *eLife, 6*, e19599.

Ronovsky, M., Berger, S., Zambon, A., Reisinger, S. N., Horvath, O., Pollak, A., et al. (2017). Maternal immune activation transgenerationally modulates maternal care and offspring depression-like behavior. *Brain, Behavior, and Immunity, 63*, 127−136.

Ronzoni, G., Del Arco, A., Mora, F., & Segovia, G. (2016). Enhanced noradrenergic activity in the amygdala contributes to hyperarousal in an animal model of PTSD. *Psychoneuroendocrinology, 70*, 1−9.

Roodveldt, C., Labrador-Garrido, A., Gonzalez-Rey, E., Lachaud, C. C., Guilliams, T., et al. (2013). Preconditioning of microglia by α-synuclein strongly affects the response induced by toll-like receptor (TLR) stimulation. *PLoS One, 8*(11), e79160.

Rook, G. A., Lowry, C. A., & Raison, C. L. (2015). Hygiene and other early childhood influences on the subsequent function of the immune system. *Brain Research, 1617*, 47−62.

Roomruangwong, C., Anderson, G., Berk, M., Stoyanov, D., Carvalho, A. F., et al. (2018). A neuro-immune, neuro-oxidative and neuro-nitrosative model of prenatal and postpartum depression. *Progress in Neuropsychopharmacology & Biological Psychiatry, 81*, 262−274.

Roosen, D. A., & Cookson, M. R. (2016). LRRK2 at the interface of autophagosomes, endosomes and lysosomes. *Molecular Neurodegeneration, 11*(1), 73.

Rosén, A., Yi, J., Kirsch, I., Kaptchuk, T. J., Ingvar, M., et al. (2017). Effects of subtle cognitive manipulations on placebo analgesia - An implicit priming study. *European Journal of Pain, 21*(4), 594−604.

Rosen, N. J., Yoshida, C. K., & Croen, L. A. (2007). Infection in the first 2 years of life and autism spectrum disorders. *Pediatrics, 119*(1), e61−69. Available from https://doi.org/10.1542/peds.2006-1788.

Rosenblat, J. D., & McIntyre, R. S. (2016). Bipolar disorder and inflammation. *The Psychiatric Clinics of North America, 39*(1), 125−137.

Rosselot, A. E., Hong, C., & Moore, S. R. (2016). Rhythm and bugs: Circadian clocks, gut microbiota, and enteric infections. *Current Opinions in Gastroenterology, 32*(1), 7−11.

Rosshart, S. P., Vassallo, B. G., Angeletti, D., Hutchinson, D. S., Morgan, A. P., et al. (2017). Wild mouse gut microbiota promotes host fitness and improves disease resistance. *Cell, 171*(5), 1015−1028.

Rossi, S., Studer, V., Motta, C., Polidoro, S., Perugini, J., Macchiarulo, G., et al. (2017). Neuroinflammation drives anxiety and depression in relapsing-remitting multiple sclerosis. *Neurology, 89*, 1338−1347.

Rossi-George, A., LeBlanc, F., Kaneta, T., Urbach, D., & Kusnecov, A. W. (2004). Effects of bacterial superantigens on behavior of mice in the elevated plus maze and light−dark box. *Brain, Behavior, and Immunity, 18*(1), 46−54.

Rosso, I. M., Crowley, D. J., Silveri, M. M., Rauch, S. L., Jensen, J. E., et al. (2017). Hippocampus glutamate and N-acetyl aspartate markers of excitotoxic neuronal compromise in posttraumatic stress disorder. *Neuropsychopharmacology, 42*(8), 1698−1705.

Roth, T. L., Lubin, F. D., Funk, A. J., & Sweatt, J. D. (2009). Lasting epigenetic influence of early- life adversity on the BDNF gene. *Biological Psychiatry, 65*(9), 760−769.

Roth, T. L., Zoladz, P. R., Sweatt, J. D., & Diamond, D. M. (2011). Epigenetic modification of hippocampal Bdnf DNA in adult rats in an animal model of post-traumatic stress disorder. *Journal of Psychiatric Research, 45*(7), 919−926.

Rothaug, M., Becker-Pauly, C., & Rose-John, S. (2016). The role of interleukin-6 signaling in nervous tissue. *Biochimica et Biophysica Acta, 1863*(6 Pt A), 1218−1227.

Rothschild, D. M., O'Grady, M., & Wecker, L. (1999). Neonatal cytomegalovirus exposure decreases prepulse inhibition in adult rats: Implications for schizophrenia. *Journal of Neuroscience Research, 57*(4), 429−434.

Rottenberg, J., Yaroslavsky, I., Carney, R. M., Freedland, K. E., George, C. J., et al. (2014). The association between major depressive disorder in childhood and risk factors for cardiovascular disease in adolescence. *Psychosomatic Medicine, 76*(2), 122−127.

Roumier, A., Pascual, O., Bechade, C., Wakselman, S., Poncer, J. C., et al. (2008). Prenatal activation of microglia induces delayed impairment of glutamatergic synaptic function. *PLoS One, 3*(7), e2595.

Rouxel, O., Da Silva, J., Beaudoin, L., Nel, I., Tard, C., et al. (2017). Cytotoxic and regulatory roles of mucosal-associated invariant T cells in type 1 diabetes. *Nature Immunology, 18*(12), 1321–1331.

Roy, B., Dunbar, M., Shelton, R. C., & Dwivedi, Y. (2016). Identification of microRNA-124-3p as a putative epigenetic signature of major depressive disorder. *Neuropsychopharmacology, 42*(4), 864–875.

Rozeske, R. R., Evans, A. K., Frank, M. G., Watkins, L. R., Lowry, C. A., et al. (2011). Uncontrollable, but not controllable, stress desensitizes 5-HT1A receptors in the dorsal raphe nucleus. *Journal of Neuroscience, 31*(40), 14107–14115.

Ruan, H. B., Dietrich, M. O., Liu, Z. W., Zimmer, M. R., Li, M. D., et al. (2014). O-GlcNAc transferase enables AgRP neurons to suppress browning of white fat. *Cell, 159*(2), 306–317.

Rubin, M., Shvil, E., Papini, S., Chhetry, B. T., Helpman, L., et al. (2016). Greater hippocampal volume is associated with PTSD treatment response. *Psychiatry Research, 252*, 36–39.

Rubio-Casillas, A., & Fernández-Guasti, A. (2016). The dose makes the poison: From glutamate-mediated neurogenesis to neuronal atrophy and depression. *Reviews of Neuroscience, 27*(6), 599–622.

Rudenko, I. N., Chia, R., & Cookson, M. R. (2012). Is inhibition of kinase activity the only therapeutic strategy for LRRK2-associated Parkinson's disease? *BMC Medicine, 10*, 20.

Rudyk, C., Litteljohn, D., Syed, S., Dwyer, Z., & Hayley, S. (2015). Paraquat and psychological stressor interactions as pertains to Parkinsonian co-morbidity. *Neurobiology of Stress, 2*, 85–93.

Rusiecki, J. A., Byrne, C., Galdzicki, Z., Srikantan, V., Chen, L., et al. (2013). PTSD and DNA methylation in select immune function gene promoter regions: A repeated measures case-control study of U.S. military service members. *Frontiers in Psychiatry, 4*, 56.

Russman Block, S., King, A. P., Sripada, R. K., Weissman, D. H., Welsh, R., et al. (2017). Behavioral and neural correlates of disrupted orienting attention in posttraumatic stress disorder. *Cognitive, Affective & Behavioral Neuroscience, 17*(2), 422–436.

Russo, E., Citraro, R., Constanti, A., & De Sarro, G. (2012). The mTOR signaling pathway in the brain: Focus on epilepsy and epileptogenesis. *Molecular Neurobiology, 46*(3), 662–681.

Russo, M. V., & McGavern, D. B. (2016). Inflammatory neuroprotection following traumatic brain injury. *Science, 353*(6301), 783–785.

Russo, R., Cristiano, C., Avagliano, C., De Caro, C., La Rana, G., et al. (2017). Gut-brain axis: Role of lipids in the regulation of inflammation, pain and CNS diseases. *Current Medicinal Chemistry*.

Russo, S. J., Murrough, J. W., Han, M. H., Charney, D. S., & Nestler, E. J. (2012). Neurobiology of resilience. *Nature Neuroscience, 15*, 1475–1484.

Rutayisire, E., Huang, K., Liu, Y., & Tao, F. (2016). The mode of delivery affects the diversity and colonization pattern of the gut microbiota during the first year of infants' life: A systematic review. *BMC Gastroenterol, 16*(1), 86.

Rutishauser, U., Mamelak, A. N., & Adolphs, R. (2015). The primate amygdala in social perception - insights from electrophysiological recordings and stimulation. *Trends in Neurosciences, 38*(5), 295–306. Available from https://doi.org/10.1016/j.tins.2015.03.001.

Ryan, B., Musazzi, L., Mallei, A., Tardito, D., Gruber, S. H., et al. (2009). Remodelling by early-life stress of NMDA receptor-dependent synaptic plasticity in a gene-environment rat model of depression. *International Journal of Neuropsychopharmacology, 12*(4), 553–559.

Rydén, M., Hrydziuszko, O., Mileti, E., Raman, A., Bornholdt, J., et al. (2016). The adipose transcriptional response to insulin is determined by obesity, not insulin sensitivity. *Cell Reports, 16*(9), 2317–2326.

Sabban, E. L., Laukova, M., Alaluf, L. G., Olsson, E., & Serova, L. I. (2015). Locus coeruleus response to single-prolonged stress and early intervention with intranasal neuropeptide Y. *Journal of Neurochemistry, 135*(5), 975–986.

Saben, J. L., Boudoures, A. L., Asghar, Z., Thompson, A., Drury, A., et al. (2016). Maternal metabolic syndrome programs mitochondrial dysfunction via germline changes across three generations. *Cell Reports, 16*(1), 1–8.

Sabihi, S., Dong, S. M., Maurer, S. D., Post, C., & Leuner, B. (2017). Oxytocin in the medial prefrontal cortex attenuates anxiety: Anatomical and receptor specificity and mechanism of action. *Neuropharmacology, 125*, 1–12.

Sacks, F. M., Lichtenstein, A. H., Wu, J. H., Appel, L. J., Creager, M. A., et al. (2017). Dietary fats and cardiovascular disease: A presidential advisory from the American Heart Association. *Circulation, 136*(3), e1–e23.

Safaiyan, S., Kannaiyan, N., Snaidero, N., Brioschi, S., Biber, K., et al. (2016). Age-related myelin degradation burdens the clearance function of microglia during aging. *Nature Neuroscience, 19*(8), 995–998.

Sah, R., Ekhator, N. N., Jefferson-Wilson, L., Horn, P. S., & Geracioti, T. D. (2014). Cerebrospinal fluid neuropeptide Y in combat veterans with and without posttraumatic stress disorder. *Psychoneuroendocrinology, 40*, 277–283.

Saint-Mezard, P., Chavagnac, C., Bosset, S., Ionescu, M., Peyron, E., et al. (2003). Psychological stress exerts an adjuvant effect on skin dendritic cell functions in vivo. *Journal of Immunology*, 171(8), 4073−4080.

Sakayori, N., Kikkawa, T., Tokuda, H., Kiryu, E., Yoshizaki, K., et al. (2016). Maternal dietary imbalance between omega-6 and omega-3 polyunsaturated fatty acids impairs neocortical development via epoxy metabolites. *Stem Cells*, 34(2), 470−482.

Salam, A. P., Borsini, A., & Zunszain, P. A. (2018). Trained innate immunity: A salient factor in the pathogenesis of neuroimmune psychiatric disorders. *Molecular Psychiatry*, 23(2), 170−176. Available from https://doi.org/10.1038/mp.2017.186, Epub 2017 Dec 12.

Salari, A. A., & Amani, M. (2017). Neonatal blockade of GABA-A receptors alters behavioral and physiological phenotypes in adult mice. *International Journal of Developmental Neuroscience*, 57, 62−71.

Salazar, A., Gonzalez-Rivera, B. L., Redus, L., Parrott, J. M., & O'Connor, J. C. (2012). Indoleamine 2, 3-dioxygenase mediates anhedonia and anxiety-like behaviors caused by peripheral lipopolysaccharide immune challenge. *Hormones and Behavior*, 62, 202−209.

Salmaso, N., Jablonska, B., Scafidi, J., Vaccarino, F. M., & Gallo, V. (2014). Neurobiology of premature brain injury. *Nature Neuroscience*, 17(3), 34134−34136.

Salter, M. W., & Stevens, B. (2017). Microglia emerge as central players in brain disease. *Nature Medicine (New York, NY, United States)*, 23(9), 1018−1027.

Salvemini, D., & Jacobson, K. A. (2017). Highly selective A3 adenosine receptor agonists relieve chronic neuropathic pain. *Expert Opinion on Therapeutic Patents*, 27(8), 967.

Samii, A., Etminan, M., Wiens, M. O., & Jafari, S. (2009). NSAID use and the risk of Parkinson's disease: Systematic review and meta-analysis of observational studies. *Drugs & Aging*, 26(9), 769−779.

Sampson, T. R., Debelius, J. W., Thron, T., Janssen, S., Shastri, G. G., et al. (2016). Gut microbiota regulate motor deficits and neuroinflammation in a model of Parkinson's disease. *Cell*, 167(6), 1469−1480.e12.

Samsam, M., Ahangari, R., & Naser, S. A. (2014). Pathophysiology of autism spectrum disorders: Revisiting gastrointestinal involvement and immune imbalance. *World Journal of Gastroenterology*, 20(29), 9942−9951. Available from https://doi.org/10.3748/wjg.v20.i29.9942.

Sanacora, G., Treccani, G., & Popoli, M. (2012). Towards a glutamate hypothesis of depression: An emerging frontier of neuropsychopharmacology for mood disorders. *Neuropharmacology*, 62(1), 63−77.

Sanchez-Ramos, J., Song, S., Sava, V., Catlow, B., & Lin, X. (2009). Granulocyte colony stimulating factor decreases brain amyloid burden and reverses cognitive impairment in Alzheimer's mice. *Neuroscience*, 163, 55−72.

Sánchez-Santed, F., Colomina, M. T., & Herrero Hernández, E. (2016). Organophosphate pesticide exposure and neurodegeneration. *Cortex; A Journal Devoted to the Study of the Nervous System and Behavior*, 74, 417−426.

Sánchez-Vidaña, D. I., Chan, N. J., Chan, A. H., Hui, K. K., Lee, S., et al. (2016). Repeated treatment with oxytocin promotes hippocampal cell proliferation, dendritic maturation and affects socio-emotional behavior. *Neuroscience*, 333, 65−77.

Sandler, R. H., Finegold, S. M., Bolte, E. R., Buchanan, C. P., Maxwell, A. P., et al. (2000). Short-term benefit from oral vancomycin treatment of regressive-onset autism. *Journal of Child Neurology*, 15(7), 429−435.

Sandman, C. A., Davis, E. P., Buss, C., & Glynn, L. M. (2011). Exposure to prenatal psychobiological stress exerts programming influences on the mother and her fetus. *Neuroendocrinology*, 95(1), 8−21.

Sando, S. B., Melquist, S., Cannon, A., Hutton, M. L., Sletvold, O., et al. (2008). APOE epsilon 4 lowers age at onset and is a high risk factor for Alzheimer's disease; a case control study from central Norway. *BMC Neurology*, 8, 9.

Sankaranarayanan, K., Ozga, A. T., Warinner, C., Tito, R. Y., Obregon-Tito, A. J., et al. (2015). Gut microbiome diversity among Cheyenne and Arapaho individuals from Western Oklahoma. *Current Biology*, 25(24), 3161−3169.

Santarelli, S., Zimmermann, C., Kalideris, G., Lesuis, S. L., Arloth, J., et al. (2017). An adverse early life environment can enhance stress resilience in adulthood. *Psychoneuroendocrinology*, 78, 213−221.

Santpere, G., Garcia-Esparcia, P., Andres-Benito, P., Lorente-Galdos, B., Navarro, A., et al. (2017). Transcriptional network analysis in frontal cortex in Lewy body diseases with focus on dementia with Lewy bodies. *Brain Pathology*. Available from https://doi.org/10.1111/bpa.12511.

Sanyal, A., Naumann, J., Hoffmann, L. S., Chabowska-Kita, A., Ehrlund, A., et al. (2017). Interplay between obesity-induced inflammation and cGMP signaling in white adipose tissue. *Cell Reports*, 18, 225−236.

Sapolsky, R. M., Romero, L. M., & Munck, A. U. (2000). How do glucocorticoids influence stress responses? Integrating permissive, suppressive, stimulatory, and preparative actions. *Endocrine Reviews*, 21, 55−89.

Saragiotto, B. T., Machado, G. C., Ferreira, M. L., Pinheiro, M. B., Abdel Shaheed, C., et al. (2016). Paracetamol for low back pain. *Cochrane Database of Systematic Reviews*, CD012230.

Saresella, M., Marventano, I., Guerini, F. R., Mancuso, R., Ceresa, L., Zanzottera, M., & Clerici, M. (2009). An autistic endophenotype results in complex immune dysfunction in healthy siblings of autistic children. *Biological Psychiatry, 66*(10), 978−984. Available from https://doi.org/10.1016/j.biopsych.2009.06.020.

Sarris, J., Murphy, J., Mischoulon, D., Papakostas, G. I., Fava, M., et al. (2016). Adjunctive nutraceuticals for depression: A systematic review and meta-analyses. *The American Journal of Psychiatry, 173*(6), 575−587.

Sasaki, A. (2017). Microglia and brain macrophages: An update. *Neuropathology, 37*, 452−464.

Sasaki, A., de Vega, W., Sivanathan, S., St-Cyr, S., & McGowan, P. O. (2014). Maternal high-fat diet alters anxiety behavior and glucocorticoid signaling in adolescent offspring. *Neuroscience, 272*, 92−101.

Sassone-Corsi, M., & Raffatellu, M. (2015). No vacancy: How beneficial microbes cooperate with immunity to provide colonization resistance to pathogens. *Journal of Immunology, 194*, 4081−4087.

Sathyanesan, M., Haiar, J. M., Watt, M. J., & Newton, S. S. (2017). Restraint stress differentially regulates inflammation and glutamate receptor gene expression in the hippocampus of C57BL/6 and BALB/c mice. *Stress, 20* (2), 197−204.

Sato, H., Zhang, L. S., Martinez, K., Chang, E. B., Yang, Q., et al. (2016). Antibiotics suppress activation of intestinal mucosal mast cells and reduce dietary lipid absorption in Sprague-Dawley rats. *Gastroenterology, 151*, 923−932.

Sattar, N., Murray, H. M., Welsh, P., Blauw, G. J., Buckley, B. M., et al. (2009). Are markers of inflammation more strongly associated with risk for fatal than for nonfatal vascular events? *PLoS Medicine, 6*, e1000099.

Saul, A. N., Oberyszyn, T. M., Daugherty, C., Kusewitt, D., Jones, S., et al. (2005). Chronic stress and susceptibility to skin cancer. *Journal of the National Cancer Institute, 97*, 1760−1767.

Savignac, H. M., Couch, Y., Stratford, M., Bannerman, D. M., Tzortzis, G., et al. (2016). Prebiotic administration normalizes lipopolysaccharide (LPS)-induced anxiety and cortical 5-HT2A receptor and IL1-β levels in male mice. *Brain, Behavior, and Immunity, 52*, 120−131.

Sawada, Y., Nishio, Y., Suzuki, K., Hirayama, K., Takeda, A., et al. (2012). Attentional set-shifting deficit in Parkinson's disease is associated with prefrontal dysfunction: An FDG-PET study. *PLoS One, 7*, e38498.

Scaini, G., Fries, G. R., Valvassori, S. S., Zeni, C. P., Zunta-Soares, G., et al. (2017). Perturbations in the apoptotic pathway and mitochondrial network dynamics in peripheral blood mononuclear cells from bipolar disorder patients. *Translational Psychiatry, 7*, e1111.

Scarante, F. F., Vila-Verde, C., Ferreira-Junior, N. C., Detoni, V. L., Guimaraes, F., et al. (2017). Cannabinoid modulation of the stressed hippocampus. *Frontiers in Molecular Neuroscience, 10*, 411.

Schafer, D. P., & Stevens, B. (2010). Synapse elimination during development and disease: Immune molecules take centre stage. *Biochemical Society Transactions, 38*, 476−481.

Schafer, M. J., White, T. A., Evans, G., Tonne, J. M., Verzosa, G. C., et al. (2016). Exercise prevents diet-induced cellular senescence in adipose tissue. *Diabetes, 65*, 1606−1615.

Scharko, A. M. (2011). The infection hypothesis of schizophrenia: A systematic review. *Behavioral and Brain Sciences, 1*, 47−56.

Schellekens, H., Finger, B. C., Dinan, T. G., & Cryan, J. F. (2012). Ghrelin signalling and obesity: At the interface of stress, mood and food reward. *Pharmacology & Therapeutics, 135*(3), 316−326.

Schenk, D., Barbour, R., Dunn, W., Gordon, G., & Grajeda, H. (1999). Immunization with amyloid-beta attenuates Alzheimer-disease-like pathology in the PDAPP mouse. *Nature, 400*, 173−177.

Schenkel, L. S., Spaulding, W. D., DiLillo, D., & Silverstein, S. M. (2005). Histories of childhood maltreatment in schizophrenia: relationships with premorbid functioning, symptomatology, and cognitive deficits. *Schizophrenia Research, 76*(2-3), 273−286. Available from https://doi.org/10.1016/j.schres.2005.03.003.

Schlenger, W. E., Corry, N. H., Williams, C. S., Kulka, R. A., Mulvaney-Day, N., et al. (2015). A prospective study of mortality and trauma-related risk factors among a nationally representative sample of Vietnam veterans. *American Journal of Epidemiology, 182*, 980−990.

Schmeltzer, S. N., Herman, J. P., & Sah, R. (2016). Neuropeptide Y (NPY) and posttraumatic stress disorder (PTSD): A translational update. *Experimental Neurology, 284*, 196−210.

Schmid, A. W., Lynch, M. A., & Herron, C. E. (2009). The effects of IL-1 receptor antagonist on beta amyloid mediated depression of LTP in the rat CA1 in vivo. *Hippocampus, 19*, 670−676.

Schmidt, E. D., Aguilera, G., Binnekade, R., & Tilders, F. J. (2003). Single administration of interleukin-1 increased corticotropin releasing hormone and corticotropin releasing hormone-receptor mRNA in the hypothalamic paraventricular nucleus which paralleled long-lasting (weeks) sensitization to emotional stressors. *Neuroscience, 116*, 275−283.

Schmidt, E. D., Binnekade, R., Janszen, A. W., & Tilders, F. J. (1996). Short stressor induced long lasting increases of vasopressin stores in hypothalamic corticotropin releasing hormone (CRH) neurons in adult rats. *Journal of Neuroendocrinology, 8*(9), 703−712.

Schmidt, K., Cowen, P. J., Harmer, C. J., Tzortzis, G., Errington, S., & Burnet, P. W. (2015). Prebiotic intake reduces the waking cortisol response and alters emotional bias in healthy volunteers. *Psychopharmacology*, *232*, 1793–1801.

Schmidt, N. B., & Keough, M. E. (2010). Treatment of panic. *Annual Review of Clinical Psychology*, *6*, 241–256.

Schmidt-Reinwald, A., Pruessner, J. C., Hellhammer, D. H., Federenko, I., Rohleder, N., et al. (1999). The cortisol response to awakening in relation to different challenge tests and a 12-hour cortisol rhythm. *Life Sciences*, *64*, 1653–1660.

Schmitz, N., Deschênes, S. S., Burns, R. J., Smith, K. J., Lesage, A., et al. (2016). Depression and risk of type 2 diabetes: The potential role of metabolic factors. *Molecular Psychiatry*, *21*, 1726–1732.

Schoenfeld, T. J., Rada, P., Pieruzzini, P. R., Hsueh, B., & Gould, E. (2013). Physical exercise prevents stress-induced activation of granule neurons and enhances local inhibitory mechanisms in the dentate gyrus. *Journal of Neuroscience*, *33*, 7770–7777.

Schormair, B., Kemlink, D., Mollenhauer, B., Fiala, O., Machetanz, G., et al. (2017). Diagnostic exome sequencing in early-onset Parkinson's disease confirms VPS13C as a rare cause of autosomal-recessive Parkinson's disease. *Clinical Genetics*, *93*(3), 603–612. Available from https://doi.org/10.1111/cge.13124.

Schreiber, S., Schreiber, F., Lockhart, S. N., Horng, A., Bejanin, A., et al. (2017). Alzheimer disease signature neurodegeneration and APOE genotype in mild cognitive impairment with suspected non-Alzheimer disease pathophysiology. *JAMA Neurology*, *74*, 650–659.

Schreier, H. M., & Wright, R. J. (2014). Stress and food allergy: Mechanistic considerations. *Annals of Allergy, Asthma & Immunol*, *112*(4), 296–301.

Schrepf, A., Markon, K., & Lutgendorf, S. K. (2014). From childhood trauma to elevated C-reactive protein in adulthood: The role of anxiety and emotional eating. *Psychosomatic Medicine*, *76*(5), 327–336.

Schroder, K., Hertzog, P. J., Ravasi, T., & Hume, D. A. (2004). Interferon-gamma: An overview of signals, mechanisms and functions. *Joural of Leukocyte Biology*, *75*(2), 163–189.

Schroder, K., & Tschopp, J. (2010). The inflammasomes. *Cell*, *140*, 821–832.

Schroeder, B. O., & Bäckhed, F. (2016). Signals from the gut microbiota to distant organs in physiology and disease. *Nature Medicine (New York, NY, United States)*, *22*, 1079–1089.

Schroeder, M., Jakovcevski, M., Polacheck, T., Lebow, M., Drori, Y., et al. (2017). A methyl-balanced diet prevents CRF-induced prenatal stress-triggered predisposition to binge eating-like phenotype. *Cell Metabolism*, *25*, 1269–1281.

Schrott, L. M., & Crnic, L. S. (1996). Increased anxiety behaviors in autoimmune mice. *Behavioral Neuroscience*, *110*(3), 492–502.

Schür, R. R., Boks, M. P., Geuze, E., Prinsen, H. C., Verhoeven-Duif, N. M., et al. (2016). Development of psychopathology in deployed armed forces in relation to plasma GABA levels. *Psychoneuroendocrinology*, *73*, 263–270.

Schwarcz, R., & Stone, T. W. (2017). The kynurenine pathway and the brain: Challenges, controversies and promises. *Neuropharmacology*, *112*, 237–247.

Schwarz, K. A., Pfister, R., & Büchel, C. (2016). Rethinking explicit expectations: Connecting placebos, social cognition, and contextual perception. *Trends in Cognitive Sciences*, *20*(6), 469–480.

Schwarzschild, M. A., Chen, J. F., & Ascherio, A. (2002). Caffeinated clues and the promise of adenosine A(2A) antagonists in PD. *Neurology*, *58*, 1154–1160.

Scott, F. I., Horton, D. B., Mamtani, R., Haynes, K., Goldberg, D. S., et al. (2016). Administration of antibiotics to children before age 2 years increases risk for childhood obesity. *Gastroenterology*, *151*, 120–129.

Scott, K. A., Ida, M., Peterson, V. L., Prenderville, J. A., Moloney, G. M., et al. (2017). Revisiting Metchnikoff: Age-related alterations in microbiota-gut-brain axis in the mouse. *Brain, Behavior, and Immunity*, *65*, 20–32.

Scott, R. A., Freitag, D. F., Li, L., Chu, A. Y., Surendran, P., et al. (2016). A genomic approach to therapeutic target validation identifies a glucose-lowering GLP1R variant protective for coronary heart disease. *Science Translational Medicine*, *8*, 341ra76.

Scotti, M. A., Carlton, E. D., Demas, G. E., & Grippo, A. J. (2015). Social isolation disrupts innate immune responses in both male and female prairie voles and enhances agonistic behavior in female prairie voles (Microtus ochrogaster). *Hormones and Behavior*, *70*, 7–13.

Sebastiani, P., Gurinovich, A., Bae, H., Andersen, S., Malovini, A., et al. (2017). Four genome-wide association studies identify new extreme longevity variants. *The Journals of Gerontology. Series A, Biological Sciences and Medical Sciences*, *72*(11), 1453–1464.

Secinti, E., Thompson, E. J., Richards, M., & Gaysina, D. (2017). Research Review: Childhood chronic physical illness and adult emotional health — A systematic review and meta-analysis. *Journal of Child Psychology and Psychiatry, and Allied Disciplines*, *58*(7), 753–769.

Segal, Z. V., Williams, J. M. G., & Teasdale, J. (2002). *Mindfulness-based cognitive therapy for depression: A new approach to preventing relapse*. New York: Guilford.

Segerstrom, S. C. (2000). Personality and the immune system: Models, methods, and mechanisms. *Annals of Behavioral Medicine*, *22*(3), 180–190.

Segerstrom, S. C., & Miller, G. E. (2004). Psychological stress and the human immune system: A meta-analytic study of 30 years of inquiry. *Psychological Bulletin*, *130*(4), 601–630.

Seifritz, E., Hatzinger, M., & Holsboer-Trachsler, E. (2016). Efficacy of hypericum extract WS 5570 compared with paroxetine in patients with a moderate major depressive episode — A subgroup analysis. *International Journal of Psychiatry in Clinical Practice, 20*(3), 126—132.

Sekar, A., Bialas, A. R., de Rivera, H., Davis, A., Hammond, T. R., et al. (2016). Schizophrenia risk from complex variation of complement component 4. *Nature, 530,* 177—183.

Sela, U., Euler, C. W., Correa da Rosa, J., & Fischetti, V. A. (2018). Strains of bacterial species induce a greatly varied acute adaptive immune response: The contribution of the accessory genome. *PLoS Pathogens, 14*(1), e1006726.

Seligman, M. E., & Csikszentmihalyi, M. (2000). Positive psychology: An introduction. *American Psychologist, 55,* 5—14.

Seligman, M. E., & Maier, S. F. (1967). Failure to escape traumatic shock. *Journal of Experimental Psychology, 74* (1), 1—9.

Selvaraj, S., Bloomfield, P. S., Cao, B., Veronese, M., Turkheimer, F., et al. (2017). Brain TSPO imaging and gray matter volume in schizophrenia patients and in people at ultra high risk of psychosis: An [(11)C]PBR28 study. *Schizophrenia Research, 169*(1—3), 508, pii: S0920-9964(17)30539-X.

Seng, J., Low, L., Sperlich, M., Ronis, D., & Liberzon, I. (2011). Post-traumatic stress disorder, child abuse history, birthweight and gestational age: A prospective cohort study. *BJOG: An International Journal of Obstetrics and Gynecology, 118*(11), 1329—1339.

Sequeira, A., Mamdani, F., Ernst, C., Vawter, M. P., Bunney, W. E., et al. (2009). Global brain gene expression analysis links glutamatergic and GABAergic alterations to suicide and major depression. *PLoS One, 4,* e6585.

Serafini, G. H., Howland, R., Rovedi, F., Girardi, P., & Amore, M. (2014). The role of ketamine in treatment-resistant depression: A systematic review. *Current Neuropharmacology, 12,* 444—461.

Serova, L., Mulhall, H., & Sabban, E. (2017). NPY1 receptor agonist modulates development of depressive-like behavior and gene expression in hypothalamus in SPS rodent PTSD model. *Frontiers in Neuroscience, 11,* 203.

Serrano-Pozo, A., Mielke, M. L., Gómez-Isla, T., Betensky, R. A., Growdon, J. H., et al. (2011). Reactive glia not only associates with plaques but also parallels tangles in Alzheimer's disease. *American Journal of Pathology, 179,* 1373—1384.

Serrats, J., Grigoleit, J. S., Alvarez-Salas, E., & Sawchenko, P. E. (2017). Pro-inflammatory immune-to-brain signaling is involved in neuroendocrine responses to acute emotional stress. *Brain, Behavior, and Immunity, 62,* 53—63.

Sessa, B. (2015). Turn on and tune in to evidence-based psychedelic research. *The Lancet. Psychiatry, 2*(1), 10—12.

Severance, E. G., Gressitt, K. L., Stallings, C. R., Katsafanas, E., Schweinfurth, L. A., et al. (2016). Candida albicans exposures, sex specificity and cognitive deficits in schizophrenia and bipolar disorder. *NPJ Schizophrenia, 2,* 16018.

Severance, E. G., Gressitt, K. L., Stallings, C. R., Katsafanas, E., Schweinfurth, L. A., et al. (2017). Probiotic normalization of Candida albicans in schizophrenia: A randomized, placebo-controlled, longitudinal pilot study. *Brain, Behavior, and Immunity, 62,* 41—45.

Shaaban, S. Y., El Gendy, Y. G., Mehanna, N. S., El-Senousy, W. M., El-Feki, H. S. A., et al. (2017). The role of probiotics in children with autism spectrum disorder: A prospective, open-label study. *Nutritional Neuroscience, 7,* 1—6.

Shackman, A. J., & Fox, A. S. (2016). Contributions of the central extended amygdala to fear and anxiety. *Journal of Neuroscience, 36,* 8050—8063.

Shackman, A. J., Stockbridge, M. D., Tillman, R. M., Kaplan, C. M., Tromp, D. P., et al. (2016). The neurobiology of dispositional negativity and attentional biases to threat: Implications for understanding anxiety disorders in adults and youth. *Journal of Experimental Psychopathology, 7*(3), 311—342.

Shahbazi, S., Kaur, J., Kuanar, A., Kar, D., Singh, S., et al. (2017). Risk of late-onset Alzheimer's disease by plasma cholesterol: Rational in silico drug investigation of pyrrole-based HMG-CoA reductase inhibitors. *ASSAY and Drug Development Technologies, 15,* 342—351.

Shalev, A., Liberzon, I., & Marmar, C. (2017). Post-traumatic stress disorder. *New England Journal of Medicine, 376,* 2459—2469.

Shalev, A. Y., Ankri, Y., Gilad, M., Israeli-Shalev, Y., Adessky, R., et al. (2016). Long-term outcome of early interventions to prevent posttraumatic stress disorder. *Journal of Clinical Psychiatry, 77,* e580—e587.

Shalev, A. Y., Ankri, Y., Israeli-Shalev, Y., Peleg, T., Adessky, R., et al. (2012). Prevention of posttraumatic stress disorder by early treatment: Results from the Jerusalem trauma outreach and prevention study. *Archives of General Psychiatry, 69,* 166—176.

Sham, H. P., Yu, E. Y., Gulen, M. F., Bhinder, G., Stahl, M., et al. (2013). SIGIRR, a negative regulator of TLR/IL-1R signalling promotes microbiota dependent resistance to colonization by enteric bacterial pathogens. *PLoS Pathogens, 9*(8), e1003539.

Shang, D. S., Yang, Y. M., Zhang, H., Tian, L., Jiang, J. S., et al. (2016). Intracerebral GM-CSF contributes to trans-endothelial monocyte migration in APP/PS1 Alzheimer's disease mice. *Journal of Cerebral Blood Flow & Metabolism, 36,* 1978—1991.

Shanks, N., Kusnecov, A., Pezzone, M., Berkun, J., & Rabin, B. S. (1997). Lactation alters the effects of conditioned stress on immune function. *American Journal of Physiology, 272*(1 Pt 2), R16–R25.

Shanks, N., & Kusnecov, A. W. (1998). Differential immune reactivity to stress in BALB/cByJ and C57BL/6J mice: In vivo dependence on macrophages. *Physiology & Behavior, 65*, 95–103.

Shanks, N., Renton, C., Zalcman, S., & Anisman, H. (1994). Influence of change from grouped to individual housing on a T-cell-dependent immune response in mice: Antagonism by diazepam. *Pharmacology, Biochemistry and Behavior, 47*, 497–502.

Shanks, N., Windle, R. J., Perks, P. A., Harbuz, M. S., Jessop, D. S., et al. (2000). Early-life exposure to endotoxin alters hypothalamic-pituitary-adrenal function and predisposition to inflammation. *Proceedings of the National Academy of Sciences of the United States of America, 97*, 5645–5650.

Shapira, M. (2016). Gut microbiotas and host evolution: Scaling up symbiosis. *Trends in Ecology & Evolution, 31* (7), 539–549.

Sharifi-Sanjani, M., Oyster, N. M., Tichy, E. D., Bedi, K. C., Jr., Harel, O., et al. (2017). Cardiomyocyte-specific telomere shortening is a distinct signature of heart failure in humans. *Journal of the American Heart Association, 6*(9), e005086.

Sharkey, K. A., & Wiley, J. W. (2016). The role of the endocannabinoid system in the gut-brain axis. *Gastroenterology, 151*, 252–266.

Sharma, A. (2016). Systems genomics support for immune and inflammation hypothesis of depression. *Current Neuropharmacology, 14*(7), 749–758.

Sharma, N., Sharma, S., & Nehru, B. (2017). Curcumin protects dopaminergic neurons against inflammation-mediated damage and improves motor dysfunction induced by single intranigral lipopolysaccharide injection. *Inflammopharmacology, 25*(3), 351–368.

Sharpe, M., Goldsmith, K. A., Johnson, A. L., Chalder, T., Walker, J., et al. (2015). Rehabilitative treatments for chronic fatigue syndrome: Long-term follow-up from the PACE trial. *The Lancet Psychiatry, 2*(12), 1067–1074.

Shavali, S., Combs, C. K., & Ebadi, M. (2006). Reactive macrophages increase oxidative stress and alpha-synuclein nitration during death of dopaminergic neuronal cells in co-culture: Relevance to Parkinson's disease. *Neurochemical Research, 31*, 85–94.

Shelton, R. C., & Miller, A. H. (2010). Eating ourselves to death (and despair): The contribution of adiposity and inflammation to depression. *Progress in Neurobiology, 91*, 275–299.

Shen, Y., Yang, L., & Li, R. (2013). What does complement do in Alzheimer's disease? Old molecules with new insights. *Translational Neurodegeneration, 2*(1), 21.

Sherer, M. L., Posillico, C. K., & Schwarz, J. M. (2017). An examination of changes in maternal neuroimmune function during pregnancy and the postpartum period. *Brain, Behavior, and Immunity, 66*, 201–209.

Sheridan, J. F. (1998). Norman Cousins Memorial Lecture 1997. Stress-induced modulation of anti-viral immunity. *Brain, Behavior, and Immunity, 12*, 1–6.

Sherman, M. P., Sherman, J., Arcinue, R., & Niklas, V. (2016). Randomized control trial of human recombinant lactoferrin: A substudy reveals effects on the fecal microbiome of very low birth weight infants. *Journal of Pediatrics, 173*, S37–S42.

Sherwin, E., Rea, K., Dinan, T. G., & Cryan, J. F. (2016). A gut (microbiome) feeling about the brain. *Current Opinion in Gastroenterology, 32*(2), 96–102.

Shi, L., Fatemi, S. H., Sidwell, R. W., & Patterson, P. H. (2003). Maternal influenza infection causes marked behavioral and pharmacological changes in the offspring. *Journal of Neuroscience, 23*, 297–302.

Shi, L., Smith, S. E., Malkova, N., Tse, D., Su, Y., et al. (2009). Activation of the maternal immune system alters cerebellar development in the offspring. *Brain, Behavior, and Immunity, 23*, 116–123.

Shi, L., Tu, N., & Patterson, P. H. (2005). Maternal influenza infection is likely to alter fetal brain development indirectly: The virus is not detected in the fetus. *International Journal of Developmental Neuroscience, 23*(2–3), 299–305.

Shi, M., Bradner, J., Hancock, A. M., Chung, K. A., Quinn, J. F., et al. (2011). Cerebrospinal fluid biomarkers for Parkinson disease diagnosis and progression. *Annals of Neurology, 69*, 570–580.

Shi, Q., Chowdhury, S., Ma, R., Le, K. X., Hong, S., et al. (2017). Complement C3 deficiency protects against neurodegeneration in aged plaque-rich APP/PS1 mice. *Science Translational Medicine, 9*(392), eaaf6295.

Shi, Q., Colodner, K. J., Matousek, S. B., Merry, K., Hong, S., et al. (2015). Complement C3-deficient mice fail to display age-related hippocampal decline. *Journal of Neuroscience, 35*, 13029–13042.

Shichita, T., Ito, M., & Yoshimura, A. (2014). Post-ischemic inflammation regulates neural damage and protection. *Frontiers in Cellular Neuroscience, 8*, 319.

Shields, G. S., Kuchenbecker, S. Y., Pressman, S. D., Sumida, K. D., & Slavich, G. M. (2016). Better cognitive control of emotional information is associated with reduced pro-inflammatory cytokine reactivity to emotional stress. *Stress, 19*(1), 63–68.

Shields, G. S., & Slavich, G. M. (2017). Lifetime stress exposure and health: A review of contemporary assessment methods and biological mechanisms. *Social and Personality Psychology Compass, 11*, pii: e12335.

Shih, I. F., Liew, Z., Krause, N., & Ritz, B. (2016). Lifetime occupational and leisure time physical activity and risk of Parkinson's disease. *Parkinsonism & Related Disordedrs, 28*, 112–117.

Shinto, L., Quinn, J., Montine, T., Dodge, H. H., Woodward, W., et al. (2014). A randomized placebo-controlled pilot trial of omega-3 fatty acids and alpha lipoic acid in Alzheimer's disease. *Journal of Alzheimers Disease, 38*(1), 111–120.

Shirahige, L., Melo, L., Nogueira, F., Rocha, S., & Monte-Silva, K. (2016). Efficacy of noninvasive brain stimulation on pain control in migraine patients: A systematic review and meta-analysis. *Headache, 56*(10), 1565–1596.

Shoaie, S., Ghaffari, P., Kovatcheva-Datchary, P., Mardinoglu, A., Sen, P., et al. (2015). Quantifying diet-induced metabolic changes of the human gut microbiome. *Cell Metabolism, 22*, 320–331.

Shonkoff, J. P., Boyce, W. T., & McEwen, B. S. (2009). Neuroscience, molecular biology, and the childhood roots of health disparities: Building a new framework for health promotion and disease prevention. *The Journal of the American Medical Association, 301*, 2252–2259.

Short, S. J., Lubach, G. R., Karasin, A. I., Olsen, C. W., Styner, M., et al. (2010). Maternal influenza infection during pregnancy impacts postnatal brain development in the rhesus monkey. *Biological Psychiatry, 67*, 965–973.

Shrestha, S., Hirvonen, J., Hines, C. S., Henter, I. D., Svenningsson, P., et al. (2011). Serotonin-1A receptors in major depression quantified using PET: Controversies, confounds, and recommendations. *NeuroImage, 59*, 3243–3251.

Shushtari, N., & Abtahi Froushani, S. M. (2017). Caffeine augments the instruction of anti-inflammatory macrophages by the conditioned medium of mesenchymal stem cells. *Cell Journal, 19*(3), 415–424.

Sibille, E., Arango, V., Galfalvy, H. C., Pavlidis, P., Erraji-Benchekroun, L., et al. (2004). Gene expression profiling of depression and suicide in human prefrontal cortex. *Neuropsychopharmacology, 29*, 351–361.

Siddiqui, S. S., Springer, S. A., Verhagen, A., Sundaramurthy, V., Alisson-Silva, F., et al. (2017). The Alzheimer's disease-protective CD33 splice variant mediates adaptive loss of function via diversion to an intracellular pool. *Journal of Biological Chemistry, 292*, 15312–15320.

Siedlik, J. A., Benedict, S. H., Landes, E. J., Weir, J. P., Vardiman, J. P., et al. (2016). Acute bouts of exercise induce a suppressive effect on lymphocyte proliferation in human subjects: A meta-analysis. *Brain, Behavior, and Immunity, 56*, 343–351.

Sijbrandij, M., Kleiboer, A., Bisson, J. I., Barbui, C., & Cuijpers, P. (2015). Pharmacological prevention of post-traumatic stress disorder and acute stress disorder: A systematic review and meta-analysis. *The Lancet Psychiatry, 2*(5), 413–421.

Silva, L. C., de Araújo, A. L., Fernandes, J. R., Matias Mde, S., Silva, P. R., et al. (2016). Moderate and intense exercise lifestyles attenuate the effects of aging on telomere length and the survival and composition of T cell subpopulations. *Age, 38*, 24.

Silveira, P. P., Gaudreau, H., Atkinson, L., Fleming, A. S., Sokolowski, M. B., et al. (2016). Genetic differential susceptibility to socioeconomic status and childhood obesogenic behavior: Why targeted prevention may be the best societal investment. *JAMA Pediatrics, 170*(4), 359–364.

Silverberg, J. I. (2016). Atopic disease and cardiovascular risk factors in US children. *Journal of Allergy and Clinical Immunology, 137*, 938–940.

Silverman, J. D., Washburne, A. D., Mukherjee, S., & David, L. A. (2017). A phylogenetic transform enhances analysis of compositional microbiota data. *eLife, 6*, e21887.

Silverman, M., Kua, L., Tanca, A., Pala, M., Palomba, A., et al. (2017). Protective major histocompatibility complex allele prevents type 1 diabetes by shaping the intestinal microbiota early in ontogeny. *Proceedings of the National Academy of Sciences of the United States of America, 114*, 9671–9676.

Silverman, M. N., Macdougall, M. G., Hu, F., Pace, T. W., Raison, C. L., et al. (2007). Endogenous glucocorticoids protect against TNF-alpha-induced increases in anxiety-like behavior in virally infected mice. *Molecular Psychiatry, 12*, 408–417.

Silverstein, A. M. (2003). Cellular versus humoral immunology: A century-long dispute. *Nature Immunology, 4*, 425–428.

Simpson, R. J., & Bosch, J. A. (2014). Special issue on exercise immunology: Current perspectives on ageing, health and extreme performance. *Brain, Behavior, and Immunity, 39*, 1–7.

Simpson, R. J., Kunz, H., Agha, N., & Graff, R. (2015). Exercise and the regulation of immune functions. *Progress in Molecular Biology and Translational Science, 135*, 355–380.

Singh, I., Sagare, A. P., Coma, M., Perlmutter, D., Gelein, R., et al. (2013). Low levels of copper disrupt brain amyloid-β homeostasis by altering its production and clearance. *Proceedings of the National Academy of Sciences of the United States of America, 110*, 14771–14776.

Singh, J. B., Fedgchin, M., Daly, E., Xi, L., Melman, C., et al. (2016). Intravenous esketamine in adult treatment-resistant depression: A double-blind, double-randomization, placebo-controlled study. *Biological Psychiatry, 80*, 424–431.

Singh, V., Roth, S., Llovera, G., Sadler, R., Garzetti, D., et al. (2016). Microbiota dysbiosis controls the neuroinflammatory response after stroke. *Journal of Neuroscience, 36*, 7428–7440.

Skilbeck, K. J., Johnston, G. A., & Hinton, T. (2010). Stress and GABA receptors. *Journal of Neurochemistry, 112*, 1115–1130.

Slater, P. G., Yarur, H. E., & Gysling, K. (2016). Corticotropin-releasing factor receptors and their interacting proteins: Functional consequences. *Molecular Pharmacology*, *90*, 627–632.

Slavich, G. M., & Irwin, M. R. (2014). From stress to inflammation and major depressive disorder: A social signal transduction theory of depression. *Psychological Bulletin*, *140*(3), 774–815.

Slavin, J. (2013). Fiber and prebiotics: mechanisms and health benefits. *Nutrients*, *5*, 1417–1435.

Slokar, G., & Hasler, G. (2015). Human endogenous retroviruses as pathogenic factors in the development of schizophrenia. *Frontiers in Psychiatry*, *6*, 183.

Slopen, N., Loucks, E. B., Appleton, A. A., Kawachi, I., Kubzansky, L. D., et al. (2015). Early origins of inflammation: An examination of prenatal and childhood social adversity in a prospective cohort study. *Psychoneuroendocrinology*, *51*, 403–413.

Ślusarczyk, J., Trojan, E., Głombik, K., Budziszewska, B., Kubera, M., et al. (2015). Prenatal stress is a vulnerability factor for altered morphology and biological activity of microglia cells. *Frontiers in Cellular Neuroscience*, *12*(9), 82.

Slyepchenko, A., Carvalho, A. F., Cha, D. S., Kasper, S., & McIntyre, R. S. (2014). Gut emotions – Mechanisms of action of probiotics as novel therapeutic targets for depression and anxiety disorders. *CNS & Neurological Disorders Drug Targets*, *13*(10), 1770–1786.

Slyepchenko, A., Maes, M., Köhler, C. A., Anderson, G., Quevedo, J., et al. (2016). T helper 17 cells may drive neuroprogression in major depressive disorder: Proposal of an integrative model. *Neuroscience & Biobehavioral Reviews*, *64*, 83–100.

Smeyne, M., Jiao, Y., Shepherd, K. R., & Smeyne, R. J. (2005). Glia cell number modulates sensitivity to MPTP in mice. *Glia*, *52*(2), 144–152.

Smith, A. K., Conneely, K. N., Kilaru, V., Mercer, K. B., Weiss, T. E., et al. (2011). Differential immune system DNA methylation and cytokine regulation in posttraumatic stress disorder. *American Journal of Medical Genetics. Part B, Neuropsychiatric Genetics*, *156B*(6), 700–708.

Smith, J. A., Zhao, W., Wang, X., Ratliff, S. M., Mukherjee, B., et al. (2017). Neighbourhood characteristics influence DNA methylation of genes involved in stress response and inflammation: The multi-ethnic study of Atherosclerosis. *Epigenetics*, *12*(8), 662–673.

Smith, M. J., & White, K. L., Jr. (2010). Establishment and comparison of delayed-type hypersensitivity models in the B(6)C(3)F(1) mouse. *Journal of Immunotoxicology*, *7*(4), 308–317.

Smith, M. L., Hostetler, C. M., Heinricher, M. M., & Ryabinin, A. E. (2016). Social transfer of pain in mice. *Science Advances*, *2*(10), e1600855.

Smith, P. M., Tuomisto, M. T., Blumenthal, J., Sherwood, A., Parkkinen, L., et al. (2013). Psychosocial correlates of atrial natriuretic peptide: A marker of vascular health. *Annals of Behavioral Medicine*, *45*(1), 99–109.

Snijders, A. M., Langley, S. A., Kim, Y. M., Brislawn, C. J., Noecker, C., et al. (2016). Influence of early life exposure, host genetics and diet on the mouse gut microbiome and metabolome. *Nature Microbiology*, *2*, 16221.

Soares, V. P., & Campos, A. C. (2017). Evidences for the anti-panic actions of cannabidiol. *Current Neuropharmacology*, *15*(2), 291–299.

Sobota, R. S., Stein, C. M., Kodaman, N., Scheinfeldt, L. B., Maro, I., et al. (2016). A locus at 5q33.3 confers resistance to tuberculosis in highly susceptible individuals. *American Journal of Human Genetics*, *98*(3), 514–524.

Söderholm, J. D., & Perdue, M. H. (2001). Stress and gastrointestinal tract. II. Stress and intestinal barrier function. *American Journal of Physiology Gastrointestinal and Liver Physiology*, *280*(1), G7–G13.

Solmi, F., Hayes, J. F., Lewis, G., & Kirkbride, J. B. (2017). Curiosity killed the cat: No evidence of an association between cat ownership and psychotic symptoms at ages 13 and 18 years in a UK general population cohort. *Psychological Medicine*, *47*(9), 1659–1667.

Sommer, I. E., de Witte, L., Begemann, M., & Kahn, R. S. (2012). Nonsteroidal anti-inflammatory drugs in schizophrenia: Ready for practice or a good start? A meta-analysis. *Journal of Clinical Psychiatry*, *73*(4), 414–419.

Sompayrac, L. (2016). *How the immune system works* (5th ed.). West Sussex, UK: Wiley-Blackwell.

Song, M., Hu, F. B., Wu, K., Must, A., Chan, A. T., et al. (2016). Trajectory of body shape in early and middle life and all cause and cause specific mortality: Results from two prospective US cohort studies. *British Medical Journal*, *353*, i2195.

Sonnenburg, E. D., Smits, S. A., Tikhonov, M., Higginbottom, S. K., Wingreen, N. S., et al. (2016). Diet-induced extinctions in the gut microbiota compound over generations. *Nature*, *529*(7585), 212–215.

Sonnenburg, J. L., & Bäckhed, F. (2016). Diet-microbiota interactions as moderators of human metabolism. *Nature*, *535*(7610), 56–64.

Sorg, R. A., Lin, L., van Doorn, G. S., Sorg, M., Olson, J., et al. (2016). Collective resistance in microbial communities by intracellular antibiotic deactivation. *PLoS Biology*, *14*(12), e2000631.

Sorge, R. E., & Totsch, S. K. (2017). Sex differences in pain. *Journal of Neuroscience Research*, *95*(6), 1271–1281.

Soriano-Tárraga, C., Jiménez-Conde, J., Giralt-Steinhauer, E., Mola-Caminal, M., Vivanco-Hidalgo, R. M., et al. (2016). Epigenome-wide association study identifies TXNIP gene associated with type 2 diabetes mellitus and sustained hyperglycemia. *Human Molecular Genetics, 25*(3), 609–619.

Sorrells, S. F., Caso, J. R., Munhoz, C. D., & Sapolsky, R. M. (2009). The stressed CNS: When glucocorticoids aggravate inflammation. *Neuron, 64*(1), 33–39.

Spahn, V., Del Vecchio, G., Labuz, D., Rodriguez-Gaztelumendi, A., Massaly, N., et al. (2017). A nontoxic pain killer designed by modeling of pathological receptor conformations. *Science, 355*(6328), 966–969.

Specht, P. G. (2007). The Peltzman effect: Do safety regulations increase unsafe behavior? *Journal of Safety, Health and Environmental Research, 4*, 1–2.

Spellberg, B. (2016). The new antibiotic mantra: "Shorter is better". *JAMA Internal Medicine, 176*(9), 1254–1255.

Spencer, S. J., Martin, S., Mouihate, A., & Pittman, Q. J. (2006). Early-life immune challenge: Defining a critical window for effects on adult responses to immune challenge. *Neuropsychopharmacology, 31*, 1910–1918.

Spencer, S. J., Xu, L., Clarke, M. A., Lemus, M., Reichenbach, A., et al. (2012). Ghrelin regulates the hypothalamic-pituitary-adrenal axis and restricts anxiety after acute stress. *Biological Psychiatry, 72*(6), 457–465.

Spencer, S. P., & Belkaid, Y. (2012). Dietary and commensal derived nutrients: Shaping mucosal and systemic immunity. *Current Opinion in Immunology, 24*(4), 379–384.

Spiegel, K., Sheridan, J. F., & Van Cauter, E. (2002). Effect of sleep deprivation on response to immunization. *The Journal of the American Medical Association, 288*(12), 1471–1472.

Spierling, S. R., & Zorrilla, E. P. (2017). Don't stress about CRF: Assessing the translational failures of CRF1 antagonists. *Psychopharmacology (Berlin, Germany), 234* (9–10), 1467–1481.

Spinler, J. K., Ross, C. L., & Savidge, T. C. (2016). Probiotics as adjunctive therapy for preventing Clostridium difficile infection — What are we waiting for? *Anaerobe, 41*, 51–57.

Spitzer, C., Barnow, S., Völzke, H., Wallaschofski, H., John, U., et al. (2010). Association of posttraumatic stress disorder with low-grade elevation of C-reactive protein: evidence from the general population. *Journal of Psychiatric Research, 44*(1), 15–21.

Spohn, S. N., & Mawe, G. M. (2017). Non-conventional features of peripheral serotonin signaling — The gut and beyond. *Nature Reviews Gastroenterology & Hepatology, 14* (7), 412–420.

Spor, A., Koren, O., & Ley, R. (2011). Unravelling the effects of the environment and host genotype on the gut microbiome. *Nature Reviews Microbiology, 9*(4), 279–290.

Stalnaker, T. A., Cooch, N. K., & Schoenbaum, G. (2015). What the orbitofrontal cortex does not do. *Nature Neuroscience, 18*(5), 620–627.

Stampanoni Bassi, M., Garofalo, S., Marfia, G. A., Gilio, L., Simonelli, I., et al. (2017). Amyloid-β homeostasis bridges inflammation, synaptic plasticity deficits and cognitive dysfunction in multiple sclerosis. *Frontiers in Molecular Neuroscience, 10*, 390.

Stanley, D., Moore, R. J., & Wong, C. H. Y. (2018). An insight into intestinal mucosal microbiota disruption after stroke. *Scientific Reports, 8*(1), 568.

Starcevic, V. (2014). The reappraisal of benzodiazepines in the treatment of anxiety and related disorders. *Expert Review of Neurotherapeutics, 14*(11), 1275–1286.

Stark, J. L., Avitsur, R., Hunzeker, J., Padgett, D. A., & Sheridan, J. F. (2002). Interleukin-6 and the development of social disruption-induced glucocorticoid resistance. *Journal of Neuroimmunology, 124*(1–2), 9–15.

Stark, J. L., Avitsur, R., Padgett, D. A., Campbell, K. A., Beck, F. M., et al. (2001). Social stress induces glucocorticoid resistance in macrophages. *American Journal of Physiology Regulatory Integrative and Comparative Physiology, 280*, R1799–1805.

Steenen, S. A., van Wijk, A. J., van der Heijden, G. J., van Westrhenen, R., de Lange, J., et al. (2016). Propranolol for the treatment of anxiety disorders: Systematic review and meta-analysis. *Journal of Psychopharmacology, 30*(2), 128–139.

Steffens, N. K., Cruwys, T., Haslam, C., Jetten, J., & Haslam, S. A. (2016). Social group memberships in retirement are associated with reduced risk of premature death: Evidence from a longitudinal cohort study. *BMJ Open, 6*(2), e010164.

Steiger, V. R., Brühl, A. B., Weidt, S., Delsignore, A., Rufer, M., et al. (2016). Pattern of structural brain changes in social anxiety disorder after cognitive behavioral group therapy: A longitudinal multimodal MRI study. *Molecular Psychiatry, 22*(8), 1164–1171.

Stein, M., Steckler, T., Lightfoot, J. D., Hay, E., & Goddard, A. W. (2010). Pharmacologic treatment of panic disorder. *Current Topics in Behavioral Neurosciences, 2*, 469–478.

Stein, M. B., Chen, C. Y., Ursano, R. J., Cai, T., Gelernter, J., et al. (2016). Genome-wide association studies of posttraumatic stress disorder in 2 cohorts of US Army soldiers. *JAMA Psychiatry, 73*, 695–704.

Steiner, J., Walter, M., Glanz, W., Sarnyai, Z., Bernstein, H. G., et al. (2013). Increased prevalence of diverse N-methyl-D-aspartate glutamate receptor antibodies in patients with an initial diagnosis of schizophrenia: Specific relevance of IgG NR1a antibodies for distinction from N-methyl-D-aspartate glutamate receptor encephalitis. *JAMA Psychiatry, 70*(3), 271–278.

Steinman, L. (2015). No quiet surrender: Molecular guardians in multiple sclerosis brain. *Journal of Clinical Investigation, 125*, 1371–1378.

Steinman, M. Q., Duque-Wilckens, N., Greenberg, G. D., Hao, R., Campi, K. L., et al. (2016). Sex-specific effects of stress on oxytocin neurons correspond with responses to intranasal oxytocin. *Biological Psychiatry*, *80*(5), 406−414.

Stelmashook, E. V., Isaev, N. K., Genrikhs, E. E., Amelkina, G. A., Khaspekov, L. G., et al. (2014). Role of zinc and copper ions in the pathogenetic mechanisms of Alzheimer's and Parkinson's diseases. *Biochemestry, Biokhimiia*, *79*(5), 391−396.

Stephan, A. H., Barres, B. A., & Stevens, B. (2012). The complement system: An unexpected role in synaptic pruning during development and disease. *Annual Review of Neuroscience*, *35*, 369−389.

Stephan, A. H., Madison, D. V., Mateos, J. M., Fraser, D. A., Lovelett, E. A., et al. (2013). A dramatic increase of C1q protein in the CNS during normal aging. *The Journal of Neuroscience*, *33*(33), 13460−13474.

Stephens, G., Derry, S., & Moore, R. A. (2016). Paracetamol (acetaminophen) for acute treatment of episodic tension-type headache in adults. *Cochrane Database of Systematic Reviews*, CD011889.

Steptoe, A., Hamer, M., & Chida, Y. (2007). The effects of acute psychological stress on circulating inflammatory factors in humans: A review and meta-analysis. *Brain, Behavior, and Immunity*, *21*(7), 901−912.

Steptoe, A., & Kivimäki, M. (2013). Stress and cardiovascular disease: An update on current knowledge. *Annual Review of Public Health*, *34*, 337−354.

Steptoe, A., Shamaei-Tousi, A., Gylfe, Å., Henderson, B., Bergström, S., et al. (2008). Socioeconomic status, pathogen burden and cardiovascular disease risk. *Heart (British Cardiac Society)*, *93*, 1567−1570.

Sternberg, E. M. (2006). Neural regulation of innate immunity: A coordinated nonspecific host response to pathogens. *Nature Reviews Immunology*, *6*(4), 318−328.

Stevens, J. S., Kim, Y. J., Galatzer-Levy, I. R., Reddy, R., Ely, T. D., et al. (2017). Amygdala reactivity and anterior cingulate habituation predict posttraumatic stress disorder symptom maintenance after acute civilian trauma. *Biological Psychiatry*, *81*(12), 1023−1029.

Stewart, R. A. H., Colquhoun, D. M., Marschner, S. L., Kirby, A. C., Simes, J., et al. (2017). Persistent psychological distress and mortality in patients with stable coronary artery disease. *Heart (British Cardiac Society)*, *103*(23), 1860−1866.

Stieg, M. R., Sievers, C., Farr, O., Stalla, G. K., & Mantzoros, C. S. (2014). Leptin: A hormone linking activation of neuroendocrine axes with neuropathology. *Psychoneuroendocrinology*, *51*, 47−57.

Stilling, R. M., Dinan, T. G., & Cryan, J. F. (2014). Microbial genes, brain & behaviour − Epigenetic regulation of the gut-brain axis. *Genes, Brain and Behavior*, *13*(1), 69−86.

Stilling, R. M., Ryan, F. J., Hoban, A. E., Shanahan, F., Clarke, G., et al. (2015). Microbes & neurodevelopment−Absence of microbiota during early life increases activity-related transcriptional pathways in the amygdala. *Brain, Behavior, and Immunity*, *50*, 209−220.

Stilling, R. M., van de Wouw, M., Clarke, G., Stanton, C., Dinan, T. G., et al. (2016). The neuropharmacology of butyrate: The bread and butter of the microbiota-gut-brain axis? *Neurochemistry International*, *99*, 110−132.

Stockmeier, C. A. (2003). Involvement of serotonin in depression: Evidence from postmortem and imaging studies of serotonin receptors and the serotonin transporter. *Journal of Psychiatric Research*, *37*(5), 357−373.

Stokes, A., & Preston, S. H. (2016). Revealing the burden of obesity using weight histories. *Proceedings of the National Academy of Sciences of the United States of America*, *113*(3), 572−577.

Stokes, J. M., MacNair, C. R., Ilyas, B., French, S., Côté, J. P., et al. (2017). Pentamidine sensitizes gram-negative pathogens to antibiotics and overcomes acquired colistin resistance. *Nature Microbiology*, *2*, 17028.

Stokes, L., Layhadi, J. A., Bibic, L., Dhuna, K., & Fountain, S. J. (2017). P2X4 receptor function in the nervous system and current breakthroughs in pharmacology. *Frontiers in Pharmacology*, *23*, 291.

Stokowska, A., Atkins, A. L., Morán, J., Pekny, T., Bulmer, L., et al. (2016). Complement peptide C3a stimulates neural plasticity after experimental brain ischaemia. *Brain*, *140*(2), 353−369.

Stouffer, M. A., Woods, C. A., Patel, J. C., Lee, C. R., Witkovsky, P., et al. (2015). Insulin enhances striatal dopamine release by activating cholinergic interneurons and thereby signals reward. *Nature Communications*, *6*, 8543.

Strachan, D. P. (2000). Family size, infection and atopy: The first decade of the "hygiene hypothesis". *Thorax*, *55* (Suppl 1), S2.

Strati, F., Cavalieri, D., Albanese, D., De Felice, C., Donati, C., et al. (2017). New evidences on the altered gut microbiota in autism spectrum disorders. *Microbiome*, *5*, 24.

Straub, R. H., Dhabhar, F. S., Bijlsma, J. W., & Cutolo, M. (2005). How psychological stress via hormones and nerve fibers may exacerbate rheumatoid arthritis. *Arthritis & Rheumatology*, *52*(1), 16−26.

Strissel, K. J., Denis, G. V., & Nikolajczyk, B. S. (2014). Immune regulators of inflammation in obesity-associated type 2 diabetes and coronary artery disease. *Current Opinion in Endocrinology, Diabetes, and Obesity*, *21*(5), 330−338.

Stuart, M. J., & Baune, B. T. (2012). Depression and type 2 diabetes: Inflammatory mechanisms of a psychoneuroendocrine co-morbidity. *Neuroscience & Biobehavioral Reviews*, *36*(1), 658−676.

Stubbs, E. G., Ritvo, E. R., & Mason-Brothers, A. (1985). Autism and shared parental HLA antigens. *Journal of the American Academy of Child Psychiatry, 24*(2), 182–185.

Stubbs, T. M., Bonder, M. J., Stark, A. K., Krueger, F., BI Ageing Clock Team., et al. (2017). Multi-tissue DNA methylation age predictor in mouse. *Genome Biology, 18*, 68.

Suárez-Zamorano, N., Fabbiano, S., Chevalier, C., Stojanović, O., Colin, D. J., et al. (2015). Microbiota depletion promotes browning of white adipose tissue and reduces obesity. *Nature Medicine (New York, NY, United States), 21*, 1497–1501.

Sublette, M. E., Galfalvy, H. C., Fuchs, D., Lapidus, M., Grunebaum, M. F., et al. (2011). Plasma kynurenine levels are elevated in suicide attempters with major depressive disorder. *Brain, Behavior, and Immunity, 25* (6), 1272–1278.

Subramaniam, S. R., & Federoff, H. J. (2017). Targeting microglial activation states as a therapeutic avenue in Parkinson's disease. *Frontiers in Aging Neuroscience, 9*, 176.

Sudo, N. (2014). Microbiome, HPA axis and production of endocrine hormones in the gut. *Microbial Endocrinology: The Microbiota-Gut-Brain Axis in Health and Disease* (pp. 177–194). New York: Springer.

Suderman, M., Borghol, N., Pappas, J. J., Pinto Pereira, S. M., Pembrey, M., et al. (2014). Childhood abuse is associated with methylation of multiple loci in adult DNA. *BMC Medical Genomics, 7*, 13.

Sugama, S., & Kakinuma, Y. (2016). Loss of dopaminergic neurons occurs in the ventral tegmental area and hypothalamus of rats following chronic stress: Possible pathogenetic loci for depression involved in Parkinson's disease. *Neurosciences Research, 111*, 48–55.

Sugama, S., Sekiyama, K., Kodama, T., Takamatsu, Y., Takenouchi, T., et al. (2016). Chronic restraint stress triggers dopaminergic and noradrenergic neurodegeneration: Possible role of chronic stress in the onset of Parkinson's disease. *Brain, Behavior, and Immunity, 51*, 39–46.

Sujan, A. C., Rickert, M. E., Öberg, A. S., Quinn, P. D., Hernández-Díaz, S., et al. (2017). Associations of maternal antidepressant use during the first trimester of pregnancy with preterm birth, small for gestational age, autism spectrum disorder, and attention-deficit/hyperactivity disorder in offspring. *The Journal of the American Medical Association, 317*(15), 1553–1562.

Sulakhiya, K., Keshavlal, G. P., Bezbaruah, B. B., Dwivedi, S., Gurjar, S. S., et al. (2016). Lipopolysaccharide induced anxiety-and depressive-like behaviour in mice are prevented by chronic pre-treatment of esculetin. *Neuroscience Letters, 611*, 106–111.

Summers, P. M., Hartmann, D. A., Hui, E. S., Nie, X., Deardorff, R. L., et al. (2017). Functional deficits induced by cortical microinfarcts. *Journal of Cerebral Blood Flow & Metabolism, 37*(11), 3599–3614.

Sun, J., Ma, H., Yu, C., Lv, J., Guo, Y., et al. (2016). Association of major depressive episodes with stroke risk in a prospective study of 0.5 million Chinese adults. *Stroke, 47*(9), 2203–2208.

Sun, J., Wang, F., Hong, G., Pang, M., Xu, H., et al. (2016). Antidepressant-like effects of sodium butyrate and its possible mechanisms of action in mice exposed to chronic unpredictable mild stress. *Neuroscience Letters, 618*, 159–166.

Sun, L., Li, Y., Jia, X., Wang, Q., Li, Y., et al. (2017). Neuroprotection by IFN-gamma via astrocyte-secreted IL-6 in acute neuroinflammation. *Oncotarget, 8*(25), 40065–40078.

Sun, Q., Xie, N., Tang, B., Li, R., & Shen, Y. (2017). Alzheimer's disease: From genetic variants to the distinct pathological mechanisms. *Frontiers in Molecular Neuroscience, 10*, 319.

Sundaram, J. R., Poore, C. P., Sulaimee, N. H. B., Pareek, T., Cheong, W. F., et al. (2017). Curcumin ameliorates neuroinflammation, neurodegeneration, and memory deficits in p25 transgenic mouse model that bears hallmarks of Alzheimer's disease. *Journal of Alzheimer's Disease, 60* (4), 1429–1442.

Sundquist, J., Palmér, K., Johansson, L. M., & Sundquist, K. (2017). The effect of mindfulness group therapy on a broad range of psychiatric symptoms: A randomised controlled trial in primary health care. *European Psychiatry, 43*, 19–27.

Surwit, R. S., Van Tilburg, M. A., Zucker, N., McCaskill, C. C., Parekh, P., et al. (2002). Stress management improves long-term glycemic control in type 2 diabetes. *Diabetes Care, 25*, 30–34.

Sutton, L. P., Ostrovskaya, O., Dao, M., Xie, K., Orlandi, C., et al. (2016). Regulator of G-protein signaling 7 regulates reward behavior by controlling opioid signaling in the striatum. *Biological Psychiatry, 80*(3), 235–245.

Svenningsson, P., Kim, Y., Warner-Schmidt, J., Oh, Y. S., & Greengard, P. (2013). p11 and its role in depression and therapeutic responses to antidepressants. *Nature Reviews Neuroscience, 14*(10), 673–680.

Svensson, M., Lexell, J., & Deierborg, T. (2015). Effects of physical exercise on neuroinflammation, neuroplasticity, neurodegeneration, and behavior: What we can learn from animal models in clinical settings. *Neurorehabilitation and Neural Repair, 29*(6), 577–589.

Swartz, J. R., Prather, A. A., Di Iorio, C. R., Bogdan, R., & Hariri, A. R. (2016). A functional interleukin-18 haplotype predicts depression and anxiety through increased threat-related amygdala reactivity in women but not men. *Neuropsychopharmacology, 42*(2), 419–426.

Swartz, J. R., Prather, A. A., & Hariri, A. R. (2017). Threat-related amygdala activity is associated with peripheral CRP concentrations in men but not women. *Psychoneuroendocrinology, 78*, 93–96.

Szaruga, M., Munteanu, B., Lismont, S., Veugelen, S., Horré, K., et al. (2017). Alzheimer's-causing mutations shift Aβ length by destabilizing γ-secretase-Aβn interactions. *Cell, 170*(3), 443–456.e14.

Szasz, T. (1973). *The second sin* (p. 113) New York: Doubleday.

Szasz, T. (1974). *The myth of mental illness: Foundations of a theory of personal conduct*. New York: Harper & Row.

Szyf, M. (2009). Epigenetics, DNA methylation, and chromatin modifying drugs. *Annual Review of Pharmacology and Toxicology, 49*, 243–263.

Szyf, M. (2011). The early life social environment and DNA methylation: DNA methylation mediating the long-term impact of social environments early in life. *Epigenetics, 6*(8), 971–978.

Szyf, M. (2015). Epigenetics, a key for unlocking complex CNS disorders? Therapeutic implications. *European Neuropsychopharmacology, 25*(5), 682–702.

Szyf, M., & Bick, J. (2013). DNA methylation: A mechanism for embedding early life experiences in the genome. *Child Development, 84*(1), 49–57.

Szyf, M., Tang, Y. Y., Hill, K. G., & Musci, R. (2016). The dynamic epigenome and its implications for behavioral interventions: A role for epigenetics to inform disorder prevention and health promotion. *Translational and Behavioral Medicine, 6*(1), 55–62.

Tabak, B. A., Meyer, M. L., Castle, E., Dutcher, J. M., Irwin, M. R., et al. (2015). Vasopressin, but not oxytocin, increases empathic concern among individuals who received higher levels of paternal warmth: A randomized controlled trial. *Psychoneuroendocrinology, 51*, 253–261.

Tabak, B. A., Meyer, M. L., Dutcher, J. M., Castle, E., Irwin, M. R., et al. (2016). Oxytocin, but not vasopressin, impairs social cognitive ability among individuals with higher levels of social anxiety: A randomized controlled trial. *Social Cognitive and Affective Neuroscience, 11*(8), 1272–1279.

Tabak, B. A., Vrshek-Schallhorn, S., Zinbarg, R. E., Prenoveau, J. M., Mineka, S., et al. (2016). Interaction of CD38 variant and chronic interpersonal stress prospectively predicts social anxiety and depression symptoms over six years. *Clinical Psychological Science, 4*(1), 17–27.

Taché, Y., Larauche, M., Yuan, P. Q., & Million, M. (2017). Brain and gut CRF signaling: Biological actions and role in the gastrointestinal tract. *Current Molecular Pharmacol, 11*(1), 51–71.

Taha, S. A., Matheson, K., & Anisman, H. (2014). H1N1 was not all that scary: Uncertainty and stressor appraisals predict anxiety related to a coming viral threat. *Stress and Health: Journal of the International Society for the Investigation of Stress., 30*(2), 149–157.

Tak, C. R., Job, K. M., Schoen-Gentry, K., Campbell, S. C., Carroll, P., et al. (2017). The impact of exposure to antidepressant medications during pregnancy on neonatal outcomes: A review of retrospective database cohort studies. *European Journal of Clinical Pharmacology, 73*(9), 1055–1069.

Takahashi, A., Chung, J. R., Zhang, S., Zhang, H., Grossman, Y., et al. (2017). Establishment of a repeated social defeat stress model in female mice. *Scientific Reports, 7*(1), 12838.

Talge, N. M., Neal, C., & Glover, V. (2007). Antenatal maternal stress and long-term effects on child neurodevelopment: How and why? *Journal of Child Psychology and Psychiatry, and Allied Disciplines, 48*(3–4), 245–261.

Talukdar, S., Zhou, Y., Li, D., Rossulek, M., Dong, J., et al. (2016). A long-acting FGF21 molecule, PF-05231023, decreases body weight and improves lipid profile in non-human primates and type 2 diabetic subjects. *Cell Metabolism, 23*, 427–440.

Tan, L. L., Pelzer, P., Heinl, C., Tang, W., Gangadharan, V., et al. (2017). A pathway from midcingulate cortex to posterior insula gates nociceptive hypersensitivity. *Nature Neuroscience, 20*(11), 1591–1601.

Tang, W. H., & Hazen, S. L. (2017). Microbiome, trimethylamine N-oxide, and cardiometabolic disease. *Translational Research, 179*, 108–115.

Tang, W. H., Wang, Z., Fan, Y., Levison, B., Hazen, J. E., et al. (2014). Prognostic value of elevated levels of intestinal microbe-generated metabolite trimethylamine-N-oxide in patients with heart failure: Refining the gut hypothesis. *Journal of the American College of Cardiology, 64*(18), 1908–1914.

Tang, W. W., Wang, Z., Li, X. S., Fan, Y., Li, D. S., et al. (2017). Increased trimethylamine N-oxide portends high mortality risk independent of glycemic control in patients with type 2 diabetes mellitus. *Clinical Chemist, 63*(1), 297–306.

Tang, Y., & Le, W. (2016). Differential roles of M1 and M2 microglia in neurodegenerative diseases. *Molecular Neurobiology, 53*(2), 1181–1194.

Tang, Z., Ye, G., Chen, X., Pan, M., Fu, J., et al. (2018). Peripheral proinflammatory cytokines in Chinese patients with generalised anxiety disorder. *Journal of Affective Disorder, 225*, 593–598.

Tanner, C. M., Kamel, F., Ross, G. W., Hoppin, J. A., Goldman, S. M., et al. (2011). Rotenone, paraquat, and Parkinson's disease. *Environmental Health Perspectives, 119*(6), 866–872.

Tapia-Rojas, C., Aranguiz, F., Varela-Nallar, L., & Inestrosa, N. C. (2016). Voluntary running attenuates memory loss, decreases neuropathological changes and induces neurogenesis in a mouse model of Alzheimer's disease. *Brain Pathology*, *26*(1), 62−74.

Tarr, A. J., Galley, J. D., Fisher, S. E., Chichlowski, M., Berg, B. M., et al. (2015). The prebiotics 3′Sialyllactose and 6′Sialyllactose diminish stressor-induced anxiety-like behavior and colonic microbiota alterations: Evidence for effects on the gut-brain axis. *Brain, Behavior, and Immunity*, *50*, 166−177.

Tawakol, A., Ishai, A., Takx, R. A., Figueroa, A. L., Ali, A., et al. (2017). Relation between resting amygdalar activity and cardiovascular events: A longitudinal and cohort study. *Lancet*, *389*(10071), 834−845.

Taylor, A. M., Mehrabani, S., Liu, S., Taylor, A. J., & Cahill, C. M. (2017). Topography of microglial activation in sensory- and affect-related brain regions in chronic pain. *Journal of Neuroscience Research*, *95*(6), 1330−1335.

Taylor, S. E., Seeman, T. E., Eisenberger, N. I., Kozanian, T. A., Moore, A. N., & Moons, W. G. (2010). Effects of a supportive or an unsupportive audience on biological and psychological responses to stress. *Journal of Personality and Social Psychology*, *98*, 487−492.

Taylor, B. K., & Westlund, K. N. (2017). The noradrenergic locus coeruleus as a chronic pain generator. *Journal of Neuroscience Research*, *95*(6), 1336−1346.

Taylor, F., Huffman, M. D., Macedo, A. F., Moore, T. H., Burke, M., et al. (2013). Statins for the primary prevention of cardiovascular disease. *Cochrane Database of Systematic Reviews*, (1), CD004816.

Taylor, S. E., Burklund, L. J., Eisenberger, N. I., Lehman, B. J., Hilmert, C. J., et al. (2008). Neural bases of moderation of cortisol stress responses by psychosocial resources. *Journal of Personality and Social Psycholology*, *95*(1), 197−211.

Taylor, S. E., Klein, L. C., Lewis, B. P., Gruenewald, T. L., Gurung, R. A., et al. (2000). Biobehavioral responses to stress in females: Tend-and-befriend, not fight-or-flight. *Psychological Review*, *107*(3), 411.

Teche, S. P., Rovaris, D. L., Aguiar, B. W., Hauck, S., Vitola, E. S., et al. (2017). Resilience to traumatic events related to urban violence and increased IL10 serum levels. *Psychiatry Research*, *250*, 136−140.

Tellez, L. A., Han, W., Zhang, X., Ferreira, T. L., Perez, I. O., et al. (2016). Separate circuitries encode the hedonic and nutritional values of sugar. *Nature Neuroscience*, *19*(3), 465−470.

Tétreault, P., Mansour, A., Vachon-Presseau, E., Schnitzer, T. J., Apkarian, A. V., et al. (2016). Brain connectivity predicts placebo response across chronic pain clinical trials (2016). *PLoS Biology*, *14*(10), e1002570.

Thaiss, C. A., Levy, M., Korem, T., Dohnalová, L., Shapiro, H., et al. (2016). Microbiota diurnal rhythmicity programs host transcriptome oscillations. *Cell*, *167*(6), 1495−1510.

Thaiss, C. A., Zmora, N., Levy, M., & Elinav, E. (2016). The microbiome and innate immunity. *Nature*, *535*(7610), 65−74.

Thakur, P., Breger, L. S., Lundblad, M., Wan, O. W., Mattsson, B., et al. (2017). Modeling Parkinson's disease pathology by combination of fibril seeds and α-synuclein overexpression in the rat brain. *Proceedings of the National Academy of Sciences of the United States of America*, *114*(39), E8284−E8293.

Thaler, R., & Sunstein, C. R. (2008). Nudge: Improving decisions about health, wealth, and happiness. New Haven, Conn: Yale University Press.

The Global BMI Mortality Collaboration., Di Angelantonio, E., Bhupathiraju, Sh. N., Wormser, D., Gao, P., et al. (2016). Body-mass index and all-cause mortality: Individual-participant-data meta-analysis of 239 prospective studies in four continents. *Lancet*, *388*, 776−786.

The Lancet. (2017). The health inequalities and ill-health of children in the UK. *Lancet*, *389*(10068), 477.

Theoharides, T. C., Tsilioni, I., Patel, A. B., & Doyle, R. (2016). Atopic diseases and inflammation of the brain in the pathogenesis of autism spectrum disorders. *Translational Psychiatry*, *6*(6), e844. Available from https://doi.org/10.1038/tp.2016.77.

Thevaranjan, N., Puchta, A., Schulz, C., Naidoo, A., Szamosi, J. C., et al. (2017). Age-associated microbial dysbiosis promotes intestinal permeability, systemic inflammation, and macrophage dysfunction. *Cell Host & Microbe*, *21*(4), 455−466.

Thévenet, J., Pescini Gobert, R., Hooft van Huijsduijnen, R., Wiessner, C., & Sagot, Y. J. (2011). Regulation of LRRK2 expression points to a functional role in human monocyte maturation. *PLoS One*, *6*(6), e21519.

Thoeringer, C. K., Henes, K., Eder, M., Dahlhoff, M., Wurst, W., et al. (2012). Consolidation of remote fear memories involves corticotropin-releasing hormone (CRH) receptor type 1-mediated enhancement of AMPA receptor GluR1 signaling in the dentate gyrus. *Neuropsychopharmacology*, *37*(3), 787−796.

Thomaes, K., Engelhard, I. M., Sijbrandij, M., Cath, D. C., & Van den Heuvel, O. A. (2016). Degrading traumatic memories with eye movements: A pilot functional MRI study in PTSD. *European Journal of Psychotraumatology*, *7*, 31371.

Thompson, P. M., Vidal, C., Giedd, J. N., Gochman, P., Blumenthal, J., et al. (2001). Mapping adolescent brain change reveals dynamic wave of accelerated gray matter loss in very early-onset schizophrenia. *Proceedings of the National Academy of Sciences of the United States of America*, *98*, 11650−11655.

Thompson, R. S., Roller, R., Mika, A., Greenwood, B. N., Knight, R., et al. (2017). Dietary prebiotics and bioactive milk fractions improve NREM sleep, enhance REM sleep rebound and attenuate the stress-induced decrease in diurnal temperature and gut microbial alpha diversity. *Frontiers in Behavioral Neuroscience, 10,* 240.

Thrane, S. (2013). Effectiveness of integrative modalities for pain and anxiety in children and adolescents with cancer: A systematic review. *Journal of Pediatric Oncology Nursing, 30,* 320–333.

Tilders, F. J., & Schmidt, E. D. (1998). Interleukin-1-induced plasticity of hypothalamic CRH neurons and long-term stress hyperresponsiveness. *Annals of the New York Academy of Sciences, 840,* 65–73.

Tilders, F. J. H., & Schmidt, E. D. (1999). Cross-sensitization between immune and non-immune stressors. A role in the etiology of depression? *Advances in Experimental Medicine, 461,* 179–197.

Tillisch, K., Labus, J., Kilpatrick, L., Jiang, Z., Stains, J., et al. (2013). Consumption of fermented milk product with probiotic modulates brain activity. *Gastroenterology, 144,* 1394–1401.

Tiwari, V., Guan, Y., & Raja, S. N. (2014). Modulating the delicate glial-neuronal interactions in neuropathic pain: Promises and potential caveats. *Neuroscience & Biobehavioral Reviews, 45,* 19–27.

Tomasetti, C., & Vogelstein, B. (2015). Cancer etiology. Variation in cancer risk among tissues can be explained by the number of stem cell divisions. *Science, 347*(6217), 78–81.

Tonelli, L. H., Stiller, J., Rujescu, D., Giegling, I., Schneider, B., et al. (2008). Elevated cytokine expression in the orbitofrontal cortex of victims of suicide. *Acta Psychiatrica Scandinavica, 117*(3), 198–206.

Tonomura, S., Ihara, M., Kawano, T., Tanaka, T., Okuno, Y., et al. (2016). Intracerebral haemorrhage and deep microbleeds associated with cnm-positive Streptococcus mutans; a hospital cohort study. *Scientific Reports, 6,* 1–9.

Torkamani, A., Andersen, K. G., Steinhubl, S. R., & Topol, E. J. (2017). High-definition medicine. *Cell, 170*(5), 828–843.

Torres, A. R., Sweeten, T. L., Johnson, R. C., Odell, D., Westover, J. B., Bray-Ward, P., & Benson, M. (2016). Common genetic variants found in HLA and KIR immune genes in autism spectrum disorder. *Frontiers in Neuroscience, 10,* 463. Available from https://doi.org/10.3389/fnins.2016.00463.

Torres-Fuentes, C., Schellekens, H., Dinan, T. G., & Cryan, J. F. (2017). The microbiota–gut–brain axis in obesity. *Lancet Gastroenterol Hepatol, 2*(10), 747–756.

Torrey, E. F., Bartko, J. J., & Yolken, R. H. (2012). Toxoplasma gondii and other risk factors for schizophrenia: An update. *Schizophrenia Bulletin, 38*(3), 642–647. Available from https://doi.org/10.1093/schbul/sbs043.

Toth, M., Flandreau, E. I., Deslauriers, J., Geyer, M. A., Mansuy, I. M., et al. (2016). Overexpression of forebrain CRH during early life increases trauma susceptibility in adulthood. *Neuropsychopharmacology, 41*(6), 1681–1690.

Toufexis, D., Rivarola, M. A., Lara, H., & Viau, V. (2014). Stress and the reproductive axis. *Journal of Neuroendocrinology, 26,* 573–586.

Tournier, J. N., Mathieu, J., Mailfert, Y., Multon, E., Drouet, C., et al. (2001). Chronic restraint stress induces severe disruption of the T-cell specific response to tetanus toxin vaccine. *Immunology, 102*(1), 87–93.

Tracey, K. J. (2009). Reflex control of immunity. *Nature Reviews Immunology, 9,* 418–428.

Tran, C. M., Mukherjee, S., Ye, L., Frederick, D. W., Kissig, M., et al. (2016). Rapamycin blocks induction of the thermogenic program in white adipose tissue. *Diabetes, 65*(4), 927–941.

Trang, T., Beggs, S., & Salter, M. W. (2011). Brain-derived neurotrophic factor from microglia: A molecular substrate for neuropathic pain. *Neuron Glia Biology, 7*(1), 99–108.

Tremlett, H., Fadrosh, D. W., Faruqi, A. A., Hart, J., Roalstad, S., et al. (2016). Gut microbiota composition and relapse risk in pediatric MS: A pilot study. *Journal of the Neurological Sciences, 363,* 153–157.

Treutlein, J., Strohmaier, J., Frank, J., Witt, S. H., Rietschel, L., et al. (2017). Association between neuropeptide Y receptor Y2 promoter variant rs6857715 and major depressive disorder. *Psychiatric Genetics, 27*(1), 34–37.

Trikojat, K., Luksch, H., Rösen-Wolff, A., Plessow, F., Schmitt, J., et al. (2017). "Allergic mood" – Depressive and anxiety symptoms in patients with seasonal allergic rhinitis (SAR) and their association to inflammatory, endocrine, and allergic markers. *Brain, Behavior, and Immunity, 65,* 202–209.

Tripp, A., Oh, H., Guilloux, J. P., Martinowich, K., Lewis, D. A., et al. (2012). Brain-derived neurotrophic factor signaling and subgenual anterior cingulate cortex dysfunction in major depressive disorder. *American Journal of Psychiatry, 169,* 1194–1202.

Tromp do, P. M., Grupe, D. W., Oathes, D. J., McFarlin, D. R., Hernandez, P. J., et al. (2012). Reduced structural connectivity of a major frontolimbic pathway in generalized anxiety disorder. *Archives of General Psychiatry, 69,* 925–934.

Tsuda, M. (2017). P2 receptors, microglial cytokines and chemokines, and neuropathic pain. *Journal of Neuroscience Research, 95,* 1319–1329.

Tu, J. V., Chu, A., Maclagan, L., Austin, P. C., Johnston, S., et al. (2017). Regional variations in ambulatory care and incidence of cardiovascular events. *CMAJ*, *189*(13), E494–E501.

Tu, M. T., Lupien, S. J., & Walker, C. D. (2005). Measuring stress responses in postpartum mothers: Perspectives from studies in human and animal populations. *Stress*, *8*(1), 19–34.

Tükel, R., Arslan, B. A., Ertekin, B. A., Ertekin, E., Oflaz, S., et al. (2012). Decreased IFN-γ and IL-12 levels in panic disorder. *Journal of Psychosomatic Research*, *73*(1), 63–67.

Turecki, G., & Meaney, M. J. (2016). Effects of the social environment and stress on glucocorticoid receptor gene methylation: A systematic review. *Biological Psychiatry*, *79*, 87–96.

Turnbaugh, P. J., Ley, R. E., Mahowald, M. A., Magrini, V., Mardis, E. R., et al. (2006). An obesity-associated gut microbiome with increased capacity for energy harvest. *Nature*, *444*, 1027–1031.

Turner, J. E. (2016). Is immunosenescence influenced by our lifetime "dose" of exercise? *Biogerontology*, *17*(3), 581–602.

Tursich, M., Neufeld, R. W., Frewen, P. A., Harricharan, S., Kibler, J. L., et al. (2014). Association of trauma exposure with proinflammatory activity: A transdiagnostic meta-analysis. *Translational Psychiatry*, *4*, e413.

Tye, K. M., Mirzabekov, J. J., Warden, M. R., Ferenczi, E. A., Tsai, H. C., et al. (2013). Dopamine neurons modulate neural encoding and expression of depression-related behaviour. *Nature*, *493*(7433), 537–541.

Tylee, D. S., Chandler, S. D., Nievergelt, C. M., Liu, X., Pazol, J., et al. (2015). Blood-based gene-expression biomarkers of post-traumatic stress disorder among deployed marines: A pilot study. *Psychoneuroendocrinology*, *51*, 472–494.

Tyml, K., Swarbreck, S., Pape, C., Secor, D., Koropatnick, J., et al. (2017). Voluntary running exercise protects against sepsis-induced early inflammatory and pro-coagulant responses in aged mice. *Critical Care*, *21*(1), 210.

Tynan, R. J., Naicker, S., & Hinwood, M. (2010). Chronic stress alters the density and morphology of microglia in a subset of stress-responsive brain regions. *Brain, Behavior, and Immunity*, *24*, 1058–1068.

Tysnes, O. B., & Storstein, A. (2017). Epidemiology of Parkinson's disease. *Journal of Neural Transmission*, *124* (8), 901–905.

Uchakin, P. N., Tobin, B., Cubbage, M., Marshall, G., Jr., & Sams, C. (2001). Immune responsiveness following academic stress in first-year medical students. *Journal of Interferon & Cytokine Research*, *21*, 687–694.

Uddin, M., Aiello, A. E., Wildman, D. E., Koenen, K. C., Pawelec, G., et al. (2010). Epigenetic and immune function profiles associated with posttraumatic stress disorder. *Proceedings of the National Academy of Sciences of the United States of America*, *107*, 9470–9475.

Udina, M., Moreno-España, J., Navinés, R., Giménez, D., Langohr, K., et al. (2013). Serotonin and interleukin-6: The role of genetic polymorphisms in IFN-induced neuropsychiatric symptoms. *Psychoneuroendocrinology*, *38*, 1803–1813.

Udina, M., Navinés, R., Egmond, E., Oriolo, G., Langohr, K., et al. (2016). Glucocorticoid receptors, brain-derived neurotrophic factor, serotonin and dopamine neurotransmission are associated with interferon-induced depression. *International Journal of Neuropsychopharmacology*, *19*, 135.

Ulland, T. K., Song, W. M., Huang, S. C. C., Ulrich, J. D., Sergushichev, A., et al. (2017). TREM2 maintains microglial metabolic fitness in Alzheimer's disease. *Cell*, *170*, 649–663, e13.

Ullrich, C., & Humpel, C. (2009). Rotenone induces cell death of cholinergic neurons in an organotypic co-culture brain slice model. *Neurochemical Research*, *34*, 2147–2153.

Umehara, H., Numata, S., Watanabe, S. Y., Hatakeyama, Y., Kinoshita, M., et al. (2017). Altered KYN/TRP, Gln/Glu, and Met/methionine sulfoxide ratios in the blood plasma of medication-free patients with major depressive disorder. *Scientific Reports*, *7*(1), 4855.

Unemo, M., Bradshaw, C. S., Hocking, J. S., de Vries, H. J. C., Fancis, S. C., et al. (2017). Sexually transmitted infections: Challenges ahead. *The Lancet Infectious Diseases*, *17*(8), e235–e279.

Unternaehrer, E., Bolten, M., Nast, I., Staehli, S., Meyer, A. H., et al. (2016). Maternal adversities during pregnancy and cord blood oxytocin receptor (OXTR) DNA methylation. *Social Cognitive and Affective Neuroscience*, *11*(9), 1460–1470.

Uschold-Schmidt, N., Nyuyki, K. D., Füchsl, A. M., Neumann, I. D., & Reber, S. O. (2012). Chronic psychosocial stress results in sensitization of the HPA axis to acute heterotypic stressors despite a reduction of adrenal in vitro ACTH responsiveness. *Psychoneuroendocrinology*, *37*, 1676–1687.

Ussar, S., Griffin, N. W., Bezy, O., Fujisaka, S., Vienberg, S., et al. (2015). Interactions between gut microbiota, host genetics and diet modulate the predisposition to obesity and metabolic syndrome. *Cell Metabolism*, *22*(3), 516–530.

Uversky, V. N., Li, J., Bower, K., & Fink, A. L. (2002). Synergistic effects of pesticides and metals on the fibrillation of alpha-synuclein: Implications for Parkinson's disease. *Neurotoxicology*, *23*, 527–536.

Vaccarino, V., Goldberg, J., Rooks, C., Shah, A. J., Veledar, E., et al. (2013). Post-traumatic stress disorder and incidence of coronary heart disease: A twin study. *Journal of the American College of Cardiology, 62,* 970–978.

Vaikunthanathan, T., Safinia, N., Lombardi, G., & Lechler, R. I. (2016). Microbiota, immunity and the liver. *Immunology Letters, 171,* 36–94.

Valastro, B., Ghribi, O., Poirier, J., Krzywkowski, P., & Massicotte, G. (2001). AMPA receptor regulation and LTP in the hippocampus of young and aged apolipoprotein E-deficient mice. *Neurobiology of Aging, 22*(1), 9–15.

Valdearcos, M., Douglass, J. D., Robblee, M. M., Dorfman, M. D., Stifler, D. R., et al. (2017). Microglial inflammatory signaling orchestrates the hypothalamic immune response to dietary excess and mediates obesity susceptibility. *Cell Metabolism, 26,* 185–197.

Valenzano, D. R., Benayoun, B. A., Singh, P. P., Zhang, E., Etter, P. D., et al. (2015). The African Turquoise Killifish genome provides insights into evolution and genetic architecture of lifespan. *Cell, 163,* 1539–1554.

Valenzuela, M. J., & Sachdev, P. (2006). Brain reserve and dementia: A systematic review. *Psychological Medicine, 36*(8), 441–454.

Valero, J., Mastrella, G., Neiva, I., Sánchez, S., & Malva, J. O. (2014). Long-term effects of an acute and systemic administration of LPS on adult neurogenesis and spatial memory. *Frontiers in Neuroscience, 8,* 83.

Vallejo-Vaz, A. J., Robertson, M., Catapano, A. L., Watts, G. F., Kastelein, J. J., et al. (2017). Low-density lipoprotein cholesterol lowering for the primary prevention of cardiovascular disease among men with primary elevations of low-density lipoprotein cholesterol levels of 190 mg/dL or above: Analyses from the WOSCOPS (West of Scotland Coronary Prevention Study) 5-year randomized trial and 20-year observational follow-up. *Circulation, 136*(20), 1878–1891.

Vallières, L., Campbell, I. L., Gage, F. H., & Sawchenko, P. E. (2002). Reduced hippocampal neurogenesis in adult transgenic mice with chronic astrocytic production of interleukin-6. *Journal of Neuroscience, 22,* 486–492.

Valtorta, N. K., Kanaan, M., Gilbody, S., Ronzi, S., & Hanratty, B. (2016). Loneliness and social isolation as risk factors for coronary heart disease and stroke: Systematic review and meta-analysis of longitudinal observational studies. *Heart, 102,* 1009–1016.

van Bloemendaal, L., Ten Kulve, J. S., la Fleur, S. E., Ijzerman, R. G., & Diamant, M. (2014). Effects of glucagon-like peptide 1 on appetite and body weight: Focus on the CNS. *Journal of Endocrinology, 221*(1), T1–T16.

Van Dam, N. T., van Vugt, M. K., Vago, D. R., Schmalzl, L., Saron, C. D., et al. (2017). Mind the hype: A critical evaluation and prescriptive agenda for research on mindfulness and meditation. *Perspectives on Psychological Science, 13*(1), 36–61.

van de Giessen, E. M., de Win, M. M. L., Tanck, M. W. T., van den Brink, W., Baas, F., et al. (2009). Striatal dopamine transporter availability associated with polymorphisms in the dopamine transporter gene SLC6A3. *Journal of Nuclear Medicine, 50*(1), 45–52.

Van den Hove, D., Jakob, S. B., Schraut, K. G., Kenis, G., Schmitt, A. G., et al. (2011). Differential effects of prenatal stress in 5-Htt deficient mice: Towards molecular mechanisms of gene × environment interactions. *PLoS One, 6*(8), e22715.

van der Meer, D., Hartman, C. A., Pruim, R. H., Mennes, M., Heslenfeld, D., et al. (2017). The interaction between 5-HTTLPR and stress exposure influences connectivity of the executive control and default mode brain networks. *Brain Imaging and Behavior, 11*(5), 1486–1496.

Vanegas, H., Vazquez, E., & Tortorici, V. (2010). NSAIDs, opioids, cannabinoids and the control of pain by the central nervous system. *Pharmaceuticals, 3*(5), 1335–1347.

Vanhaecke, T., Aubert, P., Grohard, P. A., Durand, T., Hulin, P., et al. (2017). L. fermentum CECT 5716 prevents stress-induced intestinal barrier dysfunction in newborn rats. *Neurogastroenterology Motility, 29.* Available from https://doi.org/10.1111/nmo.13069.

van Honk, J., Bos, P. A., Terburg, D., Heany, S., & Stein, D. J. (2015). Neuroendocrine models of social anxiety disorder. *Dialogues in Clinical Neuroscience, 17*(3), 287–293.

van Kesteren, C. F., Gremmels, H., de Witte, L. D., Hol, E. M., Van Gool, A. R., et al. (2017). Immune involvement in the pathogenesis of schizophrenia: A meta-analysis on postmortem brain studies. *Translational Psychiatry, 7*(3), e1075.

van Langenberg, D. R., Yelland, G. W., Robinson, S. R., & Gibson, P. R. (2016). Cognitive impairment in Crohn's disease is associated with systemic inflammation, symptom burden and sleep disturbance. *United European Gastroenterology Journa, 5*(4), 579–587.

van Leeuwen, W. M., Lehto, M., Karisola, P., Lindholm, H., Luukkonen, R., et al. (2009). Sleep restriction increases the risk of developing cardiovascular diseases by augmenting proinflammatory responses through IL-17 and CRP. *PLoS One, 4*(2), e4589.

van Nuenen, B. F. L., Helmich, R. C., Ferraye, M., Thaler, A., Hendler, T., et al. (2012). Cerebral pathological and compensatory mechanisms in the premotor phase of leucine-rich repeat kinase 2 parkinsonism. *Brain:A Journal of Neurology, 135,* 3687–3698.

van Rooij, S. J., Stevens, J. S., Ely, T. D., Fani, N., Smith, A. K., et al. (2016). Childhood trauma and COMT genotype interact to increase hippocampal activation in resilient individuals. *Frontiers in Psychiatry, 7,* 156.

Van't Wout, M., Longo, S. M., Reddy, M. K., Philip, N. S., Bowker, M. T., et al. (2017). Transcranial direct current stimulation may modulate extinction memory in post-traumatic stress disorder. *Brain and Behavior*, 7(5), e00681.

Varga, Z., Csabai, D., Miseta, A., Wiborg, O., & Czéh, B. (2017). Chronic stress affects the number of GABAergic neurons in the orbitofrontal cortex of rats. *Behavioural Brain Research*, 316, 104−114.

Vargas, D. L., Nascimbene, C., Krishnan, C., Zimmerman, A. W., & Pardo, C. A. (2005). Neuroglial activation and neuroinflammation in the brain of patients with autism. *Annals of Neurology*, 57(1), 67−81. Available from https://doi.org/10.1002/ana.20315.

Veer, I. M., Oei, N. Y., van Buchem, M. A., Spinhoven, P., Elzinga, B. M., et al. (2015). Evidence for smaller right amygdala volumes in posttraumatic stress disorder following childhood trauma. *Psychiatry Research: Neuroimaging*, 233(3), 436−442.

Vegas, A. J., Veiseh, O., Gürtler, M., Millman, J. R., Pagliuca, F. W., et al. (2016). Long-term glycemic control using polymer-encapsulated human stem cell-derived beta cells in immune-competent mice. *Nature Medicine (New York, NY, United States)*, 22, 306−311.

Velazquez, P., Cribbs, D. H., Poulos, T. L., & Tenner, A. J. (1997). Aspartate residue 7 in amyloid beta-protein is critical for classical complement pathway activation: Implications for Alzheimer's disease pathogenesis. *Nature Medicine (New York, NY, United States)*, 3, 77−79.

Velmurugan, G., Ramprasath, T., Gilles, M., Swaminathan, K., & Ramasamy, S. (2017). Gut microbiota, endocrine-disrupting chemicals, and the diabetes epidemic. *Trends in Endocrinology and Metabolism*, 28(8), 612−625.

Venegas, C., & Heneka, M. T. (2017). Danger-associated molecular patterns in Alzheimer's disease. *Journal of Leukocyte Biology*, 101(1), 87−98.

Vgontzas, A. N., Fernandez-Mendoza, J., Liao, D., & Bixler, E. O. (2013). Insomnia with objective short sleep duration: The most biologically severe phenotype of the disorder. *Sleep Medicine Reviews*, 17(4), 241−254.

Vidarsson, G., Dekkers, G., & Rispens, T. (2014). IgG subclasses and allotypes: From structure to effector functions. *Frontiers in Immunology*, 5, 520.

Vieira, M. M., Ferreira, T. B., Pacheco, P. A., Barros, P. O., Almeida, C. R., et al. (2010). Enhanced Th17 phenotype in individuals with generalized anxiety disorder. *Journal of Neuroimmunology*, 229(1−2), 212−218.

Viladomiu, M., Kivolowitz, C., Abdulhamid, A., Dogan, B., Victorio, D., et al. (2017). IgA-coated E. coli enriched in Crohn's disease spondyloarthritis promote TH17-dependent inflammation. *Science Translational Medicine*, 9(376), eaaf9655.

Villafane, G., Thiriez, C., Audureau, E., Straczek, C., Kerschen, P., et al. (2017). High-dose transdermal nicotine in Parkinson's disease patients: A randomized, open-label, blinded-endpoint evaluation phase 2 study. *European Journal of Neurology*, 25(1), 120−127.

Villeda, S. A., Plambeck, K. E., Middeldorp, J., Castellano, J. M., Mosher, K. I., et al. (2014). Young blood reverses age-related impairments in cognitive function and synaptic plasticity in mice. *Nature Medicine (New York, NY, United States)*, 20, 659−663.

Vinet, E., Pineau, C. A., Clarke, A. E., Scott, S., Fombonne, E., Joseph, L., & Bernatsky, S. (2015). Increased risk of autism spectrum disorders in children born to women with systemic lupus erythematosus: Results from a large population-based cohort. *Arthritis & Rheumatology*, 67(12), 3201−3208. Available from https://doi.org/10.1002/art.39320.

Viswanathan, K., Daugherty, C., & Dhabhar, F. S. (2005). Stress as an endogenous adjuvant: Augmentation of the immunization phase of cell-mediated immunity. *International Immunology*, 17(8), 1059−1069.

Vitek, M. P., Brown, C. M., & Colton, C. A. (2009). APOE genotype-specific differences in the innate immune response. *Neurobiology of Aging*, 30, 1350−1360.

Vogelzangs, N., Beekman, A. T., de Jonge, P., & Penninx, B. W. (2013). Anxiety disorders and inflammation in a large adult cohort. *Translational Psychiatry*, 3, e249.

Vogelzangs, N., de Jonge, P., Smit, J. H., Bahn, S., & Penninx, B. W. (2016). Cytokine production capacity in depression and anxiety. *Translational Psychiatry*, 6(5), e825.

Voigt, R. M., Forsyth, C. B., Green, S. J., Engen, P. A., & Keshavarzian, A. (2016). Circadian rhythm and the gut microbiome. *Internayional Review of Neurobiology*, 131, 193−205.

Volkow, N., Benveniste, H., & McLellan, A. T. (2017). Use and misuse of opioids in chronic pain. *Annual Review of Medicine*, 69, 451−465.

Volkow, N. D., Wang, G. J., Fowler, J. S., Tomasi, D., Telang, F., et al. (2010). Addiction: Decreased reward sensitivity and increased expectation sensitivity conspire to overwhelm the brain's control circuit. *BioEssays: News and Reviews in Molecular, Cellular and Developmental Biology*, 32(9), 748−755.

Vollmer, L. L., Schmeltzer, S., Schurdak, J., Ahlbrand, R., Rush, J., et al. (2016). Neuropeptide Y impairs retrieval of extinguished fear and modulates excitability of neurons in the infralimbic prefrontal cortex. *Journal of Neuroscience*, 36(4), 1306−1315.

Volmar, C. H., Ait-Ghezala, G., Frieling, J., Paris, D., & Mullan, M. J. (2008). The granulocyte macrophage colony stimulating factor (GM-CSF) regulates amyloid beta (Abeta) production. *Cytokine*, 42(3), 336−344.

Voorhees, J. L., Tarr, A. J., Wohleb, E. S., Godbout, J. P., Mo, X., et al. (2013). Prolonged restraint stress increases IL-6, reduces IL-10, and causes persistent depressive-like behavior that is reversed by recombinant IL-10. *PLoS One, 8*(3), e58488.

Vorstman, J. A. S., Parr, J. R., Moreno-De-Luca, D., Anney, R. J. L., Nurnberger, J. I., Jr., & Hallmayer, J. F. (2017). Autism genetics: Opportunities and challenges for clinical translation. *Nature Reviews Genetics, 18*(6), 362–376. Available from https://doi.org/10.1038/nrg.2017.4.

Voss, M. W., Weng, T. B., Burzynska, A. Z., Wong, C. N., Cooke, G. E., et al. (2015). Fitness, but not physical activity, is related to functional integrity of brain networks associated with ageing. *NeuroImage, 131*, 113–125.

Vukojevic, V., Kolassa, I. T., Fastenrath, M., Gschwind, L., Spalek, K., et al. (2014). Epigenetic modification of the glucocorticoid receptor gene is linked to traumatic memory and post-traumatic stress disorder risk in genocide survivors. *Journal of Neuroscience, 34*(31), 10274–10284.

Vyas, S., Rodrigues, A. J., Silva, J. M., Tronche, F., Almeida, O. F. X., et al. (2016). Chronic stress and glucocorticoids: From neuronal plasticity to neurodegeneration. *Neural Plasticity, 2016*, 6391686.

Wager, T. D., Atlas, L. Y., & Botvinick, M. M. (2016). Pain in the ACC? *Proceedings of the National Academy of Sciences of the United States of America, 113*, E2474–2475.

Wahner, A. D., Bronstein, J. M., Bordelon, Y. M., & Ritz, B. (2007). Nonsteroidal anti-inflammatory drugs may protect against Parkinson disease. *Neurology, 69*(19), 1836–1842.

Walker, A. K., Hawkins, G., Sominsky, L., & Hodgson, D. M. (2012). Transgenerational transmission of anxiety induced by neonatal exposure to lipopolysaccharide: Implications for male and female germ lines. *Psychoneuroendocrinology, 37*(8), 1320–1335.

Walker, A. K., Nakamura, T., & Hodgson, D. M. (2010). Neonatal lipopolysaccharide exposure alters central cytokine responses to stress in adulthood in Wistar rats. *Stress, 13*, 506–515.

Walker, C. D. (2010). Maternal touch and feed as critical regulators of behavioral and stress responses in the offspring. *Developmental Psychobiology, 52*(7), 638–650.

Walker, K. A., Hoogeveen, R. C., Folsom, A. R., Ballantyne, C. M., Knopman, D. S., et al. (2017). Midlife systemic inflammatory markers are associated with late-life brain volume: The ARIC study. *Neurology, 89*(22), 2262–2270.

Wallace, C. J., & Milev, R. (2017). The effects of probiotics on depressive symptoms in humans: A systematic review. *Annals of General Psychiatry, 16*, 14–20.

Wallace, S., Nazroo, J., & Bécares, L. (2016). Cumulative effect of racial discrimination on the mental health of ethnic minorities in the United Kingdom. *American Journal of Public Health, 106*(7), 1294–1300.

Wallin, C., Sholts, S. B., Österlund, N., Luo, J., Jarvet, J., et al. (2017). Alzheimer's disease and cigarette smoke components: Effects of nicotine, PAHs, and Cd(II), Cr (III), Pb(II), Pb(IV) ions on amyloid-β peptide aggregation. *Scientific Reports, 7*(1), 14423.

Wallings, R., Manzoni, C., & Bandopadhyay, R. (2015). Cellular processes associated with LRRK2 function and dysfunction. *Federation of European Biochemical Societies Journal, 282*, 2806–2826.

Walters, E. T. (2014). Neuroinflammatory contributions to pain after SCI: Roles for central glial mechanisms and nociceptor-mediated host defense. *Experimental Neurology, 258*, 48–61.

Wan, M. T., Shin, D. B., Hubbard, R. A., Noe, M. H., Mehta, N. N., et al. (2017). Psoriasis and the risk of diabetes: A prospective population-based cohort study. *Journal of the American Academy of Dermatology, 78*, 315–322, pii: S0190-9622(17)32616-6.

Wang, A. K., & Miller, B. J. (2018). Meta-analysis of cerebrospinal fluid cytokine and tryptophan catabolite alterations in psychiatric patients: Comparisons between schizophrenia, bipolar disorder, and depression. *Schizophrenia Bulletin, 44*(1), 75–83.

Wang, D. D., Li, Y., Stampfer, M. J., Stampfer, M. J., Manson, J. E., et al. (2016). Association of specific dietary fats with total and cause-specific mortality. *Journal of the American Medical Association: Internal Medicine, 176* (8), 1134–1145.

Wang, F., Stefano, G. B., & Kream, R. M. (2014). Epigenetic modification of DRG neuronal gene expression subsequent to nerve injury: Etiological contribution to complex regional pain syndromes (Part I). *Medical Science Monitor, 20*, 1067–1077.

Wang, G. Y., Zhu, Z. M., Cui, S., & Wang, J. H. (2016). Glucocorticoid induces incoordination between glutamatergic and GABAergic neurons in the amygdala. *PLoS One, 11*(11), e0166535.

Wang, H. T., Huang, F. L., Hu, Z. L., Zhang, W. J., Qiao, X. Q., et al. (2017). Early-life social isolation-induced depressive-like behavior in rats results in microglial activation and neuronal histone methylation that are mitigated by minocycline. *Neurotoxicity Research, 31*(4), 505–520.

Wang, J., Hossain, M., Thanabalasuriar, A., Gunzer, M., Meininger, C., et al. (2017). Visualizing the function and fate of neutrophils in sterile injury and repair. *Science, 358*(6359), 111–116.

Wang, L., Christophersen, C. T., Sorich, M. J., Gerber, J. P., Angley, M. T., et al. (2012). Elevated fecal short chain fatty acid and ammonia concentrations in children with autism spectrum disorder. *Digestive Diseases and Sciences, 57*(8), 2096–2102.

Wang, M., Yan, W., Liu, Y., Hu, H., Sun, Q., et al. (2017). Erythropoietin ameliorates diabetes-associated cognitive dysfunction in vitro and in vivo. *Scientific Reports, 7*(1), 2801.

Wang, P., Guan, P. P., Yu, X., Zhang, L. C., Su, Y. N., et al. (2016). Prostaglandin I2 attenuates prostaglandin E2-stimulated expression of interferon γ in a β-amyloid protein- and NF-κB-dependent mechanism. *Scientific Reports, 6*, 20879.

Wang, Q., Ao, Y., Yang, K., Tang, H., & Chen, D. (2016). Circadian clock gene Per2 plays an important role in cell proliferation, apoptosis and cell cycle progression in human oral squamous cell carcinoma. *Oncology Reports, 35*, 3387–3394.

Wang, Q., Dong, X., Wang, Y., Liu, M., Sun, A., et al. (2017). Adolescent escitalopram prevents the effects of maternal separation on depression- and anxiety-like behaviours and regulates the levels of inflammatory cytokines in adult male mice. *International Journal of Developmental Neuroscience, 62*, 37–45.

Wang, Q. P., Lin, Y. Q., Zhang, L., Wilson, Y. A., Oyston, L. J., et al. (2016). Sucralose promotes food intake through NPY and a neuronal fasting response. *Cell Metabolism, 24*(1), 75–90.

Wang, W., Chen, L., Zhou, R., Wang, X., Song, L., et al. (2014). Increased proportions of Bifidobacterium and the Lactobacillus group and loss of butyrate-producing bacteria in inflammatory bowel disease. *Journal of Clinical Microbiology, 52*(2), 398–406.

Wang, W. Y., Liu, Y., Wang, H. F., Tan, L., Sun, F. R., et al. (2017). Impacts of CD33 genetic variations on the atrophy rates of hippocampus and parahippocampal gyrus in normal aging and mild cognitive impairment. *Molecular Neurobiology, 54*(2), 1111–1118.

Wang, X., Grace, P. M., Pham, M. N., Cheng, K., Strand, K. A., et al. (2013). Rifampin inhibits Toll-like receptor 4 signaling by targeting myeloid differentiation protein 2 and attenuates neuropathic pain. *Federation of American Societies for Experimental Biology Journal, 27*(7), 2713–2722.

Wang, X., Häring, M. F., Rathjen, T., Lockhart, S. M., Sørensen, D., et al. (2017). Insulin resistance in vascular endothelial cells promotes intestinal tumour formation. *Oncogene, 36*(35), 4987–4996.

Wang, X., Lopez, O. L., Sweet, R. A., Becker, J. T., DeKosky, S. T., et al. (2015). Genetic determinants of disease progression in Alzheimer's disease. *Journal of Alzheimers Disease, 43*(2), 649–655.

Wang, Y., Lawson, M. A., Dantzer, R., & Kelley, K. W. (2010). LPS-induced indoleamine 2,3-dioxygenase is regulated in an interferon-gamma-independent manner by a JNK signaling pathway in primary murine microglia. *Brain, Behavior, and Immunity, 24*(2), 201–209.

Wang, Y., Ulland, T. K., Ulrich, J. D., Song, W., Tzaferis, J. A., et al. (2016). TREM2-mediated early microglial response limits diffusion and toxicity of amyloid plaques. *Journal of Experimental Medicine, 213*(5), 667–675.

Wang, Y., Zhao, Z., Rege, S. V., Wang, M., Si, G., et al. (2016). 3K3A-activated protein C stimulates postischemic neuronal repair by human neural stem cells in mice. *Nature Medicine (New York, NY, United States), 22*(9), 1050–1055.

Wang, Z., Klipfell, E., Bennett, B. J., Koeth, R., Levison, B. S., et al. (2011). Gut flora metabolism of phosphatidylcholine promotes cardiovascular disease. *Nature, 472*(7341), 57–63.

Wang, Z., Roberts, A. B., Buffa, J. A., Levison, B. S., Zhu, W., et al. (2015). Non-lethal inhibition of gut microbial trimethylamine production for the treatment of atherosclerosis. *Cell, 163*(7), 1585–1595.

Wang, Z., Wei, X., Yang, J., Suo, J., Chen, J., et al. (2016). Chronic exposure to aluminum and risk of Alzheimer's disease: A meta-analysis. *Neuroscience Letters, 610*, 200–206.

Ware, M. A., Wang, T., Shapiro, S., Robinson, A., Ducruet, T., et al. (2010). Smoked cannabis for chronic neuropathic pain: A randomized controlled trial. *Canadian Medical Association Journal, 182*(14), E694–E701.

Warren, H. R., Evangelou, E., Cabrera, C. P., Gao, H., Ren, M., et al. (2017). Genome-wide association analysis identifies novel blood pressure loci and offers biological insights into cardiovascular risk. *Nature Genetics, 49*(3), 403–415.

Waszak, S. M., Delaneau, O., Gschwind, A. R., Kilpinen, H., Raghav, S. K., et al. (2015). Population variation and genetic control of modular chromatin architecture in humans. *Cell, 162*(5), 1039–1050.

Watkins, C., Stanton, C., Ryan, C. A., & Ross, R. P. (2017). Microbial therapeutics designed for infant health. *Frontiers in Nutrition, 4*, 48.

Watanabe, Y., Nunokawa, A., Kaneko, N., Muratake, T., Arinami, T., Ujike, H., … Someya, T. (2009). Two-stage case-control association study of polymorphisms in rheumatoid arthritis susceptibility genes with schizophrenia. *Journal of Human Genetics, 54*(1), 62–65. Available from https://doi.org/10.1038/jhg.2008.4.

Wattananit, S., Tornero, D., Graubardt, N., Memanishvili, T., Monni, E., et al. (2016). Monocyte-derived macrophages contribute to spontaneous long-term functional recovery after stroke in mice. *Journal of Neuroscience, 36*, 4182–4195.

Weaver, I. C., Champagne, F. A., Brown, S. E., Dymov, S., Sharma, S., et al. (2005). Reversal of maternal programming of stress responses in adult offspring through methyl supplementation: Altering epigenetic marking later in life. *Journal of Neuroscience, 25*, 11045–11054.

Weber, M. D., Frank, M. G., Tracey, K. J., Watkins, L. R., & Maier, S. F. (2015). Stress induces the danger-associated molecular pattern HMGB-1 in the hippocampus of male Sprague Dawley rats: A priming stimulus of microglia and the NLRP3 inflammasome. *Journal of Neuroscience, 35*(1), 316–324.

Weber, M. D., Godbout, J. P., & Sheridan, J. F. (2017). Repeated social defeat, neuroinflammation, and behavior: Monocytes carry the signal. *Neuropsychopharmacology, 42*(1), 46–61.

Weber-Stadlbauer, U. (2017). Epigenetic and transgenerational mechanisms in infection-mediated neurodevelopmental disorders. *Translational Psychiatry, 7*(5), e1113.

Wegman, M. P., Guo, M., Bennion, D. M., Shankar, M. N., Chrzanowski, S. M., et al. (2015). Practicality of intermittent fasting in humans and its effect on oxidative stress and genes related to ageing and metabolism. *Rejuvenation Research, 18*(2), 162–172.

Weinstock, M. (2007). Gender differences in the effects of prenatal stress on brain development and behaviour. *Neurochemical Research, 32*(10), 1730–1740.

Weintraub, M. K., Kranjac, D., Eimerbrink, M. J., Pearson, S. J., Vinson, B. T., et al. (2014). Peripheral administration of poly I:C leads to increased hippocampal amyloid-beta and cognitive deficits in a non-transgenic mouse. *Behavioural Brain Research, 266*, 183–187.

Weissman, M. M., Wickramaratne, P., Gameroff, M. J., Warner, V., Pilowsky, D., et al. (2016). Offspring of depressed parents: 30 years later. *American Journal of Psychiatry, 173*(10), 1024–1032.

Weisz, J. R., Kuppens, S., Ng, M. Y., Eckshtain, D., Ugueto, A. M., et al. (2017). What five decades of research tells us about the effects of youth psychological therapy: a multilevel meta-analysis and implications for science and practice. *American Psychologist, 72*(2), 79–117.

Weizman, R., Laor, N., Barber, Y., Hermesh, H., Notti, I., et al. (1996). Cytokine production in obsessive-compulsive disorder. *Biological Psychiatry, 40*(9), 908–912.

Wells, J. M., Brummer, R. J., Derrien, M., MacDonald, T. T., Troost, F., Cani, P. D., & Garcia-Rodenas, C. L. (2017). Homeostasis of the gut barrier and potential biomarkers. *American Journal of Physiology Gastrointestinal and Liver Physiology, 312*(3), G171–G193. Available from https://doi.org/10.1152/ajpgi.00048.2015.

Wen, D. J., Poh, J. S., Ni, S. N., Chong, Y. S., Chen, H., et al. (2017). Influences of prenatal and postnatal maternal depression on amygdala volume and microstructure in young children. *Translational Psychiatry, 7*(4), e1103.

West, A. B. (2017). Achieving neuroprotection with LRRK2 kinase inhibitors in Parkinson disease. *Experimental Neurology, 298*(Pt B), 236–245.

West, C. E., Jenmalm, M. C., Kozyrskyj, A. L., & Prescott, S. L. (2016). Probiotics for treatment and primary prevention of allergic diseases and asthma: Looking back and moving forward. *Expert Review of Clinical Immunology, 12*(6), 625–639.

Westberg, J. A., Serlachius, M., Lankila, P., Penkowa, M., Hidalgo, J., & Andersson, L. C. (2007). Hypoxic preconditioning induces neuroprotective stanniocalcin-1 in brain via IL-6 signaling. *Stroke, 38*(3), 1025–1030. Available from https://doi.org/10.1161/01.STR.0000258113.67252.fa.

Westermann, J., Lange, T., Textor, J., & Born, J. (2015). System consolidation during sleep – A common principle underlying psychological and immunological memory formation. *Trends in Neurosciences, 38*(10), 585–597.

Wettstein, A., Aeppli, J., Gautschi, K., & Peters, M. (1991). Failure to find a relationship between mnestic skills of octogenarians and aluminum in drinking water. *International Archives of Occupational and Environmental Health, 63*(2), 97–103.

Wheaton, M. G., Galfalvy, H., Steinman, S. A., Wall, M. M., Foa, E. B., et al. (2016). Patient adherence and treatment outcome with exposure and response prevention for OCD: Which components of adherence matter and who becomes well? *Behaviour Research and Therapy, 85*, 6–12.

White, J., Kivimäki, M., Jokela, M., & Batty, G. D. (2017). Association of inflammation with specific symptoms of depression in a general population of older people: The English longitudinal study of ageing. *Brain, Behavior, and Immunity, 61*, 27–30.

Whiting, P. F., Wolff, R. F., Deshpande, S., Di Nisio, M., Duffy, S., et al. (2015). Cannabinoids for medical use: A systematic review and meta-analysis. *Journal of the American Medical Association, 313*, 2456–2473.

Whitmer, R. A., Gunderson, E. P., Barrett-Connor, E., Quesenberry, C. P., & Yaffe, K. (2005). Obesity in middle age and future risk of dementia: A 27 year longitudinal population based study. *British Medical Journal, 330*(7504), 1360.

Wichers, M. C., Koek, G. H., Robaeys, G., Verkerk, R., Scharpé, S., et al. (2005). IDO and interferon-alpha-induced depressive symptoms: A shift in hypothesis from tryptophan depletion to neurotoxicity. *Molecular Psychiatry, 10*(6), 538–544.

Wilcock, D. M., Rojiani, A., Rosenthal, A., Levkowitz, G., Subbarao, S., et al. (2004). Passive amyloid immunotherapy clears amyloid and transiently activates microglia in a transgenic mouse model of amyloid deposition. *Journal of Neuroscience, 24*(27), 6144–6151.

Wiles, N. J., Thomas, L., Turner, N., Garfield, K., Kounali, D., et al. (2016). Long-term effectiveness and cost-effectiveness of cognitive behavioural therapy as an adjunct to pharmacotherapy for treatment-resistant depression in primary care: Follow-up of the CoBalT randomised controlled trial. *Lancet Psychiatry, 3*(2), 137–144.

Willette, A. A., Lubach, G. R., Knickmeyer, R. C., Short, S. J., Styner, M., et al. (2011). Brain enlargement and increased behavioral and cytokine reactivity in infant monkeys following acute prenatal endotoxemia. *Behavioural Brain Research, 219*(1), 108–115.

Williams, L. M., Debattista, C., Duchemin, A. M., Schatzberg, A. F., & Nemeroff, C. B. (2016). Childhood trauma predicts antidepressant response in adults with major depression: Data from the randomized international study to predict optimized treatment for depression. *Translational Psychiatry, 6*, e799.

Williams, S., Chen, L., Savignac, H. M., Tzortzis, G., Anthony, D. C., et al. (2016). Neonatal prebiotic (BGOS) supplementation increases the levels of synaptophysin, GluN2A-subunits and BDNF proteins in the adult rat hippocampus. *Synapses, 70*(3), 121–124.

Williams, S., Sakic, B., & Hoffman, S. A. (2010). Circulating brain-reactive autoantibodies and behavioral deficits in the MRL model of CNS lupus. *Journal of Neuroimmunology, 218*(1–2), 73–82.

Williamson, L. L., Sholar, P. W., Mistry, R. S., Smith, S. H., & Bilbo, S. D. (2011). Microglia and memory: Modulation by early-life infection. *Journal of Neuroscience, 31*(43), 15511–15521.

Willing, B. P., Vacharaksa, A., Croxen, M., Thanachayanont, T., & Finlay, B. B. (2011). Altering host resistance to infections through microbial transplantation. *PLoS One, 6*(10), e26988.

Willner, P. (2016). Reliability of the chronic mild stress model of depression: A user survey. *Neurobiology of Stress, 6*, 68–77.

Wills, S., Cabanlit, M., Bennett, J., Ashwood, P., Amaral, D. G., & Van de Water, J. (2009). Detection of autoantibodies to neural cells of the cerebellum in the plasma of subjects with autism spectrum disorders. *Brain, Behavior, and Immunity, 23*(1), 64–74. Available from https://doi.org/10.1016/j.bbi.2008.07.007.

Willyard, C. (2017). An epigenetics gold rush: New controls for gene expression. *Nature, 542*(7642), 406–408.

Wilson, C. B., McLaughlin, L. D., Nair, A., Ebenezer, P. J., Dange, R., et al. (2013). Inflammation and oxidative stress are elevated in the brain, blood, and adrenal glands during the progression of post-traumatic stress disorder in a predator exposure animal model. *PLoS One, 8*(10), e76146.

Wing, L., & Gould, J. (1979). Severe impairments of social interaction and associated abnormalities in children: Epidemiology and classification. *Journal of Autism and Developmental Disorders, 9*(1), 11–29.

Wingfield, J. C., & Sapolsky, R. M. (2003). Reproduction and resistance to stress: When and how. *Journal of Neuroendocrinology, 15*(8), 711–724.

Winglee, K., Howard, A. G., Sha, W., Gharaibeh, R. Z., Liu, J., et al. (2017). Recent urbanization in China is correlated with a Westernized microbiome encoding increased virulence and antibiotic resistance genes. *Microbiome, 5*(1), 121.

Wirdefeldt, K., Adami, H.-O., Cole, P., Trichopoulos, D., & Mandel, J. (2011). Epidemiology and etiology of Parkinson's disease: A review of the evidence. *European Journal of Epidemiology, 2*(Suppl 1), S1–S58.

Wischhof, L., Irrsack, E., Osorio, C., & Koch, M. (2015). Prenatal LPS-exposure—A neurodevelopmental rat model of schizophrenia—Differentially affects cognitive functions, myelination and parvalbumin expression in male and female offspring. *Progress in Neuro-Psychopharmacology & Biological Psychiatry, 57*, 17–30.

Wiseman, H., Barber, P., Raz, A., Yam, I., Foltz, C., et al. (2002). Parental communication of Holocaust experiences and interpersonal patterns in offspring of Holocaust survivors. *International Journal of Behavioral Development, 26*, 371–381.

Witcher, K. G., Eiferman, D. S., & Godbout, J. P. (2015). Priming the inflammatory pump of the CNS after traumatic brain injury. *Trends in Neurosciences, 38*, 609–620.

Withers, S. B., Forman, R., Meza-Perez, S., Sorobetea, D., Sitnik, K., et al. (2017). Eosinophils are key regulators of perivascular adipose tissue and vascular functionality. *Scientific Reports, 7*, 1–12.

Wohleb, E. S., Franklin, T., Iwata, M., & Duman, R. S. (2016). Integrating neuroimmune systems in the neurobiology of depression. *Nature Reviews Neuroscience, 17*(8), 497–511.

Wohleb, E. S., Hanke, M. L., Corona, A. W., Powell, N. D., Stiner, L. M., et al. (2011). β-adrenergic receptor antagonism prevents anxiety-like behavior and microglial reactivity induced by repeated social defeat. *Journal of Neuroscience, 31*(7), 6277–6288.

Wohlfert, E. A., Blader, I. J., & Wilson, E. H. (2017). Brains and brawn: Toxoplasma infections of the central nervous system and skeletal muscle. *Trends in Parasitology, 33*(7), 519–531.

Wolf, J., Rose-John, S., & Garbers, C. (2014). Interleukin-6 and its receptors: a highly regulated and dynamic system. *Cytokine, 70*(1), 11–20. Available from https://doi.org/10.1016/j.cyto.2014.05.024.

Wolf, O. T., Atsak, P., de Quervain, D. J., Roozendaal, B., & Wingenfeld, K. (2016). Stress and memory: A selective review on recent developments in the understanding of stress hormone effects on memory and their clinical relevance. *Journal of Neuroendocrinology, 28*(8). Available from https://doi.org/10.1111/jne.12353.

Wolkowitz, O. M., Wolf, J., Shelly, W., Rosser, R., Burke, H. M., et al. (2011). Serum BDNF levels before treatment predict SSRI response in depression. *Progress in Neuro-Psychopharmacology & Biological Psychiatry*, 35(7), 1623–1630.

Wong, M. L., Inserra, A., Lewis, M. D., Mastronardi, C. A., Leong, L., et al. (2016). Inflammasome signaling affects anxiety- and depressive-like behavior and gut microbiome composition. *Molecular Psychiatry*, 21(6), 797–805.

Woo, C. W., Roy, M., Buhle, J. T., & Wager, T. D. (2015). Distinct brain systems mediate the effects of nociceptive input and self-regulation on pain. *PLoS Biology*, 13(1), e1002036.

Wood, H. (2017). Parkinson disease: Caffeine and nicotine do not provide symptomatic relief in Parkinson disease. *Nature Reviews Neurology*, 13(12), 707.

Wood, P. G., Karol, M. H., Kusnecov, A. W., & Rabin, B. S. (1993). Enhancement of antigen-specific humoral and cell-mediated immunity by electric footshock stress in rats. *Brain, Behavior, and Immunity*, 7, 121–134.

Woodhams, S. G., Sagar, D. R., Burston, J. J., & Chapman, V. (2015). The role of the endocannabinoid system in pain. *Handbook Experimental Pharmacology*, 227, 119–143.

Woody, E. Z., & Szechtman, H. (2011). Adaptation to potential threat: The evolution, neurobiology, and psychopathology of the security motivation system. *Neuroscience & Biobehavioral Reviews*, 35(4), 1019–1033.

Woon, F. L., & Hedges, D. W. (2008). Hippocampal and amygdala volumes in children and adults with childhood maltreatment-related posttraumatic stress disorder: A meta-analysis. *Hippocampus*, 18(8), 729–736.

World Health Organization (2016a). *Tuberculosis: Fact sheet*. Retrieved from http://www.who.int/mediacentre/factsheets/fs104/en/. Accessed May 2016.

World Health Organization (2016b). *Investing in treatment for depression and anxiety leads to fourfold return*. Retrieved from http://www.who.int/mediacentre/news/releases/2016/depression-anxiety-treatment/en/. Accessed May 2016.

World Health Organization (2017). *Depression: Lets talk: Says WHO, as depression tops list of causes of ill health*. Retrieved from http://www.who.int/mediacentre/news/releases/2017/world-health-day/en/

Wrocklage, K. M., Averill, L. A., Cobb Scott, J., Averill, C. L., Schweinsburg, B., et al. (2017). Cortical thickness reduction in combat exposed U.S. veterans with and without PTSD. *European Neuropsychopharmacology*, 27(5), 515–525.

Wu, H., Esteve, E., Tremaroli, V., Khan, M. T., Caesar, R., et al. (2017). Metformin alters the gut microbiome of individuals with treatment-naive type 2 diabetes, contributing to the therapeutic effects of the drug. *Nature Medicine (New York, NY, United States)*, 23(7), 850–858.

Wu, M. K., Huang, T. L., Huang, K. W., Huang, Y. L., & Hung, Y. Y. (2015). Association between toll-like receptor 4 expression and symptoms of major depressive disorder. *Neuropsychiatric Disease and Treatment*, 11, 1853–1857.

Wu, S., Ding, Y., Wu, F., Li, R., Xie, G., Hou, J., & Mao, P. (2015). Family history of autoimmune diseases is associated with an increased risk of autism in children: A systematic review and meta-analysis. *Neuroscience & Biobehavioral Reviews*, 55, 322–332. Available from https://doi.org/10.1016/j.neubiorev.2015.05.004.

Wu, S. C., Cao, Z. S., Chang, K. M., & Juang, J. L. (2017). Intestinal microbial dysbiosis aggravates the progression of Alzheimer's disease in Drosophila. *Nature Communications*, 8, 24.

Wu, S. Y., Wang, T. F., Yu, L., Jen, C. J., Chuang, J. I., et al. (2011). Running exercise protects the substantia nigra dopaminergic neurons against inflammation-induced degeneration via the activaition of the BDNF signaling pathway. *Brain, Behavior, and Immunity*, 25(1), 135–146.

Wu, Z. M., Zheng, C. H., Zhu, Z. H., Wu, F. T., Ni, G. L., et al. (2016). SiRNA-mediated serotonin transporter knockdown in the dorsal raphe nucleus rescues single prolonged stress-induced hippocampal autophagy in rats. *Journal of the Neurological Sciences*, 360, 133–140.

Xia, C. Y., Zhang, S., Gao, Y., Wang, Z. Z., & Chen, N. H. (2015). Selective modulation of microglia polarization to M2 phenotype for stroke treatment. *International Immunopharmacology*, 25(2), 377–382.

Xiao, L., Priest, M. F., Nasenbeny, J., Lu, T., & Kozorovitskiy, Y. (2017). Biased oxytocinergic modulation of midbrain dopamine systems. *Neuron*, 95(2), 368–384.

Xie, P., Kranzler, H. R., Poling, J., Stein, M. B., Anton, R. F., et al. (2009). Interactive effect of stressful life events and the serotonin transporter 5-HTTLPR genotype on posttraumatic stress disorder diagnosis in 2 independent populations. *Archives of General Psychiatry*, 66(11), 1201–1209.

Xu, J., Marshall, J. J., Fernandes, H. B., Nomura, T., Copits, B. A., et al. (2017). Complete disruption of the kainate receptor gene family results in corticostriatal dysfunction in mice. *Cell Reports*, 18(8), 1848–1857.

Xu, T., Magnusson Hanson, L. L., Lange, T., Starkopf, L., Westerlund, H., et al. (2018). Workplace bullying and violence as risk factors for type 2 diabetes: A multicohort study and meta-analysis. *Diabetologia*, 61(1), 75–83.

Xu, Y., Zhou, H., & Zhu, Q. (2017). The impact of microbiota-gut-brain axis on diabetic cognition impairment. *Frontiers in Aging Neuroscience*, 9, 1–18.

Xue, K. S., Stevens-Ayers, T., Campbell, A. P., Englund, J. A., Pergam, S. A., et al. (2017). Parallel evolution of influenza across multiple spatiotemporal scales. *eLife*, 6, e26875.

Xue, M., Brimacombe, M., Chaaban, J., Zimmerman-Bier, B., & Wagner, G. C. (2008). Autism spectrum disorders: Concurrent clinical disorders. *Journal of Child Neurology*, 23(1), 6–13. Available from https://doi.org/10.1177/0883073807307102.

Xue, Y., Xu, X., Zhang, X. Q., Farokhzad, O. C., & Langer, R. (2016). Preventing diet-induced obesity in mice by adipose tissue transformation and angiogenesis using targeted nanoparticles. *Proceedings of the National Academy of Sciences of the United States of America, 113* (20), 5552–5557.

Yakushev, I., Gerhard, A., Müller, M. J., Lorscheider, M., Buchholz, H. G., et al. (2011). Relationships between hippocampal microstructure, metabolism, and function in early Alzheimer's disease. *Brain Structure & Function, 216*, 219–226.

Yamamoto, S., Morinobu, S., Takei, S., Fuchikami, M., Matsuki, A., et al. (2009). Single prolonged stress: Toward an animal model of posttraumatic stress disorder. *Depress Anxiety, 26*(12), 1110–1117.

Yamanashi, T., Iwata, M., Kamiya, N., Tsunetomi, K., Kajitani, N., et al. (2017). Beta-hydroxybutyrate, an endogenic NLRP3 inflammasome inhibitor, attenuates stress-induced behavioral and inflammatory responses. *Scientific Reports, 7*(1), 7677.

Yan, D., Zhang, Y., Liu, L., & Yan, H. (2016). Pesticide exposure and risk of Alzheimer's disease: A systematic review and meta-analysis. *Scientific Reports, 6*, 32222.

Yan, J. Q., Sun, J. C., Zhai, M. M., Cheng, L. N., Bai, X. L., et al. (2015). Lovastatin induces neuroprotection by inhibiting inflammatory cytokines in 6-hydroxydopamine treated microglia cells. *International Journal of Clinical and Experimental Medicine, 8*(6), 9030–9037.

Yan, M., Audet-Walsh, É., Manteghi, S., Dufour, C. R., Walker, B., et al. (2016). Chronic AMPK activation via loss of FLCN induces functional beige adipose tissue through PGC-1α/ERRα. *Genes & Development, 30*(9), 1034–1046.

Yan, T., Xie, Q., Zheng, Z., Zou, K., & Wang, L. (2017). Different frequency repetitive transcranial magnetic stimulation (rTMS) for posttraumatic stress disorder (PTSD): A systematic review and meta-analysis. *Journal of Psychiatric Research, 89*, 125–135.

Yang, E. V., & Glaser, R. (2002). Stress-associated immunomodulation and its implications for responses to vaccination. *Expert Review of Vaccines, 1*, 453–459.

Yang, G. R., Murray, J. D., & Wang, X. J. (2016). A dendritic disinhibitory circuit mechanism for pathway-specific gating. *Nature Communications, 7*, 12815.

Yang, J. J., Wang, N., Yang, C., Shi, J. Y., Yu, H. Y., et al. (2015). Serum interleukin-6 is a predictive biomarker for ketamine's antidepressant effect in treatment-resistant patients with major depression. *Biological Psychiatry, 77* (3), e19–e20.

Yang, Y., Tian, J., & Yang, B. (2018). Targeting gut microbiome: A novel and potential therapy for autism. *Life Sciences, 194*, 111–119.

Yao, X., Wang, D., Li, H., Shen, H., Shu, Z., et al. (2016). Erythropoietin treatment in acute ischemic stroke: A systematic review and meta-analysis. *Current Drug Delivery, 14*, 853–860.

Yasmin, F., Saxena, K., McEwen, B. S., & Chattarji, S. (2016). The delayed strengthening of synaptic connectivity in the amygdala depends on NMDA receptor activation during acute stress. *Physiological Reports, 4*(20), e13002.

Ye, K., Gao, F., Wang, D., Bar-Yosef, O., Keinan, A., et al. (2017). Dietary adaptation of FADS genes in Europe varied across time and geography. *Nature Ecology & Evolution, 1*(7), 0167.

Yehuda, R. (2002). Current status of cortisol findings in post-traumatic stress disorder. *The Psychiatric Clinics of North America, 25*(2), 341–368.

Yehuda, R., & Bierer, L. M. (2009). The relevance of epigenetics to PTSD: Implications for the DSM-V. *Journal of Traumatic Stress, 22*(5), 427–434.

Yehuda, R., Brand, S., & Yang, R. K. (2006). Plasma neuropeptide Y concentrations in combat exposed veterans: Relationship to trauma exposure, recovery from PTSD, and coping. *Biological Psychiatry, 59*(7), 660–663.

Yehuda, R., Daskalakis, N. P., Bierer, L. M., Bader, H. N., Klengel, T., et al. (2016). Holocaust exposure induced intergenerational effects on FKBP5 methylation. *Biological Psychiatry, 80*(5), 372–380.

Yehuda, R., Daskalakis, N. P., Desarnaud, F., Makotkine, I., Lehrner, A., et al. (2013). Epigenetic biomarkers as predictors and correlates of symptom improvement following psychotherapy in combat veterans with PTSD. *Frontiers in Psychiatry, 4*, 118.

Yehuda, R., Flory, J. D., Pratchett, L. C., Buxbaum, J., Ising, M., et al. (2010). Putative biological mechanisms for the association between early life adversity and the subsequent development of PTSD. *Psychopharmacology (Berlin, Germany), 212*(3), 405–417.

Yin, J., Liao, S. X., He, Y., Wang, S., Xia, G. H., et al. (2015). Dysbiosis of gut microbiota with reduced trimethylamine-N-oxide level in patients with large-artery atherosclerotic stroke of transient ischemic attack. *Journal of the American Heart Association, 4*(11), e002699.

Yirmiya, R. (2000). Depression in medical illness: The role of the immune system. *The Western Journal of Medicine, 173*(5), 333–336.

Yirmiya, R., Pollak, Y., Barak, O., Avitsur, R., Ovadia, H., et al. (2001). Effects of antidepressant drugs on the behavioral and physiological responses to lipopolysaccharide (LPS) in rodents. *Neuropsychopharmacology, 24*(5), 531–544.

Yirmiya, R., Pollak, Y., Morag, M., Reichenberg, A., & Barak, O. (2000). Illness, cytokines, and depression. *Ann NY Acad Sci, 917*, 478–487.

Yirmiya, R., Rimmerman, N., & Reshef, R. (2015). Depression as a microglial disease. *Trends in Neurosciences, 38*(10), 637–658.

Yohn, N. L., & Blendy, J. A. (2017). Adolescent chronic unpredictable stress exposure is a sensitive window for long-term changes in adult behavior in mice. *Neuropsychopharmacology, 42*(8), 1670–1678.

Yohn, S. E., Arif, Y., Haley, A., Tripodi, G., Baqi, Y., et al. (2016). Effort-related motivational effects of the pro-inflammatory cytokine interleukin-6: Pharmacological and neurochemical characterization. *Psychopharmacology (Berlin, Germany), 233*(19–20), 3575–3586.

Yolken, R., Adamos, M., Katsafanas, E., Khushalani, S., Origoni, A., et al. (2016). Individuals hospitalized with acute mania have increased exposure to antimicrobial medications. *Bipolar Disorders, 18*, 404–409.

Yoon, S., Cho, H., Kim, J., Lee, D. W., Kim, G. H., et al. (2017). Brain changes in overweight/obese and normal-weight adults with type 2 diabetes mellitus. *Diabetologia, 60*(7), 1207–1217.

Youm, Y. H., Nguyen, K. Y., Grant, R. W., Goldberg, E. L., Bodogai, M., et al. (2015). The ketone metabolite β-hydroxybutyrate blocks NLRP3 inflammasome-mediated inflammatory disease. *Nature Medicine (New York, NY, United States), 21*(3), 263–269.

Young, A. M., Campbell, E., Lynch, S., Suckling, J., & Powis, S. J. (2011). Aberrant NF-kappaB expression in autism spectrum condition: A mechanism for neuroinflammation. *Frontiers in Psychiatry, 2*, 27. Available from https://doi.org/10.3389/fpsyt.2011.00027.

Youngson, N. A., & Whitelaw, E. (2008). Transgenerational epigenetic effects. *Annual Review of Genomics and Human Genetics, 9*, 233–257.

Yousef Yengej, F. A., van Royen-Kerkhof, A., Derksen, R., & Fritsch-Stork, R. D. E. (2017). The development of offspring from mothers with systemic lupus erythematosus. A systematic review. *Autoimmunity Reviews, 16*(7), 701–711. Available from https://doi.org/10.1016/j.autrev.2017.05.005.

Ysseldyk, R., Matheson, K., & Anisman, H. (2010). Religiosity as identity: Toward an understanding of religion from a social identity perspective. *Personality and Social Psychology Review, 14*(1), 60–71.

Yu, A., & Dang, W. (2017). Regulation of stem cell aging by SIRT1 – Linking metabolic signaling to epigenetic modifications. *Molecular and Cellular Endocrinology, 455*, 75–82.

Yu, Q., Daugherty, A. M., Anderson, D. M., Nishimura, M., Brush, D., et al. (2017). Socioeconomic status and hippocampal volume in children and young adults. *Developmental Science, 21*(3), e12561. Available from https://doi.org/10.1111/desc.12561.

Yuen, E. Y., Wei, J., & Yan, Z. (2016). Estrogen in prefrontal cortex blocks stress-induced cognitive impairments in female rats. *Journal of Steroid Biochemistry and Molecular Biology, 160*, 221–226.

Yulug, B., Hanoglu, L., & Kilic, E. (2017). Does sleep disturbance affect the amyloid clearance mechanisms in Alzheimer's disease? *Psychiatry and Clinical Neurosciences, 71*(10), 673–677.

Zaganas, I., Kapetanaki, S., Mastorodemos, V., Kanavouras, K., Colosio, C., et al. (2013). Linking pesticide exposure and dementia: What is the evidence? *Toxicology, 307*, 3–11.

Zai, G., Robbins, T. W., Sahakian, B. J., & Kennedy, J. L. (2017). A review of molecular genetic studies of neurocognitive deficits in schizophrenia. *Neuroscience & Biobehavioral Reviews, 72*, 50–67.

Zalcman, S., & Anisman, H. (1993). Acute and chronic stressor effects on the antibody response to sheep red blood cells. *Pharmacology, Biochemistry and Behavior, 46*(2), 445–452.

Zalli, A., Carvalho, L. A., Lin, J., Hamer, M., Erusalimsky, J. D., et al. (2014). Shorter telomeres with high telomerase activity are associated with raised allostatic load and impoverished psychosocial resources. *Proceedings of the National Academy of Sciences of the United States of America, 111*(12), 4519–4524.

Zamanian, J. L., Xu, L., Foo, L. C., Nouri, N., Zhou, L., et al. (2012). Genomic analysis of reactive astrogliosis. *Journal of Neuroscience, 32*, 6391–6410.

Zaneveld, J. R., McMinds, R., & Vega Thurber, R. (2017). Stress and stability: Applying the Anna Karenina principle to animal microbiomes. *Nature Microbiology, 2*, 17121.

Zanoni, P., Khetarpal, S. A., Larach, D. B., Hancock-Cerutti, W. F., Millar, J. S., et al. (2016). Rare variant in scavenger receptor BI raises HDL cholesterol and increases risk of coronary heart disease. *Science, 351*, 1166–1171.

Zareie, M., Johnson-Henry, K., Jury, J., Yang, P. C., Ngan, B. Y., et al. (2006). Probiotics prevent bacterial translocation and improve intestinal barrier function in rats following chronic psychological stress. *Gut, 55*, 1553–1560.

Zass, L. J., Hart, S. A., Seedat, S., Hemmings, S. M., & Malan-Müller, S. (2017). Neuroinflammatory genes associated with post-traumatic stress disorder: Implications for comorbidity. *Psychiatric Genetics, 27*, 1–16.

Zeevi, D., Korem, T., Zmora, N., Israeli, D., Rothschild, D., et al. (2015). Personalized nutrition by prediction of glycemic responses. *Cell, 163*(5), 1079–1094.

Zeidan, F., Emerson, N. M., Farris, S. R., Ray, J. N., Jung, Y., et al. (2015). Mindfulness meditation-based pain relief employs different neural mechanisms than placebo and sham mindfulness meditation-induced analgesia. *Journal of Neuroscience, 35*, 15307–15325.

Zeineh, M. M., Kang, J., Atlas, S. W., Raman, M. M., Reiss, A. L., et al. (2014). Right arcuate fasciculus abnormality in chronic fatigue syndrome. *Radiology, 274*, 517–526.

Zenk, F., Loeser, E., Schiavo, R., Kilpert, F., Bogdanović, O., et al. (2017). Germ line–inherited H3K27me3 restricts enhancer function during maternal-to-zygotic transition. *Science, 357*(6347), 212–216.

Zerbo, O., Qian, Y., Yoshida, C., Fireman, B. H., Klein, N. P., & Croen, L. A. (2017). Association Between Influenza Infection and Vaccination During Pregnancy and Risk of Autism Spectrum Disorder. *JAMA Pediatr, 171*(1), e163609. Available from https://doi.org/10.1001/jamapediatrics.2016.3609.

Zhang, D., Kishihara, K., Wang, B., Mizobe, K., Kubo, C., et al. (1998). Restraint stress-induced immunosuppression by inhibiting leukocyte migration and Th1 cytokine expression during the intraperitoneal infection of Listeria monocytogenes. *Journal of Neuroimmunology, 92*(1–2), 139–151.

Zhang, D., Wang, X., Wang, B., Garza, J. C., Fang, X., et al. (2016). Adiponectin regulates contextual fear extinction and intrinsic excitability of dentate gyrus granule neurons through AdipoR2 receptors. *Molecular Psychiatry, 22*(7), 1044–1055.

Zhang, F., Signore, A. P., Zhou, Z., Wang, S., Cao, G., et al. (2006). Erythropoietin protects CA1 neurons against global cerebral ischemia in rat: Potential signaling mechanisms. *Journal of Neuroscience Research, 83*, 1241–1251.

Zhang, J., Lamers, F., Hickie, I. B., He, J. P., Feig, E., et al. (2013). Differentiating nonrestorative sleep from nocturnal insomnia symptoms: Demographic, clinical, inflammatory, and functional correlates. *Sleep, 36*(5), 671–679.

Zhang, J. C., Yao, W., Dong, C., Yang, C., Ren, Q., et al. (2017). Blockade of interleukin-6 receptor in the periphery promotes rapid and sustained antidepressant actions: A possible role of gut-microbiota-brain axis. *Translational Psychiatry, 7*(5), e1138.

Zhang, L., Hu, X. Z., Li, X., Li, H., Smerin, S., et al. (2014). Telomere length – A cellular aging marker for depression and post-traumatic stress disorder. *Medical Hypotheses, 83*(2), 182–185.

Zhang, W., & Rosenkranz, J. A. (2016). Effects of repeated stress on age-dependent GABAergic regulation of the lateral nucleus of the amygdala. *Neuropsychopharmacology, 41*(9), 2309–2323.

Zhang, Y., Shao, F., Wang, Q., Xie, X., & Wang, W. (2017). Neuroplastic correlates in the mPFC underlying the impairment of stress-coping ability and cognitive flexibility in adult rats exposed to chronic mild stress during adolescence. *Neural Plasticity, 2017*, 9382797.

Zhang, Z., Cai, Y. Q., Zou, F., Bie, B., & Pan, Z. Z. (2011). Epigenetic suppression of GAD65 expression mediates persistent pain. *Nature Medicine (New York, NY, United States), 17*, 1448–1455.

Zhang, Z., Fan, Q., Bai, Y., Wang, Z., Zhang, H., et al. (2016). Brain gamma-aminobutyric acid (GABA) concentration of the prefrontal lobe in unmedicated patients with obsessive-compulsive disorder: A

research of magnetic resonance spectroscopy. *Shanghai Archives of Psychiatry, 28*(5), 263–270.

Zhao, S. C., Ma, L. S., Chu, Z. H., Xu, H., Wu, W. Q., et al. (2017). Regulation of microglial activation in stroke. *Acta Pharmacologica Sinica, 38*(4), 445–458.

Zhao, W., Rasheed, A., Tikkanen, E., Lee, J. J., Butterworth, A. S., et al. (2017). Identification of new susceptibility loci for type 2 diabetes and shared etiological pathways with coronary heart disease. *Nature Genetics, 49*(10), 1450–1457.

Zheng, Y., Manson, J. E., Yuan, C., Liang, M. H., Grodstein, F., et al. (2017). Associations of weight gain from early to middle adulthood with major health outcomes later in life. *The Journal of the American Medical Association, 318*(3), 255–269.

Zhernakova, A., Kurilshikov, A., Bonder, M. J., Tigchelaar, E. F., Schirmer, M., et al. (2016). Population-based metagenomics analysis reveals markers for gut microbiome composition and diversity. *Science, 352*(6285), 565–569.

Zhou, D., Kusnecov, A. W., Shurin, M. R., DePaoli, M., & Rabin, B. S. (1993). Exposure to physical and psychological stressors elevates plasma interleukin 6: Relationship to the activation of hypothalamic-pituitary-adrenal axis. *Endocrinology, 133*(6), 2523–2530.

Zhou, J., Luo, Y., Zhang, J. T., Li, M. X., Wang, C. M., et al. (2015). Propranolol decreases retention of fear memory by modulating the stability of surface glutamate receptor GluA1 subunits in the lateral amygdala. *British Journal of Pharmacology, 172*(21), 5068–5082.

Zhou, J., Nagarkatti, P., Zhong, Y., Ginsberg, J. P., Singh, N. P., et al. (2016). Dysregulation in microRNA expression is associated with alterations in immune functions in combat veterans with post-traumatic stress disorder. *PLoS One, 9*(3), e94075.

Zhou, R., Chen, F., Feng, X., Zhou, L., Li, Y., et al. (2015). Perinatal exposure to low-dose of bisphenol A causes anxiety-like alteration in adrenal axis regulation and behaviors of rat offspring: A potential role for metabotropic glutamate 2/3 receptors. *Journal of Psychiatric Research, 64*, 121–129.

Zhou, T., Huang, Z., Sun, X., Zhu, X., Zhou, L., et al. (2017). Microglia polarization with M1/M2 phenotype changes in rd1 mouse model of retinal degeneration. *Frontiers in Neuroanatomy, 11*, 77.

Zhou, W., Lv, H., Li, M. X., Su, H., Huang, L. G., et al. (2015). Protective effects of bifidobacteria on intestines in newborn rats with necrotizing enterocolitis and its regulation on TLR2 and TLR4. *Genetics and Molecular Research, 14*, 11505–11514.

Zhu, W., Gregory, J. C., Org, E., Buffa, J. A., Gupta, N., et al. (2016). Gut microbial metabolite TMAO enhances platelet hyperreactivity and thrombosis risk. *Cell, 165* (1), 111−124.

Ziegler, C., Dannlowski, U., Bräuer, D., Stevens, S., Laeger, I., et al. (2015). Oxytocin receptor gene methylation: Converging multilevel evidence for a role in social anxiety. *Neuropsychopharmacology, 40*(6), 1528−1538.

Zieker, J., Zieker, D., Jatzko, A., Dietzsch, J., Nieselt, K., et al. (2007). Differential gene expression in peripheral blood of patients suffering from post-traumatic stress disorder. *Molecular Psychiatry, 12*(2), 116−118.

Zimmerman, A. W., Jyonouchi, H., Comi, A. M., Connors, S. L., Milstien, S., Varsou, A., & Heyes, M. P. (2005). Cerebrospinal fluid and serum markers of inflammation in autism. *Pediatric Neurology, 33*(3), 195−201. Available from https://doi.org/10.1016/j.pediatrneurol.2005.03.014.

Zimprich, A., Biskup, S., Leitner, P., Lichtner, P., Farrer, M., Lincoln, S., … Gasser, T. (2004). Mutations in LRRK2 cause autosomal-dominant parkinsonism with pleomorphic pathology. *Neuron, 44*(4), 601−607.

Zincir, S., Öztürk, P., Bilgen, A. E., İzci, F., & Yükselir, C. (2016). Levels of serum immunomodulators and alterations with electroconvulsive therapy in treatment-resistant major depression. *Neuropsychiatric Disease and Treatment, 12*, 1389−1396.

Zinman, B., Wanner, C., Lachin, J. M., Fitchett, D., Bluhmki, E., et al. (2015). Empagliflozin, cardiovascular outcomes, and mortality in type 2 diabetes. *New England Journal of Medicine, 373*(22), 2117−2128.

Zivadinov, R., Jakimovski, D., Gandhi, S., Ahmed, R., Dwyer, M. G., Horakova, D., … Bergsland, N. (2016). Clinical relevance of brain atrophy assessment in multiple sclerosis. Implications for its use in a clinical routine. *Expert Rev Neurother, 16*(7), 777−793. Available from https://doi.org/10.1080/14737175.2016.1181543.

Zobel, A. W., Nickel, T., Sonntag, A., Uhr, M., Holsboer, F., et al. (2001). Cortisol response in the combined dexamethasone/CRH test as predictor of relapse in patients with remitted depression: A prospective study. *Journal of Psychiatric Research, 35*(2), 83−94.

Zoccola, P. M., Figueroa, W. S., Rabideau, E. M., Woody, A., & Benencia, F. (2014). Differential effects of poststressor rumination and distraction on cortisol and C-reactive protein. *Health Psychology, 33*(12), 1606−1609.

Zohar, J., Fostick, L., Juven-Wetzler, A., Kaplan, Z., Shalev, H., et al. (2018). Secondary prevention of chronic PTSD by early and short-term administration of escitalopram: A prospective randomized, placebo-controlled, double-blind trial. *Journal of Clinical Psychiatry*, pii: 16m10730.

Zonis, S., Pechnick, R. N., Ljubimov, V. A., Mahgerefteh, M., Wawrowsky, K., et al. (2015). Chronic intestinal inflammation alters hippocampal neurogenesis. *Journal of Neuroinflammation, 12*, 65.

Zotova, E., Bharambe, V., Cheaveau, M., Morgan, W., Holmes, C., et al. (2013). Inflammatory components in human Alzheimer's disease and after active amyloid-β42 immunization. *Brain: A Journal of Neurology, 136*(9), 2677−2696.

Zovkic, I. B., & Sweatt, J. D. (2013). Epigenetic mechanisms in learned fear: Implications for PTSD. *Neuropsychopharmacology, 38*, 77−93.

Zucca, F. A., Segura-Aguilar, J., Ferrari, E., Muñoz, P., Paris, I., et al. (2017). Interactions of iron, dopamine and neuromelanin pathways in brain aging and Parkinson's disease. *Progress in Neurobiology, 155*, 96−119.

Index